Applied Instrumentation in the Process Industries

Volume III
Third Edition
Engineering Data and Resource Material

Applied Instrumentation in the Process Industries

Volume III
Third Edition
Engineering Data and Resource Material

W.G. Andrew
H.B. Williams

Gulf Publishing Company
Houston, London, Paris, Zurich, Tokyo

Applied Instrumentation in the Process Industries
Volume III
Engineering Data and Resource Material
Third Edition

Gulf Publishing Company
Book Division
P.O. Box 2608, Houston, Texas 77252-2608

10 9 8 7 6 5 4 3 2 1

Library of Congress Cataloging-in-Publication Data

Andrew, W. G. (William G.)
Applied instrumentation in the process industries / W. G. Andrew, H. B. Williams. — 3rd ed.
p. cm.
Includes index.
Contents: —v. 3. Engineering data and resource material.
ISBN 0-87201-047-3 (v. 3)
1. Process control. 2. Automatic control. 3. Engineering instruments. I. Williams, H. B., 1939– . II. Title.
TS156.8.A534 1993 92-35107
CIP

Printed in the United States of America.

Dedication

To my wife, Patsy,
for her patience and encouragement.

H. B. Williams

Contents

Acknowledgments

Any technical book draws material from a large number of sources. Although many of these are referenced in the text, it is not feasible to include all the contributors to whom the authors are indebted. Data and information were furnished by many industrial companies.

The authors were encouraged to undertake the work by A. C. Lederer, former president of S.I.P. Engineering, Inc. The cooperation of W. L. Hampton, President, S.I.P. Engineering, Inc., a Parsons Corporation Company, is gratefully acknowledged in producing the work on schedule.

Appreciation is extended particularly to B. J. Normand and K. G. Rhea for time spent in reviewing and criticizing many chapters and sections of the manuscript. Others who contributed in this area include L. Ashley, W. E. DeLong, D. M. Dudney, L. C. Hoffman, T. E. Lasseter, and J. G. Royle.

In addition, the authors are deeply appreciative to S.I.P. Engineering, Inc., and its staff in providing the environment and materials for producing this work.

On the second and third editions: Mr. H. B. Williams and Mr. W. J. Ruschel give their profound appreciation to all who have assisted in preparing this revised text—those who have contributed technical information; those who have typed and proofread; those who willingly provided illustrations, suggestions, and encouragement; and to all who have, by their confidence, inspired this effort.

Contributors

William G. Andrew	PE, BSEE, University of Texas
David M. Dudney	BSChE, University of Houston S.I.P. Engineering, Inc.
James L. Jackson	BSChE, Rice University Borg-Warner Corporation Marbon Division (Formerly with S.I.P Engineering, Inc.)
B. J. Normand	PE, BSChE, University of Southwestern Louisiana S.I.P. Engineering, Inc.
Wayne J. Ruschel	PE, BSEE, Louisiana State University S.I.P. Engineering, Inc.
Andrew Jackson Stockton	S.I.P. Engineering, Inc.
H. Baxter Williams	PE, BSEE, Louisiana Polytechnic Institute S.I.P. Engineering, Inc.

Preface

The third volume of *Applied Instrumentation In The Process Industries,* "Engineering Data and Resource Material," is yet another unique addition to the instrument field. It is filled with the type of information that is a necessary part of the activities of people involved in instrument application, engineering, design, and operation.

It begins with a thorough treatment of problems associated with fluid flow. Physical properties of fluids are discussed, and the nature of liquid and compressible fluid flows are treated clearly and concisely. This section alone makes the volume a worthy addition to a technical library.

Also provided are charts, tables, nomagraphs, formulas and symbols that instrument people need in performing their work. This information is arranged topically and indexed for easy reference. There is an abundance of information on the physical properties of fluids, flow data, conversion data, mathematical functions, piping information, and electrical data that are essential to instrument engineering.

Other features included in Volume 3 are: (1) a listing of formulas needed for the many calculations that must be made from time to time in engineering work. Many of them are used infrequently and may be forgotten. They are listed topically for easy reference. (2) Typical installation details for many instrument devices that many people will find useful in the preparation of standards and in daily work. (3) Calculation examples that are extremely helpful to novices in the industry. These include fluid flow problems, orifice calculations, control valve and relief valve calculations and other problems that confront the average instrument engineer. These calculations involve the use of charts, tables, etc., given in other sections of the volume.

This volume has under one cover the information that the average instrument engineer must ordinarily use many sources to obtain.

1 Fluid Flow

David M. Dudney
James L. Jackson

Introduction

This section is intended as an aid in solving a majority of fluid flow problems in the chemical processing industries. The procedures are presented for rapid and convenient solution of these types of flow problems. Two-phase liquid-vapor flow is given a limited, but sufficient, treatment for solution of many industrial problems.

For convenience, data are included for viscosities and densities of many fluids and equivalent lengths of fittings.

Nature of Fluids

Solving fluid flow problems requires an understanding of the physical nature of fluids, which undergo continuous deformation when subjected to shear stress. The physical properties of greatest interest are viscosity and density, which are directly involved in most flow problems. Other fluid characteristics of fluid flow problems include vapor pressure, melting or pour point, molecular weight, compressibility factor, isentropic expansion coefficient ($k = C_p/C_v$), critical temperature and pressure, enthalpy data, vapor-liquid equilibrium data and corrosion-erosion characteristics.

Viscosity

Viscosity is a measure of a fluid's resistance to flow – the greater the viscosity the greater the resistance. Viscosity can vary with temperature, pressure, shear stress and the fluid's history.

Increasing temperatures cause decreasing liquid viscosities and increasing vapor viscosities. Increasing pressures cause increasing viscosities for both liquids and gases. Generally, vapor viscosities are more sensitive to pressure changes than are liquids and less sensitive to temperature changes.

Newtonian fluids are those for which viscosities vary only with temperature and pressure. Examples are gases and most liquids of moderate viscosities that contain little or no suspended solids such as alcohol, water and gasoline.

Non-Newtonian fluids are those whose viscosities vary with shear stress and/or fluid history as well as temperature and pressure. This behavior is frequently exhibited by polymeric solutions and melts, emulsions and suspensions of solids. Examples are pulp slurries, catsup, mayonnaise, paints, inks, quicksand and drilling muds. (Treatment of non-Newtonian fluid flow is outside the scope of this book. A good discussion of these fluids may be found in Reference 1.)

There are two types of viscosity units commonly in use – kinematic and dynamic. Kinematic viscosity is expressed by two different methods. One method is to express the time required for a specified quantity of fluid to drain from a reservoir of standardized dimensions. An example of this is the Saybolt Universal Viscometer. The second method is to use units with dimensions of area per unit time; for example, centistokes which equals 10^{-2} cm^2/sec.

Units with dimensions of force-time/area or the equivalent of mass/length-time are used for dynamic viscosity; for

Table 1.1. Kinematic Viscosity Conversions

To convert the numerical value of a property expressed in one of the units in the left-hand column of the table to the numerical value expressed in one of the units in the top row of the table, multiply the former value by the factor in the block common to both units.

Units→ ↓	sq ft/hr	sq ft/sec	sq m/hr	sq cm/sec (Stokes)	sq cm/sec × 10² (Centistokes)
sq ft/hr	1	2.778 × 10⁻⁴	9.290 × 10⁻²	0.2581	25.81
sq ft/sec	3,600	1	3.345 × 10²	929	9.29 × 10⁴
sq m/hr	10.76	2.990 × 10⁻³	1	2.778	277.8
sq cm/sec (Stokes)	3.875	1.076 × 10⁻³	0.3600	1	100
sq cm/sec × 10² (Centistokes)	3.875 × 10⁻²	1.076 × 10⁻⁵	3.600 × 10⁻³	0.0100	1

Table 1.2. Absolute Viscosity Conversions

To convert the numerical value of a property expressed in one of the units in the left-hand column of the table to the numerical value expressed in one of the units in the top row of the table, multiply the former value by the factor in the block common to both units.

Units→ ↓	lb/sec—ft	lb/hr—ft	lb [force]—sec/sq ft	g/sec—cm × 10² (Centipoises)*	kg/hr—m
lb/sec—ft	1	3,600	0.03108	1,488	5,357
lb/hr—ft	2.778 × 10⁻⁴	1	8.634 × 10⁻⁶	0.4134	1.488
lb [force]—sec/sq ft	32.17	1.158 × 10⁵	1	47,880	1.724 × 10⁵
g/sec—cm × 10² (Centipoises)*	6.720 × 10⁻⁴	2.419	2.089 × 10⁻⁵	1	3.600
kg/hr—m	1.867 × 10⁻⁴	0.6720	5.801 × 10⁻⁶	0.2778	1

* 1 poise = 100 centipoises = 1 g/sec—cm. Kinematic viscosity, in centistokes, times density (g/cu cm) at same temperature equals centipoises.

example, centipoise (μ) which equals 10^{-2} dyne-sec./cm^2 = 10^{-2} gm/cm-sec.

The relationship between dynamic and kinematic viscosity (units of area/time) for a consistent set of units is

$$\text{kinematic vis.} = (\text{dynamic vis.})/\text{density} \tag{1.1}$$

Viscosity of liquid mixtures may be estimated by the equation

$$\mu_{\text{mix}} = (\Sigma_i\, x_i\, \mu_i^{1/3})^3 \tag{1.2}$$

This equation works best for liquids that are similar in molecular weight and character.

Viscosities for vapor mixtures of reduced pressure less than 0.6 may be estimated by the equation

$$\mu_{\text{mix}} = \frac{\Sigma_i y_i \mu_i (MW)_i}{\Sigma_i y_i (MW)_i} \tag{1.3}$$

A common problem is estimating the viscosity of a fluid at a reduced temperature and pressure different from those of a

Table 1.3. Equivalents of Kinematic and Saybolt Universal Viscosity

Kinematic Viscosity, Centistokes v	Equivalent Saybolt Universal Viscosity, Sec — At 100 F Basic Values	At 210 F
1.83	32.01	32.23
2.0	32.62	32.85
4.0	39.14	39.41
6.0	45.56	45.88
8.0	52.09	52.45
10.0	58.91	59.32
15.0	77.39	77.93
20.0	97.77	98.45
25.0	119.3	120.1
30.0	141.3	142.3
35.0	163.7	164.9
40.0	186.3	187.6
45.0	209.1	210.5
50.0	232.1	233.8
55.0	255.2	257.0
60.0	278.3	280.2
65.0	301.4	303.5
70.0	324.4	326.7
75.0	347.6	350.0
80.0	370.8	373.4
85.0	393.9	396.7
90.0	417.1	420.0
95.0	440.3	443.4
100.0	463.5	466.7
120.0	556.2	560.1
140.0	648.9	653.4
160.0	741.6	
180.0	834.2	
200.0	926.9	
220.0	1019.6	
240.0	1112.3	
260.0	1205.0	
280.0	1297.7	
300.0	1390.4	
320.0	1483.1	
340.0	1575.8	
360.0	1668.5	Saybolt Seconds equal Centistokes times 4.6673
380.0	1761.2	
400.0	1853.9	
420.0	1946.6	
440.0	2039.3	
460.0	2132.0	
480.0	2224.7	
500.0	2317.4	
Over 500	Saybolt Seconds equal Centistokes times 4.6347	

Note: To obtain the Saybolt Universal viscosity equivalent to a kinematic viscosity determined at t, multiply the equivalent Saybolt Universal viscosity at 100 F by $1 + (t - 100)\ 0.000\ 064$.
For example, 10 v at 210 F are equivalent to 58.91 multiplied by 1.0070 or 59.32 sec Saybolt Universal at 210 F.

Source: Crane Company (T.P. #410) and by permission of the American Society for Testing Materials (ASTM).

Table 1.4. Equivalents of Kinematic and Saybolt Furol Viscosity

Kinematic Viscosity, Centistokes v	Equivalent Saybolt Furol Viscosity, Sec — At 122 F	At 210 F
48	25.3	
50	26.1	25.2
60	30.6	29.8
70	35.1	34.4
80	39.6	39.0
90	44.1	43.7
100	48.6	48.3
125	60.1	60.1
150	71.7	71.8
175	83.8	83.7
200	95.0	95.6
225	106.7	107.5
250	118.4	119.4
275	130.1	131.4
300	141.8	143.5
325	153.6	155.5
350	165.3	167.6
375	177.0	179.7
400	188.8	191.8
425	200.6	204.0
450	212.4	216.1
475	224.1	228.3
500	235.9	240.5
525	247.7	252.8
550	259.5	265.0
575	271.3	277.2
600	283.1	289.5
625	294.9	301.8
650	306.7	314.1
675	318.4	326.4
700	330.2	338.7
725	342.0	351.0
750	353.8	363.4
775	365.5	375.7
800	377.4	388.1
825	389.2	400.5
850	400.9	412.9
875	412.7	425.3
900	424.5	437.7
925	436.3	450.1
950	448.1	462.5
975	459.9	474.9
1000	471.7	487.4
1025	483.5	499.8
1050	495.2	512.3
1075	507.0	524.8
1100	518.8	537.2
1125	530.6	549.7
1150	542.4	562.2
1175	554.2	574.7
1200	566.0	587.2
1225	577.8	599.7
1250	589.5	612.2
1275	601.3	624.8
1300	613.1	637.3
Over 1300	*	†

*OVER 1300 CENTISTOKES AT 122 F:
Saybolt Fluid Sec = Centistokes x 0.4717

†OVER 1300 CENTISTOKES AT 210 F:
Log (Saybolt Furol Sec − 2.87) = 1.0276 [Log (Centistokes)] − 0.3975

Source: Crane Company (T.P. #410) and by permission of the American Society for Testing Materials (ASTM).

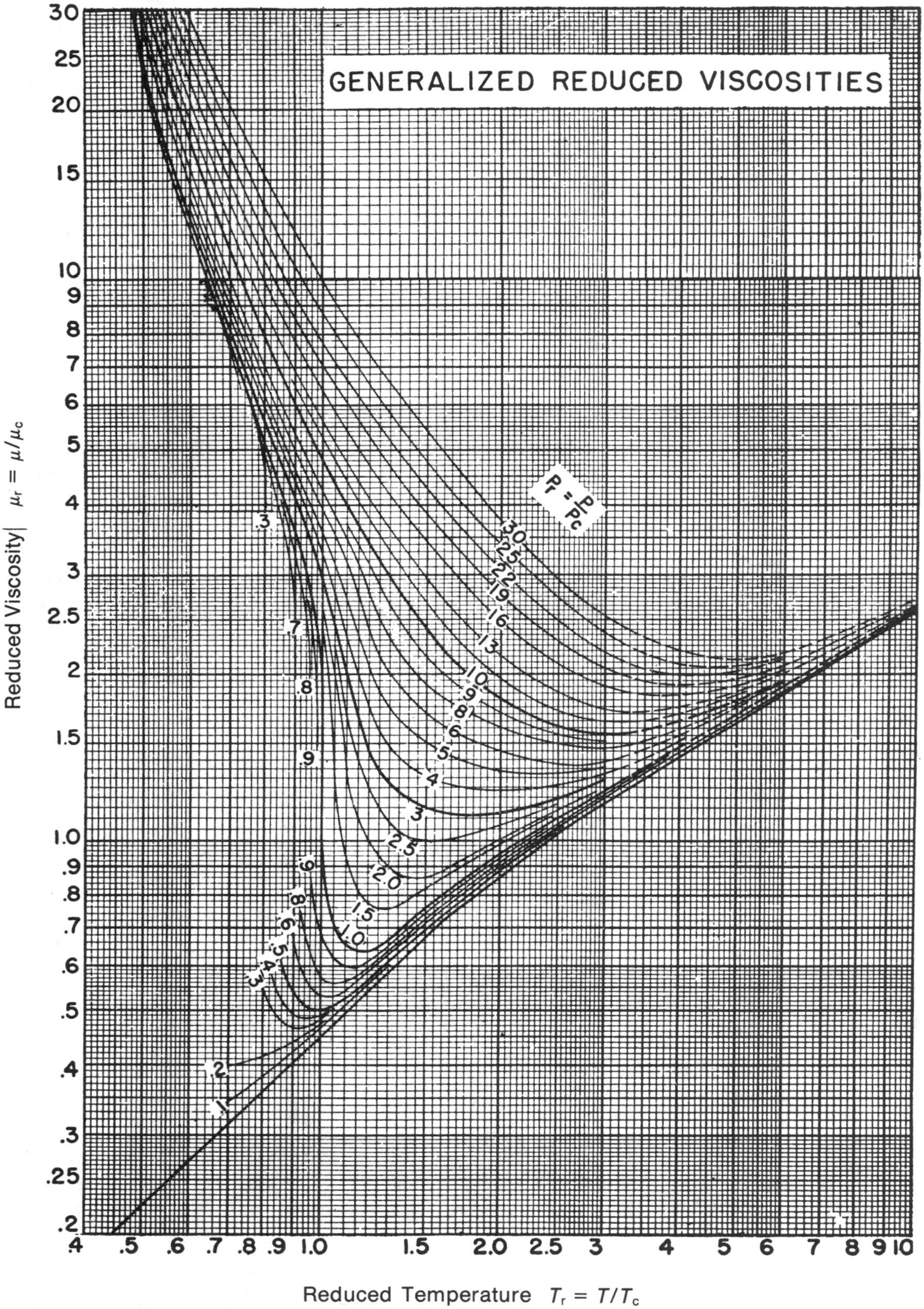

Figure 1.1. Generalized reduced viscosities. From Hougen and Watson, Chemical Process Principles, *Part III, John Wiley and Sons, Inc., reprinted by permission.*

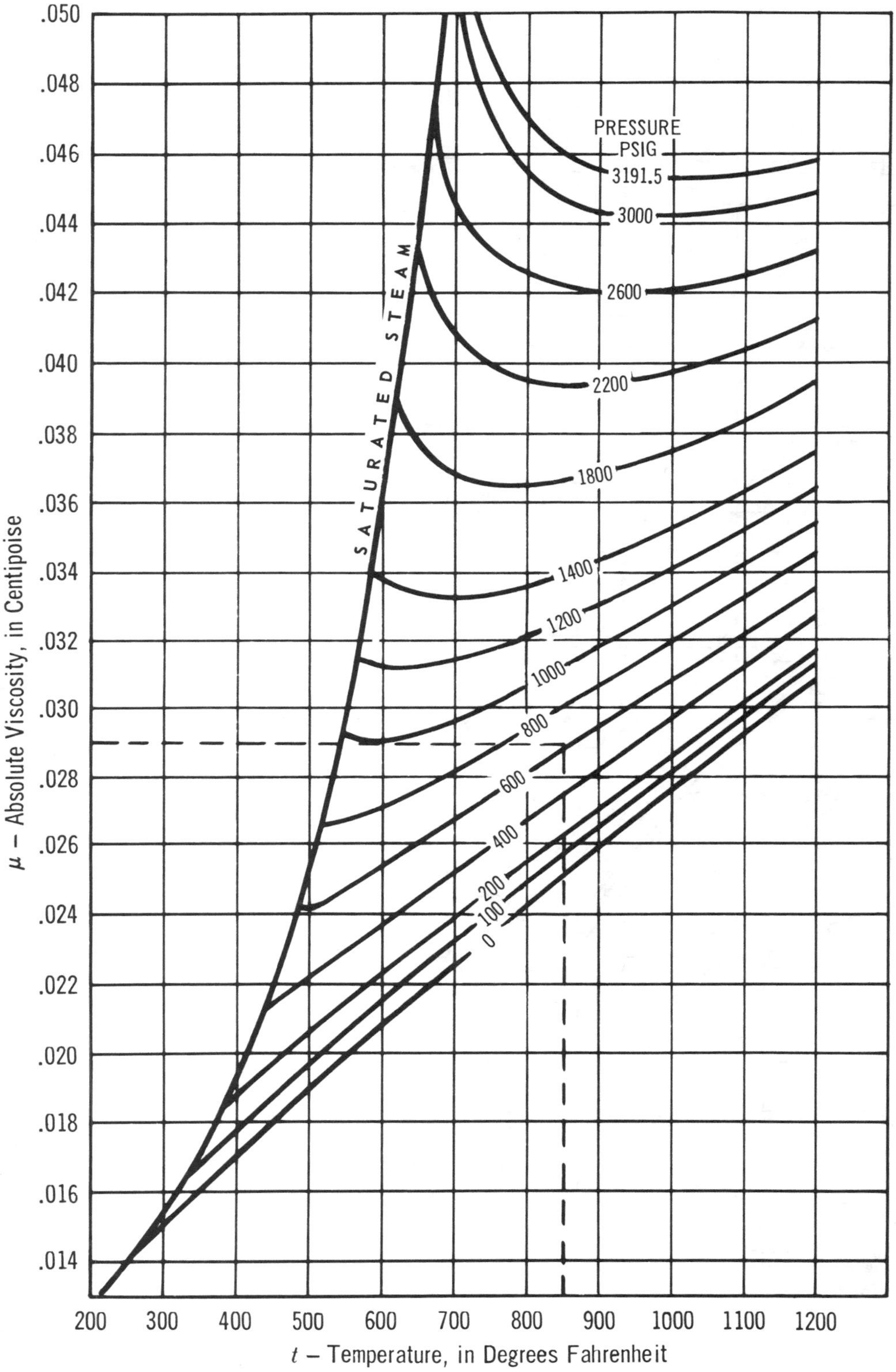

Example: Viscosity of 600 psig, 850 F steam is 0.029 centipoise.

Figure 1.2. Viscosity of steam. Courtesy Crane Co. (T.P. No. 410) and the Ronald Press Co. (adapted from Steam Power Plants, *Phillip J. Potter, copyright © 1949).*

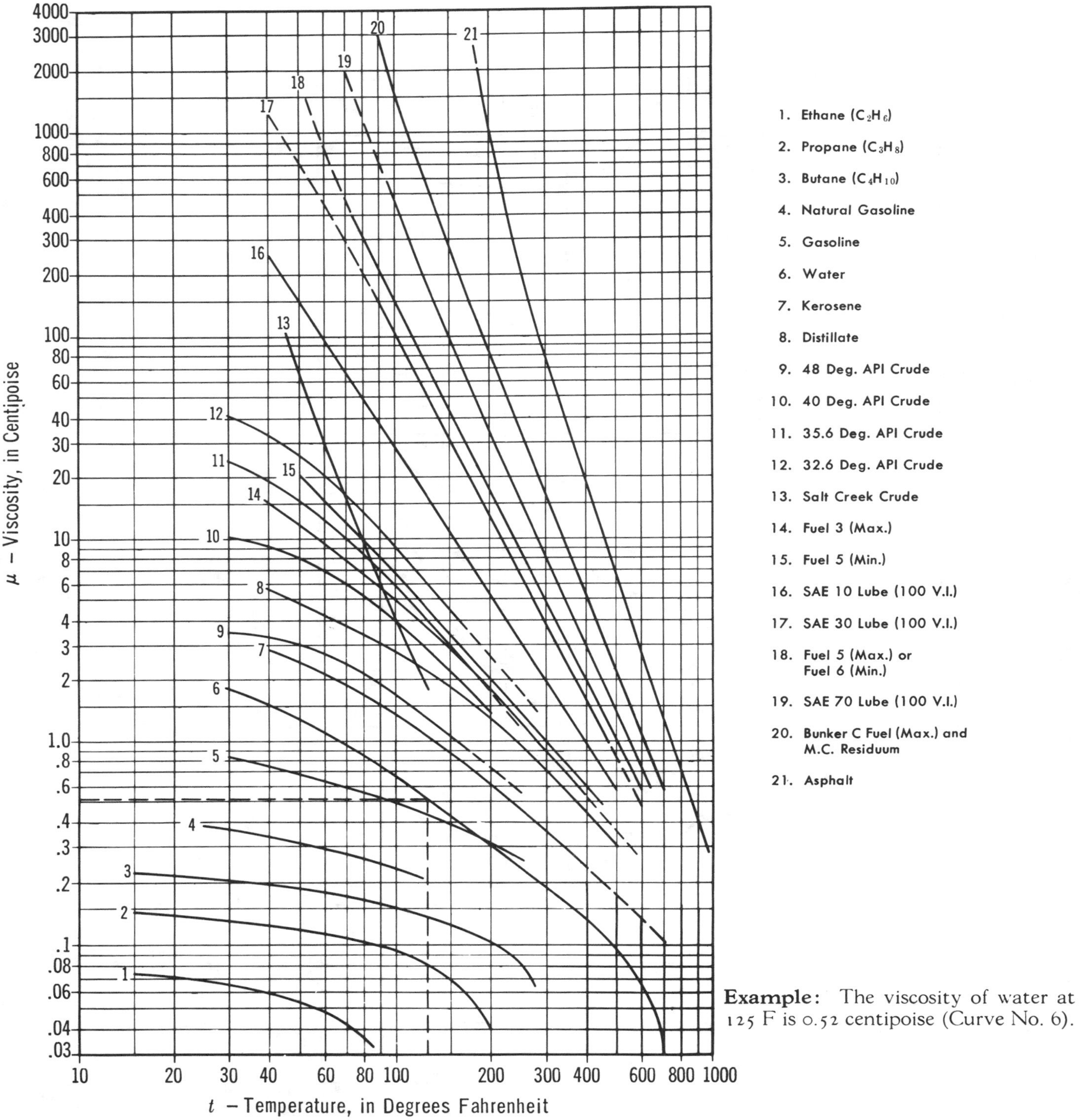

Figure 1.3. Viscosity of water and liquid petroleum products. Courtesy Crane Co. (T.P. No. 410) and by permission of the Oil and Gas Journal.

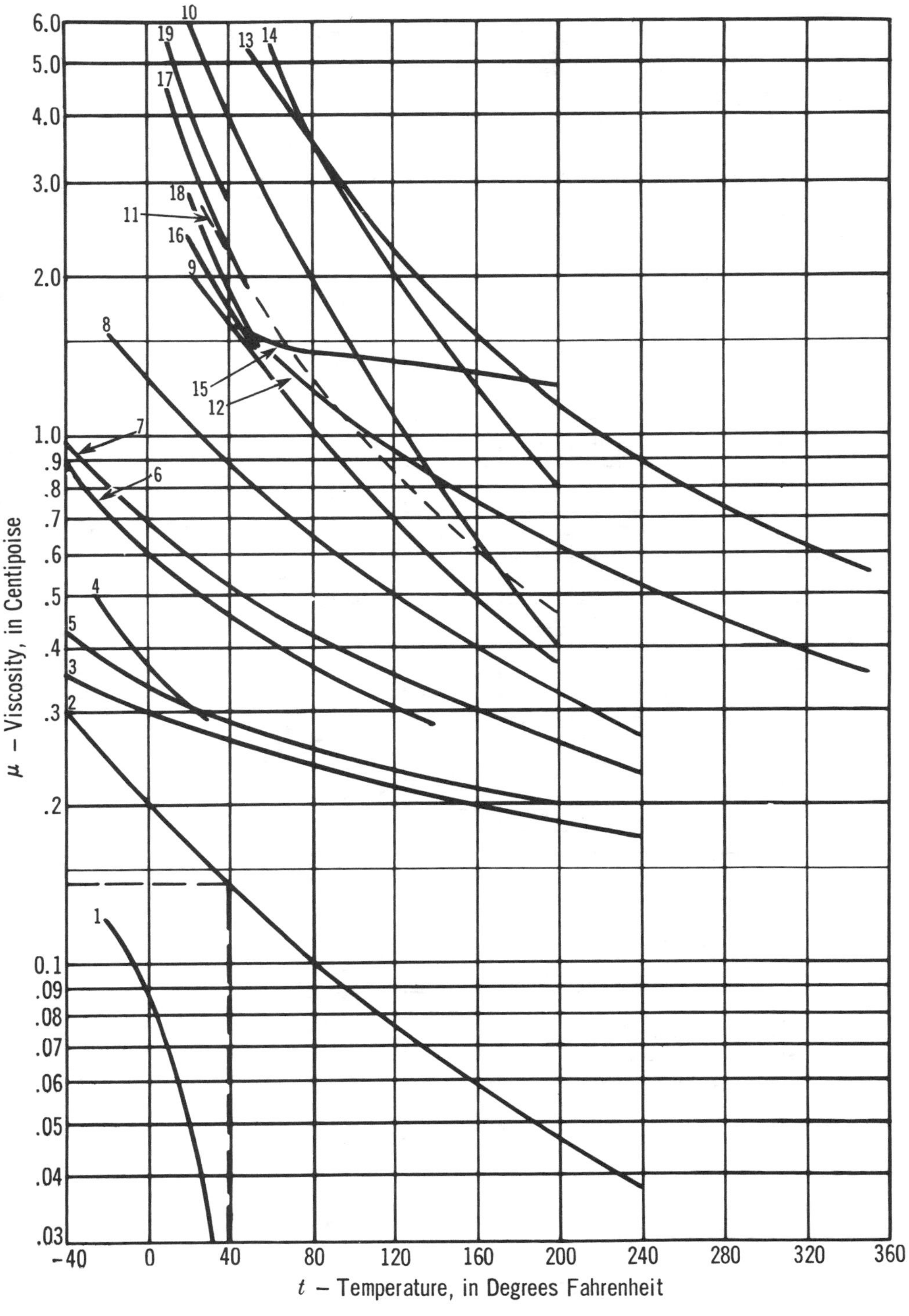

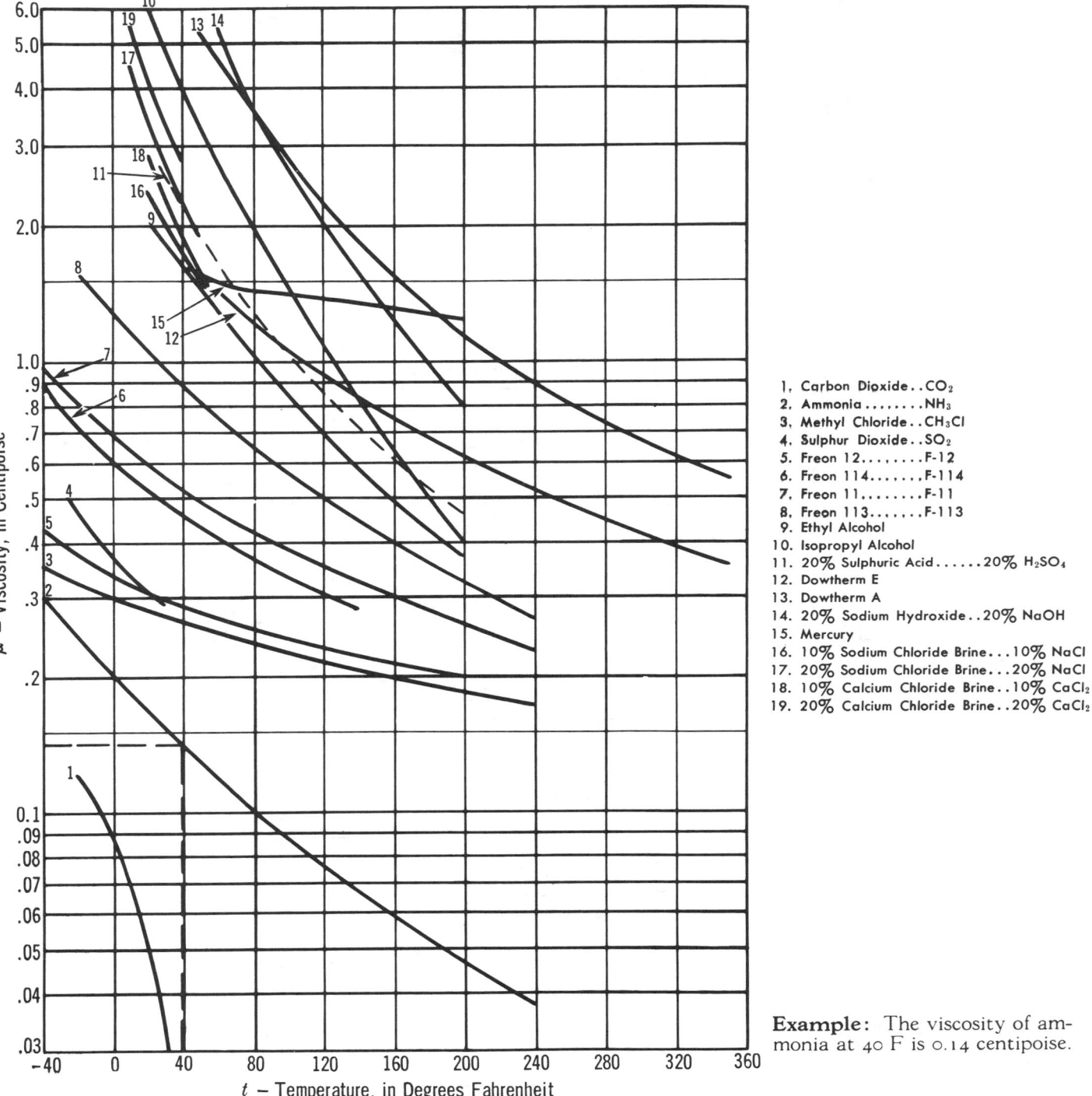

1. Carbon Dioxide..CO_2
2. Ammonia........NH_3
3. Methyl Chloride..CH_3Cl
4. Sulphur Dioxide..SO_2
5. Freon 12........F-12
6. Freon 114.......F-114
7. Freon 11........F-11
8. Freon 113.......F-113
9. Ethyl Alcohol
10. Isopropyl Alcohol
11. 20% Sulphuric Acid......20% H_2SO_4
12. Dowtherm E
13. Dowtherm A
14. 20% Sodium Hydroxide..20% NaOH
15. Mercury
16. 10% Sodium Chloride Brine...10% NaCl
17. 20% Sodium Chloride Brine...20% NaCl
18. 10% Calcium Chloride Brine..10% $CaCl_2$
19. 20% Calcium Chloride Brine..20% $CaCl_2$

Example: The viscosity of ammonia at 40 F is 0.14 centipoise.

Figure 1.4. Viscosity of various liquids. Courtesy Crane Co. (T.P. No. 410).

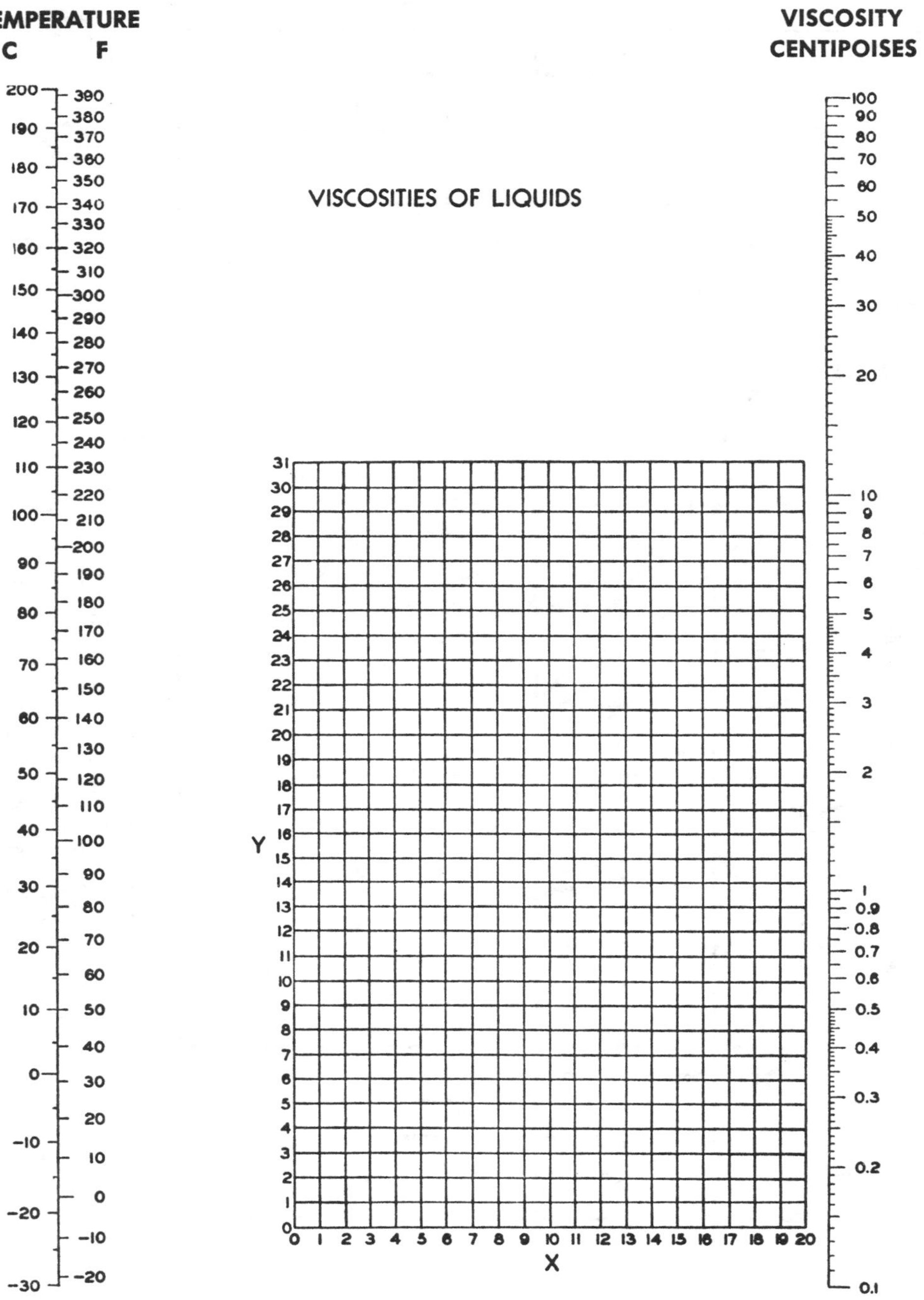

Figure 1.5. Viscosity of liquids. From Perry's Chemical Engineers' Handbook, *4th ed., copyright © 1963 by McGraw-Hill, Inc. Used with permission of McGraw-Hill Book Company.*

known viscosity. This can be done for both liquids and gases by using Figure 1.1 and the relation

$$\mu = \mu_r \, \mu_c \tag{1.4}$$

It is best if the known viscosity is selected at conditions close to those for which the estimated viscosity is to be obtained.

If no viscosity data are available, approximations can be made by employing Figure 1.1 and estimating μ_c with the Licht and Stechert equation

$$\mu_c = 1.416 \times 10^{-4} \frac{(MW)^{1/2} (P_c)^{2/3}}{T_c^{1/6}} \tag{1.5}$$

Density

Density as commonly defined in fluid flow is a measure of the mass at a temperature and pressure that can be placed in a vessel of unit volume.

Table 1.5. Viscosities of Liquids (Coordinates for Use with Figure 1.5)

Liquid	X	Y	Liquid	X	Y
Acetaldehyde	15.2	4.8	Freon–22	17.2	4.7
Acetic acid, 100%	12.1	14.2	Freon–113	12.5	11.4
Acetic acid, 70%	9.5	17.0	Freon–114	14.6	8.3
Acetic anhydride	12.7	12.8	Glycerol, 100%	2.0	30.0
Acetone, 100%	14.5	7.2	Glycerol, 50%	6.9	19.6
Acetone, 35%	7.9	15.0	Heptane	14.1	8.4
Allyl alcohol	10.2	14.3	Hexane	14.7	7.0
Ammonia, 100%	12.6	2.0	Hydrochloric acid, 31.5%	13.0	16.6
Ammonia, 26%	10.1	13.9	Isobutyl alcohol	7.1	18.0
Amyl acetate	11.8	12.5	Isobutyric acid	12.2	14.4
Amyl alcohol	7.5	18.4	Isopropyl alcohol	8.2	16.0
Aniline	8.1	18.7	Kerosene	10.2	16.9
Anisole	12.3	13.5	Linseed oil, raw	7.5	27.2
Arsenic trichloride	13.9	14.5	Mercury	18.4	16.4
Benzene	12.5	10.9	Methanol, 100%	12.4	10.5
Brine, $CaCl_2$, 25%	6.6	15.9	Methanol, 90%	12.3	11.8
Brine, NaCl, 25%	10.2	16.6	Methanol, 40%	7.8	15.5
Bromine	14.2	13.2	Methyl acetate	14.2	8.2
Bromotoluene	20.0	15.9	Methyl chloride	15.0	3.8
Butyl acetate	12.3	11.0	Methyl ethyl ketone	13.9	8.6
Butyl alcohol	8.6	17.2	Naphthalene	7.9	18.1
Butyric acid	12.1	15.3	Nitric acid, 95%	12.8	13.8
Carbon dioxide	11.6	0.3	Nitric acid, 60%	10.8	17.0
Carbon disulfide	16.1	7.5	Nitrobenzene	10.6	16.2
Carbon tetrachloride	12.7	13.1	Nitrotoluene	11.0	17.0
Chlorobenzene	12.3	12.4	Octane	13.7	10.0
Chloroform	14.4	10.2	Octyl alcohol	6.6	21.1
Chlorosulfonic acid	11.2	18.1	Pentachloroethane	10.9	17.3
Chlorotoluene, ortho	13.0	13.3	Pentane	14.9	5.2
Chlorotoluene, meta	13.3	12.5	Phenol	6.9	20.8
Chlorotoluene, para	13.3	12.5	Phosphorus tribromide	13.8	16.7
Cresol, meta	2.5	20.8	Phosphorus trichloride	16.2	10.9
Cyclohexanol	2.9	24.3	Propionic acid	12.8	13.8
Dibromoethane	12.7	15.8	Propyl alcohol	9.1	16.5
Dichloroethane	13.2	12.2	Propyl bromide	14.5	9.6
Dichloromethane	14.6	8.9	Propyl chloride	14.4	7.5
Diethyl oxalate	11.0	16.4	Propyl iodide	14.1	11.6
Dimethyl oxalate	12.3	15.8	Sodium	16.4	13.9
Diphenyl	12.0	18.3	Sodium hydroxide, 50%	3.2	25.8
Dipropyl oxalate	10.3	17.7	Stannic chloride	13.5	12.8
Ethyl acetate	13.7	9.1	Sulfur dioxide	15.2	7.1
Ethyl alcohol, 100%	10.5	13.8	Sulfuric acid, 110%	7.2	27.4
Ethyl alcohol, 95%	9.8	14.3	Sulfuric acid, 98%	7.0	24.8
Ethyl alcohol, 40%	6.5	16.6	Sulfuric acid, 60%	10.2	21.3
Ethyl benzene	13.2	11.5	Sulfuryl chloride	15.2	12.4
Ethyl bromide	14.5	8.1	Tetrachloroethane	11.9	15.7
Ethyl chloride	14.8	6.0	Tetrachloroethylene	14.2	12.7
Ethyl ether	14.5	5.3	Titanium tetrachloride	14.4	12.3
Ethyl formate	14.2	8.4	Toluene	13.7	10.4
Ethyl iodide	14.7	10.3	Trichloroethylene	14.8	10.5
Ethylene glycol	6.0	23.6	Turpentine	11.5	14.9
Formic acid	10.7	15.8	Vinyl acetate	14.0	8.8
Freon–11	14.4	9.0	Water	10.2	13.0
Freon–12	16.8	5.6	Xylene, ortho	13.5	12.1
Freon–21	15.7	7.5	Xylene, meta	13.9	10.6
			Xylene, para	13.9	10.9

Source: *Perry's Chemical Engineers' Handbook*, 4th ed., © 1963 by McGraw-Hill, Inc. Used with permission of McGraw-Hill Book Company.

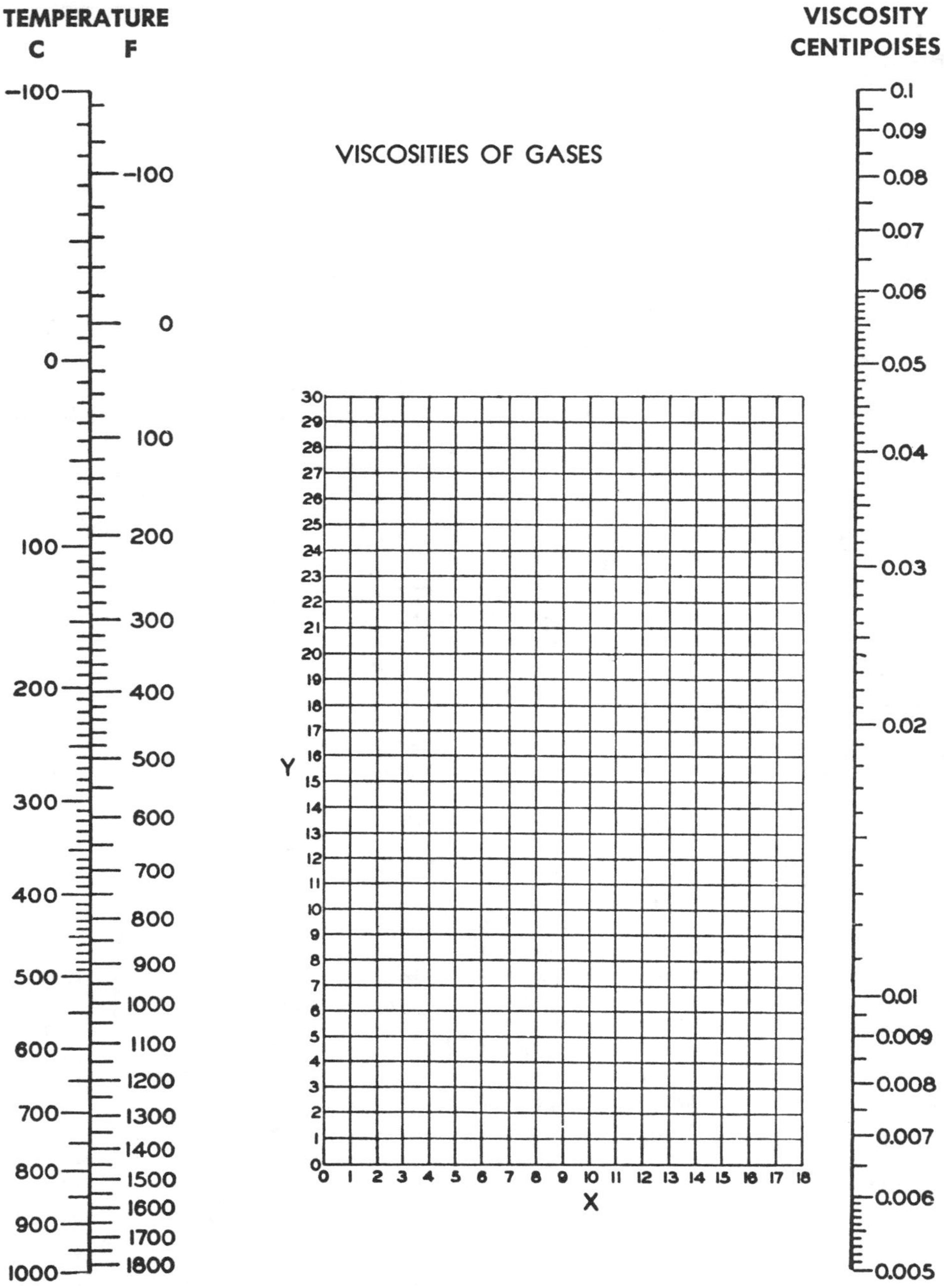

Figure 1.6. Viscosities of gases at atmospheric pressure. From Perry's Chemical Engineers' Handbook, *4th ed., copyright © 1963 by McGraw-Hill, Inc. Used with permission of McGraw-Hill Book Company.*

Liquid Density

Since liquid density data is available from a myriad of sources, this section includes little liquid density data. Two common problems in utilizing liquid density data are conversion of units and extrapolation to desired temperature and/or pressure. Table 1.7 permits conversions between various commonly used density units. Densities can be indicated by a variety of methods.

Specific Volume

Specific volume is the reciprocal of density and is defined as volume per unit mass or weight.

$$\text{density} = 1.0/(\text{specific volume}) \tag{1.6}$$

Liquid Specific Gravity (SG)

Specific gravity is the ratio of the density of a liquid at

Table 1.6. Viscosity of Gases at Atmospheric Pressure (Coordinates for Use with Figure 1.6)

No.	Gas	X	Y
1	Acetic acid	7.7	14.3
2	Acetone	8.9	13.0
3	Acetylene	9.8	14.9
4	Air	11.0	20.0
5	Ammonia	8.4	16.0
6	Argon	10.5	22.4
7	Benzene	8.5	13.2
8	Bromine	8.9	19.2
9	Butene	9.2	13.7
10	Butylene	8.9	13.0
11	Carbon dioxide	9.5	18.7
12	Carbon disulfide	8.0	16.0
13	Carbon monoxide	11.0	20.0
14	Chlorine	9.0	18.4
15	Chloroform	8.9	15.7
16	Cyanogen	9.2	15.2
17	Cyclohexane	9.2	12.0
18	Ethane	9.1	14.5
19	Ethyl acetate	8.5	13.2
20	Ethyl alcohol	9.2	14.2
21	Ethyl chloride	8.5	15.6
22	Ethyl ether	8.9	13.0
23	Ethylene	9.5	15.1
24	Fluorine	7.3	23.8
25	Freon–11	10.6	15.1
26	Freon–12	11.1	16.0
27	Freon–21	10.8	15.3
28	Freon–22	10.1	17.0
29	Freon–113	11.3	14.0
30	Helium	10.9	20.5
31	Hexane	8.6	11.8
32	Hydrogen	11.2	12.4
33	$3H_2 + 1N_2$	11.2	17.2
34	Hydrogen bromide	8.8	20.9
35	Hydrogen chloride	8.8	18.7
36	Hydrogen cyanide	9.8	14.9
37	Hydrogen iodide	9.0	21.3
38	Hydrogen sulfide	8.6	18.0
39	Iodine	9.0	18.4
40	Mercury	5.3	22.9
41	Methane	9.9	15.5
42	Methyl alcohol	8.5	15.6
43	Nitric oxide	10.9	20.5
44	Nitrogen	10.6	20.0
45	Nitrosyl chloride	8.0	17.6
46	Nitrous oxide	8.8	19.0
47	Oxygen	11.0	21.3
48	Pentane	7.0	12.8
49	Propane	9.7	12.9
50	Propyl alcohol	8.4	13.4
51	Propylene	9.0	13.8
52	Sulfur dioxide	9.6	17.0
53	Toluene	8.6	12.4
54	2, 3, 3–Trimethylbutane	9.5	10.5
55	Water	8.0	16.0
56	Xenon	9.3	23.0

Source: *Perry's Chemical Engineers' Handbook*, 4th ed., © 1963 by McGraw-Hill, Inc. Used with permission of McGraw-Hill Book Company.

Table 1.7 Density Conversions

To convert the numerical value of a property expressed in one of the units in the left-hand column of the table to the numerical value expressed in one of the units in the top row of the table, multiply the former value by the factor in the block common to both units.

Units→ ↓	g/cu cm	g/ml	lb/cu in.	lb/cu ft	lb/gal (U.S.)
g/cu cm	1	1.000028	0.036128	62.428	8.3455
g/ml	0.99997	1	0.036126	62.427	8.3452
lb/cu in.	27.680	27.681	1	1,728	231
lb/cu ft	0.016018	0.016019	5.7870×10^{-4}	1	0.13368
lb/gal (U.S.)	0.11983	0.11983	4.3290×10^{-3}	7.4805	1

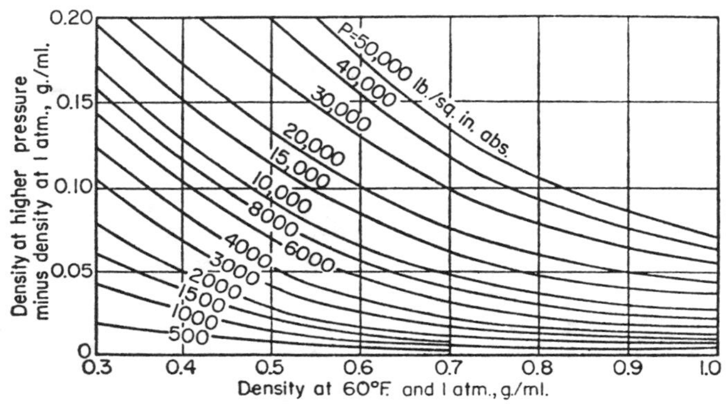

Figure 1.7. Approximate effect of pressure on liquid density. From Perry's Chemical Engineers' Handbook, *4th ed., copyright © 1963 by McGraw-Hill, Inc. Used with permission of McGraw-Hill Book Company.*

temperature T_1 to the density of water at some reference temperature T_2, written as SGT_1/T_2. Reference temperatures of 60°F, 20°C and 4°C are frequently used for water and at these temperatures, its densities at saturation pressure are 62.34, 62.30 and 62.42 pounds mass per cubic foot, respectively.

$$\rho @ T_1 = (SG\ T_1/T_2)\ (\rho \text{ of water} @ T_2) \quad (1.7)$$

Specific Weight (γ)

Specific weight is defined as weight per unit volume. Where gravity is equal to that of the earth (g = 32.17 ft/sec^2 with minor geographical variations) specific weight in lb_f/ft^3 is numerically equal to density in lb_m/ft^3.

$$\gamma = \rho\ (g/g_c) \quad (1.8)$$

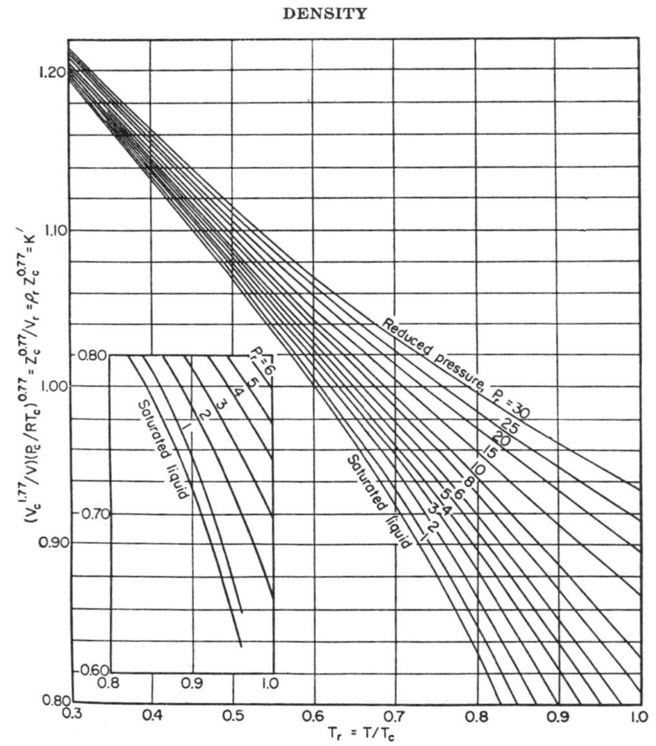

Figure 1.8. Liquid molar volume chart. From Perry's Chemical Engineers' Handbook, *4th ed., copyright © 1963 by McGraw-Hill, Inc. Used with permission of McGraw-Hill Book Company.*

Hydrometer Scales

A number of types of hydrometer scales are used for indicating liquid densities. The three most common are API for oils and two Baumé scales. The relationship between the scales and liquid specific gravities are as follows:

$$SG\ ^{60°}/_{60°}F = \frac{141.5}{131.5 + °\ API} \quad (1.9)$$

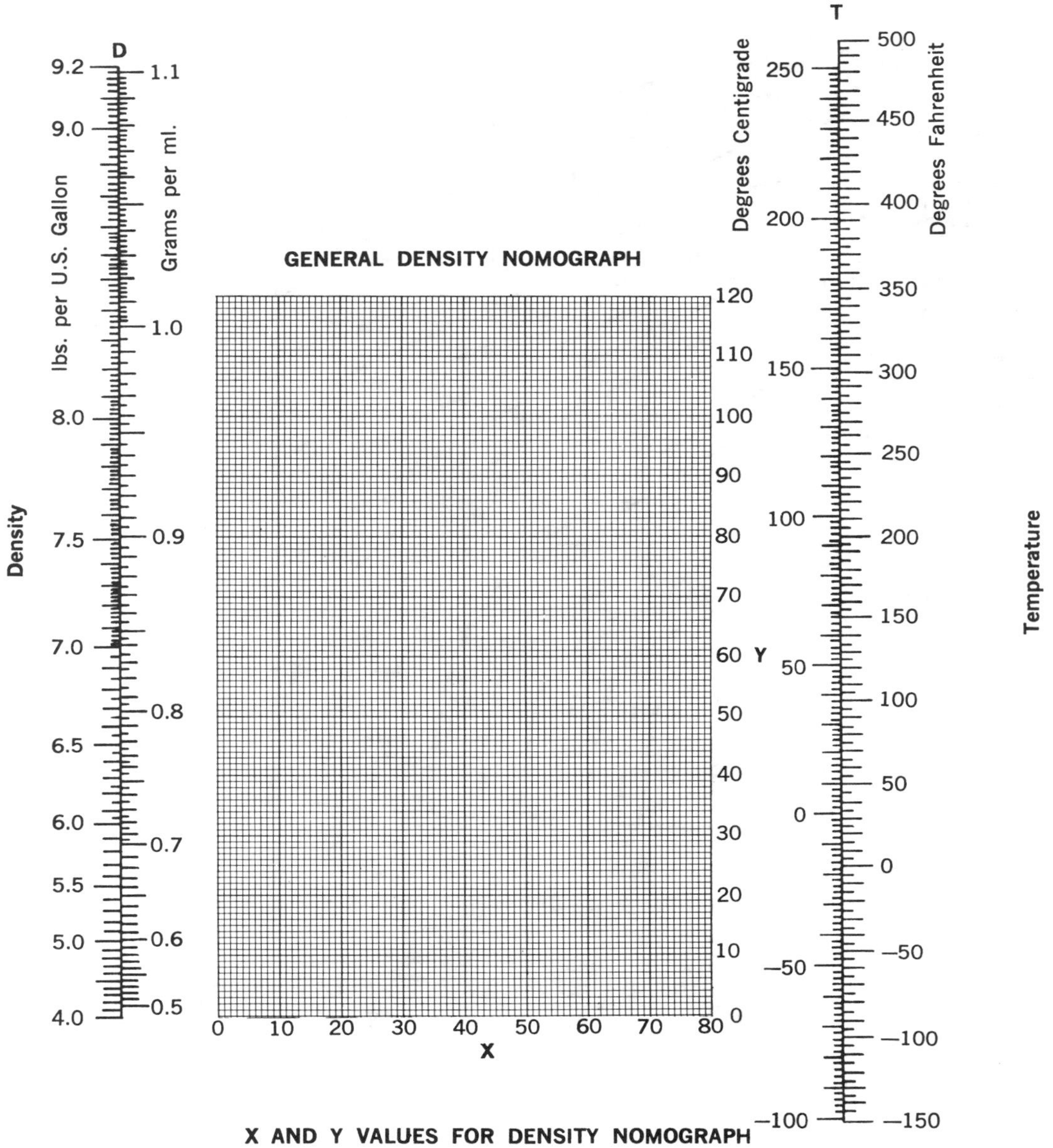

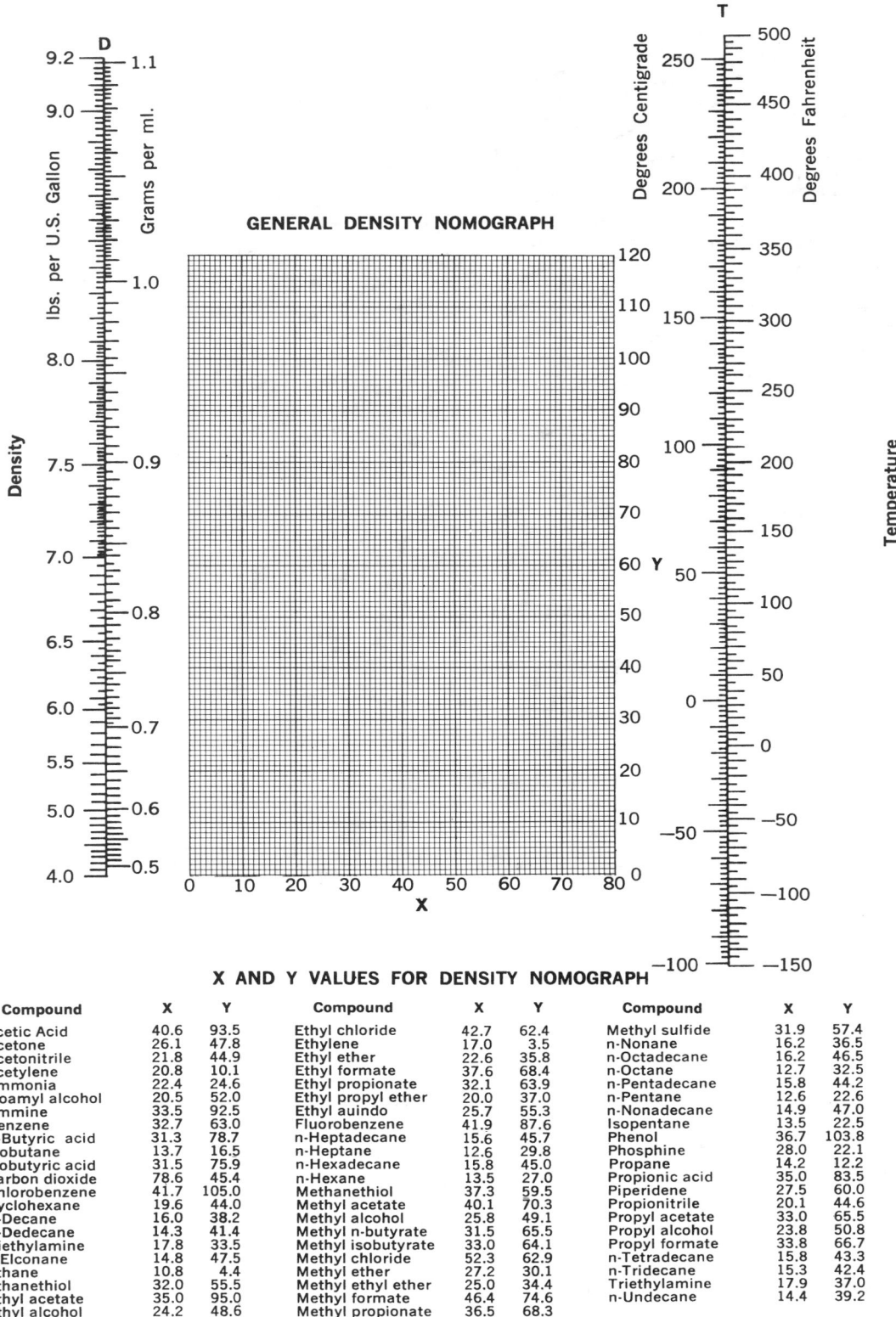

X AND Y VALUES FOR DENSITY NOMOGRAPH

Compound	X	Y	Compound	X	Y	Compound	X	Y
Acetic Acid	40.6	93.5	Ethyl chloride	42.7	62.4	Methyl sulfide	31.9	57.4
Acetone	26.1	47.8	Ethylene	17.0	3.5	n-Nonane	16.2	36.5
Acetonitrile	21.8	44.9	Ethyl ether	22.6	35.8	n-Octadecane	16.2	46.5
Acetylene	20.8	10.1	Ethyl formate	37.6	68.4	n-Octane	12.7	32.5
Ammonia	22.4	24.6	Ethyl propionate	32.1	63.9	n-Pentadecane	15.8	44.2
Isoamyl alcohol	20.5	52.0	Ethyl propyl ether	20.0	37.0	n-Pentane	12.6	22.6
Ammine	33.5	92.5	Ethyl auindo	25.7	55.3	n-Nonadecane	14.9	47.0
Benzene	32.7	63.0	Fluorobenzene	41.9	87.6	Isopentane	13.5	22.5
n-Butyric acid	31.3	78.7	n-Heptadecane	15.6	45.7	Phenol	36.7	103.8
Isobutane	13.7	16.5	n-Heptane	12.6	29.8	Phosphine	28.0	22.1
Isobutyric acid	31.5	75.9	n-Hexadecane	15.8	45.0	Propane	14.2	12.2
Carbon dioxide	78.6	45.4	n-Hexane	13.5	27.0	Propionic acid	35.0	83.5
Chlorobenzene	41.7	105.0	Methanethiol	37.3	59.5	Piperidene	27.5	60.0
Cyclohexane	19.6	44.0	Methyl acetate	40.1	70.3	Propionitrile	20.1	44.6
n-Decane	16.0	38.2	Methyl alcohol	25.8	49.1	Propyl acetate	33.0	65.5
n-Dedecane	14.3	41.4	Methyl n-butyrate	31.5	65.5	Propyl alcohol	23.8	50.8
Diethylamine	17.8	33.5	Methyl isobutyrate	33.0	64.1	Propyl formate	33.8	66.7
n-Elconane	14.8	47.5	Methyl chloride	52.3	62.9	n-Tetradecane	15.8	43.3
Ethane	10.8	4.4	Methyl ether	27.2	30.1	n-Tridecane	15.3	42.4
Ethanethiol	32.0	55.5	Methyl ethyl ether	25.0	34.4	Triethylamine	17.9	37.0
Ethyl acetate	35.0	95.0	Methyl formate	46.4	74.6	n-Undecane	14.4	39.2
Ethyl alcohol	24.2	48.6	Methyl propionate	36.5	68.3			

Figure 1.9. General liquid density nomograph. Reprinted by permission of Tubular Exchanger Manufacturers Association (TEMA).

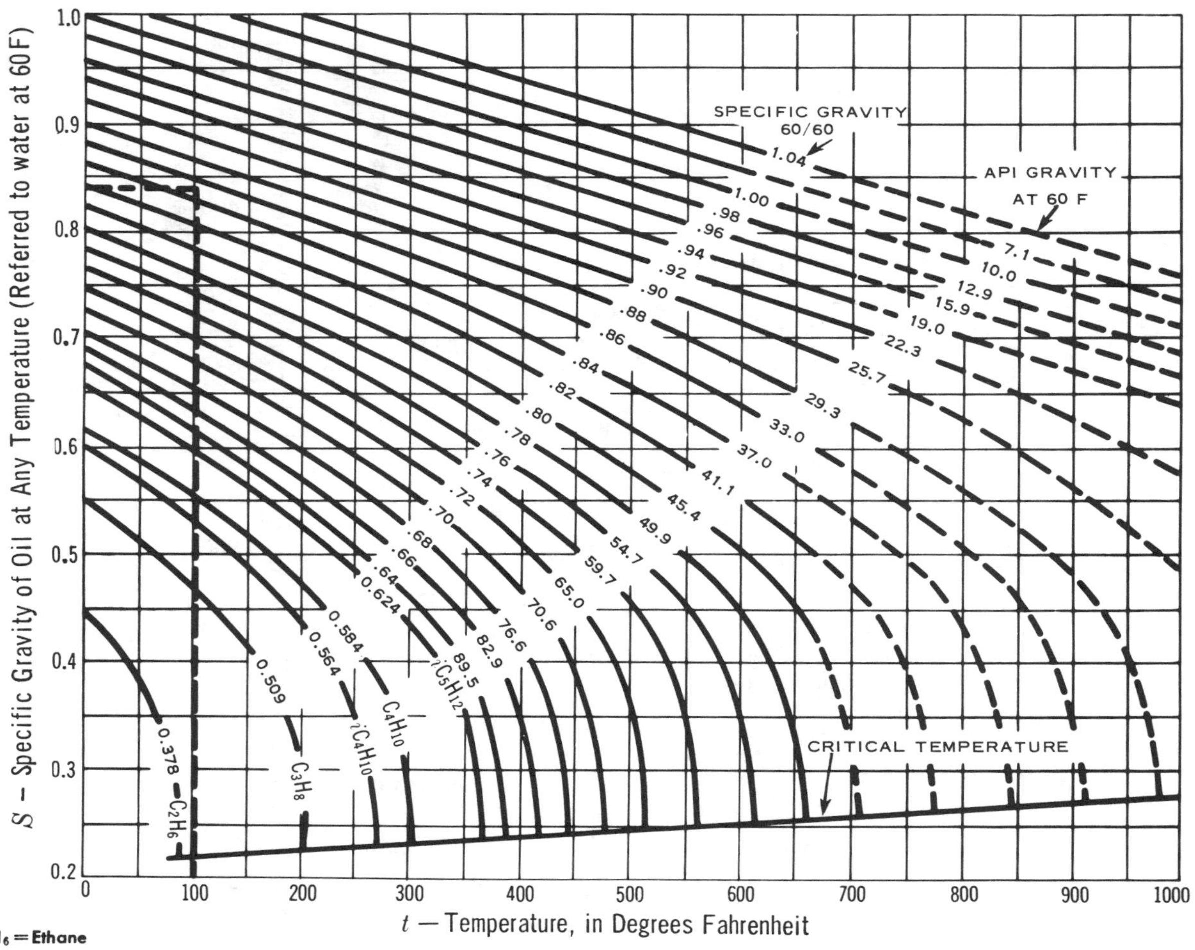

C_2H_6 = Ethane
C_3H_8 = Propane iC_4H_{10} = Isobutane
C_4H_{10} = Butane iC_5H_{12} = Isopentane

Example: The specific gravity of an oil at 60 F is 0.85 The specific gravity at 100 F = 0.83.

To find the weight density of a petroleum oil at its flowing temperature when the specific gravity at 60 F/60 F is known, multiply the specific gravity of the oil at flowing temperature (see chart above) by 62.4, the density of water at 60 F.

Figure 1.10. Specific gravity/temperature relationship for petroleum oils. Courtesy Crane Co. (T.P. No. 410).

For liquids more dense than water

$$SG\ ^{60°}/_{60°}\text{F} = \frac{145}{145 - °\text{Baumé}} \tag{1.10}$$

For liquids less dense than water

$$SG\ ^{60°}/_{60°}\text{F} = \frac{140}{130 + °\text{Baumé}} \tag{1.11}$$

The weak influence of pressure on liquid density is approximately shown by Figure 1.7. Lu's generalized liquid molar volume chart (Figure 1.8) may be used in two methods for estimating variation of liquid density with both temperature and pressure if $Z_c \geqslant 0.24$. One method is if T_c, P_c and $\overline{V}_c$ or Z_c is known for a given fluid, a value of K' from Figure 1.8 may be obtained at the T_r and P_r of interest and density calculated as

$$\rho\ @\ T_r P_r = \rho_c\ K'/Z_c^{0.77} \tag{1.12}$$

where

$$\rho_c = (MW)/\overline{V}_c \tag{1.13}$$

and the relationship between the critical properties are

$$P_c\ \overline{V}_c = RZ_c\ T_c \tag{1.14}$$

Table 1.8. Weight Density and Specific Gravity of Various Liquids

Liquid	Temp. t Deg. Fahr.	Weight Density ρ Lbs. per Cu. Ft.	Specific Gravity S	Liquid	Temp. t Deg. Fahr.	Weight Density ρ Lbs. per Cu. Ft.	Specific Gravity S
Acetone	60	49.4	0.792	Mercury	20	849.74	...
Ammonia, Saturated	10	40.9	...	Mercury	40	848.03	...
Benzene	32	56.1	...	Mercury	60	846.32	13.570
Brine, 10% Ca Cl	32	68.05	...	Mercury	80	844.62	...
Brine, 10% Na Cl	32	67.24	...	Mercury	100	842.93	...
Bunkers C Fuel Max.	60	63.25	1.014	Milk	...	†	...
Carbon Disulphide	32	80.6	...	Olive Oil	59	57.3	0.919
Distillate	60	52.99	0.850	Pentane	59	38.9	0.624
Fuel 3 Max.	60	56.02	0.898	SAE 10 Lube‡	60	54.64	0.876
Fuel 5 Min.	60	60.23	0.966	SAE 30 Lube‡	60	56.02	0.898
Fuel 5 Max.	60	61.92	0.993	SAE 70 Lube‡	60	57.12	0.916
Fuel 6 Min.	60	61.92	0.993	Salt Creek Crude	60	52.56	0.843
Gasoline	60	46.81	0.751	32.6° API Crude	60	53.77	0.862
Gasoline, Natural	60	42.42	0.680	35.6° API Crude	60	52.81	0.847
Kerosene	60	50.85	0.815	40° API Crude	60	51.45	0.825
M. C. Residuum	60	58.32	0.935	48° API Crude	60	49.16	0.788

*Liquid at 60 F referred to water at 60 F.

†Milk has a weight density of 64.2 to 64.6.

‡100 Viscosity Index.

Source: Courtesy Crane Co. (T.P. No. 410).

A second method is if a value of ρ is known for a given T_r and P_r, then Figure 1.8 can be used to obtain values of ρ at other conditions by obtaining K_1', at the known density ρ_1, and K_2' at the desired conditions and using the relationship

$$\rho_2 = \rho_1 \,(K_1'/K_2') \tag{1.15}$$

Best results are obtained by the second method, and greatest deviations are in the critical region.

Densities of ideal liquid mixtures can be estimated by:

$$\rho_{mix} = \Sigma_i \,(\rho_i \,\phi_i) \tag{1.16}$$

Equation 1.16 applies only if there is no interaction between any of the mixture components, but it often works satisfactorily in cases of mild interaction.

Gas Density

Gas density data are often presented as lines of constant specific volume on Mollier diagrams or as tabular data for specific conditions, such as "standard temperature and pressure" or saturation pressure at varying temperatures. Since a variety of "standard" conditions are in common use, one should always determine what "standard" conditions are when using such data.

Gas densities may be calculated by means of the gas equation

$$\rho = \frac{P(MW)}{RTZ} \tag{1.17}$$

Care must be taken to use absolute temperature and pressure units as shown below for the units in this section.

$$\rho,\ \text{lb}_m/\text{ft}^3 = \frac{(P,\ \text{psia})\,(MW)}{(R = 10.73)\,(T,\ ^\circ\text{R})\,(Z,\ \text{dim.})} \tag{1.18}$$

For many problems the conditions are often such that the compressibility factor, Z, can be taken as 1.0 with negligible error. An ideal gas is by definition one for which $Z = 1.0$. High temperature and low pressure favor ideal gas behavior. A common industrial practice is to assume $Z = 1.0$ for all gases at standard conditions. This is especially true for material balance calculations; then all standard gas volumes are additive and the standard molar volumes of all gases are the same.

There are many correlations for estimating values of Z. Figure 1.11 shows one that expresses Z as a function of reduced temperature and reduced pressure. It may be used for both pure and mixed component gases.

Gas mixture densities are calculated by determining mixture molecular weight, reduced temperature and reduced pressure and then using these properties as if for a pure gas. The basic equations for doing so are as follows:

$$y_i = \text{mole fraction of component } i \tag{1.19}$$

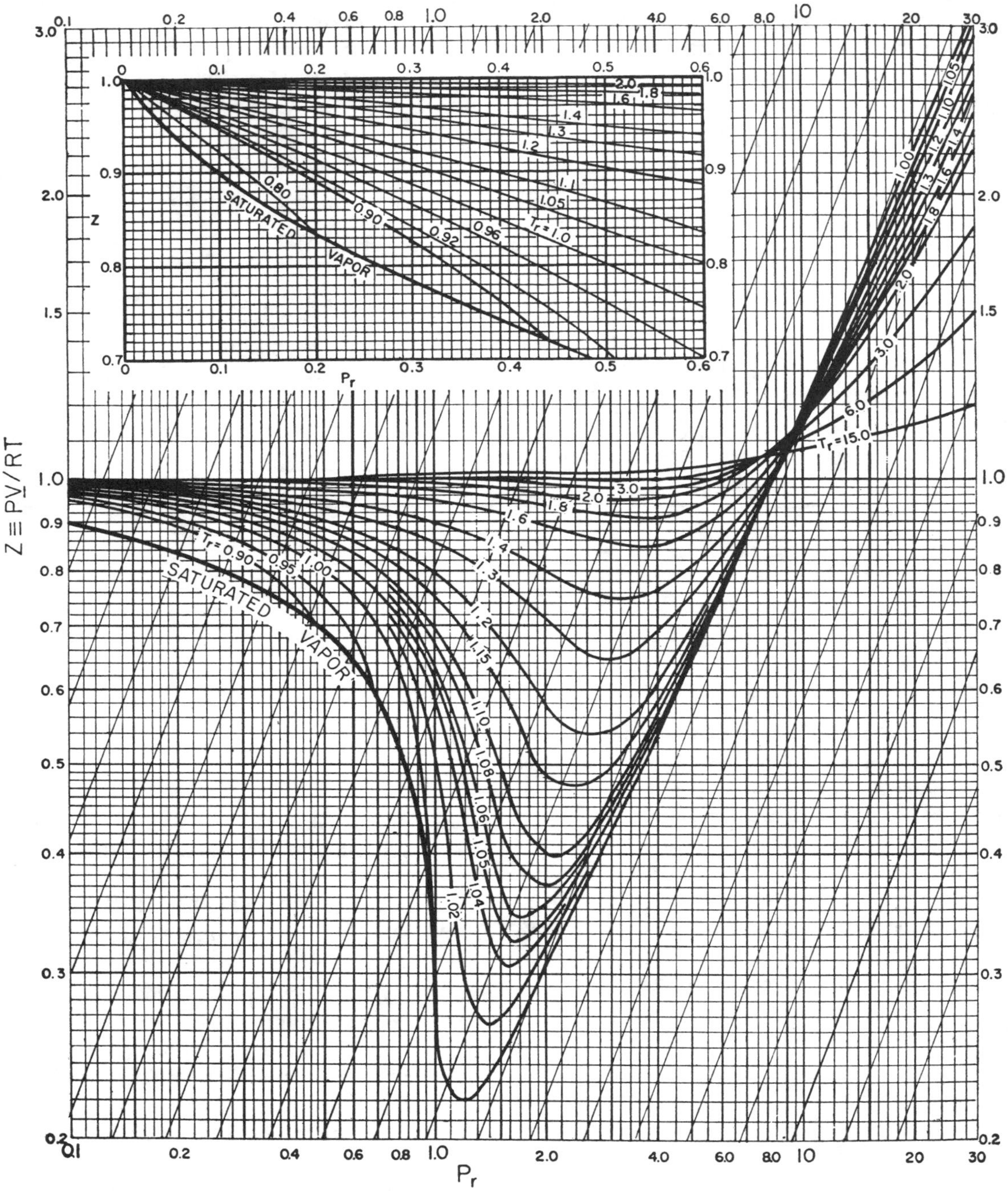

Figure 1.11. Generalized compressibility charts. From Perry's Chemical Engineers' Handbook, *4th ed., copyright © 1963 by McGraw-Hill, Inc. Used with permission of McGraw-Hill Book Company.*

Table 1.9. International Atomic Weights Based on Carbon-12

	Symbol	Atomic Number	Atomic Weight		Symbol	Atomic Number	Atomic Weight
Actinium	Ac	89	[227] *	Mercury	Hg	80	200.59
Aluminum	Al	13	26.9815	Molybdenum	Mo	42	95.94
Americium	Am	95	[243] *	Neodymium	Nd	60	144.24
Antimony	Sb	51	121.75	Neon	Ne	10	20.183
Argon	Ar	18	39.948	Neptunium	Np	93	[237] *
Arsenic	As	33	74.9216	Nickel	Ni	28	58.71
Astatine	At	85	[210] *	Niobium	Nb	41	92.906
Barium	Ba	56	137.34	Nitrogen	N	7	14.0067
Berkelium	Bk	97	[249] *	Nobelium	No	102	[254] *
Beryllium	Be	4	9.0122	Osmium	Os	76	190.2
Bismuth	Bi	83	208.980	Oxygen	O	8	15.9994 [a]
Boron	B	5	10.811 [a]	Palladium	Pd	46	106.4
Bromine	Br	35	79.909 [b]	Phosphorus	P	15	30.9738
Cadmium	Cd	48	112.40	Platinum	Pt	78	195.09
Calcium	Ca	20	40.08	Plutonium	Pu	94	[242] *
Californium	Cf	98	[251] *	Polonium	Po	84	[210] *
Carbon	C	6	12.01115 [a]	Potassium	K	19	39.102
Cerium	Ce	58	140.12	Praseodymium	Pr	59	140.907
Cesium	Cs	55	132.905	Promethium	Pm	61	[147] *
Chlorine	Cl	17	35.453 [b]	Protactinium	Pa	91	[231] *
Chromium	Cr	24	51.996 [b]	Radium	Ra	88	[226] *
Cobalt	Co	27	58.9332	Radon	Rn	86	[222] *
Copper	Cu	29	63.54	Rhenium	Re	75	186.2
Curium	Cm	96	[247] *	Rhodium	Rh	45	102.905
Dysprosium	Dy	66	162.50	Rubidium	Rb	37	85.47
Einsteinium	Es	99	[254] *	Ruthenium	Ru	44	101.07
Erbium	Er	68	167.26	Samarium	Sm	62	150.35
Europium	Eu	63	151.96	Scandium	Sc	21	44.956
Fermium	Fm	100	[253] *	Selenium	Se	34	78.96
Fluorine	F	9	18.9984	Silicon	Si	14	28.086 [a]
Francium	Fr	87	[223] *	Silver	Ag	47	107.870 [b]
Gadolinium	Gd	64	157.25	Sodium	Na	11	22.9898
Gallium	Ga	31	69.72	Strontium	Sr	38	87.62
Germanium	Ge	32	72.59	Sulfur	S	16	32.064 [a]
Gold	Au	79	196.967	Tantalum	Ta	73	180.948
Hafnium	Hf	72	178.49	Technetium	Tc	43	[99] *
Helium	He	2	4.0026	Tellurium	Te	52	127.60
Holmium	Ho	67	164.930	Terbium	Tb	65	158.924
Hydrogen	H	1	1.00797 [a]	Thallium	Tl	81	204.37
Indium	In	49	114.82	Thorium	Th	90	232.038
Iodine	I	53	126.9044	Thulium	Tm	69	168.934
Iridium	Ir	77	192.2	Tin	Sn	50	118.69
Iron	Fe	26	55.847 [b]	Titanium	Ti	22	47.90
Krypton	Kr	36	83.80	Tungsten	W	74	183.85
Lanthanum	La	57	138.91	Uranium	U	92	238.03
Lead	Pb	82	207.19	Vanadium	V	23	50.942
Lithium	Li	3	6.939	Xenon	Xe	54	131.30
Lutetium	Lu	71	174.97	Ytterbium	Yb	70	173.04
Magnesium	Mg	12	24.312	Yttrium	Y	39	88.905
Manganese	Mn	25	54.9380	Zinc	Zn	30	65.37
Mendelevium	Md	101	[256] *	Zirconium	Zr	40	91.22

* Value in brackets denotes the mass number of the isotope of longest known half life (or a better known one for Bk, Cf, Po, Pm, and Tc).

[a] Atomic weight varies because of natural variation in isotopic composition: B, ±0.003; C, ±0.00005; H, ±0.00001; O, ±0.0001; Si, ±0.001; S, ±0.003.

[b] Atomic weight is believed to have following experimental uncertainty: Br, ±0.002; Cl, ±0.001; Cr, ±0.001; Fe, ±0.003; Ag, ±0.003. For other elements, the last digit given for the atomic weight is believed reliable to ±0.5. Lawrencium, Lw, has been proposed as the name for element No. 103, nuclidic mass about 257.

Source: Perry's Chemical Engineers' Handbook, 4th ed., copyright © 1963 by McGraw-Hill, Inc. Used with permission of McGraw-Hill Book Company.

$$(MW)_{mix} = \Sigma_i[y(MW)]_i \quad (1.20)$$

$$(T_c)_{mix} = \Sigma_i\,(yT_c)_i \quad (1.21)$$

$$(P_c)_{mix} = \Sigma_i\,(yP_c)_i \quad (1.22)$$

Example Problem 1.1
Viscosity of Liquid Mixtures

Estimate the viscosity @ 180°F of a liquid mixture that is 30 weight per cent glycerol (MW = 92) and 70% ethylene glycol (MW = 62).
Use Equation 1.2

$$\mu_{mix} = (\Sigma_i x_i \mu_i^{1/3})^3 \quad (1.2)$$

which requires conversion of weight % to mole fraction (x).

	A	B	C = B/A	$x = \frac{C_i}{\Sigma C_i}$
Compound	MW	Wt.%	Moles/100 lbs.	Mole Fract.
Glycerol	92	30	0.33	0.23
Ethylene glycol	62	70	1.13	0.77
		100	Σ = 1.46	1.00

From Table 1.5 obtain coordinates (X,Y) for use in estimating viscosities of each compound at 180°F with Figure 1.5.

Compound	Viscosity Coordinates X	Viscosity Coordinates Y	μ Vis., CP @ 180°F	$\mu^{1/3}$	$x\mu^{1/3}$
Glycerol	2.0	30.0	33	3.21	0.74
Ethylene glycol	6.0	23.6	3.3	1.49	1.15
					Σ=1.89

$$\mu_{mix} = (\Sigma x_i \mu_i^{1/3})^3 = (1.89)^3 = 6.75 \text{ cp @ } 180°\text{F}.$$

Example Problem 1.2
Viscosity of a Gas Mixture

Estimate the viscosity at 100°F and 3.0 psig of a gas mixture that is composed of 79 mole % nitrogen (MW = 28) and 21% oxygen (MW = 32).

Procedure will be to estimate viscosity of each compound using Figure 1.6 and then combine the individual viscosities into a mixture viscosity using

$$\mu_{mix} = \frac{\Sigma_i\,(y_i \mu_i\,(MW)_i)}{\Sigma_i\,(y_i\,(MW)_i)} \quad (1.3)$$

Although Figure 1.6 is for viscosities at atmospheric pressure, it is also valid at low pressures in general. Viscosity coordinates for use with Figure 1.6 are obtained from Table 1.6.

Item	N_2	O_2	Sum
MW = mole wt.	28	32	
y = mole fract.	0.79	0.21	
$y_i\,(MW)_i$ –	22.12	6.72	28.8
Viscosity Coordinates			
X	10.6	11.0	
Y	20.0	21.3	
μ = viscosity, CP @ 100°F	0.0182	0.0211	
$y\,(MW)\mu$	0.403	0.142	0.545

$$\mu_{mix} = \frac{\Sigma_i y_i\,(MW)_i\,\mu_i}{\Sigma y_i\,(MW)_i} = \frac{0.545}{28.8} = 0.0189$$

Example Problem 1.3
Extrapolation of Viscosity Data Using Figure 1.1

Estimate the viscosity of nitrogen at 200°F and 3,000 psig, knowing its viscosity is 0.0182 cp. at 100°F and one atmosphere pressure. The critical temperature of nitrogen is −232.6°F and critical pressure is 492 psia.

1. Calculate reduced temperature and pressure at the conditions for the known viscosity.

$$T_r = \frac{T}{T_c} = \frac{100 + 460}{-232.6 + 460} = \frac{560°\text{R}}{227.4°\text{R}} = 2.462$$

$$P_r = P/P_c = \frac{14.7}{492} = 0.0299$$

2. Obtain μ_r at known conditions using Figure 1.1.

$$\mu_r = 1.02$$

3. Calculate μ_c using Equation 1.4.

$$\mu_c = \frac{\mu}{\mu_r} = \frac{0.0182}{1.02} = 0.0178$$

4. Calculate reduced temperature and pressure at the conditions for which the viscosity is desired.

$$T_r = T/T_c = \frac{200 + 460}{227.4} = 2.90$$

$$P_r = P/P_c = \frac{3{,}000 + 14.7}{492} = 6.13$$

5. Obtain μ_r at above condition using Figure 1.1.

$\mu_r = 1.39$

6. Estimate the desired viscosity using Equation 1.4.

$\mu = \mu_r\mu_c = 1.39\ (0.0178) = 0.0247$ cp.

Example Problem 1.4
Estimating Viscosities Knowing *MW*, T_c, P_c

Estimate the viscosity of nitrogen at 200°F and 3,000 psig knowing $MW = 28$, $T_c = 227.4$°R, $P_c = 492$ psia.

1. Estimate viscosity at critical point using Equation 1.5

$$\mu_c = 1.416 \times 10^{-4}\ \frac{(MW)^{1/2}\ (P_c)^{2/3}}{T_c^{\ 1/6}} = 0.0189 \text{ cp.}$$

(Compare with $\mu_c = 0.0178$ for nitrogen as estimated in example problem 1.3).

2. Calculate reduced temperature and pressure at the conditions for which the viscosity is to be estimated.

$$T_r = T/T_c = \frac{200 + 460}{227.4} = 2.90$$

$$P_r = P/P_c = \frac{3{,}000 + 14.7}{227.4} = 6.13$$

3. Obtain μ_r at above condition using Figure 1.1

$\mu_r = 1.39$

4. Estimate viscosity using Equation 1.4

$\mu = \mu_r\mu_c = 1.39(0.0189) = 0.0263$ cp.

Example Problem 1.5
Estimating Liquid Densities Knowing P_c, T_c, and V_c or Z_c

Estimate the density of normal butane ($MW = 58.12$) at 150°F and its vapor pressure at that temperature. The critical properties of n-butane are

$P_c = 550.7$ psia, $T_c = 765.6$°R, and $\overline{V}_c = 4.08$ cu.ft/lb-mole.

1. Calculate Z_c using Equation 1.14

$$Z_c = \frac{P_c\overline{V}_c}{RT_c} = \frac{(550.7)(4.08)}{(10.73)(765.6)} = 0.274$$

2. Calculate ρ_c using Equation 1.13

$\rho_c = (MW)/\overline{V}_c = 58.12/4.08 = 14.25\ \text{lb}_m/\text{ft}^3$

3. Calculate reduced temperature

$$T_r = T/T_c = \frac{150 + 460}{765.6} = 0.797$$

4. Read K' from Figure 1.8 for $T_r = 0.797$ using saturated liquid line.

$K' = 0.833$

5. Calculate desired density using Equation 1.12

$\rho = \rho_c K'/Zc^{0.77} = 32.2\ \text{lb}_m/\text{ft}^3$

Example Problem 1.6
Density of a Nonideal Gas Mixture

Estimate the density at 110°F and 200 psig of a gas mixture composed as follows:

Compound	y mole fract.	*MW* mole wt.	T_c critical temp. °R	P_c critical pressure psia
methane	0.05	16.0	344	673
ethane	0.28	30.1	550	710
propane	0.67	44.1	666	617

$$(MW)_{\text{mix}} = \Sigma_i(y(MW))_i \qquad (1.20)$$
$$= 0.05(16.0) + 0.28(30.1) + 0.67(44.1) = 38.8$$

$$(T_c)_{\text{mix}} = \Sigma_i(yT_c)_i \qquad (1.21)$$
$$= 0.05(344) + 0.28(550) + 0.67(666) = 617°\text{R}$$

$$(P_c)_{\text{mix}} = \Sigma_i(yP_c)_i \qquad (1.22)$$
$$= 0.05(673) + 0.28(710) + 0.67(617) = 646 \text{ psia}$$

$$T_r = T/T_c = \frac{110 + 460}{617} = 0.924$$

$$P_r = P/P_c = \frac{200 + 14.7}{646} = 0.332$$

From Figure 1.11, $Z = 0.830$ at $T_r = 0.924, P_r = 0.332$

$$\rho = \frac{(P_1 \text{ psia})(MW)}{10.73\ TZ} = \frac{(214.7)(38.8)}{10.73(570)(0.830)} \quad (1.18)$$
$$= 1.64 \text{ lb./ft.}^3$$

The Nature of Flow in Conduits and The Friction Factor

Two flow patterns are exhibited by Newtonian fluids flowing in closed ducts running full: laminar and turbulent. A third term, transitional flow, is used when flow occurs near the critical Reynolds number and flow pattern is difficult to predict. For pipes of circular cross section, laminar flow normally exists for Reynolds numbers between zero and the critical Reynolds number of approximately 2,000 to 3,000, where transition to turbulent flow occurs. Knowing the distinguishing characteristics of the two flow patterns is important for understanding how mixing, heat transfer and other transport phenomena are influenced by the two flow patterns.

Laminar Flow

The most important characteristic of laminar flow is the absence of cross mixing of fluid across the cross section of the duct, i.e., flow streamlines remain distinct and are parallel for uniform flow. Thus it is sometimes referred to as streamline flow. It is also called viscous flow because it is often associated with viscous fluids in practical applications. Since there is no cross mixing, a stream of dye injected into a laminar flow pattern does not blend into the other fluid except by the slow process of diffusion. Thus, it is relatively easy for solids to settle and fluids to stratify. Also, in laminar flow heat transfer is relatively much slower than turbulent flow heat transfer.

Turbulent Flow

Turbulent flow is characterized by a flow pattern that involves intermixing of fluid as it flows. The greater the Reynolds number above the critical, the more turbulent the flow and the more rapid the mixing.

The Friction Factor

The friction factor (f) is an empirically determined dimensionless value used in estimating pressure drops or head losses occurring for fluid flow in ducts. Experimentally it is determined by flowing at a known velocity a fluid of known density and viscosity through a pipe of known roughness, diameter or hydraulic radius, and length; measuring the head loss that occurs and then calculating f with the equation

$$f = \frac{2h_f Dg}{LV^2} \quad (1.23)$$

By conducting a large number of these experiments the plot of friction factor versus Reynolds number for parameters of relative pipe wall roughness shown in Figure 1.12 was developed. It is generally considered to be accurate to within 10% and may be used for estimating friction factors from values of Reynolds number (R_e) and relative roughness (ϵ/D). It should be noted that there are two types of friction factors commonly found in the literature: Fanning and Moody. The Moody friction factor is the one used in this section. The two are interchangeable except that Moody friction factor values are four times larger than Fanning values obtained for identical conditions.

Reynolds number is given by:

$$R_e = 1488 \frac{\rho VD}{\mu} \quad \text{(general)}$$

$$= 0.5263 \frac{W}{\mu D} \quad \text{(round pipe only)} \quad (1.24)$$

Relative roughness is the ratio of absolute pipe wall roughness to pipe diameter.

$$\text{Relative Roughness} = \epsilon/D \quad (1.25)$$

For noncircular ducts four times the hydraulic radius ($4R_h$) is substituted for diameter.

$$\text{Relative Roughness} = \epsilon/(4R_h) \quad (1.26)$$

The influence of several flow variables on the friction factor is apparent from Figure 1.12. For laminar flow the function factor is a function of Reynolds number:

$$f, \text{ Laminar Flow} = 64/R_e \quad (1.27)$$

Combining Equations 1.24 and 1.27 yields

$$f, \text{ Laminar Flow} = 0.0430 \frac{\mu}{\rho VD} \quad \text{(general)}$$

$$= \frac{121.6\ \mu D}{W} \quad \text{(round pipe only)} \quad (1.28)$$

which shows how the friction factor is influenced by several variables.

For turbulent flow, f is a function of both R_e and ϵ/D until for fully developed turbulent flow (large values of R_e) f is a

Material	ϵ, Feet
Riveted Steel	0.003 - 0.03
Concrete	0.001 - 0.01
Wood Stave	0.0006 - 0.003
Cast Iron	0.00085
Galvanized Iron	0.0005
Asphalted Cast Iron	0.0004
Commercial Steel or Wrought Iron	0.00015
Drawn Tubing	0.000005

Table 1.10 Values of Absolute Roughness For Various Materials of Construction

function of ϵ/D only. The turbulent region of Figure 1.12 is accurrately represented by the Colebrook equation:

$$\frac{1}{\sqrt{f}} = -2 \log_{10} \left[\frac{\epsilon}{3.7D} + \frac{2.51}{R_e\sqrt{f}}\right] \quad (1.29)$$

Table 1.10 has values of absolute pipe roughness for a variety of materials of construction.

Bernoulli's Equation

Bernoulli's equation expresses the law of conservation of energy as it applies in fluid flow. The following equations do not rigorously express conservation of energy because they assume negligible heat effects due to friction and no extraneous energy effects, such as chemical reactions. Also, flow is assumed to be steady and one dimensional to permit direct application in the most common fluid flow problems of the chemical processing industries, the flow of incompressible fluids of moderate or low viscosities in conduits. The two equations differ only in the units of the additive terms. For term units of feet of flowing fluid

$$\frac{V_1^2}{2g} + \frac{114P_1}{\gamma} + E_1 + (TDH) = \frac{V_2^2}{2g} = \frac{114P_2}{\gamma} + E_2 + h_L \quad (1.30)$$

For term units of pounds per square inch the above equation is multiplied by $\gamma/144$.

$$\frac{\gamma V_1^2}{288g} + P_1 + \frac{E_1\gamma}{144} + \frac{(TDH)\gamma}{144}$$

$$= \frac{\gamma V_2^2}{288g} + P_2 + \frac{E_2\gamma}{144} + \Sigma\Delta P \quad (1.31)$$

Note that

$$\Sigma\Delta P = \frac{h_L\gamma}{144} \quad (1.32)$$

Each of the terms in the Bernoulli equations have been given names. The term $V^2/2g$ is called the velocity head, 144 P/γ is the static pressure head, E is the potential or elevation head, TDH is the total differential head across a pump (TDH is positive) or hydraulic turbine (TDH is negative) and h_L is the sum of all the head losses due to flow through pipe, fittings, equipment, control valves, orifices, etc. Estimation of losses due to flow through pipe and fittings will be discussed in detail later in this section. The terms "total head" or "total pressure" are often used to indicate the sum of the heads for velocity, pressure and elevation.

Incompressible Flow in Conduits

By definition, the density of an incompressible fluid does not vary with pressure. No real fluids are incompressible, but for many cases the assumption of incompressible flow results in only negligible error in estimating head or pressure losses. Density changes of only a few percent over the length of the flow path may be ignored in most cases. Where density changes cannot be ignored, the flow path may be broken into segments and each treated separately, or the more accurate procedures for compressible flow described later in this section may be used.

Forms of the so-called Darcy equation are used for estimating head losses (h_f, ft. fluid) or pressure losses (ΔP, psi) for incompressible flow in conduits.

$$h_f = \frac{fLV^2}{2Dg} = \frac{fLV^2}{64.34D} \quad (1.33)$$

$$\Delta P = \frac{fL}{D}\frac{V^2}{288g} = \frac{fL}{D}\frac{\gamma V^2}{9{,}265} \quad (1.34)$$

Common engineering practice is to calculate a head gradient (100 h_f/L, ft. fluid per 100 ft. pipe) or pressure gradient (100 $\Delta P/L$, psi per 100 ft. pipe) and then obtain h_f or ΔP by multiplying the appropriate gradient by equivalent length of the flow path expressed as $L/100$. Gradients are estimated by using rearrangements of the above equations.

$$100h_f/L = \frac{1.554fV^2}{D} \quad (1.35)$$

$$100\Delta P/L = \frac{f\gamma V^2}{92.65D} \quad (1.36)$$

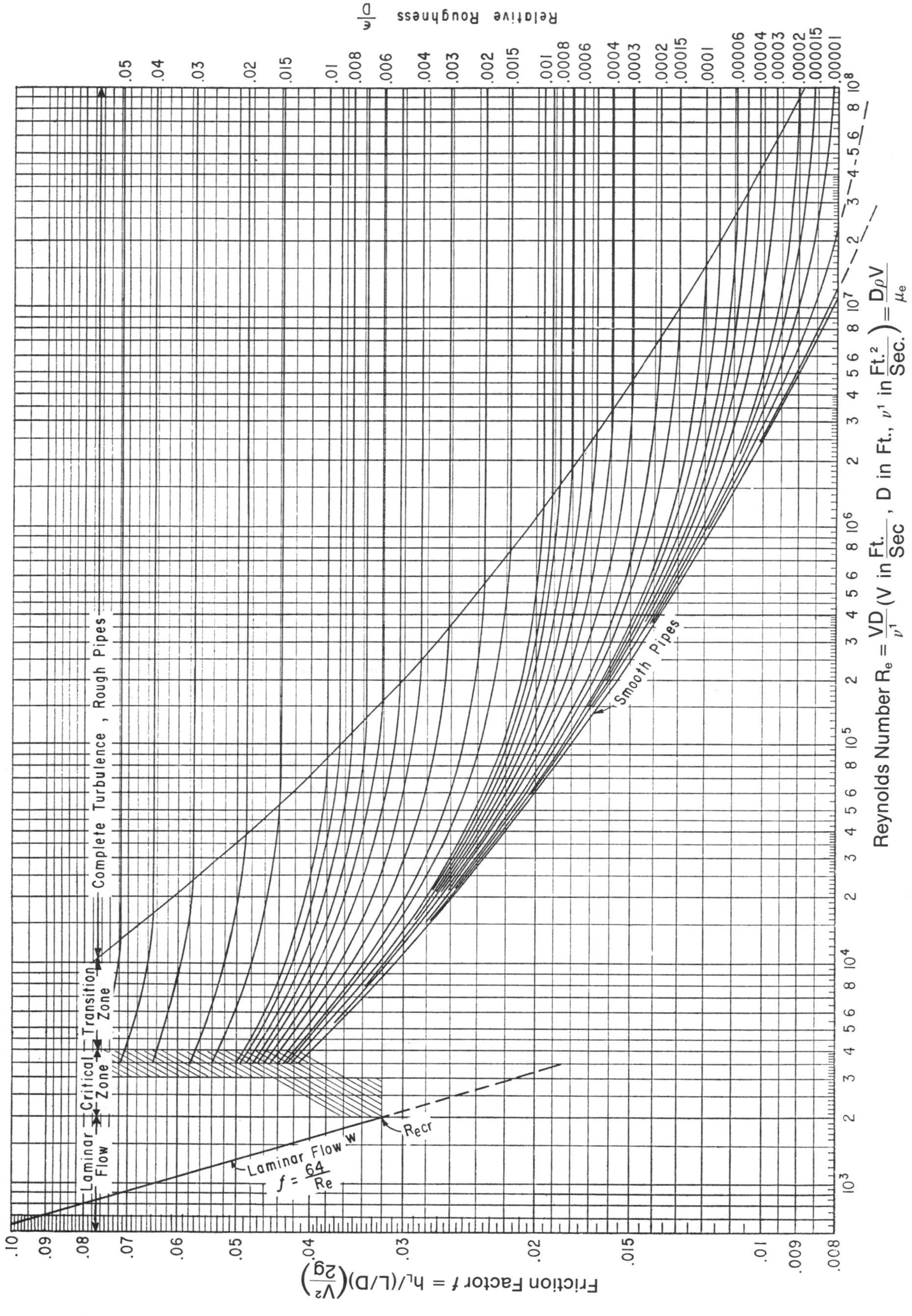

Figure 1.12. Friction factors for any kind and size of pipe. Reprinted from Pipe Friction Manual, copyright © 1954 by the Hydraulic Institute.

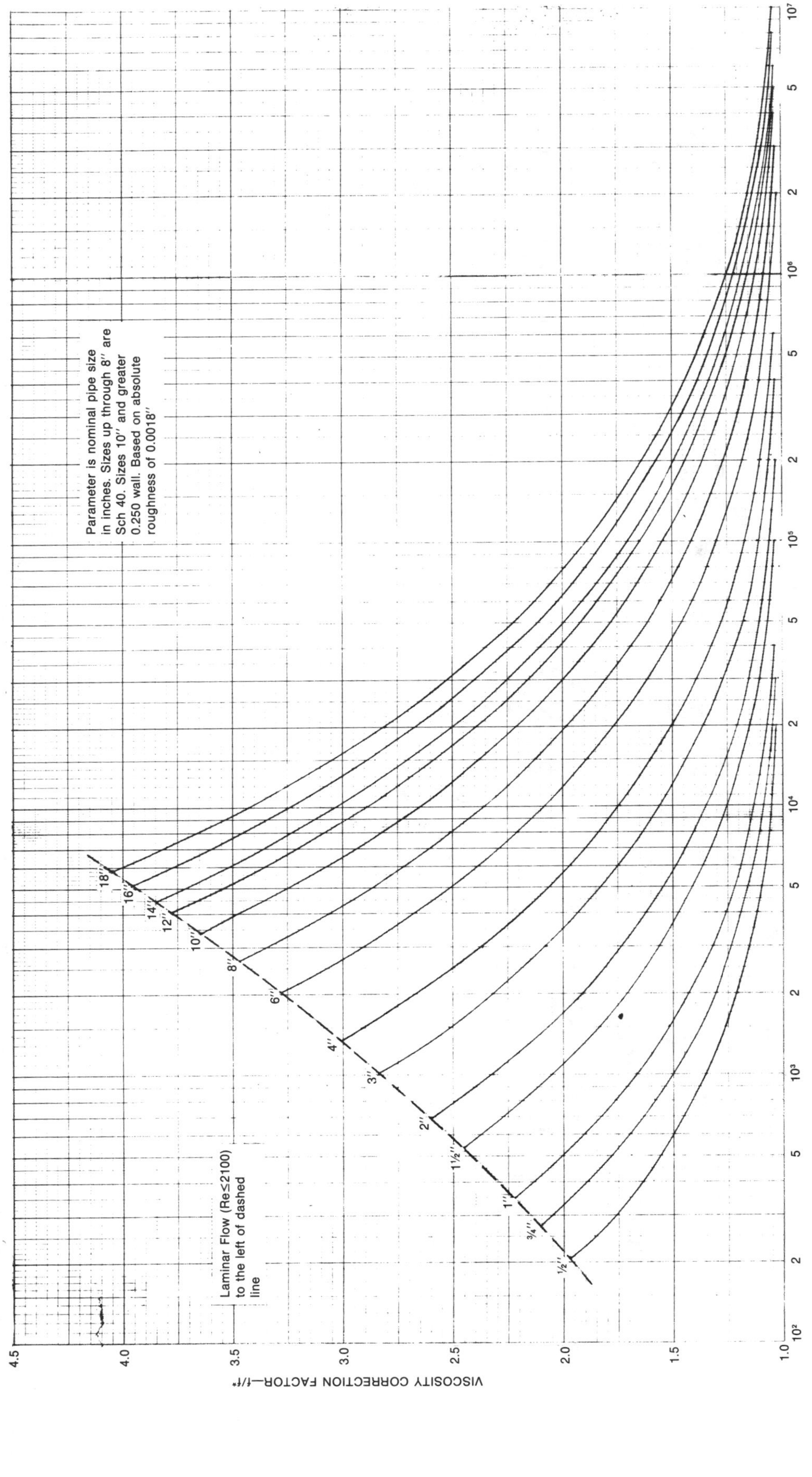

Figure 1.13. Viscosity corrections for steel pipe.

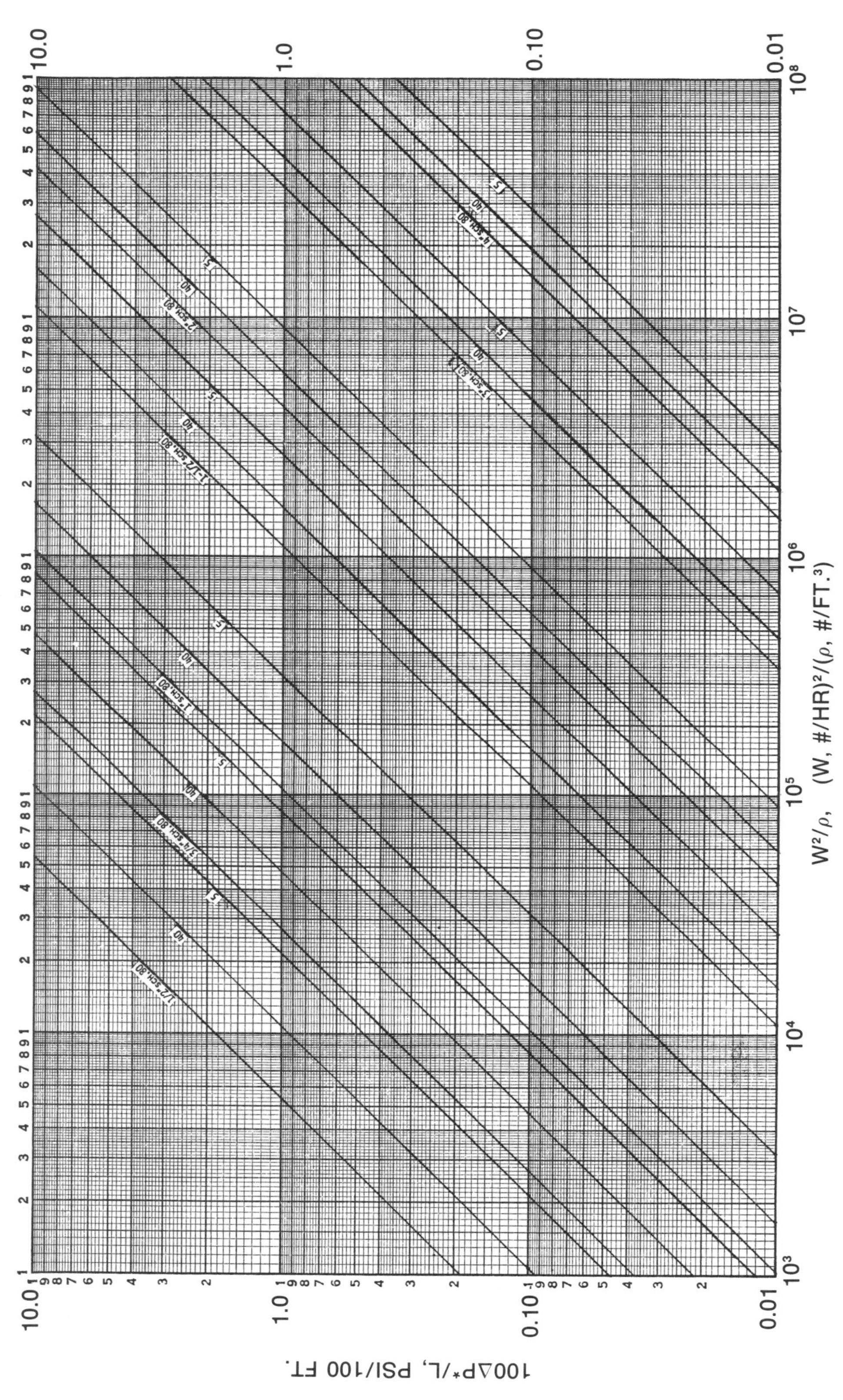

Figure 1.14. Frictional pressure loss in steel pipe for fully developed turbulent flow.

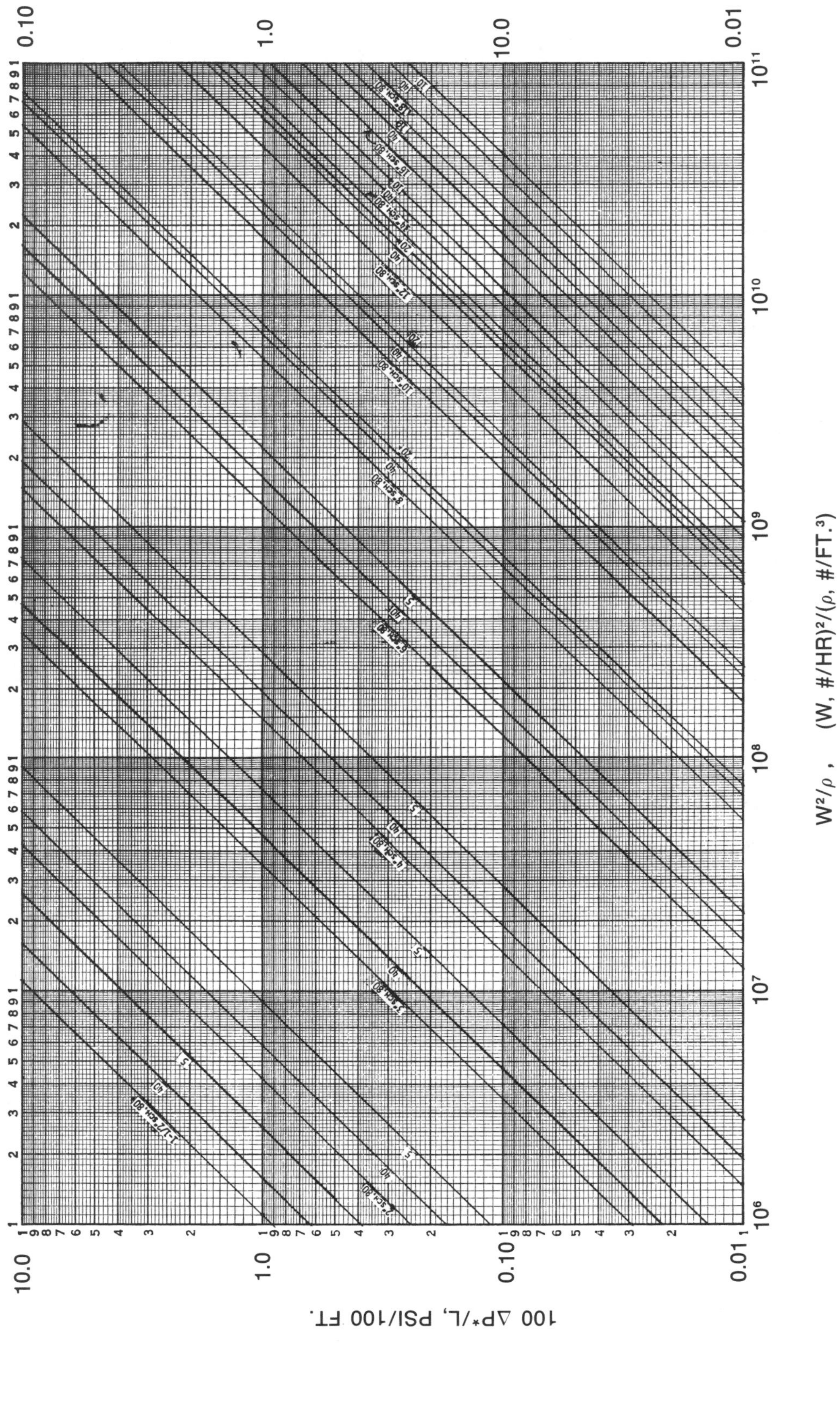

Figure 1.15. Frictional pressure loss in steel pipe for fully developed turbulent flow.

Table 1.11a. Data for Commercial Steel Pipe

Schedule	Nominal Pipe Size	Outside Diameter	Wall Thickness	Inside Diameter d	Inside Diameter D	Flow Area	Inside Diameter Functions D^4	Inside Diameter Functions D^5	Limiting Friction Factor f^*	$(1.35E\text{-}7)\, f^*/D^5 = \frac{(100\Delta P^*/L)}{(W^2/\rho)}$
	Inches	Inches	Inches	Inches	Feet	Sq. Ft.			f^*	
Schedule 10	14	14.000	.250	13.500	1.1250	.99402	1.602E+00	1.802E+00	.0127	9.487E−12
	16	16.000	.250	15.500	1.2917	1.31036	2.784E+00	3.595E+00	.0123	4.629E−12
	18	18.000	.250	17.500	1.4583	1.67033	4.523E+00	6.596E+00	.0120	2.465E−12
	20	20.000	.250	19.500	1.6250	2.07394	6.973E+00	1.133E+01	.0118	1.406E−12
	24	24.000	.250	23.500	1.9583	3.01206	1.471E+01	2.880E+01	.0114	5.341E−13
	30	30.000	.312	29.376	2.4480	4.70666	3.591E+01	8.791E+01	.0109	1.680E−13
Schedule 20	8	8.625	.250	8.125	.6771	.36006	2.102E−01	1.423E−01	.0140	1.330E−10
	10	10.750	.250	10.250	.8542	.57303	5.323E−01	4.547E−01	.0134	3.971E−11
	12	12.750	.250	12.250	1.0208	.81846	1.086E+00	1.109E+00	.0129	1.572E−11
	14	14.000	.312	13.376	1.1147	.97584	1.544E+00	1.721E+00	.0127	9.953E−12
	16	16.000	.312	15.376	1.2813	1.28948	2.696E-00	3.454E+00	.0123	4.826E−12
	18	18.000	.312	17.376	1.4480	1.64675	4.396E+00	6.366E+00	.0121	2.558E−12
	20	20.000	.375	19.250	1.6042	2.02110	6.622E+00	1.062E+01	.0118	1.503E−12
	24	24.000	.375	23.250	1.9375	2.94831	1.409E+01	2.730E+01	.0114	5.646E−13
Schedule 30	8	8.625	.277	8.071	.6726	.35529	2.046E−01	1.376E−01	.0140	1.377E−10
	10	10.750	.307	10.136	.8447	.56035	5.090E−01	4.300E−01	.0134	4.209E−11
	12	12.750	.330	12.090	1.0075	.79722	1.030E+00	1.038E+00	.0129	1.683E−11
	14	14.000	.375	13.250	1.1042	.95754	1.486E+00	1.641E+00	.0127	1.046E−11
	16	16.000	.375	15.250	1.2708	1.26843	2.608E+00	3.315E+00	.0124	5.037E−12
	18	18.000	.438	17.124	1.4270	1.59933	4.147E+00	5.917E+00	.0121	2.760E−12
	20	20.000	.500	19.000	1.5833	1.96895	6.285E+00	9.951E+00	.0119	1.609E−12
	24	24.000	.562	22.876	1.9063	2.85422	1.321E+01	2.518E+01	.0115	6.141E−13
	30	30.000	.625	28.750	2.3958	4.50820	3.295E+01	7.894E+01	.0110	1.878E−13
Schedule 40	1/8	.405	.068	.269	.0224	.00039	2.525E−07	5.661E−09	.0332	7.927E−03
	1/4	.540	.088	.364	.0303	.00072	8.466E−07	2.568E−08	.0303	1.591E−03
	3/8	.675	.091	.493	.0411	.00133	2.849E−06	1.170E−07	.0277	3.192E−04
	1/2	.840	.109	.622	.0518	.00211	7.218E−06	3.741E−07	.0259	9.347E−05
	3/4	1.050	.113	.824	.0687	.00370	2.223E−05	1.527E−06	.0240	2.121E−05
	1	1.315	.133	1.049	.0874	.00600	5.840E−05	5.105E−06	.0225	5.950E−06
	1 1/4	1.660	.140	1.380	.1150	.01039	1.749E−04	2.011E−05	.0210	1.408E−06
	1 1/2	1.900	.145	1.610	.1342	.01414	3.240E−04	4.347E−05	.0202	6.267E−07
	2	2.375	.154	2.067	.1722	.02330	8.803E−04	1.516E−04	.0190	1.691E−07
	2 1/2	2.875	.203	2.469	.2057	.03325	1.792E−03	3.687E−04	.0182	6.667E−08
	3	3.500	.216	3.068	.2557	.05134	4.273E−03	1.092E−03	.0173	2.140E−08
	3 1/2	4.000	.226	3.548	.2957	.06866	7.642E−03	2.259E−03	.0168	1.001E−08
	4	4.500	.237	4.026	.3355	.08840	1.267E−02	4.251E−03	.0163	5.173E−09
	5	5.563	.258	5.047	.4206	.13893	3.129E−02	1.316E−02	.0155	1.590E−09
	6	6.625	.280	6.065	.5054	.20063	6.525E−02	3.298E−02	.0149	6.101E−10
	8	8.625	.322	7.981	.6651	.34741	1.957E−01	1.301E−01	.0141	1.460E−10
	10	10.750	.365	10.020	.8350	.54760	4.861E−01	4.059E−01	.0134	4.469E−11
	12	12.750	.406	11.938	.9948	.77730	9.795E−01	9.744E−01	.0130	1.797E−11
	14	14.000	.438	13.124	1.0937	.93942	1.431E+00	1.565E+00	.0127	1.099E−11
	16	16.000	.500	15.000	1.2500	1.22718	2.441E+00	3.052E+00	.0124	5.489E−12
	18	18.000	.562	16.876	1.4063	1.55334	3.912E+00	5.501E+00	.0121	2.977E−12
	20	20.000	.593	18.814	1.5678	1.93059	6.042E+00	9.473E+00	.0119	1.693E−12
	24	24.000	.687	22.626	1.8855	2.79218	1.264E+01	2.383E+01	.0115	6.501E−13
Schedule 60	8	8.625	.406	7.813	.6511	.33294	1.797E−01	1.170E−01	.0141	1.631E−10
	10	10.750	.500	9.750	.8125	.51849	4.358E−01	3.541E−01	.0135	5.151E−11
	12	12.750	.562	11.626	.9688	.73720	8.810E−01	8.536E−01	.0130	2.063E−11
	14	14.000	.593	12.814	1.0678	.89556	1.300E+00	1.388E+00	.0128	1.244E−11
	16	16.000	.656	14.688	1.2240	1.17666	2.245E+00	2.747E+00	.0125	6.122E−12
	18	18.000	.750	16.500	1.3750	1.48489	3.574E+00	4.915E+00	.0122	3.346E−12
	20	20.000	.812	18.376	1.5313	1.84174	5.499E+00	8.421E+00	.0119	1.913E−12
	24	24.000	.968	22.064	1.8387	2.65519	1.143E+01	2.101E+01	.0115	7.407E−13
Schedule 80	1/8	.405	.095	.215	.0179	.00025	1.030E−07	1.846E−09	.0357	2.613E−02
	1/4	.540	.119	.302	.0252	.00050	4.011E−07	1.010E−08	.0320	4,286E−03
	3/8	.675	.126	.423	.0352	.00098	1.544E−06	5.442E−08	.0289	7.179E−04
	1/2	.840	.147	.546	.0455	.00163	4.286E−06	1.950E−07	.0269	1.860E−04
	3/4	1.050	.154	.742	.0618	.00300	1.462E−05	9.039E−07	.0247	3.685E−05
	1	1.315	.179	.957	.0797	.00500	4.045E−05	3.226E−06	.0230	9.644E−06
	1 1/4	1.660	.191	1.278	.1065	.00891	1.286E−04	1.370E−05	.0214	2.107E−06

Table 1.11b Data for Commercial Steel Pipe

Schedule	Nominal Pipe Size	Outside Diameter	Wall Thickness	Inside Diameter d	Inside Diameter D	Flow Area	Inside Diameter Functions D^4	Inside Diameter Functions D^5	Limiting Friction Factor f^*	$(1.35E\text{-}7)\, f^*/D^5 = \frac{(100\Delta P^*/L)}{(W^2/\rho)}$
	Inches	Inches	Inches	Inches	Feet	Sq. Ft.				
Schedule 80—Cont.	1½	1.900	.200	1.500	.1250	.01227	2.441E−04	3.052E−05	.0205	9.086E−07
	2	2.375	.218	1.939	.1616	.02051	6.817E−04	1.101E−04	.0193	2.364E−07
	2½	2.875	.276	2.323	.1936	.02943	1.404E−03	2.719E−04	.0185	9.173E−08
	3	3.500	.300	2.900	.2417	.04587	3.411E−03	8.243E−04	.0175	2.873E−08
	3½	4.000	.318	3.364	.2803	.06172	6.176E−03	1.731E−03	.0170	1.322E−08
	4	4.500	.337	3.826	.3188	.07984	1.033E−02	3.295E−03	.0165	6.750E−09
	5	5.563	.375	4.813	.4011	.12635	2.588E−02	1.038E−02	.0157	2.037E−09
	6	6.625	.432	5.761	.4801	.18102	5.312E−02	2.550E−02	.0151	7.976E−10
	8	8.625	.500	7.625	.6354	.31711	1.630E−01	1.036E−01	.0142	1.851E−10
	10	10.750	.593	9.564	.7970	.49889	4.035E−01	3.216E−01	.0136	5.694E−11
	12	12.750	.687	11.376	.9480	.70584	8.077E−01	7.657E−01	.0131	2.310E−11
	14	14.000	.750	12.500	1.0417	.85221	1.177E+00	1.226E+00	.0129	1.415E−11
	16	16.000	.843	14.314	1.1928	1.11750	2.025E+00	2.415E+00	.0125	6.999E−12
	18	18.000	.937	16.126	1.3438	1.41834	3.261E+00	4.383E+00	.0122	3.769E−12
	20	20.000	1.031	17.938	1.4948	1.75499	4.993E+00	7.464E+00	.0120	2.168E−12
	24	24.000	1.218	21.564	1.7970	2.53621	1.043E+01	1.874E+01	.0116	8.342E−13
Schedule 100	8	8.625	.593	7.439	.6199	.30183	1.477E−01	9.155E−02	.0143	2.106E−10
	10	10.750	.718	9.314	.7762	.47315	3.629E−01	2.817E−01	.0136	6.535E−11
	12	12.750	.843	11.064	.9220	.66765	7.226E−01	6.663E−01	.0132	2.669E−11
	14	14.000	.937	12.126	1.0105	.80198	1.043E+00	1.054E+00	.0129	1.657E−11
	16	16.000	1.031	13.938	1.1615	1.05957	1.820E+00	2.114E+00	.0126	8.037E−12
	18	18.000	1.156	15.688	1.3073	1.34234	2.921E+00	3.819E+00	.0123	4.348E−12
	20	20.000	1.281	17.438	1.4532	1.65852	4.459E+00	6.480E+00	.0121	2.511E−12
	24	24.000	1.531	20.938	1.7448	2.39110	9.269E+00	1.617E+01	.0116	9.720E−13
Schedule 120	4	4.500	.438	3.624	.3020	.07163	8.318E−03	2.512E−03	.0167	8.962E−09
	5	5.563	.500	4.563	.3802	.11356	2.091E−02	7.950E−03	.0158	2.691E−09
	6	6.625	.562	5.501	.4584	.16505	4.416E−02	2.024E−02	.0152	1.015E−09
	8	8.625	.718	7.189	.5991	.28188	1.288E−01	7.717E−02	.0144	2.516E−10
	10	10.750	.843	9.064	.7553	.44809	3.255E−01	2.459E−01	.0137	7.529E−11
	12	12.750	1.000	10.750	.8958	.63030	6.440E−01	5.769E−01	.0132	3.100E−11
	14	14.000	1.093	11.814	.9845	.76124	9.394E−01	9.249E−01	.0130	1.898E−11
	16	16.000	1.218	13.564	1.1303	1.00347	1.632E+00	1.845E+00	.0127	9.257E−12
	18	18.000	1.375	15.250	1.2708	1.26843	2.608E+00	3.315E+00	.0124	5.037E−12
	20	20.000	1.500	17.000	1.4167	1.57625	4.028E+00	5.706E+00	.0121	2.866E−12
	24	24.000	1.812	20.376	1.6980	2.26446	8.313E+00	1.412E+01	.0117	1.119E−12
Schedule 140	8	8.625	.812	7.001	.5834	.26733	1.159E−01	6.759E−02	.0145	2.888E−10
	10	10.750	1.000	8.750	.7292	.41758	2.827E−01	2.061E−01	.0138	9.045E−11
	12	12.750	1.125	10.500	.8750	.60132	5.862E−01	5.129E−01	.0133	3.503E−11
	14	14.000	1.250	11.500	.9583	.72131	8.435E−01	8.083E−01	.0131	2.183E−11
	16	16.000	1.438	13.124	1.0937	.93942	1.431E+00	1.565E+00	.0127	1.099E−11
	18	18.000	1.562	14.876	1.2397	1.20698	2.362E+00	2.928E+00	.0124	5.730E−12
	20	20.000	1.750	16.500	1.3750	1.48489	3.574E+00	4.915E+00	.0122	3.346E−12
	24	24.000	2.062	19.876	1.6563	2.15469	7.526E+00	1.247E+01	.0118	1.273E−12
Schedule 160	½	.840	.187	.466	.0388	.00118	2.274E−06	8.831E−08	.0281	4.300E−04
	¾	1.050	.218	.614	.0512	.00206	6.854E−06	3.507E−07	.0260	1.001E−04
	1	1.315	.250	.815	.0679	.00362	2.128E−05	1.445E−06	.0241	2.247E−05
	1¼	1.660	.250	1.160	.0967	.00734	8.732E−05	8.441E−06	.0219	3.506E−06
	1½	1.900	.281	1.338	.1115	.00976	1.546E−04	1.723E−05	.0211	1.656E−06
	2	2.375	.343	1.689	.1407	.01556	3.925E−04	5.524E−05	.0199	4.874E−07
	2½	2.875	.375	2.125	.1771	.02463	9.834E−04	1.741E−04	.0189	1.463E−07
	3	3.500	.438	2.624	.2187	.03755	2.286E−03	4.999E−04	.0186	4.848E−08
	4	4.500	.531	3.438	.2865	.06447	6.737E−03	1.930E−03	.0169	1.180E−08
	5	5.563	.625	4.313	.3594	.10146	1.669E−02	5.998E−03	.0160	3.611E−09
	6	6.625	.718	5.189	.4324	.14686	3.496E−02	1.512E−02	.0154	1.376E−09
	8	8.625	.906	6.813	.5677	.25317	1.039E−01	5.899E−02	.0145	3.328E−10
	10	10.750	1.125	8.500	.7083	.39406	2.517E−01	1.783E−01	.0139	1.052E−10
	12	12.750	1.310	10.130	.8442	.55969	5.078E−01	4.287E−01	.0134	4.222E−11
	14	14.000	1.400	11.200	.9333	.68417	7.588E−01	7.082E−01	.0131	2.505E−11
	16	16.000	1.590	12.820	1.0683	.89640	1.303E+00	1.392E+00	.0128	1.241E−11
	18	18.000	1.780	14.440	1.2033	1.13727	2.097E+00	2.523E+00	.0125	6.688E−12
	20	20.000	1.960	16.080	1.3400	1.41026	3.224E+00	4.320E+00	.0122	3.825E−12
	24	24.000	2.340	19.320	1.6100	2.03583	6.719E+00	1.082E+01	.0118	1.475E−12

Table 1.11c. Data for Commercial Steel Pipe

Nominal Pipe Size	Outside Diameter	Wall Thickness	Inside Diameter d	Inside Diameter D	Flow Area	Inside Diameter Functions D^4	Inside Diameter Functions D^5	Limiting Friction Factor f^*	$(1.35E\text{-}7)\, f^*/D^5 = \frac{(100\Delta P^*/L)}{(W^2/\rho)}$
Inches	Inches	Inches	Inches	Feet	Sq. Ft.				
Standard Wall Pipe									
⅛	.405	.068	.269	.0224	.00039	2.525E−07	5.661E−09	.0332	7.927E−03
¼	.540	.088	.364	.0303	.00072	8.466E−07	2.568E−08	.0303	1.591E−03
⅜	.675	.091	.493	.0411	.00133	2.849E−06	1.170E−07	.0277	3.192E−04
½	.840	.109	.622	.0518	.00211	7.218E−06	3.741E−07	.0259	9.347E−05
¾	1.050	.113	.824	.0687	.00370	2.223E−05	1.527E−06	.0240	2.121E−05
1	1.315	.133	1.049	.0874	.00600	5.840E−05	5.105E−06	.0225	5.950E−06
1¼	1.660	.140	1.380	.1150	.01039	1.749E−04	2.011E−05	.0210	1.408E−06
1½	1.900	.145	1.610	.1342	.01414	3.240E−04	4.347E−05	.0202	6.267E−07
2	2.375	.154	2.067	.1722	.02330	8.803E−04	1.516E−04	.0190	1.691E−07
2½	2.875	.203	2.469	.2057	.03325	1.792E−03	3.687E−04	.0182	6.667E−08
3	3.500	.216	3.068	.2557	.05134	4.273E−03	1.092E−03	.0173	2.140E−08
3½	4.000	.226	3.548	.2957	.06866	7.642E−03	2.259E−03	.0168	1.001E−08
4	4.500	.237	4.026	.3355	.08840	1.267E−02	4.251E−03	.0163	5.173E−09
5	5.563	.258	5.047	.4206	.13893	3.129E−02	1.316E−02	.0155	1.590E−09
6	6.625	.280	6.065	.5054	.20063	6.525E−02	3.298E−02	.0149	6.101E−10
8	8.625	.277	8.071	.6726	.35529	2.046E−01	1.376E−01	.0140	1.377E−10
8S	8.625	.322	7.981	.6651	.34741	1.957E−01	1.301E−01	.0141	1.460E−10
10	10.750	.279	10.192	.8493	.56656	5.204E−01	4.420E−01	.0134	4.090E−11
10	10.750	.307	10.136	.8447	.56035	5.090E−01	4.300E−01	.0134	4.209E−11
10S	10.750	.365	10.020	.8350	.54760	4.861E−01	4.059E−01	.0134	4.469E−11
12	12.750	.330	12.090	1.0075	.79722	1.030E+00	1.038E+00	.0129	1.683E−11
12S	12.750	.375	12.000	1.0000	.78540	1.000E+00	1.000E+00	.0130	1.750E−11
Extra Strong Pipe									
⅛	.405	.095	.215	.0179	.00025	1.030E−07	1.846E−09	.0357	2.613E−02
¼	.540	.119	.302	.0252	.00050	4.011E−07	1.010E−08	.0320	4.286E−03
⅜	.675	.126	.423	.0352	.00098	1.544E−06	5.442E−08	.0289	7.179E−04
½	.840	.147	.546	.0455	.00163	4.286E−06	1.950E−07	.0269	1.860E−04
¾	1.050	.154	.742	.0618	.00300	1.462E−05	9.039E−07	.0247	3.685E−05
1	1.315	.179	.957	.0797	.00500	4.045E−05	3.226E−06	.0230	9.644E−06
1¼	1.660	.191	1.278	.1065	.00891	1.286E−04	1.370E−05	.0214	2.107E−06
1½	1.900	.200	1.500	.1250	.01227	2.441E−04	3.052E−05	.0205	9.086E−07
2	2.375	.218	1.939	.1616	.02051	6.817E−04	1.101E−04	.0193	2.364E−07
2½	2.875	.276	2.323	.1936	.02943	1.404E−03	2.719E−04	.0185	9.173E−08
3	3.500	.300	2.900	.2417	.04587	3.411E−03	8.243E−04	.0175	2.873E−08
3½	4.000	.318	3.364	.2803	.06172	6.176E−03	1.731E−03	.0170	1.322E−08
4	4.500	.337	3.826	.3188	.07984	1.033E−02	3.295E−03	.0165	6.750E−09
5	5.563	.375	4.813	.4011	.12635	2.588E−02	1.038E−02	.0157	2.037E−09
6	6.625	.432	5.761	.4801	.18102	5.312E−02	2.550E−02	.0151	7.976E−10
8	8.625	.500	7.625	.6354	.31711	1.630E−01	1.036E−01	.0142	1.851E−10
10	10.750	.500	9.750	.8125	.51849	4.358E−01	3.541E−01	.0135	5.151E−11
12	12.750	.500	11.750	.9792	.75301	9.192E−01	9.001E−01	.0130	1.952E−11
Double Extra Strong Pipe									
½	.840	.294	.252	.0210	.00035	1.945E−07	4.084E−09	.0339	1.122E−02
¾	1.050	.308	.434	.0362	.00103	1.711E−06	6.188E−08	.0287	6.266E−04
1	1.315	.358	.599	.0499	.00196	6.208E−06	3.099E−07	.0262	1.140E−04
1¼	1.660	.382	.896	.0747	.00438	3.108E−05	2.321E−06	.0234	1.364E−05
1½	1.900	.400	1.100	.0917	.00660	7.061E−05	6.472E−06	.0222	4.635E−06
2	2.375	.436	1.503	.1252	.01232	2.461E−04	3.082E−05	.0205	8.991E−07
2½	2.875	.552	1.771	.1476	.01711	4.744E−04	7.001E−05	.0197	3.801E−07
3	3.500	.600	2.300	.1917	.02885	1.350E−03	2.587E−04	.0185	9.664E−08
3½	4.000	.636	2.728	.2273	.04059	2.671E−03	6.072E−04	.0178	3.956E−08
4	4.500	.674	3.152	.2627	.05419	4.760E−03	1.250E−03	.0172	1.858E−08
5	5.563	.750	4.063	.3386	.09004	1.314E−02	4.450E−03	.0163	4.932E−09
6	6.625	.864	4.897	.4081	.13079	2.773E−02	1.132E−02	.0156	1.861E−09
8	8.625	.875	6.875	.5729	.25779	1.077E−01	6.172E−02	.0145	3.175E−10

Table 1.11d. Data for Commercial Steel Pipe

Nominal Pipe Size	Outside Diameter	Wall Thickness	Inside Diameter d	Inside Diameter D	Flow Area	Inside Diameter Functions D^4	Inside Diameter Functions D^5	Limiting Friction Factor f^*	$(1.35E\text{-}7)\, f^*/D^5 = \frac{(100\Delta P^*/L)}{(W^2/\rho)}$
Inches	Inches	Inches	Inches	Feet	Sq. Ft.				
Schedule 5S									
½	.840	.065	.710	.0592	.00275	1.225E−05	7.251E−07	.0250	4.649E−05
¾	1.050	.065	.920	.0767	.00462	3.455E−05	2.649E−06	.0233	1.187E−05
1	1.315	.065	1.185	.0987	.00766	9.509E−05	9.390E−06	.0218	3.134E−06
1¼	1.660	.065	1.530	.1275	.01277	2.643E−04	3.369E−05	.0204	8.189E−07
1½	1.900	.065	1.770	.1475	.01709	4.733E−04	6.982E−05	.0197	3.813E−07
2	2.375	.065	2.245	.1871	.02749	1.225E−03	2.292E−04	.0186	1.097E−07
2½	2.875	.083	2.709	.2257	.04003	2.597E−03	5.863E−04	.0178	4.103E−08
3	3.500	.083	3.334	.2778	.06063	5.959E−03	1.655E−03	.0170	1.386E−08
3½	4.000	.083	3.834	.3195	.08017	1.042E−02	3.329E−03	.0165	6.677E−09
4	4.500	.083	4.334	.3612	.10245	1.701E−02	6.145E−03	.0160	3.521E−09
5	5.563	.109	5.345	.4454	.15582	3.936E−02	1.753E−02	.0153	1.179E−09
6	6.625	.109	6.407	.5339	.22389	8.126E−02	4.339E−02	.0147	4.584E−10
8	8.625	.109	8.407	.7006	.38549	2.409E−01	1.688E−01	.0139	1.114E−10
10	10.750	.134	10.482	.8735	.59926	5.822E−01	5.085E−01	.0133	3.535E−11
12	12.750	.156	12.438	1.0365	.84378	1.154E+00	1.196E+00	.0129	1.452E−11
Schedule 10S									
⅛	.405	.049	.307	.0256	.00051	4.284E−07	1.096E−08	.0319	3.928E−03
¼	.540	.065	.410	.0342	.00092	1.363E−06	4.656E−08	.0292	8.469E−04
⅜	.675	.065	.545	.0454	.00162	4.255E−06	1.932E−07	.0269	1.879E−04
½	.840	.083	.674	.0562	.00248	9.952E−06	5.590E−07	.0253	6.118E−05
¾	1.050	.083	.884	.0737	.00426	2.945E−05	2.169E−06	.0235	1.465E−05
1	1.315	.109	1.097	.0914	.00656	6.984E−05	6.385E−06	.0222	4.702E−06
1¼	1.660	.109	1.442	.1202	.01134	2.085E−04	2.506E−05	.0207	1.118E−06
1½	1.900	.109	1.682	.1402	.01543	3.860E−04	5.410E−05	.0200	4.982E−07
2	2.375	.109	2.157	.1797	.02538	1.044E−03	1.876E−04	.0188	1.353E−07
2½	2.875	.120	2.635	.2196	.03787	2.325E−03	5.105E−04	.0179	4.743E−08
3	3.500	.120	3.260	.2717	.05796	5.447E−03	1.480E−03	.0171	1.558E−08
3½	4.000	.120	3.760	.3133	.07711	9.639E−03	3.020E−03	.0165	7.393E−09
4	4.500	.120	4.260	.3550	.09898	1.588E−02	5.638E−03	.0161	3.852E−09
5	5.563	.134	5.295	.4412	.15292	3.791E−02	1.673E−02	.0153	1.238E−09
6	6.625	.134	6.357	.5297	.22041	7.876E−02	4.172E−02	.0148	4.775E−10
8	8.625	.148	8.329	.6941	.37837	2.321E−01	1.611E−01	.0139	1.169E−10
10	10.750	.165	10.420	.8683	.59219	5.685E−01	4.937E−01	.0133	3.645E−11
12	12.750	.180	12.390	1.0325	.83728	1.136E+00	1.173E+00	.0129	1.482E−11

Values of the friction factor, f, may be estimated using Figure 1.12.

A graphical procedure for estimating pressure loss gradients by the Darcy equation is shown in Figures 1.13, 1.14 and 1.15. The graphical procedure is specific for incompressible flow of any Newtonian fluid in steel pipes of circular cross section and involves no simplifying assumptions. It is intended for rapid estimation of pressure losses or selection of pipe diameter. Primary advantages are speed of application and minimization of calculations. Where warranted by frequency of use, the graphical solution may be extended for pipe materials other than steel.

Development of the procedure is based on two factors:

1. Absolute wall roughness for a given pipe material is a constant and friction factors for fully developed turbulent flow, designated as f^* for the procedure, are a function of relative pipe wall roughness, ϵ/D, only. This means that for a given pipe material f^* is a function of diameter only. Values of f^* for steel pipe are listed in Table 1.11 or they may be calculated for any pipe from the Colebrook Equation for R_e = infinity which reduces to

$$\frac{1}{\sqrt{f^*}} = -2 \log_{10} \left(\frac{\epsilon}{3.7D}\right) \tag{1.37}$$

2. The Darcy equation (Equation 1.34) can be rearranged for ducts with circular cross section to

$$100\ \Delta P/L = 1.350 \times 10^{-7} \frac{f^*}{D^5} \frac{f}{f^*} \frac{W^2}{\rho} \tag{1.38}$$

noting that

$$f = f^* \, (f/f^*) \qquad (1.39)$$

The form of Equation 1.38 suggests that $100\Delta P/L$ values may be calculated as the product of three terms.

The first term, $1.35 \times 10^{-7} f^*/D^5$, is unique to each pipe of a given material and diameter and calculated values for steel pipe are listed in Table 1.11.

The second term, f/f^*, may be considered a viscosity correction factor. Its value approaches 1.0 for large values of R_e. For a given pipe material, values of f/f^* can be plotted versus the ratio W/μ for parameters of pipe diameter. The ratio W/μ serves as an effective Reynolds number, but has the advantage of being independent of diameter allowing various trial diameters to be studied without recalculation. Figure 1.13 is such a plot for a number of commonly used steel pipes.

The third term, W^2/ρ, is also conveniently independent of pipe diameter. It is the principal flow variant in pressure loss calculations. For many preliminary calculations involving gases or low viscosity liquids the viscosity correction factor, f/f^*, may be neglected or approximated until refined calculations are made.

Figures 1.14 and 1.15 provide for steel pipe graphical solutions for $100\Delta P^*/L$, the pressure loss gradient uncorrected for viscosity. These are straight line plots in accordance with

$$100\,\Delta P^*/L = 1.35 \times 10^{-7} \frac{f^*}{D^5} (W^2/\rho) \qquad (1.40)$$

which is Equation 1.38 with the f/f^* term omitted. Values of $100\Delta P/L$ for steel pipe may be rapidly calculated using Figures 1.13, 1.14, and 1.15 and the relation

$$100\Delta P/L = 100\Delta P^*/L \, (f/f^*) \qquad (1.41)$$

Example Problem 1.7

Estimation of Pressure Loss Gradients for Incompressible Flow

Estimate $100\,\Delta P/L$ for the flow of 62,500 lb./hr. of fluid through a 3″ sch. 40 commercial steel pipe. Fluid density is 57.5 lb_m/ft^3 and viscosity is 1.3 centipoise.

1. $100\,\Delta P/L$ using Equation 1.36

   For 3″ sch. 40 pipe

   D = 0.2557 ft., A = 0.0513 sq. ft. (Table 1.11)

$$R_e = 0.5236 \frac{W}{\mu D} = 0.5263 \frac{62{,}500}{(1.3)(0.2557)} \qquad (1.24)$$

$$= 9.90 \times 10^4$$

ϵ = 0.00015 feet (Table 1.10)

$$\epsilon/D = 0.00015/0.2557 = 0.000587 \qquad (1.25)$$

f = 0.021 @ R_e, ϵ/D (Fig. 1.12)

$$\text{Flow, cu. ft./sec.} = \frac{W}{3600\rho} = \frac{62{,}500}{3600(57.5)} = 0.302$$

V = $0.302/A$ = 5.89 ft./sec.

$$100\,\Delta P/L = \frac{f\gamma V^2}{92.65D} = \frac{0.021(57.5)(5.89)^2}{92.65(0.2557)} \qquad (1.36)$$

$$= 1.77 \text{ psi}/100'$$

2. $100\Delta P/L$ using Figures 1.13 and 1.14

$W^2/\rho = (62{,}500)^2/57.5 = 6.79 \times 10^7$

$W/\mu = 62{,}500/1.3 = 4.81 \times 10^4$

f/f^* = 1.18 for 3″ sch. 40 pipe @ W/μ (Fig. 1.13)

$100\,\Delta P^*/L$ = 1.48 psi/100′ for 3″ sch. 40 pipe @ W^2/ρ (Fig. 1.14)

$$100\,\Delta P/L = (f/f^*)\, 100\,\Delta P^*/L = 1.18\,(1.48) = 1.75 \text{ psi}/100' \qquad (1.41)$$

Compressible Flow In Conduits

As a compressible fluid flows subsonically, with friction, through pipe, its density decreases because of decreasing pressure, resulting in an increase in fluid velocity and kinetic energy. When pressure loss is great enough to cause significant variation (>10%) in fluid density, the fluid flow calculation method must take into account variable fluid properties, therefore, use of the Darcy equation for incompressible flow, Equation 1.34, may result in serious error. Crane Co. (2) suggests that the Darcy equation may be used with reasonable accuracy in compressible flow calculations under the following conditions:

1. If pressure loss is less than 10% of upstream pressure, with fluid properties based on either upstream or downstream conditions
2. If pressure loss is greater than 10%, but less than 40% of upstream pressure, with fluid properties based on average of upstream and downstream conditions
3. If pressure loss is greater than 40% of upstream pressure, or if a high degree of accuracy is required, more rigorous methods of calculation must be used

These guidelines must be qualified, however, to apply when Mach number of the fluid is less than ≈0.3. At higher values of Mach number, significant error in predicted pressure loss

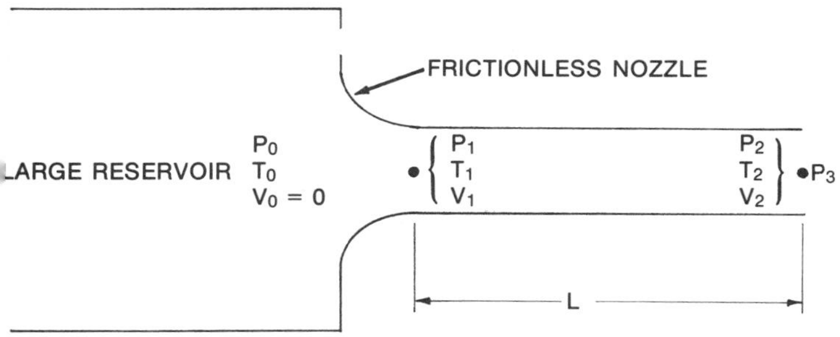

Figure 1.16. Nomenclature for compressible flow conditions.

(>20%) will result through use of the Darcy equation for compressible flow calculations even though pressure loss is within the limits stated above.

Compressible flow calculations are usually classified as *isothermal flow,* in which heat transfer occurs between the fluid and the surroundings to maintain constant fluid temperature, or as *adiabatic flow,* in which no heat transfer takes place between the fluid and the surroundings. The greater the value of specific heat ratio ($k = C_p/C_v$) for the fluid, or the closer the approach to sonic flow (Mach number = 1), the greater the differences between isothermal and adiabatic flow. As specific heat ratio approaches 1.0, or the length of pipe through which the fluid is flowing increases, the smaller the difference between the two types of flow. In real systems, flow is neither truly adiabatic nor isothermal, and the choice of calculation method is a matter of judgment.

From a practical design standpoint, if resistance coefficient, K, is greater than 10, and final Mach number is less than 0.75, the difference in predicted pressure loss for isothermal versus adiabatic flow is well within the limits of precision with which flow conditions, flow path, and fluid properties are known, and the most convenient method of calculation should be used. Assumption of isothermal flow will result in larger predicted values of pressure loss for a given mass flow rate than adiabatic flow but will result in predicted downstream pressure no less than 80% of that predicted for adiabatic flow for any given inlet conditions or pipe configuration. However, pressure loss differences predicted by the two procedures for the same inlet conditions may vary widely. (See example problems 1.10 and 1.11.)

Isothermal Compressible Flow

The equation for isothermal compressible flow relating pressure to distance along a pipe may be expressed in terms of inlet conditions:

$$f\frac{L}{D} = \frac{1}{M_{t1}^2}\left[1 - \left(\frac{P_2}{P_1}\right)\right]^2 - 2\ln\frac{P_1}{P_2} \qquad (1.42)$$

or outlet conditions:

$$f\frac{L}{D} = \frac{1}{M_{t2}^2}\left[\left(\frac{P_1}{P_2}\right)^2 - 1\right] - 2\ln\frac{P_1}{P_2} \qquad (1.43)$$

where

$$M_t = \frac{V}{V_t} \qquad (1.44)$$

and V_t is the limiting velocity for isothermal flow:

$$V_t = \frac{V_s}{\sqrt{k}} \qquad (1.45)$$

For a gas:

$$V_t = 68.1\sqrt{\frac{P}{\rho}} = 223.1\sqrt{\frac{ZT}{MW}} \qquad (1.46)$$

and

$$M_t = 1.336 \times 10^{-5}\,\frac{W}{AP}\sqrt{\frac{ZT}{MW}} \qquad (1.47)$$

The term $f(L/D)$ is the familiar velocity head count, (or resistance coefficient) K, but in compressible flow calculations, it may be applied only for a pipe segment having no restrictions which produce critical flow prior to the end of the segment. If a restriction exists, the restriction and pipe segments on either side must be treated separately.

Another useful expression relating pressure and Mach number is

$$\frac{M_1}{M_2} = \frac{P_2}{P_1} \qquad (1.48)$$

It can be shown that this ratio of Mach numbers is also equal to M_{t1}/M_{t2} and ρ_2/ρ_1.

The assumptions made in the isothermal flow equations are:

1. Flow is steady state, one dimensional, and subsonic
2. Flow path is horizontal, and of constant cross section
3. Compressibility factor of the fluid is constant
4. Friction factor is constant throughout the length of the flow path

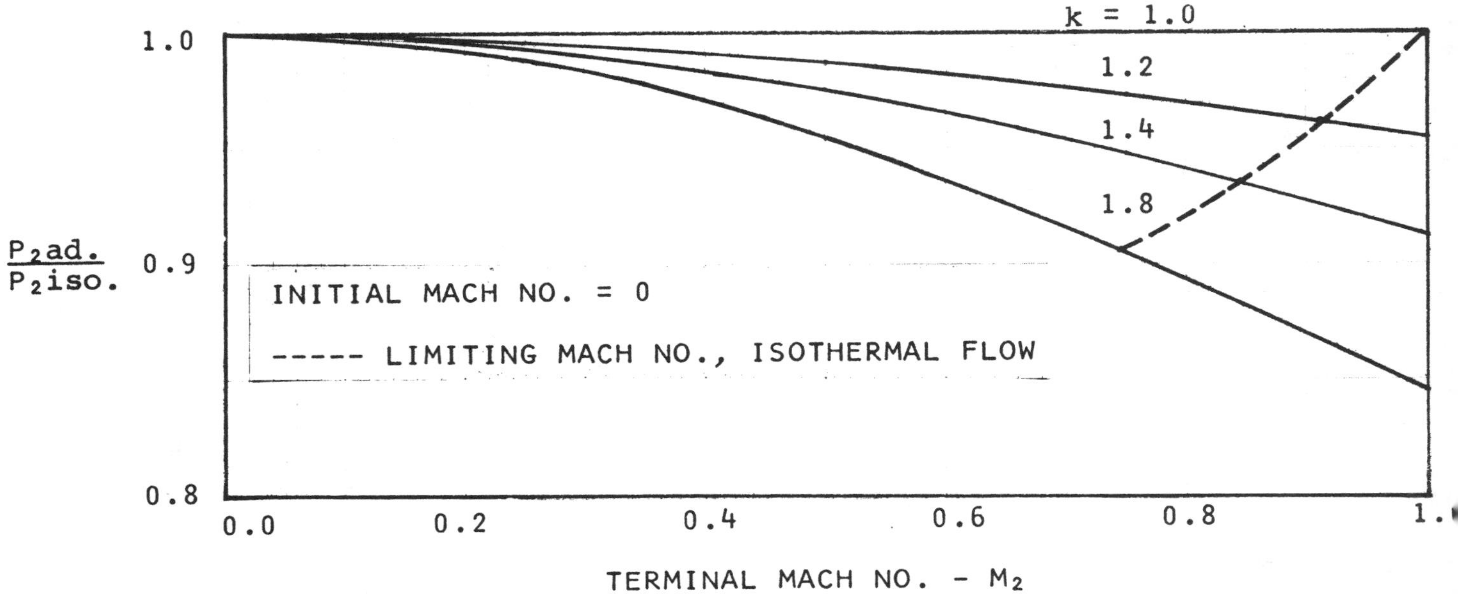

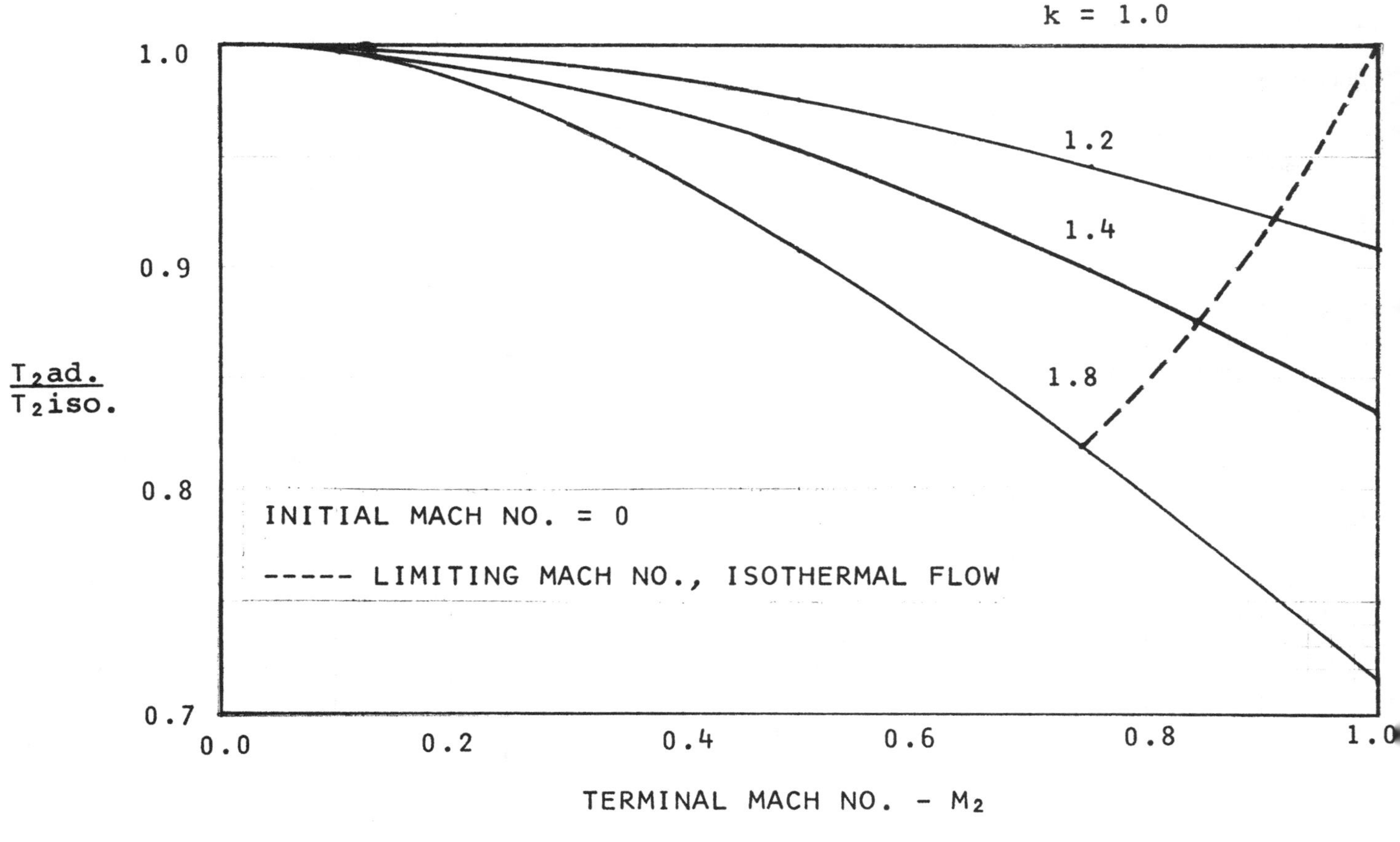

Figure 1.17. Comparison of adiabatic and isothermal compressible flow for given inlet conditions.

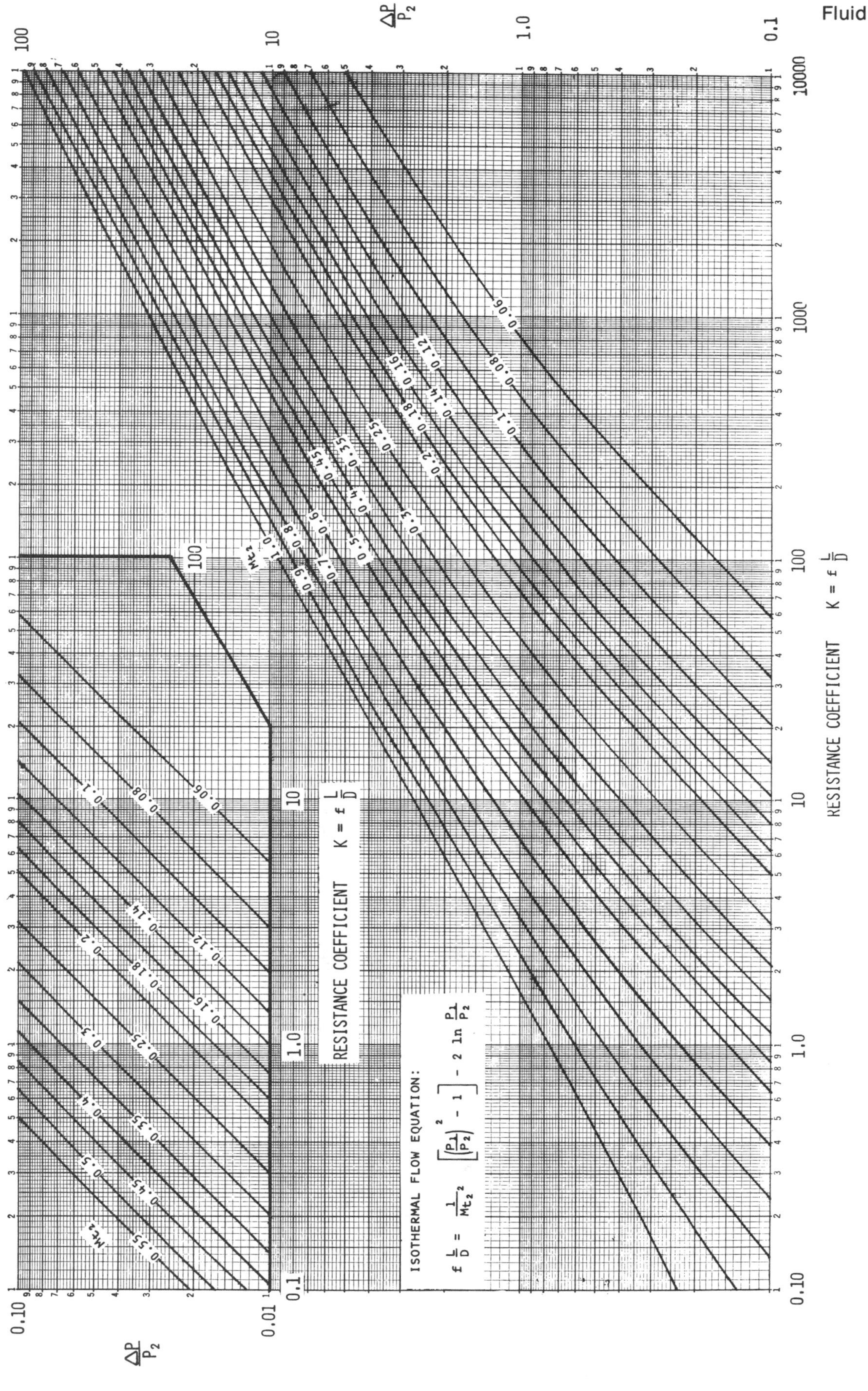

Figure 1.18. Isothermal compressible flow chart based on downstream conditions.

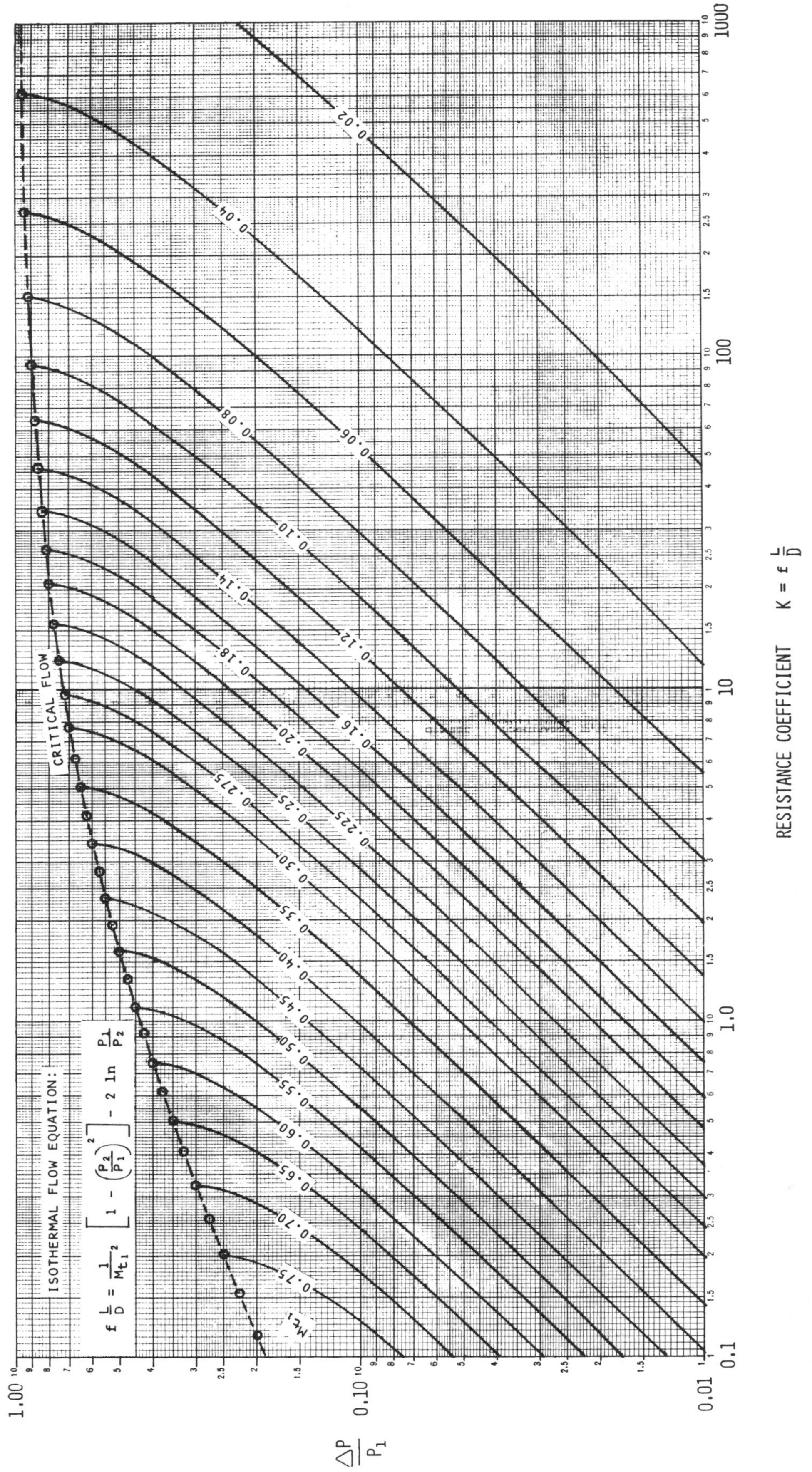

Figure 1.19. Isothermal compressible flow chart based on inlet conditions.

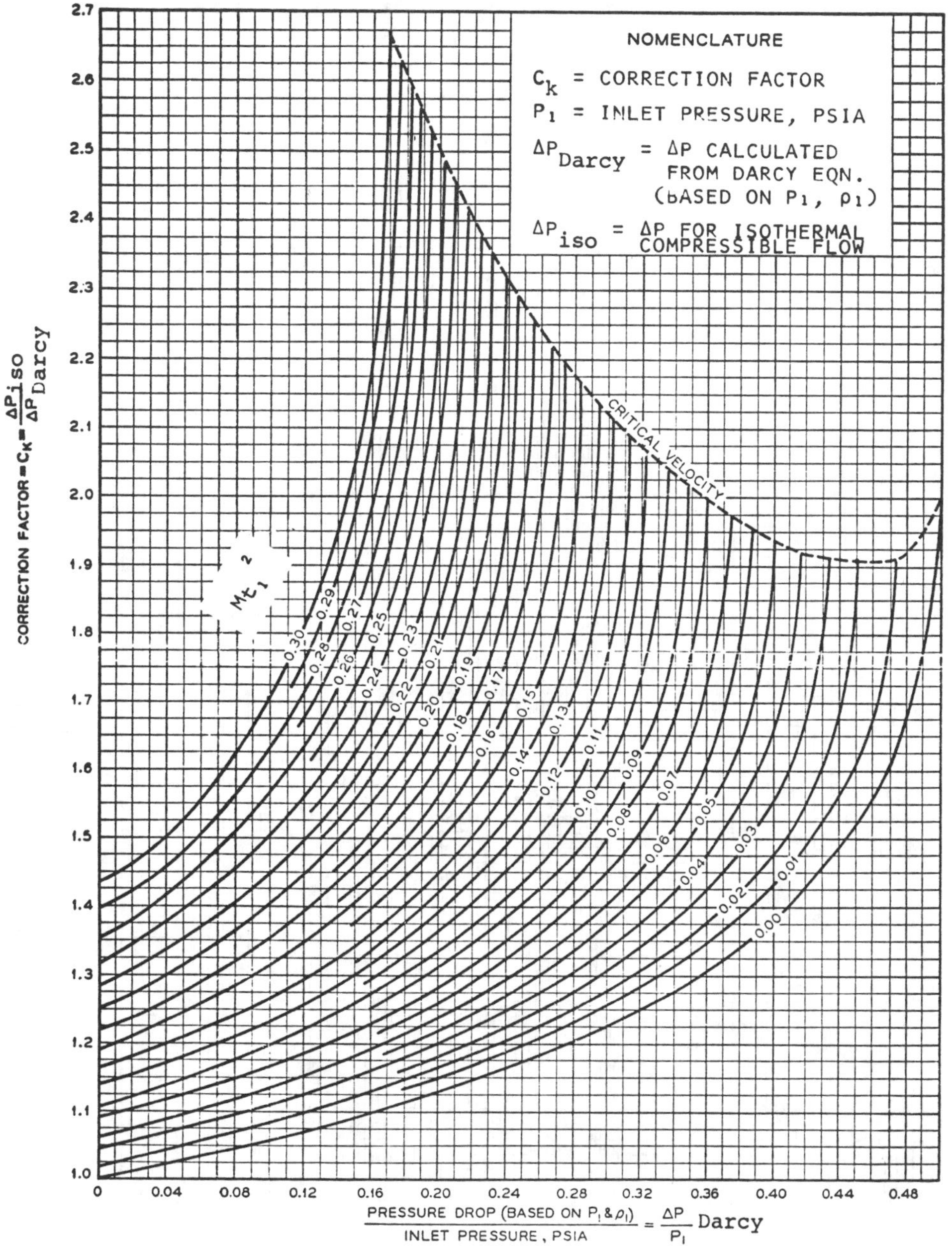

Figure 1.20. Isothermal compressible-flow/correction chart for Darcy equation. Reprinted from API RP520 by permission of the American Society of Mechanical Engineers (ASME).

Procedure for Calculation of ΔP_{iso}:

1. Calculate ΔP_{Darcy} based on P_1, T_1, W, L & D. (See section on non-compressible flow, p. 21.)

2. Calculate $M_{t1}{}^2$.

$$M_{t1} = 1.336 \times 10^{-3} \frac{W}{AP_1} \sqrt{\frac{ZT_1}{MW}}$$

3. From chart, determine C_k at calculated values of $\Delta P_{Darcy}/P_1$ and $M_{t1}{}^2$.

4. Calculate $\Delta P_{iso} = C_k \ \Delta P_{Darcy}$.

5. If intersection of coordinates for $\Delta P_{Darcy}/P_1$ and $M_{t1}{}^2$ falls above dashed line, critical flow exists at the outlet, indicating that W exceeds capacity of the line for given conditions. To avoid this situation, W or L must be reduced, or P_1 or D increased, and pressure drop recalculated.

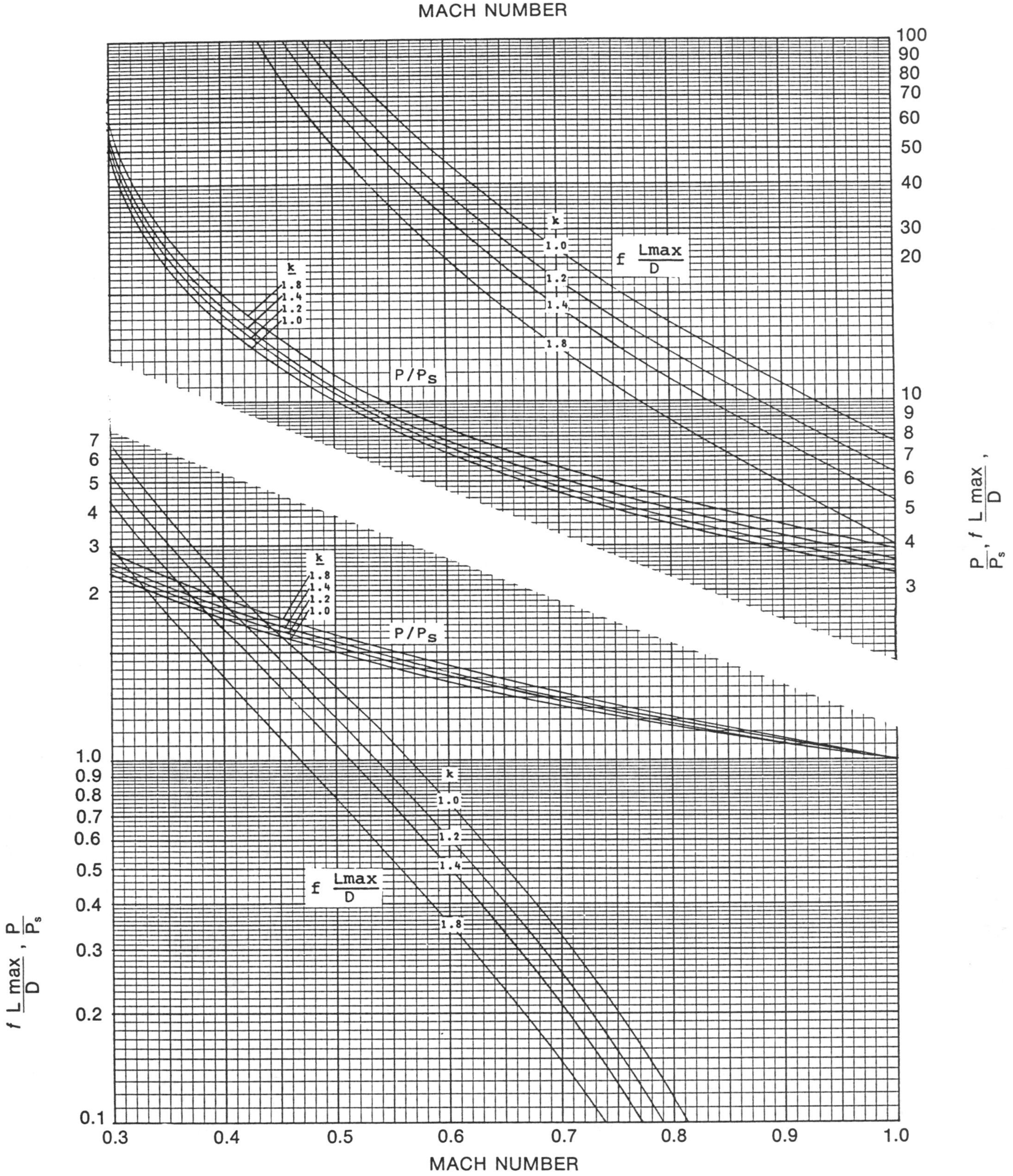

Figure 1.21. Adiabatic compressible flow chart.

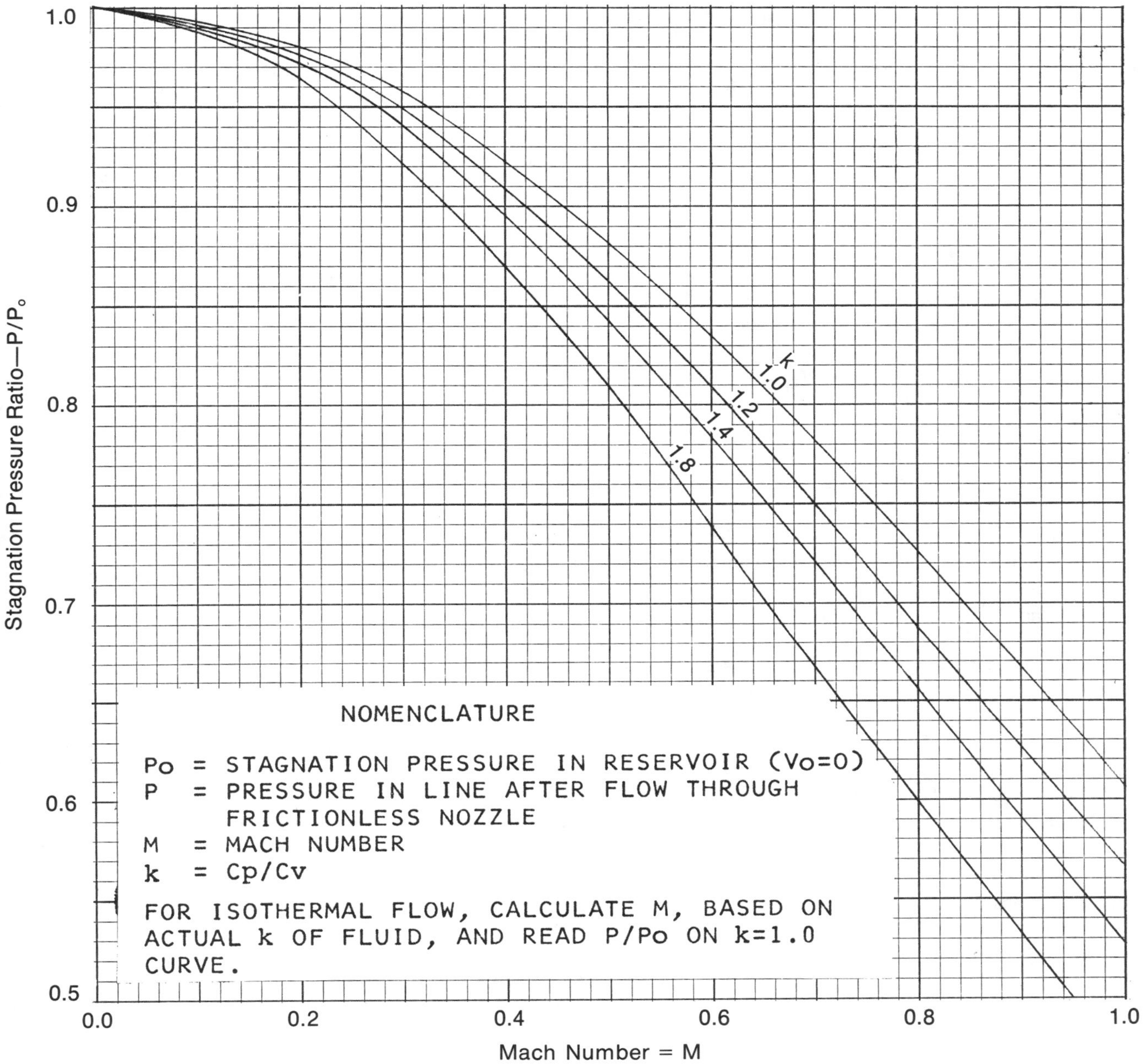

Figure 1.22. Stagnation pressure chart.

For any inlet condition, there is a maximum length for which continuous isothermal flow is possible, dependent only on initial Mach number:

$$\left(f\frac{L_{max}}{D}\right)_1 = \frac{1}{M_{t1}^2} - 1 + \ln M_{t1}^2 \qquad (1.49)$$

For a given inlet Mach number, the following expression must be satisfied:

$$\left(f\frac{L}{D}\right)_{design} \leqslant \left(f\frac{L_{max}}{D}\right)_1 \qquad (1.50)$$

Thus, for a given piping configuration, there is a maximum attainable inlet Mach number which occurs when:

$$\left(f\frac{L}{D}\right)_{design} = \left(f\frac{L_{max}}{D}\right)_1 \qquad (1.51)$$

When this condition is reached, $M_{t_2} = 1$ and $P_2 = P_3$ (see Figure 1.16). Equation 1.48 shows that P_2/P_1 becomes constant, equal to M_{t_1}, and it follows that:

$$\left(\frac{\Delta P}{P_1}\right)_{\text{critical}} = 1 - M_{t_1 \text{ critical}} \tag{1.52}$$

$$\left(\frac{\Delta P}{P_2}\right)_{\text{critical}} = \frac{1}{M_{t_1 \text{ critical}}} - 1 \tag{1.53}$$

As a mass flow is increased beyond this point, P_1 and P_2 must increase accordingly, and P_2 becomes greater than P_3, a condition called choke flow.

The relationship of $[f(L_{max}/D)]_1$ to M_{t_1} is shown graphically in Figure 1.19 by the dashed line at the top of the chart. At this limiting condition, since M_{t_2} is known, P_2 may be calculated from Equation 1.47 for a given mass flow rate.

In calculations where flow rate is unknown, it is necessary to calculate K without knowing the friction factor, f. A good first approximation for f is f^*, the friction factor at fully developed turbulent flow. Values of f^* for commercial steel pipe are listed in Table 1.11. After estimating flow rate, friction factor may be calculated and flow rate corrected accordingly.

Graphical solutions to Equations 1.42 and 1.43 are presented in Figures 1.18 and 1.19, respectively. With these charts, direct solutions to most isothermal compressible flow problems may be obtained. However, when pipe diameter is unknown, trial and error is required. See example problems 1.8 and 1.9 for illustration of how charts are used.

Another procedure for calculating isothermal compressible pressure loss, by using a factor to correct pressure loss calculated with the Darcy equation, is presented in Figure 1.20. Inlet conditions must be known to use this procedure conveniently. Other graphical procedures have been developed by Lapple (3) and Smith (4), also based on inlet conditions. All of these procedures give essentially the same results, so the procedure used is a matter of personal preference.

Adiabatic Compressible Flow

Since adiabatic flow calculations are more complex to handle than isothermal, a common practice is to treat any compressible flow problem as being isothermal, unless a high degree of precision is required. Each such problem should be analyzed to determine which calculation method best approximates the real situation.

In general, it is necessary to consider adiabatic flow only in short lines ($K<10$), or at high Mach numbers (>0.75). A comparison of final pressure and temperature for isothermal flow and adiabatic flow for given inlet conditions is shown in Figure 1.17. It can be seen that differences in final pressure are reasonably small up to the limiting Mach number for isothermal flow. It should be noted, however, that the P_2 ratios do not correspond to equal values of $f(L/D)$, thus, differences in final pressures do not reflect equivalent flow paths.

Comparison of adiabatic and isothermal flow for identical flow paths and inlet conditions are shown in example problems 1.10 and 1.11.

The equations necessary for solving most adiabatic compressible flow problems are as follows:

$$\bar{f}\frac{L}{D} = \left(\bar{f}\frac{L_{max}}{D}\right)_1 - \left(\bar{f}\frac{L_{max}}{D}\right)_2 \tag{1.54}$$

$$\left(f\frac{L_{max}}{D}\right) = \frac{1 - M^2}{kM^2} + \frac{k+1}{2k} \ln \left[\frac{(k+1)M^2}{2 + (k-1)M^2}\right] \tag{1.55}$$

$$\frac{P}{P_s} = \frac{1}{M}\sqrt{\frac{k+1}{2 + (k-1)M^2}} \tag{1.56}$$

$$P_2 = P_1 \frac{(P_2/P_s)}{(P_1/P_s)} \tag{1.57}$$

$$M = 1.336 \times 10^{-5} \frac{W}{AP}\sqrt{\frac{ZT}{kMW}} \tag{1.58}$$

In the above equations f is an average friction factor determined from values calculated at initial and final conditions. As a first assumption for $\bar{f}$, f^* may be used (see Table 1.11 for values of f^* for commercial steel pipe.) P/P_s is the ratio of static pressure to pressure at sonic velocity, and is a function of Mach number and specific heat ratio (Equation 1.56).

Solution of adiabatic compressible flow problems involves a degree of trial and error, depending on the problem. Kirkpatrick has developed a graphical procedure (5) which eliminates trial and error for many problems. A chart for this procedure is presented in Figure 1.21. Use of the chart is explained below.

When pipe configuration, flow rate, and inlet conditions are known, downstream pressure may be calculated as follows:

1. Approximate $\bar{f}\ (L/D)$, using f^* for $\bar{f}$
2. Calculate $(f\,(L_{max}/D)_1$, from Equation 1.55.
3. Calculate $(f\,(L_{max}/D)_2$, using Equation 1.54.
4. Determine M_2 by trial and error solution (or graphical, see Figure 1.21) of Equation 1.55.
5. Calculate P_1/P_s and P_2/P_s, using Equation 1.56, or read values from Figure 1.21.
6. Determine P_2 from Equation 1.57.
7. Determine f_1 and f_2, using Figure 1.12 or Equation 1.39 and Figure 1.13, then calculate $\bar{f} = (f_1 + f_2/2)$.
8. Recalculate P_2 using new value of $\bar{f}$.

If $\bar{f}$ (L/D) is greater than $[f(L_{max}/D)]_1$, the given mass flow rate exceeds the capacity of the line for the given inlet conditions. In this situation, M_1 must be reduced by increasing upstream pressure, or decreasing flow rate until $[f(L_{max}/D)] \geqslant \bar{f}(L/D)$, or $\bar{f}(L/D)$ reduced by increasing pipe diameter.

For the case where $\bar{f}(L/D) = [\bar{f}(L_{max}/D)]_1$, $P_2 \geqslant P_3$ and $P_2/P_s = 1$. Since M is known, P_1/P_s may be determined from Equation 1.56, and P_2 from Equation 1.57. If desired, $\bar{f}$ may be calculated as given above, and used to recalculate a more precise value of P_2. See example problems 1.10 and 1.11 for illustration of adiabatic flow calculation procedures.

Problems in which pipe diameter is unknown involve somewhat more complex trial and error procedures, but in general, are handled in a similar manner.

Other graphical procedures for solving adiabatic compressible flow problems have been developed by Lapple (3) and Crane Co. (2).

Stagnation Presssure

Stagnation pressure, P_o, is the pressure reached isentropically at a point of zero velocity ($V_o = 0$). It is also called reservoir pressure. From a practical standpoint, it may be used to estimate changes in static pressure at a surge vessel or knockout pot in which velocity is much lower than that in the line. Since the inlet and outlet paths from such vessels are not frictionless nozzles, entrance and exit losses must be taken into consideration by using an approximate value of K to evaluate the total pressure change. A chart showing values of P/P_o versus M at various values of K, presented in Figure 1.22, is useful in evaluating these pressure changes. The use of this chart is illustrated in Example Problem 1.9.

Example Problems

Example Problem 1.8
Isothermal Compressible Flow, Downstream Conditions Known

A 12″ sch. 20 flare header discharging to atmosphere having an equivalent length of 1000 ft. must handle 400,000 lb./hr. of hydrocarbon gas at 270°F. $MW = 74$, $\mu = 0.012$ cp, $k = 1.2$, $Z = 1.0$.

(a) What pressure is developed at the inlet to the header?

(b) What flow rate would the header handle at an inlet pressure of 60 psia?

Solution

(a) Assume $P_2 = P_3$ and check for critical flow at outlet. From Table 1.11: 12″ Sch. 20 pipe, $d = 12.25''$, $A = 0.8185$ ft.^{2}, $f^* = 0.0129$

From Equation 1.47:

$$M_{t_2} = \frac{(1.336 \times 10^{-5})(400{,}000)}{(0.8185)(14.7)} \sqrt{\frac{(1)(730)}{74}}$$

$$M_{t_2} = 1.395$$

This indicates critical flow at the outlet, and setting $M_{t_2} = 1$:

$$P_2 = (1.395)(14.7) = 20.5 \text{ psia}$$

Calculate f:

$$\frac{W}{\mu} = \frac{400{,}000}{0.012} = 3.3 \times 10^7$$

From chart, Figure 1.13, read $f/f^* = 1.01$
Calculate $f = f^*(f/f^*) = (0.0129)(1.01) = 0.0130$

Calculate: $K = f\dfrac{L}{D} = \dfrac{(0.0130 \times 1000)(12)}{(12.25)} = 12.7$

From Chart, Figure 1.18, @ $M_{t_2} = 1$, $K = 12.7$, read

$$\frac{\Delta P}{P_2} = 3.08$$

$$\Delta P = (3.08)(20.5) = 63.1 \text{ psi}$$

$$P_1 = 63.1 + 20.5 = 83.6 \text{ psia}$$

(b) Assume $P_2 = P_3$ and check for critical flow at outlet:

$$\frac{\Delta P}{P_2} = \frac{60 - 14.7}{14.7} = 3.08$$

From Chart, Figure 1.18, @ $K = 12.7$, $\dfrac{\Delta P}{P_2} = 3.08$, read

$$M_{t_2} = 0.98$$

Thus, flow is below critical and $P_2 = 14.7$ psia
From Equation 1.47,

$$W = \frac{(0.98)(0.8185)(14.7)}{1.336 \times 10^{-5}} \sqrt{\frac{74}{(1)(730)}}$$

$$W = 281{,}000 \text{ lb./hr.}$$

This solution assumes that f is the same as in part (a). In this case the error incurred is negligible. If desired, f could be recalculated for the new flow rate, and flow rate recalculated. f^* could also have been used for f in the first calculation.

Example Problem 1.9
Isothermal Compressible Flow, Inlet Conditions Known

An 8″ sch. 20 line of 2,000 ft. equivalent length must carry a flow of 150,000 lb./hr. of hydrocarbon gas at 300°F. MW = 80, μ = 0.012 cp. Inlet pressure = 150 psig, k = 1.2.

(a) At what pressure will a knock-out vessel operate that is located an equivalent length of 1000 ft. from the inlet?

(b) At what pressure must the pipe outlet be controlled to maintain the desired inlet pressure?

Solution

(a) From Table 1.11, 8″ Sch. 20 pipe:

A = 0.3601 ft.^{2}
d = 8.125 in.
f^* = 0.0140

Calculate f:

From chart, Figure 1.13, read $f/f^* = 1.01$
Calculate: $f = f^* \, (f/f^*) = (0.0140)(1.01) = 0.0141$

$$\text{Calculate: } K = f\frac{L}{D} = \frac{(0.0141)(1000)(12)}{(8.125)} = 20.8$$

K for entrance to vessel (exist loss) = 1
$K_{total} = 20.8 + 1 = 21.8$

From Equation 1.47, calculate M_{t_1}

$$M_{t_1} = \frac{(1.336 \times 10^{-5})(150{,}000)}{(0.3601)(164.7)} \sqrt{\frac{(1)(760)}{80}}$$

$M_{t_1} = 0.104$

From chart, Figure 1.19, @ K = 21.8, M_{t_1} = 0.104, read

$$\frac{\Delta P}{P_1} = 0.120$$

$\Delta P = (0.120 \times 164.7) = 19.8$ psi
P_2 = 164.7 - 19.8 = 144.9 psia or 130.2 psig = P_3

(b) Assuming that velocity in the knockout vessel is very low, determine the pressure at the inlet (point 4 of calculations below) of the vessel outlet line based on stagnation pressure ratio P/P_o from Figure 1.22:
Trial and error is necessary: Assume $M_2 = M_4$

$$M_{t_2} = M_{t_1}\frac{P_1}{P_2} = 0.104\left(\frac{164.7}{144.9}\right) = 0.118$$

$$M_2 = \frac{0.118}{\sqrt{1.2}} = 0.108 = M_4$$

From Figure 1.22, @ M = 0.108, k = 1.0, read P/P_o = 0.993
P_4 = (144.9) (0.993) = 143.9 psia
Then, using P_4, recalculate M_4:

$$M_4 = \frac{0.104}{\sqrt{1.2}} = \frac{(164.7)}{(143.9)} = 0.109$$

From Figure 1.22, P/P_o = 0.992
P_4 = (0.992) (144.9) = 143.7 psia (Close enough to previous value)
Since M_4 is low (<0.3), P/P_o almost = 1, and the assumption that $M_4 = M_2$ is sufficiently precise for most calculations, making trial and error unnecessary.

$$\text{Calculate: } K = f\frac{L}{D} = \frac{(0.0141)(1000)(12)}{(8.125)} = 20.8$$

K for exit from vessel (entrance loss) = 0.5
K_{total} = 20.8 + 0.5 = 21.3
$M_{t_4} = M_4\sqrt{k} = 0.109\sqrt{1.2} = 0.119$

From Figure 11.9, @ M_{t_1} = 0.119, K = 21.3 read

$$\frac{\Delta P}{P_1} = 0.165$$

ΔP = (0.165) (143.7) = 23.7 psia
P_5 = 143.7 - 23.7 = 120.0 psia or 105.3 psig

Example Problem 1.10
Comparison of Isothermal and Adiabatic Flow (*K*<10, *M*$_2$>0.75)

A gas (MW = 30) is flowing through an 8″ sch. 20 line at inlet Mach number (M_1) = 0.75. P_1 = 250 psia, T_1 = 560°R, k = 1.2, Z = 1.0.

(a) Calculate maximum length of the line and pressure drop for isothermal flow.

(b) For line of length calculated in (a), calculate pressure drop for adiabatic flow at same inlet conditions.

Solution

(a) Calculate $\left(f\frac{L_{max}}{D}\right)_1$ from Equation 1.49:

$$\left(f\frac{L_{max}}{D}\right)_1 = \frac{1}{(1.2)(0.75)^2} - 1 + \ln\left[(1.2)(0.75)^2\right] = 0.088$$

$$L_{max} = K\frac{D}{f}$$

From Table 1.11, 8″ sch. 20 pipe

d = 8.125″
A = 0.3601 ft.²
f^* = 0.0140

$$L_{max} = \frac{(0.088)(8.125)}{(0.0140)(12)} = 4.3 \text{ ft.}$$

The use of f^* for f in this calculation results in negligible error since flow is fully developed. If desired, f may be calculated from procedure given for Figure 1.12.

By definition at $(f\frac{L_{max}}{D})_1$:

$$M_2 = \frac{1}{\sqrt{k}} = \frac{1}{\sqrt{1.2}} = 0.913 \text{ and,}$$

$$P_2 = P_1\frac{M_1}{M_2} = 250\frac{(0.75)}{(0.913)} = 205.4 \text{ psia}$$

$\Delta P = 250 - 205.4 = 44.6$ psia

(b) Assume friction factor same as in (a). Then $K = 0.088$

Calculate $\left(f\frac{L_{max}}{D}\right)_1$ from Equation 1.55:

$$\left(f\frac{L_{max}}{D}\right)_1 = \frac{1-(0.75)^2}{(1.2 \times 0.75)^2} = \frac{2.2}{2.4}\ln\left[\frac{2.2(0.75)^2}{2+(0.2)(0.75)^2}\right]$$

$$\left(f\frac{L_{max}}{D}\right)_2 = 0.58 - 0.088 = 0.070$$

Solve for M_2 in Equation 1.55 by trial and error.

Try $M_2 = 0.85$

$$\left(f\frac{L_{max}}{D}\right)_2 = 0.046$$ – This is too far below 0.070.

Try $M_2 = 0.82$

$$\left(f\frac{L_{max}}{D}\right)_2 = 0.070$$ – Close enough!

Calculate P_1/P_s from Equation 1.56 (or read from Figure 1.21):

$P_1/P_s = 1.36$

Calculate P_2/P_s from same equation (or read from Figure 1.21):

$P_2/P_s = 1.23$

Calculate P_2 from Equation 1.57:

$$P_2 = 250\frac{(1.23)}{(1.36} = 226.1 \text{ psia}$$

$\Delta P = 250 - 226.1 = 23.9$ psi

Comparing adiabatic and isothermal flow at these conditions ($K = 0.088$, $M_1 = 0.75$):

$$\frac{P_{2\,ad}}{P_{2\,iso}} = \frac{226.1}{205.4} = 1.10$$

$$\frac{\Delta P_{ad}}{\Delta P_{iso}} = \frac{23.9}{44.6} = 0.54$$

Thus, although downstream pressures are reasonably close, there is a very large difference in pressure loss between adiabatic and isothermal flow for this set of conditions.

Example Problem 1.11
Comparison of Isothermal and Adiabatic Flow (K = 10, M_2<0.75)

A gas (MW = 30) is flowing through an 8″ sch. 20 line at M_1 = 0.24, P_1 = 250 psia, T_1 = 560°R, k = 1.2, Z = 1.0, K = 10.

(a) Calculate downstream pressure, pressure loss and exit Mach number for isothermal flow.

(b) Calculate same for adiabatic flow.

Solution

(a) Calculate $M_{t_1} = M_1 \quad k = 0.24 \quad 1.2 = 0.263$
From chart, Figure 1.19, @ $K = 10$, $M_{t_1} = 0.263$, read

$$\frac{\Delta P}{P_1} = 0.57$$

$\Delta P = (0.57)(250) = 142.5$ psi
$P_2 = 250 - 142.5 = 107.5$ psia

$$M_2 = M_1\frac{P_1}{P_2} = 0.24\frac{(250)}{(107.5)} = 0.56$$

(b) Calculate $\left(f\frac{L_{max}}{D}\right)_1$ from Equation 1.55, or read from chart, Figure 1.21:

$$\left(f\frac{L_{max}}{D}\right)_1 = 10.93$$

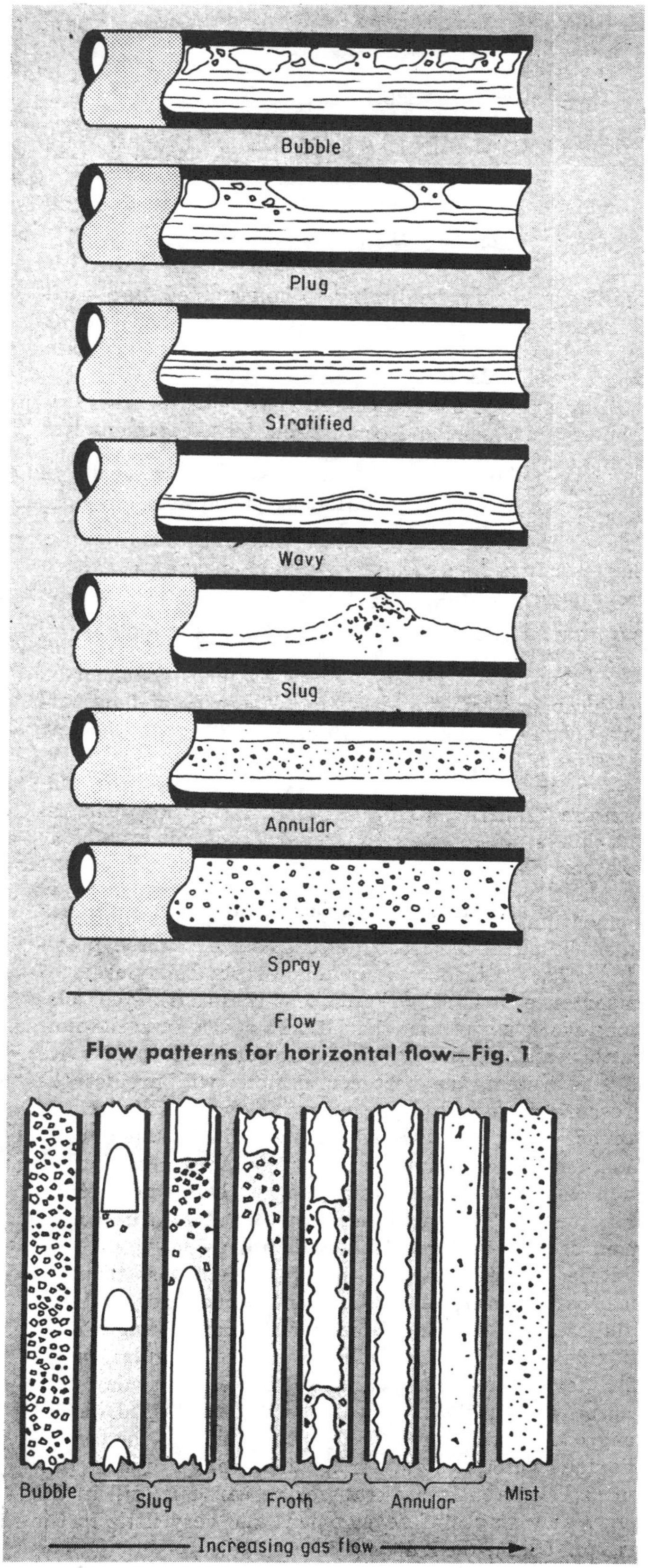

Figure 1.23. Two-phase flow patterns. Reprinted from Two-Phase Flow Patterns, Chemical Engineering, *December 6, 1965, Anderson, R.J. and Russell, T.W.F. by permission.*

Calculate $\left(f \dfrac{L_{max}}{D}\right)_2$ from Equation 1.54:

$$\left(f \frac{L_{max}}{D}\right)_2 = 10.93 - 10 = 0.93$$

Determine M_2 by trial and error solution of Equation 1.55, or read from Figure 1.21 @ $\left(f \dfrac{L_{max}}{D}\right)_2 = 0.93$, $k = 1.2$:

$$M_2 = 0.543$$

Calculate P_1/P_s from Equation 1.56, or read from Figure 1.21:

$$P_1/P_s = 4.36$$

Calculate P_2/P_s from the same equation, or read from Figure 1.21:

$$P_2/P_s = 1.90$$

Calculate P_2 from Equation 1.57:

$$P_2 = 250 \frac{(1.90)}{(4.36)} = 108.9 \text{ psia}$$

$$P = 250 - 108.9 = 141.1 \text{ psi}$$

Comparing adiabatic and isothermal flow at these conditions ($K = 10, M_1 = 0.24$):

$$\frac{P_{2\ ad}}{P_{2\ iso}} = \frac{108.9}{107.5} = 1.01 \text{ and } \frac{\Delta P_{ad}}{\Delta P_{iso}} = \frac{141.1}{142.5} = 0.99$$

Thus, there is very little difference in results whether isothermal or adiabatic flow is used as the calculation basis for this set of conditions.

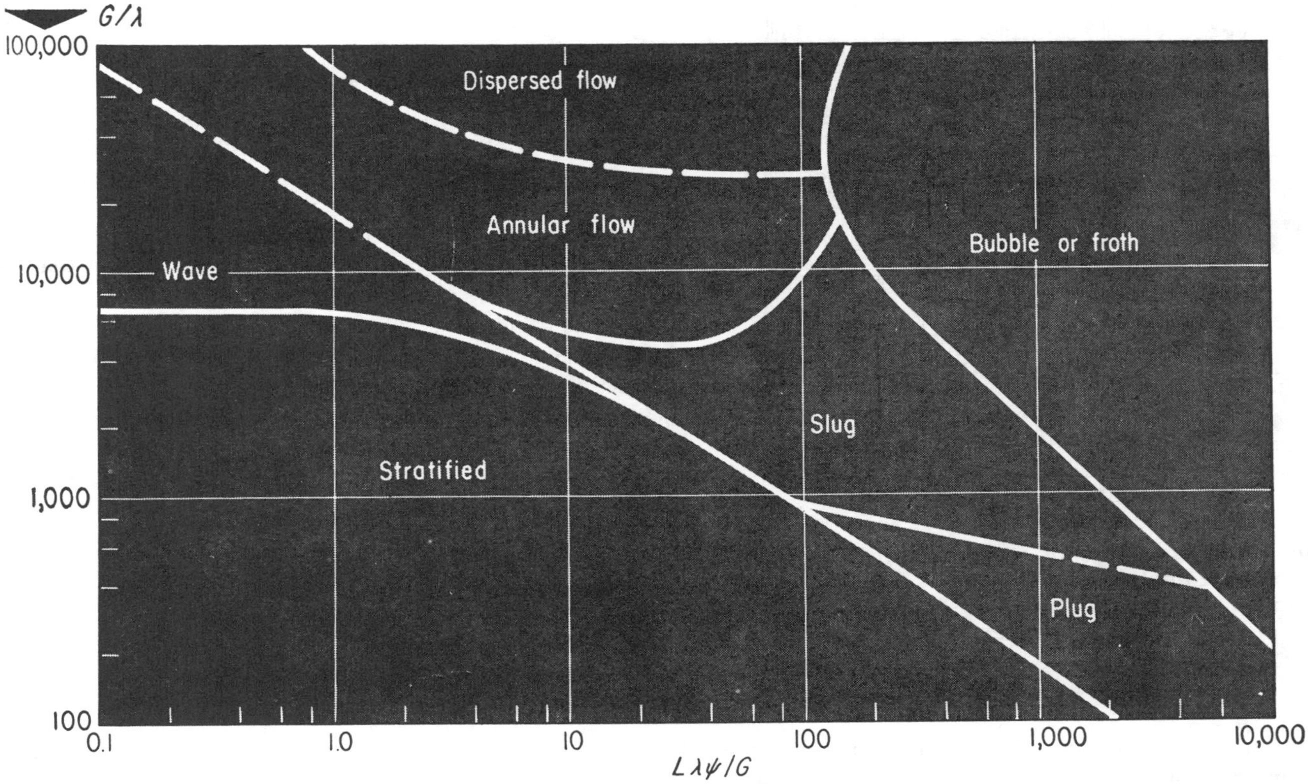

Figure 1.24. Flow pattern prediction chart for horizontal flow. Reprinted from article by O. Barker in Oil and Gas Journal, *November 10, 1958, per permission.*

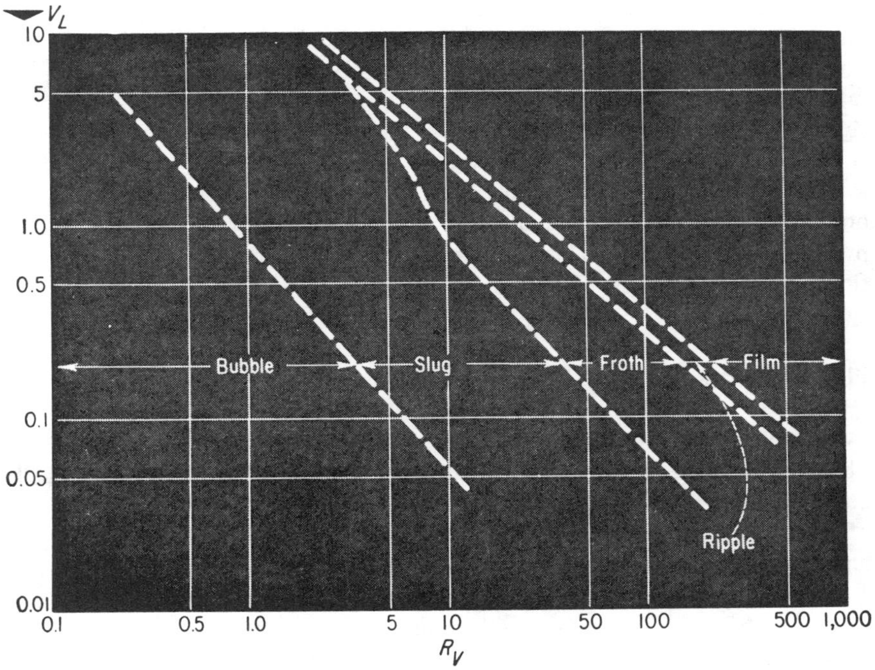

Figure 1.25. Flow pattern prediction chart for vertical flow. Reprinted from article by G.W. Govier in Chemical Engineering *in 1957 by permission.*

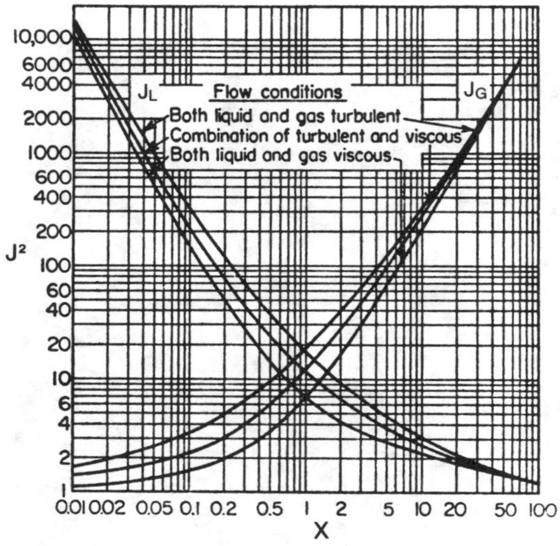

Figure 1.26. Parameters for pressure drop in liquid-gas flow through horizontal pipes. From Perry's Chemical Engineers' Handbook, *4th ed., copyright © 1963 by McGraw-Hill, Inc. Used with permission of McGraw-Hill Book Company.*

Two-Phase Vapor-Liquid Flow

No attempt is made here to treat two-phase vapor-liquid flow in great detail. It is a complex subject and, although much work has been done in the field, many questions remain unanswered. The objective of this section is to help the reader understand and approximately predict the flow patterns that may be present in any two-phase pipeline, and to provide a means for estimating two-phase pressure drop. For more detailed treatment consult the references cited.

In comparing two-phase flow to single-phase flow, the main difficulty is the variety of flow patterns in a gas-liquid system. The pattern encountered depends on fluid properties, flow rates, and equipment geometry. Unit pressure loss may vary significantly between patterns. Currently, there are no general correlations that accurately predict the behavior of two-phase systems over the range of possible flow patterns. There are, however, correlations for specific flow patterns that predict pressure loss with reasonable accuracy (~± 40%), when the flow pattern and physical and geometric properties of the system are known.

Flow Patterns

Two-phase flow patterns have been classified by visual observation by Alves for horizontal flow (6), and by Nicklin and Davidson for vertical flow (7). Illustration and description of each of the seven basic flow patterns for horizontal flow, and five basic patterns for vertical flow are shown in Figure 1.23.

Although implications of the various flow regimes on pressure loss, heat transfer, and mass transfer are not fully understood, the pattern of slug flow must be avoided in normal commercial design, because it can cause severe vibration in piping and equipment that could possibly result in eventual mechanical failure, in addition to complicating process measurement and control. Thus, it is important to be able to predict with reasonable assurance the flow pattern that will occur for a given design problem to prevent unsafe, rough operation due to slug flow, to be able to choose the best correlation for process calculations and to permit translation of satisfactory performance associated with a particular flow regime from one scale to another.

A number of charts have been prepared to predict flow patterns, based on flow conditions, system geometry, and fluid properties. Baker (8) presents such a chart for horizontal flow (see Figure 1.24). The various flow pattern regions are plotted as a function of Baker parameters, B_y and B_x which are expressed as follows:

$$B_y = G_G/\lambda \tag{1.59}$$

where

$$G_G = \frac{W_G}{A} \tag{1.60}$$

and

$$\lambda = \left[\left(\frac{\rho_G}{0.075}\right)\left(\frac{\rho_L}{62.3}\right)\right]^{1/2} = 0.463\sqrt{\rho_G \rho_L} \tag{1.61}$$

thus

$$B_y = 2.16\,\frac{W_g}{A\sqrt{\rho_G \rho_L}} \tag{1.62}$$

Also

$$B_x = \frac{G_L\,\lambda\,\psi}{G_G} \tag{1.63}$$

Where

$$G_L = \frac{W_L}{A} \tag{1.64}$$

and

$$\psi = \left[\frac{73}{\sigma_L}\right]\mu_L\left[\left(\frac{62.3}{\rho_L}\right)^2\right]^{1/3} \tag{1.65}$$

Thus

$$B_x = 531\,\frac{W_L}{W_G}\,\frac{\rho_G{}^{1/2}}{\rho_L{}^{1/6}}\,\frac{\mu_L{}^{1/3}}{\sigma_L} \tag{1.66}$$

With these two parameters, the type of flow may be approximated from Figure 1.24. It must be emphasized that the borders of flow pattern regions are not sharply defined but are zones of transition between flow patterns. Thus, if coordinates of the parameters intersect near the border of a region, consequences of flow patterns in adjacent regions must also be considered.

A chart for vertical two-phase flow similar to that of Baker's for horizontal flow has been developed by Govier, Radford, and Dunn (9). The parameters used to prepare the chart (see Figure 1.25) are superficial liquid velocity:

$$V_L = 2.78 \times 10^{-4} \frac{W_L}{\rho_L A} \tag{1.67}$$

and volumetric ratio of gas to liquid entering the line:

$$R_V = \frac{Q_G}{Q_L} \tag{1.68}$$

As with the Baker charts, the flow patterns predicted by this chart are approximate, and good engineering judgment must be applied in using the results.

Pressure Loss Prediction

Pressure loss due to friction for co-current vapor-liquid flow in horizontal pipes may be estimated by the correlation of Lockhart and Martinelli (10). To use this correlation, it is necessary to estimate the pressure loss each phase would have if it were flowing alone. The single phase pressure loss for the gas phase is ΔP_G, and for the liquid phase, ΔP_L. The parameter X (see Figure 1.26) is calculated as follows:

$$X = \sqrt{\frac{\Delta P_L}{\Delta P_G}} \tag{1.69}$$

Single phase flow is classified into four categories:

1. Both liquid and gas flow are turbulent
2. Liquid flow is viscous, gas flow is turbulent
3. Liquid flow is turbulent, gas flow is viscous
4. Both liquid and gas flow are viscous

Each of these regimes is shown as a separate curve in Figure 1.26, except that only one curve is given for liquid viscous-gas turbulent and liquid turbulent-gas viscous because the difference between the experimental curves is small compared to the uncertainty of the correlation. The correlating parameter J, is related to the two-phase pressure drop as follows:

$$\Delta P_{TP} = (J_L)^2\, \Delta P_L = (J_G)^2\, \Delta P_G \tag{1.70}$$

The transition point between viscous and turbulent flow for this correlation is not definitely known. However, for design purposes, single phase flow may be considered turbulent for $R_e > 2{,}000$, and viscous for $R_e \leqslant 2{,}000$.

Thus, when X has been determined, and single phase flow characterized, $(J_L)^2$ or $(J_G)^2$ may be read from the chart (Figure 1.26) and the two-phase pressure drop calculated by Equation 1.70.

Davis (11) has pointed out that vertical two-phase up-flow results in higher frictional pressure losses than those experienced in horizontal flow. He suggested a modification to the Lockhart-Martinelli modulus, X, which is valid for all flow regions, with an expected accuracy of ±20% at Reynolds numbers above 8,000 for the liquid phase, and above 2,100 for the gas phase. The Davis correlation is as follows:

$$X_D = 0.19\, X\, (F_r)^{0.185} \tag{1.71}$$

Where F_r is the Froude number:

$$F_r = 3.11 \times 10^{-2} \frac{V^2}{D} \tag{1.72}$$

X_D is used in place of X in Figure 1.26 to determine a value of $(J_L)^2$ or $(J_G)^2$ for calculating pressure loss in two-phase vertical up-flow. Only the curves for liquid turbulent-gas turbulent flow apply when using this correlation.

A more complex but perhaps more accurate correlation for predicting two-phase pressure losses has been developed by Dukler, et al (12). Discussions of this method by Anderson and Russel (13), and Simpson (14) may also be helpful.

Example Problem 1.12
Use of Baker Flow-Pattern Prediction Chart

A liquid is flowing in a horizontal 1" sch. 40 pipe (flow area = 0.00600 ft^2) along with 25 bl/hr of a gas. Average physical properties of liquid and gas at flowing conditions are:

$\mu_L = 0.5$ cp., $\rho_L = 55.0$ lb/ft^3, $\sigma_L = 17$ dynes/cm.
$\rho_G = 0.095$ $\overline{\text{lb}}_m$/ft^3 $\mu_G = 0.014$ cp

Find the expected flow pattern using the method of Baker.

Solution

1. Calculate B_Y using Equation 1.62

$$B_Y = \frac{(2.16)(25)}{(0.00600)\,[(0.095)(55.0)]^{1/2}} = 3{,}937$$

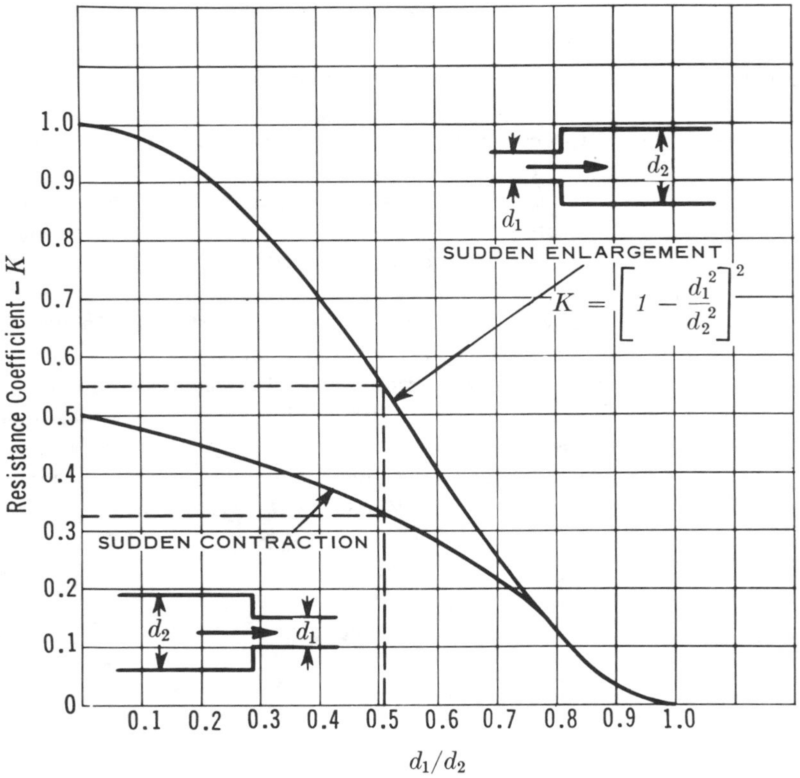

Sudden enlargement: The resistance coefficient K for a sudden enlargement from 6-inch Schedule 40 pipe to 12-inch Schedule 40 pipe is **0.55,** based on the 6-inch pipe size.

$$\frac{d_1}{d_2} = \frac{6.065}{11.938} = 0.51$$

Sudden contraction: The resistance coefficient K for a sudden contraction from 12-inch Schedule 40 pipe to 6-inch Schedule 40 pipe is **0.33,** based on the 6-inch pipe size.

$$\frac{d_1}{d_2} = \frac{6.065}{11.938} = 0.51$$

Note: The values for the resistance coefficient, K, are based on velocity in the small pipe. To determine K values in terms of the greater diameter, multiply the chart values by $(d_2/d_1)^4$.

Figure 1.27. Resistance in pipe due to sudden enlargements and contractions. Courtesy Crane Co. (T.P. #410).

2. Calculate B_X from Equation 1.66

$$B_X = \frac{(531)(2{,}500)(0.095)^{1/2}(0.5)^{1/3}}{(25)(55.0)^{1/6}(17)} = 392$$

From Figure 1.24, pattern is slug flow.

Example Problem 1.13
Use of Govier Chart for Prediction of Vertical Flow Patterns

Using same flow rates and physical properties as in Example Problem 1.12, determine expected pattern in vertical flow by the method of Govier, et al.

Solution

1. Calculate V_L from Equation 1.67

$$V_L = \frac{2.78 \times 10^{-4}\,(2{,}500)}{(55.0)(0.00600)} = 2.1 \text{ ft/sec.}$$

2. Calculate R_V using Equation 1.68

$$R_V = \frac{Q_G}{Q_L}$$

$$Q_G = \frac{25}{0.095} = 263.2 \text{ ft}^3\text{/hr.}$$

$$Q_L = \frac{2{,}500}{55.0} = 45.5 \text{ ft}^3\text{/hr.}$$

$$R_V = \frac{263.2}{45.5} = 5.79$$

From Figure 1.25, expected flow pattern falls into the slug flow region.

Example Problem 1.14
Illustration of Lockhart-Martinelli Method for Prediction of Two-Phase Pressure Drop in Horizontal Flow

A mixture of 5,000 lb/hr of liquid and 40 lb/hr of a gas are flowing in a horizontal 1″ sch 40 pipe having an equivalent length of 250 ft. Average physical properties at flowing conditions are same as in Example Problem 1.12.

Determine expected flow regime and pressure drop.

Solution

1. Use Baker Method to determine flow pattern.

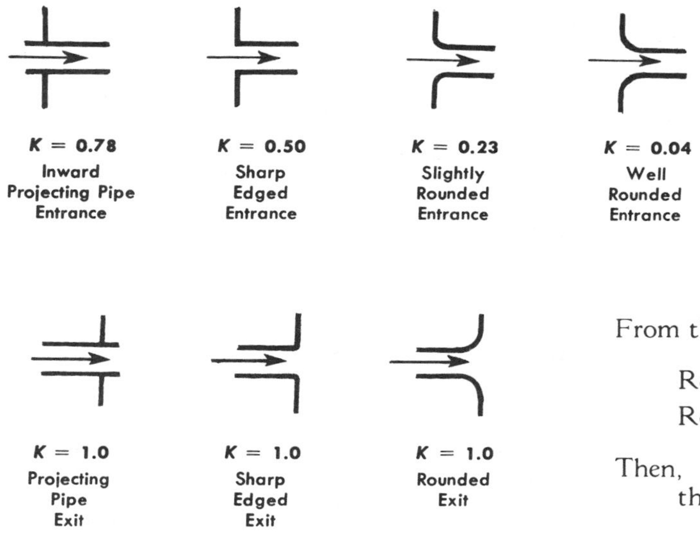

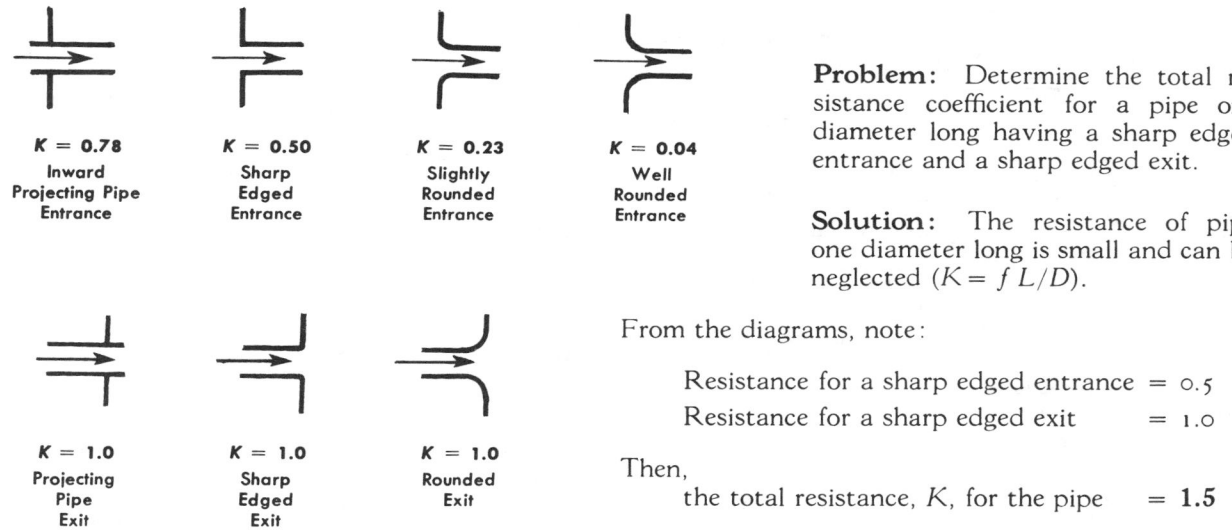

Problem: Determine the total resistance coefficient for a pipe one diameter long having a sharp edged entrance and a sharp edged exit.

Solution: The resistance of pipe one diameter long is small and can be neglected ($K = f\,L/D$).

From the diagrams, note:

Resistance for a sharp edged entrance = 0.5
Resistance for a sharp edged exit = 1.0

Then,
the total resistance, K, for the pipe = **1.5**

Figure 1.28. Resistance due to pipe entrance and exit. Courtesy Crane Co. (T.P. No. 410).

$$B_Y = \frac{(2.16)(40)}{(0.00600)\,[(0.095)(55.0)]^{1/2}} = 6{,}300$$

$$B_X = \frac{(531)(5{,}000)(0.095)^{1/2}(0.5)^{1/3}}{(40)(55)^{1/6}(17)} = 490$$

From Figure 1.24. Flow is bubble or froth flow.

2. Calculate ΔP_L according to section on Incompressible Flow.

$$W^2/\rho = \frac{(5{,}000)^2}{55.0} = 4.5 \times 10^5$$

From Figure 1.14, $\Delta P_L{}^*/L = 2.7$ psi/100′

$$\frac{W}{\mu} = \frac{5{,}000}{0.5} = 1.0 \times 10^4$$

From Figure 1.13 $f/f^* = 1.14$
$\Delta P_L/L = (2.7)(1.14) = 3.1$ psi/100′

3. Calculate $\Delta \rho_G$ by same procedure

$$\frac{W^2}{\rho} = \frac{(40)^2}{(0.095)} = 1.7 \times 10^4$$

From Figure 1.14, $\Delta P_G{}^*/L = 0.10$ psi/100′

$$\frac{W}{\mu} = \frac{40}{0.0.4} = 2.9 \times 10^3$$

From Figure 1.13, $f/f^* = 1.34$

$\Delta P_G/L = (0.10)(1.34) = 0.13$ psi/100′

4. Calculate parameter X from Equation 1.69

$$X = \left[\frac{3.1}{0.13}\right]^{1/2} = 4.88$$

5. Read parameter $(J)^2$ from Figure 1.26, on line for both phases turbulent flow.

$(J_L)^2 = 5$
$(J_G)^2 = 120$

6. Calculate ΔP_{TP} using Equation 1.70

$\Delta P_{TP}/L = (5)(3.1) = 15.5$ psi/100′
$\Delta P_{TP}/L = (120)(0.13) = 15.6$ psi/100′

$$\Delta P_{TP} = 15.5 \times \frac{250}{100} = 38.8 \text{ psi}$$

Equivalent Line Lengths for Pipe

An equivalent line length, L, is the fictitious length of straight pipe that would cause the same head loss in a given flow situation as a real flow path composed of a combination of lengths of straight pipe, fittings, entrances and exits. Equivalent lengths of many piping configurations found in industry are often much greater then the sum of their straight lengths of pipe, especially for larger sizes of pipe.

Table 1.12. Representative Equivalent Length in Pipe Diameters (L/D) of Valves and Fittings for Turbulent Flow

Description of Product				Equivalent Length In Pipe Diameters (*L/D*)
Globe Valves	Stem Perpendicular to Run	With no obstruction in flat, bevel, or plug type seat	Fully open	340
		With wing or pin guided disc	Fully open	450
	Y-Pattern	(No obstruction in flat, bevel, or plug type seat)		
		– With stem 60 degrees from run of pipe line	Fully open	175
		– With stem 45 degrees from run of pipe line	Fully open	145
Angle Valves		With no obstruction in flat, bevel, or plug type seat	Fully open	145
		With wing or pin guided disc	Fully open	200
Gate Valves	Wedge, Disc, Double Disc, or Plug Disc		Fully open	13
			Three-quarters open	35
			One-half open	160
			One-quarter open	900
	Pulp Stock		Fully open	17
			Three-quarters open	50
			One-half open	260
			One-quarter open	1200
Conduit Pipe Line Gate, Ball, and Plug Valves			Fully open	3**
Check Valves	Conventional Swing		0.5†. . . Fully open	135
	Clearway Swing		0.5†. . . Fully open	50
	Globe Lift or Stop; Stem Perpendicular to Run or Y-Pattern		2.0†. . . Fully open	Same as Globe
	Angle Lift or Stop		2.0†. . . Fully open	Same as Angle
	In-Line Ball		2.5 vertical and 0.25 horizontal†. . . Fully open	150
Foot Valves with Strainer		With poppet lift-type disc	0.3†. . . Fully open	420
		With leather-hinged disc	0.4†. . . Fully open	75
Butterfly Valves (8-inch and larger)			Fully open	40
Cocks	Straight-Through	Rectangular plug port area equal to 100% of pipe area	Fully open	18
	Three-Way	Rectangular plug port area equal to 80% of pipe area (fully open)	Flow straight through	44
			Flow through branch	140
Fittings	90 Degree Standard Elbow			30
	45 Degree Standard Elbow			16
	90 Degree Long Radius Elbow			20
	90 Degree Street Elbow			50
	45 Degree Street Elbow			26
	Square Corner Elbow			57
	Standard Tee	With flow through run		20
		With flow through branch		60
	Close Pattern Return Bend			50
Pipe	90 Degree Pipe Bends			
	Miter Bends			
	Sudden Enlargements and Contractions			
	Entrance and Exit Losses			

**Exact equivalent length is equal to the length between flange faces or welding ends.

†Minimum calculated pressure drop (psi) across valve to provide sufficient flow to lift disc fully.

Source: Crane Company (T.P. #410).

Since the equivalent length of a straight pipe is simply its actual length, equivalent length data is required only for fittings, entrances and exits. Methods of presentation for equivalent length data vary.

One method is use of an equivalent length ratio, L/D, for each fitting. The equivalent length of the fitting is then

$$L = D(L/D) \tag{1.73}$$

Another method is the use of a velocity head count, (also called resistance coefficient), K, for each fitting. The relation between K and L/D is

The chart at the right shows the resistance of 90 degree bends to the flow of fluids in terms of equivalent lengths of straight pipe.

Resistance of bends greater than 90 degrees is found using the formula:

$$\frac{L}{D} = R_t + (n - 1)\left(R_l + \frac{R_b}{2}\right)$$

n = total number of 90° bends in coil
R_t = total resistance due to one 90° bend, in L/D
R_l = resistance due to length of one 90° bend, in L/D
R_b = bend resistance due to one 90° bend, in L/D

Problem: Determine the equivalent lengths in pipe diameters of a 90 degree bend and a 270 degree bend having a relative radius of 12.

Solution: Referring to the "Total Resistance" curve, the equivalent length for a 90 degree bend is **34.5** pipe diameters.

The equivalent length of a 270 degree bend is:

$$L/D = 34.5 + (3 - 1)\,[18.7 + (15.8 \div 2)]$$
$$L/D = \mathbf{87.7} \text{ pipe diameters}$$

Note: This loss is less than the sum of losses through three 90 degree bends separated by tangents.

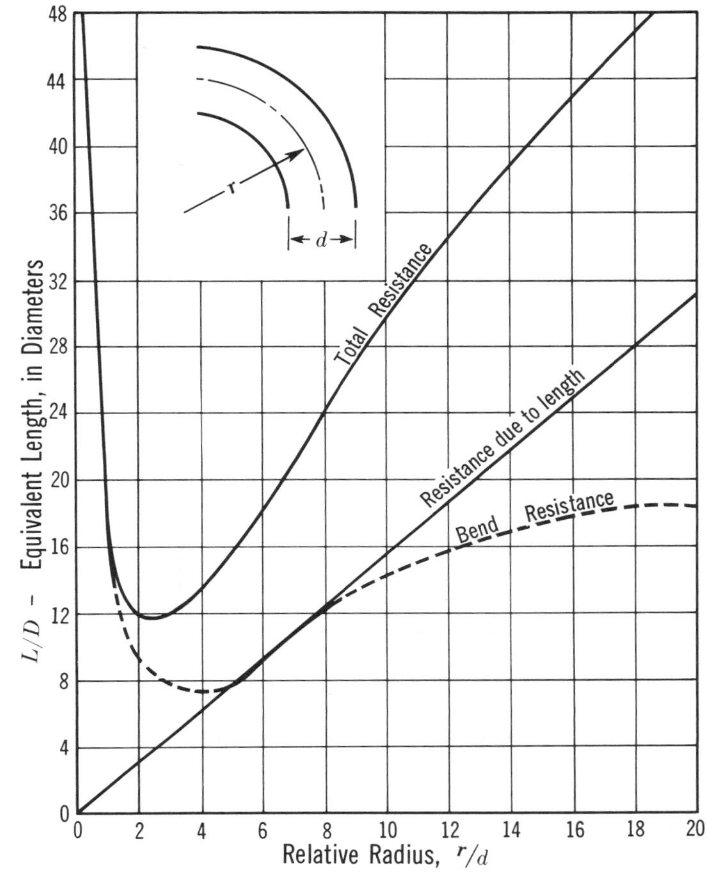

Figure 1.29. Resistance of 90° bends. Courtesy Crane Co. (T.P. #410) and National Bureau of Standards.

$$K = f(L/D) \tag{1.74}$$

A third method of presenting data on the flow resistance of fittings, especially valves, is use of a flow coefficient, C_v, defined as the flow of water in gallons per minute at 60°F that occurs through a fitting at a pressure loss of one psi. The relationship between C_v, L/D, and K is

$$C_V = \frac{4{,}295\,D^2}{K} = \frac{4{,}295\,D^2}{fL/D} \tag{1.75}$$

or

$$L = 1.845 \times 10^7 \frac{D^5}{fC_V^{\,2}} \tag{1.76}$$

When a flow path is composed of components of two or more pipe sizes it is necessary to adjust their equivalent lengths to a common pipe size before the component equivalent lengths may be totaled to get the total equivalent line length.

Adjustment of (L/D) ratios for pipe size "a" to pipe size "b" is made using the ratio (D_b/D_a) to the fourth power.

$$(L/D)_b = (L/D)_a\,(D_b/D_a)^4 \tag{1.77}$$

Equivalent lengths or lengths of straight pipe are adjusted by the ratio (D_b/D_a) to the fifth power.

$$L_b = L_a\,(D_b/D_a)^5 \tag{1.78}$$

Adjustment of K values for fittings and pipe involve ratios of friction factors and inside pipe diameters.

$$K_b = K_a\,(f_b/f_a)\,(D_b/D_a)^4 \tag{1.79}$$

Adjustment of K values for sudden contractions and expansions, entrances, and exits is made without the friction factor correction term.

Accurate estimation of the equivalent lengths of fittings is difficult due to insufficient data on the many different types and sizes and the wide variety of flow conditions in which they may be involved. Exact data should be used if available. The data presented in Table 1.12 and Figures 1.27 through 1.31 should be sufficiently accurate for most industrial fluid flow calculations.

A basic assumption in the presented data is that (L/D) ratios are a constant for a given type of fitting regardless of size and are independent of flow conditions for Reynolds number greater than 1,000. This assumption is not completely true. Geometric similarity is rarely preserved between fittings of the same type, but different size. True values of K and (L/D) ratios do vary somewhat due to flow conditions where R_e greater than 1,000. Crane Company (2) considers (L/D) ratios to vary less than K values with variation in flow conditions.

For Reynolds numbers less than 1,000, (L/D) ratios may be estimated by

$$(L/D)_L = \frac{R_e}{1{,}000} (L/D)_T \tag{1.80}$$

where the subscript "L" denotes laminar flow for R_e less than 1000 and $(L/D)_T$ is the turbulent flow (L/D) ratio such as those listed in Table 1.12.

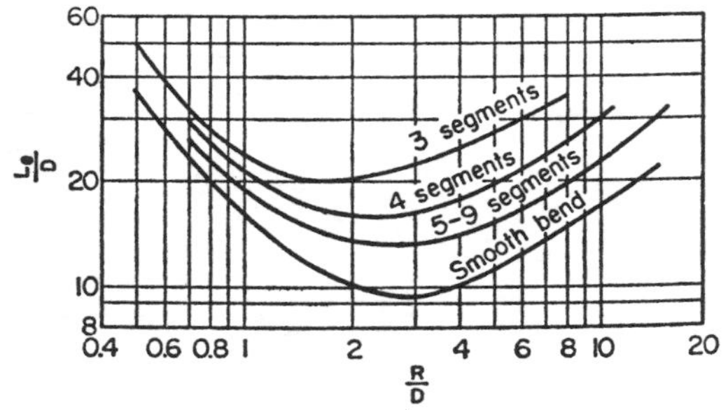

Figure 1.31. Total friction loss in 90° bends. From Perry's Chemical Engineers' Handbook, *4th ed., copyright © 1963 by McGraw-Hill, Inc. Used with permission of McGraw-Hill Book Company.*

Terminology

Numerous words and phrases used in fluid flow are either unique to the field or have specialized meanings. The terms defined below are a few of those basic to fluid flow.

Absolute units. Units of measurement that are of zero value at the minimum possible or potential value. Thus, there can be no negative values of quantities measured in absolute units.

Adiabatic. Without heat transfer. Compressible fluids having a ratio of specific heats greater than 1 and subjected to adiabatic flow will decrease in temperature in the direction of flow.

The chart at the lower right shows the resistance of miter bends to the flow of fluids. The chart is based on data published by the American Society of Mechanical Engineers (ASME).

Problem: Determine the equivalent length in pipe diameters of a 40 degree miter bend.

Solution: Referring to the "Total Resistance" curve in the chart, the equivalent length is **12** pipe diameters.

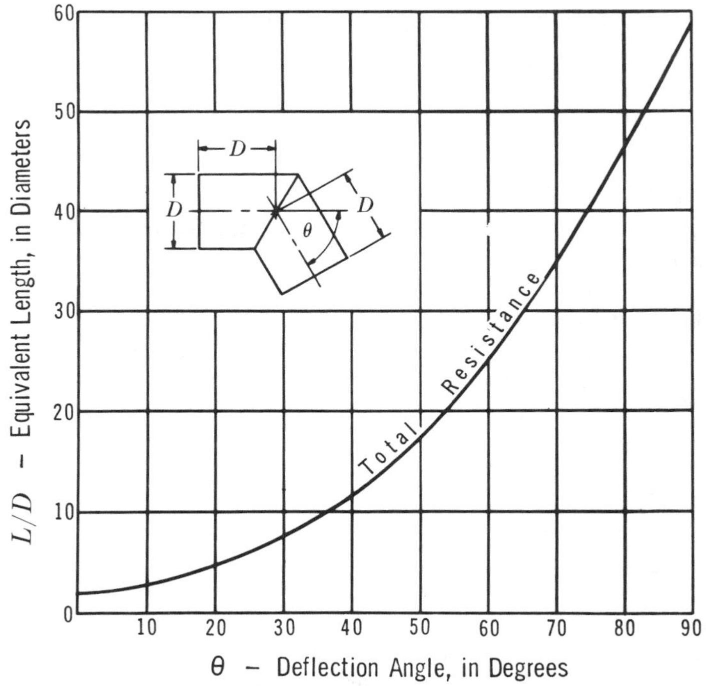

Figure 1.30. Resistance in miter bends. Courtesy Crane Co. (T.P. No. 410).

Choke flow. A condition in the compressible flow field where critical flow exists within a flow path. If the flow path is uniform over its entire length, the point of critical flow will be the downstream terminus. The significance of choke flow is that for this condition and a given flow path, the mass flow rate can be increased only by increasing the upstream pressure.

Critical properties. For any pure fluid there is a maximum temperature at which the fluid may occur as a liquid. This temperature is called the critical temperature, T_c. The vapor pressure at the critical temperature is called the critical pressure, P_c. Critical properties are fluid physical properties at conditions of critical temperature and pressure. Liquid and vapor are indistinguishable at critical conditions.

Critical Reynolds Number. The Reynolds number at which transition between laminar and turbulent flow occurs.

Fully Developed Turbulent Flow. Flow is very turbulent and the friction factor no longer changes with a change in Reynolds number. It is represented by the extreme right hand portion of Figure 1.12. Values of the friction factor at fully developed turbulent flow, f^*, have been tabulated for steel pipe in Table 1.12.

Hydraulic Radius (R_H). Defined by the equation

$$R_H = \frac{\text{flow area}}{\text{wetted perimeter}} \tag{1.81}$$

Where flow area is that portion of the duct cross-sectional area occupied by the flowing fluid and wetted perimeter is that portion wetted by flowing fluid. R_H is used in flow problems involving ducts of noncircular cross-section or ducts not filled with flowing fluid. For a round pipe flowing full

$$R_H = D/4 \tag{1.82}$$

and $4R_H$ can be used as a replacement for D in a number, but not all, of fluid flow equations.

$$\epsilon/D = \epsilon/(4R_H) \tag{1.83}$$

$$R_e = 1{,}488\ \rho VD/\mu = 1{,}488\ \rho\ \frac{V(4R_H)}{\mu} \tag{1.84}$$

$$\Delta P/L = \frac{f\rho V^2}{9{,}265D} = \frac{f\rho V^2}{9{,}265\ (4R_H)} \tag{1.85}$$

Results of Equations 1.82, 1.83, and 1.84 can be used in conjunction with Figure 1.12 for obtaining friction factors in the turbulent region.

Isothermal. Constant temperature.

Limiting velocity ($V_t = V_s/\ k$). The maximum velocity that can potentially occur for isothermal subsonic compressible flow. At this velocity the pressure loss gradient becomes infinite.

Mach number (M). A dimensionless number equal to the ratio of superficial velocity to the velocity of sound for a fluid at the conditions of the fluid. It is an important parameter in adiabatic compressible flow.

Mach number, isothermal (M_t). A dimensionless number equal to Mach number divided by ratio of specific heat (k). It is an important parameter in isothermal compressible flow.

One-dimensional flow. Flow in which all fluid and flow parameters (velocity, pressure, density, viscosity, etc.) are constant over any cross section normal to the flow.

Reduced properties. Physical properties expressed as dimensionless numbers. This is done by dividing the value of a property measured in absolute units for a given fluid at given conditions by the value of the property measured in the same units for the same fluid at critical conditions. Principle application is in estimating physical properties because many fluids exhibit identical reduced properties at identical reduced conditions.

Relative radius (R/D). A characteristic of pipe bends, it is the ratio of the radius of the bend axis to the inside diameter of the pipe comprising the bend.

Reynolds number (R_e). A dimensionless number proportional to the ratio of inertial force to viscous force of flow. It is used as one of the parameters for determining friction factors and for distinguishing between laminar and turbulent flow.

Sonic velocity (V_s). The velocity of sound through a material. Subsonic flow occurs when actual velocity is less than sonic velocity. Supersonic flow occurs when actual velocity is greater than sonic velocity.

Steady flow. Invariant with time, i.e., flow, temperature, pressure, duct area, etc., are independent of time. The converse of unsteady flow.

Streamline. A line that lies in the direction of flow at every point in a given instant.

Superficial velocity (V). The volumetric flow rate divided by the flow area. It is used to indicate a mean velocity as distinguished from local velocity, which will vary from point-to-point over the flow area.

Uniform. A flow path is uniform if the size and shape of its cross section is the same for the length of the flow path.

Nomenclature

A = Area for flow, square feet.
B_x = Baker parameter in flow pattern prediction method (Equation 1.62).
B_Y = Baker parameter in flow pattern prediction method (Equation 1.58).
C_p = Specific heat at constant pressure, btu/lb.-°F
C_v = Specific heat at constant volume, btu/lb.-°F
C_V = Valve flow coefficient = $GPM/\sqrt{\Delta P/SG}$
d = Pipe inside diameter, inches.
D = Pipe inside diameter, feet.
E = Elevation referred to any arbitrary reference plane, feet.
f = Moody friction factor, dimensionless
$\overline{f}$ = Average friction factor based on values at initial and final conditions.
g = Acceleration of gravity = 32.17 ft./sec.
G = Fluid mass velocity - $lb_m/hr\text{-}ft^2$
g_c = A constant equal to 32.17 (lb_m-ft.)/(lb_f-sec.^{2})
h_f = Head loss over a flow path due to flow through pipe and fittings, feet fluid
h_L = Sum of all head losses over a flow path due to flow, feet fluid = h_f + equipment head losses + instrument head losses
J = Correlating parameter in Lockhart-Martinelli method for prediction of two-phase flow pressure loss (Equation 1.69).
k = Specific heat ratio = C_p/C_v, dimensionless.
K = Resistance coefficient, dimensionless.
L = Equivalent length, feet.
L_{max} = Maximum equivalent length for continuous compressible flow and a given set of conditions, feet.
M = Mach number = V/V_s, dimensionless.
M_t = Isothermal Mach number = V/V_t, dimensionless.
MW = Molecular weight, lb./mole.
P = Absolute pressure = gauge pressure + atmospheric pressure, psia
Q = Volumetric flow rate, cu. ft./sec.
R = Universal gas constant = 10.73 (psia - ft.^{3})/(lb. mole- °R).
R_e = Reynolds number, dimensionless.
R_H = Hydraulic radius, feet.
R_V = Volumetric ratio of gas to liquid entering pipe (Equation 1.67).
SG = Specific gravity, dimensionless
T = Absolute temperature = 460 + °F, °R
TDH = Total Differential Head, feet of fluid
V = Superficial velocity, ft./sec.
$\overline{V}$ = Molar volume, ft^3/(lb-mole)
V_s = Sonic velocity, ft./sec.
V_t = Limiting velocity for isothermal compressible flow, ft./sec.
W = Flow rate, pounds mass per hour.
x = Mole fraction of liquid phase, dimensionless.
X = Parameter in Lockhart-Martinelli method prediction of two-phase flow pressure drop.
y = Mole fraction of vapor phase, dimensionless.
Z = Compressibility factor = $P\overline{V}/(RT)$, dimensionless.

Greek Symbols

γ = Specific weight = $\rho g/g_c$, lb_f/cu. ft.
Δ = Finite difference.
ϵ = Absolute roughness of pipe inner wall, feet.
λ = Parameter in Baker's flow pattern prediction method (Equation 1.60), dimensionless
μ = Dynamic viscosity, centipoise
ρ = Density, lb_m/cu. ft.
Σ = Mathematical symbol for summation
σ = Surface tension, dynes/cm.
ϕ = Volume fraction, dimensionless.
ψ = Parameter in Baker's flow pattern prediction method (Equation 1.64).

Mixed Symbols

ΔP = Pressure loss over a flow path due to flow through pipe and fittings, psi
$\Sigma\Delta P$ = Sum of all pressure losses over a flow path due to flow, psi

Subscripts

c = critical property
G = gas phase
i = belonging to component i
L = liquid phase
mix = mixture property or condition
o = stagnation condition
r = reduced property or condition
s = condition for sonic velocity
TP = two-phase
1 = upstream condition
2 = downstream condition

Superscripts

*properties at fully developed turbulent flow.

References

1. Perry, John H., Chilton and Kirkpatrick, eds. *Chemical Engineer's Handbook*, 4th ed. New York: McGraw-Hill Book Co., 1969.

2. Engineering Division of Crane Co. *Flow of Fluids Through Valves, Fittings, and Pipe,* Technical Paper No. 410. Chicago: Crane Co., 1965.
3. Lapple, C.E. *Trans. A.I.Ch.E.* 39, (1943) pp 385-432.
4. Smith, Benjamin. "Charts Used for Easier Pipe Sizing," *Hydrocarbon Processing,* December, 1969, pp. 135-138.
5. Kirkpatrick, D.M. "Simpler Sizing of Gas Piping," *Hydrocarbon Processing,* December, 1969, pp. 135-138.
6. Alves, G.E. *Chemical Engineering Progress* 50, 1954, pp. 449.
7. Nicklin, D.J. and Davison, J.R. "Symposium on Two-Phase Flow," Inst. Mech. Eng., Paper No. 4, London, February, 1962.
8. Baker, Ovid. "Multiphase Flow in Pipelines," *Oil and Gas Journal,* November 10, 1958, pp. 156-167.
9. Govier, G.W., Radford, B.A. and Dunn, J.S. "The Upward Vertical Flow of Air Water Mixtures," *Canadian Journal of Chemical Engineering,* 35, August 1957, p. 58-70.
10. Lockhart, R.W. and Martinelli, R.C. "Proposed Correlation of Data for Isothermac Two-Phase, Two-Component Flow in Pipes," *Chemical Engineering Process,* 45, No. 1, January, 1949, pp. 39-48.
11. Davis, W.J. *British Chemical Engineering,* 8, July, 1963.
12. Dukler, A.E., Wicks, Moye III and Cleveland, R.G. "Frictional Pressure Drop in Two-Phase Flow: B. An Approach Through Similarity Analysis," *A.I.Ch.E. Journal,* 10, No. 1, January 1949, pp. 44-51.
13. Anderson, R.J., Russel, T.W.F. "Designing for Two-Phase Flow," *Chemical Engineering,* Part I, December 6, 1965; Part II, December 20, 1966; Part III, January 3, 1966.
14. Simpson, Larry L. "Sizing Piping for Process Plants," *Chemical Engineering,* June 17, 1968, pp. 192-214.
15. Olson, Ruben M. *Essentials of Engineering Fluid Mechanics.* Scranton: International Textbook Co., 1961, pp. 230-241.
16. Kern, Robert. "How to Size Process Piping for Two-Phase Flow," *Hydrocarbon Processing,* October, 1969, pp. 105-116.
17. Paige, Peter M. "How to Estimate the Pressure Drop of Flashing Mixtures," *Chemical Engineering,* August 14, 1967, pp. 159-164.
18. Ludwig, E.E. *Applied Process Design for Chemical and Petrochemical Plants,* Volume 1, Gulf Publishing Co., 1964.
19. American Petroleum Institute RP 520, "Recommended Practice for the Design and Installation of Pressure Relieving Systems in Refineries," Part 1, 3rd ed., 1967.
20. American Petroleum Institute RP 521, "Guide for Pressure Relief and Depressuring Systems," 1969.

2 Engineering Graphical Symbols

William G. Andrew

Introduction

This section on Engineering Symbols presents graphic symbols used in piping, control systems, architectural, civil and structural disciplines to convey information efficiently among the users of engineering data in these disciplines. The symbols are often used as standards by manufacturers, processors and other users of this type of communication. Electrical symbols may be found in Chapter 9.

Flow Sheet Symbols

A wide variety of equipment, piping and instrument symbols is used on process, mechanical and utility flow diagrams to convey quickly and clearly information to users of these drawings. This is accomplished by standardizing the presentation.

In 1961 a set of standard symbols was published by the American Society of Mechanical Engineers titled "Graphical Symbols for Process Flow Diagrams in Petroleum and Chemistry Industries." Variations and additions to this basic standard are now in common use. Many of these symbols are shown in Figure 2.1 along with some common variations of these symbols.

Figure 2.1. American National Standards Institute (ANSI) process flow sheet symbols. Reprinted in part (and by permission) from ANSI Y32.11, published by the American Society of Mechanical Engineers (ASME).

LINES

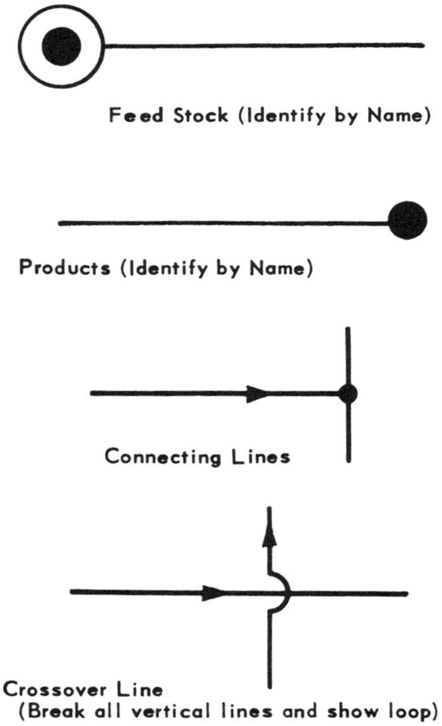

Figure 2.1 continued

Figure 2.1 continued

VALVES

Gate

Globe

Plug (Cock)

Check

REMARKS: Show valves only where necessary to clarify Process Flow.
If valve is closed during normal operation write "CLOSED" directly above the valve.

FURNACES & BOILERS

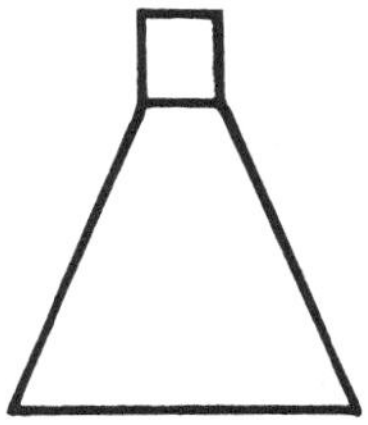

"A" Frame

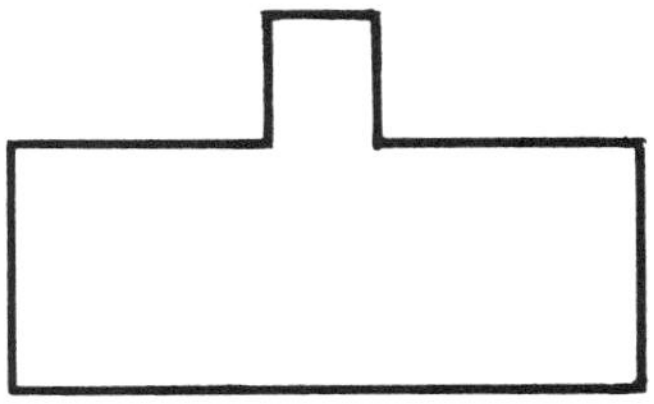

Box Type

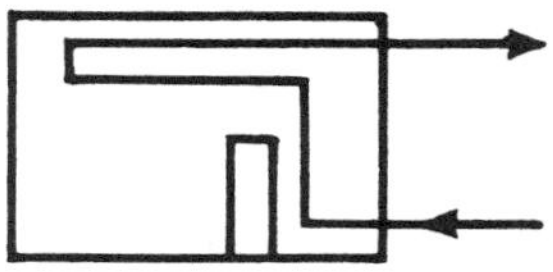

Radiant Type (Single Coil)

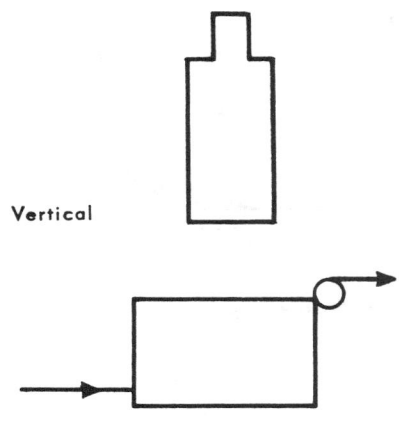

Vertical

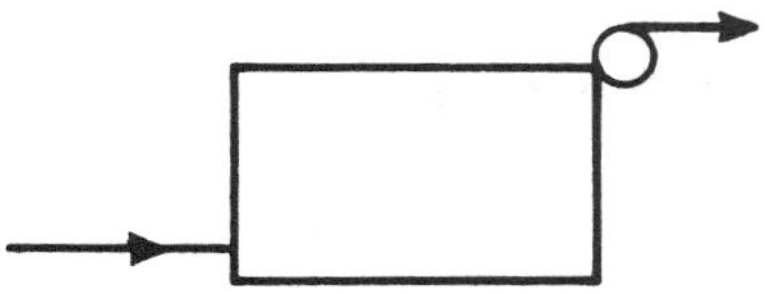

Boiler Fired or Waste Heat

REMARKS:
Indicate approximate position of inlet and outlet.
If dual coil indicate path of both streams.
Do not indicate type of fuel or firing position.

HEAT TRANSFER

Water Cooled Exchanger

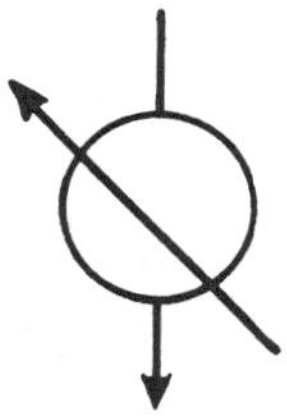

Water Cooled Condenser

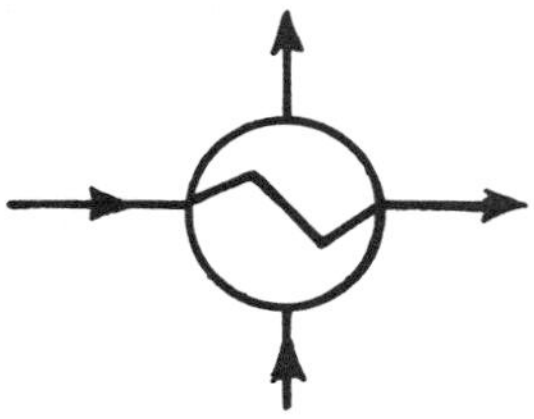

Shell & Tube Exchanger

Cooling Tower

Figure 2.1 continued

Figure 2.1 continued

HEAT TRANSFER (Cont'd)

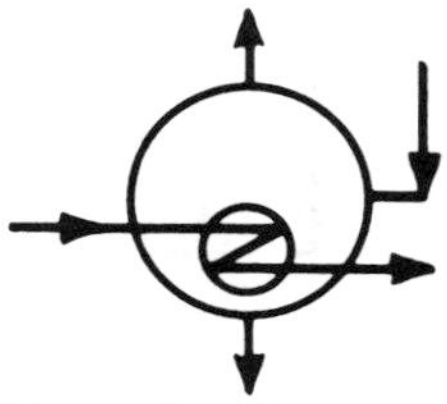

Reboiler (Kettle Type)

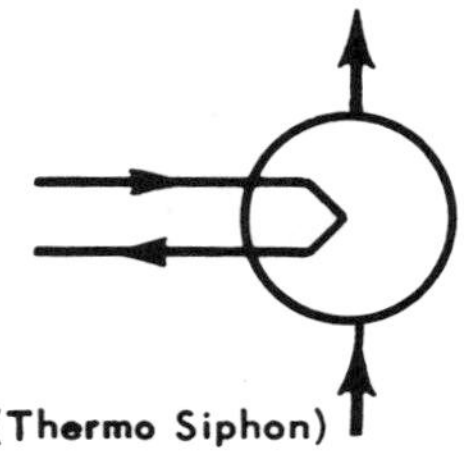

Reboiler (Thermo Siphon)

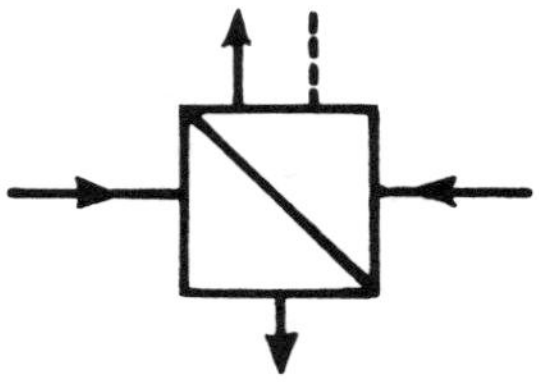

Superheater or Reheater

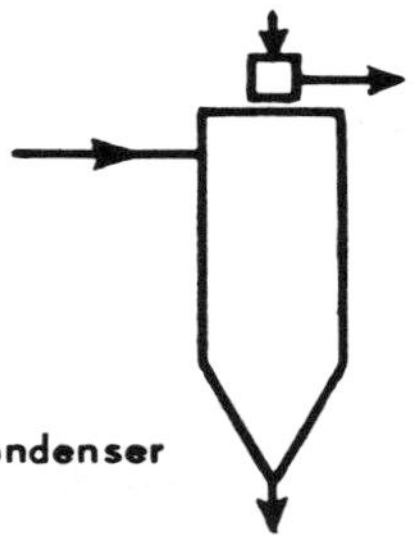

Barometric Condenser

PUMPS & COMPRESSORS

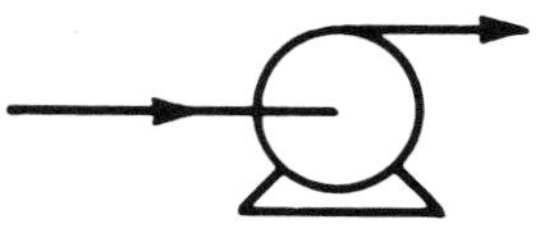

Centrifugal

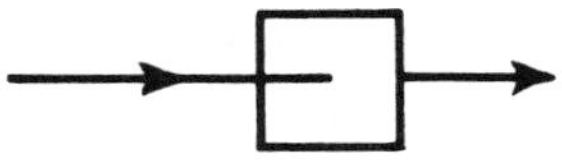

Reciprocating

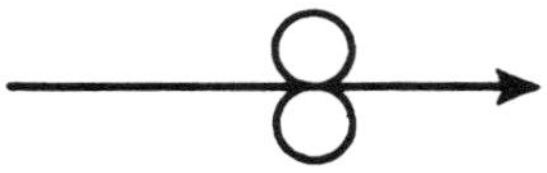

Rotary

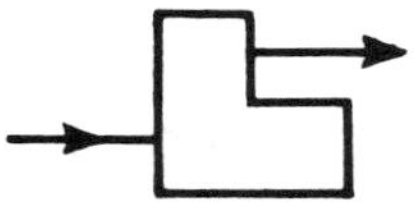

Proportioning

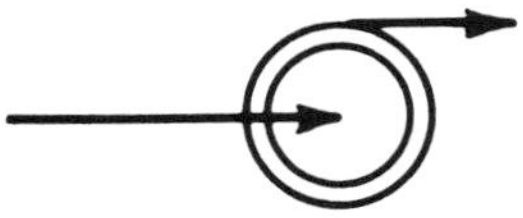

Blower or Fan (Centrifugal)

DRIVERS

Motor

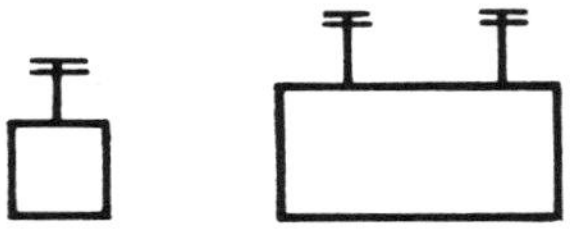

Engine — Single Drive — Dual Drive

Turbine

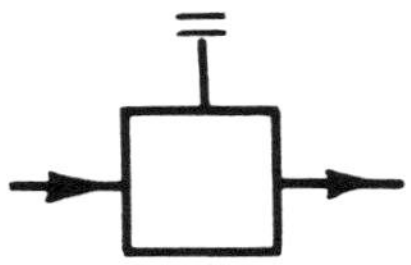

Steam Piston

REMARKS: Drivers may be shown, if desired, attached to driven equip.

Figure 2.1 continued

Figure 2.1 continued

PROCESS PRESSURE VESSELS

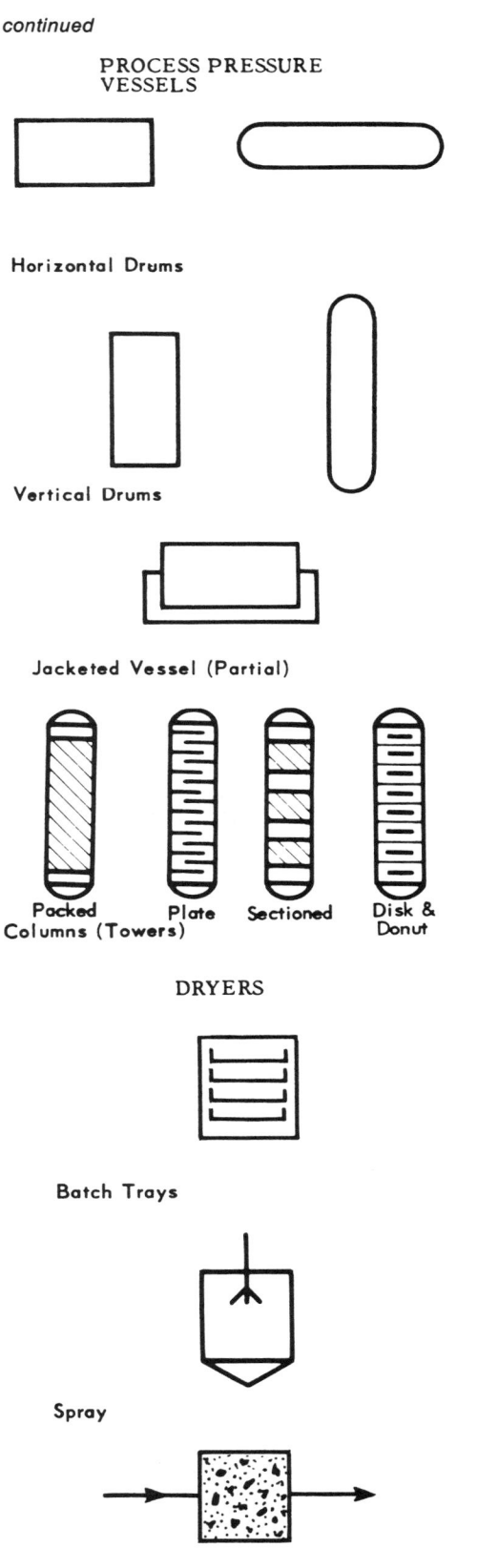

Horizontal Drums

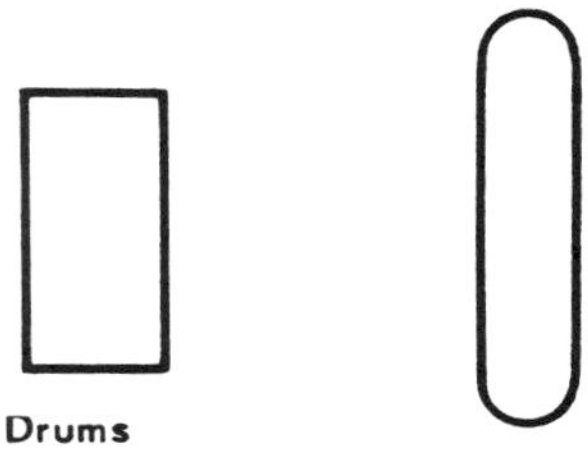

Vertical Drums

Jacketed Vessel (Partial)

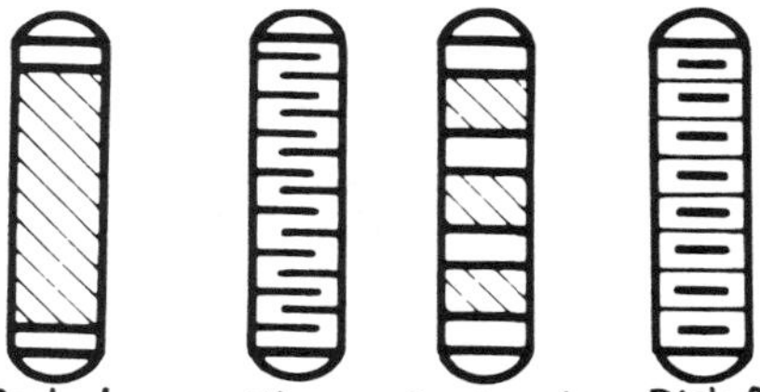

Packed Plate Sectioned Disk & Donut
Columns (Towers)

DRYERS

Batch Trays

Spray

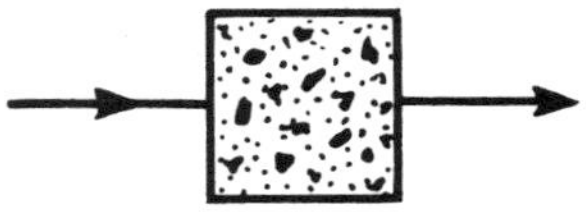

Desiccant

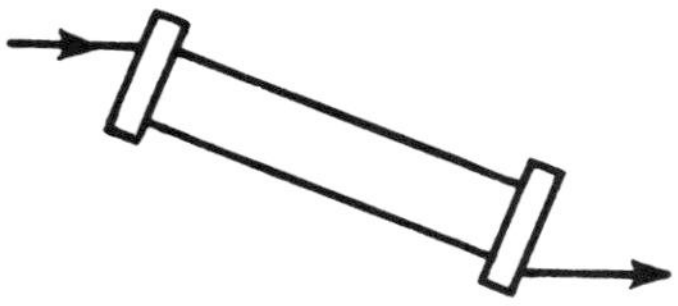

Rotary Drum Dryer or Kiln

MATERIAL HANDLING EQUIPMENT

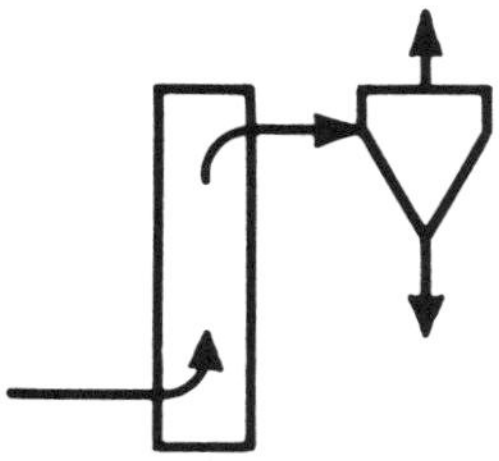

Air Lift

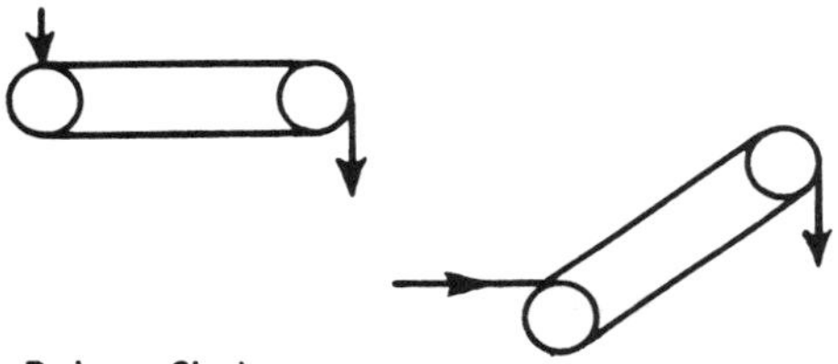

Belt or Shaker

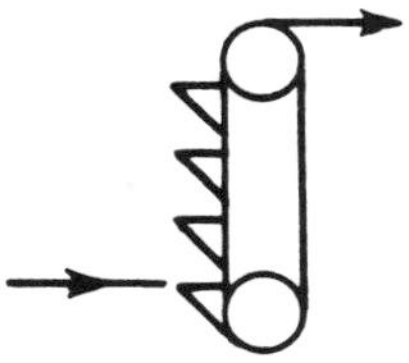

Bucket or Flight Conveyor

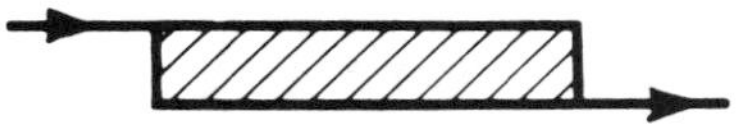

Screw Conveyor

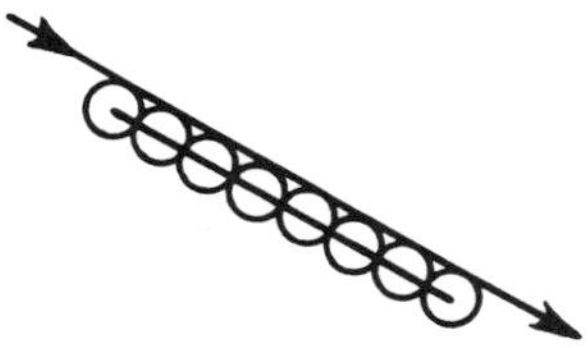

Roller Conveyor

Figure 2.1 continued

Figure 2.1 continued

MATERIAL HANDLING EQUIPMENT (Cont'd)

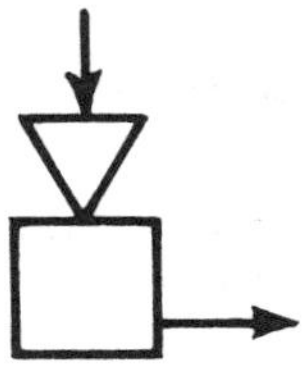

Feeder & Hopper

Rotary Feeder

SIZE REDUCING EQUIPMENT

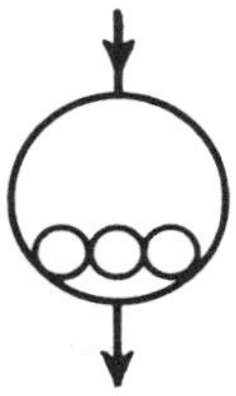

Ball Mill

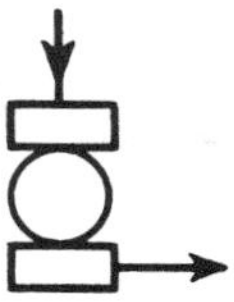

Grinder

Roller Crusher

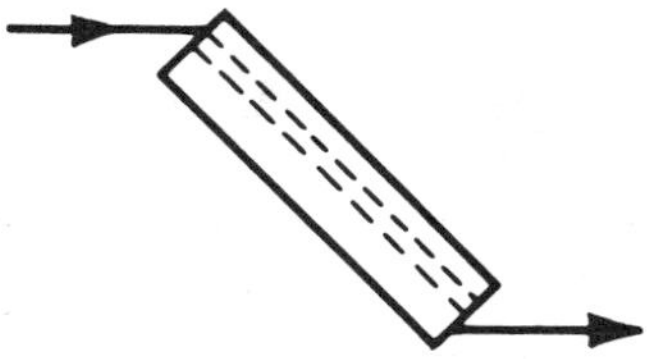

Screener

PROCESSING EQUIPMENT

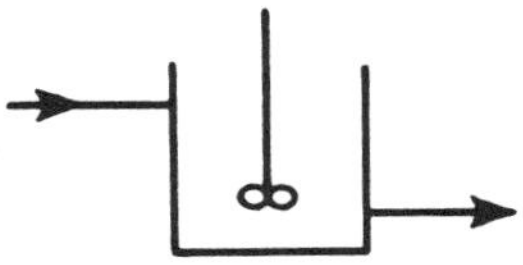

Mixer

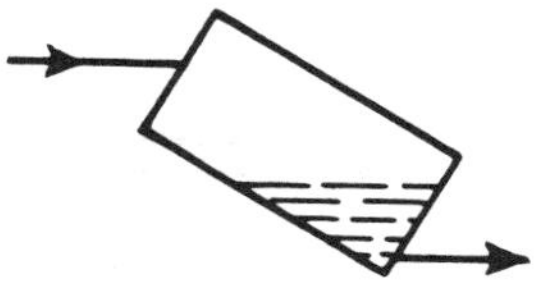

Settler

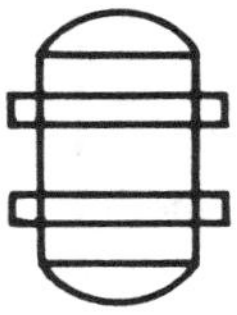

Autoclave

Kettle–Jacketed

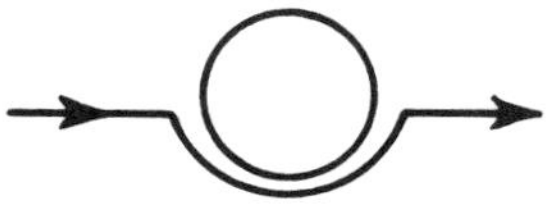

Rotary Film Dryer or Flaker

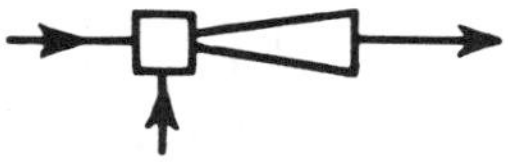

Jet Mixer
Injector, Ejector, Eductor

Figure 2.1 continued

Figure 2.1 continued

SEPARATORS

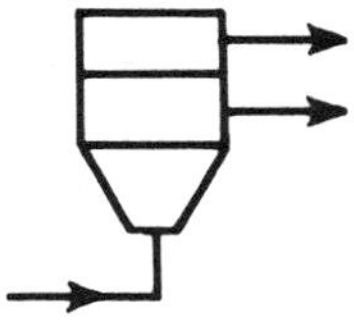

Centrifuge

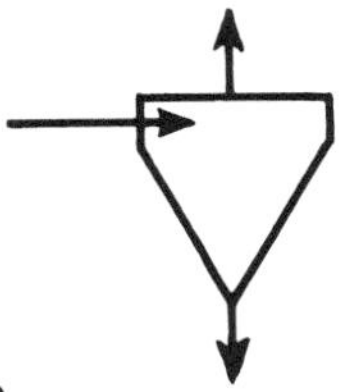

Cyclone

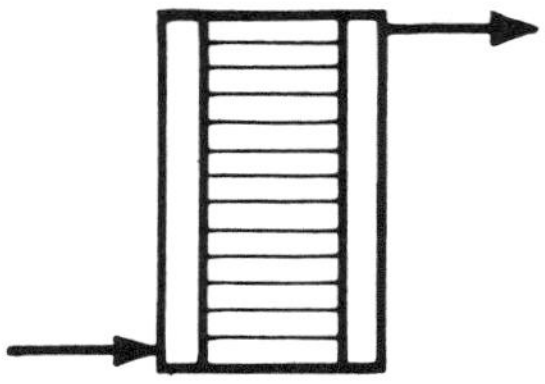

Electrical Precipitator

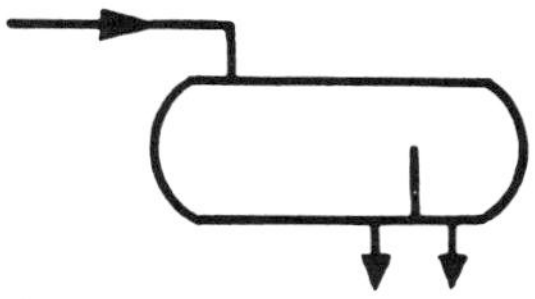

Drum Settler

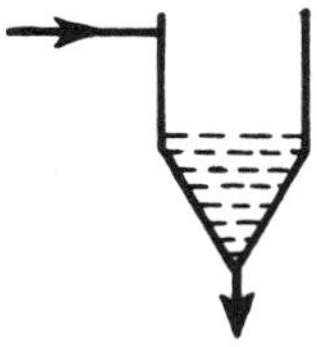

Open Settling Tank

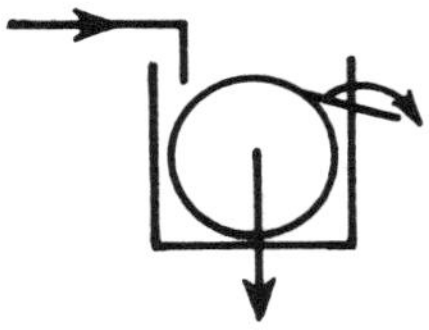

Rotary Vacuum Filter

STORAGE VESSELS

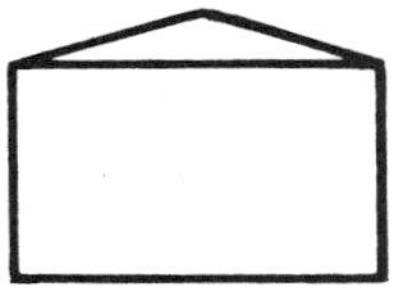

Atmospheric (Cone Roof) Tank

Floating Roof Tank

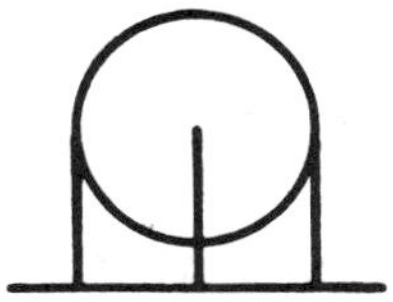

Pressure Storage (Sphere or Spheroid)

Cone Bottom Bin (Bulk Storage)

Open Top

Closed Top

Bulk Storage (Non-pressure)

Flow Sheet Codes and Line Symbols

Flow Diagram Line Codes

Code Letter Service

A	Acid
AL	Alcohol
CA	Catalyst
CWS	Cooling Water Supply
CWR	Cooling Water Return
D	Drain and Overflow Lines
DW	Domestic Water
FG	Fuel Gas
FW	Fire Water
H	Hydrogen
IA	Instrument Air
IW	Industrial Water
K	Caustic
LO	Lube Oil
N	Nitrogen
NG	Natural Gas
O	Oil, Hydrocarbon Liquids
PA	Plant Air
PP	Plastic Powder and Pellets
PV	Process Vent
S	Steam
SC	Steam Condensate
SL	Slurry
SR	Safety Release
TW	Tempered Water
V	Vapor, Hydrocarbon

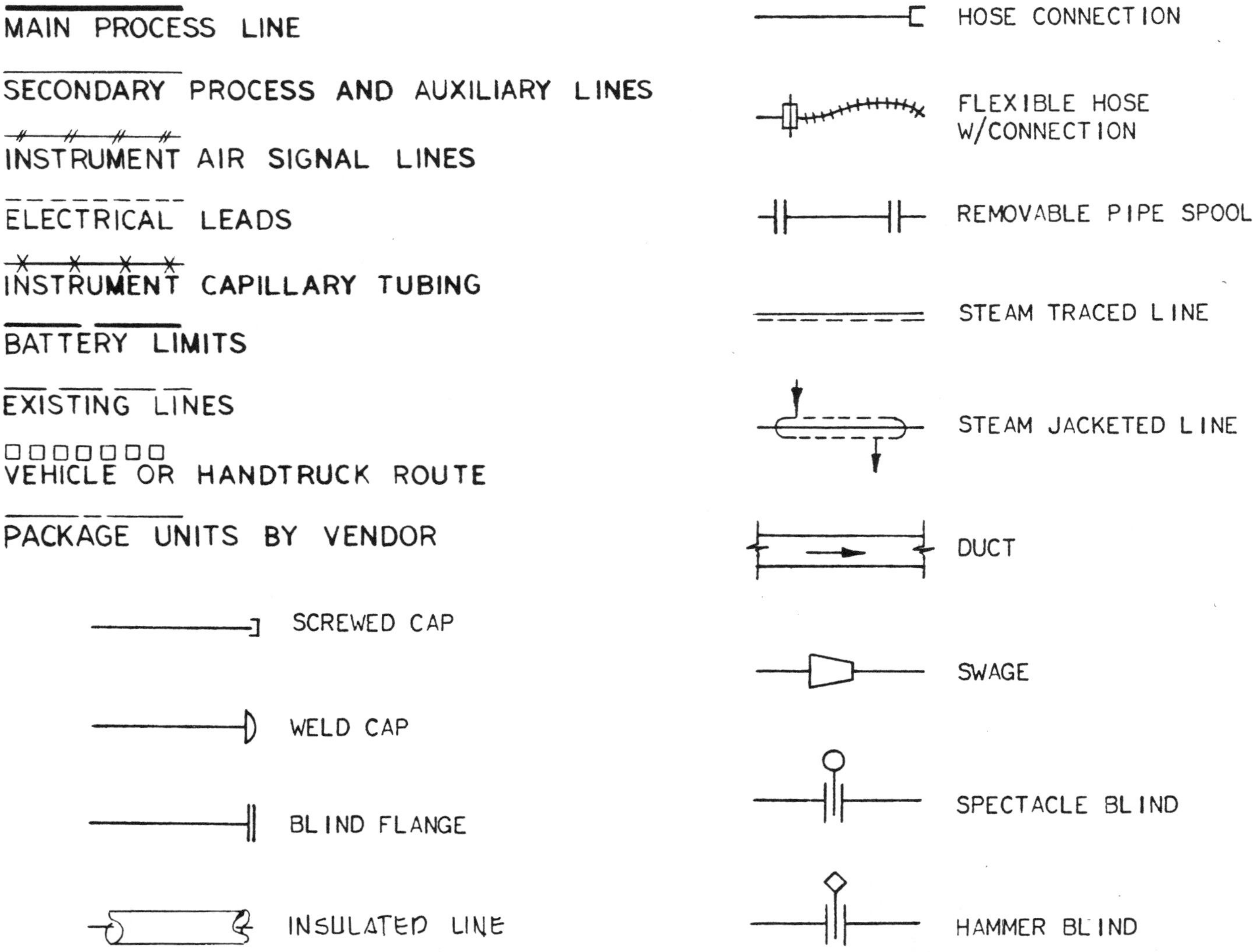

Figure 2.2. Flow sheet codes and line symbols. Courtesy of S.I.P. Engineering, Inc. *Figure 2.2 continued*

Figure 2.2 continued

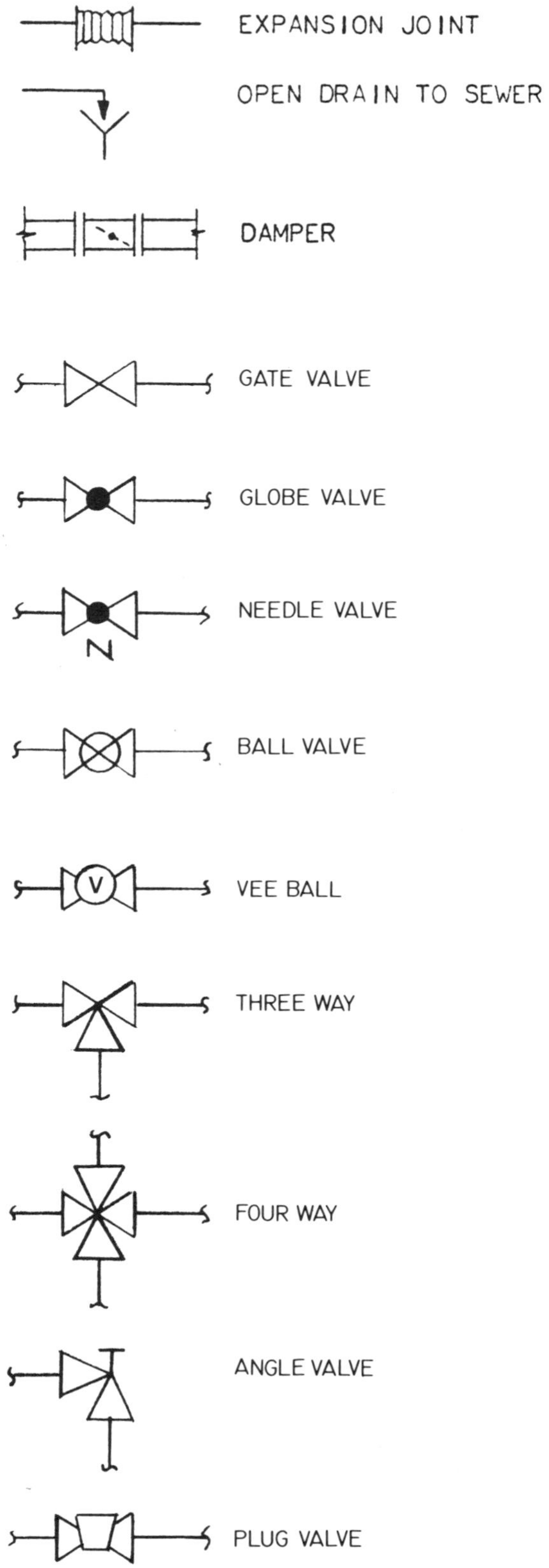
EXPANSION JOINT
OPEN DRAIN TO SEWER
DAMPER
GATE VALVE
GLOBE VALVE
NEEDLE VALVE
N
BALL VALVE
V
VEE BALL
THREE WAY
FOUR WAY
ANGLE VALVE
PLUG VALVE

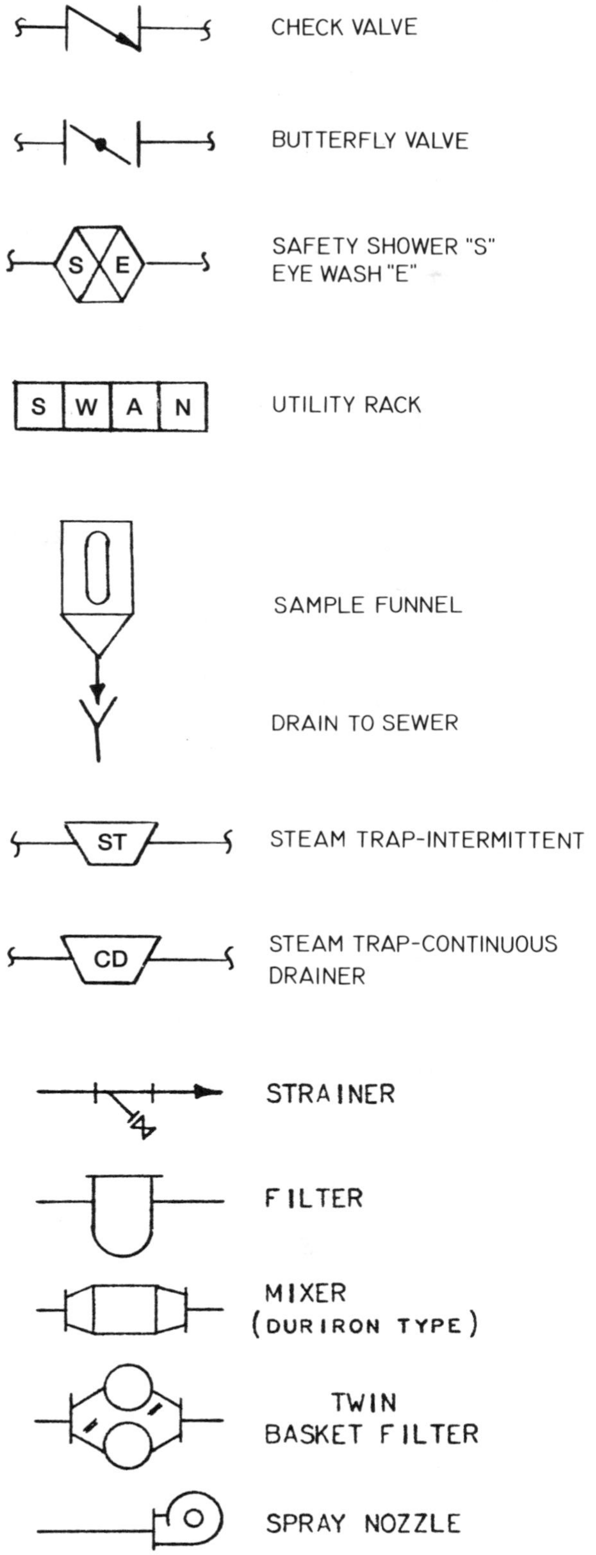
CHECK VALVE
BUTTERFLY VALVE
S
E
SAFETY SHOWER "S"
EYE WASH "E"
S
W
A
N
UTILITY RACK
SAMPLE FUNNEL
DRAIN TO SEWER
ST
STEAM TRAP-INTERMITTENT
CD
STEAM TRAP-CONTINUOUS DRAINER
STRAINER
FILTER
MIXER
(DURIRON TYPE)
TWIN
BASKET FILTER
SPRAY NOZZLE

Figure 2.2 continued

Figure 2.2 continued

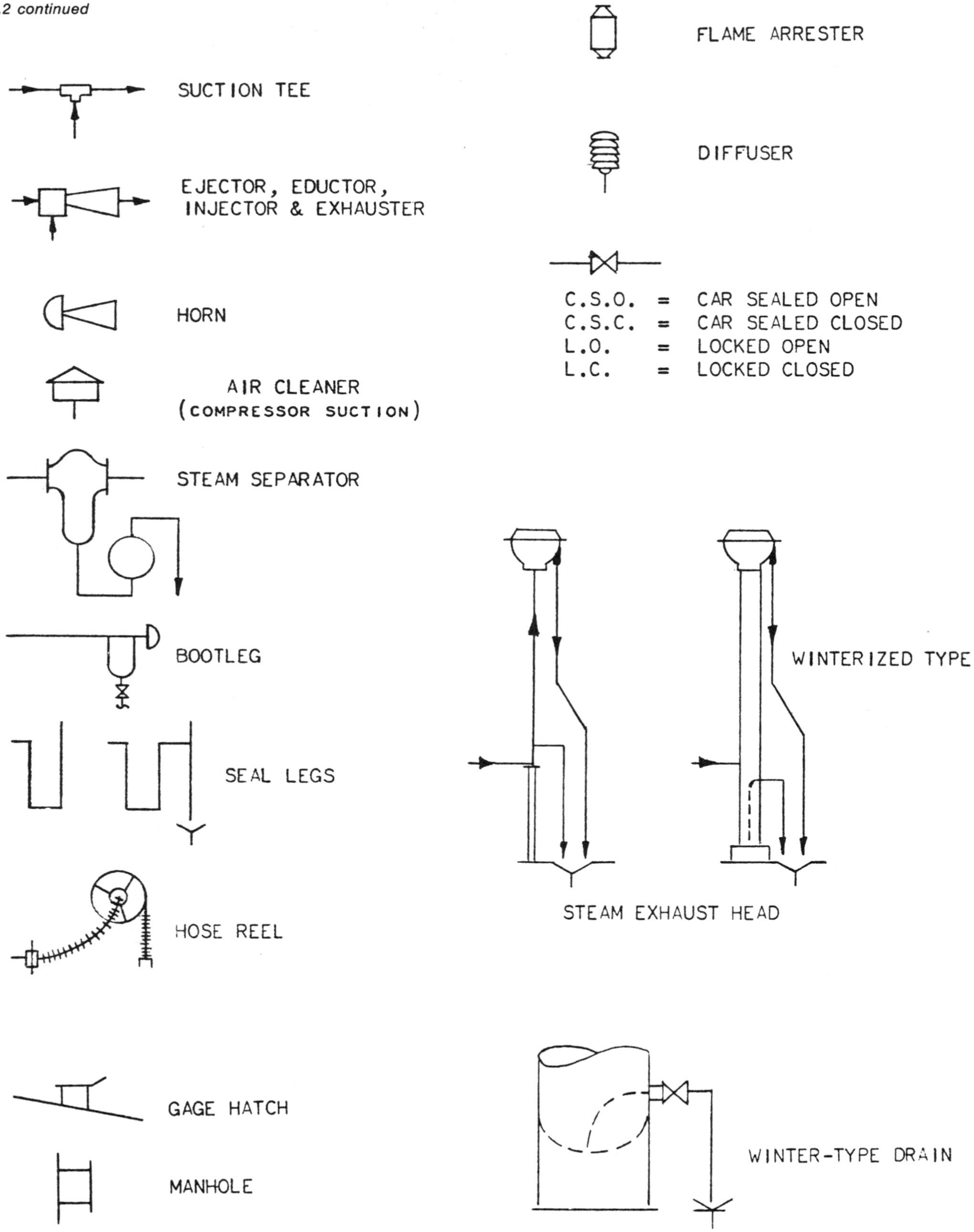
SUCTION TEE
EJECTOR, EDUCTOR, INJECTOR & EXHAUSTER
HORN
AIR CLEANER (COMPRESSOR SUCTION)
STEAM SEPARATOR
BOOTLEG
SEAL LEGS
HOSE REEL
GAGE HATCH
MANHOLE
FLAME ARRESTER
DIFFUSER
C.S.O. = CAR SEALED OPEN
C.S.C. = CAR SEALED CLOSED
L.O. = LOCKED OPEN
L.C. = LOCKED CLOSED
WINTERIZED TYPE
STEAM EXHAUST HEAD
WINTER-TYPE DRAIN

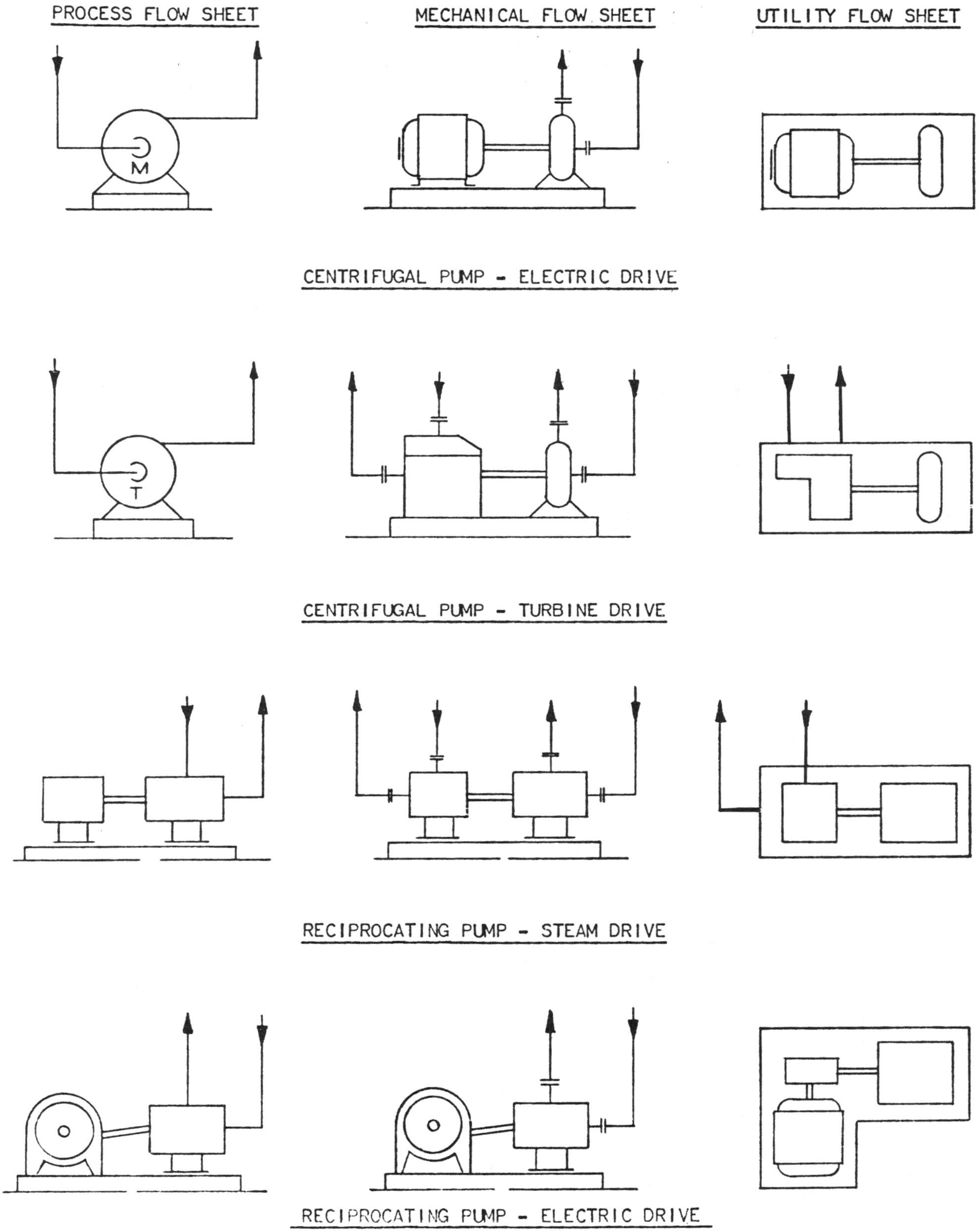

Figure 2.3 continued

Figure 2.3. Miscellaneous equipment symbols. Courtesy of S.I.P. Engineering, Inc.

Figure 2.3 continued

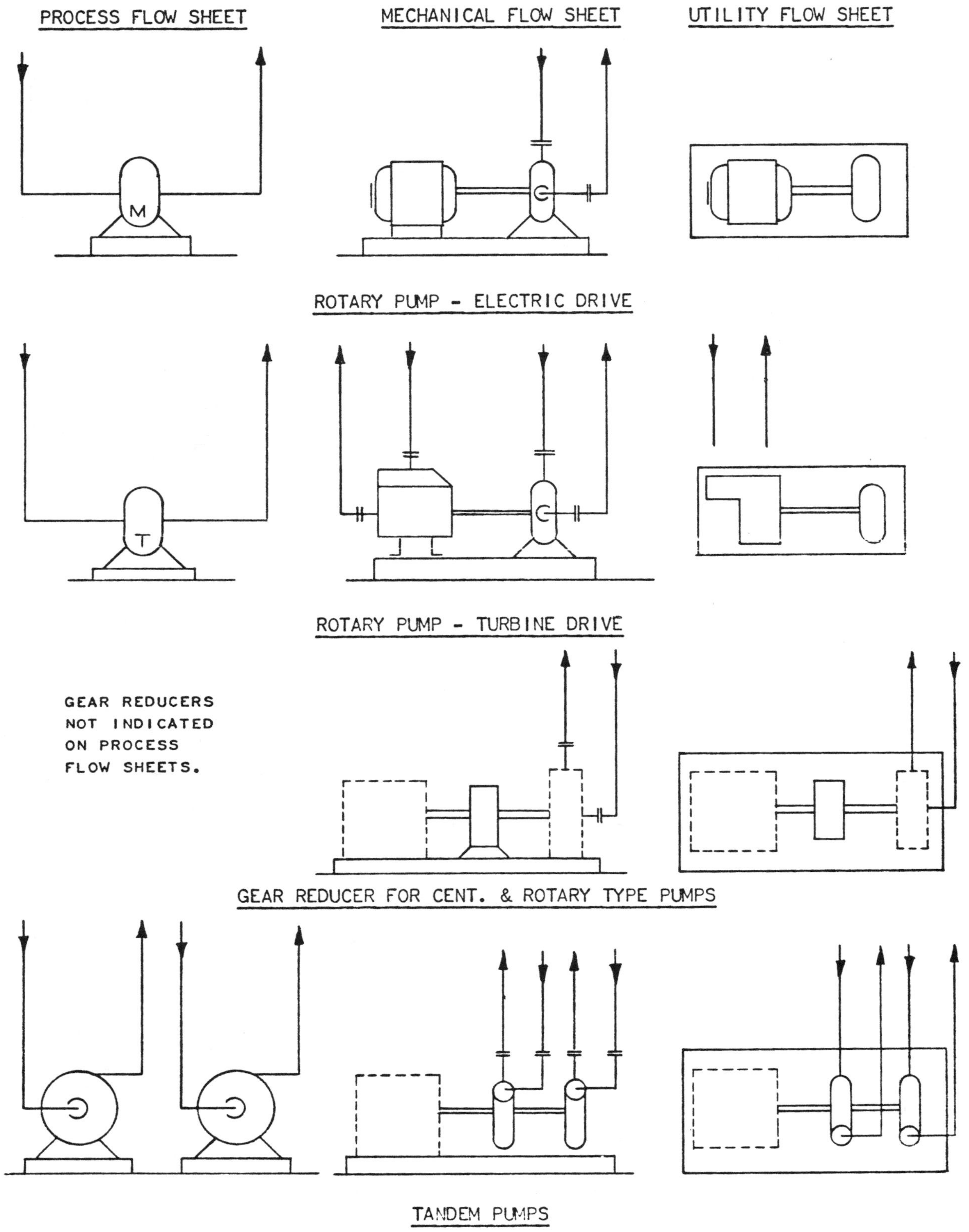

Figure 2.3 continued

Figure 2.3 continued

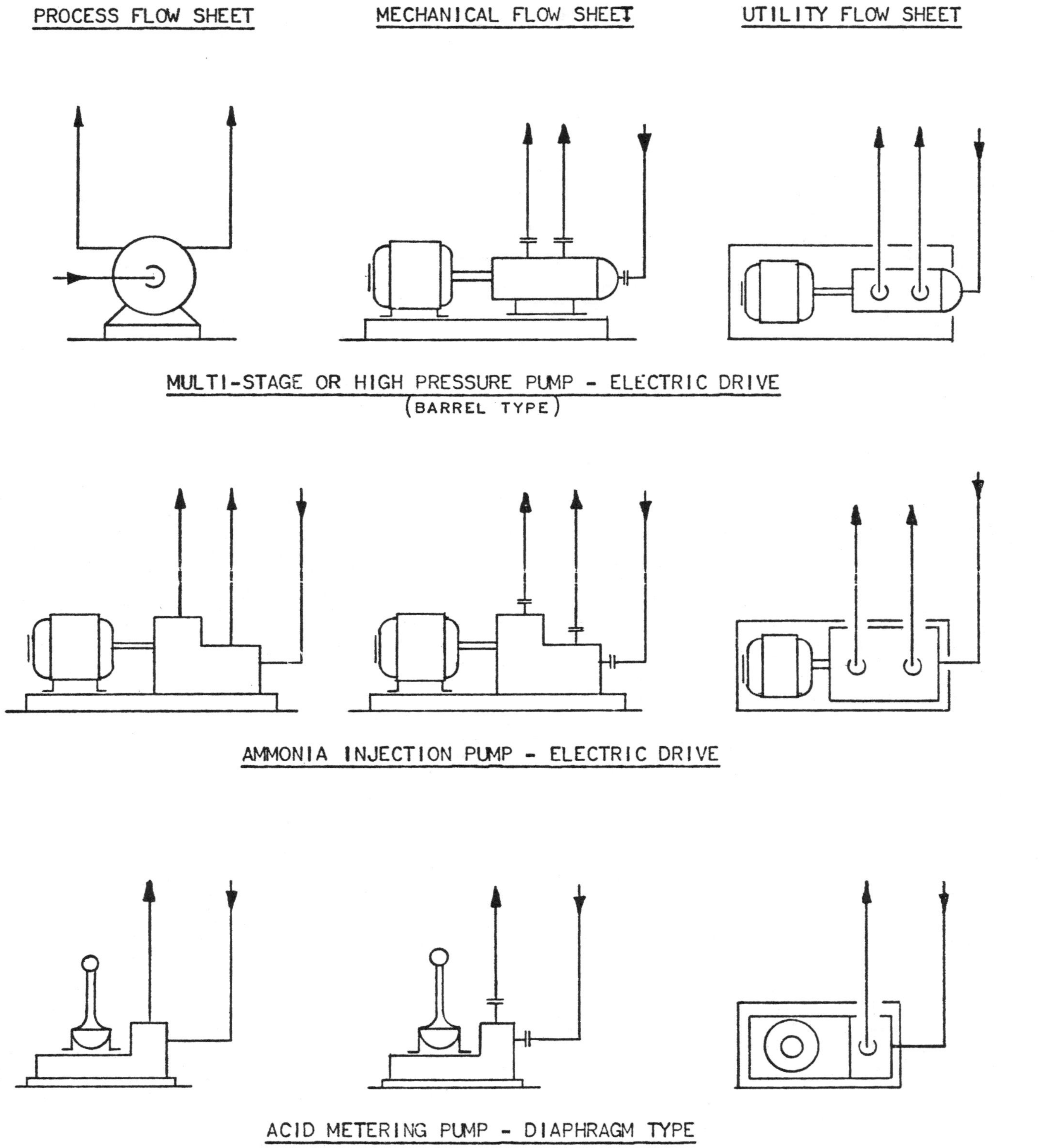
PROCESS FLOW SHEET
MECHANICAL FLOW SHEET
UTILITY FLOW SHEET
MULTI-STAGE OR HIGH PRESSURE PUMP - ELECTRIC DRIVE
(BARREL TYPE)
AMMONIA INJECTION PUMP - ELECTRIC DRIVE
ACID METERING PUMP - DIAPHRAGM TYPE

Figure 2.3 continued

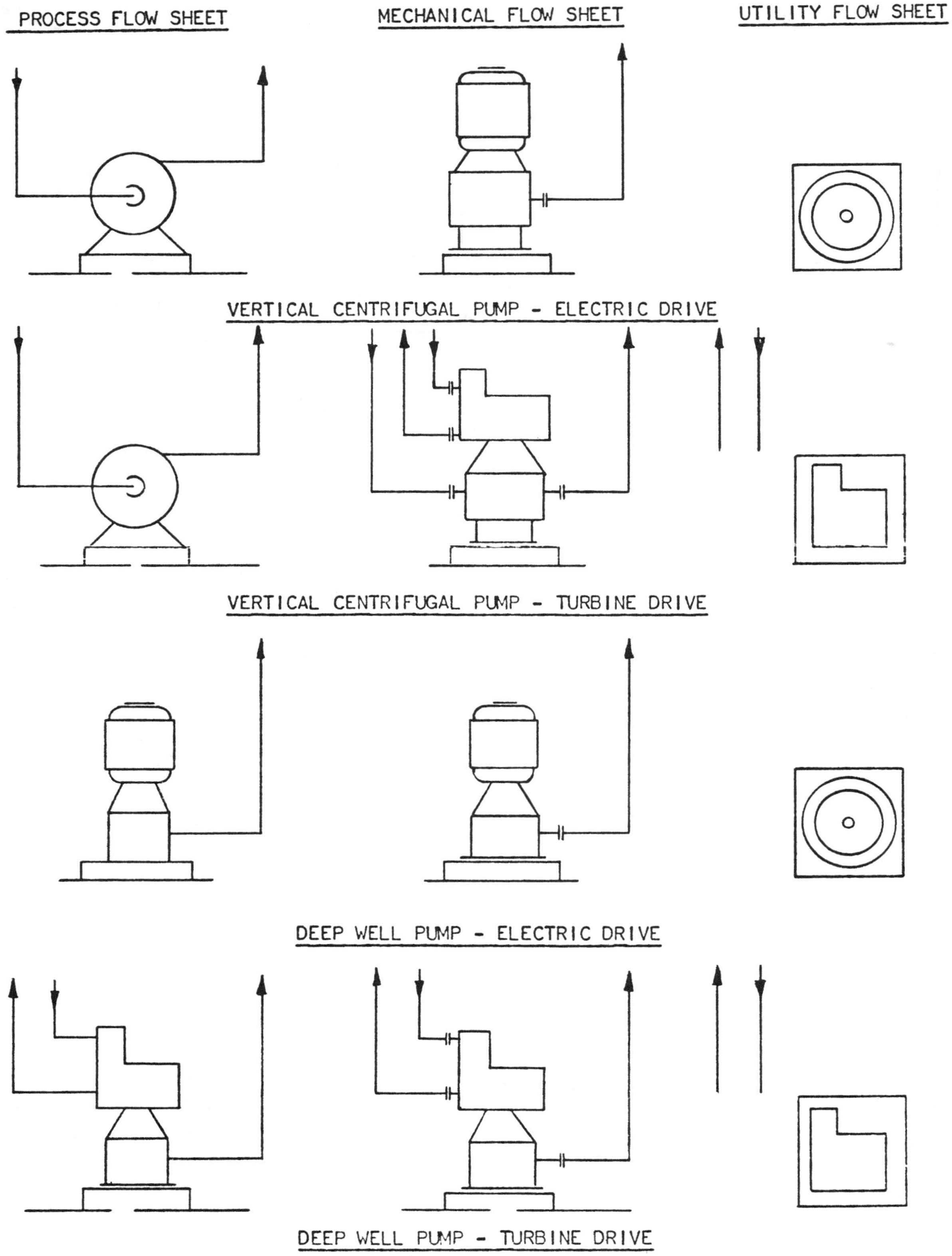

Figure 2.3 continued

Figure 2.3 continued

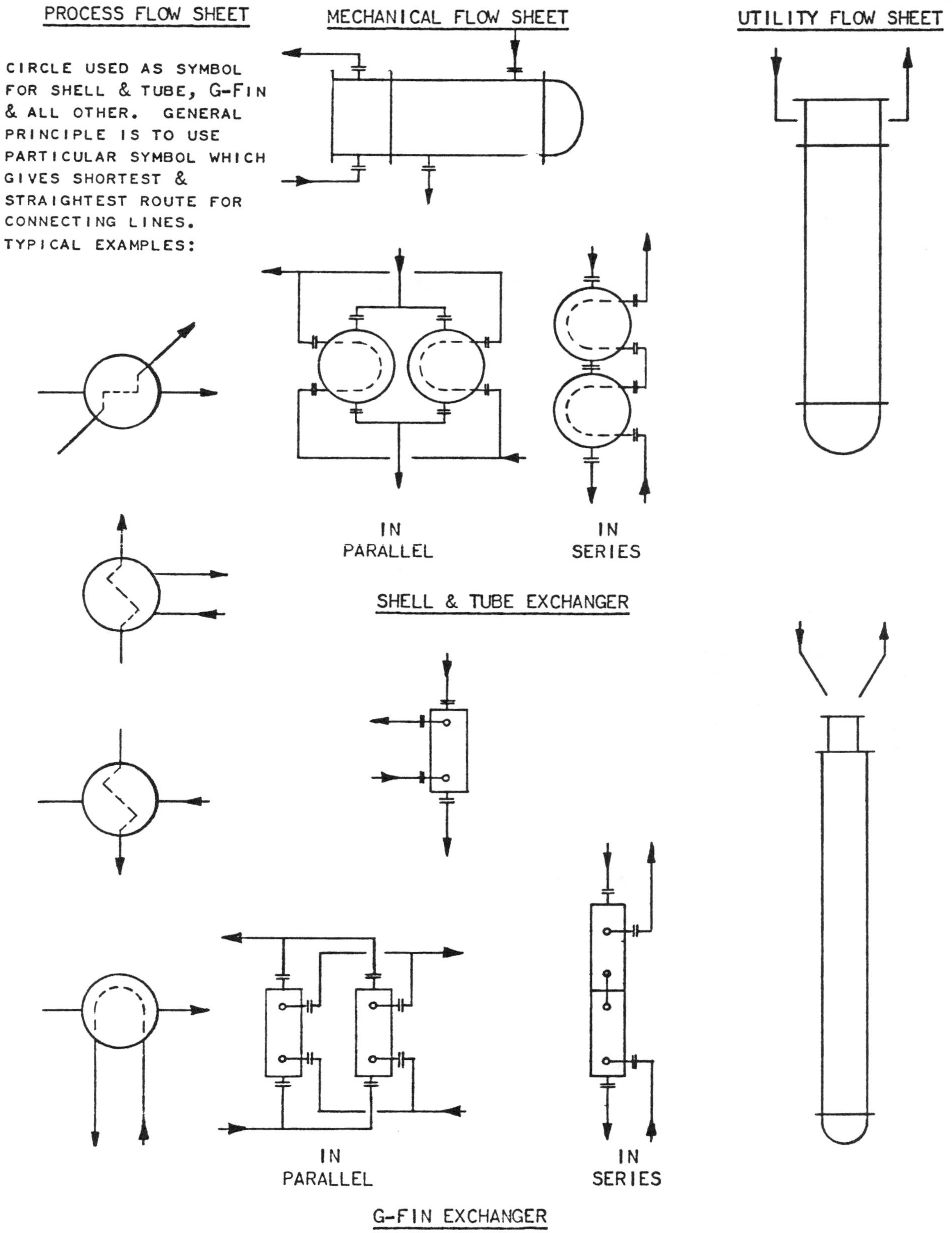

Figure 2.3 continued

Figure 2.3 continued

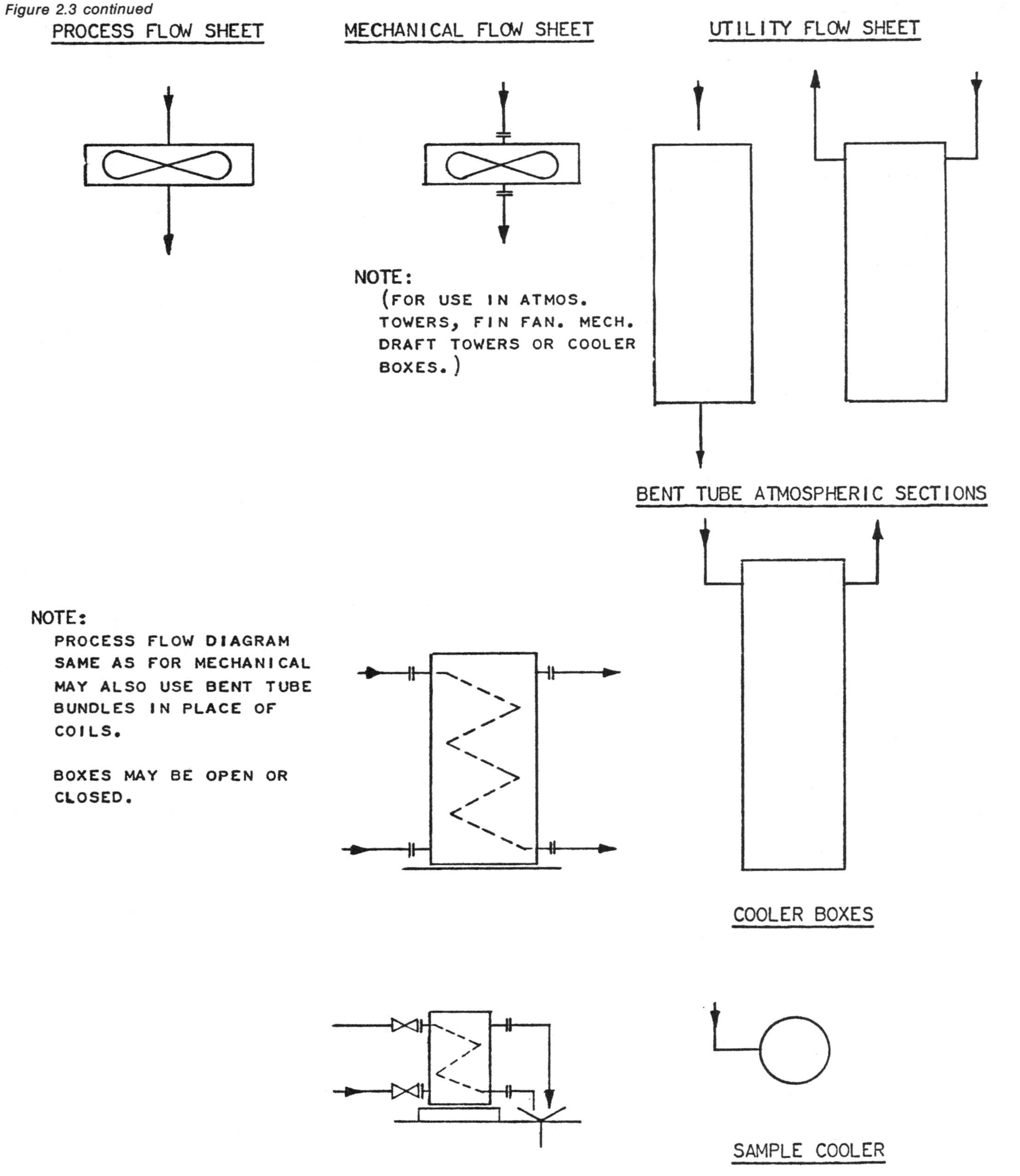

Figure 2.3 continued

Figure 2.3 continued

PROCESS FLOW SHEET | MECHANICAL FLOW SHEET | UTILITY FLOW SHEET

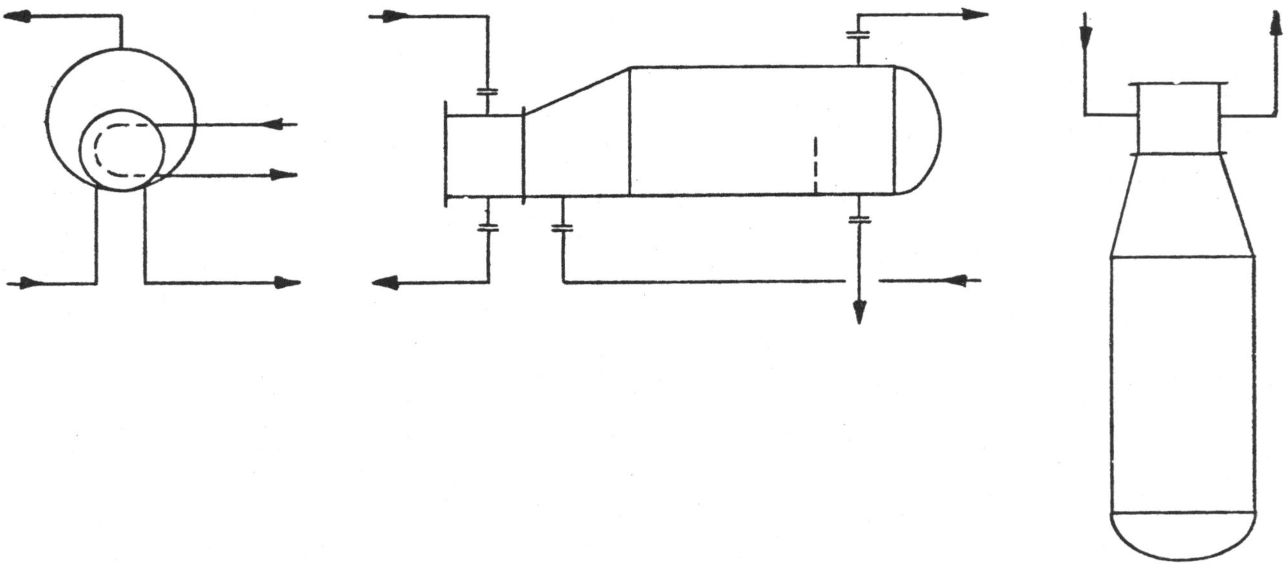

REBOILER - KETTLE TYPE

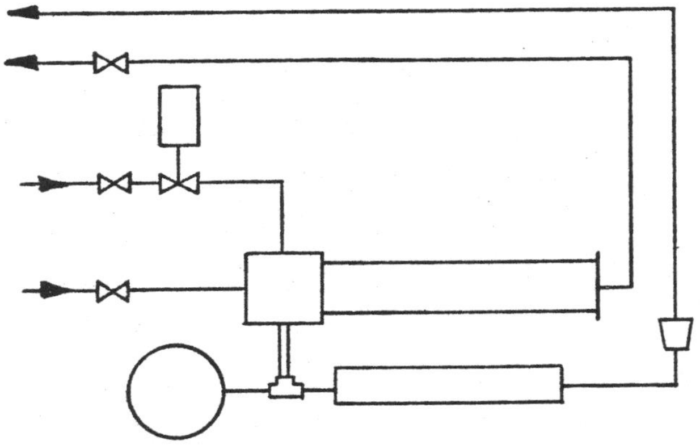

TYPICAL STEAM WATER HEATER

Figure 2.3 continued

Figure 2.3 continued

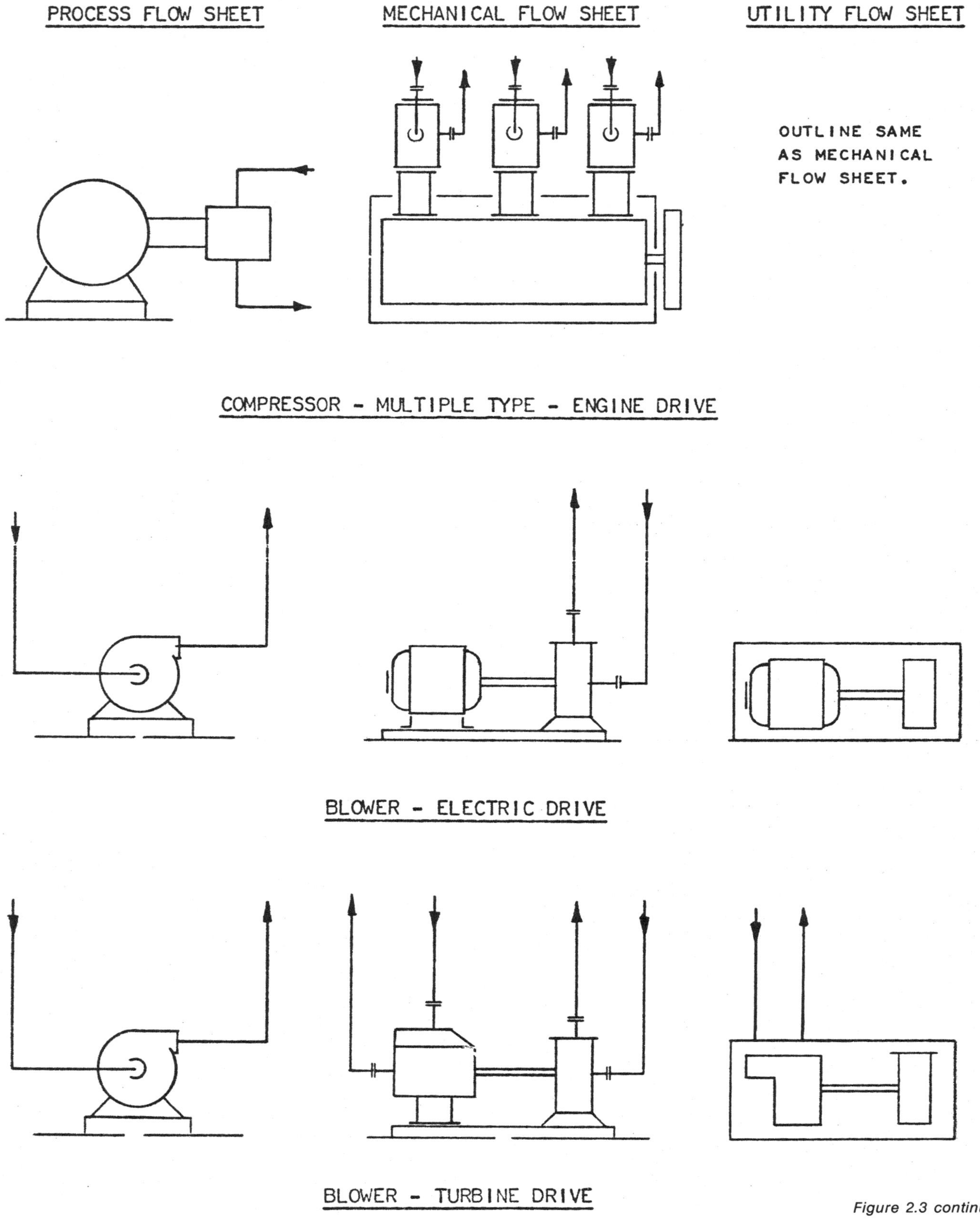

Figure 2.3 continued

Figure 2.3 continued

PROCESS FLOW SHEET | MECHANICAL FLOW SHEET

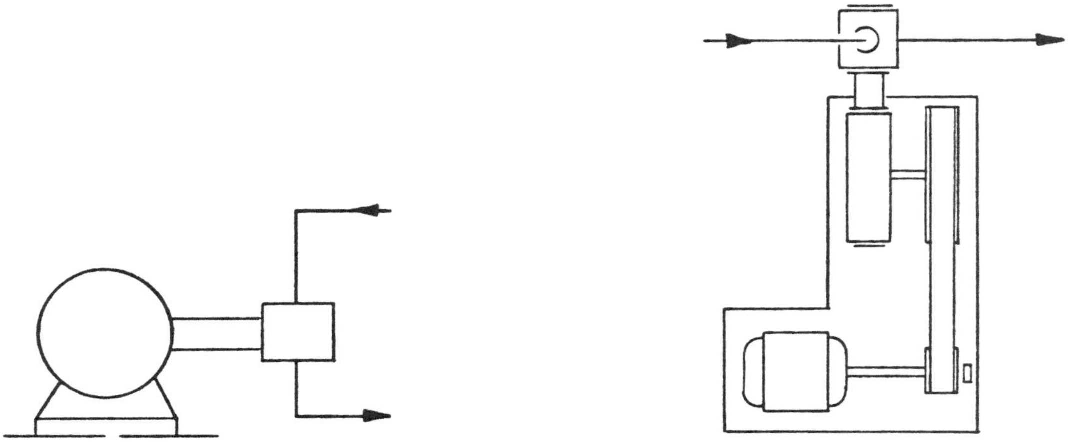

COMPRESSOR - SINGLE TYPE - ELECTRIC DRIVE

COMPRESSOR - SINGLE TYPE - ENGINE DRIVE

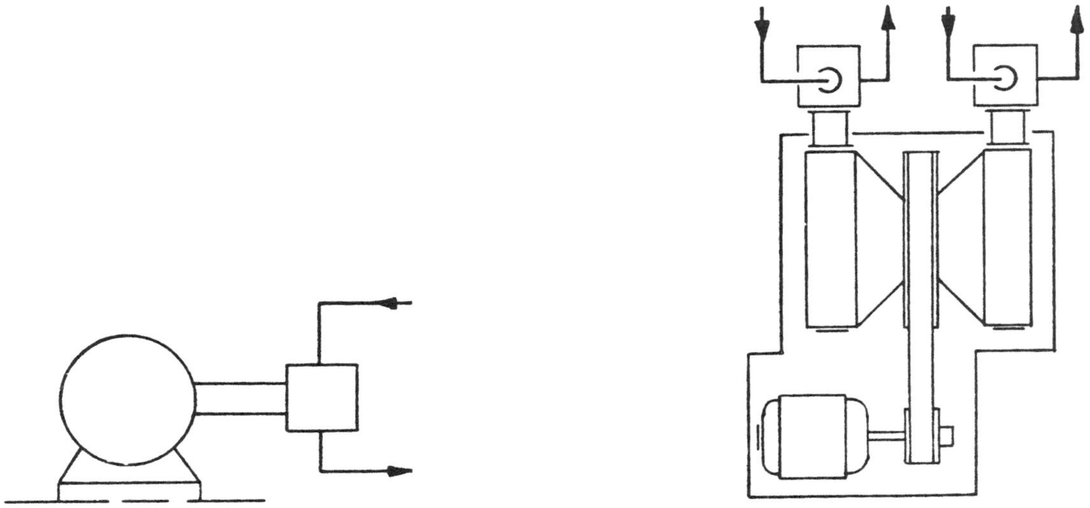

COMPRESSOR - TWIN TYPE - ELECTRIC DRIVE

Figure 2.3 continued

Figure 2.3 continued

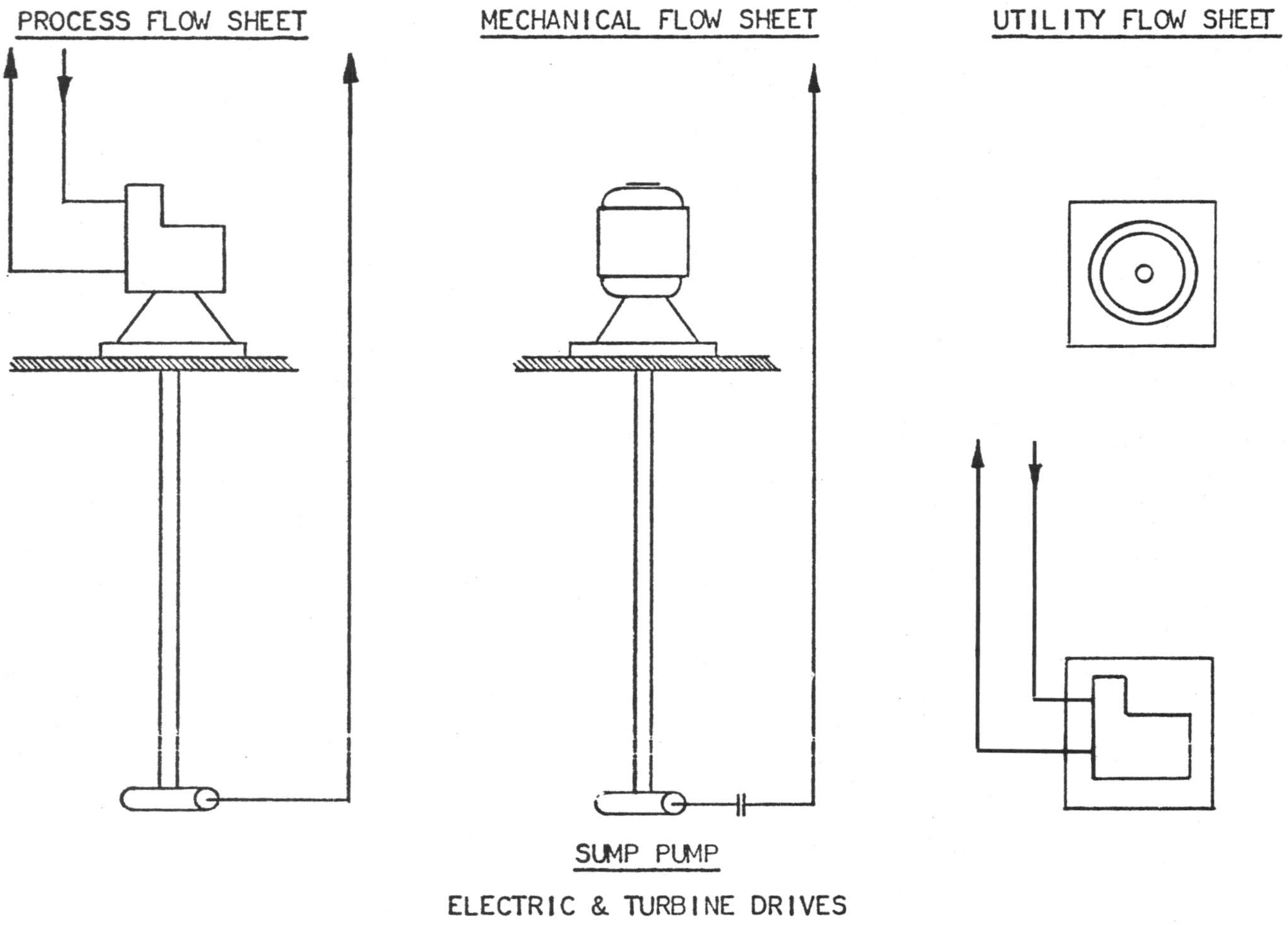

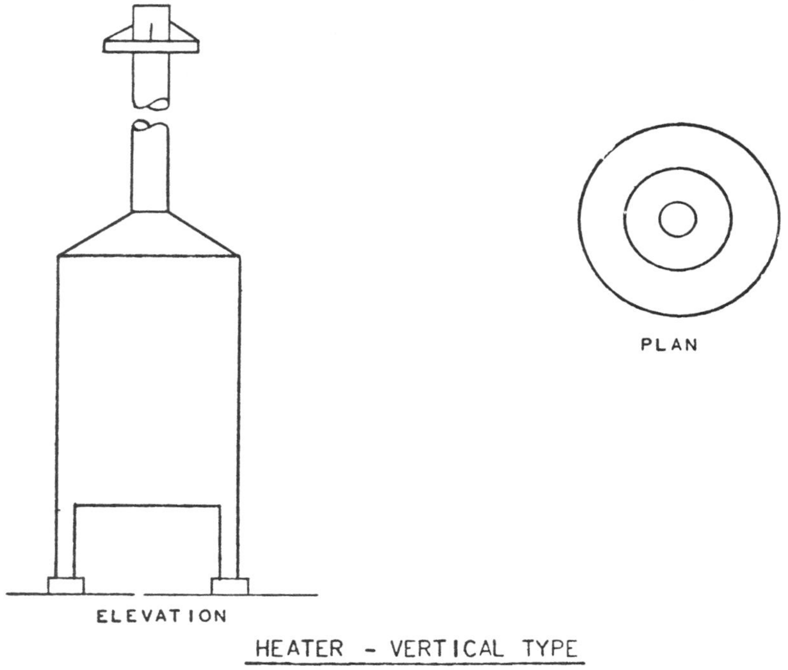

Figure 2.3 continued

Figure 2.3 continued

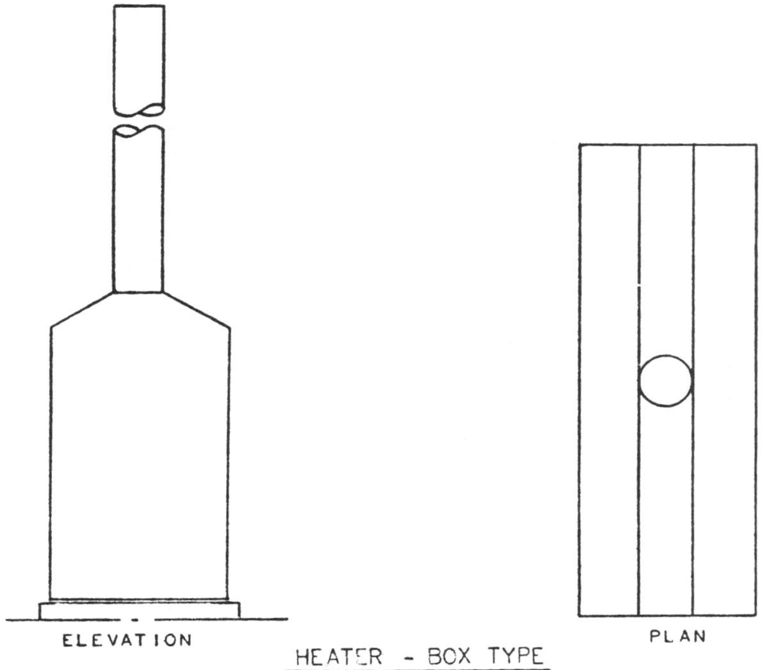

HEATER - BOX TYPE

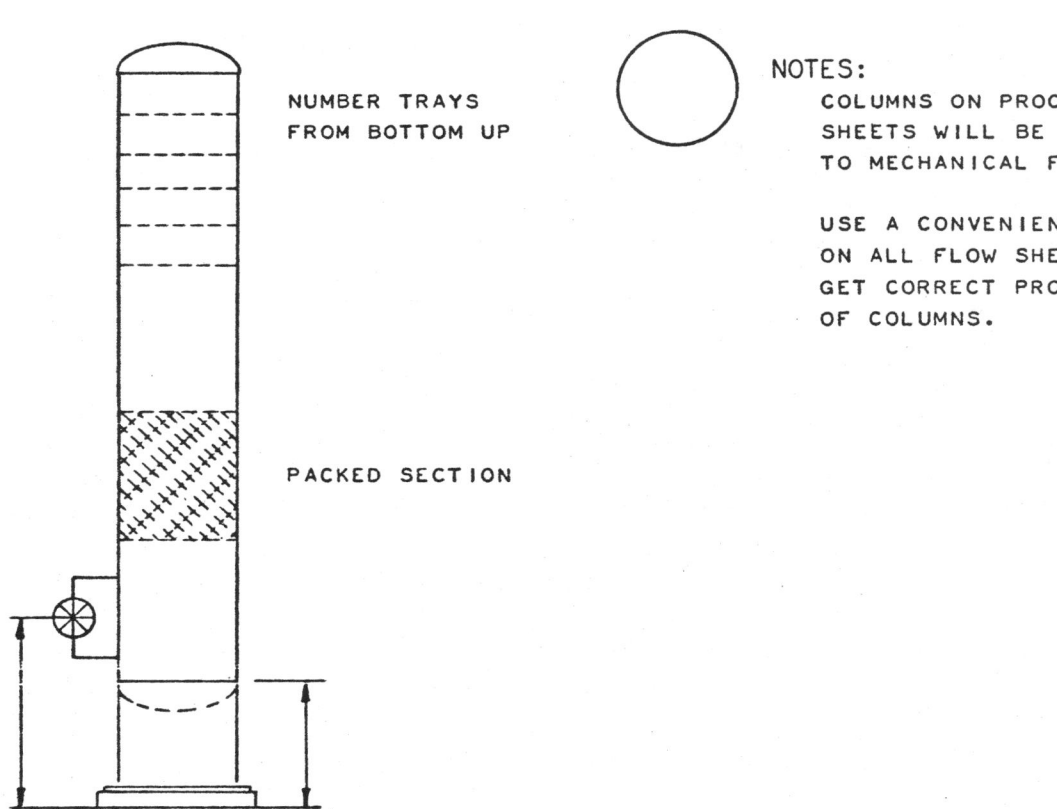

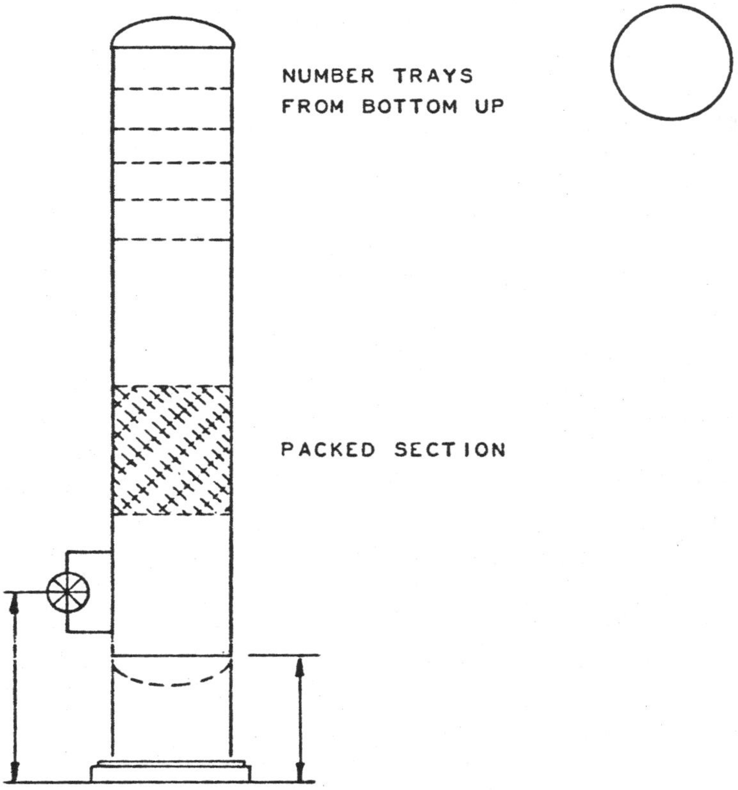

NOTES:

COLUMNS ON PROCESS FLOW SHEETS WILL BE SIMILAR TO MECHANICAL FLOW SHEET.

USE A CONVENIENT SCALE ON ALL FLOW SHEETS TO GET CORRECT PROPORTIONS OF COLUMNS.

COLUMNS

Instrument Symbols and Identification

General

The Instrument Society of America published "Recommended Practice RP5.1" in 1949 to standardize the designation of instruments and instrument systems used for measurement and control. As processes increased in complexity and as specialized instrument hardware was developed, additional ways were needed to identify and describe these new functions. In 1968 RP5.1 was replaced with Standard ISA S5.1, which establishes a uniform method of designating instruments and control systems for the processing industries.

The identification system described below is based entirely on ISA S5.1, and tables and symbols presented are taken directly from that standard.

Identification

Instruments are identified by a system of letters and numbers as shown in Table 2.1. The number is generally common to all instruments of the loop of which it is a part.

The first two letters identify the function of the instrument and are selected from Table 2.2. The succeeding numbers and letters identify the particular loop.

The first letter designates the measured or initiating variable such as temperature, level, flow, etc. Modifying letters such as *D* for differential, *F* for ratio, and *Q* for totalizing may follow the first letter. For example, a *TDI* is a differential temperature indicator and a *FQR* is a flow recorder with an integrator in the loop.

The succeeding letters designate one or more functions of the loop such as readout, passive function, or output.

The loop identification method assigns a number to each loop. It may begin with 1, 201, or 1201, which may include a plant area coding system.

Prefix numbers may be assigned to a number to designate plant areas. For example, 6-TRC-2 may indicate an instrument located in Plant Area 6.

If an instrument is common to more than one loop it may be assigned a separate number.

Table 2.1

T	R C	- 2	A
First Letter	Succeeding Letters	Loop Number	Suffix
Functional Identification		Loop Identification	
Instrument Identification or Tag Number			

For loops that have more than one instrument with the same functional identification, suffixes should be added to the loop number; e.g., FV-2A, FV-2B, FV-2C, etc., or TE-25-1, TE-25-2, TE-25-3, etc.

Generally items such as steam traps, pressure gages, and temperature wells that are purchased in bulk quantities are not identified as loops.

Instrument Line Symbols

Instrument line symbols used are shown in Table 2.3, and lines are drawn lightly in relation to process piping lines.

Typical Instrument Symbol Drawings

Typical drawings that illustrate symbols and identification of instruments on flow diagrams, engineering drawings, and other literature pertinent to instrumentation are shown in Figures 2.4 through 2.12. Generally one signal line represents all interconnections between two devices. In most cases, minor components such as auxiliary relays, pressure regulators, etc., are not shown.

Table 2.2. Meanings of Identification Letters

This table applies only to the functional identification of instruments. Numbers in table refer to notes following.

	FIRST-LETTER (4)		SUCCEEDING-LETTERS (3)		
	MEASURED OR INITIATING VARIABLE	MODIFIER	READOUT OR PASSIVE FUNCTION	OUTPUT FUNCTION	MODIFIER
A	Analysis(5,19)		Alarm		
B	Burner, Combustion		User's Choice(1)	User's Choice(1)	User's Choice(1)
C	User's Choice(1)			Control(13)	
D	User's Choice(1)	Differential(4)			
E	Voltage		Sensor (Primary Element)		
F	Flow Rate	Ratio (Fraction)(4)			
G	User's Choice(1)		Glass, Viewing Device(9)		
H	Hand				High(7,15,16)
I	Current (Electrical)		Indicate(10)		
J	Power	Scan(7)			
K	Time, Time Schedule	Time Rate of Change(4,21)		Control Station (22)	
L	Level		Light(11)		Low(7,15,16)
M	User's Choice(1)	Momentary(4)			Middle, Intermediate(7,15)
N	User's Choice(1)		User's Choice(1)	User's Choice(1)	User's Choice(1)
O	User's Choice(1)		Orifice, Restriction		
P	Pressure, Vacuum		Point (Test) Connection		
Q	Quantity	Integrate, Totalize(4)			
R	Radiation		Record(17)		
S	Speed, Frequency	Safety(8)		Switch(13)	
T	Temperature			Transmit(18)	
U	Multivariable(6)		Multifunction(12)	Multifunction(12)	Multifunction(12)
V	Vibration, Mechanical Analysis(19)			Valve, Damper, Louver(13)	
W	Weight, Force		Well		
X	Unclassified(2)	X Axis	Unclassified(2)	Unclassified(2)	Unclassified(2)
Y	Event, State or Presence(20)	Y Axis		Relay, Compute, Convert(13,14,18)	
Z	Position, Dimension	Z Axis		Driver, Actuator, Unclassified Final Control Element	

Source: Instrument Society of America

(1) A "user's choice" letter is intended to cover unlisted meanings that will be used repetitively in a particular project. If used, the letter may have one meaning as a first-letter and another meaning as a succeeding-letter. The meanings need to be defined only once in a legend, or other place, for that project. For example, the letter *N* may be defined as "modulus of elasticity" as a first-letter and "oscilloscope" as a succeeding-letter.

(2) The unclassified letter *X* is intended to cover unlisted meanings that will be used only once or used to a limited extent. If used, the letter may have any number of meanings as a first-letter and any number of meanings as a succeeding-letter. Except for its use with distinctive symbols, it is expected that the meanings will be defined outside a tagging bubble on a flow diagram. For example, *XR-2* may be a stress recorder and *XX-4* may be a stress oscilloscope.

(3) The grammatical form of the succeeding-letter meanings may be modified as required. For example, "indicate" may be applied as "indicator" or "indicating," "transmit" as "transmitter" or "transmitting," etc.

(4) Any first-letter, if used in combination with modifying letters *D* (differential), *F* (ratio), *M* (momentary), *K* (time rate of change), *Q* (integrate or totalize), or any combination of these is intended to represent a new and separate measured variable, and the combination is treated as a first-letter entity. Thus, instruments *TDI* and *TI* indicate two different variables, namely, differential-temperature and temperature. Modifying letters are used when applicable.

(5) First-letter *A* (analysis) covers all analyses not described by a "user's choice" letter. It is expected that the type of analysis will be defined outside a tagging bubble.

(6) Use of first-letter *U* for "multivariable" in lieu of a combination of first-letters is optional. It is recommended that nonspecific variable designators such as *U* be used sparingly.

(7) The use of modifying terms "high," "low," "middle" or "intermediate," and "scan" is optional.

(8) The term "safety" applies to emergency protective primary elements and emergency protective final control elements only. Thus, a self-actuated valve that prevents operation of a fluid system at a higher-than-desired pressure by bleeding fluid from the system is a back-pressure-type *PCV,* even if the valve is not intended to be used normally. However, this valve is designated as a *PSV* if it is intended to protect against emergency conditions, *i.e.,* conditions that are hazardous to personnel and/or equipment and that are not expected to arise normally.
The designation *PSV* applies to all valves intended to protect against emergency pressure conditions regardless of whether the valve construction and mode of operation place them in the category of the safety valve, relief valve, or safety relief valve. A rupture disc is designated *PSE.*

(9) The passive function *G* applies to instruments or devices that provide an uncalibrated view, such as sight glasses and television monitors.

(10) "Indicate" normally applies to the readout—analog or digital—of an actual measurement. In the case of a manual loader, it may be used for the dial or setting indication, *i.e.,* for the value of the initiating variable.

(11) A pilot light that is part of an instrument loop should be designated by a first-letter followed by the succeeding-letter *L.* For example, a pilot light that indicates an expired time period should be tagged *KQL.* If it is desired to tag a pilot light that is not part of an instrument loop, the light is designated in the same way. For example, a running light for an electric motor may be tagged *EL,* assuming voltage to be the appropriate measured variable, or *YL,* assuming the operating status is being monitored. The unclassified variable *X* should be used only for applications which are limited in extent. The designation *XL* should not be used for motor running lights, as these are commonly numerous. It is permissible to use the user's choice letters *M, N* or *O* for a motor running light when the meaning is previously defined. If *M* is used, it must be clear that the letter does not stand for the word "motor," but for a monitored state.

(12) Use of a succeeding-letter *U* for "multifunction" instead of a combination of other functional letters is optional. This nonspecific function designator should be used sparingly.

(13) A device that connects, disconnects, or transfers one or more circuits may be either a switch, a relay, an ON-OFF controller, or a control valve, depending on the application.
If the device manipulates a fluid process stream and is not a hand-actuated ON-OFF block valve, it is designated as a control valve. It is incorrect to use the succeeding-letters *CV* for anything other than a self-actuated control valve. For all applications other than fluid process streams, the device is designated as follows:

A switch, if it is actuated by hand.

A switch or an ON-OFF controller, if it is automatic and is the first such device in a loop. The term "switch" is generally used if the device is used for alarm, pilot light, selection, interlock, or safety.

The term "controller" is generally used if the device is used for normal operating control.

A relay, if it is automatic and is not the first such device in a loop, *i.e.,* it is actuated by a switch or an ON-OFF controller.

(14) It is expected that the functions associated with the use of succeeding-letter *Y* will be defined outside a bubble on a diagram when further definition is considered necessary. This definition need not be made when the function is self-evident, as for a solenoid valve in a fluid signal line.

(15) The modifying terms "high," and "low," and "middle" or "intermediate" correspond to values of the measured variable, not to values of the signal, unless otherwise noted. For example, a high-level alarm derived from a reverse-acting level transmitter signal should be an *LAH,* even though the alarm is actuated when the signal falls to a low value. The terms may be used in combinations as appropriate.

(16) The terms "high" and "low," when applied to positions of valves and other open-close devices, are defined as follows: "high" denotes that the valve is in or approaching the fully open position, and "low" denotes that it is in or approaching the fully closed position.

(17) The word "record" applies to any form of permanent storage of information that permits retrieval by any means.

(18) For use of the term "transmitter" versus "converter," see the definitions in ISA 55.1.

(19) First-letter *V,* "vibration or mechanical analysis," is intended to perform the duties in machinery monitoring that the letter *A* performs in more general analyses. Except for vibration, it is expected that the variable of interest will be defined outside the tagging bubble.

(20) First-letter *Y* is intended for use when control or monitoring responses are event-driven as opposed to time- or time schedule-driven. The letter *Y,* in this position, can also signify presence or state.

(21) Modifying-letter *K,* in combination with a first-letter such as *L, T,* or *W,* signifies a time rate of change of the measured or initiating variable. The variable *WKIC,* for instance, may represent a rate-of-weight-loss controller.

(22) Succeeding-letter *K* is a user's option for designating a control station, while the succeeding-letter *C* is used for describing automatic or manual controllers.

Table 2.3. Instrument Line Symbols

ALL LINES TO BE FINE IN RELATION TO PROCESS PIPING LINES.

(1)	INSTRUMENT SUPPLY * OR CONNECTION TO PROCESS	
(2)	UNDEFINED SIGNAL	
(3)	PNEUMATIC SIGNAL **	
(4)	ELECTRIC SIGNAL	 OR 
(5)	HYDRAULIC SIGNAL	
(6)	CAPILLARY TUBE	
(7)	ELECTROMAGNETIC OR SONIC SIGNAL *** (GUIDED)	
(8)	ELECTROMAGNETIC OR SONIC SIGNAL *** (NOT GUIDED)	
(9)	INTERNAL SYSTEM LINK (SOFTWARE OR DATA LINK)	
(10)	MECHANICAL LINK	
	OPTIONAL BINARY (ON-OFF) SYMBOLS	
(11)	PNEUMATIC BINARY SIGNAL	
(12)	ELECTRIC BINARY SIGNAL	 OR 

NOTE: 'Or' means user's choice. Consistency is recommended.

* The following abbreviations are suggested to denote the types of power supply. These designations may also be applied to purge fluid supplies.

AS - Air Supply
IA - Instrument Air } Options
PA - Plant Air
ES - Electric Supply
GS - Gas Supply
HS - Hydraulic Supply
NS - Nitrogen Supply
SS - Steam Supply
WS - Water Supply

The supply level may be added to the instrument supply line, e.g., AS-100, a 100-psig air supply; ES-24DC, a 24-volt direct current power supply.

** The pneumatic signal symbol applies to a signal using any gas as the signal medium. If a gas other than air is used, the gas may be identified by a note on the signal symbol or otherwise.

*** Electromagnetic phenomena include heat, radio waves, nuclear radiation, and light.

Source: Instrument Society of America

Table 2.4. Function Designations for Relays

The function designations associated with relays may be used as follows, individually or in combination (see Table 2.3, note 14). The use of a box enclosing a symbol is optional; the box is intended to avoid confusion by setting off the symbol from other markings on a diagram.

SYMBOL	FUNCTION
1. 1-0 or ON-OFF	Automatically connect, disconnect, or transfer one or more circuits provided that this is not the first such device in a loop (see Table 1, note 13).
2. Σ or ADD	Add or totalize (add and subtract) †
3. Δ or DIFF.	Subtract †
4. ± + [–]	Bias*
5. AVG.	Average
6. % or 1:3 or 2:1 (typical)	Gain or attenuate (input:output)*
7. [x]	Multiply †
8. ÷	Divide †
9. [√] or SQ. RT.	Extract square root
10. x^n or $x^{1/n}$	Raise to power
11. f (x)	Characterize
12. 1:1	Boost
13. [>] or HIGHEST (MEASURED VARIABLE)	High-select. Select highest (higher) measured variable (not signal, unless so noted).
14. [<] or LOWEST (MEASURED VARIABLE)	Low-select. Select lowest (lower) measured variable (not signal, unless so noted).
15. REV.	Reverse
16.	Convert
a. E/P or P/I (typical)	For input/output sequences of the following: **Designation** — **Signal** E — Voltage H — Hydraulic I — Current (electrical) O — Electromagnetic or sonic P — Pneumatic R — Resistance (electrical)
b. A/D or D/A	For input/output sequences of the following: A — Analog D — Digital
17. ∫	Integrate (time integral)
18. D or d/dt	Derivative or rate
19. 1/D	Inverse derivative
20. As required	Unclassified

* Used for single-input relay.
† Used for relay with two or more inputs.

Source: Instrument Society of America

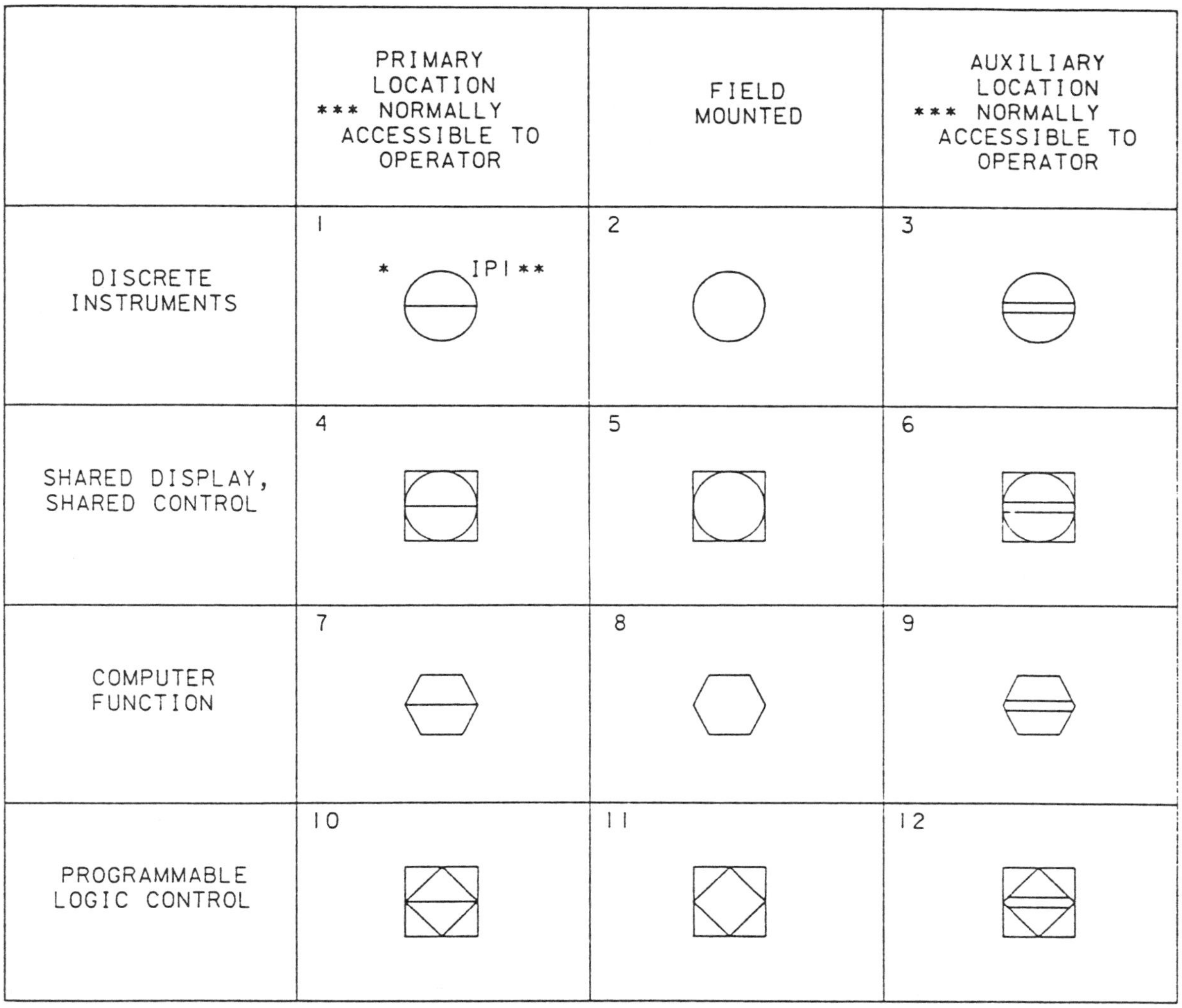
PRIMARY LOCATION *** NORMALLY ACCESSIBLE TO OPERATOR
FIELD MOUNTED
AUXILIARY LOCATION *** NORMALLY ACCESSIBLE TO OPERATOR
DISCRETE INSTRUMENTS
SHARED DISPLAY, SHARED CONTROL
COMPUTER FUNCTION
PROGRAMMABLE LOGIC CONTROL
1
* IPI **
2
3
4
5
6
7
8
9
10
11
12

* Symbol size may vary according to the user's needs and the type of document. A suggested square and circle size for large diagrams is shown above. Consistency is recommended.

** Abbreviations of the user's choice such as IP1 (Instrument Panel #1), IC2 (Instrument Console #2), CC3 (Computer Console #3), etc., may be used when it is necessary to specify instrument or function location.

*** Normally inaccessible or behind-the-panel devices or functions may be depicted by using the same symbols but with dashed horizontal bars, i.e.

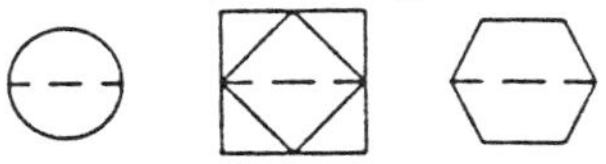

Figure 2.4. General instrument or function symbols. Courtesy of Instrument Society of America.

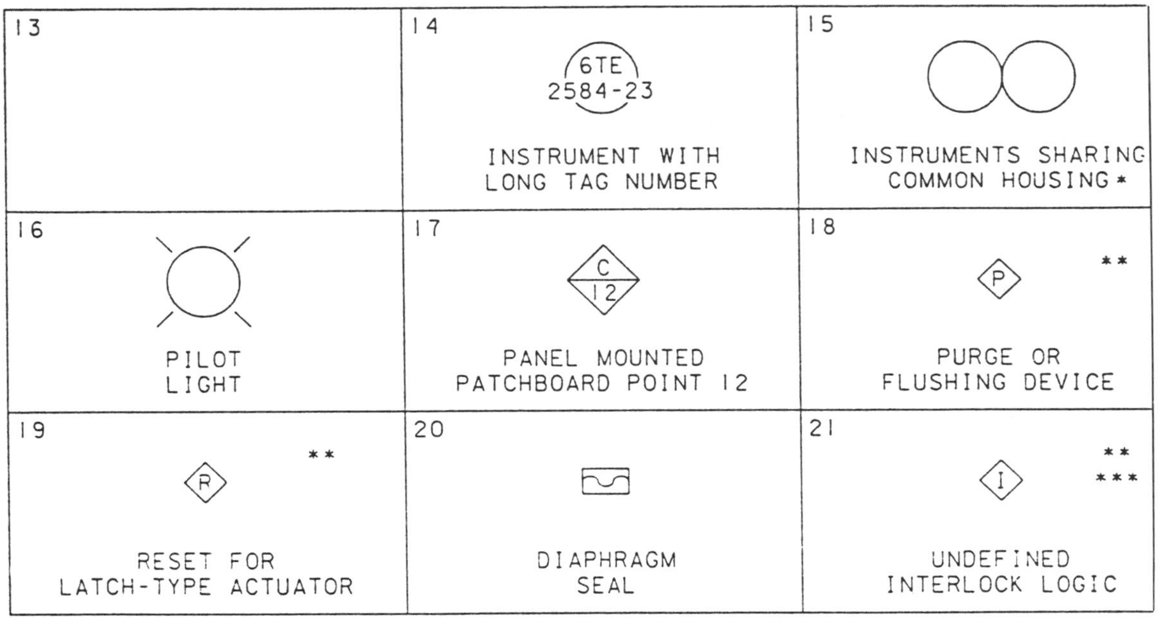

* It is not mandatory to show a common housing.
** These diamonds are approximately half the size of the larger ones.
*** For specific logic symbols, see ANSI/ISA Standard S5.2.

Figure 2.4 continued

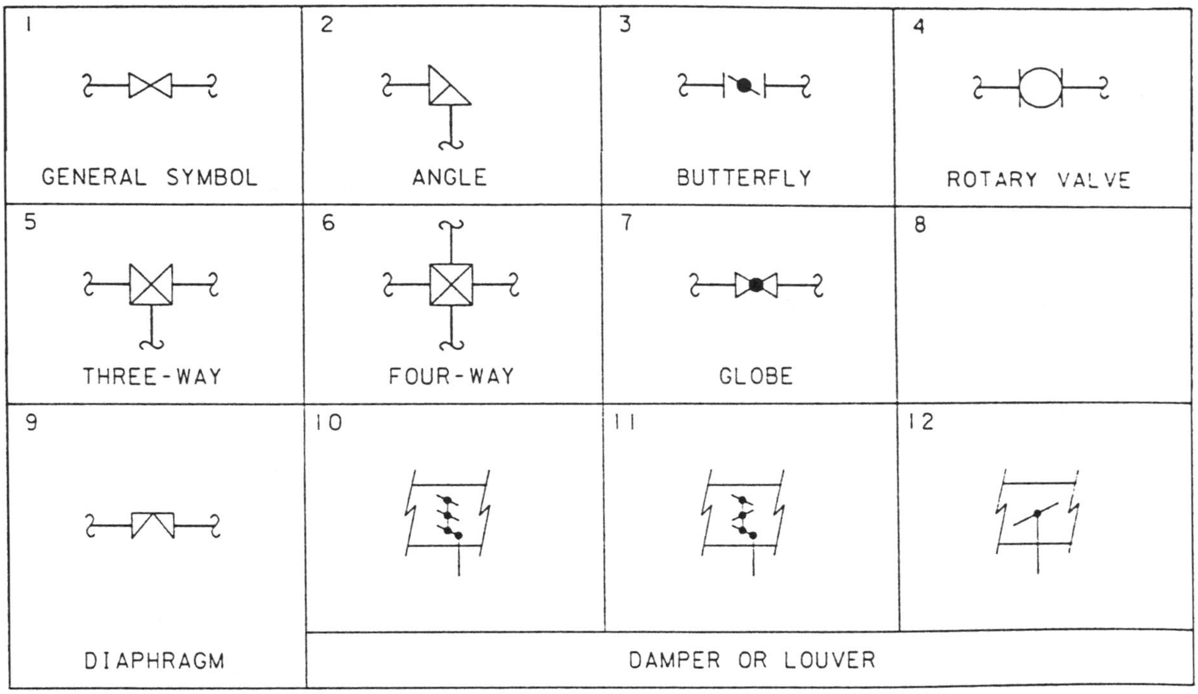

Further information may be added adjacent to the body symbol either by note or code number.

Figure 2.5. Control valve body symbols and damper symbols. Courtesy of Instrument Society of America.

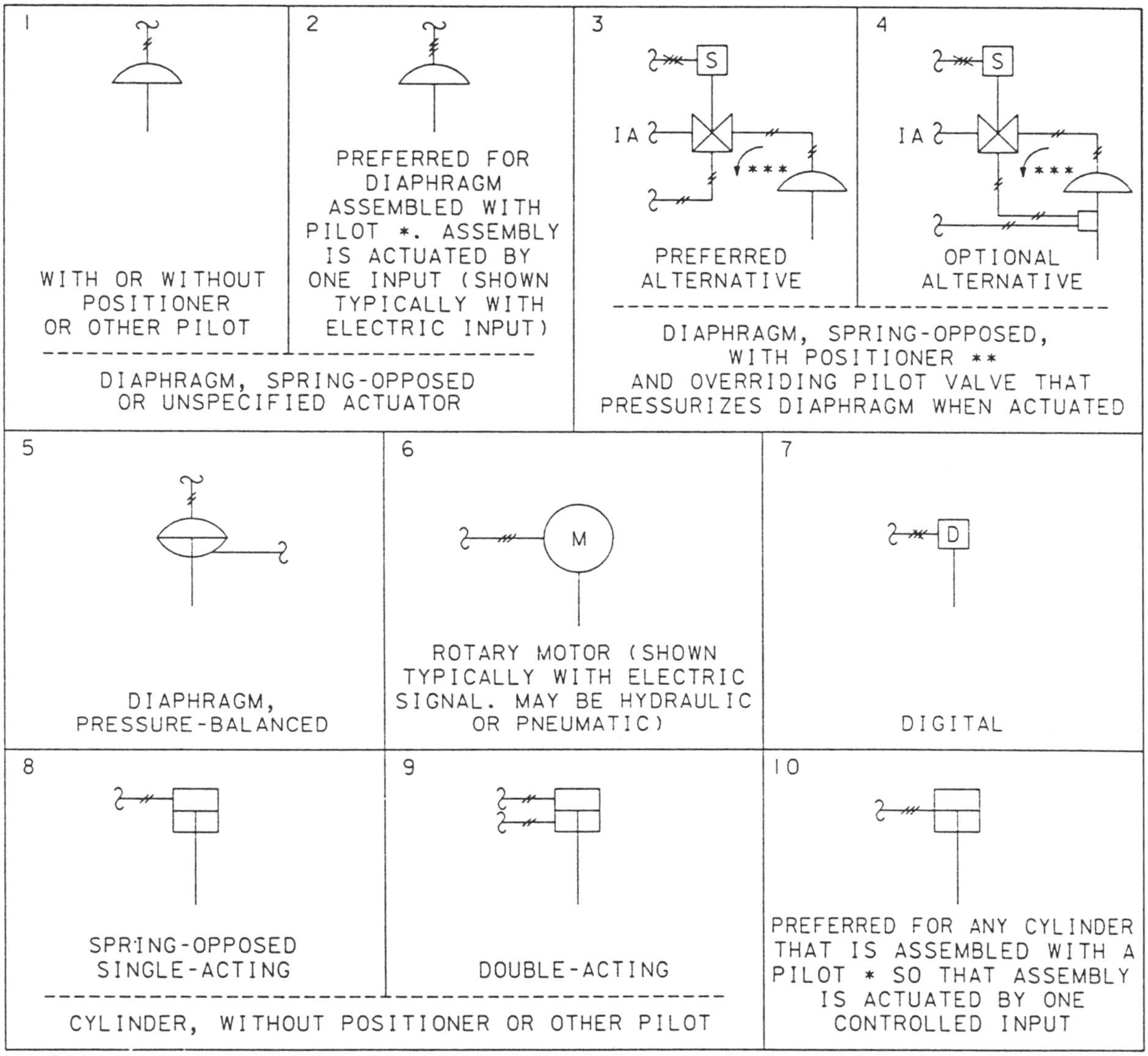

* Pilot may be positioner, solenoid valve, signal converter, etc.

** The positioner need not be shown unless an intermediate device is on its output. The positioner tagging, ZC, need not be used even if the positioner is shown. The positioner symbol, a box drawn on the actuator shaft, is the same for all types of actuators. When the symbol is used, the type of instrument signal, i.e., pneumatic, electric, etc., is drawn as appropriate. If the positioner symbol is used and there is no intermediate device on its output, then the positioner output signal need not be shown.

*** The arrow represents the path from a common to a fail open port. It does not correspond necessarily to the direction of fluid flow.

Figure 2.6. Actuator symbols. Courtesy of Instrument Society of America.

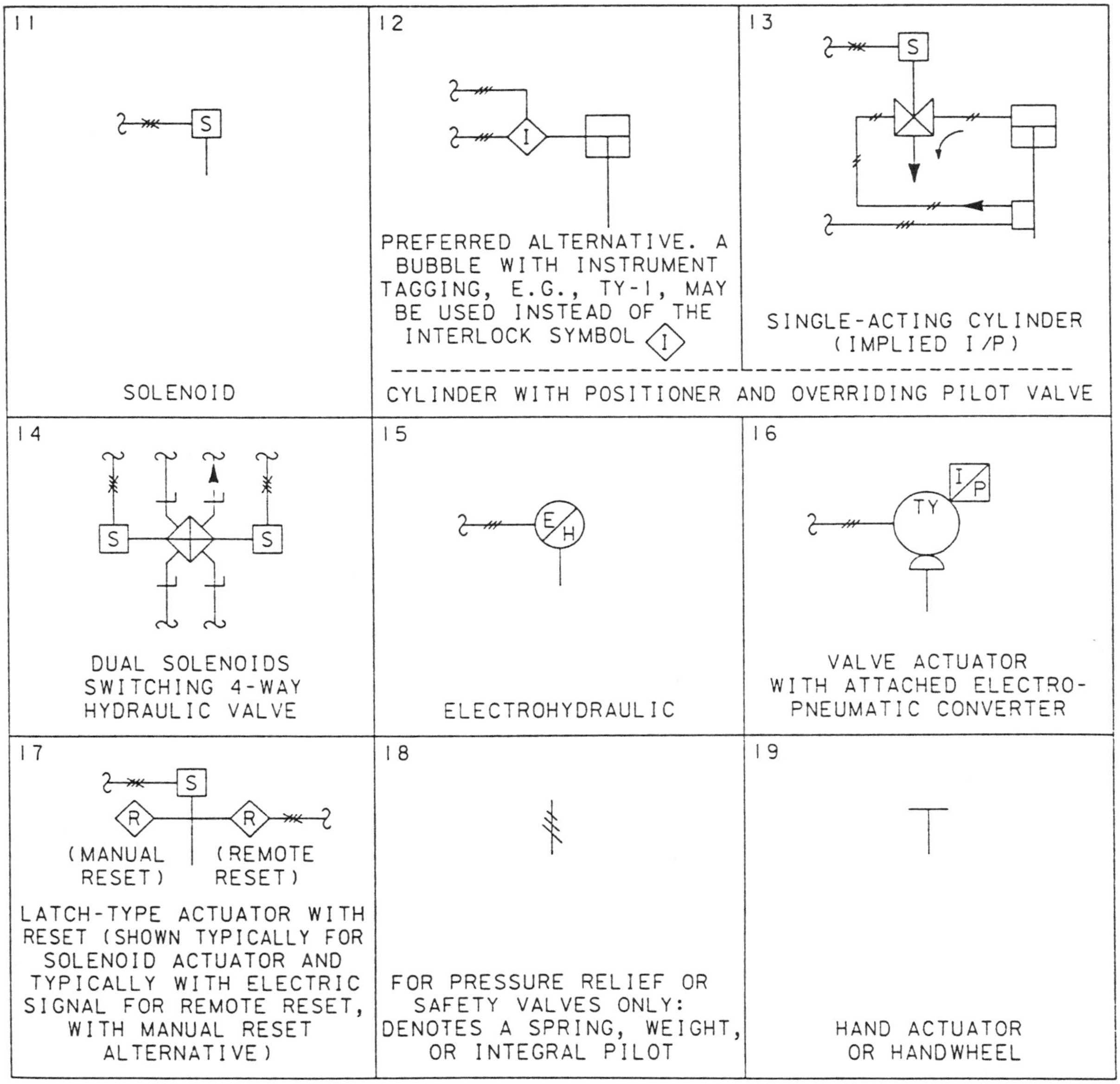

Figure 2.6 continued

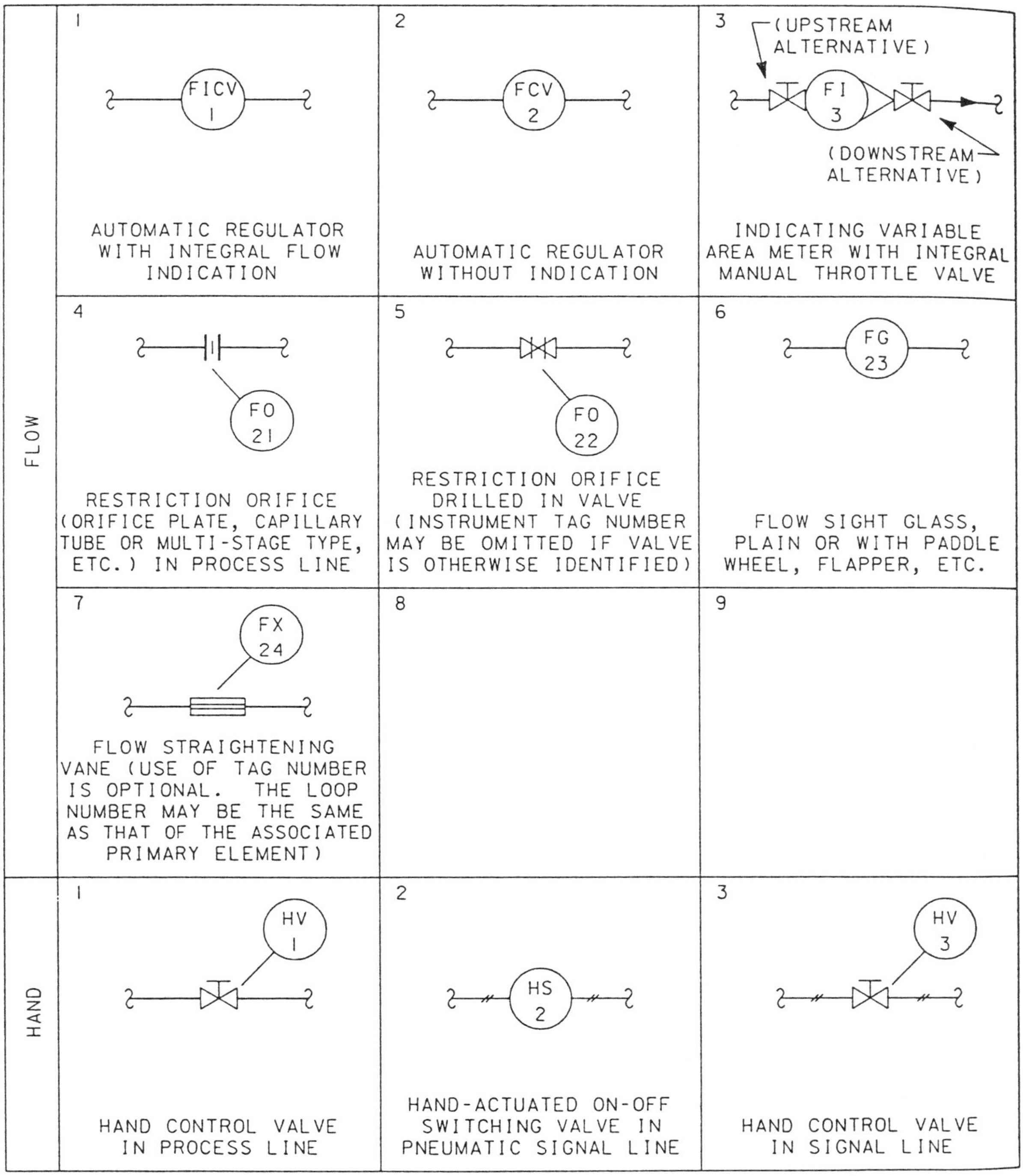

Figure 2.7. Symbols for self-actuated regulators, valves and other devices. Courtesy of Instrument Society of America.

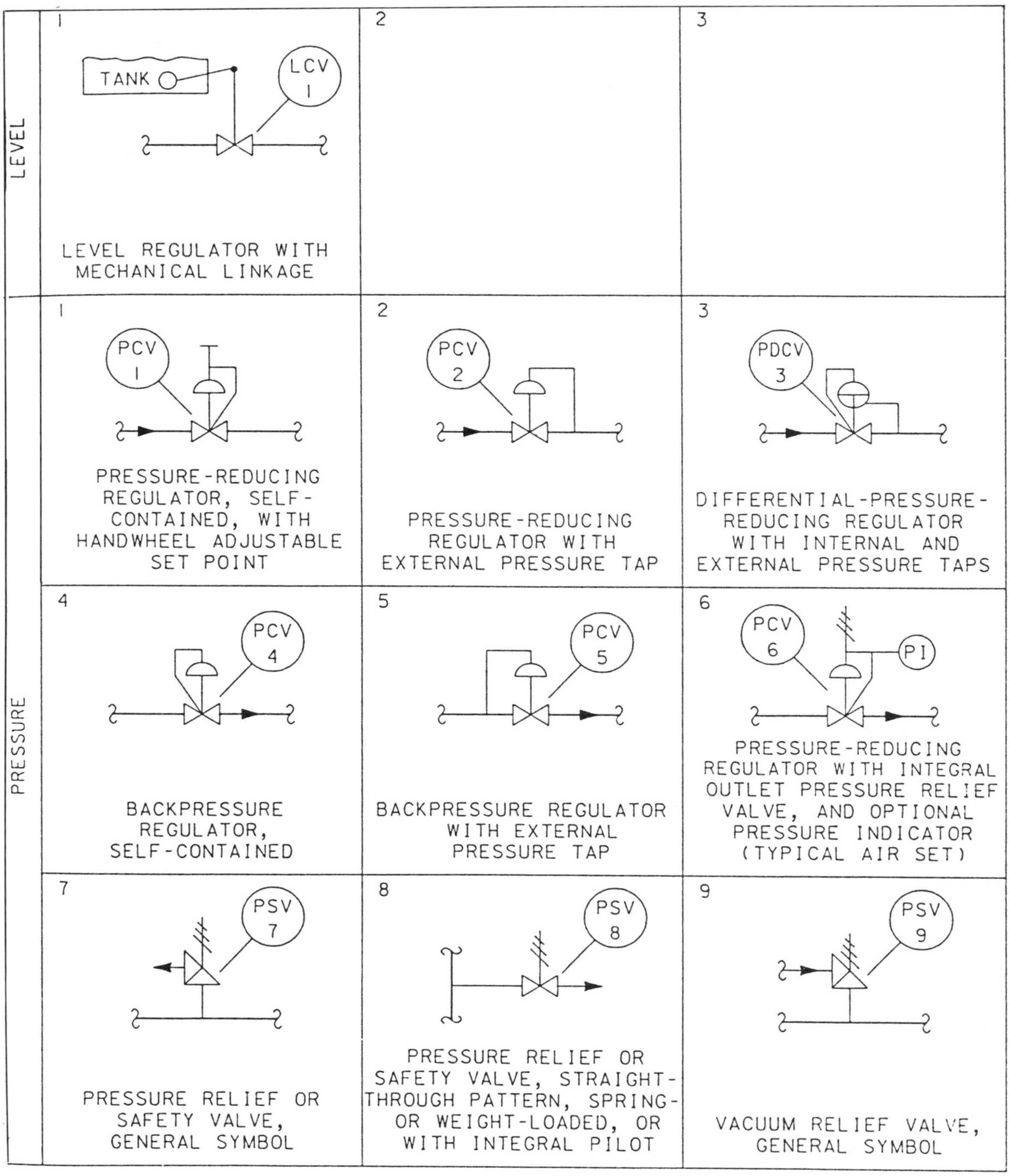

Figure 2.7 continued

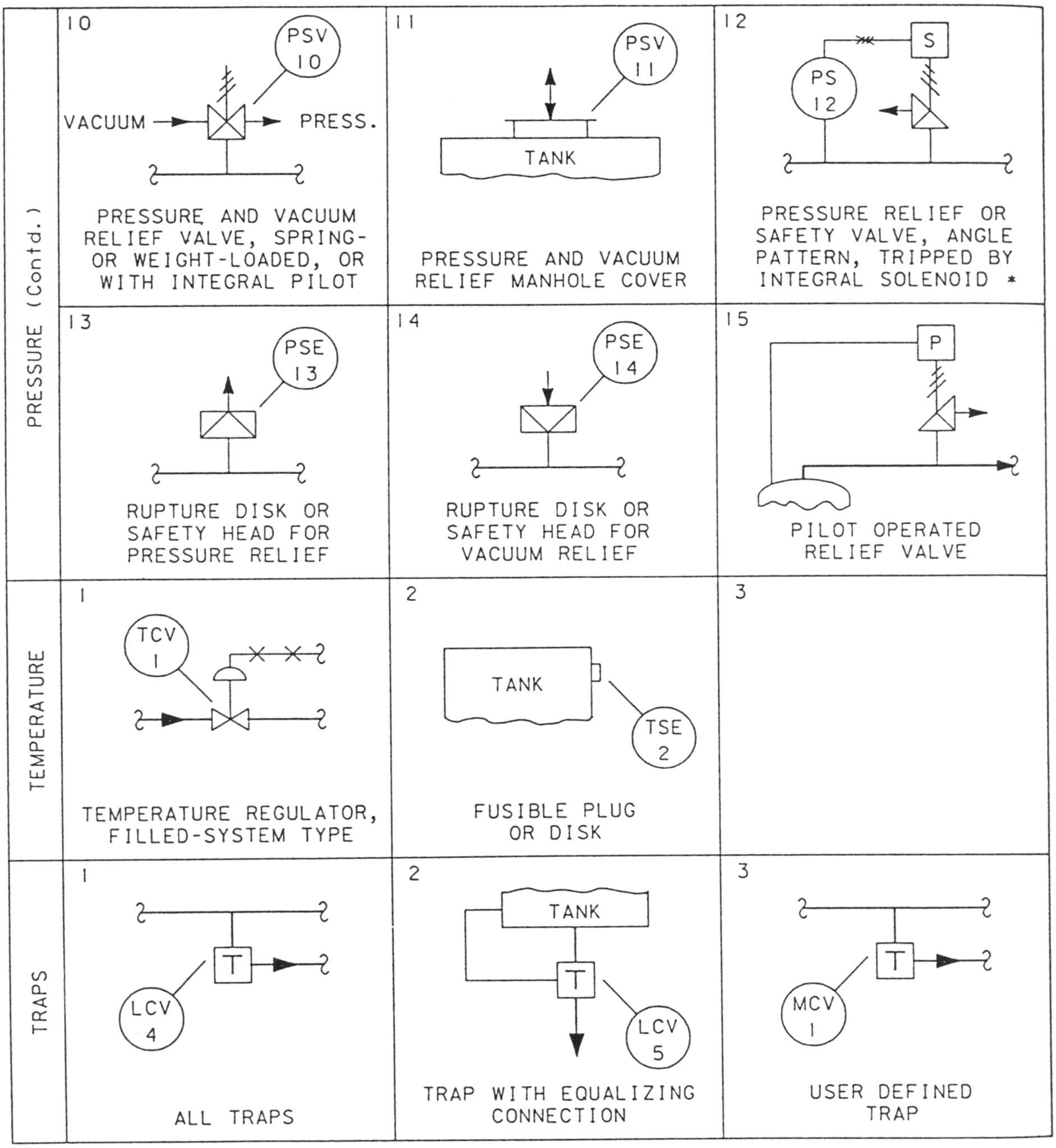

* The solenoid-tripped pressure relief valve is one of the class of power-actuated relief valves and is grouped with the other types of relief valves even though it is not entirely a self-actuated device.

Figure 2.7 continued

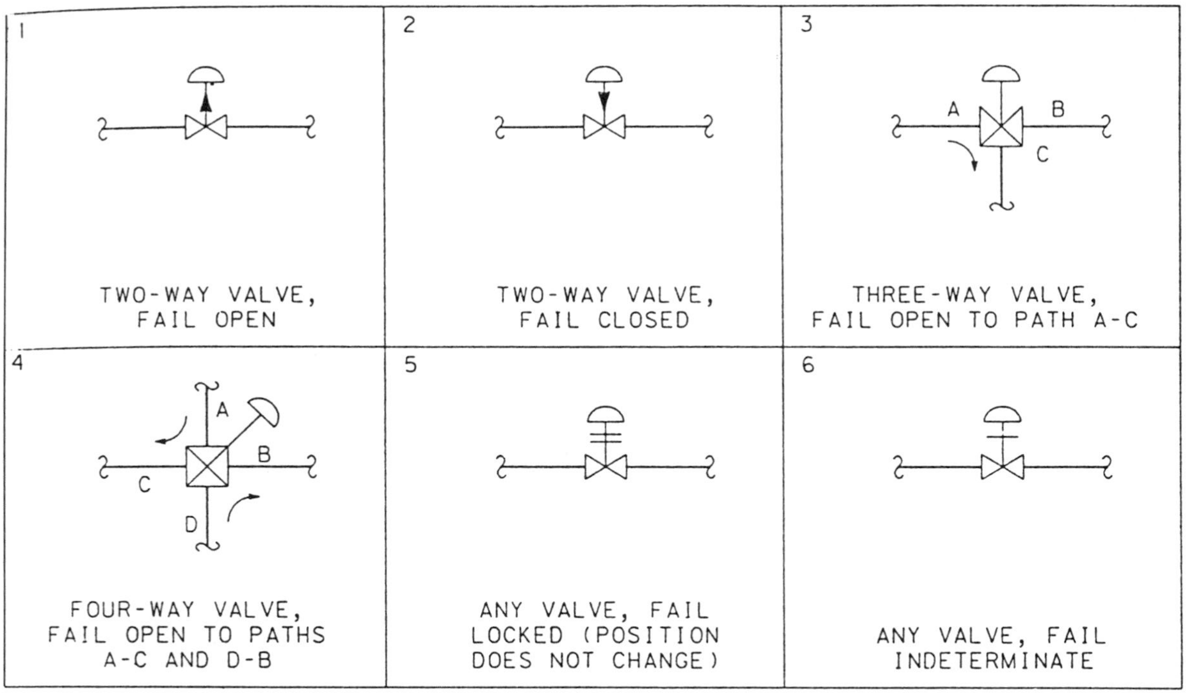

The failure modes indicated are those commonly defined by the term, "shelf-position". As an alternative to the arrows and bars, the following abbreviations may be employed:

FO - Fail Open
FC - Fail Closed
FL - Fail Locked (last position)
FI - Fail Indeterminate

Figure 2.8. Symbols for actuator action in the event of power failure. Courtesy of Instrument Society of America.

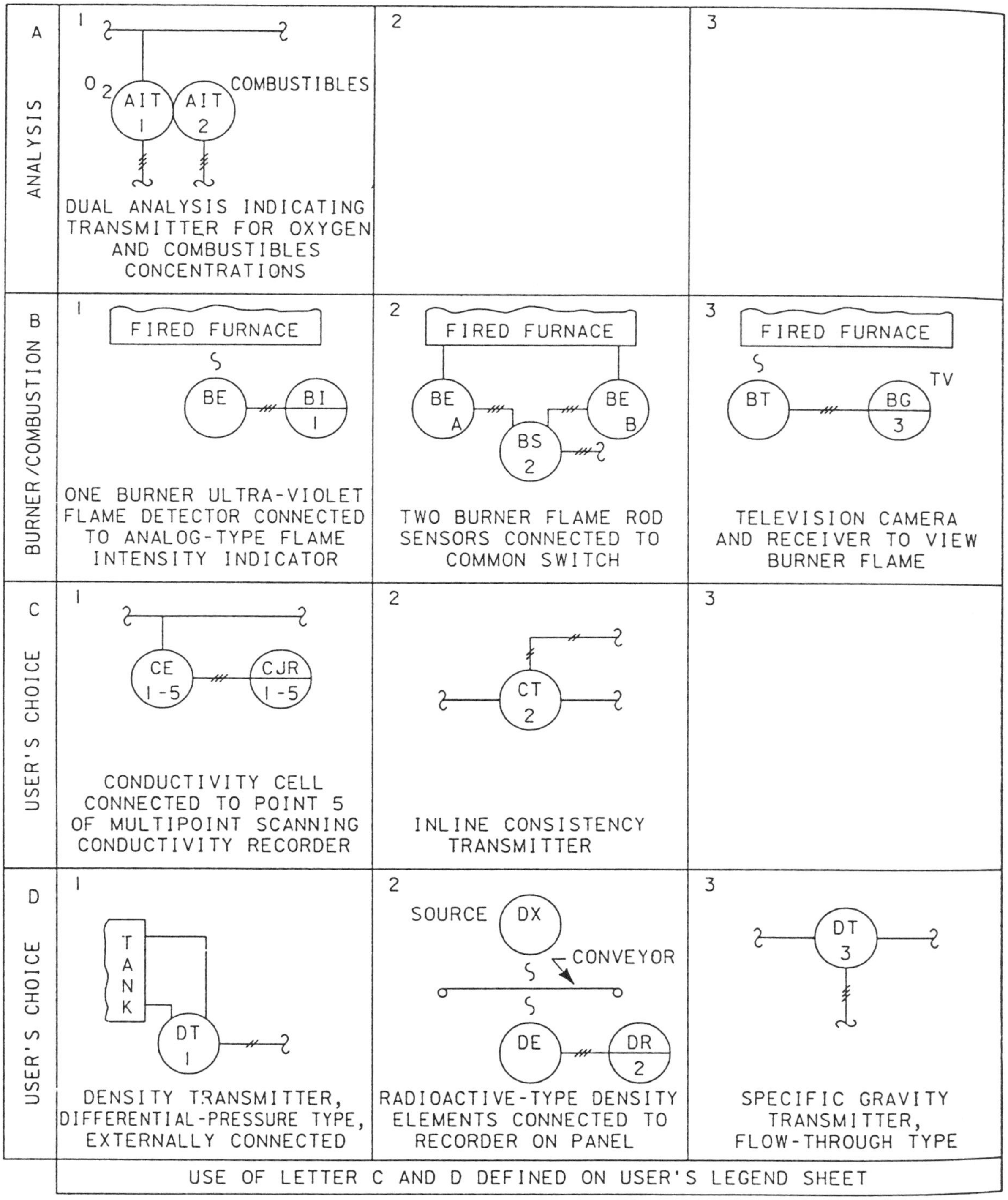

Figure 2.9. Primary element symbols. Courtesy of Instrument Society of America.

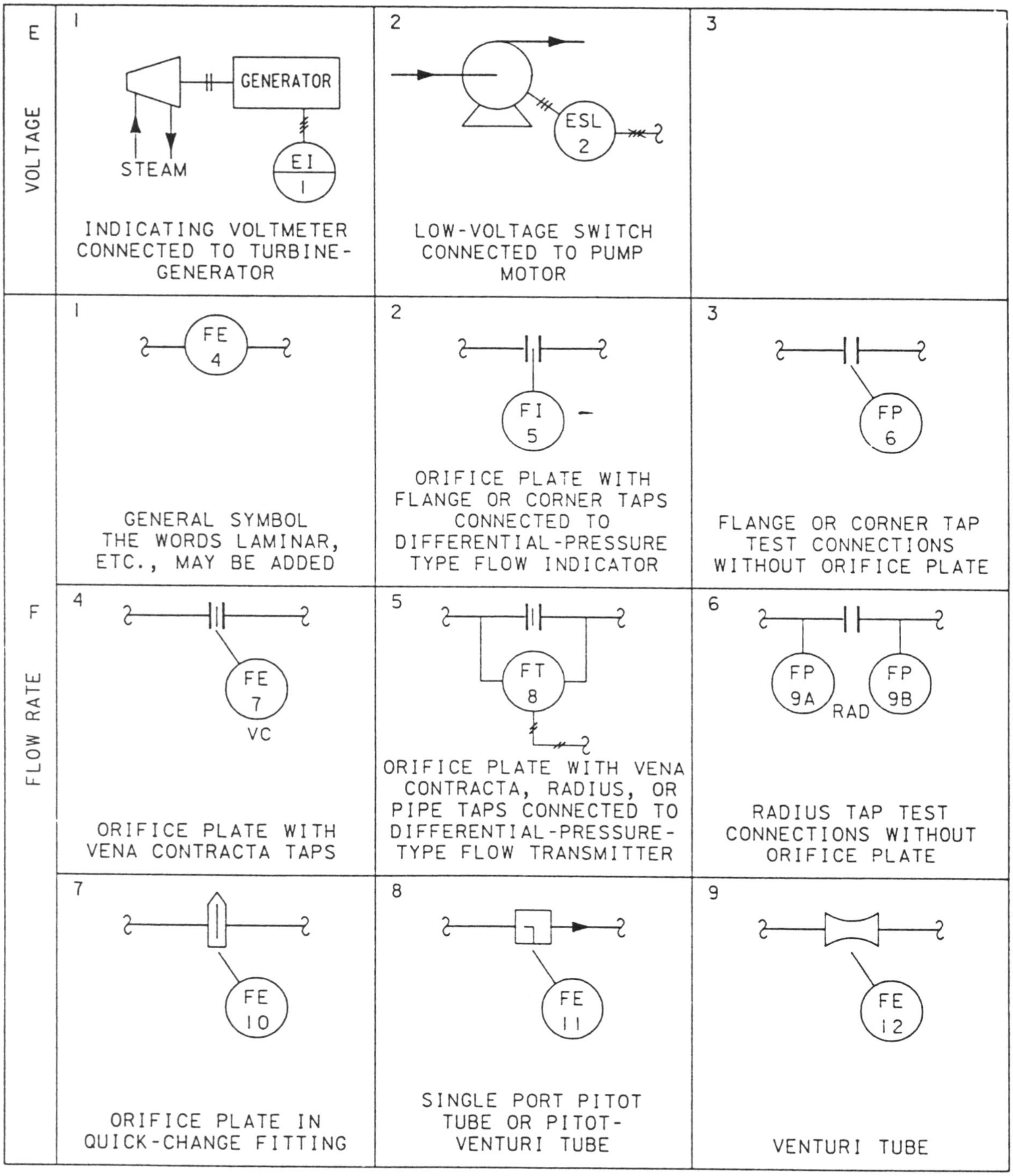

Figure 2.9 continued

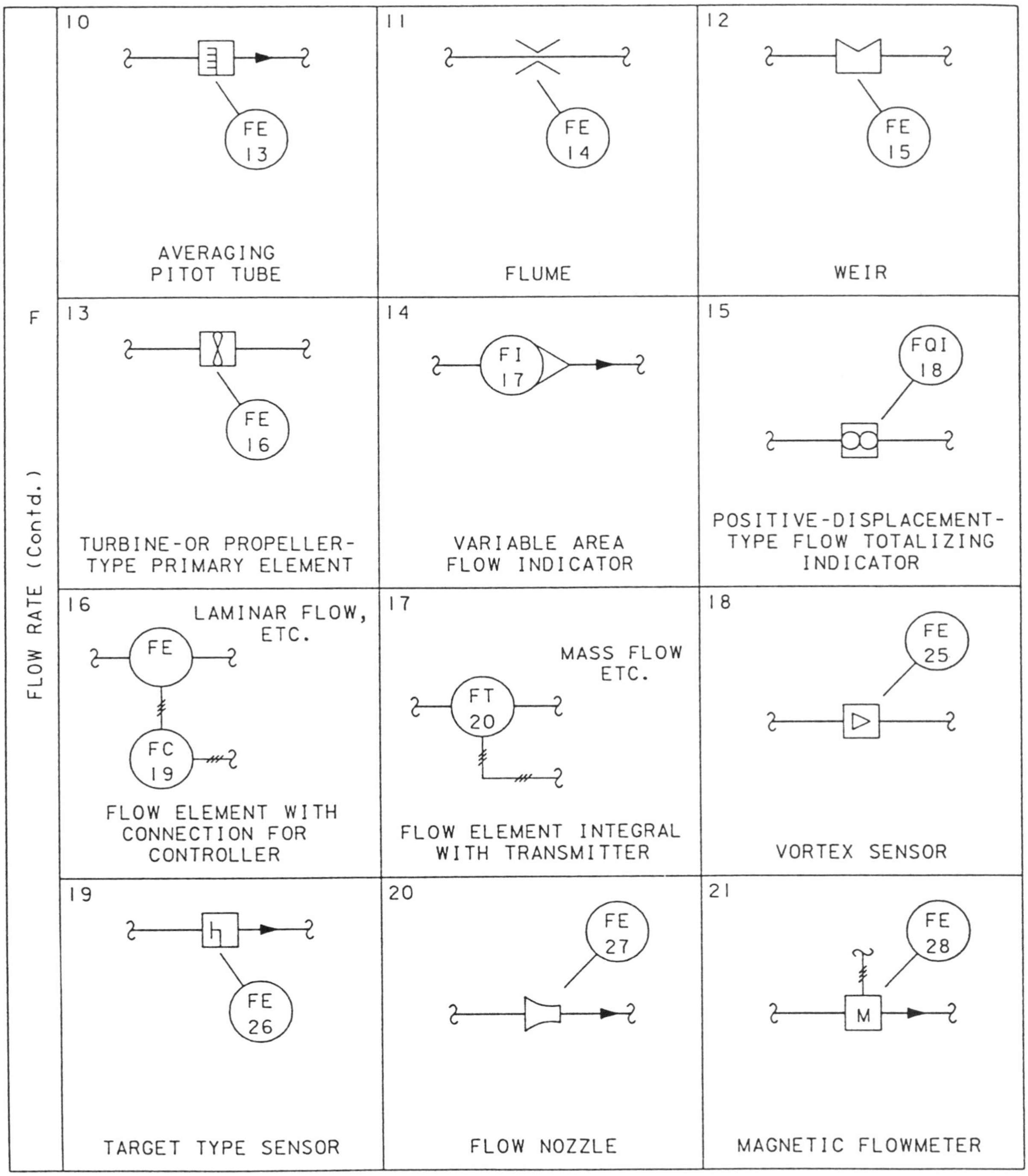

Figure 2.9 continued

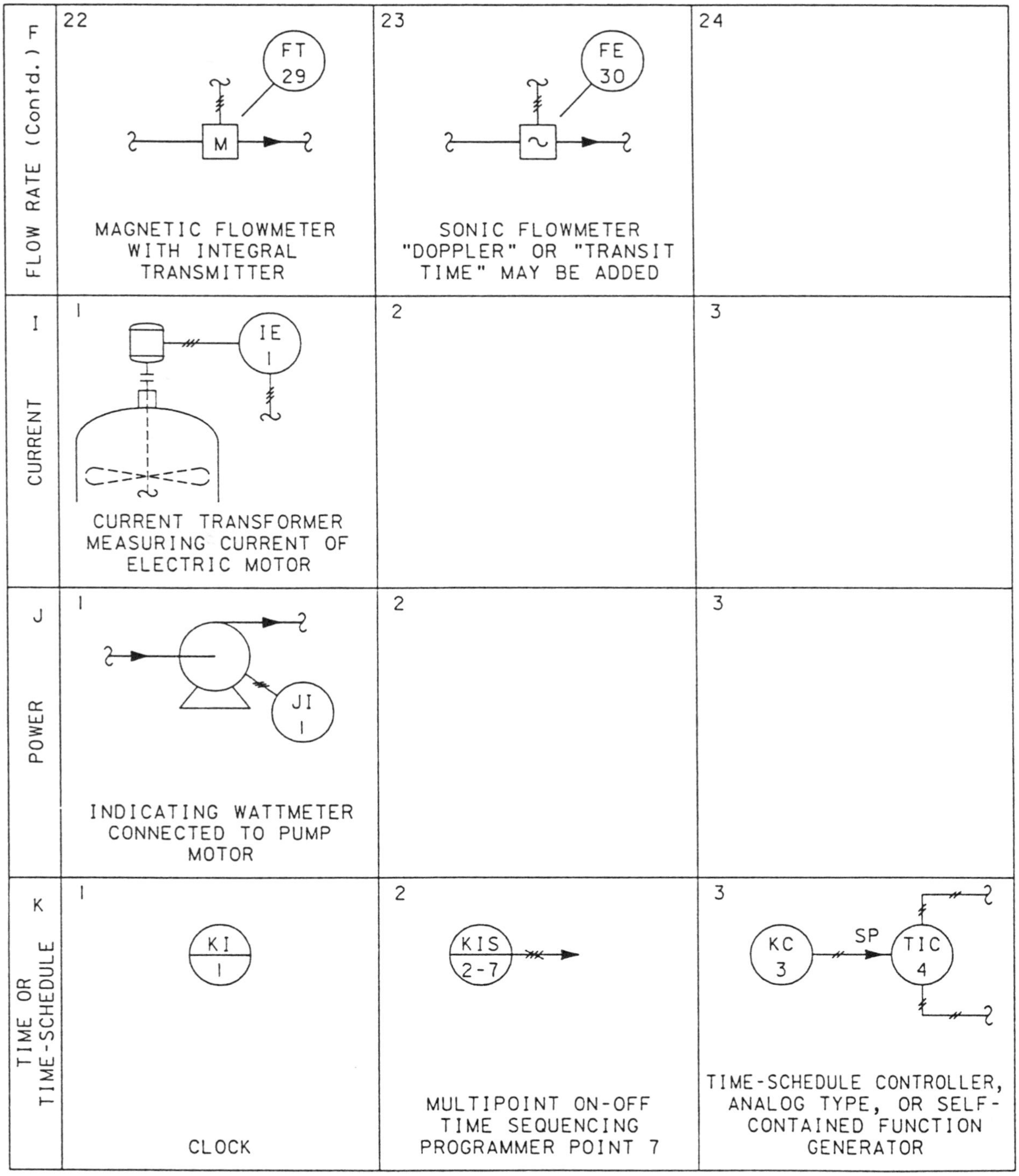

Figure 2.9 continued

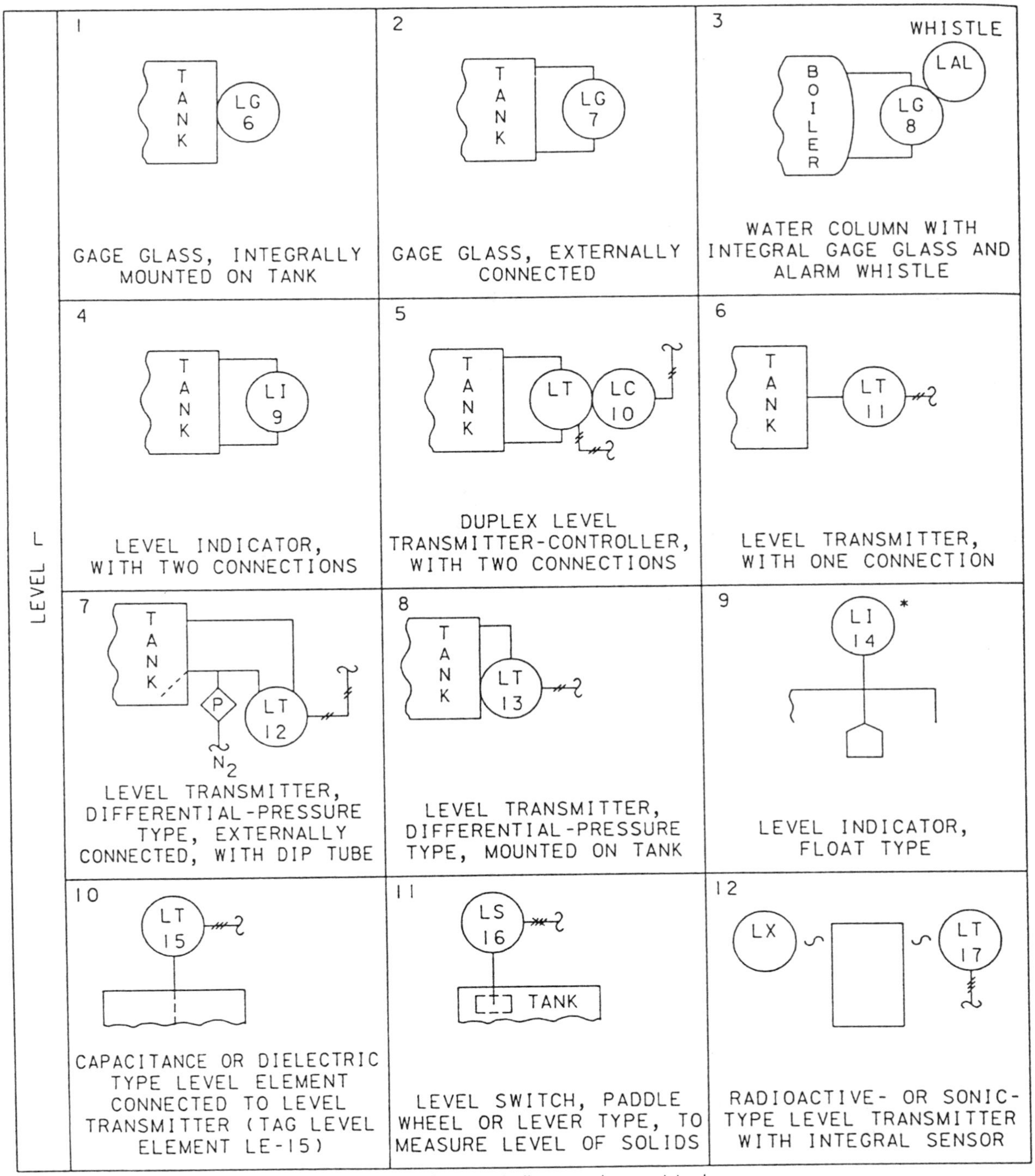
LEVEL L
1
TANK
LG 6
GAGE GLASS, INTEGRALLY MOUNTED ON TANK
2
TANK
LG 7
GAGE GLASS, EXTERNALLY CONNECTED
3
WHISTLE
LAL
BOILER
LG 8
WATER COLUMN WITH INTEGRAL GAGE GLASS AND ALARM WHISTLE
4
TANK
LI 9
LEVEL INDICATOR, WITH TWO CONNECTIONS
5
TANK
LT
LC 10
DUPLEX LEVEL TRANSMITTER-CONTROLLER, WITH TWO CONNECTIONS
6
TANK
LT 11
LEVEL TRANSMITTER, WITH ONE CONNECTION
7
TANK
P
LT 12
N2
LEVEL TRANSMITTER, DIFFERENTIAL-PRESSURE TYPE, EXTERNALLY CONNECTED, WITH DIP TUBE
8
TANK
LT 13
LEVEL TRANSMITTER, DIFFERENTIAL-PRESSURE TYPE, MOUNTED ON TANK
9
LI 14
*
LEVEL INDICATOR, FLOAT TYPE
10
LT 15
CAPACITANCE OR DIELECTRIC TYPE LEVEL ELEMENT CONNECTED TO LEVEL TRANSMITTER (TAG LEVEL ELEMENT LE-15)
11
LS 16
TANK
LEVEL SWITCH, PADDLE WHEEL OR LEVER TYPE, TO MEASURE LEVEL OF SOLIDS
12
LX
LT 17
RADIOACTIVE- OR SONIC-TYPE LEVEL TRANSMITTER WITH INTEGRAL SENSOR

* Notations such as "mounted at grade" may be added.

Figure 2.9 continued

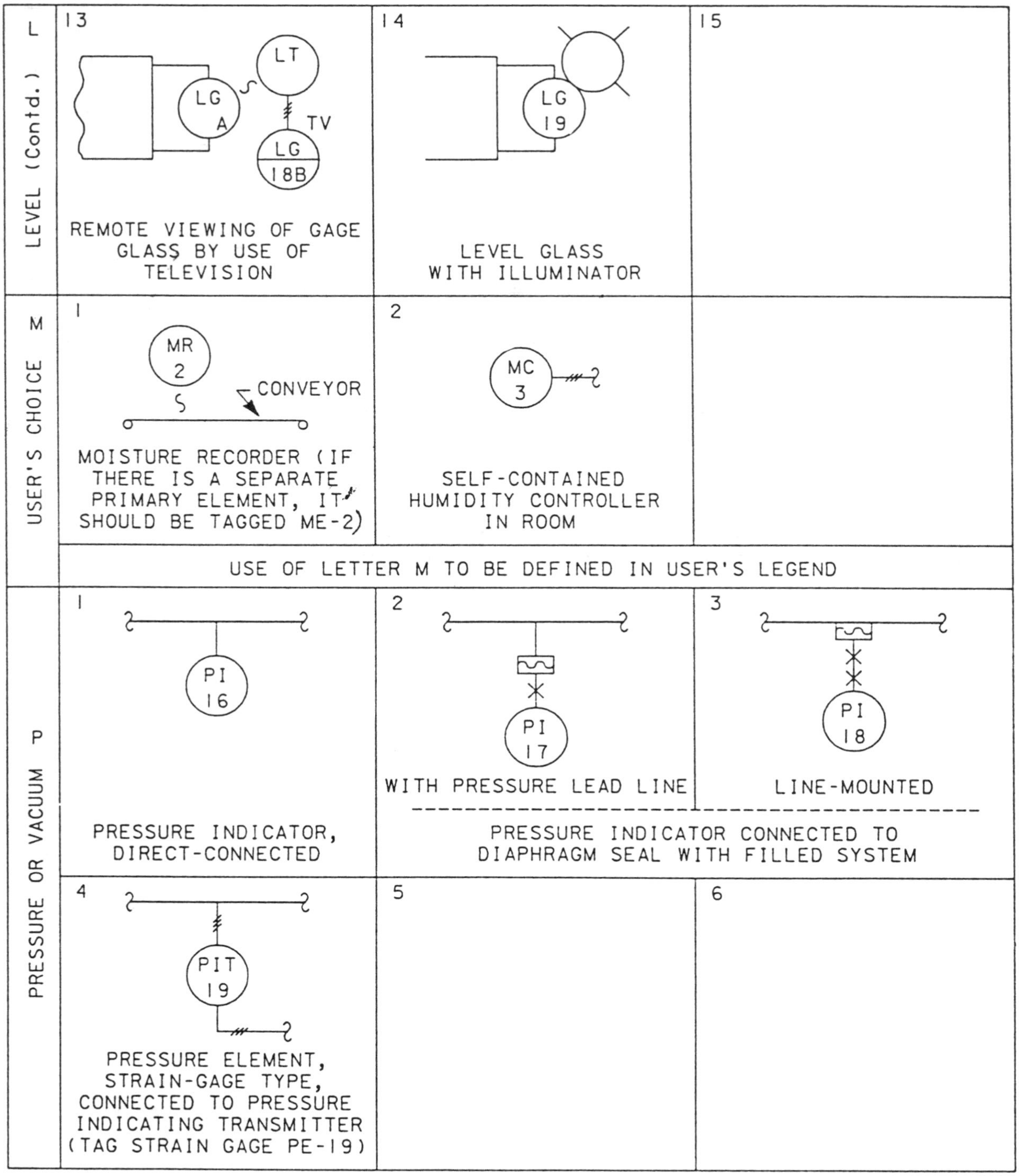

Figure 2.9 continued

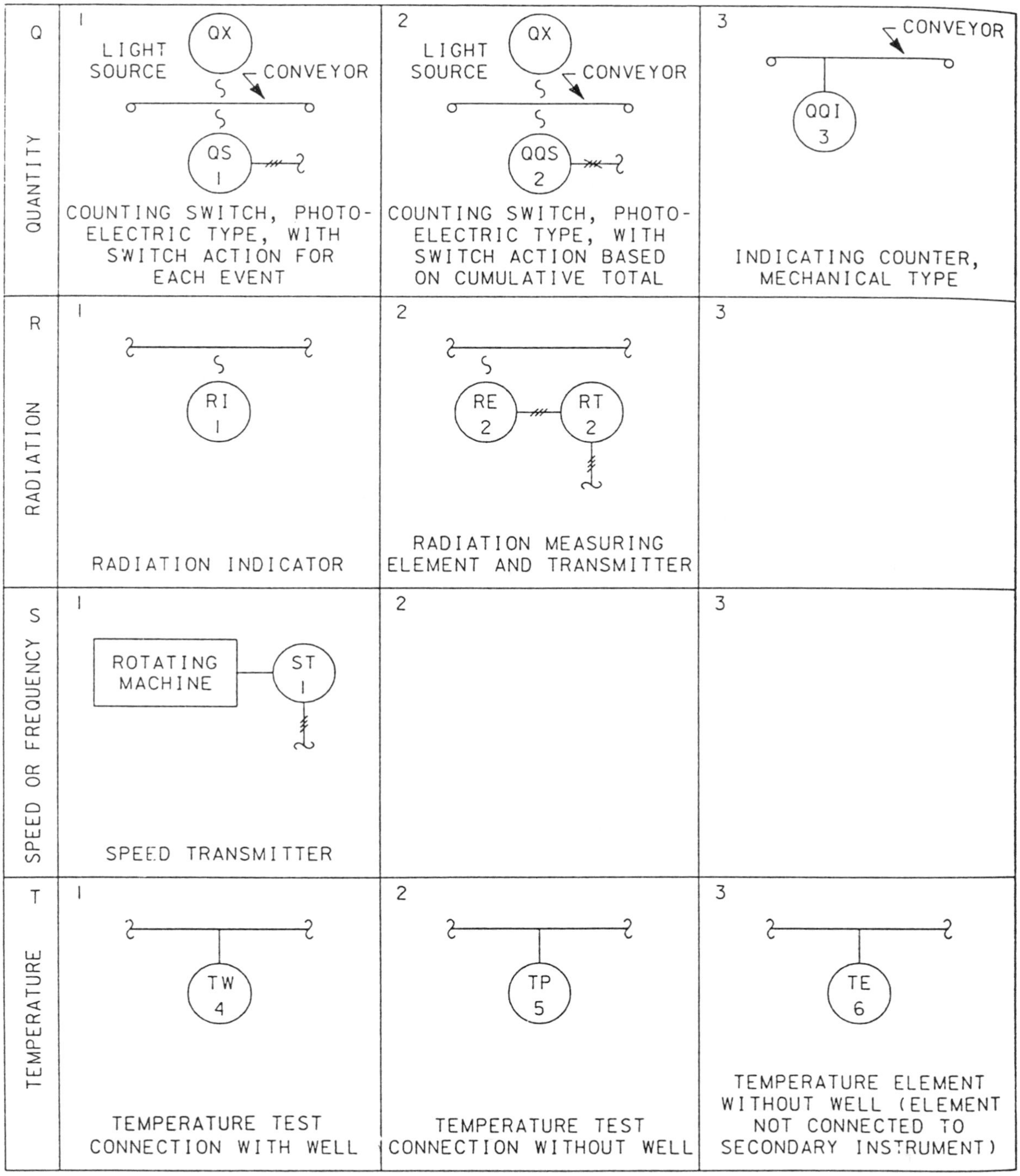

Figure 2.9 continued

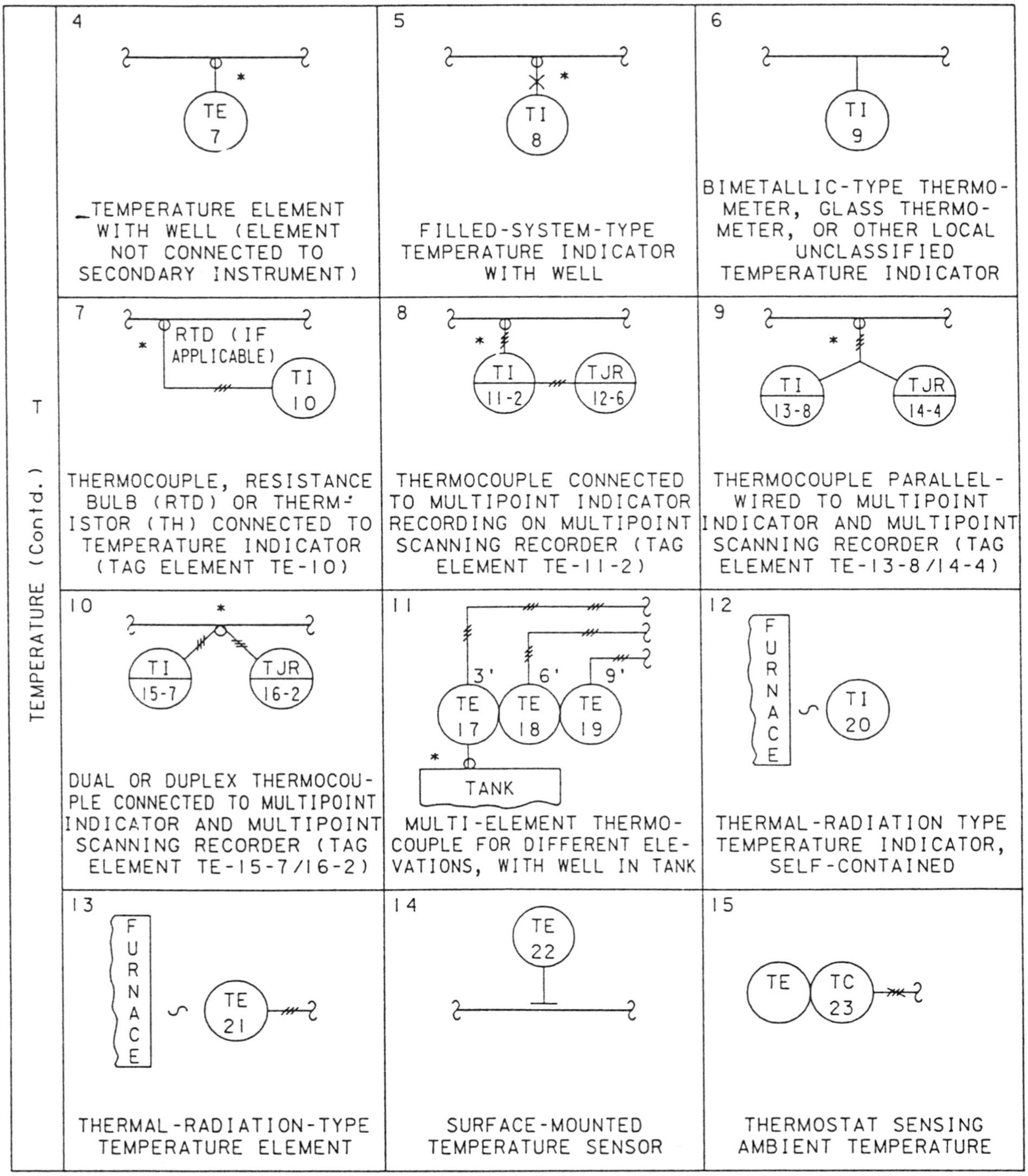

* Use of the thermowell symbol is optional. However, use or omission of the symbol should be consistent throughout a project.

Figure 2.9 continued

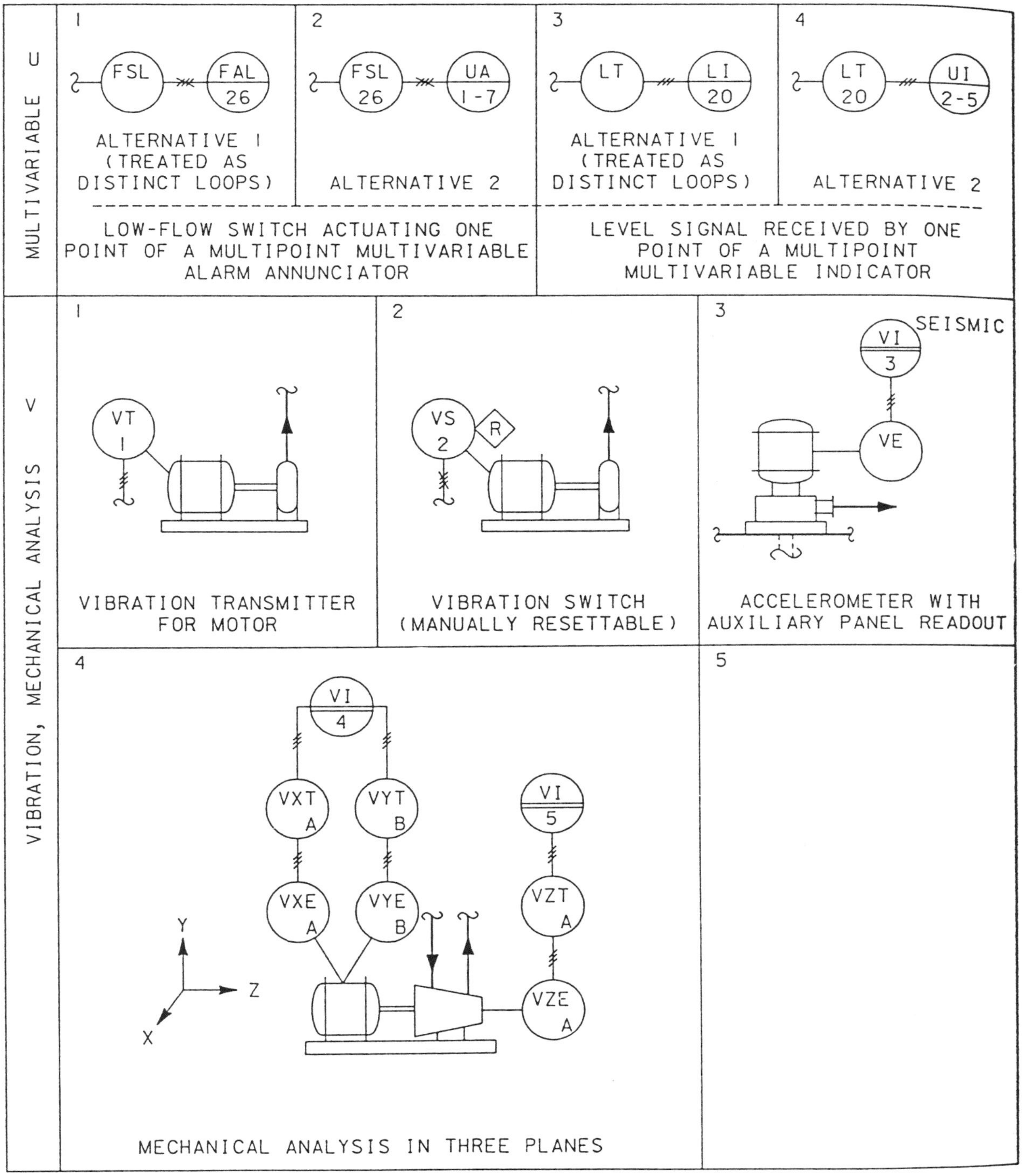

Figure 2.9 continued

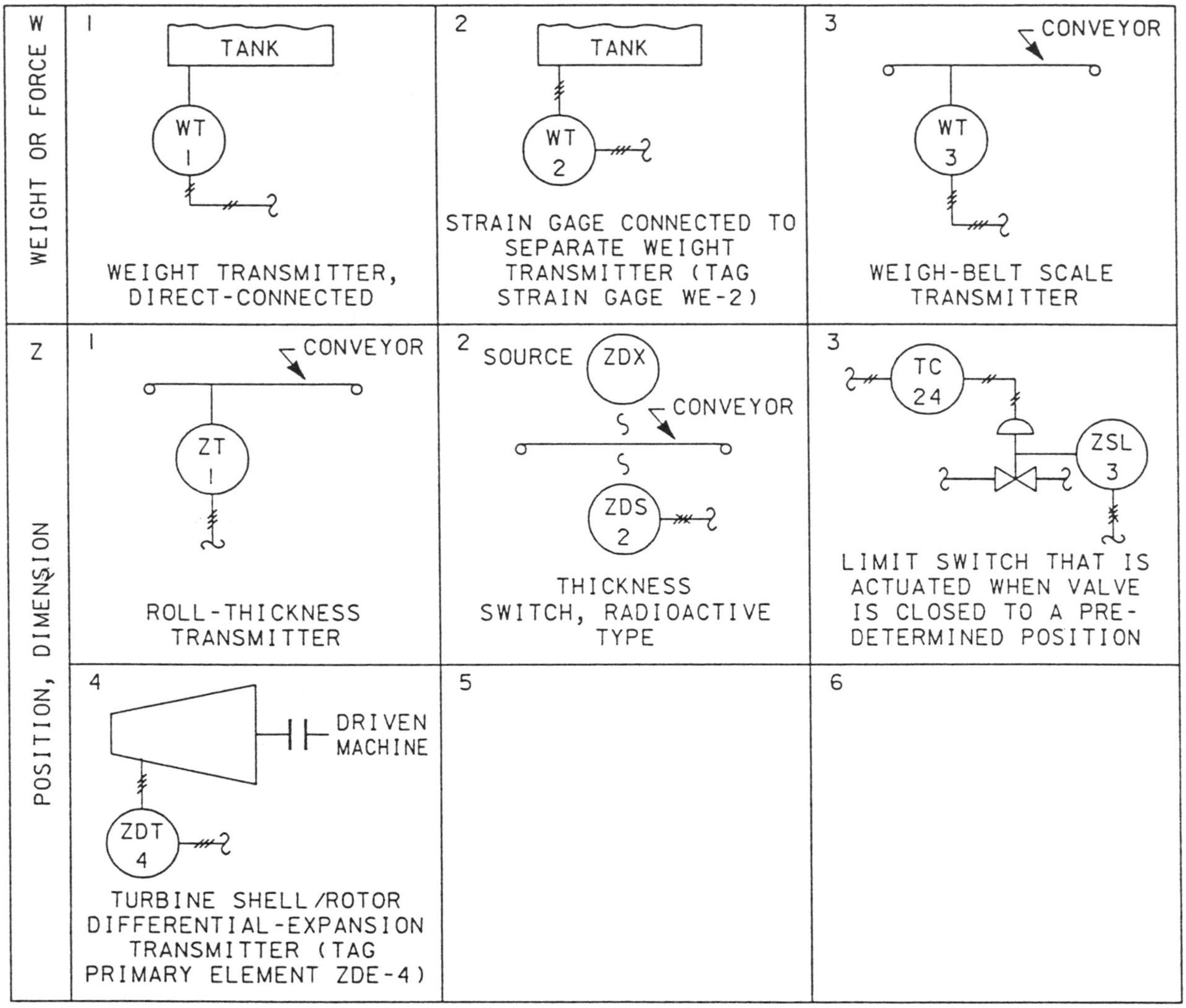

Figure 2.9 continued

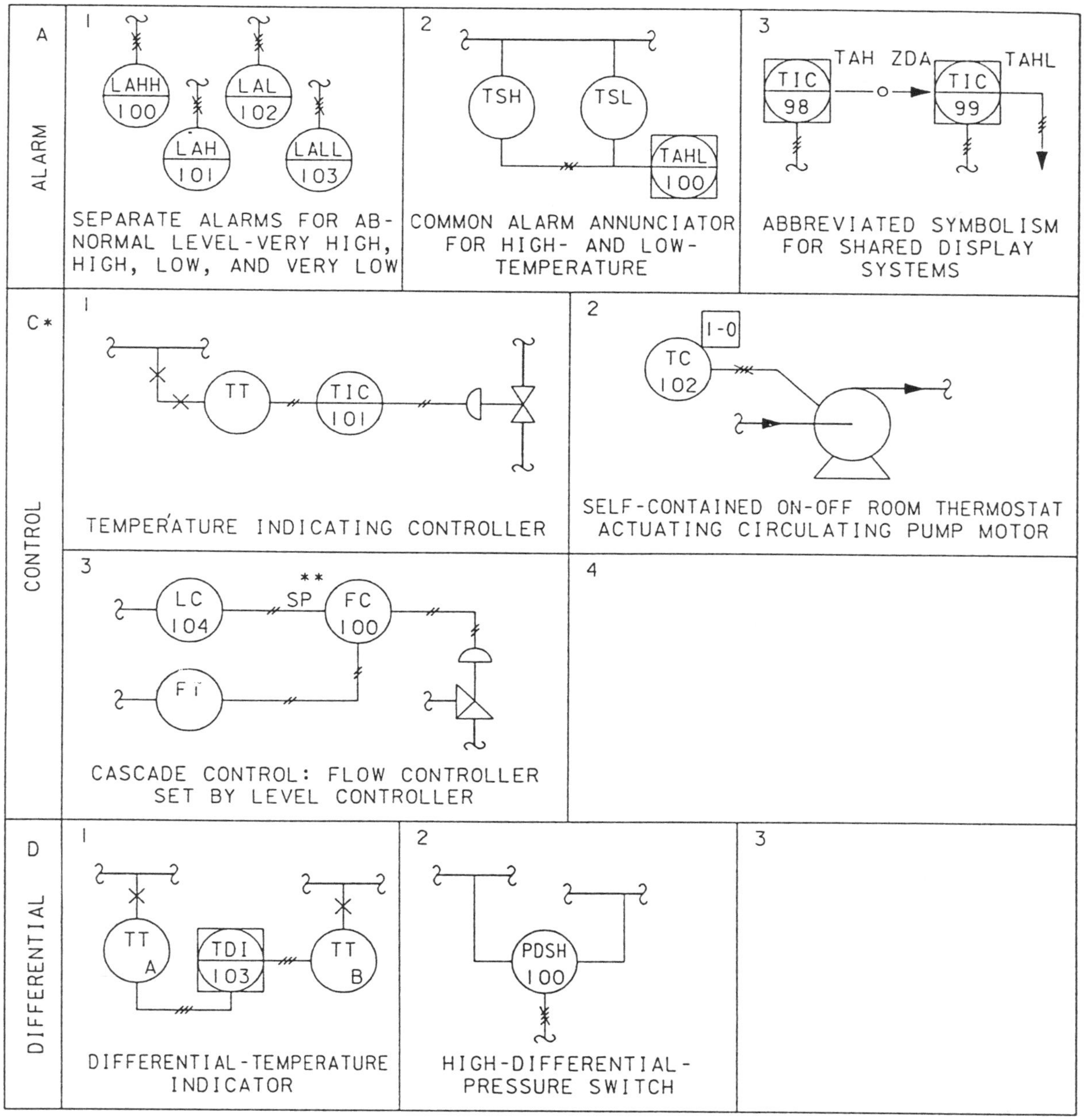

* It is expected that control modes will not be designated on a diagram. However, designations may be used outside the controller symbol, if desired, in combinations such as [%], [∫], [1-0].

** A controller is understood to have integral manual set-point adjustment unless means of remote adjustment is indicated. The remote set-point designation is SP.

Figure 2.10. Function symbols. Courtesy of Instrument Society of America.

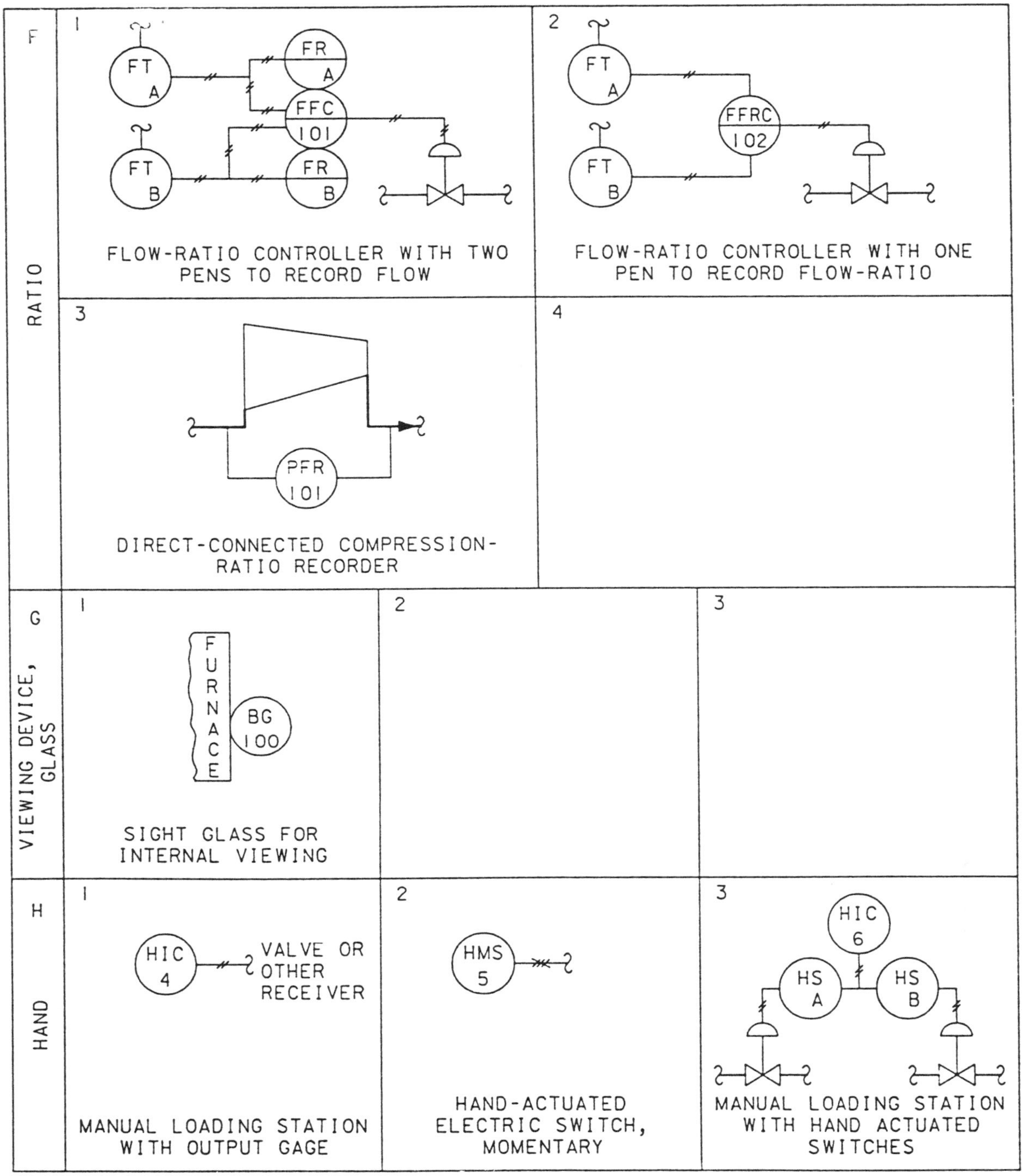

Figure 2.10 continued

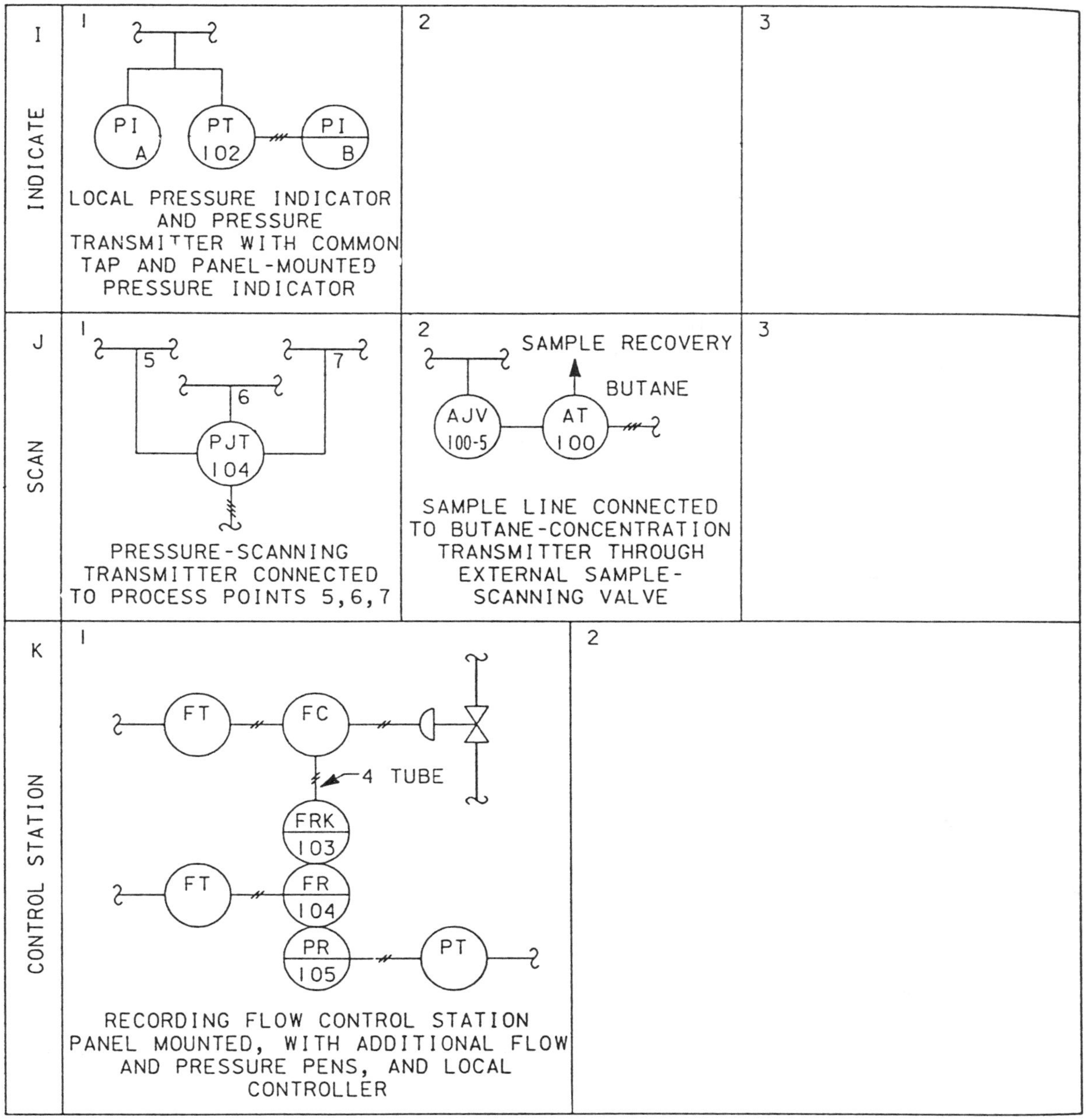
I
INDICATE
1
PI
A
PT
102
PI
B
LOCAL PRESSURE INDICATOR AND PRESSURE TRANSMITTER WITH COMMON TAP AND PANEL-MOUNTED PRESSURE INDICATOR
2
3
J
SCAN
1
5
6
7
PJT
104
PRESSURE-SCANNING TRANSMITTER CONNECTED TO PROCESS POINTS 5,6,7
2
SAMPLE RECOVERY
BUTANE
AJV
100-5
AT
100
SAMPLE LINE CONNECTED TO BUTANE-CONCENTRATION TRANSMITTER THROUGH EXTERNAL SAMPLE-SCANNING VALVE
3
K
CONTROL STATION
1
FT
FC
4 TUBE
FRK
103
FT
FR
104
PR
105
PT
RECORDING FLOW CONTROL STATION PANEL MOUNTED, WITH ADDITIONAL FLOW AND PRESSURE PENS, AND LOCAL CONTROLLER
2

Figure 2.10 continued

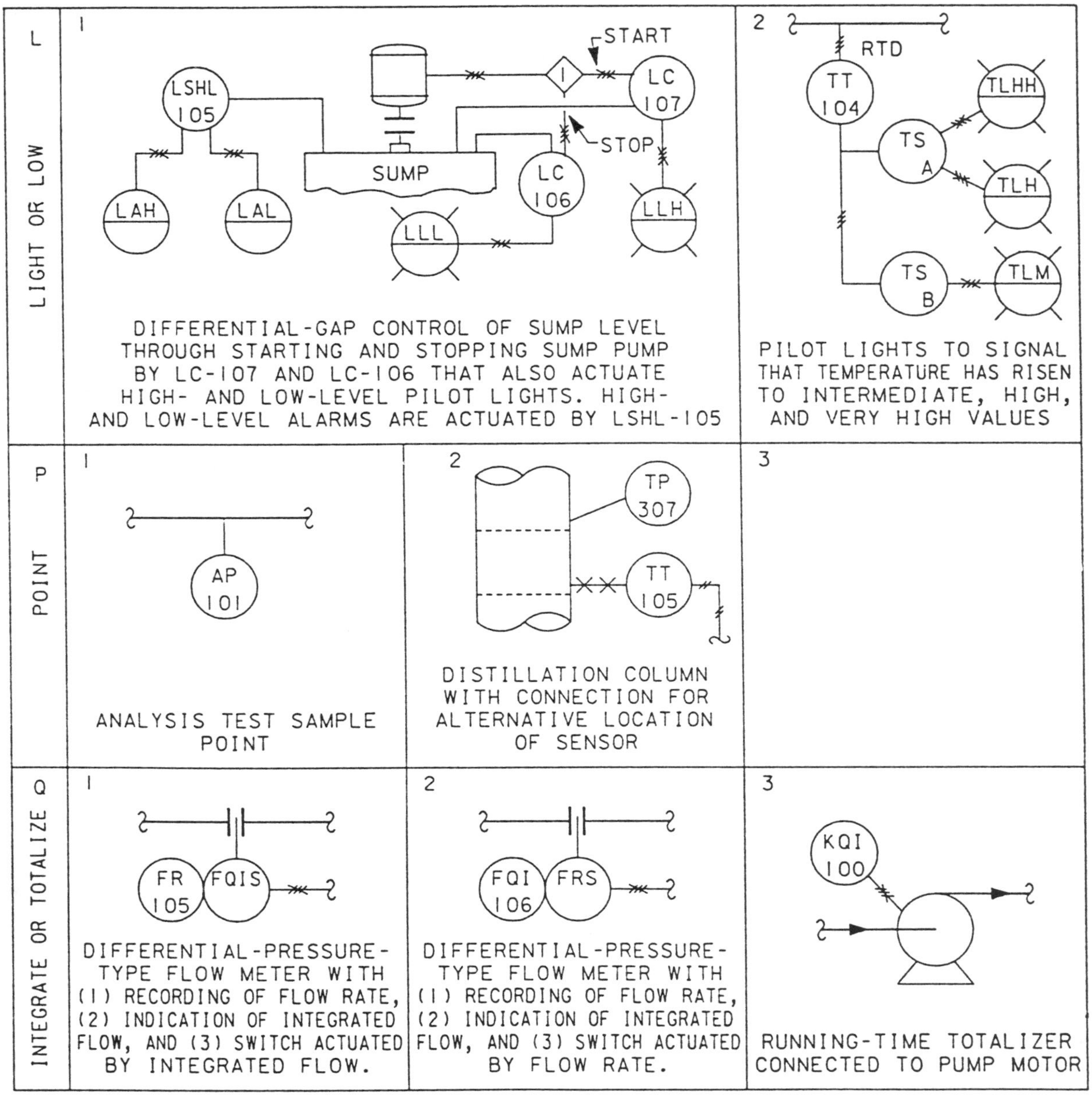

Figure 2.10 continued

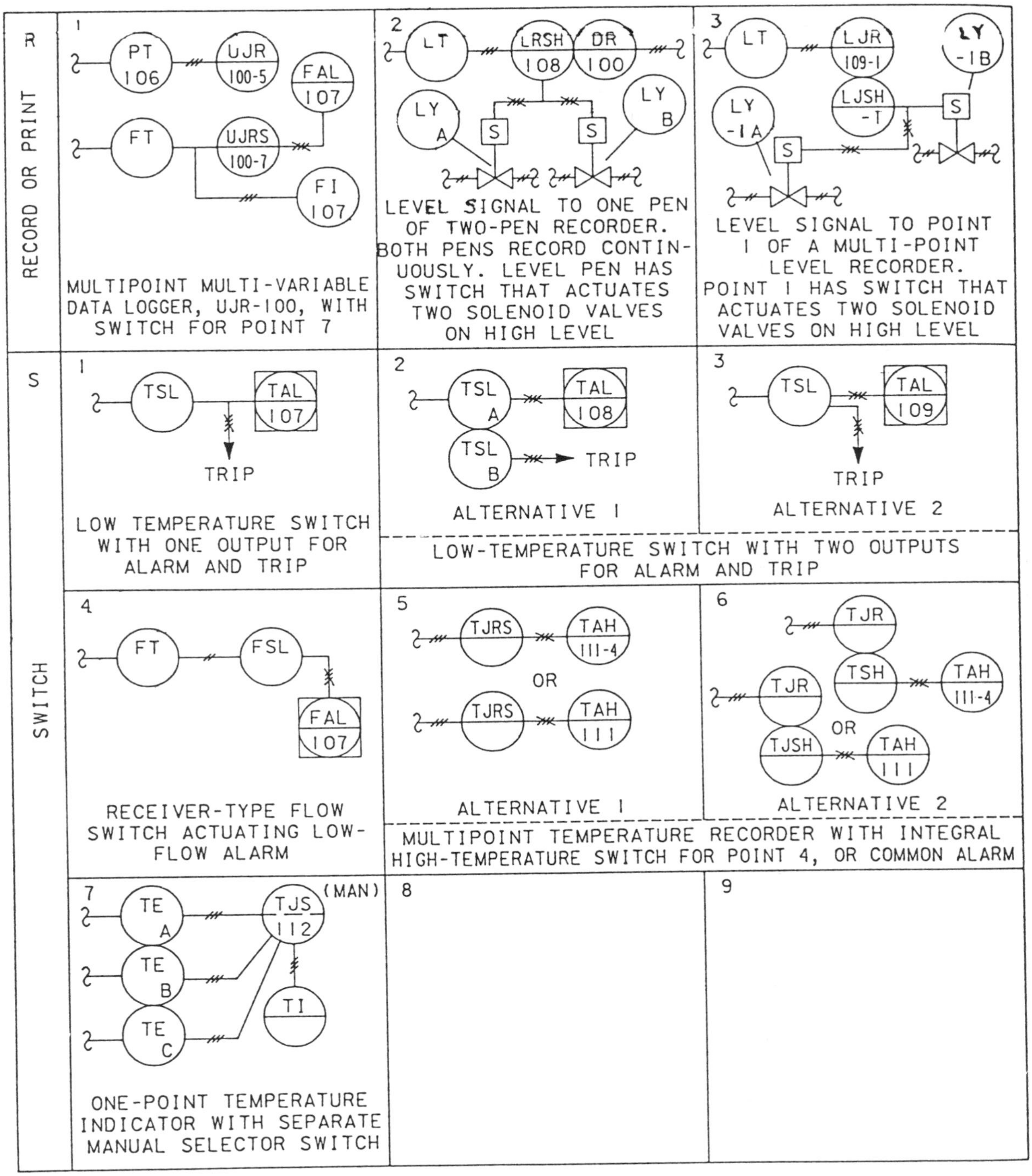

Figure 2.10 continued

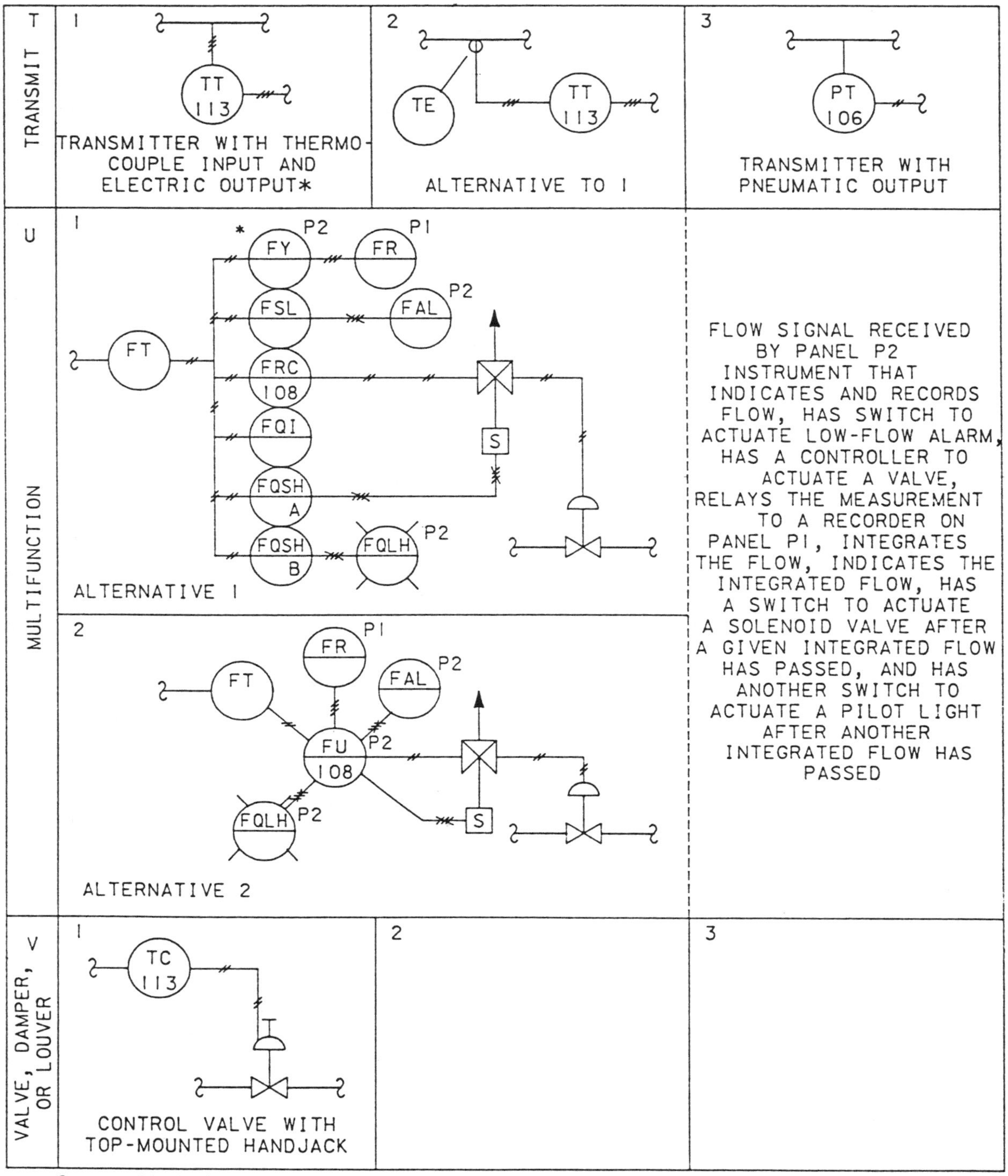
TRANSMIT T
1
TT
113
TRANSMITTER WITH THERMO-
COUPLE INPUT AND
ELECTRIC OUTPUT*
2
TE
TT
113
ALTERNATIVE TO 1
3
PT
106
TRANSMITTER WITH
PNEUMATIC OUTPUT
MULTIFUNCTION U
1
*
FY
P2
FR
P1
FSL
FAL
P2
FT
FRC
108
FQI
S
FQSH
A
FQSH
B
FQLH
P2
ALTERNATIVE 1
2
FR
P1
FT
FAL
P2
FU
108
P2
FQLH
P2
S
ALTERNATIVE 2
FLOW SIGNAL RECEIVED BY PANEL P2 INSTRUMENT THAT INDICATES AND RECORDS FLOW, HAS SWITCH TO ACTUATE LOW-FLOW ALARM, HAS A CONTROLLER TO ACTUATE A VALVE, RELAYS THE MEASUREMENT TO A RECORDER ON PANEL P1, INTEGRATES THE FLOW, INDICATES THE INTEGRATED FLOW, HAS A SWITCH TO ACTUATE A SOLENOID VALVE AFTER A GIVEN INTEGRATED FLOW HAS PASSED, AND HAS ANOTHER SWITCH TO ACTUATE A PILOT LIGHT AFTER ANOTHER INTEGRATED FLOW HAS PASSED
VALVE, DAMPER, OR LOUVER V
1
TC
113
CONTROL VALVE WITH
TOP-MOUNTED HANDJACK
2
3

* See definition of converter versus transmitter.

Figure 2.10 continued

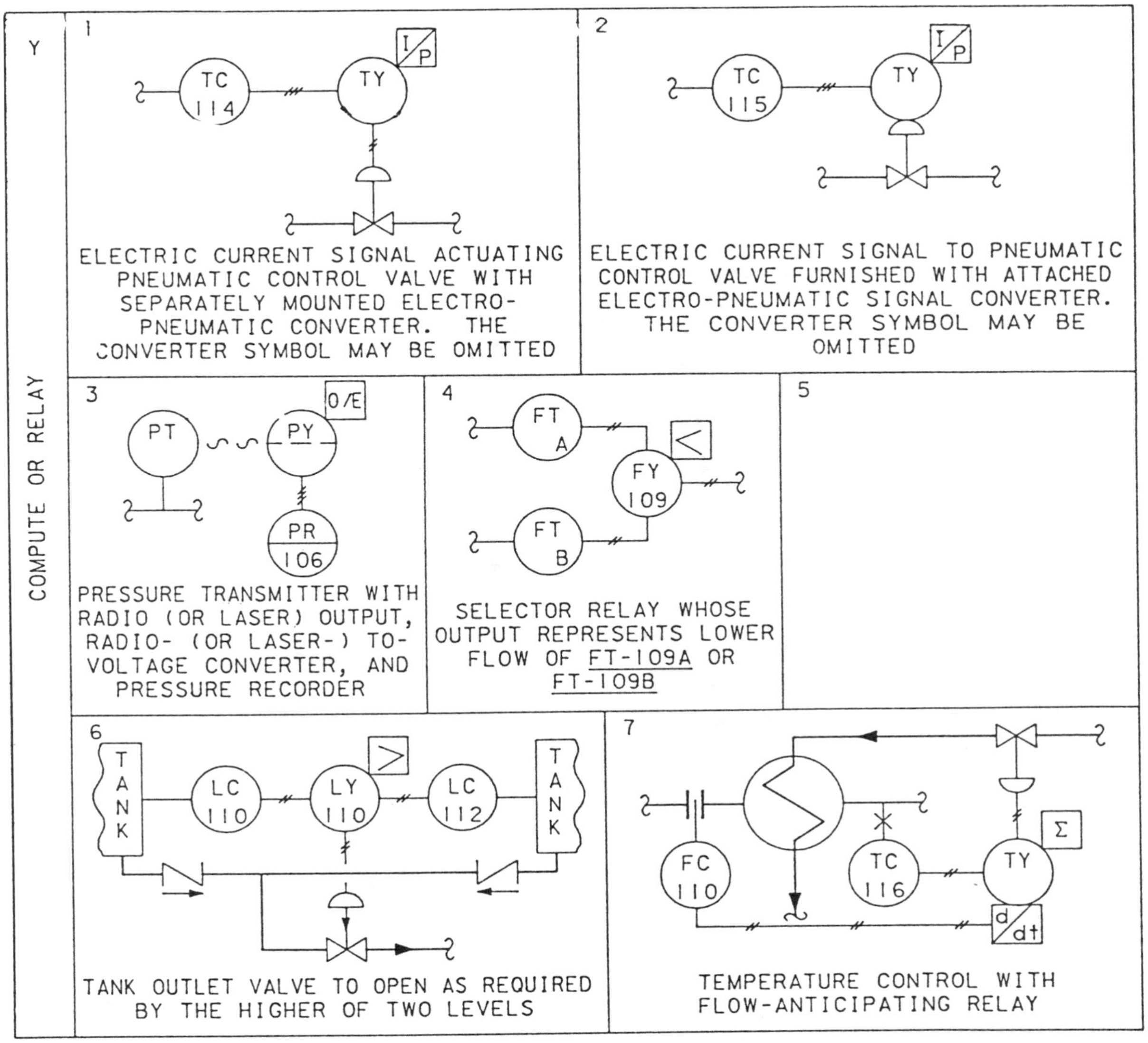

Figure 2.10 continued

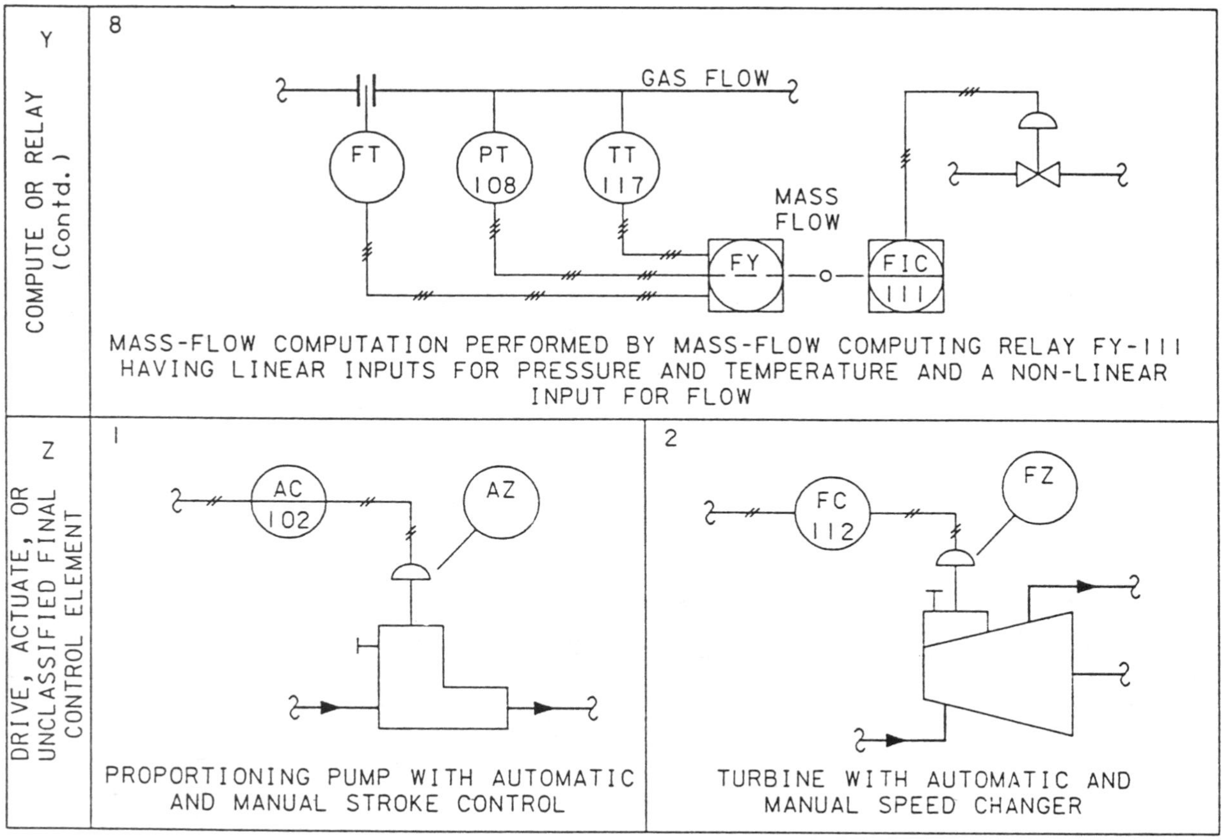
Y
COMPUTE OR RELAY (Contd.)
8
GAS FLOW
FT
PT 108
TT 117
MASS FLOW
FY
FIC 111
MASS-FLOW COMPUTATION PERFORMED BY MASS-FLOW COMPUTING RELAY FY-111 HAVING LINEAR INPUTS FOR PRESSURE AND TEMPERATURE AND A NON-LINEAR INPUT FOR FLOW
Z
DRIVE, ACTUATE, OR UNCLASSIFIED FINAL CONTROL ELEMENT
1
AC 102
AZ
PROPORTIONING PUMP WITH AUTOMATIC AND MANUAL STROKE CONTROL
2
FC 112
FZ
TURBINE WITH AUTOMATIC AND MANUAL SPEED CHANGER

Figure 2.10 continued

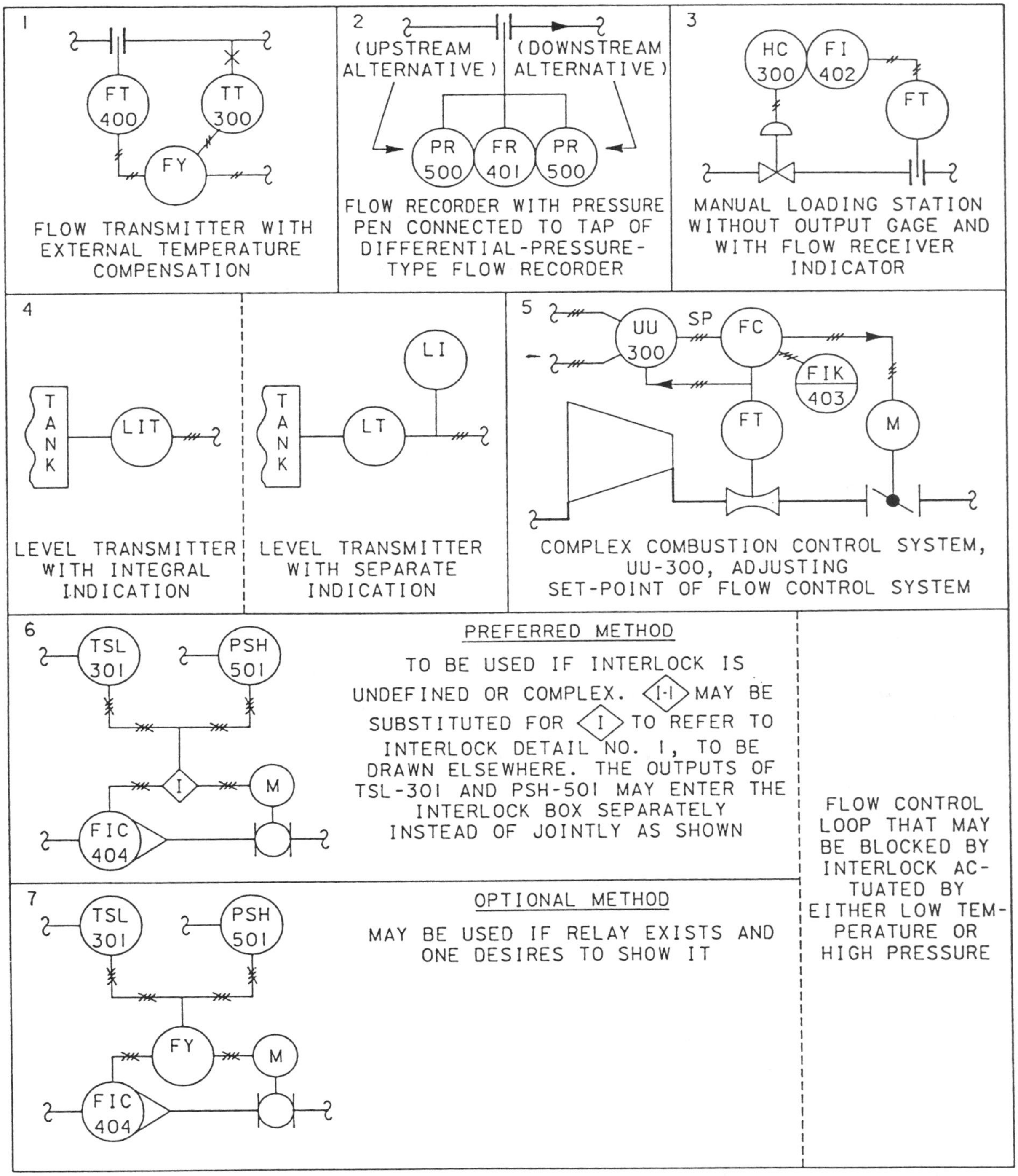

Figure 2.11. Miscellaneous systems. Courtesy of Instrument Society of America.

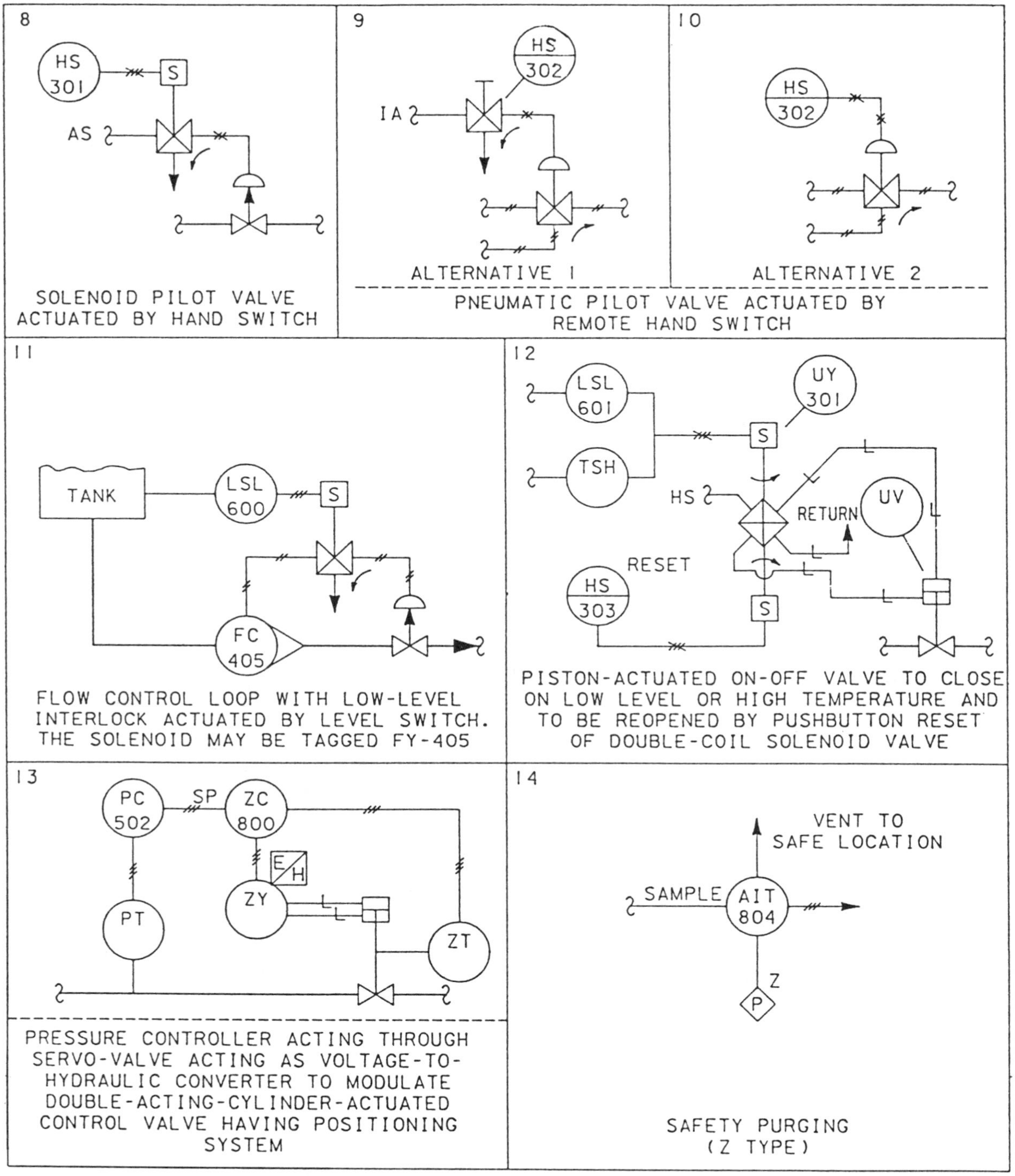

8
HS 301
S
AS
SOLENOID PILOT VALVE ACTUATED BY HAND SWITCH
9
HS 302
IA
ALTERNATIVE 1
10
HS 302
ALTERNATIVE 2
PNEUMATIC PILOT VALVE ACTUATED BY REMOTE HAND SWITCH
11
TANK
LSL 600
S
FC 405
FLOW CONTROL LOOP WITH LOW-LEVEL INTERLOCK ACTUATED BY LEVEL SWITCH. THE SOLENOID MAY BE TAGGED FY-405
12
LSL 601
TSH
UY 301
S
HS
RETURN
UV
L
RESET
HS 303
S
PISTON-ACTUATED ON-OFF VALVE TO CLOSE ON LOW LEVEL OR HIGH TEMPERATURE AND TO BE REOPENED BY PUSHBUTTON RESET OF DOUBLE-COIL SOLENOID VALVE
13
PC 502
SP
ZC 800
E H
ZY
L
PT
ZT
PRESSURE CONTROLLER ACTING THROUGH SERVO-VALVE ACTING AS VOLTAGE-TO-HYDRAULIC CONVERTER TO MODULATE DOUBLE-ACTING-CYLINDER-ACTUATED CONTROL VALVE HAVING POSITIONING SYSTEM
14
VENT TO SAFE LOCATION
SAMPLE
AIT 804
Z
P
SAFETY PURGING (Z TYPE)

Figure 2.11 continued

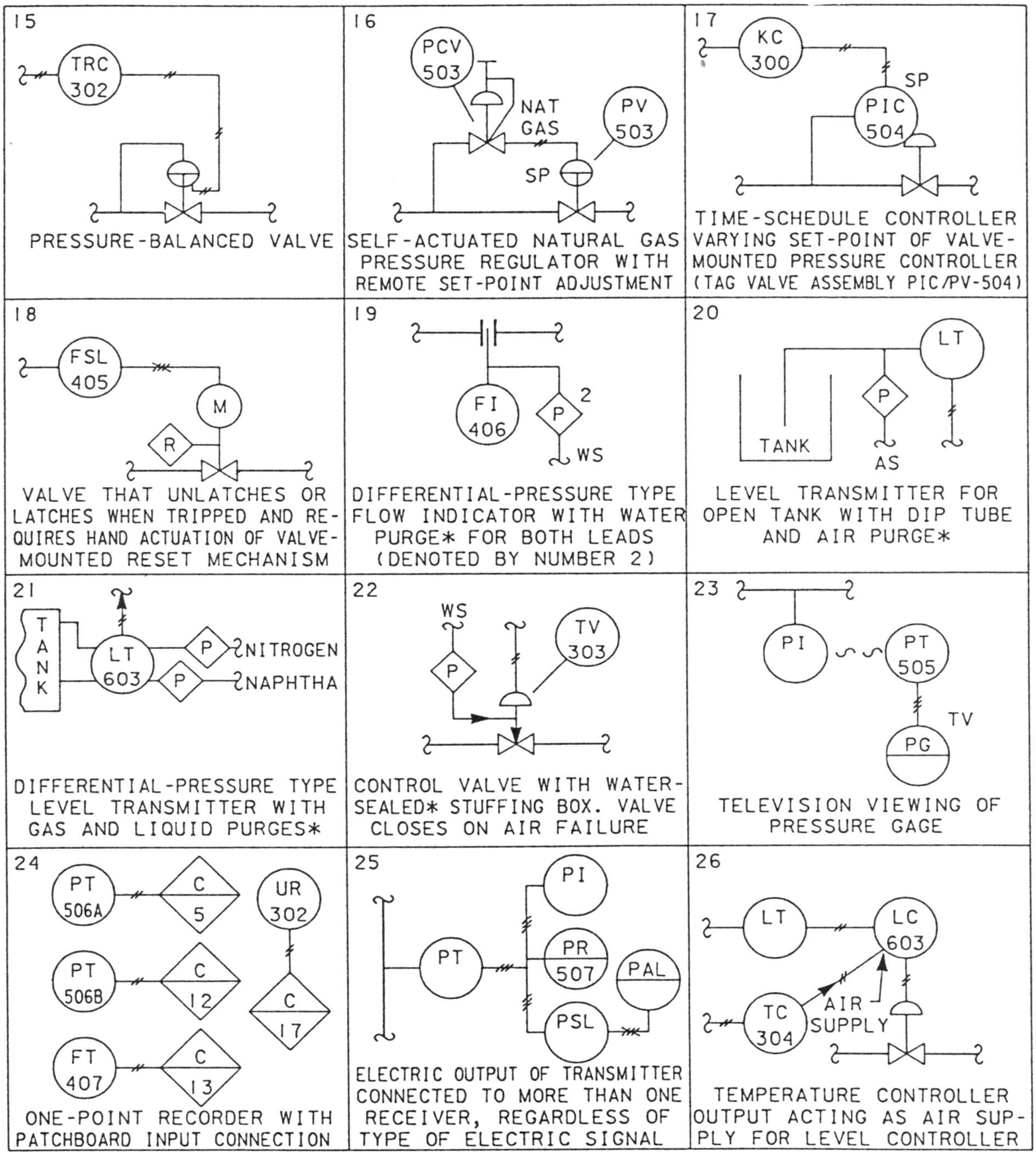
15
TRC
302
PRESSURE-BALANCED VALVE
16
PCV
503
NAT
GAS
PV
503
SP
SELF-ACTUATED NATURAL GAS
PRESSURE REGULATOR WITH
REMOTE SET-POINT ADJUSTMENT
17
KC
300
SP
PIC
504
TIME-SCHEDULE CONTROLLER
VARYING SET-POINT OF VALVE-
MOUNTED PRESSURE CONTROLLER
(TAG VALVE ASSEMBLY PIC/PV-504)
18
FSL
405
M
R
VALVE THAT UNLATCHES OR
LATCHES WHEN TRIPPED AND RE-
QUIRES HAND ACTUATION OF VALVE-
MOUNTED RESET MECHANISM
19
FI
406
P
2
WS
DIFFERENTIAL-PRESSURE TYPE
FLOW INDICATOR WITH WATER
PURGE* FOR BOTH LEADS
(DENOTED BY NUMBER 2)
20
LT
P
TANK
AS
LEVEL TRANSMITTER FOR
OPEN TANK WITH DIP TUBE
AND AIR PURGE*
21
TANK
LT
603
P
NITROGEN
P
NAPHTHA
DIFFERENTIAL-PRESSURE TYPE
LEVEL TRANSMITTER WITH
GAS AND LIQUID PURGES*
22
WS
P
TV
303
CONTROL VALVE WITH WATER-
SEALED* STUFFING BOX. VALVE
CLOSES ON AIR FAILURE
23
PI
PT
505
TV
PG
TELEVISION VIEWING OF
PRESSURE GAGE
24
PT
506A
C
5
UR
302
PT
506B
C
12
C
17
FT
407
C
13
ONE-POINT RECORDER WITH
PATCHBOARD INPUT CONNECTION
25
PI
PT
PR
507
PAL
PSL
ELECTRIC OUTPUT OF TRANSMITTER
CONNECTED TO MORE THAN ONE
RECEIVER, REGARDLESS OF
TYPE OF ELECTRIC SIGNAL
26
LT
LC
603
TC
304
AIR
SUPPLY
TEMPERATURE CONTROLLER
OUTPUT ACTING AS AIR SUP-
PLY FOR LEVEL CONTROLLER

* The purge fluid supplies may use the same abbreviations as the instrument power supplies.

Figure 2.11 continued

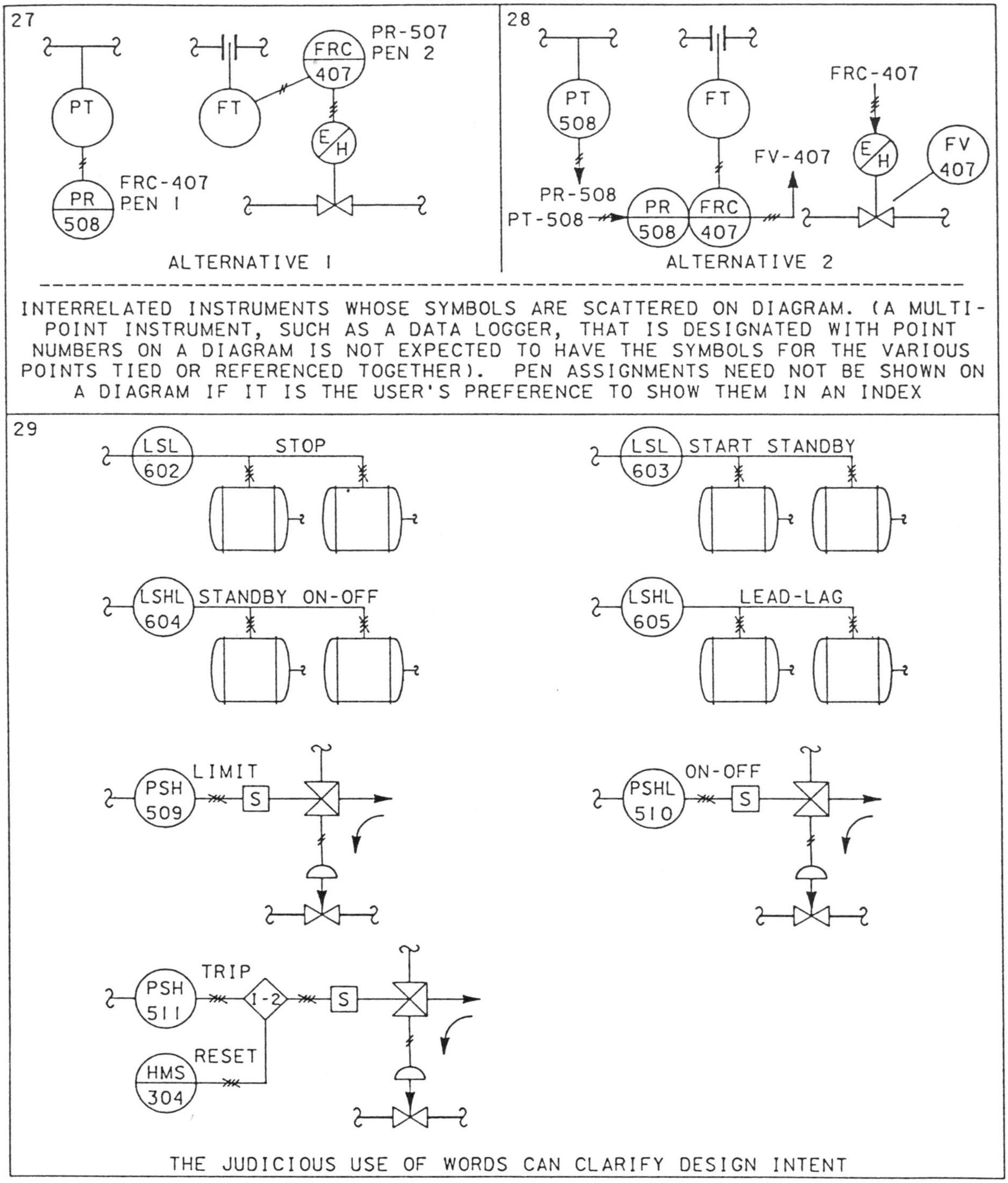
27
PT
PR
508
FRC-407
PEN 1
FT
FRC
407
PR-507
PEN 2
E/H
ALTERNATIVE 1
28
PT
508
PR-508
PT-508
PR
508
FT
FRC
407
FV-407
FRC-407
E/H
FV
407
ALTERNATIVE 2
INTERRELATED INSTRUMENTS WHOSE SYMBOLS ARE SCATTERED ON DIAGRAM. (A MULTI-POINT INSTRUMENT, SUCH AS A DATA LOGGER, THAT IS DESIGNATED WITH POINT NUMBERS ON A DIAGRAM IS NOT EXPECTED TO HAVE THE SYMBOLS FOR THE VARIOUS POINTS TIED OR REFERENCED TOGETHER). PEN ASSIGNMENTS NEED NOT BE SHOWN ON A DIAGRAM IF IT IS THE USER'S PREFERENCE TO SHOW THEM IN AN INDEX
29
LSL
602
STOP
LSL
603
START STANDBY
LSHL
604
STANDBY ON-OFF
LSHL
605
LEAD-LAG
PSH
509
LIMIT
S
PSHL
510
ON-OFF
S
PSH
511
TRIP
I-2
S
HMS
304
RESET
THE JUDICIOUS USE OF WORDS CAN CLARIFY DESIGN INTENT

Figure 2.11 continued

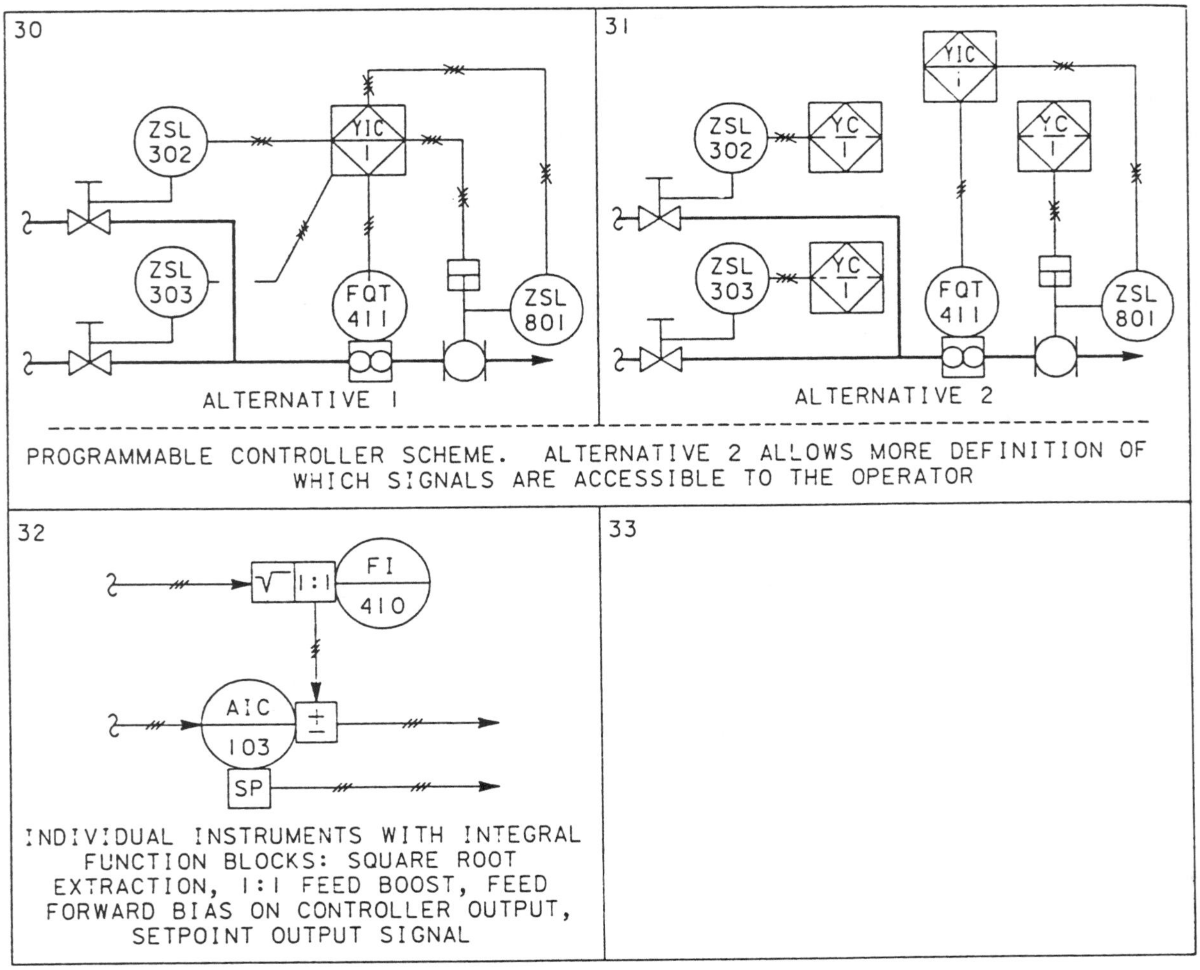
30
ZSL 302
ZSL 303
YIC I
FQT 411
ZSL 801
ALTERNATIVE 1
31
YIC I
ZSL 302
YC I
ZSL 303
YC I
FQT 411
ZSL 801
ALTERNATIVE 2
PROGRAMMABLE CONTROLLER SCHEME. ALTERNATIVE 2 ALLOWS MORE DEFINITION OF WHICH SIGNALS ARE ACCESSIBLE TO THE OPERATOR
32
√ 1:1
FI 410
AIC 103
±
SP
INDIVIDUAL INSTRUMENTS WITH INTEGRAL FUNCTION BLOCKS: SQUARE ROOT EXTRACTION, 1:1 FEED BOOST, FEED FORWARD BIAS ON CONTROLLER OUTPUT, SETPOINT OUTPUT SIGNAL
33

Figure 2-11 continued

Graphic Symbols for Distributed Control/Shared Display Instrumentation, Logic, and Computer Systems*

DEFINITIONS AND ABBREVIATIONS

Accessible - A system feature that is viewable by and interactive with the operator, and allows the operator to perform user permissible control actions, e.g. set point changes, auto-manual transfers, or on-off actions.
Assignable - A system feature that permits an operator to channel (or direct) a signal from one device to another, without the need for changes in wiring, either by means of switches or via keyboard commands to the system.
Communication Link - Computer control is a device in which control and/or display actions are generated for use by other system devices. When used with other control devices on the communication link the computer normally performs or functions in a hierarchical relationship to the other control devices.
Computer Control System - A system in which all control action takes place within the control computer. Single or redundant computers may be used.
Configurable - A system feature that permits selection through entry of keyboard commands of the basic structure and characteristics of a device or system, such as control algorithms, display formats, or input/output terminations.
Distributed Control System - That class of instrumentation (input/output devices, control devices and operator interface devices) which in addition to executing the state control functions also permits transmission of control, measurement, and operating information to and from a single or a plurality of user specifiable locations, connected by a communication link.
I/O - Input/Output
Shared Controller - A control device that contains a plurality of pre-programmed algorithms which are user retrievable, configurable, and connectable, and allows user defined control strategies or functions to be implemented. Control of multiple process variables can be implemented by sharing the capabilities of a single device of this kind.
Shared Display - The operator interface device used to display signals and/or data on a time shared basis. The signals and/or data, i.e. alphanumeric and/or graphic, reside in a data base from where selective accessibility for display is at the command of a user.
Software - Digital programs, procedures, rules, and associated documentation required for the operation and/or maintenance of a digital system.
Software Link - The interconnection of system components or functions via software or keyboard instruction.
Supervisory Set Point Control System - The generation of set point and/or other control information by a computer control system for use by shared control, shared display or other regulatory control devices.

*Source: Instrument Society of America

SYMBOLS

Distributed Control/Shared Display Symbols

Advances in control systems brought about by microprocessor based instrumentation permit shared functions such as display, control and signal lines. Therefore, the symbology defined here should be "Shared Instruments," which means shared display and/or shared control. The square portion of this symbol has the meaning of shared type instrument.

Normally Accessible to Operator

Indicator/Controller/Recorder or Alarm Points—usually used to indicate video display.

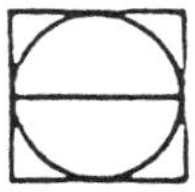

(1) Shared display.
(2) Shared display and shared control.
(3) Access limited to communication link.
(4) Operator Interface on communication link.

Auxiliary Operator's Interface Device

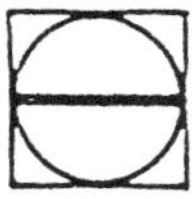

(1) Panel mounted-normally having an analog faceplate—not normally mounted on main operator console.
(2) Can be a backup controller or manual station.
(3) Access may be limited to communication link.
(4) Operator interface via the communication link.

Not Normally Accessible to Operator

(1) Shared blind controller.
(2) Shared display installed in field.
(3) Computation, signal conditioning in shared controller.
(4) May be on communication link.
(5) Normally blind operation.
(6) May be altered by configuration

Computer Symbols

The following symbols should be used where systems include components identified as "computers," as distinct from an integral processor, which drive the various functions of a "distributed control system." The computer component may be integrated with the system via the data link, or it may be a stand-alone computer.

Normally Accessible to Operator

Indicator/Controller/Recorder or Alarm Point—usually used to indicate video display.

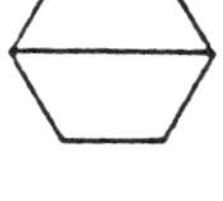

Figure 2.12. Graphic symbols. Courtesy of Instrument Society of America.

Not Normally Accessible to Operator

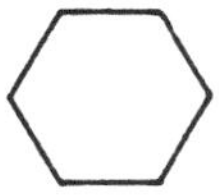

(1) Input/Output interface.
(2) Computation/Signal conditioning within a computer.
(3) May be used as a blind controller or a software calculation module.

Logic and Sequential Control Symbols

General Symbol - For undefined complex interconnecting logic or sequence control.

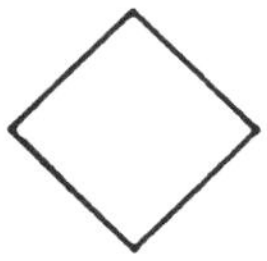

Distributed control interconnecting logic controller with binary or sequential logic functions.

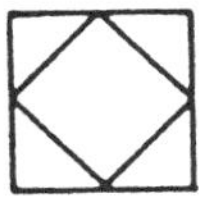

(1) Packaged programmable logic controller, or digital logic controls integral to the distributed control equipment.
(2) Not normally accessible to the operator.

Distributed control interconnecting logic controller with binary or sequential logic functions.

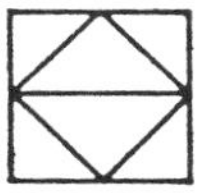

(1) Packaged programmable logic controller, or digital logic controls integral to the distributed control equipment.
(2) Normally accessible to the operator.

Internal System Function Symbols

Computation/Signal Conditioning

(1) For block identification refer to Table 2.3 "Function Designations for Relays."
(2) For extensive computational requirements, use designation "C". Explain on supplementary documentation.
(3) Used in conjunction with function relay bubbles

Common Symbols

System Link

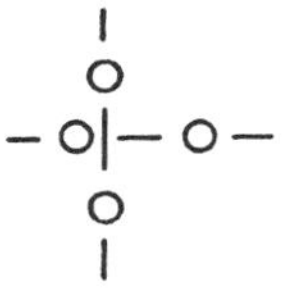

(1) Used to indicate either a software link or manufacturer's system supplied connections between functions.
(2) Alternatively, link can be implicitly shown by contiguous symbols.
(3) May be used to indicate a communication link at the user's option.

IDENTIFICATION

Software Alarms

Software alarms may be identified by placing letter designators on the input or output signal lines of the controls, or other specific integral system component.

Contiguity of Symbols

Two or more symbols can adjoin to express the following

(1) Communication among the associated instruments, e.g.
  - Hard wiring
  - Internal system link
  - Backup

(2) Instrument integrated with multiple functions, e.g.
  - Multipoint recorder
  - Control valve with integrally mounted controller.

The application of contiguous symbols is a user option.

Instrument System Alarms

Multiple alarm capability is provided in most systems.

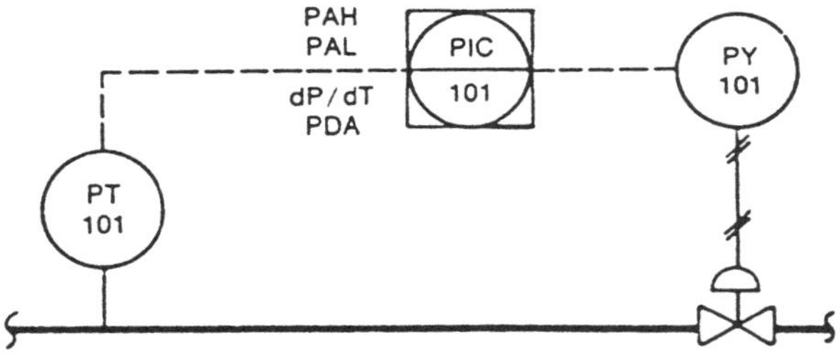

Alarms on measured variables shall include the variable identifiers, i.e.:

Pressure:	PAH	(High)
	PAL	(Low)
	dP/dt	(Rate of change)
	PDA	(deviation from set point)

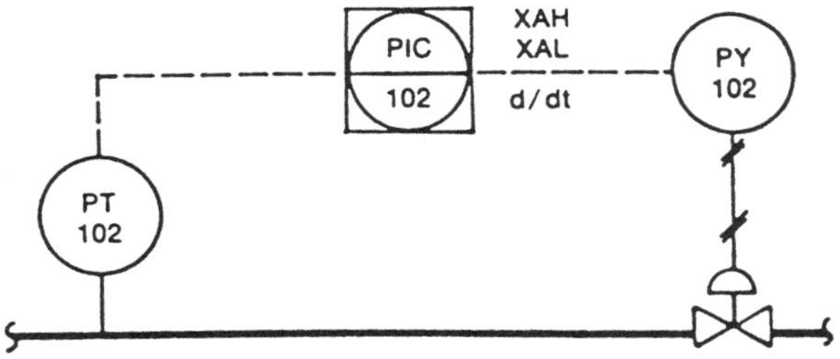

Alarms on controller output shall use the undefined variable identifier X, i.e.:

XAH	(High)
XAL	(Low)
d/dt	(Rate of change)

Figure 2.12 continued

EXAMPLES

Examples of Use

The following figures illustrate some of the various combinations of symbols. These symbols may be combined as necessary to fulfill the needs of the user.

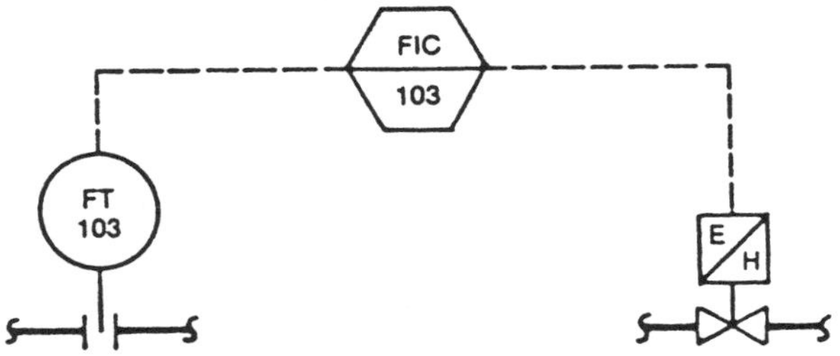

Computer Control—No Backup - Shared Display

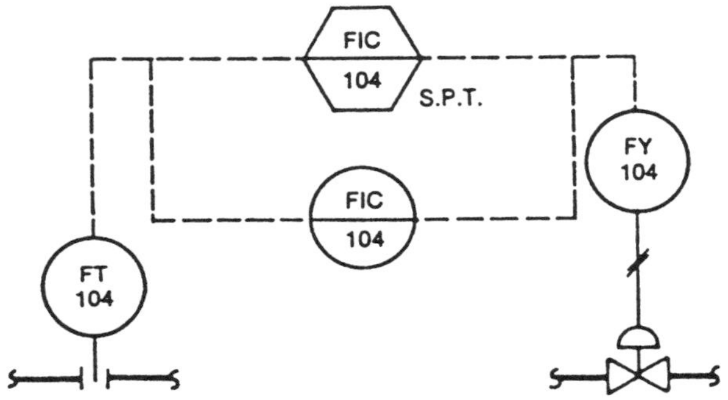

Computer Control—Full Analog Backup Through Set Point Tracking (SPT)

Controllers located in the diagram main information line are to be considered the primary controllers. All devices outside the main line provide a backup or secondary function.

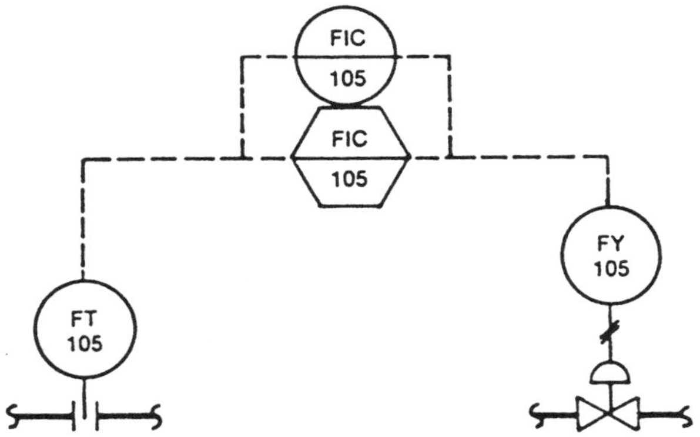

Computer Control—With Analog Backup

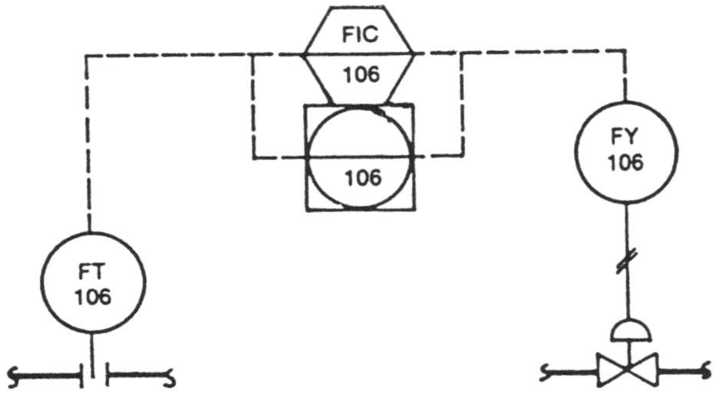

Computer Control-Full Backup from Distributed Control Instrumentation. Computer Uses Instrument System Communication Link

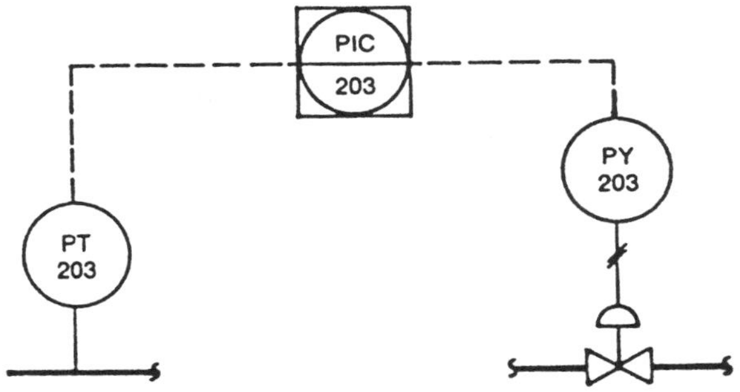

Shared Display/Shared Control—No Backup

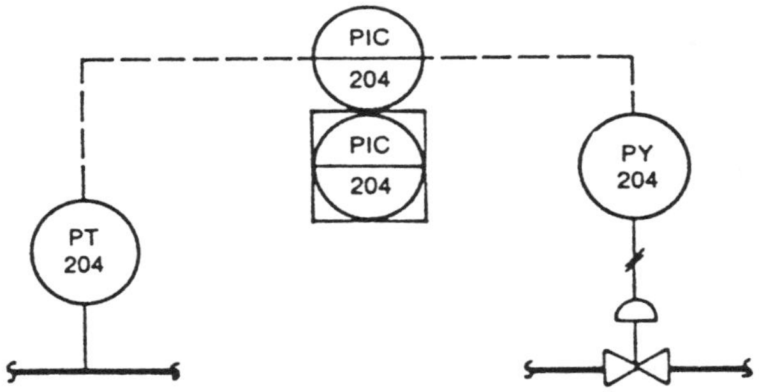

Analog Control—Interfaced with Shared Display. Shared Control Backup

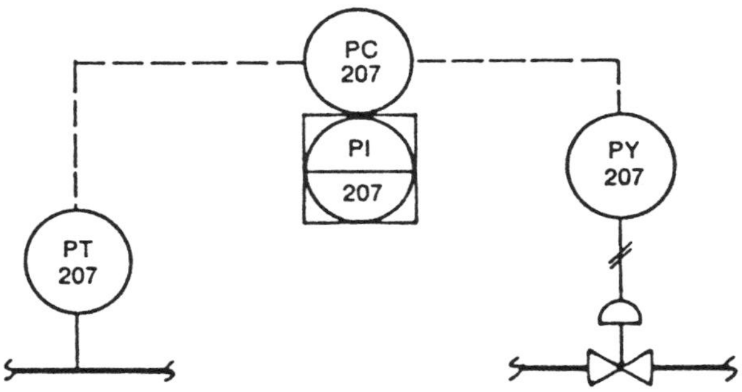

Analog Control—Blind Controller. Shared Display

Figure 2.12 continued

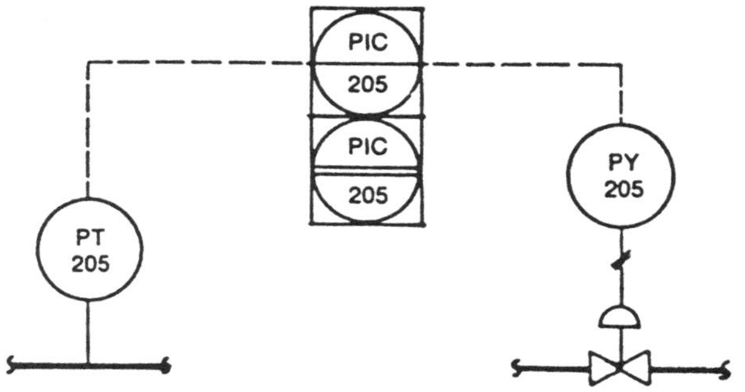

Shared Display/Shared Control—With Auxiliary Operator's Interface Device

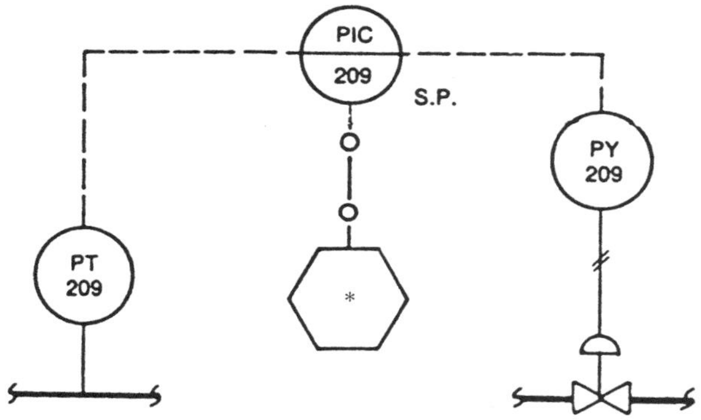

Supervisory Set Point Control—Analog Con troller with Conventional Faceplate. Computer Supervisory Set Point via Communication Link

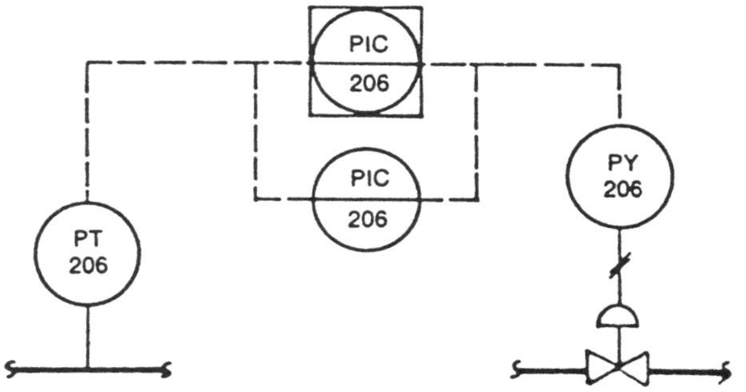

Shared Display/Shared Control—With Analog Controller Backup

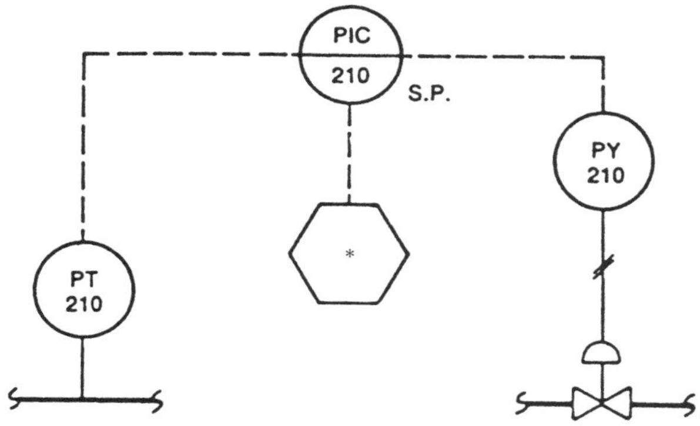

Supervisory Set Point Control—Analog Con troller Complete with Conventional Faceplate. Computer Supervisory Set Point Hardwired.

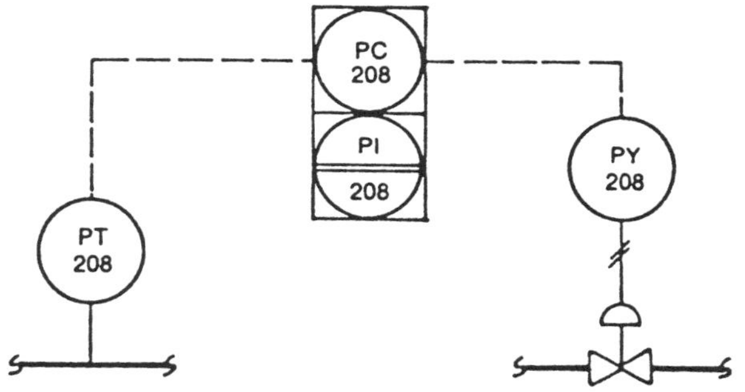

Blind Shared Control—With Auxiliary Operators Interface Backup

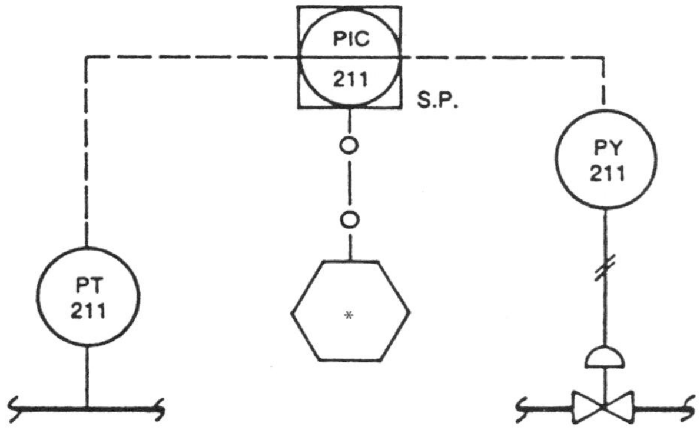

Supervisory Set Point Control—Shared Dis play Shared Control with Full Computer Access via the Communication Link

*User identification is optional

Figure 2.12 continued

Typical Flow Diagrams

The following simplified drawing (Figure 2.13) combines the basic symbols of this standard. It is intended to provide a hypothetical example and to stimulate the user's imagination in the application of symbolism to this equipment. It is arranged in the following manner:

(1) Volumetric fuel and air flows provide inputs for combustion system firing rate and fuel air ratio via distributed control instrumentation. Set points for both rate and ratio can be computer generated.

(2) Combustion air and gas pressures are monitored by pressure switches which control the gas safety shut-off valve via UC-600 "distributed control interconnecting logic."

(3) Material moisture content is measured, dry weight of the input material is calculated, and feed rate is controlled by MT-300 and WC-301. Discharged material moisture content is read by MT-302. At this point, firing rate and/or feed rate could be controlled by the Distributed Control System (DCS) instrumentation or by the computer taking other process variables into consideration.

(4) British thermal unit (Btu) analysis (AT-97) is input to the computer system to generate feed forward control adjusting firing rate, in Btu/hr. The set point is calculated by the computer, based on feed rate, weight, and moisture content.

(5) Internal system links are shown for selected computer input/output, while the firing rate and ratio set points are implied. Shown in the same manner, the links between the calculation modules and the controllers are implied by contiguous symbols, while the wild flow to the ratio control is shown in the system link symbol.

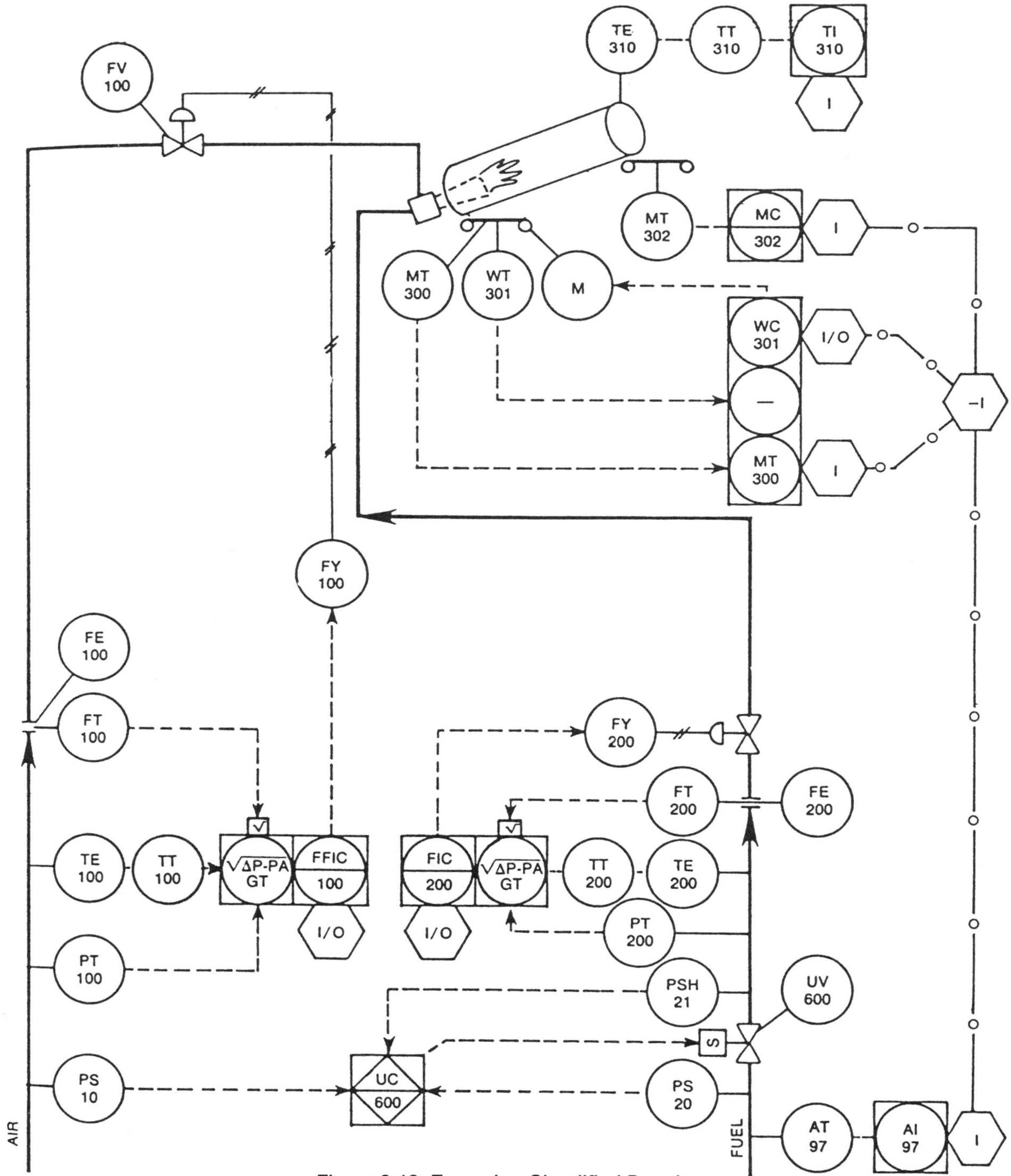

Figure 2.13. Example—Simplified Drawing

This typical flow diagram of a cascade control loop combines the symbols to depict a cascade loop with alarms. Notes are added on the diagram itself for clarification purposes only.

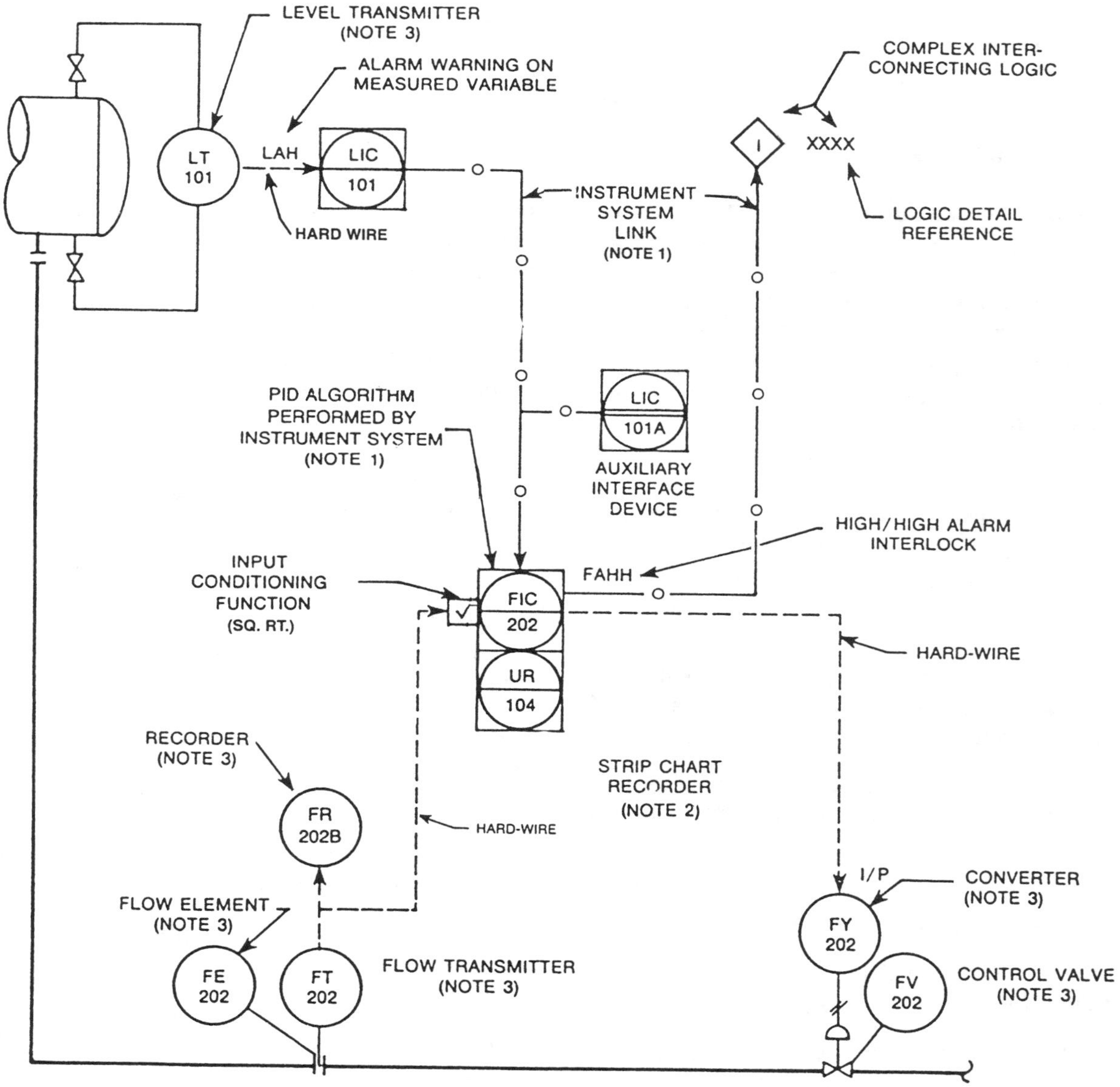

Figure 2.14. Typical Flow Diagram—Cascade Control Loop

Notes: Shared Display
1. Display/adjustments on console. Communication via data link.
2. Located in console. Signal selected from instrument system data base.
3. Field mounted.

ICS Part 1-102
GRAPHIC SYMBOLS FOR LOGIC DIAGRAMS *

The standards in this part apply to all other parts in the NEMA Standards Publications for Industrial Controls and Systems unless otherwise specified.

Part 1-102 establishes graphic symbols specifically designed for use in the preparation of logic diagrams for systems of two-state devices and sets forth principles governing the formation of graphic symbols for logic diagrams in which connections between symbols are generally shown with lines. See 1-101.3.

The definitions contained in Part 1-100 of this Standards Publication also apply to this part.

DEFINITIONS

1-102.1 DEFINITIONS

application. The standards of 1-102.1 through 1-102.5 are applicable to either positive or negative logic, but no provision has been made for a combination of both within a single system.

Authorized Engineering Information 1-21-1964.

binary notation. A numerical system in which the digits are coefficients of the powers of the radix 2. There are two symbols, 0 and 1.

NEMA Standard 5-21-1962.

For example, binary number 101 is the notation for:

$$1 \times 2^2 + 0 \times 2^1 + 1 \times 2^0$$

Authorized Engineering Information 5-21-1962.

elements.

FLIP-FLOP. A bistable device which assumes a given stable state depending upon the history of its input or inputs.

NAND. An element which produces an output under all input conditions except when all inputs are energized.

NOR. An element which produces an output if and only if all inputs are deenergized.

NOT. An element which produces an output if and only if its input is deenergized.

OR. An element which produces an output if and only if one or more inputs is energized.

OR, EXCLUSIVE. An element with two inputs which produces an output if and only if either one but not both of its inputs is energized.

NEMA Standard 5-21-1962.

logic function. A storage, delay, or sequential function expressing a relationship between signal input(s) to an element or device and the resultant output(s).

NEMA Standard 1-21-1964.

logic symbol. The graphical representation in diagrammatic form of a logic function.

NEMA Standard 1-21-1964.

negative logic. If the less positive potential (current) is consistently selected as the 1-state, the resultant system or device is said to have negative logic.

NEMA Standard 1-21-1964.

positive logic. If all signal line terminals in a logic diagram of a system or device have the same pair of physical states, and if both are electrical potentials (currents), and if the more positive potential (current) is consistently selected as the 1-state, the resultant system or device is said to have positive logic.

NEMA Standard 1-21-1964.

states in binary logic. The two physical states on each terminal of each signal line shall be referred to as the "0-state" and the "1-state." The 0-state may be called the deenergized (reference) state, and the 1-state the energized (significant) state. This must not be construed as implying that the 1-state requires more power, contains more energy, or is at a higher potential than the 0-state. The state designations are purely arbitrary as far as the physical interpretation is concerned.

NEMA Standard 1-21-1964.

1-102.2 ASSIGNMENT OF LOGIC LEVELS TO BINARY LOGIC ELEMENTS

1-102.2.1 Boolean Truth Tables

Consider a circuit whose output (F) is a function of two variables (A, B), and whose output and input levels are capable of assuming only +2 volts and −3 volts independent of time. Assume that the circuit behaves according to the following table of combinations:

Inputs		Output
A	B	F
−3 volts	−3 volts	−3 volts
−3 volts	+2 volts	−3 volts
+2 volts	−3 volts	−3 volts
+2 volts	+2 volts	+2 volts

In positive logic, the −3 volt level is the 0-state and the +2 volt level is the 1-state. (See 1-102.1.4.) Substitution of the logic values for the voltage levels results in the following Boolean truth table:

Inputs		Output
A	B	F
0	0	0
0	1	0
1	0	0
1	1	1

This is the truth table for the AND function.

In negative logic, the −3 volt level is the 1-state and +2 volt level is the 0-state. (See 1-102.1.5.) Substitution of the logic values for the voltage levels results in the following Boolean truth table:

Inputs		Output
A	B	F
1	1	1
1	0	1
0	1	1
0	0	0

This is the truth table for the OR function.

Each row of the Boolean truth table uniquely describes the behavior of the device for a particular combination of input states. The truth table must include all possible combinations of input states, with one row for each combination. Each row is a complete and independent statement of the behavior of the device and, therefore, need not be read in any particular sequence. The Boolean truth table does not take into account previous history or time dependence of the device.

1-102.2.2 Sequential Truth Tables

Sequential truth tables are extensions of Boolean truth tables, in which time and history are accounted for. These are used to describe the operation of functions, such as FLIP-FLOPS, TIME DELAY and others, in which time or history, or both, are necessary to obtain an understanding of the operation.

Sequential truth tables are read with each horizontal line independent of the others, and the values at a given instant of time are those shown in corresponding vertical columns in each group of multiple columns. The first digit of the group under each input or output shows the state of that input or output during the first time increment. Digits further to the right in the same line show the states at later time increments. A horizontal bar between digits represents time delay in changing from one state to the other state. It is assumed that input signals are maintained long enough to effect complete reset or time delay functions.

Two digits are generally sufficient to describe a TIME DELAY while three or more are required for other functions, such as FLIP-FLOPS.

The following example shows a portion of a FLIP-FLOP truth table:

Inputs			Outputs	
S (Set)	C (Clear)	T (Trigger)	1	0
010	000	000	011	100
010	000	000	111	000

During the first time increment, the first horizontal row in the table shows that the FLIP-FLOP is in the clear (reset) state by showing that the 1 and 0 outputs are respectively in the 0 and 1 states. At the start of the second time increment, the table shows that the FLIP-FLOP changes to the set state by showing that the 1 and the 0 outputs change respectively to the 1 and 0 state. During the third time increment the truth table shows the FLIP-FLOP remains in the set state by showing that the 1 and 0 outputs do not change even though the set input changes to a 0 state.

The apparent anomaly in the table shown here, in which the same input conditions during all time increments result in different output conditions during the first time increment, is basic to any FLIP-FLOP. This indicates the necessity for considering history. It is important to realize that the first column of each group is the result of the history of the element or its starting point for the sequence being considered on any given row.

NEMA Standard 1-21-1964.

1-102.3 PRESENTATION TECHNIQUES

1-102.3.1 Symbol Shape

Symbol Shape shall be rectangular and preferably square.

1-102.3.2 Symbol Size

Symbol size shall not affect the meaning of the symbol. An effort shall be made to keep all basic symbols of like shape and size.

1-102.3.3 Symbol Line Thickness

Symbol line thickness shall not affect the meaning of the symbol.

1-102.3.5 Symbol Orientation

Symbol orientation shall not affect the meaning of the symbol.

1-102.3.5 Inputs

All inputs shall enter the basic symbols at the leading edge opposite the output only. If more inputs have to enter a symbol than the symbol provides room for, the leading edge shall be extended to provide entry space for these inputs. The extended leading edge shall be permitted to be shown even though not required for this purpose.

1-102.3.6 Signal Flow Direction

When logic diagrams do not completely indicate direction of signal flow by the symbols themselves, the leading edge of the logic symbols shall be extended to show the signal flow direction. Arrowheads superimposed on lines shall be permitted to be used as an additional aid if required.

1-102.3.7 Functional Identification

Functional identification is part of the symbol and shall be located inside and at the top center of the symbol. It shall be larger than any other character contained within the symbol.

When internal input and output identifications are part of the symbol, they shall be permitted to be placed above the identification of the logic function, depending upon symbol orientation.

1-102.3.8 Supplementary Information

Supplementary information, when required, shall be located inside the symbol below the functional identification on tagging lines. Typical supplementary information shall be: physical location, characteristics, and other pertinent data.

NEMA Standard 1-24-1964.

1-102.4 COMBINATION SYMBOLS

NOR, NOT, and NAND, which are actually combinations of basic logic functions rather than being basic logic functions themselves, are frequently implemented as specific elements.

This requires the use of two types of graphic symbology: a unique symbol and a combination symbol. The unique symbol is an application of the general symbol and the combination symbol is an application of the basic symbols of this standard, such as AND and LOGIC NEGATION. (See Table 1-102-1.) Either the unique or the combination symbol shall be used on a single drawing; they shall not be mixed.

In the case of the combination symbols, the choice of whether to use the AND or the OR approach will be determined in each case by whether negation occurs on the input or the output of the basic element.

It is expected that more complex logic will evolve in the future which will use both negative and positive logic in combination. In such cases the specialized symbols do not lend themselves well to diagram needs.

NEMA Standard 1-21-1964.

Table 1-102-1
GRAPHIC SYMBOLS

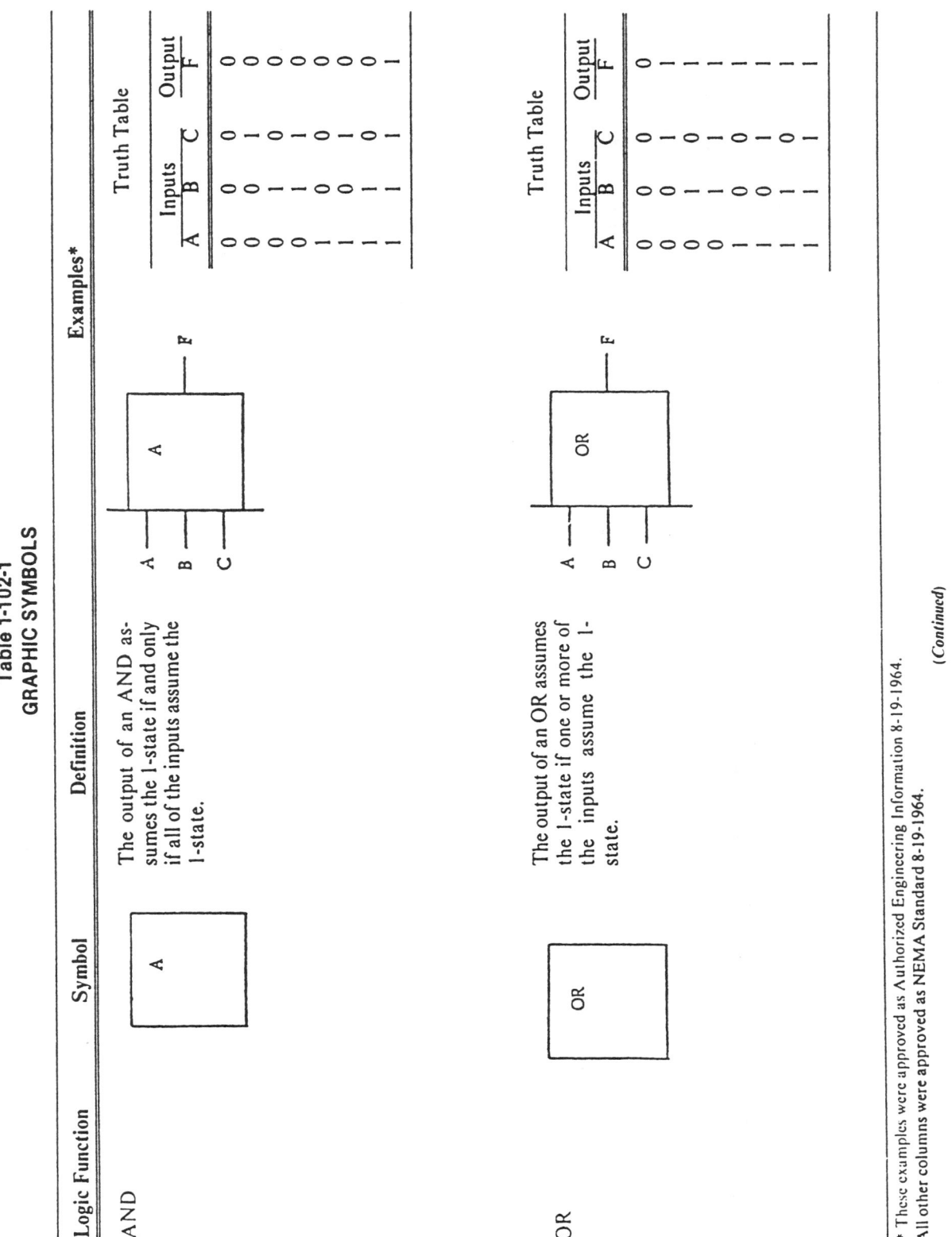

Logic Function	Symbol	Definition	Examples*
AND	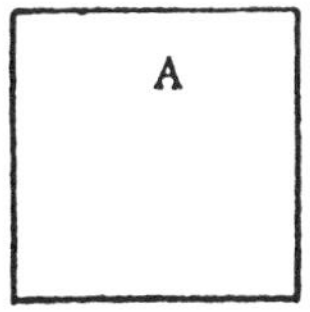	The output of an AND assumes the 1-state if and only if all of the inputs assume the 1-state.	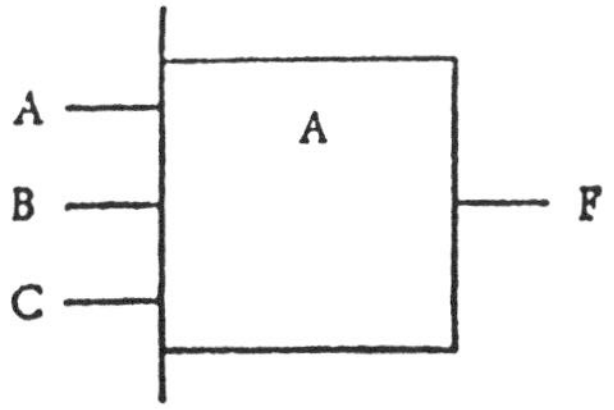
OR	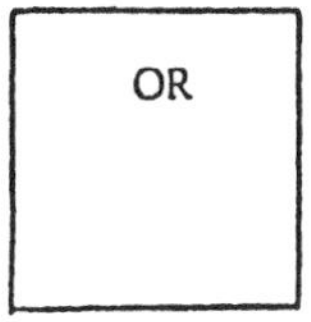	The output of an OR assumes the 1-state if one or more of the inputs assume the 1-state.	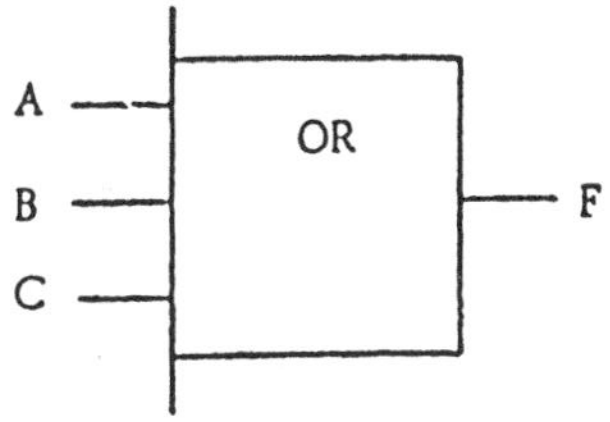

AND — Truth Table

Inputs A	Inputs B	Inputs C	Output F
0	0	0	0
0	0	1	0
0	1	0	0
0	1	1	0
1	0	0	0
1	0	1	0
1	1	0	0
1	1	1	1

OR — Truth Table

Inputs A	Inputs B	Inputs C	Output F
0	0	0	0
0	0	1	1
0	1	0	1
0	1	1	1
1	0	0	1
1	0	1	1
1	1	0	1
1	1	1	1

* These examples were approved as Authorized Engineering Information 8-19-1964.
All other columns were approved as NEMA Standard 8-19-1964.

(Continued)

Table 2.5. Graphic Symbols. This material is reproduced by permission of the National Electrical Manufacturers Association from NEMA Standards Publication ICS 6-1988 "ENCLOSURES FOR INDUSTRIAL CONTROL AND SYSTEMS," copyright 1988 by NEMA.

Table 1-102-1 (*Cont.*)
GRAPHIC SYMBOLS

Logic Function	Symbol	Definition	Examples*
EXCLUSIVE OR	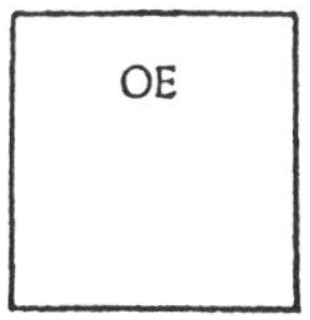	The output of an EXCLUSIVE OR with two inputs assumes the 1-state if one and only one input assumes the 1-state.	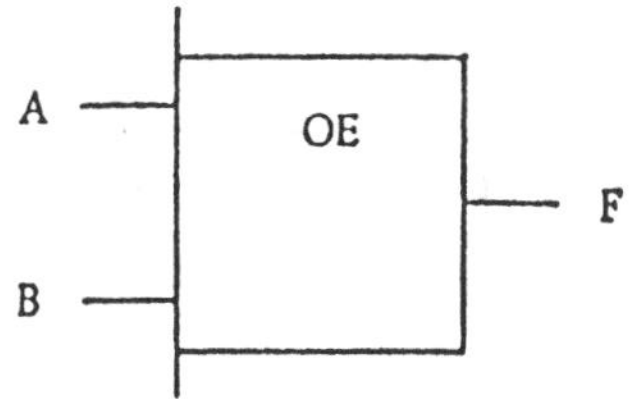 Truth Table (below)
LOGIC NEGATION		The output of LOGIC NEGATION operation takes on the 1-state if and only if the input does not take on the 1-state. It is an auxiliary symbol to be used only with one of the regular symbols to indicate logic negation. A small circle drawn at the point where a signal line joins a logic symbol indicates logic negation.	

Truth Table

Inputs A	Inputs B	Output F
0	0	0
1	0	1
0	1	1
1	1	0

Table 2.5 continued

NOT

(NOT is not a basic logic function. It can be represented as shown in the examples at the right. See 1-102.4.)

The output of a NOT assumes the 1-state if and only if the input assumes the 0-state.

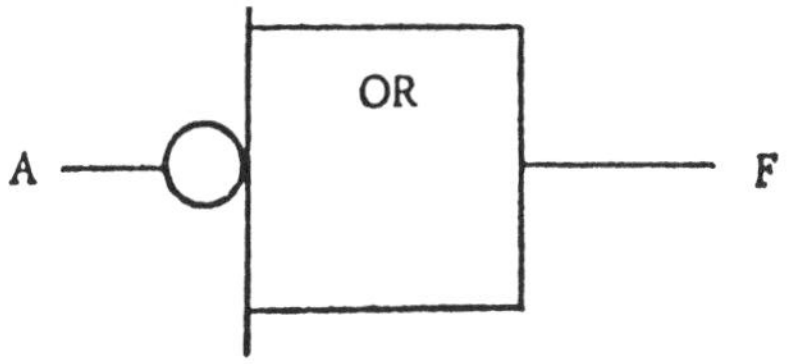

Truth Table

Input A	Output F
1	0
0	1

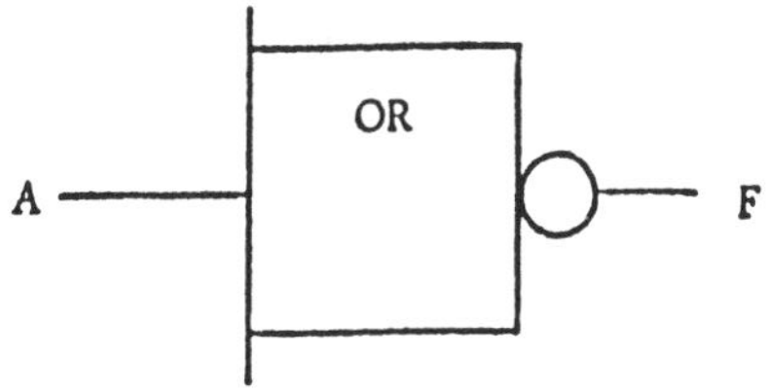

Combinational logic representations of the NOT function, such as these first four examples, should be used on drawings made up of basic logic functions.

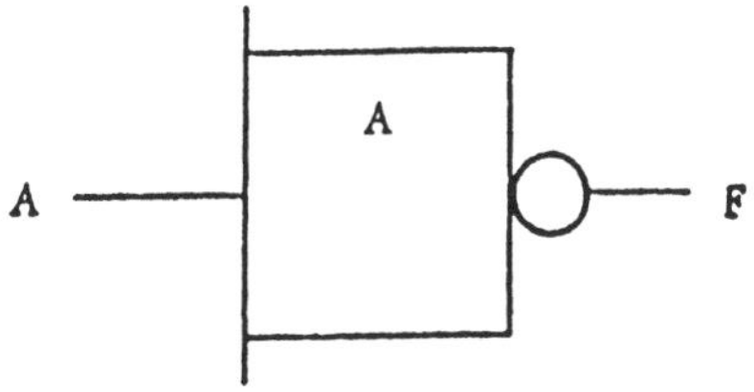

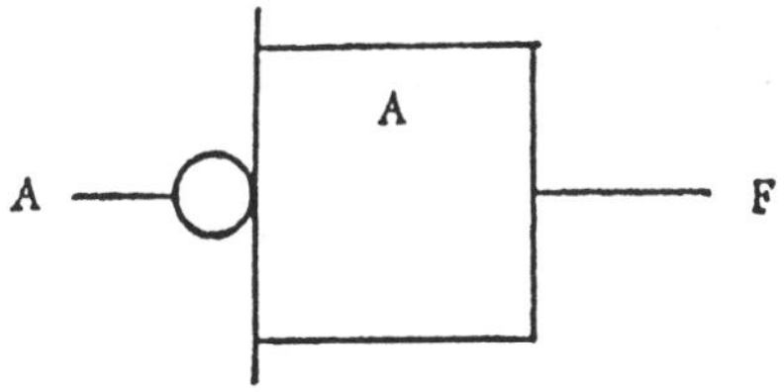

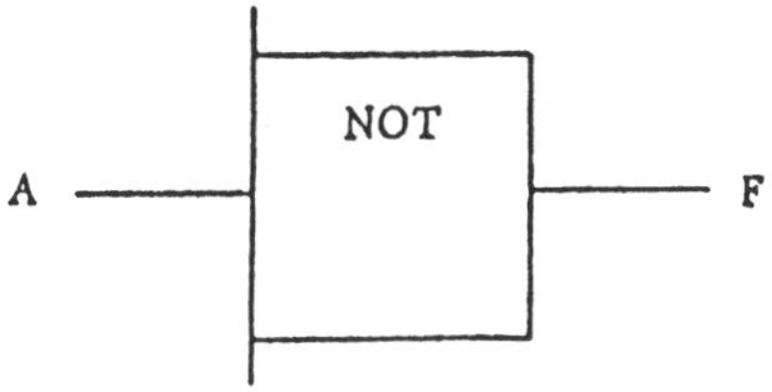

This last example is a unique representation for the NOT function. Combinational forms shown in the preceding four examples are preferred.

* These examples were approved as Authorized Engineering Information 8-19-1964.
All other columns were approved as a NEMA Standard 8-19-1964.

(*Continued*)

Table 2.5 continued

Table 1-102-1 (*Cont.*)
GRAPHIC SYMBOLS

Logic Function	Symbol	Definition	Examples*
NOR	(NOR is not a basic logic function. It can be represented as shown in the examples at the right. See 1-102.4.)	The output of a NOR assumes the 1-state if and only if all inputs assume the 0-state.	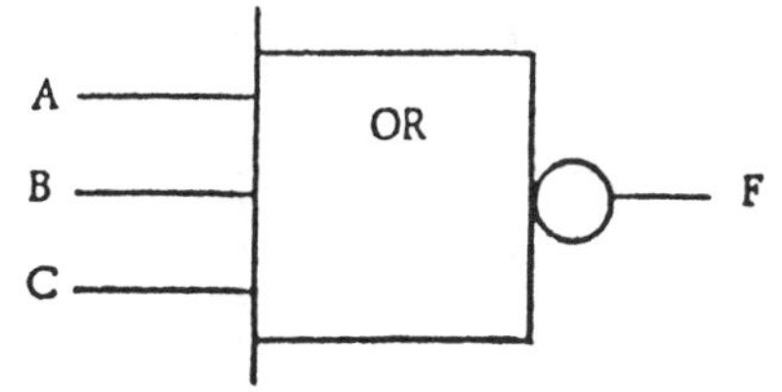 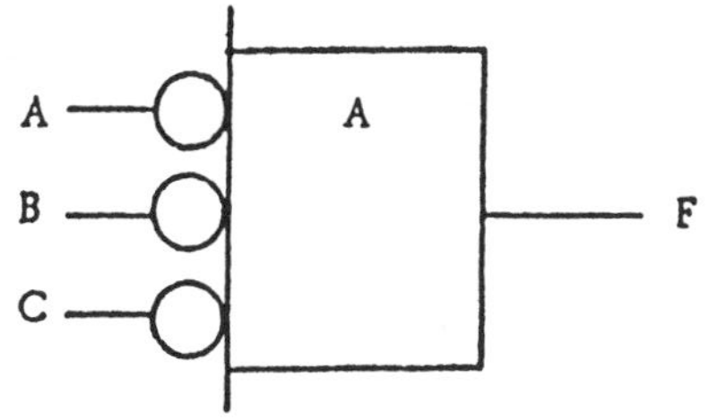 Truth Table (Three-input) — see below. Combinational logic representations of the NOR function, such as these first two examples, should be used on drawings made up of basic logic functions. 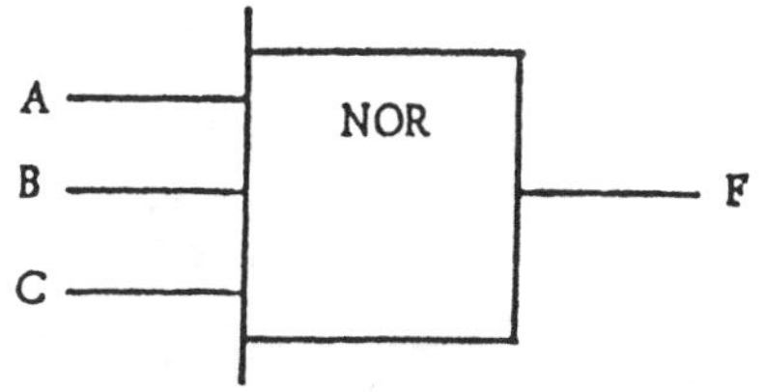 This last example is a unique logic representation for the NOR function and may be used on drawings for systems made up of NOR elements.

Truth Table (Three-input)

Inputs			Output
A	B	C	F
0	0	0	1
0	0	1	0
0	1	0	0
0	1	1	0
1	0	0	0
1	0	1	0
1	1	0	0
1	1	1	0

Table 2.5 continued

NAND

(NAND is not a basic logic function. It can be represented as shown in the examples at the right. See 1-102.4.)

The output of a NAND assumes the 0-state if and only if all inputs assume the 1-state.

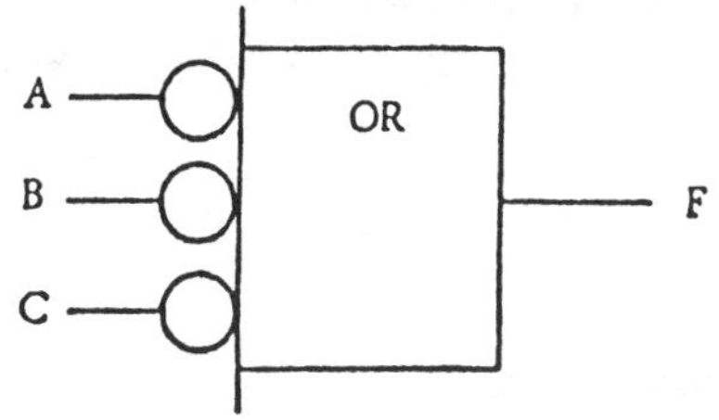

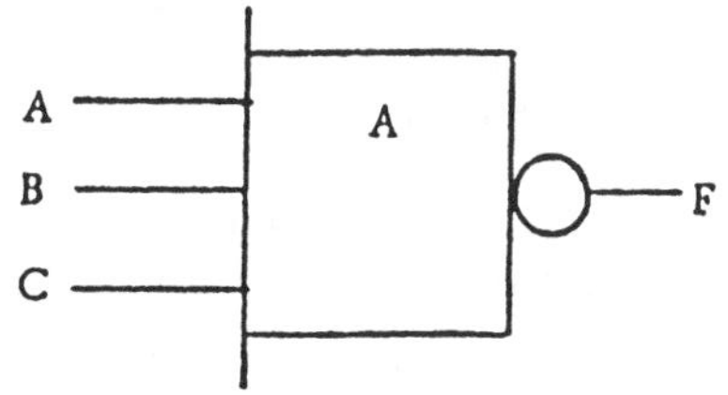

Truth Table (Three-element)

Inputs A	Inputs B	Inputs C	Output F
0	0	0	1
0	0	1	1
0	1	0	1
0	1	1	1
1	0	0	1
1	0	1	1
1	1	0	1
1	1	1	0

Combinational logic representations of the NAND function should be used on drawings made up of basic logic functions.

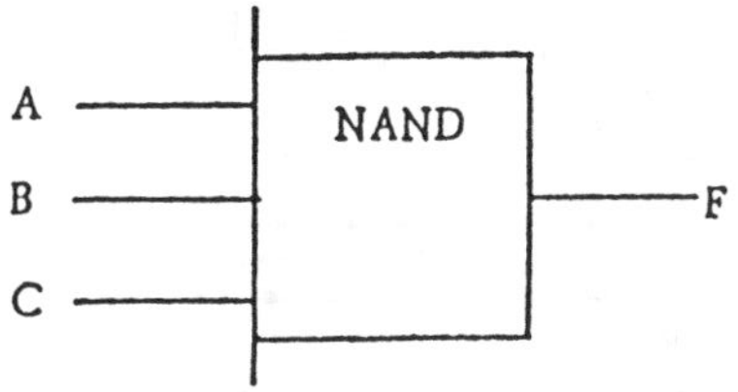

This last example is a unique logic representation for the NAND function and may be used on drawings for systems made up of NAND elements.

* These examples were approved as Authorized Engineering Information 8-19-1964.
All other columns were approved as a NEMA Standard 8-19-1964.

(Continued)

Table 2.5 continued

Table 1-102-1 (*Cont.*)
GRAPHIC SYMBOLS

Logic Function	Symbol	Definition	Examples*
FLIP-FLOP	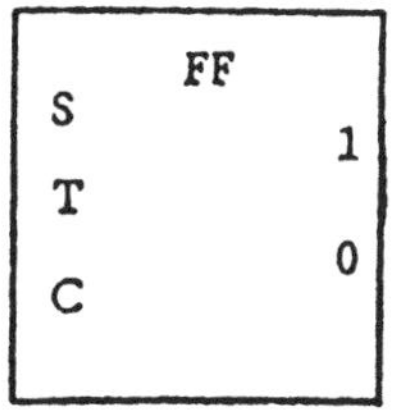 [The internal designations S (Set), T (Trigger), C (Clear), 1 and 0, when present, are part of the symbol. The set input shall be in proximity to the 1-output; the clear input shall be in proximity to the 0-output.]	A FLIP-FLOP assumes one of two stable states depending upon the history of the input or inputs.	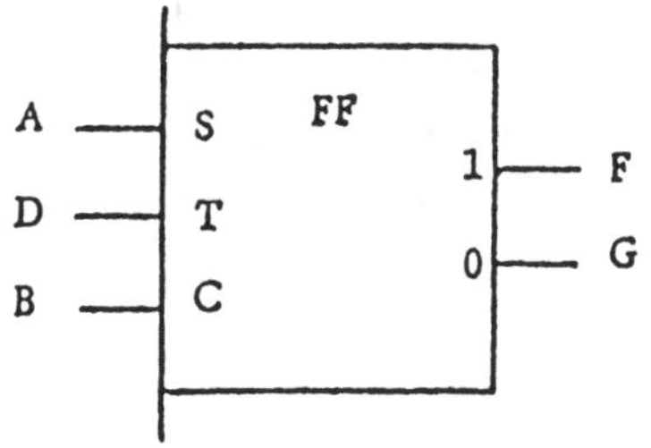

Truth Table

Inputs A (S)	Inputs B (C)	Inputs D (T)	Outputs F (1)	Outputs G (0)
000	000	010	100	011
000	000	010	011	100
010	000	000	111	000
010	000	000	011	100
000	010	000	100	011
000	010	000	000	111

There are many forms of a FLIP-FLOP. It frequently has two inputs (set S and clear C) and may have a third input [toggle (trigger) T] and complementary outputs.

A 1 is stored in a FLIP-FLOP when a 1 is first applied to the set input. The 1 (normal) output assumes the 1-state, and the 0 (complementary) output assumes the 0-state.

A 0 is stored in the FLIP-FLOP when a 1 is first applied to the clear input. The outputs then are in the opposite states from what they were when the FLIP-FLOP stored a 1.

The application of a 1-state signal to the toggle reverses the states of the FLIP-FLOP.

When necessary, the results of applying simultaneous signals to the inputs should be explained by means of an accompanying truth table.

Table 2.5 continued

Table 2.5 continued

SINGLE SHOT

A SINGLE SHOT produces a 1-state output of significant shape and duration when its input assumes a 1-state.

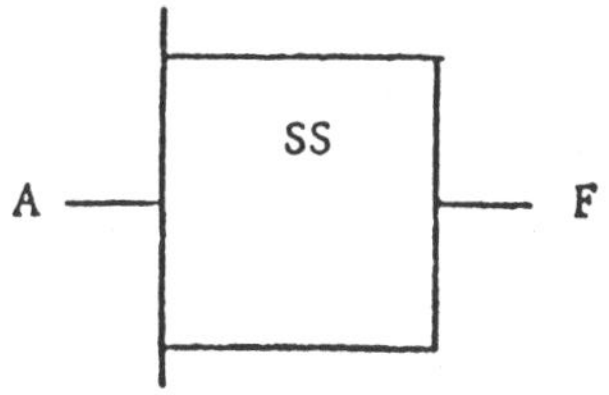

Truth Table

Input A	Output F
0 to 1 to 0	0 to 1 to 0
0 to 1 to 1	0 to 1 to 0
1 to 1 to 1	0 to 0 to 0
1 to 0 to 0	0 to 0 to 0

Output signal shape, amplitude, duration and polarity are determined by the circuit characteristics of the SINGLE SHOT, not by the input signal.

The normal (inactive) state of the SINGLE SHOT output is the 0-state. When activated by a 1-state input, its output changes to the indicated 1-state, remains there for the characteristic time of the device, and returns to the 0-state.

The duration of the on-time of the SINGLE SHOT will normally need to be included on a tag line inside the symbol. When required, waveforms indicating duration, amplitude, and rise and fall time may be used.

There may be a complementary output in addition to the output shown.

LEVEL TRIGGER (SCHMITT TRIGGER)

The output of a LEVEL TRIGGER whose inactive state is the 0-state, assumes the 1-state when the input exceeds a specified turn-on level, going in the direction of the 1-state, and the output continues until the input falls below a specified turn-off value going in the direction of the 0-state.

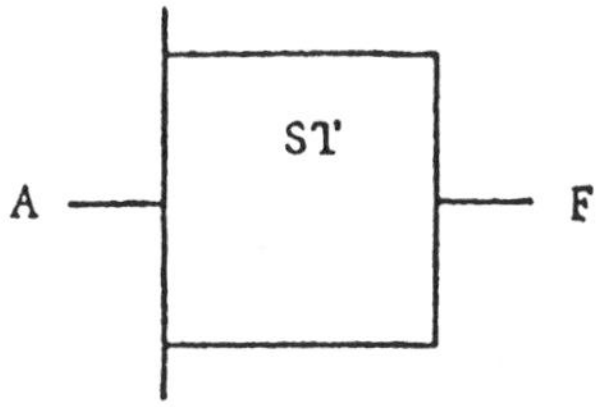

There may be a complementary output in addition to the output shown.

* These examples were approved as Authorized Engineering Information 8-19-1964.
All other columns were approved as a NEMA Standard 8-19-1964.

(Continued)

Table 1-102-1 (*Cont.*)
GRAPHIC SYMBOLS

Logic Function	Symbol	Definition	Examples*	
SWITCHING AMPLIFIER	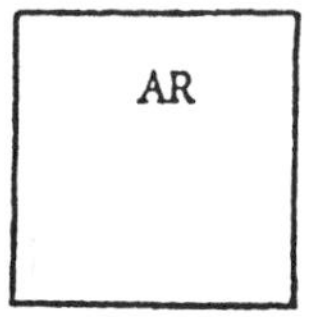 (The SWITCHING AMPLIFIER shall be permitted to be either linear or nonlinear. This symbol shall not be used for a proportional amplifier.)	The output of a SWITCHING AMPLIFIER assumes the 1-state if and only if the input assumes the 1-state.	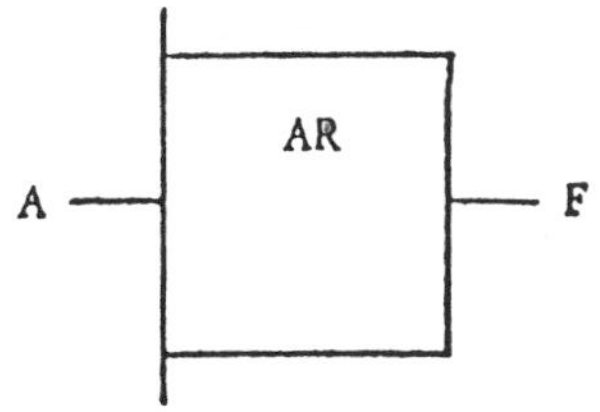	Truth Table Input A: 0, 1 Output F: 0, 1
SIGNAL (TIME) DELAY	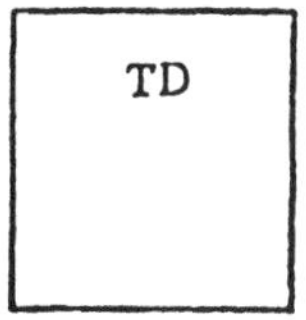 (This symbol can be used to represent what is called a delay line.)	In a SIGNAL (TIME) DELAY element, the output reproduces the input after a predetermined time delay.	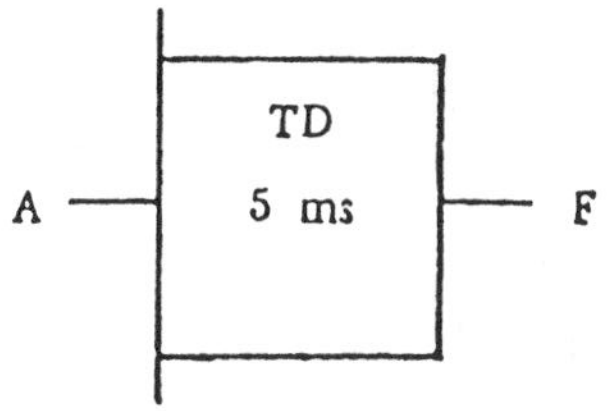	The duration of the SIGNAL (TIME) DELAY is included with the symbol on a tagging line.
			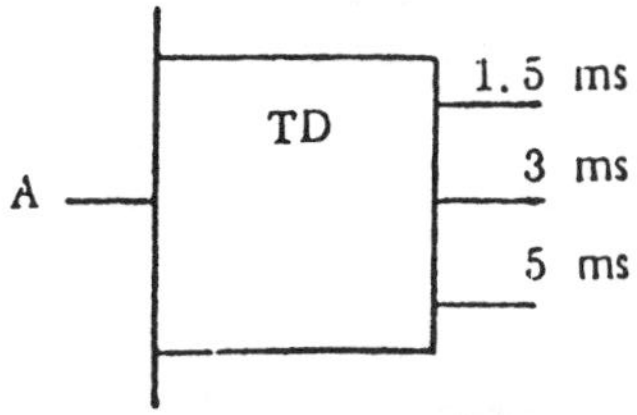	If the element is tapped, the delays with respect to the input should be included :at the tap output.

Truth Table (Switching Amplifier):

Input A	Output F
0	0
1	1

TIME DELAY, ON

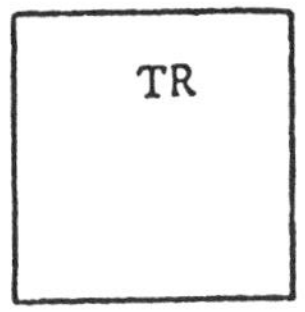

(Tag lines shall show the amount of delay or range if adjustable.)

(There shall be permitted to be a complementary output in addition to the one shown in the examples.)

The output of a TIME DELAY, ON assumes the 1-state following a definite intentional time interval after its input assumes the 1-state. The output returns to the 0-state with no intentional delay after the input returns to the 0-state. It is assumed that the input signal states are maintained long enough to effect complete time delay or reset functions.

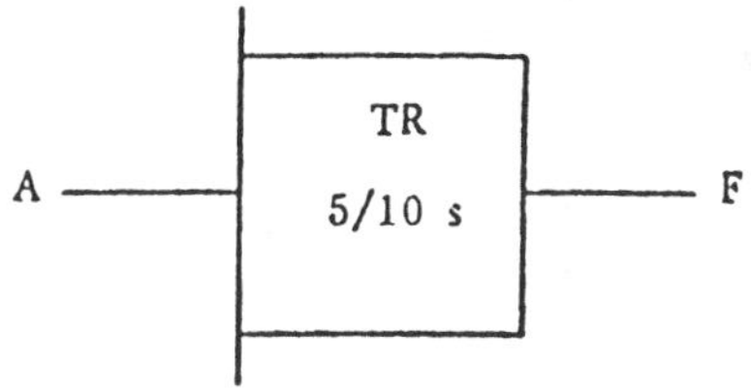

(either 5-second or 10-second delay)

Truth Table

Input A	Output F
0 to 0	0 to 0
0 to 1	0 — 1
1 to 1	1 to 1
1 to 0	1 to 0

The bar (—) denotes time delay.

The following examples of logic negation in conjunction with the TIME DELAY, ON symbol provide combinational symbols for TIME DELAY, OFF and TIME DELAY, ON AND OFF.

TIME DELAY, OFF

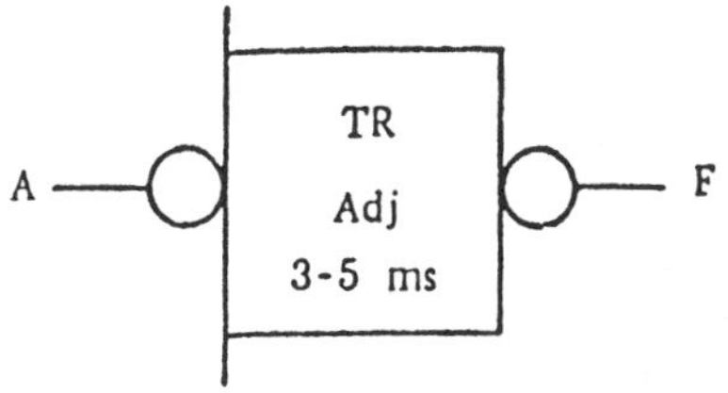

(time delay adjustable from 3 to 5 milliseconds)

Truth Table

Input A	Output F
1 to 1	1 to 1
1 to 0	1 — 0
0 to 0	0 to 0
0 to 1	0 to 1

The bar (—) denotes time delay.

* These examples were approved as Authorized Engineering Information 8-19-1964.
All other columns were approved as NEMA Standard 8-19-1964.

(*Continued*)

Table 2.5 continued

Table 1-102-1 (*Cont.*)
GRAPHIC SYMBOLS

Logic Function	Symbol	Definition	Examples*
TIME DELAY, ON AND OFF			Truth Table (see below) 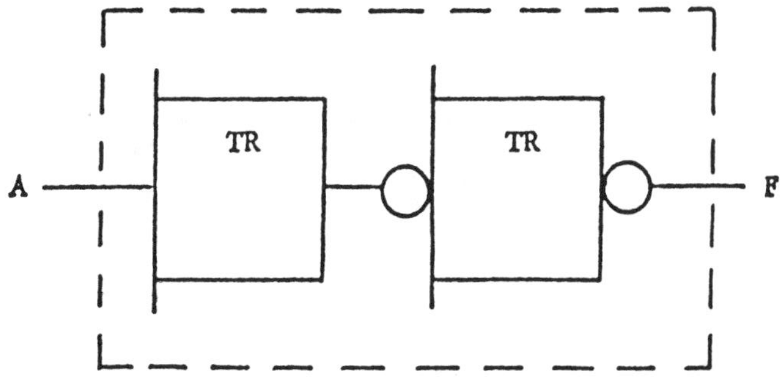
OSCILLATOR		An OSCILLATOR produces an alternating or fluctuating output of a frequency which is self-determined.	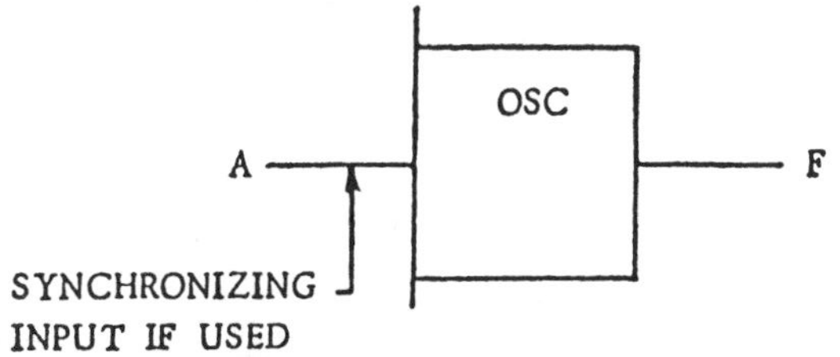 (On tagging lines, indicate the frequency and range of adjustments if applicable.)

Truth Table

Input A	Output F
0 to 0	0 to 0
0 to 1	0 — 1
1 to 1	1 to 1
1 to 0	1 — 0

The bar (—) denotes time delay.

Table 2.5 continued

TRANSMISSION GATE

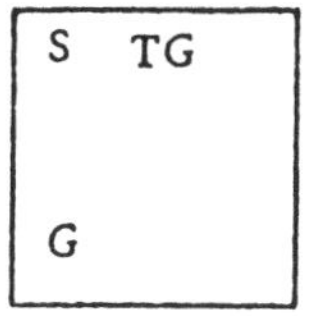

The input signal (X) to a TRANSMISSION GATE appears at the output only if the gate assumes the 1-state.

S = Signal G = Gate

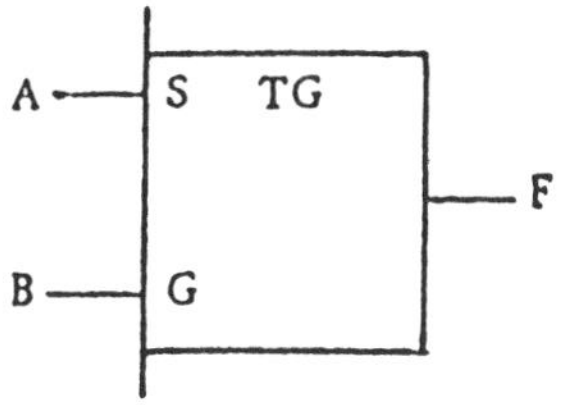

Truth Table

Inputs		Output
A	B	F
X	0	0
X	1	X

GENERAL SYMBOL NOT PREVIOUSLY IDENTIFIED

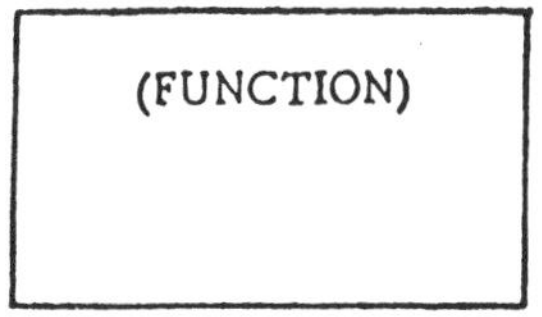

Symbol for functions not elsewhere specified. The symbol shall be adequately labeled to identify the function performed. This shall not be used to describe a function which can be conveniently expressed by a combination of basic logic symbols in this publication.

The longer side of the rectangle shall be permitted to be in either the horizontal or vertical plane.

Inputs and outputs shall not be on the same side and should perferably be on opposite sides.

Engineering data including a truth table, if applicable, should be proivided.

* These examples were approved as Authorized Engineering Information 8-19-1964.

All other columns were approved as NEMA Standard 8-19-1964.

Table 2.5 continued

ICS Part 1-103
STATIC SWITCHING CONTROL DEVICES*

The standards in this part apply to all other parts in the NEMA Standards Publications for Industrial Controls and Systems unless otherwise specified.

The definitions contained in Part 1-100 of this Standards Publication also apply to this part.

1-103.1 GENERAL

Static switching control is a method of switching electrical circuits without the use of contacts (the control primarily includes magnetic amplifiers and solid-state devices).

Static switching control performs functions similar to those of conventionally used relays but without switching contacts. Relays convert a single input to a coil into various outputs by means of its moving armature closing or opening contacts. Static switching control elements convert a single or combination of inputs into outputs by controlling the circuit's impedance or modulation, or both.

The logic terminology associated with static switching control is included herein to illustrate some operating logic functions. The same type of device may perform different control functions in the same equipment depending upon the required functions.

NEMA Standard 11-17-1960.

1-103.2 LOGIC FUNCTIONS, SYMBOLS, AND DEFINITIONS

Device logic functions and symbols are illustrated and defined in Table 1-103-1. They are intended for use on electrical elementary and schematic diagrams to indicate the operating functions. The symbols are a shorthand graphical representation and are not necessarily intended to represent the physical likeness of the device nor certain complete electrical connections.

The symbols may be combined into various circuit adaptations as required by the particular application.

Input terminals on symbols and truth tables are designated by the letters A, B and C, and the output by the letter X. These markings are only for purposes of reference, to explain the functions, and are not terminal or wire markings.

Input terminals are shown on the left side of symbols and output terminals are shown on the right side of symbols.

The number of the symbol inputs may vary because of different designs. They represent typical examples and are not all-inclusive.

With reference to MEMORY, RETENTIVE MEMORY and OFF RETURN MEMORY, the memory input "on" terminals are the upper lines on the symbols. The "off" input terminals (and output terminals, when required) are on the lower lines.

For GATE symbols, the control input is the top left terminal. The power circuit terminals are the right and lower left terminals. (See GATE, NONISOLATED and ISOLATED.)

Return circuit terminals are not shown when directly connected to common or ground. When return circuits need to be shown, the control circuit shall be connected as a top vertical terminal. (For example, see GATES AND SIGNAL CONVERTER WITH RETURN CIRCUITS.)

The term "input" or "output" denotes a signal which is useful or significant to elements of the system.

The terms "on" and "off" denote the presence or absence of an output, respectively. The "off" condition does not necessarily mean that there is no signal.

NEMA Standard 7-12-1961.

1-103.3 DEVICE IDENTIFICATION

Markings on static units shall include the following:

(a) Catalog number.
(b) Manufacturer's identification.
(c) Functional marking (does not apply to multi-function units; e.g., AND, OR, NOT, etc.) as required by the particular application.

NEMA Standard 11-17-1960.

1-103.4 VOLTAGE VARIATION

The control system shall operate on a supply voltage from minus 10 percent to plus 10 percent of the rated control system voltage.

NEMA Standard 9-29-1960.

*This material is reproduced by permission of the National Electrical Manufacturers Association from NEMA Standards Publication ICS 6-1988 "ENCLOSURES FOR INDUSTRIAL CONTROL AND SYSTEMS," copyright 1988 by NEMA.

Table 1-103-1
DEVICE LOGIC FUNCTIONS AND SYMBOLS

Logic Function	Symbol	Definitions
AND	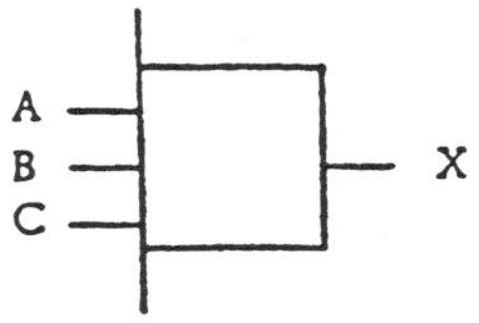	A device which produces an output only when every input is energized.
NOT	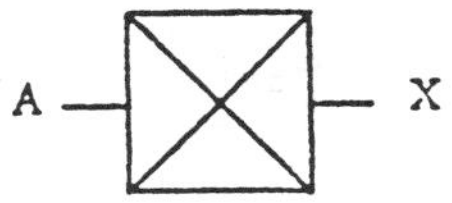	A device which produces an output only when the input is not energized.
NOR	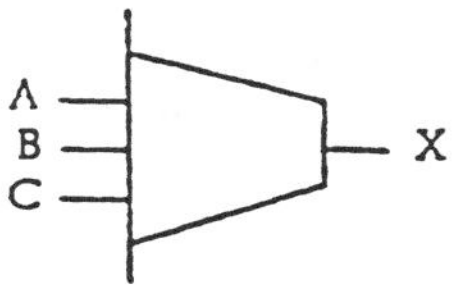	A device which produces an output only when every input is not energized.
OR	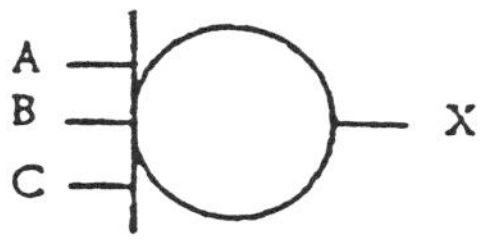	A device which produces an output when one input (or more) is energized.

Truth Tables

Truth Table—AND

A	B	C	X
0	0	0	0
0	0	1	0
0	1	0	0
0	1	1	0
1	0	0	0
1	0	1	0
1	1	0	0
1	1	1	1

Truth Table—NOT

A	X
0	1
1	0

Truth Table—NOR

A	B	C	X
0	0	0	1
0	0	1	0
0	1	0	0
0	1	1	0
1	0	0	0
1	0	1	0
1	1	0	0
1	1	1	0

Truth Table—OR

A	B	C	X
0	0	0	0
0	0	1	1
0	1	0	1
0	1	1	1
1	0	0	1
1	0	1	1
1	1	0	1
1	1	1	1

Logic Function	Symbol	Definitions
MEMORY	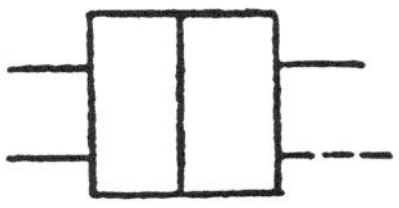	A device which retains the condition of output corresponding to the input last energized as long as power is maintained.

Table 2.6. Device Logic Functions and Symbols. This material is reproduced by permission of the National Electrical Manufacturers Association from NEMA Standards Publication ICS 6-1988 "ENCLOSURES FOR INDUSTRIAL CONTROL AND SYSTEMS," copyright 1988 by NEMA.

(*Continued*)

Table 1-103-1 (*Cont.*)
DEVICE LOGIC FUNCTIONS AND SYMBOLS

Logic Function	Symbol	Definitions
RETENTIVE MEMORY	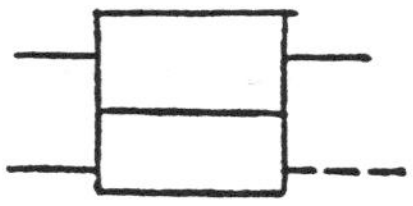	A device which retains the conditions of output corresponding to the input last energized.
OFF RETURN MEMORY	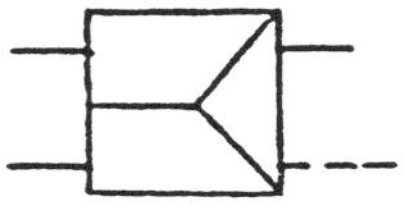	A device which retains the condition of output corresponding to the input last energized, except upon interruption of power it returns to the off condition.
TIME DELAY ENERGIZING	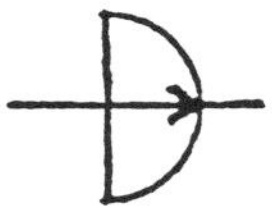	A device which produces an output following definite intentional time delay after its input is energized.
TIME DELAY DEENERGIZING	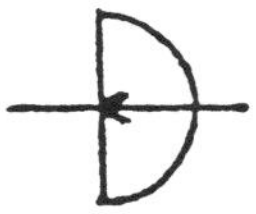	A device whose output is deenergized following a definite intentional time delay after its input is deenergized.
TIME DELAY ENERGIZING AND DEENERGIZING	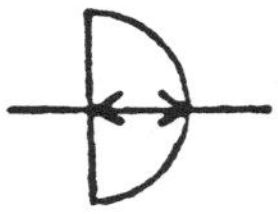	A device which produces an output following a definite intentional time delay after its input is energized and whose output is deenergized following a definite intentional time delay after its input is deenergized.
ADJUSTABLE TIME DELAY		Otherwise similar to time delay energizing, time delay deenergizing, and time delay energizing and deenergizing.
AMPLIFIER	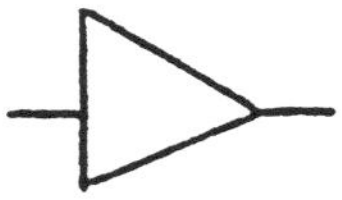	A device in which an input signal controls a local source of power to produce an output enlarged relative to the input.
GATE (NON-ISOLATED)		A device which permits current flow in a non-isolated circuit when the control input is energized.
GATE (ISOLATED)	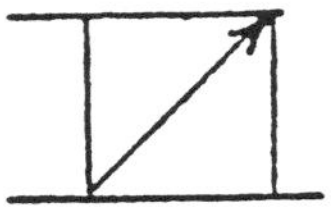	A device which permits current flow in an isolated circuit when the control input is energized.
SIGNAL CONVERTER		A device for changing a pilot signal to a logic input.
GATES AND SIGNAL CONVERTER WITH RETURN CIRCUITS		
GATE	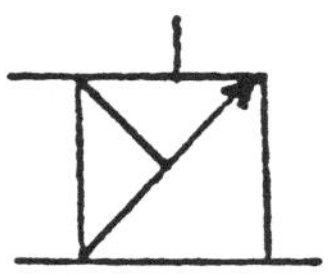	A device which permits current flow in a non-isolated circuit when the control input is energized.

Table 2.6 continued

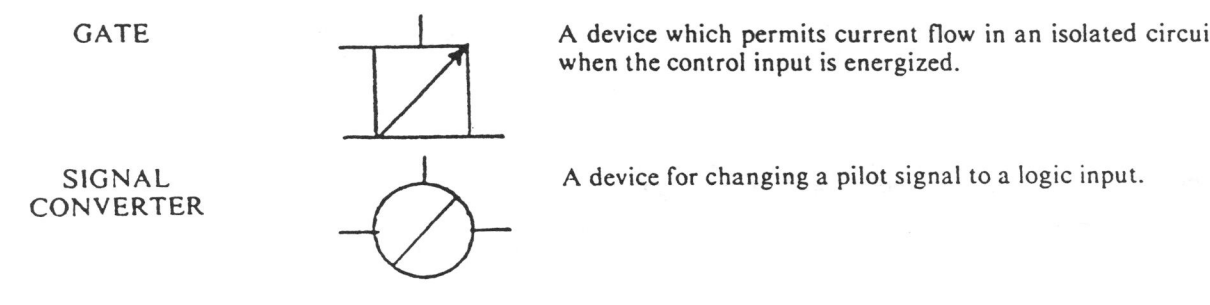

GATE	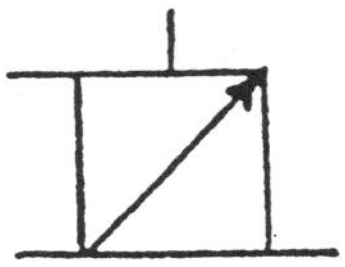	A device which permits current flow in an isolated circuit when the control input is energized.
SIGNAL CONVERTER	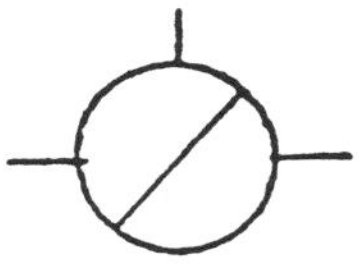	A device for changing a pilot signal to a logic input.

Table 2.6 continued

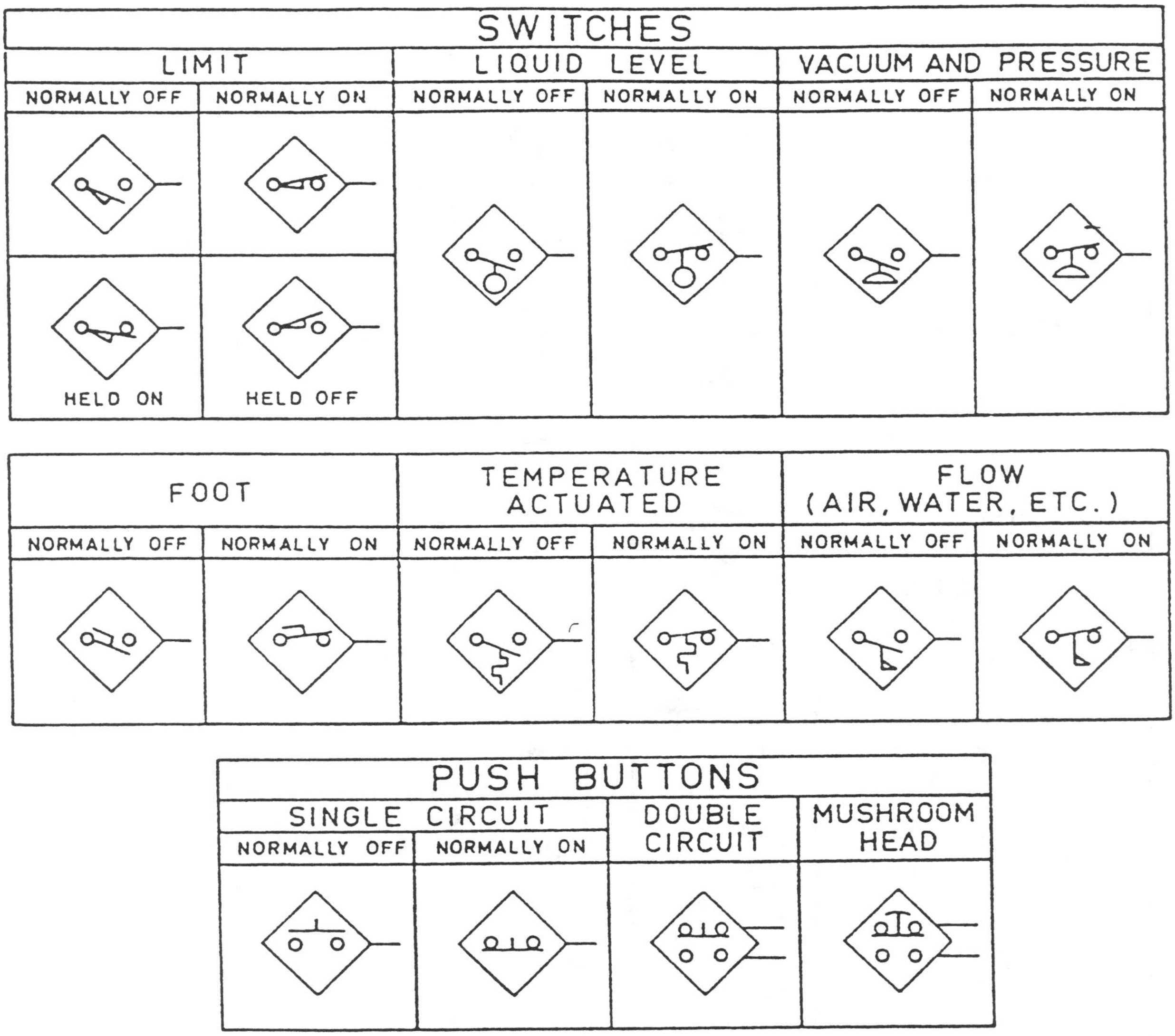
SWITCHES
LIMIT
LIQUID LEVEL
VACUUM AND PRESSURE
NORMALLY OFF
NORMALLY ON
NORMALLY OFF
NORMALLY ON
NORMALLY OFF
NORMALLY ON
HELD ON
HELD OFF
FOOT
TEMPERATURE ACTUATED
FLOW (AIR, WATER, ETC.)
NORMALLY OFF
NORMALLY ON
NORMALLY OFF
NORMALLY ON
NORMALLY OFF
NORMALLY ON
PUSH BUTTONS
SINGLE CIRCUIT
DOUBLE CIRCUIT
MUSHROOM HEAD
NORMALLY OFF
NORMALLY ON

Figure 2.15. Symbols for Static Pilot Devices. This material is reproduced by permission of the National Electrical Manufacturers Association from NEMA Standards Publication ICS 6-1988 "ENCLOSURES FOR INDUSTRIAL CONTROL AND SYSTEMS," copyright 1988 by NEMA.

Graphical Symbols for Pipe Fittings, Valves, and Piping

The piping symbols in Figure 2.16 are reprinted in part from ANSI Z32.2.3 1949 (R 1953) by permission of the ASME.

Figure 2.16. Graphical symbols for pipe fittings, valves and piping. Reprinted in part from ANSI Z.32.2.3. 1949 (R 1953) published by the American Society of Mechanical Engineers (ASME).

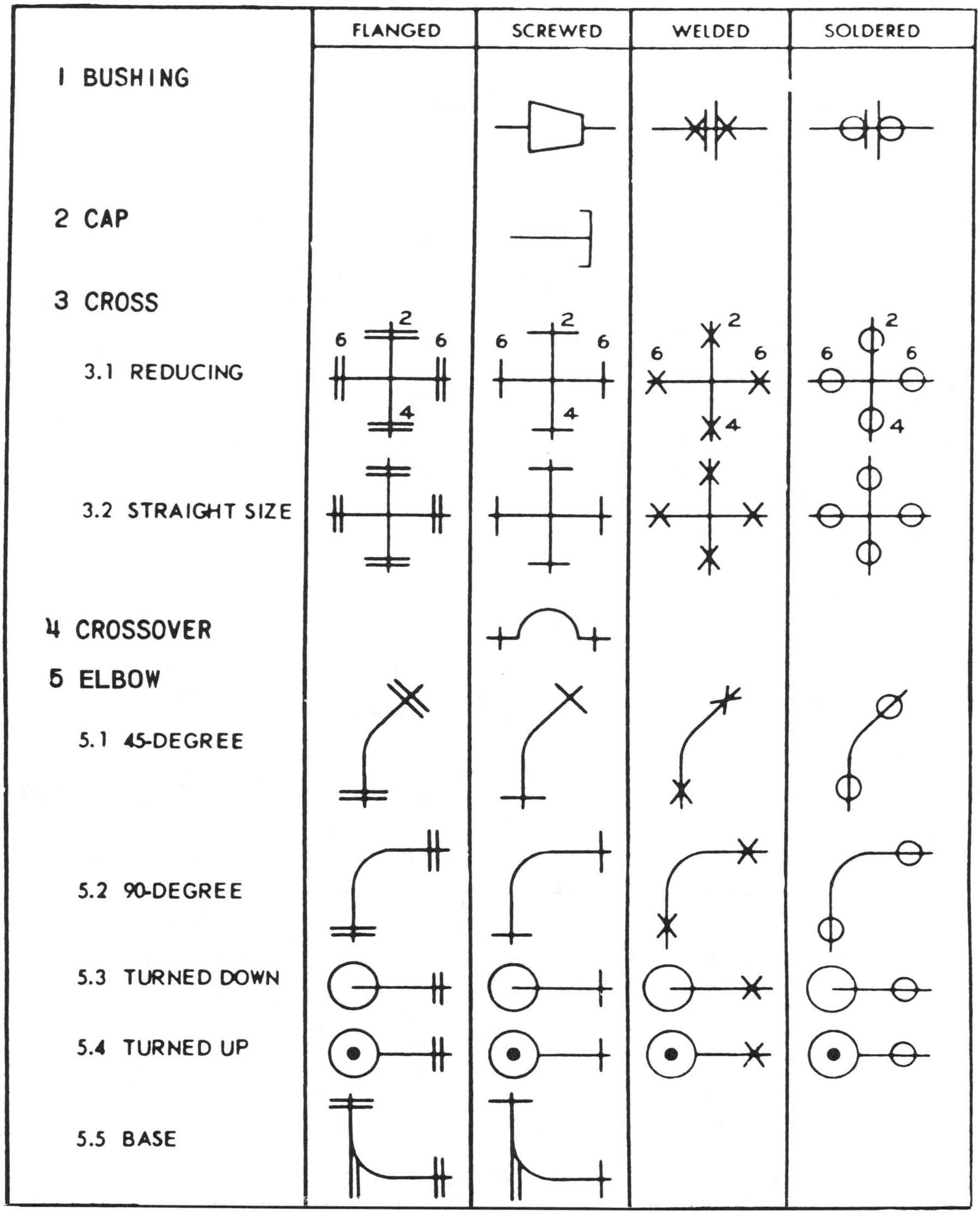
FLANGED
SCREWED
WELDED
SOLDERED
1 BUSHING
2 CAP
3 CROSS
3.1 REDUCING
3.2 STRAIGHT SIZE
4 CROSSOVER
5 ELBOW
5.1 45-DEGREE
5.2 90-DEGREE
5.3 TURNED DOWN
5.4 TURNED UP
5.5 BASE

Figure 2.16 continued

Figure 2.16 continued

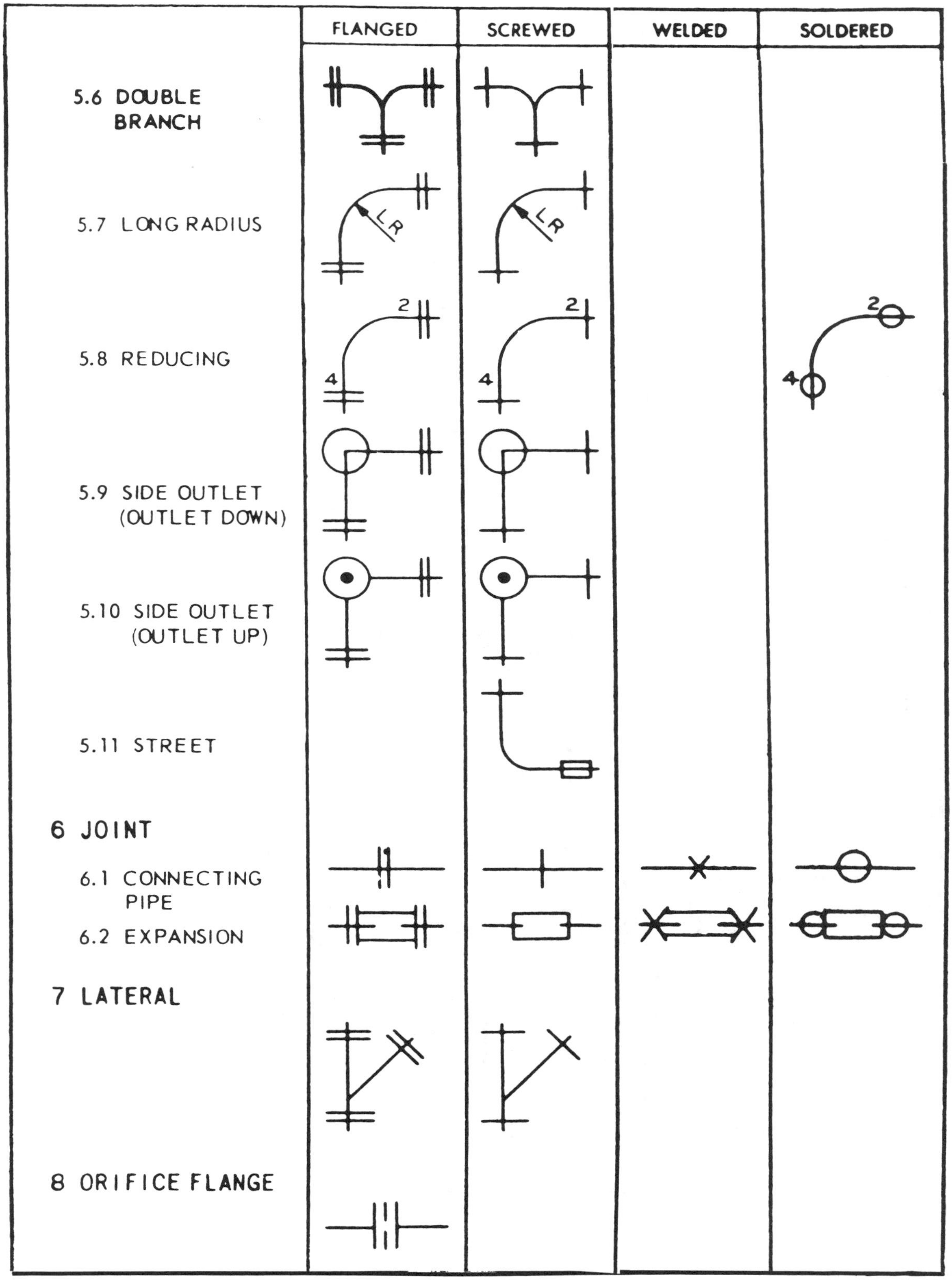

FLANGED
SCREWED
WELDED
SOLDERED
5.6 DOUBLE BRANCH
5.7 LONG RADIUS
LR
5.8 REDUCING
2
4
5.9 SIDE OUTLET (OUTLET DOWN)
5.10 SIDE OUTLET (OUTLET UP)
5.11 STREET
6 JOINT
6.1 CONNECTING PIPE
6.2 EXPANSION
7 LATERAL
8 ORIFICE FLANGE

Figure 2.16 continued

Figure 2.16 continued

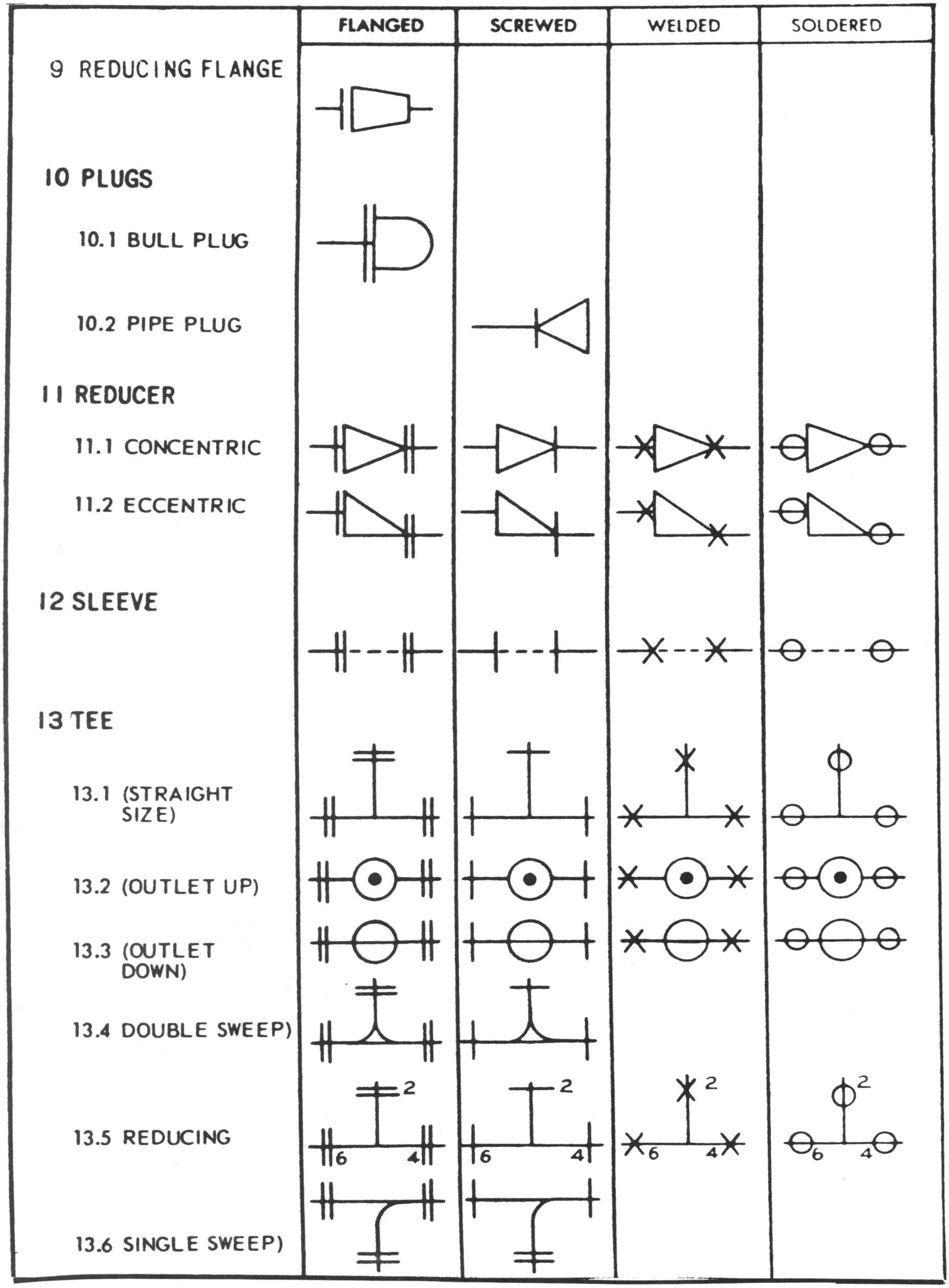
FLANGED
SCREWED
WELDED
SOLDERED
9 REDUCING FLANGE
10 PLUGS
10.1 BULL PLUG
10.2 PIPE PLUG
11 REDUCER
11.1 CONCENTRIC
11.2 ECCENTRIC
12 SLEEVE
13 TEE
13.1 (STRAIGHT SIZE)
13.2 (OUTLET UP)
13.3 (OUTLET DOWN)
13.4 DOUBLE SWEEP)
13.5 REDUCING
2
6
4
13.6 SINGLE SWEEP)

Figure 2.16 continued

Figure 2.16 continued

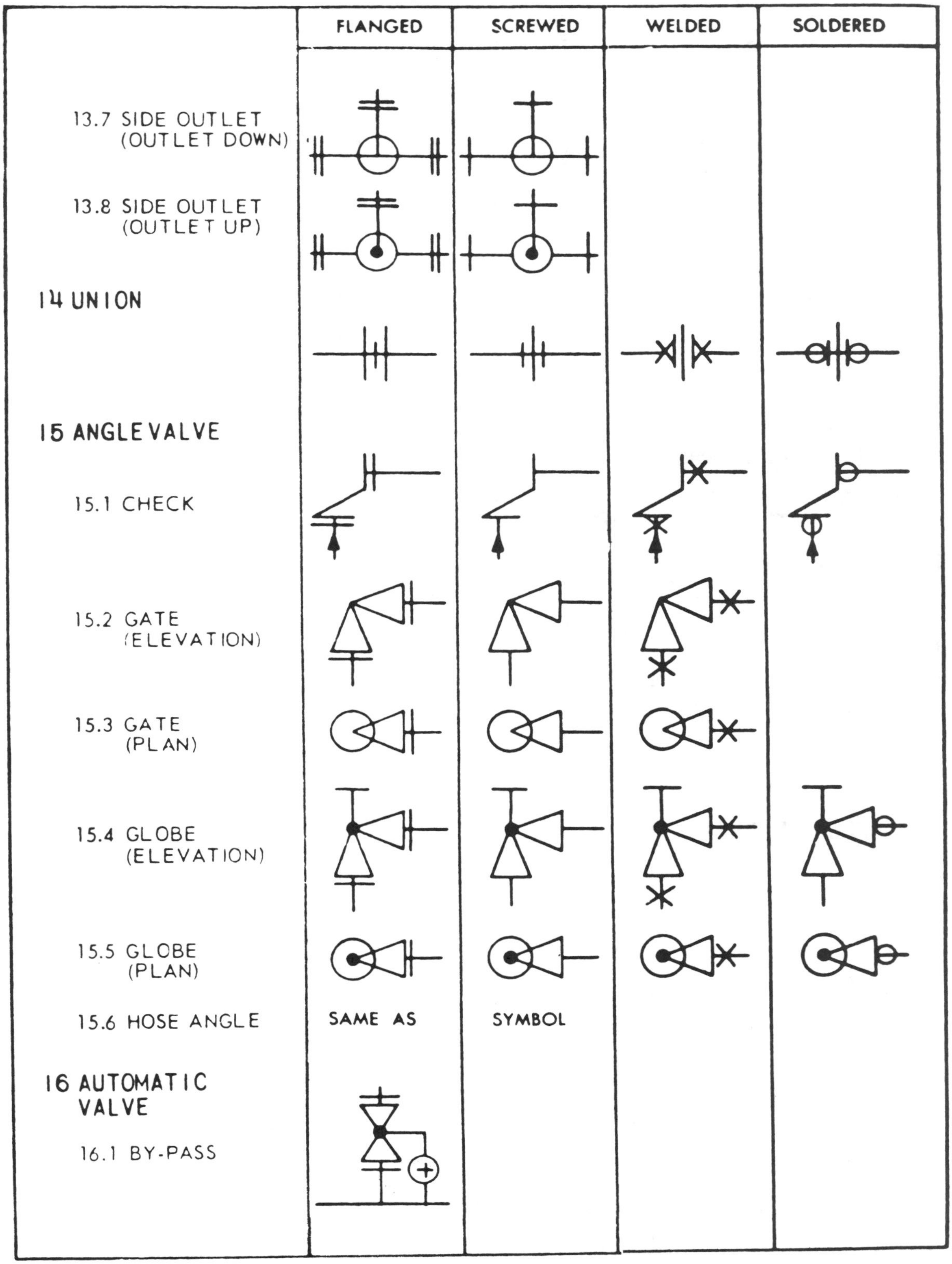

Figure 2.16 continued

Figure 2.16 continued

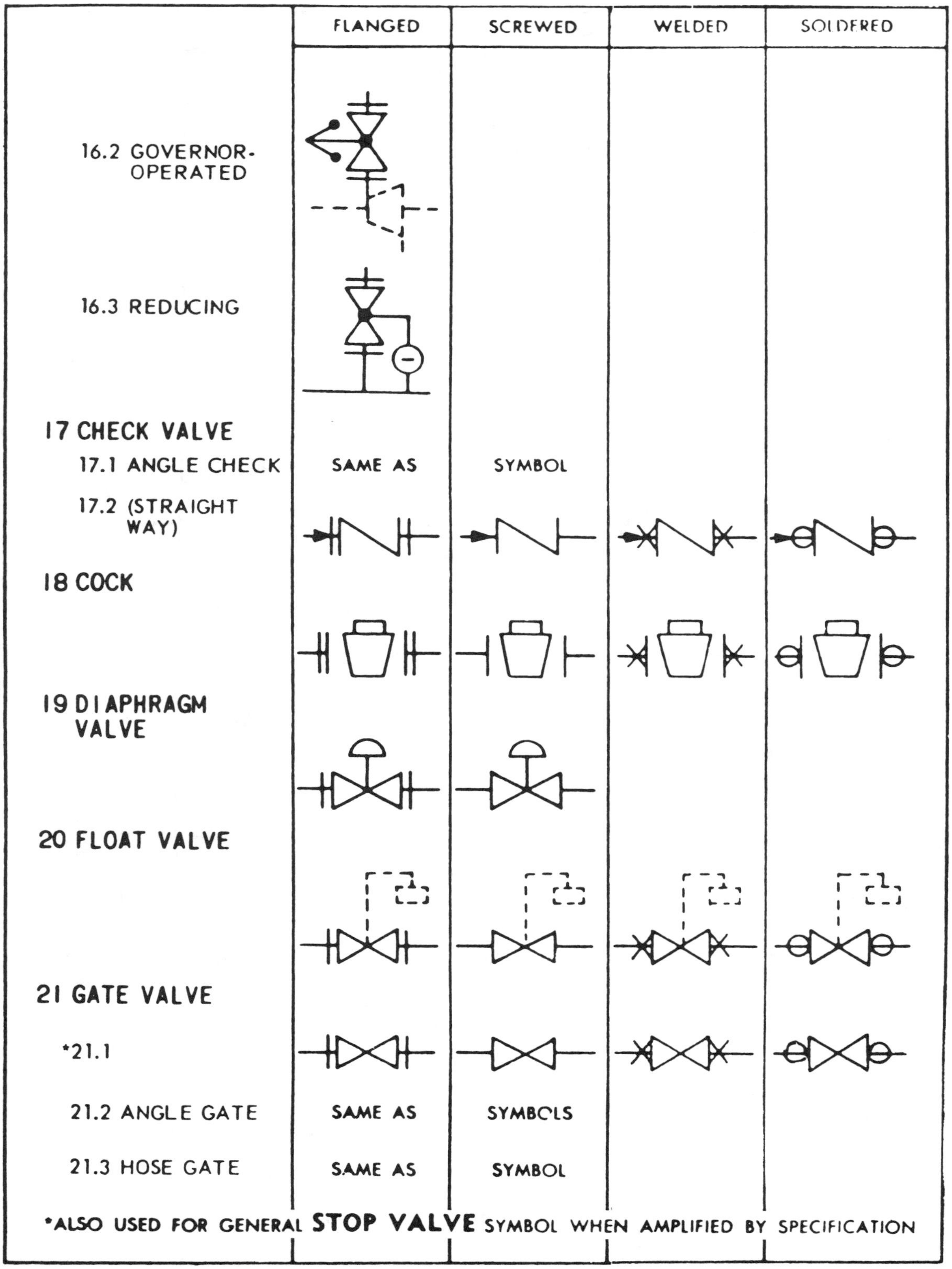
FLANGED
SCREWED
WELDED
SOLDERED
16.2 GOVERNOR-OPERATED
16.3 REDUCING
17 CHECK VALVE
17.1 ANGLE CHECK
SAME AS
SYMBOL
17.2 (STRAIGHT WAY)
18 COCK
19 DIAPHRAGM VALVE
20 FLOAT VALVE
21 GATE VALVE
*21.1
21.2 ANGLE GATE
SAME AS
SYMBOLS
21.3 HOSE GATE
SAME AS
SYMBOL
*ALSO USED FOR GENERAL STOP VALVE SYMBOL WHEN AMPLIFIED BY SPECIFICATION

Figure 2.16 continued

Figure 2.16 continued

	FLANGED	SCREWED	WELDED	SOLDERED
21.4 MOTOR-OPERATED	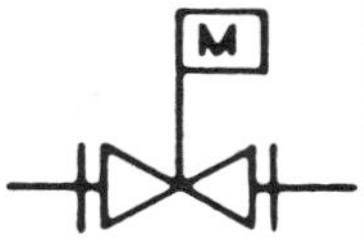	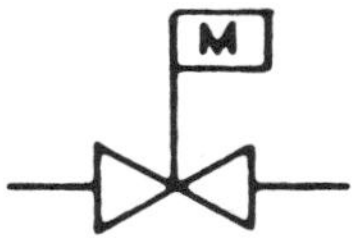	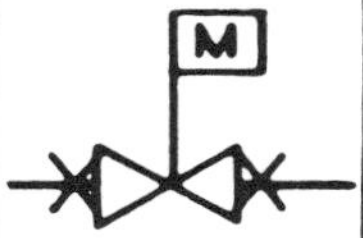	
22 GLOBE VALVE				
22.1				
22.2 ANGLE GLOBE	SAME AS	SYMBOLS		
22.3 HOSE GLOBE	SAME AS	SYMBOL		
22.4 MOTOR-OPERATED	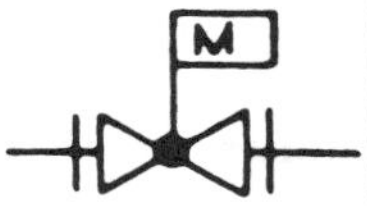	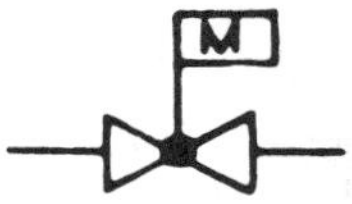	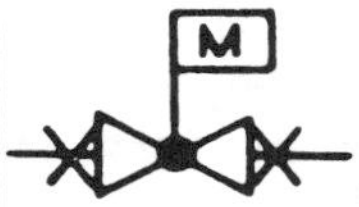	
23 HOSE VALVE				
23.1 ANGLE	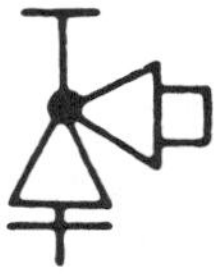			
23.2 GATE				
23.3 GLOBE				
24 LOCKSHIELD VALVE				
25 QUICK OPENING VALVE				
26 SAFETY VALVE				
27 STOP VALVE	SAME AS	SYMBOL		

Graphic Symbols for Fluid Power Diagrams

The symbols in Figure 2.17 are commonly used in circuit diagrams for fluid power systems. They are reprinted in part from ANSI Y-32.10-1967 by permission of the ASME.

1. Introduction

1.1 **General**

Fluid power systems are those that transmit and control power through use of a pressurized fluid within an enclosed circuit.

Types of symbols commonly used in drawing circuit diagrams for fluid power systems are Pictorial, Cutaway, and Graphic.

Pictorial symbols are useful for showing the interconnection of components. They are difficult to standardize from a functional basis.

Cutaway symbols emphasize construction. These are complex to draw and the functions are not readily apparent.

Graphic symbols emphasize the function and methods of operation of components. These are simple to draw. Component functions and methods of operation are obvious. Graphic symbols are capable of crossing language barriers, and promote a universal understanding of fluid power systems.

Graphic symbols for fluid power systems should be used in conjunction with the graphic symbols for other systems published by the USA Standards Institute.

Complete graphic symbols are those which give symbolic representation of the component and all of its features pertinent to the circuit diagram.

Simplified graphic symbols are stylized versions of the complete symbols.

Composite graphic symbols are an organization of simplified or complete symbols. Composite symbols usually represent a complex component.

1.2 **Scope and Purpose**

Scope

This standard presents a system of graphic symbols for fluid power diagrams.

Elementary forms of symbols are:

Circles	Triangles	Lines
Squares	Arcs	Dots
Rectangles	Arrows	Crosses

Symbols using words or their abbreviations are avoided. Symbols capable of crossing language barriers are presented herein.

Component function rather than construction is emphasized by the symbol.

The means of operating fluid power components are shown as part of the symbol (where applicable).

This standard shows the basic symbols, describes the principles on which the symbols are based, and illustrates some representative composite symbols. Composite symbols can be devised for any fluid power component by combining basic symbols.

Simplified symbols are shown for commonly used components.

This standard provides basic symbols which differentiate between hydraulic and pneumatic fluid power media.

Purpose

The purpose of this standard is:

To provide a system of fluid power graphic symbols for industrial and educational purposes.

To simplify design, fabrication, analysis, and service of fluid power circuits.

To provide fluid power graphic symbols which are internationally recognized.

To promote universal understanding of fluid power systems.

1.3 **Terms and Definitions**

Terms and corresponding definitionss found in this standard are listed in **Ref. 8**.

2. Symbol Rules (See Section 10)

2.1 Symbols show connections, flow paths, and functions of components represented. They indicate conditions occurring during transition from one flow path arrangement to another. Symbols do not indicate construction, nor do they indicate values, such as pressure, flow rate, and other component settings.

2.2 Symbols do not indicate locations of ports, direction of shifting of spools, or positions of actuators on actual component.

2.3 Symbols may be rotated or reversed without altering their meaning except in the cases of: a.) Lines to Reservoir, 4.1.1; b.) Vented Manifold, 4.1.2.3; c.) Accumulator, 4.2.

2.4 Line Technique
Keep line widths approximately equal. Line width does not alter meaning of symbols.

2.4.1 Solid Line

(Main line conductor, outline, and shaft)

2.4.2 Dash Line

(Pilot line for control)

2.4.3 Dotted Line

(Exhaust or Drain Line)

2.4.4 Center Line

(Enclosure outline)

2.4.5 Lines Crossing
(The intersection is not necessarily at a 90 deg angle.)

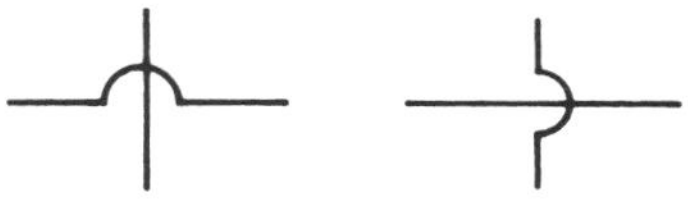

or

2.4.6 Lines Joining

or

2.5 Basic symbols may be shown any suitable size. Size may be varied for emphasis or clarity. Relative sizes should be maintained. (As in the following example.)

2.5.1 Circle and Semi-Circle

2.5.1.1 Large and small circles may be used to signify that one component is the "main" and the other the auxiliary.

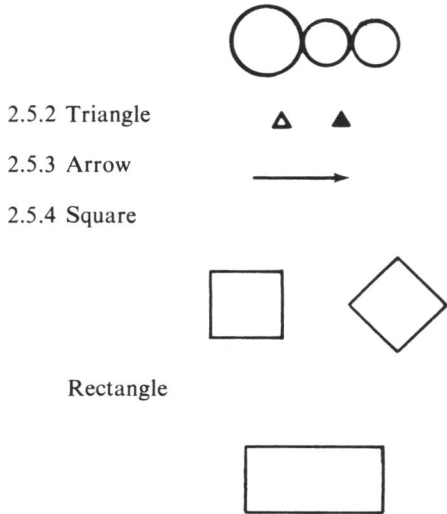

2.5.2 Triangle

2.5.3 Arrow

2.5.4 Square

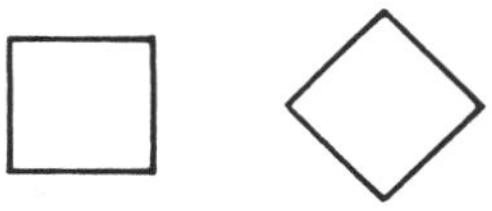

Rectangle

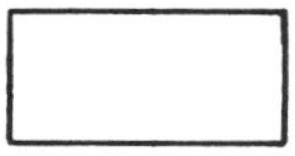

2.6 Letter combinations used as parts of graphic symbols are not necessarily abbreviations.

2.7 In multiple envelope symbols, the flow condition shown nearest an actuator symbol takes place when that control is caused or permitted to actuate.

2.8 Each symbol is drawn to show normal, at-rest, or neutral condition of component unless multiple diagrams are furnished showing various phases of circuit operation. Show an actuator symbol for each flow path condition possessed by the component.

2.9 An arrow through a symbol at approximately 45 degrees indicates that the component can be adjusted or varied.

2.10 An arrow parallel to the short side of a symbol, within the symbol, indicates that the component is pressure compensated.

2.11 A line terminating in a dot to represent a thermometer is the symbol for temperature cause or effect.

See Temperature Controls 7.9, Temperature Indicators and Recorders 9.1.2, and Temperature Compensation 10.16.3 and 4.

Figure 2.17. Graphic symbols for fluid power diagrams. Reprinted from the ANSI Y.32. 10-1967 published by the American Society of Mechanical Engineers (ASME).

Figure 2.17 continued

Figure 2.17 continued

2.12 External ports are located where flow lines connect to basic symbol, except where component enclosure symbol is used.

External ports are located at intersections of flow lines and component enclosure symbol when enclosure is used.

2.13 Rotating shafts are symbolized by an arrow which indicates direction of rotation (assume arrow on near side of shaft).

3. Conductor, Fluid

3.1 Line, Working (main)

3.2 Line, Pilot (for control)

3.3 Line, Exhaust and Liquid Drain

3.4 Line, sensing, etc. such as gage lines shall be drawn the same as the line to which it connects.

3.5 Flow, Direction of

3.5.1 Pneumatic

3.5.2 Hydraulic

3.6 Line, Pneumatic
Outlet to Atmosphere

3.6.1 Plain orifice, unconnectable

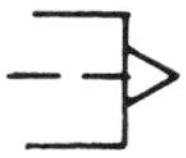

3.6.2 Connectable orifice (e.g. Thread)

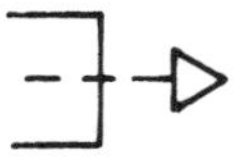

3.7 Line with Fixed Restriction

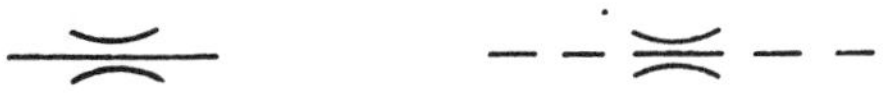

3.8 Line, Flexible

3.9 Station, Testing, measurement, or power take-off

3.9.1 Plugged part

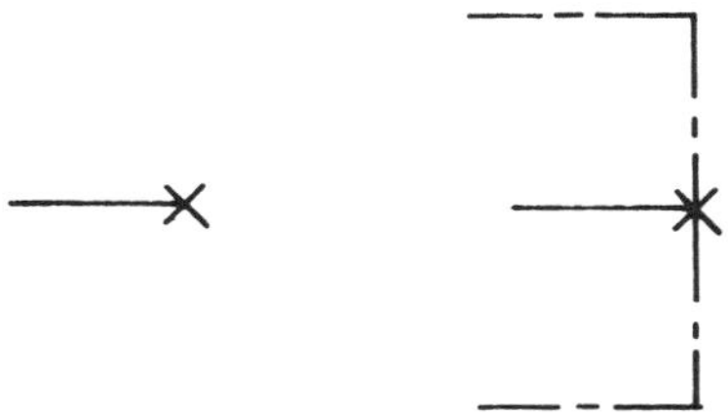

3.10 Quick Disconnect

3.10.1 Without Checks

Connected

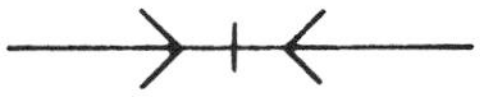

Disconnected

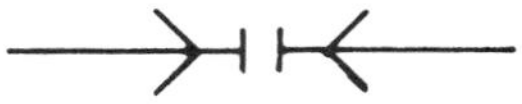

3.10.2 With Two Checks

Connected

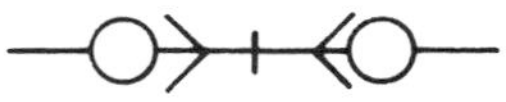

Disconnected

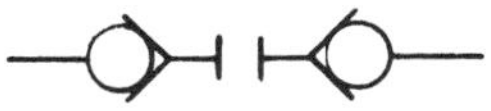

3.10.3 With One Check

Connected

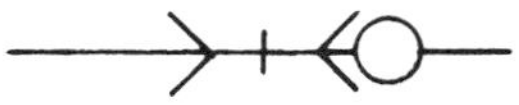

Disconnected

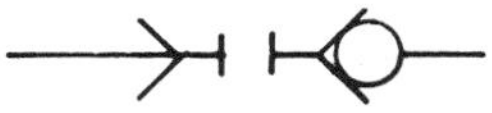

3.11 Rotating Coupling

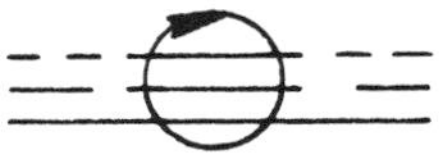

Figure 2.17 continued

Figure 2.17 continued

4. Energy Storage and Fluid Storage

4.1 Reservoir

Vented

Pressurized

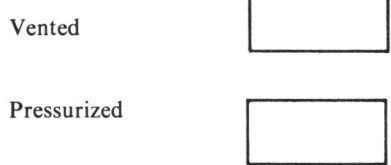

Note: Reservoirs are conventionally drawn in the horizontal plane. All lines enter and leave from above. Examples:

4.1.1 Reservoir with Connecting Lines

Above Fluid Level

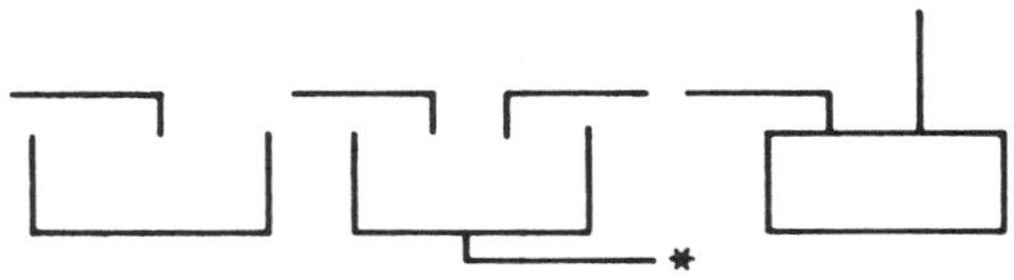

Below Fluid Level

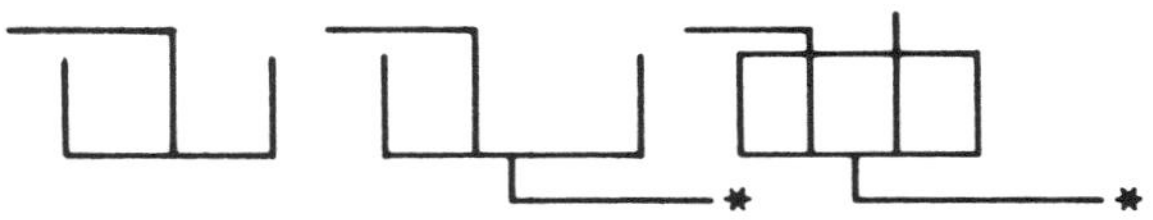

4.1.2 Simplified symbol

The symbols are used as part of a complete circuit. They are analogous to the ground symbol of electrical diagrams. 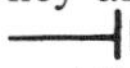 Several such symbols may be used in one diagram to represent the same reservoir.

4.1.2.1 Below Fluid Level

4.1.2.2 Above Fluid Level

(The return line is drawn to terminate at the upright legs of the tank symbol.)

* Show line entering or leaving below reservoir only when such bottom connection is essential to circuit function.

4.1.2.3 Vented Manifold

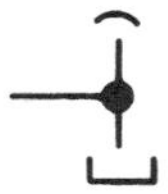

4.2 Accumulator

4.2.1 Accumulator, Spring Loaded

4.2.2 Accumulator, Gas Charged

4.2.3 Accumulator, Weighted

4.3 Receiver, for Air or Other Gases

4.4 Energy Source
(Pump, Compressor, Accumulator, etc.)

This symbol may be used to represent a fluid power source which may be a pump, compressor, or another associated system.

Hydraulic 

Pneumatic 

Simplified Symbol

Example:

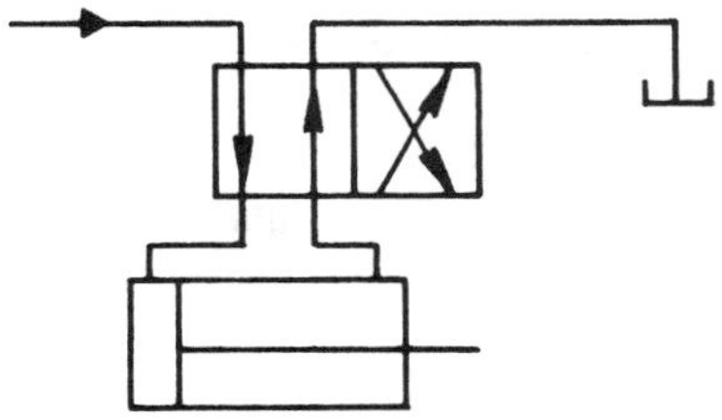

Figure 2.17 continued

Figure 2.17 continued

5. Fluid Conditioners

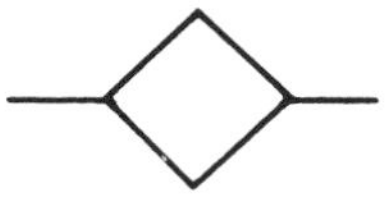

Devices which control the physical characteristics of the fluid.

5.1 Heat Exchanger

5.1.1 Heater

Inside triangles indicate the introduction of heat.

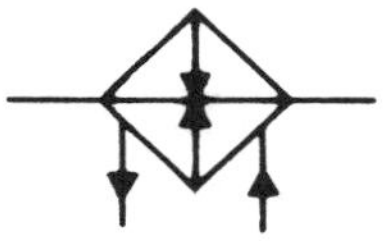

Outside triangles show the heating medium is liquid.

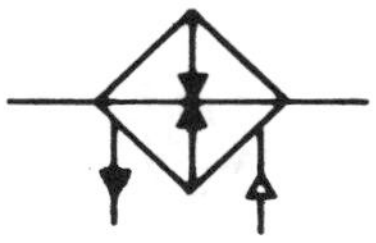

Outside triangles show the heating medium is gaseous.

5.1.2 Cooler

or

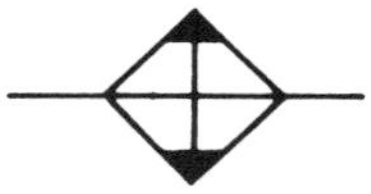

Inside triangles indicate heat dissipation

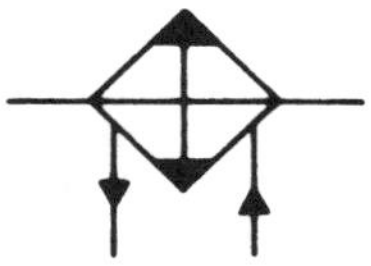

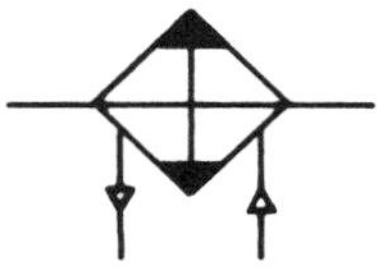

(Corners may be filled in to represent triangles.)

5.1.3 Temperature Controller
(The temperature is to be maintained between two predetermined limits.)

or

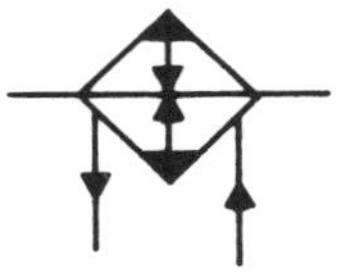

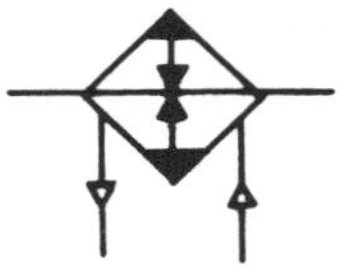

5.2 Filter—Strainer

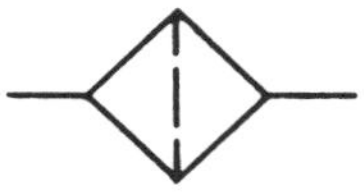

5.3 Separator

5.3.1 With Manual Drain

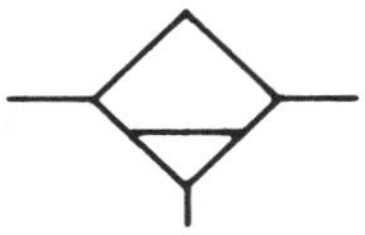

5.3.2 With Automatic Drain

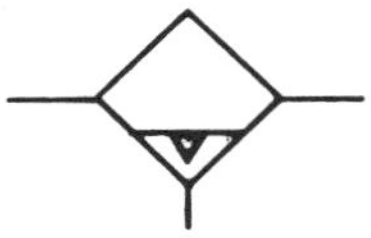

5.4 Filter—Separator

5.4.1 With Manual Drain

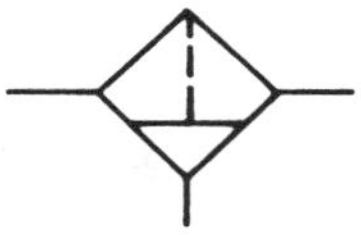

5.4.2 With Automatic Drain

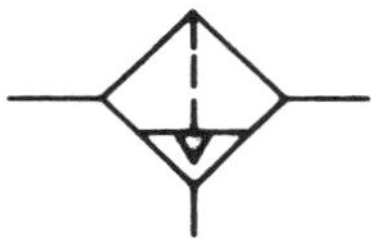

5.5 Dessicator (Chemical Dryer)

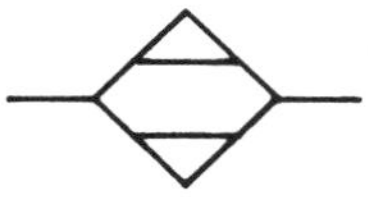

5.6 Lubricator

5.6.1 Less Drain

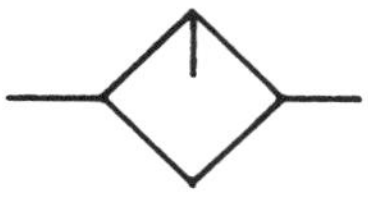

Figure 2.17 continued

Figure 2.17 continued

5.6.2 With Manual Drain

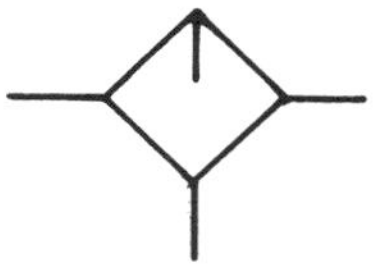

6. Linear Devices

6.1 Cylinders, Hydraulic & Pneumatic

6.1.1 Single Acting

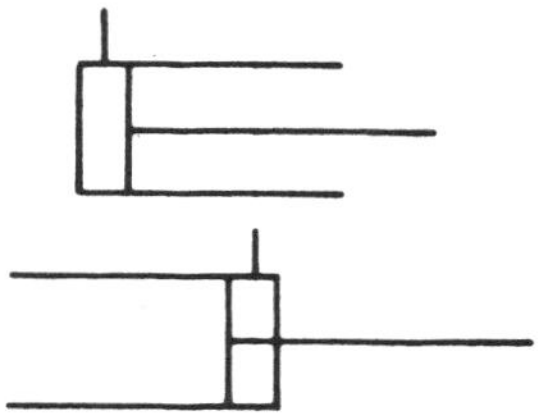

6.1.2 Double Acting

6.1.2.1 Single End Rod

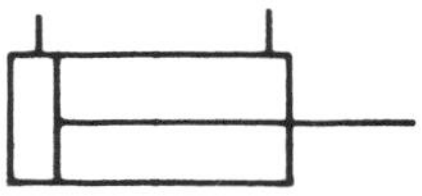

6.1.2.2 Double End Rod

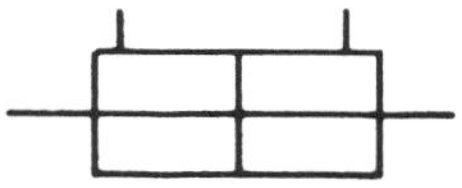

6.1.2.3 Fixed Cushion, Advance & Retract

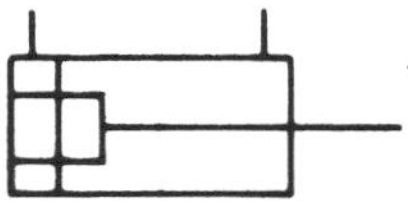

6.1.2.4 Adjustable Cushion, Advance Only

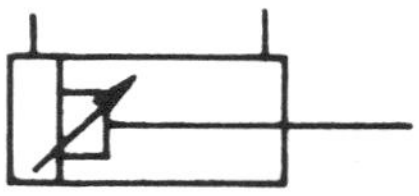

6.1.2.5 Use these symbols when diameter of rod compared to diameter of bore is significant to circuit function.

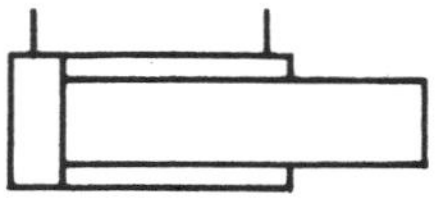

(Non-Cushion)

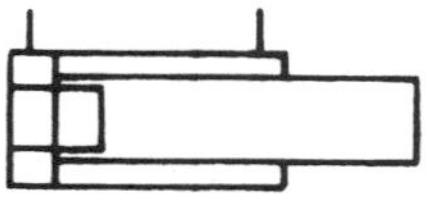

(Cushion, Advance & Retract)

6.2 Pressure Intensifier

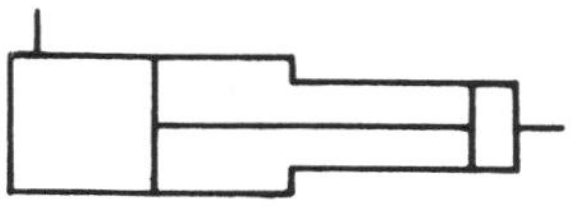

6.3 Servo Positioner (Simplified)

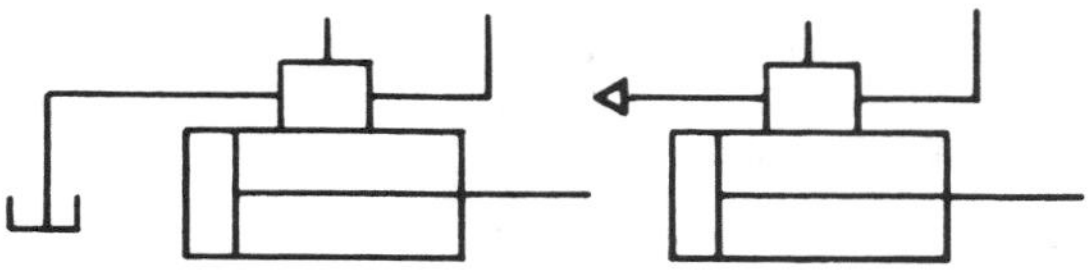

Hydraulic Pneumatic

6.4 Discrete Positioner

Combine two or more basic cylinder symbols.

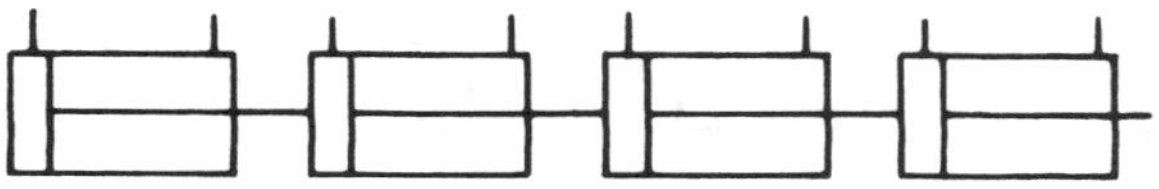

7. Actuators and Controls

7.1 Spring

7.2 Manual

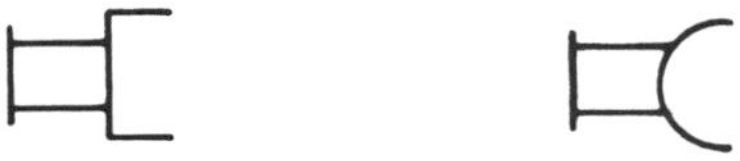

(Use as general symbol without indication of specific type; i.e., foot, hand, leg, arm.)

7.2.1 Push Button

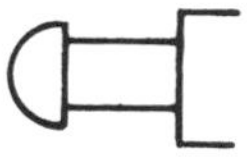

7.2.2 Lever

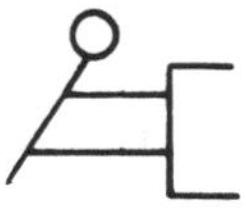

7.2.3 Pedal or Treadle

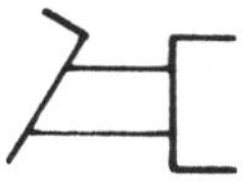

Figure 2.17 continued

Figure 2.17 continued

7.3 Mechanical

7.4 Detent

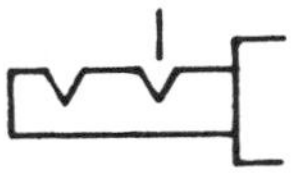

(Show a notch for each detent in the actual component being symbolized. A short line indicates which detent is in use.) Detent may, for convenience, be positioned on either end of symbol.

7.5 Pressure Compensated

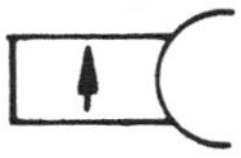

7.6 Electrical

7.6.1 Solenoid (Single Winding)

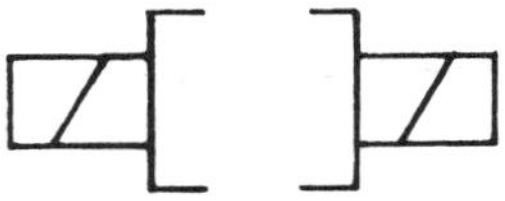

7.6.2 Reversing Motor

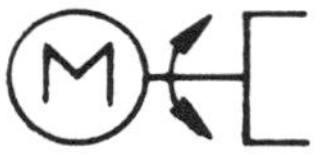

7.7 Pilot Pressure

7.7.1

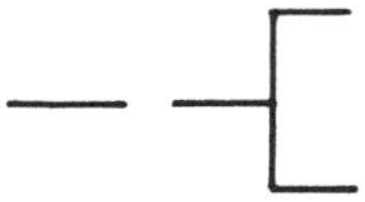

Remote Supply

7.7.2

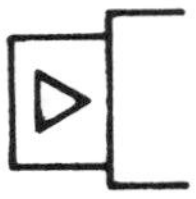

Internal Supply

7.7.3 Actuation by Released Pressure

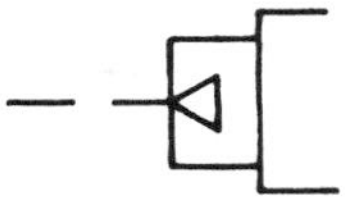

by Remote Exhaust

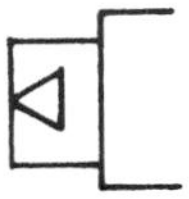

by Internal Return

7.7.4 Pilot Controlled, Spring Centered

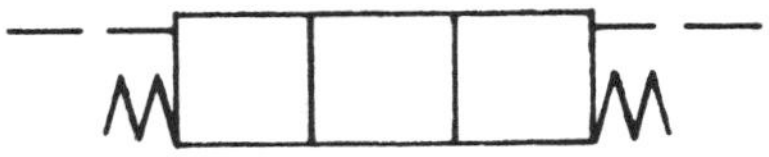

Simplified Symbol

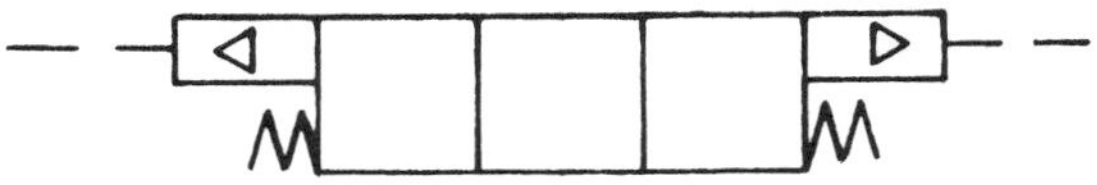

Complete Symbol

7.7.5 Pilot Differential

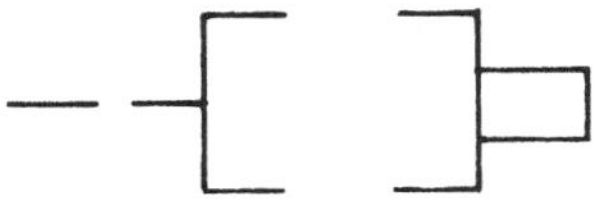

Simplified Symbol

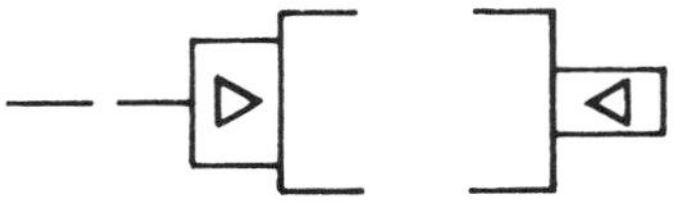

Complete Symbol

7.8 Solenoid Pilot

7.8.1 Solenoid *or* Pilot

External Pilot Supply

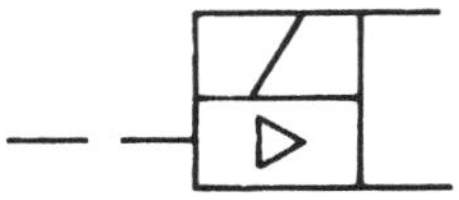

Internal Pilot Supply and Exhaust

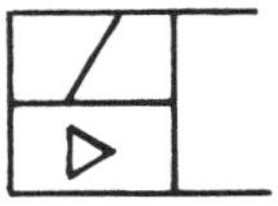

7.8.2 Solenoid *and* Pilot

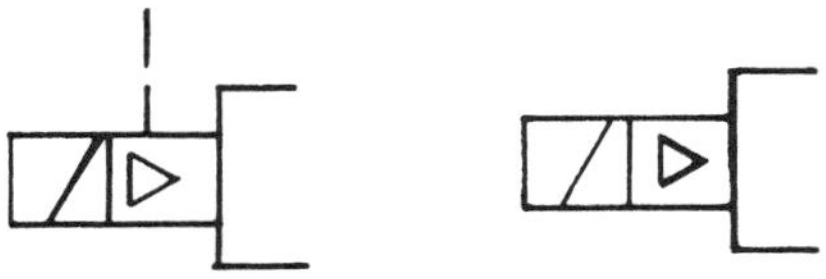

7.9 Thermal

A mechanical device responding to thermal change.

7.9.1 Local Sensing

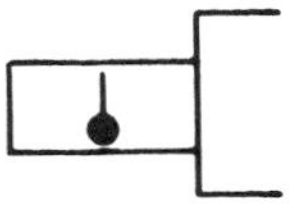

Figure 2.17 continued

Figure 2.17 continued

7.9.2 With Bulb for Remote Sensing

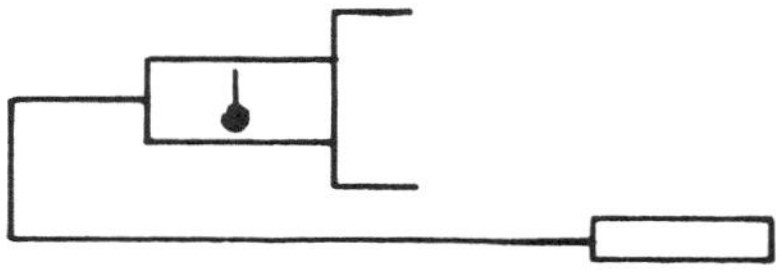

7.10 Servo

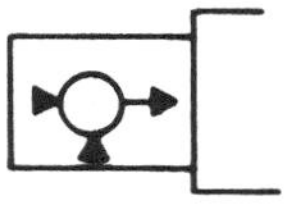

(This symbol contains representation for energy input, command input, and resultant output.)

7.11 Composite Actuators (and, or, and/or)

Basic One Signal only causes the device to operate.

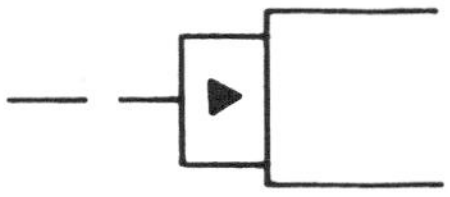

And One signal *and* a second signal both cause the device to operate.

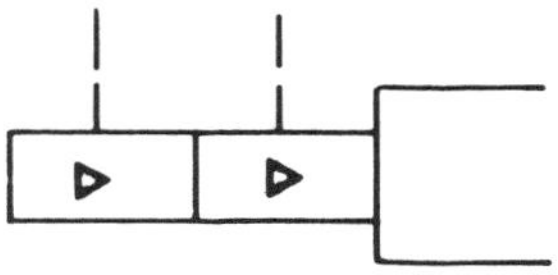

Or One signal *or* the other signal causes the device to operate.

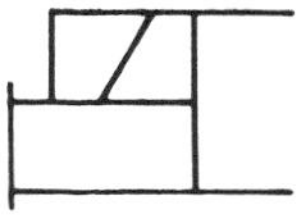

And/Or The solenoid *and* the pilot *or* the manual override alone causes the device to operate.

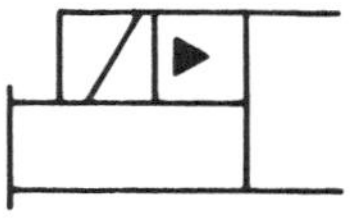

The solenoid *and* the pilot *or* the manual override *and* the pilot.

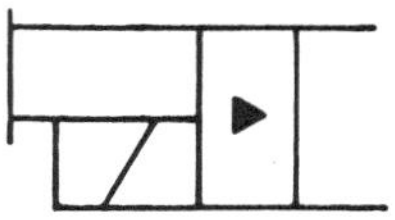

The solenoid *and* the pilot *or* a manual override *and* the pilot *or* a manual override alone.

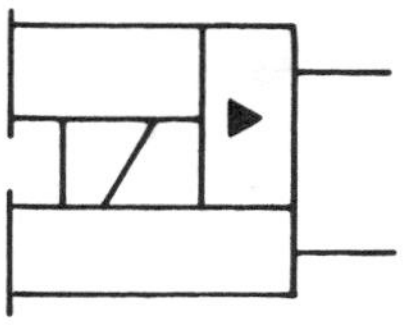

8. Rotary Devices

8.1 Basic Symbol

8.1.1 With Ports

8.1.2 With Rotating Shaft, with control, and with Drain

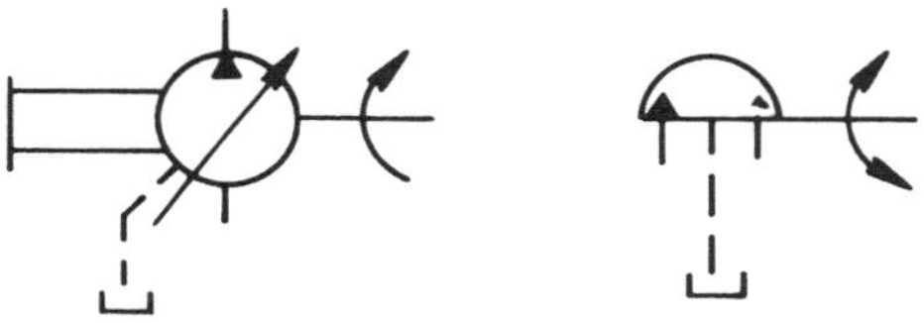

8.2 Hydraulic Pump

8.2.1 Fixed Displacement

8.2.1.1 Unidirectional

8.2.1.2 Bidirectional

8.2.2 Variable Displacement, Non-Compensated

8.2.2.1 Unidirectional

Simplified

Figure 2.17 continued

Figure 2.17 continued

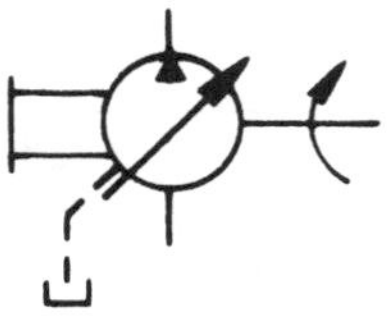

Complete

8.2.2.2 Bidirectional

Simplified

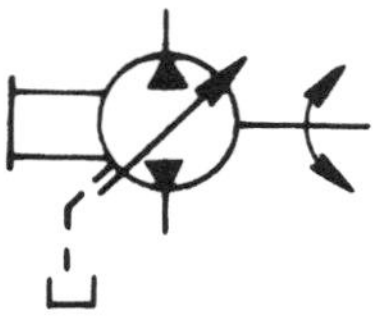

Complete

8.2.3 Variable Displacement, Préssure Compensated

8.2.3.1 Unidirectional

Simplified

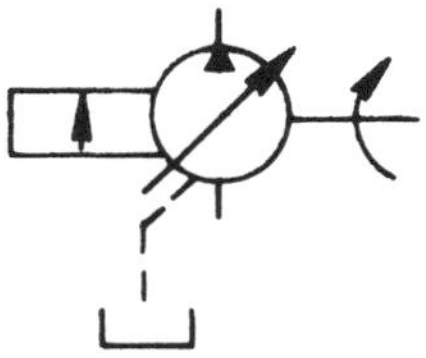

Complete

8.2.3.2 Bidirectional

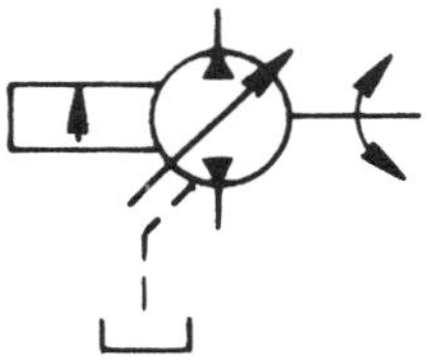

Simplified Complete

8.3 Hydraulic Motor

8.3.1 Fixed Displacement

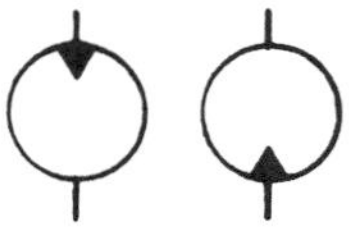

8.3.1.2 Bidirectional

8.3.2 Variable Displacement

8.3.2.1 Unidirectional

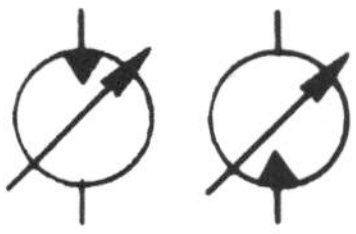

8.3.2.2 Bidirectional

8.4 Pump-Motor, Hydraulic

8.4.1 Operating in one direction as a pump. Operating in the other direction as a motor.

8.4.1.1 Complete Symbol

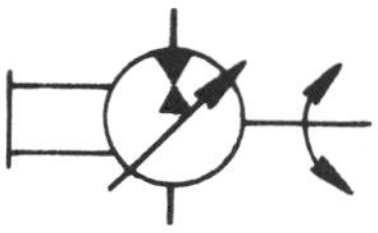

8.4.1.2 Simplified Symbol

8.4.2 Operating one direction of flow as either a pump or as a pump. Operating in the other direction as a motor.

8.4.2.1 Complete Symbol

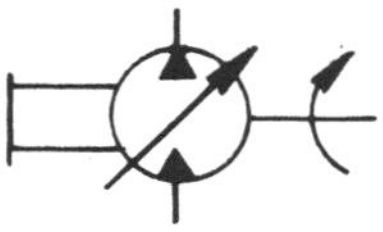

8.4.2.2 Simplified Symbol

8.4.3. Operating one direction of flow as either a pump or as a motor.

8.4.3.1 Complete Symbol

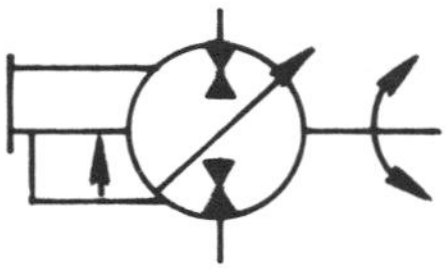

Figure 2.17 continued

Figure 2.17 continued

8.4.3.2 Simplified Symbol

8.5 Pump, Pneumatic

8.5.1 Compressor, Fixed Displacement

8.5.2 Vacuum Pump, Fixed Displacement

8.6 Motor, Pneumatic

8.6.1 Unidirectional

8.6.2 Bidirectional

8.7 Oscillator

8.7.1 Hydraulic

8.7.2 Pneumatic

8.8 Motors, Engines

8.8.1 Electric Motor

8.8.2 Heat Engine (E.G. internal combustion engine.)

9. Instruments and Accessories

9.1 Indicating and Recording

9.1.1 Pressure

9.1.2 Temperature

9.1.3 Flow Meter

9.1.3.1 Flow Rate

9.1.3.2 Totalizing

9.2 Sensing

9.2.1 Venturi

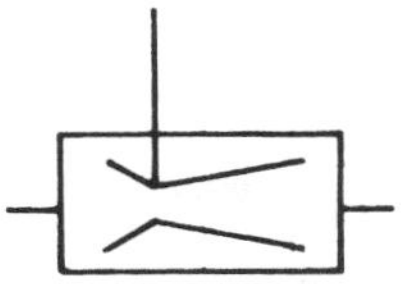

9.2.2 Orifice Plate

9.2.3 Pitot Tube

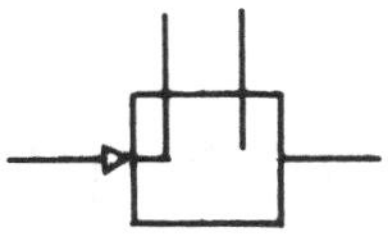

Figure 2.17 continued

Figure 2.17 continued

9.2.4 Nozzle

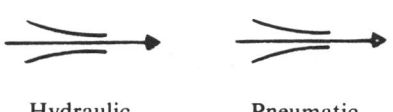

Hydraulic Pneumatic

9.3 Accessories

9.3.1 Pressure Switch

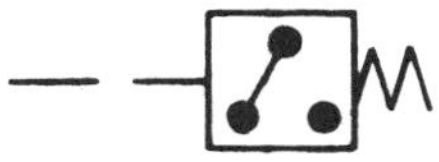

9.3.2 Muffler

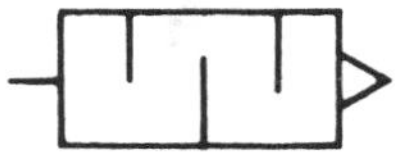

10. Valves

A basic valve symbol is composed of one or more envelopes with lines inside the envelope to represent flow paths and flow conditions between ports. Three symbol systems are used to represent valve types: single envelope, both finite and infinite position; multiple envelope, finite position; and multiple envelope, infinite position.

10.1 In infinite position single envelope valves, the envelope is imagined to move to illustrate how pressure or flow conditions are controlled as the valve is actuated.

10.2 Multiple envelopes symbolize valves providing more than one finite flow path option for the fluid. The multiple envelope moves to represent how flow paths change when the valving element within the component is shifted to its finite positions.

10.3 Multiple envelope valves capable of infinite positioning between certain limits are symbolized as in 10.2 above with the addition of horizontal bars which are drawn parallel to the envelope. The horizontal bars are the clues to the infinite positioning function possessed by the valve re-represented.

10.4 Envelopes

10.5 Ports

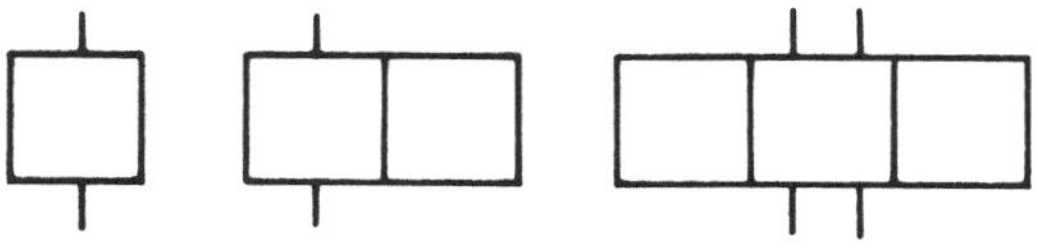

10.6 Ports, Internally Blocked

Symbol System 10.1

Symbol System 10.2

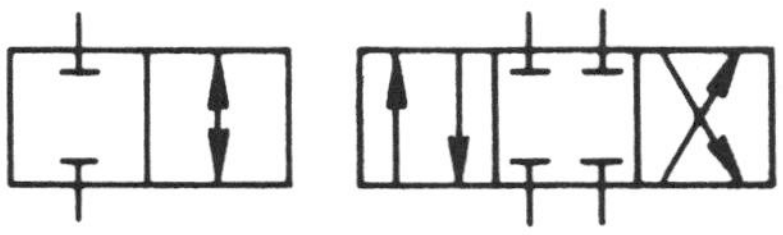

10.7 Flow Paths, Internally Open (Symbol System 10.1 and 10.2)

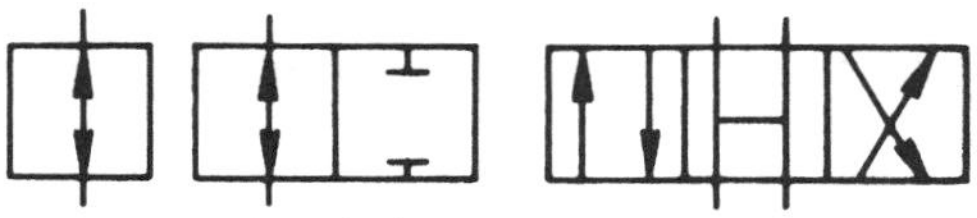

Symbol System 10.1 Symbol System 10.2

10.8 Flow Paths, Internally Open (Symbol System 10.3)

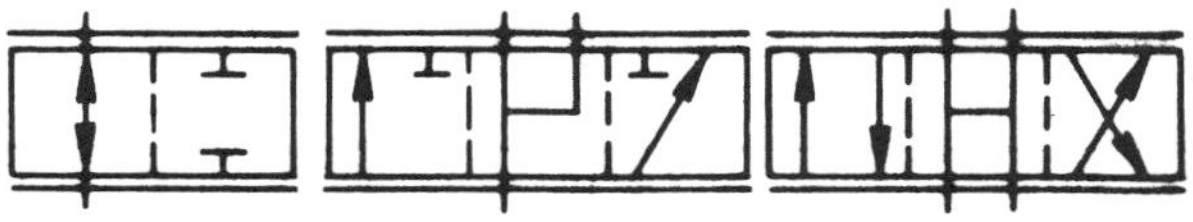

10.9 Two-Way Valves (2 Ported Valves)

10.9.1 On-Off (Manual Shut-Off)

Simplified

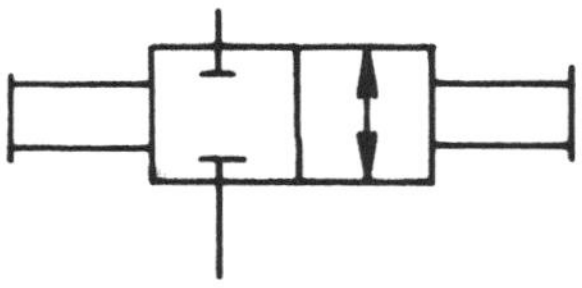

Off

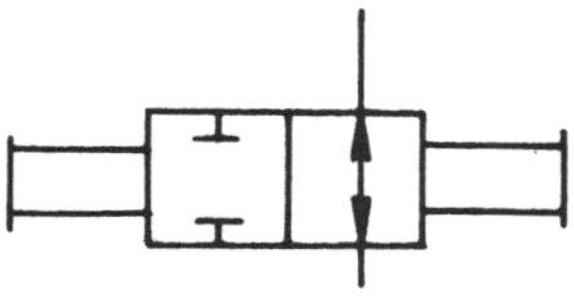

On

Figure 2.17 continued

Figure 2.17 continued

10.9.2 Check

Simplified Symbol

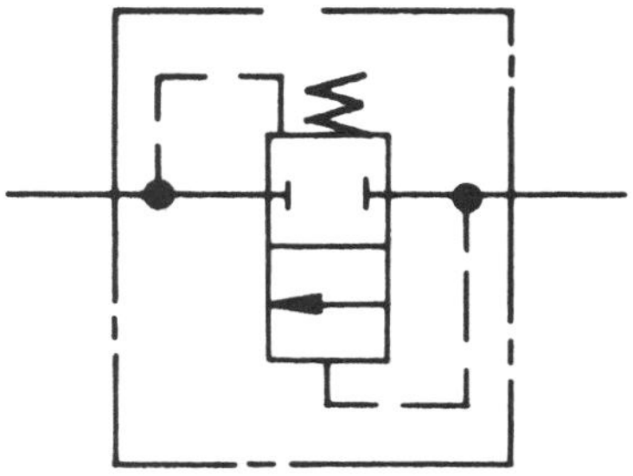

Flow to the right is blocked. Flow to the left is permitted.

(Composite Symbol)

10.9.3 Check, Pilot-Operated to Open

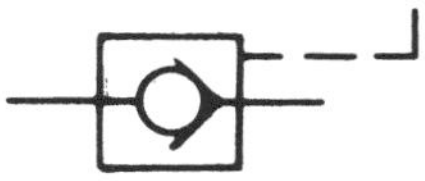

10.9.4 Check, Pilot-Operated to Close

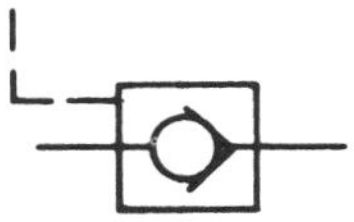

10.9.5 Two-Way Valves

10.9.5.1 Two-Position

Normally Closed

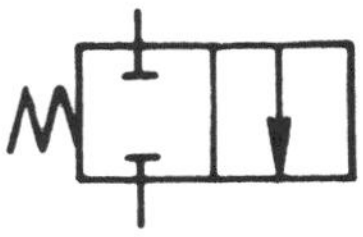

Normally Open

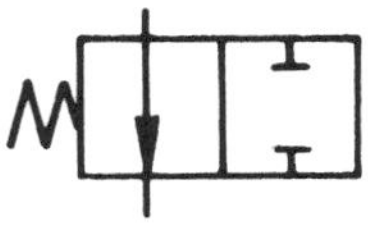

10.9.5.2 Infinite Position

Normally Closed

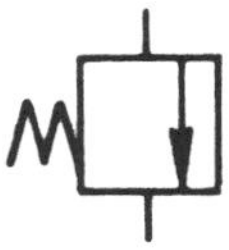

Normally Open

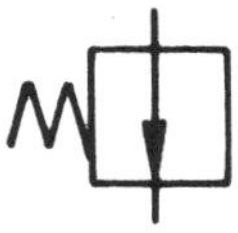

10.10 Three-Way Valves

10.10.1 Two-Position

10.10.1.1 Normally Open

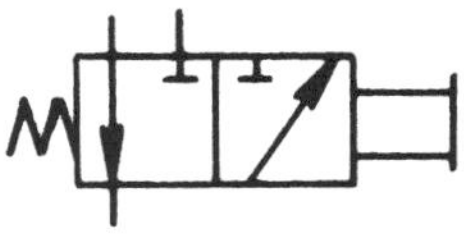

10.10.1.2 Normally Closed

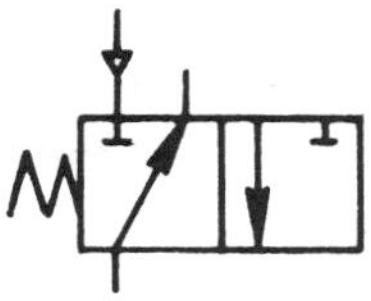

10.10.1.3 Distributor (Pressure is distributed first to one port, then the other)

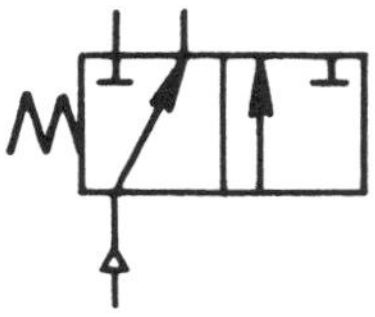

10.10.1.4 Two-Pressure

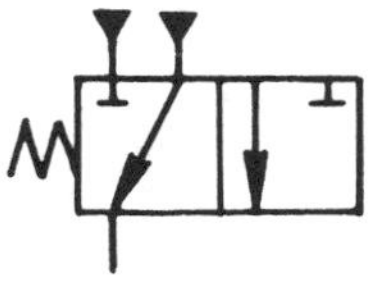

10.10.2 Double Check Valve

Double check valves can be built with and without "cross bleed". Such valves with two poppets do not usually allow pressure to momentarily "cross bleed" to return during transition. Valves with one poppet may allow "cross bleed" as these symbols illustrate.

10.10.2.1 Without Cross Bleed (One Way Flow)

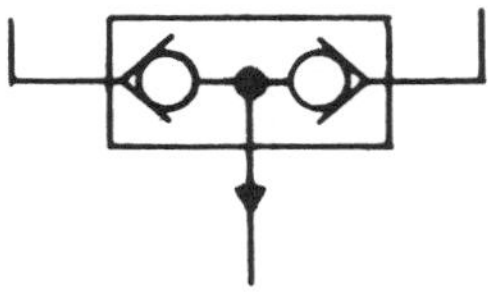

10.10.2.2 With Cross Bleed (Reverse Flow Permitted)

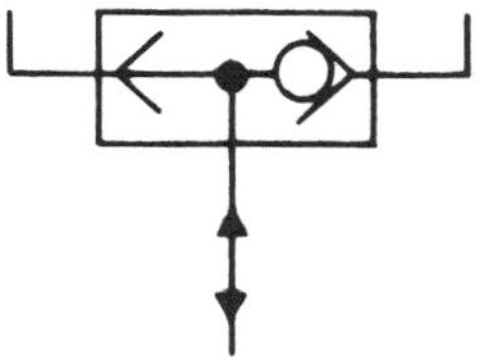

10.11 Four-Way Valves

10.11.1 Two Position

Normal

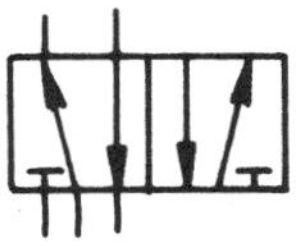

Figure 2.17 continued

Figure 2.17 continued

Actuated

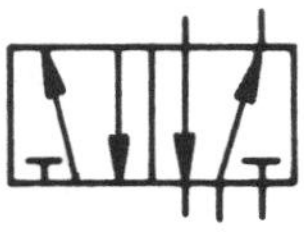

10.11.2 Three Position

(a) Normal

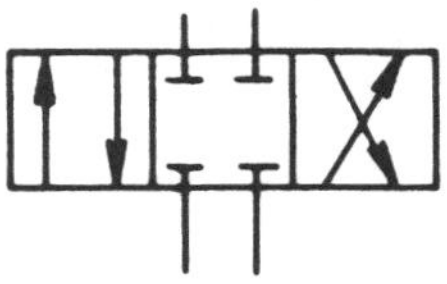

(b) Actuated Left

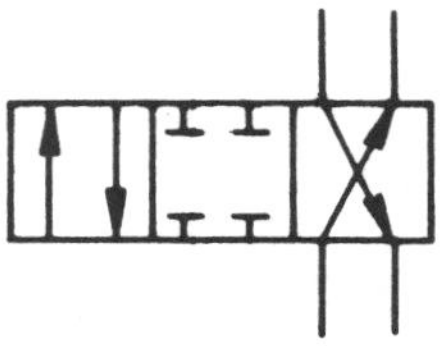

(c) Actuated Right

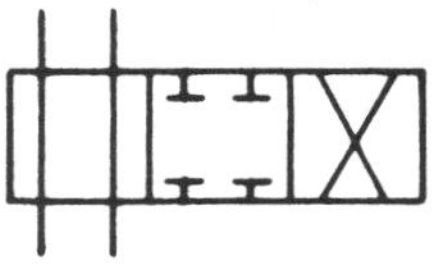

10.11.3 Typical Flow Paths for Center Condition of Three Position Valves

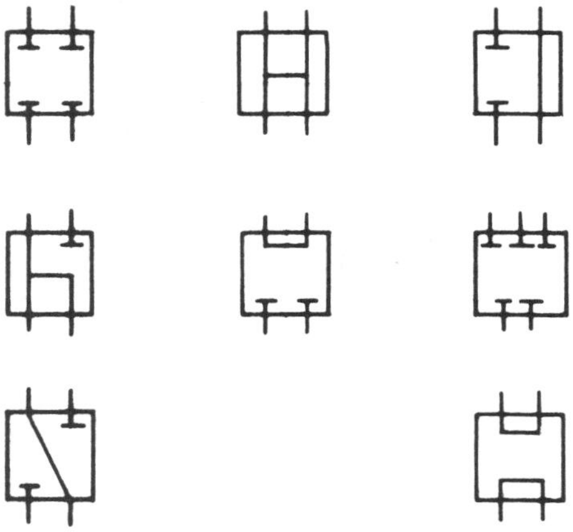

10.11.4 Two-Position, Snap Action with Transition.

As the valve element shifts from one position to the other, it passes through an intermediate position. If it is essential to circuit function to symbolize this "in transit" condition, it can be shown in the center position, enclosed by dashed lines.

Typical Transition Symbol

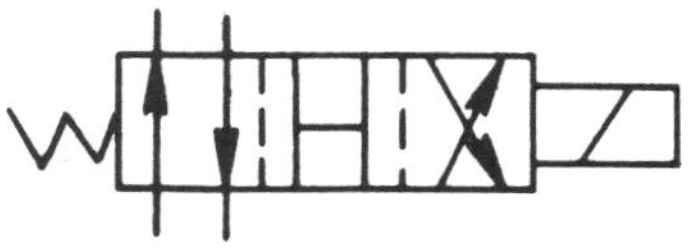

10.12 Infinite Positioning (Between Open & Closed)

10.12.1 Normally Closed

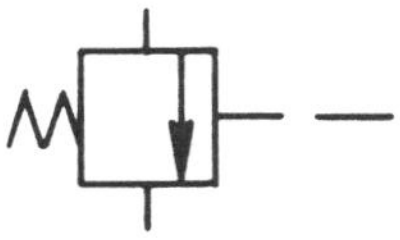

10.12.2 Normally Open

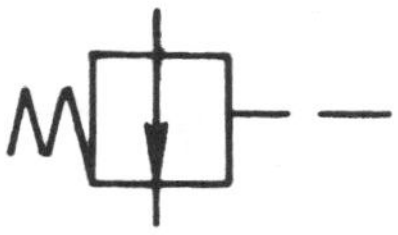

10.13 Pressure Control Valves

10.13.1 Pressure Relief

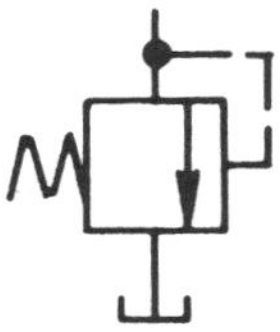

Simplified Symbol Denotes

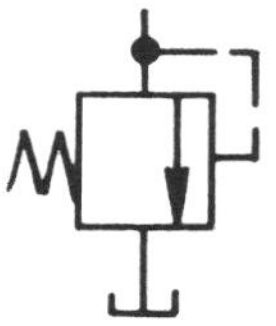

Normal

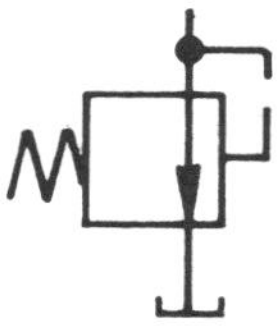

Actuated (Relieving)

10.13.2 Sequence

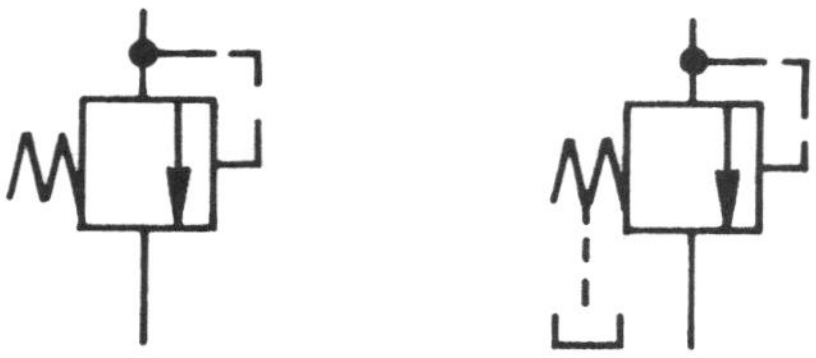

10.13.3 Pressure Reducing

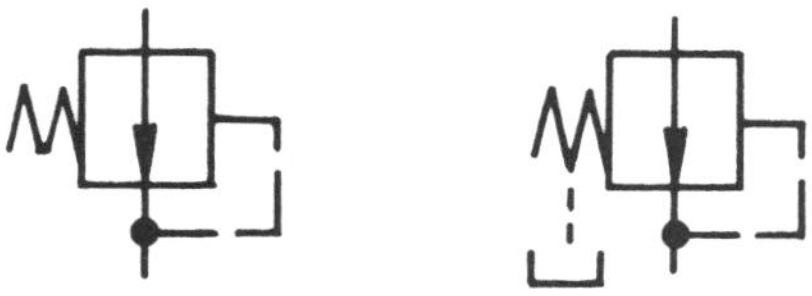

Figure 2.17 continued

Figure 2.17 continued

10.13.4 Pressure Reducing and Relieving

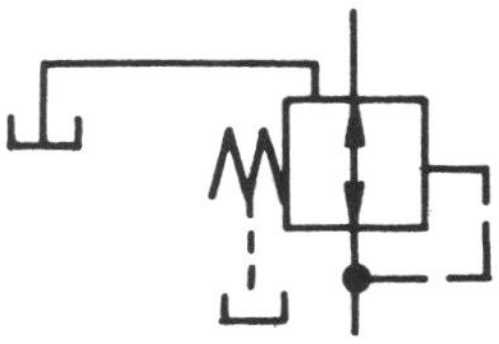

10.13.5 Airline Pressure Regulator (Adjustable, Relieving)

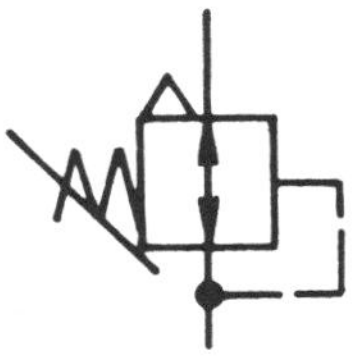

10.14 Infinite Positioning Three-Way Valves

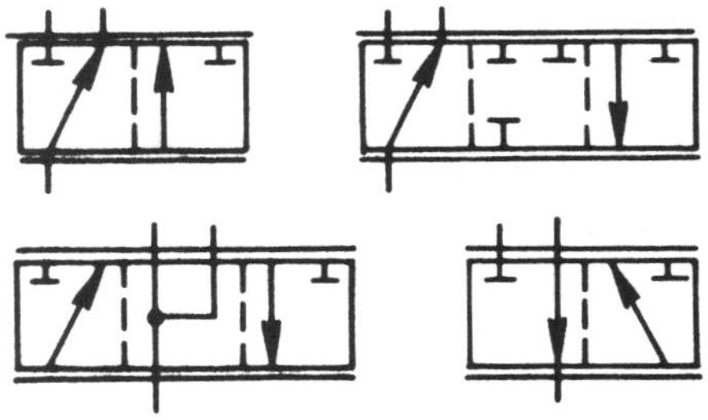

10.15 Infinite Positioning Four-Way Valves

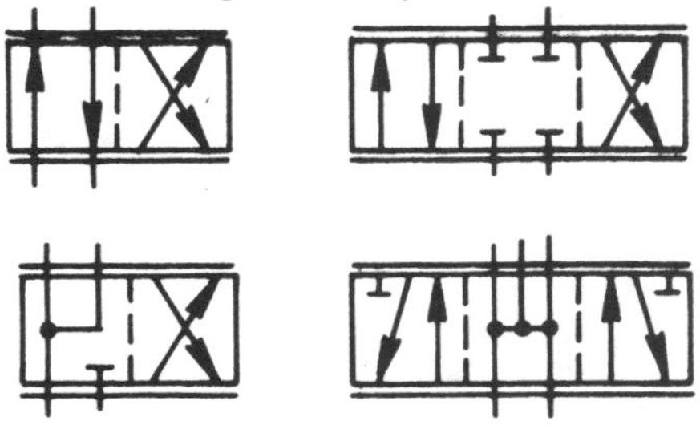

10.16 Flow Control Valves (See 3.7)

10.16.1 Adjustable, Non-Compensated (Flow control in each direction)

10.16.2 Adjustable with Bypass

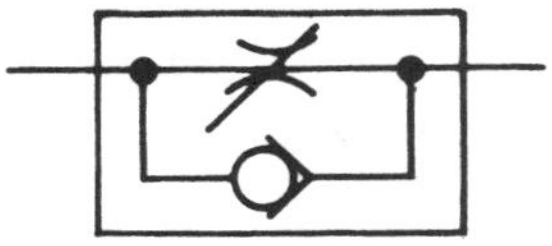

Flow is controlled to the right
Flow to the left bypasses control

10.16.3 Adjustable and Pressure Compensated with Bypass

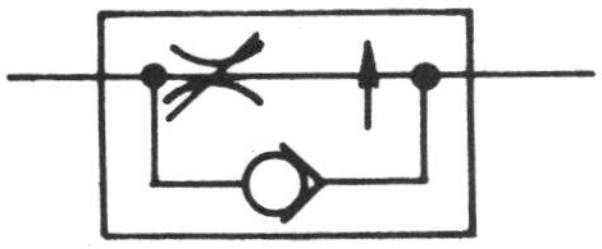

10.16.4 Adjustable, Temperature & Pressure Compensated

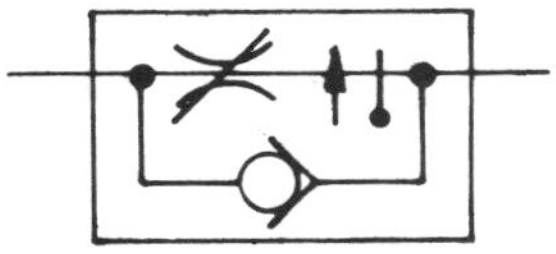

3 Charts

William G. Andrew

Introduction

A chart is defined as a map, a sheet of facts in graphical or tabular form, or a graph showing changes or variations of some quantity in relation to one or more other quantities.

It is in the latter text that most of the following graphs or charts apply. The information contained in many of them can also be obtained in other sections of this volume under *Tables* or *Formulas*. Electrical charts may be found in chapters 9 and 10.

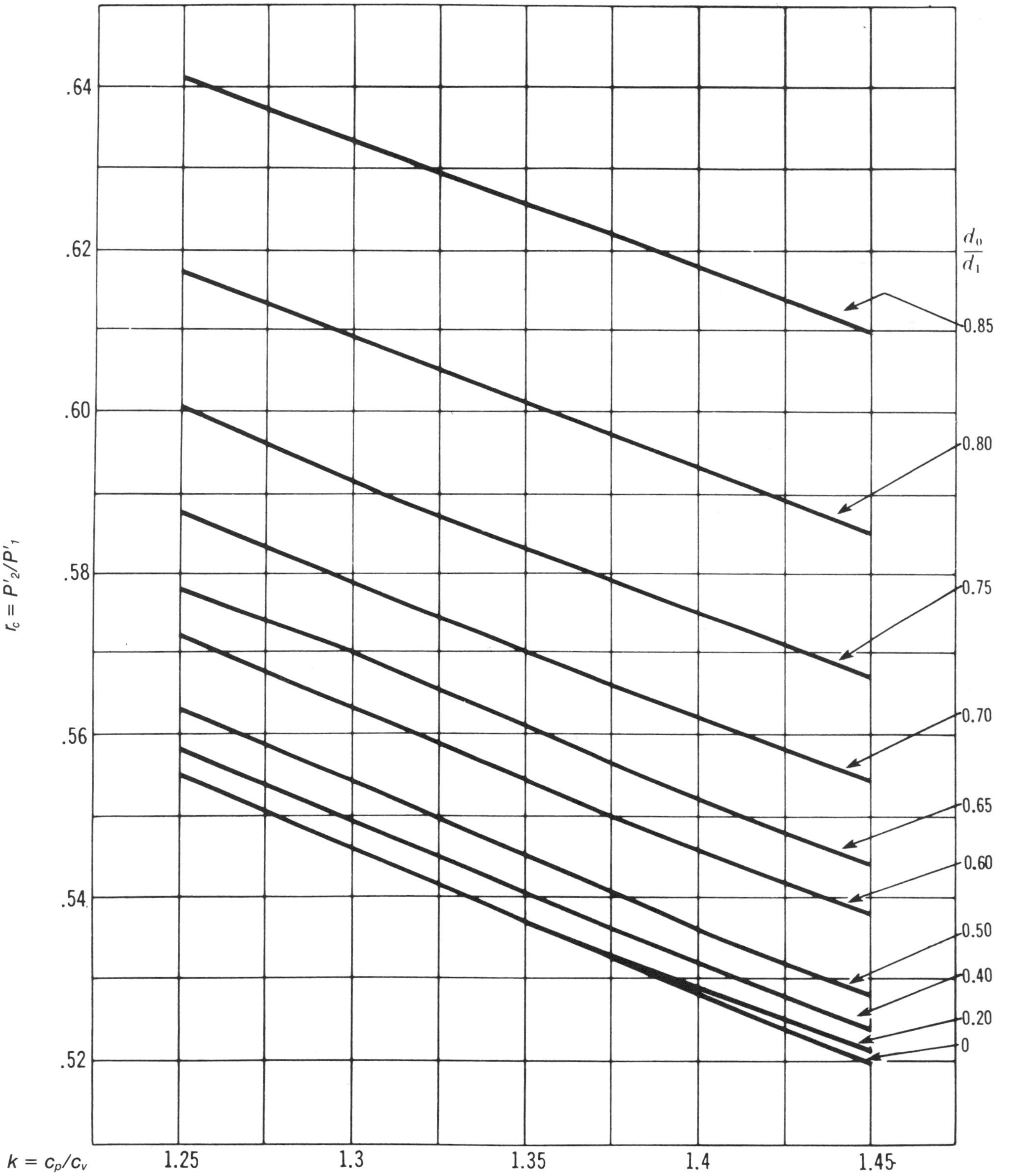

where

c_p = specific heat at constant pressure
c_v = specific heat at constant volume
k = ratio of specific heats

d_o = inside diameter of nozzle or venturi
d_i = inside diameter of pipe
P'_1 = upstream pressure, pounds per square inch absolute
P'_2 = downstream pressure, pounds per square inch absolute

Chart 3.1. Critical pressure ratio, r_c, *for compressible flow through nozzles and venturi tubes. Courtesy Crane Co. (T.P. No. 410).*

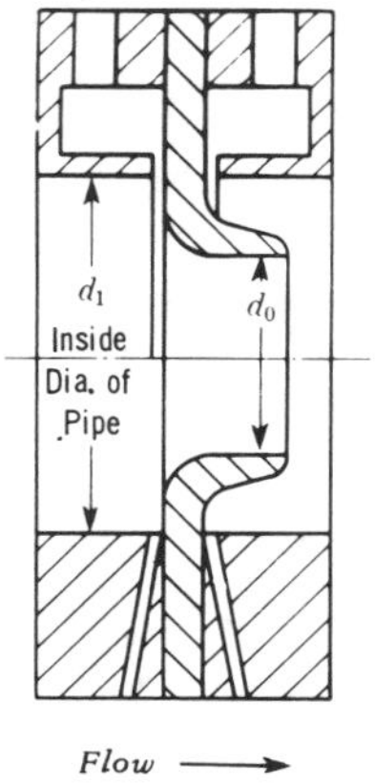

$$C = \frac{C_d}{\sqrt{1 - \left(\frac{d_0}{d_1}\right)^4}}$$

Example: The flow coefficient C for a diameter ratio d_0/d_1 of 0.60 at a Reynolds number of 20,000 (2×10^4) equals 1.01.

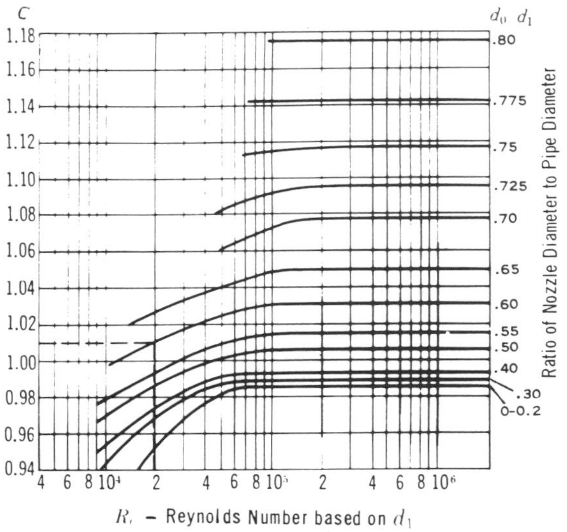

Chart 3.2. Flow coefficient C *for nozzles. Courtesy Crane Co. (T.P. No. 410).*

$$C = \frac{C_d}{\sqrt{1 - \left(\frac{d_0}{d_1}\right)^4}}$$

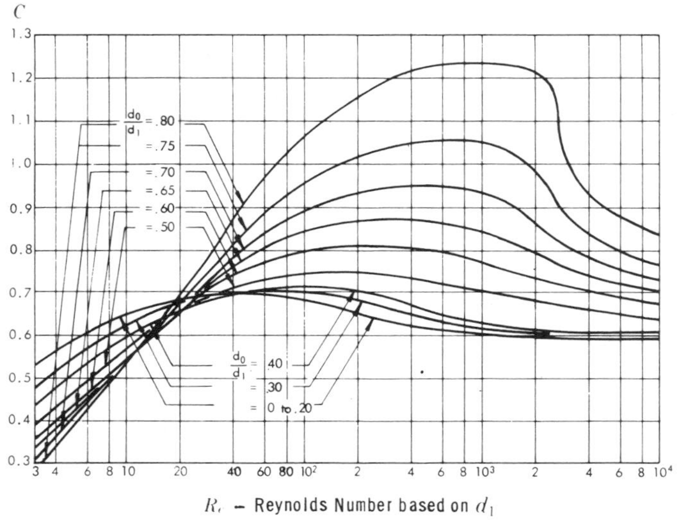

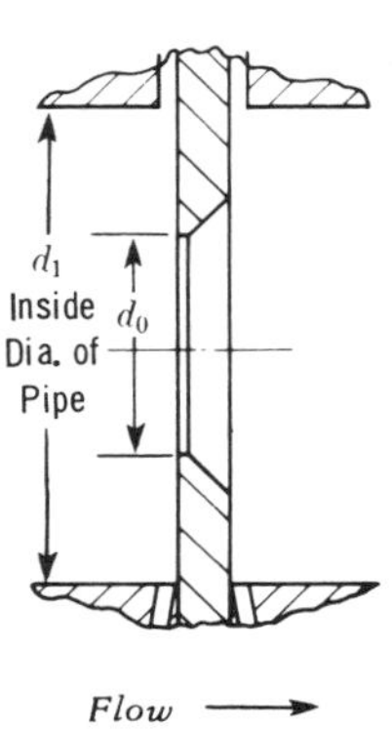

$$C = \frac{C_d}{\sqrt{1 - \left(\frac{d_0}{d_1}\right)^4}}$$

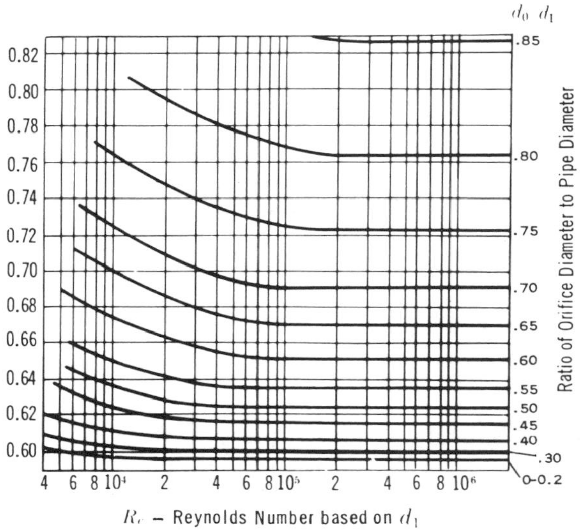

Chart 3.3. Flow coefficient C *for square edged orifices. Courtesy Crane Co. (T.P. No. 410).*

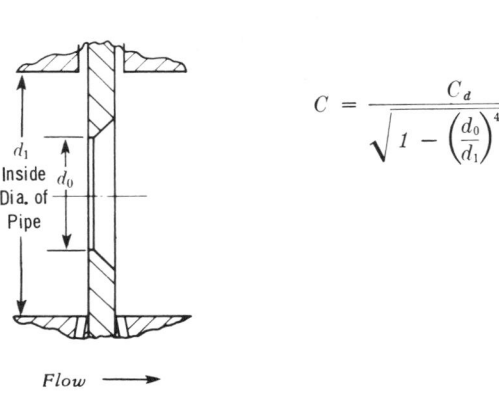

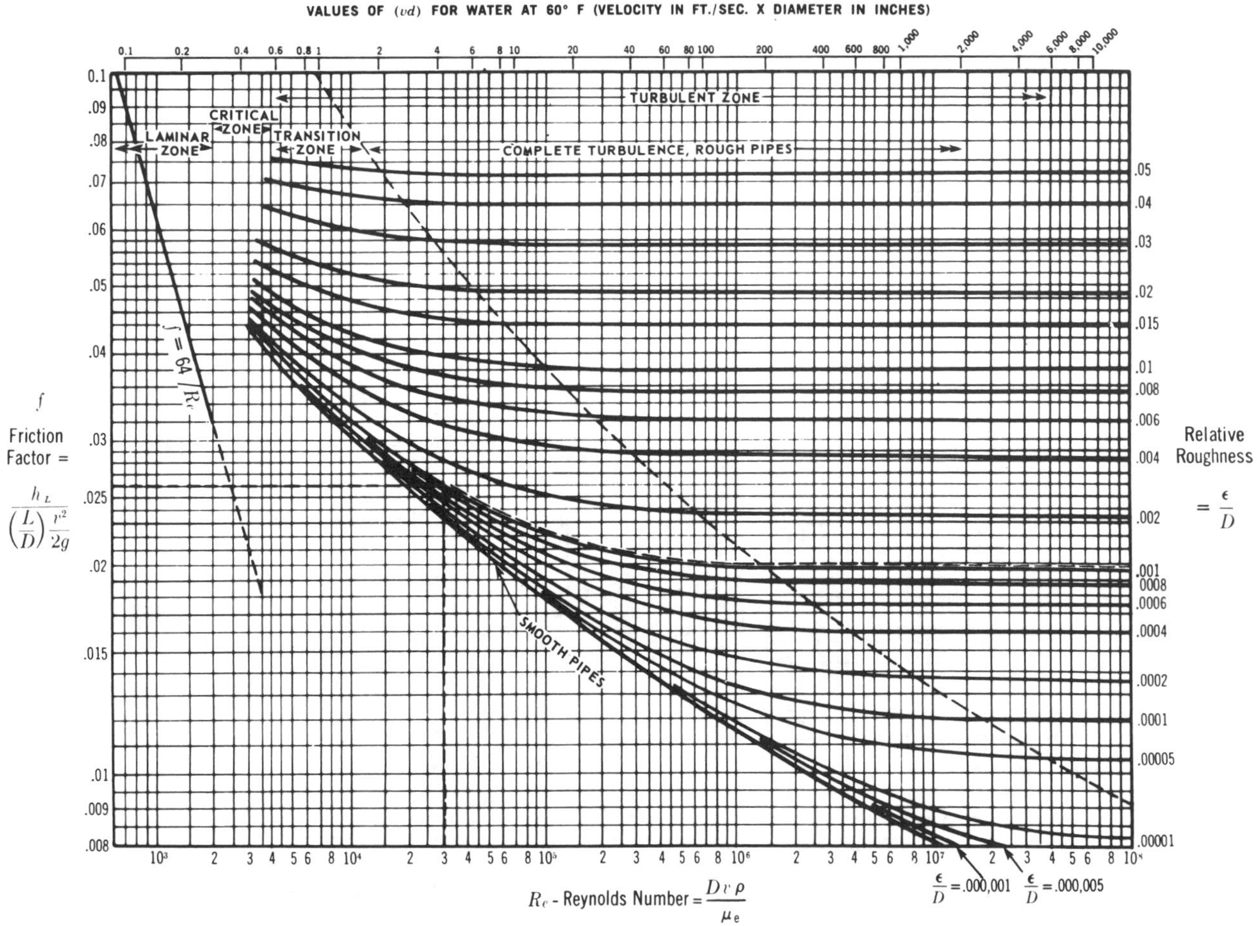

Problem:
Determine the friction factor for 10-inch cast iron pipe (10.16″ I.D.) at a Reynolds number flow of 30,000.

Solution: The relative roughness (see page A-23) is 0.001. Then, the friction factor (f) equals 0.026.

where

f = friction factor in formula $h_L = fLv^2/D2g$
h_L = loss of static pressure head due to fluid flow, in feet of fluid

L = length of pipe, in feet
ϵ = absolute roughness or effective height of pipe wall irregularities in feet
D = internal diameter of pipe, in feet
γ = mean velocity of flow, in feet per second
ρ = weight density of fluid, pounds per cubic feet
μ_e = absolute viscosity, in pounds mass per foot second or poundal seconds per square foot

Chart 3.4. Friction factors for any type of commercial pipe. Courtesy Crane Co. (T.P. No. 410) and the ASME.

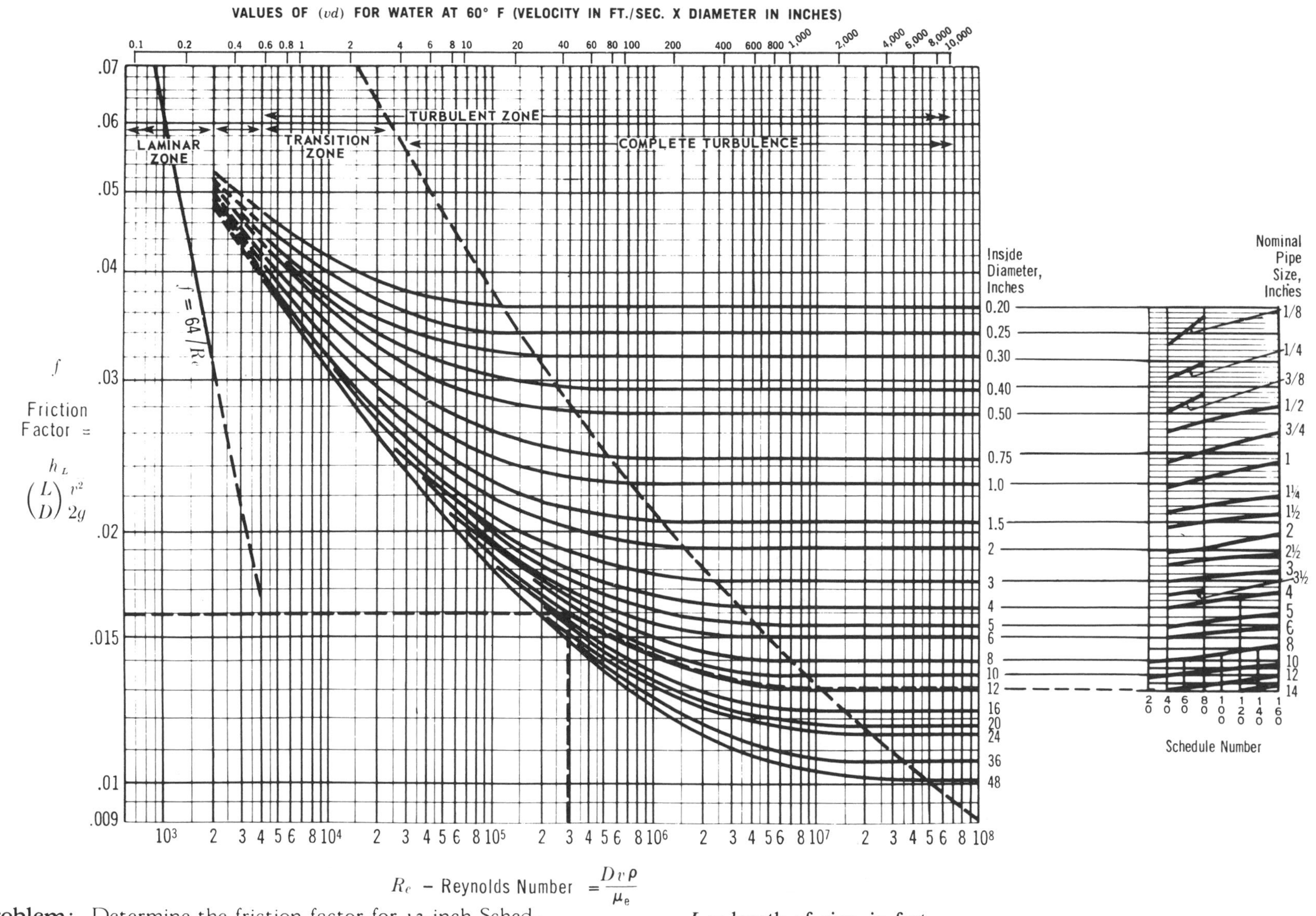

$$f = \frac{h_L}{\left(\frac{L}{D}\right)\frac{v^2}{2g}} \qquad R_e = \frac{D v \rho}{\mu_e}$$

Problem: Determine the friction factor for 12-inch Schedule 40 pipe at a flow having a Reynolds number of 300,000.

Solution:
The friction factor (f) equals 0.016.

where

h_L = loss of static pressure head due to fluid flow, in feet of fluid
L = length of pipe, in feet
D = internal diameter of pipe, in feet
v = mean velocity of flow, in feet per second
ρ = weight density of fluid, pounds per cubic feet
μ_e = absolute viscosity, in pounds mass per foot second or poundal seconds per square foot
f = friction factor in formula $h_L = fL\, v^2/D2_g$

Chart 3.5. Friction factors for clean commercial steel and wrought iron pipe. Courtesy Crane Co. (T.P. No. 410).

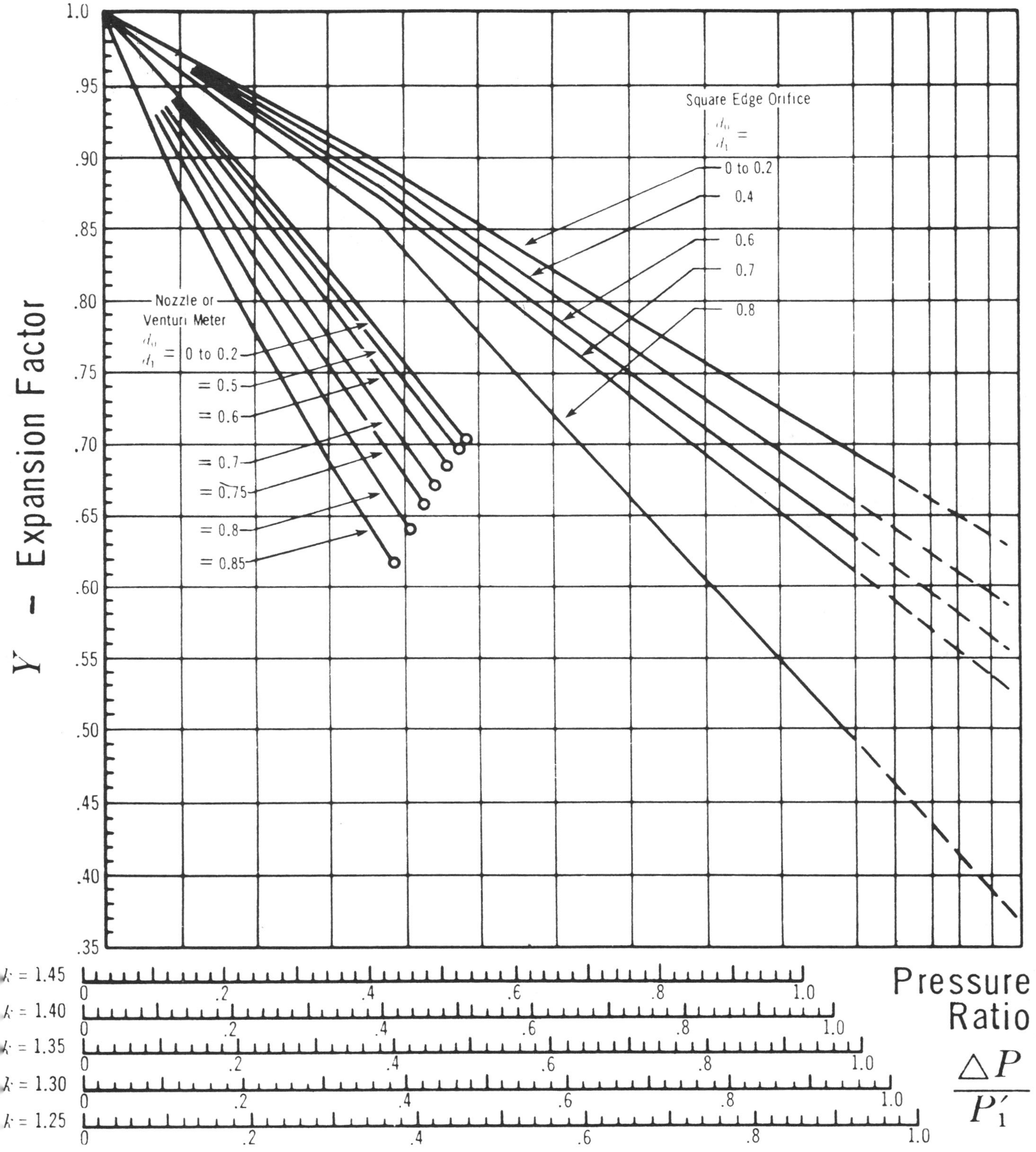

where

d_i = inside diameter of pipe
d_o = inside diameter of orifice

Y = net expansion factor for compressible flow through nozzles or pipe
P'_1 = upstream pressure in pounds per square inch absolute

Chart 3.6. Net expansion factor, Y, for compressible flow through nozzles and orifices. Courtesy Crane Co. (T.P. No. 410) and the ASME.

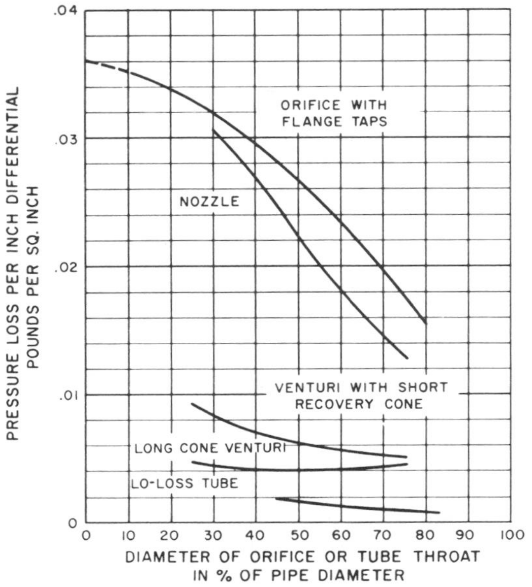

Chart 3.7. Pressure drop vs. beta ratio for various primary elements. Courtesy The Foxboro Co.

Pressure Drop vs. Beta Ratio for Various Primary Elements

The differential range of a flow meter depends on the required maximum rate of flow through the orifice, or other primary device. Standard practice is to select a range that places normal flow rate at about two-thirds of full-scale flow, equivalent to approximately one-half of differential range.

For accurate flow measurements using orifice plates, the ratio of the diameter of the primary device to the interior diameter of the pipe must be less than 0.75—best practice indicates it should be between 0.35 and 0.65. Within the limits of these diameter ratios, the differential range corresponding to full-scale flow rate should then be selected.

The differential pressure range is also limited by the static pressure of the fluid being metered. Liquids usually have ample pressure for a 100-inch range manometer. When gas, steam or vapor is involved, good practice limits the differential range in inches of water to values not exceeding the line static pressure in psi absolute. If the water pressure is low, the differential should not be great enough to lower the static pressure substantially below the vapor pressure of the liquid.

Choice of the primary device is a matter of selecting the most economical type that will meet the requirements of the installation. A practical consideration is the pressure loss due to the restriction. As indicated on the graph, this varies with the type of device and its diameter ratio. These curves show pressure loss in pounds per square inch, per inch of differential. For installations where liquid contacts the mercury in the manometer, values of differential indicated on the meter should be multiplied by a factor of about 0.93.

The Pitot Tube

The Pitot tube measures velocity at one point in the conduit. If quantity rate measurement is desired, it must be calculated from the ratio of average velocity to the velocity at the point of measurement.

The graph, Figure 3.8, shows the normal ratio of average-to-center velocity for smooth iron pipe.

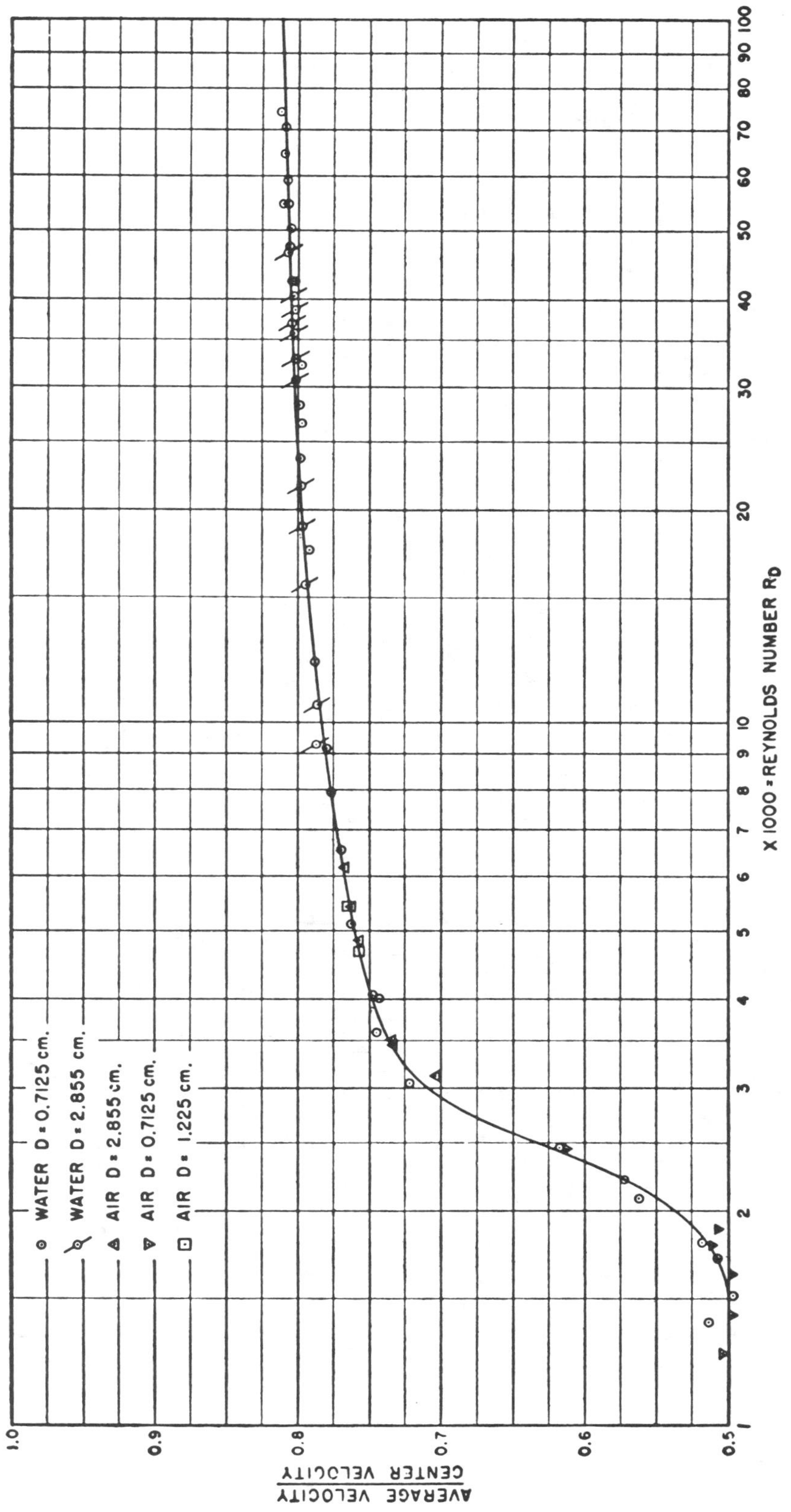

Chart 3.8. Ratio of average to center velocity in straight smooth pipe. Courtesy The Foxboro Co.

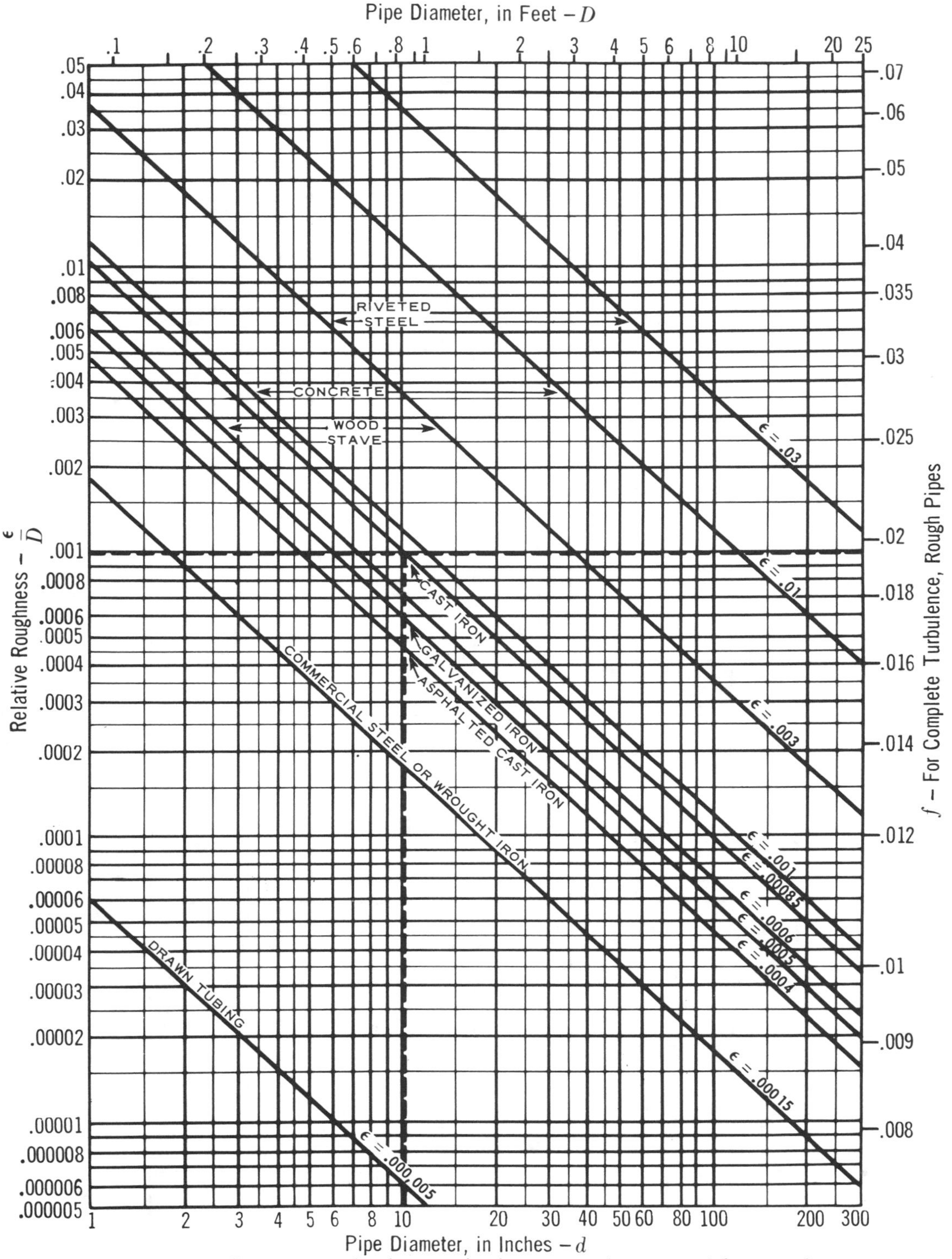

Problem: Determine absolute and relative roughness, and friction factor, for fully turbulent flow in 10-inch cast iron pipe (I.D. = 10.16″).

Solution: Absolute roughness (ϵ) = 0.00085 . . . . . Relative roughness (ϵ/D) = 0.001 . . . . . Friction factor at fully turbulent flow (f) = 0.0196.

f = friction factor in formula $h_L = fLv^2/D2g$
ϵ = absolute roughness or effective height of pipe wall irregularities, in feet
D = internal diameter of pipe, in feet
L = length of pipe in feet
v = mean velocity of flow in feet per second
d = internal diameter of pipe, in inches.

Chart 3.9. Relative roughness of pipe materials and friction factors for complete turbulence. Courtesy Crane Co. (T.P. No. 410) and the ASME.

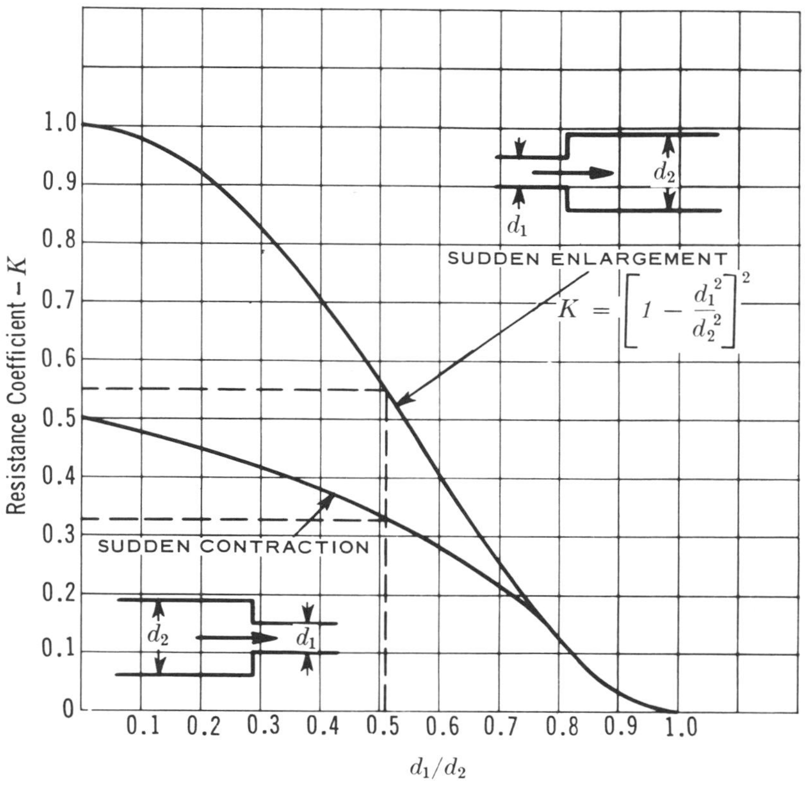

a.

Sudden enlargement: The resistance coefficient K for a sudden enlargement from 6-inch Schedule 40 pipe to 12-inch Schedule 40 pipe is **0.55,** based on the 6-inch pipe size.

$$\frac{d_1}{d_2} = \frac{6.065}{11.938} = 0.51$$

Sudden contraction: The resistance coefficient K for a sudden contraction from 12-inch Schedule 40 pipe to 6-inch Schedule 40 pipe is **0.33,** based on the 6-inch pipe size.

$$\frac{d_1}{d_2} = \frac{6.065}{11.938} = 0.51$$

Note: The values for the resistance coefficient, K, are based on velocity in the small pipe. To determine K values in terms of the greater diameter, multiply the chart values by $(d_2/d_1)^4$.

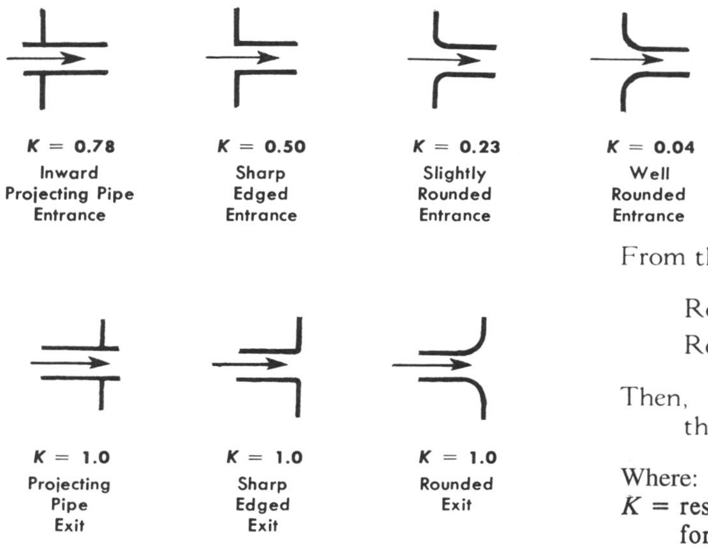

b.

Chart 3.10. Resistance in pipe (a) enlargements and contractions, (b) entrance and exit, (c) bends. Courtesy Crane Co. (T.P. No. 410) and the National Bureau of Standards.

Problem: Determine the total resistance coefficient for a pipe one diameter long having a sharp edged entrance and a sharp edged exit.

Solution: The resistance of pipe one diameter long is small and can be neglected ($K = f\,L/D$).

From the diagrams, note:

Resistance for a sharp edged entrance = 0.5
Resistance for a sharp edged exit = 1.0

Then,
the total resistance, K, for the pipe = **1.5**

Where:
K = resistance coefficient or velocity head loss in the formula $h_L = Kv^2/2g$
v = mean velocity of flow in feet per second
L = length of pipe in feet
d_1 & d_2 are as shown
D = internal diameter in feet
r, d, and Θ are as shown.

Chart 3.10 continued

Chart 3.10 continued

The chart at the right shows the resistance of 90 degree bends to the flow of fluids in terms of equivalent lengths of straight pipe.

Resistance of bends greater than 90 degrees is found using the formula:

$$\frac{L}{D} = R_t + (n-1)\left(R_l + \frac{R_b}{2}\right)$$

n = total number of 90° bends in coil
R_t = total resistance due to one 90° bend, in L/D
R_l = resistance due to length of one 90° bend, in L/D
R_b = bend resistance due to one 90° bend, in L/D

Problem: Determine the equivalent lengths in pipe diameters of a 90 degree bend and a 270 degree bend having a relative radius of 12.

Solution: Referring to the "Total Resistance" curve, the equivalent length for a 90 degree bend is **34.5** pipe diameters.

The equivalent length of a 270 degree bend is:

$L/D = 34.5 + (3-1)\,[18.7 + (15.8 \div 2)]$
$L/D =$ **87.7** pipe diameters

Note: This loss is less than the sum of losses through three 90 degree bends separated by tangents.

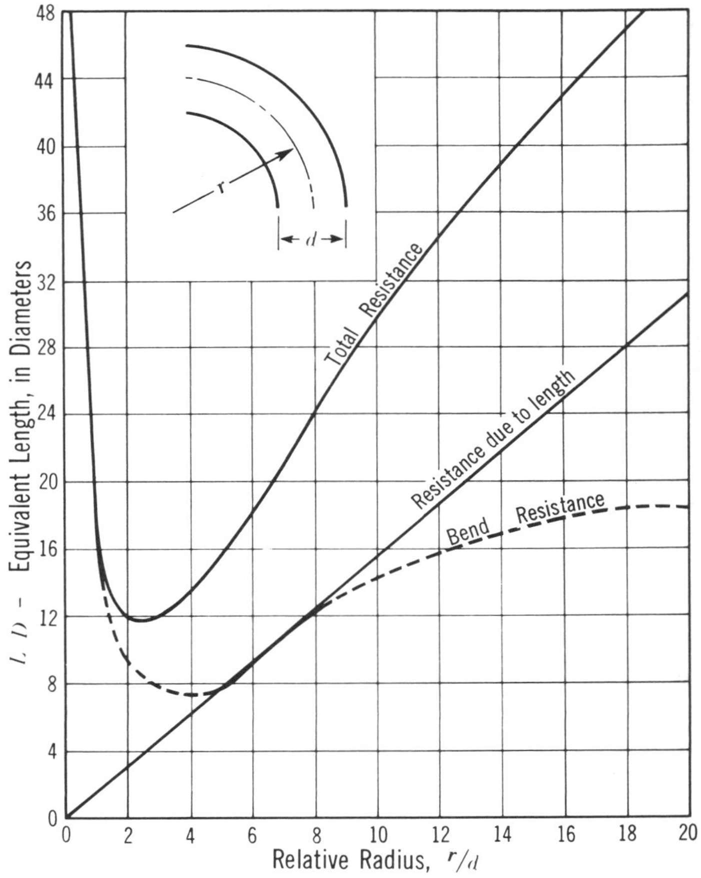

C.

The chart at the lower right shows the resistance of miter bends to the flow of fluids. The chart is based on data published by the American Society of Mechanical Engineers (ASME).

Problem: Determine the equivalent length in pipe diameters of a 40 degree miter bend.

Solution: Referring to the "Total Resistance" curve in the chart, the equivalent length is **12** pipe diameters.

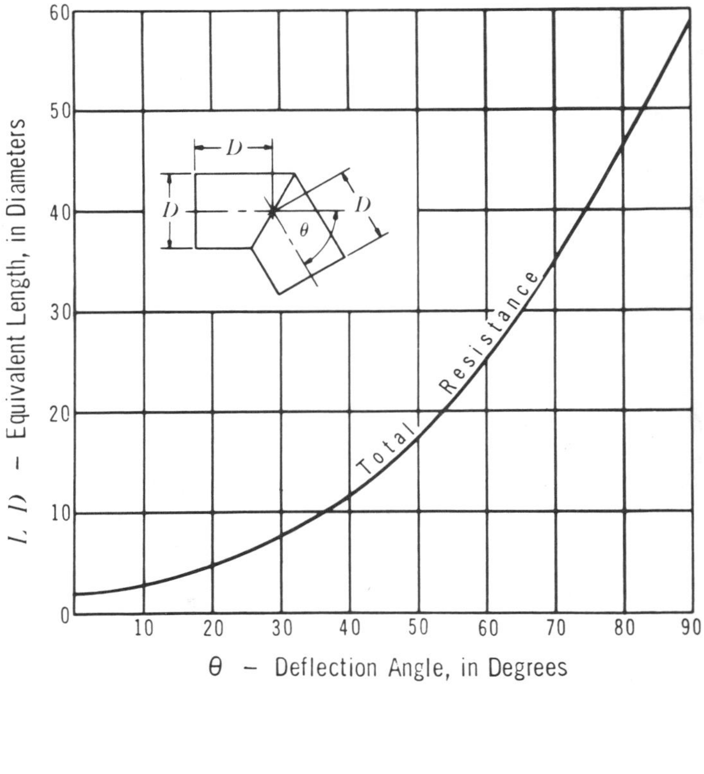

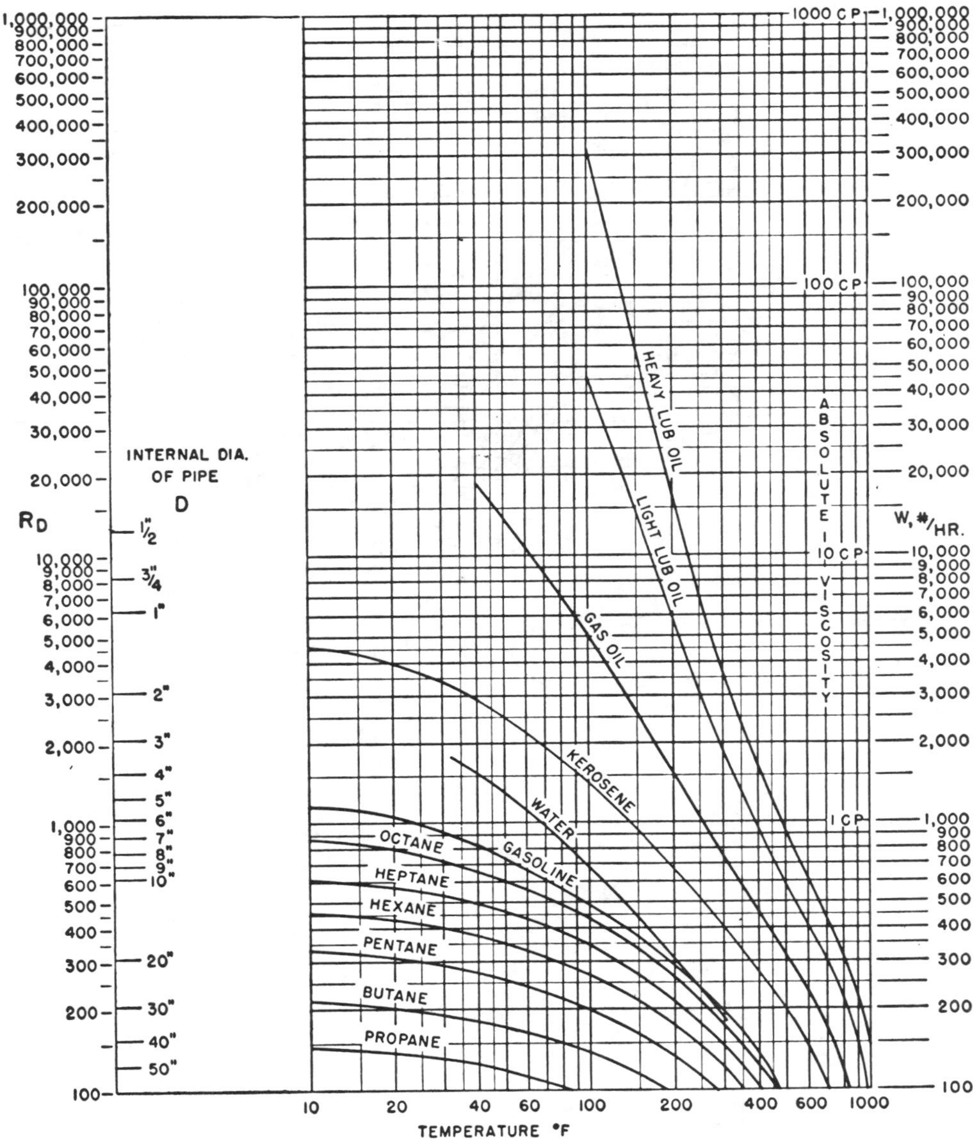

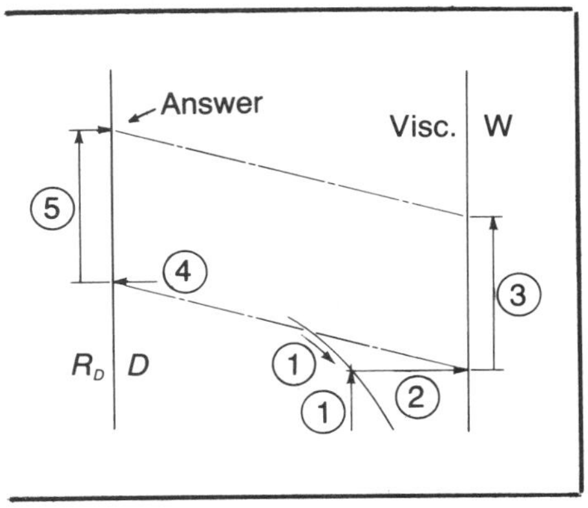

To use Chart 3.11, follow the steps shown on the inset in the numbered sequence:

Step 1. Locate the point on the fluid curve corresponding to product temperature.
Step 2. Read the value absolute viscosity in centipoise on the right.
Step 3. Measure the vertical distance from absolute viscosity to the value of flow rate in pounds per hour.
Step 4. Locate the pipe size on the left-hand scale.
Step 5. Measure from the pipe-size intercept along the left-hand scale a vertical distance that is the same length as that in Step 3, and in the same direction; it will end at the value of the Reynolds number.

Chart 3.11. Reynolds number, R_D, for liquids. Courtesy The Foxboro Co.

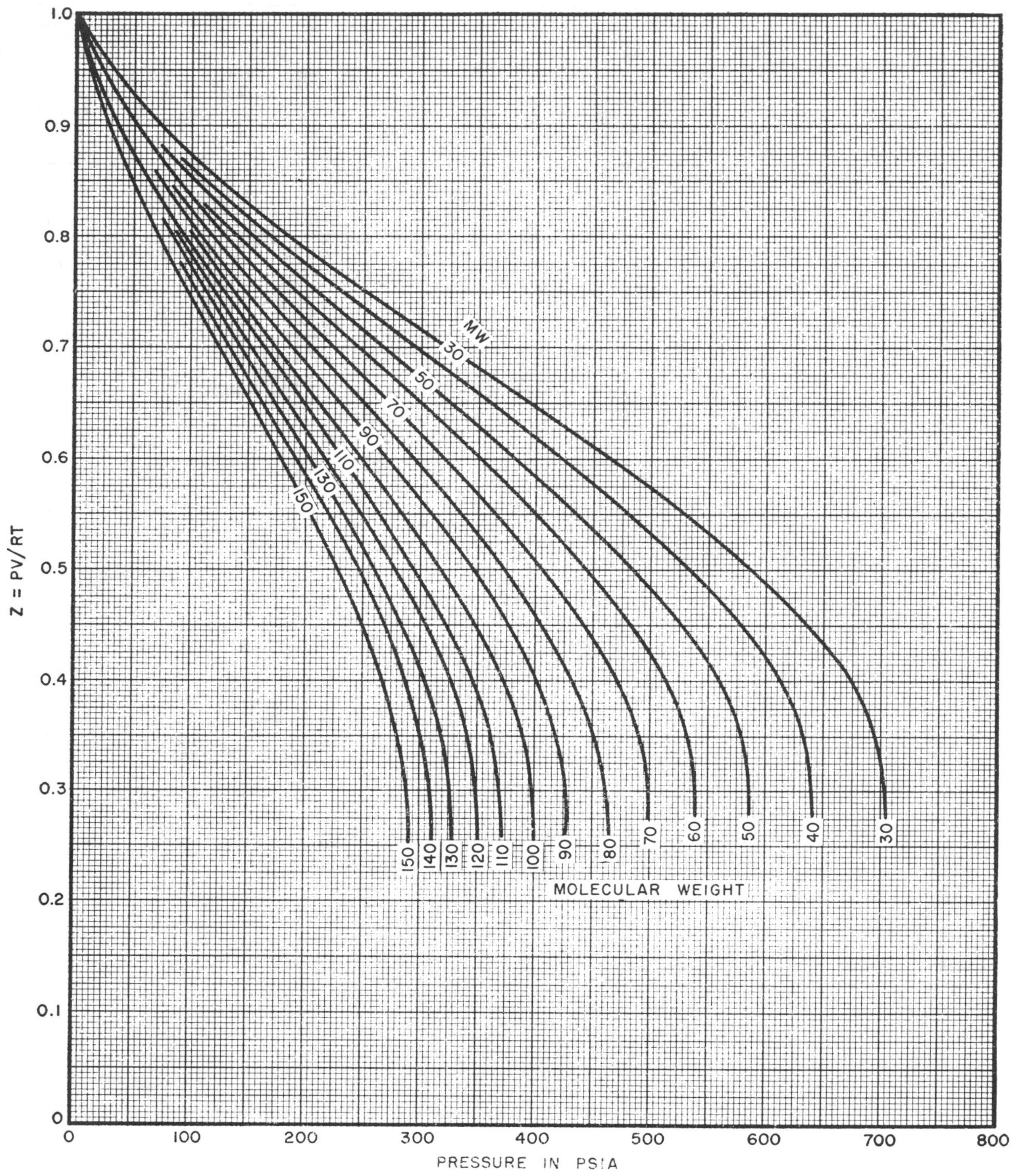

Z = compressibility factor for the deviation of the actual gas from a perfect gas, evaluated at inlet conditions.
P = upstream pressure in pounds per square inch absolute
V = volume in cubic feet
R = gas constant: 10.73 per pound mole
T = temperature absolute (460 + °F)

Chart 3.12. Compressibility factor, Z, for paraffin hydrocarbons at saturated vapor conditions. Courtesy American Petroleum Institute.

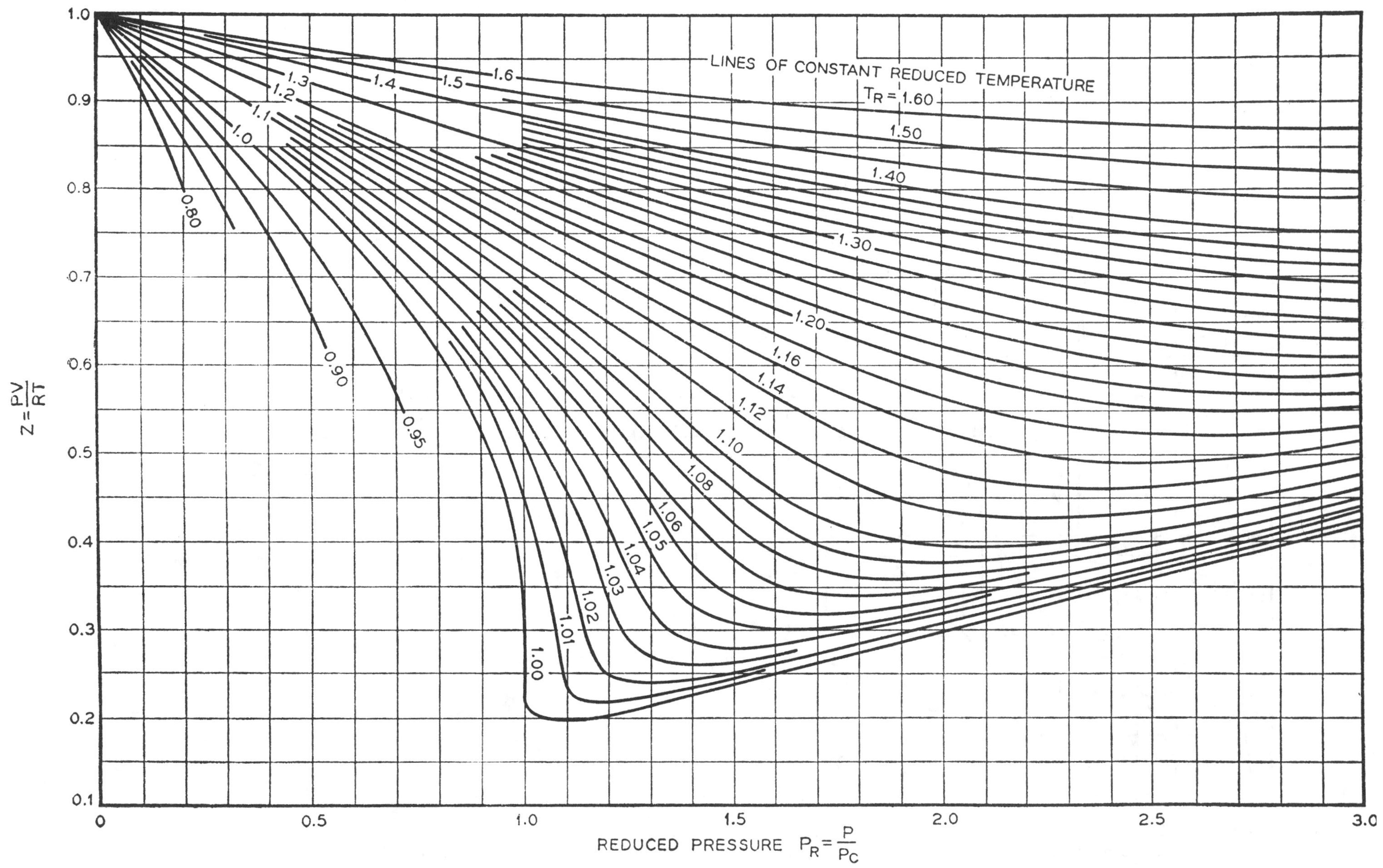

P = upstream pressure, pounds per square inch absolute
V = volume in cubic feet
R = gas constant, 10.73
T = absolute temperature (460 + °F)

P_r = reduced pressure
P_c = critical pressure
T_r = reduced temperature
Z = compressibility factor

Chart 3.13. Compressibility factor, Z, vs. reduced pressure, P , at various reduced temperature, T , values. Courtesy American Petroleum Institute.

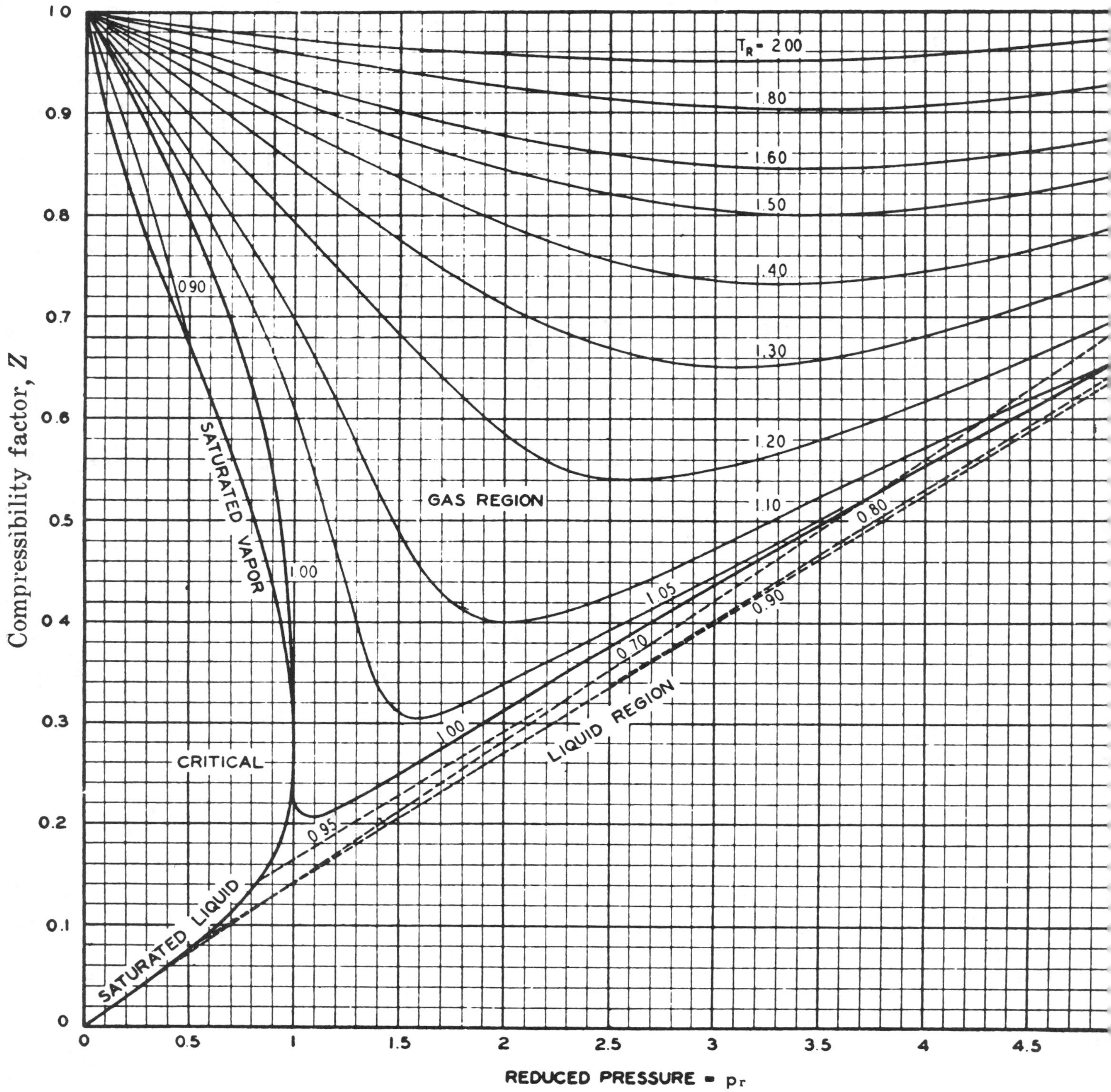

Z = compressibility factor
T_r = Reduced temperature = flowing absolute temperature divided by critical absolute temperature
P_r = Reduced pressure = flowing absolute pressure divided by the critical absolute pressure

Chart 3.14. Gas compressibility factor, Z, vs. reduced pressure, P , at various values of T . Courtesy Fischer Porter Co.

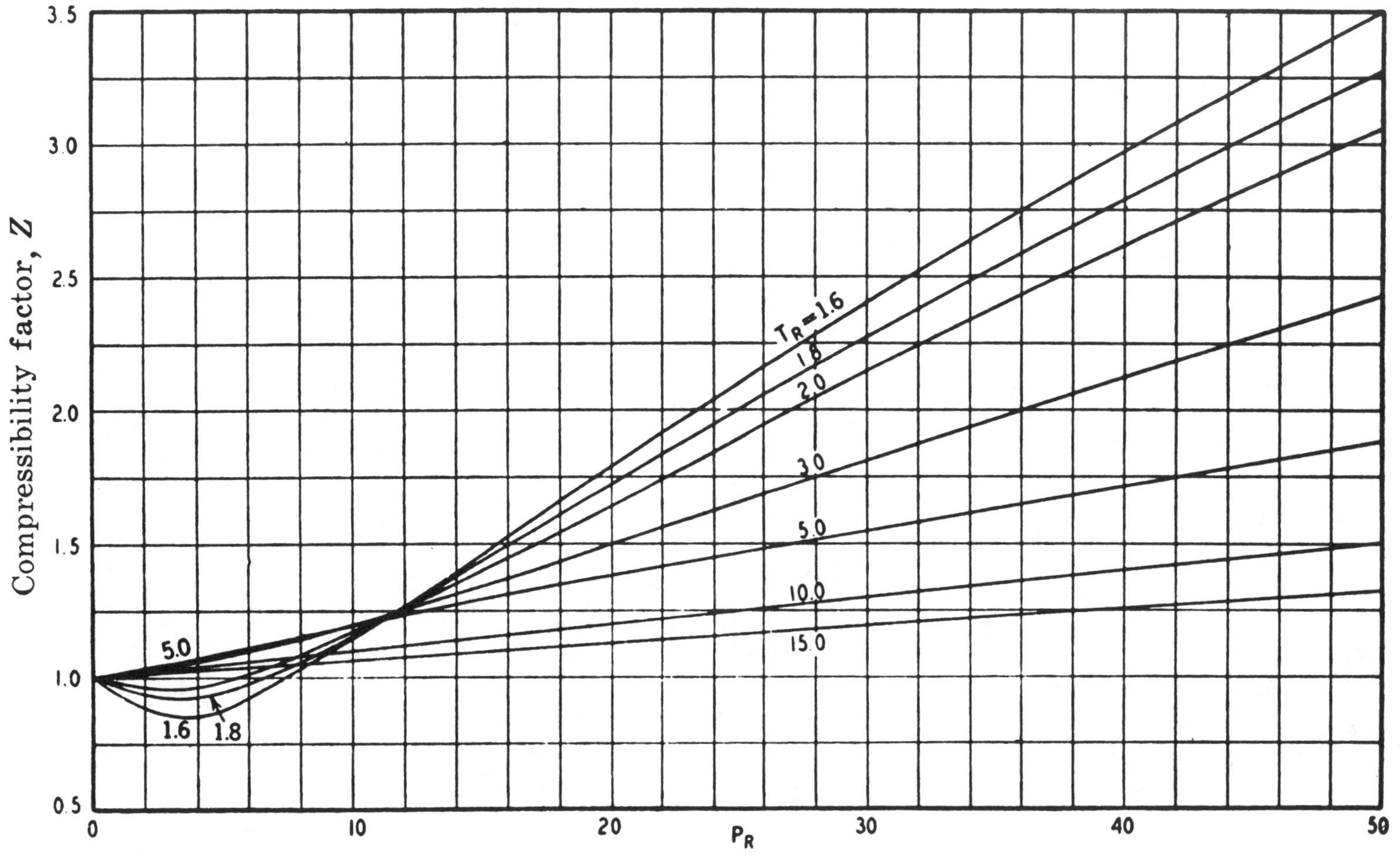

Z = Compressibility Factor
T_r = Reduced temperature = flowing absolute temperature divided by critical absolute temperature
P_r = Reduced pressure = flowing absolute pressure divided by the critical absolute pressure

Chart 3.15. Gas compressibility factor, Z, at high pressures. Courtesy Fischer Porter Co.

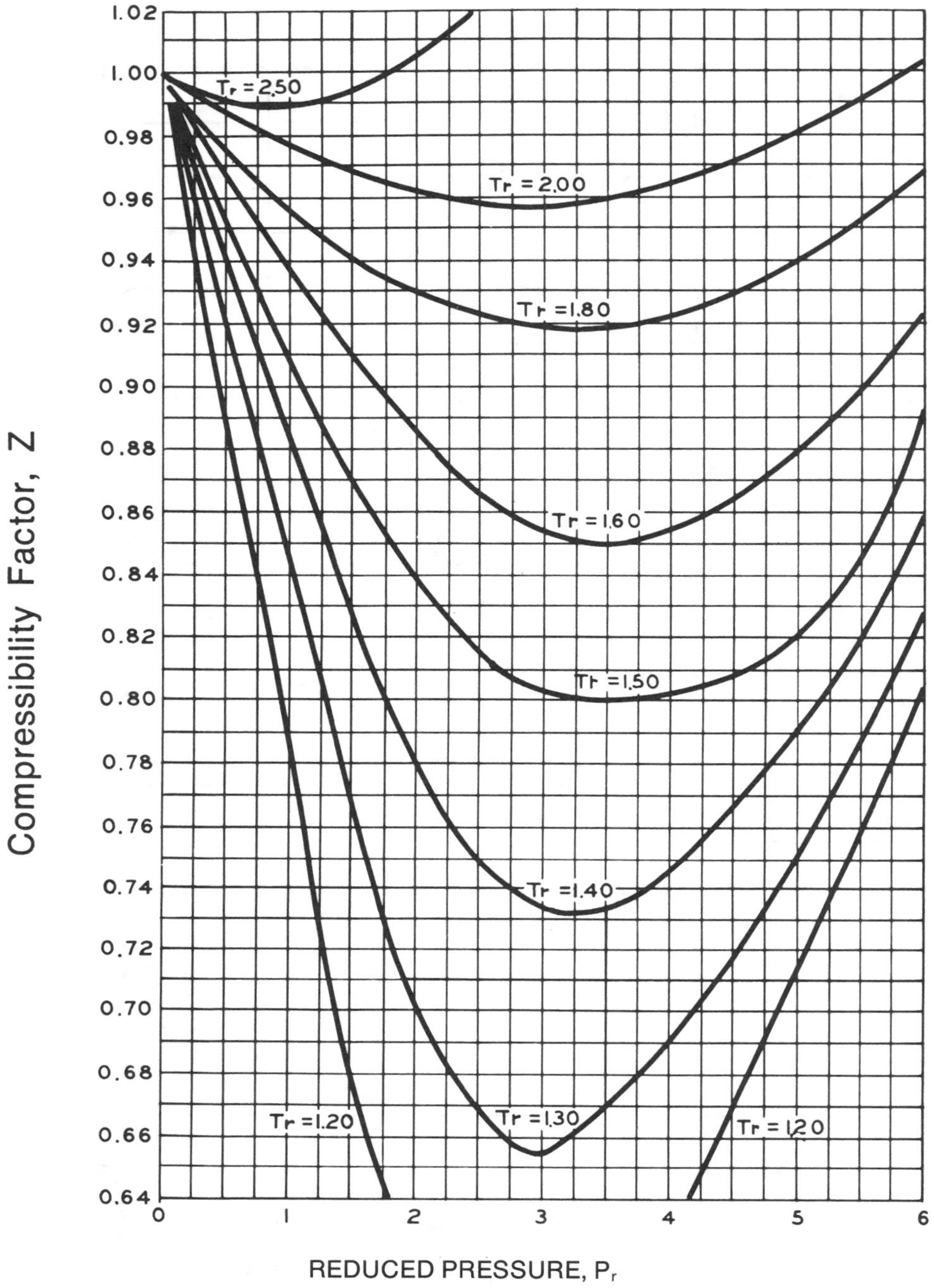

Z = compressibility factor
T_r = reduced temperature
P_r = reduced pressure

Chart 3.16. Compressibility factors for gases with reduced pressures from 0 to 6. Courtesy Masoneilan International, Inc.

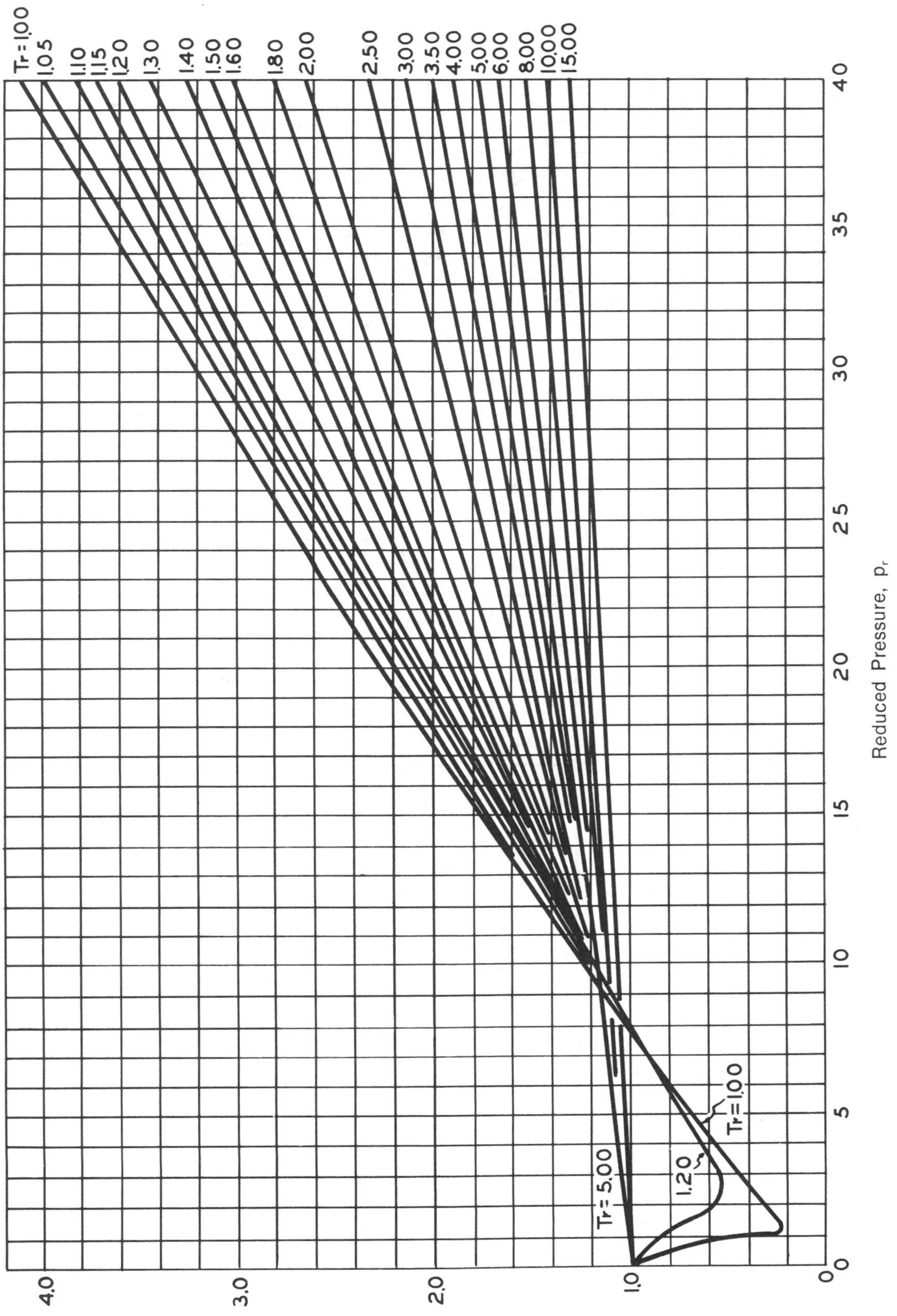

$P_r = \frac{\text{inlet pressure}}{\text{critical pressure}}$

$t_r = \frac{\text{inlet temperature}}{\text{critical temperature}}$

Chart 3.17. Compressibility factors for gases with reduced pressures from 0 to 40. Courtesy Masoneilan International, Inc.

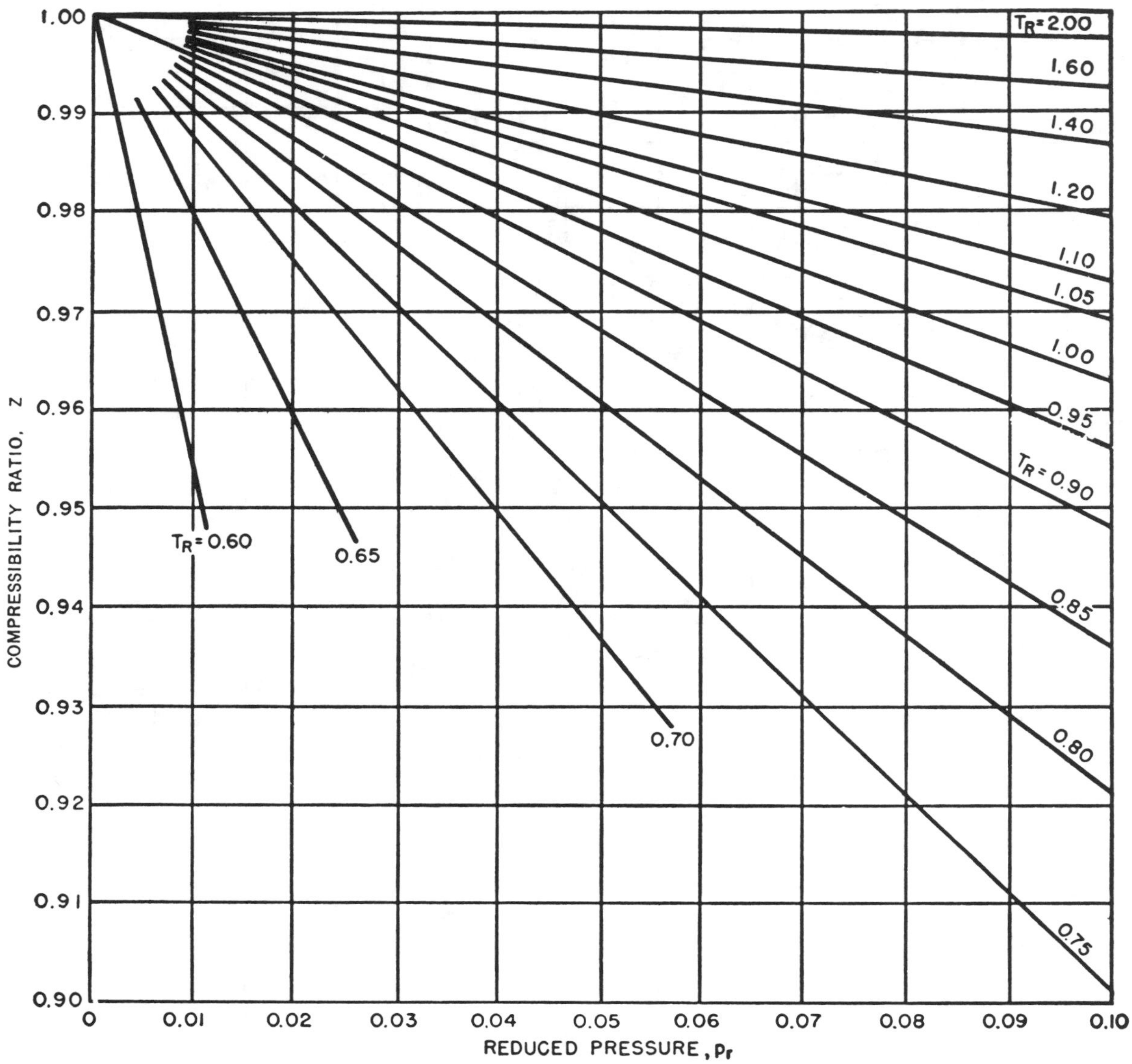

Note: Do not use for ammonia. Maximum deviation for water, hydrogen, or helium ≈ 1½ to 2%. Maximum deviation for 26 other listed gases is less than 1%.

Chart 3.18. Nelson-Obert compressibility chart (from pr = 0 to 0.10). Courtesy The Foxboro Co.

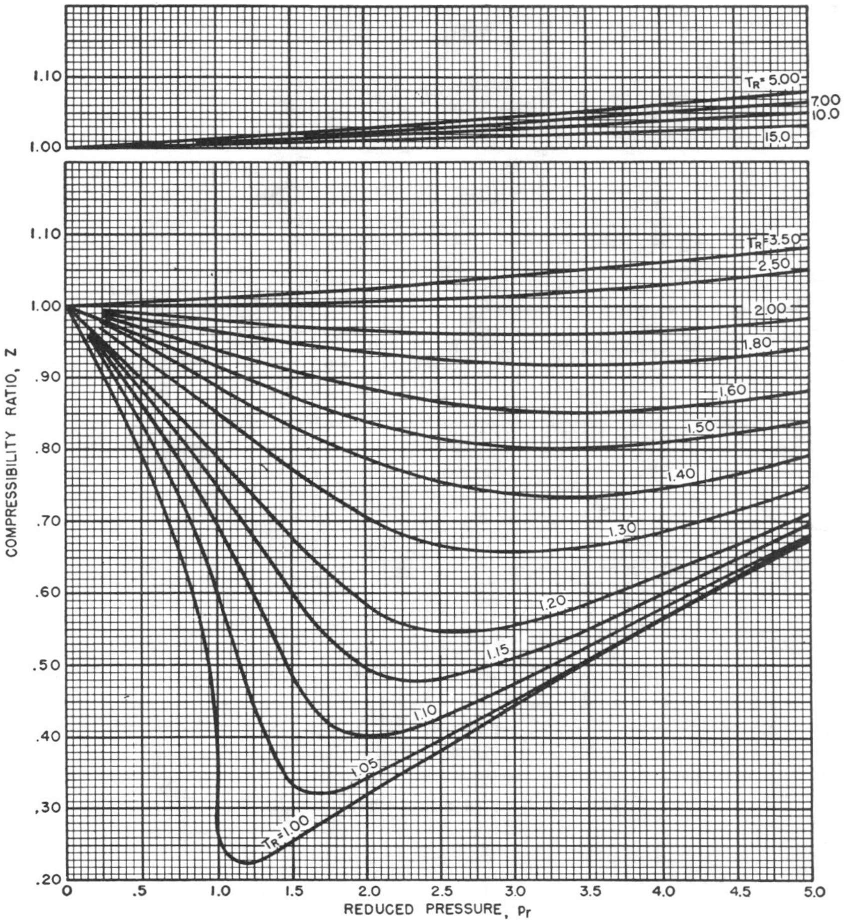

Note: Do not use for ammonia. Do not use for methyl fluoride below $T_r = 1.3$ or for hydrogen or helium below $T_r = 2.5$. For hydrogen or helium above $T_r = 2.5$, use adjusted constants: $T_{ca} + 14.4°$ F. and $p_{ca} + 117.6$ psi. Maximum deviation, when used as above, is less than $2\frac{1}{2}\%$ except near $T_r = 1.0$ and $p_r = 1.0$.

Chart 3.19. Nelson-Obert compressibility chart (from pr = 0 to 10.0). Courtesy the Foxboro Co.

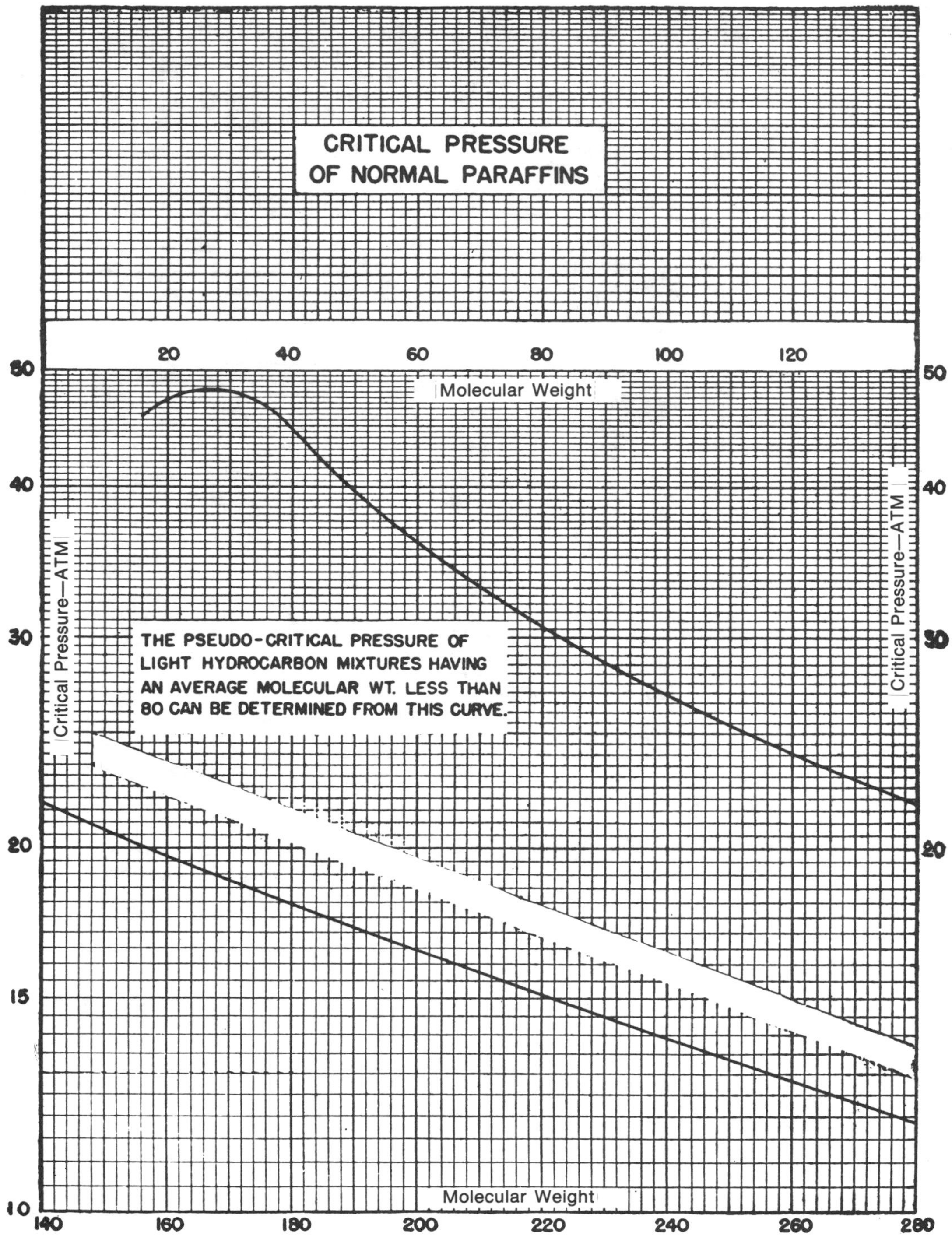

Chart 3.20. Critical pressure of normal paraffins. Data Book on Hydrocarbons, *J.B. Maxwell, copyright © 1950 by Litton Educational Publishing, Inc. Reprinted by permission of Van Nostrand Reinhold Co.*

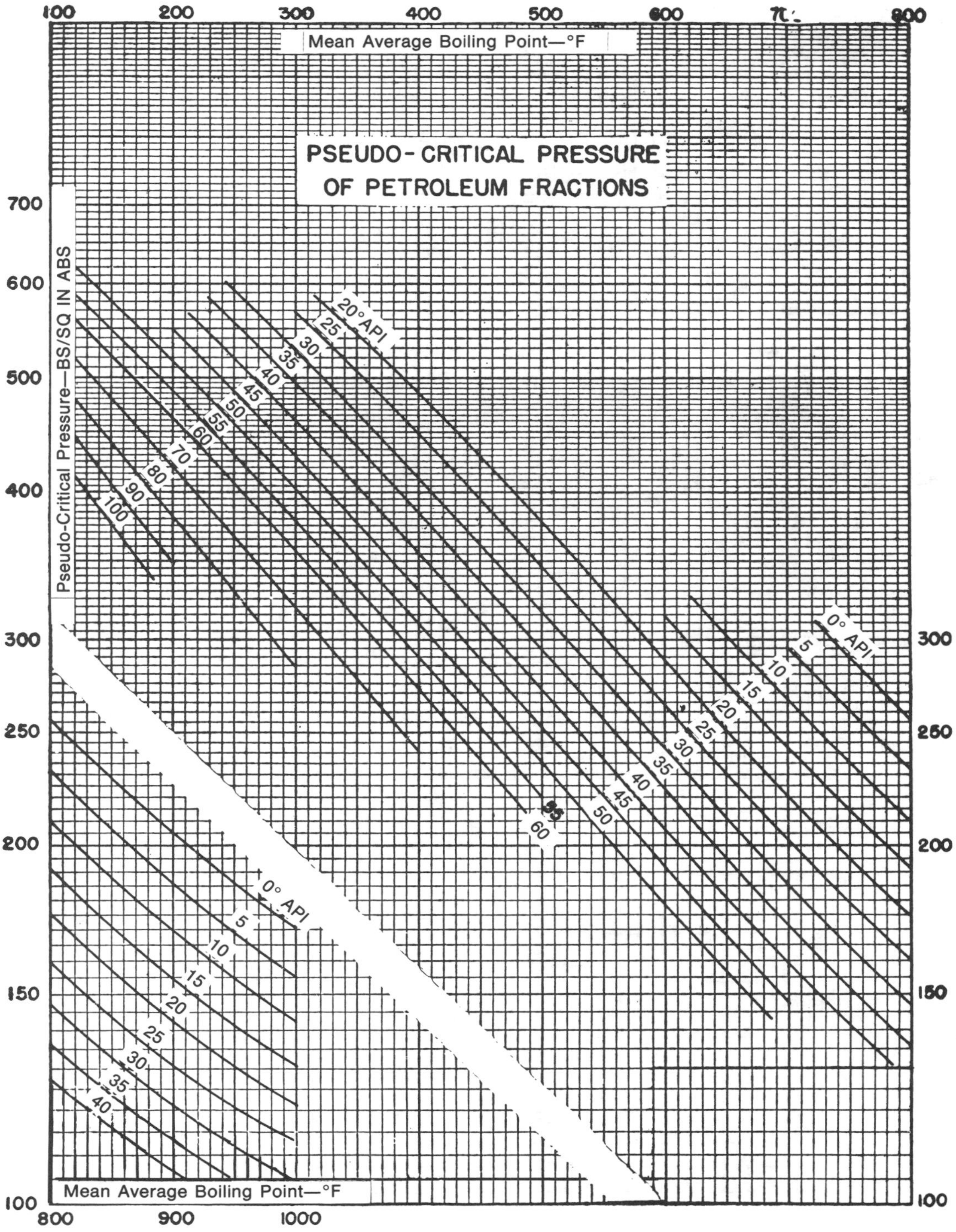

Chart 3.21. Pseudo-critical pressure of petroleum fractions. Data Book on Hydrocarbons, *J.B. Maxwell, copyright © 1950 by Litton Educational Publishing, Inc. Reprinted by permission of Van Nostrand Reinhold Co.*

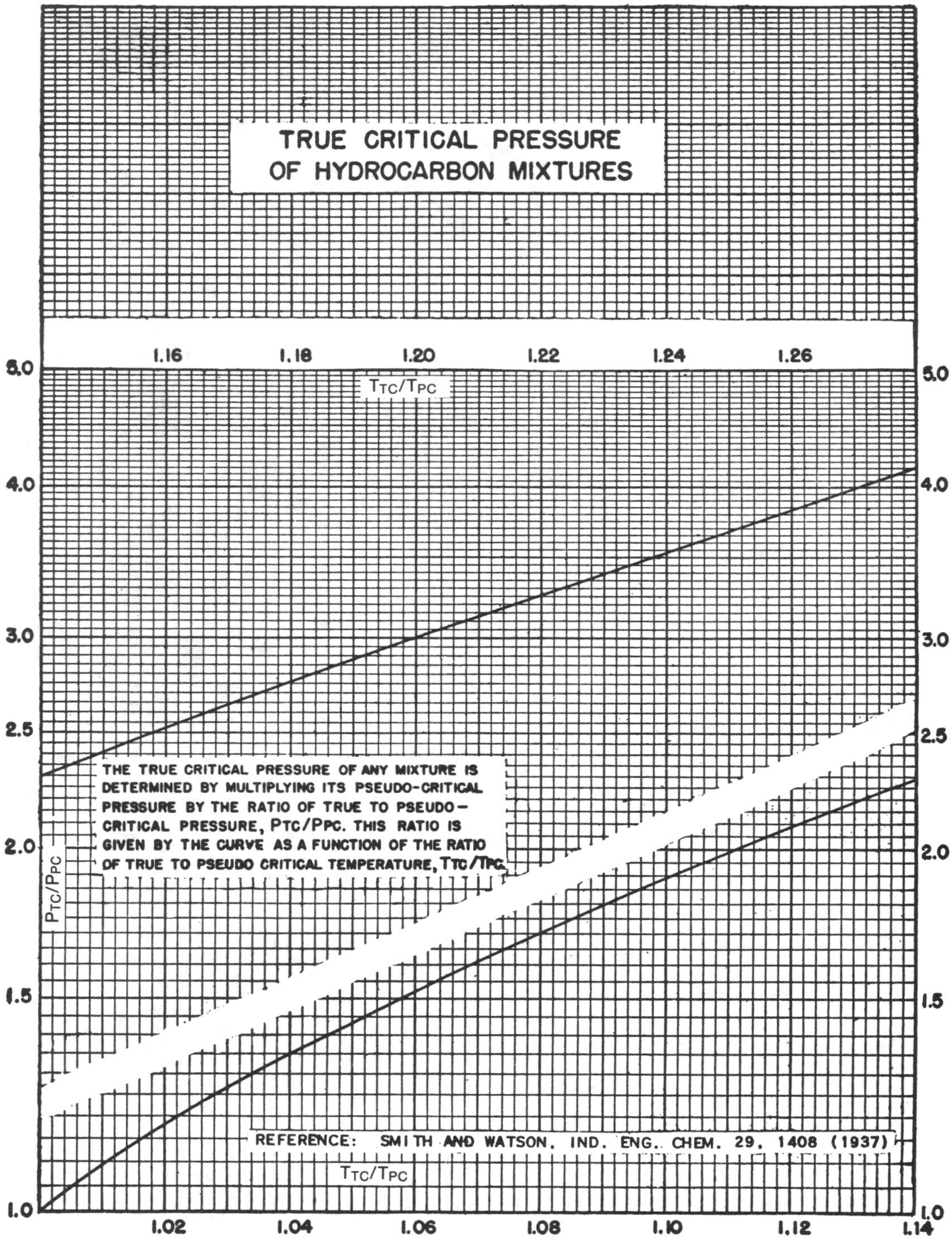
TRUE CRITICAL PRESSURE
OF HYDROCARBON MIXTURES
1.16
1.18
1.20
1.22
1.24
1.26
TTC/TPC
5.0
4.0
3.0
2.5
2.0
1.5
1.0
THE TRUE CRITICAL PRESSURE OF ANY MIXTURE IS DETERMINED BY MULTIPLYING ITS PSEUDO-CRITICAL PRESSURE BY THE RATIO OF TRUE TO PSEUDO-CRITICAL PRESSURE, PTC/PPC. THIS RATIO IS GIVEN BY THE CURVE AS A FUNCTION OF THE RATIO OF TRUE TO PSEUDO CRITICAL TEMPERATURE, TTC/TPC.
PTC/PPC
REFERENCE: SMITH AND WATSON, IND. ENG. CHEM. 29, 1408 (1937)
TTC/TPC
1.02
1.04
1.06
1.08
1.10
1.12
1.14

Chart 3.22. True critical pressure of hydrocarbon mixtures. Data Book on Hydrocarbons, *J.B. Maxwell, copyright © 1950 by Litton Educational Publishing, Inc. Reprinted by permission of Van Nostrand Reinhold Co.*

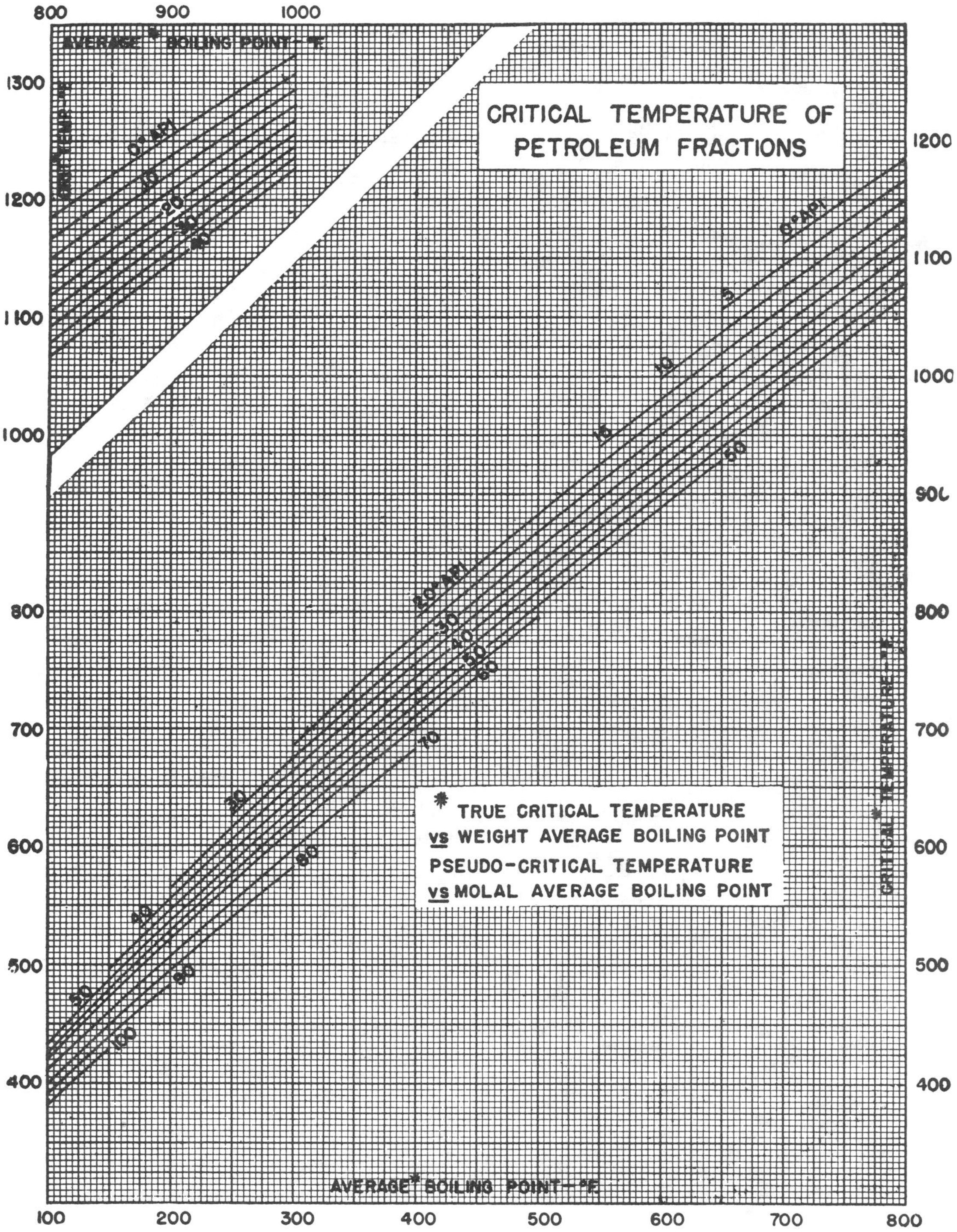

Chart 3.23. *Critical temperature of petroleum fractions.* Data Book on Hydrocarbons, *J.B. Maxwell, copyright © 1950 by Litton Educational Publishing, Inc. Reprinted by permission of Van Nostrand Reinhold Co.*

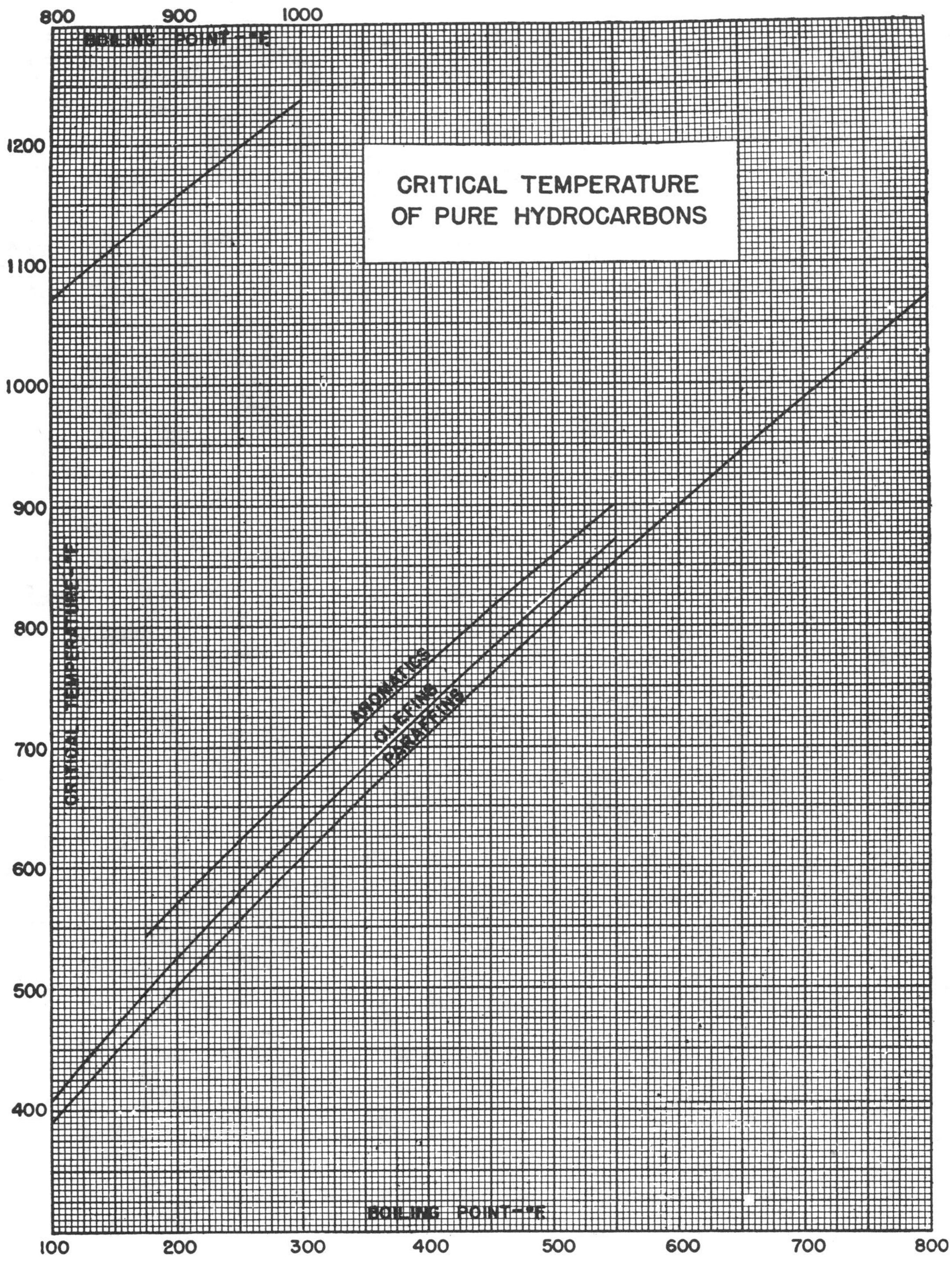

Chart 3.24. Critical temperature of pure hydrocarbons. Data Book on Hydrocarbons, *J.B. Maxwell, copyright © 1950 by Litton Educational Publishing, Inc. Reprinted by permission of Van Nostrand Reinhold Co.*

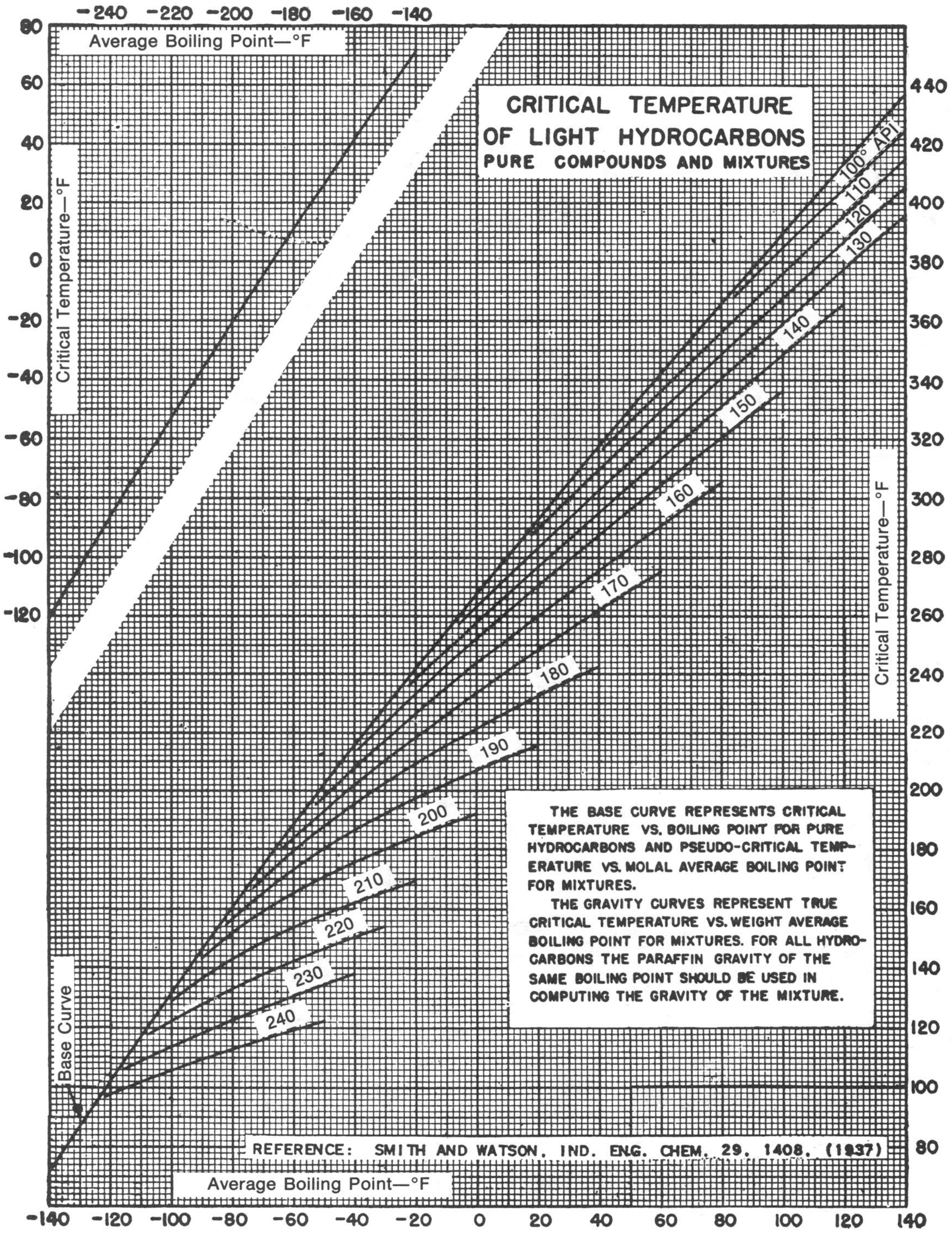

Chart 3.25. Critical temperature of light hydrocarbons. Data Book on Hydrocarbons, *J.B. Maxwell, copyright © 1950 by Litton Educational Publishing, Inc. Reprinted by permission of Van Nostrand Reinhold Co.*

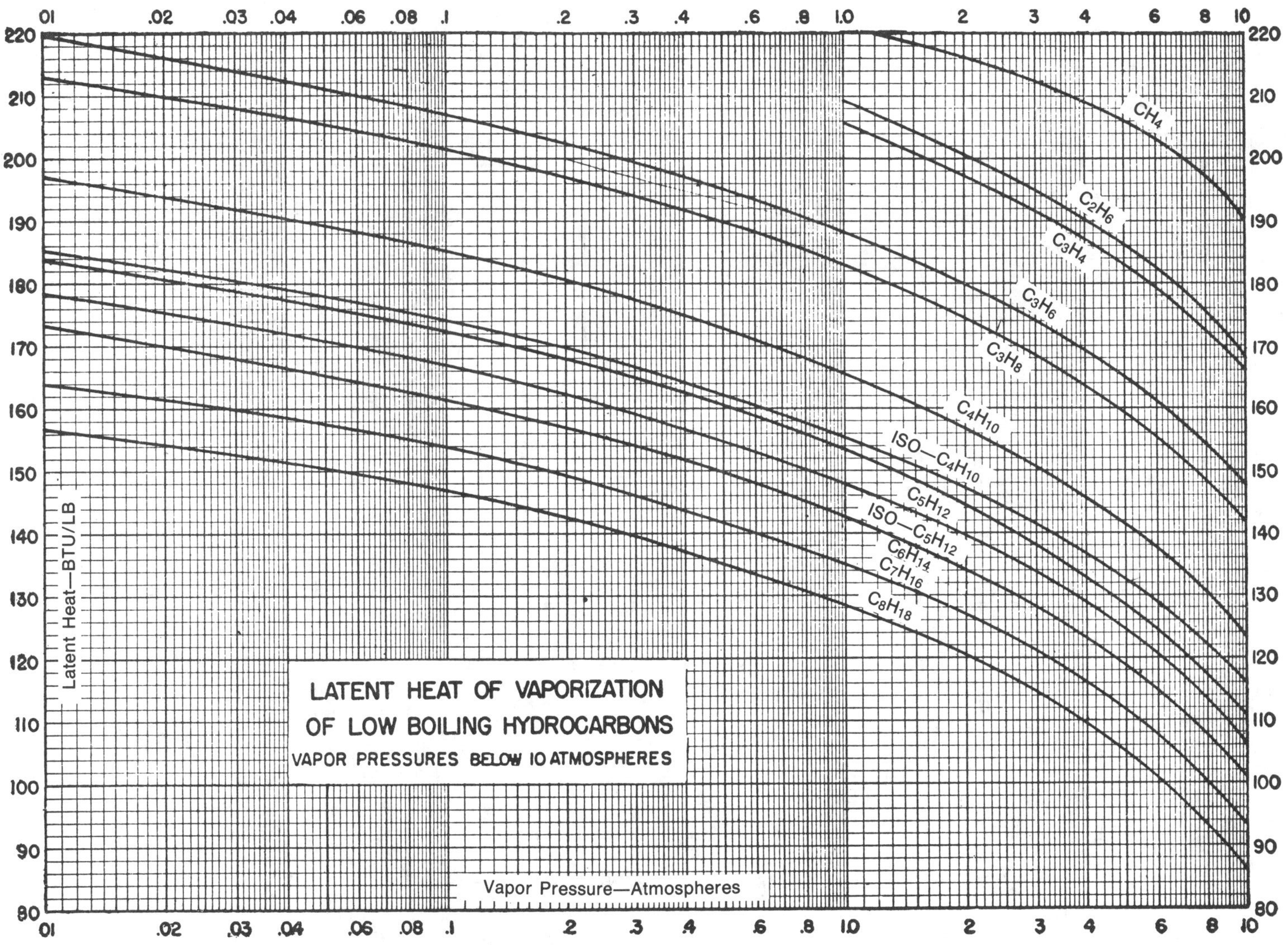

Chart 3.26. Latent heat of vaporization of low boiling hydrocarbons below 10 atmospheres. Data Book on Hydrocarbons, *J.B. Maxwell, copyright © 1950 by Litton Educational Publishing, Inc. Reprinted by permission of Van Nostrand Reinhold Co.*

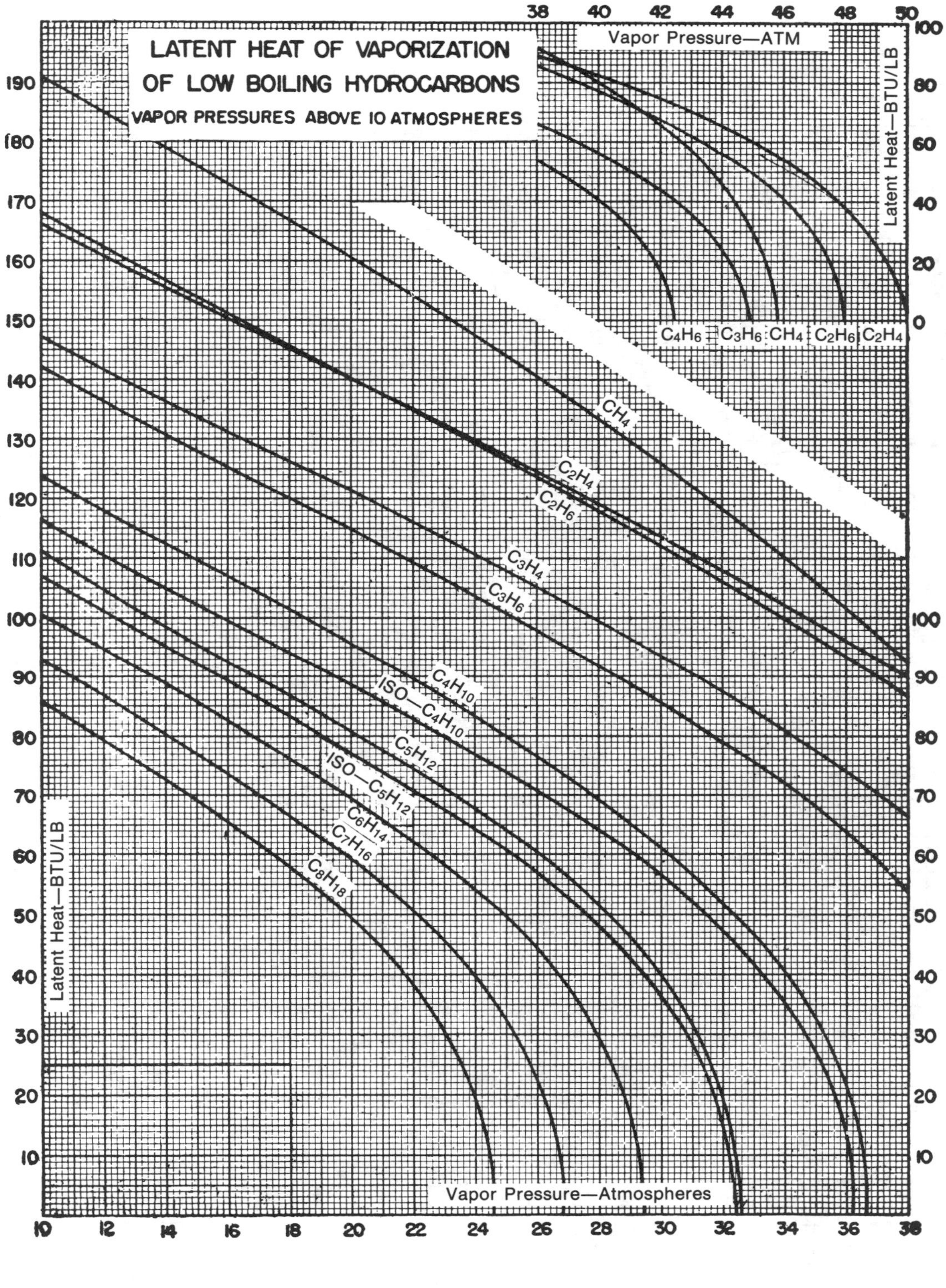

Chart 3.27. Latent heat of vaporization of low boiling hydrocarbons above 10 atmospheres. Data Book on Hydrocarbons, *J.B. Maxwell, copyright © 1950 by Litton Educational Publishing, Inc. Reprinted by permission of Van Nostrand Reinhold Co.*

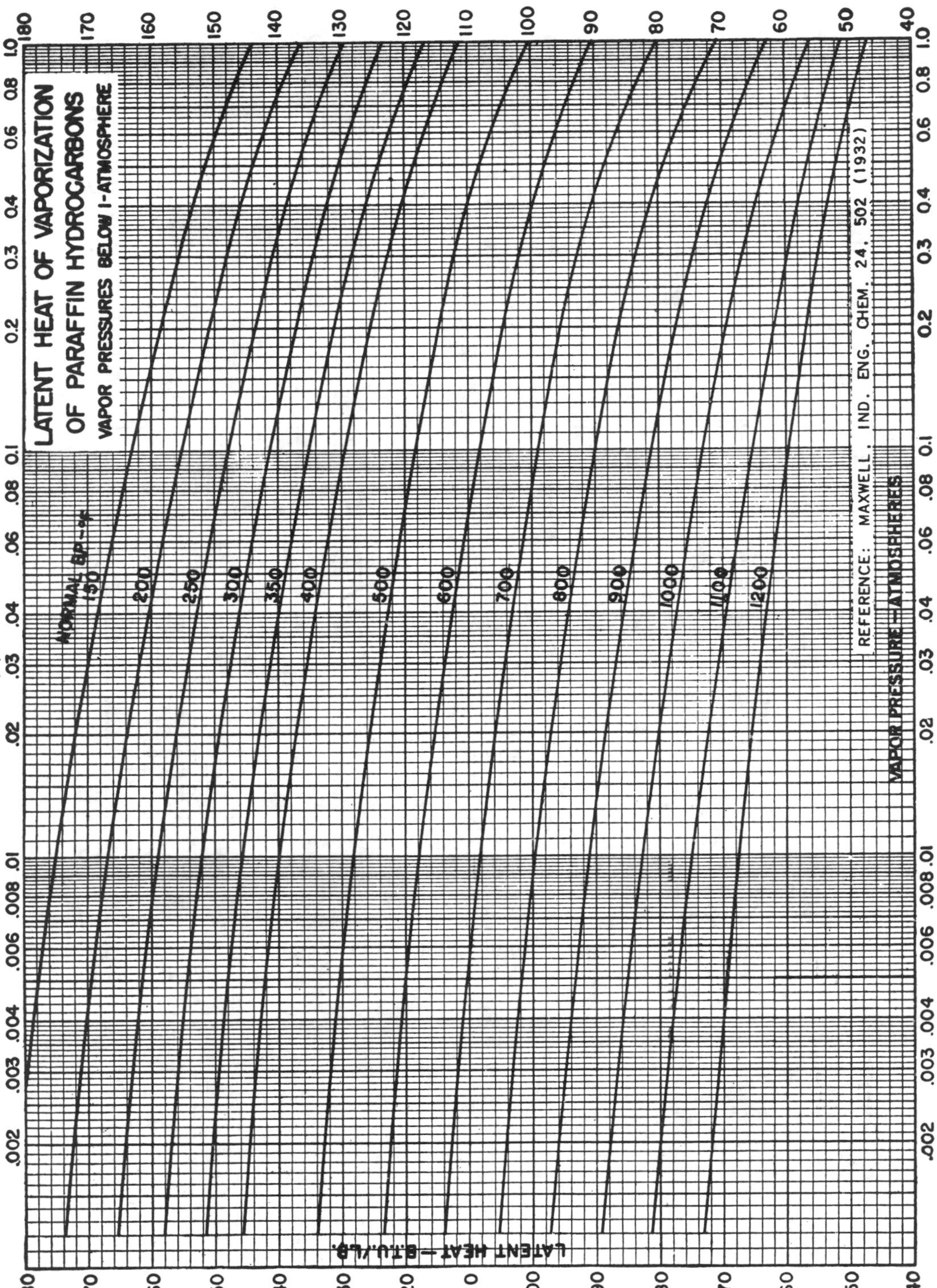

Chart 3.28. Latent heat of vaporization of paraffin hydrocarbons at vapor pressure below 1 atmosphere. Data Book on Hydrocarbons, *J.B. Maxwell, copyright © 1950 by Litton Educational Publishing, Inc. Reprinted by permission of Van Nostrand Reinhold Co.*

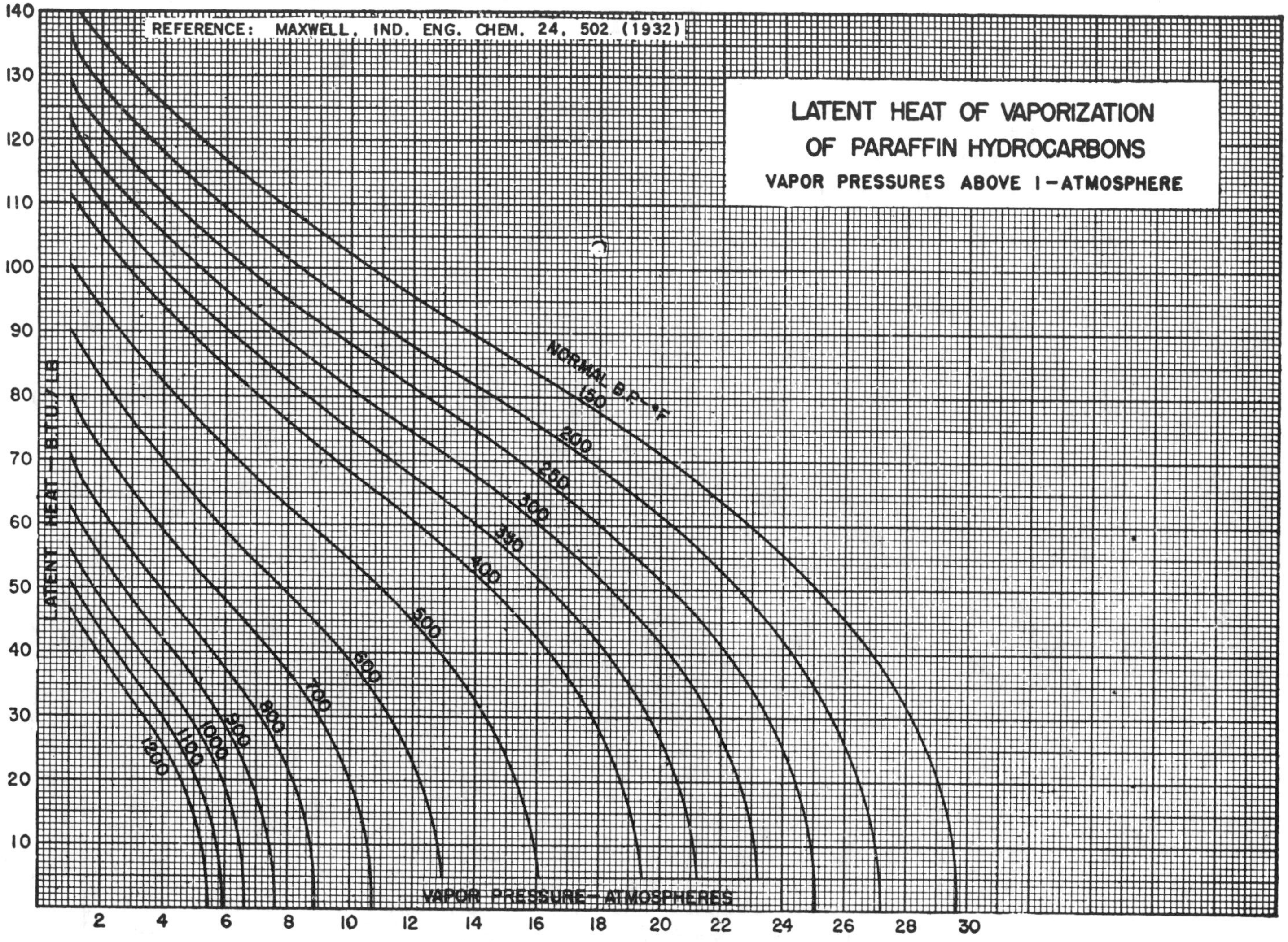

Chart 3.29. Latent heat of vaporization of paraffin hydrocarbons at vapor pressures above 1 atmosphere. Data Book on Hydrocarbons, *J.B. Maxwell, copyright © 1950 by Litton Educational Publishing, Inc. Reprinted by permission of Van Nostrand Reinhold Co.*

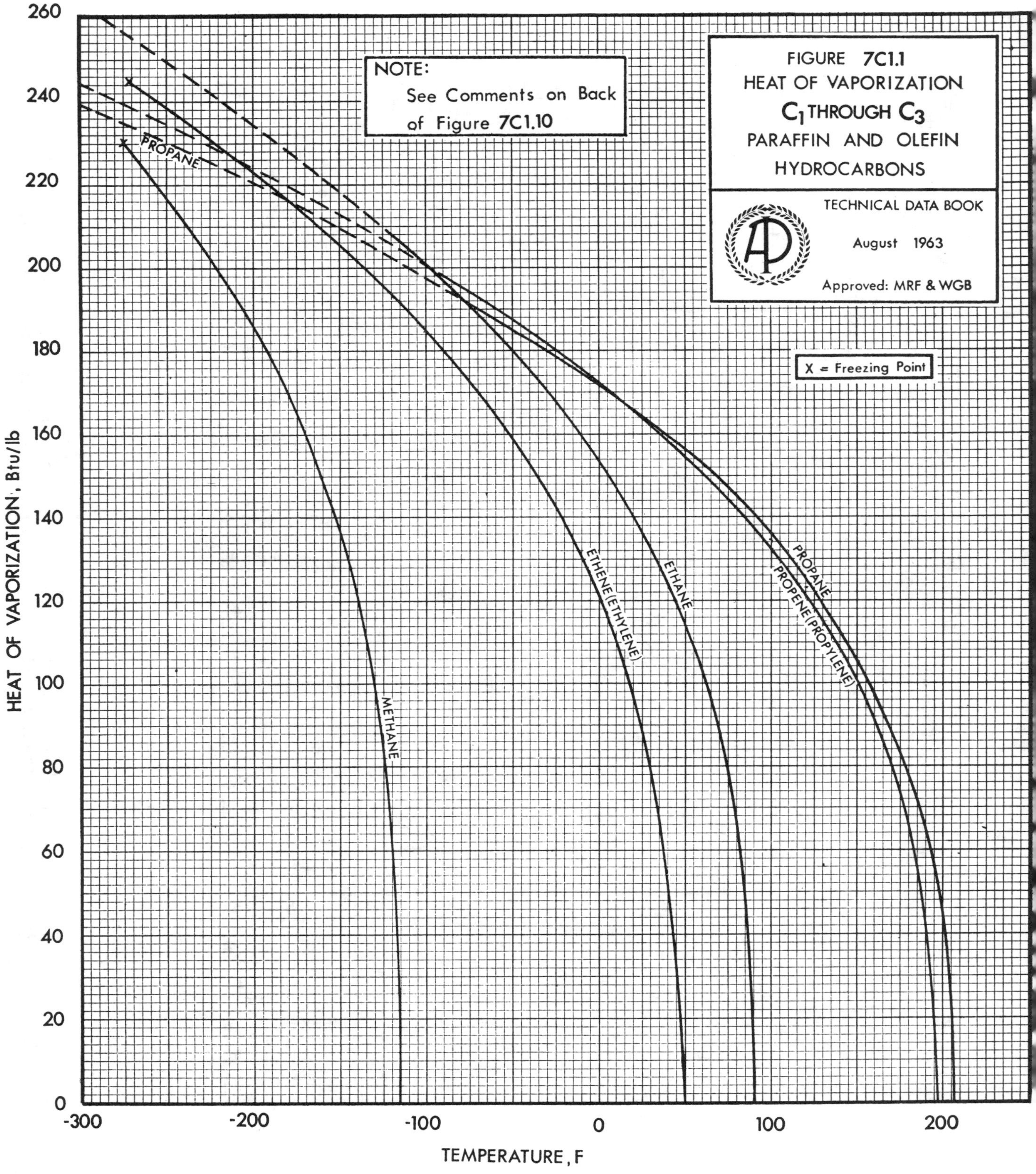

Chart 3.30. Heat of vaporization, C_1 through C_3, paraffin and olefin hydrocarbons. Courtesy American Petroleum Institute.

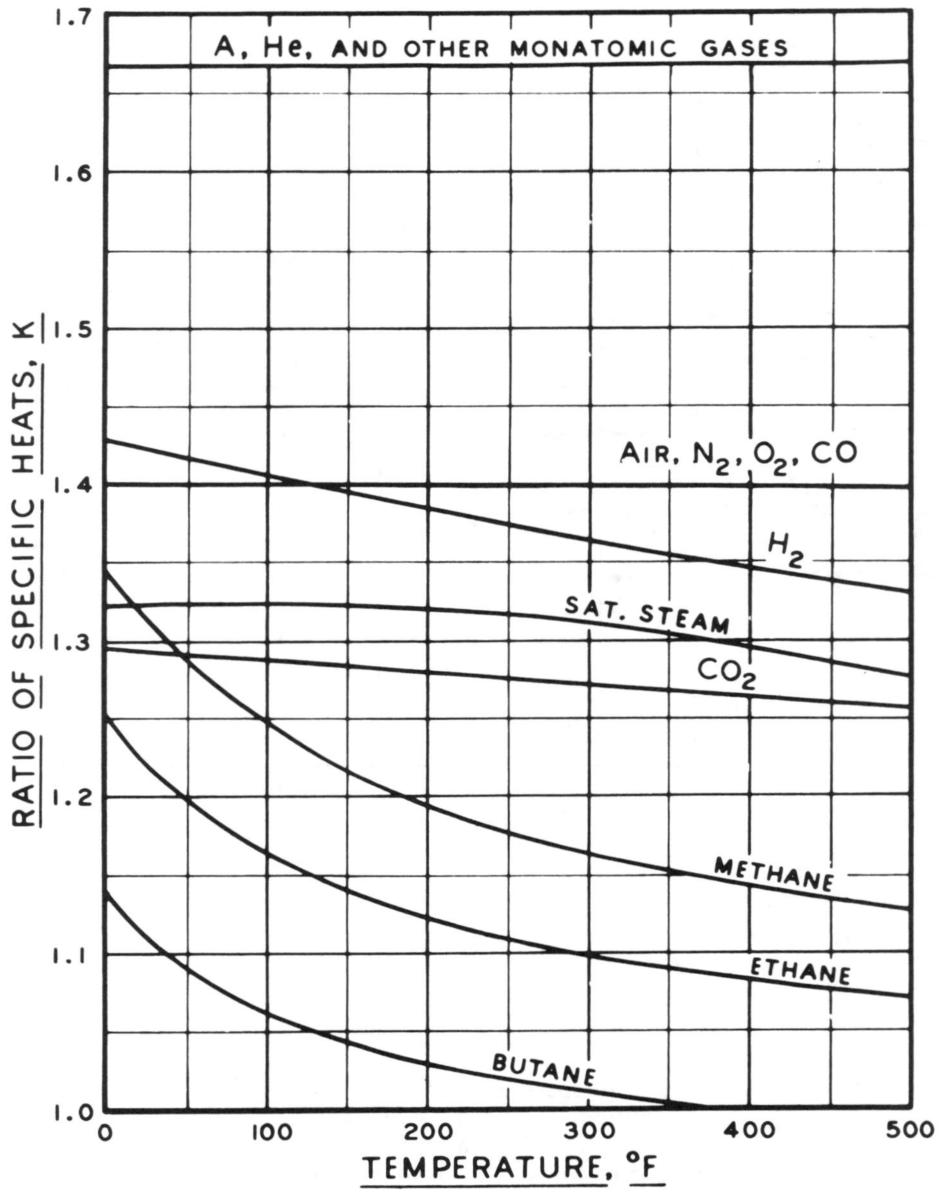

k = ratio of specific heats
c_p = specific heat at constant pressure
c_v = specific heat at constant volume

Chart 3.31. Ratio of specific heats, K, *vs. temperature.* Thermodynamics for Chemical Engineers, *Harold C. Weber, published by John Wiley & Sons, Inc., New York, New York.*

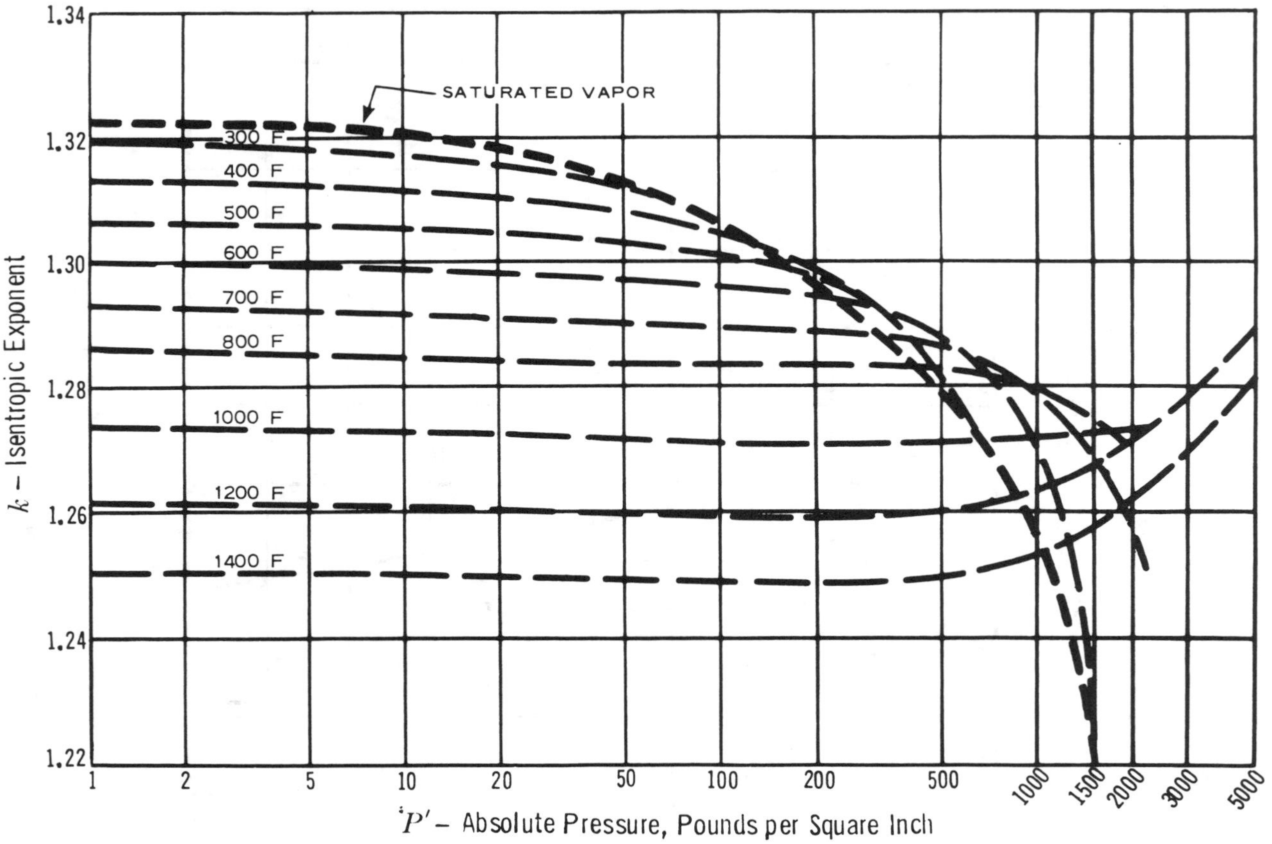

$K = C_p/C_v$

For small changes in pressure (or volume) along an isentropic, pv^k = constant

k = ratio of specific heats
c_p = specific heat at constant pressure
c_v = specific heat at constant volume
P' = absolute pressure, pounds per square inch

Chart 3.32. Ratio of specific heat, K, *for steam. Courtesy Crane Co. (T.P. No. 410) and Keenan & Keyes:* Thermodynamic Properties of Steam, *published by John Wiley & Sons, Inc., New York, New York.*

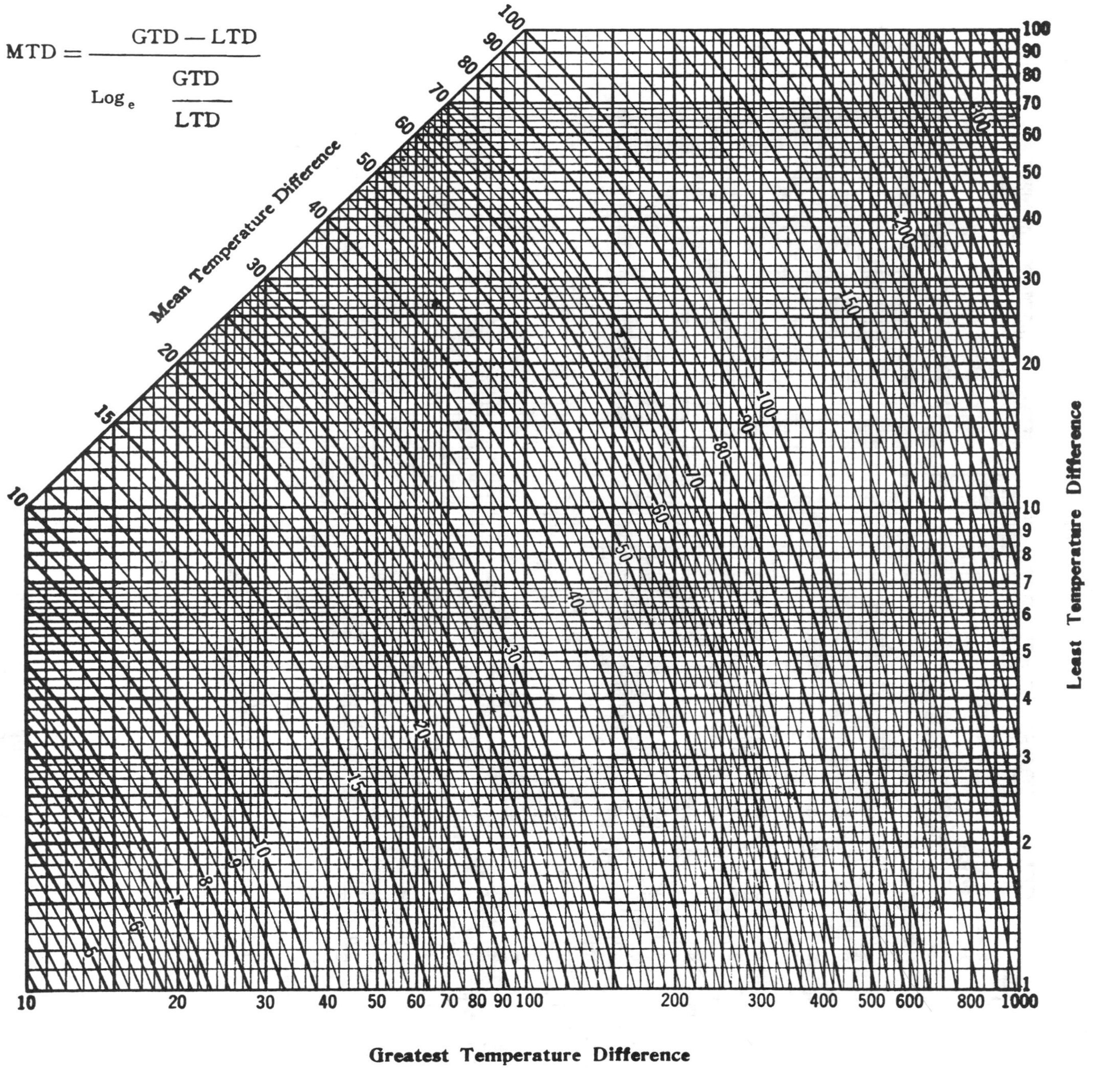

Example: GTD = 110 LTD = 12
From intersection these lines, read upward on curving lines to 44 on MTD scale.

Chart 3.33. Solving for mean temperature difference. Courtesy Natural Gas Processors Suppliers Association.

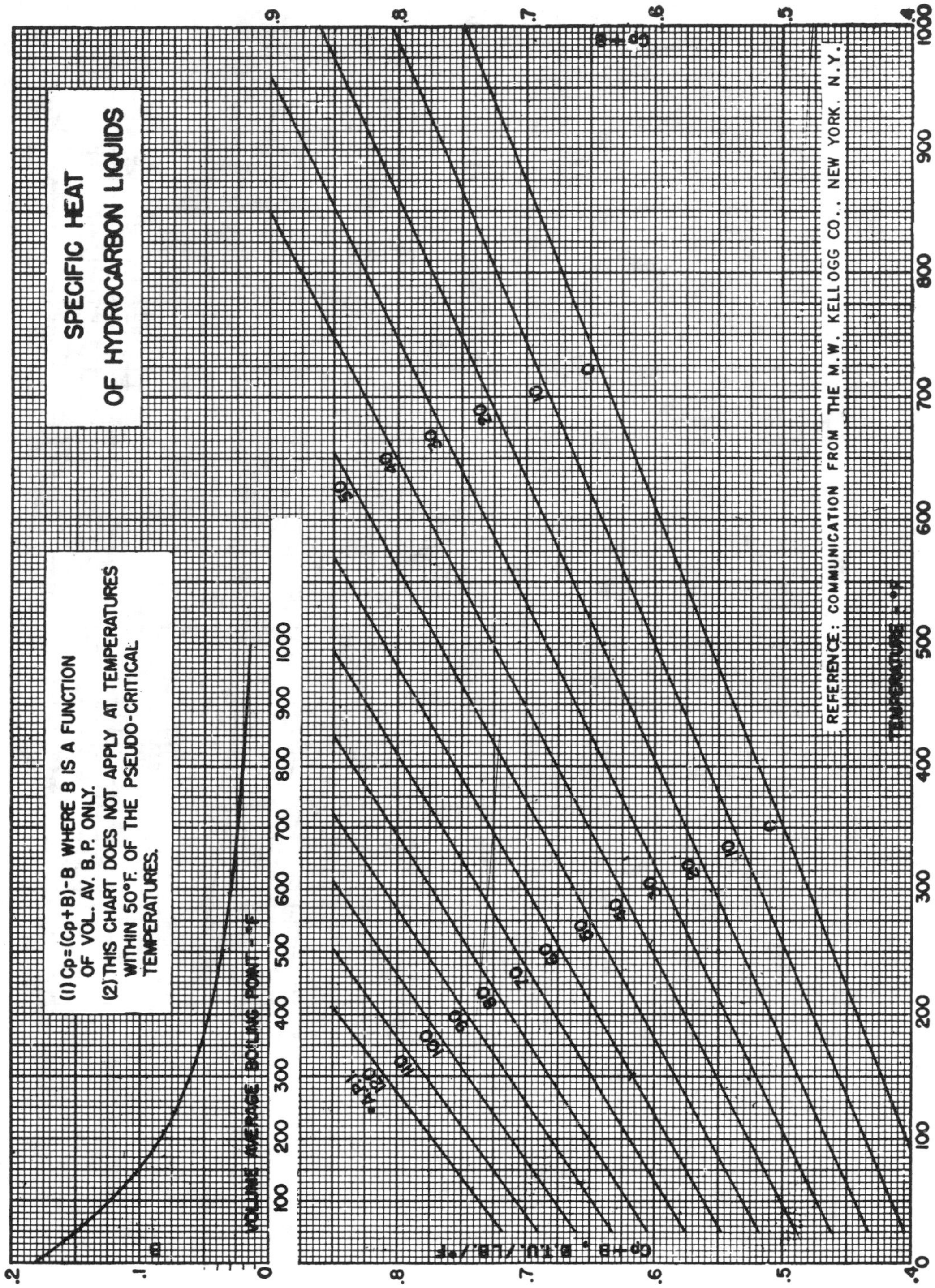

Chart 3.34. Specific heat of hydrocarbon liquids. Data Book on Hydrocarbons, *J.B. Maxwell, copyright © 1950 by Litton Educational Publishing, Inc. Reprinted by permission of Van Nostrand Reinhold Co.*

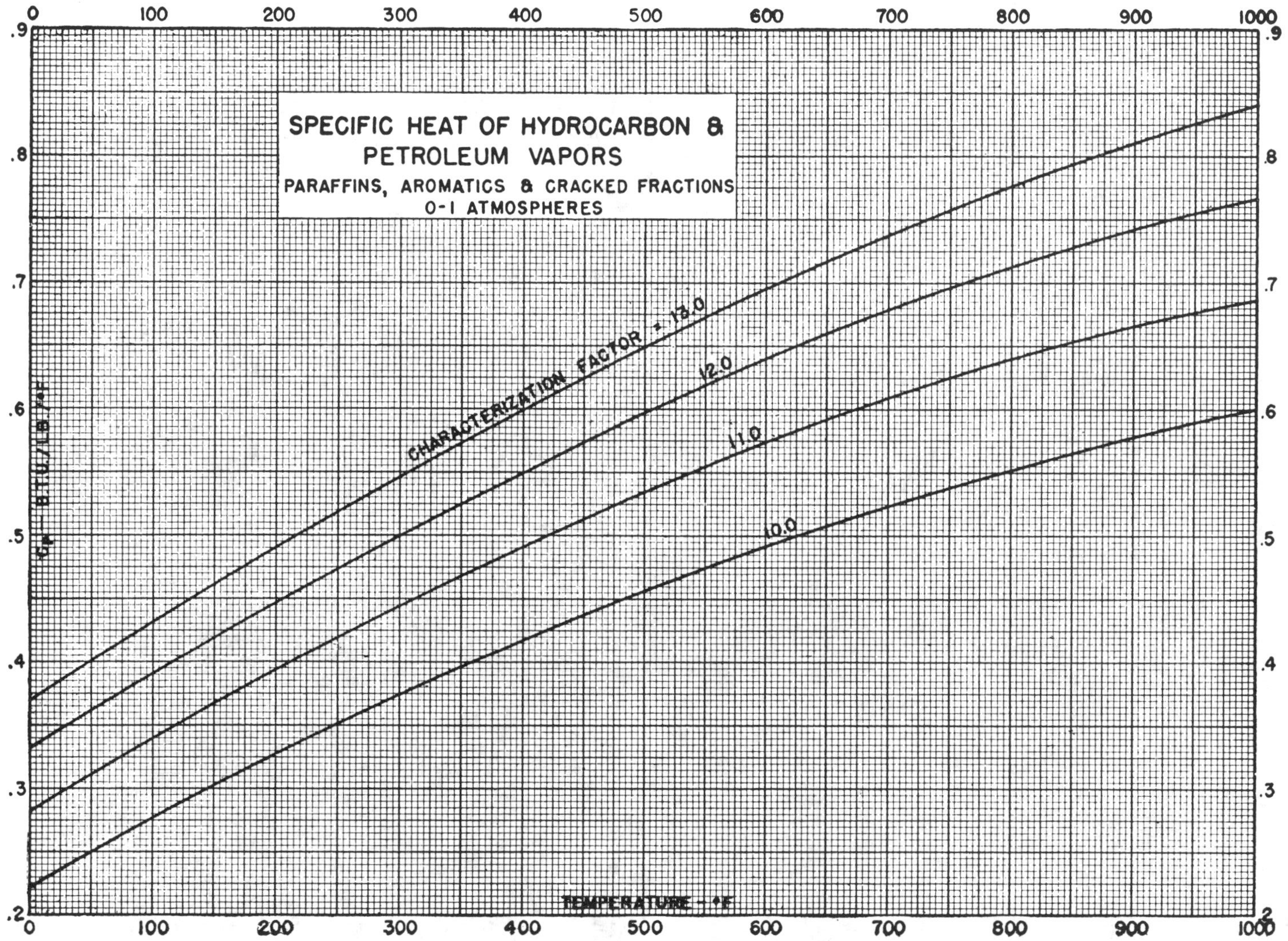

Chart 3.35. Specific heat of hydrocarbon and petroleum vapors. Data Book on Hydrocarbons, *J.B. Maxwell, copyright © 1950 by Litton Educational Publishing, Inc. Reprinted by permission of Van Nostrand Reinhold Co.*

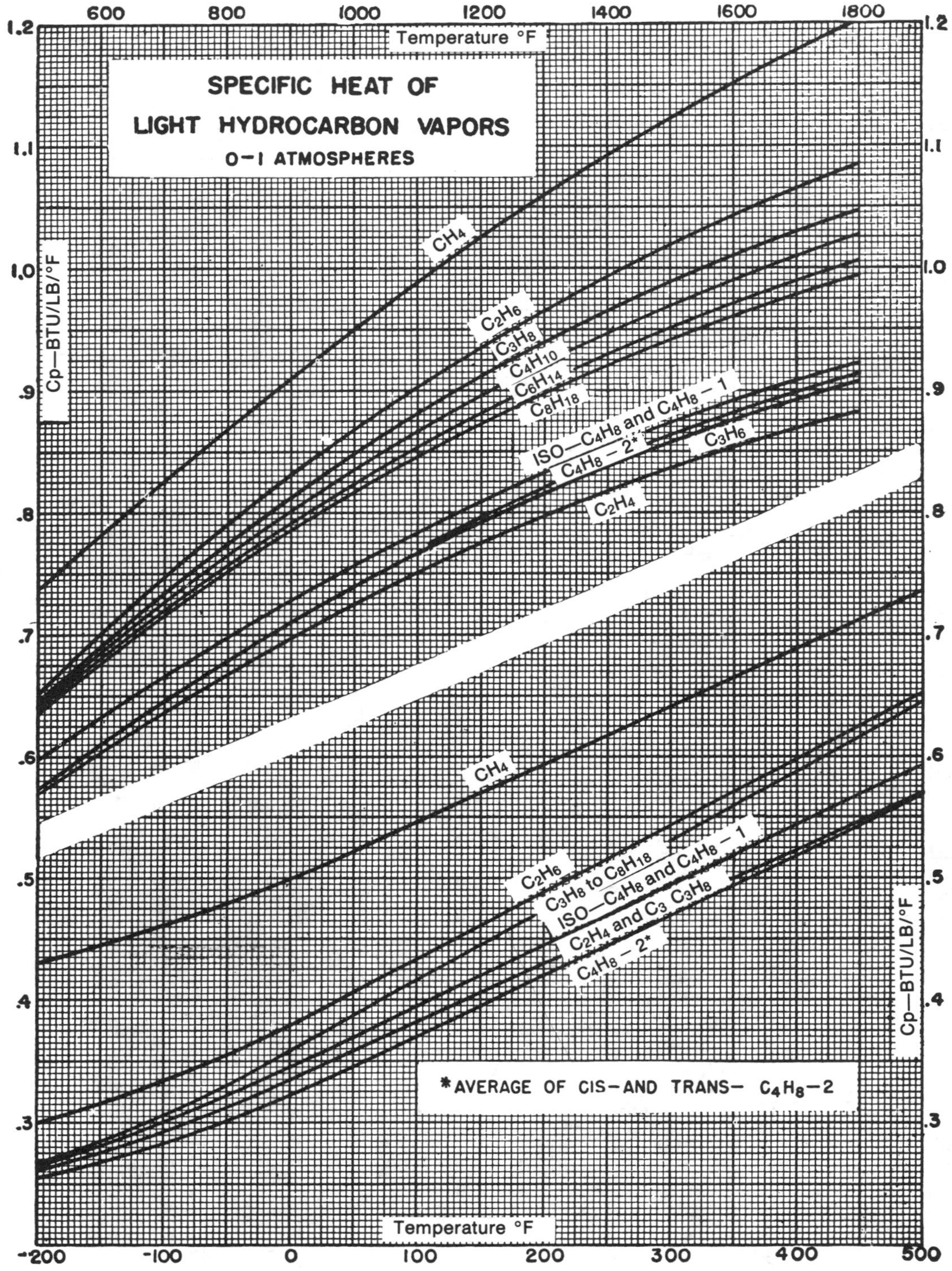

Chart 3.36. Specific heat of light hydrocarbon vapors—0 to 1 atmospheres. Data Book on Hydrocarbons, *J.B. Maxwell,* *copyright © 1950 by Litton Educational Publishing, Inc. Reprinted by permission of Van Nostrand Reinhold Co.*

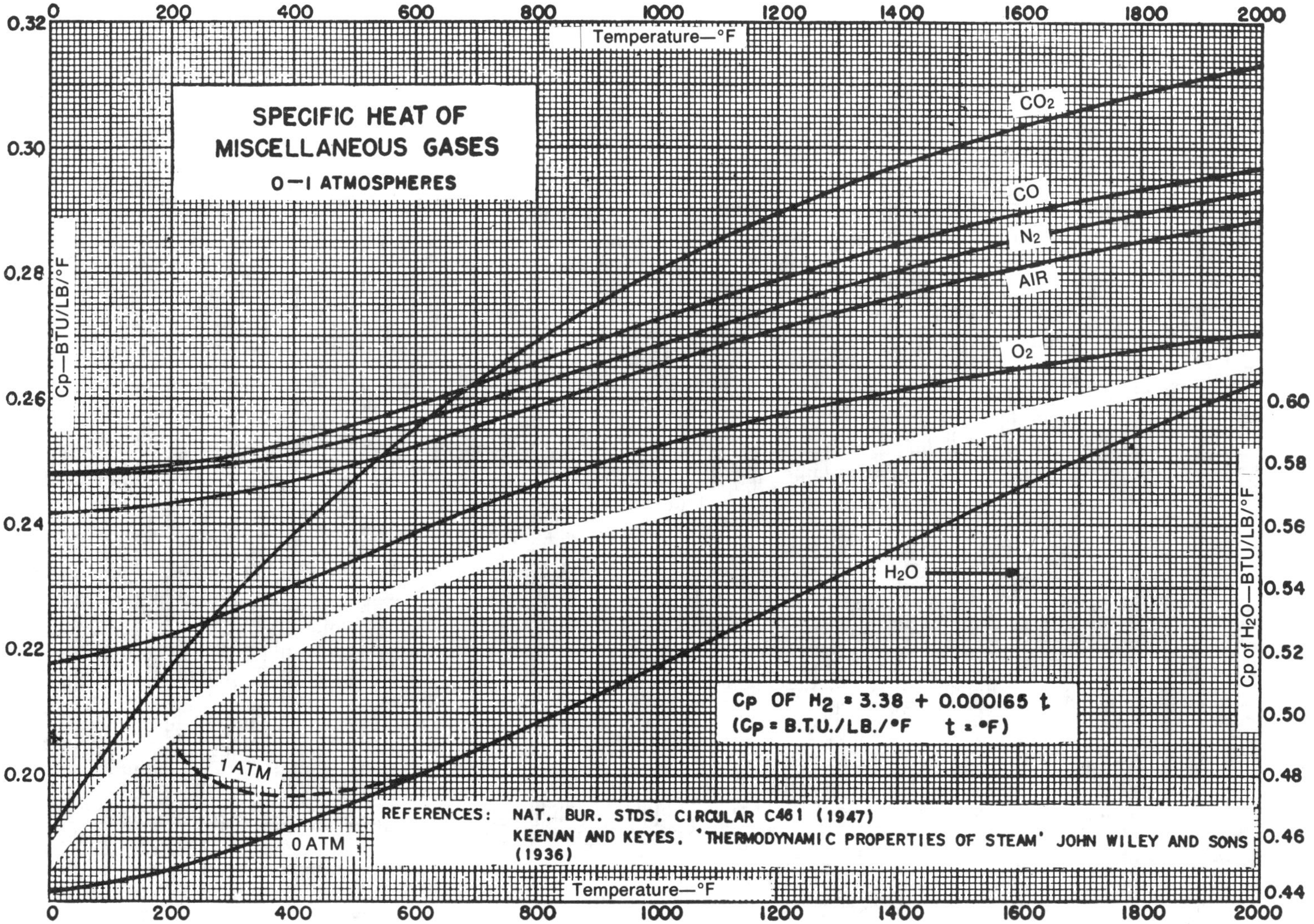

Chart 3.37. Specific heat of miscellaneous gases—0 to 1 atmospheres. Data Book on Hydrocarbons, *J.B. Maxwell, copyright © 1950 by Litton Educational Publishing, Inc. Reprinted by permission of Van Nostrand Reinhold Co.*

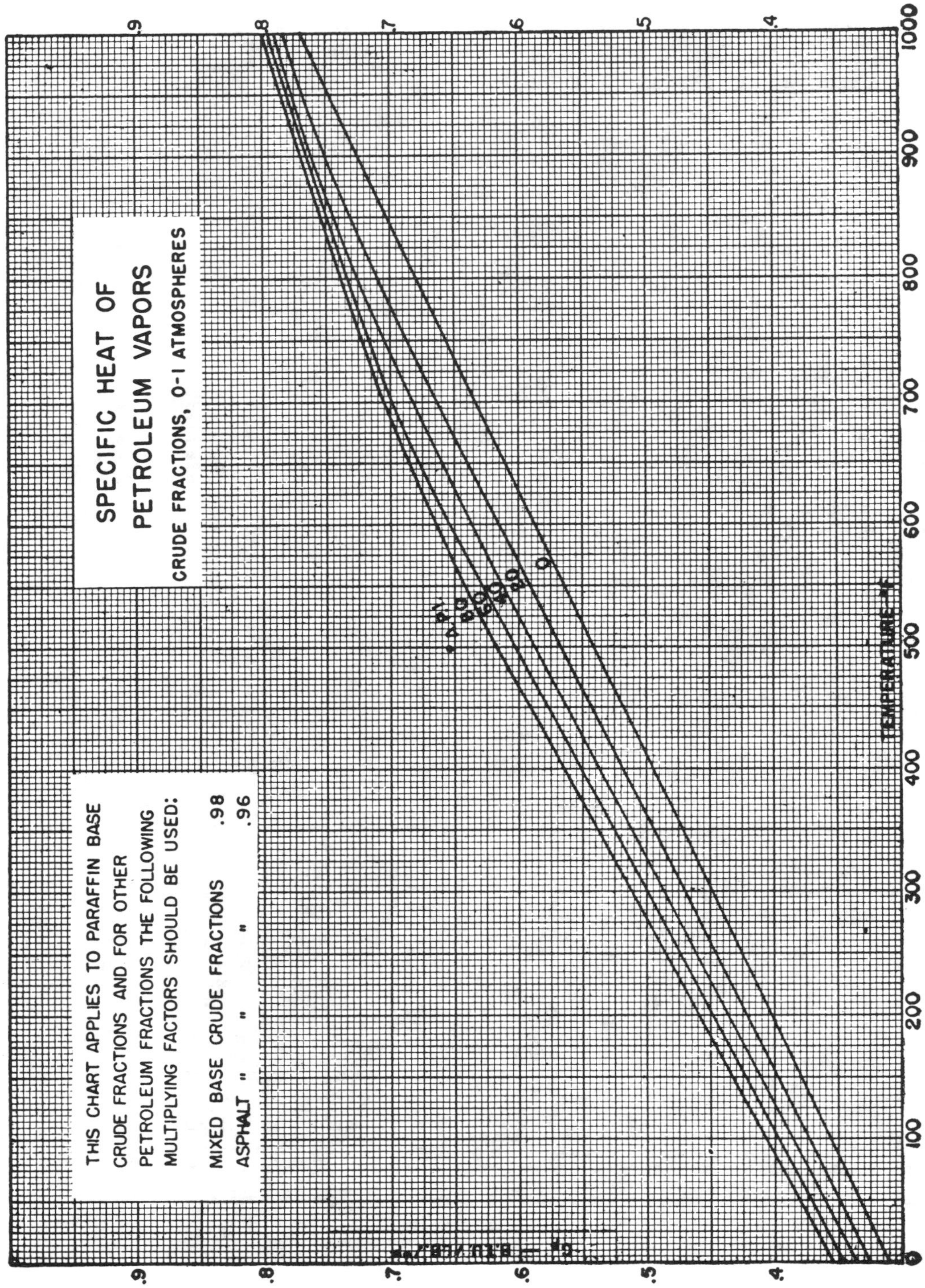

Chart 3.38. Specific heat of petroleum vapors, crude fractions—0 to 1 atmospheres. Data Book on Hydrocarbons, J.B. Maxwell, *copyright © 1950 by Litton Educational Publishing, Inc. Reprinted by permission of Van Nostrand Reinhold Co.*

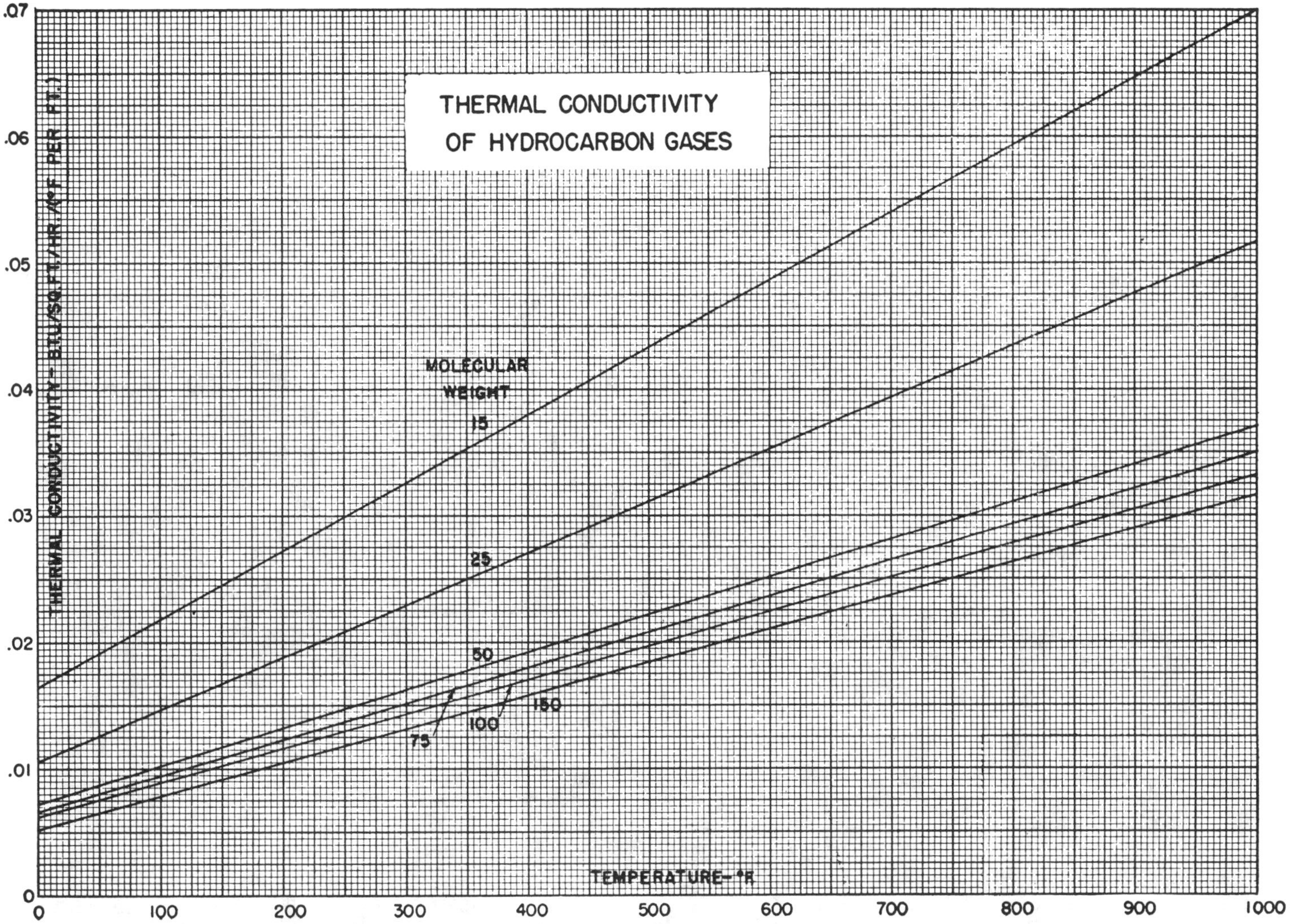

Chart 3.39. Thermal conductivity of hydrocarbon gases. Data Book on Hydrocarbons, *J.B. Maxwell, copyright © 1950 by Litton Educational Publishing, Inc. Reprinted by permission of Van Nostrand Reinhold Co.*

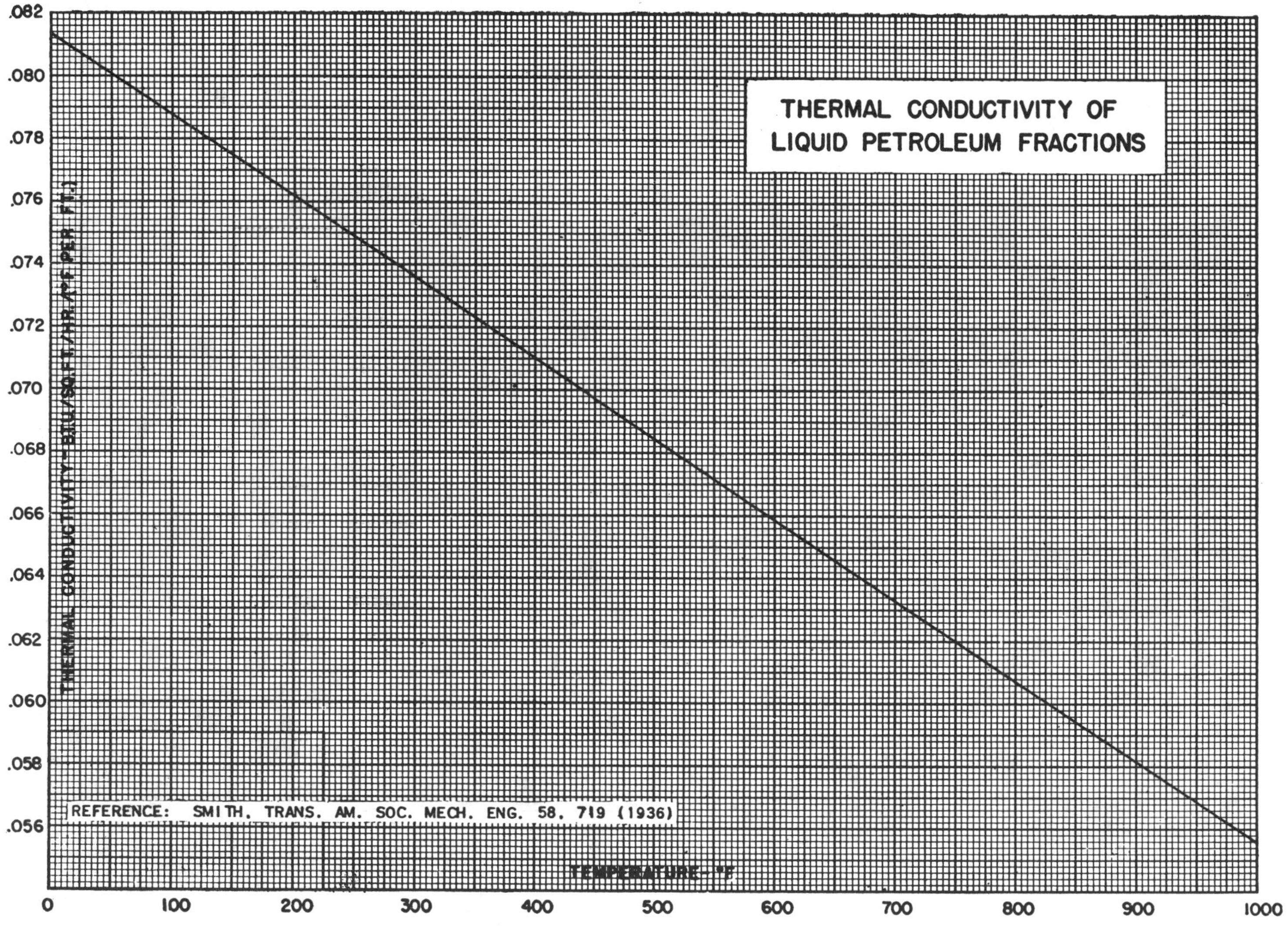

Chart 3.40. Thermal conductivity of liquid petroleum fractions. Data Book on Hydrocarbons, *J.B. Maxwell, copyright © 1950 by Litton Educational Publishing, Inc. Reprinted by permission of Van Nostrand Reinhold Co.*

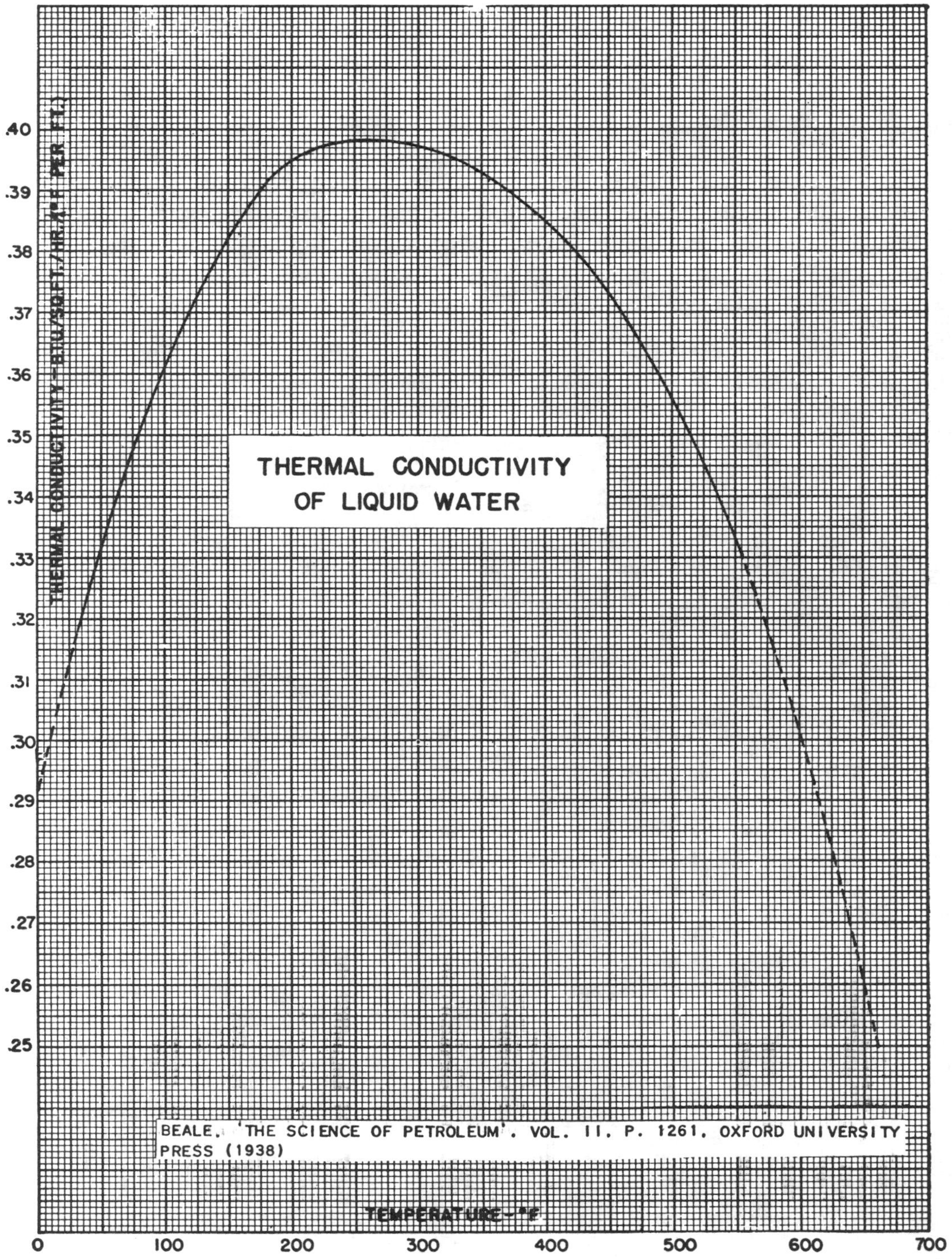

Chart 3.41. Thermal conductivity of liquid water. Data Book on Hydrocarbons, *J.B. Maxwell, copyright © 1950 by Litton Educational Publishing, Inc. Reprinted by permission of Van Nostrand Reinhold Co.*

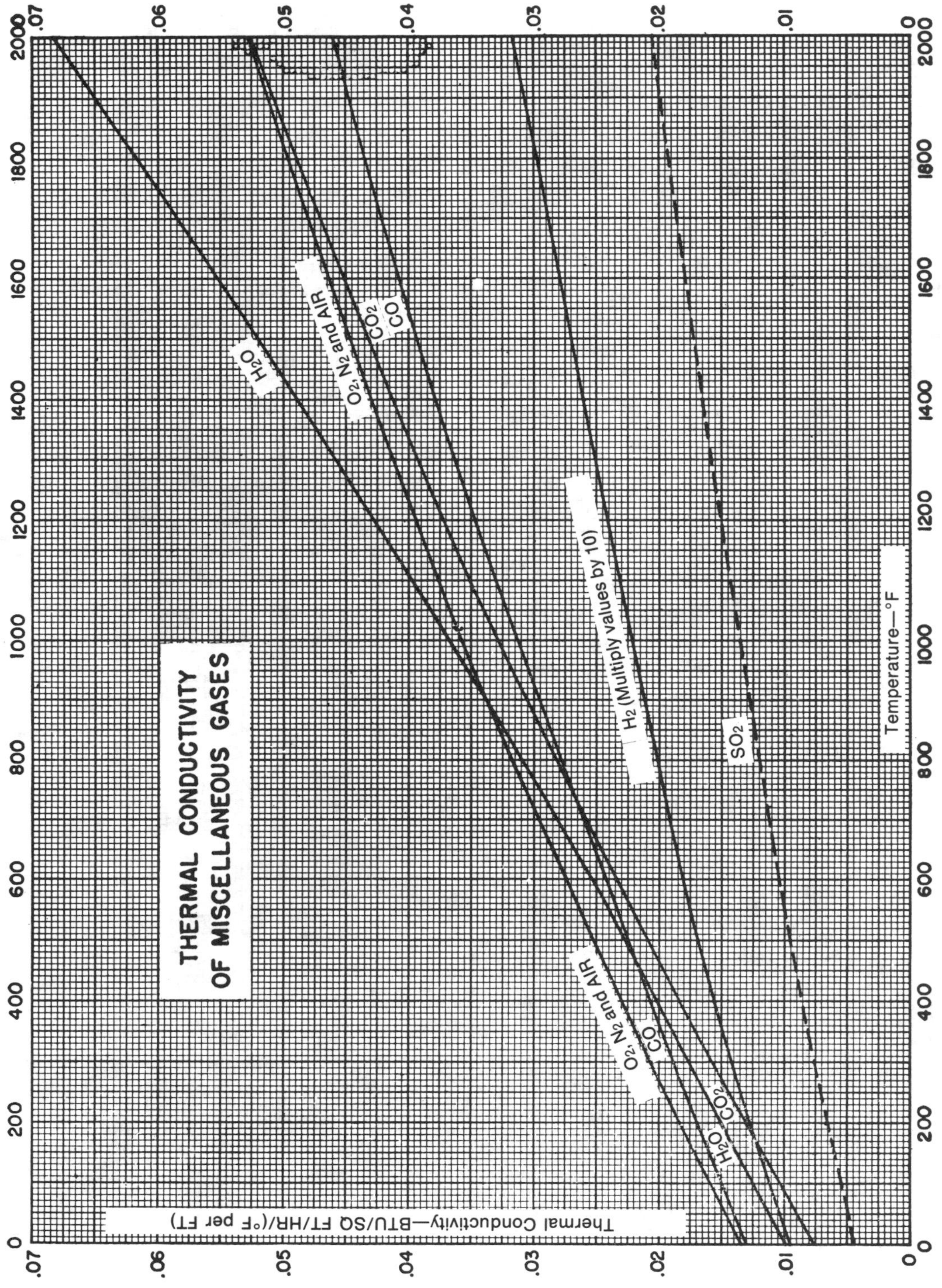

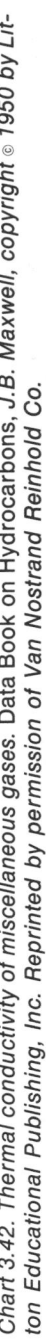

Chart 3.42. Thermal conductivity of miscellaneous gases. Data Book on Hydrocarbons, *J.B. Maxwell, copyright © 1950 by Litton Educational Publishing, Inc. Reprinted by permission of Van Nostrand Reinhold Co.*

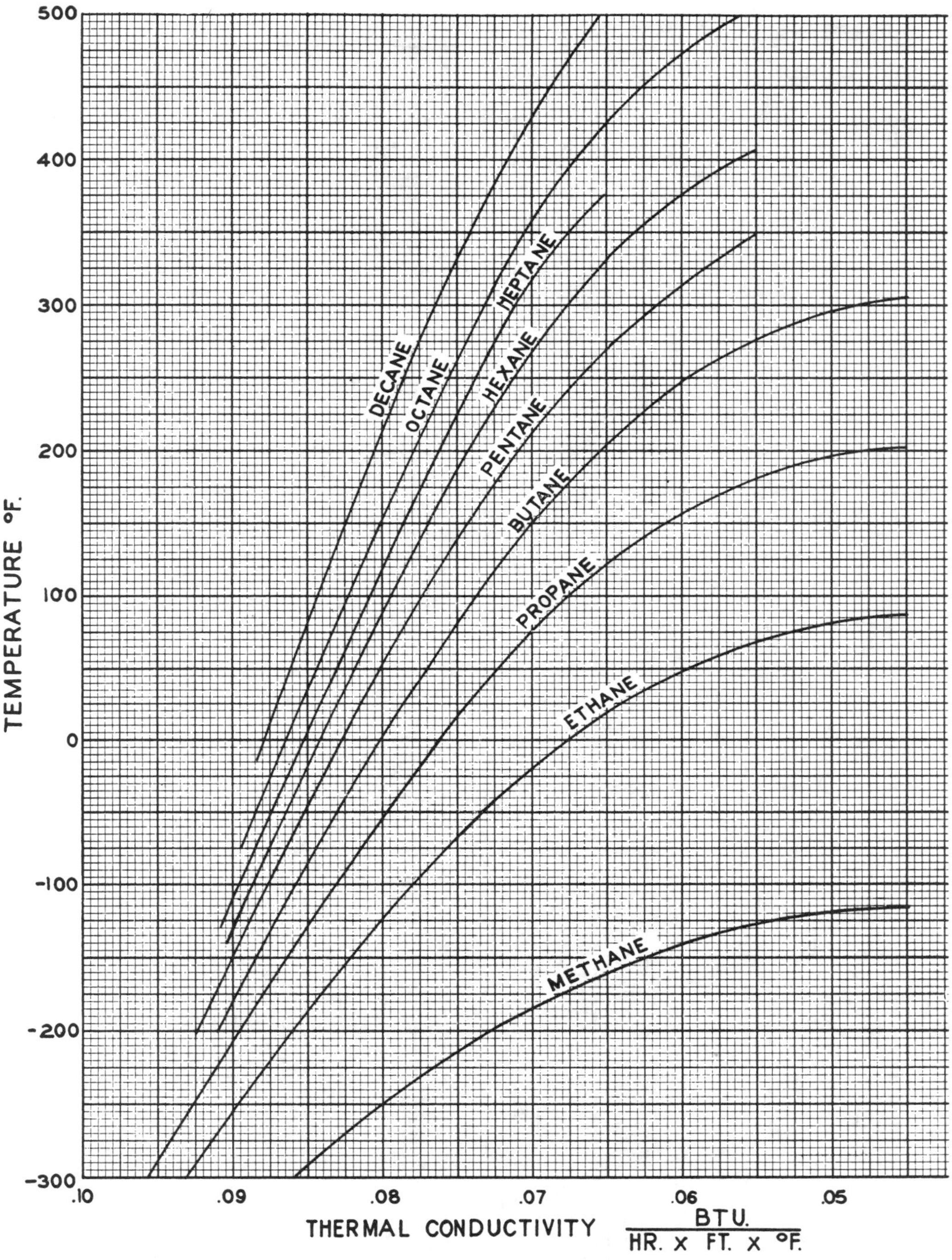

Chart 3.43. Thermal conductivity of normal paraffinic hydrocarbon liquids. Courtesy Natural Gas Processors Suppliers Association.

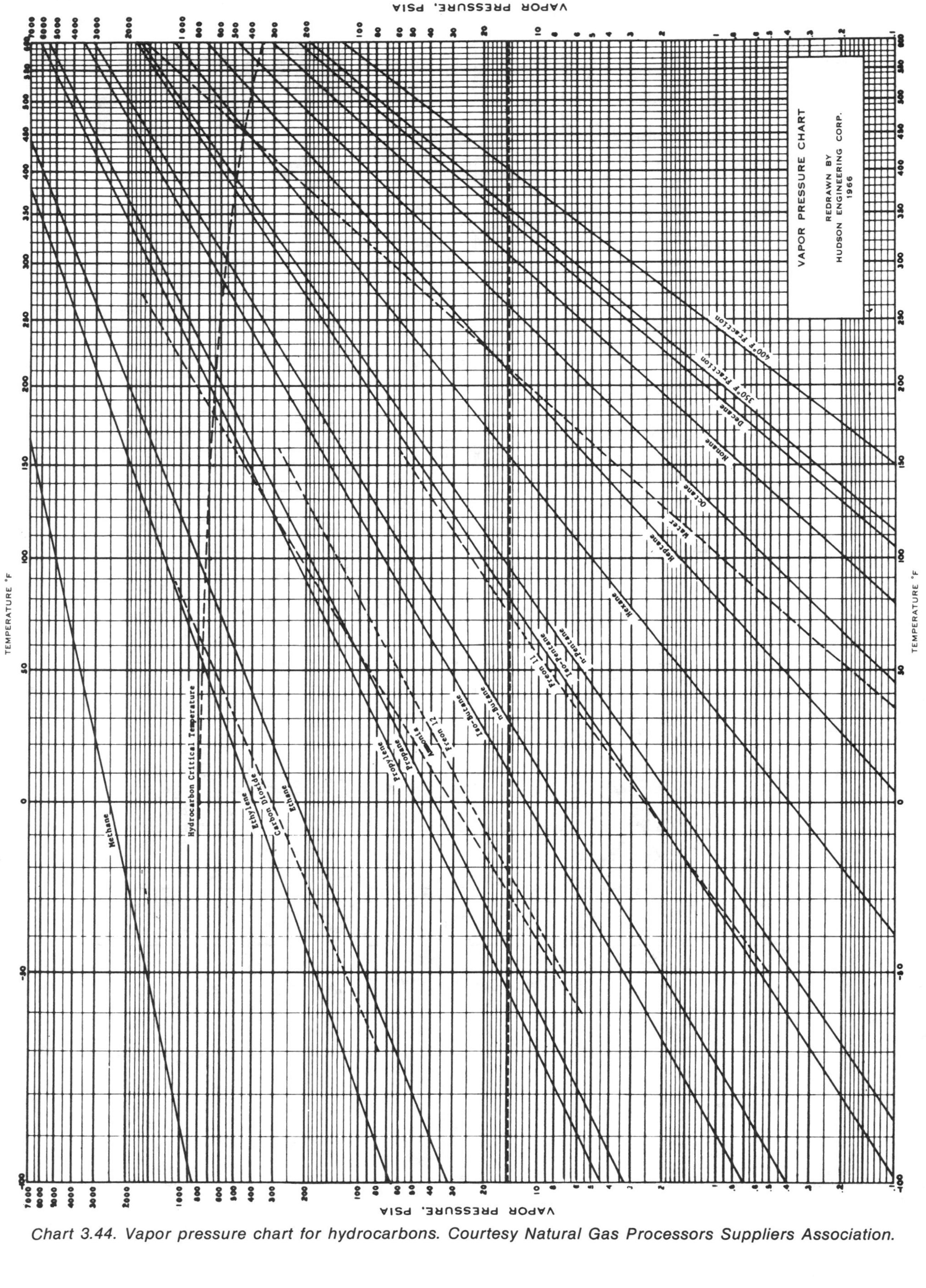

Chart 3.44. Vapor pressure chart for hydrocarbons. Courtesy Natural Gas Processors Suppliers Association.

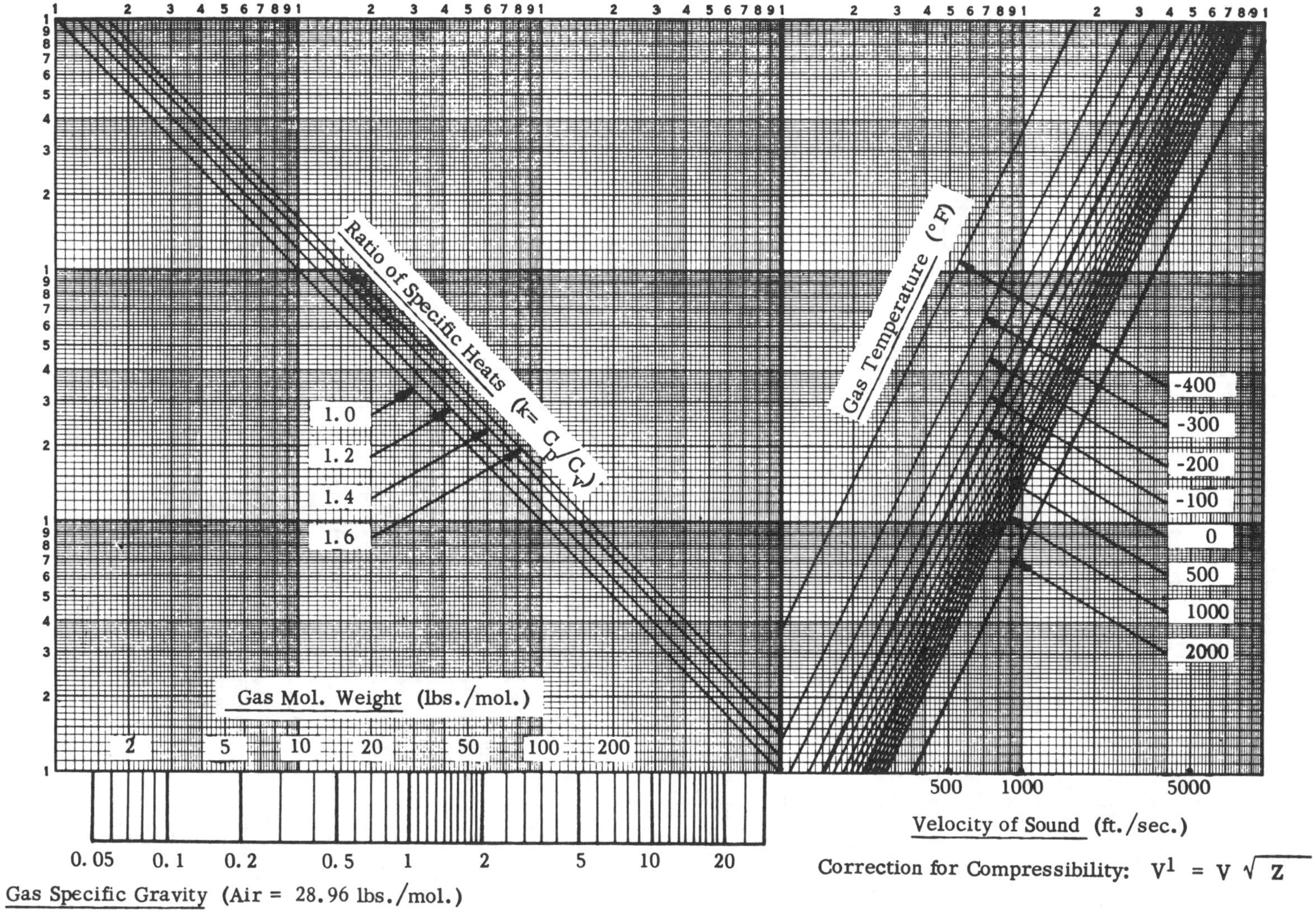

V = velocity, feet per second
V' = velocity, feet per second, corrected for compressibility
Z = compressibility factor
C_p = specific heat at constant pressure
C_v = specific heat at constant volume

Chart 3.45. Velocity of sound in gases. Courtesy Natural Gas Processors Suppliers Association.

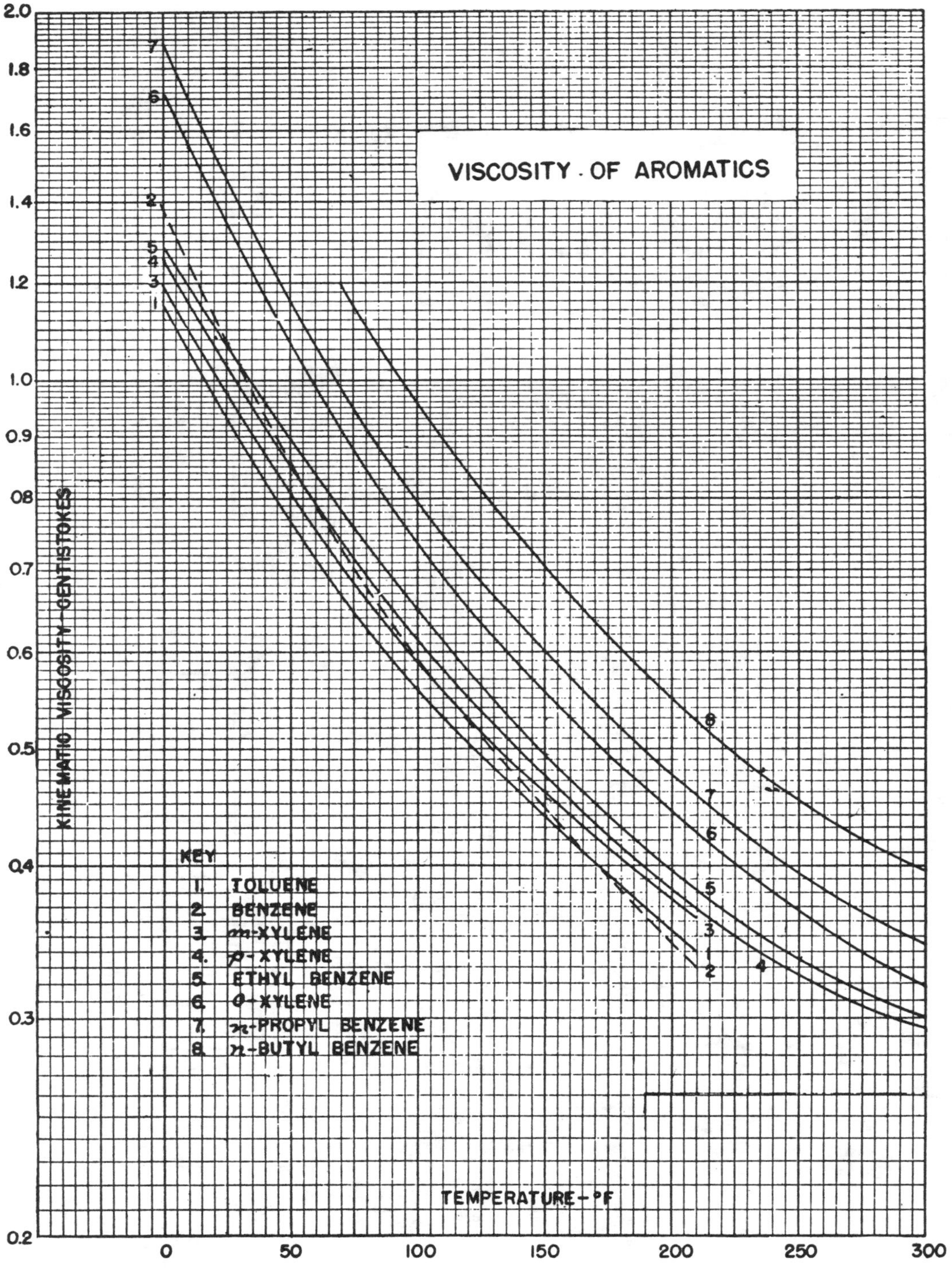

Chart 3.46. Viscosity of aromatics. Data Book on Hydrocarbons, *J.B. Maxwell, copyright © 1950 by Litton Educational Publishing, Inc. Reprinted by permission of Van Nostrand Reinhold Co.*

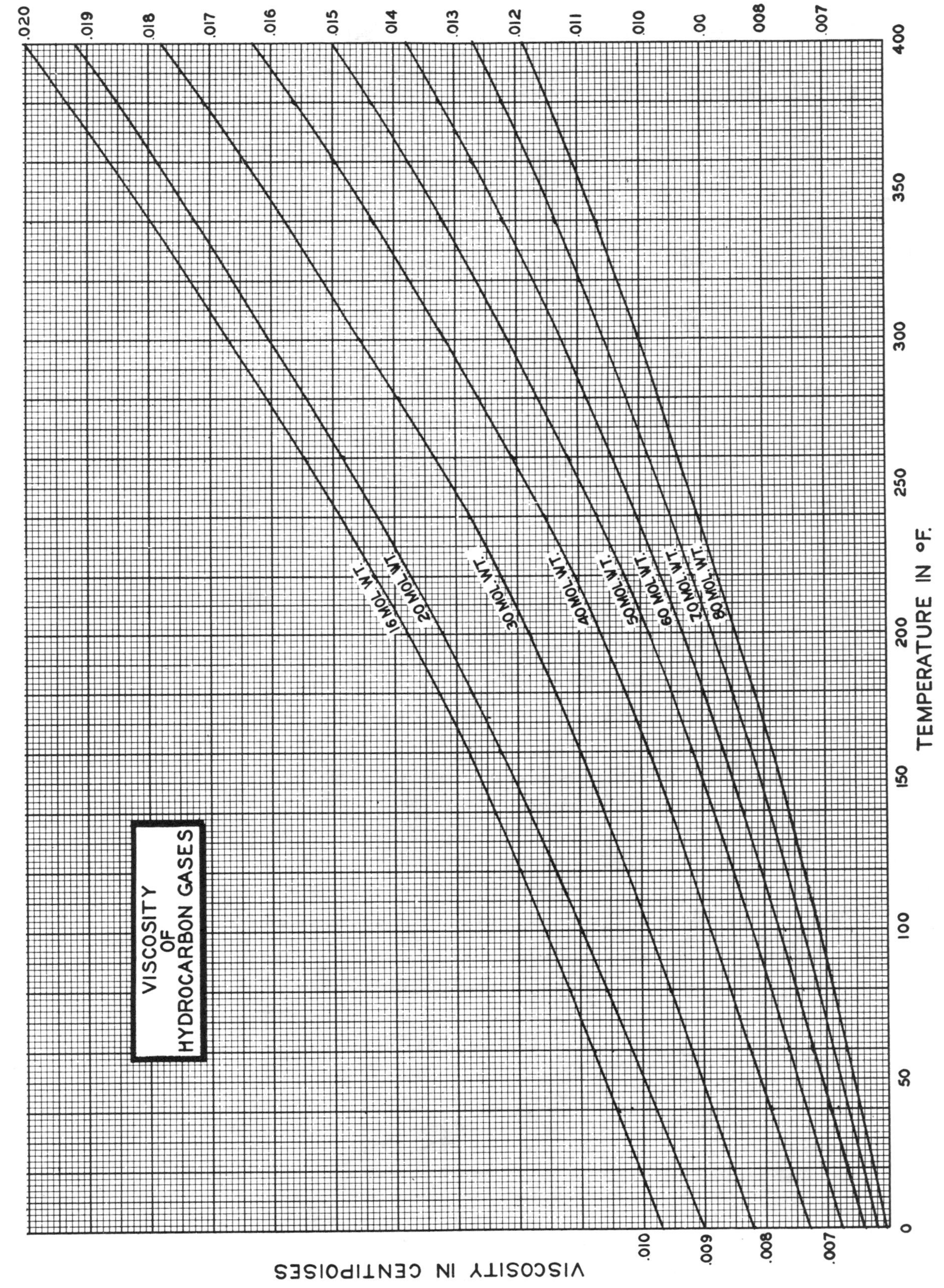

Chart 3.47. Viscosity of hydrocarbon gases. Courtesy Natural Gas Processors Suppliers Association.

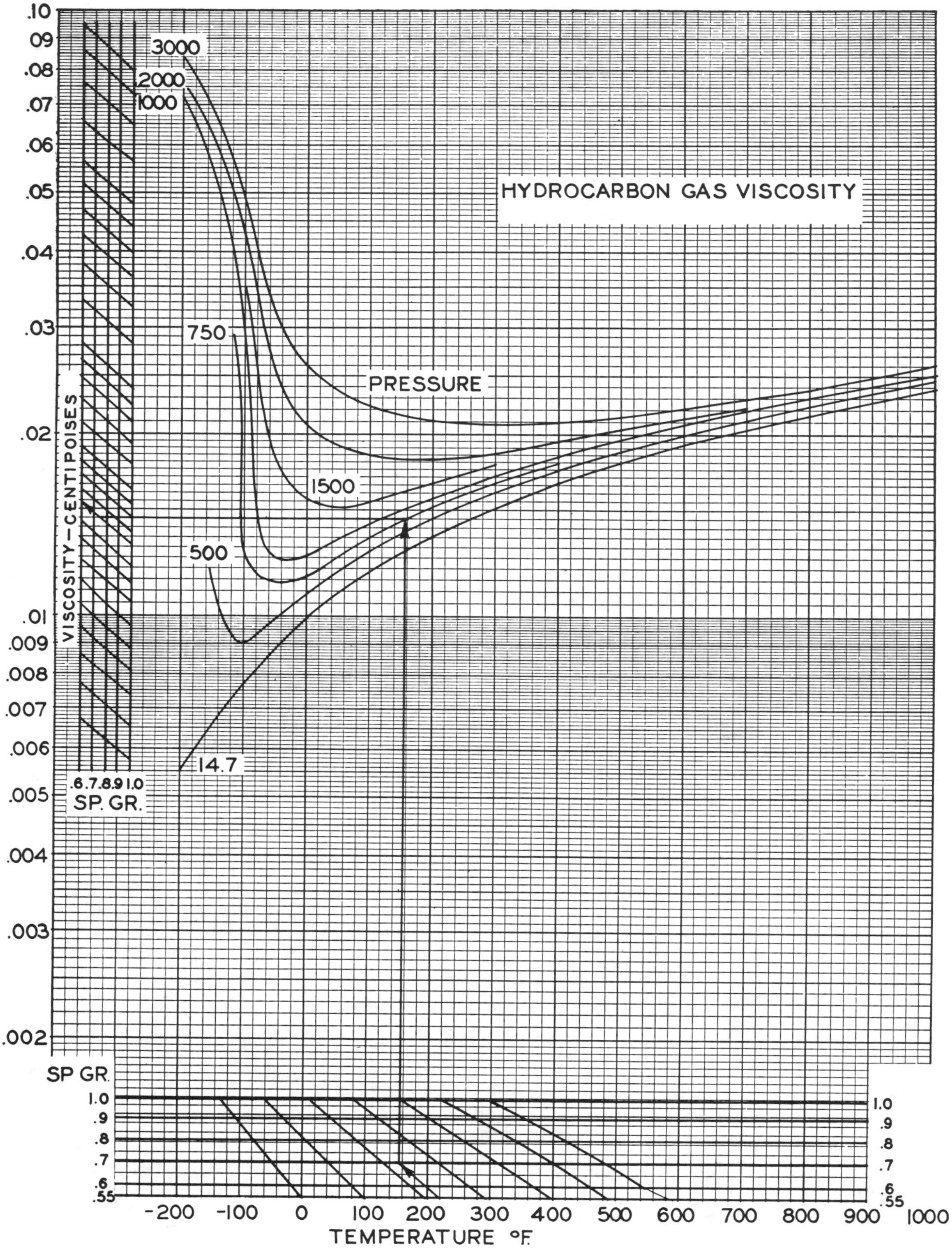

Chart 3.48. Hydrocarbon gas viscosity. Courtesy Natural Gas Processors Suppliers Association.

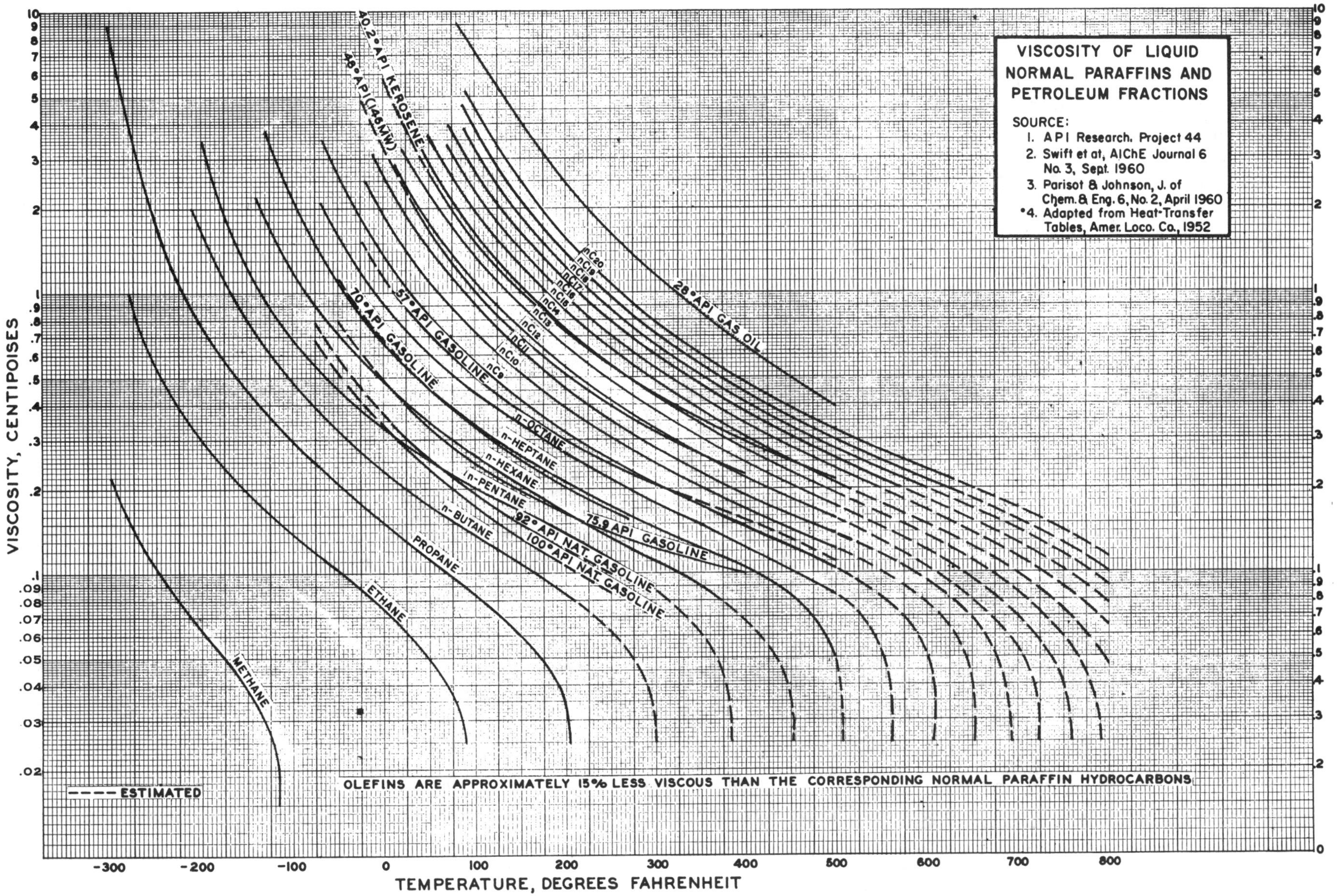

Chart 3.49. Viscosity of liquid normal paraffins and petroleum fractions. Courtesy Natural Gas Processors Suppliers Association.

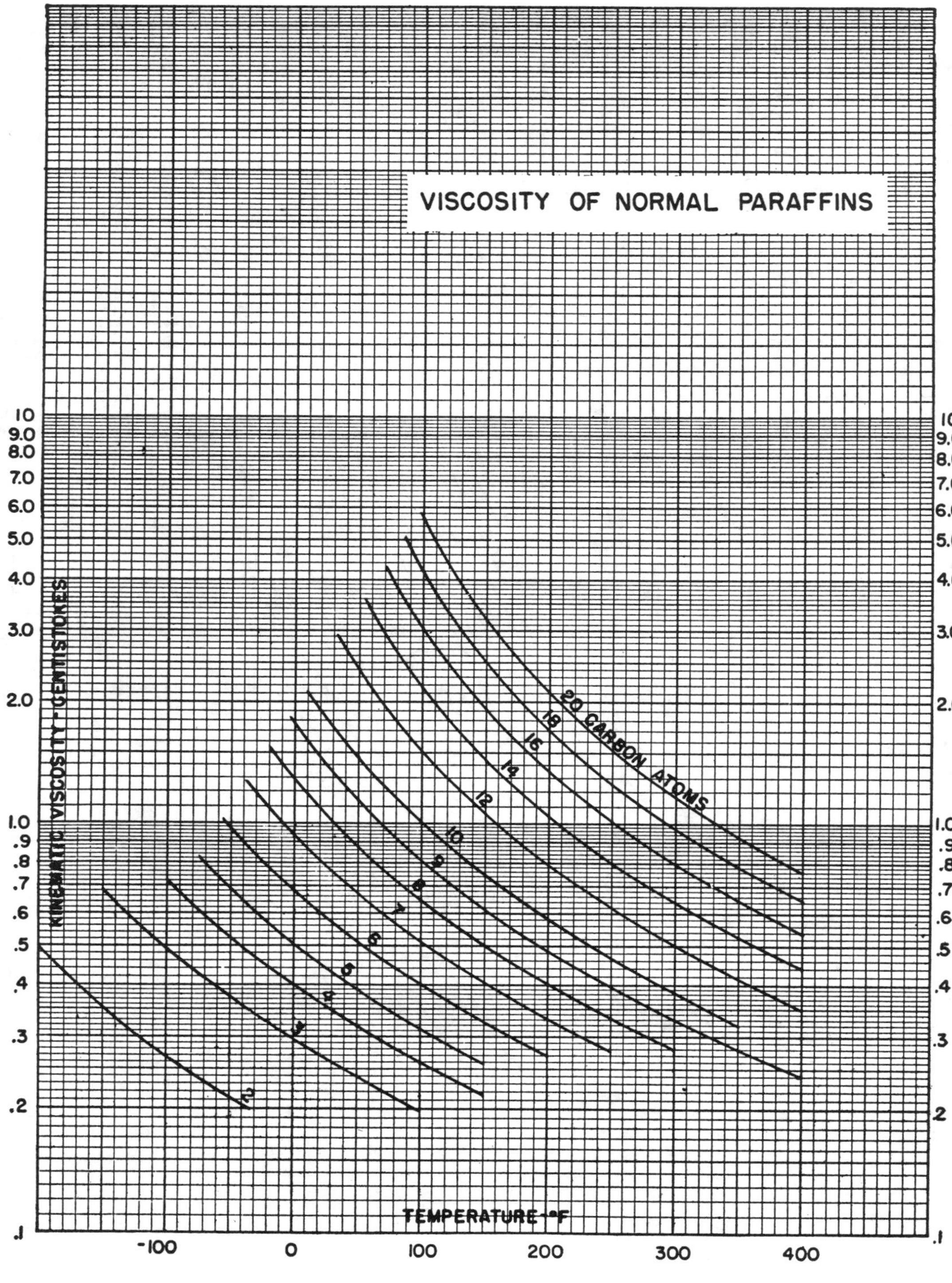

Chart 3.50. Viscosity of normal paraffins. Data Book on Hydrocarbons, *J.B. Maxwell, copyright © 1950 by Litton Educational Publishing, Inc. Reprinted by permission of Van Nostrand Reinhold Co.*

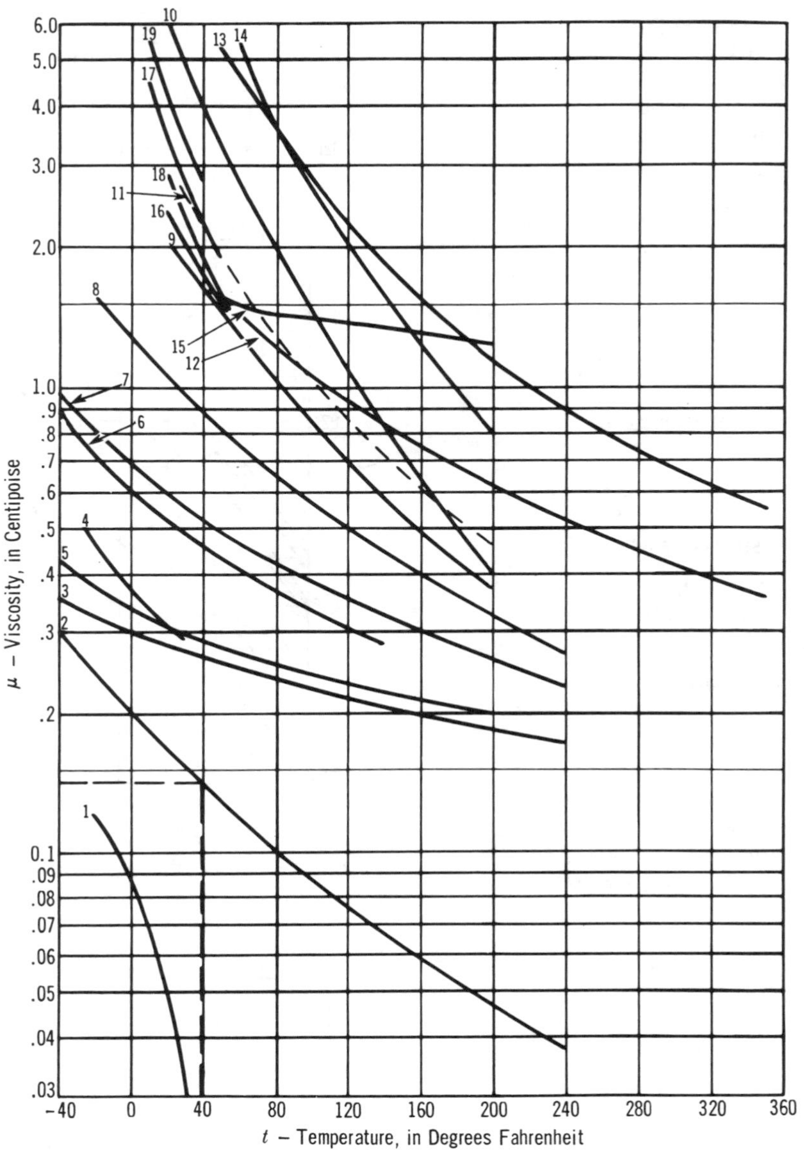

1. Carbon Dioxide . . CO_2
2. Ammonia . . . . NH_3
3. Methyl Chloride . . CH_3Cl
4. Sulphur Dioxide . . SO_2
5. Freon 12 . . . . F-12
6. Freon 114 . . . . F-114
7. Freon 11 . . . . F-11
8. Freon 113 . . . . F-113
9. Ethyl Alcohol
10. Isopropyl Alcohol
11. 20% Sulphuric Acid . . . . . . 20% H_2SO_4
12. Dowtherm E
13. Dowtherm A
14. 20% Sodium Hydroxide . . 20% NaOH
15. Mercury
16. 10% Sodium Chloride Brine . . . 10% NaCl
17. 20% Sodium Chloride Brine . . . 20% NaCl
18. 10% Calcium Chloride Brine . . 10% $CaCl_2$
19. 20% Calcium Chloride Brine . . 20% $CaCl_2$

Example: The viscosity of ammonia at 40 F is 0.14 centipoise.

Chart 3.51. Viscosity of various liquids. Courtesy Crane Co. (T.P. No. 410).

The curves for hydrocarbon vapors and natural gases in the chart at the upper right are taken from Maxwell ; the curves for all other gases (except helium) in the chart are based upon Sutherland's formula, as follows:

$$\mu = \mu_0 \left(\frac{0.555\ T_0 + C}{0.555\ T + C}\right)\left(\frac{T}{T_0}\right)^{3/2}$$

where:

μ = viscosity, in centipoise at temperature T.

μ_0 = viscosity, in centipoise at temperature T_0.

T = absolute temperature, in degrees Rankine (460 + deg. F) for which viscosity is desired.

T_0 = absolute temperature, in degrees Rankine, for which viscosity is known.

C = Sutherland's constant.

Note: The variation of viscosity with pressure is small for most gases. For gases given on this page, the correction of viscosity for pressure is less than 10 per cent for pressures up to 500 pounds per square inch.

Fluid	Approximate Values of "C"
O_2	127
Air	120
N_2	111
CO_2	240
CO	118
SO_2	416
NH_3	370
H_2	72

Upper chart example: The viscosity of sulphur dioxide gas (SO_2) at 200 F is 0.016 centipoise.

Lower chart example: The viscosity of carbon dioxide gas (CO_2) at about 80 F is 0.015 centipoise.

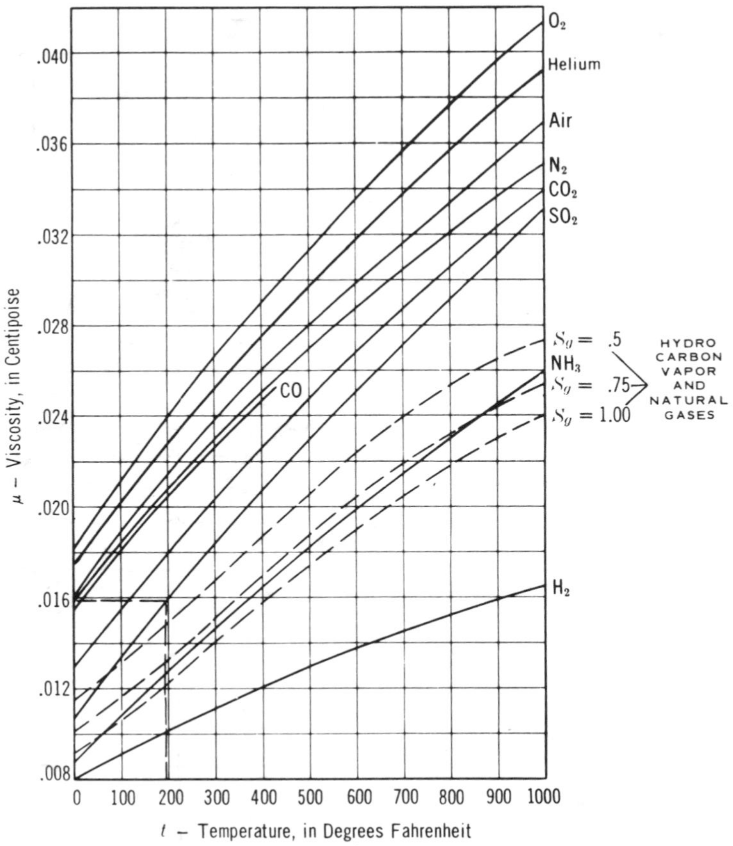

Chart 3.52. Viscosity of various gases. Courtesy Crane Co. (T.P. No. 410).

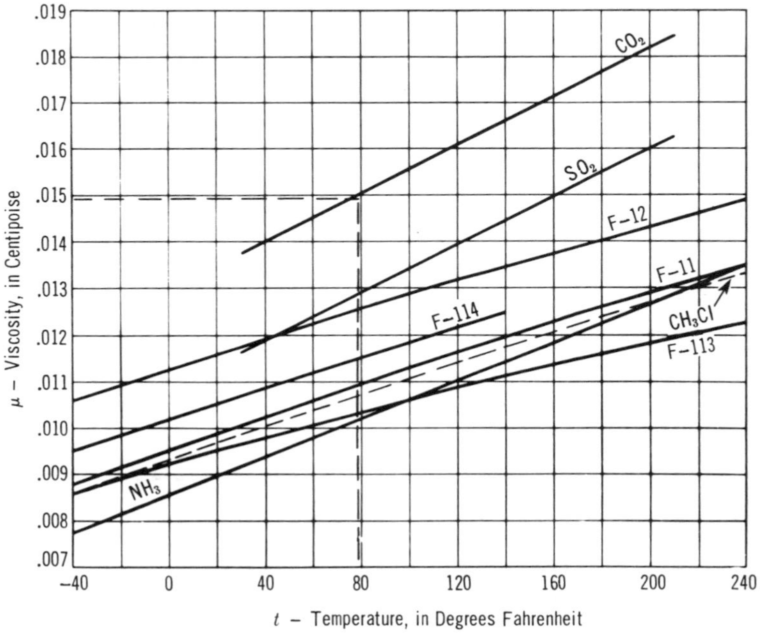

Chart 3.53. Viscosity of refrigerant vapors. Courtesy Crane Co. (T.P. No. 410).

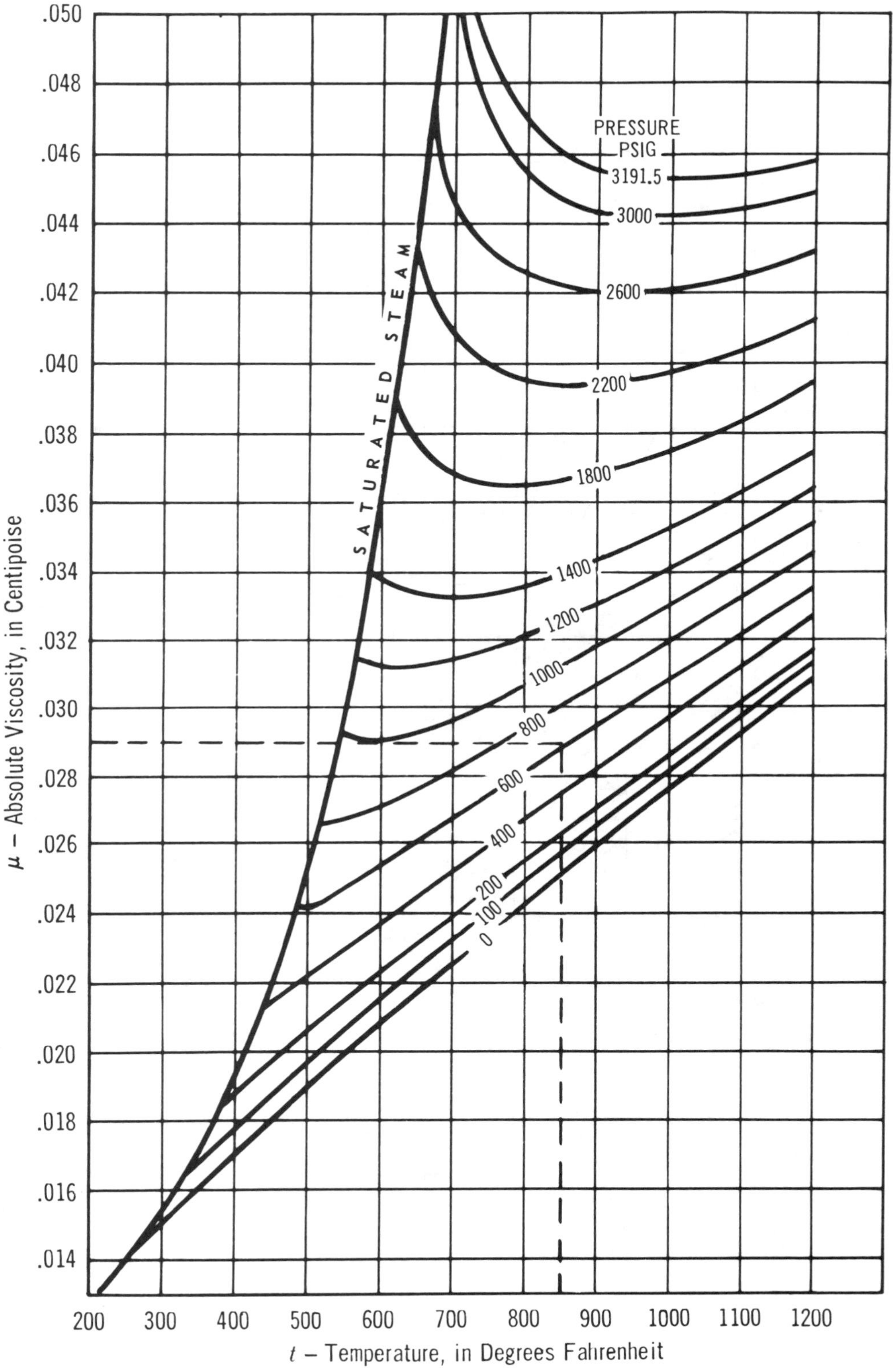

Example: Viscosity of 600 psig, 850 F steam is 0.029 centipoise.

Chart 3.54. Viscosity of steam. Courtesy Crane Co. (T.P. No. 410) and adapted from Steam Power Plants, *Phillip J. Potter, copyright © 1949 by The Ronald Press Co.*

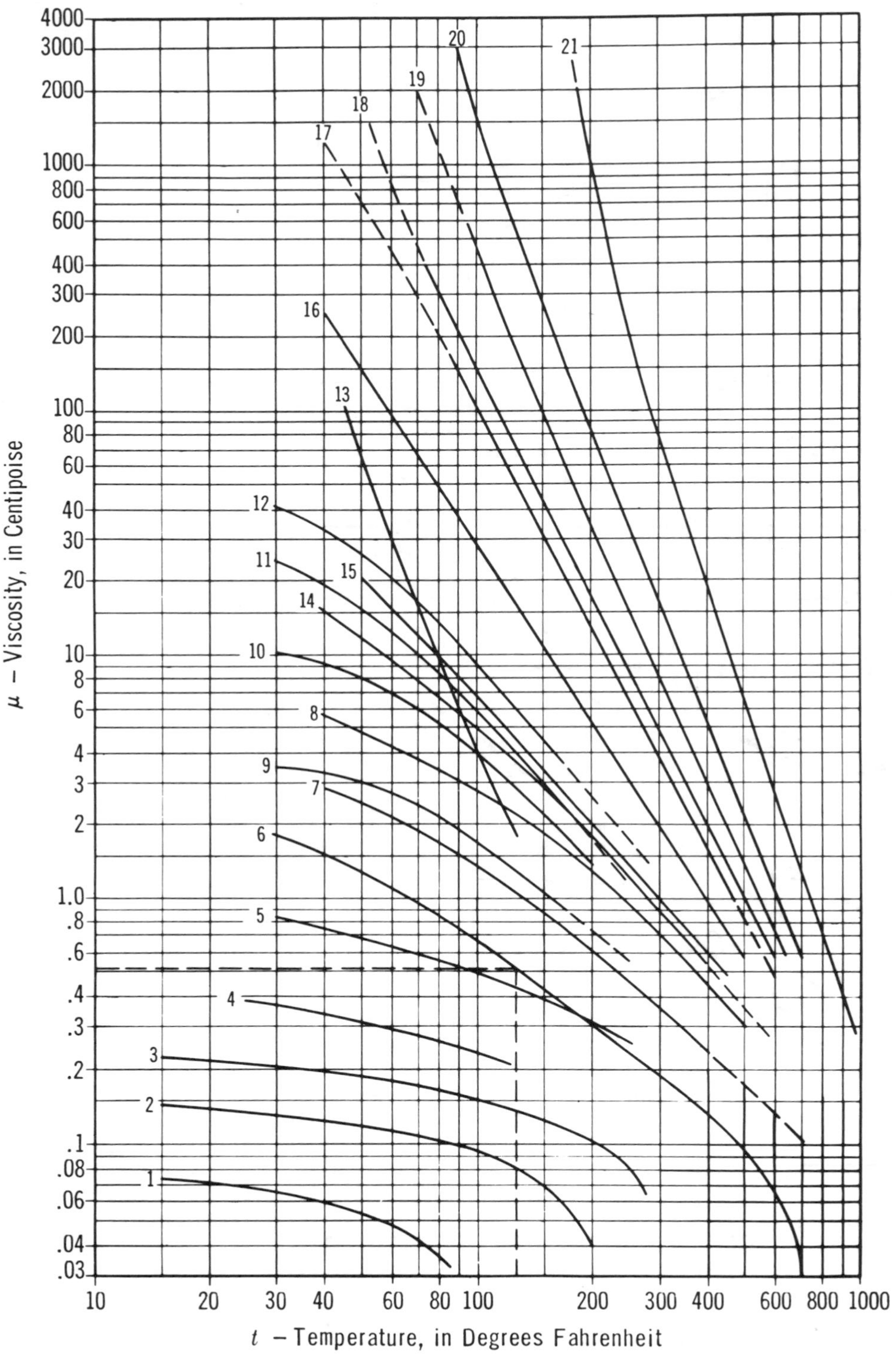

1. Ethane (C_2H_6)
2. Propane (C_3H_8)
3. Butane (C_4H_{10})
4. Natural Gasoline
5. Gasoline
6. Water
7. Kerosene
8. Distillate
9. 48 Deg. API Crude
10. 40 Deg. API Crude
11. 35.6 Deg. API Crude
12. 32.6 Deg. API Crude
13. Salt Creek Crude
14. Fuel 3 (Max.)
15. Fuel 5 (Min.)
16. SAE 10 Lube (100 V.I.)
17. SAE 30 Lube (100 V.I.)
18. Fuel 5 (Max.) or Fuel 6 (Min.)
19. SAE 70 Lube (100 V.I.)
20. Bunker C Fuel (Max.) and M.C. Residuum
21. Asphalt

Example: The viscosity of water at 125 F is 0.52 centipoise (Curve No. 6).

Chart 3.55. Viscosity of water and liquid petroleum products. Courtesy Crane Co. (T.P. No. 410) and data extracted in part from the Oil and Gas Journal.

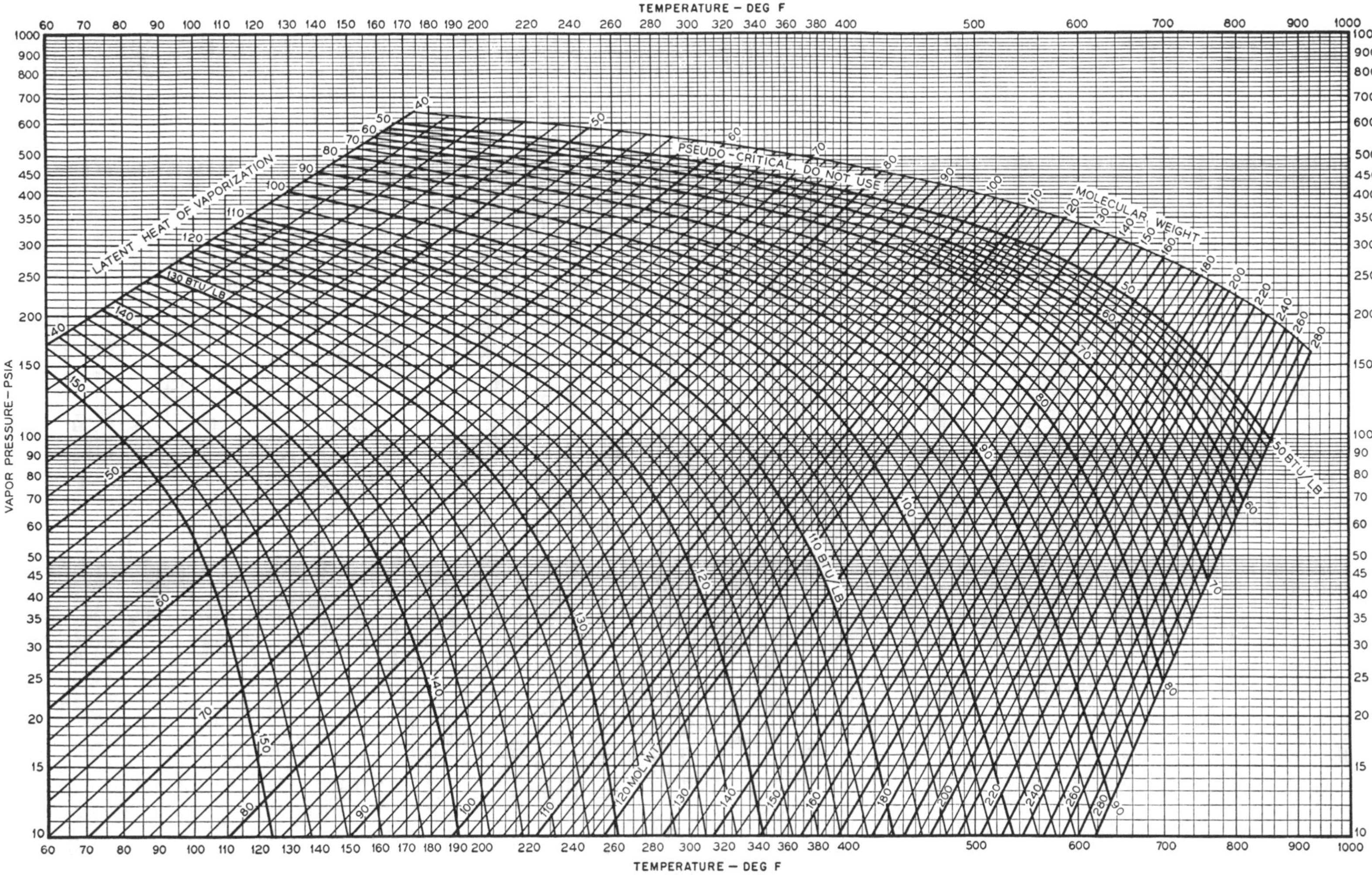

Note: This chart is directly applicable only to pure single-component paraffin hydrocarbon liquids. Consult Par. 6.5 of the text before using this chart.

Chart 3.56. Vapor pressure and heat of vaporization of pure single-component paraffin hydrocarbon liquids. Courtesy American Petroleum Institute.

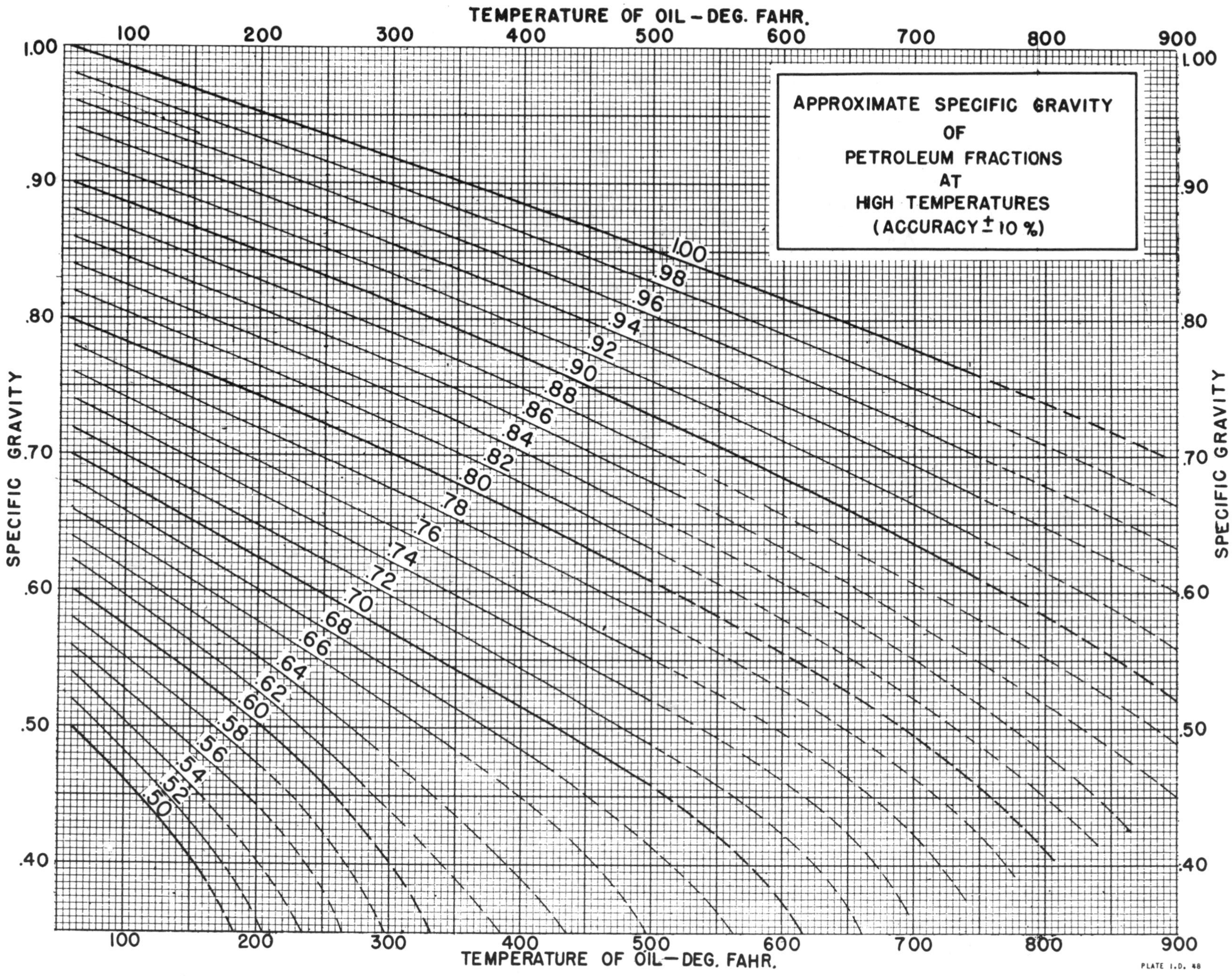

Chart 3.57. Approximate specific gravity of petroleum fractions at high temperatures. Courtesy The Foxboro Co.

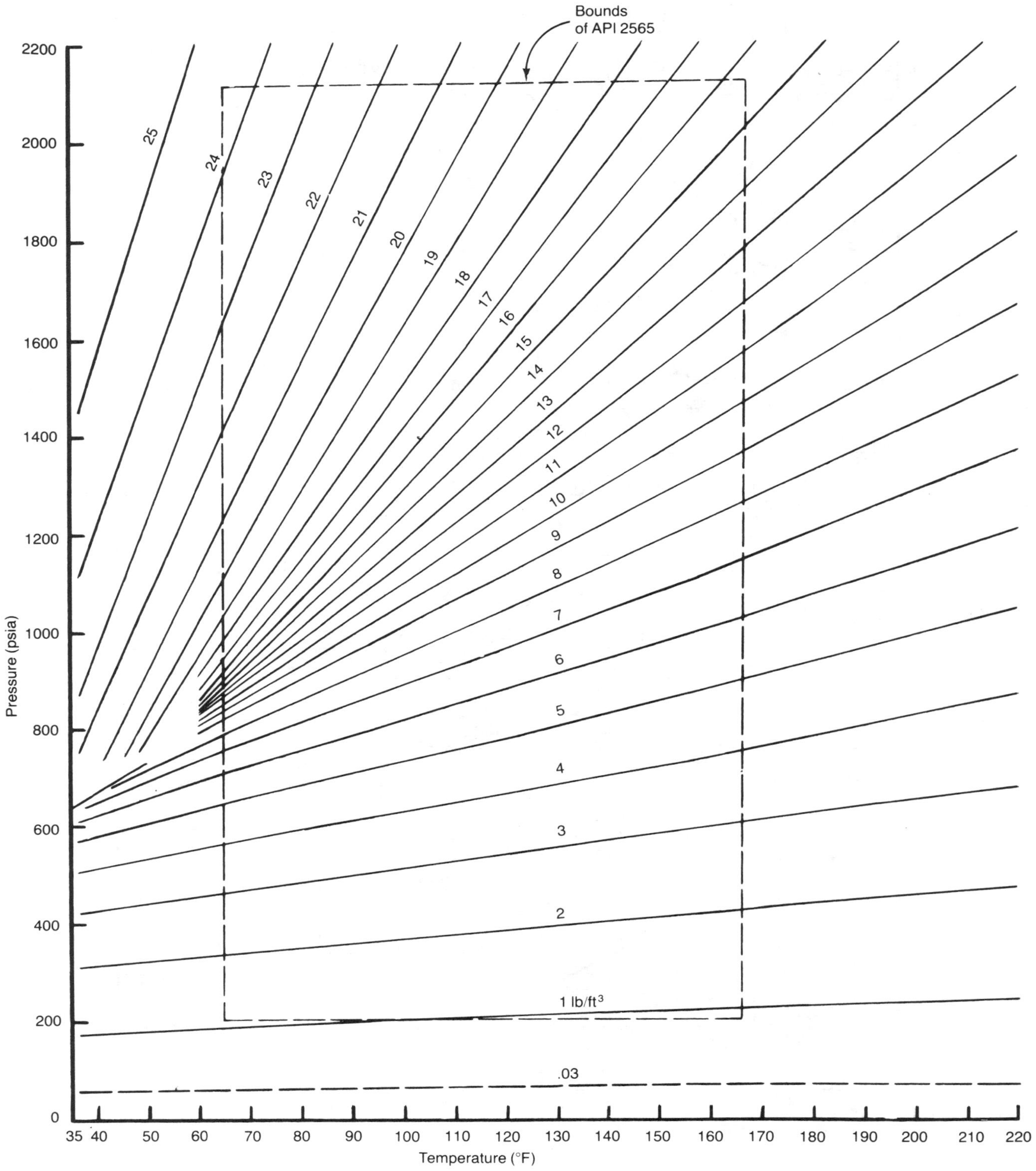

Chart 3.58. Ethylene density in 1 lb./ft^3 increments. Courtesy Daniel Industries, Inc.

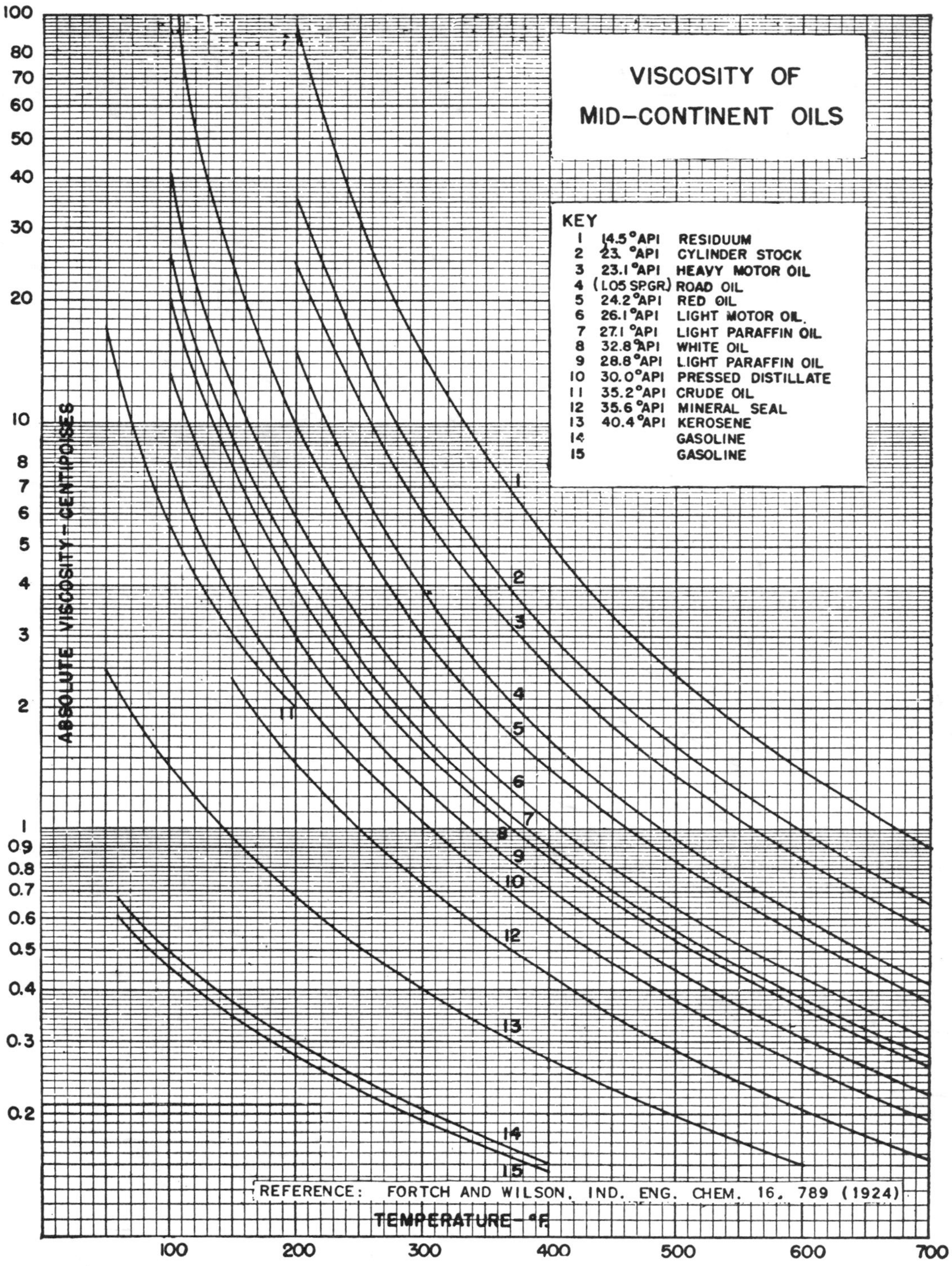

Chart 3.59. Vicosity of mid-continent oils. Data Book on Hydrocarbons, *J. B. Maxwell, copyright © 1950 Litton Educational Publishing, Inc. Reprinted by permission of Van Nostrand Reinhold Company.*

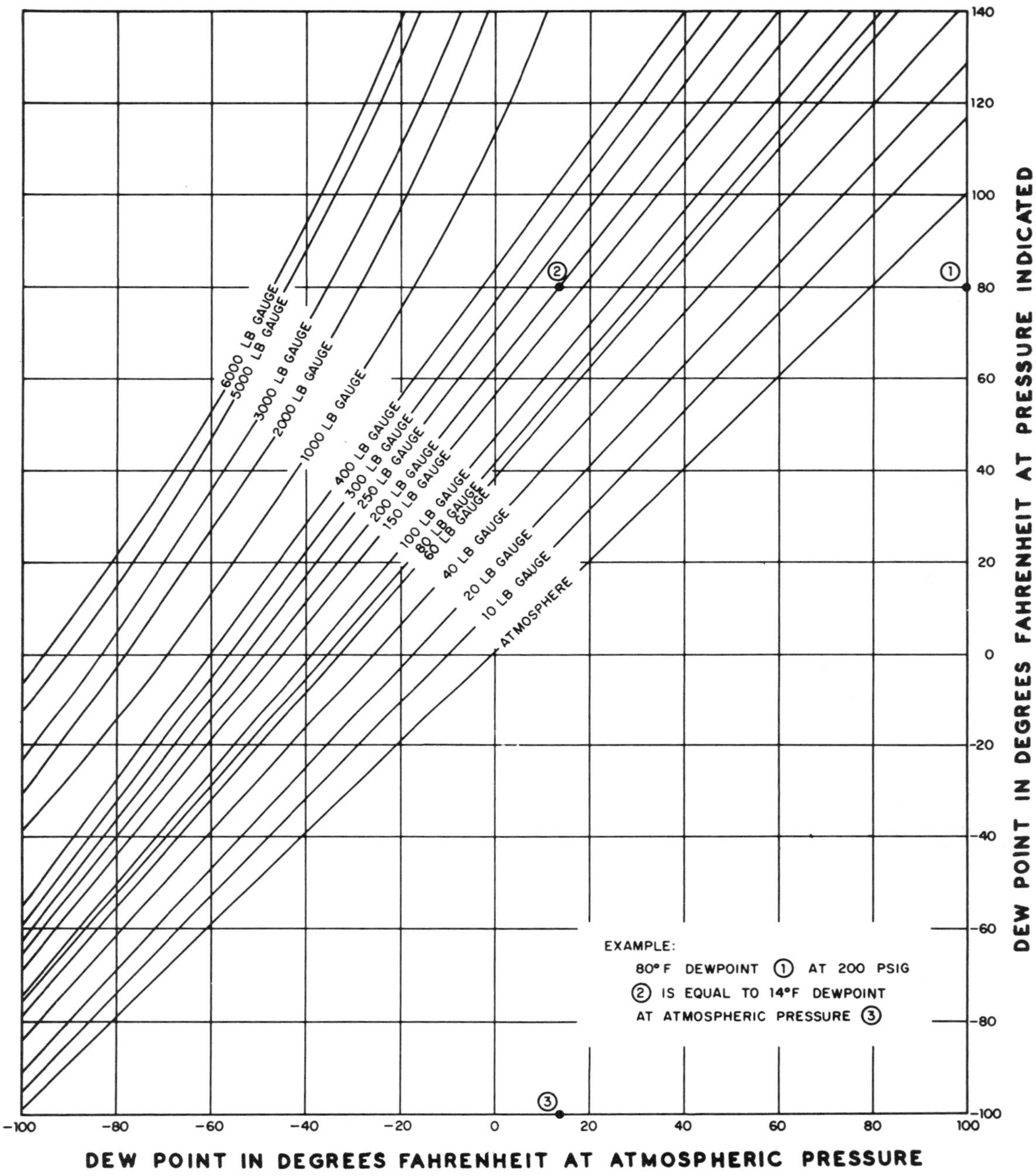

Chart 3.60. Conversion from atmospheric dew points to dew points at other pressures. Courtesy J. H. Jones Company.

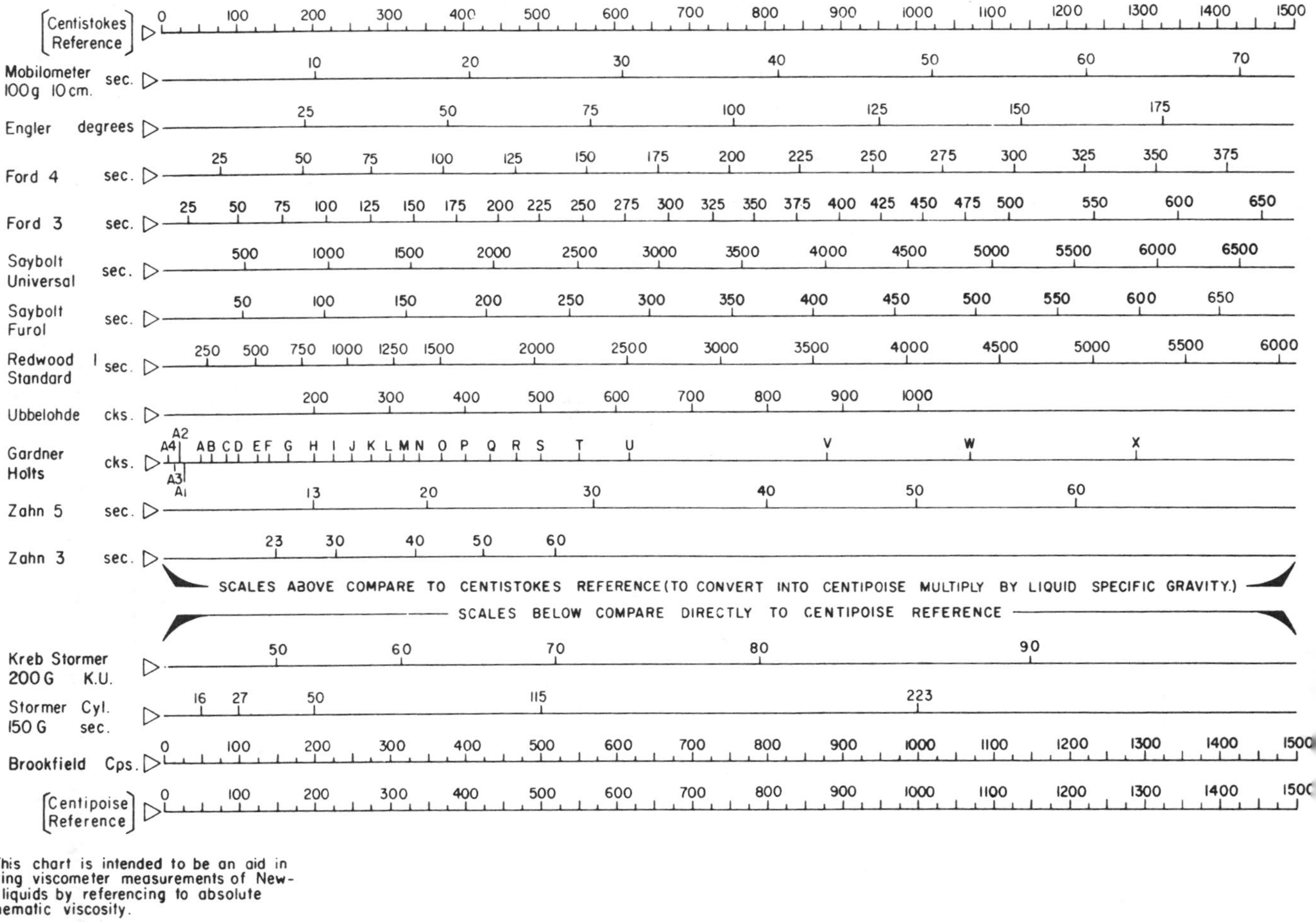

Chart 3.61. Viscosometer comparison chart for Newtonian liquids. Courtesy The Foxboro Company.

Ells Preceding and Following Meter Tube

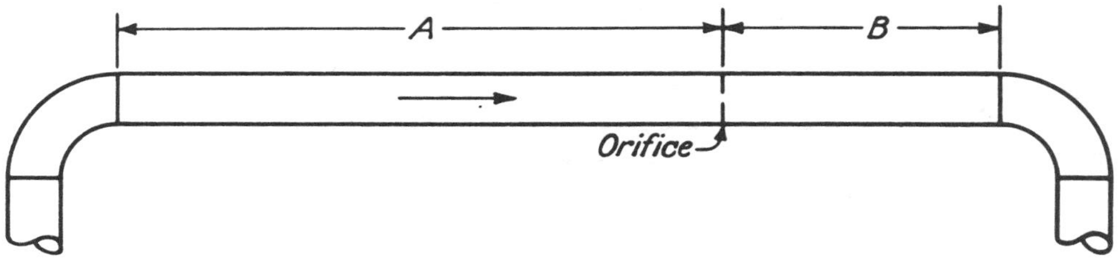

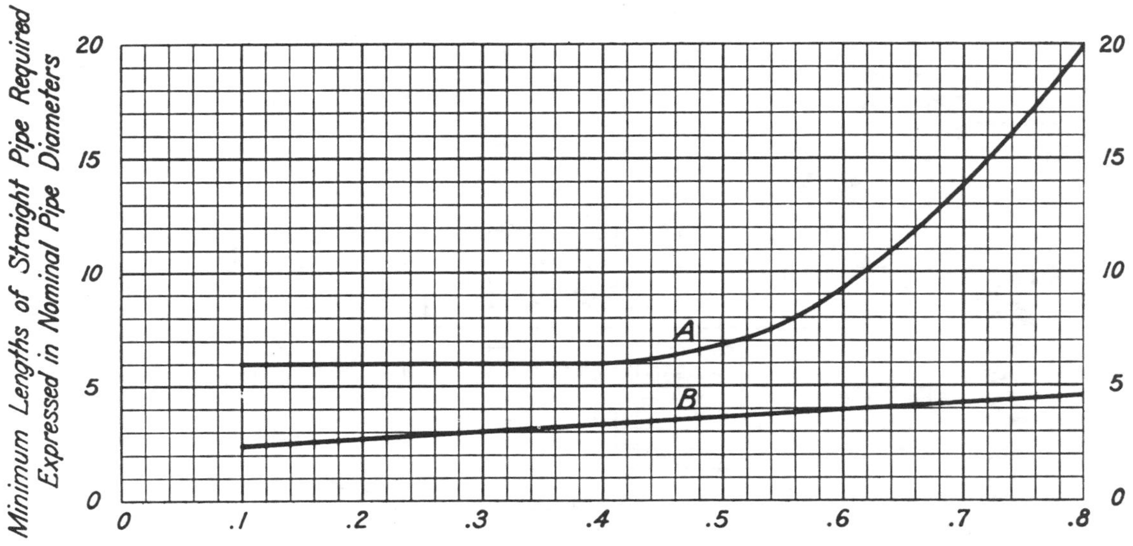

NOTE 1 - When "Pipe Taps" are used, A should be increased by 2 pipe diameters and B by 8 pipe diameters.

NOTE 2 - When the diameter of the orifice may require changing to meet different conditions, the lengths of straight pipe should be those required for the maximum orifice to pipe diameter ratio that may be used.

Chart 3.62. Beta ratio vs. pipe length for various piping configurations. Courtesy American Gas Association.

Chart 3.62 continued

Chart 3.62 continued

Two Ells or Bends Preceding Meter Tube (Bends in Same Plane)

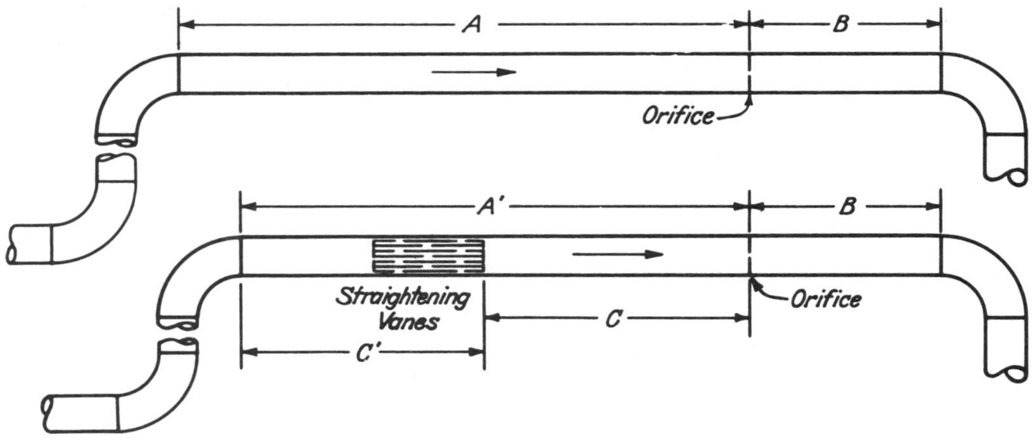

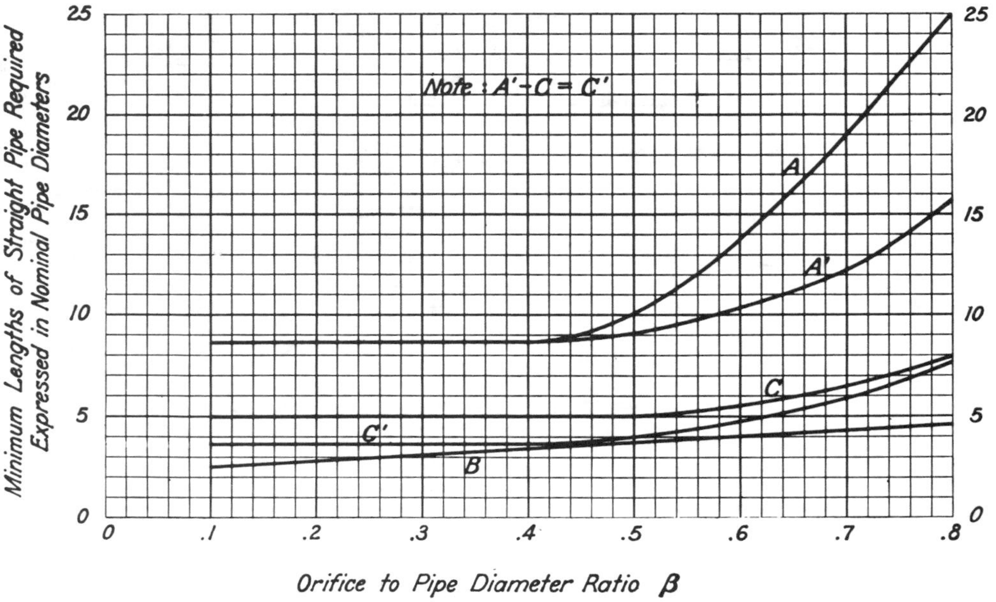

NOTE 1 - When "Pipe Taps" are used, A, A', and C should be increased by 2 pipe diameters, and B by 8 pipe diameters.

NOTE 2 - When the diameter of the orifice may require changing to meet different conditions, the lengths of straight pipe should be those required for the maximum orifice to pipe diameter ratio that may be used.

Chart 3.62 continued

Chart 3.62 continued

Two Ells or Bends Preceding Meter Tube (Bends Not in Same Plane)

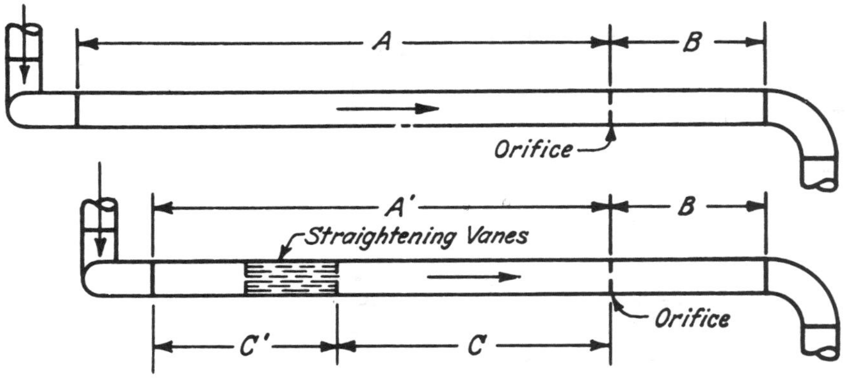

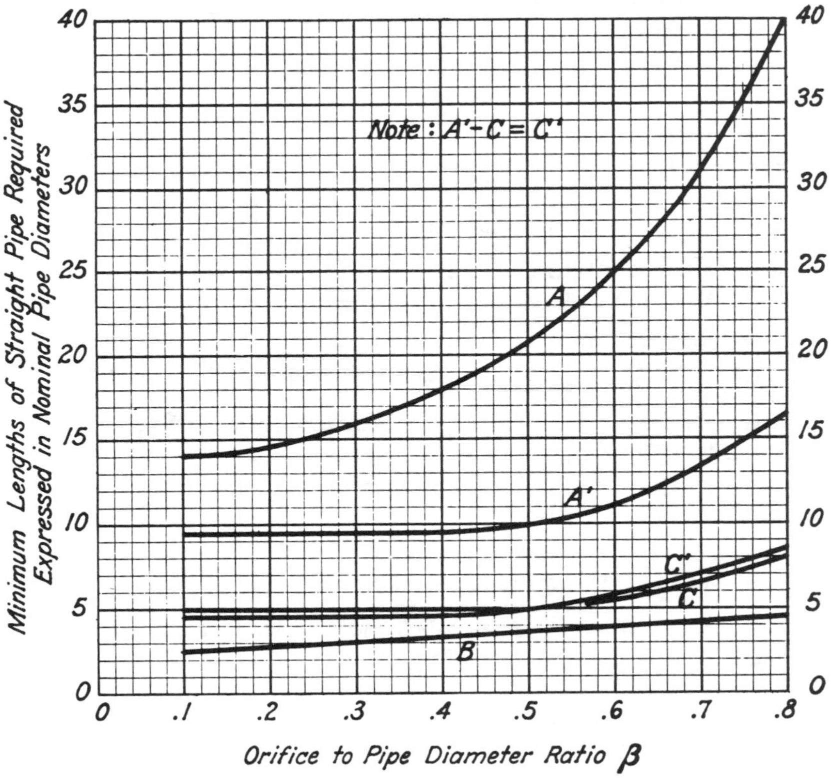

NOTE 1 - When "Pipe Taps" are used, A, A', and C should be increased by 2 pipe diameters and B by 8 pipe diameters.

NOTE 2 - When the 2 ells shown in the above sketches are closely preceded by a third which is not in the same plane as the middle or second ell, the piping requirements shown by A should be doubled.

NOTE 3 - When the diameter of the orifice may require changing to meet different conditions, the lengths of straight pipe should be those required for the maximum orifice to pipe diameter ratio that may be used.

Chart 3.62 continued

Reducer Preceding Meter Tube

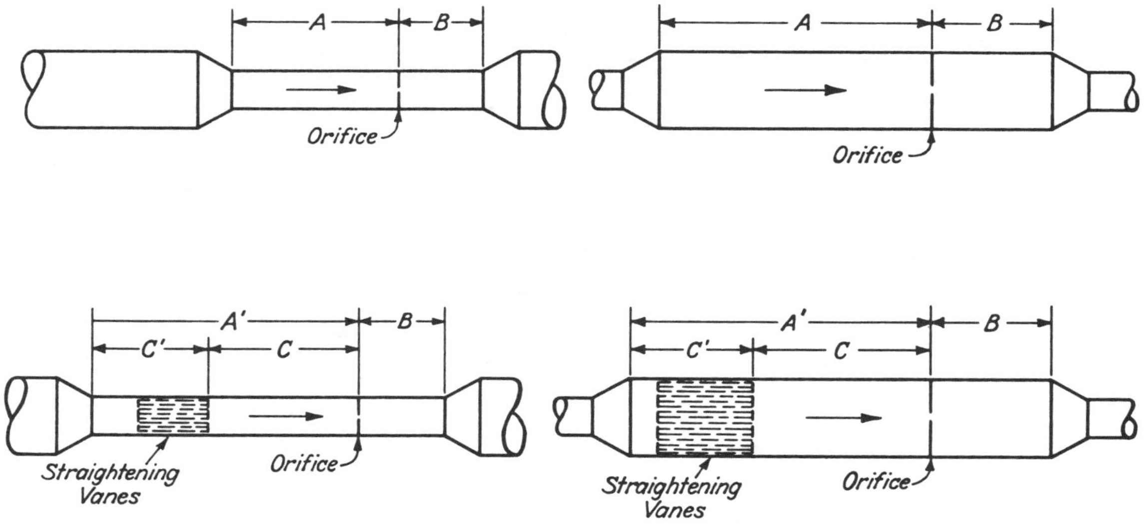

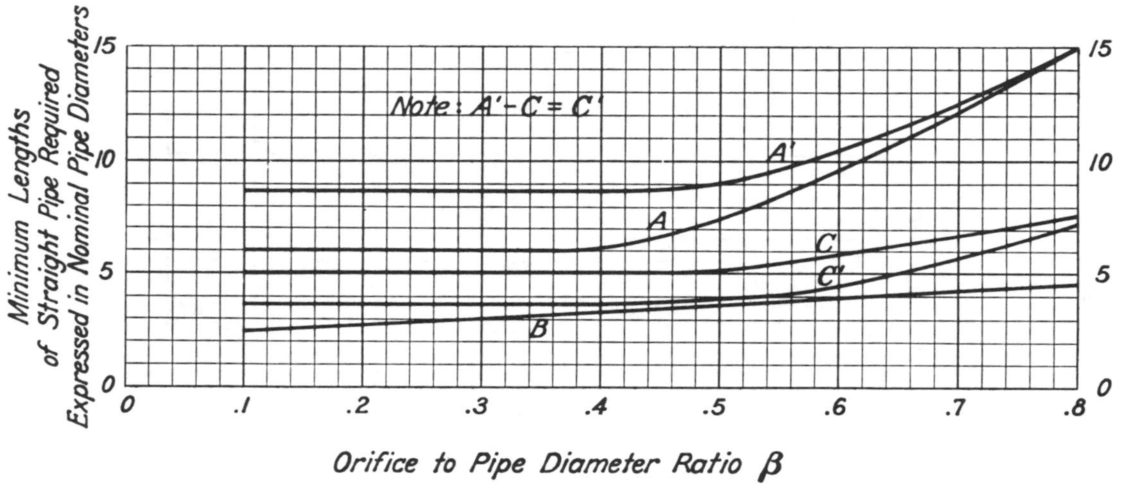

NOTE 1 - When "Pipe Taps" are used, length A, A', and C should be increased by 2 pipe diameters, and B by 8 pipe diameters.

NOTE 2 - Straightening vanes will not reduce lengths of straight pipe A. Straightening vanes are not required because of the reducers, they are required because of other fittings which precede the reducer. Length A is to be increased by an amount equal to the length of the straightening vanes whenever they are used (see bottom sketches).

NOTE 3 - When the diameter of the orifice may require changing to meet different conditions, the lengths of straight pipe should be those required for the maximum orifice to pipe diameter ratio that may be used.

Chart 3.62 continued

Chart 3.62 continued

Valve or Regulator Preceding Meter Tube

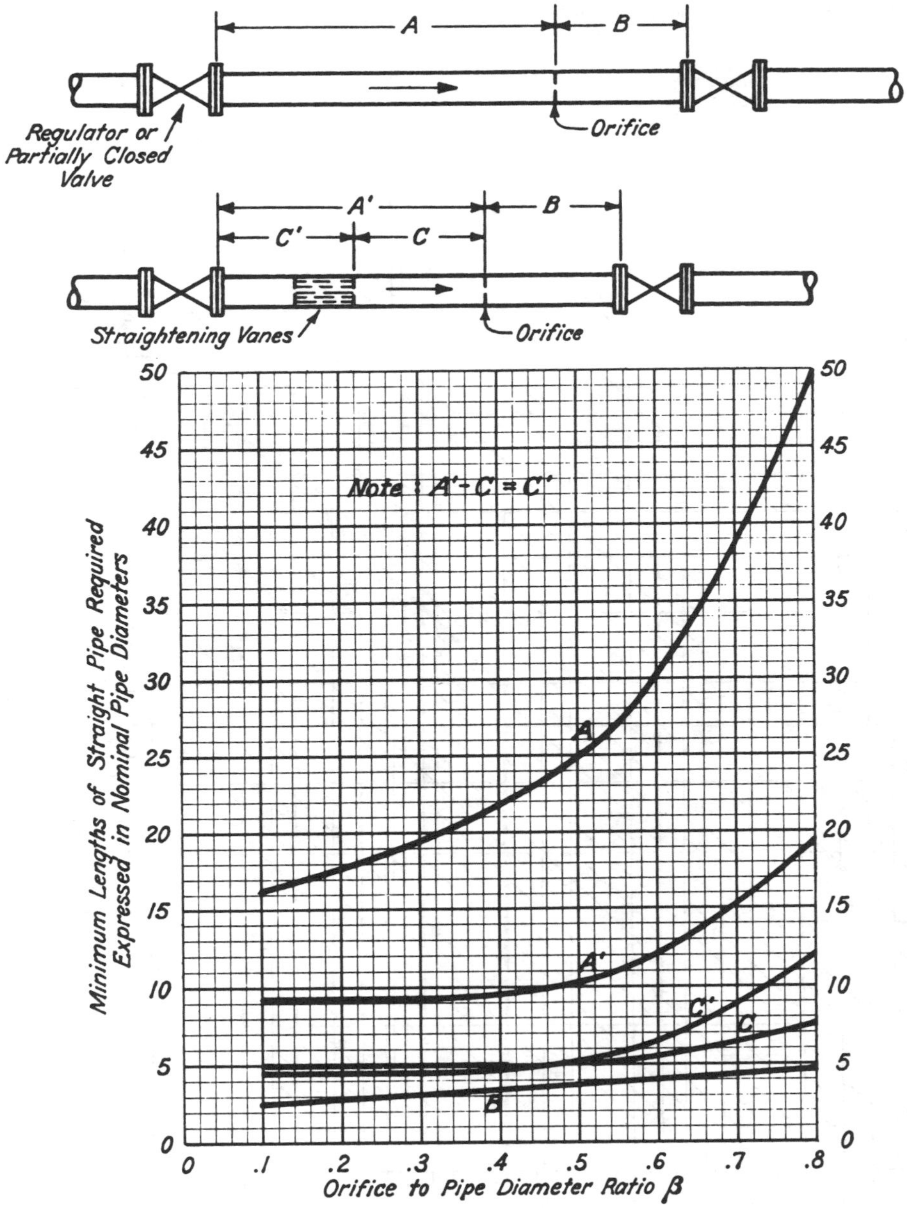

NOTE 1 - When "Pipe Taps" are used, length A, A', and C should be increased by 2 pipe diameters, and B by 8 pipe diameters.

NOTE 2 - Line A, A', C, and C' apply to regulators or gate valves, globe valves, and plug valves which are used for throttling the flow (partially closed).

NOTE 3 - When the diameter of orifice may require changing to meet different conditions, the lengths of straight pipe should be those required for the maximum orifice to pipe diameter ratio that may be used.

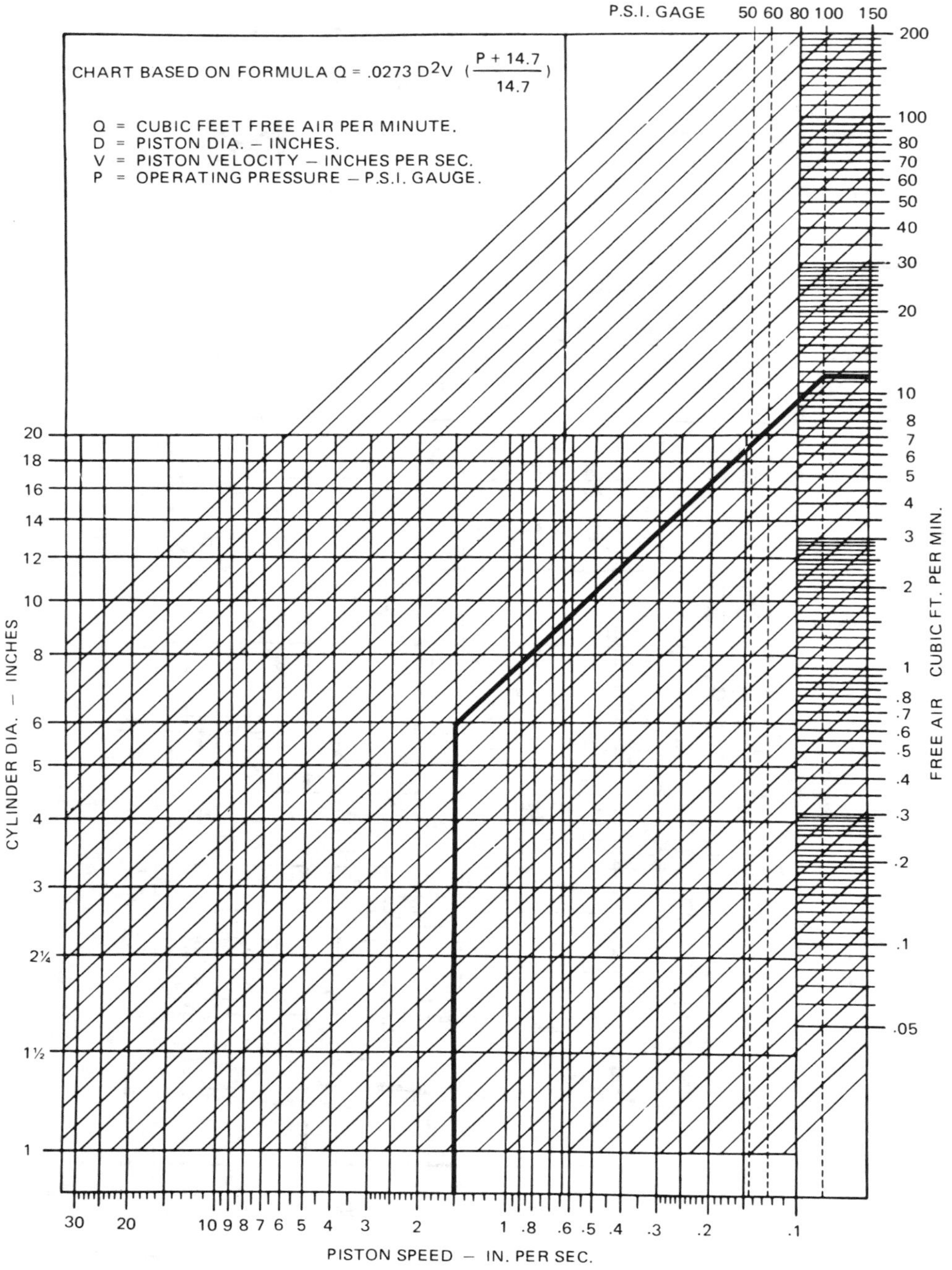

Chart 3.63. Cylinder air consumption chart. Fluid Power Handbook & Directory, *1972/1973.*

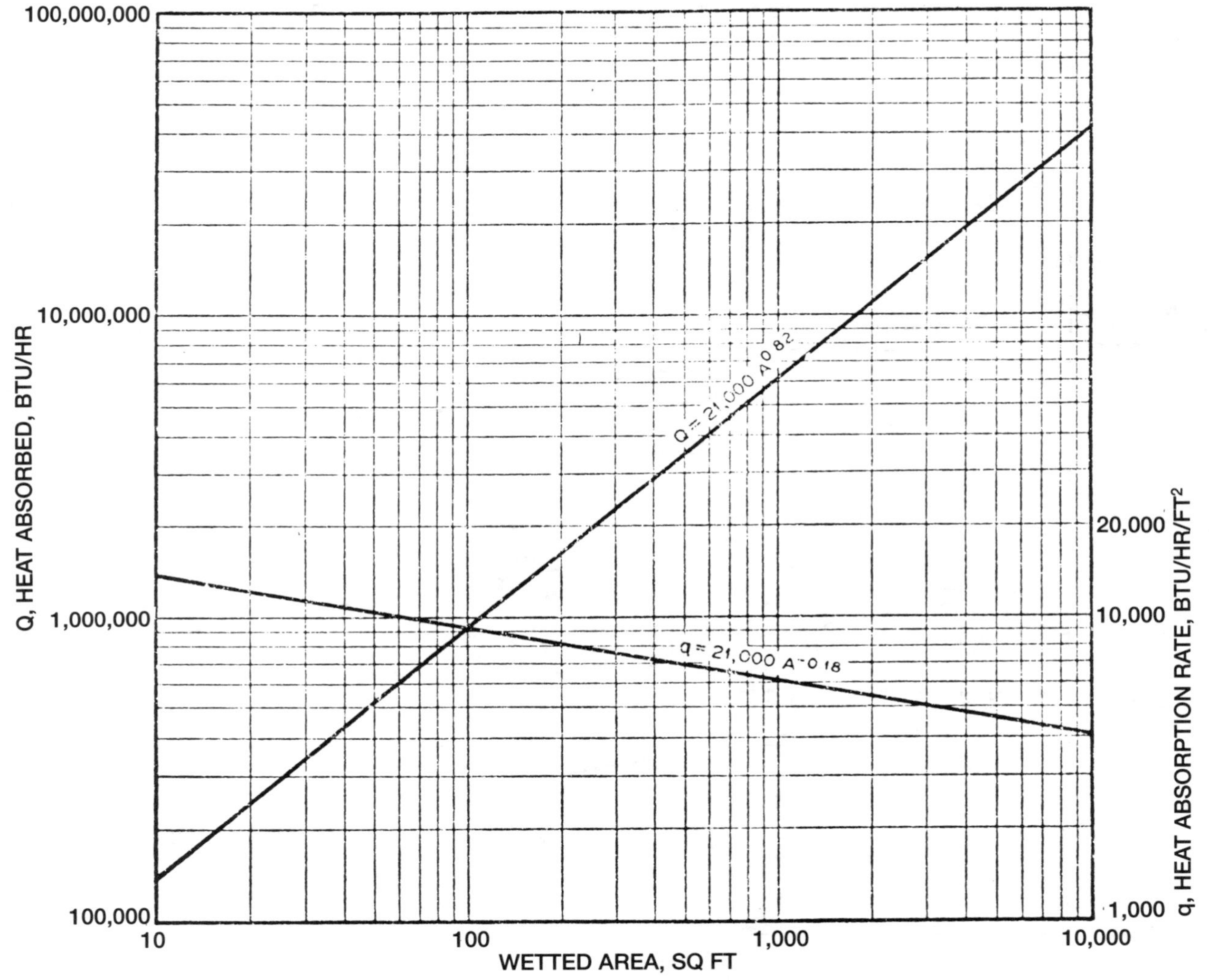

Chart 3.64. Vessel heat absorption formula for computing safety relief valve capacity. Courtesy American Petroleum Institute.

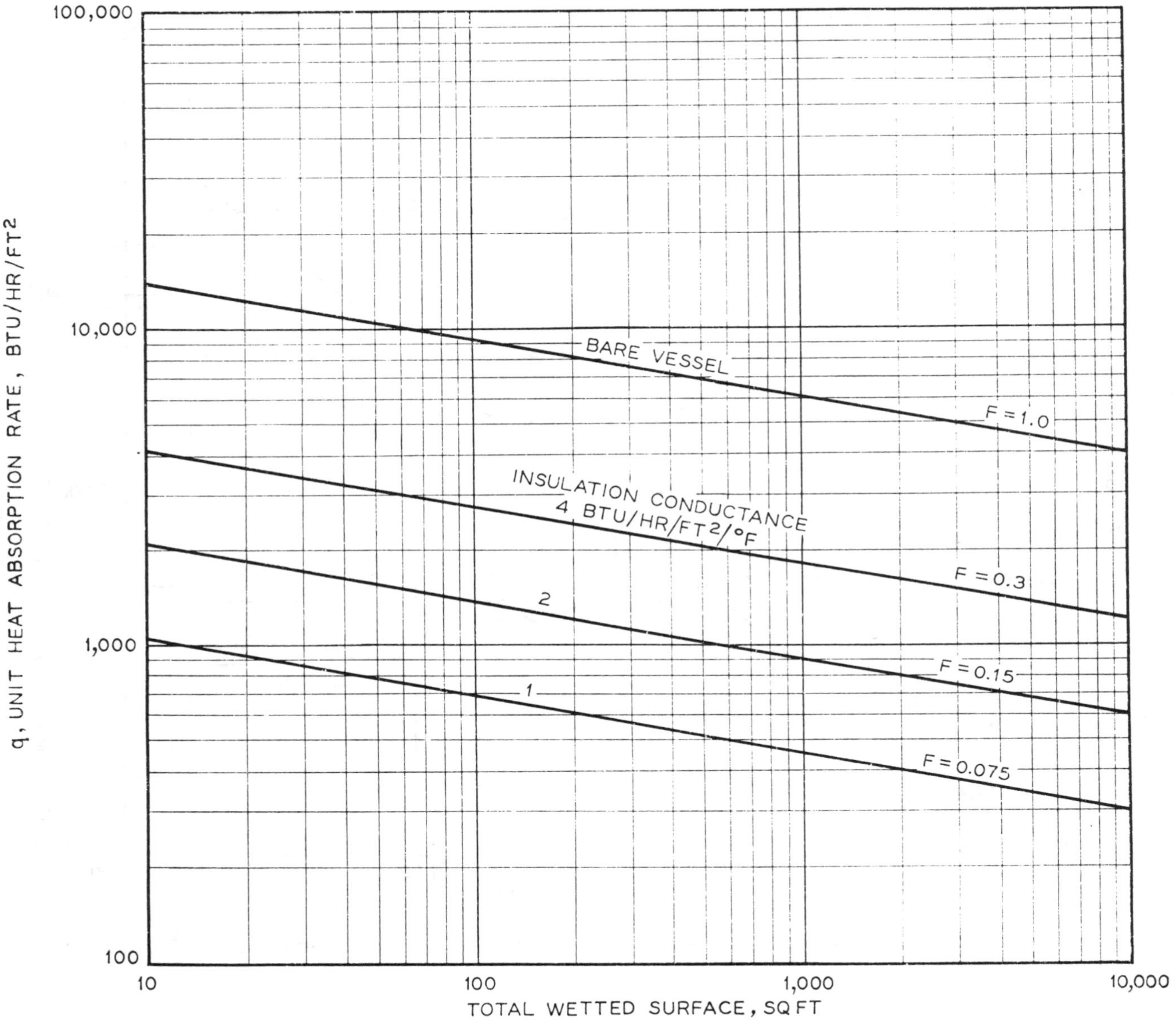

Chart 3.65. Average vessel heat absorption per square foot based on API formula. Courtesy American Petroleum Institute.

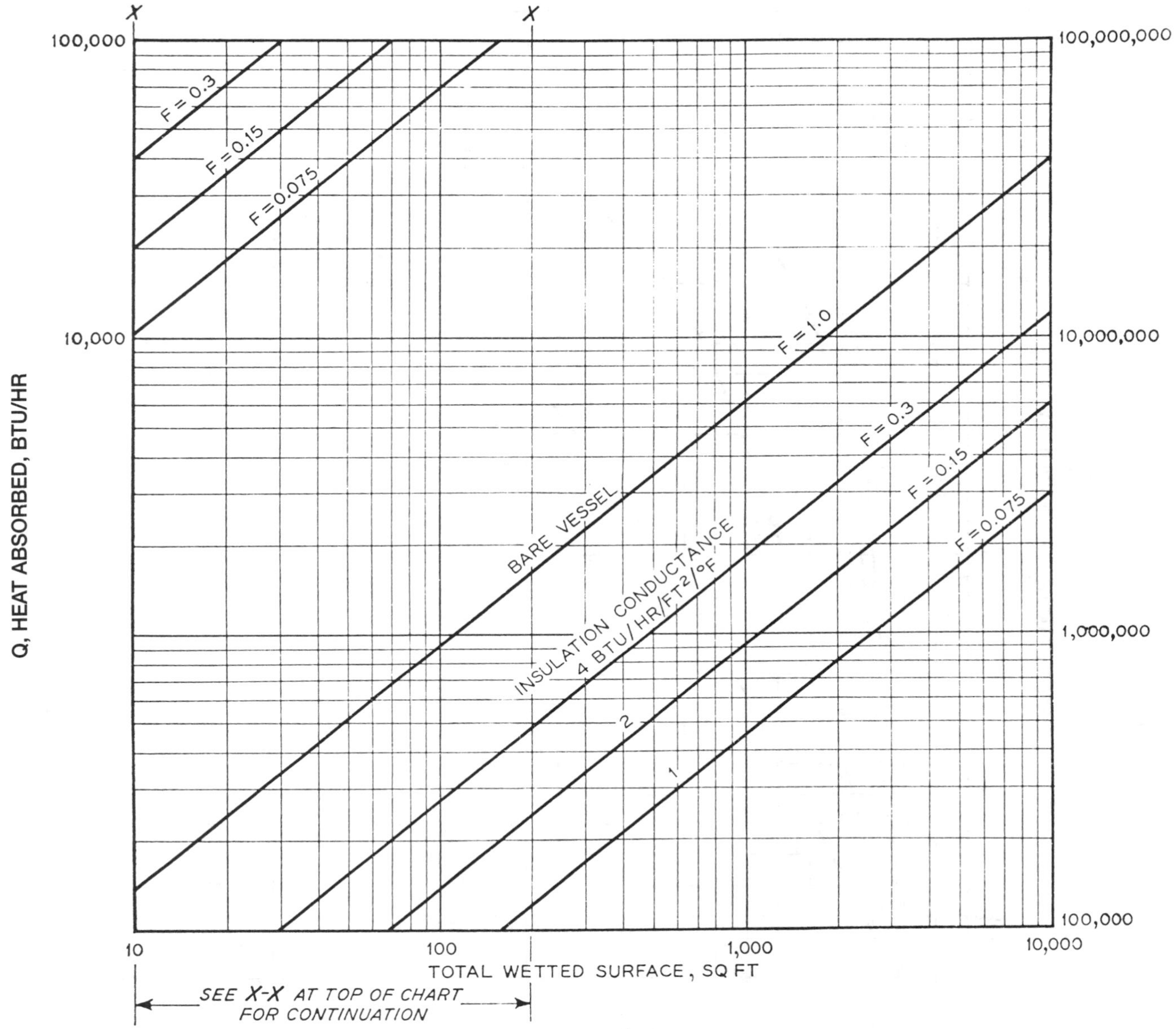

Chart 3.66. API formula for heat absorbed from fire on wetted surface of pressure vessel. Courtesy American Petroleum Institute.

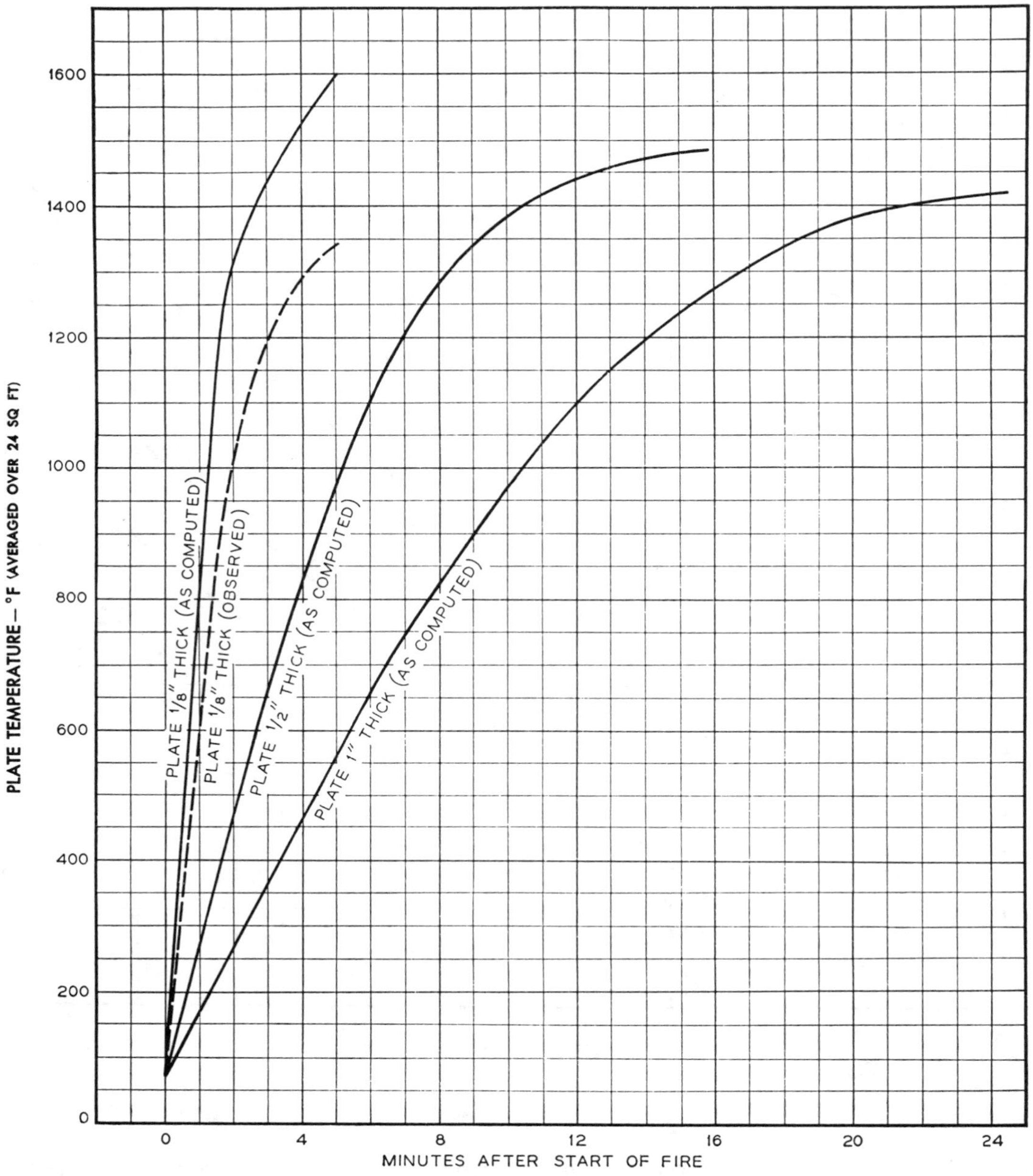

Chart 3.67. Average rate of heating of steel plates exposed to open gasoline fire on one side. Courtesy American Petroleum Institute.

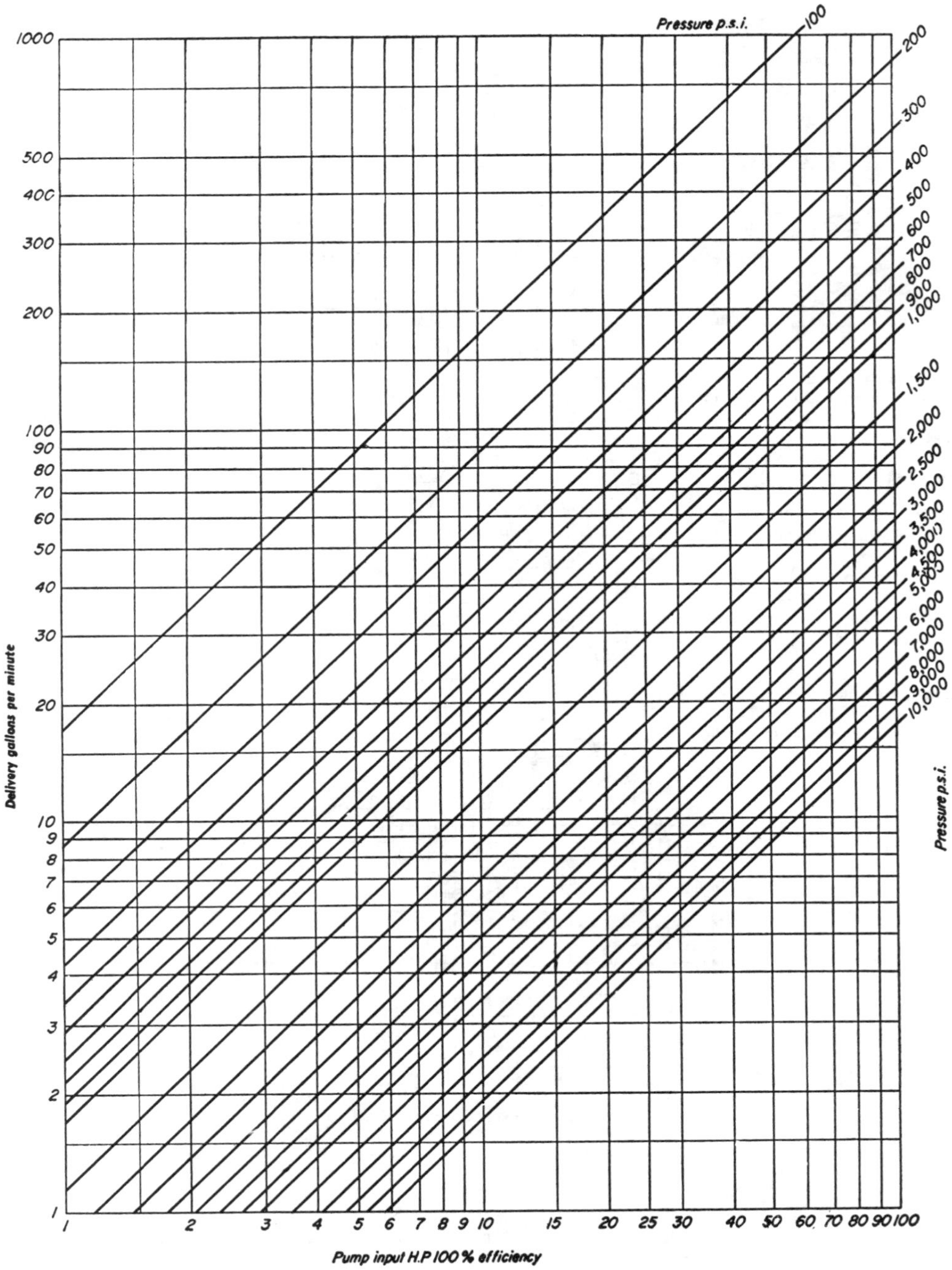

Chart 3.68. Pump input horsepower versus delivery. Fluid Power Handbook & Directory, *1972/73.*

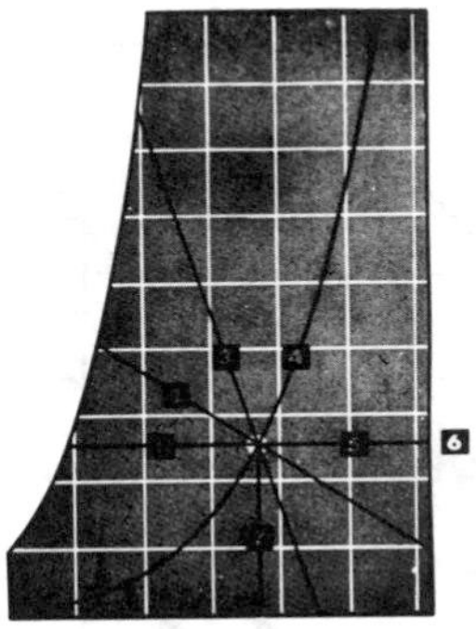

LEGEND

1 Dew Point Temperature
2 Wet Bulb Temperature
3 Cubic Feet Per Pound
4 Relative Humidity
5 Grains of Moisture
6 Pounds of Water
7 Dry Bulb Temperature

HOW TO USE CHART

Locate point on chart at intersection of any two values. Obtain other values by following chart lines.

Example: at a dry bulb temperature 7 of 172 F and a wet bulb temperature 2 of 102 F, the following values may be obtained from the chart: dew point temperature 1 87.5 F; cubic feet per pound of dry air 3 16.7; relative humidity 4 10.3 percent; grains of moisture per pound of dry air 5 201; pounds of water per pound of dry air 6 .0287.

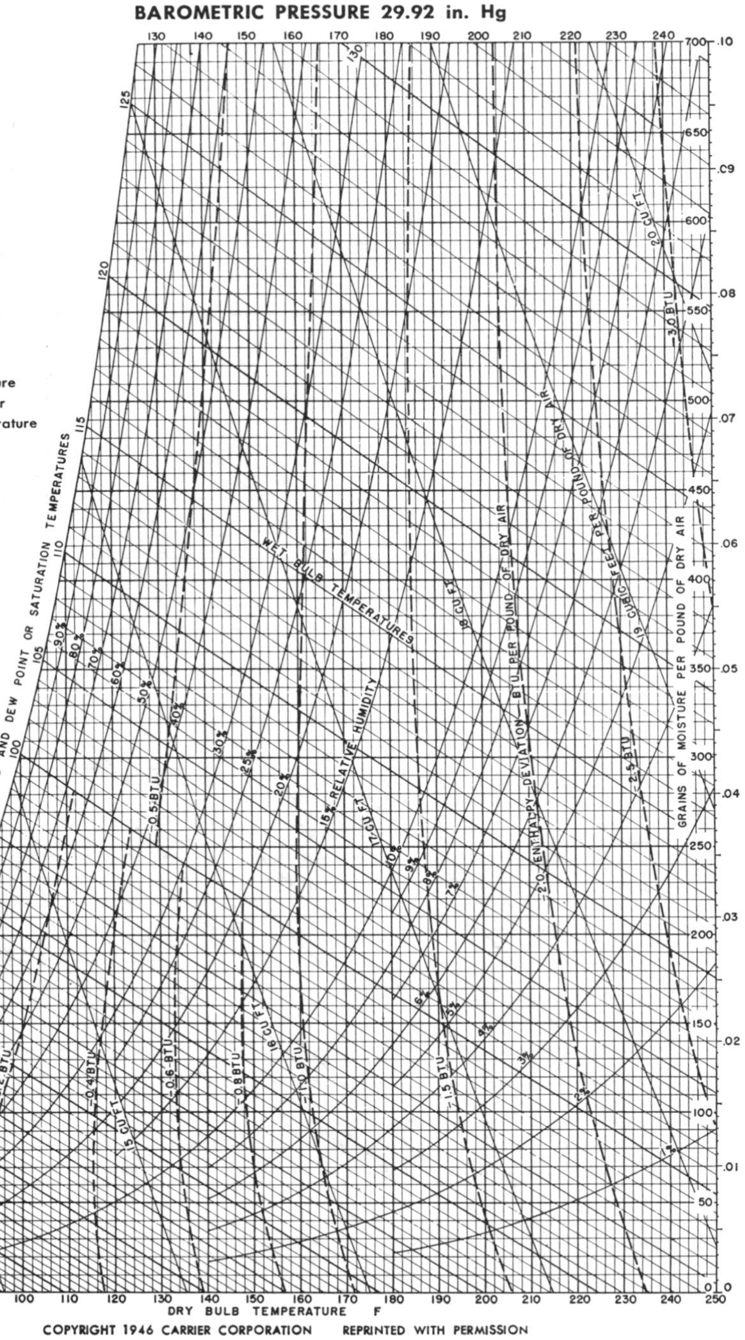

Chart 3.69. Psychrometric chart. Courtesy The Foxboro Company.

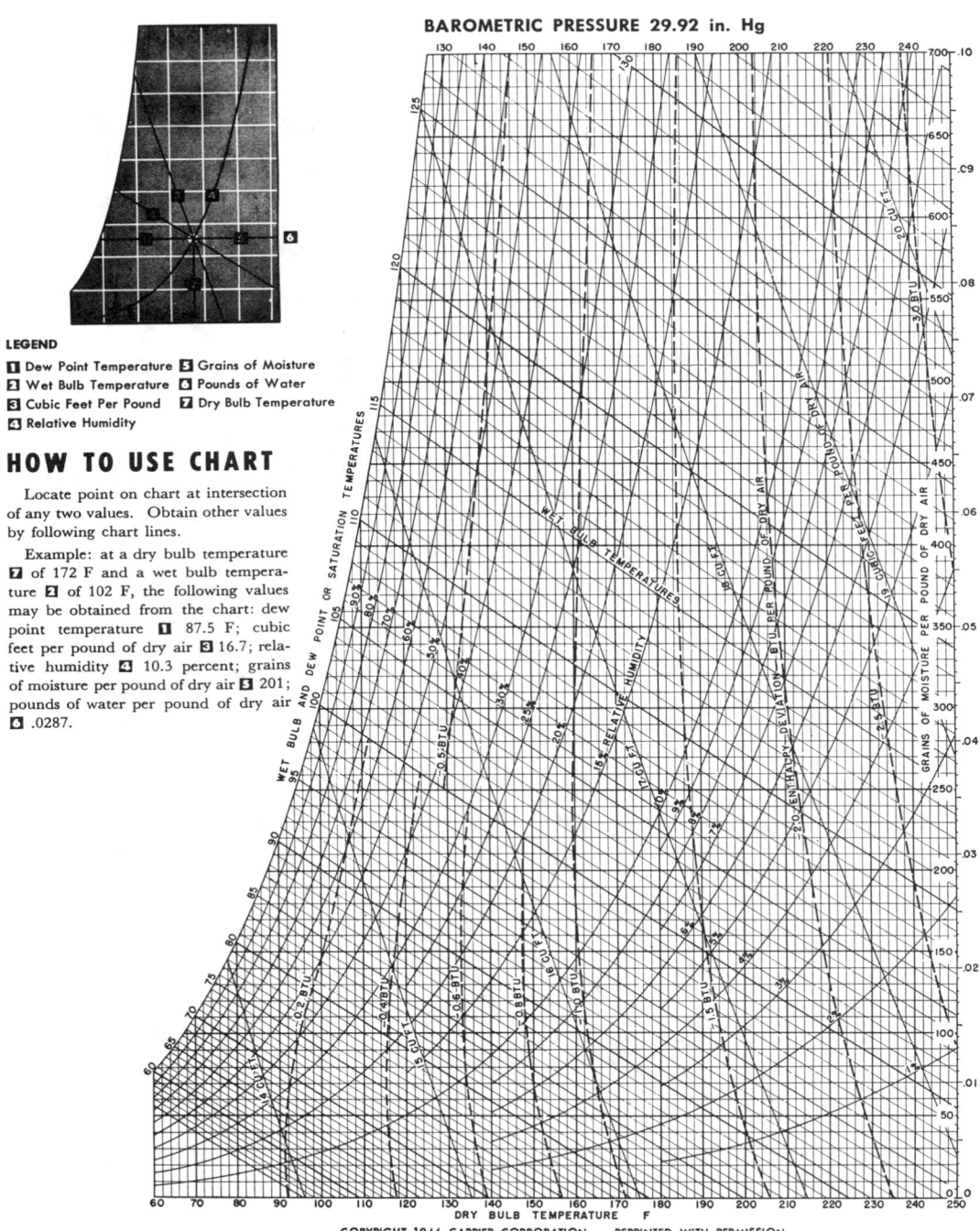

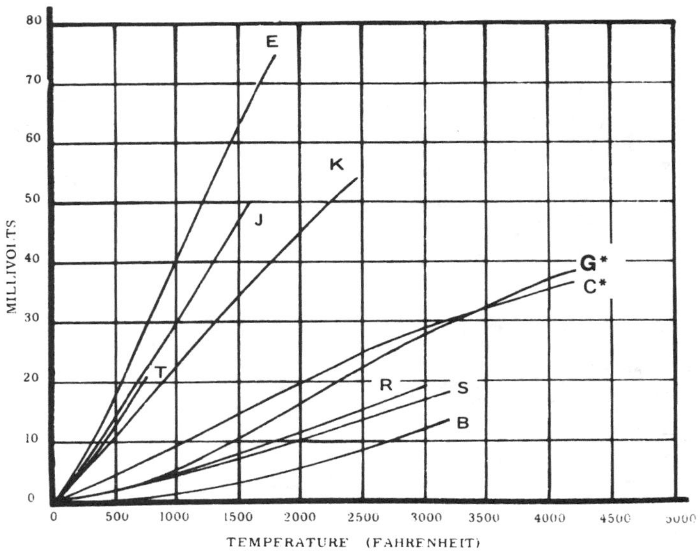

T	Copper vs. Constantan
E	Chromel vs. Constantan
J	Iron vs. Constantan
K	Chromel vs. Alumel
G*	Tungsten vs. Tungsten 26% Rhenium
C*	Tungsten 5% Rhenium vs. Tungsten 26% Rhenium
R	Platinum vs. Platinum 13% Rhodium
S	Platinum vs. Platinum 10% Rhodium
B	Platinum 6% Rhodium vs. Platinum 30% Rhodium

*Not ANSI Symbol

Chart 3.70. Temperature—millivolts graph for thermocouples. Courtesy Omega Engineering, Inc.

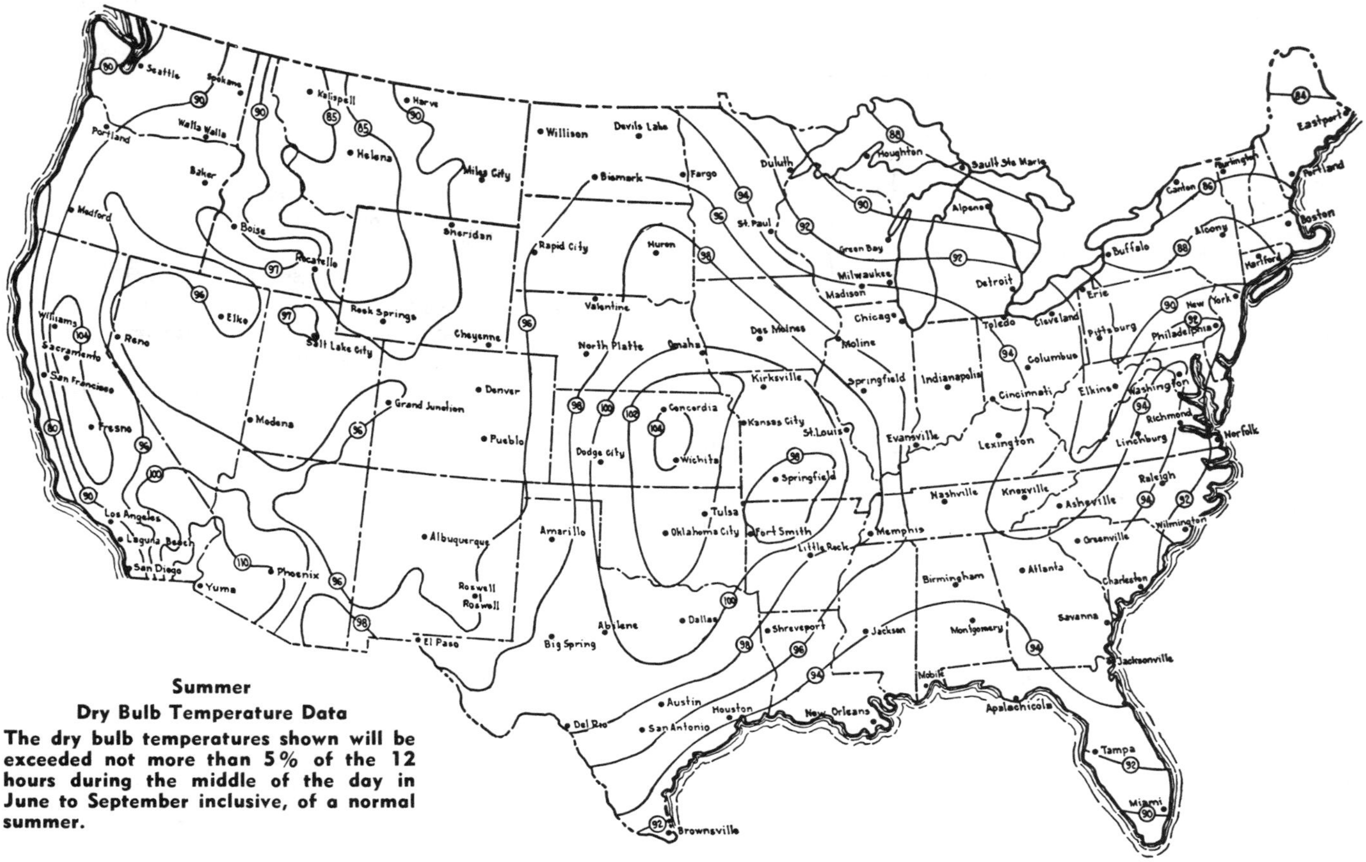

Summer
Dry Bulb Temperature Data
The dry bulb temperatures shown will be exceeded not more than 5% of the 12 hours during the middle of the day in June to September inclusive, of a normal summer.

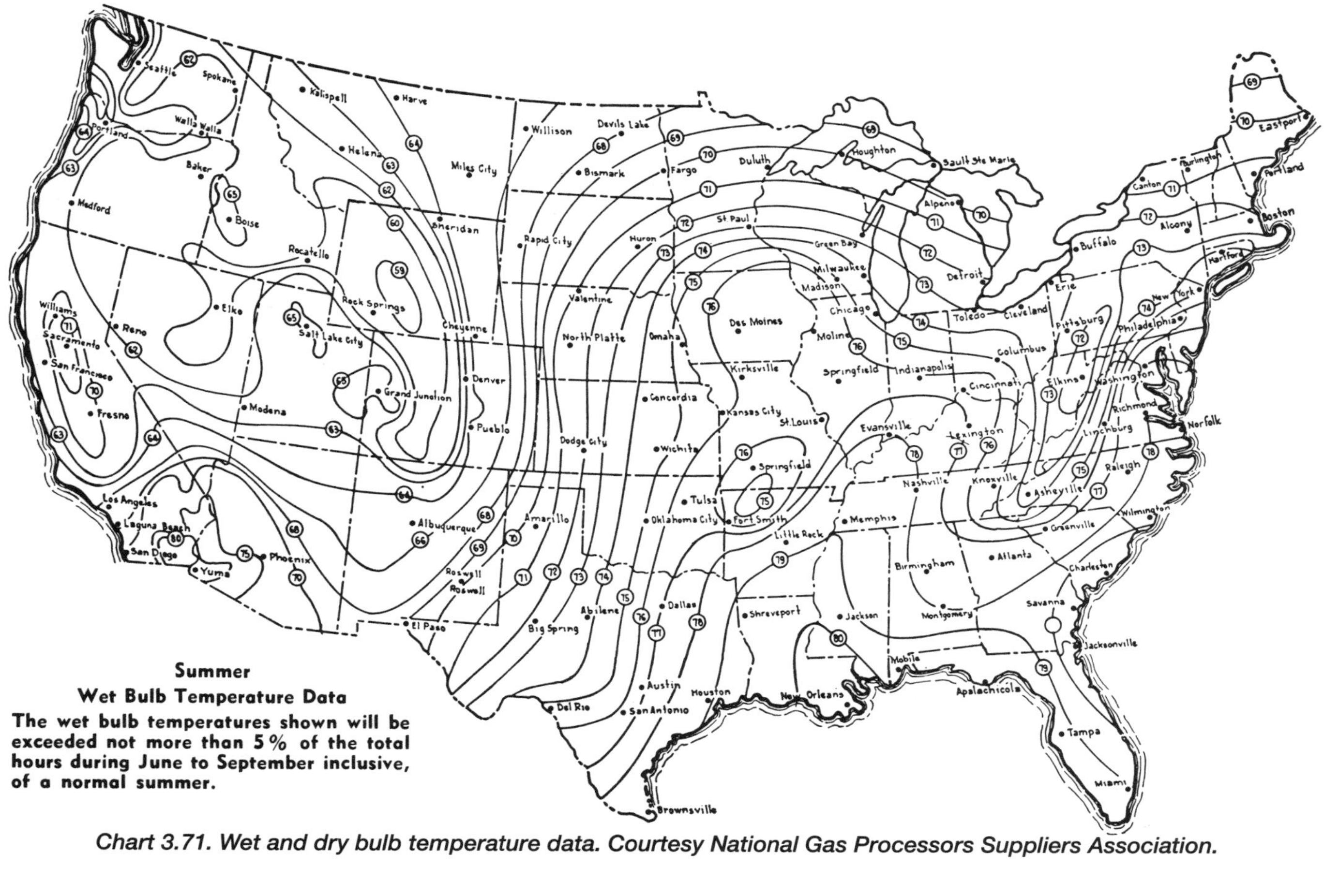

Summer
Wet Bulb Temperature Data
The wet bulb temperatures shown will be exceeded not more than 5% of the total hours during June to September inclusive, of a normal summer.

Chart 3.71. Wet and dry bulb temperature data. Courtesy National Gas Processors Suppliers Association.

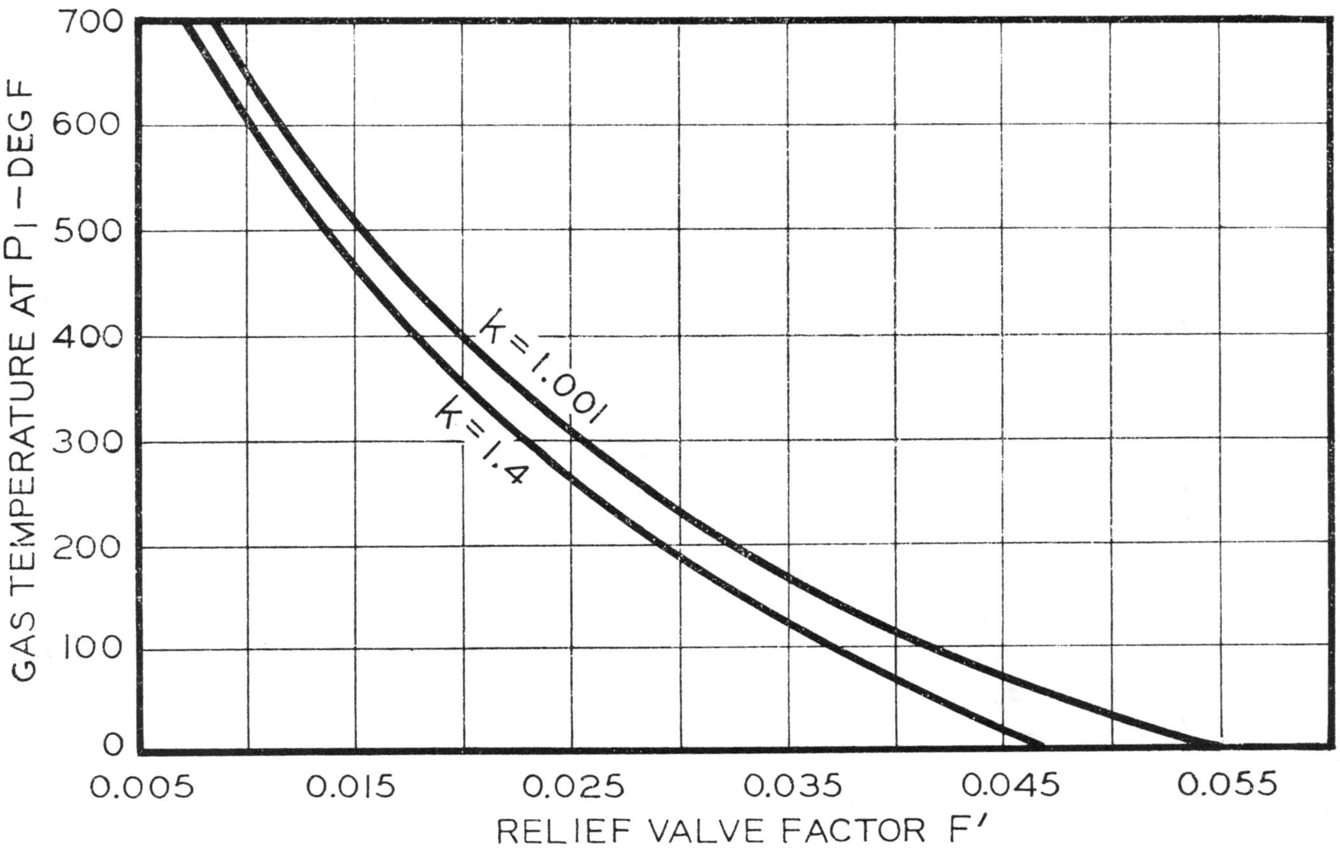

Notes:

1. Table 1 gives k values for some gases; for others, the values can be determined from the properties of gases as presented in any acceptable reference work.
2. These curves are for vessels of carbon steel.
3. These curves conform to the relationship $F' = \left(\frac{0.1406}{CK}\right)\left(\frac{\Delta T^{1.25}}{T_n^{0.6506}}\right)$

Where:

C = coefficient which is determined by the ratio of the specific heats of the gas at standard conditions. This can be obtained from Chart 3.76 or 3.77.

K = coefficient of discharge, which value is obtainable from the valve manufacturer. The K for a number of nozzle-type valves is 0.975.

T_1 = gas temperature, absolute, in degrees fahrenheit + 460, at the upstream pressure, and is determined from the relationship:

$$T_1 = \left(\frac{P_1}{P_n}\right) T_n$$

T_n = normal operating gas temperature, in degrees fahrenheit + 460.

P_n = normal operating gas pressure, in pounds per square inch absolute.

P_1 = upstream pressure, in pounds per square inch absolute. This is the set pressure multiplied by 1.20 (or less, depending on the overpressure permissible) plus the atmospheric pressure, in pounds per square inch absolute.

ΔT = $T_w - T_1$. Difference between wall temperature and the temperature of the gas at P_1.

T_w = vessel wall temperature, in degrees fahrenheit + 460.

The curves are drawn using 1,100 F as the vessel wall temperature. This value is a recommended maximum temperature for the usual carbon steel plate materials whose physical properties at temperatures in excess of 1,100 F show signs of undesirable tendencies. Where vessels are fabricated from alloy materials, the value for T_w should be changed to a more proper recommended maximum.

It is recommended that the minimum value of $F' = 0.01$.

Chart 3.72. Relief valve factors for non-insulated vessels in gas service exposed to open fires. Courtesy American Petroleum Institute.

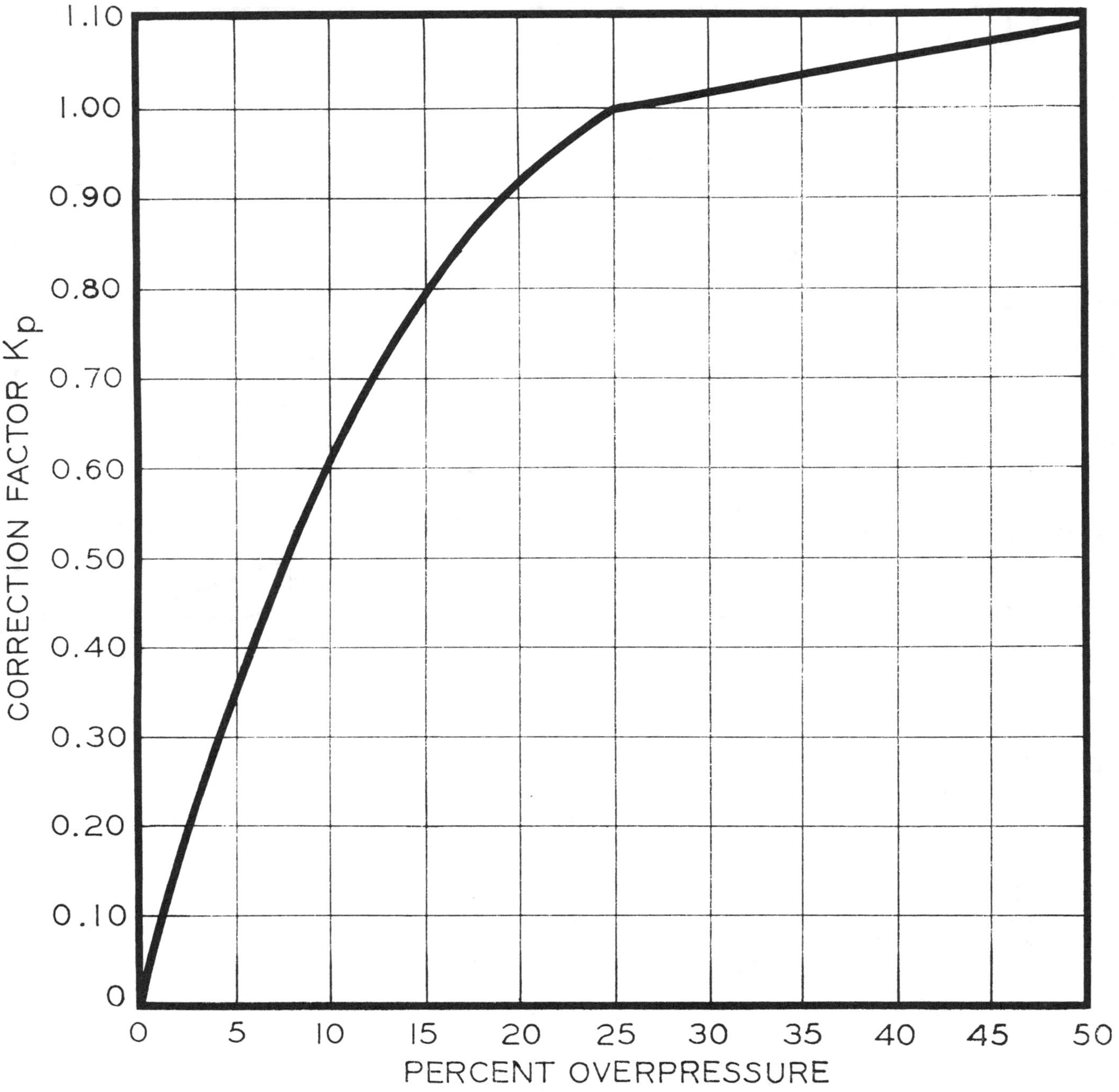

Note: The above curve shows that up to and including 25 percent overpressure, capacity is affected by the change in lift, the change in orifice discharge coefficient, and the change in overpressure. Above 25 percent, capacity is affected only by the change in overpressure.

Valves operating at low overpressures tend to "chatter"; therefore, overpressures of less than 10 percent should be avoided.

Chart 3.73. Capacity correction factors due to over-pressure for relief and safety relief valves in liquid service. Courtesy American Petroleum Institute.

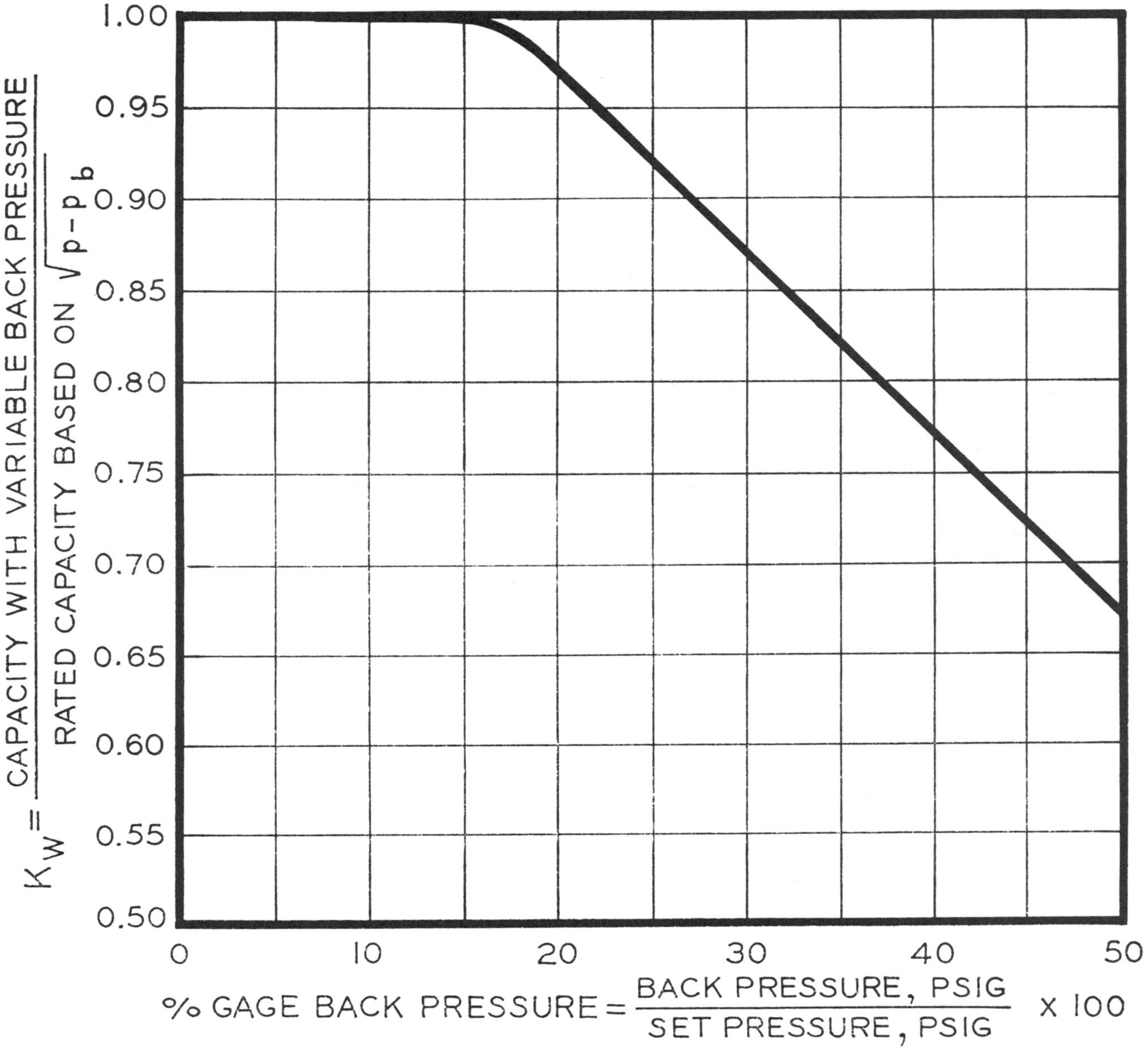

Note: The above curve represents a compromise of the values recommended by a number of relief valve manufacturers. This curve may be used when the make of the valve is not known. When the make is known, the manufacturer should be consulted for the correction factor.

Chart 3.74. Variable or constant back-pressure sizing factor, Kw, *for 25% over-pressure on balanced bellows safety relief valves (liquids only). Courtesy American Petroleum Institute.*

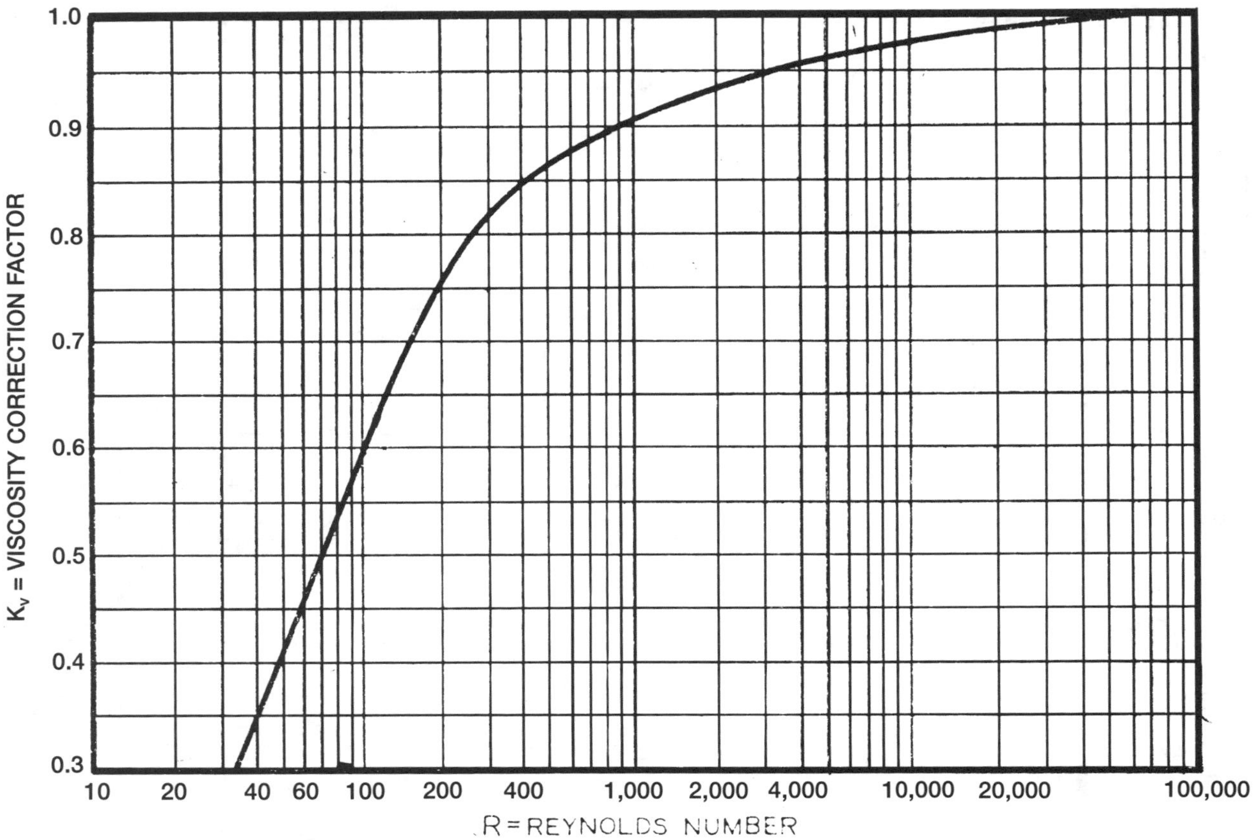

Chart 3.75. Relief valve capacity correction factor due to viscosity, K. *Courtesy American Petroleum Institute.*

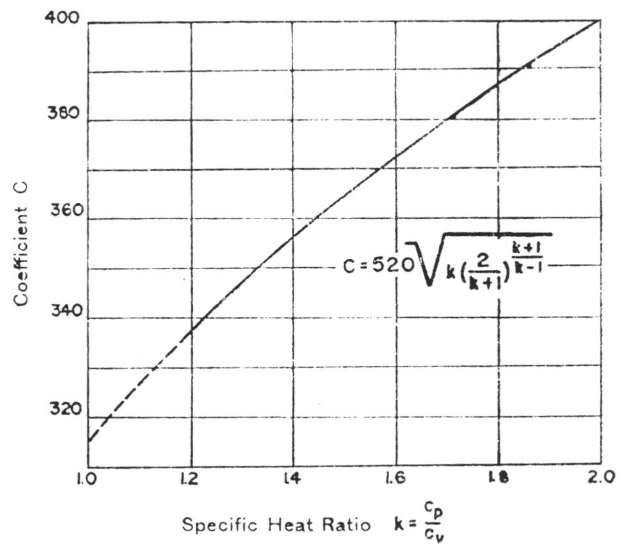

Values of Coefficient *C*

k, n	*C*	*k, n*	*C*	*k, n*	*C*	*k, n*	*C*	*k, n*	*C*	*k, n*	*C*
0.41	219.28	0.71	276.09	1.01	316.56*	1.31	347.91	1.61	373.32	1.91	394.56
0.42	221.59	0.72	277.64	1.02	317.74	1.32	348.84	1.62	374.09	1.92	395.21
0.43	223.86	0.73	279.18	1.03	318.90	1.33	349.77	1.63	374.85	1.93	395.86
0.44	226.10	0.74	280.70	1.04	320.05	1.34	350.68	1.64	375.61	1.94	396.50
0.45	228.30	0.75	282.20	1.05	321.19	1.35	351.60	1.65	376.37	1.95	397.14
0.46	230.47	0.76	283.69	1.06	322.32	1.36	352.50	1.66	377.12	1.96	397.78
0.47	232.61	0.77	285.16	1.07	323.44	1.37	353.40	1.67	377.86	1.97	398.41
0.48	234.71	0.78	286.62	1.08	324.55	1.38	354.29	1.68	378.61	1.98	399.05
0.49	236.78	0.79	288.07	1.09	325.65	1.39	355.18	1.69	379.34	1.99	399.67
0.50	238.83	0.80	289.49	1.10	326.75	1.40	356.06	1.70	380.08	2.00	400.30
0.51	240.84	0.81	290.91	1.11	327.83	1.41	356.94	1.71	380.80	2.01	400.92
0.52	242.82	0.82	292.31	1.12	328.91	1.42	357.81	1.72	381.53	2.02	401.53
0.53	244.78	0.83	293.70	1.13	329.98	1.43	358.67	1.73	382.25	2.03	402.15
0.54	246.72	0.84	295.07	1.14	331.04	1.44	359.53	1.74	382.97	2.04	402.76
0.55	248.62	0.85	296.43	1.15	332.09	1.45	360.38	1.75	383.68	2.05	403.37
0.56	250.50	0.86	297.78	1.16	333.14	1.46	361.23	1.76	384.39	2.06	403.97
0.57	252.36	0.87	299.11	1.17	334.17	1.47	362.07	1.77	385.09	2.07	404.58
0.58	254.19	0.88	300.43	1.18	335.20	1.48	362.91	1.78	385.79	2.08	405.18
0.59	256.00	0.89	301.74	1.19	336.22	1.49	363.74	1.79	386.49	2.09	405.77
0.60	257.79	0.90	303.04	1.20	337.24	1.50	364.56	1.80	387.18	2.10	406.37
0.61	259.55	0.91	304.33	1.21	338.24	1.51	365.39	1.81	387.87	2.11	406.96
0.62	261.29	0.92	305.60	1.22	339.24	1.52	366.20	1.82	388.56	2.12	407.55
0.63	263.01	0.93	306.86	1.23	340.23	1.53	367.01	1.83	389.24	2.13	408.13
0.64	264.72	0.94	308.11	1.24	341.22	1.54	367.82	1.84	389.92	2.14	408.71
0.65	266.40	0.95	309.35	1.25	342.19	1.55	368.62	1.85	390.59	2.15	409.29
0.66	268.06	0.96	310.58	1.26	343.16	1.56	369.41	1.86	391.26	2.16	409.87
0.67	269.70	0.97	311.80	1.27	344.13	1.57	370.21	1.87	391.93	2.17	410.44
0.68	271.33	0.98	313.01	1.28	345.08	1.58	370.99	1.88	392.59	2.18	411.01
0.69	272.93	0.99	314.19*	1.29	346.03	1.59	371.77	1.89	393.25	2.19	411.58
0.70	274.52	1.00	315.38*	1.30	346.98	1.60	372.55	1.90	393.91	2.20	412.15

* Interpolated values, since *C* becomes indeterminate as either *n* or *k* approaches 1.00.

Chart 3.76. Curve for evaluating coefficient, C, in flow, from specific heat ratio. Courtesy American Petroleum Institute.

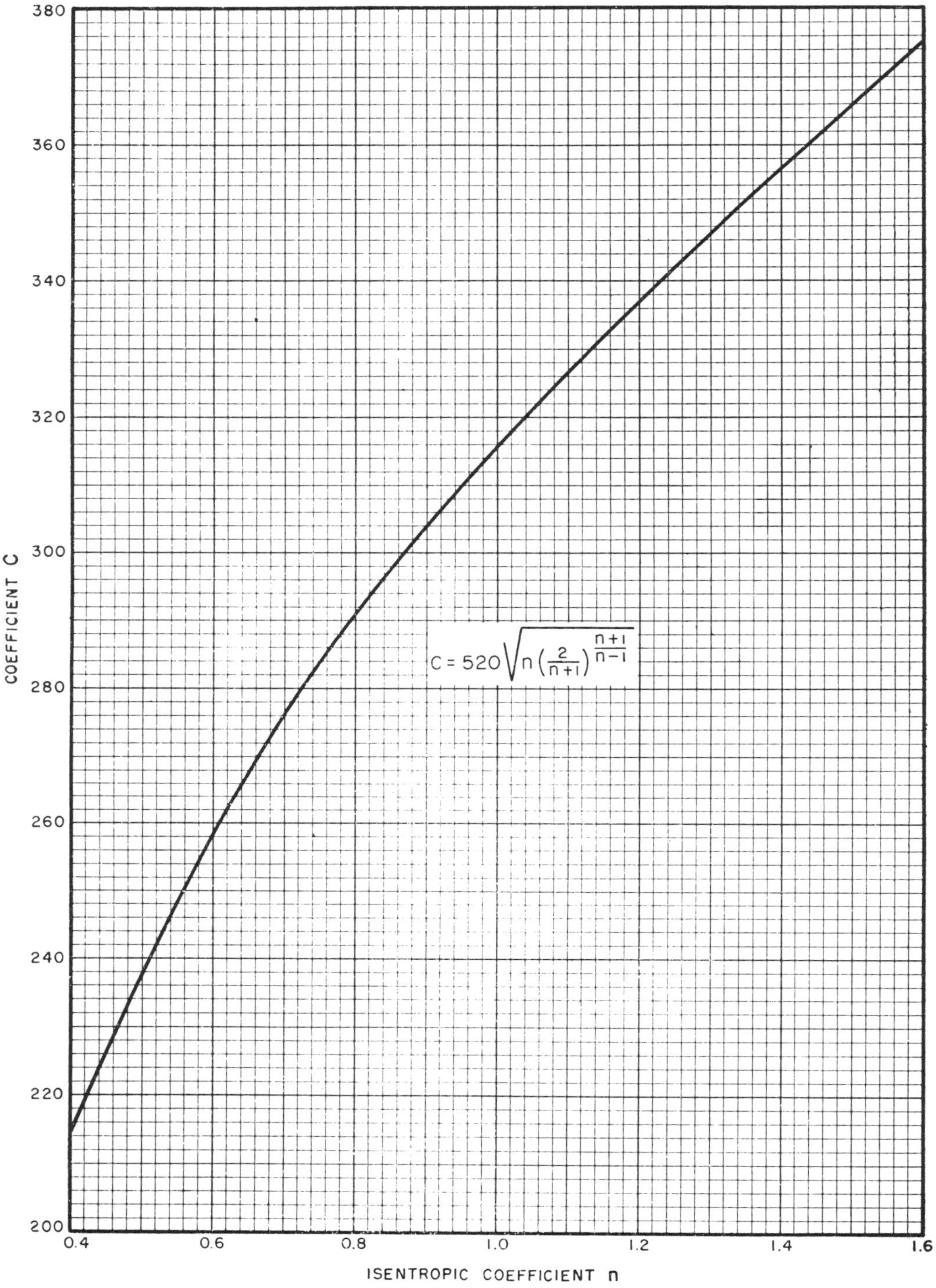

Chart 3.77. Curve for evaluating coefficient, C, *in safety relief valve flow formula from isentropic coefficient,* n. *Courtesy American Petroleum Institute.*

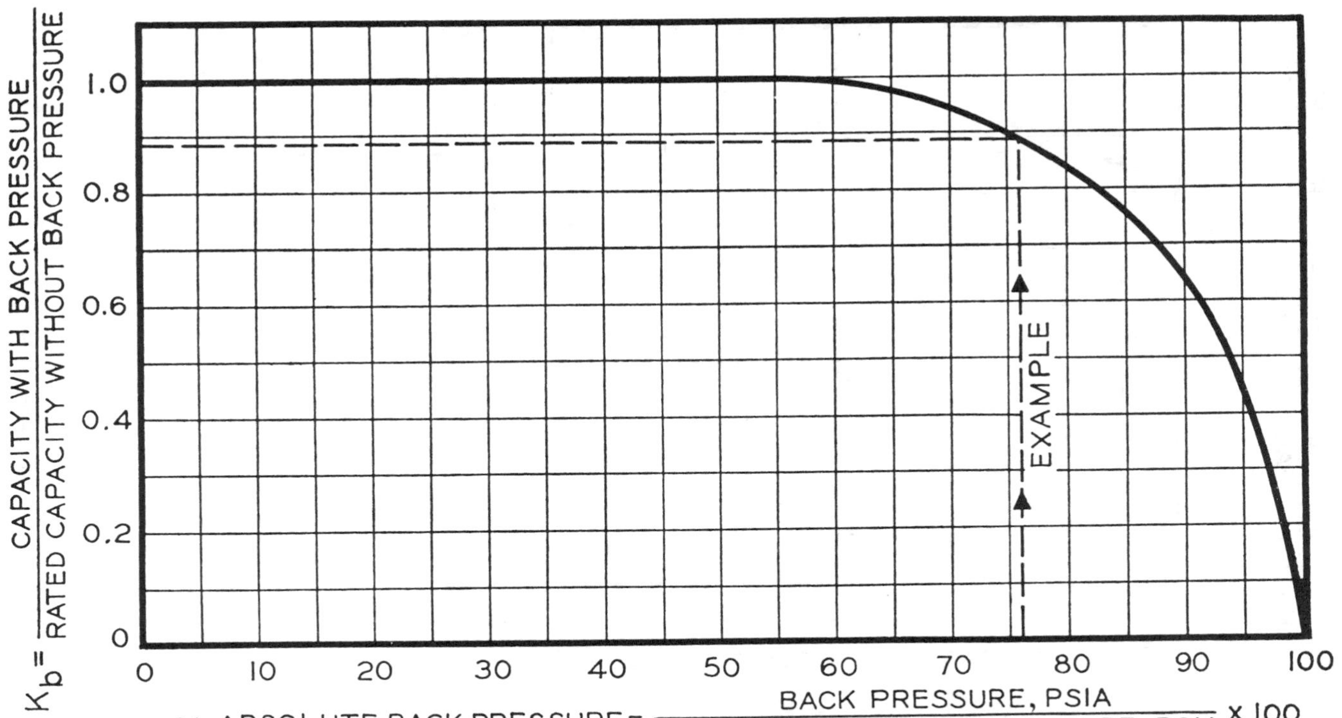

$$\% \text{ ABSOLUTE BACK PRESSURE} = \frac{\text{BACK PRESSURE, PSIA}}{\text{SET PRESSURE + OVERPRESSURE, PSIA}} \times 100$$

Example:

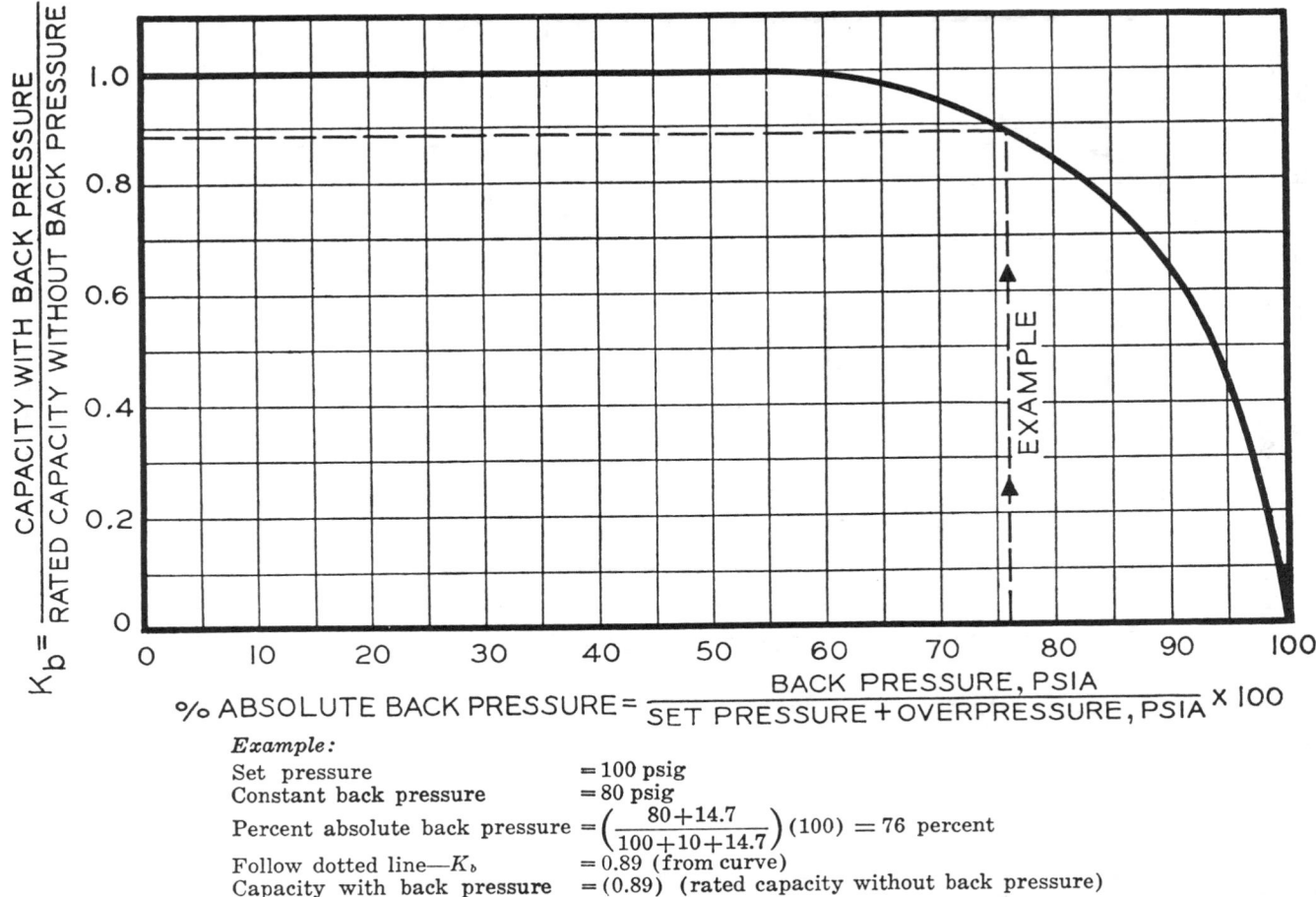

Set pressure = 100 psig

Constant back pressure = 80 psig

Percent absolute back pressure $= \left(\frac{80+14.7}{100+10+14.7}\right)(100) = 76$ percent

Follow dotted line—K_b = 0.89 (from curve)

Capacity with back pressure = (0.89) (rated capacity without back pressure)

Note: This chart is typical and suitable for use only when the make of valve or the actual critical flow pressure point for the vapor or gas is unknown; otherwise, the valve manufacturer should be consulted for specific data.

Chart 3.78. Constant back-pressure sizing factors, Kb, *for conventional safety relief valves (vapors and gases). Courtesy American Petroleum Institute.*

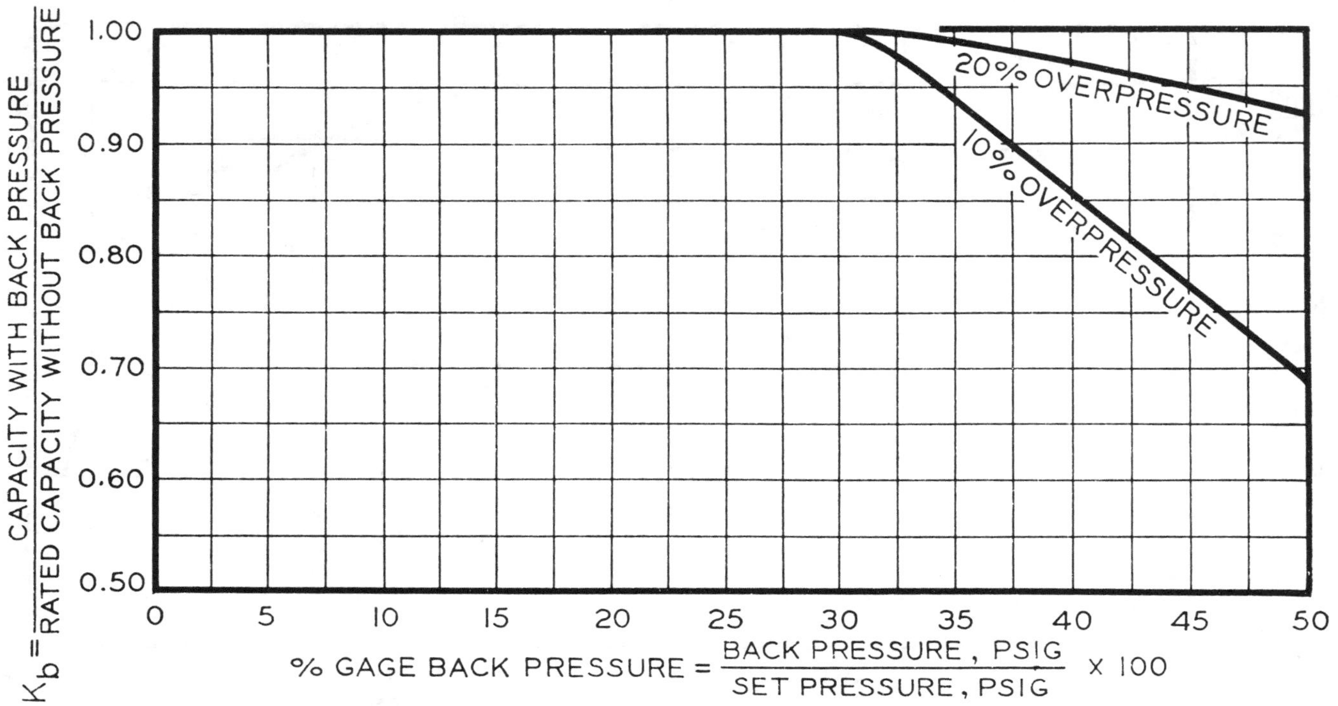

Note: The above curves represent a compromise of the values recommended by a number of relief valve manufacturers and may be used when the make of valve or the actual critical flow pressure point for the vapor or gas is unknown. When the make is known, the manufacturer should be consulted for the correction factor.

These curves are for set pressures of 50 psig and above; for set pressures lower than 50 psig, the manufacturer should be consulted for the values of K_b.

Chart 3.79. Variable or constant back-pressure sizing factor, Kb, *for balanced bellows safety relief valves (vapors and gases). Courtesy American Petroleum Institute.*

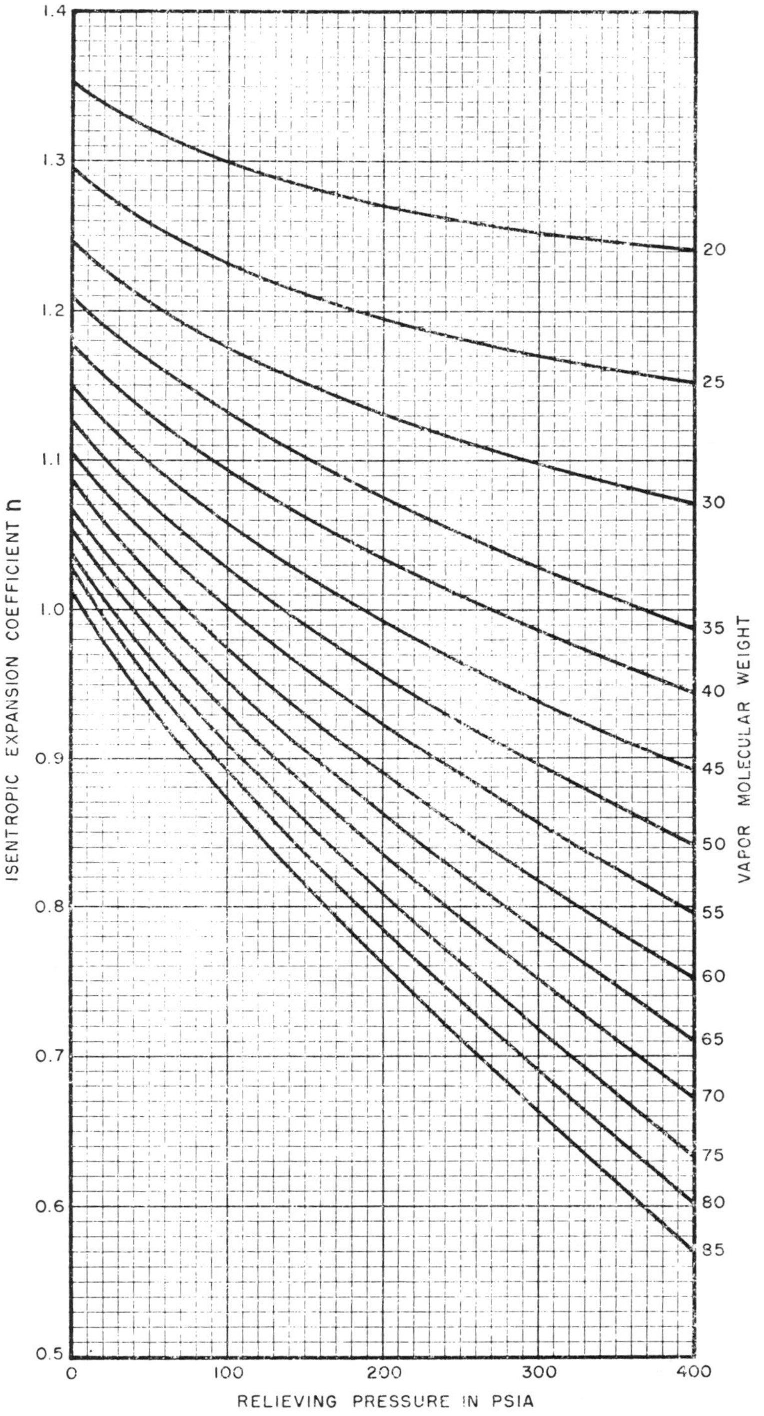

Chart 3.80. Curves for paraffin hydrocarbons expanding from relieving pressures at or near saturation to critical flow pressures. Courtesy American Petroleum Institute.

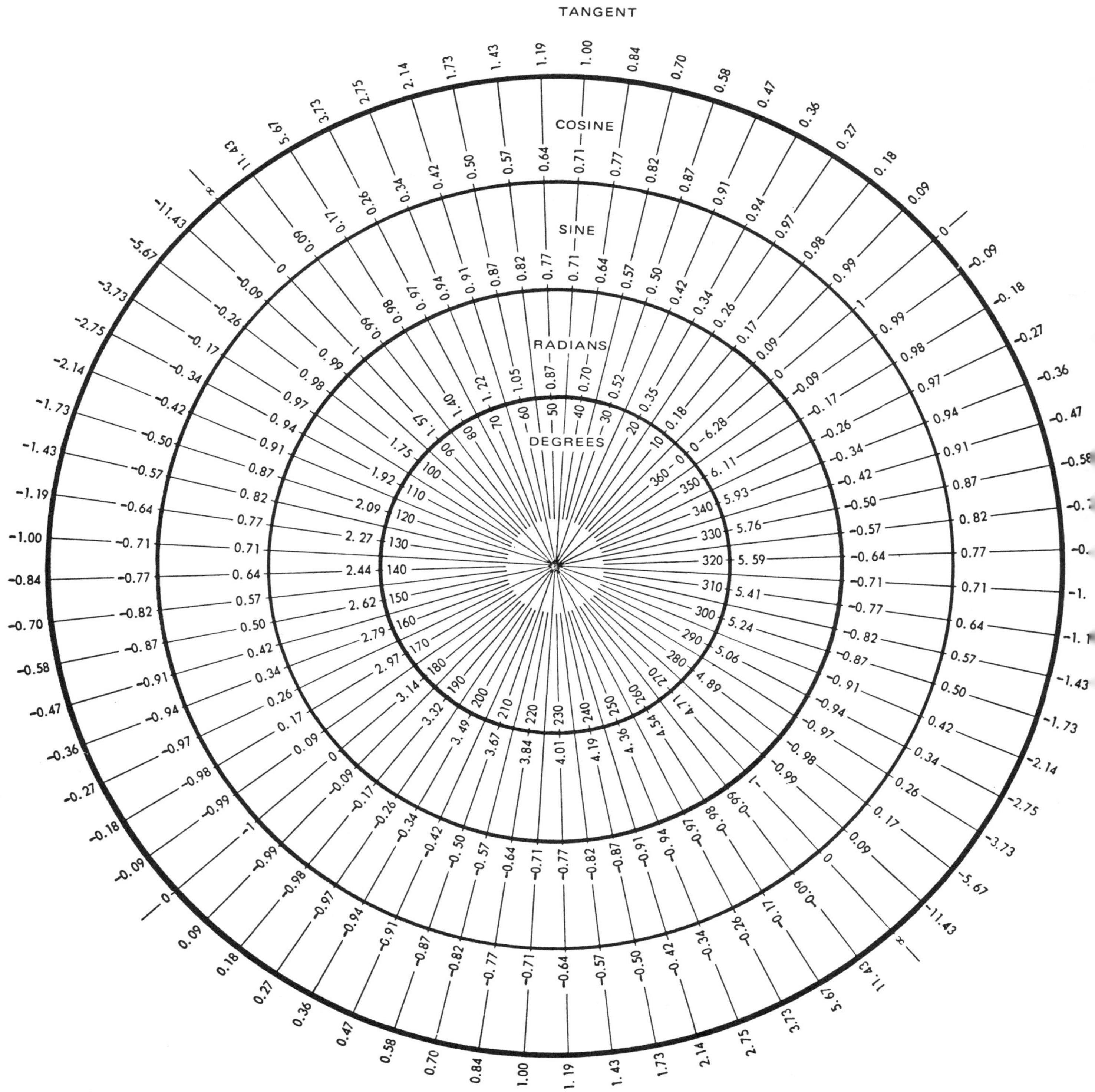

Chart 3.81. Trigonometric shortcut. Design News, *1973.*

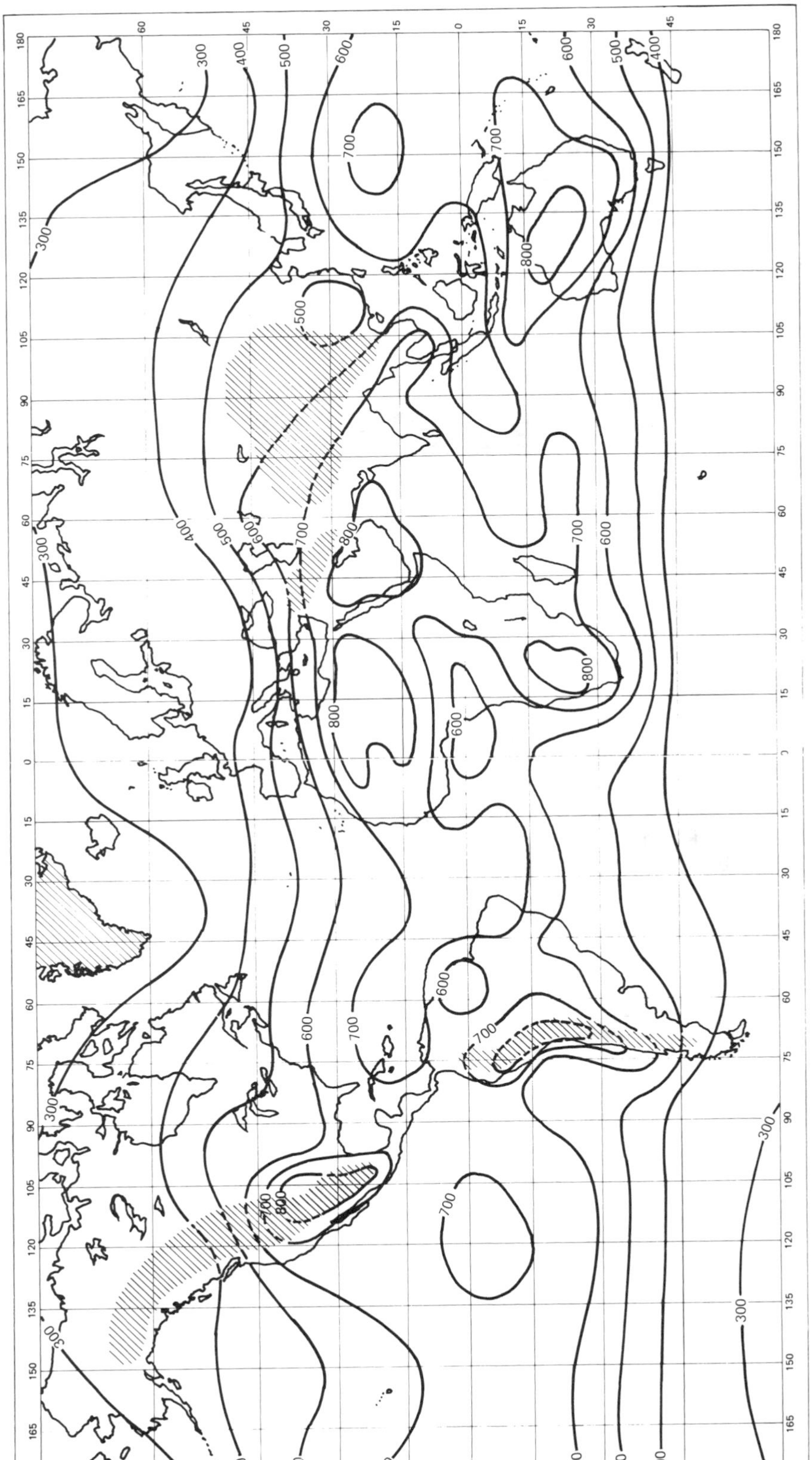

Chart 3.82. Annual solar radiation in kJ/cm². Courtesy University of Wisconsin, Solar Energy Lab, Report No. 21.

4 Tables

William G. Andrew

Introduction

This section of Engineering Data and Resource Materials covers "Tables" that engineers find helpful and time-saving in their normal duties. Included are conversion tables and factors from units of one system to another; thermocouple conversion tables; mathematical tables; physical properties of fluids, piping information tables and many others. Electrical tables may be found in chapters 9 and 10.

BRINELL INDENTATION DIAMETER, MM.	BRINELL HARDNESS NO.—10-MM. BALL. 3000-KG. LOAD: STANDARD OR TUNGSTEN CARBIDE BALL	DIAMOND PYRAMID HARDNESS NUMBER. 50-KG. LOAD	ROCKWELL HARDNESS NUMBER: B-SCALE 100-KG. LOAD 1/16 IN. DIA. BALL	ROCKWELL HARDNESS NUMBER: C-SCALE 150-KG. LOAD BRALE PENETRATOR	ROCKWELL SUPERFICIAL HARDNESS NUMBER SUPERFICIAL BRALE PENETRATOR: 15-N SCALE 15-KG. LOAD	30-N SCALE 30-KG. LOAD	45-N SCALE 45-KG. LOAD	SHORE SCLEROSCOPE HARDNESS NUMBER.	TENSILE STRENGTH (APPROX.) 1000 PSI.
2.95	429	455	—	45.7	83.4	64.6	49.9	61	217
3.00	415	440	—	44.5	82.8	63.5	48.4	59	210
3.05	401	425	—	43.1	82.0	62.3	46.9	58	202
3.10	388	410	—	41.8	81.4	61.1	45.3	56	195
3.15	375	396	—	40.4	80.6	59.9	43.6	54	188
3.20	363	383	—	39.1	80.0	58.7	42.0	52	182
3.25	352	372	(110.0)	37.9	79.3	57.6	40.5	51	176
3.30	341	360	(109.0)	36.6	78.6	56.4	39.1	50	170
3.35	331	350	(108.5)	35.5	78.0	55.4	37.8	48	166
3.40	321	339	(108.0)	34.3	77.3	54.3	36.4	47	160
3.45	311	328	(107.5)	33.1	76.7	53.3	34.4	46	155
3.50	302	319	(107.0)	32.1	76.1	52.2	33.8	45	150
3.55	293	309	(106.0)	30.9	75.5	51.2	32.4	43	145
3.60	285	301	(105.5)	29.9	75.0	50.3	31.2	—	141
3.65	277	292	(104.5)	28.8	74.4	49.3	29.9	41	137
3.70	269	284	(104.0)	27.6	73.7	48.3	28.5	40	133
3.75	262	276	(103.0)	26.6	73.1	47.3	27.3	39	129
3.80	255	269	(102.0)	25.4	72.5	46.2	26.0	38	126
3.85	248	261	(101.0)	24.2	71.7	45.1	24.5	37	122
3.90	241	253	100.0	22.8	70.9	43.9	22.8	36	118
3.95	235	247	99.0	21.7	70.3	42.9	21.5	35	115
4.00	229	241	98.2	20.5	69.7	41.9	20.1	34	111
4.05	223	234	97.3	(18.8)	—	—	—	—	—
4.10	217	228	96.4	(17.5)	—	—	—	33	105
4.15	212	222	95.5	(16.0)	—	—	—	—	102
4.20	207	218	94.6	(15.2)	—	—	—	32	100
4.25	201	212	93.8	(13.8)	—	—	—	31	98
4.30	197	207	92.8	(12.7)	—	—	—	30	95
4.35	192	202	91.9	(11.5)	—	—	—	29	93
4.40	187	196	90.7	(10.0)	—	—	—	—	90
4.45	183	192	90.0	(9.0)	—	—	—	28	89
4.50	179	188	89.0	(8.0)	—	—	—	27	87
4.55	174	182	87.8	(6.4)	—	—	—	—	85
4.60	170	178	86.8	(5.4)	—	—	—	26	83
4.65	167	175	86.0	(4.4)	—	—	—	—	81
4.70	163	171	85.0	(3.3)	—	—	—	25	79
4.80	156	163	82.9	(0.9)	—	—	—	—	76
4.90	149	156	80.8	—	—	—	—	23	73
5.00	143	150	78.7	—	—	—	—	22	71
5.10	137	143	76.4	—	—	—	—	21	67
5.20	131	137	74.0	—	—	—	—	—	65
5.30	126	132	72.0	—	—	—	—	20	63
5.40	121	127	69.8	—	—	—	—	19	60
5.50	116	122	67.6	—	—	—	—	18	58
5.60	111	117	65.7	—	—	—	—	15	56

NOTE: Values in ( ) are beyond normal range; given for information only.

Table 4.1. Hardness comparison scales for various steels.

Metric Conversion Table

To convert from	To	Multiply by	To convert from	To	Multiply by
Length			**Length**		
mm	inches	.03937	inches	mm	25.40
cm	inches	.3937	inches	cm	2.540
meters	inches	39.37	inches	meters	.0254
meters	feet	3.281	feet	meters	.3048
meters	yards	1.0936	feet	km	.0003048
km	feet	3280.8	yards	meters	.9144
km	yards	1093.6	yards	km	.0009144
km	miles	.6214	miles	km	1.609
Area			**Area**		
sq mm	sq inches	.00155	sq inches	sq mm	645.2
sq cm	sq inches	.155	sq inches	sq cm	6.452
sq meters	sq feet	10.764	sq feet	sq meters	.09290
sq meters	sq yards	1.196	sq yards	sq meters	.8361
sq km	sq miles	.3861	sq miles	sq km	2.590
hectares	acres	2.471	acres	hectares	.4047
Volume			**Volume**		
cu cm	cu inches	.06102	cu inches	cu cm	16.387
cu cm	fl ounces	.03381	cu inches	liters	.01639
cu meters	cu feet	35.314	cu feet	cu meters	.02832
cu meters	cu yards	1.308	cu feet	liters	28.317
cu meters	US gal	264.2	cu yards	cu meters	.7646
liters	cu inches	61.023	fl ounces	cu cm	29.57
liters	cu feet	.03531	US gal	cu meters	.003785
liters	US gal	.2642	US gal	liters	3.785
Weight			**Weight**		
grams	grains	15.432	grains	grams	.0648
grams	ounces†	.0353	ounces†	grams	28.350
kg	ounces†	35.27	ounces†	kg	.02835
kg	pounds†	2.2046	pounds†	kg	.4536
kg	US tons	.001102	pounds†	tonnes	.000454
kg	long tons	.000984	US tons	kg	907.2
tonnes	pounds†	2204.6	US tons	tonnes	.9072
tonnes	US tons	1.1023	long tons	kg	1016
tonnes	long tons	.9842	long tons	tonnes	1.0160

†avoirdupois pounds and ounces.

To convert from	To	Multiply by	To convert from	To	Multiply by
Unit weight			**Unit weight**		
gr/sq cm	lb/sq in	.01422	lb/ft	kg/m	1.4881
gr/cu cm	lb/cu in	.0361	lb/sq in	gr/sq cm	70.31
kg/sq cm	lb/sq in	14.22	lb/sq in	kg/sq cm	.07031
kg/cu m	lb/cu ft	.0624	lb/cu in	gr/cu cm	27.68
kg/m	lb/ft	.6720	lb/cu ft	kg/cu m	16.018
Unit volume			**Unit volume**		
liters/min	US gpm	.2642	US gpm	liters/min	3.785
liters/min	cfm	.03531	US gpm	liters/hr	227.1
liters/hr	US gpm	.0044	US gpm	cu m/hr	.2271
cu m/min	cfm	35.314	cfm	liters/min	28.317
cu m/hr	cfm	.5886	cfm	cu m/min	.02832
cu m/hr	US gpm	4.4028	cfm	cu m/hr	1.6992
Power			**Power**		
watts	ft-lb/sec	.7376	ft-lb/sec	watts	1.356
watts	hp	.00134	hp	watts	745.7
kw	hp	1.3410	hp	kw	.7457
cheval-vap	hp	.9863	hp	cheval-vap	1.0139
Heat			**Heat**		
gr-cal	Btu	.003969	Btu	gr-cal	252
kg-cal	Btu	3.9693	Btu	kg-cal	.252
kg-cal/kg	Btu/lb	1.800	Btu/lb	kg-cal/kg	.5556
gr-cal/sq cm	Btu/sq ft	3.687	Btu/sq ft	gr-cal/sq cm	.2713
kg-cal/cu m	But/cu ft	.1124	Btu/cu ft	kg-cal/cu m	8.899
Work or energy			**Work or energy**		
joule	ft-lb	.7376	ft-lb	joule	1.356
meter-kg	ft-lb	7.2330	ft-lb	meter-kg	.1383
gr-cal	ft-lb	3.087	ft-lb	gr-cal	.3239
kg-cal	ft-lb	3087	ft-lb	kg-cal	.000324
hp-hr	ft-lb	1,980,000	ft-lb	hp-hr	5.05×10^{-7}
kwhr	ft-lb	2,655,000	ft-lb	kwhr	3.766×10^{-7}
Btu	ft-lb	778.0	ft-lb	Btu	.001285

Table 4.2. Metric conversions—metric to English, English to metric. Courtesy Cameron Hydraulic Data Book, *Ingersoll-Rand Company.*

NOTE: The numbers in boldface refer to the temperature in degrees, either Centigrade or Fahrenheit, which it is desired to convert into the other scale. If converting from Fahrenheit to Centigrade degrees, the equivalent temperature will be found in the left column; while if converting from degrees Centigrade to degrees Fahrenheit, the answer will be found in the column on the right.

Centigrade		*Fahrenheit*	*Centigrade*		*Fahrenheit*	*Centigrade*		*Fahrenheit*	*Centigrade*		*Fahrenheit*
-73.3	**-100**	-148.0	2.8	**37**	98.6	33.3	**92**	197.6	293	**560**	1040
-67.8	**-90**	-130.0	3.3	**38**	100.4	33.9	**93**	199.4	299	**570**	1058
-62.2	**-80**	-112.0	3.9	**39**	102.2	34.4	**94**	201.2	304	**580**	1076
-59.4	**-75**	-103.0	4.4	**40**	104.0	35.0	**95**	203.0	310	**590**	1094
-56.7	**-70**	-94.0	5.0	**41**	105.8	35.6	**96**	204.8	316	**600**	1112
-53.9	**-65**	-85.0	5.6	**42**	107.6	36.1	**97**	206.6	321	**610**	1130
-51.1	**-60**	-76.0	6.1	**43**	109.4	36.7	**98**	208.4	327	**620**	1148
-48.3	**-55**	-67.0	6.7	**44**	111.2	37.2	**99**	210.2	332	**630**	1166
-45.6	**-50**	-58.0	7.2	**45**	113.0	37.8	**100**	212.0	338	**640**	1184
-42.8	**-45**	-49.0	7.8	**46**	114.8	43	**110**	230	343	**650**	1202
-40.0	**-40**	-40.0	8.3	**47**	116.6	49	**120**	248	349	**660**	1220
-37.2	**-35**	-31.0	8.9	**48**	118.4	54	**130**	266	354	**670**	1238
-34.4	**-30**	-22.0	9.4	**49**	120.2	60	**140**	284	360	**680**	1256
-31.7	**-25**	-13.0	10.0	**50**	122.0	66	**150**	302	366	**690**	1274
-28.9	**-20**	-4.0	10.6	**51**	123.8	71	**160**	320	371	**700**	1292
-26.1	**-15**	5.0	11.1	**52**	125.6	77	**170**	338	377	**710**	1310
-23.3	**-10**	14.0	11.7	**53**	127.4	82	**180**	356	382	**720**	1328
-20.6	**-5**	23.0	12.2	**54**	129.2	88	**190**	374	388	**730**	1346
-17.8	**0**	32.0	12.8	**55**	131.0	93	**200**	392	393	**740**	1364
-17.2	**1**	33.8	13.3	**56**	132.8	99	**210**	410	399	**750**	1382
-16.7	**2**	35.6	13.9	**57**	134.6	100	**212**	414	404	**760**	1400
-16.1	**3**	37.4	14.4	**58**	136.4	104	**220**	428	410	**770**	1418
-15.6	**4**	39.2	15.0	**59**	138.2	110	**230**	446	416	**780**	1436
-15.0	**5**	41.0	15.6	**60**	140.0	116	**240**	464	421	**790**	1454
-14.4	**6**	42.8	16.1	**61**	141.8	121	**250**	482	427	**800**	1472
-13.9	**7**	44.6	16.7	**62**	143.6	127	**260**	500	432	**810**	1490
-13.3	**8**	46.4	17.2	**63**	145.4	132	**270**	518	438	**820**	1508
-12.8	**9**	48.2	17.8	**64**	147.2	138	**280**	536	443	**830**	1526
-12.2	**10**	50.0	18.3	**65**	149.0	143	**290**	554	449	**840**	1544
-11.7	**11**	51.8	18.9	**66**	150.8	149	**300**	572	454	**850**	1562
-11.1	**12**	53.6	19.4	**67**	152.6	154	**310**	590	460	**860**	1580
-10.6	**13**	55.4	20.0	**68**	154.4	160	**320**	608	466	**870**	1598
-10.0	**14**	57.2	20.6	**69**	156.2	166	**330**	626	471	**880**	1616
-9.4	**15**	59.0	21.1	**70**	158.0	171	**340**	644	477	**890**	1634
-8.9	**16**	60.8	21.7	**71**	159.8	177	**350**	662	482	**900**	1652
-8.3	**17**	62.6	22.2	**72**	161.6	182	**360**	680	488	**910**	1670
-7.8	**18**	64.4	22.8	**73**	163.4	188	**370**	698	493	**920**	1688
-7.2	**19**	66.2	23.3	**74**	165.2	193	**380**	716	499	**930**	1706
-6.7	**20**	68.0	23.9	**75**	167.0	199	**390**	734	504	**940**	1724
-6.1	**21**	69.8	24.4	**76**	168.8	204	**400**	752	510	**950**	1742
-5.6	**22**	71.6	25.0	**77**	170.6	210	**410**	770	516	**960**	1760
-5.0	**23**	73.4	25.6	**78**	172.4	216	**420**	788	521	**970**	1778
-4.4	**24**	75.2	26.1	**79**	174.2	221	**430**	806	527	**980**	1796
-3.9	**25**	77.0	26.7	**80**	176.0	227	**440**	824	532	**990**	1814
-3.3	**26**	78.8	27.2	**81**	177.8	232	**450**	842	538	**1000**	1832
-2.8	**27**	80.6	27.8	**82**	179.6	238	**460**	860	566	**1050**	1922
-2.2	**28**	82.4	28.3	**83**	181.4	243	**470**	878	593	**1100**	2012
-1.7	**29**	84.2	28.9	**84**	183.2	249	**480**	896	621	**1150**	2102
-1.1	**30**	86.0	29.4	**85**	185.0	254	**490**	914	649	**1200**	2192
-0.6	**31**	87.8	30.0	**86**	186.8	260	**500**	932	677	**1250**	2282
0.0	**32**	89.6	30.6	**87**	188.6	266	**510**	950	704	**1300**	2372
0.6	**33**	91.4	31.1	**88**	190.4	271	**520**	968	732	**1350**	2462
1.1	**34**	93.2	31.7	**89**	192.2	277	**530**	986	760	**1400**	2552
1.7	**35**	95.0	32.2	**90**	194.0	282	**540**	1004	788	**1450**	2642
2.2	**36**	96.8	32.8	**91**	195.8	288	**550**	1022	816	**1500**	2732

The formulas at the right may also be used for converting Centigrade or Fahrenheit degrees into the other scale.

Degrees Cent., $C° = \frac{5}{9}(F° - 32)$ Degrees Fahr., $F° = \frac{9}{5}(C° + 32)$

Table 4.3. Temperature conversion—Centigrade to Fahrenheit or Fahrenheit to Centigrade. Courtesy Natural Gas Processors Suppliers Association.

TO CONVERT	INTO	MULTIPLY BY
	A	
Abcoulomb	Statcoulombs	2.998×10^{10}
Acre	Sq. chain (Gunters)	10
Acre	Rods	160
Acre	Square links (Gunters)	1×10^{5}
Acre	Hectare or sq. hectometer	.4047
acres	sq feet	43,560.0
acres	sq meters	4,047.
acres	sq miles	1.562×10^{-3}
acres	sq yards	4,840.
acre-feet	cu feet	43,560.0
acre-feet	gallons	3.259×10^{5}
Angstrom unit	Inch	3937×10^{-9}
Angstrom unit	Meter	1×10^{-10}
Angstrom unit	Micron or (Mu)	1×10^{-4}
Are	Acre (US)	.02471
Ares	sq. yards	119.60
ares	acres	0.02471
ares	sq meters	100.0
Astronomical Unit	Kilometers	1.495×10^{8}
Atmospheres	Ton/sq. inch	.007348
atmospheres	cms of mercury	76.0
atmospheres	ft of water (at 4°C)	33.90
atmospheres	in. of mercury (at 0°C)	29.92
atmospheres	kgs/sq cm	1.0333
atmospheres	kgs/sq meter	10,332.
atmospheres	pounds/sq in.	14.70
atmospheres	tons/sq ft	1.058
	B	
Barrels (U.S., dry)	cu. inches	7056.
Barrels (U.S., dry)	quarts (dry)	105.0
Barrels (U.S., liquid)	gallons	31.5
barrels (oil)	gallons (oil)	42.0
bars	atmospheres	0.9869
bars	dynes/sq cm	10^{6}
bars	kgs/sq meter	1.020×10^{4}
bars	pounds/sq ft	2,089.
bars	pounds/sq in.	14.50
Baryl	Dyne/sq. cm.	1.000
Bolt (US Cloth)	Meters	36.576
BTU	Liter—Atmosphere	10.409
Btu	ergs	1.0550×10^{10}
Btu	foot-lbs	778.3
Btu	gram-calories	252.0
Btu	horsepower-hrs	3.931×10^{-4}
Btu	joules	1,054.8
Btu	kilogram-calories	0.2520
Btu	kilogram-meters	107.5
Btu	kilowatt-hrs	2.928×10^{-4}
Btu/hr	foot-pounds/sec	0.2162
Btu/hr	gram-cal/sec	0.0700
Btu/hr	horsepower-hrs	3.929×10^{-4}
Btu/hr	watts	0.2931
Btu/min	foot-lbs/sec	12.96
Btu/min	horsepower	0.02356
Btu/min	kilowatts	0.01757
Btu/min	watts	17.57
Btu/sq ft/min	watts/sq in.	0.1221
Bucket (Br. dry)	Cubic Cm.	1.818×10^{4}
bushels	cu ft	1.2445
bushels	cu in.	2,150.4
bushels	cu meters	0.03524
bushels	liters	35.24
bushels	pecks	4.0
bushels	pints (dry)	64.0
bushels	quarts (dry)	32.0
	C	
Calories, gram (mean)	B.T.U. (mean)	3.9685×10^{-3}
Candle/sq. cm	Lamberts	3.142
Candle/sq. inch	Lamberts	.4870
centares (centiares)	sq meters	1.0
Centigrade	Fahrenheit	(C°x9/5)+32
centigrams	grams	0.01
Centiliter	Ounce fluid (US)	.3382
Centiliter	Cubic inch	.6103
Centiliter	drams	2.705
centiliters	liters	0.01
centimeters	feet	3.281×10^{-2}
centimeters	inches	0.3937
centimeters	kilometers	10^{-5}
centimeters	meters	0.01
centimeters	miles	6.214×10^{-6}
centimeters	millimeters	10.0
centimeters	mils	393.7
centimeters	yards	1.094×10^{-2}
centimeter-dynes	cm-grams	1.020×10^{-3}
centimeter-dynes	meter-kgs	1.020×10^{-8}
centimeter-dynes	pound-feet	7.376×10^{-8}
centimeter-grams	cm-dynes	980.7
centimeter-grams	meter-kgs	10^{-5}
centimeter-grams	pound-feet	7.233×10^{-5}
centimeters of mercury	atmospheres	0.01316
centimeters of mercury	feet of water	0.4461
centimeters of mercury	kgs/sq meter	136.0
centimeters of mercury	pounds/sq ft	27.85
centimeters of mercury	pounds/sq in.	0.1934
centimeters/sec	feet/min	1.1969
centimeters/sec	feet/sec	0.03281
centimeters/sec	kilometers/hr	0.036
centimeters/sec	knots	0.1943
centimeters/sec	meters/min	0.6
centimeters/sec	miles/hr	0.02237
centimeters/sec	miles/min	3.728×10^{-4}
centimeters/sec/sec	feet/sec/sec	0.03281
centimeters/sec/sec	kms/hr/sec	0.036
centimeters/sec/sec	meters/sec/sec	0.01
centimeters/sec/sec	miles/hr/sec	0.02237
Chain	Inches	792.00
Chain	meters	20.12
Chains (surveyors' or Gunter's)	yards	22.00
circular mils	sq cms	5.067×10^{-6}
circular mils	sq mils	0.7854
Circumference	Radians	6.283
circular mils	sq inches	7.854×10^{-7}
Cords	cord feet	8
Cord feet	cu. feet	16
cubic centimeters	cu feet	3.531×10^{-5}
cubic centimeters	cu inches	0.06102
cubic centimeters	cu meters	10^{-6}
cubic centimeters	cu yards	1.308×10^{-6}
cubic centimeters	gallons (U. S. liq.)	2.642×10^{-4}
cubic centimeters	liters	0.001
cubic centimeters	pints (U.S. liq.)	2.113×10^{-3}
cubic centimeters	quarts (U.S. liq.)	1.057×10^{-3}
cubic feet	bushels (dry)	0.8036
cubic feet	cu cms	— 28,320.0
cubic feet	cu inches	1,728.0
cubic feet	cu meters	0.02832
cubic feet	cu yards	0.03704
cubic feet	gallons (U.S. liq.)	7.48052
cubic feet	liters	28.32
cubic feet	pints (U.S. liq.)	59.84
cubic feet	quarts (U.S. liq.)	29.92

Table 4.4 continued

Table 4.4. Units of measurement from one system to another.

Table 4.4 continued

TO CONVERT	INTO	MULTIPLY BY
cubic feet/min	cu cms/sec	472.0
cubic feet/min	gallons/sec	0.1247
cubic feet/min	liters/sec	0.4720
cubic feet/min	pounds of water/min	62.43
cubic feet/sec	million gals/day	0.646317
cubic feet/sec	gallons/min	448.831
cubic inches	cu cms	16.39
cubic inches	cu feet	5.787×10^{-4}
cubic inches	cu meters	1.639×10^{-5}
cubic inches	cu yards	2.143×10^{-5}
cubic inches	gallons	4.329×10^{-3}
cubic inches	liters	0.01639
cubic inches	mil-feet	1.061×10^{5}
cubic inches	pints (U.S. liq.)	0.03463
cubic inches	quarts (U.S. liq.)	0.01732
cubic meters	bushels (dry)	28.38
cubic meters	cu cms	10^{6}
cubic meters	cu feet	35.31
cubic meters	cu inches	61,023.0
cubic meters	cu yards	1.308
cubic meters	gallons (U.S. liq.)	264.2
cubic meters	liters	1,000.0
cubic meters	pints (U.S. liq.)	2,113.0
cubic meters	quarts (U.S. liq.)	1,057.
cubic yards	cu cms	7.646×10^{5}
cubic yards	cu feet	27.0
cubic yards	cu inches	46,656.0
cubic yards	cu meters	0.7646
cubic yards	gallons (U.S. liq.)	202.0
cubic yards	liters	764.6
cubic yards	pints (U.S. liq.)	1,615.9
cubic yards	quarts (U.S. liq.)	807.9
cubic yards/min	cubic ft/sec	0.45
cubic yards/min	gallons/sec	3.367
cubic yards/min	liters/sec	12.74
D		
Dalton	Gram	1.650×10^{-24}
days	seconds	86,400.0
decigrams	grams	0.1
deciliters	liters	0.1
decimeters	meters	0.1
degrees (angle)	quadrants	0.01111
degrees (angle)	radians	0.01745
degrees (angle)	seconds	3,600.0
degrees/sec	radians/sec	0.01745
degrees/sec	revolutions/min	0.1667
degrees/sec	revolutions/sec	2.778×10^{-3}
dekagrams	grams	10.0
dekaliters	liters	10.0
dekameters	meters	10.0
Drams (apothecaries' or troy)	ounces (avoidupois)	0.1371429
Drams (apothecaries' or troy)	ounces (troy)	0.125
Drams (U.S., fluid or apoth.)	cubic cm.	3.6967
drams	grams	1.7718
drams	grains	27.3437
drams	ounces	0.0625
Dyne/cm	Erg/sq. millimeter	.01
Dyne/sq. cm.	Atmospheres	9.869×10^{-7}
Dyne/sq. cm.	Inch of Mercury at 0°C	2.953×10^{-5}
Dyne/sq. cm.	Inch of Water at 4°C	4.015×10^{-4}
dynes	grams	1.020×10^{-3}
dynes	joules/cm	10^{-7}
dynes	joules/meter (newtons)	10^{-5}
dynes	kilograms	1.020×10^{-6}
dynes	poundals	7.233×10^{-5}
dynes	pounds	2.248×10^{-6}
dynes/sq cm	bars	10^{-6}

TO CONVERT	INTO	MULTIPLY BY
E		
Ell	Cm.	114.30
Ell	Inches	45
Em, Pica	Inch	.167
Em, Pica	Cm.	.4233
Erg/sec	Dyne – cm/sec	1.000
ergs	Btu	9.480×10^{-11}
ergs	dyne-centimeters	1.0
ergs	foot-pounds	7.367×10^{-8}
ergs	gram-calories	0.2389×10^{-7}
ergs	gram-cms	1.020×10^{-3}
ergs	horsepower-hrs	3.7250×10^{-14}
ergs/sec	Btu/min	$5,688 \times 10^{-9}$
ergs/sec	ft-lbs/min	4.427×10^{-6}
ergs/sec	ft-lbs/sec	7.3756×10^{-8}
ergs/sec	horsepower	1.341×10^{-10}
ergs/sec	kg-calories/min	1.433×10^{-9}
F		
Fathom	Meter	1.828804
fathoms	feet	6.0
feet	centimeters	30.48
feet	kilometers	3.048×10^{-4}
feet	meters	0.3048
feet	miles (naut.)	1.645×10^{-4}
feet	miles (stat.)	1.894×10^{-4}
feet	millimeters	304.8
feet	mils	1.2×10^{4}
feet of water	atmospheres	0.02950
feet of water	in. of mercury	0.8826
feet of water	kgs/sq cm	0.03048
feet of water	kgs/sq meter	304.8
feet of water	pounds/sq ft	62.43
feet of water	pounds/sq in.	0.4335
feet/min	cms/sec	0.5080
feet/min	feet/sec	0.01667
feet/min	kms/hr	0.01829
feet/min	meters/min	0.3048
feet/min	miles/hr	0.01136
feet/sec	cms/sec	30.48
feet/sec	kms/hr	1.097
feet/sec	knots	0.5921
feet/sec	meters/min	18.29
feet/sec	miles/hr	0.6818
feet/sec	miles/min	0.01136
feet/sec/sec	cms/sec/sec	30.48
feet/sec/sec	kms/hr/sec	1.097
feet/sec/sec	meters/sec/sec	0.3048
feet/sec/sec	miles/hr/sec	0.6818
feet/100 feet	per cent grade	1.0
Foot – candle	Lumen/sq. meter	10.764
foot-pounds	Btu	1.286×10^{-3}
foot-pounds	ergs	1.356×10^{7}
foot-pounds	gram-calories	0.3238
foot-pounds	hp-hrs	5.050×10^{-7}
foot-pounds	joules	1.356
foot-pounds	kg-calories	3.24×10^{-4}
foot-pounds	kg-meters	0.1383
foot-pounds/min	Btu/min	1.286×10^{-3}
foot-pounds/min	foot-pounds/sec	0.01667
foot-pounds/min	horsepower	3.030×10^{-5}
foot-pounds/min	kg-calories/min	3.24×10^{-4}
foot-pounds/sec	Btu/hr	4.6263
foot-pounds/sec	Btu/min	0.07717
foot-pounds/sec	horsepower	1.818×10^{-3}
foot-pounds/sec	kg-calories/min	0.01945
foot-pounds/sec	kilowatts	1.356×10^{-3}
Furlongs	miles (U.S.)	0.125
furlongs	rods	40.0
furlongs	feet	660.0

Table 4.4 continued

Table 4.4 continued

TO CONVERT	INTO	MULTIPLY BY
	G	
gallons	cu cms	3,785.0
gallons	cu feet	0.1337
gallons	cu inches	231.0
gallons	cu meters	3.785×10^{-3}
gallons	cu yards	4.951×10^{-3}
gallons	liters	3.785
gallons (liq. Br. Imp.)	gallons (U.S. liq.)	1.20095
gallons (U.S.)	gallons (Imp.)	0.83267
gallons of water	pounds of water	8.3453
gallons/min	cu ft/sec	2.228×10^{-3}
gallons/min	liters/sec	0.06308
gallons/min	cu ft/hr	8.0208
Gills (British)	cubic cm.	142.07
gills	liters	0.1183
gills	pints (liq.)	0.25
Grade	Radian	.01571
Grains	drams (avoirdupois)	0.03657143
grains (troy)	grains (avdp)	1.0
grains (troy)	grams	0.06480
grains (troy)	ounces (avdp)	2.0833×10^{-3}
grains (troy)	pennyweight (troy)	0.04167
grains/U.S. gal	parts/million	17.118
grains/U.S. gal	pounds/million gal	142.86
grains/Imp. gal	parts/million	14.286
grams	dynes	980.7
grams	grains	15.43
grams	joules/cm	9.807×10^{-5}
grams	joules/meter (newtons)	9.807×10^{-3}
grams	kilograms	0.001
grams	milligrams	1,000.
grams	ounces (avdp)	0.03527
grams	ounces (troy)	0.03215
grams	poundals	0.07093
grams	pounds	2.205×10^{-3}
grams/cm	pounds/inch	5.600×10^{-3}
grams/cu cm	pounds/cu ft	62.43
grams/cu cm	pounds/cu in	0.03613
grams/cu cm	pounds/mil-foot	3.405×10^{-7}
grams/liter	grains/gal	58.417
grams/liter	pounds/1,000 gal	8.345
grams/liter	pounds/cu ft	0.062427
grams/liter	parts/million	1,000.0
grams/sq cm	pounds/sq ft	2.0481
gram-calories	Btu	3.9683×10^{-3}
gram-calories	ergs	4.1868×10^{7}
gram-calories	foot-pounds	3.0880
gram-calories	horsepower-hrs	1.5596×10^{-6}
gram-calories	kilowatt-hrs	1.1630×10^{-6}
gram-calories	watt-hrs	1.1630×10^{-3}
gram-calories/sec	Btu/hr	14.286
gram-centimeters	Btu	9.297×10^{-8}
gram-centimeters	ergs	980.7
gram-centimeters	joules	9.807×10^{-5}
gram-centimeters	kg-cal	2.343×10^{-8}
gram-centimeters	kg-meters	10^{-5}
	H	
Hand	Cm.	10.16
hectares	acres	2.471
hectares	sq feet	1.076×10^{5}
hectograms	grams	100.0
hectoliters	liters	100.0
hectometers	meters	100.0
Hogsheads (British)	cubic ft.	10.114
Hogsheads (U.S.)	cubic ft.	8.42184
Hogsheads (U.S.)	gallons (U.S.)	63
horsepower	Btu/min	42.44
horsepower	foot-lbs/min	33,000.
horsepower	foot-lbs/sec	550.0
horsepower (metric) (542.5 ft lb/sec)	horsepower (550 ft lb/sec)	0.9863
horsepower (550 ft lb/sec)	horsepower (metric) (542.5 ft lb/sec)	1.014
horsepower	kg-calories/min	10.68
horsepower (boiler)	Btu/hr	33.479
horsepower-hrs	Btu	2,547.
horsepower-hrs	ergs	2.6845×10^{13}
horsepower-hrs	foot-lbs	1.98×10^{6}
horsepower-hrs	gram-calories	641,190.
horsepower-hrs	joules	2.684×10^{6}
horsepower-hrs	kg-calories	641.1
horsepower-hrs	kg-meters	2.737×10^{5}
hours	days	4.167×10^{-2}
hours	weeks	5.952×10^{-3}
Hundredweights (long)	pounds	112
Hundredweights (long)	tons (long)	0.05
Hundredweights (short)	ounces (avoirdupois)	1600
Hundredweights (short)	pounds	100
Hundredweights (short)	tons (metric)	0.0453592
Hundredweights (short)	tons (long)	0.0446429
	I	
inches	centimeters	2.540
inches	meters	2.540×10^{-2}
inches	miles	1.578×10^{-5}
inches	millimeters	25.40
inches	mils	1,000.0
inches	yards	2.778×10^{-2}
inches of mercury	atmospheres	0.03342
inches of mercury	feet of water	1.133
inches of mercury	kgs/sq cm	0.03453
inches of mercury	kgs/sq meter	345.3
inches of mercury	pounds/sq ft	70.73
inches of mercury	pounds/sq in.	0.4912
inches of water (at 4°C)	atmospheres	2.458×10^{-3}
inches of water (at 4°C)	inches of mercury	0.07355
inches of water (at 4°C)	kgs/sq cm	2.540×10^{-3}
inches of water (at 4°C)	ounces/sq in.	0.5781
inches of water (at 4°C)	pounds/sq ft	5.204
inches of water (at 4°C)	pounds/sq in.	0.03613
	J	
joules	Btu	9.480×10^{-4}
joules	ergs	10^{7}
joules	foot-pounds	0.7376
joules	kg-calories	2.389×10^{-4}
joules	kg-meters	0.1020
joules/cm	grams	1.020×10^{4}
joules/cm	dynes	10^{7}
joules/cm	joules/meter (newtons)	100.0
joules/cm	poundals	723.3
joules/cm	pounds	22.48
	K	
kilograms	dynes	980,665.
kilograms	grams	1,000.0
kilograms	joules/cm	0.09807
kilograms	joules/meter (newtons)	9.807
kilograms	poundals	70.93
kilograms	pounds	2.205
kilograms	tons (long)	9.842×10^{-4}
kilograms	tons (short)	1.102×10^{-3}
kilograms/cu meter	grams/cu cm	0.001
kilograms/cu meter	pounds/cu ft	0.06243
kilograms/cu meter	pounds/cu in.	3.613×10^{-5}

Table 4.4 continued

Table 4.4 continued

TO CONVERT	INTO	MULTIPLY BY
kilograms/cu meter	pounds/mil-foot	3.405×10^{-10}
kilograms/meter	pounds/ft	0.6720
Kilogram/sq. cm.	Dynes	980,665
kilograms/sq cm	atmospheres	0.9678
kilograms/sq cm	feet of water	32.81
kilograms/sq cm	inches of mercury	28.96
kilograms/sq cm	pounds/sq ft	2,048.
kilograms/sq cm	pounds/sq in.	14.22
kilograms/sq meter	atmospheres	9.678×10^{-5}
kilograms/sq meter	bars	98.07×10^{-6}
kilograms/sq meter	feet of water	3.281×10^{-3}
kilograms/sq meter	inches of mercury	2.896×10^{-3}
kilograms/sq meter	pounds/sq ft	0.2048
kilograms/sq meter	pounds/sq in.	1.422×10^{-3}
kilograms/sq mm	kgs/sq meter	10^6
kilogram-calories	Btu	3.968
kilogram-calories	foot-pounds	3,088.
kilogram-calories	hp-hrs	1.560×10^{-3}
kilogram-calories	joules	4,186.
kilogram-calories	kg-meters	426.9
kilogram-calories	kilojoules	4.186
kilogram-calories	kilowatt-hrs	1.163×10^{-3}
kilogram meters	Btu	9.294×10^{-3}
kilogram meters	ergs	9.804×10^7
kilogram meters	foot-pounds	7.233
kilogram meters	joules	9.804
kilogram meters	kg-calories	2.342×10^{-3}
kilogram meters	kilowatt-hrs	2.723×10^{-6}
kilolines	maxwells	1,000.0
kiloliters	liters	1,000.0
kilometers	centimeters	10^5
kilometers	feet	3,281.
kilometers	inches	3.937×10^4
kilometers	meters	1,000.0
kilometers	miles	0.6214
kilometers	millimeters	10^6
kilometers	yards	1,094.
kilometers/hr	cms/sec	27.78
kilometers/hr	feet/min	54.68
kilometers/hr	feet/sec	0.9113
kilometers/hr	knots	0.5396
kilometers/hr	meters/min	16.67
kilometers/hr	miles/hr	0.6214
kilometers/hr/sec	cms/sec/sec	27.78
kilometers/hr/sec	ft/sec/sec	0.9113
kilometers/hr/sec	meters/sec/sec	0.2778
kilometers/hr/sec	miles/hr/sec	0.6214
knots	feet/hr	6,080.
knots	kilometers/hr	1.8532
knots	nautical miles/hr	1.0
knots	statute miles/hr	1.151
knots	yards/hr	2,027.
knots	feet/sec	1.689
L		
league	miles (approx.)	3.0
Light year	Miles	5.9×10^{12}
Light year	Kilometers	9.46091×10^{12}
lines/sq cm	gausses	1.0
lines/sq in.	gausses	0.1550
lines/sq in.	webers/sq cm	1.550×10^{-9}
lines/sq in.	webers/sq in.	10^{-8}
lines/sq in.	webers/sq meter	1.550×10^{-5}
links (engineer's)	inches	12.0
links (surveyor's)	inches	7.92
liters	bushels (U.S. dry)	0.02838

TO CONVERT	INTO	MULTIPLY BY
liters	cu cm	1,000.0
liters	cu feet	0.03531
liters	cu inches	61.02
liters	cu meters	0.001
liters	cu yards	1.308×10^{-3}
liters	gallons (U.S. liq.)	0.2642
liters	pints (U.S. liq.)	2.113
liters	quarts (U.S. liq.)	1.057
liters/min	cu ft/sec	5.886×10^{-4}
liters/min	gals/sec	4.403×10^{-3}
lumens/sq ft	foot-candles	1.0
Lumen	Spherical candle power	.07958
Lumen/sq. ft.	Lumen/sq. meter	10.76
lux	foot-candles	0.0929
M		
meters	centimeters	100.0
meters	feet	3.281
meters	inches	39.37
meters	kilometers	0.001
meters	miles (naut.)	5.396×10^{-4}
meters	miles (stat.)	6.214×10^{-4}
meters	millimeters	1,000.0
meters	yards	1.094
meters	varas	1.179
meters/min	cms/sec	1.667
meters/min	feet/min	3.281
meters/min	feet/sec	0.05468
meters/min	kms/hr	0.06
meters/min	knots	0.03238
meters/min	miles/hr	0.03728
meters/sec	feet/min	196.8
meters/sec	feet/sec	3.281
meters/sec	kilometers/hr	3.6
meters/sec	kilometers/min	0.06
meters/sec	miles/hr	2.237
meters/sec	miles/min	0.03728
meters/sec/sec	cms/sec/sec	100.0
meters/sec/sec	ft/sec/sec	3.281
meters/sec/sec	kms/hr/sec	3.6
meters/sec/sec	miles/hr/sec	2.237
meter-kilograms	cm-dynes	9.807×10^7
meter-kilograms	cm-grams	10^5
meter-kilograms	pound-feet	7.233
micrograms	grams	10^{-6}
microhms	megohms	10^{-12}
microhms	ohms	10^{-6}
microliters	liters	10^{-6}
Microns	meters	1×10^{-6}
miles (naut.)	feet	6,080.27
miles (naut.)	kilometers	1.853
miles (naut.)	meters	1,853.
miles (naut.)	miles (statute)	1.1516
miles (naut.)	yards	2,027.
miles (statute)	centimeters	1.609×10^5
miles (statute)	feet	5,280.
miles (statute)	inches	6.336×10^4
miles (statute)	kilometers	1.609
miles (statute)	meters	1,609.
miles (statute)	miles (naut.)	0.8684
miles (statute)	yards	1,760.
miles/hr	cms/sec	44.70
miles/hr	feet/min	88.
miles/hr	feet/sec	1.467
miles/hr	kms/hr	1.609
miles/hr	kms/min	0.02682
miles/hr	knots	0.8684
miles/hr	meters/min	26.82
miles/hr	miles/min	0.1667
miles/hr/sec	cms/sec/sec	44.70
miles/hr/sec	feet/sec/sec	1.467

Table 4.4 continued

Table 4.4 continued

TO CONVERT	INTO	MULTIPLY BY
miles/hr/sec	kms/hr/sec	1.609
miles/hr/sec	meters/sec/sec	0.4470
miles/min	cms/sec	2,682.
miles/min	feet/sec	88.
miles/min	kms/min	1.609
miles/min	knots/min	0.8684
miles/min	miles/hr	60.0
mil-feet	cu inches	9.425×10^{-6}
milliers	kilograms	1,000.
Millimicrons	meters	1×10^{-9}
Milligrams	grains	0.01543236
milligrams	grams	0.001
milligrams/liter	parts/million	1.0
millihenries	henries	0.001
milliliters	liters	0.001
millimeters	centimeters	0.1
millimeters	feet	3.281×10^{-3}
millimeters	inches	0.03937
millimeters	kilometers	10^{-6}
millimeters	meters	0.001
millimeters	miles	6.214×10^{-7}
millimeters	mils	39.37
millimeters	yards	1.094×10^{-3}
million gals/day	cu ft/sec	1.54723
mils	centimeters	2.540×10^{-3}
mils	feet	8.333×10^{-5}
mils	inches	0.001
mils	kilometers	2.540×10^{-8}
mils	yards	2.778×10^{-5}
miner's inches	cu ft/min	1.5
Minims (British)	cubic cm.	0.059192
Minims (U.S., fluid)	cubic cm.	0.061612
minutes (angles)	degrees	0.01667
minutes (angles)	quadrants	1.852×10^{-4}
minutes (angles)	radians	2.909×10^{-4}
minutes (angles)	seconds	60.0
myriagrams	kilograms	10.0
myriameters	kilometers	10.0
	N	
nepers	decibels	8.686
Newton	Dynes	1×10^{5}
	O	
ounces	drams	16.0
ounces	grains	437.5
ounces	grams	28.349527
ounces	pounds	0.0625
ounces	ounces (troy)	0.9115
ounces	tons (long)	2.790×10^{-5}
ounces	tons (metric)	2.835×10^{-5}
ounces (fluid)	cu inches	1.805
ounces (fluid)	liters	0.02957
ounces (troy)	grains	480.0
ounces (troy)	grams	31.103481
ounces (troy)	ounces (avdp.)	1.09714
ounces (troy)	pennyweights (troy)	20.0
ounces (troy)	pounds (troy)	0.08333
Ounce/sq. inch	Dynes/sq. cm.	4309
ounces/sq in.	pounds/sq in.	0.0625
	P	
Parsec	Miles	19×10^{12}
Parsec	Kilometers	3.084×10^{13}
parts/million	grains/U.S. gal	0.0584
parts/million	grains/Imp. gal	0.07016
parts/million	pounds/million gal	8.345
Pecks (British)	cubic inches	554.6
Pecks (British)	liters	9.091901
Pecks (U.S.)	bushels	0.25
Pecks (U.S.)	cubic inches	537.605
Pecks (U.S.)	liters	8.809582
Pecks (U.S.)	quarts (dry)	8
pennyweights (troy)	grains	24.0
pennyweights (troy)	ounces (troy)	0.05
pennyweights (troy)	grams	1.55517
pennyweights (troy)	pounds (troy)	4.1667×10^{-3}
pints (dry)	cu inches	33.60
pints (liq.)	cu cms.	473.2
pints (liq.)	cu feet	0.01671
pints (liq.)	cu inches	28.87
pints (liq.)	cu meters	4.732×10^{-4}
pints (liq.)	cu yards	6.189×10^{-4}
pints (liq.)	gallons	0.125
pints (liq.)	liters	0.4732
pints (liq.)	quarts (liq.)	0.5
Planck's quantum	Erg – second	6.624×10^{-27}
Poise	Gram/cm. sec.	1.00
Pounds (avoirdupois)	ounces (troy)	14.5833
poundals	dynes	13,826.
poundals	grams	14.10
poundals	joules/cm	1.383×10^{-3}
poundals	joules/meter (newtons)	0.1383
poundals	kilograms	0.01410
poundals	pounds	0.03108
pounds	drams	256.
pounds	dynes	44.4823×10^{4}
pounds	grains	7,000.
pounds	grams	453.5924
pounds	joules/cm	0.04448
pounds	joules/meter (newtons)	4.448
pounds	kilograms	0.4536
pounds	ounces	16.0
pounds	ounces (troy)	14.5833
pounds	poundals	32.17
pounds	pounds (troy)	1.21528
pounds	tons (short)	0.0005
pounds (troy)	grains	5,760.
pounds (troy)	grams	373.24177
pounds (troy)	ounces (avdp.)	13.1657
pounds (troy)	ounces (troy)	12.0
pounds (troy)	pennyweights (troy)	240.0
pounds (troy)	pounds (avdp.)	0.822857
pounds (troy)	tons (long)	3.6735×10^{-4}
pounds (troy)	tons (metric)	3.7324×10^{-4}
pounds (troy)	tons (short)	4.1143×10^{-4}
pounds of water	cu feet	0.01602
pounds of water	cu inches	27.68
pounds of water	gallons	0.1198
pounds of water/min	cu ft/sec	2.670×10^{-4}
pound-feet	cm-dynes	1.356×10^{7}
pound-feet	cm-grams	13,825.
pound-feet	meter-kgs	0.1383
pounds/cu ft	grams/cu cm	0.01602
pounds/cu ft	kgs/cu meter	16.02
pounds/cu ft	pounds/cu in.	5.787×10^{-4}
pounds/cu ft	pounds/mil-foot	5.456×10^{-9}
pounds/cu in.	gms/cu cm	27.68
pounds/cu in.	kgs/cu meter	2.768×10^{4}
pounds/cu in.	pounds/cu ft	1,728.
pounds/cu in.	pounds/mil-foot	9.425×10^{-6}
pounds/ft	kgs/meter	1.488
pounds/in.	gms/cm	178.6
pounds/mil-foot	gms/cu cm	2.306×10^{6}
pounds/sq ft	atmospheres	4.725×10^{-4}
pounds/sq ft	feet of water	0.01602
pounds/sq ft	inches of mercury	0.01414
pounds/sq ft	kgs/sq meter	4.882
pounds/sq ft	pounds/sq in.	6.944×10^{-3}
pounds/sq in.	atmospheres	0.06804

Table 4.4 continued

Table 4.4 continued

TO CONVERT	INTO	MULTIPLY BY
pounds/sq in.	feet of water	2.307
pounds/sq in.	inches of mercury	2.036
pounds/sq in.	kgs/sq meter	703.1
pounds/sq in.	pounds/sq ft	144.0
	Q	
quadrants (angle)	degrees	90.0
quadrants (angle)	minutes	5,400.0
quadrants (angle)	radians	1.571
quadrants (angle)	seconds	3.24×10^5
quarts (dry)	cu inches	67.20
quarts (liq.)	cu cms	946.4
quarts (liq.)	cu feet	0.03342
quarts (liq.)	cu inches	57.75
quarts (liq.)	cu meters	9.464×10^{-4}
quarts (liq.)	cu yards	1.238×10^{-3}
quarts (liq.)	gallons	0.25
quarts (liq.)	liters	0.9463
	R	
radians	degrees	57.30
radians	minutes	3,438.
radians	quadrants	0.6366
radians	seconds	2.063×10^5
radians/sec	degrees/sec	57.30
radians/sec	revolutions/min	9.549
radians/sec	revolutions/sec	0.1592
radians/sec/sec	revs/min/min	573.0
radians/sec/sec	revs/min/sec	9.549
radians/sec/sec	revs/sec/sec	0.1592
revolutions	degrees	360.0
revolutions	quadrants	4.0
revolutions	radians	6.283
revolutions/min	degrees/sec	6.0
revolutions/min	radians/sec	0.1047
revolutions/min	revs/sec	0.01667
revolutions/min/min	radians/sec/sec	1.745×10^{-3}
revolutions/min/min	revs/min/sec	0.01667
revolutions/min/min	revs/sec/sec	2.778×10^{-4}
revolutions/sec	degrees/sec	360.0
revolutions/sec	radians/sec	6.283
revolutions/sec	revs/min	60.0
revolutions/sec/sec	radians/sec/sec	6.283
revolutions/sec/sec	revs/min/min	3,600.0
revolutions/sec/sec	revs/min/sec	60.0
Rod	Chain (Gunters)	.25
Rod	Meters	5.029
Rods (Surveyors' meas.)	yards	5.5
rods	feet	16.5
	S	
Scruples	grains	20
seconds (angle)	degrees	2.778×10^{-4}
seconds (angle)	minutes	0.01667
seconds (angle)	quadrants	3.087×10^{-6}
seconds (angle)	radians	4.848×10^{-6}
Slug	Kilogram	14.59
Slug	Pounds	32.17
Sphere	Steradians	12.57
square centimeters	circular mils	1.973×10^5
square centimeters	sq feet	1.076×10^{-3}
square centimeters	sq inches	0.1550
square centimeters	sq meters	0.0001
square centimeters	sq miles	3.861×10^{-11}
square centimeters	sq millimeters	100.0
square centimeters	sq yards	1.196×10^{-4}
square feet	acres	2.296×10^{-5}
square feet	circular mils	1.833×10^8
square feet	sq cms	929.0
square feet	sq inches	144.0

TO CONVERT	INTO	MULTIPLY BY
square feet	sq meters	0.09290
square feet	sq miles	3.587×10^{-8}
square feet	sq millimeters	9.290×10^4
square feet	sq yards	0.1111
square inches	circular mils	1.273×10^6
square inches	sq cms	6.452
square inches	sq feet	6.944×10^{-3}
square inches	sq millimeters	645.2
square inches	sq mils	10^6
square inches	sq yards	7.716×10^{-4}
square kilometers	acres	247.1
square kilometers	sq cms	10^{10}
square kilometers	sq ft	10.76×10^6
square kilometers	sq inches	1.550×10^9
square kilometers	sq meters	10^6
square kilometers	sq miles	0.3861
square kilometers	sq yards	1.196×10^6
square meters	acres	2.471×10^{-4}
square meters	sq cms	10^4
square meters	sq feet	10.76
square meters	sq inches	1,550.
square meters	sq miles	3.861×10^{-7}
square meters	sq millimeters	10^6
square meters	sq yards	1.196
square miles	acres	640.0
square miles	sq feet	27.88×10^6
square miles	sq kms	2.590
square miles	sq meters	2.590×10^6
square miles	sq yards	3.098×10^6
square millimeters	circular mils	1,973.
square millimeters	sq cms	0.01
square millimeters	sq feet	1.076×10^{-5}
square millimeters	sq inches	1.550×10^{-3}
square mils	circular mils	1.273
square mils	sq cms	6.452×10^{-6}
square mils	sq inches	10^{-6}
square yards	acres	2.066×10^{-4}
square yards	sq cms	8,361.
square yards	sq feet	9.0
square yards	sq inches	1,296.
square yards	sq meters	0.8361
square yards	sq miles	3.228×10^{-7}
square yards	sq millimeters	8.361×10^5
	T	
temperature (°C) +273	absolute temperature (°C)	1.0
temperature (°C) +17.78	temperature (°F)	1.8
temperature (°F) +460	absolute temperature (°F)	1.0
temperature (°F) −32	temperature (°C)	5/9
tons (long)	kilograms	1,016.
tons (long)	pounds	2,240.
tons (long)	tons (short)	1.120
tons (metric)	kilograms	1,000.
tons (metric)	pounds	2,205.
tons (short)	kilograms	907.1848
tons (short)	ounces	32,000.
tons (short)	ounces (troy)	29,166.66
tons (short)	pounds	2,000.
tons (short)	pounds (troy)	2,430.56
tons (short)	tons (long)	0.89287
tons (short)	tons (metric)	0.9078
tons (short)/sq ft	kgs/sq meter	9,765.
tons (short)/sq ft	pounds/sq in.	2,000.
tons of water/24 hrs	pounds of water/hr	83.333
tons of water/24 hrs	gallons/min	0.16643
tons of water/24 hrs	cu ft/hr	1.3349
	Y	
yards	centimeters	91.44
yards	kilometers	9.144×10^{-4}
yards	meters	0.9144
yards	miles (naut.)	4.934×10^{-4}
yards	miles (stat.)	5.682×10^{-4}
yards	millimeters	914.4

Absolute or Dynamic Viscosity		Centipoise (μ)	Poise; Gram/Cm Sec; Dyne Sec/Cm^2 ($100\ \mu$)	Slugs/Ft Sec; *$Pound_f$ Sec/Ft^2 (μ'_e)	†$Pound_m$/Ft Sec; Poundal Sec/Ft^2 (μ_e)
Centipoise	(μ)	1	0.01	2.09 (10^{-5})	6.72 (10^{-4})
Poise; Gram/Cm Sec; Dyne Sec/Cm^2	($100\ \mu$)	100	1	2.09 (10^{-3})	0.0672
Slugs/Ft Sec; *$Pound_f$ Sec/Ft^2	(μ'_e)	47 900	479	1	g or 32.2
†$Pound_m$/Ft Sec; Poundal Sec/Ft^2	(μ_e)	1487	14.87	$\frac{1}{g}$ or .0311	1

*$Pound_f$ = Pound of Force   †$Pound_m$ = Pound of Mass

To convert absolute or dynamic viscosity from one set of units to another, locate the given set of units in the left hand column and multiply the numerical value by the factor shown horizontally to the right under the set of units desired.

As an example, suppose a given absolute viscosity of 2 poise is to be converted to slugs/foot second. By referring to the table, we find the conversion factor to be 2.09 (10^{-3}). Then, 2 (poise) times 2.09 (10^{-3}) = 4.18 (10^{-3}) = 0.00418 slugs/foot second.

Table 4.5. Viscosity—equivalents of absolute viscosity. Courtesy Crane Company (T. P. #410).

Kinematic Viscosity		Centistokes (ν)	Stokes; Cm^2/Sec ($100\ \nu$)	Ft^2/Sec (ν')
Centistokes	(ν)	1	0.01	1.076 (10^{-5})
Stokes; Cm^2/Sec	($100\ \nu$)	100	1	1.076 (10^{-3})
Ft^2/Sec	(ν')	92 900	929	1

To convert kinematic viscosity from one set of units to another, locate the given set of units in the left hand column and multiply the numerical value by the factor shown horizontally to the right, under the set of units desired.

As an example, suppose a given kinematic viscosity of 0.5 square foot/second is to be converted to centistokes. By referring to the table, we find the conversion factor to be 92,900. Then, 0.5 (sq ft/sec) times 92,900 = 46,450 centistokes.

Table 4.6. Viscosity—equivalents of kinematic viscosity. Courtesy Crane Company (T. P. #410).

Kinematic Viscosity, Centistokes v	Equivalent Saybolt Universal Viscosity, Sec	
	At 100 F Basic Values	At 210 F
1.83	32.01	32.23
2.0	32.62	32.85
4.0	39.14	39.41
6.0	45.56	45.88
8.0	52.09	52.45
10.0	58.91	59.32
15.0	77.39	77.93
20.0	97.77	98.45
25.0	119.3	120.1
30.0	141.3	142.3
35.0	163.7	164.9
40.0	186.3	187.6
45.0	209.1	210.5
50.0	232.1	233.8
55.0	255.2	257.0
60.0	278.3	280.2
65.0	301.4	303.5
70.0	324.4	326.7
75.0	347.6	350.0
80.0	370.8	373.4
85.0	393.9	396.7
90.0	417.1	420.0
95.0	440.3	443.4
100.0	463.5	466.7
120.0	556.2	560.1
140.0	648.9	653.4
160.0	741.6	
180.0	834.2	
200.0	926.9	
220.0	1019.6	
240.0	1112.3	
260.0	1205.0	
280.0	1297.7	
300.0	1390.4	
320.0	1483.1	
340.0	1575.8	
360.0	1668.5	Saybolt Seconds equal Centistokes times 4.6673
380.0	1761.2	
400.0	1853.9	
420.0	1946.6	
440.0	2039.3	
460.0	2132.0	
480.0	2224.7	
500.0	2317.4	
Over 500	Saybolt Seconds equal Centistokes times 4.6347	

Note: To obtain the Saybolt Universal viscosity equivalent to a kinematic viscosity determined at t, multiply the equivalent Saybolt Universal viscosity at 100 F by $1 + (t - 100)\ 0.000\ 064$.
For example, 10 v at 210 F are equivalent to 58.91 multiplied by 1.0070 or 59.32 sec Saybolt Universal at 210 F.

Above. ***Table 4.7. Viscosity—equivalents of kinematic and Saybolt Universal Viscosity. Courtesy Crane Company and American Society for Testing and Materials.***

Right. ***Table 4.8. Viscosity—equivalents of kinematic and Saybolt Furol Viscosity at 122°F. Courtesy Crane Company and American Society for Testing and Materials.***

Kinematic Viscosity, Centistokes v	Equivalent Saybolt Furol Viscosity, Sec	
	At 122 F	At 210 F
48	25.3	
50	26.1	25.2
60	30.6	29.8
70	35.1	34.4
80	39.6	39.0
90	44.1	43.7
100	48.6	48.3
125	60.1	60.1
150	71.7	71.8
175	83.8	83.7
200	95.0	95.6
225	106.7	107.5
250	118.4	119.4
275	130.1	131.4
300	141.8	143.5
325	153.6	155.5
350	165.3	167.6
375	177.0	179.7
400	188.8	191.8
425	200.6	204.0
450	212.4	216.1
475	224.1	228.3
500	235.9	240.5
525	247.7	252.8
550	259.5	265.0
575	271.3	277.2
600	283.1	289.5
625	294.9	301.8
650	306.7	314.1
675	318.4	326.4
700	330.2	338.7
725	342.0	351.0
750	353.8	363.4
775	365.5	375.7
800	377.4	388.1
825	389.2	400.5
850	400.9	412.9
875	412.7	425.3
900	424.5	437.7
925	436.3	450.1
950	448.1	462.5
975	459.9	474.9
1000	471.7	487.4
1025	483.5	499.8
1050	495.2	512.3
1075	507.0	524.8
1100	518.8	537.2
1125	530.6	549.7
1150	542.4	562.2
1175	554.2	574.7
1200	566.0	587.2
1225	577.8	599.7
1250	589.5	612.2
1275	601.3	624.8
1300	613.1	637.3
Over 1300	*	†

*OVER 1300 CENTISTOKES AT 122 F:
Saybolt Fluid Sec = Centistokes x 0.4717

†OVER 1300 CENTISTOKES AT 210 F:
Log (Saybolt Furol Sec − 2.87) = 1.0276 [Log (Centistokes)] − 0.3975

APPROXIMATE CONVERSION TABLE

SECONDS SAYBOLT UNIVERSAL	SECONDS SAYBOLT FUROL	ENGLER DEGREES	ENGLER TIME	BARBEY	REDWOOD STANDARD	REDWOOD ADMIRALTY	KINEMATIC VISCOSITY STOKES
35		1.18	60	2,800	32		0.026
50		1.60	82	880	44		0.074
75		2.30	102	460	65		0.141
100	15	3.00	153	320	88		0.202
150	19	4.40	230	205	128		0.318
200	23	5.90	305	148	170	18	0.431
250	28	7.60	375	118	212	23	0.543
300	33	8.90	450	98	254	27	0.651
400	42	11.80	550	72	338	36	0.876
500	52	14.50	750	59	423	45	1.10
600	61	17.50	900	48	518	53	1.32
700	71	20.6	1050	41	592	62	1.54
800	81	23	1,200	36.5	677	71	1.76
900	91	27	1,300	32.0	762	78	1.98
1,000	100	29	1,500	29.5	846	87	2.20
1,500	150	42	2,300	19.5	1,270	135	3.30
2,000	200	59	3,000	14.5	1,695	175	4.40
2,500	250	73	3,750	11.5	2,120	230	5.50
3,000	300	87	4,500	9.6	2,540	260	6.60
4,000	400	117	6,000	7.4	3,380	350	8.80
5,000	500	145	7,500	6.0	4,230	435	11.00
6,000	600	175	9,000	5.2	5,080	530	13.20
7,000	700	205	10,500	4.1	5,925	610	15.40
8,000	800	230	12,000	3.7	6,770	700	17.60
9,000	900	260	13,500	3.2	7,620	780	19.8
10,000	1,000	290	15,000	2.9	8,460	870	22
20,000	2,000	590	30,000	1.4	16,920	1,760	44
40,000	4,000	1,170	60,000		33,850	3,600	88
60,000	6,000	1,750	90,000		50,800	5,300	132
80,000	8,000	2,300	120,000		67,700	7,000	176
100,000	10,000	2,900	150,000		84,600	8,700	220
200,000	20,000	5,900	300,000		169,200	17,600	440
400,000	40,000	11,700	600,000		338,500	36,000	880
600,000	60,000	17,500	900,000		508,000	53,000	1,320
800,000	80,000	23,000	1,200,000		677,000	70,000	1,760
1,000,000	100,000	29,000	1,500,000		846,000	87,000	2,200

ABSOLUTE VISCOSITY is the force required to move a unit plane surface over another plane surface at a unit velocity when surfaces are separated by a layer of fluid of unit thickness

KINEMATIC VISCOSITY is the absolute viscosity divided by the density of the fluid.

SAE VISCOSITY NUMBERS are arbitrary numbers which have been assigned to classify lubricating oils and do not have any functional relationship with other systems.

VISCOSITY INDEX is an arbitrary system relating the effect of temperature change to viscosity. It is the slope of a plot of viscosity versus temperature for a given fluid.

1 Stoke	= 1 Poise/Density in Grams	1 Centistoke	= 1/100 Stoke	1 Centipoise	= 0.0000209 Slugs/(Ft)(Se
1 Centistoke	= 1.076×10^{-5} Ft^2/Sec	1 Centipoise	= 0.000672 Lbs/(Ft)(Sec)	1 Centipoise	= 2.42 Lbs/(Ft)(Hr)
1 Poise	= 1 Gram/(Cm)(Sec)	1 Centipoise	= 0.000672 (Poundals)(Sec)/Ft^2	1 Slug/(Ft)(Sec)	= 1/0.002089 Poise
1 Centipoise	= 1/100 Poise	1 Centipoise	= 0.0000209 (Lbs)(Sec)/Ft^2	1 Ft^2/Sec	= 92903 Centistokes

Table 4.9. Viscosity—general conversions. Courtesy J.H. Jones Company.

A.P.I. gravity	Baume gravity	Specific gravity	Lb/ U.S. gal	U.S. gal/ lb
0	10.247	1.0760	8.962	0.1116
1	9.223	1.0679	8.895	0.1124
2	8.198	1.0599	8.828	0.1133
3	7.173	1.0520	8.762	0.1141
4	6.148	1.0443	8.698	0.1150
5	5.124	1.0366	8.634	0.1158
6	4.099	1.0291	8.571	0.1167
7	3.074	1.0217	8.509	0.1175
8	2.049	1.0143	8.448	0.1184
9	1.025	1.0071	8.388	0.1192
10	10.00	1.0000	8.328	0.1201
11	10.99	0.9930	8.270	0.1209
12	11.98	0.9861	8.212	0.1218
13	12.97	0.9792	8.155	0.1226
14	13.96	0.9725	8.099	0.1235
15	14.95	0.9659	8.044	0.1243
16	15.94	0.9593	7.989	0.1252
17	16.93	0.9529	7.935	0.1260
18	17.92	0.9465	7.882	0.1269
19	18.90	0.9402	7.830	0.1277
20	19.89	0.9340	7.778	0.1286
21	20.88	0.9279	7.727	0.1294
22	21.87	0.9218	7.676	0.1303
23	22.86	0.9159	7.627	0.1311
24	23.85	0.9100	7.578	0.1320
25	24.84	0.9042	7.529	0.1328
26	25.83	0.8984	7.481	0.1337
27	26.82	0.8927	7.434	0.1345
28	27.81	0.8871	7.387	0.1354
29	28.80	0.8816	7.341	0.1362
30	29.79	0.8762	7.296	0.1371
31	30.78	0.8708	7.251	0.1379
32	31.77	0.8654	7.206	0.1388
33	32.76	0.8602	7.163	0.1396
34	33.75	0.8550	7.119	0.1405
35	34.73	0.8498	7.076	0.1413
36	35.72	0.8448	7.034	0.1422
37	36.71	0.8398	6.993	0.1430
38	37.70	0.8348	6.951	0.1439
39	38.69	0.8299	6.910	0.1447
40	39.68	0.8251	6.870	0.1456
41	40.67	0.8203	6.830	0.1464
42	41.66	0.8155	6.790	0.1473
43	42.65	0.8109	6.752	0.1481
44	43.64	0.8063	6.713	0.1490
45	44.63	0.8017	6.675	0.1498
46	45.62	0.7972	6.637	0.1507
47	50.61	0.7927	6.600	0.1515
48	50.60	0.7883	6.563	0.1524
49	50.59	0.7839	6.526	0.1532
50	50.58	0.7796	6.490	0.1541

A.P.I. gravity	Baume gravity	Specific gravity	Lb/ U.S. gal	U.S. gal/ lb
51	50.57	0.7753	6.455	0.1549
52	51.55	0.7711	6.420	0.1558
53	52.54	0.7669	6.385	0.1566
54	53.53	0.7628	6.350	0.1575
55	54.52	0.7587	6.316	0.1583
56	55.51	0.7547	6.283	0.1592
57	56.50	0.7507	6.249	0.1600
58	57.49	0.7467	6.216	0.1609
59	58.48	0.7428	6.184	0.1617
60	59.47	0.7389	6.151	0.1626
61	60.46	0.7351	6.119	0.1634
62	61.45	0.7313	6.087	0.1643
63	62.44	0.7275	6.056	0.1651
64	63.43	0.7238	6.025	0.1660
65	64.42	0.7201	5.994	0.1668
66	65.41	0.7165	5.964	0.1677
67	66.40	0.7128	5.934	0.1685
68	67.39	0.7093	5.904	0.1694
69	68.37	0.7057	5.874	0.1702
70	69.36	0.7022	5.845	0.1711
71	70.35	0.6988	5.817	0.1719
72	71.34	0.6953	5.788	0.1728
73	72.33	0.6919	5.759	0.1736
74	73.32	0.6886	5.731	0.1745
75	74.31	0.6852	5.703	0.1753
76	75.30	0.6819	5.676	0.1762
77	76.29	0.6787	5.649	0.1770
78	77.28	0.6754	5.622	0.1779
79	78.27	0.6722	5.595	0.1787
80	79.26	0.6690	5.568	0.1796
81	80.25	0.6659	5.542	0.1804
82	81.24	0.6628	5.516	0.1813
83	82.23	0.6597	5.491	0.1821
84	83.22	0.6566	5.465	0.1830
85	84.20	0.6536	5.440	0.1838
86	85.19	0.6506	5.415	0.1847
87	86.18	0.6476	5.390	0.1855
88	87.17	0.6446	5.365	0.1864
89	88.16	0.6417	5.341	0.1872
90	89.15	0.6388	5.316	0.1881
91	90.14	0.6360	5.293	0.1889
92	91.13	0.6331	5.269	0.1898
93	92.12	0.6303	5.246	0.1906
94	93.11	0.6275	5.222	0.1915
95	94.10	0.6247	5.199	0.1924
96	95.09	0.6220	5.176	0.1932
97	96.08	0.6193	5.154	0.1940
98	97.07	0.6166	5.131	0.1949
99	98.06	0.6139	5.109	0.1957
100	99.05	0.6112	5.086	0.1966

The relation of Degrees Baumé or A.P.I. to Specific Gravity is expressed by the following formulas:

For liquids lighter than water:

$$\text{Degrees Baumé} = \frac{140}{G} - 130, \quad G = \frac{140}{130 + \text{Degrees Baumé}}$$

$$\text{Degrees A.P.I.} = \frac{141.5}{G} - 131.5, \quad G = \frac{141.5}{131.5 + \text{Degrees A.P.I.}}$$

For liquids heavier than water:

$$\text{Degrees Baumé} = 145 - \frac{145}{G}, \quad G = \frac{145}{145 - \text{Degrees Baumé}}$$

G = Specific Gravity = ratio of the weight of a given volume of oil at 60° Fahrenheit to the weight of the same volume of water at 60° Fahrenheit.

The above tables are based on the weight of 1 gallon (U. S.) of oil with a volume of 231 cubic inches at 60 degrees Fahrenheit in air at 760 m.m. pressure and 50% humidity. Assumed weight of 1 gallon of water at 60° Fahrenheit in air is 8.32828 pounds.

To determine the resulting gravity by mixing oils of different gravities:

$$D = \frac{md_1 + nd_2}{m + n}$$

D = Density or Specific Gravity of mixture
m = Proportion of oil of d_1 density
n = Proportion of oil of d_2 density
d_1 = Specific Gravity of m oil
d_2 = Specific Gravity of n oil

Table 4.10. API and Baume gravity tables and weight factors. Courtesy Gas Processors Suppliers Association.

Table 4.11
Decimal numbers versus digital numbers through decimal 256 (2^8).

Decimal	Binary	Octal	Hexadecimal
0	0000	0	0
1	0001	1	1
2	0010	2	2
3	0011	3	3
4	0100	4	4
5	0101	5	5
6	0110	6	6
7	0111	7	7
8	1000	10	8
9	1001	11	9
10	1010	12	A
11	1011	13	B
12	1100	14	C
13	1101	15	D
14	1110	16	E
15	1111	17	F
16	10000	20	10
17	10001	21	11
18	10010	22	12
19	10011	23	13
20	10100	24	14
21	10101	25	15
22	10110	26	16
23	10111	27	17
24	11000	30	18
25	11001	31	19
26	11010	32	1A
27	11011	33	1B
28	11100	34	1C
29	11101	35	1D
30	11110	36	1E
31	11111	37	1F
32	100000	40	20
↓	↓	↓	↓
255	11111111	317	FF
256	100000000	320	100

Table 4.11. Decimal numbers versus digital numbers through decimal 256 (= 2^8).

Dia.	Area	Circum.	Dia.	Area	Circum.	Dia.	Area	Circum.	Dia.	Area	Circum.	Dia.	Area	Circum.	Dia.	Area	Circum.
0			2	3.1416	6.28319	5	19.635	15.7080	8	50.265	25.1327	14	153.94	43.9823	20	314.16	62.8319
1/32	0.00077	0.098175	1/16	3.3410	6.47953	1/16	20.129	15.9043	1/8	51.849	25.5254	1/8	156.70	44.3750	1/8	318.10	63.2246
1/16	0.00307	0.196350	1/8	3.5466	6.67588	1/8	20.629	16.1007	1/4	53.456	25.9181	1/4	159.48	44.7677	1/4	322.06	63.6173
3/32	0.00690	0.294524	3/16	3.7583	6.87223	3/16	21.135	16.2970	3/8	55.088	26.3108	3/8	162.30	45.1604	3/8	326.05	64.0100
1/8	0.01227	0.392699	1/4	3.9761	7.06858	1/4	21.648	16.4934	1/2	56.745	26.7035	1/2	165.13	45.5531	1/2	330.06	64.4026
5/32	0.01917	0.490874	5/16	4.2000	7.26493	5/16	22.166	16.6897	5/8	58.426	27.0962	5/8	167.99	45.9458	5/8	334.10	64.7953
3/16	0.02761	0.589049	3/8	4.4301	7.46128	3/8	22.691	16.8861	3/4	60.132	27.4889	3/4	170.87	46.3385	3/4	338.16	65.1880
7/32	0.03758	0.687223	7/16	4.6664	7.65763	7/16	23.221	17.0824	7/8	61.862	27.8816	7/8	173.78	46.7312	7/8	342.25	65.5807
1/4	0.04909	0.785398	1/2	4.9087	7.85398	1/2	23.758	17.2788	9	63.617	28.2743	15	176.71	47.1239	21	346.36	65.9734
9/32	0.06213	0.883573	9/16	5.1572	8.05033	9/16	24.301	17.4751	1/8	65.397	28.6670	1/8	179.67	47.5166	1/8	350.50	66.3661
5/16	0.07670	0.981748	5/8	5.4119	8.24668	5/8	24.850	17.6715	1/4	67.201	29.0597	1/4	182.65	47.9093	1/4	354.66	66.7588
11/32	0.09281	1.07992	11/16	5.6727	8.44303	11/16	25.406	17.8678	3/8	69.029	29.4524	3/8	185.66	48.3020	3/8	358.84	67.1515
3/8	0.11045	1.17810	3/4	5.9396	8.63938	3/4	25.967	18.0642	1/2	70.882	29.8451	1/2	188.69	48.6947	1/2	363.05	67.5442
13/32	0.12962	1.27627	13/16	6.2126	8.83573	13/16	26.535	18.2605	5/8	72.760	30.2378	5/8	191.75	49.0874	5/8	367.28	67.9369
7/16	0.15033	1.37445	7/8	6.4918	9.03208	7/8	27.109	18.4569	3/4	74.662	30.6305	3/4	194.83	49.4801	3/4	371.54	68.3296
15/32	0.17257	1.47262	15/16	6.7771	9.22843	15/16	27.688	18.6532	7/8	76.589	31.0232	7/8	197.93	49.8728	7/8	375.83	68.7223
1/2	0.19635	1.57080	3	7.0686	9.42478	6	28.274	18.8496	10	78.540	31.4159	16	201.06	50.2655	22	380.13	69.1150
17/32	0.22166	1.66897	1/16	7.3662	9.62113	1/16	28.867	19.0460	1/8	80.516	31.8086	1/8	204.22	50.6582	1/8	384.46	69.5077
9/16	0.24850	1.76715	1/8	7.6699	9.81748	1/8	29.465	19.2423	1/4	82.516	32.2013	1/4	207.39	51.0509	1/4	388.82	69.9004
19/32	0.27688	1.86532	3/16	7.9769	10.0138	3/16	30.069	19.4387	3/8	84.541	32.5940	3/8	210.60	51.4436	3/8	393.20	70.2931
5/8	0.30680	1.96350	1/4	8.2958	10.2102	1/4	30.680	19.6350	1/2	86.590	32.9867	1/2	213.82	51.8363	1/2	397.61	70.6858
21/32	0.33824	2.06167	5/16	8.6179	10.4065	5/16	31.296	19.8314	5/8	88.664	33.3794	5/8	217.08	52.2290	5/8	402.04	71.0785
11/16	0.37122	2.15984	3/8	8.9462	10.6029	3/8	31.919	20.0277	3/4	90.763	33.7721	3/4	220.35	52.6217	3/4	406.49	71.4712
23/32	0.40574	2.25802	7/16	9.2806	10.7992	7/16	32.548	20.2241	7/8	92.886	34.1648	7/8	223.65	53.0144	7/8	410.97	71.8639
3/4	0.44179	2.35619	1/2	9.6211	10.9956	1/2	33.183	20.4204	11	95.033	34.5575	17	226.98	53.4071	23	415.48	72.2566
25/32	0.47937	2.45437	9/16	9.9678	11.1919	9/16	33.824	20.6168	1/8	97.205	34.9502	1/8	230.33	53.7998	1/8	420.00	72.6493
13/16	0.51849	2.55254	5/8	10.321	11.3883	5/8	34.471	20.8131	1/4	99.402	35.3429	1/4	233.71	54.1925	1/4	424.56	73.0420
27/32	0.55914	2.65072	11/16	10.680	11.5846	11/16	35.125	21.0095	3/8	101.62	35.7356	3/8	237.10	54.5852	3/8	429.13	73.4347
7/8	0.60132	2.74889	3/4	11.045	11.7810	3/4	35.785	21.2058	1/2	103.87	36.1283	1/2	240.53	54.9779	1/2	433.74	73.8274
29/32	0.64504	2.84707	13/16	11.416	11.9773	13/16	36.451	21.4022	5/8	106.14	36.5210	5/8	243.98	55.3706	5/8	438.36	74.2201
15/16	0.69029	2.94524	7/8	11.793	12.1737	7/8	37.122	21.5984	3/4	108.43	36.9137	3/4	247.45	55.7633	3/4	443.01	74.6128
31/32	0.73708	3.04342	15/16	12.177	12.3700	15/16	37.800	21.7949	7/8	110.75	37.3064	7/8	250.95	56.1560	7/8	447.69	75.0055
1	0.78540	3.14159	4	12.566	12.5664	7	38.485	21.9911	12	113.10	37.6991	18	254.47	56.5487	24	452.39	75.3982
1/16	0.88664	3.33794	1/16	12.962	12.7627	1/16	39.175	22.1876	1/8	115.47	38.0918	1/8	258.02	56.9414	1/8	457.11	75.7909
1/8	0.99402	3.53429	1/8	13.364	12.9591	1/8	39.871	22.3838	1/4	117.86	38.4845	1/4	261.59	57.3341	1/4	461.86	76.1836
3/16	1.1075	3.73064	3/16	13.772	13.1554	3/16	40.574	22.5803	3/8	120.28	38.8772	3/8	265.18	57.7268	3/8	466.64	76.5783
1/4	1.2272	3.92699	1/4	14.186	13.3518	1/4	41.282	22.7765	1/2	122.72	39.2699	1/2	268.80	58.1195	1/2	471.44	76.9690
5/16	1.3530	4.12334	5/16	14.607	13.5481	5/16	41.997	22.9730	5/8	125.19	39.6626	5/8	272.45	58.5122	5/8	476.26	77.3617
3/8	1.4849	4.31969	3/8	15.033	13.7445	3/8	42.718	23.1692	3/4	127.68	40.0553	3/4	276.12	58.9049	3/4	481.11	77.7544
7/16	1.6230	4.51604	7/16	15.466	13.9408	7/16	43.446	23.3657	7/8	130.19	40.4480	7/8	279.81	59.2976	7/8	485.98	78.1471
1/2	1.7671	4.71239	1/2	15.904	14.1372	1/2	44.179	23.5619	13	132.73	40.8407	19	283.53	59.6903	25	490.87	78.5398
9/16	1.9175	4.90874	9/16	16.349	14.3335	9/16	44.918	23.7584	1/8	135.30	41.2334	1/8	287.27	60.0830	1/8	495.79	78.9325
5/8	2.0739	5.10509	5/8	16.800	14.5299	5/8	45.664	23.9546	1/4	137.89	41.6261	1/4	291.04	60.4757	1/4	500.74	79.3252
11/16	2.2365	5.30144	11/16	17.257	14.7262	11/16	46.415	24.1511	3/8	140.50	42.0188	3/8	294.83	60.8684	3/8	505.71	79.7179
3/4	2.4053	5.49779	3/4	17.721	14.9226	3/4	47.173	24.3473	1/2	143.14	42.4115	1/2	298.65	61.2611	1/2	510.71	80.1106
1/16	2.5802	5.69414	13/16	18.190	15.1189	13/16	47.937	24.5428	5/8	145.80	42.8042	5/8	302.49	61.6538	5/8	515.72	80.5033
7/8	2.7612	5.89049	7/8	18.665	15.3153	7/8	48.707	24.7400	3/4	148.49	43.1969	3/4	306.35	62.0465	3/4	520.77	80.8960
15/16	2.9483	6.08684	15/16	19.147	15.5116	15/16	49.483	24.9364	7/8	151.20	43.5896	7/8	310.24	62.4392	7/8	525.84	81.2887

Table 4.12 continued

Table 4.12. Circles, areas, and circumferences for diameters in units and fractions.

Table 4.12 continued

Dia.	Area	Circum.	Dia.	Area	Circum.	Dia.	Area	Circum.	Dia.	Area	Circum.	Dia.	Area	Circum.	Dia.	Area	Circum.
26	530.93	81.6814	32	804.25	100.531	38	1134.1	119.381	44	1520.5	138.230	50	1963.5	157.080	56	2463.0	175.929
⅛	536.05	82.0741	⅛	810.54	100.924	⅛	1141.6	119.773	⅛	1529.2	138.623	⅛	1973.3	157.472	⅛	2474.0	176.322
¼	541.19	82.4668	¼	816.86	101.316	¼	1149.1	120.166	¼	1537.9	139.015	¼	1983.2	157.865	¼	2485.0	176.715
⅜	546.35	82.8595	⅜	823.21	101.709	⅜	1156.6	120.559	⅜	1546.6	139.408	⅜	1993.1	158.258	⅜	2496.1	177.107
½	551.55	83.2522	½	829.58	102.102	½	1164.2	120.951	½	1555.3	139.801	½	2003.0	158.650	½	2507.2	177.500
⅝	556.76	83.6449	⅝	835.97	102.494	⅝	1171.7	121.344	⅝	1564.0	140.194	⅝	2012.9	159.043	⅝	2518.3	177.893
¾	562.00	84.0376	¾	842.39	102.887	¾	1179.3	121.737	¾	1572.8	140.586	¾	2022.8	159.436	¾	2529.4	178.285
⅞	567.27	84.4303	⅞	848.83	103.280	⅞	1186.9	122.129	⅞	1581.6	140.979	⅞	2032.8	159.829	⅞	2540.6	178.678
27	572.56	84.8230	33	855.30	103.673	39	1194.6	122.522	45	1590.4	141.372	51	2042.8	160.221	57	2551.8	179.071
⅛	577.87	85.2157	⅛	861.79	104.065	⅛	1202.3	122.915	⅛	1599.3	141.764	⅛	2052.8	160.614	⅛	2563.0	179.463
¼	583.21	85.6084	¼	868.31	104.458	¼	1210.0	123.308	¼	1608.2	142.157	¼	2062.9	161.007	¼	2574.2	179.856
⅜	588.57	86.0011	⅜	874.85	104.851	⅜	1217.7	123.700	⅜	1617.0	142.550	⅜	2073.0	161.399	⅜	2585.4	180.249
½	593.96	86.3938	½	881.41	105.243	½	1225.4	124.093	½	1626.0	142.942	½	2083.1	161.792	½	2596.7	180.642
⅝	599.37	86.7865	⅝	888.00	105.636	⅝	1233.2	124.486	⅝	1634.9	143.335	⅝	2093.2	162.185	⅝	2608.0	181.034
¾	604.81	87.1792	¾	894.62	106.029	¾	1241.0	124.878	¾	1643.9	143.728	¾	2103.3	162.577	¾	2619.4	181.427
⅞	610.27	87.5719	⅞	901.26	106.421	⅞	1248.8	125.271	⅞	1652.9	144.121	⅞	2113.5	162.970	⅞	2630.7	181.820
28	615.75	87.9646	34	907.92	106.814	40	1256.6	125.664	46	1661.9	144.513	52	2123.7	163.363	58	2642.1	182.212
⅛	621.26	88.3573	⅛	914.61	107.207	⅛	1264.5	126.056	⅛	1670.9	144.906	⅛	2133.9	163.756	⅛	2653.5	182.605
¼	626.80	88.7500	¼	921.32	107.600	¼	1272.4	126.449	¼	1680.0	145.299	¼	2144.2	164.148	¼	2664.9	182.998
⅜	632.36	89.1427	⅜	928.06	107.992	⅜	1280.3	126.842	⅜	1689.1	145.691	⅜	2154.5	164.541	⅜	2676.4	183.390
½	637.94	89.5354	½	934.82	108.385	½	1288.2	127.235	½	1698.2	146.084	½	2164.8	164.934	½	2687.8	183.783
⅝	643.55	89.9281	⅝	941.61	108.788	⅝	1296.2	127.627	⅝	1707.4	146.477	⅝	2175.1	165.326	⅝	2699.3	184.176
¾	649.18	90.3208	¾	948.42	109.170	¾	1304.2	128.020	¾	1716.5	146.869	¾	2185.4	165.719	¾	2710.9	184.569
⅞	654.85	90.7135	⅞	955.25	109.563	⅞	1312.2	128.413	⅞	1725.7	147.262	⅞	2195.8	166.112	⅞	2722.4	184.961
29	660.52	91.1062	35	962.11	109.956	41	1320.3	128.805	47	1734.9	147.655	53	2206.2	166.504	59	2734.0	185.354
⅛	666.23	91.4989	⅛	969.00	110.348	⅛	1328.3	129.198	⅛	1744.2	148.048	⅛	2216.6	166.897	⅛	2745.6	185.747
¼	671.96	91.8916	¼	975.91	110.741	¼	1336.4	129.591	¼	1753.5	148.440	¼	2227.0	167.290	¼	2757.2	186.139
⅜	677.71	92.2843	⅜	982.84	111.134	⅜	1344.5	129.983	⅜	1762.7	148.833	⅜	2237.5	167.683	⅜	2768.8	186.532
½	683.49	92.6770	½	989.80	111.527	½	1352.7	130.376	½	1772.1	149.226	½	2248.0	168.075	½	2780.5	186.925
⅝	689.30	93.0697	⅝	996.78	111.919	⅝	1360.8	130.769	⅝	1781.4	149.618	⅝	2258.5	168.468	⅝	2792.2	187.317
¾	695.13	93.4624	¾	1003.8	112.312	¾	1369.0	131.161	¾	1790.8	150.011	¾	2269.1	168.861	¾	2803.9	187.710
⅞	700.98	93.8551	⅞	1010.8	112.705	⅞	1377.2	131.554	⅞	1800.1	150.404	⅞	2279.6	169.253	⅞	2815.7	188.103
30	706.86	94.2478	36	1017.9	113.097	42	1385.4	131.947	48	1809.6	150.796	54	2290.2	169.646	60	2827.4	188.496
⅛	712.76	94.6405	⅛	1025.0	113.490	⅛	1393.7	132.340	⅛	1819.0	151.189	⅛	2300.8	170.039	⅛	2839.2	188.888
¼	718.69	95.0332	¼	1032.1	113.883	¼	1402.0	132.732	¼	1828.5	151.582	¼	2311.5	170.431	¼	2851.0	189.281
⅜	724.64	95.4259	⅜	1039.2	114.275	⅜	1410.3	133.125	⅜	1837.9	151.975	⅜	2322.1	170.824	⅜	2862.9	189.674
½	730.62	95.8186	½	1046.3	114.668	½	1418.6	133.518	½	1847.5	152.367	½	2332.8	171.217	½	2874.8	190.066
⅝	736.62	96.2113	⅝	1053.5	115.061	⅝	1427.0	133.910	⅝	1857.0	152.760	⅝	2343.5	171.609	⅝	2886.6	190.459
¾	742.64	96.6040	¾	1060.7	115.454	¾	1435.4	134.303	¾	1866.5	153.153	¾	2354.3	172.002	¾	2898.6	190.852
⅞	748.69	96.9967	⅞	1068.0	115.846	⅞	1443.8	134.696	⅞	1876.1	153.545	⅞	2365.0	172.395	⅞	2910.5	191.244
31	754.77	97.3894	37	1075.2	116.239	43	1452.2	135.088	49	1885.7	153.938	55	2375.8	172.788	61	2922.5	191.637
⅛	760.87	97.7821	⅛	1082.5	116.632	⅛	1460.7	135.481	⅛	1895.4	154.331	⅛	2386.6	173.180	⅛	2934.5	192.030
¼	766.99	98.1748	¼	1089.8	117.024	¼	1469.1	135.874	¼	1905.0	154.723	¼	2397.5	173.573	¼	2946.5	192.423
⅜	773.14	98.5675	⅜	1097.1	117.417	⅜	1477.6	136.267	⅜	1914.7	155.116	⅜	2408.3	173.966	⅜	2958.5	192.815
½	779.31	98.9602	½	1104.5	117.810	½	1486.2	136.659	½	1924.2	155.509	½	2419.2	174.358	½	2970.6	193.208
⅝	785.51	99.3529	⅝	1111.8	118.202	⅝	1494.7	137.052	⅝	1934.2	155.902	⅝	2430.1	174.751	⅝	2982.7	193.601
¾	791.73	99.7456	¾	1119.2	118.595	¾	1503.3	137.445	¾	1943.9	156.294	¾	2441.1	175.144	¾	2994.8	193.993
⅞	797.98	100.138	⅞	1126.7	118.988	⅞	1511.9	137.837	⅞	1953.7	156.687	⅞	2452.0	175.536	⅞	3006.9	194.386

Table 4.12 continued

Table 4.12 continued

Dia.	Area	Circum.	Dia.	Area	Circum.	Dia.	Area	Circum.	Dia.	Area	Circum.	Dia.	Area	Circum.	Dia.	Area	Circum.
62	3019.1	194.779	68	3631.7	213.628	74	4300.8	232.478	80	5026.5	251.327	86	5808.8	270.177	92	6647.6	289.027
⅛	3031.3	195.171	⅛	3645.0	214.021	⅛	4315.4	232.871	⅛	5042.3	251.720	⅛	5825.7	270.570	⅛	6665.7	289.419
¼	3043.5	195.564	¼	3658.4	214.414	¼	4329.9	233.263	¼	5058.0	252.113	¼	5842.6	270.962	¼	6683.8	289.812
⅜	3055.7	195.957	⅜	3671.8	214.806	⅜	4344.5	233.656	⅜	5073.8	252.506	⅜	5859.6	271.355	⅜	6701.9	290.205
½	3068.0	196.350	½	3685.3	215.199	½	4359.2	234.049	½	5089.6	252.898	½	5876.5	271.748	½	6720.1	290.597
⅝	3080.3	196.742	⅝	3698.7	215.592	⅝	4373.8	234.441	⅝	5105.4	253.291	⅝	5893.5	272.140	⅝	6738.2	290.990
¾	3092.6	197.135	¾	3712.2	215.984	¾	4388.5	234.834	¾	5121.2	253.684	¾	5910.6	272.533	¾	6756.4	291.383
⅞	3104.9	197.528	⅞	3725.7	216.337	⅞	4403.1	235.227	⅞	5137.1	254.076	⅞	5927.6	272.926	⅞	6774.7	291.775
63	3117.2	197.920	69	3739.3	216.770	75	4417.9	235.619	81	5153.0	254.469	87	5944.7	273.319	93	6792.9	292.168
⅛	3129.6	198.313	⅛	3752.8	217.163	⅛	4432.6	236.012	⅛	5168.9	254.862	⅛	5961.8	273.711	⅛	6811.2	292.561
¼	3142.0	198.706	¼	3766.4	217.555	¼	4447.4	236.405	¼	5184.9	255.254	¼	5978.9	274.104	¼	6829.5	292.954
⅜	3154.5	199.098	⅜	3780.0	217.948	⅜	4462.2	236.798	⅜	5200.8	255.647	⅜	5996.0	274.497	⅜	6847.8	293.346
½	3166.9	199.491	½	3793.7	218.341	½	4477.0	237.190	½	5216.8	256.040	½	6013.2	274.889	½	6866.1	293.739
⅝	3179.4	199.884	⅝	3807.3	218.733	⅝	4491.8	237.583	⅝	5232.8	256.433	⅝	6030.4	275.282	⅝	6884.5	294.132
¾	3191.9	200.277	¾	3821.0	219.120	¾	4506.7	237.976	¾	5248.9	256.825	¾	6047.6	275.675	¾	6902.9	294.524
⅞	3204.4	200.669	⅞	3834.7	219.519	⅞	4521.5	238.368	⅞	5264.9	257.218	⅞	6064.9	276.067	⅞	6921.3	294.917
64	3217.0	201.062	70	3848.5	219.911	76	4536.5	238.761	82	5281.0	257.611	88	6082.1	276.460	94	6939.8	295.310
⅛	3229.6	201.455	⅛	3862.2	220.304	⅛	4551.4	239.154	⅛	5297.1	258.003	⅛	6099.4	276.853	⅛	6958.2	295.702
¼	3242.2	201.847	¼	3876.0	220.697	¼	4566.4	239.546	¼	5313.3	258.396	¼	6116.7	277.246	¼	6976.7	296.095
⅜	3254.8	202.240	⅜	3889.8	221.090	⅜	4581.3	239.939	⅜	5329.4	258.789	⅜	6134.1	277.638	⅜	6995.3	296.488
½	3267.5	202.633	½	3903.6	221.482	½	4596.3	240.332	½	5345.6	259.181	½	6151.4	278.031	½	7013.8	296.881
⅝	3280.1	203.025	⅝	3917.5	221.875	⅝	4611.4	240.725	⅝	5361.8	259.574	⅝	6168.8	278.424	⅝	7032.4	297.273
¾	3292.8	203.418	¾	3931.4	222.268	¾	4626.4	241.117	¾	5378.1	259.967	¾	6186.2	278.816	¾	7051.0	297.666
⅞	3305.6	203.811	⅞	3945.3	222.660	⅞	4641.5	241.510	⅞	5394.3	260.359	⅞	6203.7	279.209	⅞	7069.6	298.059
65	3318.3	204.204	71	3959.2	223.053	77	4656.6	241.903	83	5410.6	260.752	89	6221.1	279.602	95	7088.2	298.451
⅛	3331.1	204.596	⅛	3973.1	223.446	⅛	4671.8	242.295	⅛	5426.9	261.145	⅛	6238.6	279.994	⅛	7106.9	298.844
¼	3343.9	204.989	¼	3987.1	223.838	¼	4686.9	242.688	¼	5443.3	261.538	¼	6256.1	280.387	¼	7125.6	299.237
⅜	3356.7	205.382	⅜	4001.1	224.231	⅜	4702.1	243.081	⅜	5459.6	261.930	⅜	6273.7	280.780	⅜	7144.3	299.629
½	3369.6	205.774	½	4015.2	224.625	½	4717.3	243.473	½	5476.0	262.323	½	6291.2	281.173	½	7163.0	300.022
⅝	3382.4	206.167	⅝	4029.2	225.017	⅝	4732.5	243.866	⅝	5492.4	262.716	⅝	6308.8	281.565	⅝	7181.8	300.415
¾	3395.3	206.560	¾	4043.3	225.409	¾	4747.8	244.259	¾	5508.8	263.103	¾	6326.4	281.958	¾	7200.6	300.807
⅞	3408.2	206.952	⅞	4057.4	225.802	⅞	4763.1	244.652	⅞	5525.3	263.501	⅞	6344.1	282.351	⅞	7219.4	301.200
66	3421.2	207.345	72	4071.5	226.195	78	4778.4	245.044	84	5541.8	263.894	90	6361.7	282.743	96	7238.2	301.593
⅛	3434.3	207.738	⅛	4085.7	226.587	⅛	4793.7	245.437	⅛	5558.3	264.286	⅛	6379.4	283.136	¼	7276.0	302.378
¼	3447.2	208.131	¼	4099.8	226.980	¼	4809.0	245.830	¼	5574.8	264.679	¼	6397.1	283.529	½	7313.8	303.164
⅜	3460.2	208.523	⅜	4114.0	227.373	⅜	4824.4	246.222	⅜	5591.4	265.072	⅜	6414.9	283.921	¾	7351.8	303.949
½	3473.2	208.916	½	4128.2	227.765	½	4839.8	246.615	½	5607.9	265.465	½	6432.6	284.314	97	7389.8	304.734
⅝	3486.3	209.309	⅝	4142.5	228.158	⅝	4855.2	247.008	⅝	5624.5	265.857	⅝	6450.4	284.707	¼	7428.0	305.520
¾	3499.4	209.701	¾	4156.8	228.551	¾	4870.7	247.400	¾	5641.2	266.250	¾	6468.2	285.100	½	7466.2	306.305
⅞	3512.5	210.094	⅞	4171.1	228.944	⅞	4886.2	247.793	⅞	5657.8	266.643	⅞	6486.0	285.492	¾	7504.5	307.091
67	3525.7	210.487	73	4185.4	229.336	79	4901.7	248.186	85	5674.5	267.035	91	6503.9	285.885	98	7543.0	307.876
⅛	3538.8	210.879	⅛	4199.7	229.729	⅛	4917.2	248.579	⅛	5691.2	267.428	⅛	6521.8	286.278	¼	7581.5	308.661
¼	3552.0	211.272	¼	4214.1	230.122	¼	4932.7	248.971	¼	5707.9	267.821	¼	6539.7	286.670	½	7620.1	309.447
⅜	3565.2	211.665	⅜	4228.5	230.514	⅜	4948.3	249.364	⅜	5724.7	268.213	⅜	6557.6	287.063	¾	7658.9	310.232
½	3578.5	212.058	½	4242.9	230.907	½	4963.9	249.757	½	5741.5	268.606	½	6575.5	287.456	99	7697.7	311.018
⅝	3591.7	212.450	⅝	4257.4	231.300	⅝	4979.5	250.149	⅝	5758.3	268.999	⅝	6593.5	287.848	¼	7736.6	311.803
¾	3605.0	212.843	¾	4271.8	231.692	¾	4995.2	250.542	¾	5775.1	269.392	¾	6611.5	288.241	½	7775.6	312.588
⅞	3618.3	213.236	⅞	4286.3	232.085	⅞	5010.9	250.935	⅞	5791.9	269.784	⅞	6629.6	288.634	¾	7814.8	313.374

Diam. in Feet	Circumference		Area of Circle		Volume of Cylinder Per Foot of Height			Diam. in Feet
	Feet	Meters	Sq. Feet	Sq. Meters	U. S. Gals.	Imperial Gals.	U. S. Bbls. (42 Gals.)	
1	3.14	0.9576	0.785	.0730	5.9	4.9	0.140	1
2	6.28	1.9151	3.142	.2919	23.5	19.6	0.560	2
3	9.42	2.8727	7.069	.6567	52.9	44.0	1.259	3
4	12.57	3.8302	12.566	1.1675	94.0	78.3	2.238	4
5	15.71	4.7878	19.635	1.8241	146.9	122.3	3.497	5
6	18.85	5.7454	28.274	2.6268	211.5	176.1	5.04	6
7	21.99	6.7029	38.485	3.5753	287.9	239.7	6.85	7
8	25.13	7.6605	50.266	4.6698	376.0	313.1	8.95	8
9	28.27	8.6180	63.617	5.9102	475.9	396.3	11.33	9
10	31.42	9.5756	78.540	7.2966	587.5	489.2	13.99	10
11	34.56	10.5332	95.033	8.8289	710.9	591.9	16.93	11
12	37.70	11.4907	113.097	10.5071	846.0	704.5	20.14	12
13	40.84	12.4483	132.732	12.3312	992.9	826.8	23.64	13
14	43.98	13.4059	153.938	14.3013	1,151.5	958.9	27.42	14
15	47.12	14.3634	176.715	16.4173	1,321.9	1,100.7	31.47	15
16	50.27	15.3210	201.062	18.6792	1,504.0	1,252.4	35.81	16
17	53.41	16.2785	226.980	21.0871	1,697.9	1,413.8	40.43	17
18	56.55	17.2361	254.469	23.6409	1,903.6	1,585.1	45.32	18
19	59.69	18.1937	283.529	26.3407	2,120.9	1,766.1	50.50	19
20	62.83	19.1512	314.159	29.1863	2,350.1	1,956.9	55.95	20
21	65.97	20.1088	346.361	32.1779	2,591.0	2,157.4	61.69	21
22	69.12	21.0663	380.133	35.3154	2,843.6	2,367.8	67.70	22
23	72.26	22.0239	415.476	38.5989	3,108.0	2,587.9	74.00	23
24	75.40	22.9815	452.389	42.0283	3,384.1	2,817.9	80.57	24
25	78.54	23.9390	490.874	45.6036	3,672.0	3,057.6	87.43	25
26	81.68	24.8966	530.929	49.3249	3,971.6	3,307.1	94.56	26
27	84.82	25.8541	572.555	53.1921	4,283.0	3,566.4	101.98	27
28	87.97	26.8117	615.752	57.2052	4,606.1	3,835.4	109.67	28
29	91.11	27.7693	660.520	61.3642	4,941.0	4,114.3	117.64	29
30	94.25	28.7268	706.858	65.6692	5,287.7	4,402.9	125.90	30
31	97.39	29.6844	754.768	70.1201	5,646.1	4,701.4	134.43	31
32	100.53	30.6420	804.248	74.7170	6,016.2	5,009.6	143.24	32
33	103.67	31.5995	855.299	79.4598	6,398.1	5,327.5	152.34	33
34	106.81	32.5571	907.920	84.3485	6,791.7	5,655.3	161.71	34
35	109.96	33.5146	962.113	89.3831	7,197.1	5,992.9	171.36	35
36	113.10	34.4722	1,017.88	94.5637	7,614.2	6,340.2	181.29	36
37	116.24	35.4298	1,075.21	99.8902	8,043.1	6,697.4	191.50	37
38	119.38	36.3873	1,134.11	105.3626	8,483.8	7,064.3	201.99	38
39	122.52	37.3449	1,194.59	110.9810	8,936.2	7,441.0	212.77	39
40	125.66	38.3024	1,256.64	116.7453	9,400.3	7,827.4	223.82	40
41	128.81	39.2600	1,320.25	122.6555	9,876.2	8,223.7	235.15	41
42	131.95	40.2176	1,385.44	128.7117	10,363.8	8,629.7	246.76	42
43	135.09	41.1751	1,452.20	134.9138	10,863.2	9,045.6	258.65	43
44	138.23	42.1327	1,520.53	141.2618	11,374.4	9,471.2	270.82	44
45	141.37	43.0902	1,590.43	147.7557	11,897.2	9,906.6	283.27	45
46	144.51	44.0478	1,661.90	154.3956	12,431.9	10,351.8	296.00	46
47	147.65	45.0054	1,734.94	161.1815	12,978.3	10,806.8	309.01	47
48	150.80	45.9629	1,809.56	168.1132	13,536.4	11,271.5	322.30	48
49	153.94	46.9205	1,885.74	175.1909	14,106.3	11,746.0	335.86	49
50	157.08	47.8781	1,963.50	182.4145	14,688.0	12,230.4	349.71	50
51	160.22	48.8356	2,042.82	189.7840	15,281.4	12,724.5	363.84	51
52	163.36	49.7932	2,123.72	197.2995	15,886.5	13,228.4	378.25	52
53	166.50	50.7507	2,206.18	204.9609	16,503.4	13,742.0	392.94	53
54	169.65	51.7083	2,290.22	212.7683	17,132.0	14,265.5	407.91	54
55	172.79	52.6659	2,375.83	220.7215	17,772.4	14,798.7	423.15	55
56	175.93	53.6234	2,463.01	228.8207	18,424.6	15,341.8	438.68	56
57	179.07	54.5810	2,551.76	237.0659	19,088.5	15,894.6	454.49	57
58	182.21	55.5386	2,642.08	245.4570	19,764.1	16,457.2	470.57	58
59	185.35	56.4961	2,733.97	253.9939	20,451.5	17,029.6	486.94	59
60	188.50	57.4537	2,827.43	262.6769	21,150.7	17,611.7	503.59	60

Notes:
1. If diameters are assumed as meters, values in columns "Circumference Feet" and "Area of Circle Square Feet" will represent circumference in meters and area of circle in square meters respectively. 2. If diameters are assumed as meters, values in column "Area of Circle Square Feet" will represent volume of cylinder in cubic meters per vertical meter of height.

Formula to determine capacity per foot of vertical height of cylinder.
D = Diameter in Feet.
$0.1398854D^2$ = Barrels of 42 U.S. Gallons per vertical foot.
$5.875185\ D^2$ = U.S. Gallons per vertical foot.
$4.892148\ D^2$ = Imperial Gallons per vertical foot.
$0.022240\ D^2$ = Cubic Meters per vertical foot.
$0.785398\ D^2$ = (D—meters) = Cubic meters per vertical meter.

Table 4.13 continued

Table 4.13. Cylinder volumes, circumferences, and area of circles. Courtesy Natural Gas Processors Suppliers Association.

Table 4.13 continued

Diam. in Feet	Circumference		Area of Circle		Volume of Cylinder Per Foot of Height			Diam. in Feet
	Feet	Meters	Sq. Feet	Sq. Meters	U. S. Gals.	Imperial Gals.	U. S. Bbls. (42 Gals.)	
61	191.64	58.4112	2,922.47	271.5057	21,861.6	18,203.7	520.51	61
62	194.78	59.3688	3,019.07	280.4805	22,584.2	18,805.4	537.72	62
63	197.92	60.3263	3,117.25	289.6013	23,318.6	19,416.9	555.21	63
64	201.06	61.2839	3,216.99	298.8679	24,064.8	20,038.2	572.97	64
65	204.20	62.2415	3,318.31	308.2805	24,822.7	20,669.3	591.02	65
66	207.35	63.1990	3,421.19	317.8390	25,592.3	21,310.2	609.34	66
67	210.49	64.1566	3,525.65	327.5435	26,373.7	21,960.9	627.95	67
68	213.63	65.1141	3,631.68	337.3939	27,166.9	22,621.3	646.83	68
69	216.77	66.0717	3,739.28	347.3902	27,971.8	23,291.5	665.99	69
70	219.91	67.0293	3,848.45	357.5324	28,788.4	23,971.5	685.44	70
71	223.05	67.9868	3,959.19	367.8206	29,616.8	24,661.3	705.16	71
72	226.19	68.9444	4,071.50	378.2547	30,457.0	25,360.9	725.17	72
73	229.34	69.9020	4,185.39	388.8347	31,308.9	26,070.3	745.45	73
74	232.48	70.8600	4,300.84	399.5607	32,172.5	26,789.4	766.01	74
75	235.62	71.8171	4,417.86	410.4326	33,047.9	27,518.3	786.86	75
76	238.76	72.7746	4,536.46	421.4505	33,935.1	28,257.0	807.98	76
77	241.90	73.7322	4,656.63	432.6142	34,834.0	29,005.5	829.38	77
78	245.04	74.6898	4,778.36	443.9239	35,744.6	29,763.8	851.06	78
79	248.19	75.6473	4,901.67	455.3796	36,667.0	30,531.9	873.02	79
80	251.33	76.6049	5,026.55	466.9811	37,601.2	31,309.7	895.27	80
81	254.47	77.5624	5,153.00	478.7286	38,547.1	32,097.4	917.79	81
82	257.61	78.5200	5,281.02	490.6220	39,504.7	32,894.8	940.59	82
83	260.75	79.4776	5,410.61	502.6614	40,474.1	33,702.0	963.67	83
84	263.89	80.4351	5,541.77	514.8467	41,455.3	34,519.0	987.03	84
85	267.04	81.3927	5,674.50	527.1779	42,448.2	35,345.8	1,010.67	85
86	270.18	82.3502	5,808.80	539.6551	43,452.9	36,182.3	1,034.59	86
87	273.32	83.3078	5,944.68	552.2781	44,469.3	37,028.7	1,058.79	87
88	276.46	84.2654	6,082.12	565.0472	45,497.4	37,884.8	1,083.27	88
89	279.60	85.2229	6,221.14	577.9621	46,537.3	38,750.7	1,108.03	89
90	282.74	86.1805	6,361.73	591.0230	47,589.0	39,626.4	1,133.07	90
91	285.88	87.1381	6,503.88	604.2298	48,652.4	40,511.9	1,158.39	91
92	289.03	88.0956	6,647.61	617.5825	49,727.6	41,407.1	1,183.99	92
93	292.17	89.0532	6,792.91	631.0812	50,814.5	42,312.2	1,209.87	93
94	295.31	90.0107	6,939.78	644.7258	51,913.1	43,227.0	1,236.03	94
95	298.45	90.9683	7,088.22	658.5163	53,023.5	44,151.6	1,262.47	95
96	301.59	91.9259	7,238.23	672.4528	54,145.7	45,086.0	1,289.18	96
97	304.73	92.8834	7,389.81	686.5352	55,279.6	46,030.2	1,316.18	97
98	307.88	93.8410	7,542.96	700.7635	56,425.3	46,984.2	1,343.46	98
99	311.02	94.7985	7,697.69	715.1378	57,582.7	47,947.9	1,371.02	99
100	314.16	95.7561	7,853.98	729.6580	58,751.9	48,921.5	1,398.85	100
101	317.30	96.7137	8,011.85	744.3241	59,932.8	49,904.8	1,426.97	101
102	320.44	97.6712	8,171.28	759.1362	61,125.4	50,897.9	1,455.37	102
103	323.58	98.6288	8,332.29	774.0942	62,329.8	51,900.8	1,484.04	103
104	326.73	99.5863	8,494.87	789.1981	63,546.0	52,913.5	1,513.00	104
105	329.87	100.5439	8,659.01	804.4479	64,773.9	53,935.9	1,542.24	105
106	333.01	101.5015	8,824.73	819.8437	66,013.6	54,968.2	1,571.75	106
107	336.15	102.4590	8,992.02	835.3854	67,265.0	56,010.2	1,601.55	107
108	339.29	103.4166	9,160.88	851.0731	68,528.2	57,062.0	1,631.62	108
109	342.43	104.3741	9,331.32	866.9067	69,803.1	58,123.6	1,661.98	109
110	345.58	105.3317	9,503.32	882.8862	71,089.7	59,195.0	1,692.61	110
111	348.72	106.2893	9,676.89	899.0116	72,388.2	60,276.2	1,723.53	111
112	351.86	107.2468	9,852.03	915.2830	73,698.3	61,367.1	1,754.72	112
113	355.00	108.2044	10,028.75	931.7003	75,020.2	62,467.8	1,786.20	113
114	358.14	109.1620	10,207.03	948.2635	76,353.9	63,578.4	1,817.95	114
115	361.28	110.1195	10,386.89	964.9727	77,699.3	64,698.7	1,849.98	115
116	364.42	111.0771	10,568.32	981.8278	79,056.5	65,828.7	1,882.30	116
117	367.57	112.0346	10,751.31	998.8288	80,425.4	66,968.6	1,914.89	117
118	370.71	112.9922	10,935.88	1,015.9758	81,806.1	68,118.3	1,947.76	118
119	373.85	113.9498	11,122.02	1,033.2687	83,198.5	69,277.7	1,980.92	119
120	376.99	114.9073	11,309.73	1,050.7075	84,602.7	70,446.9	2,014.35	120
121	380.13	115.8649	11,499.01	1,068.2923	86,018.6	71,625.9	2,048.06	121
122	383.27	116.8224	11,689.86	1,086.0230	87,446.3	72,814.7	2,082.05	122
123	386.42	117.7800	11,882.29	1,103.8996	88,885.7	74,013.3	2,116.33	123
124	389.56	118.7376	12,076.28	1,121.9221	90,336.8	75,221.7	2,150.88	124
125	392.70	119.6951	12,271.84	1,140.0906	91,799.8	76,439.8	2,185.71	125
126	395.84	120.6527	12,468.98	1,158.4050	93,274.4	77,667.7	2,220.82	126
127	398.98	121.6102	12,667.68	1,176.8654	94,760.9	78,905.5	2,256.21	127
128	402.12	122.5678	12,867.96	1,195.4717	96,259.0	80,153.0	2,291.88	128
129	405.27	123.5254	13,069.81	1,214.2239	97,769.0	81,410.2	2,327.83	129
130	408.41	124.4829	13,273.23	1,233.1220	99,290.6	82,677.3	2,364.06	130

Table 4.13 continued

Table 4.13 continued

Diam. in Feet	Circumference		Area of Circle		Volume of Cylinder Per Foot of Height			Diam. in Feet
	Feet	Meters	Sq. Feet	Sq. Meters	U. S. Gals.	Imperial Gals.	U. S. Bbls. (42 Gals.)	
131	411.55	125.4405	13,478.22	1,252.1661	100,824.0	83,954.2	2,400.57	131
132	414.69	126.3981	13,684.77	1,271.3561	102,369.2	85,240.8	2,437.36	132
133	417.83	127.3556	13,892.91	1,290.6920	103,926.1	86,537.2	2,474.43	133
134	420.97	128.3132	14,102.61	1,310.1739	105,494.8	87,843.4	2,511.78	134
135	424.12	129.2707	14,313.88	1,329.8017	107,075.2	89,159.4	2,549.41	135
136	427.26	130.2283	14,526.72	1,349.5754	108,667.4	90,485.2	2,587.32	136
137	430.40	131.1859	14,741.14	1,369.4951	110,271.3	91,820.7	2,625.51	137
138	433.54	132.1434	14,957.12	1,389.5607	111,887.0	93,166.1	2,663.98	138
139	436.68	133.1010	15,174.67	1,409.7722	113,514.4	94,521.2	2,702.73	139
140	439.82	134.0585	15,393.80	1,430.1297	115,153.6	95,886.1	2,741.75	140
141	442.96	135.0161	15,614.50	1,450.6331	116,804.6	97,260.8	2,781.06	141
142	446.11	135.9737	15,836.77	1,471.2824	118,467.2	98,645.3	2,820.65	142
143	449.25	136.9312	16,060.60	1,492.0776	120,141.7	100,039.5	2,860.52	143
144	452.39	137.8888	16,286.01	1,513.0188	121,827.8	101,443.6	2,900.66	144
145	455.53	138.8463	16,512.99	1,534.1059	123,525.8	102,857.4	2,941.09	145
146	458.67	139.8039	16,741.54	1,555.3390	125,235.4	104,281.0	2,981.80	146
147	461.81	140.7615	16,971.67	1,576.7180	126,956.9	105,714.4	3,022.78	147
148	464.96	141.7190	17,203.36	1,598.2429	128,690.1	107,157.6	3,064.05	148
149	468.10	142.6766	17,436.62	1,619.9137	130,435.0	108,610.6	3,105.60	149
150	471.24	143.6342	17,671.46	1,641.7305	132,191.7	110,073.3	3,147.42	150
151	474.38	144.5917	17,907.86	1,663.6932	133,960.1	111,545.9	3,189.53	151
152	477.52	145.5493	18,145.84	1,685.8018	135,740.3	113,028.2	3,231.91	152
153	480.66	146.5068	18,385.38	1,708.0564	137,532.2	114,520.3	3,274.58	153
154	483.81	147.4644	18,626.50	1,730.4569	139,335.9	116,022.2	3,317.52	154
155	486.95	148.4220	18,869.19	1,753.0033	141,151.3	117,533.9	3,360.75	155
156	490.09	149.3795	19,113.45	1,775.6957	142,978.5	119,055.3	3,404.25	156
157	493.23	150.3371	19,359.28	1,798.5340	144,817.4	120,586.6	3,448.04	157
158	496.37	151.2946	19,606.68	1,821.5182	146,668.1	122,127.6	3,492.10	158
159	499.51	152.2522	19,855.65	1,844.6484	148,530.6	123,678.4	3,536.44	159
160	502.65	153.2098	20,106.19	1,867.9245	150,404.7	125,239.0	3,581.07	160
161	505.80	154.1673	20,358.30	1,891.3465	152,290.7	126,809.4	3,625.97	161
162	508.94	155.1249	20,611.99	1,914.9145	154,188.4	128,389.5	3,671.15	162
163	512.08	156.0824	20,867.24	1,938.6283	156,097.8	129,979.5	3,716.62	163
164	515.22	157.0400	21,124.06	1,962.4882	158,019.0	131,579.2	3,762.36	164
165	518.36	157.9976	21,382.46	1,986.4939	159,951.9	133,188.7	3,808.38	165
166	521.50	158.9551	21,642.43	2,010.6456	161,896.6	134,808.0	3,854.68	166
167	524.65	159.9127	21,903.96	2,034.9432	163,853.0	136,437.1	3,901.26	167
168	527.79	160.8702	22,167.07	2,059.3867	165,821.2	138,076.0	3,948.13	168
169	530.93	161.8278	22,431.75	2,083.9762	167,801.2	139,724.6	3,995.27	169
170	534.07	162.7854	22,698.00	2,108.7116	169,792.8	141,383.1	4,042.69	170
171	537.21	163.7429	22,965.82	2,133.5930	171,796.3	143,051.3	4,090.39	171
172	540.35	164.7005	23,235.21	2,158.6202	173,811.5	144,729.3	4,138.37	172
173	543.50	165.6581	23,506.18	2,183.7934	175,838.4	146,417.1	4,186.63	173
174	546.64	166.6156	23,778.71	2,209.1126	177,877.1	148,114.7	4,235.17	174
175	549.78	167.5732	24,052.81	2,234.5776	179,927.5	149,822.0	4,283.99	175
176	552.92	168.5307	24,328.49	2,260.1886	181,989.7	151,539.2	4,333.09	176
177	556.06	169.4883	24,605.73	2,285.9455	184,063.7	153,266.1	4,382.47	177
178	559.20	170.4459	24,884.55	2,311.8484	186,149.4	155,002.8	4,432.13	178
179	562.35	171.4034	25,164.94	2,337.8972	188,246.8	156,749.3	4,482.07	179
180	565.49	172.3610	25,446.90	2,364.0919	190,356.0	158,505.6	4,532.29	180
181	568.63	173.3185	25,730.42	2,390.4326	192,476.9	160,271.7	4,582.79	181
182	571.77	174.2761	26,015.52	2,416.9192	194,609.6	162,047.5	4,633.56	182
183	574.91	175.2337	26,302.19	2,443.5517	196,754.1	163,833.1	4,684.62	183
184	578.05	176.1912	26,590.43	2,470.3301	198,910.3	165,628.6	4,735.96	184
185	581.19	177.1488	26,880.25	2,497.2545	201,078.2	167,433.8	4,787.58	185
186	584.34	178.1063	27,171.63	2,524.3248	203,257.9	169,248.8	4,839.48	186
187	587.48	179.0639	27,464.58	2,551.5411	205,449.3	171,073.5	4,891.65	187
188	590.62	180.0215	27,759.11	2,578.9032	207,652.5	172,908.1	4,944.11	188
189	593.76	180.9790	28,055.20	2,606.4113	209,867.5	174,752.4	4,996.85	189
190	596.90	181.9366	28,352.87	2,634.0654	212,094.2	176,606.5	5,049.86	190
191	600.04	182.8942	28,652.10	2,661.8653	214,332.6	178,470.5	5,103.16	191
192	603.19	183.8517	28,952.91	2,689.8113	216,582.8	180,344.1	5,156.74	192
193	606.33	184.8093	29,255.29	2,717.9031	218,844.8	182,227.6	5,210.59	193
194	609.47	185.7668	29,559.24	2,746.1408	221,118.5	184,120.9	5,264.73	194
195	612.61	186.7244	29,864.76	2,774.5245	223,403.9	186,023.9	5,319.14	195
196	615.75	187.6820	30,171.85	2,803.0542	225,701.1	187,936.8	5,373.84	196
197	618.89	188.6395	30,480.51	2,831.7297	228,010.1	189,859.4	5,428.81	197
198	622.04	189.5971	30,790.74	2,860.5512	230,330.8	191,791.8	5,484.07	198
199	625.18	190.5546	31,102.55	2,889.5186	232,663.2	193,734.0	5,539.60	199
200	628.32	191.5122	31,415.93	2,918.6320	235,007.4	195,685.9	5,595.42	200

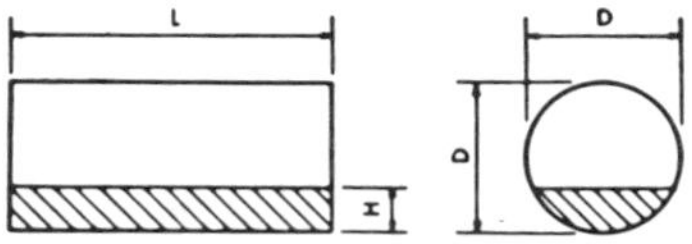

Partial volume in height (H) = Cylindrical coefficient for H/D × total volume.

$$\text{Total volume} = \frac{\pi L D^2}{4}$$

COEFFICIENTS FOR PARTIAL VOLUMES OF HORIZONTAL CYLINDERS

H/D	0	1	2	3	4	5	6	7	8	9
.00	.000000	.000053	.000151	.000279	.000429	.000600	.000788	.000992	.001212	.001445
.01	.001692	.001952	.002223	.002507	.002800	.003104	.003419	.003743	.004077	.004421
.02	.004773	.005134	.005503	.005881	.006267	.006660	.007061	.007470	.007886	.008310
.03	.008742	.009179	.009625	.010076	.010534	.010999	.011470	.011947	.012432	.012920
.04	.013417	.013919	.014427	.014940	.015459	.015985	.016515	.017052	.017593	.018141
.05	.018692	.019250	.019813	.020382	.020955	.021533	.022115	.022703	.023296	.023894
.06	.024496	.025103	.025715	.026331	.026952	.027578	.028208	.028842	.029481	.030124
.07	.030772	.031424	.032081	.032740	.033405	.034073	.034747	.035423	.036104	.036789
.08	.037478	.038171	.038867	.039569	.040273	.040981	.041694	.042410	.043129	.043852
.09	.044579	.045310	.046043	.046782	.047523	.048268	.049017	.049768	.050524	.051283
.10	.052044	.052810	.053579	.054351	.055126	.055905	.056688	.057474	.058262	.059054
.11	.059850	.060648	.061449	.062253	.063062	.063872	.064687	.065503	.066323	.067147
.12	.067972	.068802	.069633	.070469	.071307	.072147	.072991	.073836	.074686	.075539
.13	.076393	.077251	.078112	.078975	.079841	.080709	.081581	.082456	.083332	.084212
.14	.085094	.085979	.086866	.087756	.088650	.089545	.090443	.091343	.092246	.093153
.15	.094061	.094971	.095884	.096799	.097717	.098638	.099560	.100486	.101414	.102343
.16	.103275	.104211	.105147	.106087	.107029	.107973	.108920	.109869	.110820	.111773
.17	.112728	.113686	.114646	.115607	.116572	.117538	.118506	.119477	.120450	.121425
.18	.122403	.123382	.124364	.125347	.126333	.127321	.128310	.129302	.130296	.131292
.19	.132290	.133291	.134292	.135296	.136302	.137310	.138320	.139332	.140345	.141361
.20	.142378	.143398	.144419	.145443	.146468	.147494	.148524	.149554	.150587	.151622
.21	.152659	.153697	.154737	.155779	.156822	.157867	.158915	.159963	.161013	.162066
.22	.163120	.164176	.165233	.166292	.167353	.168416	.169480	.170546	.171613	.172682
.23	.173753	.174825	.175900	.176976	.178053	.179131	.180212	.181294	.182378	.183463
.24	.184550	.185639	.186729	.187820	.188912	.190007	.191102	.192200	.193299	.194400
.25	.195501	.196604	.197709	.198814	.199922	.201031	.202141	.203253	.204368	.205483
.26	.206600	.207718	.208837	.209957	.211079	.212202	.213326	.214453	.215580	.216708
.27	.217839	.218970	.220102	.221235	.222371	.223507	.224645	.225783	.226924	.228065
.28	.229209	.230352	.231498	.232644	.233791	.234941	.236091	.237242	.238395	.239548
.29	.240703	.241859	.243016	.244173	.245333	.246494	.247655	.248819	.249983	.251148
.30	.252315	.253483	.254652	.255822	.256992	.258165	.259338	.260512	.261687	.262863
.31	.264039	.265218	.266397	.267578	.268760	.269942	.271126	.272310	.273495	.274682
.32	.275869	.277058	.278247	.279437	.280627	.281820	.283013	.284207	.285401	.286598
.33	.287795	.288992	.290191	.291390	.292591	.293793	.294995	.296198	.297403	.298605
.34	.299814	.301021	.302228	.303438	.304646	.305857	.307068	.308280	.309492	.310705
.35	.311918	.313134	.314350	.315566	.316783	.318001	.319219	.320439	.321660	.322881
.36	.324104	.325326	.326550	.327774	.328999	.330225	.331451	.332678	.333905	.335134
.37	.336363	.337593	.338823	.340054	.341286	.342519	.343751	.344985	.346220	.347455
.38	.348690	.349926	.351164	.352402	.353640	.354879	.356119	.357359	.358599	.359840
.39	.361082	.362325	.363568	.364811	.366056	.367300	.368545	.369790	.371036	.372282
.40	.373530	.374778	.376026	.377275	.378524	.379774	.381024	.382274	.383526	.384778
.41	.386030	.387283	.388537	.389790	.391044	.392298	.393553	.394808	.396063	.397320
.42	.398577	.399834	.401092	.402350	.403608	.404866	.406125	.407384	.408645	.409904
.43	.411165	.412426	.413687	.414949	.416211	.417473	.418736	.419998	.421261	.422525
.44	.423788	.425052	.426316	.427582	.428846	.430112	.431378	.432645	.433911	.435178
.45	.436445	.437712	.438979	.440246	.441514	.442782	.444050	.445318	.446587	.447857
.46	.449125	.450394	.451663	.452932	.454201	.455472	.456741	.458012	.459283	.460554
.47	.461825	.463096	.464367	.465638	.466910	.468182	.469453	.470725	.471997	.473269
.48	.474541	.475814	.477086	.478358	.479631	.480903	.482176	.483449	.484722	.485995
.49	.487269	.488542	.489814	.491087	.492360	.493633	.494906	.496179	.497452	.498726
.50	.500000	.501274	.502548	.503821	.505094	.506367	.507640	.508913	.510186	.511458
.51	.512731	.514005	.515278	.516551	.517824	.519097	.520369	.521642	.522914	.524186
52	.525459	.526731	.528003	.529275	.530547	.531818	.533090	.534362	.535633	.536904
.53	.538175	.539446	.540717	.541988	.543259	.544528	.545799	.547068	.548337	.549606
.54	.550875	.552143	.553413	.554682	.555950	.557218	.558486	.559754	.561021	.562288
.55	.563555	.564822	.566089	.567355	.568622	.569888	.571154	.572418	.573684	.574948
.56	.576212	.577475	.578739	.580002	.581264	.582527	.583789	.585051	.586313	.587574
.57	.588835	.590096	.591355	.592616	.593875	.595134	.596392	.597650	.598908	.600166
.58	.601423	.602680	.603937	.605192	.606447	.607702	.608956	.610210	.611463	.612717
.59	.613970	.615222	.616474	.617726	.618976	.620226	.621476	.622725	.623974	.625222

Table 4.14. continued.

Table 4.14. Cylinder volumes, partial and horizontal cylinders. Courtesy Natural Gas Processors Suppliers Association.

Table 4.14. continued.

COEFFICIENTS FOR PARTIAL VOLUMES OF HORIZONTAL CYLINDERS

H/D	0	1	2	3	4	5	6	7	8	9
.60	.626470	.627718	.628964	.630210	.631455	.632700	.633944	.635189	.636432	.637675
.61	.638918	.640160	.641401	.642641	.643881	.645121	.646360	.647598	.648836	.650074
.62	.651310	.652545	.653780	.655015	.656249	.657481	.658714	.659946	.661177	.662407
.63	.663637	.664866	.666095	.667322	.668549	.669775	.671001	.672226	.673450	.674674
.64	.675896	.677119	.678340	.679561	.680781	.681999	.683217	.684434	.685650	.686866
.65	.688082	.689295	.690508	.691720	.692932	.694143	.695354	.696562	.697772	.698979
.66	.700186	.701392	.702597	.703802	.705005	.706207	.707409	.708610	.709809	.711008
.67	.712205	.713402	.714599	.715793	.716987	.718180	.719373	.720563	.721753	.722942
.68	.724131	.725318	.726505	.727690	.728874	.730058	.731240	.732422	.733603	.734782
.69	.735961	.737137	.738313	.739488	.740662	.741835	.743008	.744178	.745348	.746517
.70	.747685	.748852	.750017	.751181	.752345	.753506	.754667	.755827	.756984	.758141
.71	.759297	.760452	.761605	.762758	.763909	.765059	.766209	.767356	.768502	.769648
.72	.770791	.771935	.773076	.774217	.775355	.776493	.777629	.778765	.779898	.781030
.73	.782161	.783292	.784420	.785547	.786674	.787798	.788921	.790043	.791163	.792282
.74	.793400	.794517	.795632	.796747	.797859	.798969	.800078	.801186	.802291	.803396
.75	.804499	.805600	.806701	.807800	.808898	.809993	.811088	.812180	.813271	.814361
.76	.815450	.816537	.817622	.818706	.819788	.820869	.821947	.823024	.824100	.825175
.77	.826247	.827318	.828387	.829454	.830520	.831584	.832647	.833708	.834767	.835824
.78	.836880	.837934	.838987	.840037	.841085	.842133	.843178	.844221	.845263	.846303
.79	.847341	.848378	.849413	.850446	.851476	.852506	.853532	.854557	.855581	.856602
.80	.857622	.858639	.859655	.860668	.861680	.862690	.863698	.864704	.865708	.866709
.81	.867710	.868708	.869704	.870698	.871690	.872679	.873667	.874653	.875636	.876618
.82	.877597	.878575	.879550	.880523	.881494	.882462	.883428	.884393	.885354	.886314
.83	.887272	.888227	.889180	.890131	.891080	.892027	.892971	.893913	.894853	.895789
.84	.896725	.897657	.898586	.899514	.900440	.901362	.902283	.903201	.904116	.905029
.85	.905939	.906847	.907754	.908657	.909557	.910455	.911350	.912244	.913134	.914021
.86	.914906	.915788	.916668	.917544	.918419	.919291	.920159	.921025	.921888	.922749
.87	.923607	.924461	.925314	.926164	.927009	.927853	.928693	.929531	.930367	.931198
.88	.932028	.932853	.933677	.934497	.935313	.936128	.936938	.937747	.938551	.939352
.89	.940150	.940946	.941738	.942526	.943312	.944095	.944874	.945649	.946421	.947190
.90	.947956	.948717	.949476	.950232	.950983	.951732	.952477	.953218	.953957	.954690
.91	.955421	.956148	.956871	.957590	.958306	.959019	.959727	.960431	.961133	.961829
.92	.962522	.963211	.963896	.964577	.965253	.965927	.966595	.967260	.967919	.968576
.93	.969228	.969876	.970519	.971158	.971792	.972422	.973048	.973669	.974285	.974897
.94	.975504	.976106	.976704	.977297	.977885	.978467	.979045	.979618	.980187	.980750
.95	.981308	.981859	.982407	.982948	.983485	.984015	.984541	.985060	.985573	.986081
.96	.986583	.987080	.987568	.988053	.988530	.989001	.989466	.989924	.990375	.990821
.97	.991258	.991690	.992114	.992530	.992939	.993340	.993733	.994119	.994497	.994866
.98	.995227	.995579	.995923	.996257	.996581	.996896	.997200	.997493	.997777	.998048
.99	.998308	.998555	.998788	.999008	.999212	.999400	.999571	.999721	.999849	.999947
1.00	1.000000									

PARTIAL VOLUME IN HORIZONTAL AND VERTICAL STORAGE TANKS WITH ELLIPSOIDAL OR HEMISPHERICAL HEADS

(A)

When liquid level is in bottom head: Partial volume (shaded area) = (ellipsoidal coefficient for H/D) × (total volume in two heads)

$0 \leq H \leq D/2$

(B)

When liquid level is in shell: Partial volume (shaded area) = (h/L) × (Volume in cylinder) + (Volume in one head)

$D/2 \leq (D/2 + h) \leq (D/2 + L)$

(C)

When liquid is in top head: Partial volume (shaded area) = (ellipsoidal coefficient for H/D) × (total volume in two heads) + (volume in cylinder)

$(L + D/2) \leq (H + L) \leq (L + D)$

Vertical Tanks

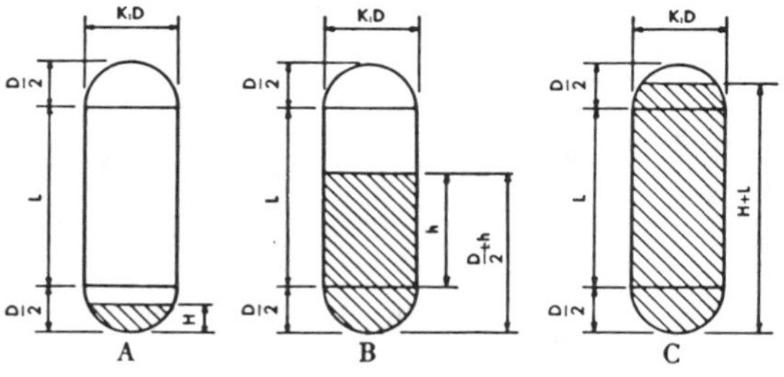

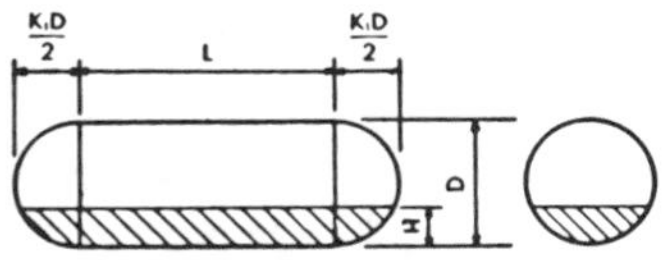

Horizontal Tanks

Partial volume in height (H) = (Cylindrical coefficient for H/D) × (total volume in cylinder) + (ellipsoidal coefficient for H/D) × (total volume in two heads)

The following formulas and table of coefficients provide a convenient method of determining partial volumes in ellipsoids and spheres. To determine the partial volume of two ellipsoidal heads for horizontal tanks, proceed as follows:

D = 20 ft. $K_1 = .5$ H = 6.5 ft.

$$\text{Total volume} = \frac{\pi}{6} K_1 D^3 =$$

$$\frac{3.1416}{6} \times .5 \times 20^3 = 2{,}094.4 \text{ cu. ft.}$$

$$\frac{H}{D} = \frac{6.5}{20} = .325$$

Refer to the first two figures (.32) in the column headed "H/D" in the table below. Proceed to the right until the coefficient is found under the column headed "5" which is the third digit. The coefficient of .325 is found to be .248219. For more than three digits it is necessary to interpolate. Partial volume in height (H) = ellipsoidal coefficient × total volume = .248219 × 2,094.4 = 519.87 cu. ft.

Two Ellipsoidal Heads for Horizontal Tanks

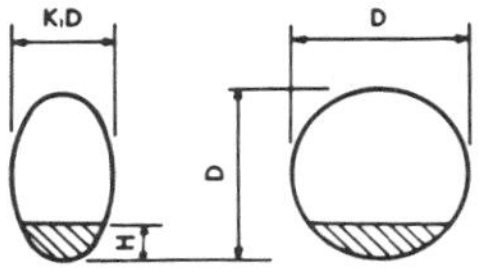

$$\text{Total Volume} = \frac{\pi}{6} K_1 D^3$$

Two Ellipsoidal Heads for Vertical Tanks

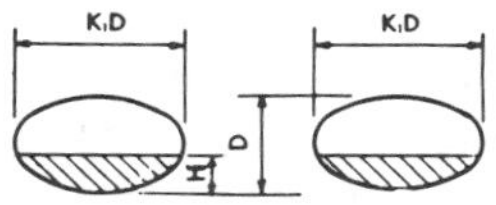

$$\text{Total Volume} = \frac{\pi}{6} K_1^2 D^3$$

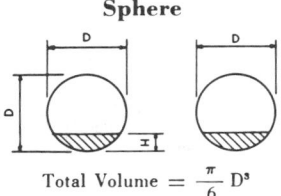

General Ellipsoid

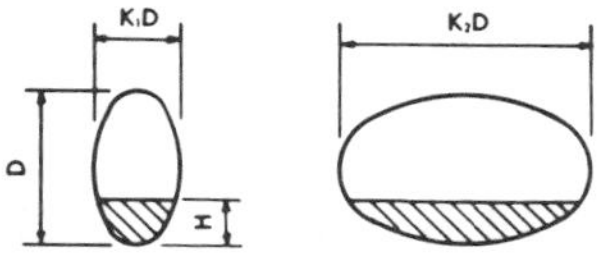

$$\text{Total Volume} = \frac{\pi}{6} K_1 K_2 D^3$$

Sphere

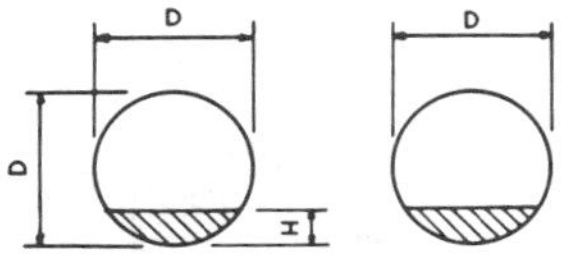

$$\text{Total Volume} = \frac{\pi}{6} D^3$$

For all true ellipsoids:

Partial volume in height (H) = Ellipsoidal coefficient for H/D × total volume. Where D is the vertical height.

COEFFICIENTS FOR PARTIAL VOLUMES OF ELLIPSOIDS (SPHERES)

H/D	0	1	2	3	4	5	6	7	8	9
.00	.000000	.000003	.000012	.000027	.000048	.000075	.000108	.000146	.000191	.000242
.01	.000298	.000360	.000429	.000503	.000583	.000668	.000760	.000857	.000960	.001069
.02	.001184	.001304	.001431	.001563	.001700	.001844	.001993	.002148	.002308	.002474
.03	.002646	.002823	.003006	.003195	.003389	.003589	.003795	.004006	.004222	.004444
.04	.004672	.004905	.005144	.005388	.005638	.005893	.006153	.006419	.006691	.006968
.05	.007250	.007538	.007831	.008129	.008433	.008742	.009057	.009377	.009702	.010032
.06	.010368	.010709	.011055	.011407	.011764	.012126	.012493	.012865	.013243	.013626
.07	.014014	.014407	.014806	.015209	.015618	.016031	.016450	.016874	.017303	.017737
.08	.018176	.018620	.019069	.019523	.019983	.020447	.020916	.021390	.021869	.022353
.09	.022842	.023336	.023835	.024338	.024847	.025360	.025879	.026402	.026930	.027462
.10	.028000	.028542	.029090	.029642	.030198	.030760	.031326	.031897	.032473	.033053
.11	.033638	.034228	.034822	.035421	.036025	.036633	.037246	.037864	.038486	.039113
.12	.039744	.040380	.041020	.041665	.042315	.042969	.043627	.044290	.044958	.045630
.13	.046306	.046987	.047672	.048362	.049056	.049754	.050457	.051164	.051876	.052592
.14	.053312	.054037	.054765	.055499	.056236	.056978	.057724	.058474	.059228	.059987
.15	.060750	.061517	.062288	.063064	.063843	.064627	.065415	.066207	.067003	.067804
.16	.068608	.069416	.070229	.071046	.071866	.072691	.073519	.074352	.075189	.076029
.17	.076874	.077723	.078575	.079432	.080292	.081156	.082024	.082897	.083772	.084652
.18	.085536	.086424	.087315	.088210	.089109	.090012	.090918	.091829	.092743	.093660
.19	.094582	.095507	.096436	.097369	.098305	.099245	.100189	.101136	.102087	.103042
.20	.104000	.104962	.105927	.106896	.107869	.108845	.109824	.110808	.111794	.112784
.21	.113778	.114775	.115776	.116780	.117787	.118798	.119813	.120830	.121852	.122876
.22	.123904	.124935	.125970	.127008	.128049	.129094	.130142	.131193	.132247	.133305
.23	.134366	.135430	.136498	.137568	.138642	.139719	.140799	.141883	.142969	.144059
.24	.145152	.146248	.147347	.148449	.149554	.150663	.151774	.152889	.154006	.155127
.25	.156250	.157376	.158506	.159638	.160774	.161912	.163054	.164198	.165345	.166495
.26	.167684	.168804	.169963	.171124	.172289	.173456	.174626	.175799	.176974	.178153
.27	.179334	.180518	.181705	.182894	.184086	.185281	.186479	.187679	.188882	.190088
.28	.191296	.192507	.193720	.194937	.196155	.197377	.198601	.199827	.201056	.202288
.29	.203522	.204759	.205998	.207239	.208484	.209730	.210979	.212231	.213485	.214741
.30	.216000	.217261	.218526	.219792	.221060	.222331	.223604	.224879	.226157	.227437
.31	.228718	.230003	.231289	.232578	.233870	.235163	.236459	.237757	.239057	.240359
.32	.241664	.242971	.244280	.245590	.246904	.248219	.249536	.250855	.252177	.253500
.33	.254826	.256154	.257483	.258815	.260149	.261484	.262822	.264161	.265503	.266847
.34	.268192	.269539	.270889	.272240	.273593	.274948	.276305	.277663	.279024	.280386

Table 4.15 continued

Table 4.15. Ellipsoids and spheres, partial volumes. Courtesy National Gas Processors Suppliers Association.

Table 4.15 continued

COEFFICIENTS FOR PARTIAL VOLUMES OF ELLIPSOIDS

H/D	0	1	2	3	4	5	6	7	8	9
.35	.281750	.283116	.284484	.285853	.287224	.288597	.289972	.291348	.292727	.294106
.36	.295488	.296871	.298256	.299643	.201031	.302421	.303812	.305205	.306600	.307996
.37	.309394	.310793	.312194	.313597	.315001	.316406	.317813	.319222	.320632	.322043
.38	.323456	.324870	.326286	.327703	.329122	.330542	.331963	.333386	334810	.336235
.39	.337662	.339090	.340519	.341950	.343382	.344815	.346250	.347685	.349122	.350561
.40	.352000	.353441	.354882	.356325	.357769	.359215	.360661	.362109	.363557	.365007
.41	.366458	.367910	.369363	.370817	.372272	.373728	.375185	.376644	.378103	.379563
.42	.381024	.382486	.383949	.395413	.386878	.388344	.389810	.391278	.392746	.394216
.43	.395686	.397157	.398629	.400102	.401575	.403049	.404524	.406000	.407477	.408954
.44	.410432	.411911	.413390	.414870	.416351	.417833	.419315	.420798	422281	.423765
.45	.425250	.426735	.428221	.429708	.431195	.432682	.434170	.435659	.437148	.438638
.46	.440128	.441619	.443110	.444601	.446093	.447586	.449079	.450572	.452066	.453560
.47	.455054	.456549	.458044	.459539	.461035	.462531	.464028	.465524	.467021	.468519
48	.470016	.471514	.473012	.474510	.476008	.477507	.479005	.480504	.482003	.483593
49	.485002	.486501	.488001	.489501	.491000	.492500	.494000	.495500	.497000	.498500
.50	500000	.501500	.503000	.504500	.506000	.507500	.509000	.510499	.511999	.513499
.51	.514998	.516497	.517997	.519496	.520995	.522493	.523992	525490	.526988	.528486
.52	.529984	.531481	.532979	.534476	.535972	.537469	.538965	.540461	.541956	.543451
.53	.544946	.546440	.547934	.549428	.550921	.552414	.553907	.555399	.556890	.558381
54	559872	.561362	.562852	.564341	.565830	.567318	.568805	.570292	.571779	573265
.55	.574750	.576235	.577719	.579202	.580685	.582167	.583649	.585130	.586610	.588089
.56	.589568	.591046	.592523	.594000	.595476	.596951	.598425	.599898	.601371	.602843
.57	.604314	.605784	.607254	.608722	.610190	.611656	.613122	.614587	.616051	.617514
.58	.618976	.620437	.621897	.623356	.624815	.626272	.627728	.629183	.630637	.632090
.59	.633542	.634993	.636443	.637891	.639339	.640785	.642231	.643675	.645118	.646559
.60	.648000	.649439	.650878	.652315	.653750	.655185	.656618	.658050	.659481	.660910
.61	.662338	.663765	.665190	.666614	.668037	.669458	.670878	.672297	.673714	.675130
.62	.676544	.677957	.679368	.680778	.682187	.683594	.684999	.686403	.687806	.689207
.63	.690606	.692004	.693400	.694795	.696188	.697579	.698969	.700357	.701744	.703129
64	.704512	.705894	.707273	.708652	.710028	.711403	.712776	.714147	.715516	.716884
.65	.718250	.719614	.720976	.722337	.723695	.725052	.726407	.727760	.729111	.730461
.66	.731808	.733153	.734497	.735839	.737178	.738516	.739851	.741185	.742517	.743846
.67	.745174	.746500	.747823	.749145	.750464	.751781	.753096	.754410	.755720	.757029
.68	.758336	.759641	.760943	.762243	.763541	.764837	.766130	.767422	.768711	.769997
.69	.771282	.772563	.773843	.775121	.776396	.777669	.778940	.780208	.781474	.782739
.70	.784000	.785259	.786515	.787769	.789021	.790270	.791516	.792761	.794002	.795241
.71	.796478	.797712	.798944	.800173	.801399	.802623	.803845	.805063	.806280	.807493
.72	.808704	.809912	.811118	.812321	.813521	.814719	.815914	.817106	.818295	.819482
.73	.820666	.821847	.823026	824201	.825374	.826544	.827711	.828876	.830037	.831196
.74	.832352	.833505	.834655	.835802	.836946	.838088	839226	.840362	.841494	.842624
.75	.843750	.844873	.845994	.847111	.848226	.849337	.850446	.851551	.852653	.853752
76	.854848	.855941	.857031	.858117	.859201	.860281	.861358	.862432	.863502	.864570
.77	.865634	.866695	.867753	.868807	.869858	.870906	.871951	.872992	.874030	875065
.78	.876096	.877124	.878148	.879170	880187	.881202	.882213	.883220	.884224	.885225
.79	.886222	.887216	.888206	.889192	.890176	.891155	.892131	.893104	.894073	.895038
.80	.896000	.896958	.897913	.898864	.899811	.900755	.901695	.902631	.903564	.904493
81	.905418	.906340	.907257	.908171	.909082	.909988	.910891	.911790	.912685	.913576
.82	.914464	.915348	.916228	.917103	.917976	.918844	.919708	.920568	.921425	922277
.83	.923126	.923971	.924811	.925648	.926481	.927309	.928134	.928954	.929771	.930584
.84	.931392	.932196	.932997	.933793	.934585	.935373	.936157	.936936	.937712	.938483
.85	.939250	.940013	.940772	.941526	.942276	.943022	.943764	.944501	.945235	.945963
.86	.946688	.947408	.948124	.948836	.949543	.950246	.950944	.951638	.952328	.95[illegible]013
.87	.953694	954370	.955042	.955710	.956373	.957031	.957685	.958335	.958980	.959620
.88	.960256	.960887	.961514	.962136	.962754	.963367	.963975	.964579	.965178	.965772
89	.966362	.966947	.967527	.968103	.968674	.969240	.969802	.970358	.970910	.971458
.90	972000	.972538	.973070	.973598	.974121	.974640	.975153	.975662	.976165	.976664
.91	.977158	.977647	.978131	.978610	.979084	.979553	.980017	.980477	.980931	.981380
.92	.981824	.982263	.982697	.983126	.983550	.983969	.984382	.984791	.985194	.985593
.93	.985986	.986374	.986757	.987135	.987507	.987874	.988236	.988593	.988945	.989291
.94	.989632	.989968	.990298	.990623	.990943	.991258	.991567	.991871	.992169	.992462
.95	.992750	.993032	.993309	.993581	.993847	.994107	.994362	.994612	.994856	.995095
.96	.995328	.995556	.995778	.995994	.996205	.996411	.996611	.996805	.996994	.997177
.97	.997354	.997526	.997692	.997852	.998007	.998156	.998300	.998437	.998569	.998696
.98	.998816	.998931	.999040	.999143	.999240	.999332	.999417	.999497	.999571	.999640
.99	.999702	.999758	.999809	.999854	.999892	.999925	.999952	.999973	.999988	.999997
1.00	1.000000									

No.	Square	Cube	Square Root	Cube Root	No.	Square	Cube	Square Root	Cube Root	No.	Square	Cube	Square Root	Cube Root
0	0000	00000	0.0000	0.0000	60	3,600	216,000	7.7460	3.9149	120	14,400	1,728,000	10.9545	4.9324
1	1	1	1.0000	1.0000	61	3,721	226,981	7.8102	3.9365	121	14,641	1,771,561	11.0000	4.9461
2	4	8	1.4142	1.2599	62	3,844	238,328	7.8740	3.9579	122	14,884	1,815,848	11.0454	4.9597
3	9	27	1.7321	1.4422	63	3,969	250,047	7.9373	3.9791	123	15,129	1,860,867	11.0905	4.9732
4	16	64	2.0000	1.5874	64	4,096	262,144	8.0000	4.0000	124	15,376	1,906,624	11.1355	4.9866
5	25	125	2.2361	1.7100	65	4,225	274,625	8.0623	4.0207	125	15,625	1,953,125	11.1803	5.0000
6	36	216	2.4495	1.8171	66	4,356	287,496	8.1240	4.0412	126	15,876	2,000,376	11.2250	5.0133
7	49	343	2.6458	1.9129	67	4,489	300,763	8.1854	4.0615	127	16,129	2,048,383	11.2694	5.0265
8	64	512	2.8284	2.0000	68	4,624	314,432	8.2462	4.0817	128	16,384	2,097,152	11.3137	5.0397
9	81	729	3.0000	2.0801	69	4,761	328,509	8.3066	4.1016	129	16,641	2,146,689	11.3578	5.0528
10	100	1,000	3.1623	2.1544	70	4,900	343,000	8.3666	4.1213	130	16,900	2,197,000	11.4018	5.0658
11	121	1,331	3.3166	2.2240	71	5,041	357,911	8.4261	4.1408	131	17,161	2,248,091	11.4455	5.0788
12	144	1,728	3.4641	2.2894	72	5,184	373,248	8.4853	4.1602	132	17,424	2,299,968	11.4891	5.0916
13	169	2,197	3.6056	2.3513	73	5,329	389,017	8.5440	4.1793	133	17,689	2,352,637	11.5326	5.1045
14	196	2,744	3.7417	2.4101	74	5,476	405,224	8.6023	4.1983	134	17,956	2,406,104	11.5758	5.1172
15	225	3,375	3.8730	2.4662	75	5,625	421,875	8.6603	4.2172	135	18,225	2,460,375	11.6190	5.1290
16	256	4,096	4.0000	2.5198	76	5,776	438,976	8.7178	4.2358	136	18,496	2,515,456	11.6619	5.1426
17	289	4,913	4.1231	2.5713	77	5,929	456,533	8.7750	4.2543	137	18,769	2,571,353	11.7047	5.1551
18	324	5,832	4.2426	2.6207	78	6,084	474,552	8.8318	4.2727	138	19,044	2,628,072	11.7473	5.1676
19	361	6,859	4.3589	2.6684	79	6,241	493,039	8.8882	4.2908	139	19,321	2,685,619	11.7898	5.1801
20	400	8,000	4.4721	2.7144	80	6,400	512,000	8.9443	4.3089	140	19,600	2,744,000	11.8322	5.1925
21	441	9,261	4.5826	2.7589	81	6,561	531,441	9.0000	4.3267	141	19,881	2,803,221	11.8743	5.2048
22	484	10,648	4.6904	2.8020	82	6,724	551,368	9.0554	4.3445	142	20,164	2,863,288	11.9164	5.2171
23	529	12,167	4.7958	2.8439	83	6,889	571,787	9.1104	4.3621	143	20,449	2,924,207	11.9583	5.2293
24	576	13,824	4.8990	2.8845	84	7,056	592,704	9.1652	4.3795	144	20,736	2,985,984	12.0000	5.2415
25	625	15,625	5.0000	2.9240	85	7,225	614,125	9.2195	4.3968	145	21,025	3,048,625	12.0416	5.2536
26	676	17,576	5.0990	2.9625	86	7,396	636,056	9.2736	4.4140	146	21,316	3,112,136	12.0830	5.2656
27	729	19,683	5.1962	3.0000	87	7,569	658,503	9.3274	4.4310	147	21,609	3,176,523	12.1244	5.2776
28	784	21,952	5.2915	3.0366	88	7,744	681,472	9.3808	4.4480	148	21,904	3,241,792	12.1655	5.2896
29	841	24,389	5.3852	3.0723	89	7,921	704,969	9.4340	4.4647	149	22,201	3,307,949	12.2066	5.3015
30	900	27,000	5.4772	3.1072	90	8,100	729,000	9.4868	4.4814	150	22,500	3,375,000	12.2474	5.3133
31	961	29,791	5.5678	3.1414	91	8,281	753,571	9.5394	4.4979	151	22,801	3,442,951	12.2882	5.3251
32	1,024	32,768	5.6569	3.1748	92	8,464	778,688	9.5917	4.5144	152	23,104	3,511,808	12.3288	5.3368
33	1,089	35,937	5.7446	3.2075	93	8,649	804,357	9.6437	4.5307	153	23,409	3,581,577	12.3693	5.3485
34	1,156	39,304	5.8310	3.2396	94	8,836	830,584	9.6954	4.5468	154	23,716	3,652,264	12.4097	5.3601
35	1,225	42,875	5.9161	3.2711	95	9,025	857,375	9.7468	4.5629	155	24,025	3,723,875	12.4499	5.3717
36	1,296	46,656	6.0000	3.3019	96	9,216	884,736	9.7980	4.5789	156	24,336	3,796,416	12.4900	5.3832
37	1,369	50,653	6.0828	3.3322	97	9,409	912,673	9.8489	4.5947	157	24,649	3,869,893	12.5300	5.3947
38	1,444	54,872	6.1644	3.3620	98	9,604	941,192	9.8995	4.6104	158	24,964	3,944,312	12.5698	5.4061
39	1,521	59,319	6.2450	3.3912	99	9,801	970,299	9.9499	4.6261	159	25,281	4,019,679	12.6095	5.4175
40	1,600	64,000	6.3246	3.4200	100	10,000	1,000,000	10.0000	4.6416	160	25,600	4,096,000	12.6491	5.4288
41	1,681	68,921	6.4031	3.4482	101	10,201	1,030,301	10.0499	4.6570	161	25,921	4,173,281	12.6886	5.4401
42	1,764	74,088	6.4807	3.4760	102	10,404	1,061,208	10.0995	4.6723	162	26,244	4,251,528	12.7279	5.4514
43	1,849	79,507	6.5574	3.5034	103	10,609	1,092,727	10.1489	4.6875	163	26,569	4,330,747	12.7671	5.4626
44	1,936	85,184	6.6332	3.5303	104	10,816	1,124,864	10.1980	4.7027	164	26,896	4,410,944	12.8062	5.4737
45	2,025	91,125	6.7082	3.5569	105	11,025	1,157,625	10.2470	4.7177	165	27,225	4,492,125	12.8452	5.4848
46	2,116	97,336	6.7823	3.5830	106	11,236	1,191,016	10.2956	4.7326	166	27,556	4,574,296	12.8841	5.4959
47	2,209	103,823	6.8557	3.6088	107	11,449	1,225,043	10.3441	4.7475	167	27,889	4,657,463	12.9228	5.5069
48	2,304	110,592	6.9282	3.6342	108	11,664	1,259,712	10.3923	4.7622	168	28,224	4,741,632	12.9615	5.5178
49	2,401	117,649	7.0000	3.6593	109	11,881	1,295,029	10.4403	4.7769	169	28,561	4,826,809	13.0000	5.5288
50	2,500	125,000	7.0711	3.6840	110	12,100	1,331,000	10.4881	4.7914	170	28,900	4,913,000	13.0384	5.5397
51	2,601	132,651	7.1414	3.7084	111	12,321	1,367,631	10.5357	4.8059	171	29,241	5,000,211	13.0767	5.5505
52	2,704	140,608	7.2111	3.7325	112	12,544	1,404,928	10.5830	4.8203	172	29,584	5,088,448	13.1149	5.5613
53	2,809	148,877	7.2801	3.7563	113	12,769	1,442,897	10.6301	4.8346	173	29,929	5,177,717	13.1529	5.5721
54	2,916	157,464	7.3485	3.7798	114	12,996	1,481,544	10.6771	4.8488	174	30,276	5,268,024	13.1909	5.5828
55	3,025	166,375	7.4162	3.8030	115	13,225	1,520,875	10.7238	4.8629	175	30,625	5,359,375	13.2288	5.5934
56	3,136	175,616	7.4833	3.8259	116	13,456	1,560,896	10.7703	4.8770	176	30,976	5,451,776	13.2665	5.6041
57	3,249	185,193	7.5498	3.8485	117	13,689	1,601,613	10.8167	4.8910	177	31,329	5,545,233	13.3041	5.6147
58	3,364	195,112	7.6158	3.8709	118	13,924	1,643,032	10.8628	4.9049	178	31,684	5,639,752	13.3417	5.6252
59	3,481	205,379	7.6811	3.8930	119	14,161	1,685,159	10.9087	4.9187	179	32,041	5,735,339	13.3791	5.6357

Table 4.16 continued

Table 4.16. Square, cube, square root, and cube root of numbers.

Table 4.16 continued

No.	Square	Cube	Square Root	Cube Root	No.	Square	Cube	Square Root	Cube Root	No.	Square	Cube	Square Root	Cube Root
180	32,400	5,832,000	13.4164	5.6462	240	57,600	13,824,000	15.4919	6.2145	300	90,000	27,000,000	17.3205	6.6943
181	32,761	5,929,741	13.4536	5.6567	241	58,081	13,997,521	15.5242	6.2231	301	90,601	27,270,901	17.3494	6.7018
182	33,124	6,028,568	13.4907	5.6671	242	58,564	14,172,488	15.5563	6.2317	302	91,204	27,543,608	17.3781	6.7092
183	33,489	6,128,487	13.5277	5.6774	243	59,049	14,348,907	15.5885	6.2403	303	91,809	27,818,127	17.4069	6.7166
184	33,856	6,229,504	13.5647	5.6877	244	59,536	14,526,784	15.6205	6.2488	304	92,416	28,094,464	17.4356	6.7240
185	34,225	6,331,625	13.6015	5.6980	245	60,025	14,706,125	15.6525	6.2573	305	93,025	28,372,625	17.4642	6.7313
186	34,596	6,434,856	13.6382	5.7083	246	60,516	14,886,936	15.6844	6.2658	306	93,636	28,652,616	17.4929	6.7387
187	34,969	6,539,203	13.6748	5.7185	247	61,009	15,069,223	15.7162	6.2743	307	94,249	28,934,443	17.5214	6.7460
188	35,344	6,644,672	13.7113	5.7287	248	61,504	15,252,992	15.7480	6.2828	308	94,864	29,218,112	17.5499	6.7533
189	35,721	6,751,269	13.7477	5.7388	249	62,001	15,438,249	15.7797	6.2912	309	95,481	29,503,629	17.5784	6.7606
190	36,100	6,859,000	13.7840	5.7489	250	62,500	15,625,000	15.8114	6.2996	310	96,100	29,791,000	17.6068	6.7679
191	36,481	6,967,871	13.8203	5.7590	251	63,001	15,813,251	15.8430	6.3080	311	96,721	30,080,231	17.6352	6.7752
192	36,864	7,077,888	13.8564	5.7690	252	63,504	16,003,008	15.8745	6.3164	312	97,344	30,371,328	17.6635	6.7824
193	37,249	7,189,057	13.8924	5.7790	253	64,009	16,194,277	15.9060	6.3247	313	97,969	30,664,297	17.6918	6.7897
194	37,636	7,301,384	13.9284	5.7890	254	64,516	16,387,064	15.9374	6.3330	314	98,596	30,959,144	17.7200	6.7969
195	38,025	7,414,875	13.9642	5.7989	255	65,025	16,581,375	15.9687	6.3413	315	99,225	31,255,875	17.7482	6.8041
196	38,416	7,529,536	14.0000	5.8088	256	65,536	16,777,216	16.0000	6.3496	316	99,856	31,554,496	17.7764	6.8113
197	38,809	7,645,373	14.0357	5.8186	257	66,049	16,974,593	16.0312	6.3579	317	100,489	31,855,013	17.8045	6.8185
198	39,204	7,762,392	14.0712	5.8285	258	66,564	17,173,512	16.0624	6.3661	318	101,124	32,157,432	17.8326	6.8256
199	39,601	7,880,599	14.1067	5.8383	259	67,081	17,373,979	16.0935	6.3743	319	101,761	32,461,759	17.8606	6.8328
200	40,000	8,000,000	14.1421	5.8480	260	67,600	17,576,000	16.1245	6.3825	320	102,400	32,768,000	17.8885	6.8399
201	40,401	8,120,601	14.1774	5.8578	261	68,121	17,779,581	16.1555	6.3907	321	103,041	33,076,161	17.9165	6.8470
202	40,804	8,242,408	14.2127	5.8675	262	68,644	17,984,728	16.1864	6.3988	322	103,684	33,386,248	17.9444	6.8541
203	41,209	8,365,427	14.2478	5.8771	263	69,169	18,191,447	16.2173	6.4070	323	104,329	33,698,267	17.9722	6.8612
204	41,616	8,489,664	14.2829	5.8868	264	69,696	18,399,744	16.2481	6.4151	324	104,976	34,012,224	18.0000	6.8683
205	42,025	8,615,125	14.3178	5.8964	265	70,225	18,609,625	16.2788	6.4232	325	105,625	34,328,125	18.0278	6.8753
206	42,436	8,741,816	14.3527	5.9059	266	70,756	18,821,096	16.3095	6.4312	326	106,276	34,645,976	18.0555	6.8824
207	42,849	8,869,743	14.3875	5.9155	267	71,289	19,034,163	16.3401	6.4393	327	106,929	34,965,783	18.0831	6.8894
208	43,264	8,998,912	14.4222	5.9250	268	71,824	19,248,832	16.3707	6.4473	328	107,584	35,287,552	18.1108	6.8964
209	43,681	9,129,329	14.4568	5.9345	269	72,361	19,465,109	16.4012	6.4553	329	108,241	35,611,289	18.1384	6.9034
210	44,100	9,261,000	14.4914	5.9439	270	72,900	19,683,000	16.4317	6.4633	330	108,900	35,937,000	18.1659	6.9104
211	44,521	9,393,931	14.5258	5.9533	271	73,441	19,902,511	16.4621	6.4713	331	109,561	36,264,691	18.1934	6.9174
212	44,944	9,528,128	14.5602	5.9627	272	73,984	20,123,648	16.4924	6.4793	332	110,224	36,594,368	18.2209	6.9244
213	45,369	9,663,597	14.5945	5.9721	273	74,529	20,346,417	16.5227	6.4872	333	110,889	36,926,037	18.2483	6.9313
214	45,796	9,800,344	14.6287	5.9814	274	75,076	20,570,824	16.5529	6.4951	334	111,556	37,259,704	18.2757	6.9382
215	46,225	9,938,375	14.6629	5.9907	275	75,625	20,796,875	16.5831	6.5030	335	112,225	37,595,375	18.3030	6.9451
216	46,656	10,077,696	14.6969	6.0000	276	76,176	21,024,576	16.6132	6.5108	336	112,896	37,933,056	18.3303	6.9521
217	47,089	10,218,313	14.7309	6.0092	277	76,729	21,253,933	16.6433	6.5187	337	113,569	38,272,753	18.3576	6.9589
218	47,524	10,360,232	14.7648	6.0185	278	77,284	21,484,952	16.6733	6.5265	338	114,244	38,614,472	18.3848	6.9658
219	47,961	10,503,459	14.7986	6.0277	279	77,841	21,717,639	16.7033	6.5343	339	114,921	38,958,219	18.4120	6.9727
220	48,400	10,648,000	14.8324	6.0368	280	78,400	21,952,000	16.7332	6.5421	340	115,600	39,304,000	18.4391	6.9795
221	48,841	10,793,861	14.8661	6.0459	281	78,961	22,188,041	16.7631	6.5499	341	116,281	39,651,821	18.4662	6.9864
222	49,284	10,941,048	14.8997	6.0550	282	79,524	22,425,768	16.7929	6.5577	342	116,964	40,001,688	18.4932	6.9932
223	49,729	11,089,567	14.9332	6.0641	283	80,089	22,665,187	16.8226	6.5654	343	117,649	40,353,607	18.5203	7.0000
224	50,176	11,239,424	14.9666	6.0732	284	80,656	22,906,304	16.8523	6.5731	344	118,336	40,707,584	18.5472	7.0068
225	50,625	11,390,625	15.0000	6.0822	285	81,225	23,149,125	16.8819	6.5808	345	119,025	41,063,625	18.5742	7.0136
226	51,076	11,543,176	15.0333	6.0912	286	81,796	23,393,656	16.9115	6.5885	346	119,716	41,421,736	18.6011	7.0203
227	51,529	11,697,083	15.0665	6.1002	287	82,369	23,639,903	16.9411	6.5962	347	120,409	41,781,923	18.6279	7.0271
228	51,984	11,852,352	15.0997	6.1091	288	82,944	23,887,872	16.9706	6.6039	348	121,104	42,144,192	18.6548	7.0338
229	52,441	12,008,989	15.1327	6.1180	289	83,521	24,137,569	17.0000	6.6115	349	121,801	42,508,549	18.6815	7.0406
230	52,900	12,167,000	15.1658	6.1269	290	84,100	24,389,000	17.0294	6.6191	350	122,500	42,875,000	18.7083	7.0473
231	53,361	12,326,391	15.1987	6.1358	291	84,681	24,642,171	17.0587	6.6267	351	123,201	43,243,551	18.7350	7.0540
232	53,824	12,487,168	15.2315	6.1446	292	85,264	24,897,088	17.0880	6.6343	352	123,904	43,614,208	18.7617	7.0607
233	54,289	12,649,337	15.2643	6.1534	293	85,849	25,153,757	17.1172	6.6419	353	124,609	43,986,977	18.7883	7.0674
234	54,756	12,812,904	15.2971	6.1622	294	86,436	25,412,184	17.1464	6.6494	354	125,316	44,361,864	18.8149	7.0740
235	55,225	12,977,875	15.3297	6.1710	295	87,025	25,672,375	17.1756	6.6569	355	126,025	44,738,875	18.8414	7.0807
236	55,696	13,144,256	15.3623	6.1797	296	87,616	25,934,336	17.2047	6.6644	356	126,736	45,118,016	18.8680	7.0873
237	56,169	13,312,053	15.3948	6.1885	297	88,209	26,198,073	17.2337	6.6719	357	127,449	45,499,293	18.8944	7.0940
238	56,644	13,481,272	15.4272	6.1972	298	88,804	26,463,592	17.2627	6.6794	358	128,164	45,882,712	18.9209	7.1006
239	57,121	13,651,919	15.4596	6.2058	299	89,401	26,730,899	17.2916	6.6869	359	128,881	46,268,279	18.9473	7.1072

Table 4.16 continued

Table 4.16 continued

No.	Square	Cube	Square Root	Cube Root	No.	Square	Cube	Square Root	Cube Root	No.	Square	Cube	Square Root	Cube Root
360	129,600	46,656,000	18.9737	7.1138	420	176,400	74,088,000	20.4939	7.4889	480	230,400	110,592,000	21.9089	7.8297
361	130,321	47,045,881	19.0000	7.1204	421	177,241	74,618,461	20.5183	7.4948	481	231,361	111,284,641	21.9317	7.8352
362	131,044	47,437,928	19.0263	7.1269	422	178,084	75,151,448	20.5426	7.5007	482	232,324	111,980,168	21.9545	7.8406
363	131,769	47,832,147	19.0526	7.1335	423	178,929	75,686,967	20.5670	7.5067	483	233,289	112,678,587	21.9773	7.8460
364	132,496	48,228,544	19.0788	7.1400	424	179,776	76,225,024	20.5913	7.5126	484	234,256	113,379,904	22.0000	7.8514
365	133,225	48,627,125	19.1050	7.1466	425	180,625	76,765,625	20.6155	7.5185	485	235,225	114,084,125	22.0227	7.8568
366	133,956	49,027,896	19.1311	7.1531	426	181,476	77,308,776	20.6398	7.5244	486	236,196	114,791,256	22.0454	7.8622
367	134,689	49,430,863	19.1572	7.1596	427	182,329	77,854,483	20.6640	7.5302	487	237,169	115,501,303	22.0681	7.8676
368	135,424	49,836,032	19.1833	7.1661	428	183,184	78,402,752	20.6882	7.5361	488	238,144	116,214,272	22.0907	7.8730
369	136,161	50,243,409	19.2094	7.1726	429	184,041	78,953,589	20.7123	7.5420	489	239,121	116,930,169	22.1133	7.8784
370	136,900	50,653,000	19.2354	7.1791	430	184,900	79,507,000	20.7364	7.5478	490	240,100	117,649,000	22.1359	7.8837
371	137,641	51,064,811	19.2614	7.1855	431	185,761	80,062,991	20.7605	7.5537	491	241,081	118,370,771	22.1585	7.8891
372	138,384	51,478,848	19.2873	7.1920	432	186,624	80,621,568	20.7846	7.5595	492	242,064	119,095,488	22.1811	7.8944
373	139,129	51,895,117	19.3132	7.1984	433	187,489	81,182,737	20.8087	7.5654	493	243,049	119,823,157	22.2036	7.8998
374	139,876	52,313,624	19.3391	7.2048	434	188,356	81,746,504	20.8327	7.5712	494	244,036	120,553,784	22.2261	7.9051
375	140,625	52,734,375	19.3649	7.2112	435	189,225	82,312,875	20.8567	7.5770	495	245,025	121,287,375	22.2486	7.9105
376	141,376	53,157,376	19.3907	7.2177	436	190,096	82,881,856	20.8806	7.5828	496	246,016	122,023,936	22.2711	7.9158
377	142,129	53,582,633	19.4165	7.2240	437	190,969	83,453,453	20.9045	7.5886	497	247,009	122,763,473	22.2935	7.9211
378	142,884	54,010,152	19.4422	7.2304	438	191,844	84,027,672	20.9284	7.5944	498	248,004	123,505,992	22.3159	7.9264
379	143,641	54,439,939	19.4679	7.2368	439	192,721	84,604,519	20.9523	7.6001	499	249,001	124,251,499	22.3383	7.9317
380	144,400	54,872,000	19.4936	7.2432	440	193,600	85,184,000	20.9762	7.6059	500	250,000	125,000,000	22.3607	7.9370
381	145,161	55,306,341	19.5192	7.2495	441	194,481	85,766,121	21.0000	7.6117	501	251,001	125,751,501	22.3830	7.9423
382	145,924	55,742,968	19.5448	7.2558	442	195,364	86,350,888	21.0238	7.6174	502	252,004	126,506,008	22.4054	7.9476
383	146,689	56,181,887	19.5704	7.2622	443	196,249	86,938,307	21.0476	7.6232	503	253,009	127,263,527	22.4277	7.9528
384	147,456	56,623,104	19.5959	7.2685	444	197,136	87,528,384	21.0713	7.6289	504	254,016	128,024,064	22.4499	7.9581
385	148,225	57,066,625	19.6214	7.2748	445	198,025	88,121,125	21.0950	7.6346	505	255,025	128,787,625	22.4722	7.9634
386	148,996	57,512,456	19.6469	7.2811	446	198,916	88,716,536	21.1187	7.6403	506	256,036	129,554,216	22.4944	7.9686
387	149,769	57,960,603	19.6723	7.2874	447	199,809	89,314,623	21.1424	7.6460	507	257,049	130,323,843	22.5167	7.9739
388	150,544	58,411,072	19.6977	7.2936	448	200,704	89,915,392	21.1660	7.6517	508	258,064	131,096,512	22.5389	7.9791
389	151,321	58,863,869	19.7231	7.2999	449	201,601	90,518,849	21.1896	7.6574	509	259,081	131,872,229	22.5610	7.9843
390	152,100	59,319,000	19.7484	7.3061	450	202,500	91,125,000	21.2132	7.6631	510	260,100	132,651,000	22.5832	7.9896
391	152,881	59,776,471	19.7737	7.3124	451	203,401	91,733,851	21.2368	7.6688	511	261,121	133,432,831	22.6053	7.9948
392	153,664	60,236,288	19.7990	7.3186	452	204,304	92,345,408	21.2603	7.6744	512	262,144	134,217,728	22.6274	8.0000
393	154,449	60,698,457	19.8242	7.3248	453	205,209	92,959,677	21.2838	7.6801	513	263,169	135,005,697	22.6495	8.0052
394	155,236	61,162,984	19.8494	7.3310	454	206,116	93,576,664	21.3073	7.6857	514	264,196	135,796,744	22.6716	8.0104
395	156,025	61,629,875	19.8746	7.3372	455	207,025	94,196,375	21.3307	7.6914	515	265,225	136,590,875	22.6936	8.0156
396	156,816	62,099,136	19.8997	7.3434	456	207,936	94,818,816	21.3542	7.6970	516	266,256	137,388,096	22.7156	8.0208
397	157,609	62,570,773	19.9249	7.3496	457	208,849	95,443,993	21.3776	7.7026	517	267,289	138,188,413	22.7376	8.0260
398	158,404	63,044,792	19.9499	7.3558	458	209,764	96,071,912	21.4009	7.7082	518	268,324	138,991,832	22.7596	8.0311
399	159,201	63,521,199	19.9750	7.3619	459	210,681	96,702,579	21.4243	7.7138	519	269,361	139,798,359	22.7816	8.0363
400	160,000	64,000,000	20.0000	7.3681	460	211,600	97,336,000	21.4476	7.7194	520	270,400	140,608,000	22.8035	8.0415
401	160,801	64,481,201	20.0250	7.3742	461	212,521	97,972,181	21.4709	7.7250	521	271,441	141,420,761	22.8254	8.0466
402	161,604	64,964,808	20.0499	7.3803	462	213,444	98,611,128	21.4942	7.7306	522	272,484	142,236,648	22.8473	8.0517
403	162,409	65,450,827	20.0749	7.3864	463	214,369	99,252,847	21.5174	7.7362	523	273,529	143,055,667	22.8692	8.0569
404	163,216	65,939,264	20.0998	7.3925	464	215,296	99,897,344	21.5407	7.7418	524	274,576	143,877,824	22.8910	8.0620
405	164,025	66,430,125	20.1246	7.3986	465	216,225	100,544,625	21.5639	7.7473	525	275,625	144,703,125	22.9129	8.0671
406	164,836	66,923,416	20.1494	7.4047	466	217,156	101,194,696	21.5870	7.7529	526	276,676	145,531,576	22.9347	8.0723
407	165,649	67,419,143	20.1742	7.4108	467	218,089	101,847,563	21.6102	7.7584	527	277,729	146,363,183	22.9565	8.0774
408	166,464	67,917,312	20.1990	7.4169	468	219,024	102,503,232	21.6333	7.7639	528	278,784	147,197,952	22.9783	8.0825
409	167,281	68,417,929	20.2237	7.4229	469	219,961	103,161,709	21.6564	7.7695	529	279,841	148,035,889	23.0000	8.0876
410	168,100	68,921,000	20.2485	7.4290	470	220,900	103,823,000	21.6795	7.7750	530	280,900	148,877,000	23.0217	8.0927
411	168,921	69,426,531	20.2731	7.4350	471	221,841	104,487,111	21.7025	7.7805	531	281,961	149,721,291	23.0434	8.0978
412	169,744	69,934,528	20.2978	7.4410	472	222,784	105,154,048	21.7256	7.7860	532	283,024	150,568,768	23.0651	8.1028
413	170,569	70,444,997	20.3224	7.4470	473	223,729	105,823,817	21.7486	7.7915	533	284,089	151,419,437	23.0868	8.1079
414	171,396	70,957,944	20.3470	7.4530	474	224,676	106,496,424	21.7715	7.7970	534	285,156	152,273,304	23.1084	8.1130
415	172,225	71,473,375	20.3715	7.4590	475	225,625	107,171,875	21.7945	7.8025	535	286,225	153,130,375	23.1301	8.1180
416	173,056	71,991,296	20.3961	7.4650	476	226,576	107,850,176	21.8174	7.8079	536	287,296	153,990,656	23.1517	8.1231
417	173,889	72,511,713	20.4206	7.4710	477	227,529	108,531,333	21.8403	7.8134	537	288,369	154,854,153	23.1733	8.1281
418	174,724	73,034,632	20.4450	7.4770	478	228,484	109,215,352	21.8632	7.8188	538	289,444	155,720,872	23.1948	8.1332
419	175,561	73,560,059	20.4695	7.4829	479	229,441	109,902,239	21.8861	7.8243	539	290,521	156,590,819	23.2164	8.1382

Table 4.16 continued

Table 4.16 continued

No.	Square	Cube	Square Root	Cube Root	No.	Square	Cube	Square Root	Cube Root	No.	Square	Cube	Square Root	Cube Root
540	291,600	157,464,000	23.2379	8.1433	600	360,000	216,000,000	24.4949	8.4343	660	435,600	287,496,000	25.6905	8.7066
541	292,681	158,340,421	23.2594	8.1483	601	361,201	217,081,801	24.5153	8.4390	661	436,921	288,804,781	25.7099	8.7110
542	293,764	159,220,088	23.2809	8.1533	602	362,404	218,167,208	24.5357	8.4437	662	438,244	290,117,528	25.7294	8.7154
543	294,849	160,103,007	23.3024	8.1583	603	363,609	219,256,227	24.5561	8.4484	663	439,569	291,434,247	25.7488	8.7198
544	295,936	160,989,184	23.3238	8.1633	604	364,816	220,348,864	24.5764	8.4530	664	440,896	292,754,944	25.7682	8.7241
545	297,025	161,878,625	23.3452	8.1683	605	366,025	221,445,125	24.5967	8.4577	665	442,225	294,079,625	25.7876	8.7285
546	298,116	162,771,336	23.3666	8.1733	606	367,236	222,545,016	24.6171	8.4623	666	443,556	295,408,296	25.8070	8.7329
547	299,209	163,667,323	23.3880	8.1783	607	368,449	223,648,543	24.6374	8.4670	667	444,889	296,740,963	25.8263	8.7373
548	300,304	164,566,592	23.4094	8.1833	608	369,664	224,755,712	24.6577	8.4716	668	446,224	298,077,632	25.8457	8.7416
549	301,401	165,469,149	23.4307	8.1882	609	370,881	225,866,529	24.6779	8.4763	669	447,561	299,418,309	25.8650	8.7460
550	302,500	166,375,000	23.4521	8.1932	610	372,100	226,981,000	24.6982	8.4809	670	448,900	300,763,000	25.8844	8.7503
551	303,601	167,284,151	23.4734	8.1982	611	373,321	228,099,131	24.7184	8.4856	671	450,241	302,111,711	25.9037	8.7547
552	304,704	168,196,608	23.4947	8.2031	612	374,544	229,220,928	24.7386	8.4902	672	451,584	303,464,448	25.9230	8.7590
553	305,809	169,112,377	23.5160	8.2081	613	375,769	230,346,397	24.7588	8.4948	673	452,929	304,821,217	25.9422	8.7634
554	306,916	170,031,464	23.5372	8.2130	614	376,996	231,475,544	24.7790	8.4994	674	454,276	306,182,024	25.9615	8.7677
555	308,025	170,953,875	23.5584	8.2180	615	378,225	232,608,375	24.7992	8.5040	675	455,625	307,546,875	25.9808	8.7721
556	309,136	171,879,616	23.5797	8.2229	616	379,456	233,744,896	24.8193	8.5086	676	456,976	308,915,776	26.0000	8.7764
557	310,249	172,808,693	23.6008	8.2278	617	380,689	234,885,113	24.8395	8.5132	677	458,329	310,288,733	26.0192	8.7807
558	311,364	173,741,112	23.6220	8.2327	618	381,924	236,029,032	24.8596	8.5178	678	459,684	311,665,752	26.0384	8.7850
559	312,481	174,676,879	23.6432	8.2377	619	383,161	237,176,659	24.8797	8.5224	679	461,041	313,046,839	26.0576	8.7893
560	313,600	175,616,000	23.6643	8.2426	620	384,400	238,328,000	24.8998	8.5270	680	462,400	314,432,000	26.0768	8.7937
561	314,721	176,558,481	23.6854	8.2475	621	385,641	239,483,061	24.9199	8.5316	681	463,761	315,821,241	26.0960	8.7980
562	315,844	177,504,328	23.7065	8.2524	622	386,884	240,641,848	24.9399	8.5362	682	465,124	317,214,568	26.1151	8.8023
563	316,969	178,453,547	23.7276	8.2573	623	388,129	241,804,367	24.9600	8.5408	683	466,489	318,611,987	26.1343	8.8066
564	318,096	179,406,144	23.7487	8.2621	624	389,376	242,970,624	24.9800	8.5453	684	467,856	320,013,504	26.1534	8.8109
565	319,225	180,362,125	23.7697	8.2670	625	390,625	244,140,625	25.0000	8.5499	685	469,225	321,419,125	26.1725	8.8152
566	320,356	181,321,496	23.7908	8.2719	626	391,876	245,314,376	25.0200	8.5544	686	470,596	322,828,856	26.1916	8.8194
567	321,489	182,284,263	23.8118	8.2763	627	393,129	246,491,883	25.0400	8.5590	687	471,969	324,242,703	26.2107	8.8237
568	322,624	183,250,432	23.8328	8.2816	628	394,384	247,673,152	25.0599	8.5635	688	473,344	325,660,672	26.2298	8.8280
569	323,761	184,220,009	23.8537	8.2865	629	395,641	248,858,189	25.0799	8.5681	689	474,721	327,082,769	26.2488	8.8323
570	324,900	185,193,000	23.8747	8.2913	630	396,900	250,047,000	25.0998	8.5726	690	476,100	328,509,000	26.2679	8.8366
571	326,041	186,169,411	23.8956	8.2962	631	398,161	251,239,591	25.1197	8.5772	691	477,481	329,939,371	26.2869	8.8408
572	327,184	187,149,248	23.9165	8.3010	632	399,424	252,435,968	25.1396	8.5817	692	478,864	331,373,888	26.3059	8.8451
573	328,329	188,132,517	23.9374	8.3059	633	400,689	253,636,137	25.1595	8.5862	693	480,249	332,812,557	26.3249	8.8493
574	329,476	189,119,224	23.9583	8.3107	634	401,956	254,840,104	25.1794	8.5907	694	481,636	334,255,384	26.3439	8.8536
575	330,625	190,109,375	23.9792	8.3155	635	403,225	256,047,875	25.1992	8.5952	695	483,025	335,702,375	26.3629	8.8578
576	331,776	191,102,976	24.0000	8.3203	636	404,496	257,259,456	25.2190	8.5997	696	484,416	337,153,536	26.3818	8.8621
577	332,929	192,100,033	24.0208	8.3251	637	405,769	258,474,853	25.2389	8.6043	697	485,809	338,608,873	26.4008	8.8663
578	334,084	193,100,552	24.0416	8.3300	638	407,044	259,694,072	25.2587	8.6088	698	487,204	340,068,392	26.4197	8.8706
579	335,241	194,104,539	24.0624	8.3348	639	408,321	260,917,119	25.2784	8.6132	699	488,601	341,532,099	26.4386	8.8748
580	336,400	195,112,000	24.0832	8.3396	640	409,600	262,144,000	25.2982	8.6177	700	490,000	343,000,000	26.4575	8.8790
581	337,561	196,122,941	24.1039	8.3443	641	410,881	263,374,721	25.3180	8.6222	701	491,401	344,472,101	26.4764	8.8833
582	338,724	197,137,368	24.1247	8.3491	642	412,164	264,609,288	25.3377	8.6267	702	492,804	345,948,408	26.4953	8.8875
583	339,889	198,155,287	24.1454	8.3539	643	413,449	265,847,707	25.3574	8.6312	703	494,209	347,428,927	26.5141	8.8917
584	341,056	199,176,704	24.1661	8.3587	644	414,736	267,089,984	25.3772	8.6357	704	495,616	348,913,664	26.5330	8.8959
585	342,225	200,201,625	24.1868	8.3634	645	416,025	268,336,125	25.3969	8.6401	705	497,025	350,402,625	26.5518	8.9001
586	343,396	201,230,056	24.2074	8.3682	646	417,316	269,586,136	25.4165	8.6446	706	498,436	351,895,816	26.5707	8.9043
587	344,569	202,262,003	24.2281	8.3730	647	418,609	270,840,023	25.4362	8.6490	707	499,849	353,393,243	26.5895	8.9085
588	345,744	203,297,472	24.2487	8.3777	648	419,904	272,097,792	25.4558	8.6535	708	501,264	354,894,912	26.6083	8.9127
589	346,921	204,336,469	24.2693	8.3825	649	421,201	273,359,449	25.4755	8.6579	709	502,681	356,400,829	26.6271	8.9169
590	348,100	205,379,000	24.2899	8.3872	650	422,500	274,625,000	25.4951	8.6624	710	504,100	357,911,000	26.6458	8.9211
591	349,281	206,425,071	24.3105	8.3919	651	423,801	275,894,451	25.5147	8.6668	711	505,521	359,425,431	26.6646	8.9253
592	350,464	207,474,688	24.3311	8.3967	652	425,104	277,167,808	25.5343	8.6713	712	506,944	360,944,128	26.6833	8.9295
593	351,649	208,527,857	24.3516	8.4014	653	426,409	278,445,077	25.5539	8.6757	713	508,369	362,467,097	26.7021	8.9337
594	352,836	209,584,584	24.3721	8.4061	654	427,716	279,726,264	25.5734	8.6801	714	509,796	363,994,344	26.7208	8.9378
595	354,025	210,644,875	24.3926	8.4108	655	429,025	281,011,375	25.5930	8.6845	715	511,225	365,525,875	26.7395	8.9420
596	355,216	211,708,736	24.4131	8.4155	656	430,336	282,300,416	25.6125	8.6890	716	512,656	367,061,696	26.7582	8.9462
597	356,409	212,776,173	24.4336	8.4202	657	431,649	283,593,393	25.6320	8.6934	717	514,089	368,601,813	26.7769	8.9503
598	357,604	213,847,192	24.4540	8.4249	658	432,964	284,890,312	25.6515	8.6978	718	515,524	370,146,232	26.7955	8.9515
599	358,801	214,921,799	24.4745	8.4296	659	434,281	286,191,179	25.6710	8.7022	719	516,961	371,694,959	26.8142	8.9587

Table 4.16 continued

Table 4.16 continued

No.	Square	Cube	Square Root	Cube Root
720	518,400	373,248,000	26.8328	8.9628
721	519,841	374,805,361	26.8514	8.9670
722	521,284	376,367,048	26.8701	8.9711
723	522,729	377,933,067	26.8887	8.9752
724	524,176	379,503,424	26.9072	8.9794
725	525,625	381,078,125	26.9258	8.9835
726	527,076	382,657,176	26.9444	8.9876
727	528,529	384,240,583	26.9629	8.9918
728	529,984	385,828,352	26.9815	8.9959
729	531,441	387,420,489	27.0000	9.0000
730	532,900	389,017,000	27.0185	9.0041
731	534,361	390,617,891	27.0370	9.0082
732	535,824	392,223,168	27.0555	9.0123
733	537,289	393,832,837	27.0740	9.0164
734	538,756	395,446,904	27.0924	9.0205
735	540,225	397,065,375	27.1109	9.0246
736	541,696	398,688,256	27.1293	9.0287
737	543,169	400,315,553	27.1477	9.0328
738	544,644	401,947,272	27.1662	9.0369
739	546,121	403,583,419	27.1846	9.0410
740	547,600	405,224,000	27.2029	9.0450
741	549,081	406,869,021	27.2213	9.0491
742	550,564	408,518,488	27.2397	9.0532
743	552,049	410,172,407	27.2580	9.0572
744	553,536	411,830,784	27.2764	9.0613
745	555,025	413,493,625	27.2947	9.0654
746	556,516	415,160,936	27.3130	9.0694
747	558,009	416,832,723	27.3313	9.0735
748	559,504	418,508,992	27.3496	9.0775
749	561,001	420,189,749	27.3679	9.0816
750	562,500	421,875,000	27.3861	9.0856
751	564,001	423,564,751	27.4044	9.0896
752	565,504	425,259,008	27.4226	9.0937
753	567,009	426,957,777	27.4408	9.0977
754	568,516	428,661,064	27.4591	9.1017
755	570,025	430,368,875	27.4773	9.1057
756	571,536	432,081,216	27.4955	9.1098
757	573,049	433,798,093	27.5136	9.1138
758	574,564	435,519,512	27.5318	9.1178
759	576,081	437,245,479	27.5500	9.1218
760	577,600	438,976,000	27.5681	9.1258
761	579,121	440,711,081	27.5862	9.1298
762	580,644	442,450,728	27.6043	9.1338
763	582,169	444,194,947	27.6225	9.1378
764	583,696	445,943,744	27.6405	9.1418
765	585,225	447,697,125	27.6586	9.1458
766	586,756	449,455,096	27.6767	9.1498
767	588,289	451,217,663	27.6948	9.1537
768	589,824	452,984,832	27.7128	9.1577
769	591,361	454,756,609	27.7308	9.1617
770	592,900	456,533,000	27.7489	9.1657
771	594,441	458,314,011	27.7669	9.1696
772	595,984	460,099,648	27.7849	9.1736
773	597,529	461,889,917	27.8029	9.1775
774	599,076	463,684,824	27.8209	9.1815
775	600,625	465,484,375	27.8388	9.1855
776	602,176	467,288,576	27.8568	9.1894
777	603,729	469,097,433	27.8747	9.1933
778	605,284	470,910,952	27.8927	9.1973
779	606,841	472,729,139	27.9106	9.2012

No.	Square	Cube	Square Root	Cube Root
780	608,400	474,552,000	27.9285	9.2052
781	609,961	476,379,541	27.9464	9.2091
782	611,524	478,211,768	27.9643	9.2130
783	613,089	480,048,687	27.9821	9.2170
784	614,656	481,890,304	28.0000	9.2209
785	616,225	483,736,625	28.0179	9.2248
786	617,796	485,587,656	28.0357	9.2287
787	619,369	487,443,403	28.0535	9.2326
788	620,944	489,303,872	28.0713	9.2365
789	622,521	491,169,069	28.0891	9.2404
790	624,100	493,039,000	28.1069	9.2443
791	625,681	494,913,671	28.1247	9.2482
792	627,264	496,793,088	28.1425	9.2521
793	628,849	498,677,257	28.1603	9.2560
794	630,436	500,566,184	28.1780	9.2599
795	632,025	502,459,875	28.1957	9.2638
796	633,616	504,358,336	28.2135	9.2677
797	635,209	506,261,573	28.2312	9.2716
798	636,804	508,169,592	28.2489	9.2754
799	638,401	510,082,399	28.2666	9.2793
800	640,000	512,000,000	28.2843	9.2832
801	641,601	513,922,401	28.3019	9.2870
802	643,204	515,849,608	28.3196	9.2909
803	644,809	517,781,627	28.3373	9.2948
804	646,416	519,718,464	28.3549	9.2986
805	648,025	521,660,125	28.3725	9.3025
806	649,636	523,606,616	28.3901	9.3063
807	651,249	525,557,943	28.4077	9.3102
808	652,864	527,514,112	28.4253	9.3140
809	654,481	529,475,129	28.4429	9.3179
810	656,100	531,441,000	28.4605	9.3217
811	657,721	533,411,731	28.4781	9.3255
812	659,344	535,387,328	28.4956	9.3294
813	660,969	537,367,797	28.5132	9.3332
814	662,596	539,353,144	28.5307	9.3370
815	664,225	541,343,375	28.5482	9.3408
816	665,856	543,338,496	28.5657	9.3447
817	667,489	545,338,513	28.5832	9.3485
818	669,124	547,343,432	28.6007	9.3523
819	670,761	549,353,259	28.6182	9.3561
820	672,400	551,368,000	28.6356	9.3599
821	674,041	553,387,661	28.6531	9.3637
822	675,684	555,412,248	28.6705	9.3675
823	677,329	557,441,767	28.6880	9.3713
824	678,976	559,476,224	28.7054	9.3751
825	680,625	561,515,625	28.7228	9.3789
826	682,276	563,559,976	28.7402	9.3827
827	683,929	565,609,283	28.7576	9.3865
828	685,584	567,663,552	28.7750	9.3902
829	687,241	569,722,789	28.7924	9.3940
830	688,900	571,787,000	28.8097	9.3978
831	690,561	573,856,191	28.8271	9.4016
832	692,224	575,930,368	28.8444	9.4053
833	693,889	578,009,537	28.8617	9.4091
834	695,556	580,093,704	28.8791	9.4129
835	697,225	582,182,875	28.8964	9.4166
836	698,896	584,277,056	28.9137	9.4204
837	700,569	586,376,253	28.9310	9.4241
838	702,244	588,480,472	28.9482	9.4279
839	703,921	590,589,719	28.9655	9.4316

No.	Square	Cube	Square Root	Cube Root
840	705,600	592,704,000	28.9828	9.4354
841	707,281	594,823,321	29.0000	9.4391
842	708,964	596,947,688	29.0172	9.4429
843	710,649	599,077,107	29.0345	9.4466
844	712,336	601,211,584	29.0517	9.4503
845	714,025	603,351,125	29.0689	9.4541
846	715,716	605,495,736	29.0861	9.4578
847	717,409	607,645,423	29.1033	9.4615
848	719,104	609,800,192	29.1204	9.4652
849	720,801	611,960,049	29.1376	9.4690
850	722,500	614,125,000	29.1548	9.4727
851	724,201	616,295,051	29.1719	9.4764
852	725,904	618,470,208	29.1890	9.4801
853	727,609	620,650,477	29.2062	9.4838
854	729,316	622,835,864	29.2233	9.4875
855	731,025	625,026,375	29.2404	9.4912
856	732,736	627,222,016	29.2575	9.4949
857	734,449	629,422,793	29.2746	9.4986
858	736,164	631,628,712	29.2916	9.5023
859	737,881	633,839,779	29.3087	9.5060
860	739,600	636,056,000	29.3258	9.5097
861	741,321	638,277,381	29.3428	9.5134
862	743,044	640,503,928	29.3598	9.5171
863	744,769	642,735,647	29.3769	9.5207
864	746,496	644,972,544	29.3939	9.5244
865	748,225	647,214,625	29.4109	9.5281
866	749,956	649,461,896	29.4279	9.5317
867	751,689	651,714,363	29.4449	9.5354
868	753,424	653,972,032	29.4618	9.5391
869	755,161	656,234,909	29.4788	9.5427
870	756,900	658,503,000	29.4958	9.5464
871	758,641	660,776,311	29.5127	9.5501
872	760,384	663,054,848	29.5296	9.5537
873	762,129	665,338,617	29.5466	9.5574
874	763,876	667,627,624	29.5635	9.5610
875	765,625	669,921,875	29.5804	9.5647
876	767,376	672,221,376	29.5973	9.5683
877	769,129	674,526,133	29.6142	9.5719
878	770,884	676,836,152	29.6311	9.5756
879	772,641	679,151,439	29.6479	9.5792
880	774,400	681,472,000	29.6648	9.5828
881	776,161	683,797,841	29.6816	9.5865
882	777,924	686,128,968	29.6985	9.5901
883	779,689	688,465,387	29.7153	9.5937
884	781,456	690,807,104	29.7321	9.5973
885	783,225	693,154,125	29.7489	9.6010
886	784,996	695,506,456	29.7658	9.6046
887	786,769	697,864,103	29.7825	9.6082
888	788,544	700,227,072	29.7993	9.6118
889	790,321	702,595,369	29.8161	9.6154
890	792,100	704,969,000	29.8329	9.6190
891	793,881	707,347,971	29.8496	9.6226
892	795,664	709,732,288	29.8664	9.6262
893	797,449	712,121,957	29.8831	9.6298
894	799,236	714,516,984	29.8998	9.6334
895	801,025	716,917,375	29.9166	9.6370
896	802,816	719,323,136	29.9333	9.6406
897	804,609	721,734,273	29.9500	9.6442
898	806,404	724,150,792	29.9666	9.6477
899	808,201	726,572,699	29.9833	9.6513

Table 4.16 continued

Table 4.16 continued

No.	Square	Cube	Square Root	Cube Root	No.	Square	Cube	Square Root	Cube Root	No.	Square	Cube	Square Root	Cube Root
900	810,000	729,000,000	30.0000	9.6549	934	872,356	814,780,504	30.5614	9.7750	967	935,089	904,231,063	31.0966	9.8888
901	811,801	731,432,701	30.0167	9.6585	935	874,225	817,400,375	30.5778	9.7785	968	937,024	907,039,232	31.1127	9.8922
902	813,604	733,870,808	30.0333	9.6620	936	876,096	820,025,856	30.5941	9.7819	969	938,961	909,853,209	31.1288	9.8956
903	815,409	736,314,327	30.0500	9.6656	937	877,969	822,656,953	30.6105	9.7854	970	940,900	912,673,000	31.1448	9.8990
904	817,216	738,763,264	30.0666	9.6692	938	879,844	825,293,672	30.6268	9.7889	971	942,841	915,498,611	31.1609	9.9024
905	819,025	741,217,625	30.0832	9.6727	939	881,721	827,936,019	30.6431	9.7924	972	944,784	918,330,048	31.1769	9.9058
906	820,836	743,677,416	30.0998	9.6763	940	883,600	830,584,000	30.6594	9.7959	973	946,729	921,167,317	31.1929	9.9092
907	822,649	746,142,643	30.1164	9.6799	941	885,481	833,237,621	30.6757	9.7993	974	948,676	924,010,424	31.2090	9.9126
908	824,464	748,613,312	30.1330	9.6834	942	887,364	835,896,888	30.6920	9.8028	975	950,625	926,859,375	31.2250	9.9160
909	826,281	751,089,429	30.1496	9.6870	943	889,249	838,561,807	30.7083	9.8063	976	952,576	929,714,176	31.2410	9.9194
910	828,100	753,571,000	30.1662	9.6905	944	891,136	841,232,384	30.7246	9.8097	977	954,529	932,574,833	31.2570	9.9227
911	829,921	756,058,031	30.1828	9.6941	945	893,025	843,908,625	30.7409	9.8132	978	956,484	935,441,352	31.2730	9.9261
912	831,744	758,550,528	30.1993	9.6976	946	894,916	846,590,536	30.7571	9.8167	979	958,441	938,313,739	31.2890	9.9295
913	833,569	761,048,497	30.2159	9.7012	947	896,809	849,278,123	30.7734	9.8201	980	960,400	941,192,000	31.3050	9.9329
914	835,396	763,551,944	30.2324	9.7047	948	898,704	851,971,392	30.7896	9.8236	981	962,361	944,076,141	31.3209	9.9363
915	837,225	766,060,875	30.2490	9.7082	949	900,601	854,670,349	30.8058	9.8270	982	964,324	946,966,168	31.3369	9.9396
916	839,056	768,575,296	30.2655	9.7118	950	902,500	857,375,000	30.8221	9.8305	983	966,289	949,862,087	31.3528	9.9430
917	840,889	771,095,213	30.2820	9.7153	951	904,401	860,085,351	30.8383	9.8339	984	968,256	952,763,904	31.3688	9.9464
918	842,724	773,620,632	30.2985	9.7188	952	906,304	862,801,408	30.8545	9.8374	985	970,225	955,671,625	31.3847	9.9497
919	844,561	776,151,559	30.3150	9.7224	953	908,209	865,523,177	30.8707	9.8408	986	972,196	958,585,256	31.4006	9.9531
920	846,400	778,688,000	30.3315	9.7259	954	910,116	868,250,664	30.8869	9.8443	987	974,169	961,504,803	31.4166	9.9565
921	848,241	781,229,961	30.3480	9.7294	955	912,025	870,983,875	30.9031	9.8477	988	976,144	964,430,272	31.4235	9.9598
922	850,084	783,777,448	30.3645	9.7329	956	913,936	873,722,816	30.9192	9.8511	989	978,121	967,361,669	31.4484	9.9632
923	851,929	786,330,467	30.3809	9.7364	957	915,849	876,467,493	30.9354	9.8546	990	980,100	970,299,000	31.4643	9.9666
924	853,776	788,889,024	30.3974	9.7400	958	917,764	879,217,912	30.9516	9.8580	991	982,081	973,242,271	31.4802	9.9699
925	855,625	791,453,125	30.4138	9.7435	959	919,681	881,974,079	30.9677	9.8614	992	984,064	976,191,488	31.4960	9.9733
926	857,476	794,022,776	30.4302	9.7470	960	921,600	884,736,000	30.9839	9.8648	993	986,049	979,146,657	31.5119	9.9766
927	859,329	796,597,983	30.4467	9.7505	961	923,521	887,503,681	31.0000	9.8683	994	988,036	982,107,784	31.5278	9.9800
928	861,184	799,178,752	30.4631	9.7540	962	925,444	890,277,128	31.0161	9.8717	995	990,025	985,074,875	31.5436	9.9833
929	863,041	801,765,089	30.4795	9.7575	963	927,369	893,056,347	31.0322	9.8751	996	992,016	988,047,936	31.5595	9.9866
930	864,900	804,357,000	30.4959	9.7610	964	929,296	895,841,344	31.0483	9.8785	997	994,009	991,026,973	31.5753	9.9900
931	866,761	806,954,491	30.5123	9.7645	965	931,225	898,632,125	31.0644	9.8819	998	996,004	994,011,992	31.5911	9.9933
932	868,624	809,557,568	30.5287	9.7680	966	933,156	901,428,696	31.0805	9.8854	999	998,001	997,002,999	31.6070	9.9967
933	870,489	812,166,237	30.5450	9.7715										

Degrees	SINES 0′	10′	20′	30′	40′	50′	60′	Cosines
0	0.00000	0.00291	0.00582	0.00873	0.01164	0.01454	0.01745	89
1	0.01745	0.02036	0.02327	0.02618	0.02908	0.03199	0.03490	88
2	0.03490	0.03781	0.04071	0.04362	0.04653	0.04943	0.05234	87
3	0.05234	0.05524	0.05814	0.06105	0.06395	0.06685	0.06976	86
4	0.06976	0.07266	0.07556	0.07846	0.08136	0.08426	0.08716	85
5	0.08716	0.09005	0.09295	0.09585	0.09874	0.10164	0.10453	84
6	0.10453	0.10742	0.11031	0.11320	0.11609	0.11898	0.12187	83
7	0.12187	0.12476	0.12764	0.13053	0.13341	0.13629	0.13917	82
8	0.13917	0.14205	0.14493	0.14781	0.15069	0.15356	0.15643	81
9	0.15643	0.15931	0.16218	0.16505	0.16792	0.17078	0.17365	80
10	0.17365	0.17651	0.17937	0.18224	0.18509	0.18795	0.19081	79
11	0.19081	0.19366	0.19652	0.19937	0.20222	0.20507	0.20791	78
12	0.20791	0.21076	0.21360	0.21644	0.21928	0.22212	0.22495	77
13	0.22495	0.22778	0.23062	0.23345	0.23627	0.23910	0.24192	76
14	0.24192	0.24474	0.24756	0.25038	0.25320	0.25601	0.25882	75
15	0.25882	0.26163	0.26443	0.26724	0.27004	0.27284	0.27564	74
16	0.27564	0.27843	0.28123	0.28402	0.28680	0.28959	0.29237	73
17	0.29237	0.29515	0.29793	0.30071	0.30348	0.30625	0.30902	72
18	0.30902	0.31178	0.31454	0.31730	0.32006	0.32282	0.32557	71
19	0.32557	0.32832	0.33106	0.33381	0.33655	0.33929	0.34202	70
20	0.34202	0.34475	0.34748	0.35021	0.35293	0.35565	0.35837	69
21	0.35837	0.36108	0.36379	0.36650	0.36921	0.37191	0.37461	68
22	0.37461	0.37730	0.37999	0.38268	0.38537	0.38805	0.39073	67
23	0.39073	0.39341	0.39608	0.39875	0.40142	0.40408	0.40674	66
24	0.40674	0.40939	0.41204	0.41469	0.41734	0.41998	0.42262	65
25	0.42262	0.42525	0.42788	0.43051	0.43313	0.43575	0.43837	64
26	0.43837	0.44098	0.44359	0.44620	0.44880	0.45140	0.45399	63
27	0.45399	0.45658	0.45917	0.46175	0.46433	0.46690	0.46947	62
28	0.46947	0.47204	0.47460	0.47716	0.47971	0.48226	0.48481	61
29	0.48481	0.48735	0.48989	0.49242	0.49495	0.49748	0.50000	60
30	0.50000	0.50252	0.50503	0.50754	0.51004	0.51254	0.51504	59
31	0.51504	0.51753	0.52002	0.52250	0.52498	0.52745	0.52992	58
32	0.52992	0.53238	0.54484	0.53730	0.53975	0.54220	0.54464	57
33	0.54464	0.54708	0.54951	0.55194	0.55436	0.55678	0.55919	56
34	0.55919	0.56160	0.56401	0.56641	0.56880	0.57119	0.57358	55
35	0.57358	0.57596	0.57833	0.58070	0.58307	0.58543	0.58779	54
36	0.58779	0.59014	0.59248	0.59482	0.59716	0.59949	0.60182	53
37	0.60182	0.60414	0.60645	0.60876	0.61107	0.61337	0.61566	52
38	0.61566	0.61795	0.62024	0.62251	0.62479	0.62706	0.62932	51
39	0.62932	0.63158	0.63383	0.63608	0.63832	0.64056	0.64279	50
40	0.64279	0.64501	0.64723	0.64945	0.65166	0.65386	0.65606	49
41	0.65606	0.65825	0.66044	0.66262	0.66480	0.66697	0.66913	48
42	0.66913	0.67129	0.67344	0.67559	0.67773	0.67987	0.68200	47
43	0.68200	0.68412	0.68624	0.68835	0.69046	0.69256	0.69466	46
44	0.69466	0.69675	0.69883	0.70091	0.70298	0.70505	0.70711	45
Sines	60′	50′	40′	30′ COSINES	20′	10′	0′	Degrees

Degrees	COSINES 0′	10′	20′	30′	40′	50′	60′	Sines
0	1.00000	1.00000	0.99998	0.99996	0.99993	0.99989	0.99985	89
1	0.99985	0.99979	0.99973	0.99966	0.99958	0.99949	0.99939	88
2	0.99939	0.99929	0.99917	0.99905	0.99892	0.99878	0.99863	87
3	0.99863	0.99847	0.99831	0.99813	0.99795	0.99776	0.99756	86
4	0.99756	0.99736	0.99714	0.99692	0.99668	0.99644	0.99619	85
5	0.99619	0.99594	0.99567	0.99540	0.99511	0.99482	0.99452	84
6	0.99452	0.99421	0.99390	0.99357	0.99324	0.99290	0.99255	83
7	0.99255	0.99219	0.99182	0.99144	0.99106	0.99067	0.99027	82
8	0.99027	0.98986	0.98944	0.98902	0.98858	0.98814	0.98769	81
9	0.98769	0.98723	0.98676	0.98629	0.98580	0.98531	0.98481	80
10	0.98481	0.98430	0.98378	0.98325	0.98272	0.98218	0.98163	79
11	0.98163	0.98107	0.98050	0.97992	0.97934	0.97875	0.97815	78
12	0.97815	0.97754	0.97692	0.97630	0.97566	0.97502	0.97437	77
13	0.97437	0.97371	0.97304	0.97237	0.97169	0.97100	0.97030	76
14	0.97030	0.96959	0.96887	0.96815	0.96742	0.96667	0.96593	75
15	0.96593	0.96517	0.96440	0.96363	0.96285	0.96206	0.96126	74
16	0.96126	0.96046	0.95964	0.95882	0.95799	0.95715	0.95630	73
17	0.95630	0.95545	0.95459	0.95372	0.95284	0.95195	0.95106	72
18	0.95106	0.95015	0.94924	0.94832	0.94740	0.94646	0.94552	71
19	0.94552	0.94457	0.94361	0.94264	0.94167	0.94068	0.93969	70
20	0.93969	0.93869	0.93769	0.93667	0.93565	0.93462	0.93358	69
21	0.93358	0.93253	0.93148	0.93042	0.92935	0.92827	0.92718	68
22	0.92718	0.92609	0.92499	0.92388	0.92276	0.92164	0.92050	67
23	0.92050	0.91936	0.91822	0.91706	0.91590	0.91472	0.91355	66
24	0.91355	0.91236	0.91116	0.90996	0.90875	0.90753	0.90631	65
25	0.90631	0.90507	0.90383	0.90259	0.90133	0.90007	0.89879	64
26	0.89879	0.89752	0.89623	0.89493	0.89363	0.89232	0.89101	63
27	0.89101	0.88968	0.88835	0.88701	0.88566	0.88431	0.88295	62
28	0.88295	0.88158	0.88020	0.87882	0.87743	0.87603	0.87462	61
29	0.87462	0.87321	0.87178	0.87036	0.86892	0.86748	0.86603	60
30	0.86603	0.86457	0.86310	0.86163	0.86015	0.85866	0.85717	59
31	0.85717	0.85567	0.85416	0.85264	0.85112	0.84959	0.84805	58
32	0.84805	0.84650	0.84495	0.84339	0.84182	0.84025	0.83867	57
33	0.83867	0.83708	0.83549	0.83389	0.83228	0.83066	0.82904	56
34	0.82904	0.82741	0.82577	0.82413	0.82248	0.82082	0.81915	55
35	0.81915	0.81748	0.81580	0.81412	0.81242	0.81072	0.80902	54
36	0.80902	0.80730	0.80558	0.80386	0.80212	0.80038	0.79864	53
37	0.79864	0.79688	0.79512	0.79335	0.79158	0.78980	0.78801	52
38	0.78801	0.78622	0.78442	0.78261	0.78979	0.77897	0.77715	51
39	0.77715	0.77531	0.77347	0.77162	0.76977	0.76791	0.76604	50
40	0.76604	0.76417	0.76229	0.76041	0.75851	0.75661	0.75471	49
41	0.75471	0.75280	0.75088	0.74896	0.74703	0.74509	0.74314	48
42	0.74314	0.74120	0.73924	0.73728	0.73531	0.73333	0.73135	47
43	0.73135	0.72937	0.72737	0.72537	0.72337	0.72136	0.71934	46
44	0.71934	0.71732	0.71529	0.71325	0.71121	0.70916	0.70711	45
Cosines	60′	50′	40′	30′ SINES	20′	10′	0′	Degrees

Table 4.17 continued

Table 4.17. Natural trigonometric functions. Courtesy J.H. Jones Company.

Table 4.17 continued

Degrees	TANGENTS							Cotangents
	0′	10′	20′	30′	40′	50′	60′	
0	0.00000	0.00291	0.00582	0.00873	0.01164	0.01455	0.01746	89
1	0.01746	0.02036	0.02328	0.02619	0.02910	0.03201	0.03492	88
2	0.03492	0.03783	0.04075	0.04366	0.04658	0.04949	0.05241	87
3	0.05241	0.05533	0.05824	0.06116	0.06408	0.06700	0.06993	86
4	0.06993	0.07285	0.07578	0.07870	0.08163	0.08456	0.08749	85
5	0.08749	0.09042	0.09335	0.09629	0.09923	0.10216	0.10510	84
6	0.10510	0.10805	0.11099	0.11394	0.11688	0.11983	0.12278	83
7	0.12278	0.12574	0.12869	0.13165	0.13461	0.13758	0.14054	82
8	0.14054	0.14351	0.14648	0.14945	0.15243	0.15540	0.15838	81
9	0.15838	0.16137	0.16435	0.16734	0.17033	0.17333	0.17633	80
10	0.17633	0.17933	0.18233	0.18534	0.18835	0.19136	0.19438	79
11	0.19438	0.19740	0.20042	0.20345	0.20648	0.20952	0.21256	78
12	0.21256	0.21560	0.21864	0.22169	0.22475	0.22781	0.23087	77
13	0.23087	0.23393	0.23700	0.24008	0.24316	0.24624	0.24933	76
14	0.24933	0.25242	0.25552	0.25862	0.26172	0.26483	0.26795	75
15	0.26795	0.27107	0.27419	0.27732	0.28046	0.28360	0.28675	74
16	0.28675	0.28990	0.29305	0.29621	0.29938	0.30255	0.30573	73
17	0.30573	0.30891	0.31210	0.31530	0.31850	0.32171	0.32492	72
18	0.32492	0.32814	0.33136	0.33460	0.33783	0.34108	0.34433	71
19	0.34433	0.34758	0.35085	0.35412	0.35740	0.36068	0.36397	70
20	0.36397	0.36727	0.37057	0.37388	0.37720	0.38053	0.38386	69
21	0.38386	0.38721	0.39055	0.39391	0.39727	0.40065	0.40403	68
22	0.40403	0.40741	0.41081	0.41421	0.41763	0.42105	0.42447	67
23	0.42447	0.42791	0.43136	0.43481	0.43828	0.44175	0.44523	66
24	0.44523	0.44872	0.45222	0.45573	0.45924	0.46277	0.46631	65
25	0.46631	0.46985	0.47341	0.47698	0.48055	0.48414	0.48773	64
26	0.48773	0.49134	0.49495	0.49858	0.50222	0.50587	0.50953	63
27	0.50953	0.51320	0.51688	0.52057	0.52427	0.52798	0.53171	62
28	0.53171	0.53545	0.53920	0.54296	0.54674	0.55051	0.55431	61
29	0.55431	0.55812	0.56194	0.56577	0.56962	0.57348	0.57735	60
30	0.57735	0.58124	0.58513	0.58905	0.59297	0.59691	0.60086	59
31	0.60086	0.60483	0.60881	0.61280	0.61681	0.62083	0.62487	58
32	0.62487	0.62892	0.63299	0.63707	0.64117	0.64528	0.64941	57
33	0.64941	0.65355	0.65771	0.66189	0.66608	0.67028	0.67451	56
34	0.67451	0.67875	0.68301	0.68728	0.69157	0.69588	0.70021	55
35	0.70021	0.70455	0.70891	0.71329	0.71769	0.72211	0.72654	54
36	0.72654	0.73100	0.73547	0.73996	0.74447	0.74900	0.75355	53
37	0.75355	0.75812	0.76272	0.76733	0.77196	0.77661	0.78129	52
38	0.78129	0.78598	0.79070	0.79544	0.80020	0.80498	0.80978	51
39	0.80978	0.81461	0.81946	0.82434	0.82923	0.83415	0.83910	50
40	0.83910	0.84407	0.84906	0.85408	0.85912	0.86419	0.86929	49
41	0.86929	0.87441	0.87955	0.88473	0.88992	0.89515	0.90040	48
42	0.90040	0.90569	0.91099	0.91633	0.92170	0.92709	0.93252	47
43	0.93252	0.93797	0.94345	0.94896	0.95451	0.96008	0.96569	46
44	0.96569	0.97133	0.97700	0.98270	0.98843	0.99420	1.00000	45
Tangents	60′	50′	40′	30′	20′	10′	0′	Degrees
	COTANGENTS							

Degrees	COTANGENTS							Tangents
	0′	10′	20′	30′	40′	50′	60′	
0	∞	343.77371	171.88540	114.58865	85.93979	68.75009	57.28996	89
1	57.28996	49.10388	42.96408	38.18846	34.36777	31.24158	28.63625	88
2	28.63625	26.43160	24.54176	22.90377	21.47040	20.20555	19.08114	87
3	19.08114	18.07498	17.16934	16.34986	15.60478	14.92442	14.30067	86
4	14.30067	13.72674	13.19688	12.70621	12.25051	11.82617	11.43005	85
5	11.43005	11.05943	10.71191	10.38540	10.07803	9.78817	9.51436	84
6	9.51436	9.25530	9.00983	8.77689	8.55555	8.34496	8.14435	83
7	8.14435	7.95302	7.77035	7.59575	7.42871	7.26873	7.11537	82
8	7.11537	6.96823	6.82694	6.69116	6.56055	6.43484	6.31375	81
9	6.31375	6.19703	6.08444	5.97576	5.87080	5.76937	5.67128	80
10	5.67128	5.57638	5.48451	5.39552	5.30928	5.22566	5.14455	79
11	5.14455	5.06584	4.98940	4.91516	4.84300	4.77286	4.70463	78
12	4.70463	4.63825	4.57363	4.51071	4.44942	4.38969	4.33148	77
13	4.33148	4.27471	4.21933	4.16530	4.11256	4.06107	4.01078	76
14	4.01078	3.96165	3.91364	3.86671	3.82083	3.77595	3.73205	75
15	3.73205	3.68909	3.64705	3.60588	3.56557	3.52609	3.48741	74
16	3.48741	3.44951	3.41236	3.37594	3.34023	3.30521	3.27085	73
17	3.27085	3.23714	3.20406	3.17159	3.13972	3.10842	3.07768	72
18	3.07768	3.04749	3.01783	2.98869	2.96004	2.93189	2.90421	71
19	2.90421	2.87700	2.85023	2.82391	2.79802	2.77254	2.74748	70
20	2.74748	2.72281	2.69853	2.67462	2.65109	2.62791	2.60509	69
21	2.60509	2.58261	2.56046	2.53865	2.51715	2.49597	2.47509	68
22	2.47509	2.45451	2.43422	2.41421	2.39449	2.37504	2.35585	67
23	2.35585	2.33693	2.31826	2.29984	2.28167	2.26374	2.24604	66
24	2.24604	2.22857	2.21132	2.19430	2.17749	2.16090	2.14451	65
25	2.14451	2.12832	2.11233	2.09654	2.08094	2.06553	2.05030	64
26	2.05030	2.03526	2.02039	2.00569	1.99116	1.97680	1.96261	63
27	1.96261	1.94858	1.93470	1.92098	1.97041	1.89400	1.88073	62
28	1.88073	1.86760	1.85462	1.84177	1.82907	1.81649	1.80405	61
29	1.80405	1.79174	1.77955	1.76749	1.75556	1.74375	1.73205	60
30	1.73205	1.72047	1.70901	1.69766	1.68643	1.67530	1.66428	59
31	1.66428	1.65337	1.64256	1.63185	1.62125	1.61074	1.60033	58
32	1.60033	1.59002	1.57981	1.56969	1.55966	1.54972	1.53987	57
33	1.53987	1.53010	1.52043	1.51084	1.50133	1.49190	1.48256	56
34	1.48256	1.47330	1.46411	1.45501	1.44598	1.43703	1.42815	55
35	1.42815	1.41934	1.41061	1.40195	1.39336	1.38484	1.37638	54
36	1.37638	1.36800	1.35968	1.35142	1.34323	1.33511	1.32704	53
37	1.32704	1.31904	1.31110	1.30323	1.29541	1.28764	1.27994	52
38	1.27994	1.27230	1.26471	1.25717	1.24969	1.24227	1.23490	51
39	1.23490	1.22758	1.22031	1.21310	1.20593	1.19882	1.19175	50
40	1.19175	1.18474	1.17777	1.17085	1.16398	1.15715	1.15037	49
41	1.15037	1.14363	1.13694	1.13029	1.12369	1.11713	1.11061	48
42	1.11061	1.10414	1.09770	1.09131	1.08496	1.07864	1.07237	47
43	1.07237	1.06613	1.05994	1.05378	1.04766	1.04158	1.03553	46
44	1.03553	1.02952	1.02355	1.01761	1.01170	1.00583	1.00000	45
Cotangents	60′	50′	40′	30′	20′	10′	0′	Degrees
	TANGENTS							

Table 4.17 continued

Table 4.17 continued

Degrees	SECANTS 0′	10′	20′	30′	40′	50′	60′	Cosecants
0	1.00000	1.00000	1.00002	1.00004	1.00007	1.00011	1.00015	89
1	1.00015	1.00021	1.00027	1.00034	1.00042	1.00051	1.00061	88
2	1.00061	1.00072	1.00083	1.00095	1.00108	1.00122	1.00137	87
3	1.00137	1.00153	1.00169	1.00187	1.00205	1.00224	1.00244	86
4	1.00244	1.00265	1.00287	1.00309	1.00333	1.00357	1.00382	85
5	1.00382	1.00408	1.00435	1.00463	1.00491	1.00521	1.00551	84
6	1.00551	1.00582	1.00614	1.00647	1.00681	1.00715	1.00751	83
7	1.00751	1.00787	1.00825	1.00863	1.00902	1.00942	1.00983	82
8	1.00983	1.01024	1.01067	1.01111	1.01155	1.01200	1.01247	81
9	1.01247	1.01294	1.01342	1.01391	1.01440	1.01491	1.01543	80
10	1.01543	1.01595	1.01649	1.01703	1.01758	1.01815	1.01872	79
11	1.01872	1.01930	1.01989	1.02049	1.02110	1.02171	1.02234	78
12	1.02234	1.02298	1.02362	1.02428	1.02494	1.02562	1.02630	77
13	1.02630	1.02700	1.02770	1.02842	1.02914	1.02987	1.03061	76
14	1.03061	1.03137	1.03213	1.03290	1.03368	1.03447	1.03528	75
15	1.03528	1.03609	1.03691	1.03774	1.03858	1.03944	1.04030	74
16	1.04030	1.04117	1.04206	1.04295	1.04385	1.04477	1.04569	73
17	1.04569	1.04668	1.04757	1.04853	1.04950	1.05047	1.05146	72
18	1.05146	1.05246	1.05347	1.05449	1.05552	1.05657	1.05762	71
19	1.05762	1.05869	1.05976	1.06085	1.06195	1.06306	1.06418	70
20	1.06418	1.06531	1.06645	1.06761	1.06878	1.06995	1.07115	69
21	1.07115	1.07235	1.07356	1.07479	1.07602	1.07727	1.07853	68
22	1.07853	1.07981	1.08109	1.08239	1.08370	1.08503	1.08636	67
23	1.08636	1.08771	1.08907	1.09044	1.09183	1.09323	1.09464	66
24	1.09464	1.09606	1.09750	1.09895	1.10041	1.10189	1.10338	65
25	1.10338	1.10488	1.10640	1.10793	1.10947	1.11103	1.11260	64
26	1.11260	1.11419	1.11579	1.11740	1.11903	1.12067	1.12233	63
27	1.12233	1.12400	1.12568	1.12738	1.12910	1.13083	1.13257	62
28	1.13257	1.13433	1.13610	1.13789	1.13970	1.14152	1.14335	61
29	1.14335	1.14521	1.14707	1.14896	1.15085	1.15277	1.15470	60
30	1.15470	1.15665	1.15861	1.16059	1.16259	1.16460	1.16663	59
31	1.16663	1.16868	1.17075	1.17283	1.17493	1.17704	1.17918	58
32	1.17918	1.18133	1.18350	1.18569	1.18790	1.19012	1.19236	57
33	1.19236	1.19463	1.19691	1.19920	1.20152	1.20386	1.20622	56
34	1.20622	1.20859	1.21099	1.21341	1.21584	1.21830	1.22077	55
35	1.22077	1.22327	1.22579	1.22833	1.23089	1.23347	1.23607	54
36	1.23607	1.23869	1.24134	1.24400	1.24669	1.24940	1.25214	53
37	1.25214	1.25489	1.25767	1.26047	1.26330	1.26615	1.26902	52
38	1.26902	1.27191	1.27483	1.27778	1.28075	1.28374	1.28676	51
39	1.28676	1.28980	1.29287	1.29597	1.29909	1.30223	1.30541	50
40	1.30541	1.30861	1.31183	1.31509	1.31837	1.32168	1.32501	49
41	1.32501	1.32838	1.33177	1.33519	1.33864	1.34212	1.34563	48
42	1.34563	1.34917	1.35274	1.35634	1.35997	1.36363	1.36733	47
43	1.36733	1.37105	1.37481	1.37860	1.38242	1.38628	1.39016	46
44	1.39016	1.39409	1.39804	1.40203	1.40606	1.41012	1.41421	45
Secants	60′	50′	40′	30′ COSECANTS	20′	10′	0′	Degrees

Degrees	COSECANTS 0′	10′	20′	30′	40′	50′	60′	Secants
0	∞	343.77516	171.88831	114.59301	85.94561	68.75736	57.29869	89
1	57.29869	49.11406	42.97571	38.20155	34.38232	31.25758	28.65371	88
2	28.65371	26.45051	24.56212	22.92559	21.49368	20.23028	19.10732	87
3	19.10732	18.10262	17.19843	16.38041	15.63679	14.95788	14.33559	86
4	14.33559	13.76312	13.23472	12.74550	12.29125	11.86837	11.47371	85
5	11.47371	11.10455	10.75849	10.43343	10.12752	9.83912	9.56677	84
6	9.56677	9.30917	9.06515	8.83367	8.61379	8.40466	8.20551	83
7	8.20551	8.01565	7.83443	7.66130	7.49571	7.33719	7.18530	82
8	7.18530	7.03962	6.89979	6.76547	6.63633	6.51208	6.39245	81
9	6.39245	6.27719	6.16607	6.05886	5.95536	5.85539	5.75877	80
10	5.75877	5.66533	5.57493	5.48740	5.40263	5.32049	5.24084	79
11	5.24084	5.16359	5.08863	5.01585	4.94517	4.87649	4.80973	78
12	4.80973	4.74482	4.68167	4.62023	4.56041	4.50216	4.44541	77
13	4.44541	4.39012	4.33622	4.28366	4.23239	4.18238	4.13357	76
14	4.13357	4.08591	4.03938	3.99393	3.94952	3.90613	3.86370	75
15	3.86370	3.82223	3.78166	3.74198	3.70315	3.66515	3.62796	74
16	3.62796	3.59154	3.55587	3.52094	3.48671	3.45317	3.42030	73
17	3.42030	3.38808	3.35649	3.32551	3.29512	3.26531	3.23607	72
18	3.23607	3.20737	3.17920	3.15155	3.12440	3.09774	3.07155	71
19	3.07155	3.04584	3.02057	2.99574	2.97135	2.94737	2.92380	70
20	2.92380	2.90063	2.87785	2.85545	2.83342	2.81175	2.79043	69
21	2.79043	2.76945	2.74881	2.72850	2.70851	2.68884	2.66947	68
22	2.66947	2.65040	2.63162	2.61313	2.59491	2.57698	2.55930	67
23	2.55930	2.54190	2.52474	2.50784	2.49119	2.47477	2.45859	66
24	2.45859	2.44264	2.42692	2.41142	2.39614	2.38107	2.36620	65
25	2.36620	2.34154	2.33708	2.32282	2.30875	2.29487	2.28117	64
26	2.28117	2.26766	2.25432	2.24116	2.22817	2.21535	2.20269	63
27	2.20269	2.19019	2.17786	2.16568	2.15366	2.14178	2.13005	62
28	2.13005	2.11847	2.10704	2.09574	2.08458	2.07356	2.06267	61
29	2.06267	2.05191	2.04128	2.03077	2.02039	2.01014	2.00000	60
30	2.00000	1.98998	1.98008	1.97029	1.96062	1.95106	1.94160	59
31	1.94160	1.93226	1.92302	1.91388	1.90485	1.89591	1.88709	58
32	1.88708	1.87834	1.86970	1.86116	1.85271	1.84435	1.83608	57
33	1.83608	1.82790	1.81981	1.81180	1.80388	1.79604	1.78829	56
34	1.78829	1.78062	1.77303	1.76552	1.75808	1.75073	1.74345	55
35	1.74345	1.73624	1.72911	1.72205	1.71506	1.70815	1.70130	54
36	1.70130	1.69452	1.68782	1.68117	1.67460	1.66809	1.66164	53
37	1.66164	1.65526	1.64894	1.64268	1.63648	1.63035	1.62427	52
38	1.62427	1.61825	1.61229	1.60639	1.60054	1.59475	1.58902	51
39	1.58902	1.58333	1.57771	1.57213	1.56661	1.56114	1.55572	50
40	1.55572	1.55036	1.54504	1.53977	1.53455	1.52938	1.52425	49
41	1.52425	1.51918	1.51415	1.50916	1.50422	1.49933	1.49448	48
42	1.49448	1.48967	1.48491	1.48019	1.47551	1.47087	1.46628	47
43	1.46628	1.46173	1.45721	1.45274	1.44831	1.44391	1.43956	46
44	1.43956	1.43524	1.43096	1.42672	1.42251	1.41835	1.41421	45
Cosecants	60′	50′	40′	30′ SECANTS	20′	10′	0′	Degrees

If the decimal point in the number is moved n places to the right (or left), the value of n (or $-n$) is added to the logarithm, thus:

log 3.14 = 0.4969
log 314. = 0.4969 + 2 or 2.4969
log .0314 = 0.4969 − 2, which may be written $\bar{2}.4969$ or 8.4969 − 10

If the given number has more than four significant figures, it should be reduced to four figures, since those beyond four figures will not affect the result in four-place computations.

The logarithm of a number having four significant figures must be interpolated by adding to the logarithm of the three figure number, the amount under the fourth figure, as read in the proportional parts section of the table.

Thus, the logarithm of 3.1416 is found as follows:

a. Reduce the number to four significant figures: 3.142
b. The log of 3.14 is .4969
c. The value of the proportional part under 2 (the fourth figure) is 3
d. Then, the log 3.142 = 0.4969 + .0003 or 0.4972

Natural logarithms: Many calculations make use of natural logarithms (Base e = 2.7183). To convert base 10 (common) logarithms to natural logarithms, multiply the value for the former by 2.30258.

Natural logarithms are also called Hyperbolic or Naperian logarithms.

$$\log ab = \log a + \log b \qquad \log a^n = n \log a$$

$$\log \frac{a}{b} = \log a - \log b \qquad \log \sqrt[n]{a} = \frac{\log a}{n}$$

N	0	1	2	3	4	5	6	7	8	9	Proportional Parts								
											1	2	3	4	5	6	7	8	9
1.0	0000	0043	0086	0128	0170	0212	0253	0294	0334	0374	4	8	12	17	21	25	29	33	37
1.1	0414	0453	0492	0531	0569	0607	0645	0682	0719	0755	4	8	11	15	19	23	26	30	34
1.2	0792	0828	0864	0899	0934	0969	1004	1038	1072	1106	3	7	10	14	17	21	24	28	31
1.3	1139	1173	1206	1239	1271	1303	1335	1367	1399	1430	3	6	10	13	16	19	23	26	29
1.4	1461	1492	1523	1553	1584	1614	1644	1673	1703	1732	3	6	9	12	15	18	21	24	27
1.5	1761	1790	1818	1847	1875	1903	1931	1959	1987	2014	3	6	8	11	14	17	20	22	25
1.6	2041	2068	2095	2122	2148	2175	2201	2227	2253	2279	3	5	8	11	13	16	18	21	24
1.7	2304	2330	2355	2380	2405	2430	2455	2480	2504	2529	2	5	7	10	12	15	17	20	22
1.8	2553	2577	2601	2625	2648	2672	2695	2718	2742	2765	2	5	7	9	12	14	16	19	21
1.9	2788	2810	2833	2856	2878	2900	2923	2945	2967	2989	2	4	7	9	11	13	16	18	20
2.0	3010	3032	3054	3075	3096	3118	3139	3160	3181	3201	2	4	6	8	11	13	15	17	19
2.1	3222	3243	3263	3284	3304	3324	3345	3365	3385	3404	2	4	6	8	10	12	14	16	18
2.2	3424	3444	3464	3483	3502	3522	3541	3560	3579	3598	2	4	6	8	10	12	14	15	17
2.3	3617	3636	3655	3674	3692	3711	3729	3747	3766	3784	2	4	6	7	9	11	13	15	17
2.4	3802	3820	3838	3856	3874	3892	3909	3927	3945	3962	2	4	5	7	9	11	12	14	16
2.5	3979	3997	4014	4031	4048	4065	4082	4099	4116	4133	2	3	5	7	9	10	12	14	15
2.6	4150	4166	4183	4200	4216	4232	4249	4265	4281	4298	2	3	5	7	8	10	11	13	15
2.7	4314	4330	4346	4362	4378	4393	4409	4425	4440	4456	2	3	5	6	8	9	11	13	14
2.8	4472	4487	4502	4518	4533	4548	4564	4579	4594	4609	2	3	5	6	8	9	11	12	14
2.9	4624	4639	4654	4669	4683	4698	4713	4728	4742	4757	1	3	4	6	7	9	10	12	13
3.0	4771	4786	4800	4814	4829	4843	4857	4871	4886	4900	1	3	4	6	7	9	10	11	13
3.1	4914	4928	4942	4955	4969	4983	4997	5011	5024	5038	1	3	4	6	7	8	10	11	12
3.2	5051	5065	5079	5092	5105	5119	5132	5145	5159	5172	1	3	4	5	7	8	9	11	12
3.3	5185	5198	5211	5224	5237	5250	5263	5276	5289	5302	1	3	4	5	6	8	9	10	12
3.4	5315	5328	5340	5353	5366	5378	5391	5403	5416	5428	1	3	4	5	6	8	9	10	11
3.5	5441	5453	5465	5478	5490	5502	5514	5527	5539	5551	1	2	4	5	6	7	9	10	11
3.6	5563	5575	5587	5599	5611	5623	5635	5647	5658	5670	1	2	4	5	6	7	8	10	11
3.7	5682	5694	5705	5717	5729	5740	5752	5763	5775	5786	1	2	3	5	6	7	8	9	10
3.8	5798	5809	5821	5832	5843	5855	5866	5877	5888	5899	1	2	3	5	6	7	8	9	10
3.9	5911	5922	5933	5944	5955	5966	5977	5988	5999	6010	1	2	3	4	5	7	8	9	10
N	0	1	2	3	4	5	6	7	8	9	1	2	3	4	5	6	7	8	9

Table 4.18 continued

Table 4.18. Four-place logarithms to base 10. Courtesy Crane Company (T. P. #410).

Table 4.18 continued

N	0	1	2	3	4	5	6	7	8	9	Proportional Parts								
											1	2	3	4	5	6	7	8	9
4.0	6021	6031	6042	6053	6064	6075	6085	6096	6107	6117	1	2	3	4	5	6	8	9	10
4.1	6128	6138	6149	6160	6170	6180	6191	6201	6212	6222	1	2	3	4	5	6	7	8	9
4.2	6232	6243	6253	6263	6274	6284	6294	6304	6314	6325	1	2	3	4	5	6	7	8	9
4.3	6335	6345	6355	6365	6375	6385	6395	6405	6415	6425	1	2	3	4	5	6	7	8	9
4.4	6435	6444	6454	6464	6474	6484	6493	6503	6513	6522	1	2	3	4	5	6	7	8	9
4.5	6532	6542	6551	6561	6571	6580	6590	6599	6609	6618	1	2	3	4	5	6	7	8	9
4.6	6628	6637	6646	6656	6665	6675	6684	6693	6702	6712	1	2	3	4	5	6	7	7	8
4.7	6721	6730	6739	6749	6758	6767	6776	6785	6794	6803	1	2	3	4	5	5	6	7	8
4.8	6812	6821	6830	6839	6848	6857	6866	6875	6884	6893	1	2	3	4	4	5	6	7	8
4.9	6902	6911	6920	6928	6937	6946	6955	6964	6972	6981	1	2	3	4	4	5	6	7	8
5.0	6990	6998	7007	7016	7024	7033	7042	7050	7059	7067	1	2	3	3	4	5	6	7	8
5.1	7076	7084	7093	7101	7110	7118	7126	7135	7143	7152	1	2	3	3	4	5	6	7	8
5.2	7160	7168	7177	7185	7193	7202	7210	7218	7226	7235	1	2	2	3	4	5	6	7	7
5.3	7243	7251	7259	7267	7275	7284	7292	7300	7308	7316	1	2	2	3	4	5	6	6	7
5.4	7324	7332	7340	7348	7356	7364	7372	7380	7388	7396	1	2	2	3	4	5	6	6	7
5.5	7404	7412	7419	7427	7435	7443	7451	7459	7466	7474	1	2	2	3	4	5	5	6	7
5.6	7482	7490	7497	7505	7513	7520	7528	7536	7543	7551	1	2	2	3	4	5	5	6	7
5.7	7559	7566	7574	7582	7589	7597	7604	7612	7619	7627	1	2	2	3	4	5	5	6	7
5.8	7634	7642	7649	7657	7664	7672	7679	7686	7694	7701	1	1	2	3	4	4	5	6	7
5.9	7709	7716	7723	7731	7738	7745	7752	7760	7767	7774	1	1	2	3	4	4	5	6	7
6.0	7782	7789	7796	7803	7810	7818	7825	7832	7839	7846	1	1	2	3	4	4	5	6	6
6.1	7853	7860	7868	7875	7882	7889	7896	7903	7910	7917	1	1	2	3	4	4	5	6	6
6.2	7924	7931	7938	7945	7952	7959	7966	7973	7980	7987	1	1	2	3	3	4	5	6	6
6.3	7993	8000	8007	8014	8021	8028	8035	8041	8048	8055	1	1	2	3	3	4	5	5	6
6.4	8062	8069	8075	8082	8089	8096	8102	8109	8116	8122	1	1	2	3	3	4	5	5	6
6.5	8129	8136	8142	8149	8156	8162	8169	8176	8182	8189	1	1	2	3	3	4	5	5	6
6.6	8195	8202	8209	8215	8222	8228	8235	8241	8248	8254	1	1	2	3	3	4	5	5	6
6.7	8261	8267	8274	8280	8287	8293	8299	8306	8312	8319	1	1	2	3	3	4	5	5	6
6.8	8325	8331	8338	8344	8351	8357	8363	8370	8376	8382	1	1	2	3	3	4	4	5	6
6.9	8388	8395	8401	8407	8414	8420	8426	8432	8439	8445	1	1	2	2	3	4	4	5	6
7.0	8451	8457	8463	8470	8476	8482	8488	8494	8500	8506	1	1	2	2	3	4	4	5	6
7.1	8513	8519	8525	8531	8537	8543	8549	8555	8561	8567	1	1	2	2	3	4	4	5	5
7.2	8573	8579	8585	8591	8597	8603	8609	8615	8621	8627	1	1	2	2	3	4	4	5	5
7.3	8633	8639	8645	8651	8657	8663	8669	8675	8681	8686	1	1	2	2	3	4	4	5	5
7.4	8692	8698	8704	8710	8716	8722	8727	8733	8739	8745	1	1	2	2	3	4	4	5	5
7.5	8751	8756	8762	8768	8774	8779	8785	8791	8797	8802	1	1	2	2	3	3	4	5	5
7.6	8808	8814	8820	8825	8831	8837	8842	8848	8854	8859	1	1	2	2	3	3	4	5	5
7.7	8865	8871	8876	8882	8887	8893	8899	8904	8910	8915	1	1	2	2	3	3	4	4	5
7.8	8921	8927	8932	8938	8943	8949	8954	8960	8965	8971	1	1	2	2	3	3	4	4	5
7.9	8976	8982	8987	8993	8998	9004	9009	9015	9020	9025	1	1	2	2	3	3	4	4	5
8.0	9031	9036	9042	9047	9053	9058	9063	9069	9074	9079	1	1	2	2	3	3	4	4	5
8.1	9085	9090	9096	9101	9106	9112	9117	9122	9128	9133	1	1	2	2	3	3	4	4	5
8.2	9138	9143	9149	9154	9159	9165	9170	9175	9180	9186	1	1	2	2	3	3	4	4	5
8.3	9191	9196	9201	9206	9212	9217	9222	9227	9232	9238	1	1	2	2	3	3	4	4	5
8.4	9243	9248	9253	9258	9263	9269	9274	9279	9284	9289	1	1	2	2	3	3	4	4	5
8.5	9294	9299	9304	9309	9315	9320	9325	9330	9335	9340	1	1	2	2	3	3	4	4	5
8.6	9345	9350	9355	9360	9365	9370	9375	9380	9385	9390	1	1	2	2	3	3	4	4	5
8.7	9395	9400	9405	9410	9415	9420	9425	9430	9435	9440	0	1	1	2	2	3	3	4	4
8.8	9445	9450	9455	9460	9465	9469	9474	9479	9484	9489	0	1	1	2	2	3	3	4	4
8.9	9494	9499	9504	9509	9513	9518	9523	9528	9533	9538	0	1	1	2	2	3	3	4	4
9.0	9542	9547	9552	9557	9562	9566	9571	9576	9581	9586	0	1	1	2	2	3	3	4	4
9.1	9590	9595	9600	9605	9609	9614	9619	9624	9628	9633	0	1	1	2	2	3	3	4	4
9.2	9638	9643	9647	9652	9657	9661	9666	9671	9675	9680	0	1	1	2	2	3	3	4	4
9.3	9685	9689	9694	9699	9703	9708	9713	9717	9722	9727	0	1	1	2	2	3	3	4	4
9.4	9731	9736	9741	9745	9750	9754	9759	9763	9768	9773	0	1	1	2	2	3	3	4	4
9.5	9777	9782	9786	9791	9795	9800	9805	9809	9814	9818	0	1	1	2	2	3	3	4	4
9.6	9823	9827	9832	9836	9841	9845	9850	9854	9859	9863	0	1	1	2	2	3	3	4	4
9.7	9868	9872	9877	9881	9886	9890	9894	9899	9903	9908	0	1	1	2	2	3	3	4	4
9.8	9912	9917	9921	9926	9930	9934	9939	9943	9948	9952	0	1	1	2	2	3	3	4	4
9.9	9956	9961	9965	9969	9974	9978	9983	9987	9991	9996	0	1	1	2	2	3	3	3	4
N	0	1	2	3	4	5	6	7	8	9	1	2	3	4	5	6	7	8	9

DIAMETER IN FT.	VOLUME OF SPHERE			DIAMETER IN FT.	VOLUME OF SPHERE		
	CUBIC FEET	U. S. GALS.	42 U. S. GAL. BBLS.		CUBIC FEET	U. S. GALS.	42 U. S. GAL. BBLS.
1	0.52	3.89	0.09	31	15599	116689	2778
2	4.19	31.34	0.75	32	17157	128344	3056
3	14.14	105.77	2.52	33	18817	140762	3352
4	33.51	250.67	5.97	34	20580	153950	3666
5	65.45	489.60	11.66	35	22449	167931	3998
6	113	845	20	36	24429	182742	4351
7	180	1346	32	37	26522	198399	4724
8	268	2005	48	38	28731	214924	5117
9	382	2858	68	39	31059	232338	5532
10	524	3920	93	40	33510	250673	5969
11	697	5213	124	41	36087	269950	6428
12	905	6770	161	42	38792	290184	6909
13	1150	8603	205	43	41630	311414	7415
14	1437	10750	256	44	44602	333647	7944
15	1767	13218	315	45	47713	356919	8498
16	2145	16046	382	46	50965	381245	9077
17	2572	19240	458	47	54362	406656	9682
18	3054	22846	544	48	57906	433167	10313
19	3591	26863	640	49	61601	460807	10972
20	4189	31336	746	50	65450	489600	11657
21	4849	36273	864	51	69456	519567	12371
22	5575	41704	993	52	73622	550731	13113
23	6371	47659	1135	53	77952	583121	13884
24	7238	54144	1289	54	82448	616754	14685
25	8181	61198	1457	55	87114	651658	15515
26	9203	68844	1639	56	91953	687856	16376
27	10306	77095	1836	57	96967	725363	17270
28	11494	85981	2047	58	102161	764217	18194
29	12770	95527	2275	59	107536	804425	19152
30	14137	105753	2518	60	113098	846032	20142

Surface area of sphere = $3.1416D^2$ Sq. Ft. Volume of sphere = $.5236D^3$ Cu. Ft. Volume of sphere = $.0932D^3$ Bbls. of 42 U. S. Gals. Number of Bbls. in 42 U. S. Gals. at any inch in a true sphere = $(3D-2H)H^2 \times .000053968$ where D is diameter and H is depth of liquid, both in inches.

Table 4.19. Volume of spheres.

Absolute Pressure: Lbs. per Sq. In. P′	Absolute Pressure: Inches of Hg	Vacuum Inches of Hg	Temperature t Degrees F.	Heat of the Liquid Btu/lb.	Latent Heat of Evaporation Btu/lb.	Total Heat of Steam h_g Btu/lb.	Specific Volume $\overline{V}$ Water Cu. ft. per lb.	Specific Volume $\overline{V}$ Steam Cu. ft. per lb.
0.0087	0.02	29.90	32.018	0.0003	1075.5	1075.5	0.016022	3302.4
0.10	0.20	29.72	35.023	3.026	1073.8	1076.8	0.016020	2945.5
0.15	0.31	29.61	45.453	13.498	1067.9	1081.4	0.016020	2004.7
0.20	0.41	29.51	53.160	21.217	1053.5	1084.7	0.016025	1526.3
0.25	0.51	29.41	59.323	27.382	1060.1	1087.4	0.016032	1235.5
0.30	0.61	29.31	64.484	32.541	1057.1	1089.7	0.016040	1039.7
0.35	0.71	29.21	68.939	36.992	1054.6	1091.6	0.016048	898.6
0.40	0.81	29.11	72.869	40.917	1052.4	1093.3	0.016056	792.1
0.45	0.92	29.00	76.387	44.430	1050.5	1094.9	0.016063	708.8
0.50	1.02	28.90	79.586	47.623	1048.6	1096.3	0.016071	641.5
0.60	1.22	28.70	85.218	53.245	1045.5	1098.7	0.016085	540.1
0.70	1.43	28.49	90.09	58.10	1042.7	1100.8	0.016099	466.94
0.80	1.63	28.29	94.38	62.39	1040.3	1102.6	0.016112	411.69
0.90	1.83	28.09	98.24	66.24	1038.1	1104.3	0.016124	368.43
1.0	2.04	27.88	101.74	69.73	1036.1	1105.8	0.016136	333.60
1.2	2.44	27.48	107.91	75.90	1032.6	1108.5	0.016158	280.96
1.4	2.85	27.07	113.26	81.23	1029.5	1110.7	0.016178	243.02
1.6	3.26	26.66	117.98	85.95	1026.8	1112.7	0.016196	214.33
1.8	3.66	26.26	122.22	90.18	1024.3	1114.5	0.016213	191.85
2.0	4.07	25.85	126.07	94.03	1022.1	1116.2	0.016230	173.76
2.2	4.48	25.44	129.61	97.57	1020.1	1117.6	0.016245	158.87
2.4	4.89	25.03	132.88	100.84	1018.2	1119.0	0.016260	146.40
2.6	5.29	24.63	135.93	103.88	1016.4	1120.3	0.016274	135.80
2.8	7.70	24.22	138.78	106.73	1014.7	1121.5	0.016287	126.67
3.0	6.11	23.81	141.47	109.42	1013.2	1122.6	0.016300	118.73
3.5	7.13	22.79	147.56	115.51	1009.6	1125.1	0.016331	102.74
4.0	8.14	21.78	152.96	120.92	1006.4	1127.3	0.016358	90.64
4.5	9.16	20.76	157.82	125.77	1003.5	1129.3	0.016384	83.03
5.0	10.18	19.74	162.24	130.20	1000.9	1131.1	0.016407	73.532
5.5	11.20	18.72	166.29	134.26	998.5	1132.7	0.016430	67.249
6.0	12.22	17.70	170.05	138.03	996.2	1134.2	0.016451	61.984
6.5	13.23	16.69	173.56	141.54	994.1	1135.6	0.016472	57.506
7.0	14.25	15.67	176.84	144.83	992.1	1136.9	0.016491	53.650
7.5	15.27	14.65	179.93	147.93	990.2	1138.2	0.016510	50.294
8.0	16.29	13.63	182.86	150.87	988.5	1139.3	0.016527	47.345
8.5	17.31	12.61	185.63	153.65	986.8	1140.4	0.016545	44.733
9.0	18.32	11.60	188.27	156.30	985.1	1141.4	0.016561	42.402
9.5	19.34	10.58	190.80	158.84	983.6	1142.4	0.016577	40.310
10.0	20.36	9.56	193.21	161.26	982.1	1143.3	0.016592	38.420
11.0	22.40	7.52	197.75	165.82	979.3	1145.1	0.016622	35.142
12.0	24.43	5.49	201.96	170.05	976.6	1146.7	0.016650	32.394
13.0	26.47	3.45	205.88	174.00	974.2	1148.2	0.016676	30.057
14.0	38.50	1.42	209.56	177.71	971.9	1149.6	0.016702	28.043

Pressure Lbs. per Sq. In.: Absolute P′	Pressure Lbs. per Sq. In.: Gage P	Temperature t Degrees F.	Heat of the Liquid Btu/lb.	Latent Heat of Evaporation Btu/lb.	Total Heat of Steam h_g Btu/lb.	Specific Volume $\overline{V}$ Water Cu. ft. per lb.	Specific Volume $\overline{V}$ Steam Cu. ft. per lb.
14.696	0.0	212.00	180.17	970.3	1150.5	0.016719	26.799
15.0	0.3	213.03	181.21	969.7	1150.9	0.016726	26.290
16.0	1.3	216.32	184.52	967.6	1152.1	0.016749	24.750
17.0	2.3	219.44	187.66	965.6	1153.2	0.016771	23.385
18.0	3.3	222.41	190.66	963.7	1154.3	0.016793	22.168
19.0	4.3	225.24	193.52	961.8	1155.3	0.016814	21.074
20.0	5.3	227.96	196.27	960.1	1156.3	0.016834	20.087
21.0	6.3	230.57	198.90	958.4	1157.3	0.016854	19.190
22.0	7.3	233.07	201.44	956.7	1158.1	0.016873	18.373
23.0	8.3	235.49	203.88	955.1	1159.0	0.016891	17.624
24.0	9.3	237.82	206.24	953.6	1159.8	0.016909	16.936
25.0	10.3	240.07	208.52	952.1	1160.6	0.016927	16.301
26.0	11.3	242.25	210.7	950.6	1161.4	0.016944	15.7138
27.0	12.3	244.36	212.9	949.2	1162.1	0.016961	15.1684
28.0	13.3	246.41	214.9	947.9	1162.8	0.016977	14.6607
29.0	14.3	248.40	217.0	946.5	1163.5	0.016993	14.1869
30.0	15.3	250.34	218.9	945.2	1164.1	0.017009	13.7436
31.0	16.3	252.22	220.8	943.9	1164.8	0.017024	13.3280
32.0	17.3	254.05	222.7	942.7	1165.4	0.017039	12.9376
33.0	18.3	255.84	224.5	941.5	1166.0	0.017054	12.5700
34.0	19.3	257.58	226.3	940.3	1166.6	0.017069	12.2234

*Abstracted from ASME Steam Tables (1967), with permission of the publisher, The American Society of Mechanical Engineers, 345 East 47th Street, New York, New York 10017.

Table 4.20 continued

Table 4.20. Properties of saturated steam. Courtesy Crane Company and Keenan and Keyes: Thermodynamic Properties of Steam, *John Wiley & Sons, Inc.*

Table 4.20 continued

Pressure Lbs. per Sq. In. Absolute P'	Gage P	Temperature t Degrees F.	Heat of the Liquid Btu/lb.	Latent Heat of Evaporation Btu/lb.	Total Heat of Steam h_g Btu/lb.	Specific Volume $\overline{V}$ Water Cu. ft. per lb.	Steam Cu. ft. per lb.
35.0	20.3	259.29	228.0	939.1	1167.1	0.017083	11.8959
36.0	21.3	260.95	229.7	938.0	1167.7	0.017097	11.5860
37.0	22.3	262.58	231.4	936.9	1168.2	0.017111	11.2923
38.0	23.3	264.17	233.0	935.8	1168.8	0.017124	11.0136
39.0	24.3	265.72	234.6	934.7	1169.3	0.017138	10.7487
40.0	25.3	267.25	236.1	933.6	1169.8	0.017151	10.4965
41.0	26.3	268.74	237.7	932.6	1170.2	0.017164	10.2563
42.0	27.3	270.21	239.2	931.5	1170.7	0.017177	10.0272
43.0	28.3	271.65	240.6	930.5	1171.2	0.017189	9.8083
44.0	29.3	273.06	242.1	929.5	1171.6	0.017202	9.5991
45.0	30.3	274.44	243.5	928.6	1172.0	0.017214	9.3988
46.0	31.3	275.80	244.9	927.6	1172.5	0.017226	9.2070
47.0	32.3	277.14	246.2	926.6	1172.9	0.017238	9.0231
48.0	33.3	278.45	247.6	925.7	1173.3	0.017250	8.8465
49.0	34.3	279.74	248.9	924.8	1173.7	0.017262	8.6770
50.0	35.3	281.02	250.2	923.9	1174.1	0.017274	8.5140
51.0	36.3	282.27	251.5	923.0	1174.5	0.017285	8.3571
52.0	37.3	283.50	252.8	922.1	1174.9	0.017296	8.2061
53.0	38.3	284.71	254.0	921.2	1175.2	0.017307	8.0606
54.0	39.3	285.90	255.2	920.4	1175.6	0.017319	7.9203
55.0	40.3	287.08	256.4	919.5	1175.9	0.017329	7.7850
56.0	41.3	288.24	257.6	918.7	1176.3	0.017340	7.6543
57.0	42.3	289.38	258.8	917.8	1176.6	0.017351	7.5280
58.0	43.3	290.50	259.9	917.0	1177.0	0.017362	7.4059
59.0	44.3	291.62	261.1	916.2	1177.3	0.017372	7.2879
60.0	45.3	292.71	262.2	915.4	1177.6	0.017383	7.1736
61.0	46.3	293.79	263.3	914.6	1177.9	0.017393	7.0630
62.0	47.3	294.86	264.4	913.8	1178.2	0.017403	6.9558
63.0	48.3	295.91	265.5	913.0	1178.6	0.017413	6.8519
64.0	49.3	296.95	266.6	912.3	1178.9	0.017423	6.7511
65.0	50.3	297.98	267.6	911.5	1179.1	0.017433	6.6533
66.0	51.3	298.99	268.7	910.8	1179.4	0.017443	6.5584
67.0	52.3	299.99	269.7	910.0	1179.7	0.017453	6.4662
68.0	53.3	300.99	270.7	909.3	1180.0	0.017463	6.3767
69.0	54.3	301.96	271.7	908.5	1180.3	0.017472	6.2896
70.0	55.3	302.93	272.7	907.8	1180.6	0.017482	6.2050
71.0	56.3	303.89	273.7	907.1	1180.8	0.017491	6.1226
72.0	57.3	304.83	274.7	906.4	1181.1	0.017501	6.0425
73.0	58.3	305.77	275.7	905.7	1181.4	0.017510	5.9645
74.0	59.3	306.69	276.6	905.0	1181.6	0.017519	5.8885
75.0	60.3	307.61	277.6	904.3	1181.9	0.017529	5.8144
76.0	61.3	308.51	278.5	903.6	1182.1	0.017538	5.7423
77.0	62.3	309.41	279.4	902.9	1182.4	0.017547	5.6720
78.0	63.3	310.29	280.3	902.3	1182.6	0.017556	5.6034
79.0	64.3	311.17	281.3	901.6	1182.8	0.017565	5.5364
80.0	65.3	312.04	282.1	900.9	1183.1	0.017573	5.4711
81.0	66.3	312.90	283.0	900.3	1183.3	0.017582	5.4074
82.0	67.3	313.75	283.9	899.6	1183.5	0.017591	5.3451
83.0	68.3	314.60	284.8	899.0	1183.8	0.017600	5.2843
84.0	69.3	315.43	285.7	898.3	1184.0	0.017608	5.2249
85.0	70.3	316.26	286.5	897.7	1184.2	0.017617	5.1669
86.0	71.3	317.08	287.4	897.0	1184.4	0.017625	5.1101
87.0	72.3	317.89	288.2	896.4	1184.6	0.017634	5.0546
88.0	73.3	318.69	289.0	895.8	1184.8	0.017642	5.0004
89.0	74.3	319.49	289.9	895.2	1185.0	0.017651	4.9473
90.0	75.3	320.28	290.7	894.6	1185.3	0.017659	4.8953
91.0	76.3	321.06	291.5	893.9	1185.5	0.017667	4.8445
92.0	77.3	321.84	292.3	893.3	1185.7	0.017675	4.7947
93.0	78.3	322.61	293.1	892.7	1185.9	0.017684	4.7459
94.0	79.3	323.37	293.9	892.1	1186.0	0.017692	4.6982
95.0	80.3	324.13	294.7	891.5	1186.2	0.017700	4.6514
96.0	81.3	324.88	295.5	891.0	1186.4	0.017708	4.6055
97.0	82.3	325.63	296.3	890.4	1186.6	0.017716	4.5606
98.0	83.3	326.36	297.0	889.8	1186.8	0.017724	4.5166
99.0	84.3	327.10	297.8	889.2	1187.0	0.017732	4.4734
100.0	85.3	327.82	298.5	888.6	1187.2	0.017740	4.4310
101.0	86.3	328.54	299.3	888.1	1187.3	0.01775	4.3895
102.0	87.3	329.26	300.0	887.5	1187.5	0.01776	4.3487
103.0	88.3	329.97	300.8	886.9	1187.7	0.01776	4.3087
104.0	89.3	330.67	301.5	886.4	1187.9	0.01777	4.2695
105.0	90.3	331.37	302.2	885.8	1188.0	0.01778	4.2309
106.0	91.3	332.06	303.0	885.2	1188.2	0.01779	4.1931
107.0	92.3	332.75	303.7	884.7	1188.4	0.01779	4.1560
108.0	93.3	333.44	304.4	884.1	1188.5	0.01780	4.1195
109.0	94.3	334.11	305.1	883.6	1188.7	0.01781	4.0837

Table 4.20 continued

Table 4.20 continued

Pressure Lbs. per Sq. In. Absolute P'	Gage P	Temperature t Degrees F.	Heat of the Liquid Btu/lb.	Latent Heat of Evaporation Btu/lb.	Total Heat of Steam h_g Btu/lb.	Specific Volume $\overline{V}$ Water Cu. ft. per lb.	Steam Cu. ft. per lb.
110.0	95.3	334.79	305.8	883.1	1188.9	0.01782	4.0484
111.0	96.3	335.46	306.5	882.5	1189.0	0.01782	4.0138
112.0	97.3	336.12	307.2	882.0	1189.2	0.01783	3.9798
113.0	98.3	336.78	307.9	881.4	1189.3	0.01784	3.9464
114.0	99.3	337.43	308.6	880.9	1189.5	0.01785	3.9136
115.0	100.3	338.08	309.3	880.4	1189.6	0.01785	3.8813
116.0	101.3	338.73	309.9	879.9	1189.8	0.01786	3.8495
117.0	102.3	339.37	310.6	879.3	1189.9	0.01787	3.8183
118.0	103.3	340.01	311.3	878.8	1190.1	0.01787	3.7875
119.0	104.3	340.64	311.9	878.3	1190.2	0.01788	3.7573
120.0	105.3	341.27	312.6	877.8	1190.4	0.01789	3.7275
121.0	106.3	341.89	313.2	877.3	1190.5	0.01790	3.6983
122.0	107.3	342.51	313.9	876.8	1190.7	0.01790	3.6695
123.0	108.3	343.13	314.5	876.3	1190.8	0.01791	3.6411
124.0	109.3	343.74	315.2	875.8	1190.9	0.01792	3.6132
125.0	110.3	344.35	315.8	875.3	1191.1	0.01792	3.5857
126.0	111.3	344.95	316.4	874.8	1191.2	0.01793	3.5586
127.0	112.3	345.55	317.1	874.3	1191.3	0.01794	3.5320
128.0	113.3	346.15	317.7	873.8	1191.5	0.01794	3.5057
129.0	114.3	346.74	318.3	873.3	1191.6	0.01795	3.4799
130.0	115.3	347.33	319.0	872.8	1191.7	0.01796	3.4544
131.0	116.3	347.92	319.6	872.3	1191.9	0.01797	3.4293
132.0	117.3	348.50	320.2	871.8	1192.0	0.01797	3.4046
133.0	118.3	349.08	320.8	871.3	1192.1	0.01798	3.3802
134.0	119.3	349.65	321.4	870.8	1192.2	0.01799	3.3562
135.0	120.3	350.23	322.0	870.4	1192.4	0.01799	3.3325
136.0	121.3	350.79	322.6	869.9	1192.5	0.01800	3.3091
137.0	122.3	351.36	323.2	869.4	1192.6	0.01801	3.2861
138.0	123.3	351.92	323.8	868.9	1192.7	0.01801	3.2634
139.0	124.3	352.48	324.4	868.5	1192.8	0.01802	3.2411
140.0	125.3	353.04	325.0	868.0	1193.0	0.01803	3.2190
141.0	126.3	353.59	325.5	867.5	1193.1	0.01803	3.1972
142.0	127.3	354.14	326.1	867.1	1193.2	0.01804	3.1757
143.0	128.3	354.69	326.7	866.6	1193.3	0.01805	3.1546
144.0	129.3	355.23	327.3	866.2	1193.4	0.01805	3.1337
145.0	130.3	355.77	327.8	865.7	1193.5	0.01806	3.1130
146.0	131.3	356.31	328.4	865.2	1193.6	0.01806	3.0927
147.0	132.3	356.84	329.0	864.8	1193.8	0.01807	3.0726
148.0	133.3	357.38	329.5	864.3	1193.9	0.01808	3.0528
149.0	134.3	357.91	330.1	863.9	1194.0	0.01808	3.0332
150.0	135.3	358.43	330.6	863.4	1194.1	0.01809	3.0139
152.0	137.3	359.48	331.8	862.5	1194.3	0.01810	2.9760
154.0	139.3	360.51	332.8	861.6	1194.5	0.01812	2.9391
156.0	141.3	361.53	333.9	860.8	1194.7	0.01813	2.9031
158.0	143.3	362.55	335.0	859.9	1194.9	0.01814	2.8679
160.0	145.3	363.55	336.1	859.0	1195.1	0.01815	2.8336
162.0	147.3	364.54	337.1	858.2	1195.3	0.01817	2.8001
164.0	149.3	365.53	338.2	857.3	1195.5	0.01818	2.7674
166.0	151.3	366.50	339.2	856.5	1195.7	0.01819	2.7355
168.0	153.3	367.47	340.2	855.6	1195.8	0.01820	2.7043
170.0	155.3	368.42	341.2	854.8	1196.0	0.01821	2.6738
172.0	157.3	369.37	342.2	853.9	1196.2	0.01823	2.6440
174.0	159.3	370.31	343.2	853.1	1196.4	0.01824	2.6149
176.0	161.3	371.24	344.2	852.3	1196.5	0.01825	2.5864
178.0	163.3	372.16	345.2	851.5	1196.7	0.01826	2.5585
180.0	165.3	373.08	346.2	850.7	1196.9	0.01827	2.5312
182.0	167.3	373.98	347.2	849.9	1197.0	0.01828	2.5045
184.0	169.3	374.88	348.1	849.1	1197.2	0.01830	2.4783
186.0	171.3	375.77	349.1	848.3	1197.3	0.01831	2.4527
188.0	173.3	376.65	350.0	847.5	1197.5	0.01832	2.4276
190.0	175.3	377.53	350.9	846.7	1197.6	0.01833	2.4030
192.0	177.3	378.40	351.9	845.9	1197.8	0.01834	2.3790
194.0	179.3	379.26	352.8	845.1	1197.9	0.01835	2.3554
196.0	181.3	380.12	353.7	844.4	1198.1	0.01836	2.3322
198.0	183.3	380.96	354.6	843.6	1198.2	0.01838	2.3095
200.0	185.3	381.80	355.5	842.8	1198.3	0.01839	2.28728
205.0	190.3	383.88	357.7	840.9	1198.7	0.01841	2.23349
210.0	195.3	385.91	359.9	839.1	1199.0	0.01844	2.18217
215.0	200.3	387.91	362.1	837.2	1199.3	0.01847	2.13315
220.0	205.3	389.88	364.2	835.4	1199.6	0.01850	2.08629
225.0	210.3	391.80	366.2	833.6	1199.9	0.01852	2.04143
230.0	215.3	393.70	368.3	831.8	1200.1	0.01855	1.99846
235.0	220.3	395.56	370.3	830.1	1200.4	0.01857	1.95725
240.0	225.3	397.39	372.3	828.4	1200.6	0.01860	1.91769
245.0	230.3	399.19	374.2	826.6	1200.9	0.01863	1.87970

Table 4.20 continued

Pressure Lbs. per Sq. In. Absolute p'	Gage p	Temperature t Degrees F.	Heat of the Liquid Btu/lb.	Latent Heat of Evaporation Btu/lb.	Total Heat of Steam h_g Btu/lb.	Specific Volume $\overline{V}$ Water Cu. ft. per lb.	Steam Cu. ft. per lb.
250.0	235.3	400.97	376.1	825.0	1201.1	0.01865	1.84317
255.0	240.3	402.72	378.0	823.3	1201.3	0.01868	1.80802
260.0	245.3	404.44	379.9	821.6	1201.5	0.01870	1.77418
265.0	250.3	406.13	381.7	820.0	1201.7	0.01873	1.74157
270.0	255.3	407.80	383.6	818.3	1201.9	0.01875	1.71013
275.0	260.3	409.45	385.4	816.7	1202.1	0.01878	1.67978
280.0	265.3	411.07	387.1	815.1	1202.3	0.01880	1.65049
285.0	270.3	412.67	388.9	813.6	1202.4	0.01882	1.62218
290.0	275.3	414.25	390.6	812.0	1202.6	0.01885	1.59482
295.0	280.3	415.81	392.3	810.4	1202.7	0.01887	1.56835
300.0	285.3	417.35	394.0	808.9	1202.9	0.01889	1.54274
320.0	305.3	423.31	400.5	802.9	1203.4	0.01899	1.44801
340.0	325.3	428.99	406.8	797.0	1203.8	0.01908	1.36405
360.0	345.3	434.41	412.8	791.3	1204.1	0.01917	1.28910
380.0	365.3	439.61	418.6	785.8	1204.4	0.01925	1.22177
400.0	385.3	444.60	424.2	780.4	1204.6	0.01934	1.16095
420.0	405.3	449.40	429.6	775.2	1204.7	0.01942	1.10573
440.0	425.3	454.03	434.8	770.0	1204.8	0.01950	1.05535
460.0	445.3	458.50	439.8	765.0	1204.8	0.01959	1.00921
480.0	465.3	462.82	444.7	760.0	1204.8	0.01967	0.96677
500.0	485.3	467.01	449.5	755.1	1204.7	0.01975	0.92762
520.0	505.3	471.07	454.2	750.4	1204.5	0.01982	0.89137
540.0	525.3	475.01	458.7	745.7	1204.4	0.01990	0.85771
560.0	545.3	478.84	463.1	741.0	1204.2	0.01998	0.82637
580.0	565.3	482.57	467.5	736.5	1203.9	0.02006	0.79712
600.0	585.3	486.20	471.7	732.0	1203.7	0.02013	0.76975
620.0	605.3	489.74	475.8	727.5	1203.4	0.02021	0.74408
640.0	625.3	493.19	479.9	723.1	1203.0	0.02028	0.71995
660.0	645.3	496.57	483.9	718.8	1202.7	0.02036	0.69724
680.0	665.3	499.86	487.8	714.5	1202.3	0.02043	0.67581
700.0	685.3	503.08	491.6	710.2	1201.8	0.02050	0.65556
720.0	705.3	506.23	495.4	706.0	1201.4	0.02058	0.63639
740.0	725.3	509.32	499.1	701.9	1200.9	0.02065	0.61822
760.0	745.3	512.34	502.7	697.7	1200.4	0.02072	0.60097
780.0	765.3	515.30	506.3	693.6	1199.9	0.02080	0.58457
800.0	785.3	518.21	509.8	689.6	1199.4	0.02087	0.56896
820.0	805.3	521.06	513.3	685.5	1198.8	0.02094	0.55408
840.0	825.3	523.86	516.7	681.5	1198.2	0.02101	0.53988
860.0	845.3	526.60	520.1	677.6	1197.7	0.02109	0.52631
880.0	865.3	529.30	523.4	673.6	1197.0	0.02116	0.51333
900.0	885.3	531.95	526.7	669.7	1196.4	0.02123	0.50091
920.0	905.3	534.56	530.0	665.8	1195.7	0.02130	0.48901
940.0	925.3	537.13	533.2	661.9	1195.1	0.02137	0.47759
960.0	945.3	539.65	536.3	658.0	1194.4	0.02145	0.46662
980.0	965.3	542.14	539.5	654.2	1193.7	0.02152	0.45609
1000.0	985.3	544.58	542.6	650.4	1192.9	0.02159	0.44596
1050.0	1035.3	550.53	550.1	640.9	1191.0	0.02177	0.42224
1100.0	1085.3	556.28	557.5	631.5	1189.1	0.02195	0.40058
1150.0	1135.3	561.82	564.8	622.2	1187.0	0.02214	0.38073
1200.0	1185.3	567.19	571.9	613.0	1184.8	0.02232	0.36245
1250.0	1235.3	572.38	578.8	603.8	1182.6	0.02250	0.34556
1300.0	1285.3	577.42	585.6	594.6	1180.2	0.02269	0.32991
1350.0	1335.3	582.32	592.2	585.6	1177.8	0.02288	0.31536
1400.0	1385.3	587.07	598.8	567.5	1175.3	0.02307	0.30178
1450.0	1435.3	591.70	605.3	567.6	1172.9	0.02327	0.28909
1500.0	1485.3	596.20	611.7	558.4	1170.1	0.02346	0.27719
1600.0	1585.3	604.87	624.2	540.3	1164.5	0.02387	0.25545
1700.0	1685.3	613.13	636.5	522.2	1158.6	0.02428	0.23607
1800.0	1785.3	621.02	648.5	503.8	1152.3	0.02472	0.21861
1900.0	1885.3	628.56	660.4	485.2	1145.6	0.02517	0.20278
2000.0	1985.3	635.80	672.1	466.2	1138.3	0.02565	0.18831
2100.0	2085.3	642.76	683.8	446.7	1130.5	0.02615	0.17501
2200.0	2185.3	649.45	695.5	426.7	1122.2	0.02669	0.16272
2300.0	2285.3	655.89	707.2	406.0	1113.2	0.02727	0.15133
2400.0	2385.3	662.11	719.0	384.8	1103.7	0.02790	0.14076
2500.0	2485.3	668.11	731.7	361.6	1093.3	0.02859	0.13068
2600.0	2585.3	673.91	744.5	337.6	1082.0	0.02938	0.12110
2700.0	2685.3	679.53	757.3	312.3	1069.7	0.03029	0.11194
2800.0	2785.3	684.96	770.7	285.1	1055.8	0.03134	0.10305
2900.0	2885.3	690.22	785.1	254.7	1039.8	0.03262	0.09420
3000.0	2985.3	695.33	801.8	218.4	1020.3	0.03428	0.08500
3100.0	3085.3	700.28	824.0	169.3	993.3	0.03681	0.07452
3200.0	3185.3	705.08	875.5	56.1	931.6	0.04472	0.05663
3208.2	3193.5	705.47	906.0	0.0	906.0	0.05078	0.05078

Table 4.20 continued

Table 4.20 continued

V = specific volume, cubic feet per pound
h_g = total heat of steam, Btu per pound

Pressure Lbs. per Sq. In.		Sat. Temp.		Total Temperature Degrees Fahrenheit (t)										
Abs. P'	Gage P	t	$\bar{V}$	350°	400°	500°	600°	700°	800°	900°	1000°	1100°	1300°	1500°
15.0	0.3	213.03	V	31.939	33.963	37.985	41.986	45.978	49.964	53.946	57.926	61.905	69.858	77.807
			h_g	1216.2	1239.9	1287.3	1335.2	1383.8	1433.2	1483.4	1534.5	1586.5	1693.2	1803.4
20.0	5.3	227.96	V	23.900	25.428	28.457	31.466	34.465	37.458	40.447	43.435	46.420	52.388	58.352
			h_g	1215.4	1239.2	1286.9	1334.9	1383.5	1432.9	1483.2	1534.3	1586.3	1693.1	1803.3
30.0	15.3	250.34	V	15.859	16.892	18.929	20.945	22.951	24.952	26.949	28.943	30.936	34.918	38.896
			h_g	1213.6	1237.8	1286.0	1334.2	1383.0	1432.5	1482.8	1534.0	1586.1	1692.9	1803.2
40.0	25.3	267.25	V	11.838	12.624	14.165	15.685	17.195	18.699	20.199	21.697	23.194	26.183	29.168
			h_g	1211.7	1236.4	1285.0	1333.6	1382.5	1432.1	1482.5	1533.7	1585.8	1692.7	1803.0
50.0	35.3	281.02	V	9.424	10.062	11.306	12.529	13.741	14.947	16.150	17.350	18.549	20.942	23.332
			h_g	1209.9	1234.9	1284.1	1332.9	1382.0	1431.7	1482.2	1533.4	1585.6	1692.5	1802.9
60.0	45.3	292.71	V	7.815	8.354	9.400	10.425	11.438	12.446	13.450	14.452	15.452	17.448	19.441
			h_g	1208.0	1233.5	1283.2	1332.3	1381.5	1431.3	1481.8	1533.2	1585.3	1692.4	1802.8
70.0	55.3	302.93	V	6.664	7.133	8.039	8.922	9.793	10.659	11.522	12.382	13.240	14.952	16.661
			h_g	1206.0	1232.0	1282.2	1331.6	1381.0	1430.9	1481.5	1532.9	1585.1	1692.2	1802.6
80.0	65.3	312.04	V	5.801	6.218	7.018	7.794	8.560	9.319	10.075	10.829	11.581	13.081	14.577
			h_g	1204.0	1230.5	1281.3	1330.9	1380.5	1430.5	1481.1	1532.6	1584.9	1692.0	1802.5
90.0	75.3	320.28	V	5.128	5.505	6.223	6.917	7.600	8.277	8.950	9.621	10.290	11.625	12.956
			h_g	1202.0	1228.9	1280.3	1330.2	1380.0	1430.1	1480.8	1532.3	1584.6	1691.8	1802.4
100.0	85.3	327.82	V	4.590	4.935	5.588	6.216	6.833	7.443	8.050	8.655	9.258	10.460	11.659
			h_g	1199.9	1227.4	1279.3	1329.6	1379.5	1429.7	1480.4	1532.0	1584.4	1691.6	1802.2
120.0	105.3	341.27	$\bar{V}$	3.7815	4.0786	4.6341	5.1637	5.6813	6.1928	6.7006	7.2060	7.7096	8.7130	9.7130
			h_g	1195.6	1224.1	1277.4	1328.2	1378.4	1428.8	1479.8	1531.4	1583.9	1691.3	1802.0
140.0	125.3	353.04	$\bar{V}$		3.4661	3.9526	4.4119	4.8588	5.2995	5.7364	6.1709	6.6036	7.4652	8.3233
			h_g		1220.8	1275.3	1326.8	1377.4	1428.0	1479.1	1530.8	1583.4	1690.9	1801.7
160.0	145.3	363.55	$\bar{V}$		3.0060	3.4413	3.8480	4.2420	4.6295	5.0132	5.3945	5.7741	6.5293	7.2811
			h_g		1217.4	1273.3	1325.4	1376.4	1427.2	1478.4	1530.3	1582.9	1690.5	1801.4
180.0	165.3	373.08	$\bar{V}$		2.6474	3.0433	3.4093	3.7621	4.1084	4.4508	4.7907	5.1289	5.8014	6.4704
			h_g		1213.8	1271.2	1324.0	1375.3	1426.3	1477.7	1529.7	1582.4	1690.2	1801.2
200.0	185.3	381.80	$\bar{V}$		2.3598	2.7247	3.0583	3.3783	3.6915	4.0008	4.3077	4.6128	5.2191	5.8219
			h_g		1210.1	1269.0	1322.6	1374.3	1425.5	1477.0	1529.1	1581.9	1689.8	1800.9
220.0	205.3	389.88	$\bar{V}$		2.1240	2.4638	2.7710	3.0642	3.3504	3.6327	3.9125	4.1905	4.7426	5.2913
			h_g		1206.3	1266.9	1321.2	1373.2	1424.7	1476.3	1528.5	1581.4	1689.4	1800.6
240.0	225.3	397.39	$\bar{V}$		1.9268	2.2462	2.5316	2.8024	3.0661	3.3259	3.5831	3.8385	4.3456	4.8492
			h_g		1202.4	1264.6	1319.7	1372.1	1423.8	1475.6	1527.9	1580.9	1689.1	1800.4
260.0	245.3	404.44	$\bar{V}$			2.0619	2.3289	2.5808	2.8256	3.0663	3.3044	3.5408	4.0097	4.4750
			h_g			1262.4	1318.2	1371.1	1423.0	1474.9	1527.3	1580.4	1688.7	1800.1
280.0	265.3	411.07	$\bar{V}$			1.9037	2.1551	2.3909	2.6194	2.8437	3.0655	3.2855	3.7217	4.1543
			h_g			1260.0	1316.8	1370.0	1422.1	1474.2	1526.8	1579.9	1688.4	1799.8
300.0	285.3	417.35	$\bar{V}$			1.7665	2.0044	2.2263	2.4407	2.6509	2.8585	3.0643	3.4721	3.8764
			h_g			1257.7	1315.2	1368.9	1421.3	1473.6	1526.2	1579.4	1688.0	1799.6
320.0	305.3	423.31	$\bar{V}$			1.6462	1.8725	2.0823	2.2843	2.4821	2.6774	2.8708	3.2538	3.6332
			h_g			1255.2	1313.7	1367.8	1420.5	1472.9	1525.6	1578.9	1687.6	1799.3
340.0	325.3	428.99	$\bar{V}$			1.5399	1.7561	1.9552	2.1463	2.3333	2.5175	2.7000	3.0611	3.4186
			h_g			1252.8	1312.2	1366.7	1419.6	1472.2	1525.0	1578.4	1687.3	1799.0
360.0	345.3	434.41	$\bar{V}$			1.4454	1.6525	1.8421	2.0237	2.2009	2.3755	2.5482	2.8898	3.2279
			h_g			1250.3	1310.6	1365.6	1418.7	1471.5	1542.4	1577.9	1686.9	1798.8

*Abstracted from ASME Steam Tables (1967) with permission of the publisher, the American Society of Mechanical Engineers, 345 East 47th Street, New York, N.Y. 10017.

Table 4.20 continued

$\overline{V}$ = specific volume, cubic feet per pound
h_g = total heat of steam, Btu per pound

Pressure Lbs. per Sq. In. Abs. P'	Gage P	Sat. Temp. t	$\overline{V}$	Total Temperature—Degrees Fahrenheit (t) 500°	600°	700°	800°	900°	1000°	1100°	1200°	1300°	1400°	1500°
380.0	365.3	439.61	$\overline{V}$	1.3606	1.5598	1.7410	1.9139	2.0825	2.2484	2.4124	2.5750	2.7366	2.8973	3.0572
			h_g	1247.7	1309.0	1364.5	1417.9	1470.8	1523.8	1577.4	1631.6	1686.5	1742.2	1798.5
400.0	385.3	444.60	$\overline{V}$	1.2841	1.4763	1.6499	1.8151	1.9759	2.1339	2.2901	2.4450	2.5987	2.7515	2.9037
			h_g	1245.1	1307.4	1363.4	1417.0	1470.1	1523.3	1576.9	1631.2	1686.2	1741.9	1798.2
420.0	405.3	449.40	$\overline{V}$	1.2148	1.4007	1.5676	1.7258	1.8795	2.0304	2.1795	2.3273	2.4739	2.6196	2.7647
			h_g	1242.4	1305.8	1362.3	1416.2	1469.4	1522.7	1576.4	1630.8	1685.8	1741.6	1798.0
440.0	425.3	454.03	$\overline{V}$	1.1517	1.3319	1.4926	1.6445	1.7918	1.9363	2.0790	2.2203	2.3605	2.4998	2.6384
			h_g	1239.7	1304.2	1361.1	1415.3	1468.7	1522.1	1575.9	1630.4	1685.5	1741.2	1797.7
460.0	445.3	458.50	$\overline{V}$	1.0939	1.2691	1.4242	1.5703	1.7117	1.8504	1.9872	2.1226	2.2569	2.3903	2.5230
			h_g	1236.9	1302.5	1360.0	1414.4	1468.0	1521.5	1575.4	1629.9	1685.1	1740.9	1797.4
480.0	465.3	462.82	$\overline{V}$	1.0409	1.2115	1.3615	1.5023	1.6384	1.7716	1.9030	2.0330	2.1619	2.2900	2.4173
			h_g	1234.1	1300.8	1358.8	1413.6	1467.3	1520.9	1574.9	1629.5	1684.7	1740.6	1797.2
500.0	485.3	467.01	$\overline{V}$	0.9919	1.1584	1.3037	1.4397	1.5708	1.6992	1.8256	1.9507	2.0746	2.1977	2.3200
			h_g	1231.2	1299.1	1357.7	1412.7	1466.6	1520.3	1574.4	1629.1	1684.4	1740.3	1796.9
520.0	505.3	471.07	$\overline{V}$	0.9466	1.1094	1.2504	1.3819	1.5085	1.6323	1.7542	1.8746	1.9940	2.1125	2.2302
			h_g	1228.3	1297.4	1356.5	1411.8	1465.9	1519.7	1573.9	1628.7	1684.0	1740.0	1796.7
540.0	525.3	475.01	$\overline{V}$	0.9045	1.0640	1.2010	1.3284	1.4508	1.5704	1.6880	1.8042	1.9193	2.0336	2.1471
			h_g	1225.3	1295.7	1355.3	1410.9	1465.1	1519.1	1573.4	1628.2	1683.6	1739.7	1796.4
560.0	545.3	478.84	$\overline{V}$	0.8653	1.0217	1.1552	1.2787	1.3972	1.5129	1.6266	1.7388	1.8500	1.9603	2.0699
			h_g	1222.2	1293.9	1354.2	1410.0	1464.4	1518.6	1572.9	1627.8	1683.3	1739.4	1796.1
580.0	565.3	482.57	$\overline{V}$	0.8287	0.9824	1.1125	1.2324	1.3473	1.4593	1.5693	1.6780	1.7855	1.8921	1.9980
			h_g	1219.1	1292.1	1353.0	1409.2	1463.7	1518.0	1572.4	1627.4	1682.9	1739.1	1795.9
600.0	585.3	486.20	$\overline{V}$	0.7944	0.9456	1.0726	1.1892	1.3008	1.4093	1.5160	1.6211	1.7252	1.8284	1.9309
			h_g	1215.9	1290.3	1351.8	1408.3	1463.0	1517.4	1571.9	1627.0	1682.6	1738.8	1795.6
650.0	635.3	494.89	$\overline{V}$	0.7173	0.8634	0.9835	1.0929	1.1969	1.2979	1.3969	1.4944	1.5909	1.6864	1.7813
			h_g	1207.6	1285.7	1348.7	1406.0	1461.2	1515.9	1570.7	1625.9	1681.6	1738.0	1794.9
700.0	685.3	503.08	$\overline{V}$	...	0.7928	0.9072	1.0102	1.1078	1.2023	1.2948	1.3858	1.4757	1.5647	1.6530
			h_g	...	1281.0	1345.6	1403.7	1459.4	1514.4	1569.4	1624.8	1680.7	1737.2	1794.3
750.0	735.3	510.84	$\overline{V}$	...	0.7313	0.8409	0.9386	1.0306	1.1195	1.2063	1.2916	1.3759	1.4592	1.5419
			h_g	...	1276.1	1342.5	1401.5	1457.6	1512.9	1568.2	1623.8	1679.8	1736.4	1793.6
800.0	785.3	518.21	$\overline{V}$	...	0.6774	0.7828	0.8759	0.9631	1.0470	1.1289	1.2093	1.2885	1.3669	1.4446
			h_g	...	1271.1	1339.3	1399.1	1455.8	1511.4	1566.9	1622.7	1678.9	1735.7	1792.9
850.0	835.3	525.24	$\overline{V}$	...	0.6296	0.7315	0.8205	0.9034	0.9830	1.0606	1.1366	1.2115	1.2855	1.3588
			h_g	...	1265.9	1336.0	1396.8	1454.0	1510.0	1565.7	1621.6	1678.0	1734.9	1792.3
900.0	885.3	531.95	$\overline{V}$	...	0.5869	0.6858	0.7713	0.8504	0.9262	0.9998	1.0720	1.1430	1.2131	1.2825
			h_g	...	1260.6	1332.7	1394.4	1452.2	1508.5	1564.4	1620.6	1677.1	1734.1	1791.6
950.0	935.3	538.39	$\overline{V}$	...	0.5485	0.6449	0.7272	0.8030	0.8753	0.9455	1.0142	1.0817	1.1484	1.2143
			h_g	...	1255.1	1329.3	1392.0	1450.3	1507.0	1563.2	1619.5	1676.2	1733.3	1791.0
1000.0	985.3	544.58	$\overline{V}$	...	0.5137	0.6080	0.6875	0.7603	0.8295	0.8966	0.9622	1,0266	1.0901	1.1529
			h_g	...	1249.3	1325.9	1389.6	1448.5	1505.4	1561.9	1618.4	1675.3	1732.5	1790.3
1050.0	1035.3	550.53	$\overline{V}$	...	0.4821	0.5745	0.6515	0.7216	0.7881	0.8524	0.9151	0.9767	1.0373	1.0973
			h_g	...	1243.4	1322.4	1387.2	1446.6	1503.9	1560.7	1617.4	1674.4	1731.8	1789.6
1100.0	1085.3	556.28	$\overline{V}$	...	0.4531	0.5440	0.6188	0.6865	0.7505	0.8121	0.8723	0.9313	0.9894	1.0468
			h_g	...	1237.3	1318.8	1384.7	1444.7	1502.4	1559.4	1616.3	1673.5	1731.0	1789.0
1150.0	1135.3	561.82	$\overline{V}$	...	0.4263	0.5162	0.5889	0.6544	0.7161	0.7754	0.8332	0.8899	0.9456	1.0007
			h_g	...	1230.9	1315.2	1382.2	1442.8	1500.9	1558.1	1615.2	1672.6	1730.2	1788.3

Table 4.20 continued

$\bar{V}$ = specific volume, cubic feet per pound
h_g = total heat of steam, Btu per pound

Pressure Lbs. per Sq. In. Abs. P′	Gage P	Sat. Temp. t	$\bar{V}$	650°	700°	750°	800°	900°	1000°	1100°	1200°	1300°	1400°	1500°
1200.0	1185.3	567.19	$\bar{V}$	0.4497	0.4905	0.5273	0.5615	0.6250	0.6845	0.7418	0.7974	0.8519	0.9055	0.9584
			h_g	1271.8	1311.5	1346.9	1379.7	1440.9	1499.4	1556.9	1614.2	1671.6	1729.4	1787.6
1300.0	1285.3	577.42	$\bar{V}$	0.4052	0.4451	0.4804	0.5129	0.5729	0.6287	0.6822	0.7341	0.7847	0.8345	0.8836
			h_g	1261.9	1303.9	1340.8	1374.6	1437.1	1496.3	1554.3	1612.0	1669.8	1727.9	1786.3
1400.0	1385.3	587.07	$\bar{V}$	0.3667	0.4059	0.4400	0.4712	0.5282	0.5809	0.6311	0.6798	0.7272	0.7737	0.8195
			h_g	1251.4	1296.1	1334.5	1369.3	1433.2	1493.2	1551.8	1609.9	1668.0	1726.3	1785.0
1500.0	1485.3	596.20	$\bar{V}$	0.3328	0.3717	0.4049	0.4350	0.4894	0.5394	0.5869	0.6327	0.6773	0.7210	0.7639
			h_g	1240.2	1287.9	1328.0	1364.0	1429.2	1490.1	1549.2	1607.7	1666.2	1724.8	1783.7
1600.0	1585.3	604.87	$\bar{V}$	0.3026	0.3415	0.3741	0.4032	0.4555	0.5031	0.5482	0.5915	0.6336	0.6748	0.7153
			h_g	1228.3	1279.4	1321.4	1358.5	1425.2	1486.9	1546.6	1605.6	1664.3	1723.2	1782.3
1700.0	1685.3	613.13	$\bar{V}$	0.2754	0.3147	0.3468	0.3751	0.4255	0.4711	0.5140	0.5552	0.5951	0.6341	0.6724
			h_g	1215.3	1270.5	1314.5	1352.9	1421.2	1483.8	1544.0	1603.4	1662.5	1721.7	1781.0
1800.0	1785.3	621.02	$\bar{V}$	0.2505	0.2906	0.3223	0.3500	0.3988	0.4426	0.4836	0.5229	0.5609	0.5980	0.6343
			h_g	1201.2	1261.1	1307.4	1347.2	1417.1	1480.6	1541.4	1601.2	1660.7	1720.1	1779.7
1900.0	1885.3	628.56	$\bar{V}$	0.2274	0.2687	0.3004	0.3275	0.3749	0.4171	0.4565	0.4940	0.5303	0.5656	0.6002
			h_g	1185.7	1251.3	1300.2	1341.4	1412.9	1477.4	1538.8	1599.1	1658.8	1718.6	1778.4
2000.0	1985.3	635.80	$\bar{V}$	0.2056	0.2488	0.2805	0.3072	0.3534	0.3942	0.4320	0.4680	0.5027	0.5365	0.5695
			h_g	1168.3	1240.9	1292.6	1335.4	1408.7	1474.1	1536.2	1596.9	1657.0	1717.0	1777.1
2100.0	2085.3	642.76	$\bar{V}$	0.1847	0.2304	0.2624	0.2888	0.3339	0.3734	0.4099	0.4445	0.4778	0.5101	0.5418
			h_g	1148.5	1229.8	1284.9	1329.3	1404.4	1470.9	1533.6	1594.7	1655.2	1715.4	1775.7
2200.0	2185.3	649.45	$\bar{V}$	0.1636	0.2134	0.2458	0.2720	0.3161	0.3545	0.3897	0.4231	0.4551	0.4862	0.5165
			h_g	1123.9	1218.0	1276.8	1323.1	1400.0	1467.6	1530.9	1592.5	1653.3	1713.9	1774.4
2300.0	2285.3	655.89	$\bar{V}$	...	0.1975	0.2305	0.2566	0.2999	0.3372	0.3714	0.4035	0.4344	0.4643	0.4935
			h_g	...	1205.3	1268.4	1316.7	1395.7	1464.2	1528.3	1590.3	1651.5	1712.3	1773.1
2400.0	2385.3	662.11	$\bar{V}$	...	0.1824	0.2164	0.2424	0.2850	0.3214	0.3545	0.3856	0.4155	0.4443	0.4724
			h_g	...	1191.6	1259.7	1310.1	1391.2	1460.9	1525.6	1588.1	1649.6	1710.8	1771.8
2500.0	2485.3	668.11	$\bar{V}$	...	0.1681	0.2032	0.2293	0.2712	0.3068	0.3390	0.3692	0.3980	0.4259	0.4529
			h_g	...	1176.7	1250.6	1303.4	1386.7	1457.5	1522.9	1585.9	1647.8	1709.2	1770.4
2600.0	2585.3	673.91	$\bar{V}$	...	0.1544	0.1909	0.2171	0.2585	0.2933	0.3247	0.3540	0.3819	0.4088	0.4350
			h_g	...	1160.2	1241.1	1296.5	1382.1	1454.1	1520.2	1583.7	1646.0	1707.7	1769.1
2700.0	2685.3	679.53	$\bar{V}$	...	0.1411	0.1794	0.2058	0.2468	0.2809	0.3114	0.3399	0.3670	0.3931	0.4184
			h_g	...	1142.0	1231.1	1289.5	1377.5	1450.7	1517.5	1581.5	1644.1	1706.1	1767.8
2800.0	2785.3	684.96	$\bar{V}$	...	0.1278	0.1685	0.1952	0.2358	0.2693	0.2991	0.3268	0.3532	0.3785	0.4030
			h_g	...	1121.2	1220.6	1282.2	1372.8	1447.2	1514.8	1579.3	1642.2	1704.5	1766.5
2900.0	2885.3	690.22	$\bar{V}$	...	0.1138	0.1581	0.1853	0.2256	0.2585	0.2877	0.3147	0.3403	0.3649	0.3887
			h_g	...	1095.3	1209.6	1274.7	1368.0	1443.7	1512.1	1577.0	1640.4	1703.0	1765.2
3000.0	2985.3	695.33	$\bar{V}$	...	0.0982	0.1483	0.1759	0.2161	0.2484	0.2770	0.3033	0.3282	0.3522	0.3753
			h_g	...	1060.5	1197.9	1267.0	1363.2	1440.2	1509.4	1574.8	1638.5	1701.4	1763.8
3100.0	3085.3	700.28	$\bar{V}$	...	...	0.1389	0.1671	0.2071	0.2390	0.2670	0.2927	0.3170	0.3403	0.3628
			h_g	...	...	1185.4	1259.1	1358.4	1436.7	1506.6	1572.6	1636.7	1699.8	1762.5
3200.0	3185.3	705.08	$\bar{V}$	...	...	0.1300	0.1588	0.1987	0.2301	0.2576	0.2827	0.3065	0.3291	0.3510
			h_g	...	...	1172.3	1250.9	1353.4	1433.1	1503.8	1570.3	1634.8	1698.3	1761.2
3300.0	3285.3	...	$\bar{V}$	...	...	0.1213	0.1510	0.1908	0.2218	0.2488	0.2734	0.2966	0.3187	0.3400
			h_g	...	...	1158.2	1242.5	1348.4	1429.5	1501.0	1568.1	1623.9	1696.7	1759.9
3400.0	3385.3	...	$\bar{V}$	...	...	0.1129	0.1435	0.1834	0.2140	0.2405	0.2646	0.2872	0.3088	0.3296
			h_g	...	...	1143.2	1233.7	1343.4	1425.9	1498.3	1565.8	1631.1	1695.1	1758.5

No.	Compound	Formula	Molecular Weight	Boiling Point °F., 14.696 psi., abs.	Vapor Pressure 100°F., psi., abs.	Freezing Point, °F., 14.696 psi., abs.	Critical Constants			Density of Liquid; 60°F., 14.696 psi., abs.				
							Pressure, psi., abs.	Temperature, °F.	Volume, cu. ft. per lb.	Specific Gravity 60°F./60°F. a,b	lb. per gal.* a (Wt. in Vacuum)	lb. per gal.a,c * (Wt. in Air)	Gal/Lb. Mol*	Temperature Coefficient of Density a,*
1	Methane	CH_4	16.042	−258.68	—	−296.46[d]	673.1	−115.78	0.0991	0.3[i]	2.5[i]	2.5[i]	6.4[i]	—
2	Ethane	C_2H_6	30.068	−127.53	—	−297.89[d]	709.8	+90.32	0.0788	0.3771[h]	3.144[h]	3.144[h]	9.56[h]	—
3	Propane	C_3H_8	44.094	−43.73	190	−305.84[d]	617.4	206.26	0.0728	0.5077[h]	4.233[h]	4.220[h]	10.42[h]	0.00152[h]
4	n-Butane	C_4H_{10}	58.120	+31.10	51.6	−217.03	550.7	305.62	0.0702	0.5844[h]	4.872[h]	4.865[h]	11.93[h]	0.00117[h]
5	2-Methylpropane (isobutane)	C_4H_{10}	58.120	+10.89	72.2	−255.28	529.1	274.96	0.0724	0.5631[h]	4.695[h]	4.686[h]	12.38[h]	0.00119[h]
6	n-Pentane	C_5H_{12}	72.146	96.93	15.570	−201.50	489.5	385.5	0.0690	0.6312	5.262	5.253	13.71	0.00087
7	2-Methylbutane (isopentane)	C_5H_{12}	72.146	82.13	20.44	−255.82	483.	369.0	0.0685	0.6248	5.209	5.199	13.85	0.00090
8	2,2-Dimethylpropane (neopentane)	C_5H_{12}	72.146	49.10	35.9	+2.21	464.0	321.08	0.0674	0.5967[h]	4.975[h]	4.965[h]	14.50[h]	0.00104[h]
9	n-Hexane	C_6H_{14}	86.172	155.73	4.956	−139.63	440.0	454.1	0.0685	0.6640	5.536	5.526	15.57	0.00075
10	2-Methylpentane	C_6H_{14}	86.172	140.49	6.767	−244.61	440.1	435.7	0.0681	0.6579	5.485	5.476	15.71	0.00078
11	3-Methylpentane	C_6H_{14}	86.172	145.91	6.098	—	453.1	448.2	0.0681	0.6690	5.578	5.568	15.45	0.00075
12	2,2-Dimethylbutane (neohexane)	C_6H_{14}	86.172	121.53	9.856	−147.77	450.7	420.1	0.0667	0.6540	5.453	5.443	15.80	0.00078
13	2,3-Dimethylbutane	C_6H_{14}	86.172	136.38	7.404	−199.37	455.4	440.2	0.0665	0.6664	5.556	5.546	15.51	0.00075
14	n-Heptane	C_7H_{16}	100.198	209.17	1.620	−131.10	396.8	512.62	0.0682	0.6882	5.738	5.728	17.46	0.00069
15	2-Methylhexane	C_7H_{16}	100.198	194.09	2.271	−180.90	400.	495.	0.0685	0.6830	5.694	5.685	17.60	0.00068
16	3-Methylhexane	C_7H_{16}	100.198	197.33	2.130	—	413.	504.4	0.0668	0.6915	5.765	5.758	17.38	0.00069
17	3-Ethylpentane	C_7H_{16}	100.198	200.26	2.012	−181.49	420.	513.8	0.0665	0.7026	5.858	5.849	17.10	0.00070
18	2,2-Dimethylpentane	C_7H_{16}	100.198	174.55	3.492	−190.86	417.	477.9	0.0646	0.6783	5.655	5.645	17.72	0.00072
19	2,4-Dimethylpentane	C_7H_{16}	100.198	176.90	3.292	−182.64	403.	476.9	0.0671	0.6772	5.646	5.637	17.75	0.00072
20	3,3-Dimethylpentane	C_7H_{16}	100.198	186.92	2.773	−210.03	440.	505.0	(0.067)	0.6977	5.817	5.807	17.23	0.00065
21	2,2,3-Trimethylbutane (triptane)	C_7H_{16}	100.198	177.59	3.374	−12.84	437.2	497.0	0.0631	0.6945	5.790	5.782	17.31	0.00069
22	n-Octane	C_8H_{18}	114.224	258.20	0.537	−70.23	362.1	563.7	0.0682	0.7068	5.893	5.883	19.38	0.00062
23	2,5-Dimethylhexane (diisobutyl)	C_8H_{18}	114.224	228.39	1.101	−132.16	362.	530.	0.0676	0.6980	5.819	5.810	19.63	0.00065
24	2.2.4-Trimethylpentane ("isooctane")	C_8H_{18}	114.224	210.63	1.708	−161.28	374.7	520.1	0.0676	0.6963	5.805	5.795	19.68	0.00065
25	n-Nonane	C_9H_{20}	128.250	303.44	0.179	−64.33	332.	610.5	0.0679	0.7217	6.017	6.008	21.31	0.00063
26	n-Decane	$C_{10}H_{22}$	142.276	345.42	0.073	−21.39	304.	651.9	0.0679	0.7341	6.120	6.112	23.25	0.00055
27	Cyclopentane	C_5H_{10}	70.130	120.67	9.914	−136.96	654.7	461.48	0.0593	0.7505	6.257	6.247	11.21	0.00070
28	Methylcyclopentane	C_6H_{12}	84.156	161.26	4.503	−224.42	549.0	499.30	0.0607	0.7535	6.282	6.274	13.40	0.00071
29	Cyclohexane	C_6H_{12}	84.156	177.33	3.264	+43.80	591.5(19)	535.6(19)	0.0586(17)	0.7834	6.531	6.522	12.89	0.00068
30	Methylcyclohexane	C_7H_{14}	98.182	213.68	1.609	−195.87	504.4	570.2	0.0562	0.7740	6.453	6.443	15.21	0.00063
31	Ethene (ethylene)	C_2H_4	28.052	−154.68	—	−272.47[d]	742.1	49.82	0.0706	0.35(16)	—	—	—	—
32	Propene	C_3H_6	42.078	−53.86	226.4	−301.45[d]	667.	197.4	0.0689	0.5220[h]	4.352[h]	4.343[h]	9.67[h]	0.00189[h]
33	1-Butene	C_4H_8	56.104	+20.73	63.05	−301.63[d]	583.	295.6	0.0689	0.6013[h]	5.013[h]	5.004[h]	11.19[h]	0.00116[h]
34	cis-2-Butene	C_4H_8	56.104	38.70	45.54	−218.04	600.	324.3	0.0503	0.6271[h]	5.228[h]	5.219[h]	10.73[h]	0.00098[h]
35	trans-2-Butene	C_4H_8	56.104	33.58	49.80	−157.99	600.	311.9	0.0503	0.6100[h]	5.086[h]	5.076[h]	11.03[h]	0.00107[h]
36	2-Methylpropene (isobutene)	C_4H_8	56.104	19.58	63.40	−220.63	579.8	292.5	0.0513	0.6004[h]	5.006[h]	4.996[h]	11.21[h]	0.00120[h]
37	1-Pentene	C_5H_{10}	70.130	85.94	19.12	−265.40	586.	376.9	(0.0672)	0.6457	5.383	5.374	13.03	0.00089
38	1,2-Butadiene	C_4H_6	54.088	51.53	(20)	−213.14	(653)	(339)	(0.0649)	0.658[h]	5.486[h]	5.47[h]	9.86[h]	0.00098[h]
39	1,3-Butadiene	C_4H_6	54.088	24.06	(60)	−164.05	628.	306.	(0.0654)	0.6861	5.229[h]	5.220[h]	10.34[h]	0.00113[h]
40	2-Methyl-1,3-butadiene (isoprene)	C_5H_8	68.114	93.32	16.672	−230.71	(558)	(412)	(0.0650)	.6861	5.720	5.711	11.91	0.00086
41	Ethyne (acetylene)	C_2H_2	26.036	−119[e]	—	−114[d]	905.	97.4	0.0695	0.6150[k]	—	—	—	—
42	Benzene	C_6H_6	78.108	176.18	3.224	+41.96	714.	552.0	0.0535	0.8845	7.374	7.365	10.59	0.00066
43	Toluene	C_7H_8	92.134	231.12	1.032	−138.98	590.	605.5	0.0564	0.8719	7.269	7.260	12.67	0.00060
44	Ethylbenzene	C_8H_{10}	106.160	277.13	0.371	−138.96	540.	651.2	0.0556	0.8717	7.268	7.259	14.61	0.00054
45	1,2-Dimethylbenzene (o-xylene)	C_8H_{10}	106.160	291.94	0.264	−13.33	530.	674.8	0.0584	0.8848	7.377	7.367	14.39	0.00055
46	1,3-Dimethylbenzene (m-xylene)	C_8H_{10}	106.160	282.39	0.326	−54.17	510.	649.9	0.0584	0.8687	7.243	7.234	14.66	0.00054
47	1,4-Dimethylbenzene (p-xylene)	C_8H_{10}	106.160	281.03	0.324	+55.87	500.	649.4	0.0556	0.8657	7.218	7.209	14.71	0.00054
48	Styrene (Phenyl Ethylene)	C_8H_8	104.144	293.25	(0.24)	−23.13	580.	706.0	0.0541	0.9111	7.596	7.586	13.71	0.00057
49	Isopropylbenzene (cumene)	C_9H_{12}	120.186	306.31	0.188	−140.86	460.	724.5	0.0591	0.8663	7.223	7.213	16.64	0.00054
50	Methyl Alcohol	CH_4O	32.042	148.1(2)	4.4(7)	−144.0(2)	1154.(17)	464.0(17)	0.0590(17)	0.796(3)	6.64	6.63	4.83	—
51	Ethyl Alcohol	C_2H_6O	46.069	173.3(2)	2.3(7)	−179.1(2)	926.(17)	469.(17)	0.0582(17)	0.794(3)	6.62	6.61	6.96	—
52	Carbon Monoxide	CO	28.010	−313.6(2)	—	−340.6(2)	507.(17)	−220.(17)	0.0532(17)	0.801[m](8)	—	—	—	—
53	Carbon Dioxide	CO_2	44.010	−109.3(2)	—	—	1071.(17)	87.8(17)	0.0343(17)	0.827[h](6)	6.90[h]	6.89[h]	6.38[h]	—
54	Hydrogen Sulfide	H_2S	34.076	−75.3(7)	394.0(6)	−117.2(7)	1306.(17)	212.7(17)	0.0461(17)	0.79[h](6)	6.59[h]	6.58[h]	5.17[h]	—
55	Sulfur Dioxide	SO_2	64.060	+14.0(7)	88.(7)	−103.9(7)	1143.(17)	315.5(17)	0.0304(17)	1.397[h](14)	11.65[h]	11.63[h]	5.50[h]	—
56	Ammonia	NH_3	17.032	−28.0(2)	212.(7)	−107.9(2)	1636.(17)	270.1(17)	0.0681(17)	0.6173(11)	5.15	5.14	3.31	—
57	Air	N_2O_2	28.966	−317.6(2)	—	—	547.(2)	−221.3(2)	0.0517(3)	0.856[m](8)	—	—	—	—
58	Hydrogen	H_2	2.016	−422.9(2)	—	−434.5(2)	188.1(17)	−399.8(17)	0.5159(17)	0.07[m](3)	—	—	—	—
59	Oxygen	O_2	32.000	−297.4(2)	—	−361.1(2)	736.(17)	−181.1(17)	0.0375(17)	1.14[m](3)	—	—	—	—
60	Nitrogen	N_2	28.016	−320.4(2)	—	−345.7(2)	492.(17)	−232.6(17)	0.0514(17)	0.808[m](3)	—	—	—	—
61	Chlorine	Cl_2	70.914	−30.3(2)	158.(7)	−150.9(2)	1120.(17)	291.(17)	0.0281(17)	1.414(14)	11.79	11.78	6.02	—
62	Water	H_2O	18.016	+212.0	0.9492(12)	32.0	3208.(17)	705.6(17)	0.0500(17)	1.000	8.337	8.328	2.16	—
63	Hydrogen Chloride	HCl	36.465	−121(16)	925(7)	−173.6(16)	1198.(17)	124.5(17)	0.0208(17)	0.8558(14)	7.135	7.127	5.11	0.003346

Table 4.21 continued

Table 4.21. Physical constants of hydrocarbons. Courtesy Gas Processors Suppliers Association.

Table 4.21 continued

Pitzer Accentric Factor (18)	Compressibility Factor of real gas, Z 14,696 psia, 60° F	Gas Density, 60 F., 14.696 psi., abs. Ideal Gas*: Specific Gravity Air = 1 *	Gas Density: cu. ft. Gas per lb. *	Gas Density: cu. ft. Gas per gal. Liquid *	Specific Heat; 60 F. 14.696 psi., abs. Cp Btu./lb/°F: Ideal Gas	Specific Heat Cp: Liquid	Calorific Value; 60 F*: Net, Btu. per cu. ft. Ideal Gas, 14,696 psi., abs. (20) *	Calorific Value, 60 F*: Gross, Btu. per cu. ft. Ideal Gas, 14,696 psi., abs. (20) *	Calorific Value, 60 F*: Gross, Btu. per lb. Liquid (Wt. in Vacuum) *	Calorific Value, 60 F*: Gross, Btu. per gal. Liquid *	Heat of Vaporization, 14.696 psi., abs. at Boiling Point, Btu. per lb.	Refractive Index*, nD 68° F.	Air Required for Combustion* Ideal Gas: cu. ft. per cu. ft.	Flammability Limits, Vol. % in Air Mixture: Lower	Flammability Limits: Higher	A.S.T.M. Octane Number: Motor Method D-357	A.S.T.M. Octane Number: Research Method D-908	No.
0.013	0.9981	0.5538	23.66	59.[i]	0.5271		909	1010	—	—	219.2	—	9.55	5.0	15.0	—	—	1
0.105	0.9916	1.038	12.62	39.7[h]	0.4097	0.926[h]	1618	1769	—	—	210.4	—	16.71	2.9	13.0	+0.05[f]	+1.6[f,j]	2
0.152	0.9820	1.522	8.606	36.43[h]	0.3885	0.592[h]	2316	2517	21,498	—	183.1	—	23.87	2.1	9.5	97.1	+1.8[f,j]	3
0.201	0.9667	2.006	6.529	31.81[h]	0.3908	0.564[h]	3011	3262	21,136	102,980	165.7	1.3326[h]	31.03	1.8	8.4	89.6[j]	93.8[j]	4
0.192	0.9699	2.006	6.529	30.66[h]	0.3872	0.570[h]	3001	3253	21,086	98,990	157.5	—	31.03	1.8	8.4	97.6	+0.10[f,j]	5
0.252	0.9435	2.491	5.260	27.68	0.3883	0.542	3707	4009	20,926	110,120	153.6	1.35748	38.19	1.4	8.3	62.6[j]	61.7[j]	6
0.206	0.9482	2.491	5.260	27.40	0.3827	0.5353	3698	4000	20,887	108,800	147.1	1.35373	38.19	1.4	(8.3)	90.3	92.3	7
0.195	(0.95)	2.491	5.260	26.17[h]	0.3914	0.554	3685	3987	20,836	103,650	135.6	1.342[h]	38.19	1.4	(8.3)	80.2	85.5	8
0.290	—	2.975	4.404	24.38	0.3864	0.5333	4403	4756	20,784	115,060	144.0	1.37486	45.35	1.2	7.7	26.0	24.8	9
—	—	2.975	4.404	24.16	—	0.5264	4395	4747	20,756	113,850	138.7	1.37145	45.35	1.2	(7.7)	73.5	73.4	10
—	—	2.975	4.404	24.56	—	0.507	4398	4751	20,768	115,840	140.1	1.37652	45.35	(1.2)	(7.7)	74.3	74.5	11
—	—	2.975	4.404	24.01	—	0.5165	4382	4735	20,711	112,930	131.2	1.36876	45.35	(1.2)	(7.7)	93.4	91.8	12
—	—	2.975	4.404	24.47	—	0.5127	4391	4744	20,743	115,250	136.1	1.37495	45.35	(1.2)	(7.7)	94.3	+0.3[f]	13
0.352	—	3.459	3.787	21.73	0.3853	0.5276	5100	5502	20,681	118,660	136.0	1.38764	52.51	1.0	7.0	0.0	0.0	14
—	—	3.459	3.787	21.57	—	0.522	5092	5495	20,658	117,630	131.6	1.38485	52.51	(1.0)	(7.0)	46.4	42.4	15
—	—	3.459	3.787	21.83	—	0.511	5095	5497	20,668	119,160	132.1	1.38864	52.51	(1.0)	(7.0)	55.8	52.0	16
—	—	3.459	3.787	22.19	—	0.506	5098	5500	20,679	121,130	132.8	1.39339	52.51	(1.0)	(7.0)	69.3	65.0	17
—	—	3.459	3.787	21.42	—	0.517	5079	5482	20,620	116,610	125.1	1.38215	52.51	(1.0)	(7.0)	95.6	92.8	18
—	—	3.459	3.787	21.38	—	0.522	5084	5486	20,636	116,510	126.6	1.38145	52.51	(1.0)	(7.0)	83.8	83.1	19
—	—	3.459	3.787	22.03	—	.502	5084	5487	20,638	120,050	127.2	1.39092	52.51	(1.0)	(7.0)	86.6	80.8	20
—	—	3.459	3.787	21.93	—	.498	5081	5483	20,627	119,430	124.2	1.38944	52.51	(1.0)	(7.0)	+0.1[f]	+1.8[f]	21
—	—	3.943	3.322	19.58	0.3845	0.52.30	5796	6249	20,604	121,420	129.5	1.39743	59.67	0.96	—	—	—	22
—	—	3.943	3.322	19.33	—	0.5114	5780	6233	20,564	119,670	122.8	1.39246	59.67	(0.98)	—	55.7	55.2	23
—	—	3.943	3.322	19.29	—	0.4892	5778	6231	20,569	119,390	116.7	1.39145	59.67	1.0	—	100	100	24
—	—	4.428	2.959	17.80	—	0.5220	6493	6996	20,543	123,610	126.7	1.40542	66.83	0.87[s]	2.9	—	—	25
—	—	4.912	2.667	16.32	—	0.5207	7189	7743	20,495	125,440	118.7	1.41189	73.99	0.78[s]	2.6	—	—	26
0.193	—	2.421	5.411	33.86	—	0.4216	3512	3763	20,187	126,310	167.3	1.40645	35.80	(1.4)	—	84.9[j]	+0.1[f]	27
0.234	—	2.905	4.509	28.33	—	0.4407	4198	4500	20,129	126,450	147.8	1.40970	42.96	(1.2)	8.35	80.0	91.3	28
0.186	—	2.905	4.509	29.45	—	0.4332	4180	4482	20,038	130,880	153.7	1.42623	42.96	1.3	7.8	77.2	83.0	29
—	—	3.390	3.865	24.94	—	0.4397	4864	5216	20,003	129,080	138.9	1.42312	50.12	1.2	—	71.1	74.8	30
—	0.9940	0.9684	13.53	—	0.3622	—	1499	1599	—	—	207.6	—	14.32	2.7	34.0	75.6	+0.03[f]	31
0.143	0.9839	1.453	9.019	39.25[h]	0.3541	—	2183	2334	—	—	188.2	—	21.48	2.0	10.0	84.9	+0.2[f]	32
0.203	0.9694	1.937	6.764	33.91[h]	0.3548	—	2879	3081	20,679	103,670	167.9	—	28.64	1.6	9.3	80.8[j]	97.4	33
0.273	0.9653(15)	1.937	6.764	35.36[h]	0.3269	—	2872	3073	20,613	107,770	178.9	—	28.64	(1.6)	—	83.5	100.	34
0.234	0.9654(15)	1.937	6.764	34.40[h]	0.3654	—	2868	3068	20,584	104,680	174.4	—	28.64	(1.6)	—	—	—	35
0.201	0.9687(15)	1.937	6.764	33.86[h]	0.3701	—	2860	3061	20,548	102,860	169.5	—	28.64	(1.6)	—	—	—	36
—	(0.95)	2.421	5.411	29.13	—	0.5196	3576	3827	20,550	110,630	154.5	1.37148	35.80	(1.4)	8.7	77.1	90.9	37
—	(0.97)	1.867	7.016	38.49[h]	0.3458	—	2789	2940	—	—	(181)	—	26.25	(2.0)	(12)	—	—	38
—	0.975	1.867	7.016	36.69[h]	0.3412	—	2730	2881	20,039	104,790	(174)	—	26.25	2.0	11.5	—	—	39
—	(0.96)	2.352	5.571	31.87	—	0.5245	3411	3612	19,955	114,150	(153)	1.42194	33.41	(1.5)	—	81.0	99.1	40
0.186	0.9925	0.8988	14.58	—	0.3966	—	1422	1473	—	—	—	—	11.93	2.5	80	—	—	41
0.215	0.929(15)	2.697	4.859	35.83	—	0.4098	3591	3742	17,991	132,670	169.3	1.50112	35.80	1.3[g]	7.9[g]	+2.8[f]	—	42
0.252	0.903(21)	3.181	4.119	29.94	—	0.4017	4273	4475	18,251	132,670	156.2	1.49693	42.96	1.2[g]	7.1[g]	+0.3[f]	+5.8[f]	43
—	—	3.665	3.575	25.98	—	0.4114	4970	5222	18,494	134,110	145.7	1.49588	50.12	0.99[g]	6.7[g]	+97.9	+0.8[f]	44
—	—	3.665	3.575	26.37	—	0.4418	4958	5210	18,445	136,070	149.1	1.50545	50.12	1.1[g]	6.4[g]	100.	—	45
—	—	3.665	3.575	25.89	—	0.4045	4957	5208	18,441	133,560	147.4	1.49722	50.12	1.1[g]	6.4[g]	+2.8[f]	+4.0[f]	46
—	—	3.665	3.575	25.80	—	0.4083	4957	5209	18,445	133,130	146.1	1.49582	50.12	1.1[g]	6.6[g]	+1.2[f]	+3.4[f]	47
—	—	3.595	3.644	27.68	—	0.4122	4830	5031	18,150	137,870	(151)	1.54682	47.73	1.1	6.1	+0.2[f]	>+3[f]	48
—	—	4.149	3.158	22.81	—	(0.41)	5661	5963	18,653	134,710	134.3	1.49145	57.28	0.88[g]	6.5[g]	99.3	+2.1[f]	49
—	—	1.106	11.84	78.6	—	0.594(7)	—	—	9,760(16)	64,810	473(2)	1.3288(8)	7.15(7)	6.72(5)	36.50	—	—	50
—	—	1.590	8.237	54.5	—	0.562(7)	—	—	12,780(16)	84,600	367(2)	1.3614(8)	14.30(7)	3.28(5)	18.95	—	—	51
0.041	0.9995(15)	0.9670	13.55	—	0.2484(13)	—	—	321[r](13)	—	—	92.7(14)	—	2.38(7)	12.50(5)	74.20	—	—	52
0.225	0.9943(15)	1.519	8.623	59.4[h]	0.1991(13)	—	—	—	—	—	238.2[n](14)	—	—	—	—	—	—	53
0.100	0.9903(15)	1.176	11.14	73.4[h]	0.238(4)	—	588(16)	637(16)	—	—	235.6(7)	—	7.15(7)	4.30(5)	45.50	—	—	54
—	—	2.212	5.924	69.0[h]	0.145(7)	0.325[h](7)	—	—	—	—	166.7(14)	—	—	—	—	—	—	55
—	—	0.5880	22.28	114.7	0.5002(10)	1.114[h](7)	359(16)	434(16)	—	—	587.2(14)	—	3.57(7)	15.50(5)	27.00	—	—	56
—	0.9996(15)	1.000	13.10	—	0.2400(9)	—	—	—	—	—	92.(3)	—	—	—	—	—	—	57
—	1.0006	0.0696	188.2	—	3.408(13)	—	274[r](13)	324[r](13)	—	—	193.9(14)	—	2.38(7)	4.00(5)	74.20	—	—	58
0.0213	—	1.105	11.86	—	0.2188(13)	—	—	—	—	—	91.6(14)	—	—	—	—	—	—	59
0.040	0.9997(15)	0.9672	13.55	—	0.2482(13)	—	—	—	—	—	87.8(14)	—	—	—	—	—	—	60
—	—	2.448	5.351	63.1	0.119(7)	—	—	—	—	—	123.8(14)	—	—	—	—	—	—	61
—	—	0.6220	21.06	175.6	0.4446(13)	1.0009(7)	—	—	—	—	970.3(12)	1.3330(8)	—	—	—	—	—	62
—	—	1.259	10.41	74.3	0.190(7)	—	—	—	—	—	185.5(14)	—	—	—	—	—	—	63

Table 4.21 continued

Table 4.21 continued

1. Values for hydrocarbons 1-49 were selected or calculated from data in ASTM Special Technical Publication No. 109A, "Physical Constants of Hydrocarbons C_1 - C_{10}, 1963, **American Society for Testing Materials,** 1916 Race Street, Philadelphia.
2. International Critical Tables.
3. Hodgman, Handbook of Chemistry & Physics, 31 Edition (1949).
4. West, J. R., **Chemical Engineering Progress,** 44, 287 (1948).
5. Jones, **Chemical Review,** 22, 1 (1938).
6. Sage & Lacey, API Research Project 37, Monograph (1955).
7. Perry, Chemical Engineers Handbook, 4th Edition (1963).
8. Matteson and Hanna, **Oil and Gas Journal,** 41, No. 2, 33 (1942).
9. Keenan & Keyes, Thermodynamic Properties of Air (1947).
10. Grahl, Edw. R., "Thermodynamic Properties at High Temperature and Pressure," **Petroleum Processing,** April, 1963.
11. NBS Circular No. 142, Thermodynamic Properties of Ammonia (1945).
12. Keenan & Keyes, Thermodynamic Properties of Steam (Twenty-ninth printing 1956).
13. A.P.I. Project 44.
14. Dreisbach, Physical Properties of Chemical Compounds **American Chemical Society,** 1961.
15. Institute of Gas Technology, Research Bulletin No. 32 (1961).
16. Maxwell, J. B., Data Book on Hydrocarbons, Van Nostrand Co. 1950.
17. Kobe, K. A. & Lynn, Jr., R. E., **Chemical Review,** 52, 117-236 (1953).
18. **Journal American Chemical Society,** 77, 3434 (1955).
19. Kay, W. B. & Albert, R. E., "Liquid-Vapor Equilibrium Relations in the Ethane-Cyclohexane System," **Industrial Engineering Chemistry,** 48, 422 (1956).
20. Rossini, Frederick D., Calculations for NGPA.

NOTES

a. Air saturated hydrocarbons.

b. Absolute values from weights in vacuum.

c. The apparent values from weight in air are shown for users' convenience and compliance with ASTM-IP Petroleum Measurement Tables. In the United States and Great Britain, all commercial weights are required by law to be weights in air. All other mass data are on an absolute mass (weight in vacuum) basis.

d. At saturation pressure (Triple Point).

e. Sublimation point.

f. The + sign and number following signify the octane number corresponding to that of 2, 2, 4 trimethylpentane with the indicated number of ml of TEL added.

g. Determined at .212°F.

h. Saturation pressure and 60°F.

i. Apparent value for methane at 60°F.

j. Average value from octane numbers of more than one sample.

k. Specific Gravity, 119 F/60°F (Sublimation point).

m. Density of liquid, gr/ml at normal boiling point.

n. Heat of sublimation.

p. Values in parenthesis are estimated.

q. Calculated from other properties.

r. Values at 77°F.

s. Extrapolated to room temperature from higher temperature.

t. Values for Cp in ideal gas column for i-C_5 n-C_5 n-C_6 n-C_7 and n-C_8 are for information only. For components 31 thru 64 Cp shown for either gas or liquid depending on the physical state at 60°F and 14.696 psia or as otherwise noted.

* Calculated Values

CONSTANTS FOR USE IN CALCULATIONS

Atomic Weights: Carbon - 12:01; Hydrogen - 1.008; 1 Gal = 3785.41 milliliters.

Molecular Weight of Air = 28.966 1 cu ft = 7.4805 gal.

1 Cu Ft = 28.317 Liters 1 Lb = 453.59 Grams.

Ideal gas @ 60°F and 14.696 lb/sq in abs = 379.49 cu ft/lb mol.

Ideal gas @ 32°F and 14.696 lb/sq in abs = 22.414 liters/gram mol.

760 mm Hg = 14.696 lb/sq inch = 1 Atm 0°F = 459.69° Rankine.

Density of Water @ 60°F = 8.3372 lbs/gal = 0.999015 g/cc (weight in vacuum).

Sp Gr. @ 60°F/60°F x 0.999015 = Density at 60°F in g/cc.

$$°F = 1.8\ (C° + 40) - 40 \qquad °API = \frac{141.5}{\text{Sp gr @ 60°F/60°F}} - 131.5$$

CALCULATED VALUES

DENSITY OF LIQUID @ 60°F and 14.696 psia

Lb/gal @ 60°F (weight-in-vacuum) =

Sp Gr @ 60°F (weight-in-vacuum) x 8.3372 lb/gal (weight-in-vacuum).

Lb/gal @ 60°F (weight-in-air) = See ASTM 109A Pg 61 and API Physical Constants of Hydrocarbons (1961) Pg 4.

$$\text{Gal/lb mol @ 60°F} = \frac{\text{Mol wt}}{\text{Lb/gal @ 60°F (weight-in-vacuum)}}.$$

TEMPERATURE COEFFICIENT OF DENSITY See ASTM 109A Pg 61 and API Physical Constants of Hydrocarbons (1961) Pg 5.

DENSITY OF GAS @ 60°Γ and 14.696 psia (Ideal Gas).

$$\text{Sp Gr. @ 60°F} = \frac{\text{Mol weight}}{28.966}$$

$$\text{Cu ft gas/lb} = \frac{379.49}{\text{Mol wt}}$$

$$\text{Cu ft gas/gal liq} = \frac{\text{lb/gal @ 60°F (weight-in-vacuum) x 379.49}}{\text{Mol wt}}$$

HEAT OF COMBUSTION @ 60°F

See ASTM Pgs 61, 62 and API Physical Constants of Hydrocarbons (1961) Pg 5.

AIR REQUIRED FOR COMBUSTION (Ideal Gas) — $C_a\ H_b$

$$\frac{\text{Cu ft air}}{\text{Cu ft gas}} = \frac{\left(a + \frac{b}{4}\right)}{0.2095}$$

See ASTM 109A Pgs 61, 62 and API Physical Constants of Hydrocarbons (1961) Pg 7.

Compound	Methane	Ethane	Propane	Iso-Butane	N-Butane	Iso-Pentane	N-Pentane	N-Hexane	N-Heptane	N-Octane	N-Nonane	N-Decane
Molecular Weight	16.042	30.068	44.094	58.120	58.120	72.146	72.146	86.172	100.198	114.224	128.250	142.276
Freezing Point @ 14.696 psia—°F	−296.5[d]	−297.9[d]	−305.8[d]	−255.3	−217.0	−255.8	−201.5	−139.6	−131.1	− 70.2	− 64.3	− 21.4
°C	−182.5[d]	−183.3[d]	−187.7[d]	−159.6	−138.3	−159.9	−129.7	− 95.3	− 90.6	− 56.8	− 53.5	− 29.7
Boiling Point @ 14.696 psia—°F	−258.7	−127.5	− 43.7	10.9	31.1	82.1	96.9	155.7	209.2	258.2	303.4	345.4
°C	−161.5	− 88.6	− 42.1	− 11.7	− 0.5	27.8	36.1	68.7	98.4	125.7	150.8	174.1
Density of Liquid @ 60°F and 14.696 psia												
Specific Gravity @ 60°F / 60°F[ba]	0.3[i]	0.3771[h]	0.5077[h]	0.5631[h]	0.5844[h]	0.6248	0.6312	0.6640	0.6882	0.7068	0.7217	0.7341
°API[ba]	* 340[i]	243.7[h]	147.2[h]	119.8[h]	110.6[h]	95.0	92.7	81.6	74.1	68.7	64.6	61.3
Lb/Gal @ 60°F Wt. in Vacuum	* 2.5[i]	3.14[h]	4.233[h]	4.695[h]	4.872[h]	5.209	5.262	5.536	5.738	5.893	6.017	6.120
Lb/Gal @ 60°F[c] Wt. in Air	* 2.5[i]	3.13[h]	4.220[h]	4.686[h]	4.865[h]	5.199	5.253	5.526	5.728	5.883	6.008	6.112
Gal/Lb Mol @ 60°F	* 6.4[i]	9.56[h]	10.42[h]	12.38[h]	11.93[h]	13.85	13.71	15.57	17.46	19.38	21.31	23.25
Density of Gas @ 60°F and 14.696 psia												
Specific Gravity Air = 1.00 — Ideal Gas	* 0.554	1.038	1.522	2.006	2.006	2.491	2.491	2.975	3.459	3.943	4.428	4.912
Lb/M Cu Ft — Ideal Gas	* 42.27	79.23	116.19	153.15	153.15	190.11	190.11	227.07	264.03	300.99	337.95	374.91
Cu Ft Gas/Gal Liquid — Ideal Gas	* 59[i]	39.7[h]	36.43[h]	30.66[h]	31.81[h]	27.40	27.68	24.38	21.73	19.58	17.80	16.32
Ratio, Gas Vol./Liquid Vol. — Ideal Gas	* 442.4[i]	296.8[h]	272.5[h]	229.3[h]	238.0[h]	205.0	207.0	182.4	162.6	146.5	133.2	122.1
Critical Conditions												
Temperature—°F	−115.8	90.3	206.3	275.0	305.6	369.0	385.5	454.1	512.6	563.7	610.5	651.9
Temperature—°C	− 82.1	32.4	96.8	135.0	152.0	187.2	196.4	234.5	267.0	295.4	321.4	344.4
Pressure — Atmosphere	45.8	48.3	42.0	36.0	37.5	32.9	33.3	29.9	27.0	24.6	22.6	20.7
Pressure — Psia	673	710	617	529	551	483	490	440	397	362	332	304
Gross Heat of Combustion @ 60°F												
BTU/Lb Gas — Ideal Gas[(m)] Wt.-in-Vacuum	* 23,885	22,323	21,664	21,238	21,299	21,041	21,088	20,944	20,840	20,762	20,701	20,652
BTU/Cu Ft — Ideal Gas[p(m)]	* 1,010	1,769	2,517	3,253	3,262	4,000	4,009	4,756	5,502	6,249	6,996	7,743
BTU/Gal Liquid @ 60°F	* —	69,522	91,001	98,990	102,980	108,800	110,120	115,060	118,660	121,420	123,610	125,440
Flammability Limits												
Lower % in Air	5.0	2.9	2.1	1.8	1.8	1.4	1.4	1.2	1.0	0.96	0.87[j]	0.78[j]
Upper % in Air	15.0	13.0	9.5	8.4	8.4	8.3[m]	8.3	7.7	7.0	—	2.9	2.6
Cu. Ft. of air to burn 1 cu. ft. gas—Ideal Gas	* 9.55	16.71	23.87	31.03	31.03	38.19	38.19	45.35	52.51	59.67	66.83	73.99
Heat of Vaporization @ 14.696 psia												
BTU/Lb @ Boiling Point	219.2	210.4	183.1	157.5	165.7	147.1	153.6	144.0	136.0	129.5	126.7	118.7
Specific Heat @ 60°F and 14.696 psia[n]												
C_p Gas BTU/Lb — Ideal Gas	0.5271	0.4097	0.3885	0.3872	0.3908	0.3827	0.3883	0.3864	0.3853	0.3845	—	—
C_v Gas BTU/Lb — Ideal Gas	* 0.403	0.344	0.343	0.353	0.357	0.355	0.361	0.363	0.366	0.367	—	—
$N = C_p/C_v$ BTU/Lb — Ideal Gas	* 1.307	1.192	1.131	1.097	1.096	1.077	1.076	1.064	1.054	1.047	—	—
C_p Liquid BTU/Lb	—	0.926[h]	0.592[h]	0.570[h]	0.564[h]	0.535	0.542	0.533	0.528	0.523	0.522	0.521
Vapor Pressure @ 100°F, psia	—	—	190	72.2	51.6	20.4	15.6	4.96	1.62	0.54	0.18	0.07
Aniline Point — °F	—	—	—	225.7	181.6	170.6	159.3	155.5	157.5	159.1	164.7	170.6
Refractive Index[a] $N^{20°C}/_D$ @ 68°F	—	—	—	—	1.3326[h]	1.3537	1.3575	1.3749	1.3876	1.3974	1.4054	1.4119
Octane Number (Motor)	—	+0.05[f]	97.1	97.6	89.6[k]	90.3	62.6[k]	26.0	0	—	—	—
(Research)	—	+1.6[fk]	1.8[fk]	+0.10[fk]	93.8[k]	92.3	61.7[k]	24.8	0	—	—	—

NOTES

a. Air saturated hydrocarbons.
b. Absolute values from weights in vacuum.
c. The apparent values from weight in air are shown for users' convenience and compliance with ASTM-IP Petroleum Measurement Tables. In the United States and Great Britain, all commercial weights are required by law to be weights in air. All other mass data are on an absolute (weight in vacuum) basis.
d. At saturation pressure (Triple Point).
e. Liquid density in gm/ml at normal boiling point.
f. The + sign and number following signify the octane number of the hydrocarbon in the gas phase corresponding to that of 2, 2, 4-trimethylpentane with the indicated number of ml of TEL added.
g. Heat of sublimation.
h. Saturation pressure and 60°F.
i. Apparent value for methane at 60°F.
j. Extrapolated to room temperature from higher temperature.
k. Average value from octane numbers of more than one sample.
m. Values in parenthesis are estimated.
n. Values for c_p in ideal gas column for i-C_5, n-C_5, n-C_6, n-C_7, and n-C_8 are for information only.
p. Gross heat on dry basis at 60°F and 14.696 psia. To convert to water saturation basis, multiply by 0.9825.
* Calculated values.

Table 4.22. Physical constants of paraffin hydrocarbons and other components of natural gases. Courtesy Natural Gas Processors Suppliers Association.

c_p = specific heat at constant pressure
c_v = specific heat at constant volume

Name of Gas	Chemical Formula or Symbol	Approx. Molecular Weight	Weight Density, Pounds per Cubic Foot*	Specific Gravity Relative To Air	Individual Gas Constant	Specific Heat Per Pound at Room Temperature		Heat Capacity Per Cubic Foot at Atmospheric Pressure and 68 F		k equal to c_p/c_v
		M	ρ	S_g	R	c_p	c_v	c_p	c_v	
Acetylene	C_2H_2	26.0	.06754	.897	59.4	.350	.2737	.0236	.0185	1.28
Air	—	29.0	.07528	1.000	53.3	.241	.1725	.0181	.0130	1.40
Ammonia	NH_3	17.0	.04420	.587	90.8	.523	.4064	.0231	.0179	1.29
Argon	A	40.0	.1037	1.377	38.7	.124	.0743	.0129	.0077	1.67
Carbon Dioxide	CO_2	44.0	.1142	1.516	35.1	.205	.1599	.0234	.0183	1.28
Carbon Monoxide	CO	28.0	.07269	.965	55.2	.243	.1721	.0177	.0125	1.41
Ethylene	C_2H_4	28.0	.0728	.967	55.1	.40	.3292	.0291	.0240	1.22
Helium	He	4.0	.01039	.138	386.	1.25	.754	.0130	.0078	1.66
Hydrochloric Acid	HCl	36.5	.09460	1.256	42.4	.191	.1365	.0181	.0129	1.40
Hydrogen	H_2	2.0	.005234	.0695	767.	3.42	2.435	.0179	.0127	1.40
Methane	CH_4	16.0	.04163	.553	96.4	.593	.4692	.0247	.0195	1.26
Methyl Chloride	CH_3Cl	50.5	.1309	1.738	30.6	.24	.2006	.0314	.0263	1.20
Nitrogen	N_2	28.0	.07274	.966	55.2	.247	.1761	.0179	.0128	1.40
Nitric Oxide	NO	30.0	.07788	1.034	51.5	.231	.1648	.0180	.0128	1.40
Nitrous Oxide	N_2O	44.0	.1143	1.518	35.1	.221	.1759	.0253	.0201	1.26
Oxygen	O_2	32.0	.08305	1.103	48.3	.217	.1549	.0180	.0129	1.40
Sulphur Dioxide	SO_2	64.0	.1663	2.208	24.1	.154	.1230	.0256	.0204	1.25

*Weight density values are at atmospheric pressure and 68 F.
For values at 60 F, multiply by 1.0154.

Table 4.23. Physical properties of gases. Courtesy Crane Company and L.S. Marks, Mechanical Engineers' Handbook, *copyright 1954. Used by permission of McGraw-Hill, Inc.*

Temperature of Water t Degrees Fahrenheit	Saturation Pressure P' Pounds per Square Inch Absolute	Specific Volume $\overline{V}$ Cubic Feet Per Pound	Weight Density ρ Pounds per Cubic Foot	Weight Pounds Per Gallon
32	0.08859	0.016022	62.414	8.3436
40	0.12163	0.016019	62.426	8.3451
50	0.17796	0.016023	62.410	8.3430
60	0.25611	0.016033	62.371	8.3378
70	0.36292	0.016050	62.305	8.3290
80	0.50683	0.016072	62.220	8.3176
90	0.69813	0.016099	62.116	8.3037
100	0.94924	0.016130	61.996	8.2877
110	1.2750	0.016165	61.862	8.2698
120	1.6927	0.016204	61.7132	8.2498
130	2.2230	0.016247	61.550	8.2280
140	2.8892	0.016293	61.376	8.2048
150	3.7184	0.016343	61.188	8.1797
160	4.7414	0.016395	60.994	8.1537
170	5.9926	0.016451	60.787	8.1260
180	7.5110	0.016510	60.569	8.0969
190	9.340	0.016572	60.343	8.0667
200	11.526	0.016637	60.107	8.0351
210	14.123	0.016705	59.862	8.0024
212	14.696	0.016719	59.812	7.9957
220	17.186	0.016775	59.613	7.9690
240	24.968	0.016926	59.081	7.8979
260	35.427	0.017089	58.517	7.8226
280	49.200	0.017264	57.924	7.7433
300	67.005	0.01745	57.307	7.6608
350	134.604	0.01799	55.586	7.4308
400	247.259	0.01864	53.648	7.1717
450	422.55	0.01943	51.467	6.8801
500	680.86	0.02043	48.948	6.5433
550	1045.43	0.02176	45.956	6.1434
600	1543.2	0.02364	42.301	5.6548
650	2208.4	0.02674	37.397	4.9993
700	3094.3	0.03662	27.307	3.6505

Specific gravity of water at 60 F = 1.00

Weight per gallon is based on 7.48052 gallons per cubic foot.

All data on volume and pressure are abstracted from ASME Steam Tables (1967), with permission of publisher, The American Society of Mechanical Engineers, 345 East 47th Street, New York, N.Y. 10017.

Table 4.24. Physical properties of water.

SPECIFIC GRAVITY OF LIQUIDS

LIQUID		TEMP °C	SP GR
Acetic acid		20/4	1.050
Acetone		15.6/15.6	.792
Alcohol, ethyl		20/4	.789
Alcohol, methyl		20/4	.792
Aluminum chloride	- 10%	20/4	1.073
	20%	20/4	1.154
	40%	20/4	1.342
Ammonium hydroxide	- 10%	15/4	.960
	20%	15/4	.925
	30%	15/4	.895
Benzene		20/4	.879
Chlorine		−33.6/4	1.560
Ethylene glycol	- 50%	15.6/15.6	1.070
Gasoline		15.6/15-6	.751
Glycerin		0/4	1.260
Hydrochloric acid	- 10%	20/4	1.047
	20%	20/4	1.098
	30%	20/4	1.149
Kerosene		15.6/15.6	.820
Mercury		15.6/15.6	13.550
Nitric acid	- 10%	20/4	1.054
	30%	20/4	1.180
	50%	20/4	1.310
Oil-Linseed		15/4	.942
Phosphoric acid		20/4	1.870
Potassium hydroxide	- 10%	15/4	1.092
	30%	15/4	1.291
	50%	15/4	1.514
Silicone (DC-200)		15.6/15.6	.920
Sodium chloride	- 10%	20/4	1.071
	20%	20/4	1.148
Sodium hydroxide	- 10%	20/4	1.109
	30%	20/4	1.328
	50%	20/4	1.525
Sulfur		572/4	1.78
Sulfuric acid	- 10%	20/4	1.066
	20%	20/4	1.139
	30%	20/4	1.218
	40%	20/4	1.303
Turpentine		15.6/15.6	.873
Water		15.6/15.6	1.000
Water - Sea		15/4	1.025

Table 4.25. Specific gravity of liquids.

DENSITY & SPECIFIC GRAVITY OF GASES

Gas or Vapor	Formula	Density* lb_m/ft^3	Sp Gr Air = 1
Acetylene	C_2H_2	0.07323	0.9073
Air	------	.08071	1.0000
Ammonia	NH_3	.04813	0.5963
Argon	A	.11135	1.3796
Butane (n)	C_4H_{10}	.15725	2.0854
Butane, iso-	C_4H_{10}	.1669	2.067
Carbon dioxide	CO_2	.12341	1.5290
Carbon monoxide	CO	.07806	0.9671
Carbon oxysulfide	COS	.170	2.10
Chlorine	Cl_2	.2006	2.486
Chlorine monoxide	Cl_2O	.243	3.01
Ethane	C_2H_6	.08469	1.0493
Ethylene	C_2H_4	.07868	0.9749
Fluorine	F_2	.1059	1.312
Helium	He	.01114	0.13804
Hydrogen	H_2	.00561	0.06952
Hydrogen bromide	HBr	.2275	2.8189
Hydrogen chlorine	HCl	.10233	1.2678
Hydrogen iodide	HI	.3614	4.4776
Hydrogen sulfide	H_2S	.09608	1.190
Krypton	Kr	.2315	2.868
Methane	CH_4	.4475	0.5544
Methylamine	CH_3NH_2	.08715	1.080
Methyl chloride	CH_3Cl	.1441	1.7848
Methyl ether	$(CH_3)_2O$	.1317	1.6318
Methyl fluoride	CH_3F	.09646	1.1951
Neon	Ne	.05621	0.69638
Nitric oxide	NO	.08367	1.0366
Nitrogen	N_2	.07807	0.96724
Nitrogen (atm.)	------	.07846	0.9721
Nitrous oxide	N_2O	.1235	1.5297
Nitroxyl fluoride	NO_2F	.181	2.24
Oxygen	O_2	.08921	1.10527
Ozone	O_3	.1338	1.658
Phosphine	PH_3	.09548	1.1829
Phosphorus oxyfluoride	POF_3	.30	3.7
Phosphorus pentafluoride	PF_5	.363	4.494
Propane	C_3H_8	.1254	1.554
Sulfur dioxide	SO_2	.1827	2.2638

*Density at 32°F and 1 atmosphere

Table 4.26. Specific gravity and density of gases.

VISCOSITY OF LIQUIDS

Temperature °F	Viscosity Centipoise
Acetic Acid	
59	1.31
64	1.30
77	1.155
86	1.04
106	1.00
212	.43
Acetic Anhydride	
32	1.24
59	.971
64	.90
86	.783
212	.490
Acetone	
14.0	.450
32.0	.399
59.0	.337
77.0	.316
Ammonia	
-92	.475
-58	.317
-40	.276
-28	.255
Benzene	
32	.912
50	.758
68	.652
86	.564
104	.503
122	.442
Carbon Tetrachloride	
32	1.329
59	1.038
68	.969
86	.843
104	.739
140	.585
Chlorine Liquid	
-40	.505
-20	.462
20	.400
60	.350
100	.313
Ethylbenzene	
63	.691
Ethylene Glycol	
68	19.9
104	9.13
140	4.95
176	3.02
Ethylene Oxide	
-57	.577
-37	.488
-5.8	.394
32.0	.320
Fluorobenzene	
68	.598
140	.389
212	.275
Glycerin	
32	12110
43	6260
59.0	2330
68	1490
77	954
36	629
Heptane	
32	.524
63	.461
68	.409
77	.386
104	.341
Hexane	
32	.401
68	.326
77	.294
104	.271
Hydrochloric Acid 31.5%	
0	3.4
20	2.9
40	2.5
60	2.0
80	1.8
100	1.6
140	1.2
Iodine Liquid	
241	2.27
Isoheptane	
32	.481
68	.384
104	.315
Isohexane	
32	.376
68	.306
104	.254
Isopentane	
32	.273
68	.223
Menthyl Alcohol (Menthanol)	
-48	1.98
32	.82
59	.623
68	.597
77	.547
86	.510
Methyl Chloride	
0	.25
20	.23
40	.21
60	.19
100	.16
Naphthalene	
176	.967
212	.776
Nitric Acid	
32	2.275
Nitrobenzene	
37	2.91
42	2.71
50	2.48
68	2.03
Nitromethane	
32	.853
77	.620
n-Octane	
32	.706
68	.524
104	.433
Penthane	
32	.289
68	.240
Phenol	
65	12.7
122	3.49
158	2.03
194	1.26
Phosphorus Liquid	
71	2.34
88	2.01
110	1.73
123	1.60
140	1.45
158	1.32

VISCOSITY OF LIQUIDS
(continued)

Temperature °F	Viscosity Centipoise
Sodium Hydroxide	
70	100
100	40
120	25
140	15
160	9.5
200	3.7
220	2.4
250	1.4
Sodium Liquid	
0	2.4
26	1.3
Sodium Liquid (Cont'd)	
40	1.2
60	1.1
100	1.0
140	.85
Sulfur (gas free)	
253	10.9
276	8.7
301	7.1
314	7.2
317	7.6
319	14.5
Sulfur Dioxide	
28	.5508
13	.4285
32	.3936
Sulfuric Acid	
32	48.4
59	32.8
68	25.4
86	15.7
104	11.5
122	8.82
140	7.22

VISCOSITY OF WATER

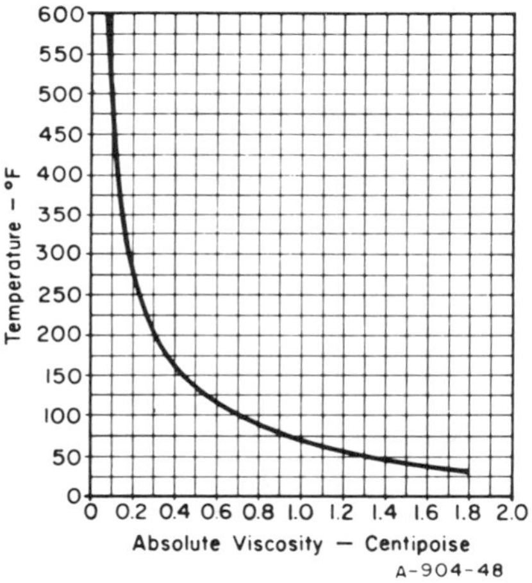

Table 4.27. Viscosity of liquids.

Temperature °F	Viscosity Centipoise
Acetylene	
32	.00935
Air	
32	.00171
104	.00190
444	.00264
633	.00312
674	.00312
768	.00341
Ammonia	
32	.0092
68	.0098
212	.0128
302	.0146
482	.0181
Argon	
32	.0210
68	.0222
212	.0269
392	.0322
Benzene	
0	.0065
40	.0070
70	.0075
100	.0080
200	.0091
Butene	
0	.0075
40	.0080
70	.0085
100	.0090
200	.0104
Butylene	
66	.0074
212	.0095
Carbon Dioxide	
144	.0090
76	.0106
32	.0139
68	.0148
86	.0153
Carbon Monoxide	
32	.0166
59	.0172
260.8	.0218
Chlorine	
122	.0147
212	.0168
302	.0187
392	.0208
Ethane	
32	.0085
63	.0090
Ethylene	
32	.0091
68	.0101
122	.0110
212	.0126
Helium	
32	.0186
68	.0194
Hydrogen	
-172	.0057
-143.5	.0062
-25	.0077
32	.0084
69	.0088
264	.0108
Hydrogen Chloride	
54.4	.0139
61.8	.0141
212	.0182
Hydrogen Sulfide	
32	.0117
62.6	.0124
212	.0159
Methane	
32	.0102
68	.0109
212	.0133
Nitrogen	
-6.7	.0156
51.6	.0171
81	.0178
261	.0219
440	.0256
Oxygen	
32	.0189
67	.0202
262	.0257
440	.0302
n-Pentane	
77	.0068
212	.0084
Propane	
64.2	.0079
213	.0101
Propylene	
62	.0083
122	.0093
Sulfur Dioxide	
32	.0116
64.4	.0124
68.9	.0125
213	.0161

Table 4.28. Viscosity of gases.

Weight density and Specific Volume of Gases and Vapors

Formula

$$\rho = \frac{144\,P'}{RT} = \frac{MP'}{10.72T} = \frac{2.70\,P'S_g}{T}$$

where: $P' = 14.7 + P$

where

ρ = weight density of fluid, pounds per cubic foot
M = molecular weight
P' = pressure, pounds per square inch absolute
R = gas constant: 10.72
T = absolute temperature, in degrees Rankine (460 + °F)
S_g = specific gravity of gas relative to air

Weight Density of Air, in Pounds per Cubic Foot For Gauge Pressures Indicated (Based on an atmospheric pressure of 14.696 and a molecular weight of 28.97)

Air Temp. Deg F.	0 psi	5 psi	10 psi	20 psi	30 psi	40 psi	50 psi	60 psi	70 psi	80 psi	90 psi	100 psi	110 psi	120 psi	130 psi	140 psi	150 psi
30°	.0811	.1087	.1363	.1915	.247	.302	.357	.412	.467	.522	.578	.633	.688	.743	.798	.853	.909
40	.0795	.1065	.1335	.1876	.242	.295	.350	.404	.458	.512	.566	.620	.674	.728	.782	.836	.890
50	.0782	.1048	.1314	.1846	.238	.291	.344	.397	.451	.504	.557	.610	.663	.717	.770	.823	.876
60	.0764	.1024	.1284	.1804	.232	.284	.336	.388	.440	.492	.544	.596	.648	.700	.752	.804	.856
70	.0750	.1005	.1260	.1770	.228	.279	.330	.381	.432	.483	.534	.585	.636	.687	.738	.789	.840
80	.0736	.0986	.1236	.1737	.224	.274	.324	.374	.424	.474	.524	.574	.624	.674	.724	.774	.824
90	.0722	.0968	.1214	.1705	.220	.269	.318	.367	.416	.465	.515	.564	.613	.662	.711	.760	.809
100	.0709	.0951	.1192	.1675	.216	.264	.312	.361	.409	.457	.505	.554	.602	.650	.698	.747	.795
110	.0697	.0934	.1171	.1645	.212	.259	.307	.354	.402	.449	.497	.544	.591	.639	.686	.734	.781
120	.0685	.0918	.1151	.1617	.208	.255	.302	.348	.395	.441	.488	.535	.581	.628	.674	.721	.768
130	.0673	.0902	.1131	.1590	.205	.251	.296	.342	.388	.434	.480	.525	.571	.617	.663	.709	.755
140	.0662	.0887	.1113	.1563	.201	.246	.291	.337	.382	.427	.472	.517	.562	.607	.652	.697	.742
150	.0651	.0873	.1094	.1537	.1981	.242	.287	.331	.375	.420	.464	.508	.553	.597	.641	.686	.730
175	.0626	.0834	.1051	.1477	.1903	.233	.275	.318	.361	.403	.446	.488	.531	.573	.616	.659	.701
200	.0602	.0807	.1011	.1421	.1831	.224	.265	.306	.347	.388	.429	.470	.511	.552	.593	.634	.675
225	.0580	.0777	.0974	.1369	.1764	.216	.255	.295	.334	.374	.413	.453	.492	.531	.571	.610	.650
250	.0559	.0750	.0940	.1321	.1702	.208	.246	.284	.322	.361	.399	.437	.475	.513	.551	.589	.627
275	.0540	.0724	.0908	.1276	.1644	.201	.238	.275	.311	.348	.385	.422	.459	.495	.532	.569	.606
300	.0523	.0700	.0878	.1234	.1590	.1945	.230	.266	.301	.337	.372	.408	.443	.479	.515	.550	.586
350	.0490	.0657	.0824	.1158	.1491	.1825	.216	.249	.283	.316	.349	.383	.416	.449	.483	.516	.550
400	.0462	.0619	.0776	.1090	.1405	.1719	.203	.235	.266	.298	.329	.360	.392	.423	.455	.486	.518
450	.0436	.0585	.0733	.1030	.1327	.1624	.1921	.222	.252	.281	.311	.341	.370	.400	.430	.459	.489
500	.0414	.0555	.0695	.0977	.1258	.1540	.1821	.210	.238	.267	.295	.323	.351	.379	.407	.436	.464
550	.0393	.0527	.0661	.0928	.1196	.1464	.1731	.1999	.227	.253	.280	.307	.334	.360	.387	.414	.441
600	.0375	.0502	.0630	.0885	.1140	.1395	.1649	.1904	.216	.241	.267	.292	.318	.343	.369	.394	.420

Air Temp. Deg F.	175 psi	200 psi	225 psi	250 psi	300 psi	400 psi	500 psi	600 psi	700 psi	800 psi	900 psi	1000 psi
30°	1.047	1.185	1.323	1.460	1.736	2.29	2.84	3.39	3.94	4.49	5.05	5.60
40	1.026	1.161	1.296	1.431	1.702	2.24	2.78	3.32	3.86	4.40	4.95	5.49
50	1.009	1.142	1.275	1.408	1.674	2.21	2.74	3.27	3.80	4.33	4.87	5.40
60	.986	1.116	1.246	1.376	1.636	2.16	2.68	3.20	3.72	4.24	4.76	5.28
70	.968	1.095	1.223	1.350	1.605	2.12	2.63	3.14	3.65	4.16	4.67	5.18
80	.950	1.075	1.200	1.325	1.575	2.08	2.58	3.08	3.58	4.08	4.58	5.08
90	.932	1.055	1.178	1.301	1.547	2.04	2.53	3.02	3.51	4.00	4.50	4.99
100	.916	1.036	1.157	1.278	1.519	2.00	2.48	2.97	3.45	3.93	4.42	4.90
110	.900	1.018	1.137	1.255	1.492	1.967	2.44	2.92	3.39	3.86	4.34	4.81
120	.884	1.001	1.117	1.234	1.467	1.933	2.40	2.86	3.33	3.80	4.26	4.73
130	.869	.984	1.098	1.213	1.442	1.900	2.36	2.82	3.27	3.73	4.19	4.65
140	.855	.967	1.080	1.193	1.418	1.868	2.32	2.77	3.22	3.67	4.12	4.57
150	.841	.951	1.062	1.173	1.395	1.838	2.28	2.72	3.17	3.61	4.05	4.50
175	.807	.914	1.020	1.127	1.340	1.765	2.19	2.62	3.04	3.47	3.89	4.32
200	.777	.879	.982	1.084	1.289	1.698	2.11	2.52	2.93	3.34	3.75	4.16
225	.749	.847	.946	1.044	1.242	1.636	2.03	2.43	2.82	3.21	3.61	4.00
250	.722	.817	.913	1.088	1.198	1.579	1.959	2.34	2.72	3.10	3.48	3.86
275	.698	.790	.881	.973	1.157	1.525	1.893	2.26	2.63	3.00	3.36	3.73
300	.675	.764	.852	.941	1.119	1.475	1.830	2.19	2.54	2.90	3.25	3.61
350	.633	.716	.800	.883	1.050	1.384	1.717	2.05	2.38	2.72	3.05	3.39
400	.596	.675	.753	.832	.989	1.303	1.618	1.932	2.25	2.56	2.87	3.19
450	.563	.638	.712	.786	.934	1.232	1.529	1.826	2.12	2.42	2.72	3.01
500	.534	.604	.675	.745	.886	1.167	1.449	1.731	2.01	2.29	2.58	2.86
550	.508	.575	.641	.708	.842	1.110	1.377	1.645	1.912	2.18	2.45	2.72
600	.484	.547	.611	.675	.802	1.057	1.312	1.567	1.822	2.08	2.33	2.59

Air Density Table

The table at the left is calculated for the perfect gas law shown at the top of the page. Correction for supercompressibility, the deviation from the perfect gas law, would be less than three percent and has not been applied.

The weight density of gases other than air can be determined from this table by multiplying the density listed for air by the specific gravity of the gas relative to air.

Table 4.29. Weight density in specific volume of gases and vapors. Courtesy Crane Company (T. P. #410).

THERMODYNAMIC CHARACTERISTICS OF VARIOUS REFRIGERANTS

For one ton of refrigeration with standard ton temperature range of 5 F to 86 F; dry compression, no liquid subcooling.

Name	Formula	Mol Wt	C_p/C_v	Weight Lb per Min	Vol cfm	Absolute Press Lb/In.² at 5 F	Absolute Press Lb/In.² at 86 F	Ratio of Compression	Coefficient of Performance	Hp	Relative % Eff
Carnot performance									5.74	.8214	100
Ammonia	NH_3	17.0	1.3	0.4214	3.44	34.28	169.2	4.93	4.85	.973	84.5
Butane	C_4H_{10}	58.1	1.11	1.619	16.14	8.2	41.6	5.07	4.63	1.017	80.7
Carbon dioxide	CO_2	44.0	1.3	3.74	.999	339	1054	3.11	2.56	1.843	44.6
Dichlorodifluoromethane; (F-12); (Freon-12)	CCl_2F_2	120.9	1.14	3.92	5.82	26.5	107.9	4.065	4.72	.997	82.2
Dichloroethylene (dieline)	$C_2H_2Cl_2$	96.9	1.14	1.768	108.4	.874	7.19	8.23	5.14	.918	89.4
Dichloromethane (methylene chloride) (Carrene No. 1)	CH_2Cl_2	84.9	1.18	1.485	74.0	1.173	10.06	8.56	4.9	.965	85.3
Methyl chloride	CH_3Cl	50.5	1.28	1.33	5.95	21.15	94.7	4.48	4.90	.963	85.4
Monofluorotrichloromethane; (Carrene No. 2); (F-11); (Freon-11)	$CFCl_3$	137.4	1.13	3.05	37.2	6.07	18.3	6.13	4.815	.980	83.8
Propane	C_3H_8	44.1	1.13	1.396	3.35	43.70	159.0	3.64	4.88	.9668	85.0
Sulfur dioxide	SO_2	64.1	1.29	1.388	9.24	11.81	66.6	5.63	4.735	.995	82.5

Table 4.30. Thermodynamic characteristics of various refrigerants.

PERIODIC TABLE OF THE ELEMENTS

KEY TO CHART

Atomic Number → 50 +2 +4 ← Oxidation States
Symbol → Sn
Atomic Weight → 118.69
18 18 4 ← Electron Configuration

Transition Elements; Group 8

1a	2a	3b	4b	5b	6b	7b	8	8	8	1b	2b	3a	4a	5a	6a	7a	0	Orbit
1 +1 −1 H 1.0079 1																	2 0 He 4.00260 2	K
3 +1 Li 6.94 2-1	4 +2 Be 9.01218 2-2											5 +3 B 10.81 2-3	6 +2 +4 −4 C 12.011 2-4	7 +1 +2 +3 +4 +5 −1 −2 −3 N 14.0067 2-5	8 −2 O 15.9994 2-6	9 −1 F 18.998403 2-7	10 0 Ne 20.17_9 2-8	K-L
11 +1 Na 22.98977 2-8-1	12 +2 Mg 24.305 2-8-2											13 +3 Al 26.98154 2-8-3	14 +2 +4 −4 Si 28.0855 2-8-4	15 +3 +5 −3 P 30.97376 2-8-5	16 +4 +6 −2 S 32.06 2-8-6	17 +1 +5 +7 −1 Cl 35.453 2-8-7	18 0 Ar 39.948 2-8-8	K-L-M
19 +1 K 39.0983 -8-8-1	20 +2 Ca 40.08 -8-8-2	21 +3 Sc 44.9559 -8-9-2	22 +2 +3 +4 Ti 47.90 -8-10-2	23 +2 +3 +4 +5 V 50.9415 -8-11-2	24 +2 +3 +6 Cr 51.996 -8-13-1	25 +2 +3 +4 +7 Mn 54.9380 -8-13-2	26 +2 +3 Fe 55.847 -8-14-2	27 +2 +3 Co 58.9332 -8-15-2	28 +2 +3 Ni 58.71 -8-16-2	29 +1 +2 Cu 63.546 -8-18-1	30 +2 Zn 65.38 -8-18-2	31 +3 Ga 69.735 -8-18-3	32 +2 +4 Ge 72.59 -8-18-4	33 +3 +5 −3 As 74.9216 -8-18-5	34 +4 +6 −2 Se 78.96 -8-18-6	35 +1 +5 −1 Br 79.904 -8-18-7	36 0 Kr 83.80 -8-18-8	-L-M-N
37 +1 Rb 85.467_8 -18-8-1	38 +2 Sr 87.62 -18-8-2	39 +3 Y 88.9059 -18-9-2	40 +4 Zr 91.22 -18-10-2	41 +3 +5 Nb 92.9064 -18-12-1	42 +6 Mo 95.94 -18-13-1	43 +4 +6 +7 Tc 98.9062 -18-13-2	44 +3 Ru 101.07 -18-15-1	45 +3 Rh 102.9055 -18-16-1	46 +2 +4 Pd 106.4 -18-18-0	47 +1 Ag 107.868 -18-18-1	48 +2 Cd 112.41 -18-18-2	49 +3 In 114.82 -18-18-3	50 +2 +4 Sn 118.69 -18-18-4	51 +3 +5 −3 Sb 121.75 -18-18-5	52 +4 +6 −2 Te 127.60 -18-18-6	53 +1 +5 +7 −1 I 126.9045 -18-18-7	54 0 Xe 131.30 -18-18-8	-M-N-O
55 +1 Cs 132.9054 -18-8-1	56 +2 Ba 137.33 -18-8-2	57* +3 La 138.9055 -18-9-2	72 +4 Hf 178.49 -32-10-2	73 +5 Ta 180.947_9 -32-11-2	74 +6 W 183.85 -32-12-2	75 +4 +6 +7 Re 186.207 -32-13-2	76 +3 +4 Os 190.2 -32-14-2	77 +3 +4 Ir 192.22 -32-15-2	78 +2 +4 Pt 195.09 -32-16-2	79 +1 +3 Au 196.9665 -32-18-1	80 +1 +2 Hg 200.59 -32-18-2	81 +1 +3 Tl 204.37 -32-18-3	82 +2 +4 Pb 207.2 -32-18-4	83 +3 +5 Bi 208.9804 -32-18-5	84 +2 +4 Po (209) -32-18-6	85 At (210) -32-18-7	86 0 Rn (222) -32-18-8	-N-O-P
87 +1 Fr (223) -18-8-1	88 +2 Ra 226.0254 -18-8-2	89** +3 Ac (227) -18-9-2	104 +4 Rf (260) -32-10-2	105 Ha (260) -32-11-2	106 (263) -32-12-2													O P Q
*Lanthanides		58 +3 +4 Ce 140.12 -20-8-2	59 +3 Pr 140.9077 -21-8-2	60 +3 Nd 144.24 -22-8-2	61 +3 Pm (145) -23-8-2	62 +2 +3 Sm 150.4 -24-8-2	63 +2 +3 Eu 151.96 -25-8-2	64 +3 Gd 157.25 -25-9-2	65 +3 Tb 158.9254 -27-8-2	66 +3 Dy 162.50 -28-8-2	67 +3 Ho 164.9304 -29-8-2	68 +3 Er 167.26 -30-8-2	69 +3 Tm 168.9342 -31-8-2	70 +2 +3 Yb 173.04 -32-8-2	71 +3 Lu 174.967 ± 0.003 -32-9-2			N O P
**Actinides		90 +4 Th 232.0381 -18-10-2	91 +5 +4 Pa 231.0359 -20-9-2	92 +3 +4 +5 +6 U 238.029 -21-9-2	93 +3 +4 +5 +6 Np 237.0482 -22-9-2	94 +3 +4 +5 +6 Pu (244) -24-8-2	95 +3 +4 +5 +6 Am (243) -25-8-2	96 +3 Cm (247) -25-9-2	97 +3 +4 Bk (247) -27-8-2	98 +3 Cf (251) -28-8-2	99 +3 Es (254) -29-8-2	100 +3 Fm (257) -30-8-2	101 +2 +3 Md (258) -31-8-2	102 +2 +3 No (259) -32-8-2	103 +3 Lr (260) -32-9-2			O P Q

Numbers in parentheses are mass numbers of most stable isotope of that element

Table 4.31. Periodic table of the elements.

Gas	Molecular Weight	Specific Heat Ratio $k=C_p/C_v$ at 60 F and 1 Atm	Critical Flow Pressure Ratio at 60 F and 1 Atm	Specific Gravity at 60 F and 1 Atm	Critical Pressure (Pounds per Square Inch Absolute)	Critical Temperature (Degrees Rankine, Fahrenheit + 460)	Condensation Temperature at 1 Atm	Flammability Limits (Percent by Volume)
Methane[3]	16	1.31	0.54	0.555	673	344	—259	5.0 to 15.0
Ethane[3]	30	1.19	0.57	1.05	708	550	—128	3.2 to 12.5
Propane[3]	44	1.13	0.58	1.55	617	666	—44	2.4 to 9.5
*n*Butane[3]	58	1.09	0.59	2.07	551	766	31	1.9 to 8.4
*n*Pentane[3]	72	1.07	0.60	2.49	490	846	97	1.4 to 7.8
*n*Hexane[3]	86	1.06 *	0.60 *	2.97	440	914	156	1.3 to 6.9
*n*Heptane[3]	100	1.05 *	0.60 *	3.49	397	973	209	1.0 to 6.0
*n*Octane[3]	114	1.05 *	0.60 *	3.94	362	1,025	258	0.84 to 3.2
*n*Nonane[3]	128	1.04 *	0.60 *	4.43	345 *	1,073 *	303	0.74 to 2.9
*n*Decane[3]	142	1.03 *	0.60 *	4.91	320 *	1,115 *	345	0.67 to 2.6
Air[4]	29	1.40	0.53	1.00	547	240	—313	... ...
Ammonia[4]	17	1.31	0.53	0.587	1,638	730	—28	16.0 to 27.0
Benzene[4]	78	1.12	0.58	2.70	700	1,011	176	1.41 to 6.75
Carbon dioxide[4]	44	1.29	0.55	1.53	1,072	548	—109	... ...
Hydrogen[4]	2	1.41	0.52	0.070	188	60	—423	4.1 to 74.2
Hydrogen sulfide[4]	34	1.32	0.53	1.19	1,306	672	—75	4.3 to 45.5
Phenol[4]	94	1.3 *	0.54 *	3.27	890	786	358	1.3* to 7.0*
Steam[4]	18	1.33	0.54	0.622	3,206	1,166	212	... ...
Sulfur dioxide[4]	64	1.29	0.55	2.26	1,141	775	14	... ...
Toluene[4]	92	1.09	0.59	3.18	611	1,069	231	1.3 to 6.8

* Estimated.

Table 4.32. Properties of gases. Courtesy American Petroleum Institute.

TABLE OF MOLAL HEAT CAPACITY MCp° (IDEAL GAS STATE), BTU PER LB. MOL PER °R

(1) Data Source: Selected Values of Properties of Hydrocarbons, API Research Project 44

Gas	Chemical Formula	Mol. Wt.	0°F	50°F	60°F	100°F	150°F	200°F	250°F	300°F
Methane	CH_4	16.042	8.233	8.414	8.456	8.651	8.947	9.277	9.638	10.01
Ethyne (Acetylene)	C_2H_2	26.036	9.683	10.23	10.33	10.71	11.13	11.54	11.88	12.22
Ethene (Ethylene)	C_2H_4	28.052	9.324	10.02	10.16	10.72	11.40	12.08	12.75	13.41
Ethane	C_2H_6	30.068	11.44	12.17	12.32	12.95	13.77	14.63	15.49	16.34
Propene (Propylene)	C_3H_6	42.078	13.63	14.69	14.90	15.75	16.80	17.85	18.87	19.89
Propane	C_3H_8	44.094	15.64	16.88	17.13	18.17	19.52	20.89	22.25	23.56
1-Butene (Butylene)	C_4H_8	56.104	17.96	19.59	19.91	21.17	22.71	24.25	25.70	27.15
Cis-2-Butene	C_4H_8	56.104	16.54	18.04	18.34	19.54	21.04	22.53	24.00	25.47
Trans-2-Butene	C_4H_8	56.104	18.84	20.22	20.50	21.61	22.99	24.37	25.72	27.06
Iso-Butane	C_4H_{10}	58.120	20.40	22.15	22.50	23.95	25.77	27.59	29.39	31.11
N-Butane	C_4H_{10}	58.120	20.80	22.38	22.71	24.07	25.81	27.54	29.23	30.90
Iso-Pentane	C_5H_{12}	72.146	24.93	27.16	27.61	29.42	31.66	33.87	36.03	38.14
N-Pentane	C_5H_{12}	72.146	25.64	27.61	28.01	29.70	31.86	33.99	36.07	38.12
Benzene	C_6H_6	78.108	16.41	18.38	18.75	20.46	22.46	24.46	27.08	29.71
N-Hexane	C_6H_{14}	86.172	30.17	32.78	33.30	35.36	37.91	40.45	42.91	45.36
N-Heptane	C_7H_{16}	100.198	34.96	37.00	38.61	41.01	43.47	46.93	49.77	52.60
Ammonia	NH_3	17.032	8.516	8.518	8.519	8.521	8.523	8.525	8.527	8.530
Air		28.966	6.944	6.951	6.952	6.960	6.973	6.990	7.009	7.033
Water	H_2O	18.016	7.983	8.006	8.010	8.033	8.075	8.116	8.171	8.226
Oxygen	O_2	32.000	6.970	6.997	7.002	7.030	7.075	7.120	7.176	7.232
Nitrogen	N_2	28.016	6.951	6.954	6.954	6.956	6.963	6.970	6.984	6.998
Hydrogen	H_2	2.016	6.782	6.856	6.871	6.905	6.929	6.953	6.965	6.977
Hydrogen-Sulfide	H_2S	34.076	8.00	8.091	8.109	8.18	8.27	8.36	8.455	8.55
Carbon Monoxide	CO	28.010	6.952	6.957	6.958	6.963	6.975	6.986	7.007	7.028
Carbon Dioxide	CO_2	44.010	8.380	8.698	8.762	9.004	8.282	9.559	10.31	10.05

(1) Exceptions: Air, Keenan and Keyes, Thermodynamic Properties of Air, Wiley, 3rd Printing 1947.
Ammonia, Edw. R. Grahl, Thermodynamic Properties of Ammonia at High Temperatures and Pressures, Petr. Processing, April 1953.
Hydrogen-Sulfide, J. R. West, Chem. Eng. Progress 44, 287, (1948).

Table 4.33. Molal heat capacity MCp° *(ideal gas state), Btu per lb. mol per °R. Courtesy Natural Gas Processors Suppliers Association.*

Schedule	Nominal Pipe Size	Outside Diameter	Thickness	Inside Diameter		Inside Diameter Functions (In Inches)				Transverse Internal Area	
				d	D	d^2	d^3	d^4	d^5	a	A
	Inches	Inches	Inches	Inches	Feet					Sq. In.	Sq. Ft.
Schedule 10	14	14	0.250	13.5	1.125	182.25	2460.4	33215.	448400.	143.14	0.994
	16	16	0.250	15.5	1.291	240.25	3723.9	57720.	894660.	188.69	1.310
	18	18	0.250	17.5	1.4583	306.25	5359.4	93789.	1641309.	240.53	1.670
	20	20	0.250	19.5	1.625	380.25	7414.9	144590.	2819500.	298.65	2.074
	24	24	0.250	23.5	1.958	552.25	12977.	304980.	7167030.	433.74	3.012
	30	30	0.312	29.376	2.448	862.95	25350.	744288.	21864218.	677.76	4.707
Schedule 20	8	8.625	0.250	8.125	0.6771	66.02	536.38	4359.3	35409.	51.85	0.3601
	10	10.75	0.250	10.25	0.8542	105.06	1076.9	11038.	113141.	82.52	0.5731
	12	12.75	0.250	12.25	1.021	150.06	1838.3	22518.	275855.	117.86	0.8185
	14	14	0.312	13.376	1.111	178.92	2393.2	32012.	428185.	140.52	0.9758
	16	16	0.312	15.376	1.281	236.42	3635.2	55894.	859442.	185.69	1.290
	18	18	0.312	17.376	1.448	301.92	5246.3	91156.	1583978.	237.13	1.647
	20	20	0.375	19.250	1.604	370.56	7133.3	137317.	2643352.	291.04	2.021
	24	24	0.375	23.25	1.937	540.56	12568.	292205.	6793832.	424.56	2.948
	30	30	0.500	29.00	2.417	841.0	24389.	707281.	20511149.	660.52	4.587
Schedule 30	8	8.625	0.277	8.071	0.6726	65.14	525.75	4243.2	34248.	51.16	0.3553
	10	10.75	0.307	10.136	0.8447	102.74	1041.4	10555.	106987.	80.69	0.5603
	12	12.75	0.330	12.09	1.0075	146.17	1767.2	21366.	258304.	114.80	0.7972
	14	14	0.375	13.25	1.1042	175.56	2326.2	30821.	408394.	137.88	0.9575
	16	16	0.375	15.25	1.2708	232.56	3546.6	54084.	824801.	182.65	1.268
	18	18	0.438	17.124	1.4270	293.23	5021.3	85984.	1472397.	230.30	1.599
	20	20	0.500	19.00	1.5833	361.00	6859.0	130321.	2476099.	283.53	1.969
	24	24	0.562	22.876	1.9063	523.31	11971.	273853.	6264703.	411.00	2.854
	30	30	0.625	28.75	2.3958	826.56	23764.	683201.	19642160.	649.18	4.508
Schedule 40	1/8	0.405	0.068	0.269	0.0224	0.0724	0.0195	0.005242	0.00141	0.057	0.00040
	1/4	0.540	0.088	0.364	0.0303	0.1325	0.0482	0.01756	0.00639	0.104	0.00072
	3/8	0.675	0.091	0.493	0.0411	0.2430	0.1198	0.05905	0.02912	0.191	0.00133
	1/2	0.840	0.109	0.622	0.0518	0.3869	0.2406	0.1497	0.09310	0.304	0.00211
	3/4	1.050	0.113	0.824	0.0687	0.679	0.5595	0.4610	0.3799	0.533	0.00371
	1	1.315	0.133	1.049	0.0874	1.100	1.154	1.210	1.270	0.864	0.00600
	1 1/4	1.660	0.140	1.380	0.1150	1.904	2.628	3.625	5.005	1.495	0.01040
	1 1/2	1.900	0.145	1.610	0.1342	2.592	4.173	6.718	10.82	2.036	0.01414
	2	2.375	0.154	2.067	0.1722	4.272	8.831	18.250	37.72	3.355	0.02330
	2 1/2	2.875	0.203	2.469	0.2057	6.096	15.051	37.161	91.75	4.788	0.03322
	3	3.500	0.216	3.068	0.2557	9.413	28.878	88.605	271.8	7.393	0.05130
	3 1/2	4.000	0.226	3.548	0.2957	12.59	44.663	158.51	562.2	9.886	0.06870
	4	4.500	0.237	4.026	0.3355	16.21	65.256	262.76	1058.	12.730	0.08840
	5	5.563	0.258	5.047	0.4206	25.47	128.56	648.72	3275.	20.006	0.1390
	6	6.625	0.280	6.065	0.5054	36.78	223.10	1352.8	8206.	28.891	0.2006
	8	8.625	0.322	7.981	0.6651	63.70	508.36	4057.7	32380.	50.027	0.3474
	10	10.75	0.365	10.02	0.8350	100.4	1006.0	10080.	101000.	78.855	0.5475
	12	12.75	0.406	11.938	0.9965	142.5	1701.3	20306.	242470.	111.93	0.7773
	14	14.0	0.438	13.124	1.0937	172.24	2260.5	29666.	389340.	135.28	0.9394
	16	16.0	0.500	15.000	1.250	225.0	3375.0	50625.	759375.	176.72	1.2272
	18	18.0	0.562	16.876	1.4063	284.8	4806.3	81111.	1368820.	223.68	1.5533
	20	20.0	0.593	18.814	1.5678	354.0	6659.5	125320.	2357244.	278.00	1.9305
	24	24.0	0.687	22.626	1.8855	511.9	11583.	262040.	5929784.	402.07	2.7921
Schedule 60	8	8.625	0.406	7.813	0.6511	61.04	476.93	3725.9	29113.	47.94	0.3329
	10	10.75	0.500	9.750	0.8125	95.06	926.86	9036.4	88110.	74.66	0.5185
	12	12.75	0.562	11.626	0.9688	135.16	1571.4	18268.	212399.	106.16	0.7372
	14	14.0	0.593	12.814	1.0678	164.20	2104.0	26962.	345480.	128.96	0.8956
	16	16.0	0.656	14.688	1.2240	215.74	3168.8	46544.	683618.	169.44	1.1766
	18	18.0	0.750	16.500	1.3750	272.25	4492.1	74120.	1222982.	213.83	1.4849
	20	20.0	0.812	18.376	1.5313	337.68	6205.2	114028.	2095342.	265.21	1.8417
	24	24.0	0.968	22.064	1.8387	486.82	10741.	236994.	5229036.	382.35	2.6552
Schedule 80	1/8	0.405	0.095	0.215	0.0179	0.0462	0.00994	0.002134	0.000459	0.036	0.00025
	1/4	0.540	0.119	0.302	0.0252	0.0912	0.0275	0.008317	0.002513	0.072	0.00050
	3/8	0.675	0.126	0.423	0.0353	0.1789	0.0757	0.03200	0.01354	0.141	0.00098
	1/2	0.840	0.147	0.546	0.0455	0.2981	0.1628	0.08886	0.04852	0.234	0.00163
	3/4	1.050	0.154	0.742	0.0618	0.5506	0.4085	0.3032	0.2249	0.433	0.00300
	1	1.315	0.179	0.957	0.0797	0.9158	0.8765	0.8387	0.8027	0.719	0.00499
	1 1/4	1.660	0.191	1.278	0.1065	1.633	2.087	2.6667	3.409	1.283	0.00891

Table 4.34. Commercial wrought steel pipe data. Courtesy Crane Company (T. P. #410).

Table 4.34 continued

Schedule	Nominal Pipe Size	Outside Diameter	Thickness	Inside Diameter		Inside Diameter Functions (In Inches)				Transverse Internal Area	
				d	D	d^2	d^3	d^4	d^5	a	A
	Inches	Inches	Inches	Inches	Feet					Sq. In.	Sq. Ft.
Schedule 80—cont.	1½	1.900	0.200	1.500	0.1250	2.250	3.375	5.062	7.594	1.767	0.01225
	2	2.375	0.218	1.939	0.1616	3.760	7.290	14.136	27.41	2.953	0.02050
	2½	2.875	0.276	2.323	0.1936	5.396	12.536	29.117	67.64	4.238	0.02942
	3	3.5	0.300	2.900	0.2417	8.410	24.389	70.728	205.1	6.605	0.04587
	3½	4.0	0.318	3.364	0.2803	11.32	38.069	128.14	430.8	8.888	0.06170
	4	4.5	0.337	3.826	0.3188	14.64	56.006	214.33	819.8	11.497	0.07986
	5	5.563	0.375	4.813	0.4011	23.16	111.49	536.38	2583.	18.194	0.1263
	6	6.625	0.432	5.761	0.4801	33.19	191.20	1101.6	6346.	26.067	0.1810
	8	8.625	0.500	7.625	0.6354	58.14	443.32	3380.3	25775.	45.663	0.3171
	10	10.75	0.593	9.564	0.7970	91.47	874.82	8366.8	80020.	71.84	0.4989
	12	12.75	0.687	11.376	0.9480	129.41	1472.2	16747.	190523.	101.64	0.7058
	14	14.0	0.750	12.500	1.0417	156.25	1953.1	24414.	305176.	122.72	0.8522
	16	16.0	0.843	14.314	1.1928	204.89	2932.8	41980.	600904.	160.92	1.1175
	18	18.0	0.937	16.126	1.3438	260.05	4193.5	67626.	1090518.	204.24	1.4183
	20	20.0	1.031	17.938	1.4948	321.77	5771.9	103536.	1857248.	252.72	1.7550
	24	24.0	1.218	21.564	1.7970	465.01	10027.	216234.	4662798.	365.22	2.5362
Schedule 100	8	8.625	0.593	7.439	0.6199	55.34	411.66	3062.	22781.	43.46	0.3018
	10	10.75	0.718	9.314	0.7762	86.75	807.99	7526.	69357.	68.13	0.4732
	12	12.75	0.843	11.064	0.9220	122.41	1354.4	14985.	165791.	96.14	0.6677
	14	14.0	0.937	12.126	1.0105	147.04	1783.0	21621.	262173.	115.49	0.8020
	16	16.0	1.031	13.938	1.1615	194.27	2707.7	37740.	526020.	152.58	1.0596
	18	18.0	1.156	15.688	1.3057	246.11	3861.0	60572.	950250.	193.30	1.3423
	20	20.0	1.281	17.438	1.4532	304.08	5302.6	92467.	1612438.	238.83	1.6585
	24	24.0	1.531	20.938	1.7448	438.40	9179.2	192195.	4024179.	344.32	2.3911
Schedule 120	4	4.50	0.438	3.624	0.302	13.133	47.595	172.49	625.1	10.315	0.07163
	5	5.563	0.500	4.563	0.3802	20.82	95.006	433.5	1978.	16.35	0.1136
	6	6.625	0.562	5.501	0.4584	30.26	166.47	915.7	5037.	23.77	0.1650
	8	8.625	0.718	7.189	0.5991	51.68	371.54	2671.	19202.	40.59	0.2819
	10	10.75	0.843	9.064	0.7553	82.16	744.66	6750.	61179.	64.53	0.4481
	12	12.75	1.000	10.750	0.8959	115.56	1242.3	13355.	143563.	90.76	0.6303
	14	14.0	1.093	11.814	0.9845	139.57	1648.9	19480.	230137.	109.62	0.7612
	16	16.0	1.218	13.564	1.1303	183.98	2495.5	33849.	459133.	144.50	1.0035
	18	18.0	1.375	15.250	1.2708	232.56	3546.6	54086.	824804.	182.66	1.2684
	20	20.0	1.500	17.000	1.4166	289.00	4913.0	83521.	1419857.	226.98	1.5762
	24	24.0	1.812	20.376	1.6980	415.18	8459.7	172375.	3512313.	326.08	2.2645
Schedule 140	8	8.625	0.812	7.001	0.5834	49.01	343.15	2402.	16819.	38.50	0.2673
	10	10.75	1.000	8.750	0.7292	76.56	669.92	5862.	51291.	60.13	0.4176
	12	12.75	1.125	10.500	0.8750	110.25	1157.6	12155.	127628.	86.59	0.6013
	14	14.0	1.250	11.500	0.9583	132.25	1520.9	17490.	201136.	103.87	0.7213
	16	16.0	1.438	13.124	1.0937	172.24	2260.5	29666.	389340.	135.28	0.9394
	18	18.0	1.562	14.876	1.2396	221.30	3292.0	48972.	728502.	173.80	1.2070
	20	20.0	1.750	16.5	1.3750	272.25	4492.1	74120.	1222981.	213.82	1.4849
	24	24.0	2.062	19.876	1.6563	395.06	7852.1	156069.	3102022.	310.28	2.1547
Schedule 160	½	0.840	0.187	0.466	0.0388	0.2172	0.1012	0.04716	0.02197	0.1706	0.00118
	¾	1.050	0.218	0.614	0.0512	0.3770	0.2315	0.1421	0.08726	0.2961	0.00206
	1	1.315	0.250	0.815	0.0679	0.6642	0.5413	0.4412	0.3596	0.5217	0.00362
	1¼	1.660	0.250	1.160	0.0966	1.346	1.561	1.811	2.100	1.057	0.00734
	1½	1.900	0.281	1.338	0.1115	1.790	2.395	3.205	4.288	1.406	0.00976
	2	2.375	0.343	1.689	0.1407	2.853	4.818	8.138	13.74	2.241	0.01556
	2½	2.875	0.375	2.125	0.1771	4.516	9.596	20.39	43.33	3.546	0.02463
	3	3.50	0.438	2.624	0.2187	6.885	18.067	47.41	124.4	5.408	0.03755
	4	4.50	0.531	3.438	0.2865	11.82	40.637	139.7	480.3	9.283	0.06447
	5	5.563	0.625	4.313	0.3594	18.60	80.230	346.0	1492.	14.61	0.1015
	6	6.625	0.718	5.189	0.4324	26.93	139.72	725.0	3762.	21.15	0.1469
	8	8.625	0.906	6.813	0.5677	46.42	316.24	2155.	14679.	36.46	0.2532
	10	10.75	1.125	8.500	0.7083	72.25	614.12	5220.	44371.	56.75	0.3941
	12	12.75	1.312	10.126	0.8438	102.54	1038.3	10514.	106461.	80.53	0.5592
	14	14.0	1.406	11.188	0.9323	125.17	1400.4	15668.	175292.	98.31	0.6827
	16	16.0	1.593	12.814	1.0678	164.20	2104.0	26961.	345482.	128.96	0.8956
	18	18.0	1.781	14.438	1.2032	208.45	3009.7	43454.	627387.	163.72	1.1369
	20	20.0	1.968	16.064	1.3387	258.05	4145.3	66590.	1069715.	202.67	1.4074
	24	24.0	2.343	19.314	1.6095	373.03	7204.7	139152.	2687582.	292.98	2.0346

Table 4.34 continued

Table 4.34 continued

Nominal Pipe Size	Outside Diameter	Thickness	Inside Diameter		Inside Diameter Functions (In Inches)				Transverse Internal Area	
			d	D					a	A
Inches	Inches	Inches	Inches	Feet	d^2	d^3	d^4	d^5	Sq. In.	Sq. Ft.
Standard Wall Pipe										
⅛	0.405	0.068	0.269	0.0224	0.0724	0.0195	0.00524	0.00141	0.057	0.00040
¼	0.540	0.088	0.364	0.0303	0.1325	0.0482	0.01756	0.00639	0.104	0.00072
⅜	0.675	0.091	0.493	0.0411	0.2430	0.1198	0.05905	0.02912	0.191	0.00133
½	0.840	0.109	0.622	0.0518	0.3869	0.2406	0.1497	0.0931	0.304	0.00211
¾	1.050	0.113	0.824	0.0687	0.679	0.5595	0.4610	0.3799	0.533	0.00371
1	1.315	0.133	1.049	0.0874	1.100	1.154	1.210	1.270	0.864	0.00600
1¼	1.660	0.140	1.380	0.1150	1.904	2.628	3.625	5.005	1.495	0.01040
1½	1.900	0.145	1.610	0.1342	2.592	4.173	6.718	10.82	2.036	0.01414
2	2.375	0.154	2.067	0.1722	4.272	8.831	18.250	37.72	3.355	0.02330
2½	2.875	0.203	2.469	0.2057	6.096	15.051	37.161	91.75	4.788	0.03322
3	3.500	0.216	3.068	0.2557	9.413	28.878	88.605	271.8	7.393	0.05130
3½	4.000	0.226	3.548	0.2957	12.59	44.663	158.51	562.2	9.886	0.06870
4	4.500	0.237	4.026	0.3355	16.21	65.256	262.76	1058.	12.730	0.08840
5	5.563	0.258	5.047	0.4206	25.47	128.56	648.72	3275.	20.006	0.1390
6	6.625	0.280	6.065	0.5054	36.78	223.10	1352.8	8206.	28.891	0.2006
8	8.625	0.277	8.071	0.6725	65.14	525.75	4243.0	34248.	51.161	0.3553
	8.625S	0.322	7.981	0.6651	63.70	508.36	4057.7	32380.	50.027	0.3474
10	10.75	0.279	10.192	0.8493	103.88	1058.7	10789.	109876.	81.585	0.5666
	10.75	0.307	10.136	0.8446	102.74	1041.4	10555.	106987.	80.691	0.5604
	10.75S	0.365	10.020	0.8350	100.4	1006.0	10080.	101000.	78.855	0.5475
12	12.75	0.330	12.090	1.0075	146.17	1767.2	21366.	258300.	114.80	0.7972
	12.75S	0.375	12.000	1.000	144.0	1728.0	20736.	248800.	113.10	0.7854
Extra Strong Pipe										
⅛	0.405	0.095	0.215	0.0179	0.0462	0.00994	0.002134	0.000459	0.036	0.00025
¼	0.540	0.119	0.302	0.0252	0.0912	0.0275	0.008317	0.002513	0.072	0.00050
⅜	0.675	0.126	0.423	0.0353	0.1789	0.0757	0.03201	0.01354	0.141	0.00098
½	0.840	0.147	0.546	0.0455	0.2981	0.1628	0.08886	0.04852	0.234	0.00163
¾	1.050	0.154	0.742	0.0618	0.5506	0.4085	0.3032	0.2249	0.433	0.00300
1	1.315	0.179	0.957	0.0797	0.9158	0.8765	0.8387	0.8027	0.719	0.00499
1¼	1.660	0.191	1.278	0.1065	1.633	2.087	2.6667	3.409	1.283	0.00891
1½	1.900	0.200	1.500	0.1250	2.250	3.375	5.062	7.594	1.767	0.01225
2	2.375	0.218	1.939	0.1616	3.760	7.290	14.136	27.41	2.953	0.02050
2½	2.875	0.276	2.323	0.1936	5.396	12.536	29.117	67.64	4.238	0.02942
3	3.500	0.300	2.900	0.2417	8.410	24.389	70.728	205.1	6.605	0.04587
3½	4.000	0.318	3.364	0.2803	11.32	38.069	128.14	430.8	8.888	0.06170
4	4.500	0.337	3.826	0.3188	14.64	56.006	214.33	819.8	11.497	0.07986
5	5.563	0.375	4.813	0.4011	23.16	111.49	536.6	2583.	18.194	0.1263
6	6.625	0.432	5.761	0.4801	33.19	191.20	1101.6	6346.	26.067	0.1810
8	8.625	0.500	7.625	0.6354	58.14	443.32	3380.3	25775.	45.663	0.3171
10	10.75	0.500	9.750	0.8125	95.06	926.86	9036.4	88110.	74.662	0.5185
12	12.75	0.500	11.750	0.9792	138.1	1622.2	19072.	223970.	108.434	0.7528
Double Extra Strong Pipe										
½	0.840	0.294	0.252	0.0210	0.0635	0.0160	0.004032	0.00102	0.050	0.00035
¾	1.050	0.308	0.434	0.0362	0.1884	0.0817	0.03549	0.01540	0.148	0.00103
1	1.315	0.358	0.599	0.0499	0.3588	0.2149	0.1287	0.07711	0.282	0.00196
1¼	1.660	0.382	0.896	0.0747	0.8028	0.7193	0.6445	0.5775	0.630	0.00438
1½	1.900	0.400	1.100	0.0917	1.210	1.331	1.4641	1.611	0.950	0.00660
2	2.375	0.436	1.503	0.1252	2.259	3.395	5.1031	7.670	1.774	0.01232
2½	2.875	0.552	1.771	0.1476	3.136	5.554	9.8345	17.42	2.464	0.01710
3	3.500	0.600	2.300	0.1917	5.290	12.167	27.984	64.36	4.155	0.02885
3½	4.000	0.636	2.728	0.2273	7.442	20.302	55.383	151.1	5.845	0.04059
4	4.500	0.674	3.152	0.2627	9.935	31.315	98.704	311.1	7.803	0.05419
5	5.563	0.750	4.063	0.3386	16.51	67.072	272.58	1107.	12.966	0.09006
6	6.625	0.864	4.897	0.4081	23.98	117.43	575.04	2816.	18.835	0.1308
8	8.625	0.875	6.875	0.5729	47.27	324.95	2234.4	15360.	37.122	0.2578

For lengths of pipe other than 100 feet, the pressure drop is proportional to the length. Thus, for 50 feet of pipe, the pressure drop is approximately one-half the value given in the table . . . for 300 feet, three times the given value, etc.

The pressure drop is also inversely proportional to the absolute pressure and directly proportional to the absolute temperature.

Therefore, to determine the pressure drop for inlet or average pressures other than 100 psi and at temperatures other than 60 F, multiply the values given in the table by the ratio:

$$\left(\frac{100 + 14.7}{P + 14.7}\right)\left(\frac{460 + t}{520}\right)$$

where:

"P" is the inlet or average gauge pressure in pounds per square inch, and,

"t" is the temperature in degrees Fahrenheit under consideration.

The cubic feet per minute of compressed air at any pressure is inversely proportional to the absolute pressure and directly proportional to the absolute temperature.

To determine the cubic feet per minute of compressed air at any temperature and pressure other than standard conditions, multiply the value of cubic feet per minute of free air by the ratio:

$$\left(\frac{14.7}{14.7 + P}\right)\left(\frac{460 + t}{520}\right)$$

Calculations for Pipe Other than Schedule 40

To determine the velocity of water, or the pressure drop of water or air, through pipe other than Schedule 40, use the following formulas:

$$v_a = v_{40}\left(\frac{d_{40}}{d_a}\right)^2$$

$$\Delta P_a = \Delta P_{40}\left(\frac{d_{40}}{d_a}\right)^5$$

Subscript "a" refers to the Schedule of pipe through which velocity or pressure drop is desired.

Subscript "40" refers to the velocity or pressure drop through Schedule 40 pipe, as given in the tables on these facing pages.

Free Air q'_m Cubic Feet Per Minute at 60 F and 14.7 psia	**Compressed Air** Cubic Feet Per Minute at 60 F and 100 psig	**Pressure Drop of Air In Pounds per Square Inch Per 100 Feet of Schedule 40 Pipe For Air at 100 Pounds per Square Inch Gauge Pressure and 60 F Temperature**								
		1/8"	1/4"	3/8"	1/2"					
1	0.128	0.361	0.083	0.018						
2	0.256	1.31	0.285	0.064	0.020					
3	0.384	3.06	0.605	0.133	0.042	3/4"				
4	0.513	4.83	1.04	0.226	0.071					
5	0.641	7.45	1.58	0.343	0.106	0.027	1"			
6	0.769	10.6	2.23	0.408	0.148	0.037				
8	1.025	18.6	3.89	0.848	0.255	0.062	0.019			
10	1.282	28.7	5.96	1.26	0.356	0.094	0.029	1 1/4"	1 1/2"	
15	1.922	. . .	13.0	2.73	0.834	0.201	0.062			
20	2.563	. . .	22.8	4.76	1.43	0.345	0.102	0.026		
25	3.204	. . .	35.6	7.34	2.21	0.526	0.156	0.039	0.019	
30	3.845	. . .	. . .	10.5	3.15	0.748	0.219	0.055	0.026	
35	4.486	. . .	. . .	14.2	4.24	1.00	0.293	0.073	0.035	
40	5.126	. . .	. . .	18.4	5.49	1.30	0.379	0.095	0.044	
45	5.767	. . .	. . .	23.1	6.90	1.62	0.474	0.116	0.055	2"
50	6.408			28.5	8.49	1.99	0.578	0.149	0.067	0.019
60	7.690	2 1/2"		40.7	12.2	2.85	0.819	0.200	0.094	0.027
70	8.971			. . .	16.5	3.83	1.10	0.270	0.126	0.036
80	10.25	0.019		. . .	21.4	4.96	1.43	0.350	0.162	0.046
90	11.53	0.023		. . .	27.0	6.25	1.80	0.437	0.203	0.058
100	12.82	0.029	3"		33.2	7.69	2.21	0.534	0.247	0.070
125	16.02	0.044			. . .	11.9	3.39	0.825	0.380	0.107
150	19.22	0.062	0.021		. . .	17.0	4.87	1.17	0.537	0.151
175	22.43	0.083	0.028		. . .	23.1	6.60	1.58	0.727	0.205
200	25.63	0.107	0.036	3 1/2"	. . .	30.0	8.54	2.05	0.937	0.264
225	28.84	0.134	0.045	0.022		37.9	10.8	2.59	1.19	0.331
250	32.04	0.164	0.055	0.027		. . .	13.3	3.18	1.45	0.404
275	35.24	0.191	0.066	0.032		. . .	16.0	3.83	1.75	0.484
300	38.45	0.232	0.078	0.037		. . .	19.0	4.56	2.07	0.573
325	41.65	0.270	0.090	0.043	4"	. . .	22.3	5.32	2.42	0.673
350	44.87	0.313	0.104	0.050		. . .	25.8	6.17	2.80	0.776
375	48.06	0.356	0.119	0.057	0.030	. . .	29.6	7.05	3.20	0.887
400	51.26	0.402	0.134	0.064	0.034	. . .	33.6	8.02	3.64	1.00
425	54.47	0.452	0.151	0.072	0.038	. . .	37.9	9.01	4.09	1.13
450	57.67	0.507	0.168	0.081	0.042	. . .	. . .	10.2	4.59	1.26
475	60.88	0.562	0.187	0.089	0.047		. . .	11.3	5.09	1.40
500	64.08	0.623	0.206	0.099	0.052		. . .	12.5	5.61	1.55
550	70.49	0.749	0.248	0.118	0.062		. . .	15.1	6.79	1.87
600	76.90	0.887	0.293	0.139	0.073		. . .	18.0	8.04	2.21
650	83.30	1.04	0.342	0.163	0.086	5"	. . .	21.1	9.43	2.60
700	89.71	1.19	0.395	0.188	0.099	0.032		24.3	10.9	3.00
750	96.12	1.36	0.451	0.214	0.113	0.036		27.9	12.6	3.44
800	102.5	1.55	0.513	0.244	0.127	0.041		31.8	14.2	3.90
850	108.9	1.74	0.576	0.274	0.144	0.046		35.9	16.0	4.40
900	115.3	1.95	0.642	0.305	0.160	0.051	6"	40.2	18.0	4.91
950	121.8	2.18	0.715	0.340	0.178	0.057	0.023	. . .	20.0	5.47
1 000	128.2	2.40	0.788	0.375	0.197	0.063	0.025	. . .	22.1	6.06
1 100	141.0	2.89	0.948	0.451	0.236	0.075	0.030	. . .	26.7	7.29
1 200	153.8	3.44	1.13	0.533	0.279	0.089	0.035	. . .	31.8	8.63
1 300	166.6	4.01	1.32	0.626	0.327	0.103	0.041	. . .	37.3	10.1
1 400	179.4	4.65	1.52	0.718	0.377	0.119	0.047			11.8
1 500	192.2	5.31	1.74	0.824	0.431	0.136	0.054			13.5
1 600	205.1	6.04	1.97	0.932	0.490	0.154	0.061	8"		15.3
1 800	230.7	7.65	2.50	1.18	0.616	0.193	0.075			19.3
2 000	256.3	9.44	3.06	1.45	0.757	0.237	0.094	0.023	10"	23.9
2 500	320.4	14.7	4.76	2.25	1.17	0.366	0.143	0.035		37.3
3 000	384.5	21.1	6.82	3.20	1.67	0.524	0.204	0.051	0.016	
3 500	448.6	28.8	9.23	4.33	2.26	0.709	0.276	0.068	0.022	
4 000	512.6	37.6	12.1	5.66	2.94	0.919	0.358	0.088	0.028	
4 500	576.7	47.6	15.3	7.16	3.69	1.16	0.450	0.111	0.035	12"
5 000	640.8	. . .	18.8	8.85	4.56	1.42	0.552	0.136	0.043	0.018
6 000	769.0	. . .	27.1	12.7	6.57	2.03	0.794	0.195	0.061	0.025
7 000	897.1	. . .	36.9	17.2	8.94	2.76	1.07	0.262	0.082	0.034
8 000	1025	. . .	. . .	22.5	11.7	3.59	1.39	0.339	0.107	0.044
9 000	1153	. . .	. . .	28.5	14.9	4.54	1.76	0.427	0.134	0.055
10 000	1282	. . .	. . .	35.2	18.4	5.60	2.16	0.526	0.164	0.067
11 000	1410	. . .	. . .	. . .	22.2	6.78	2.62	0.633	0.197	0.081
12 000	1538	. . .	. . .	. . .	26.4	8.07	3.09	0.753	0.234	0.096
13 000	1666	. . .	. . .	. . .	31.0	9.47	3.63	0.884	0.273	0.112
14 000	1794	. . .	. . .	. . .	36.0	11.0	4.21	1.02	0.316	0.129
15 000	1922	. . .	. . .	. . .	. . .	12.6	4.84	1.17	0.364	0.148
16 000	2051	. . .	. . .	. . .	. . .	14.3	5.50	1.33	0.411	0.167
18 000	2307	. . .	. . .	. . .	. . .	18.2	6.96	1.68	0.520	0.213
20 000	2563	. . .	. . .	. . .	. . .	22.4	8.60	2.01	0.642	0.260
22 000	2820	. . .	. . .	. . .	. . .	27.1	10.4	2.50	0.771	0.314
24 000	3076	. . .	. . .	. . .	. . .	32.3	12.4	2.97	0.918	0.371
26 000	3332	. . .	. . .	. . .	. . .	37.9	14.5	3.49	1.12	0.435
28 000	3588	. . .	. . .	. . .	. . .	. . .	16.9	4.04	1.25	0.505
30 000	3845	. . .	. . .	. . .	. . .	. . .	19.3	4.64	1.42	0.520

Table 4.35. Flow of air through schedule 40 steel pipe. Courtesy Crane Company (T. P. #410).

Discharge		Pressure Drop per 100 feet and Velocity in Schedule 40 Pipe for Water at 60 F.															
Gallons per Minute	Cubic Ft. per Second	Velocity Feet per Second	Press. Drop Lbs. per Sq. In.	Velocity Feet per Second	Press. Drop Lbs. per Sq. In.	Velocity Feet per Second	Press. Drop Lbs. per Sq. In.	Velocity Feet per Second	Press. Drop Lbs. per Sq. In.	Velocity Feet per Second	Press. Drop Lbs. per Sq. In.	Velocity Feet per Second	Press. Drop Lbs. per Sq. In.	Velocity Feet per Second	Press. Drop Lbs. per Sq. In.	Velocity Feet per Second	Press. Drop Lbs. per Sq. In.
		1/8"		1/4"		3/8"		1/2"									
.2	0.000446	1.13	1.86	0.616	0.359												
.3	0.000668	1.69	4.22	0.924	0.903	0.504	0.159	0.317	0.061	3/4"							
.4	0.000891	2.26	6.98	1.23	1.61	0.672	0.345	0.422	0.086								
.5	0.00111	2.82	10.5	1.54	2.39	0.840	0.539	0.528	0.167	0.301	0.033						
.6	0.00134	3.39	14.7	1.85	3.29	1.01	0.751	0.633	0.240	0.361	0.041						
.8	0.00178	4.52	25.0	2.46	5.44	1.34	1.25	0.844	0.408	0.481	0.102	1"		1¼"			
1	0.00223	5.65	37.2	3.08	8.28	1.68	1.85	1.06	0.600	0.602	0.155	0.371	0.048			1½"	
2	0.00446	11.29	134.4	6.16	30.1	3.36	6.58	2.11	2.10	1.20	0.526	0.743	0.164	0.429	0.044		
3	0.00668			9.25	64.1	5.04	13.9	3.17	4.33	1.81	1.09	1.114	0.336	0.644	0.090	0.473	0.043
4	0.00891			12.33	111.2	6.72	23.9	4.22	7.42	2.41	1.83	1.49	0.565	0.858	0.150	0.630	0.071
5	0.01114	2"				8.40	36.7	5.28	11.2	3.01	2.75	1.86	0.835	1.073	0.223	0.788	0.104
6	0.01337	0.574	0.044	2½"		10.08	51.9	6.33	15.8	3.61	3.84	2.23	1.17	1.29	0.309	0.946	0.145
8	0.01782	0.765	0.073			13.44	91.1	8.45	27.7	4.81	6.60	2.97	1.99	1.72	0.518	1.26	0.241
10	0.02228	0.956	0.108	0.670	0.046			10.56	42.4	6.02	9.99	3.71	2.99	2.15	0.774	1.58	0.361
15	0.03342	1.43	0.224	1.01	0.094	3"				9.03	21.6	5.57	6.36	3.22	1.63	2.37	0.755
20	0.04456	1.91	0.375	1.34	0.158	0.868	0.056	3½"		12.03	37.8	7.43	10.9	4.29	2.78	3.16	1.28
25	0.05570	2.39	0.561	1.68	0.234	1.09	0.083	0.812	0.041	4"		9.28	16.7	5.37	4.22	3.94	1.93
30	0.06684	2.87	0.786	2.01	0.327	1.30	0.114	0.974	0.056			11.14	23.8	6.44	5.92	4.73	2.72
35	0.07798	3.35	1.05	2.35	0.436	1.52	0.151	1.14	0.704	0.882	0.041	12.99	32.2	7.51	7.90	5.52	3.64
40	0.08912	3.83	1.35	2.68	0.556	1.74	0.192	1.30	0.095	1.01	0.052	14.85	41.5	8.59	10.24	6.30	4.65
45	0.1003	4.30	1.67	3.02	0.668	1.95	0.239	1.46	0.117	1.13	0.064			9.67	12.80	7.09	5.85
50	0.1114	4.78	2.03	3.35	0.839	2.17	0.288	1.62	0.142	1.26	0.076	5"		10.74	15.66	7.88	7.15
60	0.1337	5.74	2.87	4.02	1.18	2.60	0.406	1.95	0.204	1.51	0.107			12.89	22.2	9.47	10.21
70	0.1560	6.70	3.84	4.69	1.59	3.04	0.540	2.27	0.261	1.76	0.143	1.12	0.047			11.05	13.71
80	0.1782	7.65	4.97	5.36	2.03	3.47	0.687	2.60	0.334	2.02	0.180	1.28	0.060	6"		12.62	17.59
90	0.2005	8.60	6.20	6.03	2.53	3.91	0.861	2.92	0.416	2.27	0.224	1.44	0.074			14.20	22.0
100	0.2228	9.56	7.59	6.70	3.09	4.34	1.05	3.25	0.509	2.52	0.272	1.60	0.090	1.11	0.036	15.78	26.9
125	0.2785	11.97	11.76	8.38	4.71	5.43	1.61	4.06	0.769	3.15	0.415	2.01	0.135	1.39	0.055	19.72	41.4
150	0.3342	14.36	16.70	10.05	6.69	6.51	2.24	4.87	1.08	3.78	0.580	2.41	0.190	1.67	0.077		
175	0.3899	16.75	22.3	11.73	8.97	7.60	3.00	5.68	1.44	4.41	0.774	2.81	0.253	1.94	0.102		
200	0.4456	19.14	28.8	13.42	11.68	8.68	3.87	6.49	1.85	5.04	0.985	3.21	0.323	2.22	0.130	8"	
225	0.5013	...	...	15.09	14.63	9.77	4.83	7.30	2.32	5.67	1.23	3.61	0.401	2.50	0.162	1.44	0.043
250	0.557	...	...	...	...	10.85	5.93	8.12	2.84	6.30	1.46	4.01	0.495	2.78	0.195	1.60	0.051
275	0.6127	...	...	...	...	11.94	7.14	8.93	3.40	6.93	1.79	4.41	0.583	3.05	0.234	1.76	0.061
300	0.6684	...	...	...	...	13.00	8.36	9.74	4.02	7.56	2.11	4.81	0.683	3.33	0.275	1.92	0.072
325	0.7241	...	...	...	...	14.12	9.89	10.53	4.09	8.19	2.47	5.21	0.797	3.61	0.320	2.08	0.083
350	0.7798			...	...	...	...	11.36	5.41	8.82	2.84	5.62	0.919	3.89	0.367	2.24	0.095
375	0.8355			...	...	...	...	12.17	6.18	9.45	3.25	6.02	1.05	4.16	0.416	2.40	0.108
400	0.8912			...	...	...	...	12.98	7.03	10.08	3.68	6.42	1.19	4.44	0.471	2.56	0.121
425	0.9469			...	...	...	...	13.80	7.89	10.71	4.12	6.82	1.33	4.72	0.529	2.73	0.136
450	1.003	10"		...	...	...	...	14.61	8.80	11.34	4.60	7.22	1.48	5.00	0.590	2.89	0.151
475	1.059	1.93	0.054			...	...	...	...	11.97	5.12	7.62	1.64	5.27	0.653	3.04	0.166
500	1.114	2.03	0.059			...	...	...	...	12.60	5.65	8.02	1.81	5.55	0.720	3.21	0.182
550	1.225	2.24	0.071			...	...	...	...	13.85	6.79	8.82	2.17	6.11	0.861	3.53	0.219
600	1.337	2.44	0.083			...	...	...	...	15.12	8.04	9.63	2.55	6.66	1.02	3.85	0.258
650	1.448	2.64	0.097	12"		...	...	...	...	...	...	10.43	2.98	7.22	1.18	4.17	0.301
700	1.560	2.85	0.112	2.01	0.047			...	...	...	...	11.23	3.43	7.78	1.35	4.49	0.343
750	1.671	3.05	0.127	2.15	0.054	14"		...	...	...	...	12.03	3.92	8.33	1.55	4.81	0.392
800	1.782	3.25	0.143	2.29	0.061			...	...	...	...	12.83	4.43	8.88	1.75	5.13	0.443
850	1.894	3.46	0.160	2.44	0.068	2.02	0.042	...	...	...	...	13.64	5.00	9.44	1.96	5.45	0.497
900	2.005	3.66	0.179	2.58	0.075	2.13	0.047	...	...	...	...	14.44	5.58	9.99	2.18	5.77	0.554
950	2.117	3.86	0.198	2.72	0.083	2.25	0.052			...	...	15.24	6.21	10.55	2.42	6.09	0.613
1 000	2.228	4.07	0.218	2.87	0.091	2.37	0.057	16"		...	...	16.04	6.84	11.10	2.68	6.41	0.675
1 100	2.451	4.48	0.260	3.15	0.110	2.61	0.068			...	...	17.65	8.23	12.22	3.22	7.05	0.807
1 200	2.674	4.88	0.306	3.44	0.128	2.85	0.080	2.18	0.042	...	...	...	...	13.33	3.81	7.70	0.948
1 300	2.896	5.29	0.355	3.73	0.150	3.08	0.093	2.36	0.048	...	...	...	...	14.43	4.45	8.33	1.11
1 400	3.119	5.70	0.409	4.01	0.171	3.32	0.107	2.54	0.055					15.55	5.13	8.98	1.28
1 500	3.342	6.10	0.466	4.30	0.195	3.56	0.122	2.72	0.063	18"				16.66	5.85	9.62	1.46
1 600	3.565	6.51	0.527	4.59	0.219	3.79	0.138	2.90	0.071					17.77	6.61	10.26	1.65
1 800	4.010	7.32	0.663	5.16	0.276	4.27	0.172	3.27	0.088	2.58	0.050			19.99	8.37	11.54	2.08
2 000	4.456	8.14	0.808	5.73	0.339	4.74	0.209	3.63	0.107	2.87	0.060	20"		22.21	10.3	12.82	2.55
2 500	5.570	10.17	1.24	7.17	0.515	5.93	0.321	4.54	0.163	3.59	0.091			24"		16.03	3.94
3 000	6.684	12.20	1.76	8.60	0.731	7.11	0.451	5.45	0.232	4.30	0.129	3.46	0.075			19.24	5.59
3 500	7.798	14.24	2.38	10.03	0.982	8.30	0.607	6.35	0.312	5.02	0.173	4.04	0.101			22.44	7.56
4 000	8.912	16.27	3.08	11.47	1.27	9.48	0.787	7.26	0.401	5.74	0.222	4.62	0.129	3.19	0.052	25.65	9.80
4 500	10.03	18.31	3.87	12.90	1.60	10.67	0.990	8.17	0.503	6.46	0.280	5.20	0.162	3.59	0.065	28.87	12.2
5 000	11.14	20.35	4.71	14.33	1.95	11.85	1.21	9.08	0.617	7.17	0.340	5.77	0.199	3.99	0.079	...	...
6 000	13.37	24.41	6.74	17.20	2.77	14.23	1.71	10.89	0.877	8.61	0.483	6.93	0.280	4.79	0.111	...	...
7 000	15.60	28.49	9.11	20.07	3.74	16.60	2.31	12.71	1.18	10.04	0.652	8.08	0.376	5.59	0.150	...	...
8 000	17.82	...	...	22.93	4.84	18.96	2.99	14.52	1.51	11.47	0.839	9.23	0.488	6.38	0.192	...	...
9 000	20.05	...	...	25.79	6.09	21.34	3.76	16.34	1.90	12.91	1.05	10.39	0.608	7.18	0.242	...	...
10 000	22.28	...	...	28.66	7.46	23.71	4.61	18.15	2.34	14.34	1.28	11.54	0.739	7.98	0.294	...	...
12 000	26.74	...	...	34.40	10.7	28.45	6.59	21.79	3.33	17.21	1.83	13.85	1.06	9.58	0.416	...	...
14 000	31.19	...	...	...	...	33.19	8.89	25.42	4.49	20.08	2.45	16.16	1.43	11.17	0.562	...	...
16 000	35.65	...	...	...	...	...	...	29.05	5.83	22.95	3.18	18.47	1.85	12.77	0.723	...	...
18 000	40.10	...	...	...	...	...	...	32.68	7.31	25.82	4.03	20.77	2.32	14.36	0.907	...	...
20 000	44.56	...	...	...	...	...	...	36.31	9.03	28.69	4.93	23.08	2.86	15.96	1.12	...	...

For pipe lengths other than 100 feet, the pressure drop is proportional to the length. Thus, for 50 feet of pipe, the pressure drop is approximately one-half the value given in the table . . . for 300 feet, three times the given value, etc.

Velocity is a function of the cross sectional flow area; thus, it is constant for a given flow rate and is independent of pipe length.

Table 4.36. Flow of water through schedule 40 steel pipe. Courtesy Crane Company (T. P. #410).

Friction losses in pipe fittings

Example: The dotted line shows that the resistance of a 6-inch Standard Elbow is equivalent to approximately 16 feet of 6-inch Standard Pipe.

Note: For sudden enlargements or sudden contractions, use the smaller diameter, d, on the pipe size scale.

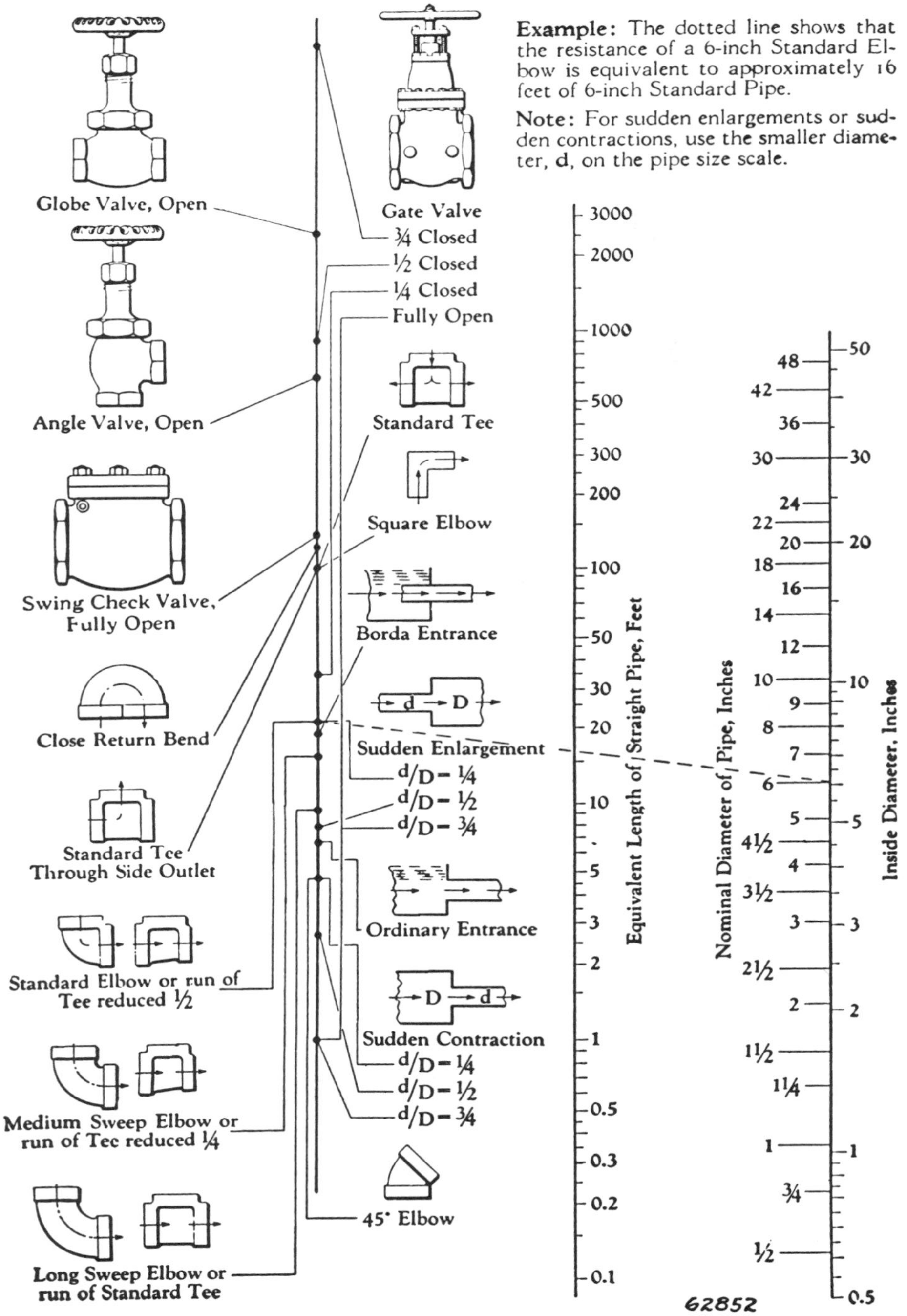

This chart may be used for any liquid or gas

Table 4.37 continued

Table 4.37. Friction loss in pipe fittings in terms of equivalent feet of straight pipe. Courtesy Cameron Hydraulic Data Book, *Ingersoll Rand Company.*

Table 4.37 continued

Nominal pipe Size D_N Std. Wt.	Actual inside diam. in.	Gate Valve FULL OPEN	45° Elbow	Long-sweep elbow or run of Std tee	Std elbow or run of tee reduced ½	Std tee thru side outlet	Close return bend	Swing check valve FULL OPEN	Angle Valve FULL OPEN	Globe Valve FULL OPEN	Equivalent Resistance of Std. Wt. Welding Elbows Length of Straight Pipe (Feet)[1]* — 90° Elbows: Short Radius $R/D_N=1$	90° Elbows: Long Radius $R/D_N=1½$	45° Elbows: Short Radius $R/D_N=1$	45° Elbows: Long Radius $R/D_N=1½$
Resistance factor		.19	.42	.6	.9	1.8	2.2	2.3	5.	10.				
½	.622	.35	.78	1.11	1.7	3.3	4.1	4.3	9.3	18.6	‡	0.68	‡	0.44
¾	.824	.44	.97	1.4	2.1	4.2	5.1	5.3	11.5	23.1	‡	0.91	‡	0.58
1	1.049	.56	1.23	1.8	2.6	5.3	6.5	6.8	14.7	29.4	1.6	1.15	1.01	0.74
1¼	1.380	.74	1.6	2.3	3.5	7.0	8.5	8.9	19.3	38.6	2.1	1.5	1.33	0.98
1½	1.610	.86	1.9	2.7	4.1	8.1	9.9	10.4	22.6	45.2	2.4	1.8	1.6	1.14
2	2.067	1.10	2.4	3.5	5.2	10.4	12.8	13.4	29	58	3.1	2.3	2.0	1.5
2½	2.469	1.32	2.9	4.2	6.2	12.4	15.2	15.9	35	69	3.7	2.7	2.4	1.7
3	3.068	1.6	3.6	5.2	7.7	15.5	18.9	19.8	43	86	4.7	3.4	3.0	2.2
4	4.026	2.1	4.7	6.8	10.2	20.3	24.8	26.0	57	113	6.1	4.4	3.9	2.9
5	5.047	2.7	5.9	8.5	12.7	25.4	31	33	71	142	7.7	5.6	4.9	3.6
6	6.065	3.2	7.1	10.2	15.3	31	37	39	85	170	9.2	6.7	5.9	4.3
7	7.024	3.7	8.3	11.8	17.7	35	43	45	98	197	‡‡	‡‡	‡‡	‡‡
8	7.981	4.3	9.4	13.4	20.2	40	49	52	112	224	12.1	8.8	7.7	5.7
10	10.020	5.3	11.8	16.9	25.3	51	62	65	141	281	15.2	11.0	9.7	7.1
12	12.000	6.4	14.1	20.2	30	61	74	77	168	336	18.2	13.2	11.6	8.5
14		7.5	16.5	23.5	35	71	86	90			20.1	14.6	12.8	9.4
16		8.5	18.8	26.9	40	81	99	104			23.1	16.8	14.7	10.8
18		9.6	21.2	30	45	91	111	116			26.2	19.0	16.7	12.2
20		10.7	23.5	34	50	101	123	129			29	21.2	18.6	13.6
24		12.8	28.2	40	61	121	148	155			35	25.6	22.5	16.5
30		16.0	35.3	50	76	151	185	193			44	32	28.3	20.7
36		19.2	42.4	61	91	181	222	232			53	39	34	25.0
42		22.4	49.4	71	106	212	259	271			63	45	40	29.2
48		25.6	57.6	81	121	242	296	310			72	52	46	33

Data on fittings based on information published by Crane Co.

*For 180° bend multiply values for 90° bend by 1.34.

Data are based on Fanning coefficient of 0.006, as taken from Chart No. 18 of Catalog 211 of Tube Turns, Inc.

‡Short Radius elbows, $R/D_N=1$, not made in this size and weight.

‡‡Not made in this size.

Representative Equivalent Length‡ in Pipe Diameters *(L/D)* Of Various Valves and Fittings

Description of Product				Equivalent Length In Pipe Diameters (L/D)
Globe Valves	Stem Perpendicular to Run	With no obstruction in flat, bevel, or plug type seat	Fully open	340
		With wing or pin guided disc	Fully open	450
	Y-Pattern	(No obstruction in flat, bevel, or plug type seat)		
		— With stem 60 degrees from run of pipe line	Fully open	175
		— With stem 45 degrees from run of pipe line	Fully open	145
Angle Valves		With no obstruction in flat, bevel, or plug type seat	Fully open	145
		With wing or pin guided disc	Fully open	200
Gate Valves	Wedge, Disc, Double Disc, or Plug Disc		Fully open	13
			Three-quarters open	35
			One-half open	160
			One-quarter open	900
	Pulp Stock		Fully open	17
			Three-quarters open	50
			One-half open	260
			One-quarter open	1200
Conduit Pipe Line Gate, Ball, and Plug Valves			Fully open	3**
Check Valves	Conventional Swing	0.5†	Fully open	135
	Clearway Swing	0.5†	Fully open	50
	Globe Lift or Stop; Stem Perpendicular to Run or Y-Pattern	2.0†	Fully open	Same as Globe
	Angle Lift or Stop	2.0†	Fully open	Same as Angle
	In-Line Ball	2.5 vertical and 0.25 horizontal†	Fully open	150
Foot Valves with Strainer		With poppet lift-type disc 0.3†	Fully open	420
		With leather-hinged disc 0.4†	Fully open	75
Butterfly Valves (8-inch and larger)			Fully open	40
Cocks	Straight-Through	Rectangular plug port area equal to 100% of pipe area	Fully open	18
	Three-Way	Rectangular plug port area equal to 80% of pipe area (fully open)	Flow straight through	44
			Flow through branch	140
Fittings	90 Degree Standard Elbow			30
	45 Degree Standard Elbow			16
	90 Degree Long Radius Elbow			20
	90 Degree Street Elbow			50
	45 Degree Street Elbow			26
	Square Corner Elbow			57
	Standard Tee	With flow through run		20
		With flow through branch		60
	Close Pattern Return Bend			50
Pipe	90 Degree Pipe Bends			Chart 3.10 p. 156-157
	Miter Bends			
	Sudden Enlargements and Contractions			
	Entrance and Exit Losses			See Page A-26

**Exact equivalent length is equal to the length between flange faces or welding ends.

†Minimum calculated pressure drop (psi) across valve to provide sufficient flow to lift disc fully.

‡For limitations, see pages 2-8 to 2-11, Crane Technical Paper 410.

Table 4.38. Representative equivalent length in pipe diameters (L/D) *of various valves and fittings. Courtesy Crane Company (T. P. #410).*

Measure

1 in.	=	25.4 mm
1 in.	=	2.54 cm
1 mm	=	0.03937 in.
1 mm	=	0.00328 ft
1 micron	=	0.000001 meter
1 torr	=	1 mm mercury
10^{-8} torr	=	1 atom mercury
1 ft	=	304.8 mm
1 ft	=	30.48 cm
1 sq. in.	=	6.4516 sq cm
1 sq cm	=	0.155 sq in.
1 sq cm	=	0.00108 sq ft
1 sq ft	=	929.03 sq cm

Circumference of a circle $= 2\pi r = \pi d$

Area of a circle $= \pi r^2 = \dfrac{\pi d^2}{4}$

Weight

1 kg = 2.205 lb
1 cu in. of water (60 F) = 0.073551 cu in. of mercury (32 F)
1 cu in. of mercury (32 F) = 13.596 cu in. of water (60 F)
1 cu in. of mercury (32 F) = 0.4905 lb

Velocity

1 ft per sec = 0.3048 m per sec
1 m per sec = 3.2808 ft per sec

Density

1 lb per cu in. = 27.68 gram per cu cm
1 gr per cu cm = 0.03613 lb per cu in.
1 lb per cu ft = 16.0184 kg per cu m
1 kg per cu m = 0.06243 lb per cu ft

Physical Constants

Base of Natural Logarithms (e) 2.718 281 828 5
Acceleration of Gravity (g) .... 32.174 ft/sec² .... (980.665 cm/sec²)
Pi (π) 3.141 592 653 6

	Degrees Kelvin	Degrees Rankine	Degrees Centigrade	Degrees Fahrenheit
Absolute Zero	0	0	− 273.16	− 459.69
Water Freezing Point (14.696 psia)	273.16	491.69	0	32
Water Boiling Point (14.696 psia)	373.16	671.69	100	212

Equivalents of Temperature

To convert degrees Centigrade to degrees Fahrenheit:

$$t = 1.8\, t_c + 32$$

To convert degrees Fahrenheit to degrees Centigrade:

$$t_c = \frac{t - 32}{1.8}$$

Where: t_c = temperature, in degrees Centigrade

Prefixes

Atto	one-quintillionth	0.000 000 000 000 000 001	10^{-18}
Femto	one-quadrillionth	0.000 000 000 000 001	10^{-15}
Pico	one-trillionth	0.000 000 000 001	10^{-12}
Nano	one-billionth	0.000 000 001	10^{-9}
Micro	one-millionth	0.000 001	10^{-6}
Milli	one-thousandth	0.001	10^{-3}
Centi	one-hundredth	0.01	10^{-2}
Deci	one-tenth	0.1	10^{-1}
Uni	one	1.0	10^{0}
Deka	ten	10.0	10^{1}
Hecto	one hundred	100.0	10^{2}
Kilo	one thousand	1 000.0	10^{3}
Mega	one million	1 000 000.0	10^{6}
Giga	one billion	1 000 000 000.0	10^{9}
Tera	one trillion	1 000 000 000 000.0	10^{12}
	one quintillion	1 000 000 000 000 000.0	10^{15}
	one quadrillion	1 000 000 000 000 000 000.0	10^{18}

Table 4.39. Equivalents and prefixes. Courtesy Crane Company (T. P. #410).

Theoretical Horsepower Required to Raise Water (at 60 F) To Different Heights

Gals. per Min.	5 feet	10 feet	15 feet	20 feet	25 feet	30 feet	35 feet	40 feet	45 feet	50 feet	60 feet	70 feet	80 feet	90 feet	100 feet
5	0.006	0.013	0.019	0.025	0.032	0.038	0.044	0.051	0.057	0.063	0.076	0.088	0.101	0.114	0.126
10	0.013	0.025	0.038	0.051	0.063	0.076	0.088	0.101	0.114	0.126	0.152	0.177	0.202	0.227	0.253
15	0.019	0.038	0.057	0.076	0.095	0.114	0.133	0.152	0.171	0.190	0.227	0.265	0.303	0.341	0.379
20	0.025	0.051	0.076	0.101	0.126	0.152	0.177	0.202	0.227	0.253	0.303	0.354	0.404	0.455	0.505
25	0.032	0.063	0.095	0.126	0.158	0.190	0.221	0.253	0.284	0.316	0.379	0.442	0.505	0.568	0.632
30	0.038	0.076	0.114	0.152	0.190	0.227	0.265	0.303	0.341	0.379	0.455	0.531	0.606	0.682	0.758
35	0.044	0.088	0.133	0.177	0.221	0.265	0.310	0.354	0.398	0.442	0.531	0.619	0.707	0.796	0.884
40	0.051	0.101	0.152	0.202	0.253	0.303	0.354	0.404	0.455	0.505	0.606	0.707	0.808	0.910	1.011
45	0.057	0.114	0.171	0.227	0.284	0.341	0.398	0.455	0.512	0.568	0.682	0.796	0.910	1.023	1.137
50	0.063	0.126	0.190	0.253	0.316	0.379	0.442	0.505	0.568	0.632	0.758	0.884	1.011	1.137	1.263
60	0.076	0.152	0.227	0.303	0.379	0.455	0.531	0.606	0.682	0.758	0.910	1.061	1.213	1.364	1.516
70	0.088	0.177	0.265	0.354	0.442	0.531	0.619	0.707	0.796	0.884	1.061	1.238	1.415	1.592	1.768
80	0.101	0.202	0.303	0.404	0.505	0.606	0.707	0.808	0.910	1.011	1.213	1.415	1.617	1.819	2.021
90	0.114	0.227	0.341	0.455	0.568	0.682	0.796	0.910	1.023	1.137	1.364	1.592	1.819	2.046	2.274
100	0.126	0.253	0.379	0.505	0.632	0.758	0.884	1.011	1.137	1.263	1.516	1.768	2.021	2.274	2.526
125	0.158	0.316	0.474	0.632	0.790	0.947	1.105	1.263	1.421	1.579	1.895	2.211	2.526	2.842	3.158
150	0.190	0.379	0.568	0.758	0.947	1.137	1.326	1.516	1.705	1.895	2.274	2.653	3.032	3.411	3.790
175	0.221	0.442	0.663	0.884	1.105	1.326	1.547	1.768	1.990	2.211	2.653	3.095	3.537	3.979	4.421
200	0.253	0.505	0.758	1.011	1.263	1.516	1.768	2.021	2.274	2.526	3.032	3.537	4.042	4.548	5.053
250	0.316	0.632	0.947	1.263	1.579	1.895	2.211	2.526	2.842	3.158	3.790	4.421	5.053	5.684	6.316
300	0.379	0.758	1.137	1.516	1.895	2.274	2.653	3.032	3.411	3.790	4.548	5.305	6.063	6.821	7.579
350	0.442	0.884	1.326	1.768	2.211	2.653	3.095	3.537	3.979	4.421	5.305	6.190	7.074	7.958	8.842
400	0.505	1.011	1.516	2.021	2.526	3.032	3.537	4.042	4.548	5.053	6.063	7.074	8.084	9.095	10.11
500	0.632	1.263	1.895	2.526	3.158	3.790	4.421	5.053	5.684	6.316	7.579	8.842	10.11	11.37	12.63

Gals. per Min.	125 feet	150 feet	175 feet	200 feet	250 feet	300 feet	350 feet	400 feet
5	0.158	0.190	0.221	0.253	0.316	0.379	0.442	0.505
10	0.316	0.379	0.442	0.505	0.632	0.758	0.884	1.011
15	0.474	0.568	0.663	0.758	0.947	1.137	1.326	1.516
20	0.632	0.758	0.884	1.011	1.263	1.516	1.768	2.021
25	0.790	0.947	1.105	1.263	1.579	1.895	2.211	2.526
30	0.947	1.137	1.326	1.516	1.895	2.274	2.653	3.032
35	1.105	1.326	1.547	1.768	2.211	2.653	3.095	3.537
40	1.263	1.516	1.768	2.021	2.526	3.032	3.537	4.042
45	1.421	1.705	1.990	2.274	2.842	3.411	3.979	4.548
50	1.579	1.895	2.211	2.526	3.158	3.790	4.421	5.053
60	1.895	2.274	2.653	3.032	3.790	4.548	5.305	6.063
70	2.211	2.653	3.095	3.537	4.421	5.305	6.190	7.074
80	2.526	3.032	3.537	4.042	5.053	6.063	7.074	8.084
90	2.842	3.411	3.979	4.548	5.684	6.821	7.958	9.095
100	3.158	3.790	4.421	5.053	6.316	7.579	8.842	10.11
125	3.948	4.737	5.527	6.316	7.895	9.474	11.05	12.63
150	4.737	5.684	6.632	7.579	9.474	11.37	13.26	15.16
175	5.527	6.632	7.737	8.842	11.05	13.26	15.47	17.68
200	6.316	7.579	8.842	10.11	12.63	15.16	17.68	20.21
250	7.895	9.474	11.05	12.63	15.79	18.95	22.11	25.26
300	9.474	11.37	13.26	15.16	18.95	22.74	26.53	30.32
350	11.05	13.26	15.47	17.68	22.11	26.53	30.95	35.37
400	12.63	15.16	17.68	20.21	25.26	30.32	35.37	40.42
500	15.79	18.95	22.11	25.26	31.58	37.90	44.21	50.53

HORSEPOWER = 33 000 . . . ft-lb/min
= 550 . . . ft-lb/sec
= 2544.48 . . . Btu/hr
= 745.7 . . . watts

$$(whp) = QH\rho \div 247\,000 = QP \div 1714$$
$$(bhp) = (whp) \div e_p = QH\rho \div 247\,000\, e_p$$
$$(e_p) = QH\rho \div 247\,000\,(bhp)$$

where: (whp) = water horsepower
H = pump head in feet
(bhp) = brake horsepower
e_p = pump efficiency

Overall efficiency (e_o) takes into account all losses in the pump and driver.

$$e_o = e_p\, e_D\, e_T$$

where: e_D = driver efficiency
e_T = transmission efficiency

e_V = volumetric efficiency

$$e_V(\%) = \frac{\text{actual pump displacement } (Q)\ (100)}{\text{theoretical pump displacement } (Q)}$$

Note: For fluids other than water, multiply table values by specific gravity. In pumping liquids with a viscosity considerably higher than that of water, the pump capacity and head are reduced. To calculate the horsepower for such fluids, pipe friction head must be added to the elevation head to obtain the total head; this value is inserted in the first horsepower equation given above.

Table 4.40. Power required for pumping. Courtesy Crane Company (T. P. #410).

%D = Percentage of Total Diameter of Tank

%C = Percentage of Total Capacity of Tank

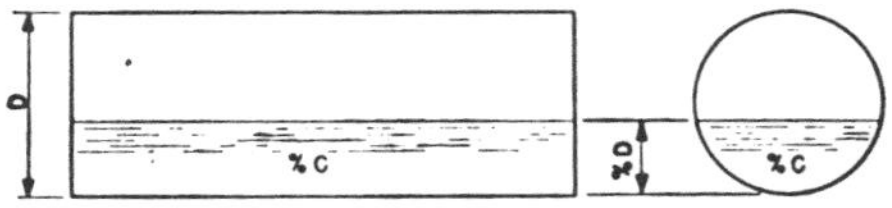

%D	%C	%D	%C	%D	%C	%D	%C	%D	%C	%D	%C
0.1	0.0053	11.2	6.1449	23.2	17.590	50.5	50.634	77.2	82.839	89.2	94.1738
0.2	0.0152	11.4	6.3060	23.4	17.806	51.0	51.271	77.4	83.051	89.4	94.3310
0.3	0.0279	11.6	6.4685	23.6	18.022	51.5	51.907	77.6	83.263	89.6	94.4874
0.4	0.0429	11.8	6.6320	23.8	18.240	52.0	52.543	77.8	83.476	89.8	94.6420
0.5	0.0600	12.0	6.7970	24.0	18.457	52.5	53.181	78.0	83.688	90.0	94.7956
0.6	0.0788	12.2	6.9630	24.2	18.675	53.0	53.812	78.2	83.899	90.2	94.9477
0.7	0.0992	12.4	7.1305	24.4	18.892	53.5	54.450	78.4	84.108	90.4	95.0985
0.8	0.1212	12.6	7.2990	24.6	19.110	54.0	55.087	78.6	84.317	90.6	95.2475
0.9	0.1446	12.8	7.4686	24.8	19.330	54.5	55.722	78.8	84.525	90.8	95.3955
1.0	0.1692	13.0	7.6390	25.0	19.551	55.0	56.356	79.0	84.733	91.0	95.5418
1.2	0.2223	13.2	7.8110	25.5	20.103	55.6	56.989	79.2	84.940	91.2	95.6869
1.4	0.2800	13.4	7.9840	26.0	20.661	56.0	57.621	79.4	85.146	91.4	95.8304
1.6	0.3419	13.6	8.1580	26.5	21.222	56.5	58.251	79.6	85.351	91.6	95.9724
1.8	0.4077	13.8	8.3330	27.0	21.785	57.0	58.884	79.8	85.556	91.8	96.1131
2.0	0.4773	14.0	8.5094	27.5	22.353	57.5	59.510	80.0	85.762	92.0	96.2520
2.2	0.5501	14.2	8.6867	28.0	22.923	58.0	60.140	80.2	85.965	92.2	96.3894
2.4	0.6267	14.4	8.8651	28.5	23.494	58.5	60.767	80.4	86.168	92.4	96.5251
2.6	0.7061	14.6	9.0440	29.0	24.072	59.0	61.396	80.6	86.370	92.6	96.6592
2.8	0.7886	14.8	9.2246	29.5	24.651	59.5	62.023	80.8	86.571	92.8	96.7918
3.0	0.8742	15.0	9.4060	30.0	25.233	60.0	62.645	81.0	86.771	93.0	96.9229
3.2	0.9625	15.2	9.588	30.5	25.818	60.5	63.268	81.2	86.970	93.2	97.0517
3.4	1.0533	15.4	9.772	31.0	26.407	61.0	63.890	81.4	87.169	93.4	97.1789
3.6	1.1470	15.6	9.956	31.5	26.996	61.5	64.509	81.6	87.367	93.6	97.3048
3.8	1.2432	15.8	10.141	32.0	27.589	62.0	65.131	81.8	87.563	93.8	97.4285
4.0	1.3418	16.0	10.327	32.5	28.184	62.5	65.746	82.0	87.760	94.0	97.5503
4.2	1.4429	16.2	10.515	33.0	28.781	63.0	66.362	82.2	87.954	94.2	97.6703
4.4	1.5461	16.4	10.703	33.5	29.380	63.5	66.975	82.4	88.149	94.4	97.7884
4.6	1.6515	16.6	10.892	34.0	29.981	64.0	67.588	82.6	88.343	94.6	97.9044
4.8	1.7594	16.8	11.082	34.5	30.587	64.5	68.198	82.8	88.535	94.8	98.1086
5.0	1.8693	17.0	11.273	35.0	31.192	65.0	68.808	83.0	88.727	95.0	98.1307
5.2	1.8914	17.2	11.465	35.5	31.802	65.5	69.413	83.2	88.918	95.2	98.2406
5.4	2.0956	17.4	11.657	36.0	32.412	66.0	70.019	83.4	89.108	95.4	98.3485
5.6	2.2116	17.6	11.851	36.5	33.025	66.5	70.620	83.6	89.297	95.6	98.4539
5.8	2.3297	17.8	12.046	37.0	33.638	67.0	71.219	83.8	89.485	95.8	98.5571
6.0	2.4497	18.0	12.240	37.5	34.254	67.5	71.816	84.0	89.673	96.0	98.6582
6.2	2.5715	18.2	12.437	38.0	34.869	68.0	72.411	84.2	89.859	96.2	98.7568
6.4	2.6952	18.4	12.633	38.5	35.491	68.5	73.004	84.4	90.044	96.4	98.8530
6.6	2.8211	18.6	12.831	39.0	36.110	69.0	73.593	84.6	90.228	96.6	98.9467
6.8	2.9483	18.8	13.030	39.5	36.732	69.5	74.182	84.8	90.412	96.8	99.0375
7.0	3.0771	19.0	13.229	40.0	37.355	70.0	74.767	85.0	90.5940	97.0	99.1258
7.2	3.2082	19.2	13.429	40.5	37.977	70.5	75.349	85.2	90.7754	97.2	99.2114
7.4	3.3408	19.4	13.630	41.0	38.604	71.0	75.928	85.4	90.9560	97.4	99.2939
7.6	3.4749	19.6	13.832	41.5	39.233	71.5	76.506	85.6	91.1349	97.6	99.3733
7.8	3.6106	19.8	14.035	42.0	39.860	72.0	77.077	85.8	91.3133	97.8	99.4499
8.0	3.7480	20.0	14.238	42.5	40.490	72.5	77.647	86.0	91.4906	98.0	99.5227
8.2	3.8869	20.2	14.444	43.0	41.116	73.0	78.215	86.2	91.6670	98.2	99.5923
8.4	4.0276	20.4	14.649	43.5	41.749	73.5	78.778	86.4	91.8420	98.4	99.6581
8.6	4.1696	20.6	14.854	44.0	42.379	74.0	79.339	86.6	92.0160	98.6	99.7200
8.8	4.3131	20.8	15.060	44.5	43.011	74.5	79.897	86.8	92.1890	98.8	99.7777
9.0	4.4582	21.0	15.267	45.0	43.644	75.0	80.449	87.0	92.3610	99.0	99.8308
9.2	4.6045	21.2	15.475	45.5	44.278	75.2	80.670	87.2	92.5314	99.1	99.8554
9.4	4.7525	21.4	15.683	46.0	44.913	75.4	80.890	87.4	92.7010	99.2	99.8788
9.6	4.9015	21.6	15.892	46.5	45.550	75.6	81.108	87.6	92.8695	99.3	99.9008
9.8	5.0523	21.8	16.101	47.0	46.188	75.8	81.325	87.8	93.0370	99.4	99.9212
10.0	5.2044	22.0	16.312	47.5	46.819	76.0	81.543	88.0	93.2030	99.5	99.9400
10.2	5.3580	22.2	16.524	48.0	47.457	76.2	81.760	88.2	93.3680	99.6	99.9508
10.4	5.5126	22.4	16.737	48.5	48.093	76.4	81.978	88.4	93.5315	99.7	99.9571
10.6	5.6690	22.6	16.949	49.0	48.729	76.6	82.194	88.6	93.6940	99.8	99.9848
10.8	5.8262	22.8	17.161	49.5	49.366	76.8	82.410	88.8	93.8551	99.9	99.9947
11.0	5.9848	23.0	17.376	50.0	50.000	77.0	82.624	89.0	94.0152	100.0	100.0000

Table 4.41. Table for gauging horizontal cylindrical tanks—flat ends.

Size	Drill Diameter	Size	Drill Diameter	Size	Drill Diameter	Size	Drill Diameter	Size	Drill Diameter
1	.2280	17	.1730	33	.1130	49	.0730	65	.0350
2	.2210	18	.1695	34	.1110	50	.0700	66	.0330
3	.2130	19	.1660	35	.1100	51	.0670	67	.0320
4	.2090	20	.1610	36	.1065	52	.0635	68	.0310
5	.2055	21	.1590	37	.1040	53	.0595	69	.0292
6	.2040	22	.1570	38	.1015	54	.0550	70	.0280
7	.2010	23	.1540	39	.0995	55	.0520	71	.0260
8	.1990	24	.1520	40	.0980	56	.0465	72	.0250
9	.1960	25	.1495	41	.0960	57	.0430	73	.0240
10	.1935	26	.1470	42	.0935	58	.0420	74	.0225
11	.1910	27	.1440	43	.0890	59	.0410	75	.0210
12	.1890	28	.1405	44	.0860	60	.0400	76	.0200
13	.1850	29	.1360	45	.0820	61	.0390	77	.0180
14	.1820	30	.1285	46	.0810	62	.0380	78	.0160
15	.1800	31	.1200	47	.0785	63	.0370	79	.0145
16	.1770	32	.1160	48	.0760	64	.0360	80	.0135

Letter Sizes

A	.234	G	.261	L	.290	Q	.332	V	.377
B	.238	H	.266	M	.295	R	.339	W	.386
C	.242	I	.272	N	.302	S	.348	X	.397
D	.246	J	.277	O	.316	T	.358	Y	.404
E	.250	K	.281	P	.323	U	.368	Z	.413
F	.257								

All dimensions are in inches.
Drills designated in common fractions are available in diameters 1/64″ to 1¾″ in 1/64″ increments, 1¾″ to 2¼″ in 1/32″ increments, and 2¼″ to 3½″ in 1/16″ increments. Drills larger than 3½″ are seldom used, and are regarded as special drills.

Table 4.42. Twist drill sizes.

Superheat Correction Factors for Safety Valves in Steam Service

Set Pressure (Pounds per Square Inch Gage)	Saturation Temperature (Degrees Fahrenheit)	Correction Factor K_{SH}											
		0.99	0.98	0.97	0.96	0.95	0.94	0.93	0.92	0.91	0.90	0.89	0.88
		Total Temperature (Degrees Fahrenheit)											
10	240	269	305	335	368	400	428	460	492	520	545	570	595
20	259	286	315	343	375	405	433	463	492	518	542	565	590
40	287	310	335	357	382	410	440	467	493	515	540	561	585
60	308	330	350	370	390	422	450	472	495	515	537	560	580
80	324	345	365	385	405	432	460	478	497	515	535	556	580
100	338	360	375	395	415	440	466	485	500	515	535	555	580
120	350	370	388	405	425	450	475	490	505	520	537	557	581
140	361	...	398	415	435	455	480	497	510	525	540	560	585
160	370	...	405	425	443	463	487	502	516	530	545	565	586
180	379	...	415	432	450	470	492	508	523	535	550	570	590
200	388	...	420	440	456	475	497	513	527	540	555	575	592
220	396	...	430	445	463	480	502	517	532	546	560	577	596
240	403	...	435	452	470	485	507	522	537	550	565	583	600
260	409	...	440	460	475	490	512	526	541	555	569	586	603
280	416	...	447	465	480	495	516	531	545	558	573	590	606
300	422	...	452	470	485	500	520	535	550	562	577	593	610
350	433	...	465	480	496	512	530	545	558	572	586	602	618
400	448	...	475	492	508	523	540	553	566	580	595	610	626
500	470	...	495	513	526	543	557	568	582	597	610	625	646
600	489	...	512	530	543	556	570	585	596	610	625	638	655
800	520	...	545	558	570	585	597	610	625	635	650	665	680
1,000	546	...	567	582	595	608	620	633	645	660	675	688	705
1,250	574	...	593	605	620	630	640	655	668	681	696	710	725
1,500	597	...	...	630	642	653	664	676	688	702	715	728	744
1,750	619	...	...	647	660	670	680	692	704	717	730	743	759
2,000	637	...	...	665	675	685	696	708	719	732	745	757	773
2,500	670	...	...	690	702	712	723	733	742	755	766	780	795
3,000	697	...	...	713	723	733	742	751	762	773	785	795	812

Table 4.43 continued

Table 4.43. Superheat correction factors for safety valves in steam service. Courtesy American Petroleum Institute.

Table 4.43 continued

Set Pressure (Pounds per Square Inch Gage)	Saturation Temperature (Degrees Fahrenheit)	Correction Factor K_{SH}											
		0.87	0.86	0.85	0.84	0.83	0.82	0.81	0.80	0.79	0.78	0.77	0.76
		Total Temperature (Degrees Fahrenheit)											
10	240	618	645	670	695	725	755	783	817	850	885	920	955
20	259	613	640	665	690	720	748	780	813	847	885	918	954
40	287	610	635	660	685	715	742	775	810	845	880	916	952
60	308	607	630	655	683	710	740	770	807	840	880	915	951
80	324	605	630	653	680	708	736	770	805	840	878	915	950
100	338	605	628	652	680	706	735	768	805	840	877	914	950
120	350	605	630	652	680	705	733	767	804	838	876	913	949
140	361	607	630	654	678	705	732	765	804	838	876	912	949
160	370	610	632	655	678	703	730	765	803	837	875	912	949
180	379	612	635	656	680	702	730	764	802	837	875	911	948
200	388	615	636	658	680	703	729	764	801	835	875	911	948
220	396	617	640	660	682	705	730	763	800	835	875	911	947
240	403	620	641	664	685	706	730	763	799	835	875	910	947
260	409	623	645	666	686	710	731	763	800	835	874	910	946
280	416	626	647	668	690	712	734	764	800	835	874	910	946
300	422	630	650	670	692	715	735	765	800	835	873	910	946
350	433	637	657	678	700	722	741	770	804	835	874	915	948
400	448	645	665	685	707	730	750	775	808	840	876	918	950
500	470	660	680	700	722	743	763	788	820	849	885	925	957
600	489	675	693	715	735	756	776	799	830	858	893	935	965
800	520	700	718	738	758	780	800	820	850	877	910	950	981
1,000	546	723	740	760	778	800	820	840	867	893	925	965	994
1,250	574	738	762	780	800	820	840	860	885	910	940	975	...
1,500	597	762	780	798	817	838	857	878	902	927	955	985	...
1,750	619	777	795	812	832	852	871	892	915	938	965	993	...
2,000	637	790	810	825	845	865	885	905	928	950	975	1,000	...
2,500	670	815	830	848	866	887	906	927	948	968	992	...	...
3,000	697	832	850	865	885	905	925	945	965	983	1,005	...	...

Note: Correction factors for pressure and temperature conditions not tabulated may be determined by interpolation if desired. However, it is practical to select the correction factor according to the next higher temperature tabulated at the closest pressure level listed.

Table 3 .... Day-long Direct Solar Irradiation in Btu/Ft² for the 21st day of Each Month at Latitudes from 24 to 64 deg North, on Surfaces Normal to the Sun's Rays; Total Irradiation on Horizontal Surfaces and South-facing Surfaces Tilted at the Following Angles above the Horizontal: $L - 10$ deg; L deg; $L + 10$ deg; $L + 20$ deg; 90 deg[a]

Date	Deg. Lat.	I_{DN}	Total Solar Irradiation, $I_{D\theta} + I_d$					
			Horiz.	$L - 10$	L	$L + 10$	$L + 20$	Vertical
Jan. 21	24	2766	1622	1984	2174	2300	2360	1766
$\delta = -20$ deg	32	2458	1288	1839	2008	2118	2166	1779
	40	2182	948	1660	1810	1906	1944	1726
	48	1710	596	1360	1478	1550	1578	1478
	56	1126	282	934	1010	1058	1074	1044
	64	306	45	268	290	302	306	304
Feb. 21	24	3036	1998	2276	2396	2446	2424	1476
$\delta = -10.6$ deg	32	2872	1724	2188	2300	2345	2322	1644
	40	2640	1414	2060	2162	2202	2176	1730
	48	2330	1080	1880	1972	2024	1978	1720
	56	1986	740	1640	1716	1792	1716	1598
	64	1432	400	1230	1286	1302	1282	1252
Mar. 21	24	3078	2270	2428	2456	2412	2298	1022
$\delta = 0.0$ deg	32	3012	2084	2378	2403	2358	2246	1276
	40	2916	1852	2308	2330	2284	2174	1484
	48	2780	1578	2208	2228	2182	2074	1632
	56	2586	1268	2066	2084	2040	1938	1700
	64	2296	932	1856	1870	1830	1736	1656
Apr. 21	24	3036	2454	2458	2374	2228	2016	488
$\delta = +11.9$ deg	32	3076	2390	2444	2356	2206	1994	764
	40	3092	2274	2412	2320	2168	1956	1022
	48	3076	2106	2358	2266	2114	1902	1262
	56	3024	1892	2282	2186	2038	1830	1450
	64	2982	1644	2776	2082	1936	1736	1594
May 21	24	3032	2556	2447	2286	2072	1800	246
$\delta = +20.3$ deg	32	3112	2582	2454	2284	2064	1788	469
	40	3160	2552	2442	2264	2040	1760	724
	48	3254	2482	2418	2234	2010	1728	982
	56	3340	2374	2374	2188	1962	1682	1218
	64	3470	2236	2312	2124	1898	1624	1436
June 21	24	2994	2574	2422	2230	1992	1700	204
$\delta = +23.45$ deg	32	3084	2634	2436	2234	1990	1690	370
	40	3180	2648	2434	2224	1974	1670	610
	48	3312	2626	2420	2204	1950	1644	874
	56	3438	2562	2388	2166	1910	1606	1120
	64	3650	2488	2342	2118	1862	1558	1356

Table 4.44. Day-long direct solar irradiation.

Table 4.44 continued

Date	Deg. Lat.	I_{DN}	Total Solar Irradiation, $I_{D\theta} + I_d$					
			Horiz.	$L - 10$	L	$L + 10$	$L + 20$	Vertical
July 21 $\delta = +20.5$ deg	24	2932	2526	2412	2250	2036	1766	246
	32	3012	2558	2442	2250	2030	1754	458
	40	3062	2534	2409	2230	2006	1728	702
	48	3158	2474	2386	2200	1974	1694	956
	56	3240	2372	2342	2152	1926	1646	1186
	64	3372	2248	2280	2090	1864	1588	1400
Aug. 21 $\delta = +12.1$ deg	24	2864	2408	2402	2316	2168	1958	470
	32	2902	2352	2388	2296	2144	1934	736
	40	2916	2244	2354	2258	2104	1894	978
	48	2898	2086	2300	2200	2046	1836	1208
	56	2850	1883	2218	2118	1966	1760	1392
	64	2808	1646	2108	1008	1860	1662	1522
Sept. 21 $\delta = 0.0$ deg	24	2878	2194	2432	2366	2322	2212	992
	32	2808	2014	2288	2308	2264	2154	1226
	40	2708	1788	2210	2228	2182	2074	1416
	48	2568	1522	2102	2118	2070	1966	1546
	56	2368	1220	1950	1962	1918	1820	1594
	64	2074	892	1726	1736	1696	1608	1532
Oct. 21 $\delta = -10.7$ deg	24	2868	1928	2198	2314	2364	2346	1442
	32	2696	1654	2100	2208	2252	2232	1588
	40	2454	1348	1962	2060	2098	2074	1654
	48	2154	1022	1774	1860	1890	1866	1626
	56	1804	688	1516	1586	1612	1588	1480
	64	1238	358	1088	1136	1152	1134	1106
Nov. 21 $\delta = -19.9$ deg	24	2706	1610	1962	2146	2268	2324	1730
	32	2405	1280	1816	1980	2084	2130	1742
	40	2128	942	1636	1778	1870	1908	1686
	48	1668	596	1336	1448	1518	1544	1442
	56	1094	284	914	986	1032	1046	1016
	64	302	46	266	286	298	302	300
Dec. 21 $\delta = -23.45$ deg	24	2624	1474	1852	2058	2204	2286	1808
	32	2348	1136	1704	1888	2016	2086	1794
	40	1978	782	1480	1634	1740	1796	1646
	48	1444	446	1136	1250	1326	1364	1304
	56	748	157	620	678	716	734	722
	64	24	2	20	22	24	24	24

5 Nomographs

William G. Andrew

Introduction

Nomographs are useful in situations where exact solutions are unnecessary and quick approximations are needed. The following nomographs include solutions relating to fluid flow data, fluid properties, conversions, volume capacities, mathematical functions and other miscellaneous data.

Air Pressure Drop In Pneumatic Tubing

The allowable length of tubing runs in pneumatic control systems is limited by the pressure drop due to friction in the lines. This limitation becomes particularly important when bleed devices are employed in the system, so that air flow is continuous. It can also be a limiting factor in the application of fluidic devices. An approximate relationship between pressure drop and flow rate for air in smooth tubing, developed from basic flow equations, lends itself to nomographic solutions as a quick check on system design.

The Darcy-Weisbach equation for frictional pressure drop of any gas can be written as in Equation (5.1).

$$\Delta P_{100} = (7.05 \times 10^{-3}) fV^2G^2/d^5\rho \qquad (5.1)$$

where ΔP_{100} is the pressure drop per 100 ft length of pipe, f is the dimensionless Darcy friction factor, V the volumetric flow rate at standard conditions (scfm at 14.7 psia and 60 deg F), G the gas specific gravity (air = 1.0), d the actual inside pipe diameter (in.), and ρ the density at flow conditions (lb_m/ft^3).

Most hydraulic and flow texts contain the experimentally derived curves for laminar and turbulent flow that relate friction factor f to Reynolds number R_e, as a function of pipe roughness. In the turbulent flow region, the curve for smooth tubing can be approximated by the Blasius Equation (5.2).

$$f = 0.3164\, R_e^{-0.25} \qquad (5.2)$$

In volumetric flow units, the Reynolds number is

$$R_e = 28.92\, VG/d\mu \qquad (5.3)$$

where μ is the gas viscosity at flow conditions (centipoise). The ASME data relating the viscosity of air to its temperature (Equation 5.3) yields a straight line when plotted on log-log paper, and can be approximated by the equation

$$\mu = (1.541 \times 10^{-4})T^{0.76} \qquad (5.4)$$

where T is the absolute temperature (deg F + 460).

Substituting Equations (5.3) and (5.4) and G = 1.0 into Equation (5.2) yields for the Darcy friction factor:

$$f = (1.520 \times 10^{-2})\, d^{0.25}T^{0.19}/V^{0.25} \qquad (5.5)$$

The air density ρ in Equation (5.1) can be replaced by the perfect gas law expression:

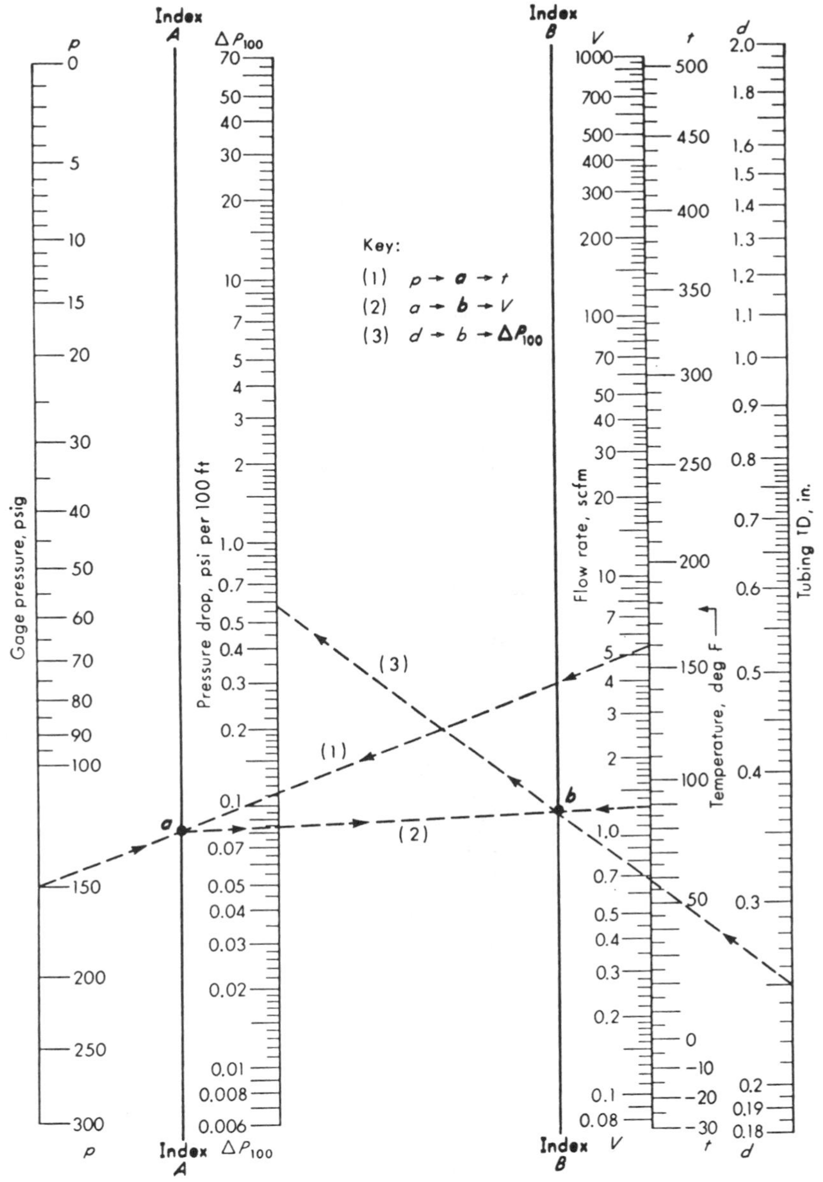

Nomograph 5.1. Air pressure drop in pneumatic tubing. Courtesy Control Engineering, copyright © *1967, Dunn Donnelley Publishing Corporation. All rights reserved.*

$$\rho = 2.70P/T \tag{5.6}$$

where P is the absolute pressure (psia), equal to the gage pressure plus ambient pressure, and T is temperature in °F.

Substituting Equations (5.5) and (5.6) into Equation (5.1) yields the desired expression for pressure drop as a function of a volumetric flow rate, temperature, pressure, and tubing size:

$$\Delta P_{100} = (3.97 \times 10^{-5})V^{1.75}T^{1.19}/d^{4.75}P \tag{5.7}$$

The nomograph for Equation (5.7) permits a rapid solution for any variable when the others are known. For example, pressure drop is found in three steps:

1. Connect temperature on the t-scale and pressure on the p-scale. Mark point a, the intersection with index scale A.
2. Connect point a with flow rate on the V-scale and mark point b, the intersection with index scale B.
3. Connect point b with tubing diameter on the d-scale.

Extend this line to read pressure drop on the ΔP_{100}-scale. This is the frictional pressure drop per 100 ft of tubing at the stated conditions.

The chart shows the nomographic solution for pressure drop of 1.30 scfm of air at 160 deg F and 150 psig in a 0.250 in. ID tube. For this example, the pressure drop is 0.58 psi per 100 ft of tubing.

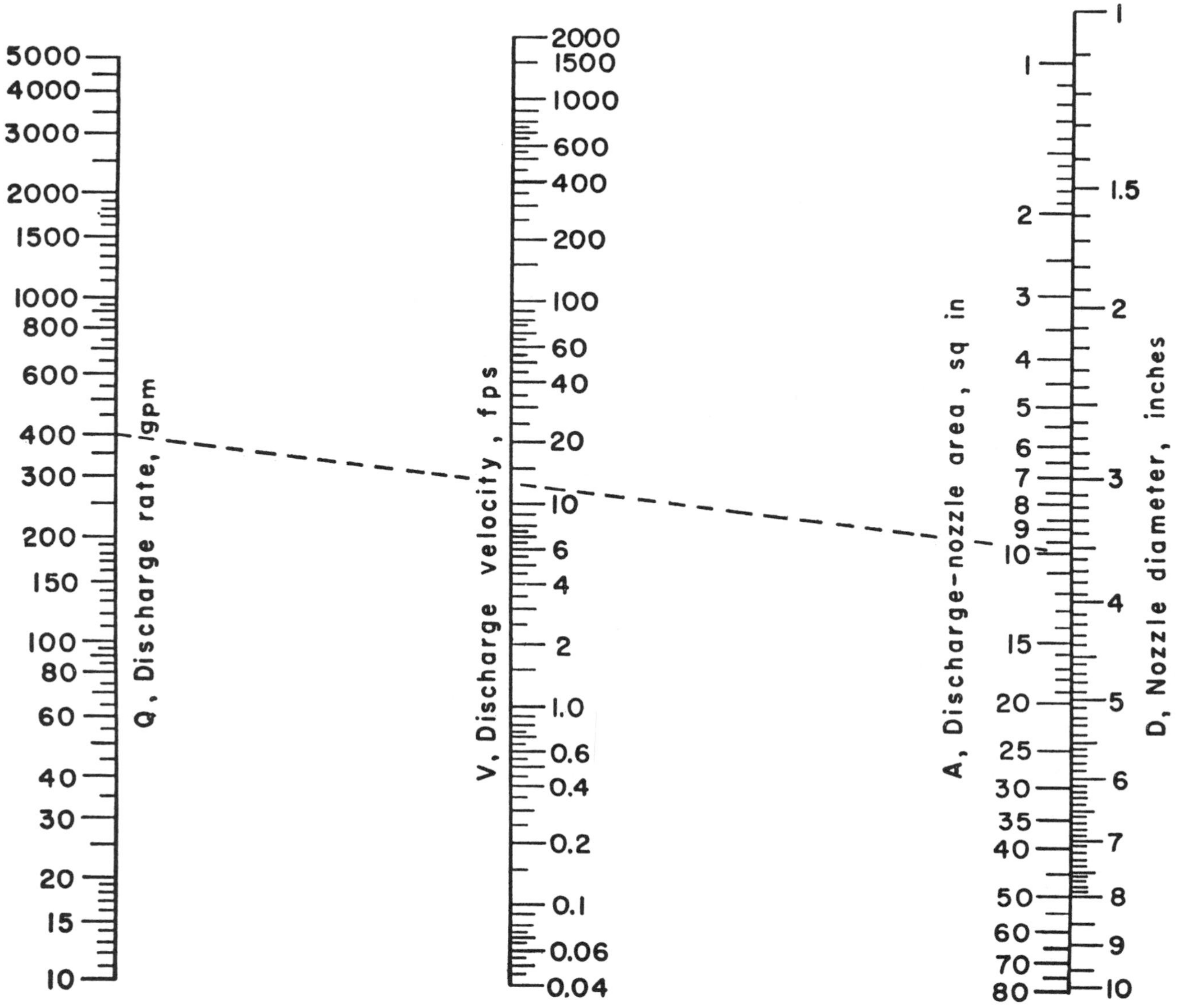

Nomograph 5.2. Discharge velocity of centrifugal pumps. Courtesy Chemical Publishing Company, Inc.

Discharge Velocity of Centrifugal Pumps

This nomograph permits quick determination of velocity of centrifugal pumps. It is based on the equation:

$$V = 0.321\ Q/A$$

where

V = velocity of discharge of a centrifugal pump, fps
Q = quantity of water discharged, gpm
A = cross-sectional area of discharge nozzle, sq in.

The chart also has a scale for converting cross-sectional area of discharge nozzles to their diameters.

Example. What will be the exit velocity if the area of the discharge nozzle is 10 sq in. and the water flow is 400 gpm?

Solution. Extend a straight line from 400 on the Q-scale to 10 on the A-scale. Read the answer where this line crosses the V-scale as 12.8 fps.

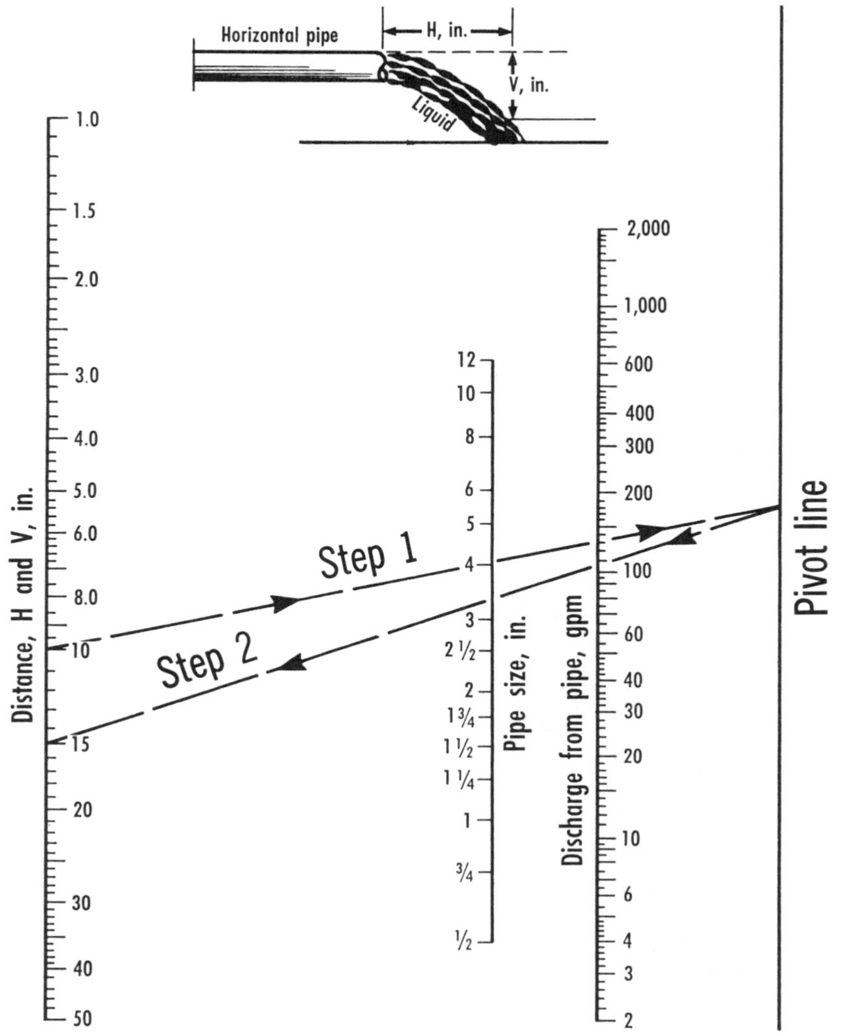

Nomograph 5.3. Volume of fluid discharged from horizontal pipe. Courtesy Oil, Gas, and Petrochem Equipment.

Volume of Fluid Discharged From Horizontal Pipe

Nomograph 5.3 provides a quick and surprisingly accurate method of determining the number of gallons per minute being discharged from an open-end horizontal pipe. Measure any convenient horizontal distance out from the open end of the pipe. This distance, in inches, will be called *H*. Now measure down (vertically) from the end of *H* to a point just touching the flowing liquid. This distance, in inches, will be called *V*. The following formula gives the liquid discharged:

$$\text{gpm} = 2.56\, Hd^2/V^{1/2}$$

where

H = distance defined above, in.
V = distance defined above, in.
d = pipe diameter, in.

See the diagram on nomograph for method of making measurements.

Example. Assume distance *H* measures 10 in. and distance *V* measures 15 in. What is the discharge from the open end of a 4-in. horizontal pipe?

Solution. Align 10 in. on *H*-and-*V* scale with 4 in. on Pipe-size scale, extend to pivot line, and mark (Step 1). Now line the point found on pivot line with 15 in. on *H*-and-*V* scale, and read 106 gpm discharge rate where line crosses Discharge-from-pipe scale (Step 2).

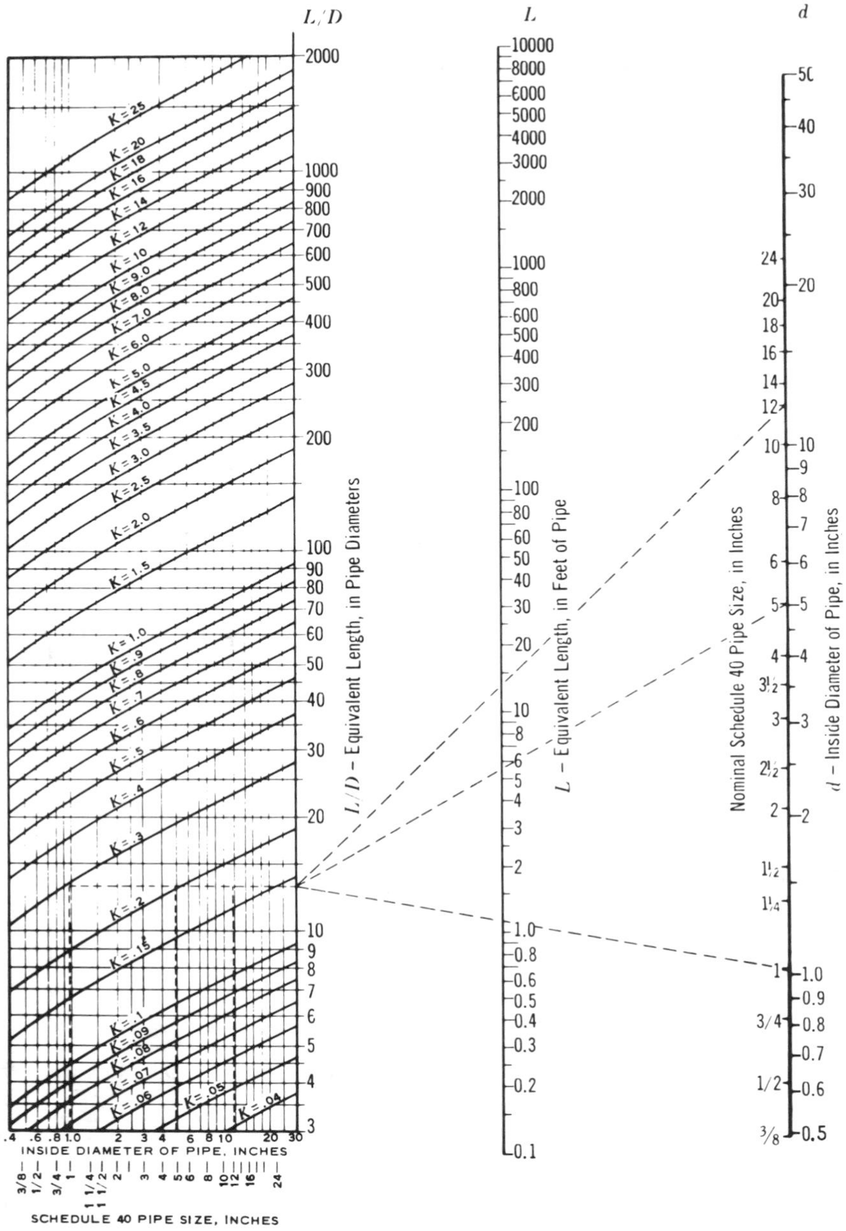

Nomograph 5.4. Equivalent lengths and L and L/D and resistance coefficient K. *Courtesy Crane Company (T.P. #410).*

*Equivalent Lengths *L* and *L/D* and Resistance Coefficient *K*

Problem. Using nomograph 5.4 find the equivalent length in pipe diameters and feet of Schedule 40 pipe, and the resistance factor *K* for 1, 5, and 12-inch fully-opened gate valves.

*Reynolds number, R_e, must be greater than 1,000.

Solution

Valve Size	1″	5″	12″	*Refer to*
Equivalent length, pipe diameters	13	13	13	Table 4.61
Equivalent length, feet of Sched. 40 pipe	1.1	5.5	13	Dotted lines on chart.
Resist. factor *K*, based on Sched. 40 pipe	0.30	0.20	0.17	

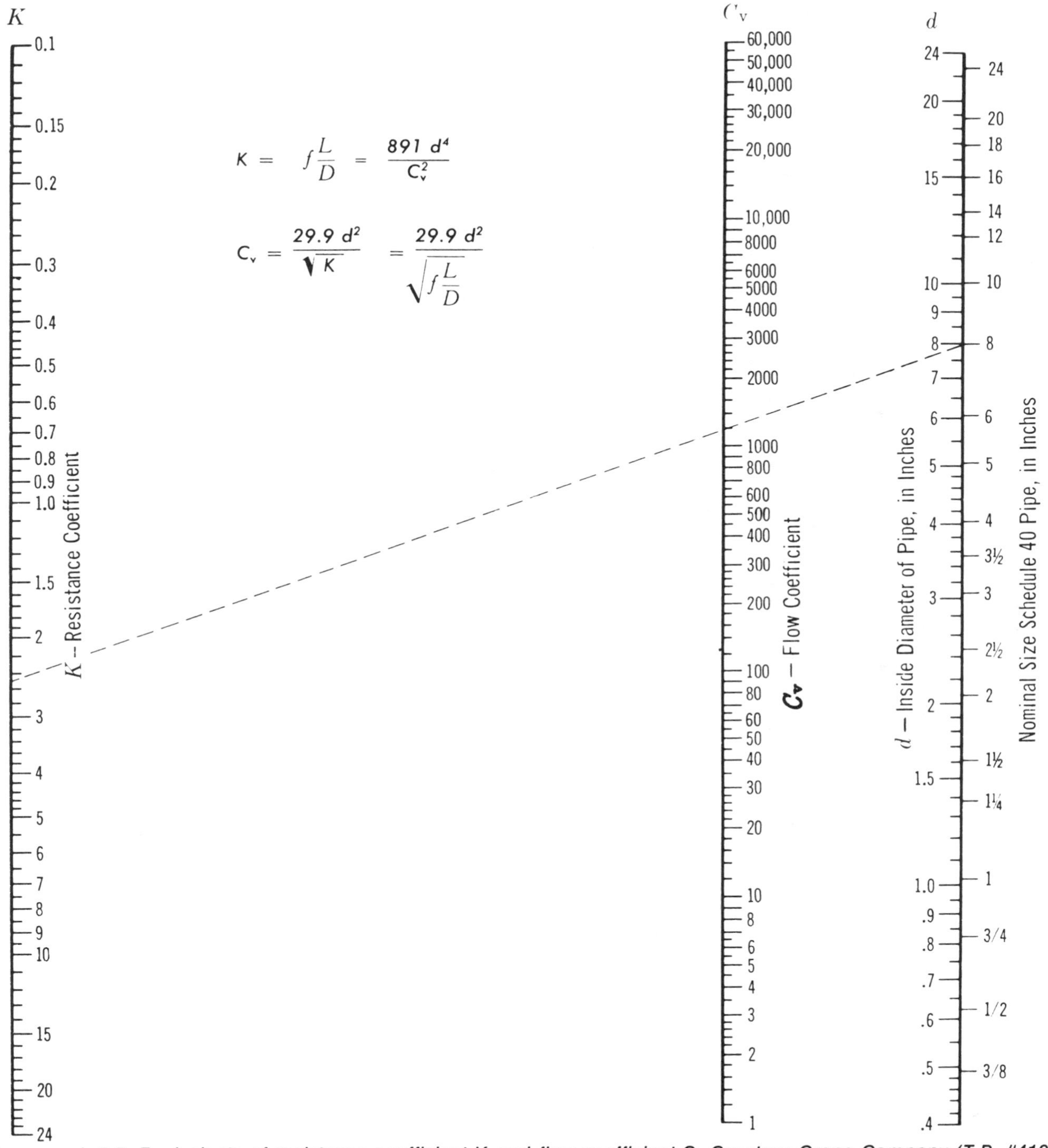

Nomograph 5.5. Equivalents of resistance coefficient K *and flow coefficient* C. *Courtesy Crane Company (T.P. #410).*

***Equivalents of Resistance Coefficient *K* And Flow Coefficient C_v**

Problem. In Nomograph 5.5 find the equivalent length in pipe diameters, the resistance coefficient K, and the flow coefficient C_v for an 8-inch, 125-pound Y-pattern globe valve with stem 60 degrees from run of valve.

Solution. Equivalent length in pipe diameters is 175 (taken from Table 4.61).

Resistance factor K based on Schedule 40 pipe is 2.5 (taken from chart shown on preceding page).

Flow coefficient C_v is 1,200 (see dotted line shown on chart above).

*Reynolds number, R_e, must be greater than 1,000.

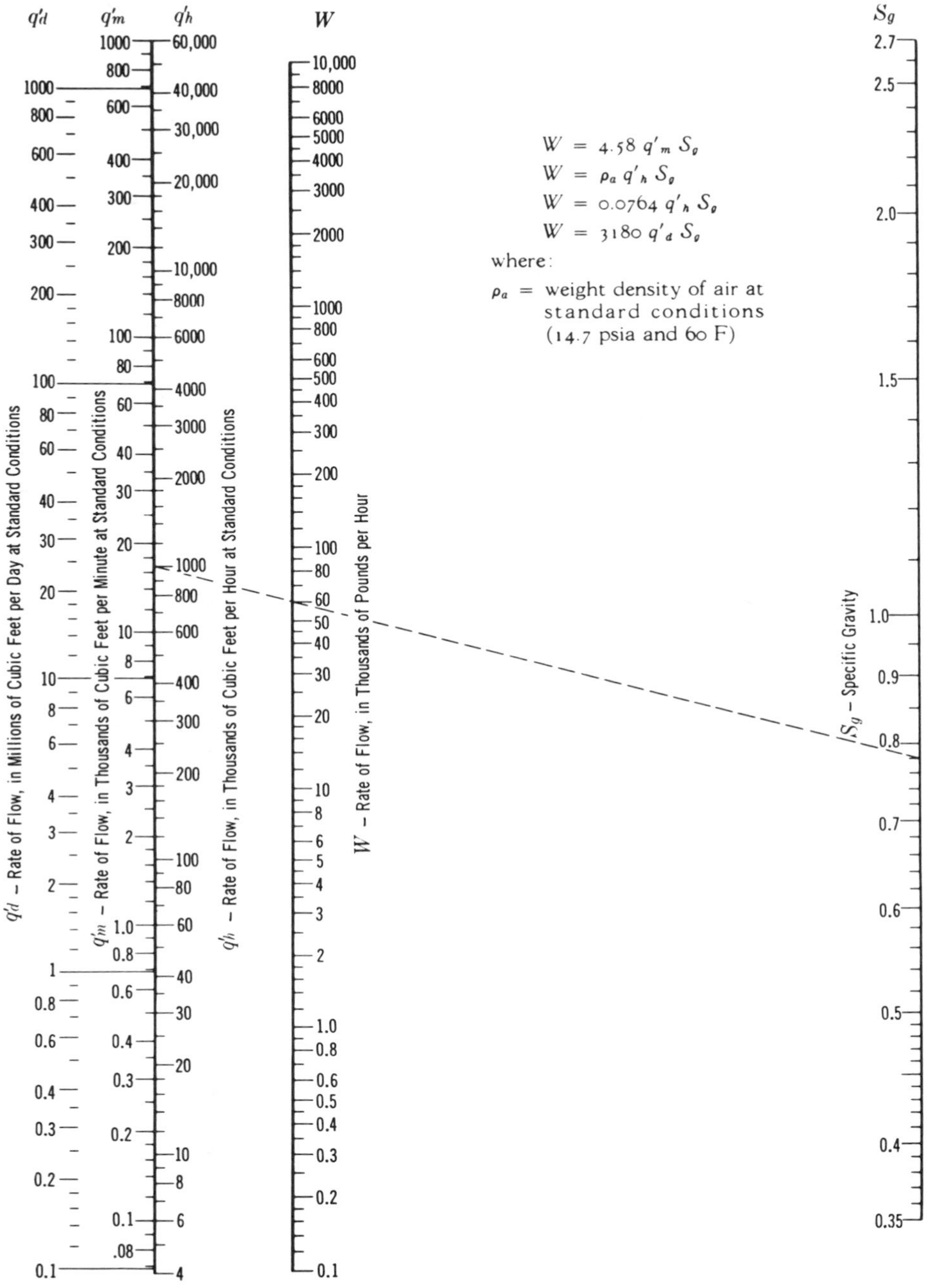

Nomograph 5.6. Equivalent volume and weight flow rates of compressible fluids. Courtesy Crane Company (T.P. #410).

Equivalent Volume and Weight Flow Rates of Compressible Fluids

Problem. In Nomograph 5.6 what is the rate of flow in pounds per hour of a gas, which has a specific gravity of 0.78, and is flowing at the rate of 1,000,000 cubic feet per hour at standard conditions?

Solution. W = 60,000 pounds per hour.

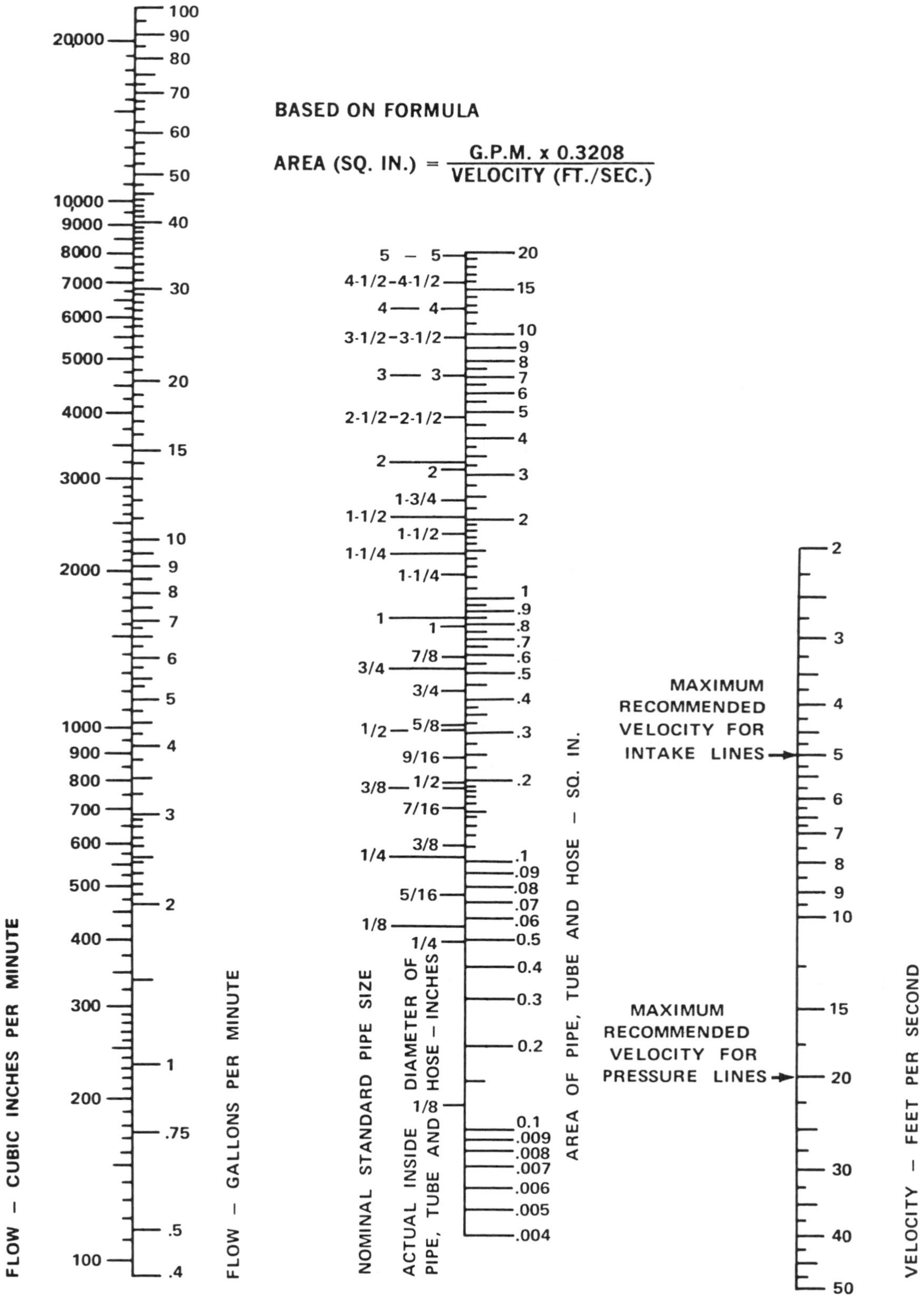

Nomograph 5.7. Flow capacities of pipe, tube and hose. Courtesy Fluid Power Handbook & Directory, *1972/1973.*

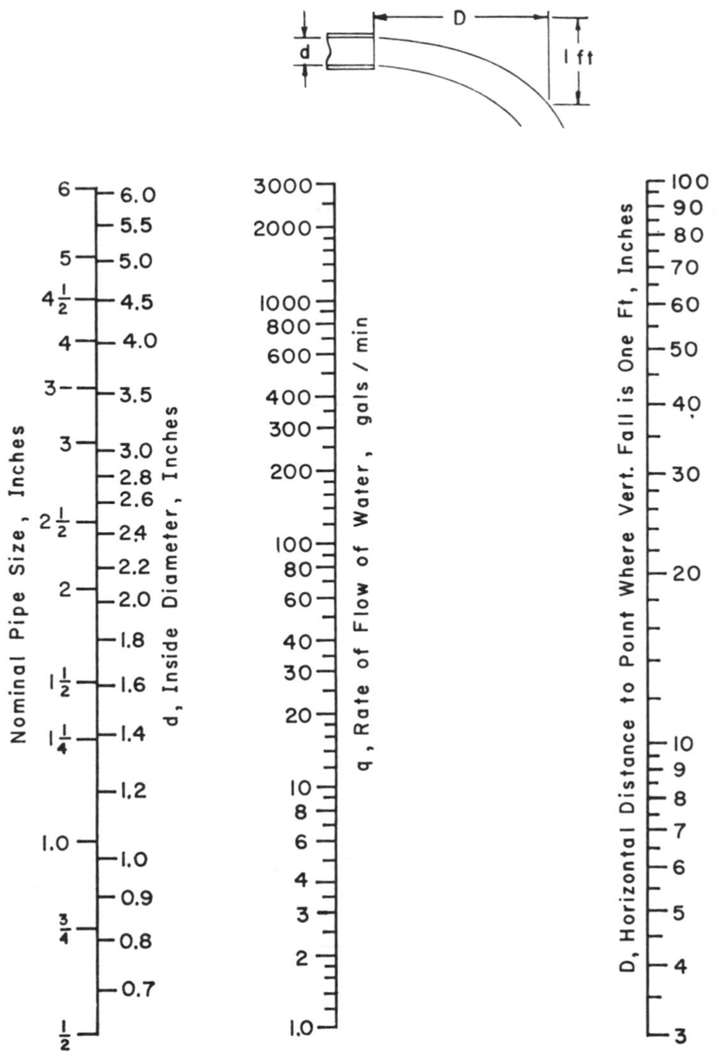

Nomograph 5.8. Flow of water from horizontal, open-end pipes. Courtesy Chemical Publishing Company, Inc.

Flow of Water from Horizontal, Open-end Pipes

Nomograph 5.8 presents a means of determining the rate of flow of water from horizontal, open-end pipes. It is based on the equation:

$$q = 1.04\ A\ D$$

where

A = internal transverse area of the pipe, sq in

D = horizontal distance from pipe opening to a point where the water stream has fallen one foot, in.

q = rate of flow of water, gpm

Example. Water is pumped out of a horizontal pipe of 4-inch nominal diameter; the stream drops one foot at a point 75 inches from pipe opening.

Connect the point of the d scale for 4-inch nominal diameter and D = 75 with a straightedge.

Read rate of flow of water as 990 gpm on q scale.

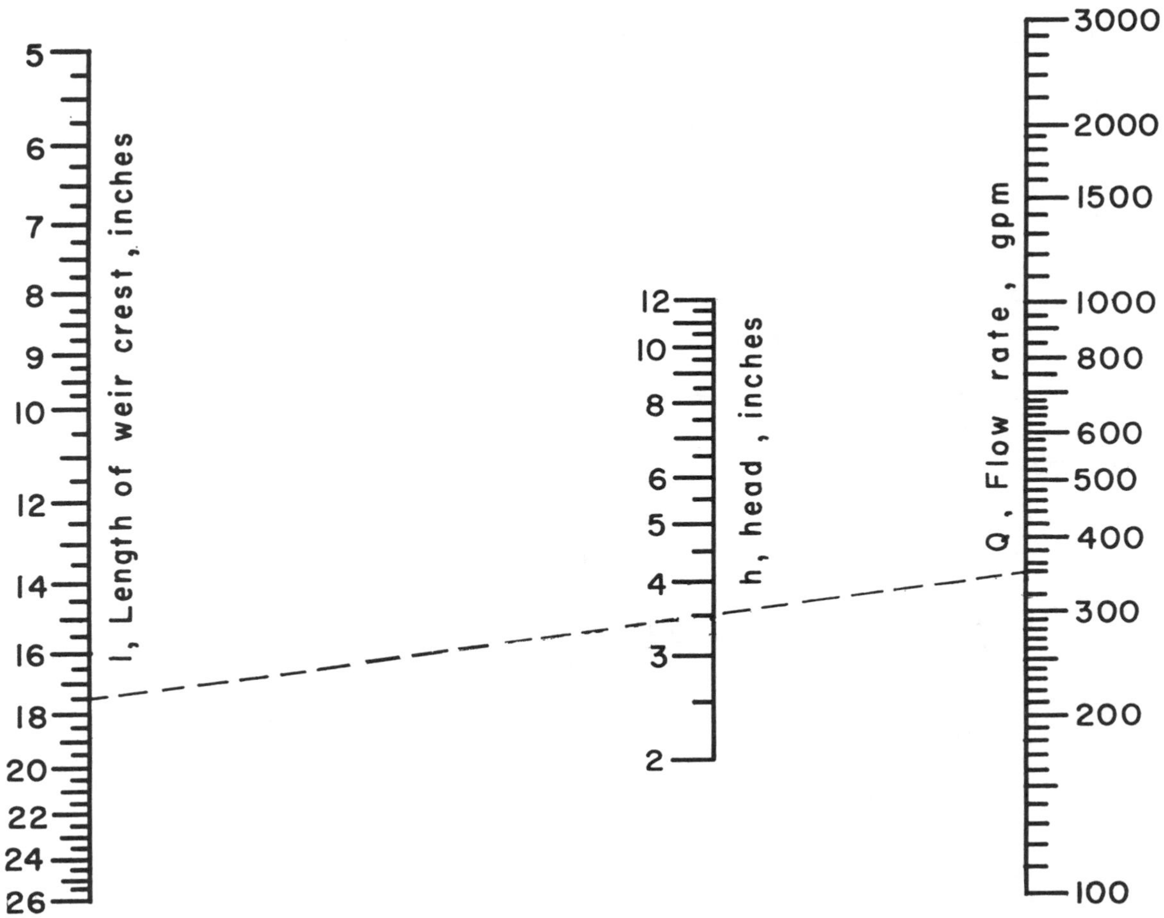

Nomograph 5.9. Flow of water over a Cipoletti weir. Courtesy Chemical Publishing Company, Inc.

Flow of Water Over a Cipolletti Weir

The flow of water over a sharp-edged Cipolletti weir—where the sides slope upward and outward—is described by the following formula:

$$Q = 3.030\ lh^{1.5}$$

where

Q = flow rate, gpm
l = length of crest, inches
h = head of water measured from weir crest, inches

The equation can be solved readily by means of Nomograph 5.9.

Example. How much water will flow over a 17.5″ Cipolletti weir when the head is 3.5″? Connect 17.5 on the *l*-scale with 3.5 on the *h*-scale. Extend the line and read 350 gpm on the *Q*-scale.

$Q\,(\text{G.P.M.}) = 24.12 \times A\,(\text{SQ. IN.}) \times \sqrt{\text{PRESSURE DROP}}\,(\text{P.S.I.})$

WHEN C = .6 AND SPECIFIC GRAVITY = .895

NO ALLOWANCE FOR VISCOSITY CHANGE

FROM:

$Q = C \times A \times \sqrt{2GH}$

WHEN: Q = FLOW IN CU. FT. / SEC.

C = ORIFICE COEFFICIENT

A = AREA IN SQ. FT.

H = PRESSURE HEAD IN FT.

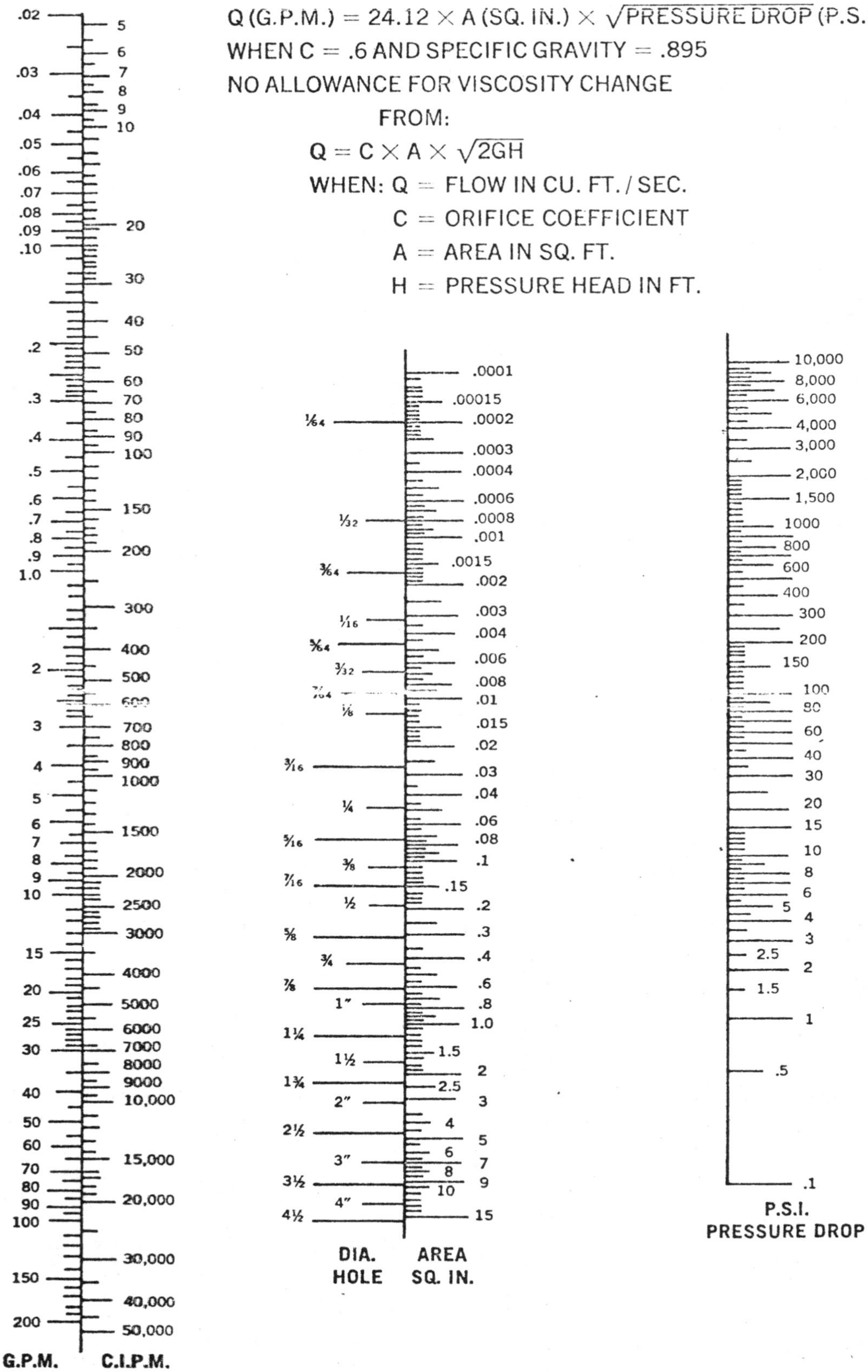

Nomograph 5.10. Orifice pressure drop.

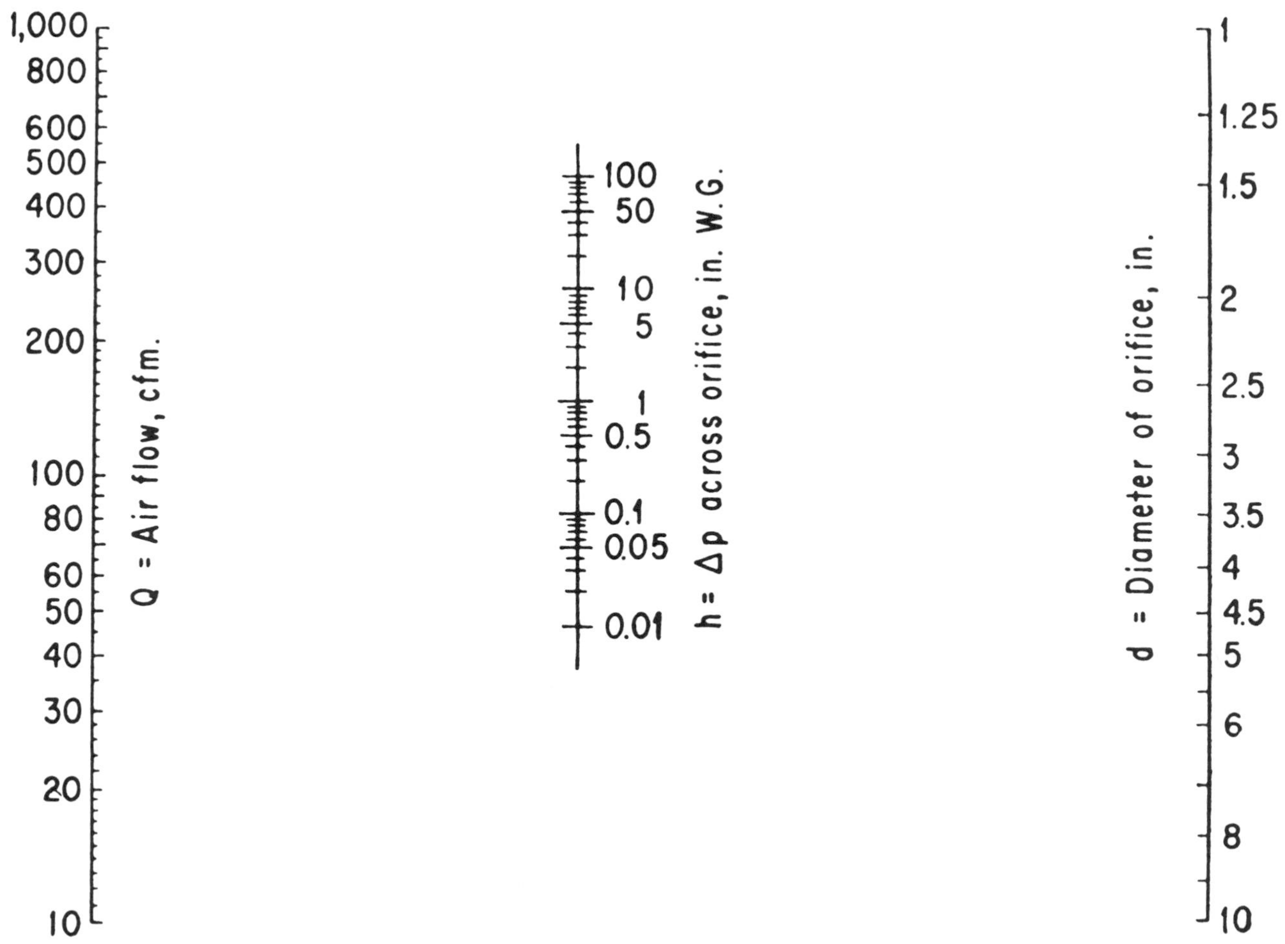

Nomograph 5.11. Air flow through thin-plate orifice. Reprinted by special permission from Chemical Engineering, *copyright* © *by McGraw-Hill, Inc. New York, N.Y. 10036.*

Air Flow Through Thin-Plate Orifice

Nomograph 5.11 gives the theoretical air flow through a thin-plate orifice if the downstream pressure is less than critical. The chart solves equation

$$Q = 21.8d^2\sqrt{h}$$

where Q is the air flow, cfm.; d is the orifice diameter, in.; h is the pressure drop across the orifice, in. water gage.

To obtain the actual air-flow value, Q must be multiplied by the coefficient of discharge for the orifice in question.

Air conditions: 100 F. and 14.7 psia.

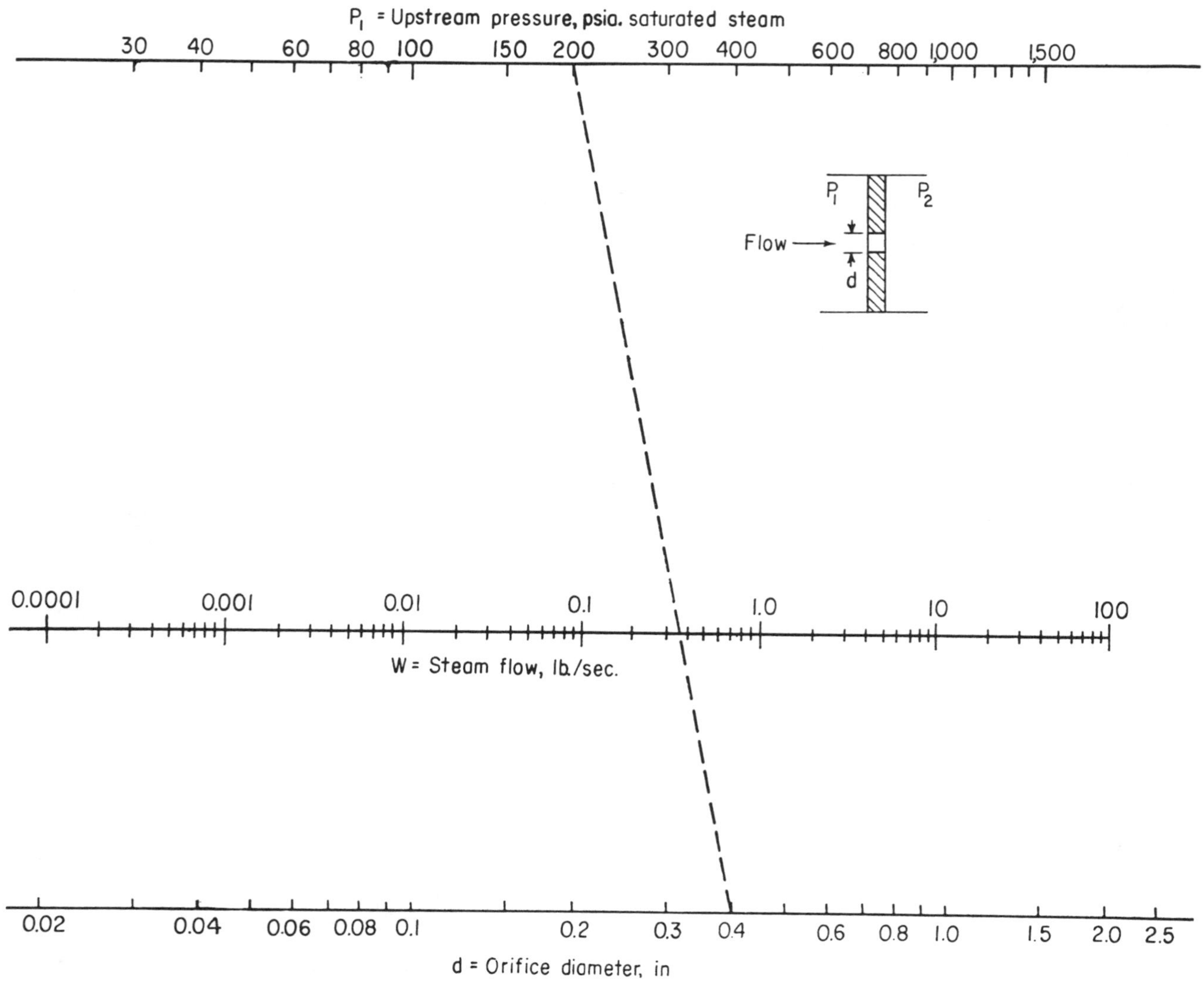

Nomograph 5.12. Steam flow through orifices. Reprinted by special permission from Chemical Engineering, *copyright © by McGraw-Hill, Inc. New York, N.Y. 10036.*

Steam Flow Through Orifices

Nomograph 5.12 is used to calculate the flow of saturated steam through an orifice. The answer obtained is the theoretical value when P_2 is less than 58% of P_1. For the actual value, multiply flow W by the coefficient of discharge for the orifice.

For wet steam, divide the answer by the square root of the quality at inlet conditions. For superheat, divide the answer by $(1 + 0.00065t)$, where t is the degrees of superheat in the steam at inlet conditions.

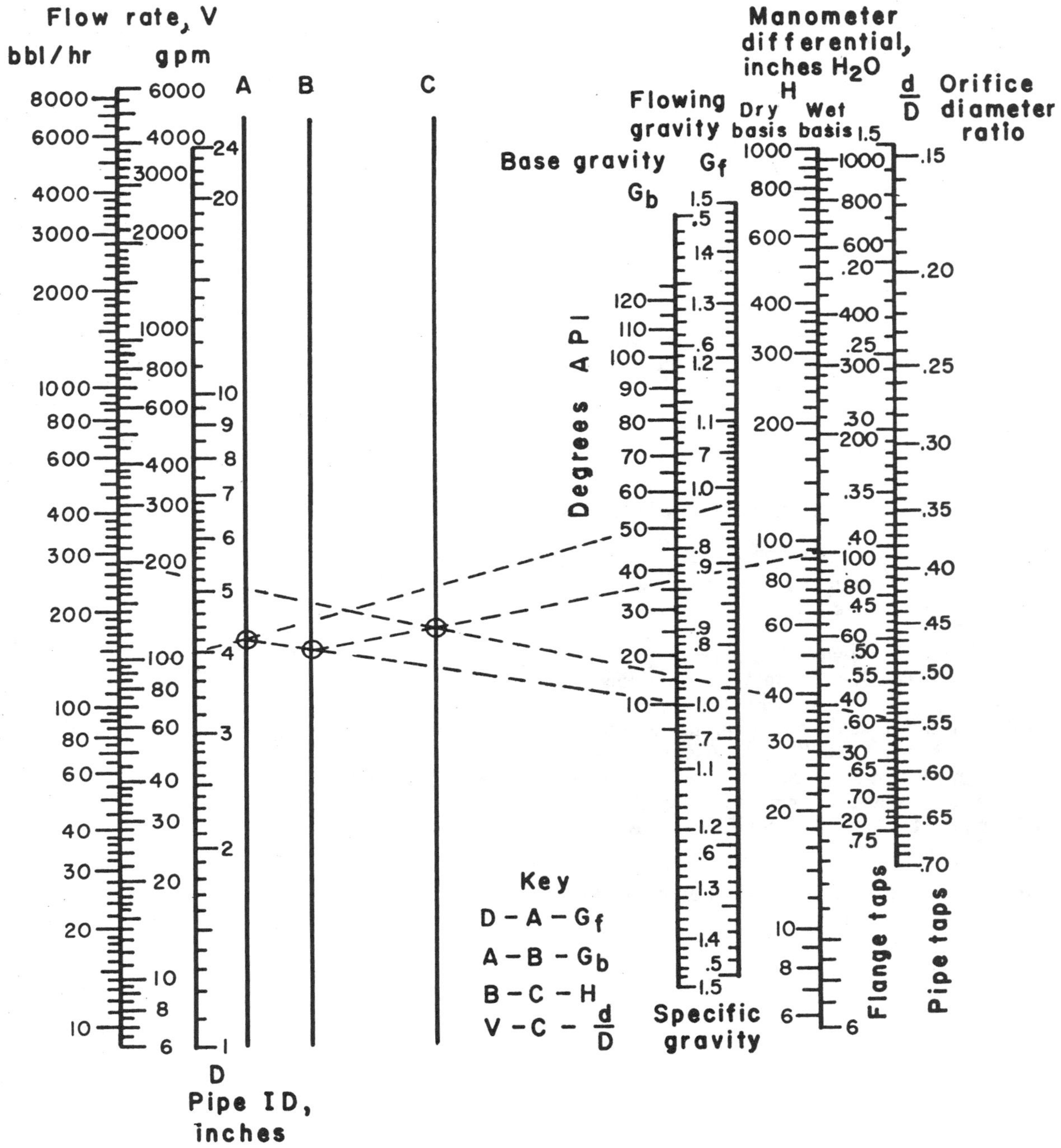

Nomograph 5.13. Orifice sizing for liquid-flow measurement. Courtesy Chemical Publishing Company, Inc.

Orifice Sizing For Liquid-Flow Measurement

Nomograph 5.13 provides a quick method of solving flow equations for orifice meters handling liquids.

Not designed for precise calculations, the nomograph does not include such factors as corrections for thermal expansion of the primary device or for Reynolds number. (These are usually negligible and seldom exceed a total of 1% correction.) However, it is particularly useful for the following:

1. Spotting gross errors in precise calculations.
2. Preliminary determination of the feasibility of metering by the orifice method when line size and flow rates are known.

3. Deciding whether flange taps or pipe taps should be employed.
4. Choosing a manometer range for a given installation.
5. Obtaining rough figures for new capacity when a change in service is contemplated for a given installation.
6. Deciding whether to change the manometer range or to bore a new orifice plate in cases where an existing installation is over-ranged or under-ranged.

The "wet-basis" side of the differential scale is used where a straight-wall mercury manometer is used with water or other liquid on the mercury surface. "Dry basis" is used when employing a spring (or spring-balance bell) meter, ring-balance or force-balance meters, or a meter with an air or gas purge to keep liquid away from the mercury.

Example

- Liquid: water at 150°F
- Line size (D): 4″ ID
- Meter: mercury manometer, water on mercury surface, 100″ range
- Maximum flow quantity (V): 200 gpm
- Determine the orifice-diameter ratio for use with pipe taps and with flange taps.

Connect 4″ on the *D*-scale with 0.98 (specific gravity of water at 150°F) on the G_f-scale. The intersection of this line with the *A*-scale is next connected with 1.0 (base specific gravity of water) on the G_b-scale. Now join the intersection of this line and the *B*-scale to 100″ on the *H*-scale. A point has now been formed on the *C*-scale. Connect this point with 200 gpm on the *V*-scale, and read an orifice-diameter ratio of 0.55 for pipe taps and 0.60 for flange taps.

Nomograph 5.14 provides a quick method of solving flow equations for orifice meters handling steam.

Obviously the nomograph is not suitable for precise calculations; for this reason it does not include such factors as corrections for thermal expansion of the primary device or Reynolds number. (These are usually negligible and seldom exceed a total of 1% correction.) However, it is particularly useful for the following:

1. Spotting gross errors in precise calculations.
2. Preliminary determination of the feasibility of metering by the orifice method when line size and flow rates are known.
3. Deciding whether flange taps or pipe taps should be employed.
4. Choosing a manometer range for a given installation.
5. Obtaining rough figures for new capacity when a change in service is contemplated for a given installation.
6. Deciding whether to change the manometer range or to bore a new orifice plate in cases where an existing installation is over-ranged or under-ranged.

The "wet basis" side of the differential scale is used where a straight-wall mercury manometer is used with water or other liquid on the mercury surface. "Dry basis" is used when employing a spring (or spring-balance bell) meter, a ring-balance meter, a force-balance meter, or a meter with an air or gas purge to keep liquid away from the mercury.

Example

- Steam conditions: 100 psia, 400°F total temperature
- Line size: 6″ ID
- Meter: Mercury manometer, steam condensate on mercury surface, 200″ range
- Maximum flow quantity: 30,000 lb/hr

The nomograph solution, indicated by dotted lines, indicates that the orifice-diameter ratio is 0.65 for pipe taps and 0.73 for flange taps.

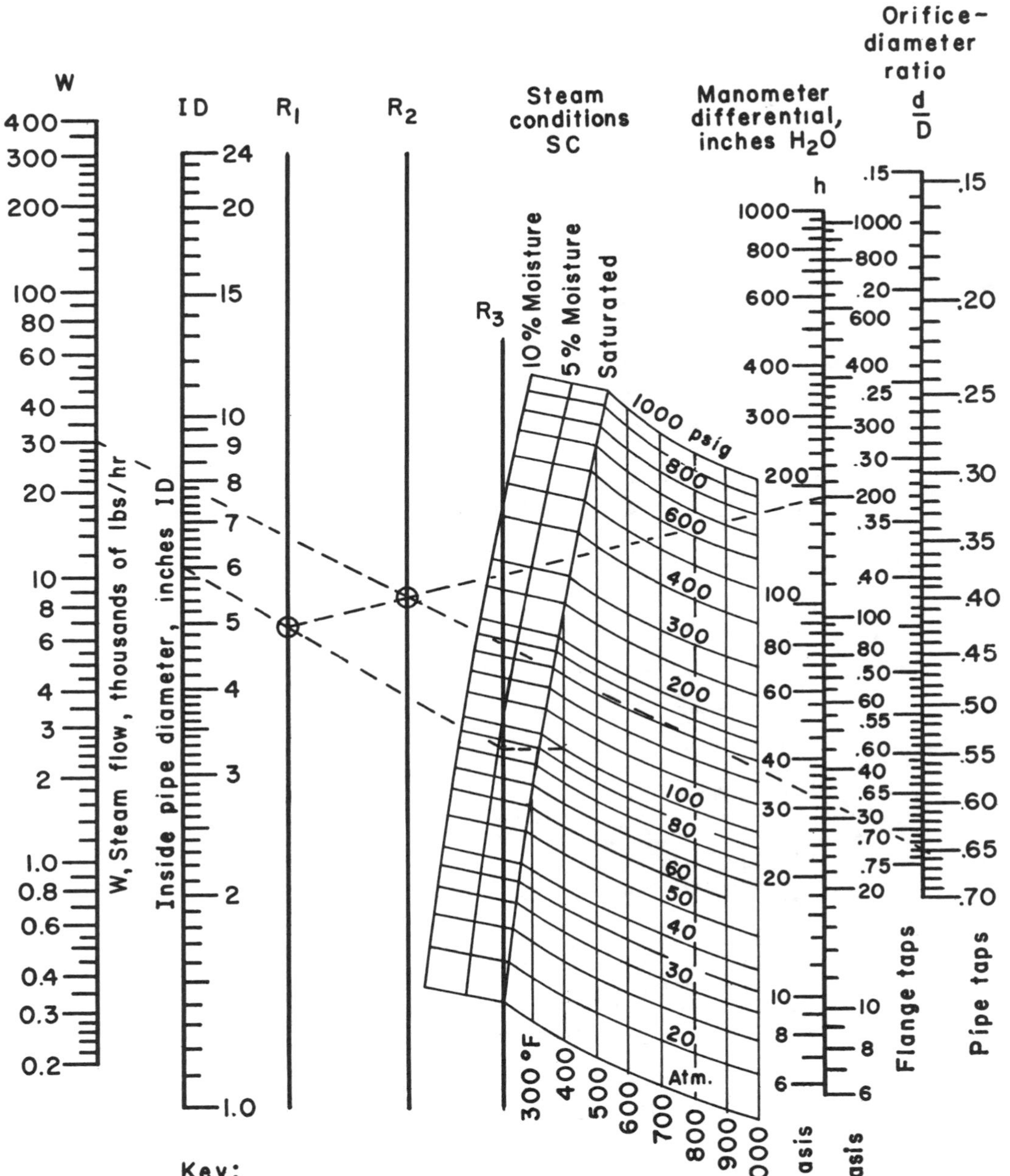

Key:
SC $-R_3$(always horizontal)
$R_3 - R_1$ - ID
$R_1 - R_2$ - h
W $-R_2 - \frac{d}{D}$

Nomograph 5.14. Orifice sizing for steam. Courtesy Chemical Publishing Company, Inc.

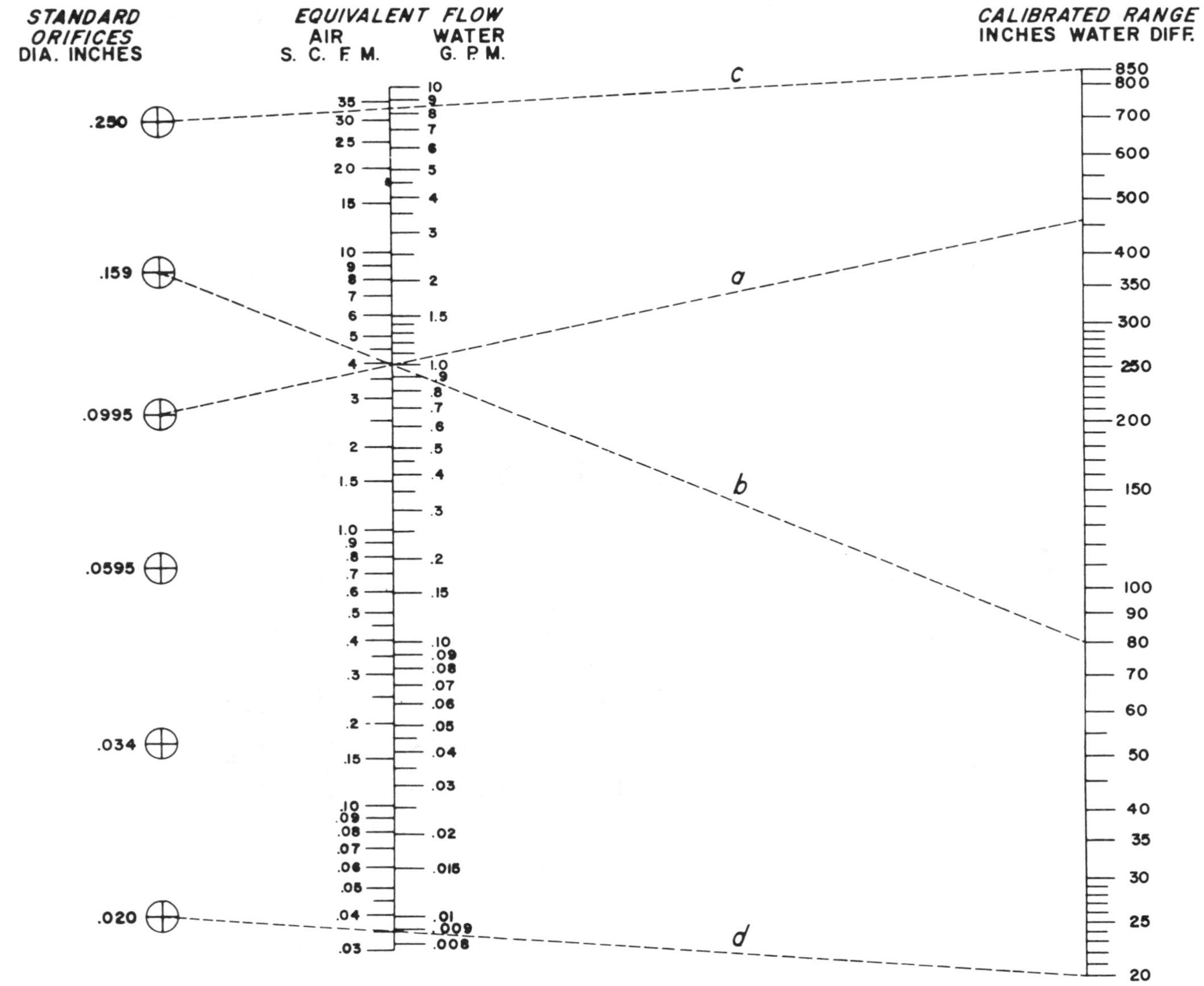

Nomograph 5.15. Integral orifice selection chart. Courtesy The Foxboro Company.

Integral Orifice Selection Chart

Do not use Nomograph 5.15 for accurate calibration. Before using the selection chart, all flows must first be converted to the equivalent flows of water or air at standard conditions. Approximate formulas for this conversion are:

Equivalent air flow (scfm) =

$$Q\sqrt{[(T/520) \times G \times (14.7/P)]} \tag{5.8}$$

where

Q = flow, scfm
T = absolute temperature of flowing gas
G = gas density compared to air
P = absolute static pressure, psia

Note: This approximate air flow formula disregards the compressibility factor, Z, which should be used if more precise calculations are desired.

Equivalent water flow (gpm) =

$$V\,G_b\sqrt{1/G_f} \tag{5.9}$$

where

V = flow, gpm at 60 F
G_b = sp gr at 60 F
G_f = sp gr at flowing temperature

NOTE: These equations disregard viscosity, since accurate data for the effect of viscosity on flow through small diameter orifices is not available. These equations and selection chart are based on empirical data which show negligible errors due to viscosity when using water and air at standard conditions. Therefore it is reasonable to assume that any fluids for which the Reynolds numbers are equal to or greater than those for water and air can be measured to the same accuracy.

For example it is required to measure 2 pounds per hour of hydrogen at 400 psi and 100 F. Dividing the given flow rate by 60 and by the density factor of 0.0056 pounds per cubic foot (weight of 1 cubic foot of hydrogen at standard conditions) gives a flow rate of 5.95 scfm of hydrogen.

Then from equation (5.8):

Equivalent air flow =

$$5.95\sqrt{(560/520 \times 0.0692 \times (14.7/414.7)}$$

$$= 0.305 \text{ scfm}$$

This flow value is then referred to the orifice selection chart to determine correct orifice size and cell range.

How To Use The Orifice Selection Chart

A straight line drawn through a flow value (on the appropriate "equivalent flow" scale) from an orifice reference point will indicate on the "calibrated range" scale the calibration required for that flow and orifice.

For example, in the Integral Orifice Selection Chart dotted lines (a) and (b) illustrate a choice of orifice sizes for an "equivalent flow" of 1.0 gpm. The 0.0995 inch orifice would be used with the higher range capsule calibrated at 460 inches. The 0.159 inch orifice would be used with the lower range capsule calibrated at 80 inches.

Dotted lines (c) and (d) indicate the maximum and minimum flow limits of the standard orifice sizes and cell ranges. If a flow rate is above line (c), the integral orifice construction cannot be provided and a conventional primary device must be used. If the flow rate is below line (d), it may be possible to provide a special orifice.

Metric Orifice Selection Chart

The Metric Orifice Selection Chart is used in the same way as the chart in English units. The orifice diameters are kept in inches (1 inch = 2.54 centimeters). The equivalent air flows are given in normal cubic meters per hour at standard conditions: zero degrees Centigrade and one atmosphere pressure (1.0333 kilograms per square centimeter). Equivalent water flows are given in liters per hour at 15 Centigrade.

Before using the selection chart, all flows must be converted to the equivalent flows of air or water at standard conditions, using the following conversion formulas:

Equivalent water flow (liters per hour) =

$$V \times G_B \times \sqrt{1/G_F} \tag{5.10}$$

where

V = flow, liters per hour at 15 C
G_B = specific gravity at 15 Centigrade
G_F = specific gravity at flowing temperature

Equivalent air flow (normal cubic meters per hour) =

$$Q\sqrt{(T/273.2) \times G \times (1.0333/P)} \tag{5.11}$$

where

Q = flow, normal cubic meters per hour
T = absolute or Kelvin temperature of flowing gas
G = gas density compared to air
P = absolute static pressure - kilograms per square centimeter

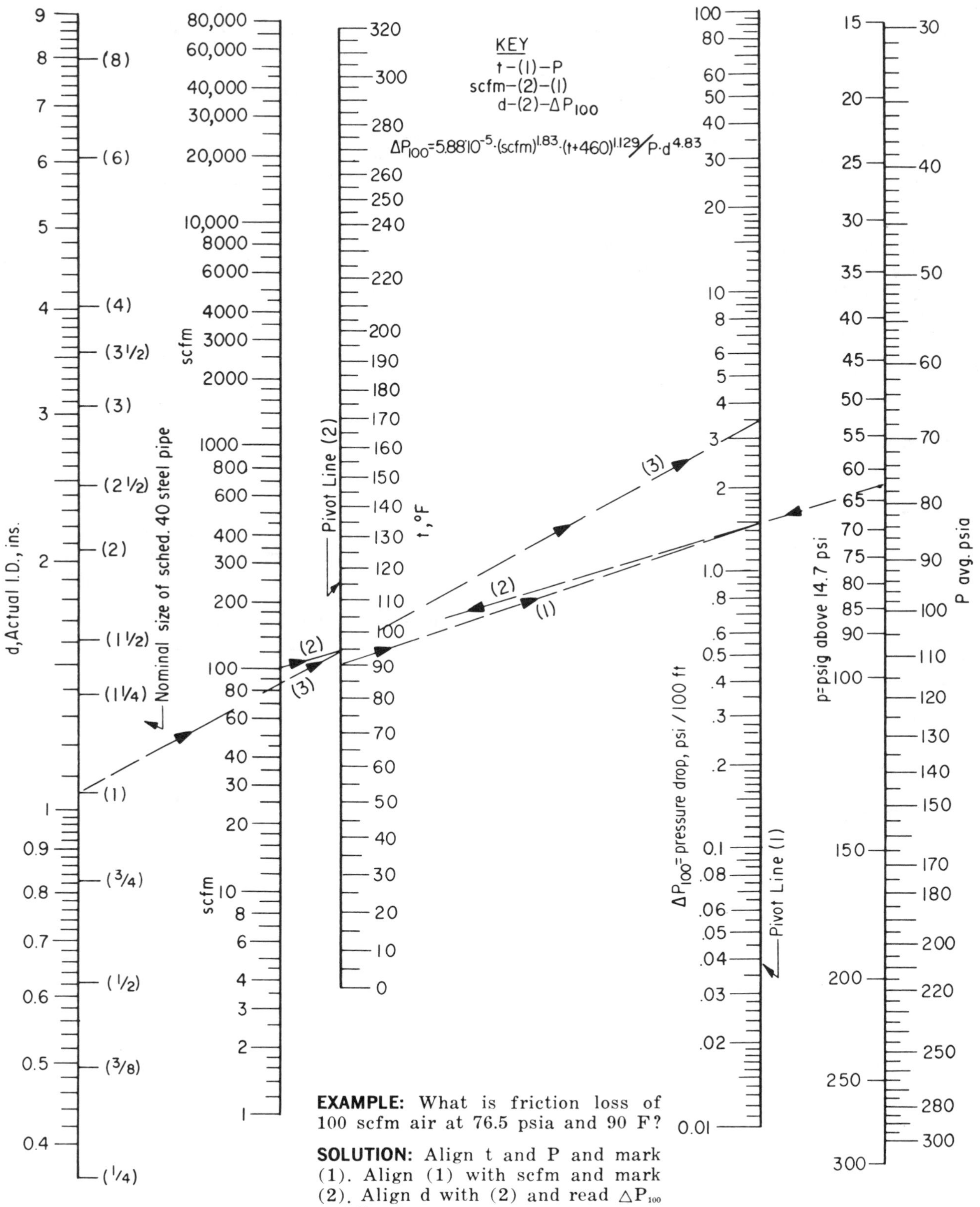

EXAMPLE: What is friction loss of 100 scfm air at 76.5 psia and 90 F?

SOLUTION: Align t and P and mark (1). Align (1) with scfm and mark (2). Align d with (2) and read ΔP_{100} = 3.5 psi.

Nomograph 5.16. Air flow in steel pipe. Courtesy The Petroleum Engineer Publishing Company.

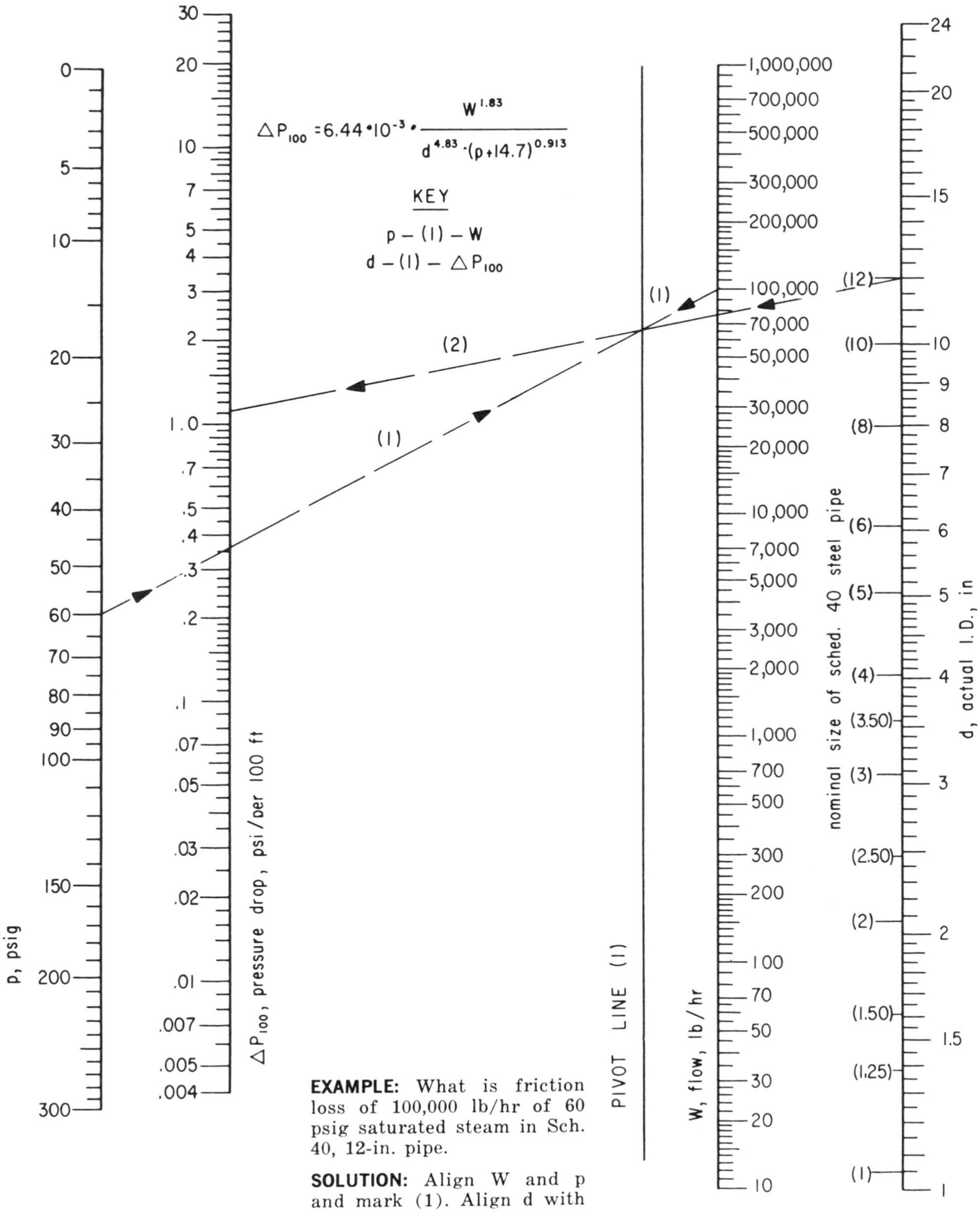

EXAMPLE: What is friction loss of 100,000 lb/hr of 60 psig saturated steam in Sch. 40, 12-in. pipe.

SOLUTION: Align W and p and mark (1). Align d with (1) and read P_{100} at 1.12.

Nomograph 5.17. Saturated steam flow in pipes. Courtesy The Petroleum Engineer Publishing Company.

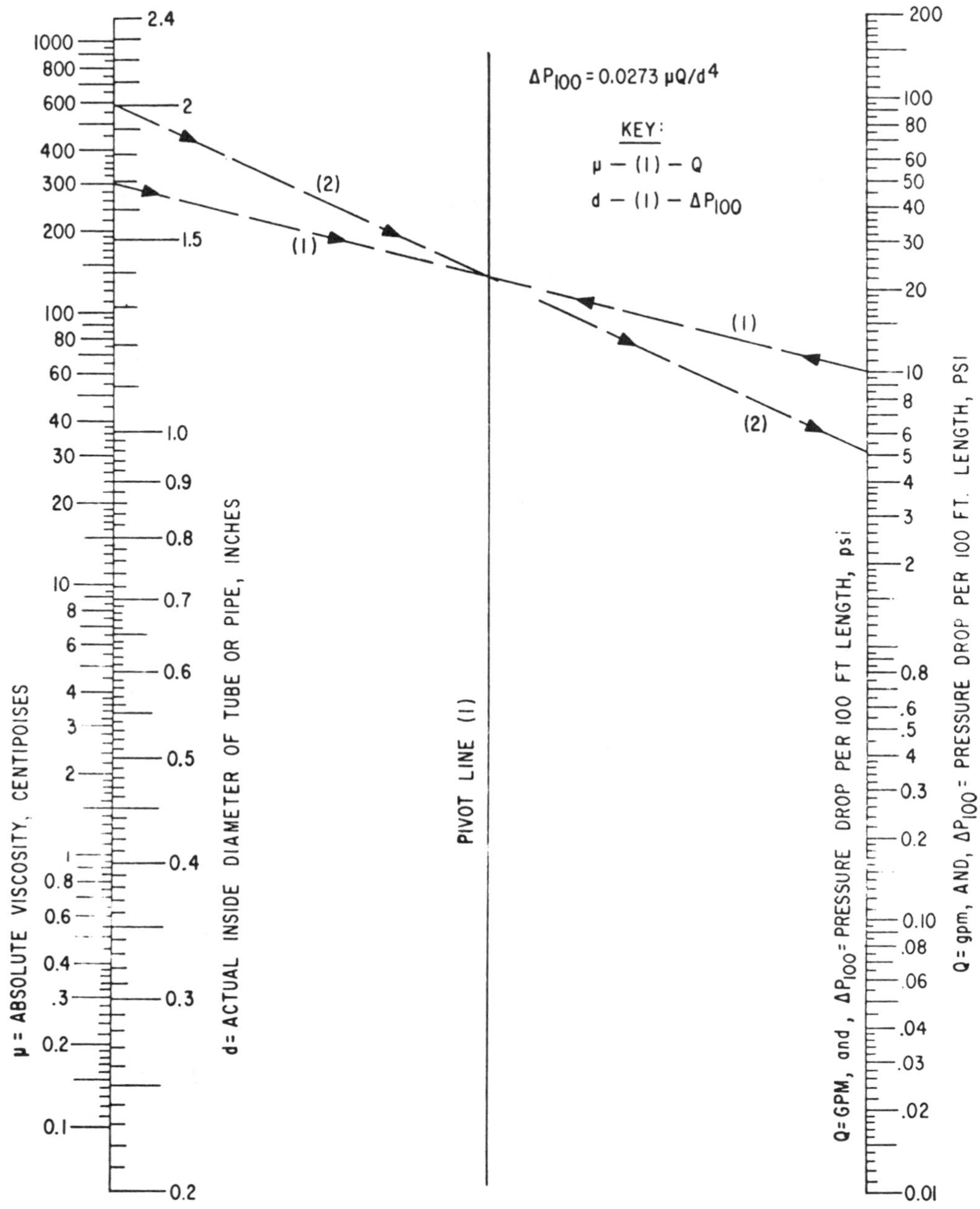

Nomograph 5.18. Pressure drop for laminar flow. Courtesy The Petroleum Engineer Publishing Company.

Pressure Drop for Laminar Flow

When the Reynolds number is less than 2,000, flow is laminar, and the equation for liquid friction loss is

$$\Delta P_{100} = 0.0273\ \mu\ Q/d^4$$

where

ΔP_{100} = pressure drop per 100 ft length, psi

μ = absolute viscosity, centipoises, of the liquid, at the flowing temperature

Q = flow, gpm

d = actual inside diameter of pipe, inches

Example. Using Nomograph 5.18 what is the pressure drop for ten gpm of hydraulic fluid (μ = 3,000 cp) in a 2-in. I.D. tube? Align μ with Q and mark pivot line (1), align d with (1) and read ΔP_{100} = 5.1 psi per 100 ft length of tube.

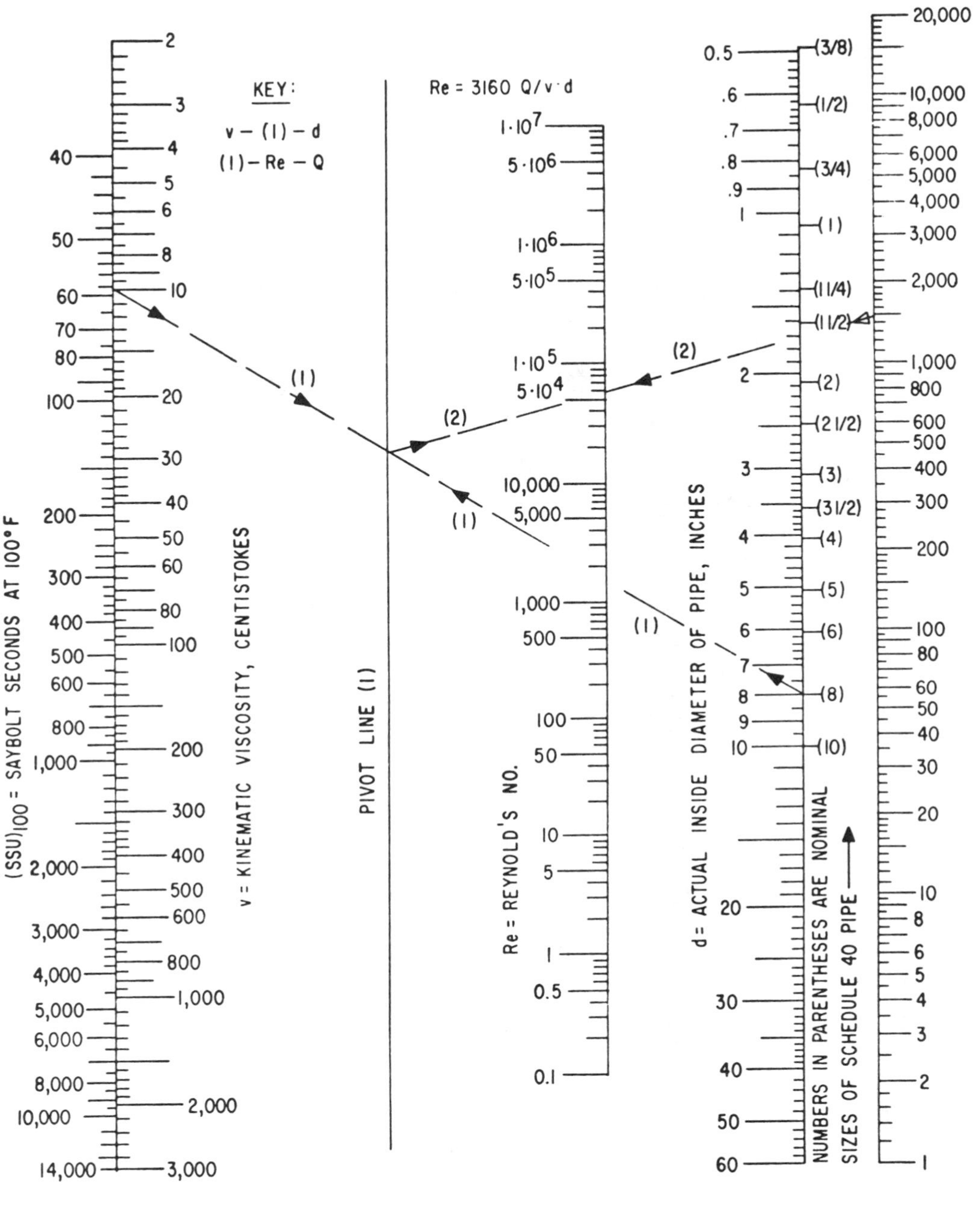

Nomograph 5.19. Reynolds number for liquid flow. Courtesy The Petroleum Engineer Publishing Company.

Reynolds Number for Liquid Flow

The equation for Reynolds number can be written: $R_e = 3{,}160\ Q/\nu d$

where

Q = flow, gpm

ν = kinematic viscosity, centistokes, at the actual temperature of the liquid

d = actual inside diameter of pipe, inches

Example. In Nomograph 5.19 what is R_e if ν = 10 centistokes (58.6 SSU at 100 F), for 1,500 gpm in an 8-in. pipe? On the nomograph, align ν with d and mark pivot line (1), align (1) with Q and read $R_e = 5.9 \times 10^4$.

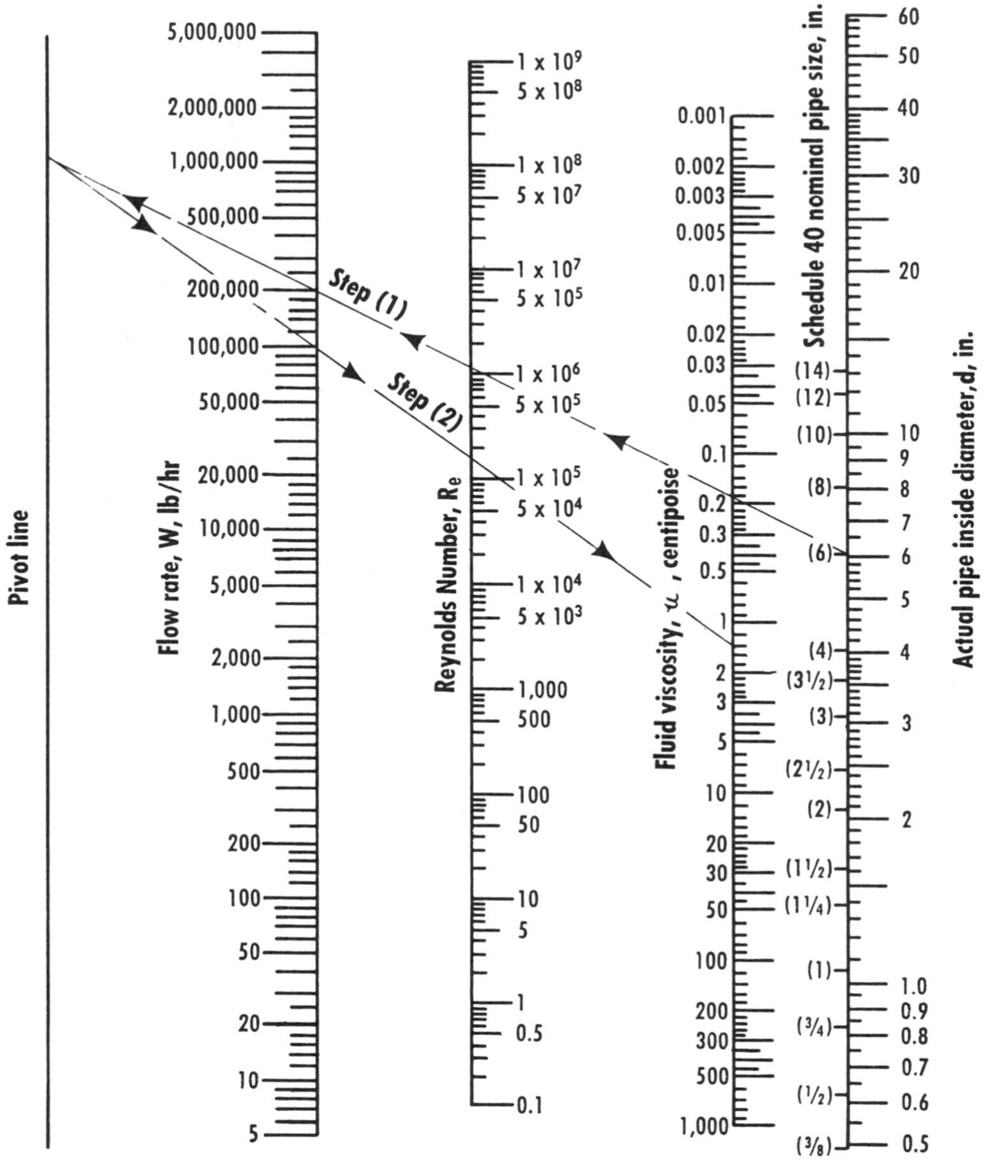

Nomograph 5.20. Reynolds number for liquids or gases. Courtesy Oil, Gas, and Petrochem Equipment.

Reynolds Number for Liquids or Gases

Nomograph 5.20 is designed to help you quickly find the Reynolds number for liquids or gases. It is based on the equation:

$$R_e = 6.32\, W/d\,\mu$$

where

R_e = Reynolds number

W = flow, lbm/hr

d = pipe inside diameter, in.

μ = fluid viscosity, centipoises

Example. What is the Reynolds number for a flow of 200,000 lb/hr in a 6-in.-ID pipe, if the fluid viscosity is 1.39 cp? To find the answer, align 6 on the pipe-inside-diameter scale with 200,000 on the flow-rate scale Step (1) and mark the intersection of a straightedge with the pivot line. Next align the mark on the pivot line with 1.39 on the fluid-viscosity scale Step (2) and read 1.5×10^5 on the Reynolds number scale.

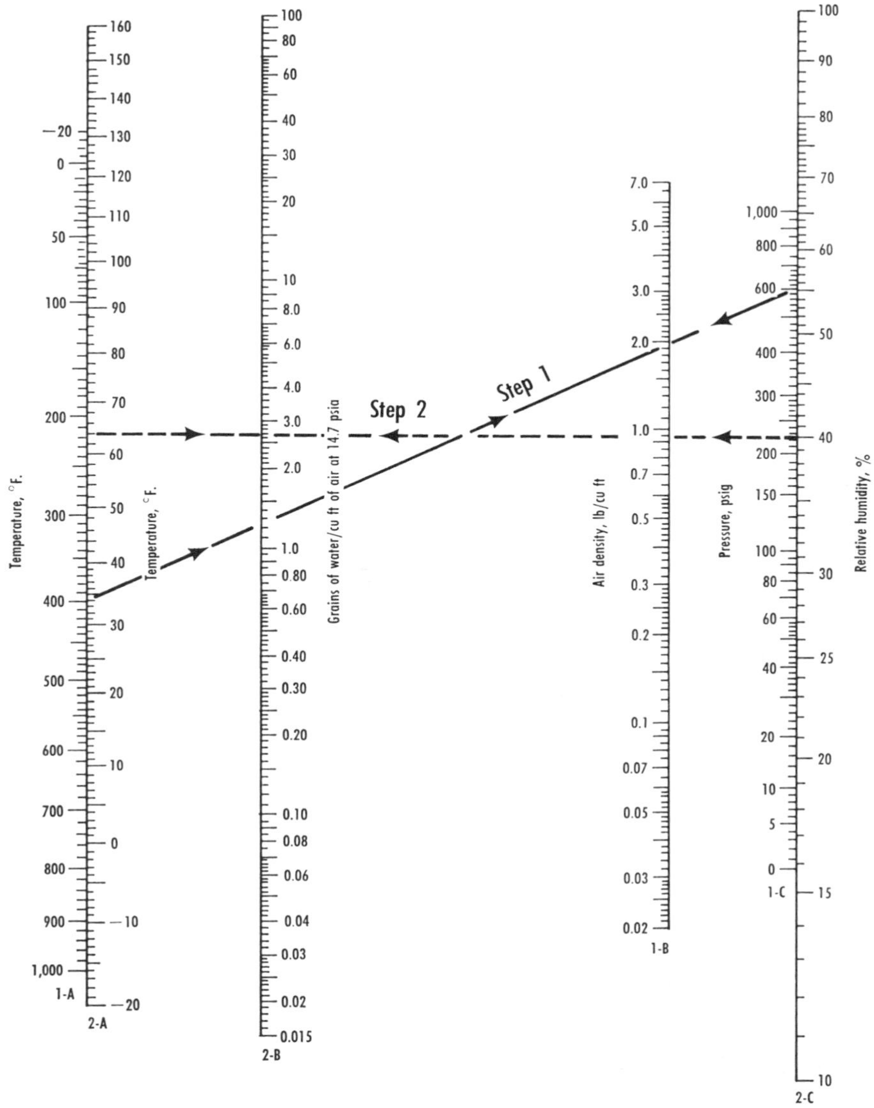

Nomograph 5.21. Density of compressed air and moisture content. Courtesy Oil, Gas, and Petrochem Equipment.

Density of Compressed Air and Moisture Content

Nomograph 5.21 provides a convenient method for finding the density, lb/cu ft, of compressed dry air over a wide range of temperatures and pressures and the grains of water/cu ft of saturated air (14.7 psia) at various temperatures and relative humidities.

The density-of-compressed-dry-air portion of the nomograph (identified by scales 1A, 1B, and 1C, and based on air at an atmospheric pressure of 14.7 psia and molecular weight of 28.97) can be useful in compressed-air flow calculations.

The grains of water/cu ft of saturated air scales (2A, 2B, and 2C) can be used in solving problems relating to air-conditioning, humidification, etc.

Example 1. What is the density of dry air at 600 psig and 400°F.?

Solution. Line 400°F. on 1-A scale with 600 psig on 1-C scale (Step 1-solid line), and read density as 1.95 lb/cu ft where line crosses 1-B scale.

Example 2. How many grains of water/cu ft of saturated air if temperature is 64°F. and relative humidity is 40%?

Solution. Line 64°F. on 2-A scale with 40% relative humidity on 2-C scale (Step 2-dashed line), and read 2.65 grains/cu ft of air where line crosses 2-B scale.

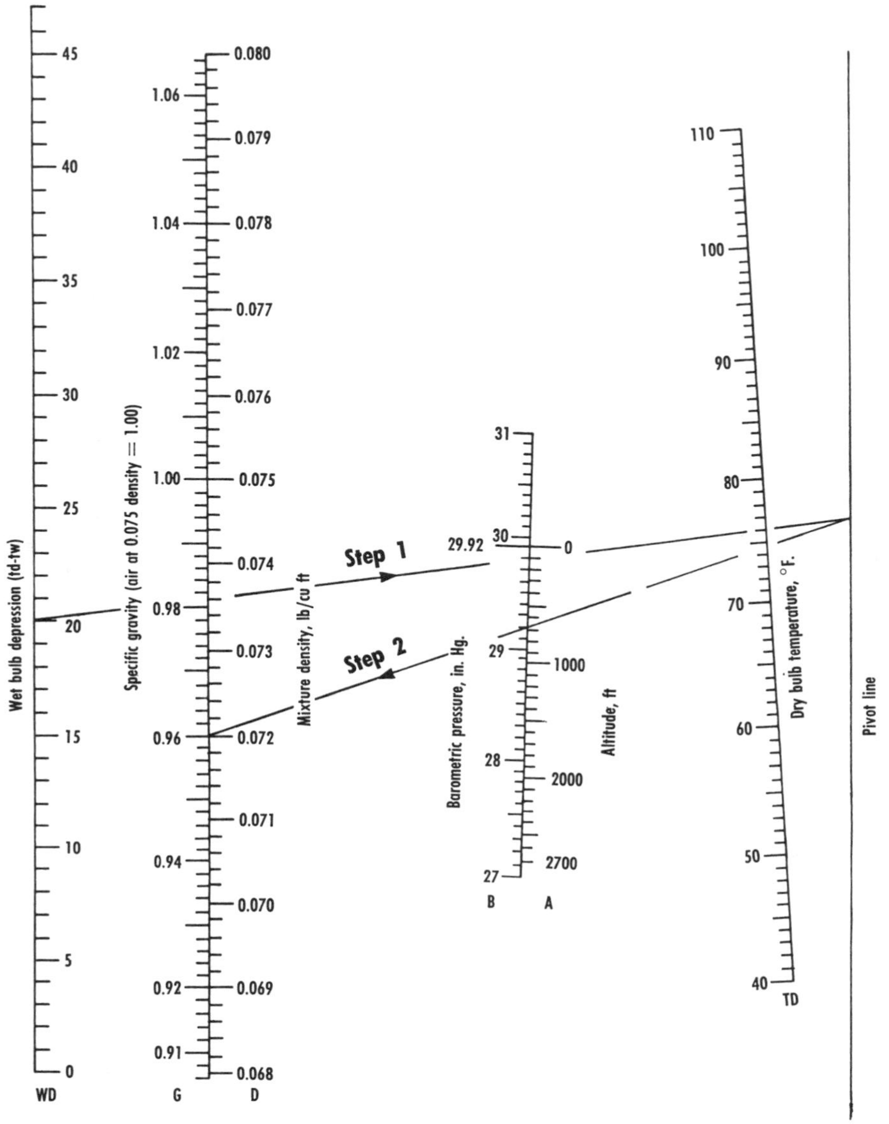

Nomograph 5.22. Air densities determined for atmospheric conditions. Courtesy Oil, Gas, and Petrochem Equipment.

Air Densities Determined for Atmospheric Conditions

Calculations relating to air systems, fans, air preheaters, air compressors, etc., require a knowledge of actual air densities at a given set of conditions.

Nomograph 5.22 based on the following formula provides a fast, accurate method for solving problems where air densities are required.

$$d_m = [B - 0.38\,pv]/[0.7541\,(td + 460)]$$

where

d_m = density of air mixture, lb_m/ft^3

pv = vapor pressure of water, in. Hg

B = barometric pressure, in. Hg

td = dry bulb temperature, °F.

Example. What is the atmospheric density and corresponding specific gravity if dry-bulb temperature is 76°F., wet bulb temperature is 56°F. and barometric pressure is 29.2 in Hg?

Solution. Line 20 ($td - tw = 76 - 56 = 20$) on *WD* scale with 76°F. on *TD* scale, extend to Pivot Line and mark (Step 1). From marked point, line with 29.2 in. Hg on *B* scale, extend to *D* and *G* scale, and read density as 0.072 lb/cu ft and specific gravity as 0.96 (Step 2).

$$\mu = \nu \rho' = \nu S$$

The empirical relation between Saybolt Universal Viscosity and Saybolt Furol Viscosity at 100 F and 122 F, respectively, and Kinematic Viscosity is taken from A.S.T.M. D2161-63T. At other temperatures, the Saybolt Viscosities vary only slightly.

Saybolt Viscosities above those shown are given by the relationships:

Saybolt Universal Seconds = Centistokes x 4.6347

Saybolt Furol Seconds = Centistokes x 0.4717

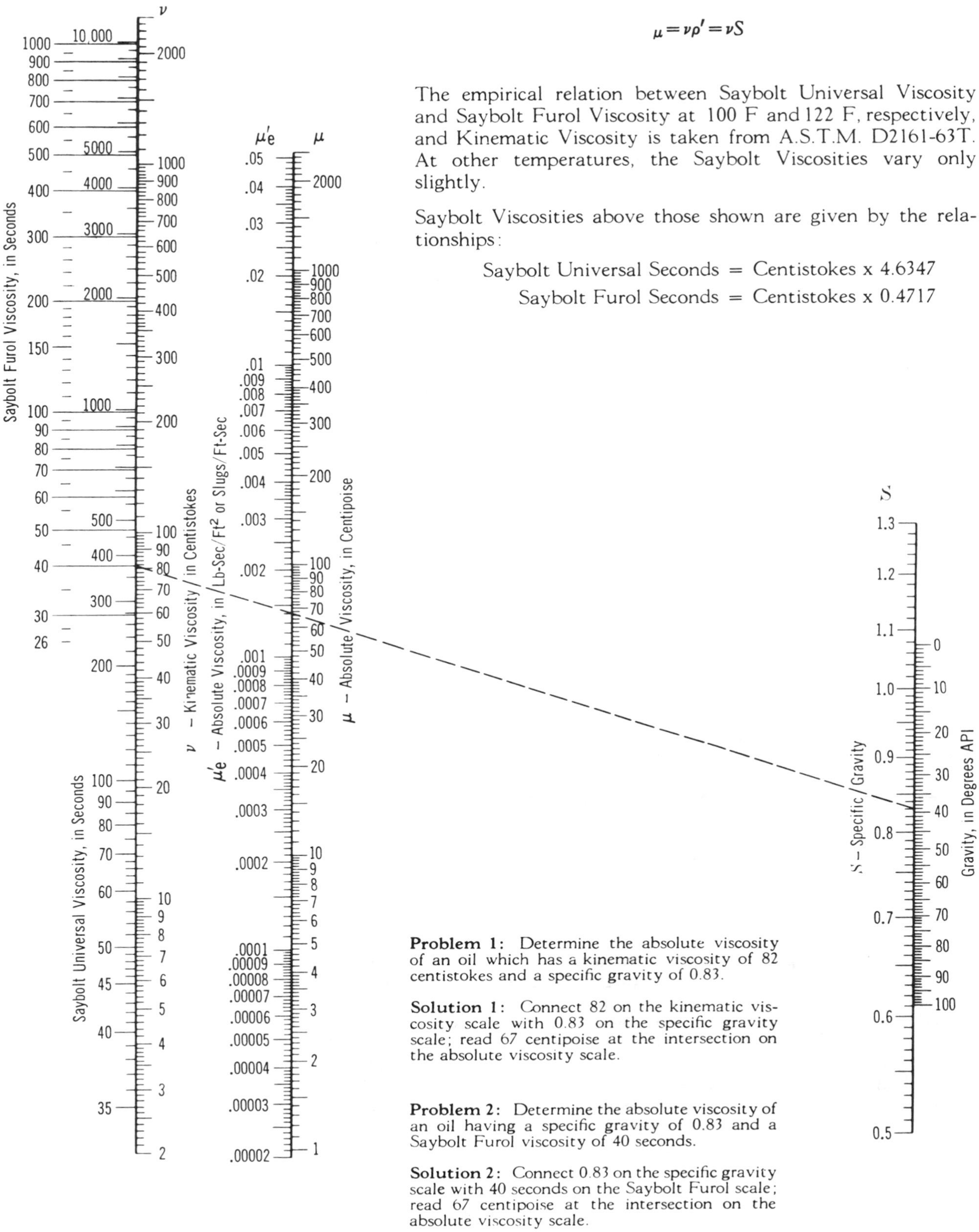

Problem 1: Determine the absolute viscosity of an oil which has a kinematic viscosity of 82 centistokes and a specific gravity of 0.83.

Solution 1: Connect 82 on the kinematic viscosity scale with 0.83 on the specific gravity scale; read 67 centipoise at the intersection on the absolute viscosity scale.

Problem 2: Determine the absolute viscosity of an oil having a specific gravity of 0.83 and a Saybolt Furol viscosity of 40 seconds.

Solution 2: Connect 0.83 on the specific gravity scale with 40 seconds on the Saybolt Furol scale; read 67 centipoise at the intersection on the absolute viscosity scale.

Nomograph 5.23. Equivalents of kinematic, Saybolt Universal, Saybolt Furol, and absolute viscosity. Courtesy Crane Company (T.P. #410).

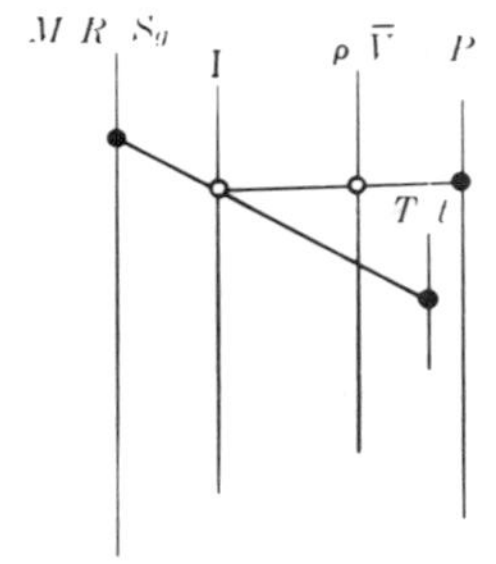

Weight Density and Specific Volume Of Gases and Vapors

Nomograph 5.24 is based on the formula:
$\rho = 144P'/RT = MP'/10.72\,T = 2.70P'\,S_g/T$

where
$P' = 14.7 + P$

Problem. What is the density of dry CH_4 if the temperature is 100°F and the gauge pressure is 15 pounds per square inch?

Solution. Refer to the table below for molecular weight, specific gravity, or individual gas constant. Connect 96.4 of the R scale with 100 on the temperature scale, t, and mark the intersection with the index scale. Connect this point with 15 on the pressure scale, P. Read the answer, 0.80 pounds per cubic foot, on the weight density scale ρ.

c_p = specific heat at constant pressure
c_v = specific heat at constant volume

Name of Gas	Chemical Formula or Symbol	Approx. Molecular Weight	Weight Density, Pounds per Cubic Foot*	Specific Gravity Relative To Air	Individual Gas Constant	Specific Heat Per Pound at Room Temperature		Heat Capacity Per Cubic Foot at Atmospheric Pressure and 68 F		k equal to c_p/c_v
		M	ρ	S_g	R	c_p	c_v	c_p	c_v	
Acetylene	C_2H_2	26.0	.06754	.897	59.4	.350	.2737	.0236	.0185	1.28
Air	—	29.0	.07528	1.000	53.3	.241	.1725	.0181	.0130	1.40
Ammonia	NH_3	17.0	.04420	.587	90.8	.523	.4064	.0231	.0179	1.29
Argon	A	40.0	.1037	1.377	38.7	.124	.0743	.0129	.0077	1.67
Carbon Dioxide	CO_2	44.0	.1142	1.516	35.1	.205	.1599	.0234	.0183	1.28
Carbon Monoxide	CO	28.0	.07269	.965	55.2	.243	.1721	.0177	.0125	1.41
Ethylene	C_2H_4	28.0	.0728	.967	55.1	.40	.3292	.0291	.0240	1.22
Helium	He	4.0	.01039	.138	386.	1.25	.754	.0130	.0078	1.66
Hydrochloric Acid	HCl	36.5	.09460	1.256	42.4	.191	.1365	.0181	.0129	1.40
Hydrogen	H_2	2.0	.005234	.0695	767.	3.42	2.435	.0179	.0127	1.40
Methane	CH_4	16.0	.04163	.553	96.4	.593	.4692	.0247	.0195	1.26
Methyl Chloride	CH_3Cl	50.5	.1309	1.738	30.6	.24	.2006	.0314	.0263	1.20
Nitrogen	N_2	28.0	.07274	.966	55.2	.247	.1761	.0179	.0128	1.40
Nitric Oxide	NO	30.0	.07788	1.034	51.5	.231	.1648	.0180	.0128	1.40
Nitrous Oxide	N_2O	44.0	.1143	1.518	35.1	.221	.1759	.0253	.0201	1.26
Oxygen	O_2	32.0	.08305	1.103	48.3	.217	.1549	.0180	.0129	1.40
Sulphur Dioxide	SO_2	64.0	.1663	2.208	24.1	.154	.1230	.0256	.0204	1.25

*Weight density values are at atmospheric pressure and 68 F.
For values at 60 F, multiply by 1.0154.

Nomograph 5.24 continued

Nomograph 5.24. Weight density and specific volume of gases and vapors. Courtesy Crane Company (T.P. #410).

Nomograph 5.24 continued

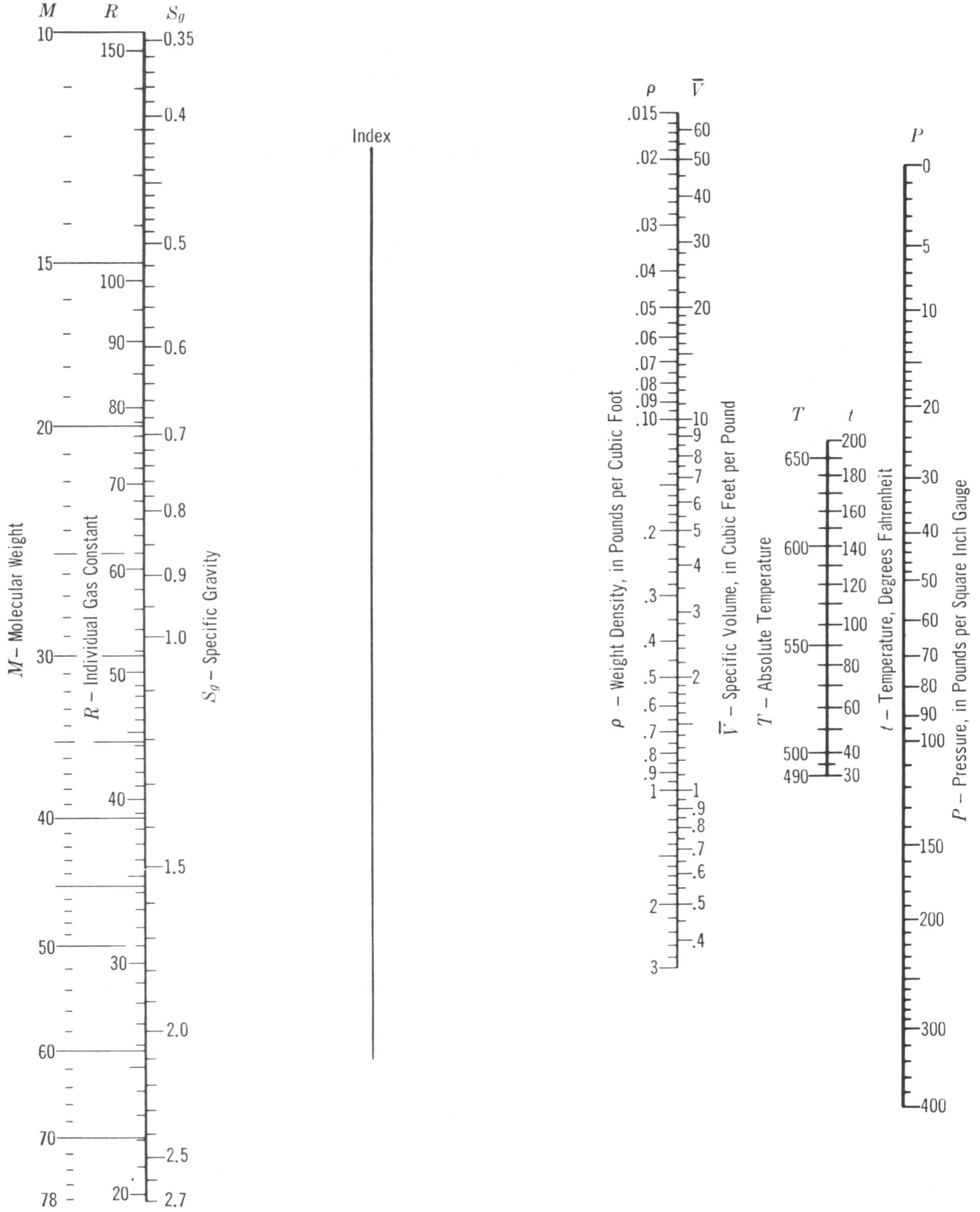

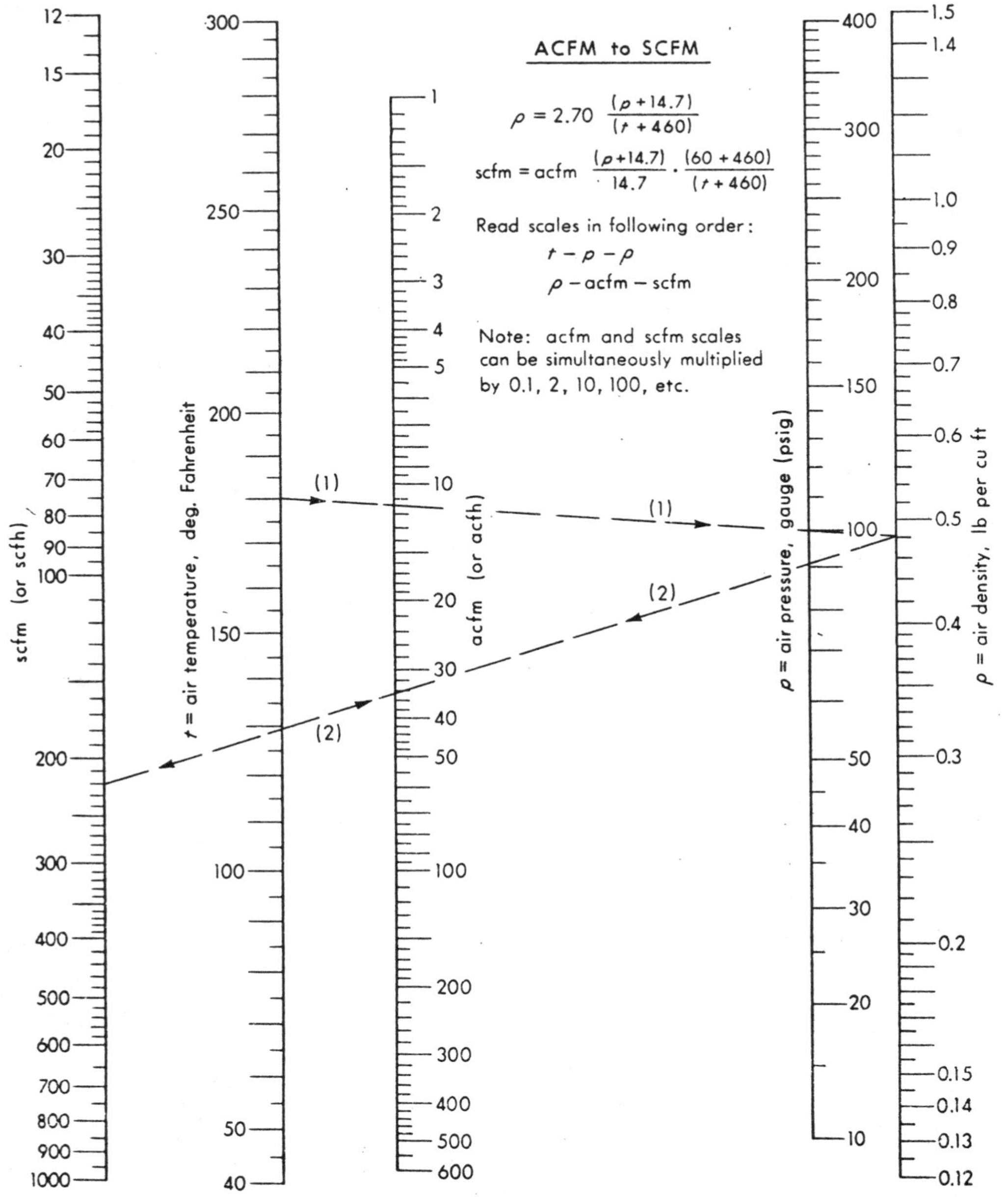

Nomograph 5.25. Airflow conversion—ACFM to SCFM. Courtesy Control Engineering, copyright © 1967, Dunn Donnelley Publishing Corporation. All rights reserved.

Airflow Conversion

Since instruments handling gases are often rated in standard cubic feet per minute (or hour), Nomograph 5.25 is presented to facilitate conversion from the volume at actual pressure and temperature (acfm) to the volume at 14.7 psia and 60 deg F (scfm). If the gas is air, the nomograph also yields actual density.

Example. A rotameter is handling 34.8 acfm at 180 deg F and 100 psig. What is the corresponding scfm? Align t = 180 with ρ = 100 and read ρ = 0.482 lb per cu ft; align ρ with acfm = 34.8 and read scfm = 220.

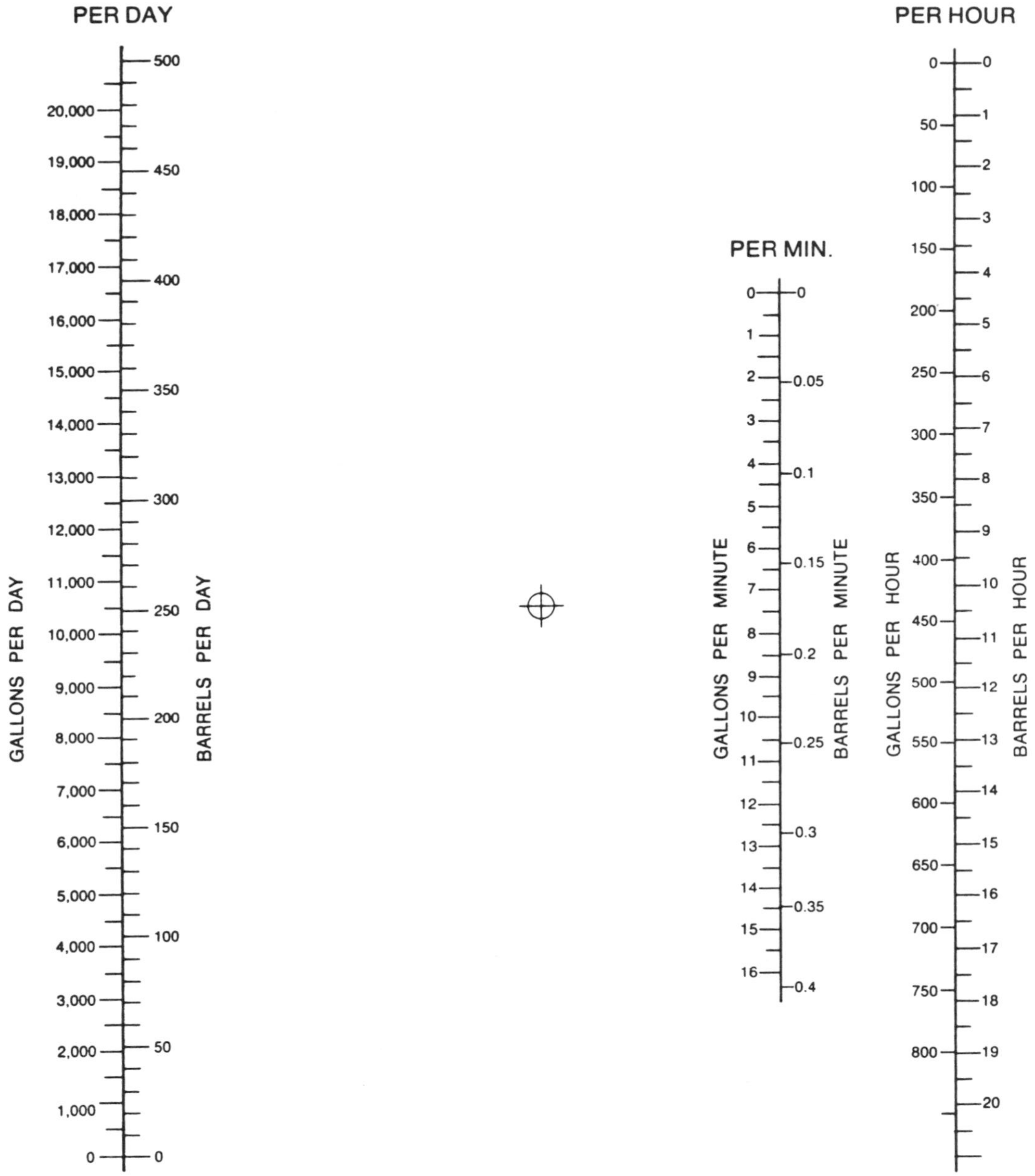

Nomograph 5.26. Flow rate vs. time.

Flow Rate vs Time

1 Barrel = 42 Gallons

Nomograph 5.26 provides a convenient method of quick conversion of barrels to gallons, gallons to barrels, and their flow rate relationship in units per minute, per hour, and per day. The nomograph is accurate to two significant figures. Relationships are linear and powers of ten may be used to extend the range.

Examples

Convert 8,000 gallons to barrels (no time base). Any of the three scales may be used; however, the left hand scale is longest and is the easiest to read. Opposite 8,000 gallons on the left hand side of the scale, estimate 190 barrels from the right hand side of the scale.

Convert 100 gallons per minute to barrels per day. Align a straightedge on the center dot and on 10 gallons per minute on the gallons per minute scale. Read 340 barrels per day on the barrels per day scale. Multiply both readings by 10 to obtain 100 gallons per minute = 2,400 barrels per day.

Nomograph 5.27. Viscosity conversion chart. Courtesy Fisher Controls.

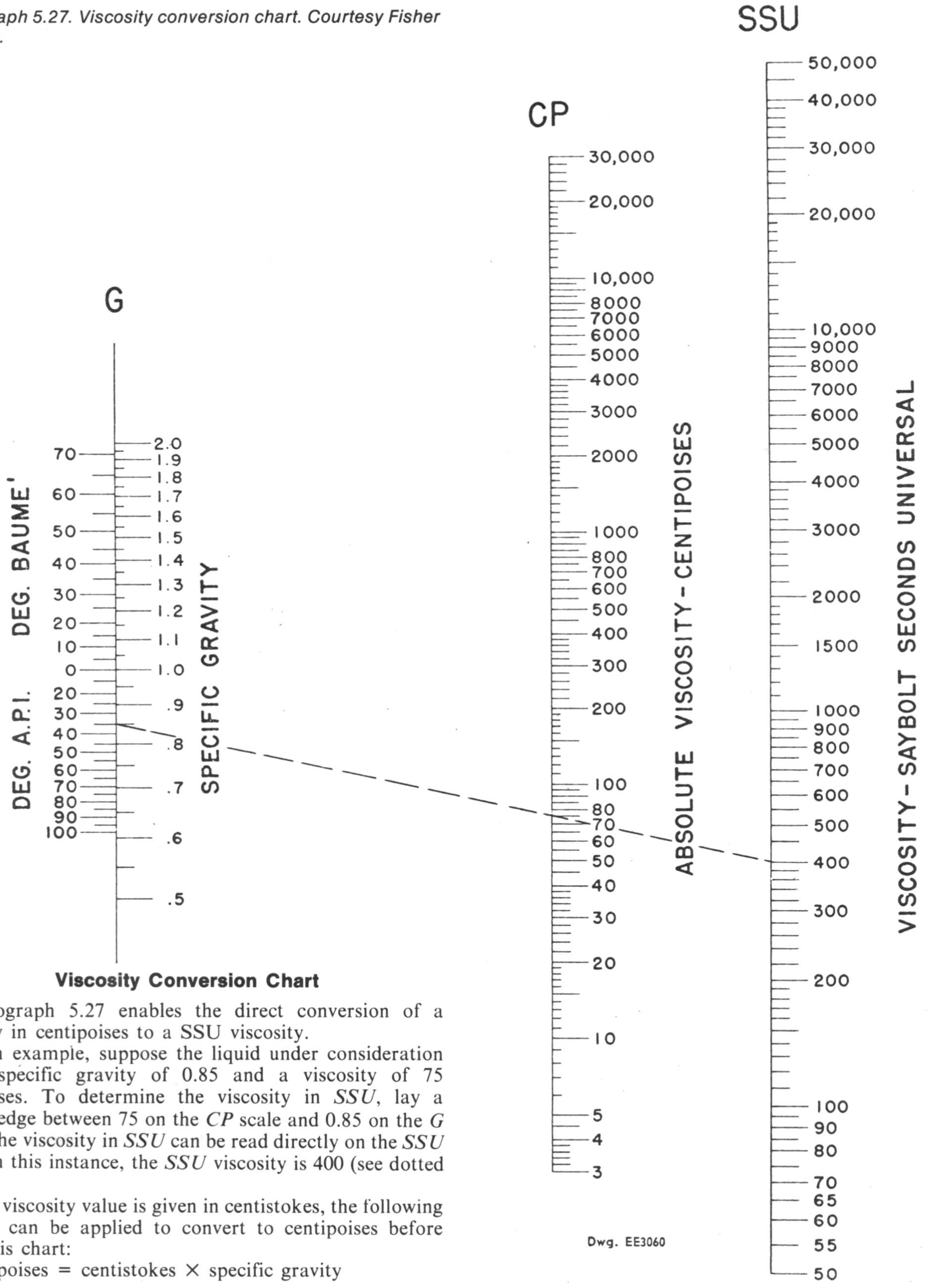

Viscosity Conversion Chart

Nomograph 5.27 enables the direct conversion of a viscosity in centipoises to a SSU viscosity.

As an example, suppose the liquid under consideration has a specific gravity of 0.85 and a viscosity of 75 centipoises. To determine the viscosity in *SSU*, lay a straightedge between 75 on the *CP* scale and 0.85 on the *G* scale. The viscosity in *SSU* can be read directly on the *SSU* scale. In this instance, the *SSU* viscosity is 400 (see dotted line).

If the viscosity value is given in centistokes, the following formula can be applied to convert to centipoises before using this chart:

Centipoises = centistokes × specific gravity

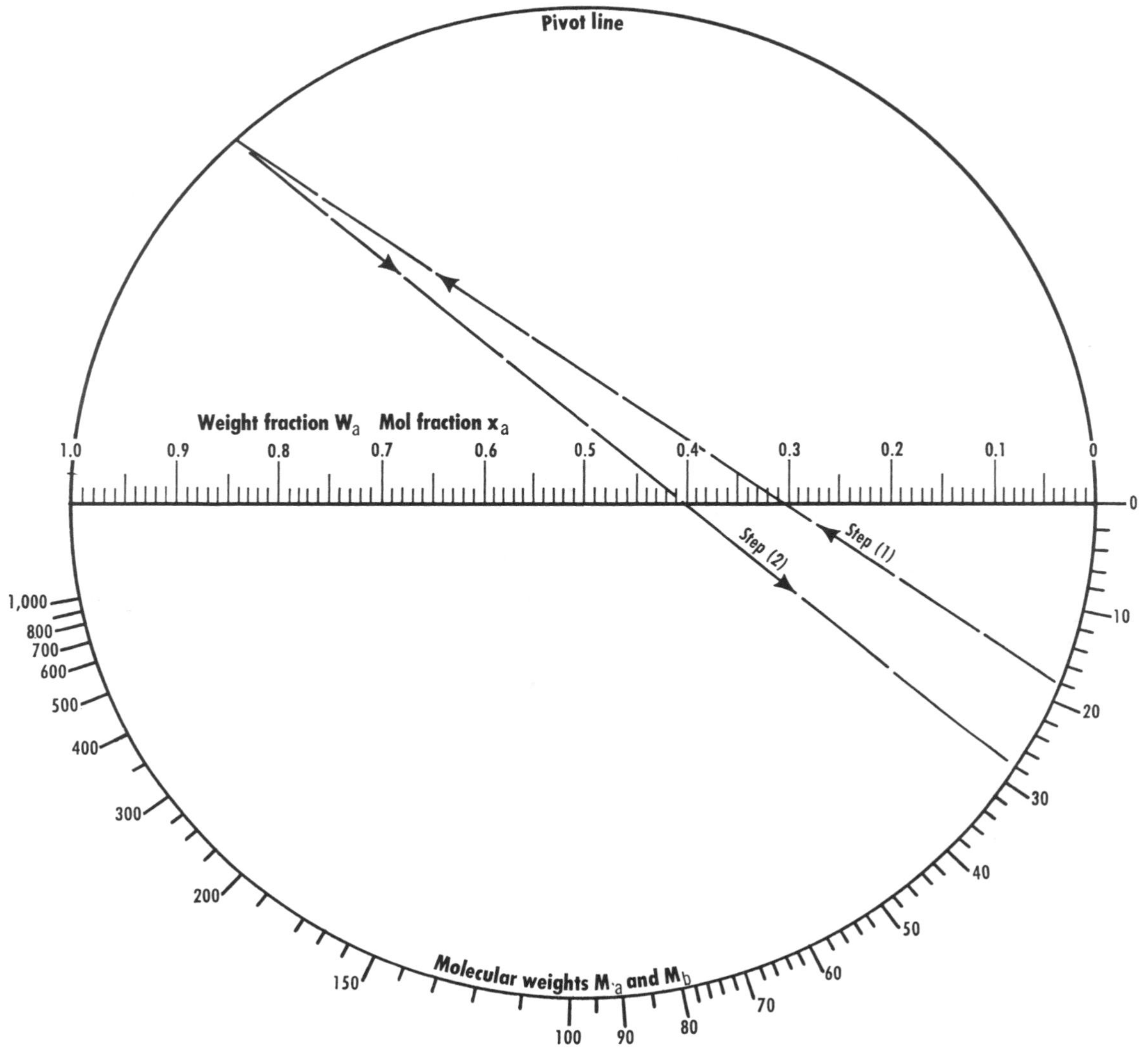

Nomograph 5.28. Convert weight fractions to mol fractions. Courtesy Chemical Publishing Company, Inc.

Convert Weight Fractions to Mol Fractions

Nomograph 5.28 can be useful in dealing with mixtures of two compounds to find the mol fraction when the two weight fractions and molecular weights are known or to find the weight fraction when the mol fractions and molecular weights are known. It provides solutions to either of these two equations:

$$x_a = \{W_a/M_a\}/\{W_a/M_a + [(1 - W_a)/M_b]\}$$

$$W_a = (x_a M_a)/[x_a M_a + (1 - x_a)M_b]$$

where

x_a = mol fraction of compound *a*
W_a = weight fraction of compound *a*
M_a = molecular weight of compound *a*
M_b = molecular weight of compound *b*

Example. What is the mol fraction of compound in a mixture of two compounds if its molecular weight is 18, its weight fraction is 0.30, and the molecular weight of compound *b* is 28? To find the answer to this question, align molecular weight, $M_a = 18$, with weight fraction $W_a = 0.30$, Step (1), and mark the intersection of a straightedge with the pivot line.

Then align the mark on the pivot line with molecular weight, Step (2), $M_b = 28$, and read mol fraction of compound *a*, $x_a = 0.40$. Note that the molecular weight values, M_a and M_b, can be simultaneously multiplied by such numbers as 0.1, 0.5, 2, 10, etc. to extend the range of the nomograph.

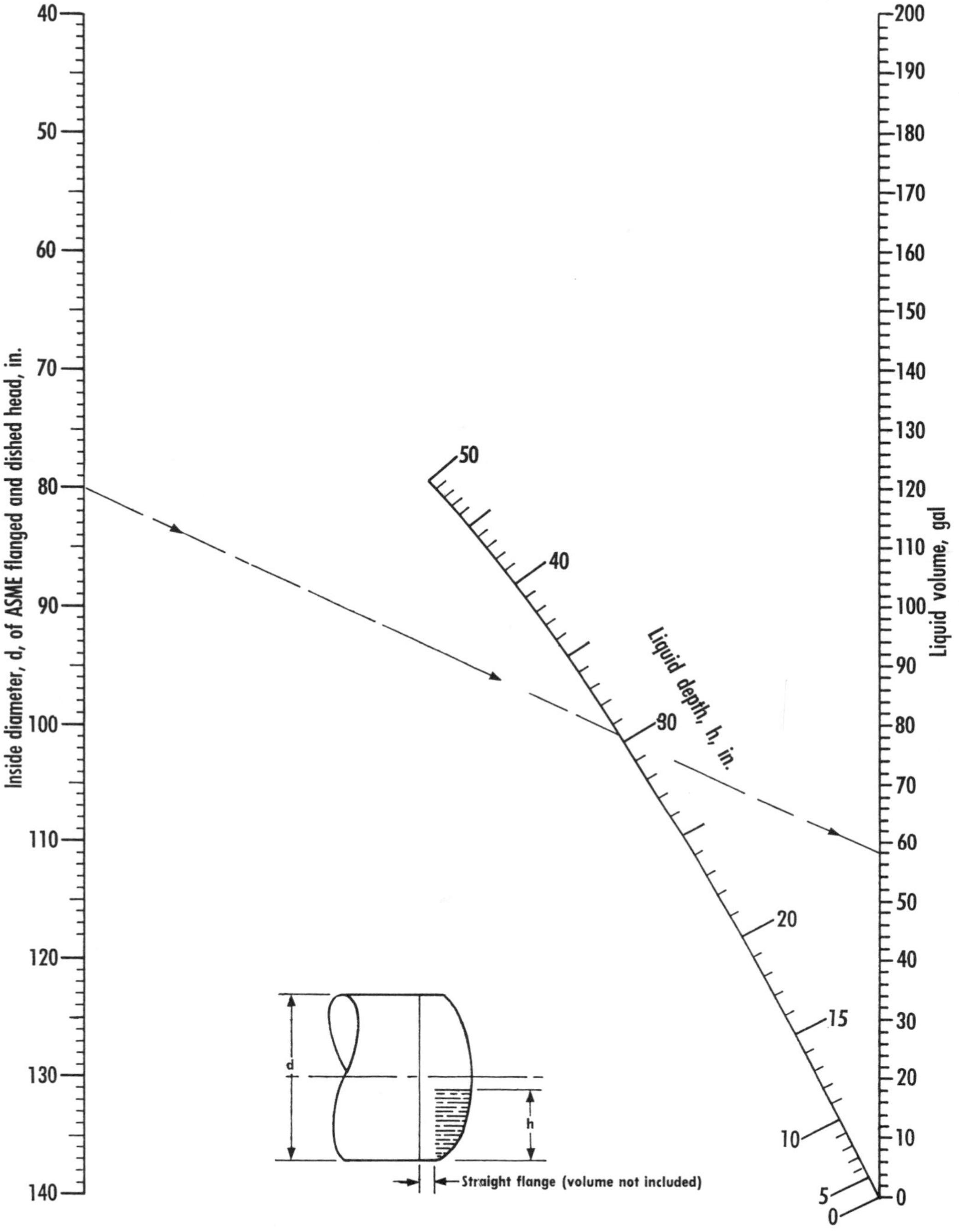

Nomograph 5.29. ASME code flanged and dished head. Courtesy Oil, Gas, and Petrochem Equipment.

ASME Code Flanged and Dished Head

Nomograph 5.29 is designed to give the partial capacity of one ASME code flanged and dished head. It is constructed from data of Buffalo Tank Corp. The following example shows how to use it to find quickly partial capacity.

Example. What is the capacity of one 80-in. ID flanged and dished head filled to a depth of 30.5 in.? To find the answer, connect, with a straightedge, 80 in. on the inside-diameter scale with 30.5 on the liquid-depth scale, and read 58.2 gal on the liquid-volume scale.

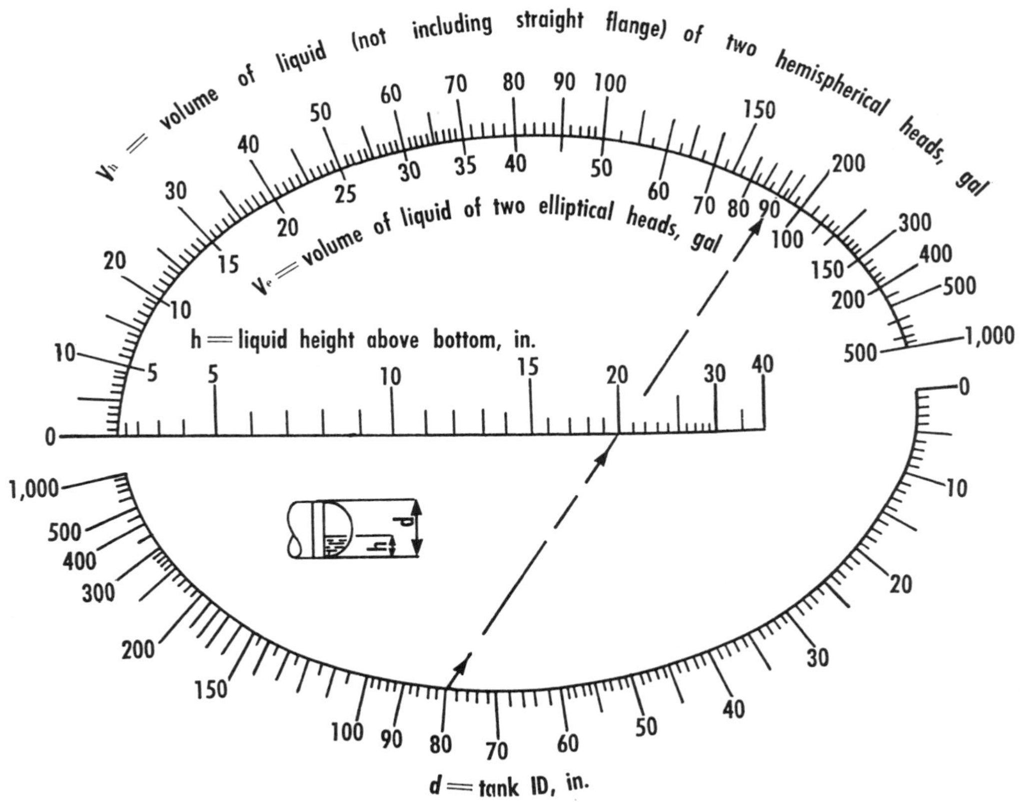

Nomograph 5.30. Vessel head content. Courtesy Oil, Gas, and Petrochem Equipment.

Vessel Head Content

Nomograph 5.30 is designed to help quickly compute the contents of two part-full hemispherical or two ASME Code elliptical (2:1) heads. The formula for the part-full contents of two hemispherical heads is this:

$V_h = 0.0068h^2[d - (h/1.5)]$
For two ASME Code elliptical (2:1) heads the formula is:
$V_e = 0.0034\,h^2[d - (h/1.5)]$

where

V_h = volume of liquid (not including straight flange) of two hemispherical heads, gal
V_e = volume of liquid of two elliptical heads, gal
h = liquid height above bottom, in.
d = tank ID, in.

Note. In using the nomograph for full head capacities, multiply the volume for $h = d/2$ by 2. If h is greater than $d/2$, subtract the difference from the full head capacity.

Example. What are the contents of two part-full hemispherical heads, if the tank ID is 80 in. and the liquid height above bottom is 20 in.? Align d = 80 with h = 20 and read V_h = 182 gal. Note that for two elliptical heads, the volume V_e = 92 gal.

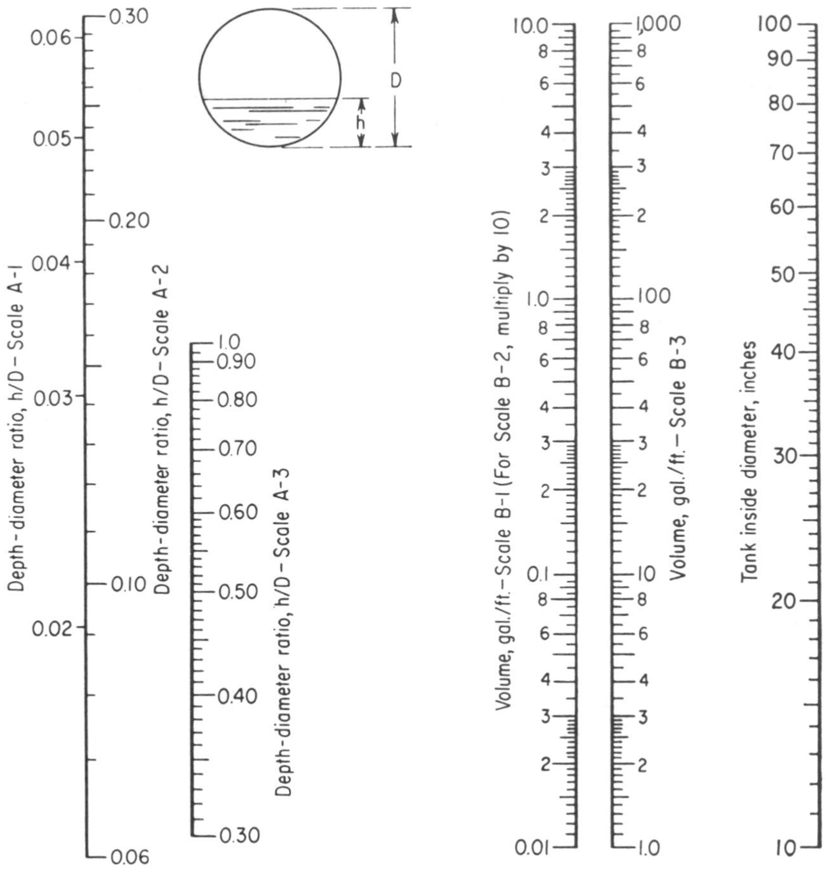

Nomograph 5.31. Calibration chart for horizontal cylindrical tanks. Reprinted by special permission from Chemical Engineering, *copyright © by McGraw-Hill, Inc. New York, N.Y. 10036.*

Calibration Chart for Horizontal Cylindrical Tanks

Chemical engineers often need to calculate the volume of liquid contained at various depths in horizontal, cylindrical tanks. Many different calculation aids have been offered to simplify this otherwise tedious task. Nomograph 5.31 is one of the most convenient as well as most accurate methods available.

First find the depth-diameter ratio, h/D, and locate its value on one of the scales, A-1, A-2 or A-3. Join the point with the tank diameter and read the volume per foot of tank length on the correspondingly numbered B-scale (i.e., Scale A-1 and Scale B-1). For gallons in two bumped ends $V = 0.000933\ h^2\ (1.5D - h)$.

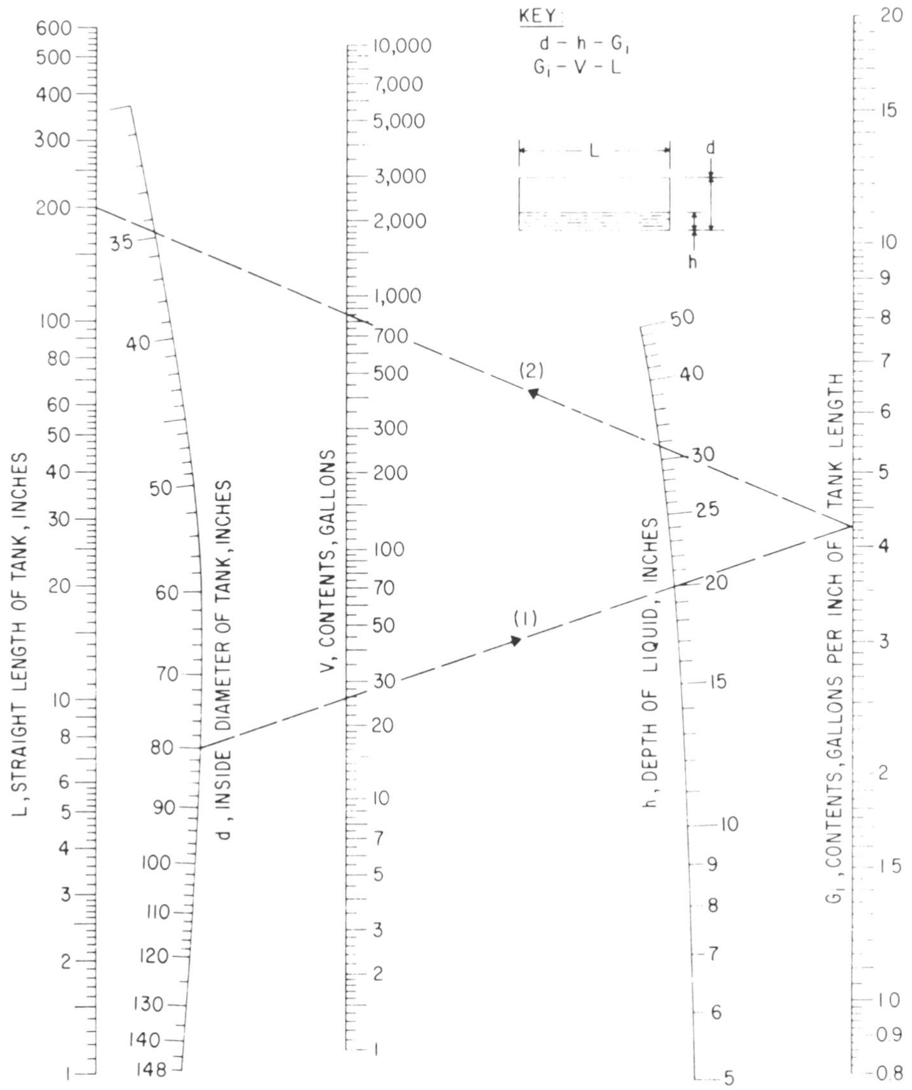

Nomograph 5.32. Volume of part-full horizontal tank. Courtesy Gulf Publishing Company.

Volume of Part-Full Horizontal Tank

The volume of part-full horizontal tanks, not considering head volumes, can be found using Nomograph 5.32 and the following equation:

$$V = 0.00433\ L \left\{ (\pi/8)d^2 - \left[(d/2 - h)\sqrt{(dh - h^2)} + d^2/4 \text{ arc Sin} \left(\frac{d/2 - h}{d/2} \right) - \pi/2 \right] \right\}$$

where

L = straight length of tank, in.
d = inside diameter of tank, in.
V = contents, gals.
h = depth of liquid, in.

Note. For full tank capacities, multiply V for $h = d/2$ by two; for $h > d/2$, subtract the void volume from the full tank.

Example. For a tank with an ID of 80 inches, depth of liquid = 20 in., and the straight length of the tank is 200 inches, what is the liquid volume content in gallons?

Solution. Follow the arrows on the nomograph and find the volume to be 26 gallons.

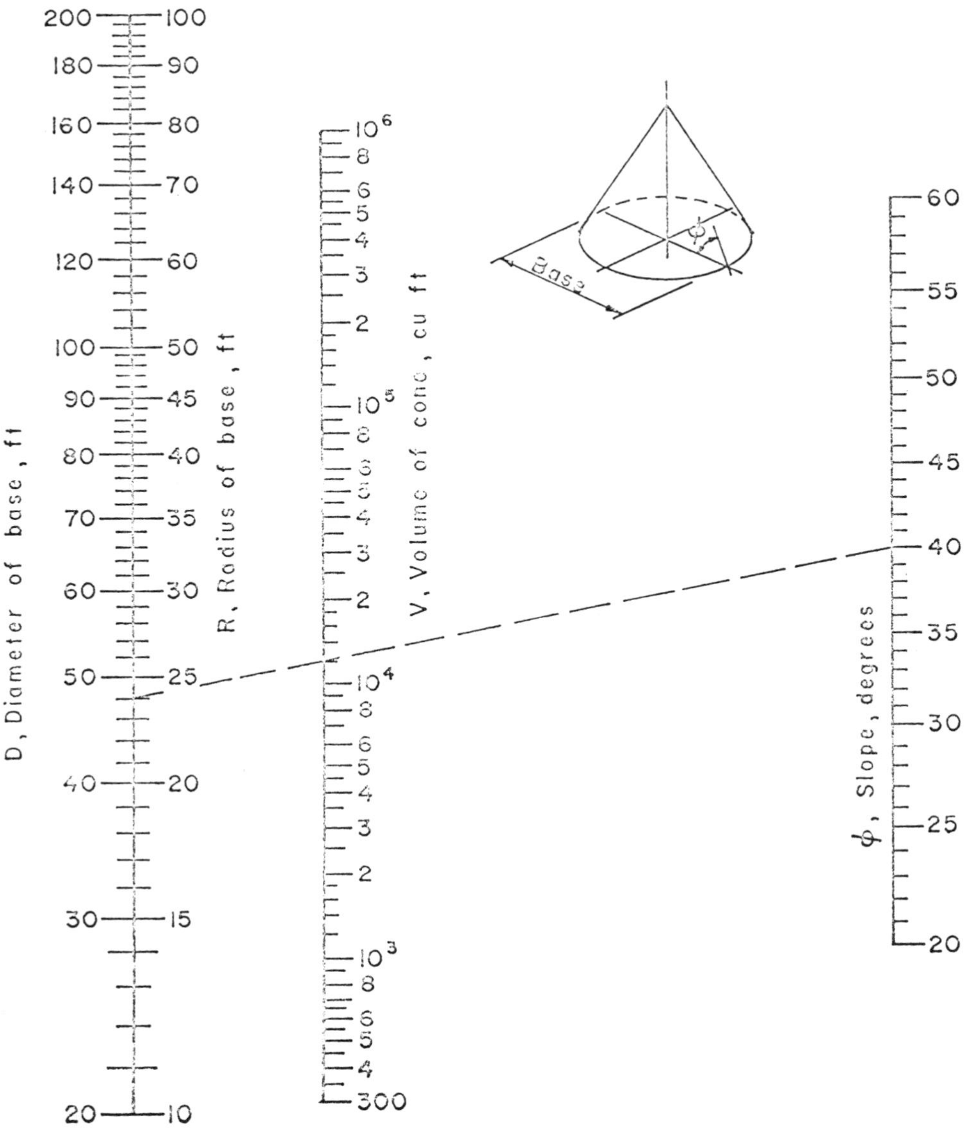

Nomograph 5.33. Volume of a cone. Courtesy Chemical Publishing Company, Inc.

Volume of a Cone

Conical shapes are encountered in pump sumps, agitators, tanks, and many stockpiles.

The volume for a right circular cone is

$$V = \pi R^2 a/3$$

where

V = volume, ft^3
R = radius of the base, ft
a = altitude, ft

When Φ is the angle between an element and the base,

$$\tan\Phi = a/R$$

so that

$$a = R \tan\Phi$$

and

$$V = \pi R^3 \tan\Phi/3$$

The last equation is the one upon which Nomograph 5.33 is based.

Example. The broken index line shows that the volume of a right circular cone with base diameter of 48′ and angle Φ of 40° is 12,100 cu ft.

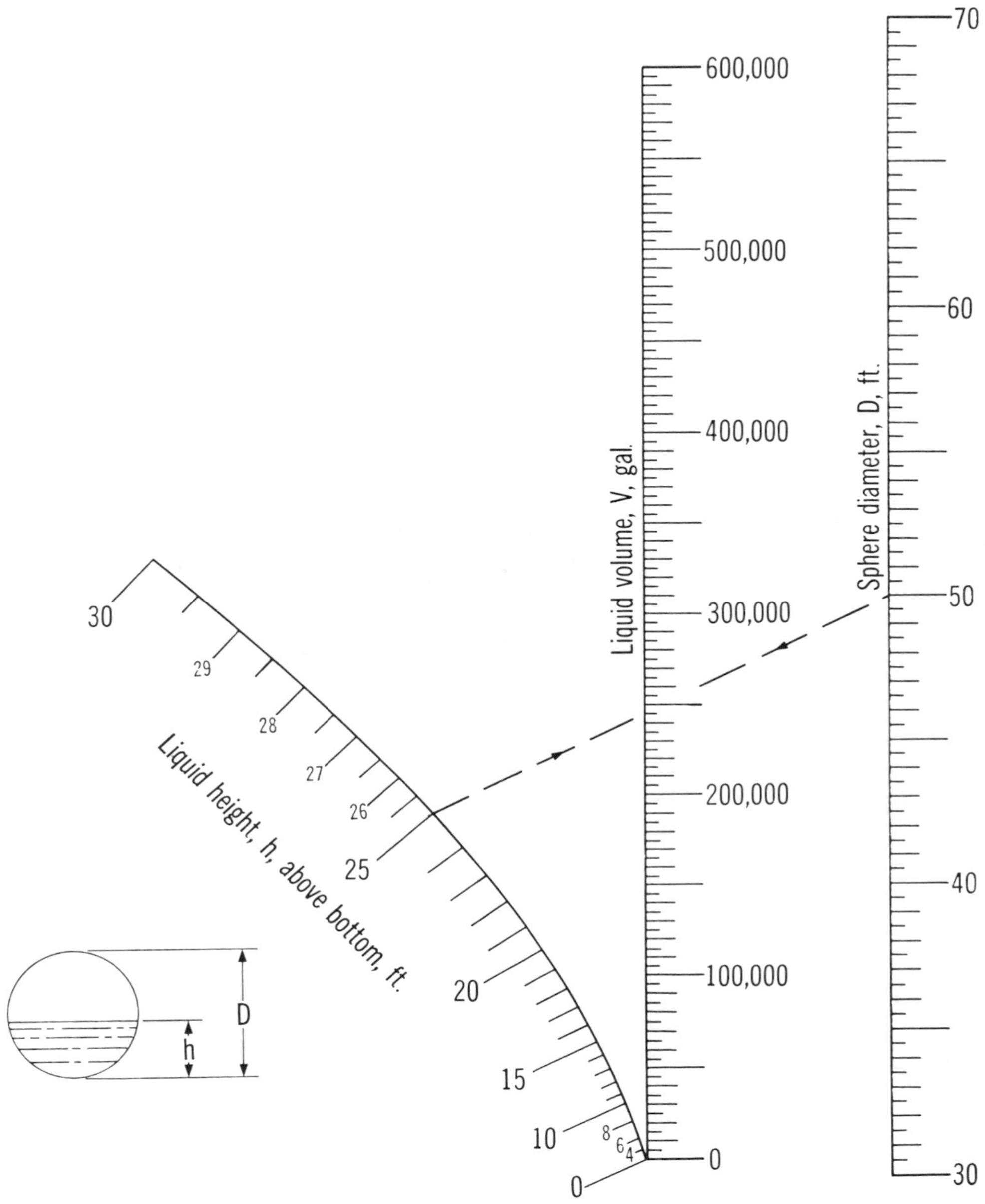

Nomograph 5.34. Partial volume of spheres. Courtesy Oil, Gas, and Petrochem Equipment.

Partial Volume of Spheres

Nomograph 5.34 is designed to give the volume of fluid in a partially filled sphere. It is based on the equation:

$$V = 7.48\ \pi h^2[(D/2) - (h/3)]$$

where

V = volume of liquid, gal
h = liquid height in sphere, ft
D = sphere diameter, ft
π = 3.1416

Example. How much liquid is in a sphere with a diameter of 50 ft and containing liquid to a height of 25 ft above bottom? To find the answer, align 50 ft on the sphere-diameter scale with 25 ft on the liquid-height scale and read 245,000 gal on the capacity scale.

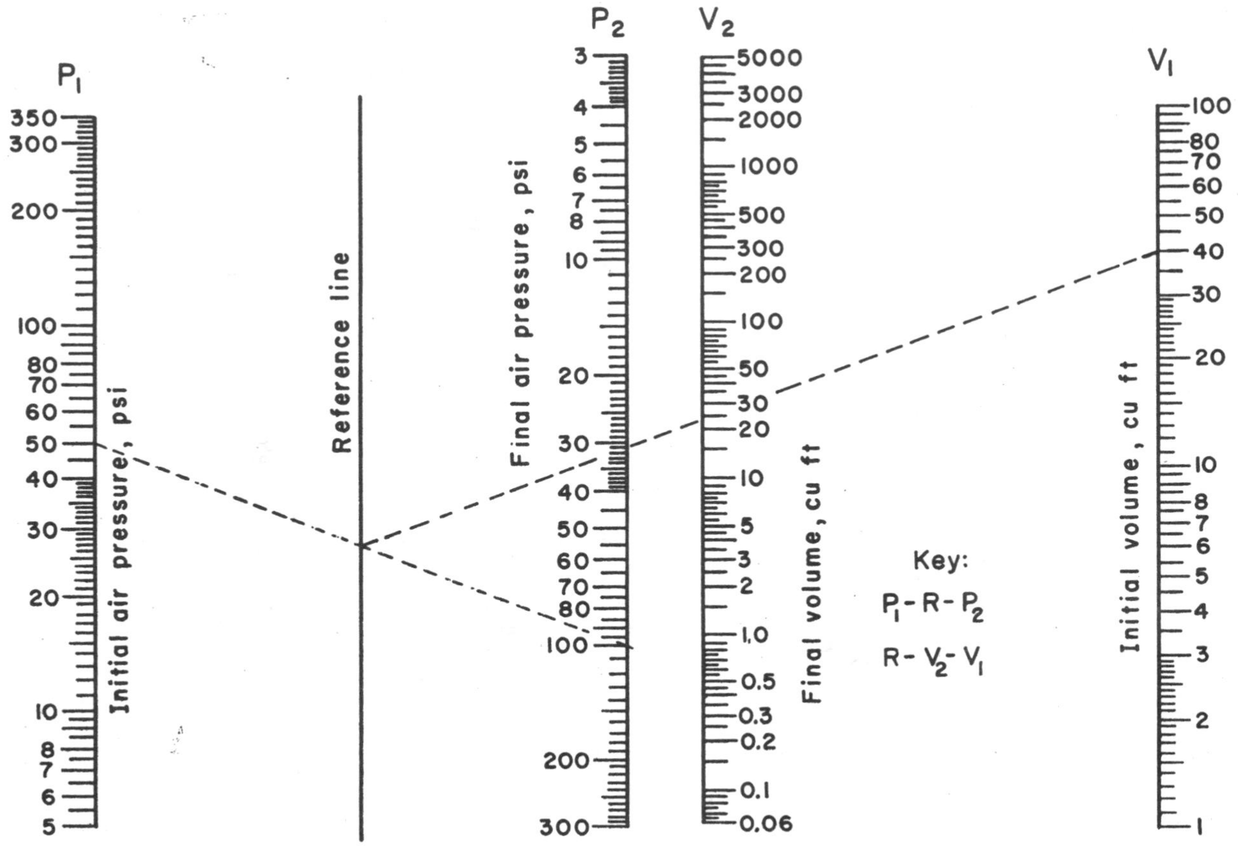

Nomograph 5.35. Adiabatic expansion of air. Courtesy Chemical Publishing Company, Inc.

Adiabatic Expansion of Air

Nomograph 5.35 permits a quick and simple determination of adiabatic expansion of air—eliminating calculations using gas-law formulas. It is based on the following equation:

$$P_1/P_2 = (V_2/V_1)^{1.41}$$

where

P_1 = initial pressure, psi
P_2 = final pressure, psi
V_1 = initial volume, cu ft
V_2 = final volume, cu ft

Example. Given an initial air pressure of 50 psi and an initial volume of 40 cu ft, what will be the final volume when a final pressure of 100 psi is applied?

Extend a line from 50 on P_1-scale to 100 on P_2-scale. The point at which this line intersects the reference scale is now used as a pivot point. Draw a line from this point to 40 on V_1-scale and read final volume—24.5 cu ft on the V_2-scale.

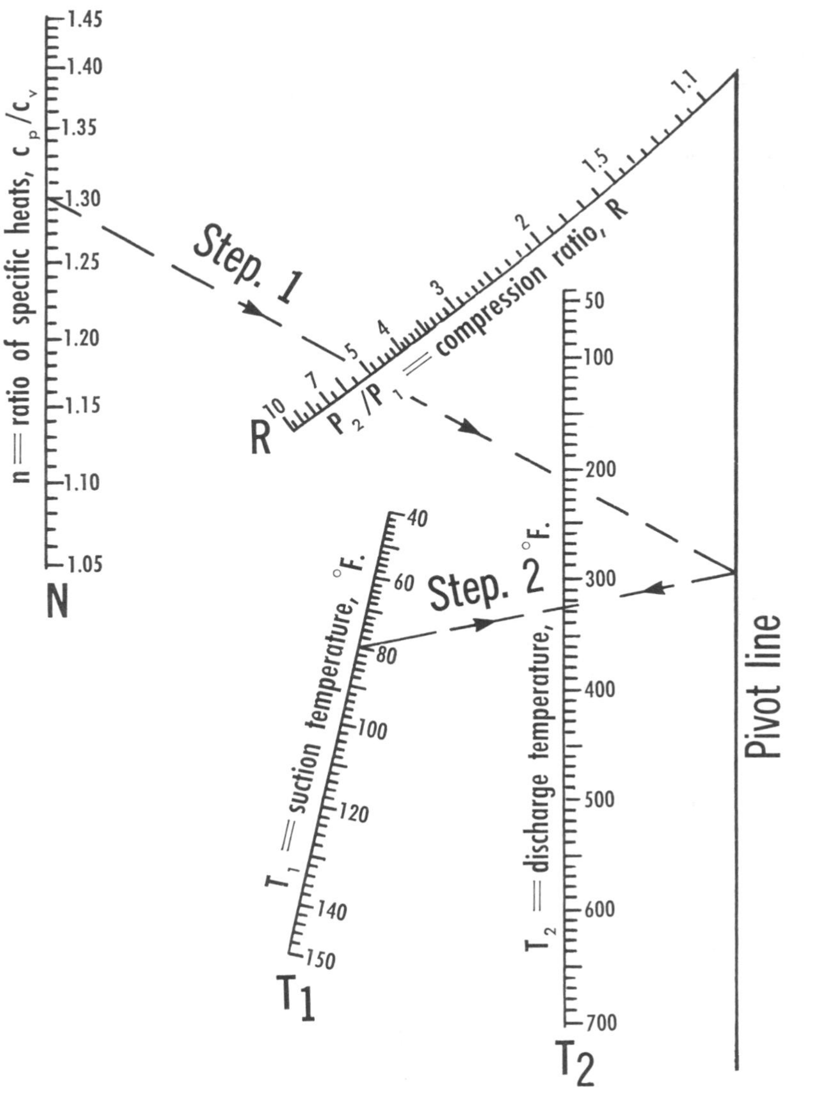

Nomograph 5.36. Determining gas compression temperature. Courtesy Oil, Gas, and Petrochem Equipment.

Determining Gas Compression Temperature

Nomograph 5.36 presented here permits a rapid determination of the compression temperatures of gases according to the equation.

$$T_2 = T_1(P_2/P_1)^{(n-1)/n}$$

where

T_2 = absolute discharge temperature, (°F.+460)
T_1 = suction temperature, (°F.+460)
P_2 = discharge pressure, psia
P_1 = suction pressure, psia
n = ratio of specific heat at constant pressure to specific heat at constant volume, (c_p/c_v)
P_2/P_1 = compression ratio, R

The ratio of specific heats, or n-value, of gases can be obtained from the known, calculated, or estimated specific gravity or percentage molecular weight of the fractions constituting the mixture of gases under consideration.

Example. What is the discharge temperature of a gas, if the compression ratio is 5, suction temperature is 80°F., and the n-value of the gas is 1.3?

Solution. On scale N align 1.3 with 5 on scale R, extend the line to the pivot line, and mark the intersection with the pivot line (Step 1). From the marked point on the pivot line align with 80°F. on scale T_1 and read the discharge temperature as 323°F. where the line intersects scale T_2 (Step 2).

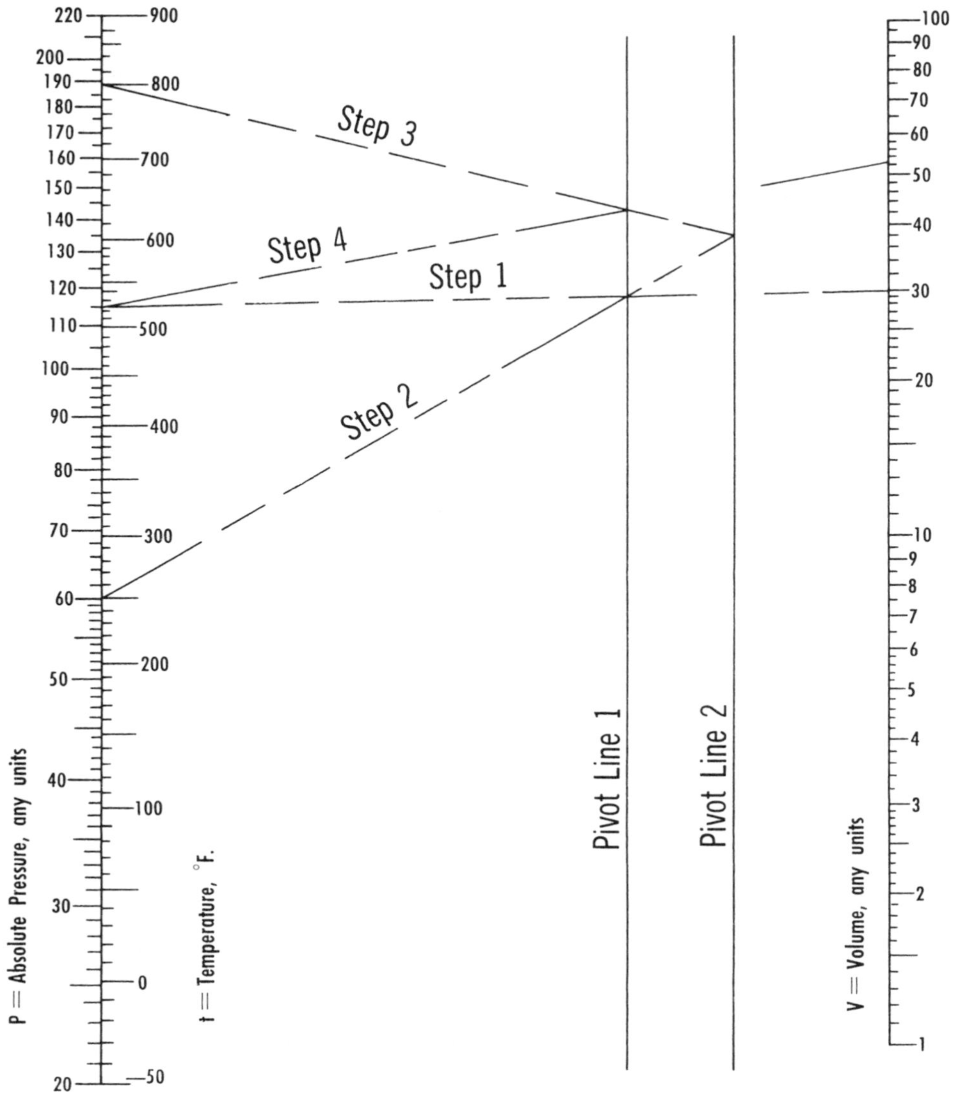

Nomograph 5.37. Solving ideal gas laws. Courtesy Oil, Gas, and Petrochem Equipment.

Solving Ideal Gas Laws

Nomograph 5.37 permits a fast solution to the equations stating the combination of Boyle's and Charles' laws for ideal gases. It quickly solves for changes in pressure and/or temperature and/or volume of a constant weight of a confined gas. The nomograph is based on these equations:

$$P_2V_2/(t_2 + 460) = P_1V_1/(t_1 + 460)$$

where

P = absolute pressure of gas
V = volume of gas
t = temperature of gas, °F.

Example. What is the final volume when nitrogen in a constant pressure accumulator is heated from 250°F. to 800°F.? Pressure is 1,150 psia and initial volume in accumulator is 30 cu in.

For Step 1, align P_1 = 115 with V_1 = 30 and mark intersection with Pivot Line 1. Align this mark with t_1 = 250 and mark intersection with pivot line 2, Step 2. For Step 3, align that mark with t_2 = 800 and mark the intersection with pivot line 1. For the fourth and final step, align the last mark on pivot line 1 with P_2 = 115 (constant pressure) and read V_2 = 53 cu in. Note that the pressure scale values can be multiplied by 2, 10, 100, etc. if the volume scale values are divided by the same number.

SPECIFIC HEATS OF PURE COMPOUNDS

Specific heat = P.c.u. / (lb.) (deg. C.) = B.t.u. / (lb.) (deg. F.) = calories / (gm.) (deg. C.)

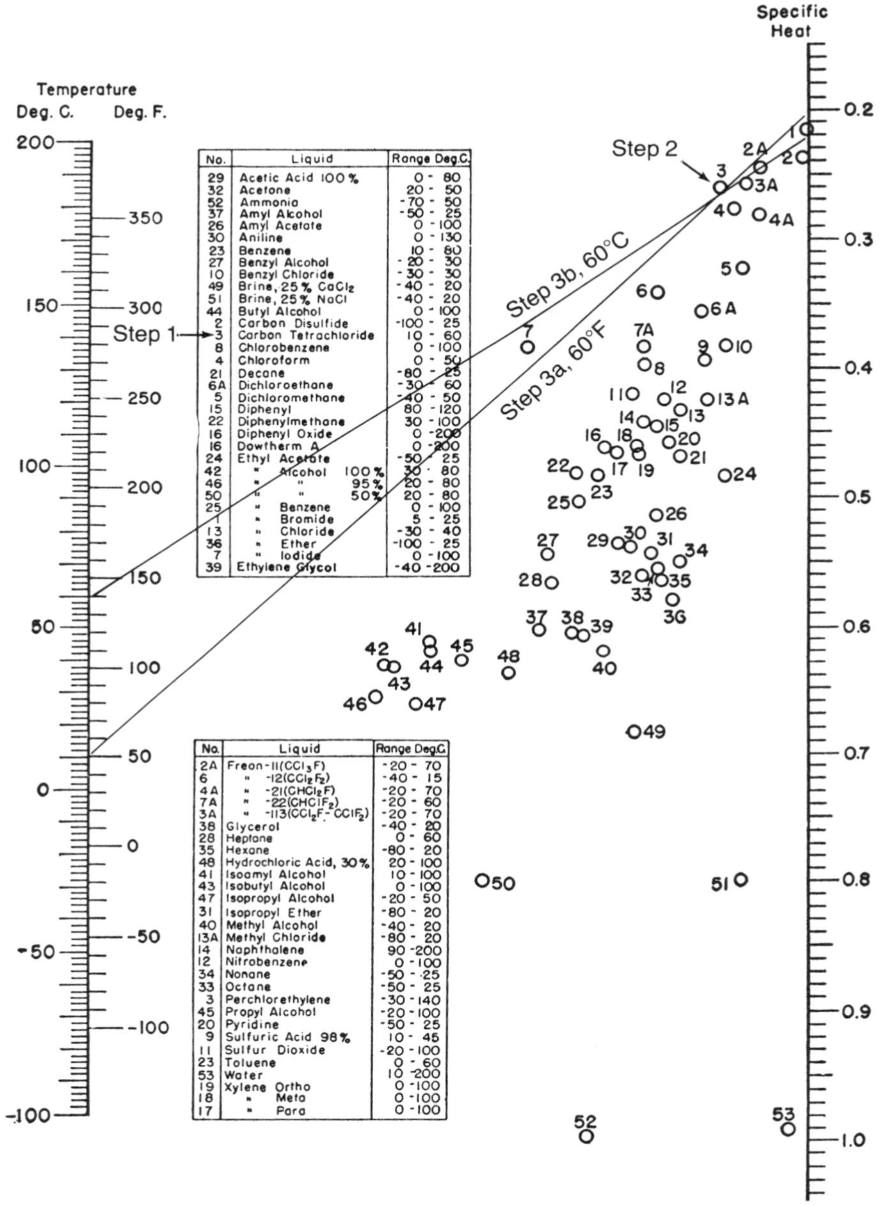

No.	Liquid	Range Deg.C.
29	Acetic Acid 100%	0 - 80
32	Acetone	20 - 50
52	Ammonia	-70 - 50
37	Amyl Alcohol	-50 - 25
26	Amyl Acetate	0 - 100
30	Aniline	0 - 130
23	Benzene	10 - 80
27	Benzyl Alcohol	-20 - 30
10	Benzyl Chloride	-30 - 30
49	Brine, 25% $CaCl_2$	-40 - 20
51	Brine, 25% NaCl	-40 - 20
44	Butyl Alcohol	0 - 100
2	Carbon Disulfide	-100 - 25
3	Carbon Tetrachloride	10 - 60
8	Chlorobenzene	0 - 100
4	Chloroform	0 - 50
21	Decane	-80 - 25
6A	Dichloroethane	-30 - 60
5	Dichloromethane	-40 - 50
15	Diphenyl	80 - 120
22	Diphenylmethane	30 - 100
16	Diphenyl Oxide	0 - 200
16	Dowtherm A	0 - 200
24	Ethyl Acetate	-50 - 25
42	" Alcohol 100%	30 - 80
46	" " 95%	20 - 80
50	" " 50%	20 - 80
25	" Benzene	0 - 100
1	" Bromide	5 - 25
13	" Chloride	-30 - 40
36	" Ether	-100 - 25
7	" Iodide	0 - 100
39	Ethylene Glycol	-40 - 200

No.	Liquid	Range Deg.C.
2A	Freon-11(CCl_3F)	-20 - 70
6	" -12(CCl_2F_2)	-40 - 15
4A	" -21($CHCl_2F$)	-20 - 70
7A	" -22($CHClF_2$)	-20 - 60
3A	" -113(CCl_2F-$CClF_2$)	-20 - 70
38	Glycerol	-40 - 20
28	Heptane	0 - 60
35	Hexane	-80 - 20
48	Hydrochloric Acid, 30%	20 - 100
41	Isoamyl Alcohol	10 - 100
43	Isobutyl Alcohol	0 - 100
47	Isopropyl Alcohol	-20 - 50
31	Isopropyl Ether	-80 - 20
40	Methyl Alcohol	-40 - 20
13A	Methyl Chloride	-80 - 20
14	Naphthalene	90 - 200
12	Nitrobenzene	0 - 100
34	Nonane	-50 - 25
33	Octane	-50 - 25
3	Perchlorethylene	-30 - 140
45	Propyl Alcohol	-20 - 100
20	Pyridine	-50 - 25
9	Sulfuric Acid 98%	10 - 45
11	Sulfur Dioxide	-20 - 100
23	Toluene	0 - 60
53	Water	10 - 200
19	Xylene Ortho	0 - 100
18	" Meta	0 - 100
17	" Para	0 - 100

Nomograph 5.38. Specific heats of liquids. From W. H. McAdams, Heat Transmission, 3rd edition. *Copyright © 1954 McGraw-Hill Book Co. Used with permission of McGraw-Hill Book Co.*

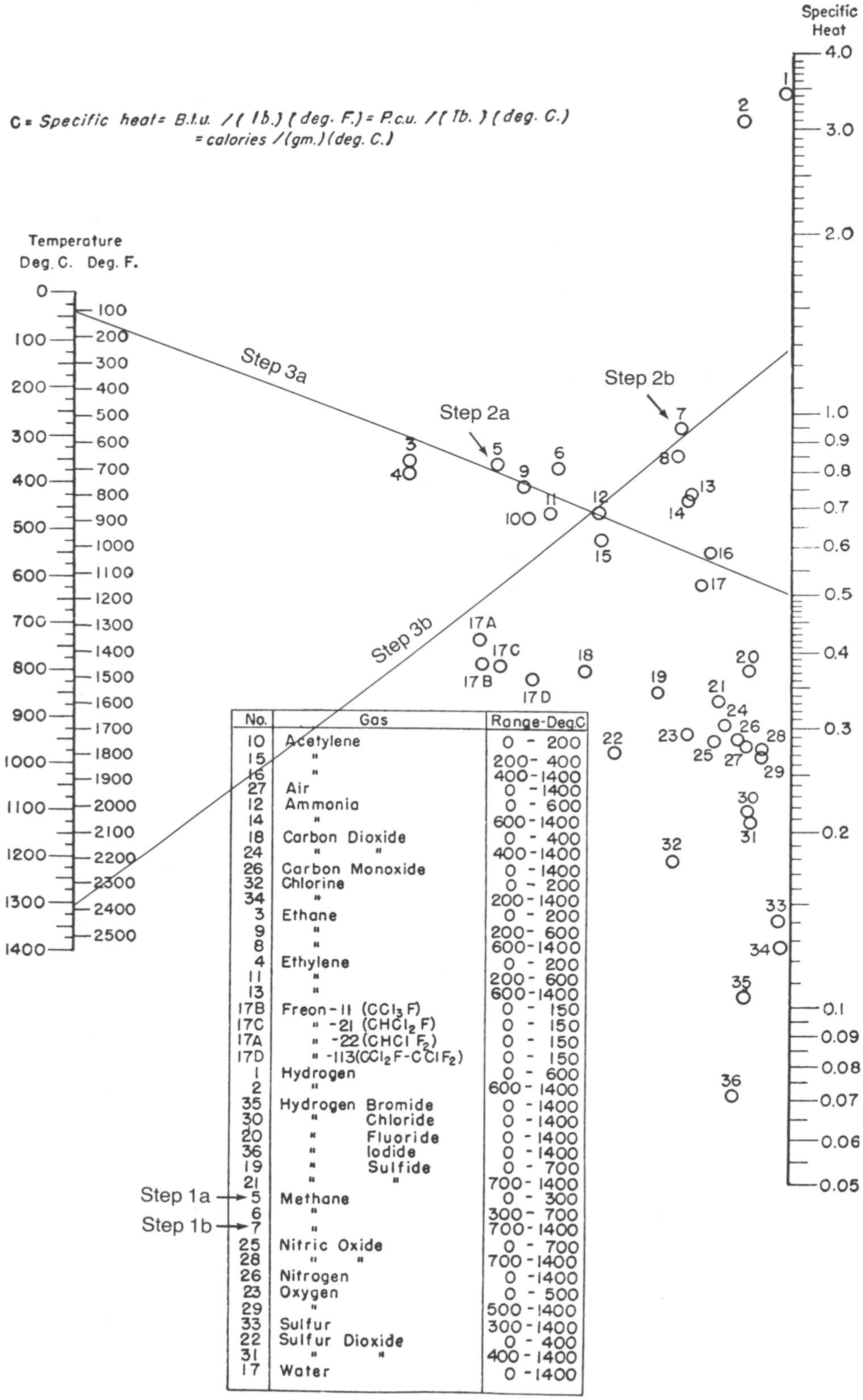

No.	Gas	Range-Deg.C
10	Acetylene	0 - 200
15	"	200 - 400
16	"	400 - 1400
27	Air	0 - 1400
12	Ammonia	0 - 600
14	"	600 - 1400
18	Carbon Dioxide	0 - 400
24	" "	400 - 1400
26	Carbon Monoxide	0 - 1400
32	Chlorine	0 - 200
34	"	200 - 1400
3	Ethane	0 - 200
9	"	200 - 600
8	"	600 - 1400
4	Ethylene	0 - 200
11	"	200 - 600
13	"	600 - 1400
17B	Freon-11 (CCl_3F)	0 - 150
17C	" -21 ($CHCl_2F$)	0 - 150
17A	" -22 ($CHClF_2$)	0 - 150
17D	" -113 (CCl_2F-$CClF_2$)	0 - 150
1	Hydrogen	0 - 600
2	"	600 - 1400
35	Hydrogen Bromide	0 - 1400
30	" Chloride	0 - 1400
20	" Fluoride	0 - 1400
36	" Iodide	0 - 1400
19	" Sulfide	0 - 700
21	" "	700 - 1400
5	Methane	0 - 300
6	"	300 - 700
7	"	700 - 1400
25	Nitric Oxide	0 - 700
28	" "	700 - 1400
26	Nitrogen	0 - 1400
23	Oxygen	0 - 500
29	"	500 - 1400
33	Sulfur	300 - 1400
22	Sulfur Dioxide	0 - 400
31	" "	400 - 1400
17	Water	0 - 1400

Nomograph 5.39. Specific heats (C_p) of gases at 1 atm pressure. From W. H. McAdams, Heat Transmission, 3rd edition. *Copyright © 1954 McGraw-Hill Book Co. Used with the permission of McGraw-Hill Book Co.*

Determining Specific Heats for Liquids and Gases

Nomographs 5.38 and 5.39 solve for specific heats of certain liquids and gases. Specific heat is defined thusly:

1. The ratio of the amount of heat required to raise a mass of material one degree in temperature to the amount of heat required to raise an equal mass of a reference substance, usually water, one degree in temperature; both measurements are made at a reference temperature, usually at constant pressure or constant volume.
2. The quantity of heat required to raise a unit mass of homogeneous material one degree in temperature in a specified way; it is assumed that during the process no phase or chemical change occurs.

The following equation defines specific heat:

$$Q = MC\,\Delta T$$

Where:

Q = heat (Btu)
M = mass (1bm)
C = specific heat (Btu/1bm°F)
ΔT = final temperature/initial temperature, (°F)

Example: What is the specific heat of liquid carbon tetrachloride which is at (a) 60°F, or at (b) 60°C?

(a) For Step 1, find carbon tetrachloride in one of the tables on Nomograph 5.38. In the left-hand column adjacent to carbon tetrachloride is the Number 3. Find the small circle associated with 3 on the nomograph, Step 2. For Step 3, locate 60°F on the temperature line; draw a line through circle Number 3 and intersecting the specific heat line; read $C = 0.20$.
(b) Repeat Steps 1 and 2 for carbon tetrachloride at 60°C. For Step 3b locate 60°C on the temperature line; draw a line through circle Number 3 and intersecting the specific heat line; read $C = 0.22$.

Example: What is the specific heat of methane which is at (a) 100°F, or at (b) 1300°C.

(a) For Step 1a, find methane in the table on Nomograph 5.39, and notice there are three entries for methane at different temperature ranges. For methane at 100°F, use the0-300°C. (same as 32-572°F) range. For Step 2a, locate the small circle associated with Number 5 on the nomograph. The third and final step is to find 100°F on the temperature scale and construct a line through circle Number 5 to the specific heat line; read $C = 0.55$.
(b) For Step 1b, locate methane on Nomograph 5.39, which is in the 700-1400°C range. Next locate the small circle associated with the Number 7, Step 2b. For Step 3b, draw a line from 1300°C on the temperature scale through circle Number 7 to the specific heat line; read $C = 1.32$.

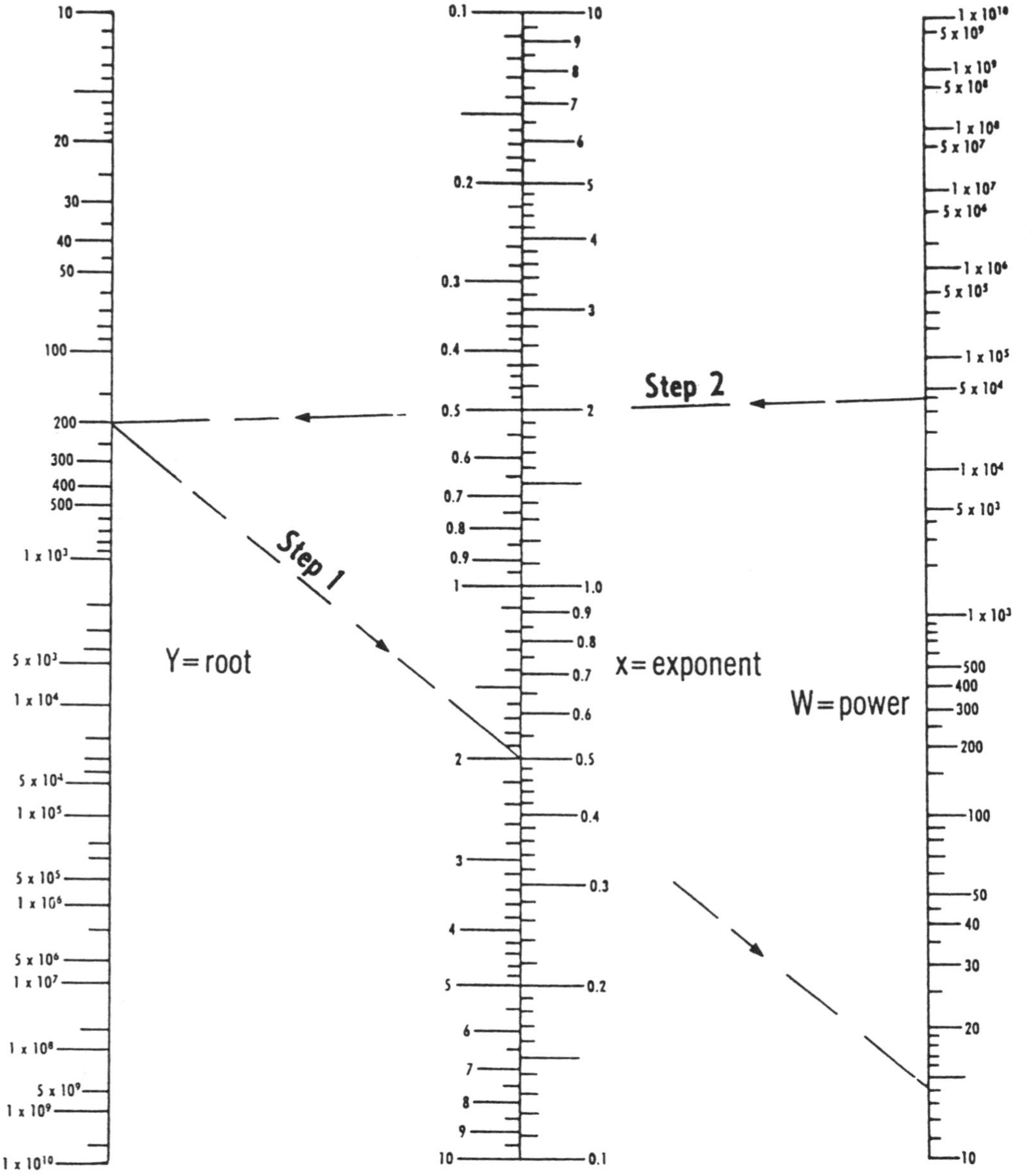

Nomograph 5.40. Determining powers and roots. Courtesy Oil, Gas, and Petrochem Equipment.

Determining Powers and Roots

Nomograph 5.38 can be used to solve for powers and roots of numbers. It is based on the equations,

$$W = Y^x$$
$$Y = W^{1/x}$$

where

W = Y raised to the x power
Y = root of W
x = power or root of number

Example. What is the square root of 200 (x = 0.5 or ½)? Step 1, align Y = 200 with x = 0.5 and read W = 14.1.

What is the value of 200 raised to the second power (squared)? Step 2, align W = 200 with x = 2 and read $Y = 4 \times 10^4$ = 40,000.

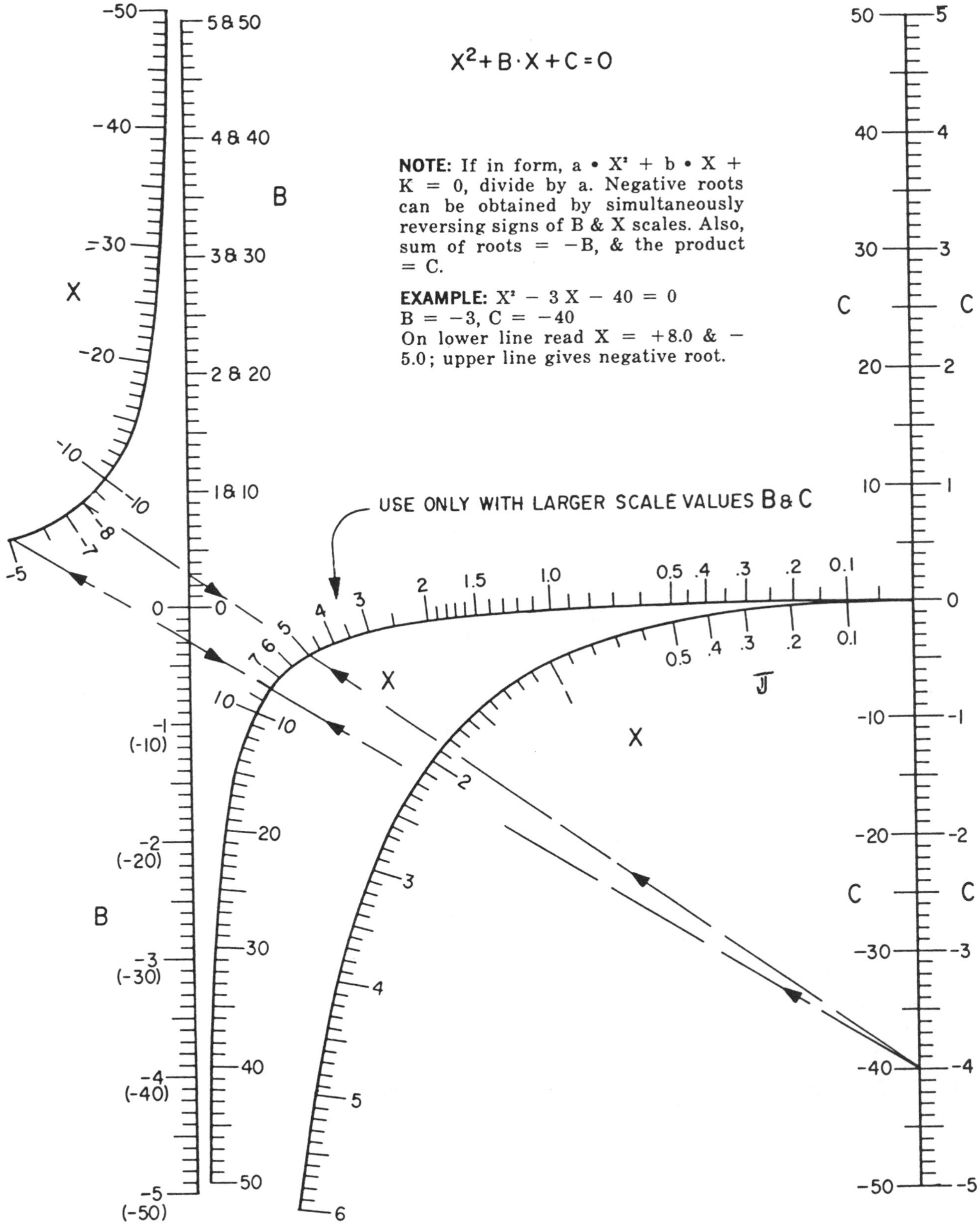

Nomograph 5.41. Solution of quadratic equations. Courtesy Chemical Publishing Company, Inc.

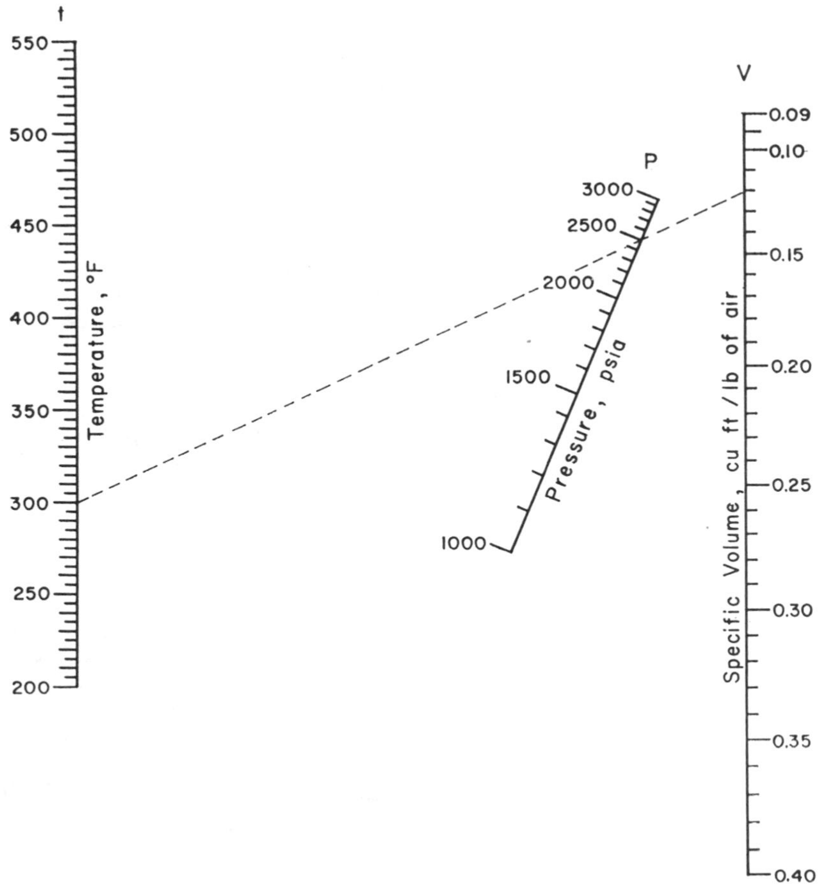

Nomograph 5.42. Determination of specific volume of air at high temperatures and pressures. Courtesy Chemical Publishing Company, Inc.

Determination of Specific Volume of Air at High Temperatures and Pressures

Nomograph 5.42 presents a convenient method for presentation and use of P-V-T data for air at high temperatures and pressures. It is based on equation:

$$0.01t + 2.795 = V^{1.25} [P/61.5]^{1.875}$$

where

t = temperature, °F
V = specific volume, cu ft/lb of air
P = pressure, psia

Example. What is the specific volume of air at 300°F and 2,500 psia? Join 300 on t-scale with 2,500 on P-scale and read value of 0.121 cu ft/lb of air on V-scale.

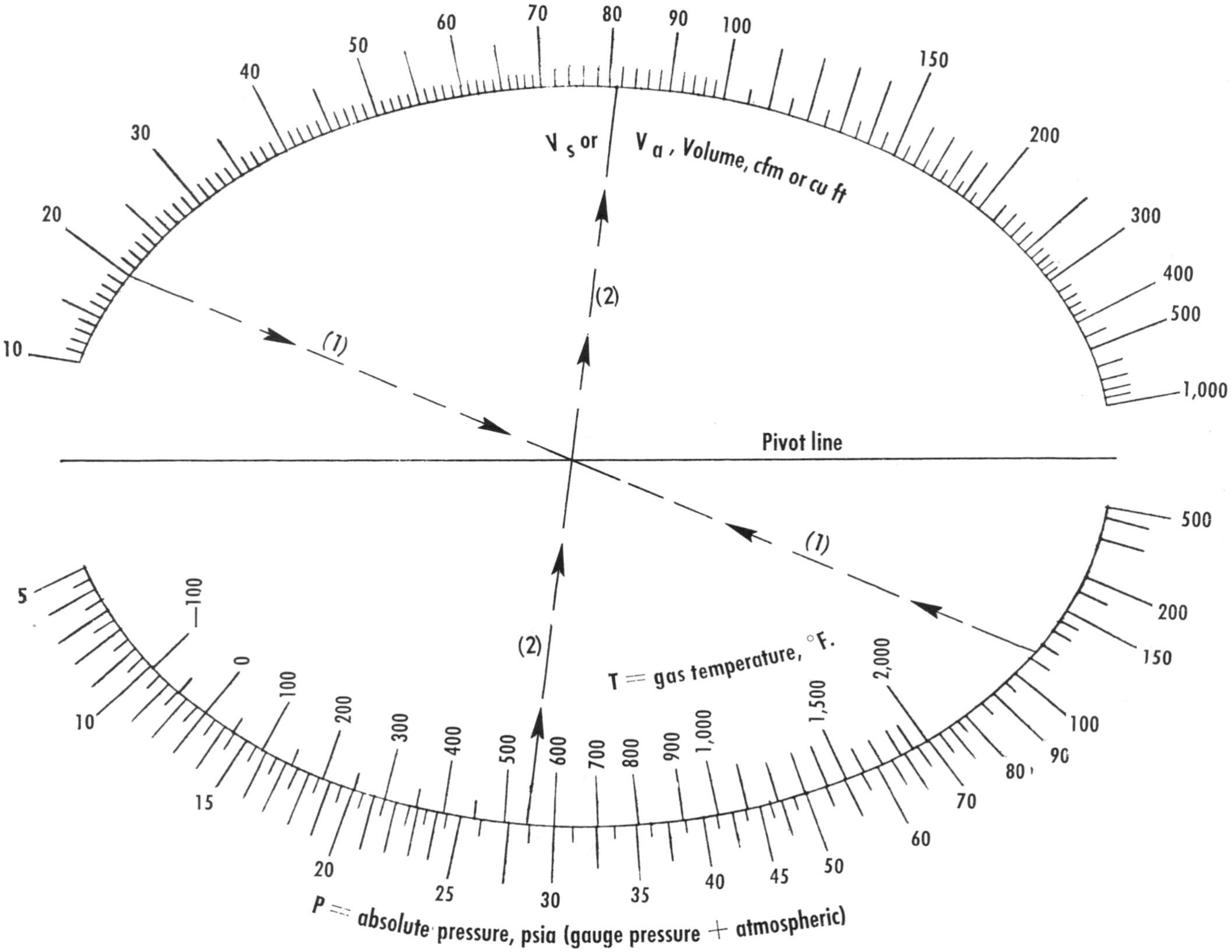

Nomograph 5.43. Conversion of actual gas volume to standard volume. Courtesy Oil, Gas, and Petrochem Equipment.

Conversion of Actual Gas Volume to Standard Volume

Nomograph 5.43. determines the volume or flow rate of a gas at standard conditions or at actual conditions. It solves the equation:

$$V_s = V_a(P/14.7)[520/(T+460)]$$

where

V_s = standard volume, cfm or cu ft, at 14.7 psia and 60°F.
V_a = volume, cfm or cu ft, at actual pressure, psia, and temperature, °F.
P = actual pressure, psia.
T = actual temperature, °F.

Example. What is the flow rate in standard cfm of a gas flowing at the rate of 20 cfm at 100 psig (114.7 psia) and 540°F.?

To determine this, use a straightedge and, Step (1), connect actual pressure, 114.7 psia, on the pressure scale with flow rate at actual conditions, 20 cfm, on the volume scale. Mark the intersection of the straightedge with the pivot line. Then, Step (2), connect the mark on the pivot line with the actual gas temperature, 540°F. on the temperature scale. Read 81 scfm (14.7 and 60°F.) on the volume scale.

By reversing this procedure, the flow rate can be determined at actual conditions. The scales may be extended by simultaneously multiplying the actual and standard volumes or flow rates by such values as 0.1, 2, 10, 100, etc.

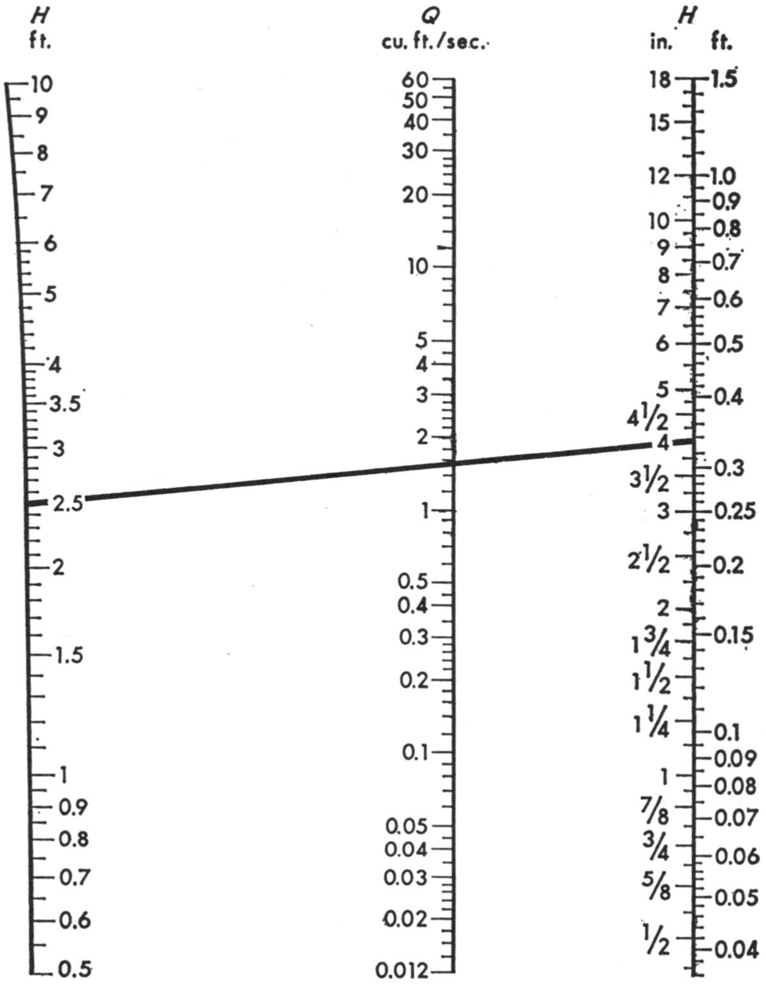

Nomograph 5.44. Calculating flow in rectangular weirs. Reprinted by special permission from Chemical Engineering, *copyright © McGraw-Hill, Inc. New York, N.Y. 10036.*

Calculating Flow in Rectangular Weirs

Nomograph 5.44 is based on the formula:

$$Q = 3.08LH^{1.46 + 0.003L}$$

where

Q = rate of flow, cu.ft./sec.;
L = length of weir crest, ft.;
H = head of inlet liquid above weir.

Example. To find the flowrate of a system where $L = 2.5$ ft. and $H = 4$ in. (0.333 ft.), connect 2.5 on the L scale with 0.333 ft. on the H scale, and read the answer, 1.57 cu.ft./sec., in the Q scale.

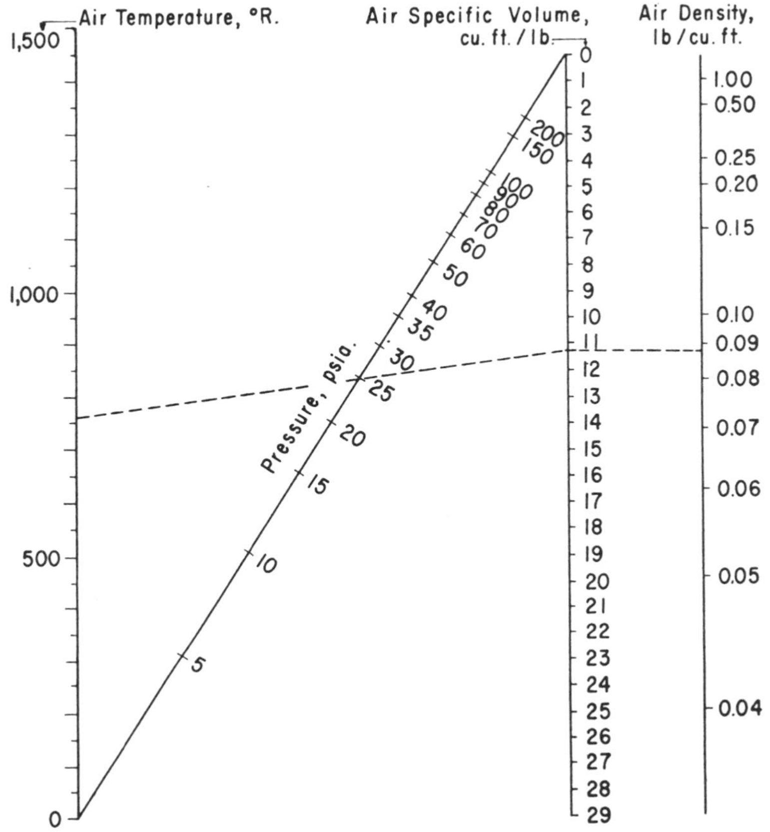

Nomograph 5.45. Determining volume or density. Reprinted by special permission from Chemical Engineering, *copyright © McGraw-Hill, Inc. New York, N.Y. 10036.*

Determining Volume or Density

Determination of the specific volume or density of air, when the pressure and temperature are known, is normally quite simple. However, as with most simple calculations, the solving of a large number of such problems can become an extremely tedious chore.

Engineers who must deal with heating and ventilation problems and compressed air production, transmission and processes requiring large quantities of air may find Nomograph 5.45 quite useful in relieving the tedium of many small, simple, time-consuming calculations.

The nomograph shows temperature in degrees R. at the left and two scales for specific volume, in lb_m/ft^3, and density, in lb_m/ft^3, at the right. Pressure, in lb_f/in^2 absolute, is depicted on a diagonal line between the temperature and specific volume scales.

Knowing the temperature and the absolute pressure, lay a straightedge on these points on the appropriate scales. Read the specific volume and/or density from the scales at right.

For example, if the temperature of the air is 300°F. and the pressure is 10 psig., convert to degrees R. and psia. Corrected temperature is then 760 R., and the pressure becomes 24.7 psia. Lay a straightedge between these points and read specific volume = 11.4 cu.f./lb. and density = 0.088 lb./cu.ft.

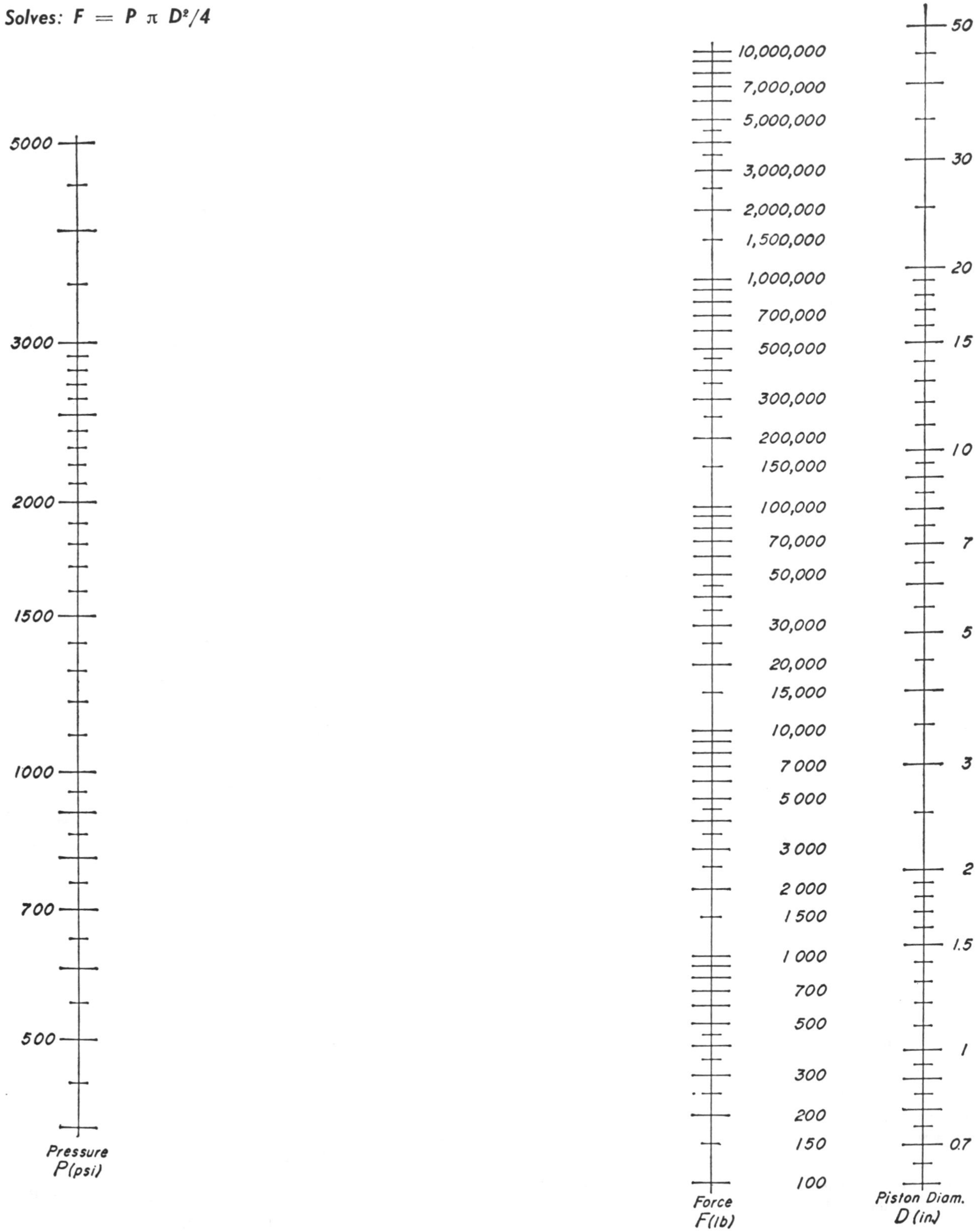

Nomograph 5.46. Cylinder force. Courtesy Fluid Power Handbook & Directory, *1972/1973.*

Solves: $V = 4Q/\pi D^2$

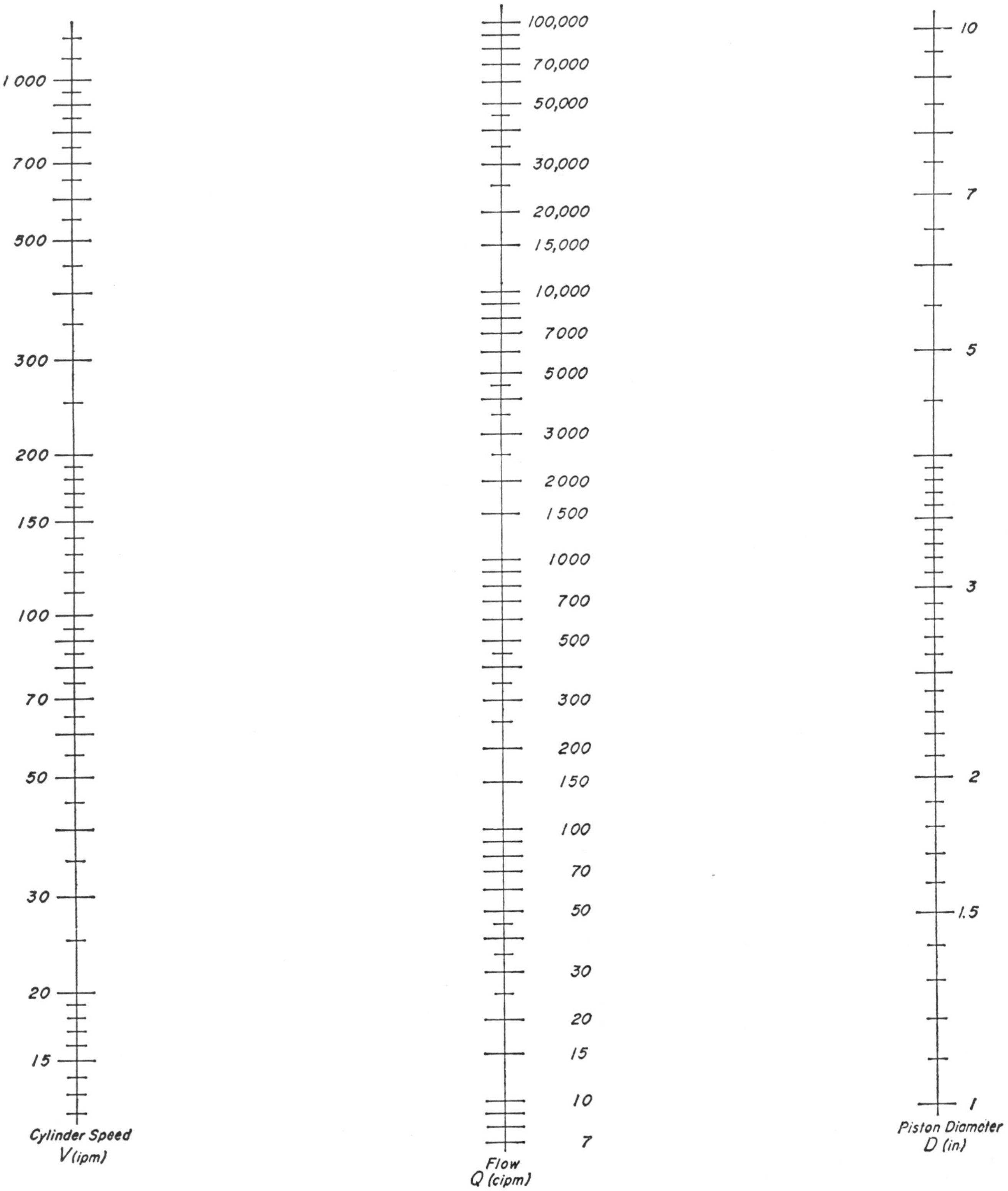

Nomograph 5.47. Cylinder speed. Courtesy Fluid Power Handbook & Directory, *1972/1973.*

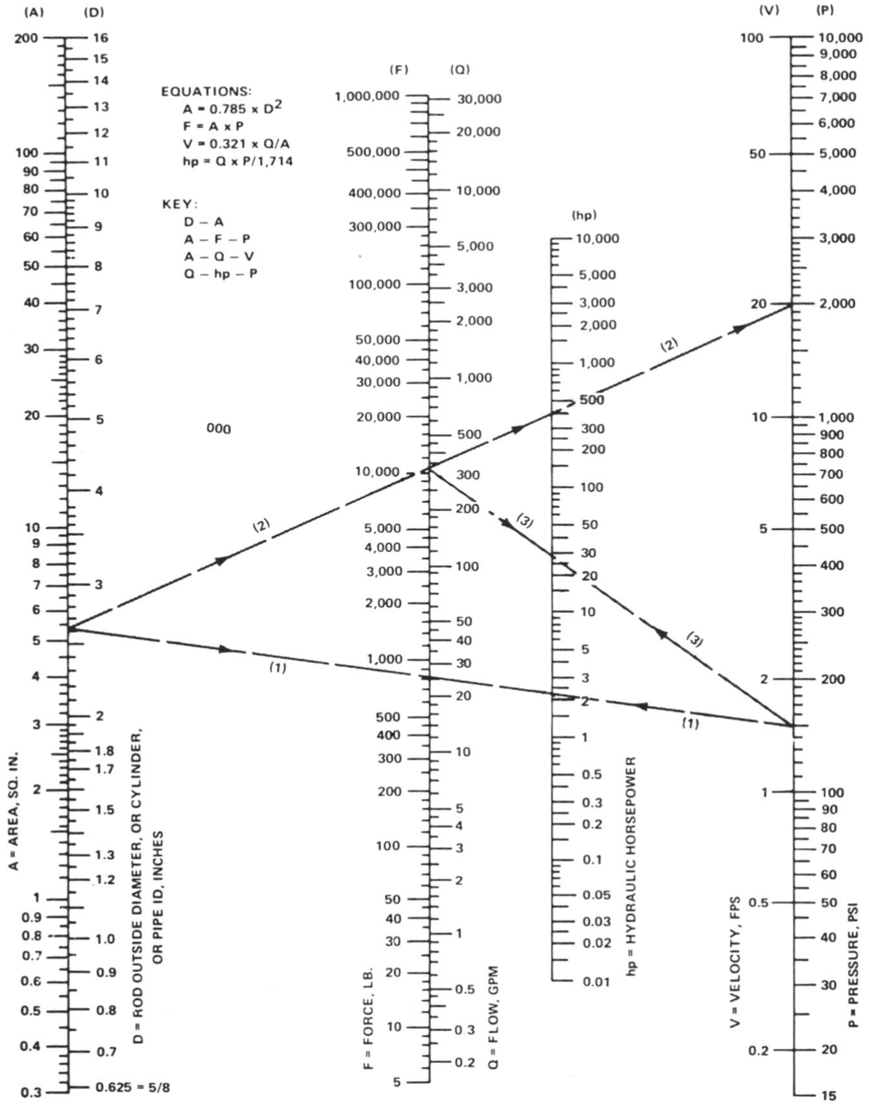

Nomograph 5.48. Sizing hydraulic cylinders. Courtesy Fluid Power Handbook & Directory, *1972/1973.*

Nomograph for Sizing Hydraulic Cylinders

This nomograph is for finding area, force, velocity and horsepower. With nomograph 5.48 determine:

- Effective area of a cylinder (bore area minus rod area).
- Force developed by a cylinder.
- Velocity of a liquid in the cylinder piston (or pipe).
- Theoretical horsepower developed by a cylinder (or required by a pump).

For example, assume that 350 gpm of fluid at 150 psi flows into a 3-inch ID cylinder with a 1.5-in. OD piston rod. Find the effective area, force, fluid velocity, and hydraulic horsepower.

To find the effective area, locate diameter 3 on scale (D) and read 7.1 sq. in. area on scale (A). Next, locate 1.5 dia. on scale (D) and read area of 1.78 sq. in. on scale (A). The effective area is $7.1 - 1.78 = 5.32$ sq. in.

To find the force developed, align 5.32 on scale (A) with 150 on scale (P), the answer is on scale (F) 800 pounds.

Fluid velocity is found by aligning 5.32 on scale (A) with 350 on scale (Q). Velocity is on scale (V): 20 fps.

To find horsepower, align 350 on scale (Q) with 150 on scale (P), the answer appears on scale (hp): 31 horsepower.

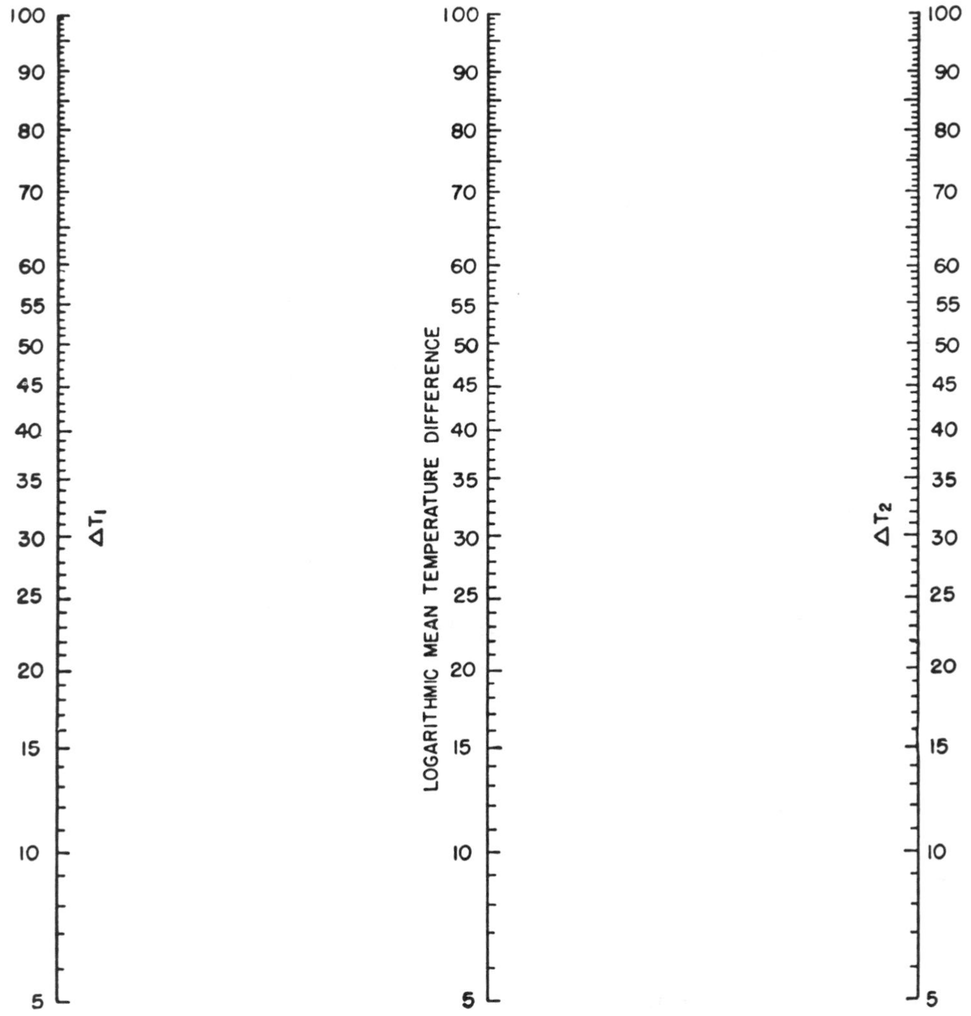

$$M.T.D. = \frac{\Delta T_1 - \Delta T_2}{LOG_E \frac{\Delta T_1}{\Delta T_2}}$$

WHEN ΔT_1 AND ΔT_2 ARE NOT WITHIN THE CHART RANGE, THESE VALUES MAY BE MULTIPLIED BY A FACTOR, AS 0.5, 2, 10, OR 100, ETC.

FOR EXAMPLE:

GIVEN ΔT_1 (ACTUAL) = 200, AND ΔT_2 (ACTUAL) = 20.

USE 0.5 AS FACTOR, AND $\Delta T_1^{\prime} = 100$, AND $\Delta T_2^{\prime} = 10$.

FROM THE CHART, $M.T.D.^{\prime} = 39.5$

OR M.T.D. (ACTUAL) = $\frac{1}{0.5} \times 39.5 = 79$.

Nomograph 5.49. Logarithmic mean temperature difference. Data Book on Hydrocarbons, *J.B. Maxwell, copyright © 1950 Litton Educational Publishing, Inc. Reprinted by permission of Van Nostrand-Reinhold Company.*

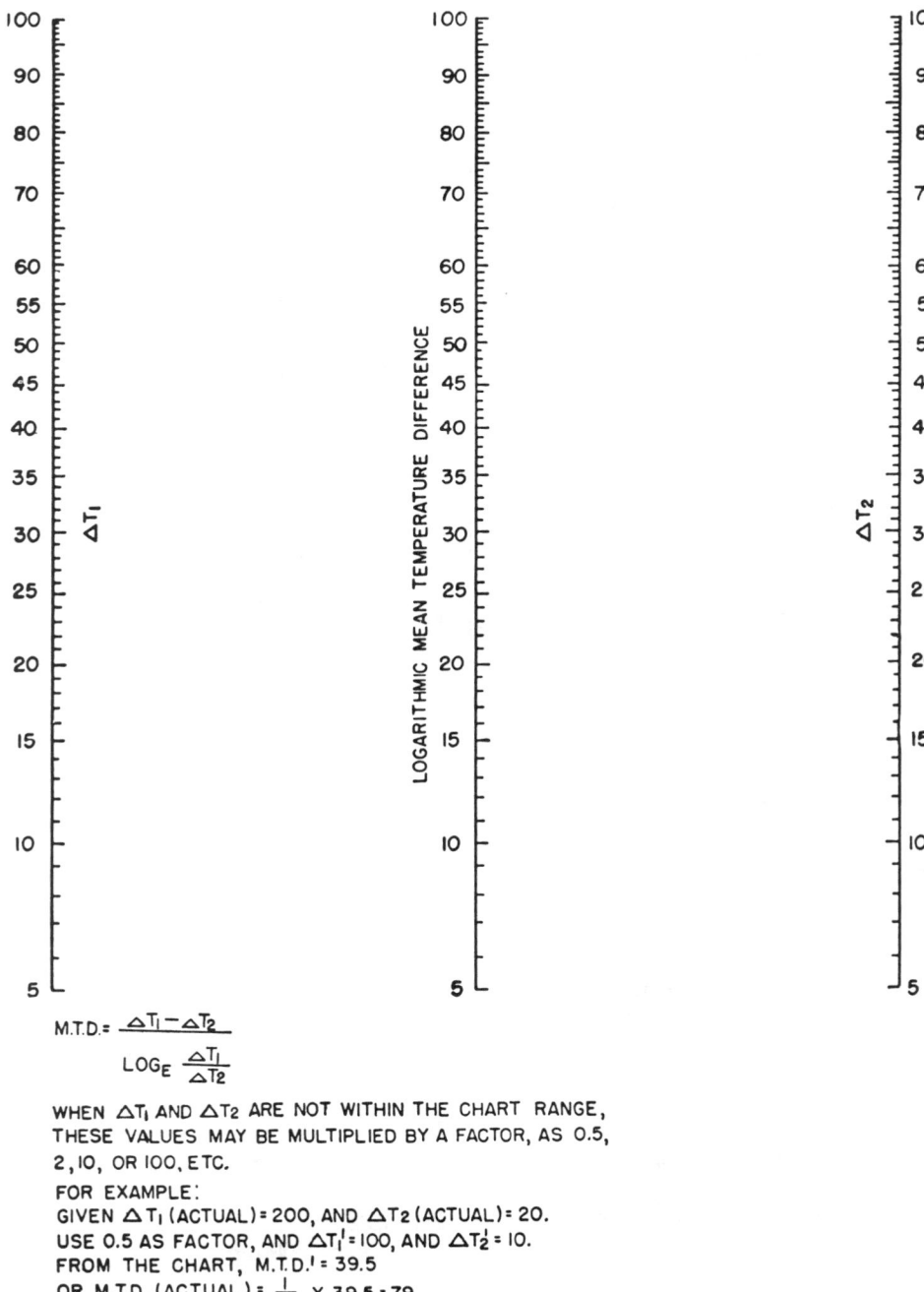

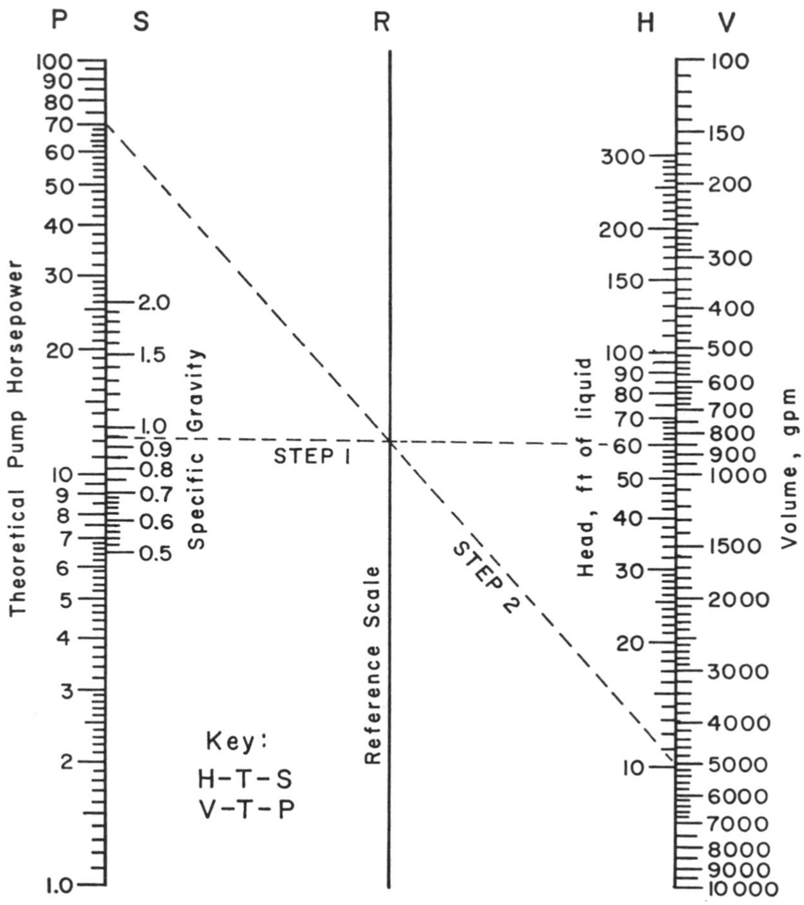

Nomograph 5.50. Theoretical pump horsepower. Courtesy Chemical Publishing Company. Inc.

Theoretical Pump Horsepower

Nomograph 5.50 provides a graphical solution for the following equation:

$P = VHS/3960$

where

P = theoretical pump horsepower
V = volume of liquid, gpm
H = head, ft of liquid
S = specific gravity of liquid

Example.
V = 5,000 gpm
S = 0.95
H = 60 ft

Connect 0.95 on the S-scale and 60 on the H-scale with a straight line. Note the point where this line intersects the turning axis. Extend a straight line from 5,000 on the V-scale through the point just found until the line intersects the P-scale at 72 horsepower.

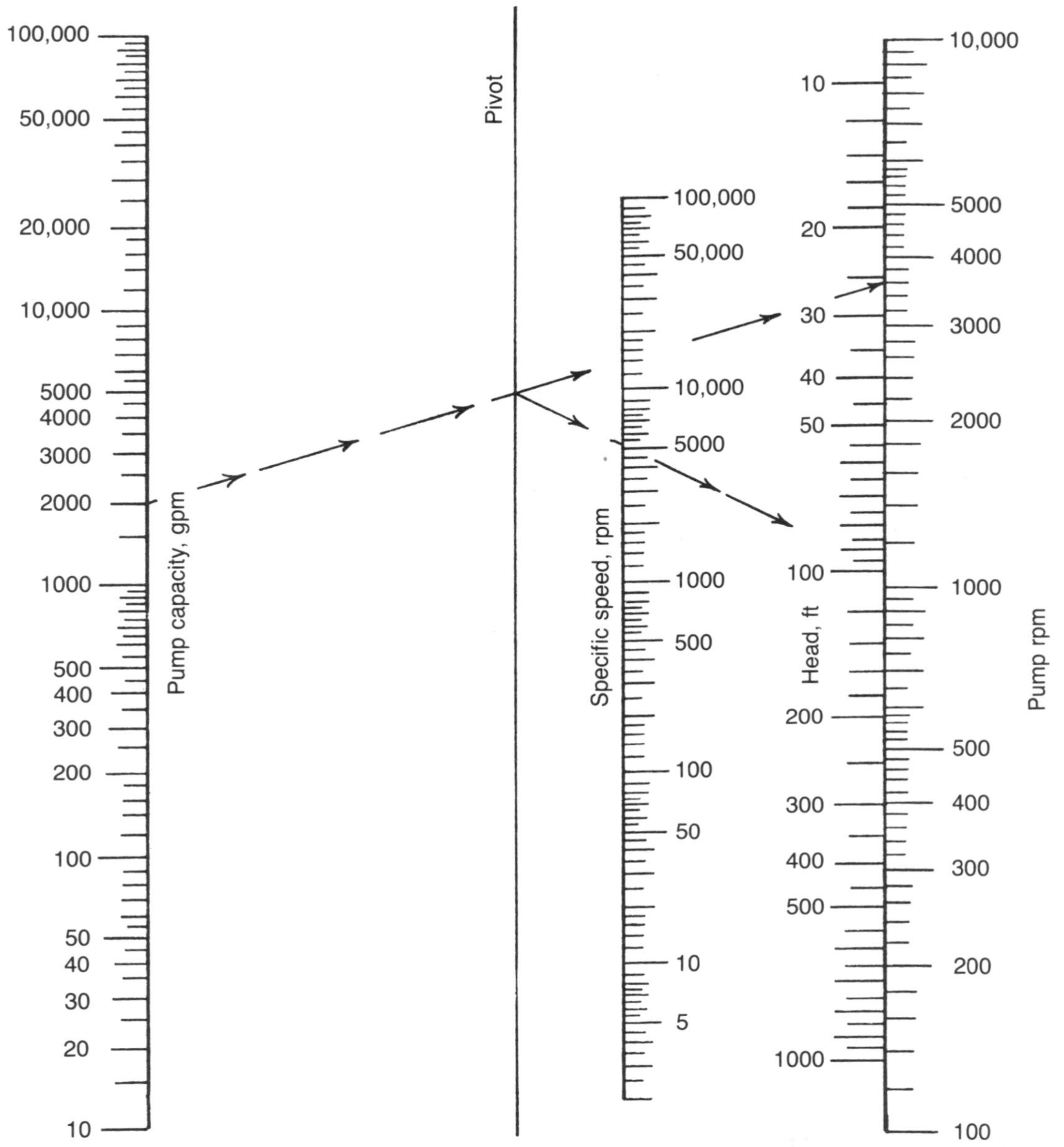

Nomograph 5.51. Determine specific speed of centrifugal pumps. Courtesy Oil, Gas, and Petrochem Equipment.

Determine Specific Speed of Centrifugal Pumps

One of the most important measures of centrifugal-pump performance is the specific speed. This quantity is widely used in the selection and application of all types of centrifugal pumps. But since computation of the specific speed of a pump involves taking the square root of the flow rate, and raising the head to the 0.75 power, the calculation can be lengthy, particularly when a number of pumps are involved. Nomograph 5.51 simplifies all specific-speed calculations.

Example. What is the specific speed of centrifugal pump handling 2,000 gpm at 3,600 rpm if the head is 100 ft. of liquid?

Solution. Enter the chart on the left at 2,000 gpm and draw a straight line through 3,600 rpm on the right. From the intersection of this line with the pivot, draw a straight line through the head, 100 ft., on the right. At the intersection with the specific-speed scale, read the result as 5,040 rpm.

This chart is based on $N = G^{1/2}\, r/H^{0.75}$, where N = specific speed, revolutions per minute; G = pump capacity, gallons per minute; r = pump speed, revolutions per minute; H = head developed by pump, feet of liquid.

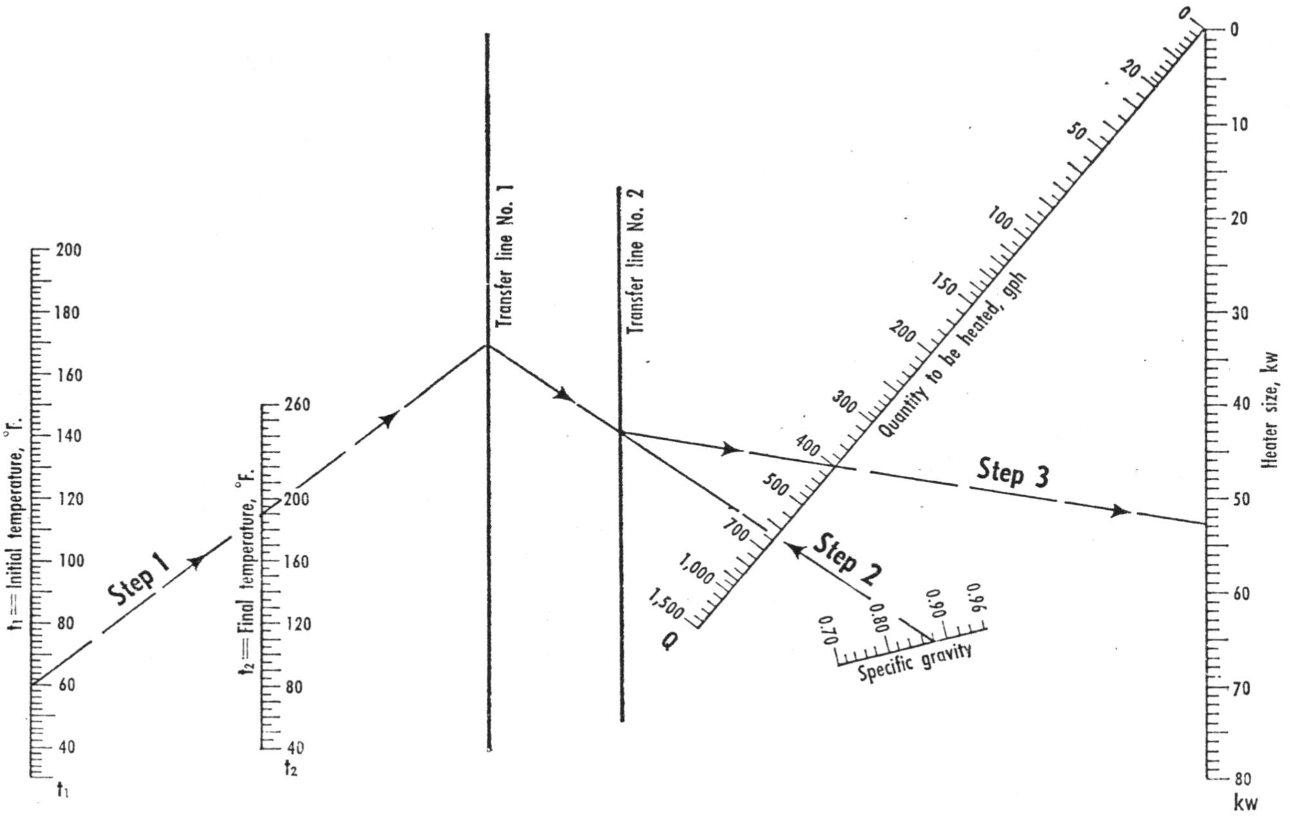

Nomograph 5.52. Size electric heaters for petroleum products. Courtesy Oil, Gas, and Petrochem Equipment.

Size Electric Heaters for Petroleum Products

Nomograph 5.52 can be of considerable assistance when it is necessary to select an electric heater for heating various petroleum products. The nomograph is constructed from the formula:

$$H = Q\, d^{1/2}[3.235 + 0.00185\,(t_2 + t_1)]\,(t_2 - t_1)/[3{,}600 \times 0.9486]$$

where

H = heater size, kw
Q = quantity to be heated, gph
t_1 = initial temperature, °F.
t_2 = final temperature, °F.
d = specific gravity (60/60 °F.)

The formula is based on the formula for variable specific heat:

$$c=(1/d^{1/2})(0.388+0.0045t) \text{ BTU/lb/°F.}$$

Example. Determine the *kw* required to heat 400 gph of oil with a specific gravity of 0.88 from an initial temperature of 60°F. to a final temperature of 190°F.

Solution. Line from 60°F. on t_1 scale with 190°F. on t_2 scale (Step 1), extend line to Transfer line no. 1 and mark. From this marked point, align with 0.88 on Sp.Gr. scale (Step 2) and mark where line intersects Transfer line no. 2. From there, line with 400 gpm on Q scale (Step 3), extend line to *kw* scale, and read 52.85 *kw* required.

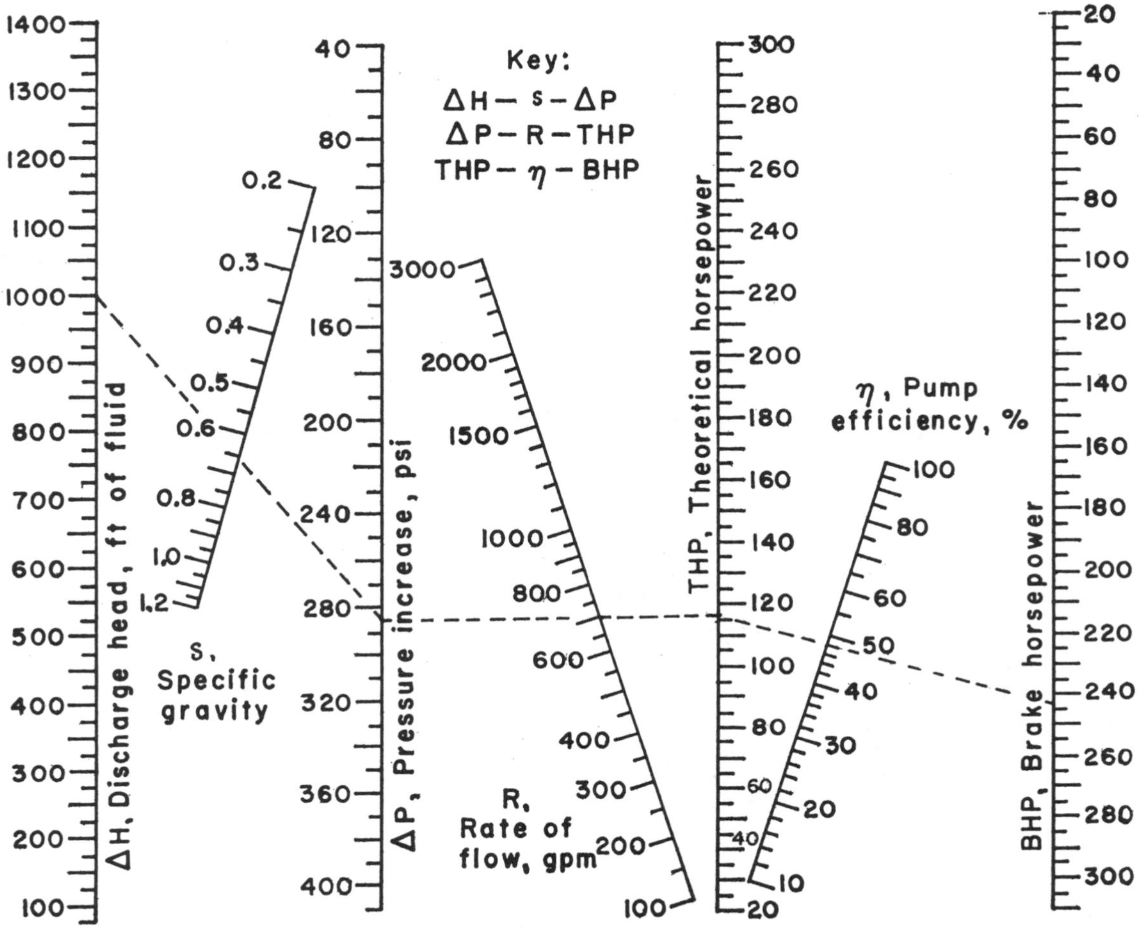

Nomograph 5.53. Sizing of pumps. Courtesy Chemical Publishing Company, Inc.

Sizing of Pumps

Nomograph 5.53 for sizing pumps quickly is based on the equations:

$$\Delta H = 2.31\ \Delta P/s$$

$$THP = (8.33\ Rs\ \Delta H/33{,}000) = 5.83(10^{-4})R\Delta P$$

$$THP = BHP/100$$

where
ΔH = discharge head, ft of fluid
ΔP = pressure increase across pump, psi
s = specific gravity of fluid
R = rate of discharge, gpm
η = pump efficiency, %

THP and *BHP* are theoretical and brake horsepowers, respectively.

Example. Hexane (specific gravity, 0.66) is pumped at 700 gpm, with a discharge head of 100′ and a pump efficiency of 48%. Find the pressure increase and the theoretical and brake horsepowers.

For convenience, multiply the head by 10, obtaining 1,000. Connect 1,000 on the ΔH-scale and 0.66 on the s-scale with a straight line. Read the pressure increase as 286 psi on the ΔP-scale. Connect this point and 700 gpm on the R-scale with a straight line. Read 117 as the theoretical horsepower on the *THP*-scale. Connect this point and 48 on the η -scale with a straight line, and find the brake horsepower to be 243 on the *BHP*-scale. Divide 243 by 10 to obtain 24.3 as the brake horsepower for 100′ of head.

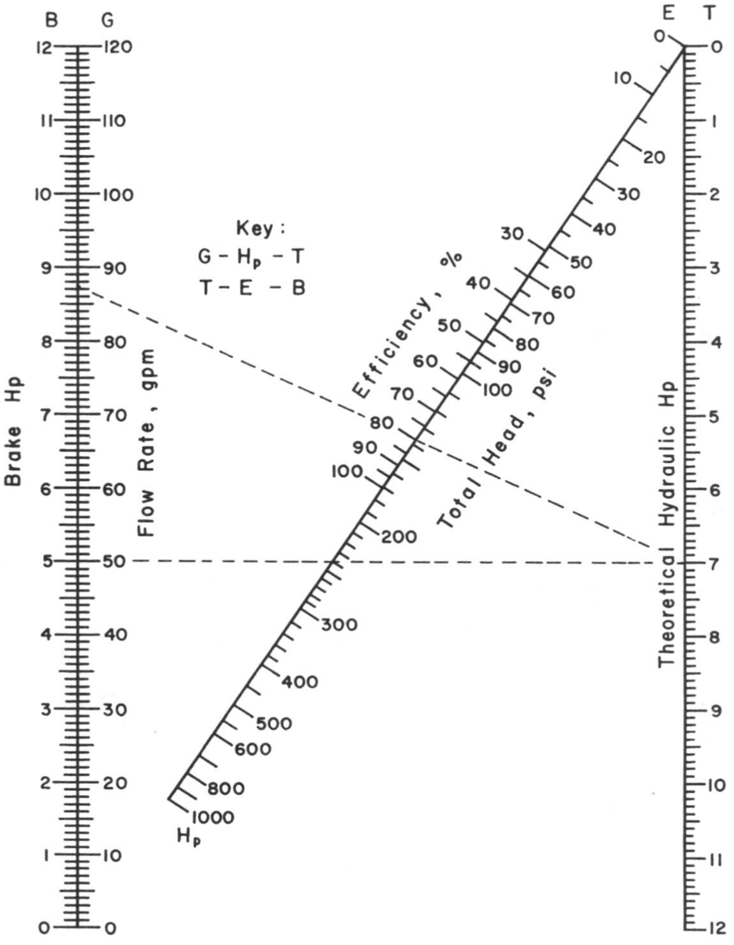

Nomograph 5.54. Sizing pump motors. Courtesy Chemical Publishing Company, Inc.

Sizing Pump Motors

The theoretical hydraulic horsepower of a 100%-efficient pump is given by the equation:

$$T = GH_p/1714$$

where

T = theoretical hydraulic horsepower
G = flow, gpm
H_p = total head, psi

The actual or brake horsepower is given by the equation:

$$B = T/E$$

where

B = brake horsepower
E = pump efficiency

Example. What bhp is required for an 80%-efficient pump to deliver 500 gpm at 240-psi head? Align $G=500=10(50)$ with $H_p=240$ and read $T=(10)7=70$; align T with $E=80$ and read bhp$=(10)$ $8.75=87.5$.

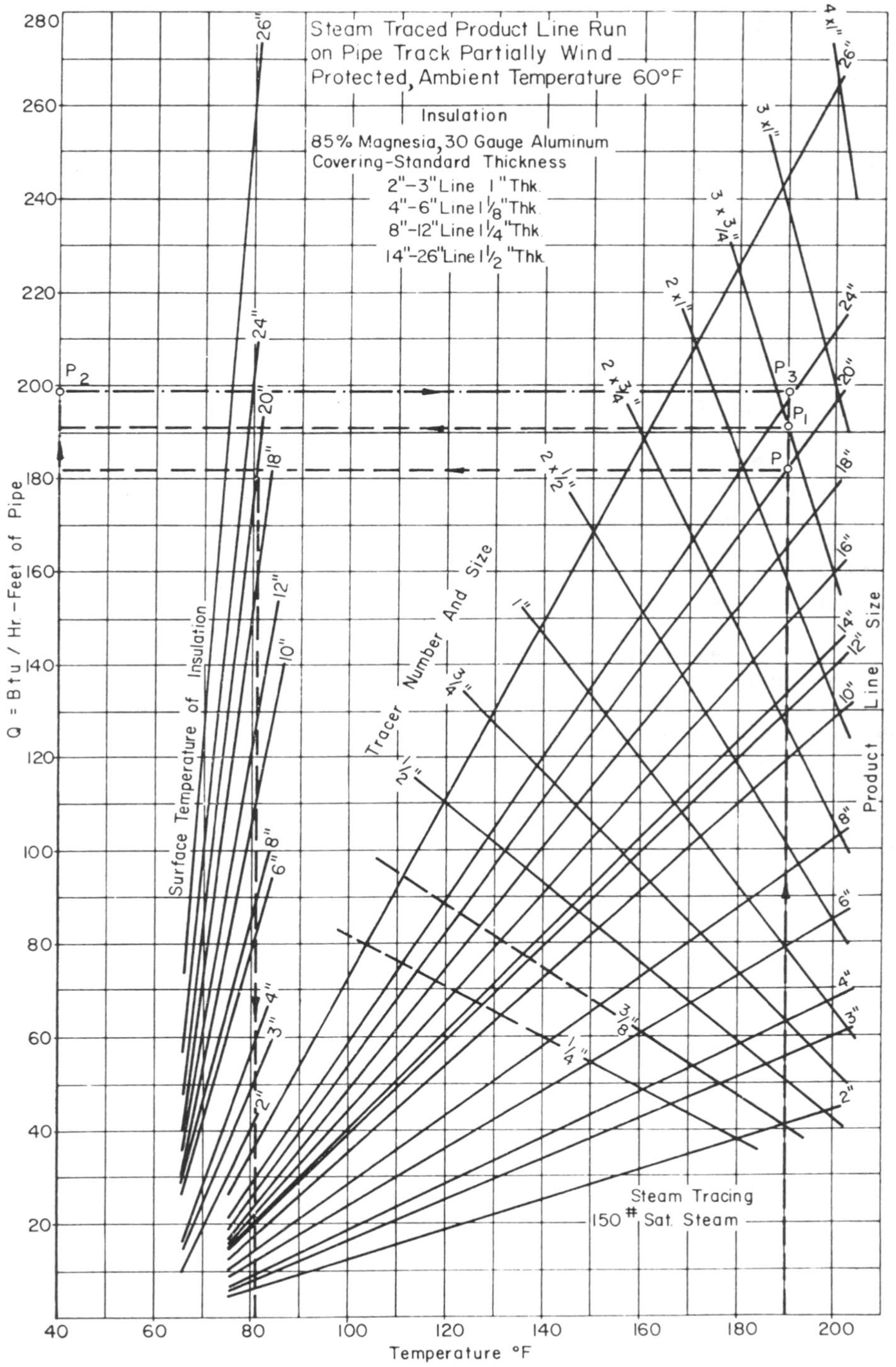

Nomograph 5.55. Sizing steam tracer lines. Courtesy Gulf Publishing Co.

Sizing Steam Tracer lines

Nomograph 5.55 may be used to determine:

- The number and size of steam tracer.
- Heat loss from a traced product line.
- Heat transfer from tracer to product.
- Outside temperature of insulation.

For Product Lines at Ground Level (Wind Protected)

Example. A product line 20 inches in size is to be held at 190°F.

To find tracer size: Enter temperature scale at 190°F and locate intersection P at 20-inch pipe size. Select tracer nearest above this intersection point, i.e., 3¾ inches at P_1.

To find heat loss from product line: From intersection P read off heat loss on Q scale, i.e., 182 Btu per hour per foot.

To find heat transfer from tracer: From intersection P_1 read off heat transfer on Q scale, i.e., 191 Btu per hour per foot.

To find outside temperature of insulation: The line from the intersection P to Q scale cuts 20-inch line on surface temperature scale at 81°F.

Correction for overhead lines (exposed 10-mph wind): Add 10 percent to heat losses derived as in "Heat Loss from product line" above (Point P_2). Extend horizontally to intersect temperature line (P_3) and read off new tracer number and size, i.e., 3-1 inch.

The diagram was checked over a number of weeks on existing installations as follows:

- On pipes in open pipeways 2 feet below ground surface, with 10-mph wind and partially protected by other pipe.
- Overhead pipes 3 to 10 feet above ground level—10-mph wind. Fully exposed.

The practical result came exceedingly close to the diagram.

The following equations were used for calculation:

$$Q_1 = U \times A$$
$$U = t_i - t_a / \{[r_2 \log (r_2/r_1)]/K + [(1/(h_c + h_r)A]\}$$
$$Q_2 = [(t_s - t_i) - 0.33\,(t_i - t_a)]d_o/1.32$$
$$T_{ins} = [Q_1/(h_c + h_r)A] + ta$$

Nomenclature

Q_1 = Heat loss, Btu per hour-foot of product line.
U = Heat loss, Btu per square foot of outer surface per hour.
A = Outer surface, square feet per feet of line.
t_1 = Product line inside temperature or holding temperature, °F.
t_a = Ambient temperature, °F.
r_2 = Radius of outer surface, inches.
r_1 = Radius of bare product line, inches.
K = Thermal conductivity of insulation material, Btu per hour per square foot per °F of one inch.
t_s = Average steam temperature, °F.
do = Outside diameter of tracer, inches.
h_c = Convection coefficient, Btu per hour per square foot per °F.
h_r = Radiation coefficient, Btu per hour per square foot per °F.
T_{ins} = Surface temperature of insulation, °F.

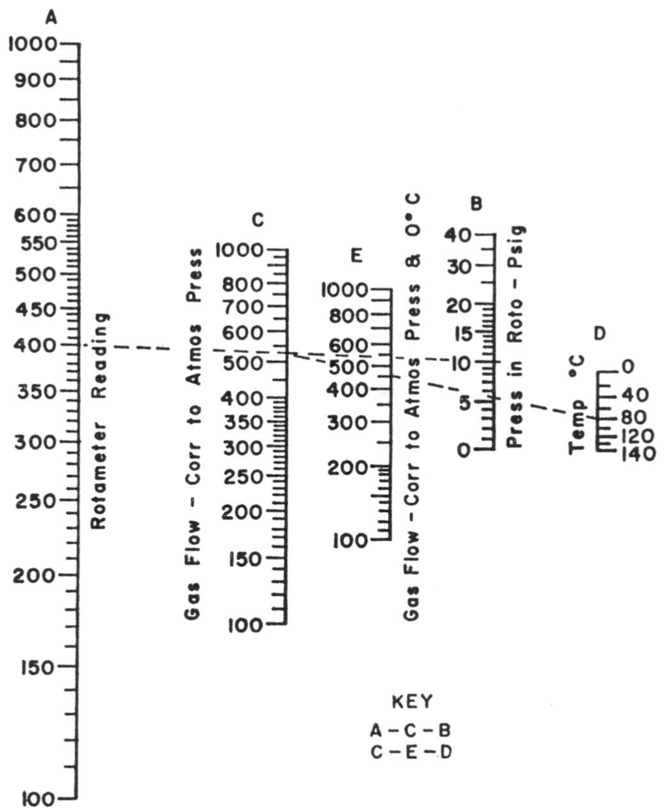

Nomograph 5.56. Temperature and pressure corrections to rotameters calibrated for gases at standard conditions. Courtesy Chemical Publishing Company, Inc.

Temperature and Pressure Corrections to Rotameters Calibrated for Gases at Standard Conditions

Quite often rotameters supplied for gas-flow measurement are calibrated for a particular gas flowing at standard conditions. In actual use, ambient conditions may be significantly different from standard conditions. In such cases correction must be applied to rotameter reading to obtain true flow.

Assuming perfect gas laws apply, the following equation may be used to correct rotameter reading:

$$Q = R\sqrt{[(P+14.7)(273)/(14.7)(t+273)]}$$

where

Q = actual gas flow, wt/vol
R = rotameter reading
P = pressure in rotameter, psig
t = temperature in rotameter, °C

Nomograph 5.56, constructed according to the procedure outlined by Davis*, solves the equation. Units of actual gas flow correspond to the units for which rotameter is calibrated. Decimal point is similarly related.

Example. Air is flowing through a rotameter calibrated for air at standard conditions. The meter indicates flow of 40 cfm. Pressure is 10 psig and the temperature is 80°C.

Connect proper values on lines A and B. Read gas flow, corrected to atmospheric pressure, on line C. Connect this value on line C with the proper temperature on line D. Read gas flow (46 cfm), corrected to standard conditions, on line E.

*Davis, D.S., *Nomography and Empirical Equations*, Chapter 6, Reinhold Publishing Corp., New York (1955).

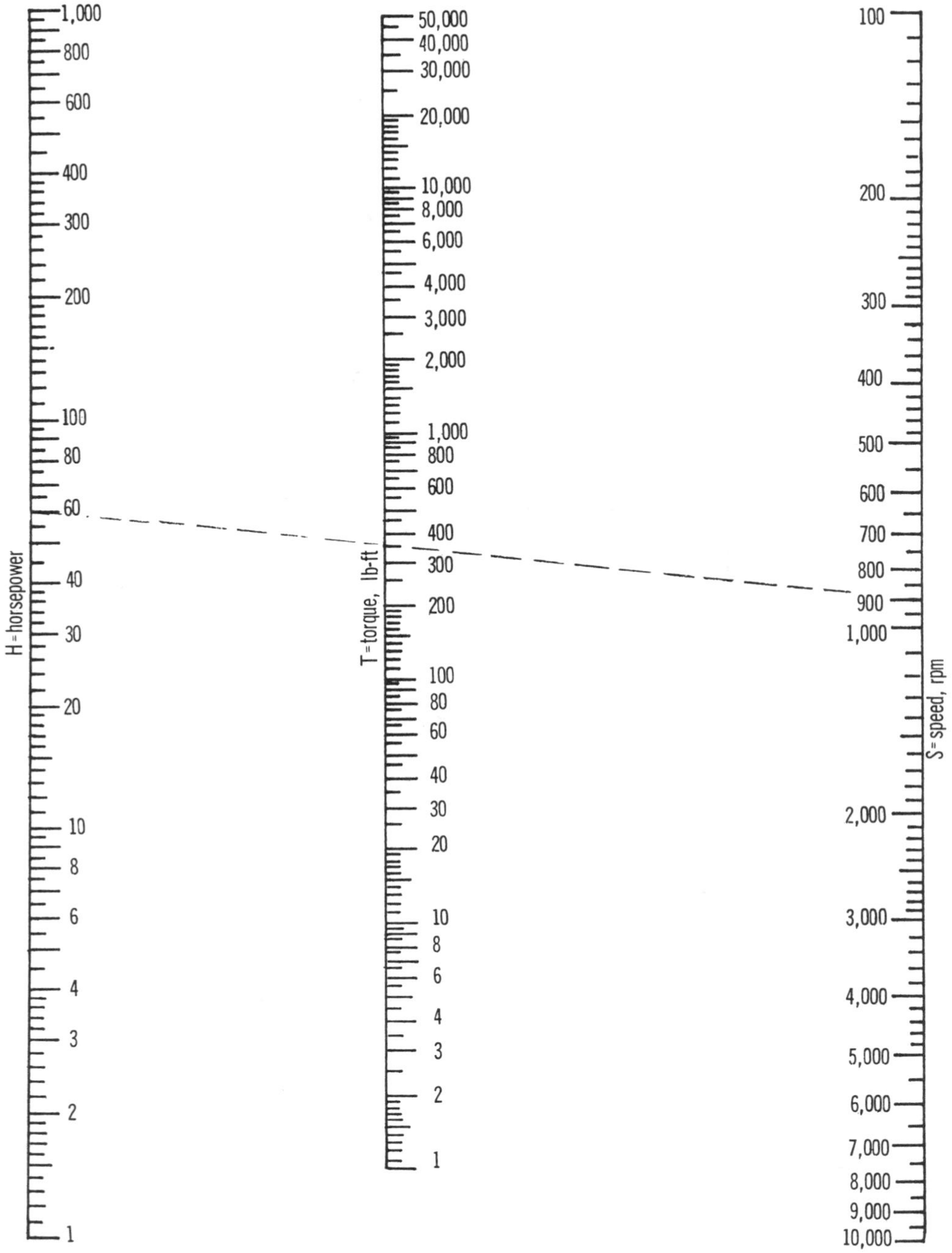

Nomograph 5.57. Torque of rotating machines. Courtesy Oil, Gas, and Petrochem Equipment.

Determine Torque of Rotating Machines

Torque of a rotating machine is directly proportional to horsepower and inversely proportional to speed. Or

$$T = 5{,}250\ H/S$$

where

T = torque, lb-ft
H = horsepower
S = speed, rpm

Nomograph 5.57 solves this equation.

Example. Find the torque when the horsepower is 60 and the speed is 900 rpm.

Align a straightedge at 60 on the H scale with 900 on the S scale, find the intersection of the straightedge on the T scale, and read the answer of 3,500 lb-ft on the T scale, as indicated by the dashed line.

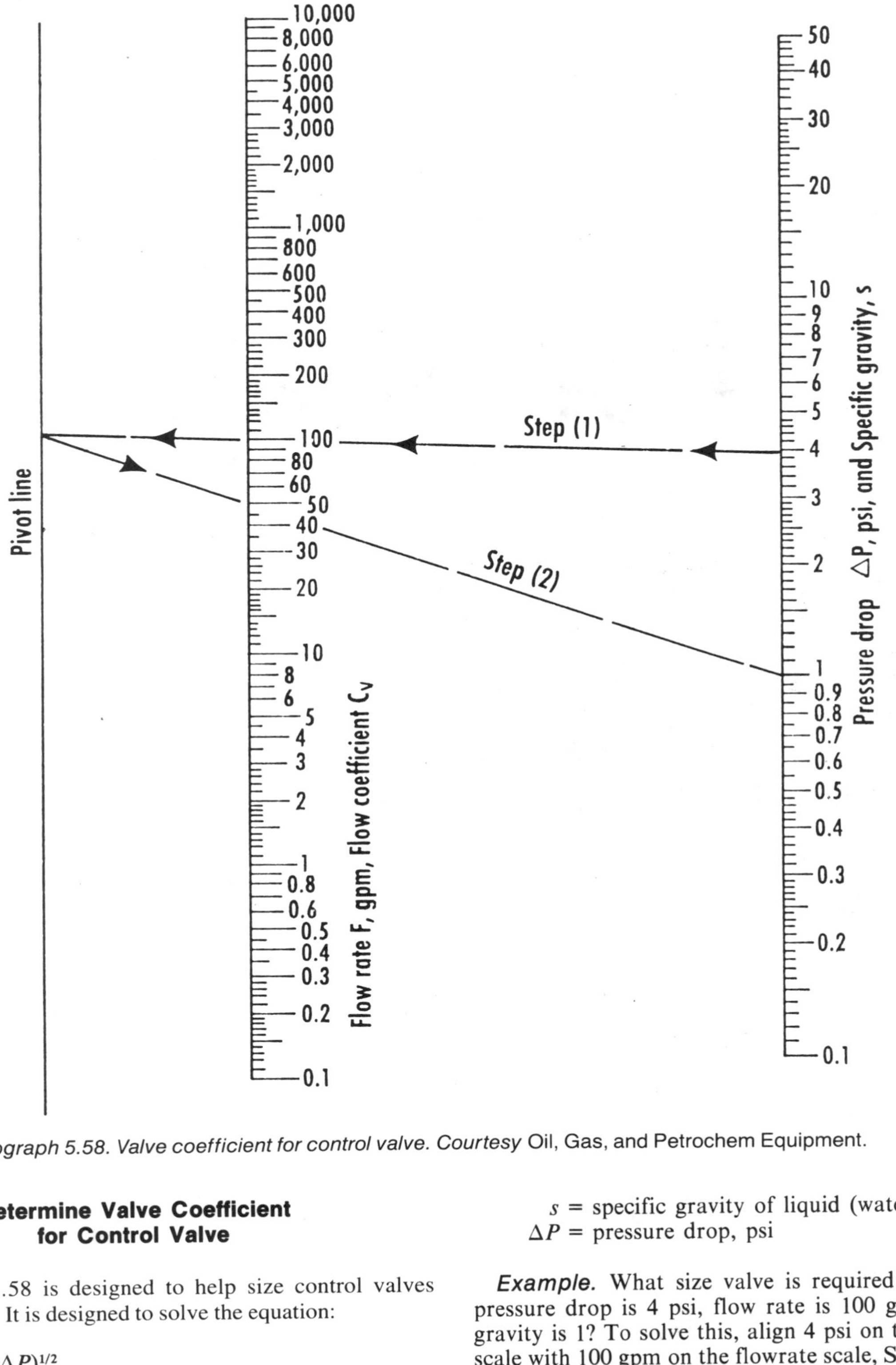

Nomograph 5.58. Valve coefficient for control valve. Courtesy Oil, Gas, and Petrochem Equipment.

Determine Valve Coefficient for Control Valve

Nomograph 5.58 is designed to help size control valves handling liquids. It is designed to solve the equation:

$$C_v = F\,(s)^{1/2}/(\Delta P)^{1/2}$$

where

C_v = valve coefficient
F = flow rate, gpm
s = specific gravity of liquid (water = 1)
ΔP = pressure drop, psi

Example. What size valve is required if the allowable pressure drop is 4 psi, flow rate is 100 gpm, and specific gravity is 1? To solve this, align 4 psi on the pressure-drop scale with 100 gpm on the flowrate scale, Step (1). Mark the intersection of the straightedge with the pivot line. Next align the mark on the pivot line with 1 on the specific-gravity scale, Step (2). Read C_v = 50 on the flow-coefficient scale.

6 Formulas

B.J. Normand

In this section consistent units must be used in formulas which do not show specific units of measurement. See Appendix A at the end of this section.

Mathematical

Areas, Volumes, Circumferences

Where units of measurement are consistent and

A = area
A_1 = surface area of solids
V = volume
C = circumference
H = height
L = length
W = width
D = diameter
R = radius
a = angle, degrees

Other letters are shown by appropriate figure.

Rectangle

$$A = W \times L$$

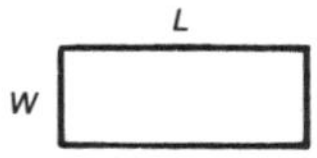

Parallelogram

$$A = H \times L$$

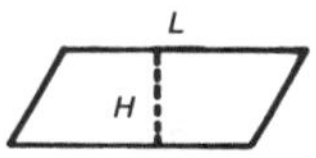

Trapezoid

$$A = H \times \frac{L_1 + L_2}{2}$$

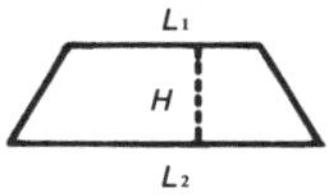

Triangle

$$A = \frac{W \times H}{2}$$

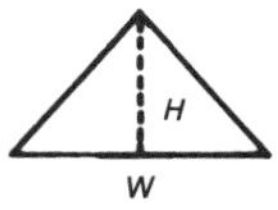

Circle

$$A = 3.142 \times R \times R$$
$$C = 3.142 \times D$$

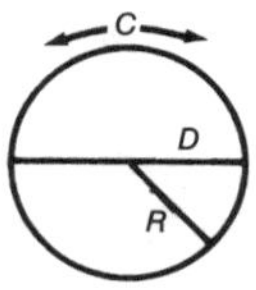

Sector of circle

$$A = \frac{3.142 \times R \times R \times a}{360}$$

$$L = 0.01745 \times R \times a$$

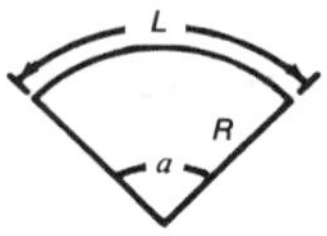

$$a = \frac{L}{0.01745 \times R}$$

$$R = \frac{L}{0.01745 \times a}$$

Ellipse

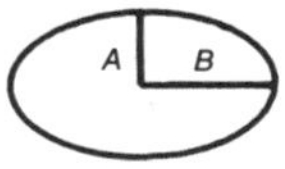

$$A = 3.142 \times A \times B$$

$$C = 6.283 \times \frac{\sqrt{A^2 + B^2}}{2}$$

Rectangular solid

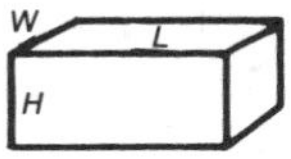

$$A_1 = 2[W \times L + L \times H + H \times W]$$
$$V = W \times L \times H$$

Cone

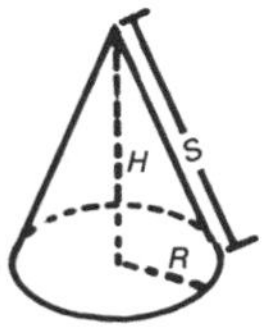

$$A_1 = 3.142 \times R \times S + 3.142 \times R \times R$$
$$V = 1.047 \times R \times R \times H$$

Cylinder

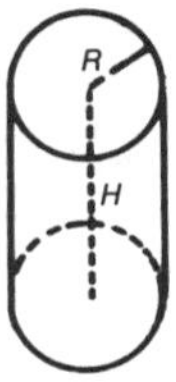

$$A_1 = 6.283 \times R \times H + 6.283 \times R \times R$$
$$V = 3.142 \times R \times R \times H$$

Elliptical Tanks

$$V = 3.142 \times A \times B \times H$$

$$A_1 = 6.283 \times \frac{\sqrt{A^2 + B^2}}{2} \times H + 6.283 \times A \times B$$

Sphere

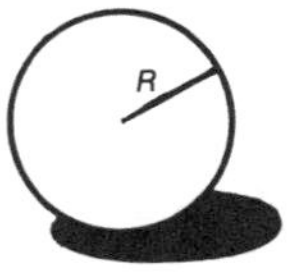

$$A_1 = 12.56 \times R \times R$$
$$V = 4.188 \times R \times R \times R$$

Frustrum

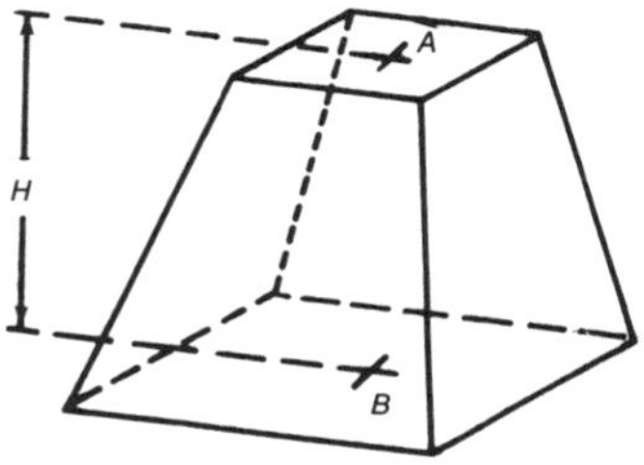

$$V = \frac{H}{3} \times (A + B + \sqrt{AB})$$

Algebraic Expressions

Exponents and Radicals

$$a^x \times a^v = a^{(x+v)} \qquad \frac{a^x}{a^v} = a^{(x-v)}$$

$$(ab)^x = a^x b^x \qquad \left(\frac{a}{b}\right)^x = \frac{a^x}{b^x}$$

$$\sqrt[x]{\frac{a}{b}} = \frac{\sqrt[x]{a}}{\sqrt[x]{b}} \qquad a^{-x} = \frac{1}{a^x}$$

$$(a^x)^v = a^{xv} \qquad \sqrt[x]{\sqrt[v]{a}} = \sqrt[xv]{a}$$

$$\sqrt[x]{ab} = \sqrt[x]{a}\ \sqrt[x]{b}$$

$$a^{\frac{x}{v}} = \sqrt[v]{a^x}$$

$$a^{\frac{1}{x}} = \sqrt[x]{a}$$

$$a^0 = 1$$

Solution of a Quadratic

Quadratic equations in the form

$$ax^2 + bx + c = 0$$

may be solved by the following:

$$x = \frac{-b \pm \sqrt{b^2 - 4ac}}{2a}$$

Transposition of Terms

If $A = \frac{B}{C}$, then $B = AC$, $C = \frac{B}{A}$.

If $\frac{A}{B} = \frac{C}{D}$, then $A = \frac{BC}{D}$,

$$B = \frac{AD}{C},\ C = \frac{AD}{B},\ D = \frac{BC}{A}.$$

If $A = \frac{1}{D\sqrt{BC}}$, then $A^2 = \frac{1}{D^2BC}$,

$$B = \frac{1}{D^2A^2C},\ C = \frac{1}{D^2A^2B},\ D = \frac{1}{A\sqrt{BC}}.$$

If $A = \sqrt{B^2 + C^2}$, then $A^2 = B^2 + C^2$,

$$B = \sqrt{A^2 - C^2},\ C = \sqrt{A^2 - B^2}.$$

Trigonometric Functions

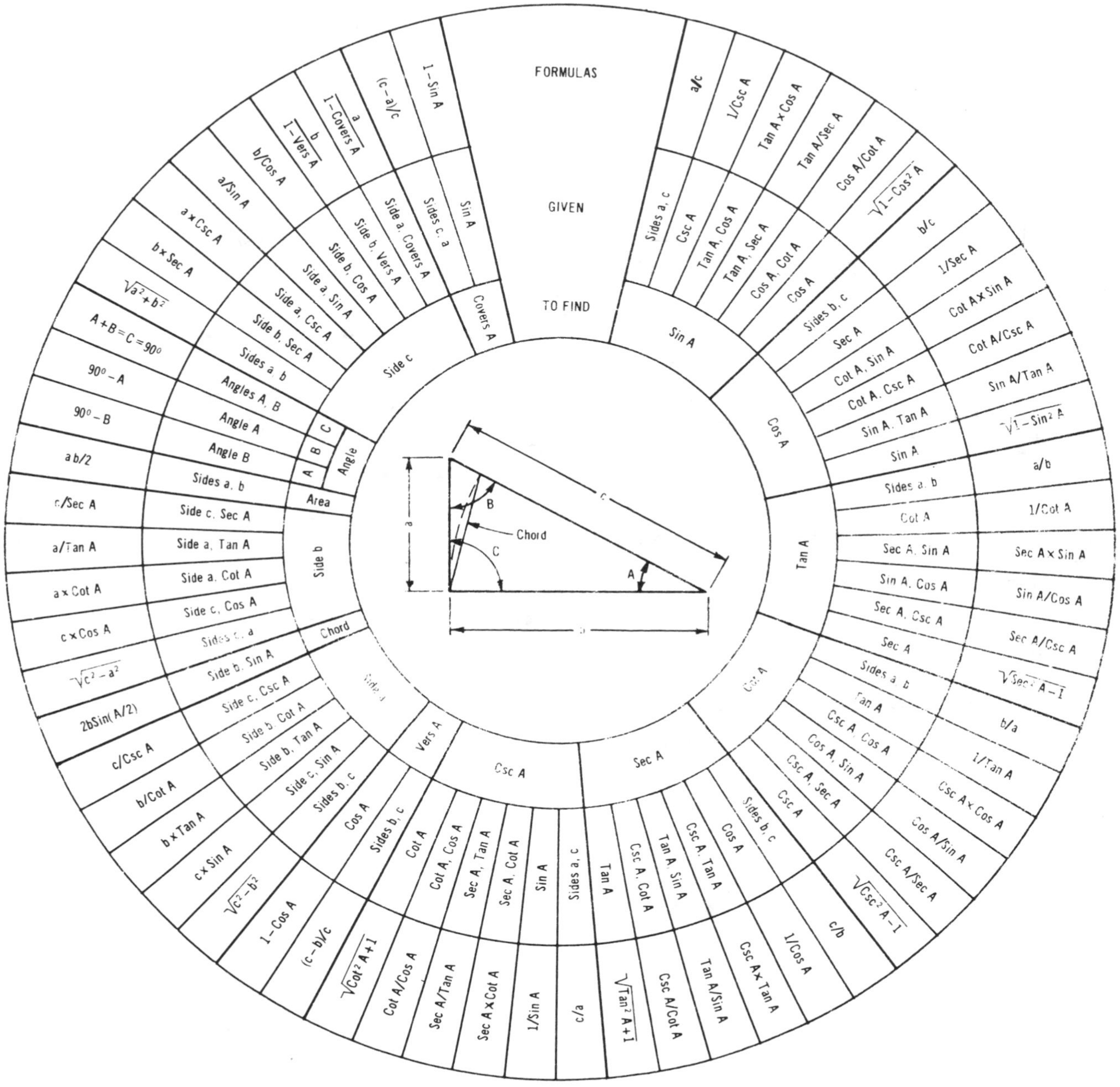

Figure 6.1. Trigonometric formulas for right triangles. Courtesy J.H. Jones Company.

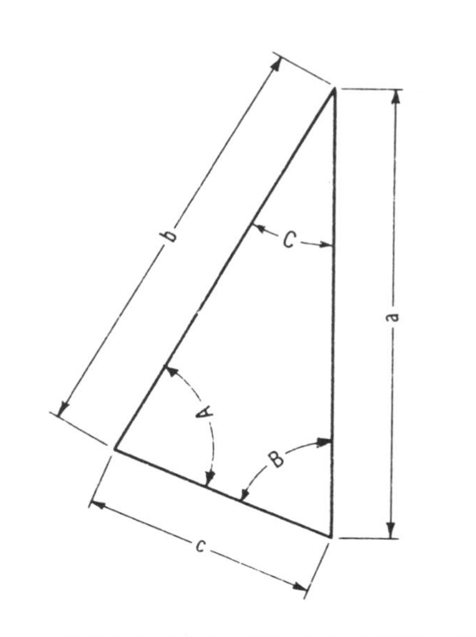

TO FIND	GIVEN	FORMULAS
A	B,C	$180^\circ - (B + C)$
Tan A	a,b,C	$\frac{a \times \text{Sin } C}{b - (a \times \text{Cos } C)}$
Cos A	a,b,c	$\frac{b^2 + c^2 - a^2}{2bc}$
Sin A	a,c,C	$\frac{a \times \text{Sin } C}{c}$
Sin A	a,b,B	$\frac{a \times \text{Sin } B}{b}$
Tan A	a,c,B	$\frac{a \times \text{Sin } B}{c - (a \times \text{Cos } B)}$
B	A,C	$180^\circ - (A + C)$
Sin B	a,b,A	$\frac{b \times \text{Sin } A}{a}$
Cos B	a,b,c	$\frac{c^2 + a^2 - b^2}{2ac}$
Tan B	b,c,A	$\frac{b \times \text{Sin } A}{c - (b \times \text{Cos } A)}$
Sin B	b,c,C	$\frac{b \times \text{Sin } C}{c}$

TO FIND	GIVEN	FORMULAS
a	c,A,C	$\frac{c \times \text{Sin } A}{\text{Sin } C}$
a	b,A,B	$\frac{b \times \text{Sin } A}{\text{Sin } B}$
a	b,c,A	$\sqrt{b^2 + c^2 - (2bc \times \text{Cos } A)}$
b	a,A,B	$\frac{a \times \text{Sin } B}{\text{Sin } A}$
b	c,B,C	$\frac{c \times \text{Sin } B}{\text{Sin } C}$
b	a,c,B	$\sqrt{c^2 + a^2 - (2ac \times \text{Cos } B)}$

FORMULAS	GIVEN	TO FIND
$\frac{a \times \text{Sin } C}{\text{Sin } A}$	a,A,C	c
$\sqrt{a^2 + b^2 - (2ab \times \text{Cos } C)}$	a,b,C	c
$\frac{b \times \text{Sin } C}{\text{Sin } B}$	b,B,C	c
$180^\circ - (A + B)$	A,B	C
$\frac{c \times \text{Sin } A}{a}$	a,c,A	Sin C
$\frac{c \times \text{Sin } A}{b - (c \times \text{Cos } A)}$	b,c,A	Tan C
$\frac{c \times \text{Sin } B}{b}$	b,c,B	Sin C
$\frac{c \times \text{Sin } B}{a - (c \times \text{Cos } B)}$	a,c,B	Tan C
$\frac{a^2 + b^2 - c^2}{2ab}$	a,b,c	Cos C
$\frac{ab \times \text{Sin } C}{2}$	a,b,C	Area
$\sqrt{S(S-a)(S-b)(S-c)}$	a,b,c	Area $S = \frac{1}{2}(a+b+c)$

Figure 6.2. Trigonometric formulas for oblique triangles. Courtesy J.H. Jones Company.

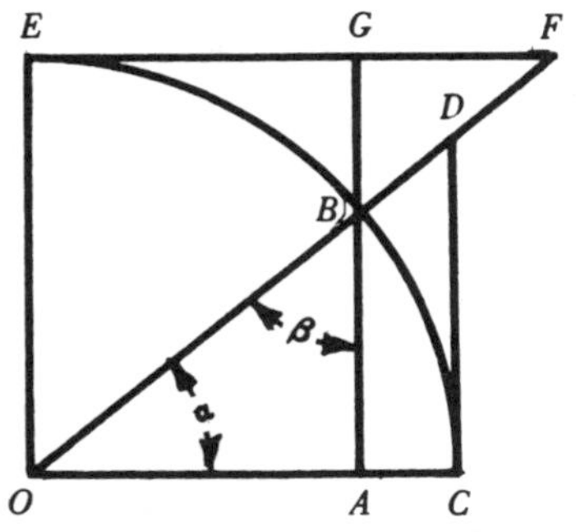

$$\text{Radius } 1=\sin^2\alpha+\cos^2\alpha=\sin\alpha\ \text{cosec}\ \alpha=\cos\alpha\ \sec\alpha=\tan\alpha\ \cot\alpha$$

$$\sin\alpha=\frac{\cos\alpha}{\cot\alpha}=\frac{1}{\text{cosec}\ \alpha}=\cos\alpha\ \tan\alpha=\sqrt{1-\cos^2\alpha}$$

$$\cos\alpha=\frac{\sin\alpha}{\tan\alpha}=\frac{1}{\sec\alpha}=\sin\alpha\ \cot\alpha=\sqrt{1-\sin^2\alpha}$$

$$\tan\alpha=\frac{\sin\alpha}{\cos\alpha}=\frac{1}{\cot\alpha}=\sin\alpha\ \sec\alpha \qquad \sec\alpha=\frac{\tan\alpha}{\sin\alpha}=\frac{1}{\cos\alpha}$$

$$\cot\alpha=\frac{\cos\alpha}{\sin\alpha}=\frac{1}{\tan\alpha}=\cos\alpha\ \text{cosec}\ \alpha \qquad \text{cosec}\ \alpha=\frac{\cot\alpha}{\cos\alpha}=\frac{1}{\sin\alpha}$$

$OC=OB=OE=1$
$AB=\text{Sin}\ \alpha$
$OA=\text{Cos}\ \alpha$
$CD=\text{Tan}\ \alpha$
$EF=\text{Cot}\ \alpha$
$OD=\text{Sec}\ \alpha$
$OF=\text{Cosec}\ \alpha$
$AC=\text{Vers}\ \alpha=1-\text{Cos}\ \alpha$
$BG=\text{Covers}\ \alpha=1-\text{Sin}\ \alpha$

$$\sin(\alpha\pm\beta)=\sin\alpha\ \cos\beta\pm\cos\alpha\ \sin\beta \qquad \tan(\alpha\pm\beta)=\frac{\tan\alpha\pm\tan\beta}{1\mp\tan\alpha\ \tan\beta}$$

$$\cos(\alpha\pm\beta)=\cos\alpha\ \cos\beta\mp\sin\alpha\ \sin\beta \qquad \cot(\alpha\pm\beta)=\frac{\cot\alpha\ \cot\beta\mp1}{\cot\beta\pm\cot\alpha}$$

$$\sin\alpha+\sin\beta=2\sin\tfrac{1}{2}(\alpha+\beta)\cos\tfrac{1}{2}(\alpha-\beta) \qquad \tan\alpha+\tan\beta=\frac{\sin(\alpha+\beta)}{\cos\alpha\cos\beta}$$

$$\sin\alpha-\sin\beta=2\cos\tfrac{1}{2}(\alpha+\beta)\sin\tfrac{1}{2}(\alpha-\beta) \qquad \tan\alpha-\tan\beta=\frac{\sin(\alpha-\beta)}{\cos\alpha\cos\beta}$$

$$\cos\alpha+\cos\beta=2\cos\tfrac{1}{2}(\alpha+\beta)\cos\tfrac{1}{2}(\alpha-\beta) \qquad \cot\alpha+\cot\beta=\frac{\sin(\beta+\alpha)}{\sin\alpha\sin\beta}$$

$$\cos\beta-\cos\alpha=2\sin\tfrac{1}{2}(\alpha+\beta)\sin\tfrac{1}{2}(\alpha-\beta) \qquad \cot\alpha-\cot\beta=\frac{\sin(\beta-\alpha)}{\sin\alpha\sin\beta}$$

$$\sin^2\alpha-\sin^2\beta=\sin(\alpha+\beta)\sin(\alpha-\beta)$$

$$\frac{\sin\alpha\pm\sin\beta}{\cos\alpha+\cos\beta}=\tan\tfrac{1}{2}(\alpha\pm\beta),$$

$$\cos^2\alpha-\sin^2\beta=\cos(\alpha+\beta)\cos(\alpha-\beta)$$

$$\frac{\sin\alpha\pm\sin\beta}{\cos\beta-\cos\alpha}=\cot\tfrac{1}{2}(\alpha\mp\beta)$$

$$\sin2\alpha=2\sin\alpha\cos\alpha \qquad \sin\tfrac{1}{2}\alpha=\sqrt{\frac{1-\cos\alpha}{2}} \qquad \sin^2\alpha=\frac{1-\cos2\alpha}{2}$$

$$\cos2\alpha=\cos^2\alpha-\sin^2\alpha \qquad \cos\tfrac{1}{2}\alpha=\sqrt{\frac{1+\cos\alpha}{2}} \qquad \cos^2\alpha=\frac{1+\cos2\alpha}{2}$$

$$\tan2\alpha=\frac{2\tan\alpha}{1-\tan^2\alpha} \qquad \tan\tfrac{1}{2}\alpha=\frac{\sin\alpha}{1+\cos\alpha} \qquad \tan^2\alpha=\frac{1-\cos2\alpha}{1+\cos2\alpha}$$

$$\cot2\alpha=\frac{\cot^2\alpha-1}{2\cot\alpha} \qquad \cot\tfrac{1}{2}\alpha=\frac{\sin\alpha}{1-\cos\alpha} \qquad \cot^2\alpha=\frac{1+\cos2\alpha}{1-\cos2\alpha}$$

Figure 6.3. General trigonometric formulas. Courtesy J.H. Jones Company.

Circumference of Circle of Diameter $1=\pi=3.14159265$
Circumference of Circle $=2\pi r=\pi d$
Diameter of Circle $=$ Circumference $\times 0.31831$
Diameter of Circle of equal periphery as square $=$ side $\times 1.27324$
Side of Square of equal periphery as circle $=$ diameter $\times 0.78540$
Diameter of Circle circumscribed about square $=$ side $\times 1.41421$
Side of Square inscribed in Circle $=$ diameter $\times 0.70711$

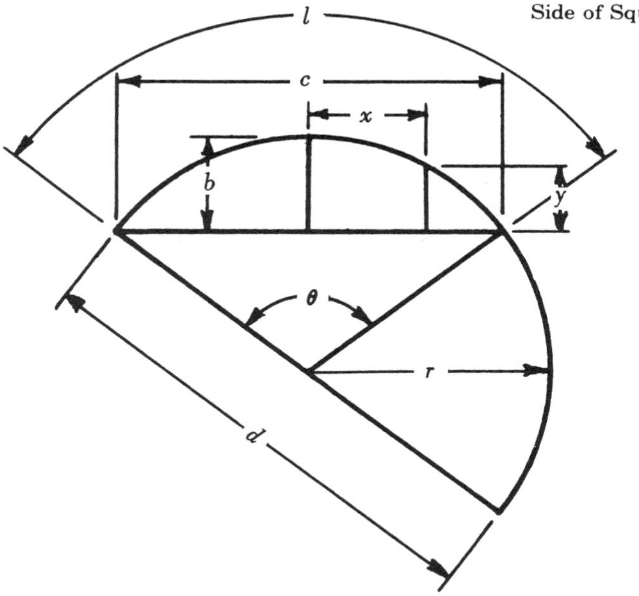

Arc, $l=\dfrac{\pi r\theta^\circ}{180}=0.017453r\theta^\circ$

Angle, $\theta=\dfrac{180^\circ l}{\pi r}=57.29578\dfrac{l}{r}$

Radius, $r=\dfrac{4b^2+c^2}{8b}$ Diameter, $d=\dfrac{4b^2+c^2}{4b}$

Chord, $c=2\sqrt{2br-b^2}=2r\sin\dfrac{\theta}{2}=d\sin\dfrac{\theta}{2}$

Rise, $b=r-\dfrac{1}{2}\sqrt{4r^2-c^2}=\dfrac{c}{2}\tan\dfrac{\theta}{4}=2r\sin^2\dfrac{\theta}{4}$

$b=r+y-\sqrt{r^2-x^2}$

$x=\sqrt{r^2=(r+y-b)^2}$

$y=b-r+\sqrt{r^2-x^2}$

$\pi=3.14159265$	log $=0.4971499$
$\dfrac{1}{\pi}=0.318310$	log $=9.502850-10$
$\dfrac{2}{\pi}=0.636620$	log $=9.803880-10$
$\dfrac{180}{\pi}=57.295780$	log $=1.758123$
$\dfrac{\pi}{180}=0.017453$	log $=8.241870-10$

$\pi^2=9.869604$	log $=0.994300$
$\pi^3=31.006277$	log $=1.491450$
$\dfrac{1}{\pi^2}=0.101321$	log $=9.005700-10$
$\dfrac{1}{\pi^3}=0.032252$	log $=8.508557-10$
$\sqrt{\pi}=1.772454$	log $=0.248575$
$1/\sqrt{\pi}=0.564190$	log $=9.751425-10$
$\sqrt[3]{\pi}=1.464592$	log $=0.165717$
$1/\sqrt[3]{\pi}=0.682784$	log $=9.834283-10$

Figure 6.4. Properties of the circle. Courtesy J.H. Jones Company.

Physics and Chemistry

Uniform Linear Motion

The *speed* of a body having uniform linear motion is defined as *the distance which the body traverses divided by the time required to traverse this distance,* or in equation form,

$$\text{Speed} = \frac{\text{distance covered}}{\text{time needed}}$$

or

$$v = \frac{s}{t}$$

The *angular speed* of a body having uniform angular motion is defined as *the angle through which the body sweeps divided by the time required to sweep through this angle*, that is

$$\text{Angular speed} = \frac{\text{angle swept through}}{\text{time needed}}$$

or

$$\omega = \frac{\theta}{t}$$

$$\text{Acceleration} = \frac{\text{final velocity} - \text{original velocity}}{\text{time interval}}$$

or

$$a = \frac{v_f - v_o}{t}$$

Units of Force

$$F = k_g ma$$

in which a is the acceleration produced in a mass m by an unbalanced force F, and k_g is a proportionality constant the value of which depends upon the units used in the expression.

Fundamentally, a unit of force is one of such magnitude as to impart unit acceleration to a unit of mass, and such a unit is styled an absolute unit since it does not involve an arbitrary constant. It can be defined, therefore, from the foregoing equation by making k_g equal to unity and writing

$$F = ma$$

In the first equation above in the British system of units, the value of k_g is $\frac{1}{32.2 \text{ ft} - \text{lb}_\text{m}/\text{sec}^2 - \text{lbf}}$ where F is in pounds force (lbf), m is in pounds mass (lbm) and a is the acceleration due to gravity in ft/sec^2.

Work

$$E = Fs$$

where E is the amount of work which an agent does on a body in exerting a push or pull F on it through a distance s, provided F and s have the same direction.

$$E = T\theta$$

where E is the work done upon a rotating body by an agent which exerts a torque T upon the body through an angle of θ radians.

Kinetic Energy

$$KE = \frac{Wv^2}{2g}$$

or

$$KE = \frac{mv^2}{2}$$

where:

W = weight of body in motion
v = velocity
m = mass
g = acceleration due to gravity

Potential Energy

$$PE = Wh$$

which shows that the amount of potential energy given to a body by raising it from one level to another depends only on the weight of the body and the vertical distance h between the two levels.

General Gas Laws for Perfect Gases

$$p'V_a = w_a RT$$

$$\rho = \frac{w_a}{V_a} = \frac{p'}{RT} = \frac{144\,P'}{RT}$$

$$R = \frac{1544}{M} = \frac{144\,P'}{\rho T}$$

$$p'V_a = n_a MRT = n_a 1544 T = \frac{W_a}{M} 1544\,T$$

$$\rho = \frac{w_a}{V_a} = \frac{p'M}{1544\,T} = \frac{P'M}{10.72\,T} = \frac{2.70\,P'S_g}{T}$$

where

- M = molecular weight
- MR = universal gas constant = 1544
- P' = pressure, pounds per square inch absolute
- p' = pressure, in pounds per square foot absolute
- R = individual gas constant = $MR/M = 1544/M$
- S_g = specific gravity of a gas relative to air = the ratio of the molecular weight of the gas to that of air
- T = absolute temperature, in degrees Rankine ($460 + t$)
- V_a = volume, in cubic feet
- w_a = weight, in pounds
- ρ = weight density of fluid, pounds per cubic ft

and where

- n_a = w_a/M = number of mols of a gas

Nonideal Gases

In dealing with gases at low pressure, the ideal gas relationship is a convenient and generally satisfactory tool. But when measurements and calculations are required for gases under high pressure the use of the ideal gas relationship may lead to appreciable errors. A convenient method of representing the pressure-volume-temperature relationships of nonideal gases is the use of the compressibility factor, Z, which serves as a multiplying correction factor to compute the correct relationship from the ideal gas equation.

- PV = $ZNRT$
- P = pressure, in lbs./sq. in. absolute
- V = volume in cu. ft.
- Z = the compressibility factor
- N = No. of pound mols, or $\dfrac{\text{pounds}}{\text{molecular wt.}}$
- R = 10.72 for all gases
- T = temperature, absolute, (460 + °F.)

The compressibility factor, Z, is dimensionless, independent of the weight of the gas, and determined by the character of the gas, its temperature and pressure.

Conversion Formulas

Gas Conversions

SCFH = Standard Cubic Feet per Hour (14.7 psia and 60°F)

$$\text{SCFH} = \frac{\text{Pounds per Hour}}{\text{Molecular Weight}} \times 379$$

$$\text{SCFH} = \frac{\text{Pounds per Hour}}{\text{Density in lbs./cu.ft. (at std. conditions)}}$$

$$\text{SCFH} = \frac{\text{Pounds per Hour}}{\text{Specific Gravity}} \times 13.1$$

$$\text{Pounds per Hour} = \frac{\text{SCFH x Molecular Weight}}{379}$$

Pounds per Hour = SCFH x Density, in lb_m/ft^3 at std. conditions.

$$\text{Pounds per Hour} = \frac{\text{SCFH x Specific Gravity}}{13.1}$$

Temperature Conversion

F = ° Fahrenheit

C = ° Celsius, or ° Centigrade

$$\frac{F - 32}{180} = \frac{C}{100}$$

$$C = \frac{5}{9}(F - 32)$$

$$F = \frac{9}{5}C + 32$$

Pressure Conversions

Multiply	By	To Obtain
Inches of Water	0.036	Pounds per Sq. In.
Inches of Mercury	0.491	Pounds per Sq. In.
Feet of Water	0.433	Pounds per Sq. In.
Ounces per Sq. In.	0.062	Pounds per Sq. In.
Atmospheres	14.7	Pounds per Sq. In.
Kilograms per Sq. Cm.	14.22	Pounds per Sq. In.
Centimeters of Mercury	0.193	Pounds per Sq. In.

Liquid Flow Conversions

Multiply	By	To Obtain
Barrels (42) per Day	0.0292	Gallons per Minute
Cu. Ft. per Second	448.8	Gallons per Minute
Cu. Ft. per Minute	7.48	Gallons per Minute
Cu. Ft. per Hour	0.1247	Gallons per Minute
Pounds per Hour of Water	0.002	Gallons per Minute
Gallons per Hour	0.0167	Gallons per Minute

Specific Gravities

Where

SG = Specific Gravity
ρ = density of fluid, pounds of mass per cubic ft.
S = specific gravity of liquids relative to water, both at standard temperature (60°F).
R = individual gas constant = $MR/M = 1544/M$
M = Molecular weight
S_g = specific gravity of a gas relative to air = the ratio of the molecular weight of the gas to that of air.

Specific Gravity of Liquids

Any Liquid:

$$SG = \frac{\rho \left(\begin{array}{c}\text{any liquid at } 60^\circ\text{F,}\\ \text{unless otherwise specified}\end{array}\right)}{\rho \quad \text{water at } 60^\circ \text{ F}}$$

Oils:

$$S\,(60^\circ\text{F}/60^\circ\text{F}) = \frac{141.5}{131.5 + \text{Deg API}}$$

Liquids lighter than water:

$$S(60^\circ\text{F}/60^\circ\text{F}) = \frac{140}{130 + \text{Deg Baume}}$$

Liquids heavier than water:

$$S(60^\circ\text{F}/60^\circ\text{F}) = \frac{145}{145 - \text{Deg Baume}}$$

Specific Gravity of gases:

$$S_g = \frac{R(\text{air})}{R(\text{gas})} = \frac{53.3}{R(\text{gas})}$$

$$S_g = \frac{M(\text{gas})}{M(\text{air})} = \frac{M(\text{gas})}{29}$$

Reynolds Number

The pipe Reynolds number, R_e, is a dimensionless ratio. If consistent units are used, the value is equal to velocity times pipe diameter times specific weight divided by the acceleration of gravity times absolute viscosity, regardless of the system of units.

The following alternative equations will give the correct value:

$$R_e = \frac{0.475 \text{ x rate of flow (lb./sec.)}}{\text{pipe diam. (inches) x abs. visc. (lb. sec./ft.}^2)}$$

$$= \frac{0.007914 \text{ x rate of flow (lb./min.)}}{\text{pipe diam. (inches) x abs. visc. (lb. sec./ft.}^2)}$$

$$= \frac{0.0001319 \text{ x rate of flow (lb./hr.)}}{\text{pipe diam. (inches) x abs. visc. (lb. sec./ft.}^2)}$$

$$= \frac{0.0632 \text{ x rate of flow (lb./hr.)}}{\text{pipe diam. (inches) x abs. visc. (poises)}}$$

$$= \frac{6.32 \text{ x rate of flow (lb./hr.)}}{\text{pipe diam. (inches) x abs. visc. (cp)}}$$

$$= \frac{0.001100 \text{ x rate of flow at base temp. (g.p.h.) x base gravit}}{\text{pipe diam. (inches) x abs. visc. (lb. sec./ft.}^2)}$$

$$= \frac{0.5271 \text{ x rate of flow at base temp. (g.p.h.) x base gravity,}}{\text{pipe diam. (inches) x abs. visc. (poises)}}$$

$$= \frac{0.0005674 \text{ x rate of flow at flowing temp. (g.p.h.)}}{\text{pipe diam. (inches) x kinematic visc. (ft.}^2\text{/sec.)}}$$

$$= \frac{0.5277 \text{ x rate of flow at flowing temp. (g.p.h.)}}{\text{pipe diam. (inches) x kinematic visc. (stokes)}}$$

$$= \frac{52.77 \text{ x rate of flow at flowing temp. (g.p.h.)}}{\text{pipe diam. (inches) x kinematic visc. (cs)}}$$

Fluid Flow

Nomenclature for all terms this section are given at the end of the section.

Head loss and pressure drop in straight pipe due to turbulent flow.

Darcy's formula:

$$h_L = f\frac{L}{D}\frac{v^2}{2g} = 0.1863\frac{fLv^2}{d}$$

$$h_L = 6260\frac{fLq^2}{d^5} = 0.0311\frac{fLQ^2}{d^5}$$

$$h_L = 0.01524\frac{fLB^2}{d^5} = 0.000483\frac{fLW^2\overline{V}}{d^5}$$

$$\Delta P = 0.001294\frac{fL\rho v^2}{d} = 0.000000359\frac{fL\rho V^2}{d}$$

$$\Delta P = 43.5\frac{fL\rho q^2}{d^5} = 0.000216\frac{fL\rho Q^2}{d^5}$$

$$\Delta P = 0.0001058\frac{fL\rho B^2}{d^5} = 0.00000336\frac{fLW^2\overline{V}}{d^5}$$

Head loss and pressure drop in straight pipe due to laminar flow (where $R_e < 2000$).

$$h_L = 0.0962\frac{\mu Lv}{d^2\rho}$$

$$h_L = 17.65\frac{\mu Lq}{d^4\rho} = 0.0393\frac{\mu LQ}{d^4\rho}$$

$$h_L = 0.0275\frac{\mu LB}{d^4\rho} = 0.004\ 90\frac{\mu LW}{d^4\rho^2}$$

$$\Delta P = 0.000668\frac{\mu Lv}{d^2} = 0.1225\frac{\mu Lq}{d^4}$$

$$\Delta P = 0.000273\frac{\mu LQ}{d^4} = 0.000191\frac{\mu LB}{d^4}$$

$$\Delta P = 0.0000340\frac{\mu LW}{d^4\rho}$$

Flow through nozzles and orifices (h_L and ΔP measured across flange taps)

Liquid

$$q = AC\sqrt{2gh_L}$$

$$q = 0.0438\ d_o^2\ C\sqrt{h_L} = 0.525\ d_o^2\ C\sqrt{\frac{\Delta P}{\rho}}$$

$$Q = 19.65\ d_o^2\ C\sqrt{h_L} = 236\ d_o^2\ C\sqrt{\frac{\Delta P}{\rho}}$$

$$w = 0.0438\ d_o^2\ C\sqrt{h_L\rho^2} = 0.525\ d_o^2\ C\sqrt{\Delta P\rho}$$

$$W = 157.6\ d_o^2\ C\sqrt{h_L\rho^2} = 1891\ d_o^2\ C\sqrt{\Delta P\rho}$$

Note: Values of C are shown on Charts 3.2 and 3.3, p. 156.

Compressible fluids

$$q'_h = 40{,}700\ Y\ d_o^2\ C\sqrt{\frac{\Delta P\ P'_1}{T_1\ S_g}}$$

$$q'_m = 678\ Y\ d_o^2\ C\sqrt{\frac{\Delta P\ P'_1}{T_1\ S_g}}$$

$$q'_m = 412\ \frac{Y\ d_o^2\ C}{S_g}\sqrt{\Delta P\rho_1}$$

$$q' = 11.30\ Y\ d_o^2\ C\sqrt{\frac{\Delta P\ P'_1}{T_1\ S_g}}$$

$$q' = 6.87\ \frac{Y\ d_o^2\ C}{S_g}\sqrt{\Delta P\rho_1}$$

$$w = 0.525\ Y\ d_o^2\ C\sqrt{\frac{\Delta P}{\overline{V}_1}}$$

$$W = 1891\ Y\ d_o^2\ C\sqrt{\frac{\Delta P}{\overline{V}_1}}$$

$$q'_h = 24{,}700\ \frac{Y\ d_o^2\ C}{S_g}\sqrt{\Delta P\rho_1}$$

Note: Values of C are shown on Charts 3.2 and 3.3. Values of Y are shown on Chart 3.6, p. 159.

Maximum (sonic) velocity of compressible fluids in pipe

The maximum possible velocity of a compressible fluid in a pipe is equivalent to the speed of sound in the fluid; this is expressed as:

$$v_s = \sqrt{kgRT}$$

$$v_s = \sqrt{kg\,144P'\,\overline{V}}$$

$$v_s = 68.1\sqrt{kP'\,\overline{V}}$$

Reynolds number of flow in pipe:

$$R_e = \frac{Dv\rho}{\mu_e} = \frac{Dv\rho}{32.2\mu'_e} = 123.9\frac{dv\rho}{\mu}$$

$$R_e = 22{,}700\frac{q\rho}{d\mu} = 50.6\frac{Q\rho}{d\mu} = 35.4\frac{B\rho}{d\mu}$$

$$R_e = 6.31\frac{W}{d\mu} = 0.482\frac{q'_h S_g}{d\mu}$$

Viscosity equivalents:

$$\nu = \frac{\mu}{\rho'} = \frac{92{,}900\,\mu}{\rho kg}$$

Nomenclature

A = cross sectional area of pipe or orifice, in square feet
B = rate of flow in barrels (42 gallons) per hour
C = flow coefficient for orifices and nozzles = discharge coefficient corrected for velocity of approach = $C_d / \sqrt{1-(d_o/d_1)^4}$
D = internal diameter of pipe, in feet
d = internal diameter of pipe, in inches
e = base of natural logarithm = 2.718
f = friction factor in formula $h_L = fLv^2/D2g$
g = acceleration of gravity = 32.2 feet per second per second
H = total head, in feet of fluid
h_L = loss of static pressure head due to fluid flow, in feet of fluid
k = ratio of specific heat at constant pressure to specific heat at constant volume = c_p/c_v
$k_g = \dfrac{1}{32.2 \text{ ft lb } m/\text{sec}^2 \text{ lb} f}$
L = length of pipe, in feet
P = pressure, in pounds per square inch gauge
P' = pressure, pounds per square inch absolute
Q = rate of flow, in gallons per minute
q = rate of flow, in cubic feet per second at flowing conditions
q'_h = rate of flow, in cubic feet per hour at standard conditions (14.7 psia and 60F), scfh
q'_m = rate of flow, in cubic feet per minute at std. conditions (14.7 psia and 60F), scfm
R = individual gas constant = MR/M = $1544/M$
R_e = Reynolds number
S = specific gravity of liquids relative to water, both at standard temperature (60F)
S_g = specific gravity of a gas relative to air = the ratio of the molecular weight of the gas to that of air
T = absolute temperature, in degrees Rankine (460 + t)
$\overline{V}$ = specific volume of fluid, in cubic feet per pound
V = mean velocity of flow, in feet per minute
v = mean velocity of flow, in feet per second
v_s = sonic (or critical) velocity of flow of a gas, in feet per second
W = rate of flow, in pounds per hour
w = rate of flow, in pounds per second
Y = net expansion factor for compressible flow through orifices, nozzles, or pipe

Subscripts

(o)...indicates orifice conditions unless otherwise specified
(1)...indicates inlet or upstream conditions unless otherwise specified
(2)...indicates outlet or downstream conditions unless otherwise specified

Greek Letters

Delta
Δ = differential between two points
Rho
ρ = weight density of fluid, pounds per cubic ft
ρ' = density of fluid, grams per cubic centimeter
Mu
μ = absolute (dynamic) viscosity, in centipoise
μ_e = absolute viscosity, in pound mass per foot second or poundal seconds per sq foot
μ'_e = absolute viscosity, in slugs per foot second or pound force seconds per square foot
Nu
ν = kinematic viscosity, in centistokes

Control Valve Sizing

Nomenclature for this group of formulas is given at the end of this section.

The value of ΔP used for sizing in the following equations should never exceed $\frac{P_1}{2}$.

(See also equations in Chapter 5, Volume 2, 2nd Ed.)

Table of ISA Formulas

REMARKS	EQUATIONS	VALUE OF N[1] U.S.	VALUE OF N[1] SI
LIQUID Turbulent and Non-Cavitating	$q_f = N_1 F_p C_v \sqrt{\dfrac{\Delta p}{G_f}}$	1.00	0.0865
	$w_f = N_6 F_p C_v \sqrt{\Delta p \gamma}$	63.3	2.73
Choked	$q_f = N_1 F_{Lp} C_v \sqrt{\dfrac{p_1 - p_{vc}}{G_f}}$	1.00	0.0865
	$w_f = N_6 F_{Lp} C_v \sqrt{\dfrac{p_1 - p_{vc}}{G_f}}$	63.3	2.73
	$p_{vc} = F_F p_v$		
	$F_F \simeq 0.96 - 0.28 \sqrt{\dfrac{p_v}{p_c}}$		
	$F_{LP} = \left[\dfrac{1}{F_L^2} + \dfrac{K_i}{N_2}\left(C_d\right)^2\right]^{-1/2}$	890	0.00214
	K_i = (See Piping Geometry Factor)		
Laminar	$q_f = N_{10} \dfrac{\Delta p}{\mu}\left(F_s F_p C_v\right)^{3/2}$	52	173
	$F_s = \left(\dfrac{F_p F_d^2}{F_{LP}}\right)^{1/3}\left[\dfrac{(F_{LP} C_v)^2}{N_2 D^4} + 1\right]^{1/6}$	890	0.00214
Transitional	$q_f = N_1 F_R F_P C_v \sqrt{\dfrac{\Delta p}{G_f}}$	1.00	0.0865

[1] U.S. units are: pounds per hour, gallons per minute, pounds per square inch absolute, pounds per cubic foot, °R, and inches.

SI units are: kilograms per hour, cubic meters per hour, kPa, kilograms per cubic meter, °K, and millimeters.

(Reprinted with permission of Instrument Society of America: from ISA Handbook of Control Valves, 2nd Edition © Instrument Society of America, 1976.)

(table continued on next page)

Table of ISA Formulas (continued)

REMARKS	EQUATIONS	VALUE OF N U.S.	SI
Gas or Vapor - (All Equations: $x \leq F_k x_T$)			
	$w_g = N_6 F_p C_v Y \sqrt{x p_1 \gamma_1}$	63.3	2.73
	$q_g = N_7 F_p C_v p_1 Y \sqrt{\frac{x}{G_g T_1 Z}}$	1360	4.17
Variations for Selected units.	$w_g = N_8 F_p C_v p_1 Y \sqrt{\frac{xM}{T_1 Z}}$	19.3	0.948
	$q_g = N_9 F_p C_v p_1 Y \sqrt{\frac{x}{M T_1 Z}}$	7320	22.4
Expansion factor lower limit = 0.667	$Y = 1 - \frac{x}{3 F_k x_T}$		
Sp. ht. ratio factor	$F_k = k/1.40$		
Mfr's. Factors	$x_T = \frac{C_1^2}{1600} = 0.84 C_f^2$		
x_T with reducers	$x_{TP} = \frac{x_T}{F_p^2}\left[\frac{x_T K_i}{N_5}\left(C_d\right)^2 + 1\right]^{-1}$ K_i = (See Piping Geometry Factor)	1000	0.0024
Steam (Dry and Saturated)			
For $x < x_{Tp}$	$w = N F_p C_v p_1 \left(3 - \frac{x}{x_{TP}}\right)\left(\sqrt{x}\right)$	1.0	0.152
For $x \geq x_{Tp}$ (Choked Flow)	$w = N F_p C_v p_1 \sqrt{x_{TP}}$	2.0	0.304

Table of ISA Formulas (continued)

REMARKS	EQUATIONS	VALUE OF N U.S.	VALUE OF N SI
Piping Geometry Factor For F_{LP} see "Liquid Choked Flow"	$F_p = \left[\frac{\Sigma K}{N_2}\left(C_d\right)^2 + 1\right]^{-1/2}$	890	0.00214
Sum of velocity head coefficients	$\Sigma K = K_1 + K_2 + K_{B1} - K_{B2}$		
Bernoulli coefficient	$K_{B1} = K_{B2} = 1 - \left(\frac{d}{D}\right)^4$		
Resistance coefficients for abrupt transitions	$K_1 = 0.5\left[1 - \left(\frac{d}{D}\right)^2\right]^2$		
Inlet fitting coefficient for F_{LP} and x_{TP}.	$K_2 = 1.0\left[1 - \left(\frac{d}{D}\right)^2\right]^2$ $K_i = K_1 + K_{B1}$		

Line Velocity	Feet/Second	Meters/Second	Range (Ft./Sec.)
Liquid	$U = \frac{q}{2.45D^2}$	$U = 354\frac{q}{D^2}$	5-10 Norm. 40-50 Max.
Gas	$U = \frac{qT}{695pD^2}$	$U = 1.24\frac{qT}{pD^2}$	250-400
Vapor	$U = \frac{w}{19.6\gamma D^2}$	$U = 354\frac{W}{\gamma D^2}$	70 Wet 300 superheated
Steam	$U = \frac{23w}{pD^2}$	$U = 685\frac{W}{pD^2}$	

Acoustic Velocity (Mach 1.0)			
Gas	$U_a = 223\sqrt{\frac{kT}{M}}$	$U_a = 91\sqrt{\frac{kT}{M}}$	< 0.3 Mach
Air	$U_a = 49\sqrt{T}$	$U_a = 20\sqrt{T}$	
Steam, Superheated	$U_a = 60\sqrt{T}$	$U_a = 24.5\sqrt{T}$	<0.15 Mach
Steam, Dry Saturated	$U_a = 1650$	$U_a = 500$	<0.10 Mach
Vapor	$U_a = 68.1\sqrt{kpv}$	$U_a = 1038\sqrt{kpv}$	

Calculated Values of F_P and x_{TP} for Valves Installed Between Short Pipe Reducers Assuming Two Reducers of the Same Size with Abrupt Change in Area

C_d	10						15						20					25					30			
x_T	.40	.50	.60	.70	.80		.40	.50	.60	.70	.80		.40	.50	.60	.70		.20	.30	.40	.50		.15	.20	.25	
d/D	x_{TP}					F_P	x_{TP}					F_P	x_{TP}				F_P	x_{TP}				F_P	x_{TP}			F_P
.80	.40	.49	.59	.69	.78	.99	.40	.49	.58	.67	.75	.98	.39	.48	.56	.64	.96	.21	.30	.39	.47	.94	.17	.21	.26	.91
.75	.40	.50	.59	.69	.78	.98	.40	.49	.58	.67	.75	.97	.40	.49	.57	.65	.94	.22	.31	.40	.48	.91	.18	.23	.27	.88
.67	.40	.50	.60	.69	.78	.98	.41	.50	.59	.68	.76	.95	.42	.51	.59	.67	.91	.24	.33	.43	.51	.87	.19	.25	.30	.83
.60	.41	.51	.60	.70	.79	.97	.42	.52	.61	.69	.78	.93	.43	.53	.61	.69	.89	.25	.36	.45	.54	.84	.21	.27	.32	.79
.50	.41	.52	.61	.70	.80	.96	.44	.53	.63	.71	.79	.91	.46	.55	.64	.72	.85	.28	.39	.49	.58	.79	.24	.30	.36	.73
.40	.42	.52	.62	.71	.80	.95	.44	.55	.65	.74	.82	.89	.49	.58	.67	.75	.82	.30	.42	.53	.62	.76	.26	.33	.40	.70
.33	.43	.53	.62	.72	.81	.94	.46	.56	.66	.75	.83	.88	.50	.60	.69	.78	.81	.31	.44	.55	.64	.74	.27	.34	.40	.69
.25	.44	.53	.63	.73	.83	.93	.48	.58	.67	.76	.85	.87	.52	.62	.71	.79	.79	.33	.46	.57	.67	.72	.27	.37	.44	.65

EXAMPLE: A 2-inch valve is rated at Cv = 80 and x_T = 0.65. Find F_p and x_{TP} if the valve is installed in a 3-inch pipe line with short reducers. $C_d = Cv/d^2$ or 20 and d/D = 2/3 or 0.67. Under the heading C_d = 20 on line d/D = 0.67 find F_p = 0.91. For x_{TP} interpolate the values given in the columns headed x_T = 0.60 and x_T = 0.70 (0.59 and 0.67 respectively). The answer is x_{TP} = 0.63.

(Reprinted with permission of Instrument Society of America; from ISA Handbook of Control Valves, 2nd Edition © Instrument Society of America, 1976.)

Valve Sizing Data
Representative Valve Factors

BODY & TRIM TYPE	FLOW DIRECTION	LINE SIZE BODY (D=d)						HALF SIZE (D=2d)			
		C_d	F_L	X_T	F_d[1]	F_s	K_c	$\frac{N_3C_v}{D^2}$	F_{LP}	x_{TP}	F_s
Single Seat Globe											
Wing Guided	Either	11	.90	.75	1.0	1.05	c	2.8	.85	.75	1.04
V-Skirt	Either	9	.90	.75	1.5	1.38	c	2.3	.86	.75	1.36
Contoured	Open	11	.90	.72	1.0	1.05	.65	2.8	.85	.73	1.04
Contoured	Close	11	.80	.55	1.0	1.09	.58	2.8	.76	.57	1.08
V-Plug	Either	9.5	.90	.75	1.0	1.05	.80	2.4	.86	.75	1.04
Cage	Open	14	.90	.75	1.0	1.06	.65	3.5	.82	.75	1.04
Cage	Close	16	.80	.70	1.0	1.11	c	4.0	.72	.71	1.08
Double Seat Globe											
Wing Guided	--	14	.90	.75	.71	0.84	c	3.5	.82	.75	0.83
V-Skirt	--	13	.90	.75	.71	0.84	c	3.3	.83	.75	0.83
Contoured	--	13	.85	.70	.71	0.85	.70	3.3	.79	.71	0.84
V-Plug	--	12.5	.90	.75	.71	0.84	.80	3.1	.83	.75	0.84
Angle											
Full Port Contour	Close	20	.80	.65	1.0	1.12	.53	5.0	.69	.68	1.08
Full Port Contour	Open	17	.90	.72	1.0	1.08	.64	4.3	.78	.73	1.04
Restricted Contour	Close	⩾6	.70	.55	1.0	1.13	c	1.5	.69	.56	1.13
Restricted Contour	Open	⩾5.5	.95	.80	1.0	1.02	c	1.3	.93	.80	1.02
2:1 Tapered Orif.	Close	12	.45	.15	1.0	1.31	c	3.0	.44	.17	1.31
Cage	Open	12[a]	.85	.65	1.0	1.08	c	3.0	.80	.66	1.06
Cage	Close	12[a]	.80	.60	1.0	1.10	c	3.0	.75	.62	1.08
Venturi	Close	22	.50	.20	1.0	1.29	.17	5.5	.46	.26	1.26
Ball											
Std. Bore[b]	--	30	.55	.15	1.0	1.28	.25	7.5	.47	.24	1.22
Characterized	--	25	.57	.25	1.0	1.25	.22	6.3	.50	.33	1.21
Butterfly											
60-Deg. Open	--	17	.68	.38	.71	0.92	.3	4.3	.63	.43	0.91
90-Deg. Open	--	>30	.55	.20	.71	1.01	c	>7.5	.45	.33	0.97

[1]The values of F_d are based on limited test data which have not been corroborated by independent laboratories. F_s is computed from F_d. Key: a = Variable, b = Orif. ≃ 0.8d, c = Unavailable.

(Reprinted with permission of Instrument Society of America; from ISA Handbook of Control Valves, 2nd Edition © Instrument Society of America, 1976.)

Reference Data for Steam and Gases

		SP. GRAVITY G	SP. HEATS RATIO k	FACTOR F_k
Acetylene		0.897	1.28	0.914
Air		1.000	1.40	1.00
Ammonia		0.587	1.29	0.921
Argon		1.377	1.67	1.19
Carbon Dioxide		1.516	1.28	0.914
Carbon Monoxide		0.965	1.41	1.01
Ethylene		0.967	1.22	0.871
Helium		0.138	1.66	1.19
Hydrogen Chloride		1.256	1.40	1.00
Hydrogen		0.0695	1.40	1.00
Methane		0.553	1.26	0.900
Methyl Chloride		1.738	1.20	0.857
Nitrogen		0.966	1.40	1.00
Nitric Oxide		1.034	1.40	1.00
Nitrous Oxide		1.518	1.26	0.900
Oxygen		1.103	1.40	1.00
Sulphur Dioxide		2.208	1.25	0.893
Steam (dry saturated)	P_1			
	0-80		1.32	0.94
	80-245		1.30	0.93
	245-475		1.29	0.92
	475-800		1.27	0.91
	800-1050		1.26	0.90
	1050-1250		1.25	0.89
	1250-1400		1.23	0.88

(Reprinted with permission of Instrument Society of America; from ISA Handbook of Control Valves, 2nd Edition © Instrument Society of America, 1976.)

Nomenclature of ISA Formulas

a	Area of orifice or valve opening, $in.^2$
C	Coefficient of discharge, dimensionless. Includes effect of jet contraction and Reynolds number, mach number (gas at high velocities), turbulence.
C_d	Relative capacity (at rated C_v) $C_d = N_3 C_v / d^2$.
c_p	Specific heat at constant pressure.
c_v	Specific heat at constant volume.
C_v	Valve coefficient, $.38\, a\overline{K}/F_L$.
d	Valve inlet diameter, inches or mm.
D	Pipe diameter, inches or mm.
F	Velocity of approach factor $= \frac{1}{\sqrt{1-m^2}}$.
F_d	Experimentally determined factor relating valve C_v to an equivalent diameter for Reynolds number.
f	Weight fraction
F_F	Liquid critical pressure ratio factor, $F_F = p_{vc}/p_v$
F_k	Ratio of specific heats factor.
F_L	Pressure recovery factor. When the valve is not choked: $F_L = \sqrt{(p_1 - p_2) / (p_1 - p_{vc})}$
F_{LP}	Combined pressure loss and piping geometry factors for valve/fitting assembly.
F_P	Correction factor for piping around valve (e.g. reducers) $F_P C_v$ = effective C_v for valve/fitting assembly.
F_R	Correction factor for Reynolds number, where $F_R C_v$ = effective C_v.
F_s	Laminar, or streamline, flow factor.
g	Acceleration due to gravity.
G_f	Specific gravity of liquids at flowing temperature relative to water at 60°F or 15°C.
G_g	Specific gravity of gas relative to air with both at standard temperature and pressure.
h	Effective differential head, height of fluid.
$\overline{K}$	Flow coefficient = CF, dimensionless.
ΣK	Sum total of effective velocity head coefficients where $K(U^2/2g) = h$.
K_B	Bernoulli coefficient $= 1 - (d/D)^4$.

(Reprinted with permission of Instrument Society of America; from ISA Handbook of Control Valves, 2nd Edition © Instrument Society of America, 1976.)

(continued on next page)

Nomenclature of ISA Formulas (continued)

K_c	Cavitation index. Actually the ratio $\Delta p/(p_1 - p_v)$ at which cavitation measurably affects the value of C_v.
K_i	Inlet velocity head coefficient, $K_1 + K_{B1}$
K_1	Resistance coefficient for inlet fitting.
K_2	Resistance coefficient for outlet fitting.
k	Ratio of specific heats of a gas = c_p/c_v, dimensionless.
M	Molecular weight.
m	Ratio of areas.
N	Numerical constant
p	Absolute static pressure.
p_c	Thermodynamic critical pressure
p_r	Reduced pressure, p/p_c
p_v	Vapor pressure of liquid at inlet.
q	Volume rate of flow.
Re_v	Reynolds number for a valve.
T	Absolute temperature.
T_c	Thermodynamic critical temperature
T_r	Reduced temperature, T/T_c
U	Average velocity.
v	Specific volume $(1/\gamma)$.
w	Weight rate of flow.
x	Ratio of differential pressure to absolute inlet static pressure, $x = (p_1 - p_2)/p_1$
x_T	Terminal or ultimate value of x, used to establish expansion factor, Y.
x_{TP}	Value of x_T for valve/fitting assembly.
Y	Expansion factor. Ratio of flow coefficient for a gas to that for a liquid at the same Reynolds number (includes radial as well as longitudinal expansion effects).

Nomenclature of ISA Fomulas (continued)

Z	Compressibility factor. (See Nelson-Obert Compressibility Charts in Appendix.)
γ (gamma)	Specific weight (1/v).
Δ (delta)	Difference, (eg., $\Delta p = p_1 - p_2$).
μ (mu)	Viscosity, centipoise.
ν (nu)	Kinematic viscosity, centistokes
Σ (sigma)	Summation

SUBSCRIPTS

1	Upstream
2	Downstream
e	Effective value
f	Liquid
g	Gas
t	Theoretical
T	Terminal or ultimate value
vc	Vena contracta (point of minimum jet stream area)

Relief Valves

Liquids

Nonviscous

The formula for sizing valves for liquid service is taken from Appendix D of API RP 520. (Charts referred to below are in section 3 of this volume).

$$gpm = \frac{27.2\, AK_pK_wK_\nu \sqrt{p - p_b}}{\sqrt{G}}$$

or

$$A = \frac{GPM}{27.2\, K_pK_wK_\nu}\sqrt{\frac{G}{p - p_b}}$$

where

GPM = Flow rate at the selected percentage overpressure, in U.S. gallons per minute.

A = Effective discharge area in square inches.

K_p = Capacity correction factor due to overpressure. Many, if not most, relief valves in liquid service are sized on the basis of 25% overpressure, in which case K_p = 1.00. The factor for other percentages of overpressure can be obtained from Chart 3.73, p. 230.

K_w = Capacity correction factor due to backpressure. If the backpressure is atmospheric, the factor can be disregarded, or K_w = 1.00. Conventional valves in backpressure service require no special correction: Kw = 1.00. Balanced bellows valves in backpressure service will require the correction factor as determined from Chart 3.74, p. 231.

K_ν = Capacity correction factor due to viscosity. For most applications, viscosity may not be significant, in which case K_ν = 1.00.

p = Set pressure at which relief valve is to begin opening, in pounds per square inch gauge.
p_b = Backpressure, in pounds per square inch gauge.
G = Specific gravity of the liquid at the flowing temperature referred to water = 1.00 at 70°F.

Viscous

When sizing relief areas for viscous liquid service the Reynolds number, R, must be determined so that the correction factor, K_v, can be determined. R is calculated from the equation below, using a preliminary area calculated from the non-viscous formula. The orifice size is then chosen as the first size greater than the calculated area. When R is found and K_v is obtained from chart 3.75, the area is then calculated on the basis of the viscous fluid. If the calculated area is larger than the chosen area used to determine R, the next larger standard orifice size is used until the correct Reynolds number is used for the orifice size eventually used.

The Reynolds number, R, is determined from either of the following relationships:

$$R = \frac{\text{GPM } (2{,}800G)}{\mu\sqrt{A}}$$

$$R = \frac{12{,}700 \text{ GPM}}{\mu\sqrt{A}}$$

where

GPM = Flow rate at the flowing temperature, in U.S. gallons per minutes.
G = Specific gravity of the liquid at the flowing temperature referred to water = 1.00 at 70°F.
μ = Absolute viscosity at the flowing temperature, in centipoises.
U = Viscosity at the flowing temperature, in Saybolt Universal Seconds.
A = Effective discharge area, in square inches (from manufacturers' standard orifice areas).

Gases or Vapors

For gas or vapor service, any one of the following formulas may be used.

$$W = \frac{CKAP_1K_b\sqrt{M}}{TZ}$$

or

$$A = \frac{W\sqrt{TZ}}{CKP_1K_b\sqrt{M}}$$

$$V = \frac{6.32\ CKAP_1K_b}{\sqrt{TZM}}$$

or

$$A = \frac{V\sqrt{TZM}}{6.32\ CKP_1K_b}$$

$$V = \frac{1.175\ CKAP_1K_b}{\sqrt{TZG}}$$

or

$$A = \frac{V\sqrt{TZG}}{1.175\ CKP_1K_b}$$

where

W = Flow through valve, in pounds per hour.
V = Flow through valve, in standard cubic feet per minute at 14.7 psia and 60°F.
C = Coefficient determined by the ratio of the specific heats of the gas or vapor at standard conditions. This can be obtained from Chart 3.76 and Chart 3.77 or Table D-1, page 42 of API RP 520, or Table 6.3 of Volume 2.
K = Coefficient of discharge, which value is obtainable from the valve manufacturer. The K for a number of nozzle-type valves is 0.975.
A = Effective discharge area of the valve, in square inches.
P_1 = Upstream pressure, in pounds per square inch absolute. This is the set pressure multiplied by 1.10 or 1.20 (depending on the amount of accumulation permissible) plus the atmospheric pressure, in pounds per square inch absolute.
K_b = Capacity correction factor due to back pressure. This can be obtained from Chart 3.78, which applies to conventional safety relief valves, or from Chart 3.79, which applies to balanced bellows valves. The correction factor value should read from the curve specifically applying to the type of valve under consideration. In connection with Chart 3.79 for set pressures lower than 50 psig, the valve manufacturer should be consulted for the proper value of correction factor K_b.
M = Molecular weight of the gas or vapor. Various handbooks carry tables of molecular weights of materials, but the composition of the flowing gas or vapor is seldom the same as that listed in the tables. This value should be obtained from the process data.

T = Absolute temperature of the inlet vapor, in degrees fahrenheit + 460.

Z = Compressibility factor for the deviation of the actual gas from a perfect gas, a ratio evaluated at inlet conditions. This can be obtained from Chart 3.13 or Chart 3.12.

G = Specific gravity of gas referred to air = 1.00 at 60°F and 14.7 psia.
= Molecular weight of the gas/28.93.

Table 6.3 of volume 2 (or Table D-1, page 42 of API RP 520) complements Chart 3.76 where

$$k = \frac{c_p}{c_v}$$

the ratio of specific heats of an ideal gas, and Chart 3.80 where n = isentropic expansion coefficient of an actual gas (such as a paraffin hydrocarbon).

When k or n cannot be determined, it is suggested that c = 315.

The discharge area for safety valves on gas-containing vessels exposed to open fires can be determined by use of the following formula:

$$A = \frac{F' A_s}{\sqrt{P_1}}$$

where

A = effective discharge area of the valve, in square inches.

F' = an operating factor determined from Chart 3.72.

A_s = exposed surface area of the vessel, in square feet.

P_1 = upstream pressure, in pounds per square inch absolute. This is the set pressure multiplied by 1.10 or 1.20 (depending on the overpressure permissible) plus the atmospheric pressure, in pounds per square inch absolute.

Steam

Safety and safety relief valves in steam service are sized by the following formula taken from API RP 520, Appendix D:

$$W = 50\, A P_1 K_{sh}$$

or

$$A = \frac{W}{50\, P_1 K_{sh}}$$

Where

W = Flow rate, in pounds per hour

A = Effective discharge area, in square inches

P_1 = Upstream pressure, in pounds per square inch absolute. This is the set pressure multiplied by 1.03 or 1.10 (depending on the permissible accumulation) plus the atmospheric pressure, in pounds per square inch absolute. The ASME power boiler code applications are permitted only 3 percent accumulation and are rated at only 90 percent of actual capacity. Other applications may need to conform to the ASME unfired pressure vessel code, which permits 10 percent accumulation.

K_{sh} = correction factor due to the amount of superheat in the steam. This can be obtained from Table 6.4 of Volume 2 or from Table C-1, page 48 of API RP 520. For saturated steam at any pressure, the factor K_{sh} = 1.0.

Appendix A

The two major systems of measurement in general use throughout the world, are usually called the *British* and the *metric*.

The British system of measurement is many centuries old; the yard and the pound, with their multiples (e.g., mile) and divisions (e.g., ounce) are basic in this system.

The metric system, established in the eighteenth century, has been adopted for general use by most countries, the notable exception being the United States. Nearly everywhere it is used for precise measurements in science. The meter and the kilogram, with their decimal multiples (e.g., kilometer) and fractions (e.g., gram) are basic in this system. The metric system is exclusively decimal, whereas British units of the same kind are related almost at random.

In the metric system, as in the British system, all but the simplest of measurements are expressed as combinations of units. For example, grams per square meter or ounces per square yard. Such combinations are called complex units.

In the metric system used in chemistry, complex quantities are measured mainly in terms of three units – the centimeter, the gram, and the second – in the mass-length-time system of dimensions. This is called the CGS System.

In problems involving units of mechanics and electromagnetic units, the MKSA or Giorgi System is used. This system is based on the meter, the kilogram, the second, and the ampere.

Units of the two metric systems and the British gravitational system are given below.

	Units for Some Basic Quantities		
	British	Metric	
Quantity	Gravitational	Mks	Cgs
Length	ft	m	cm
Time	sec	sec	sec
Velocity	ft/sec	m/sec	cm/sec
Acceleration	ft/sec^2	m/sec^2	cm/sec^2
Force	lb	newton	dyne
Mass	slug	kg	gm

Another system, the International System of units (SI), was adopted by the eleventh General Conference on Weights and Measures in 1960 and was endorsed by the International Organization for Standardization. It has been recommended by the national standardizing bodies in many countries. As a matter of information, these basic units are listed below.

Quantity	Name of Unit	Unit Symbol
length	metre	m
mass	kilogramme	kg
time	second	s
electric current	ampere	A
thermodynamic temperature	degree Kelvin	°K
luminous intensity	candela	cd
amount of substance	mole	mol
plane angle	radian	rad
solid angle	steradian	sr

7 Typical Installation Details

William G. Andrew,
Robert E. Schneider,
O.K. Payne

Introduction

Instrument installation details serve several purposes:

1. They indicate the proper mounting of instruments including supports, orientation and arrangement.
2. They detail the hook-up of instruments including power supply (air or electric), signal output (pneumatic or electronic) and connections to process lines. Hookup details may be rather complicated for special devices such as analyzers.
3. They indicate the quantity, sizes and materials of construction for the equipment which serve as a convenient takeoff for procurement of materials.
4. They serve as a convenient communications link between the instrument group and other groups (piping, vessel and electrical) so that the necessary services are provided for the instruments.

Instrument installations should be standardized as much as possible to minimize the ease and cost of installation. One drawing detail may serve as a guideline for more than one device.

Several precautions should be observed in designing installation details:

1. All materials must be suitable for the process operating conditions (pressure, temperature and corrosion properties) for normal as well as extreme conditions. Piping specifications usually designate materials for various services and instrument specifications list deviations from the piping specifications that are allowed or required. Examples of such deviations include the use of stainless steel tubing instead of standard pipe, or using 3,000 psig bar stock gauge valves instead of 600 psig forged steel gate valves. Other frequent deviations from piping specifications are the use of tubing and tubing fittings or screwed fittings instead of flanges or welding fittings that are required for the main pipe lines.

2. Where process lines contain vapors that may condense, the connecting instrument lines should be arranged to be self-draining, or seal pots should be installed to insure constant head pressure conditions at the instrument sensing elements.

3. Instrument lead lines containing dangerous fluids (i.e., toxic, high pressure, high temperature, or corrosive) must be properly vented or drained so that misoperation will not result in personnel or property damage.

4. Instrument lead lines containing fluids that may freeze at ambient conditions must be protected by heat tracing and insulating or by sealing with a non-freezing fluid compatible with the process fluid.

5. There are other requirements that need to be observed for installation of instrument devices. Reference should be made to vendors' instructions to insure proper installation of various devices, particularly if

the devices serve unusual functions or are built for special purposes.

In most of the details presented, the primary purpose is to show orientation, mounting and connections to the process. To avoid repetition where transmission devices are used, signal systems have been shown for only one type of signal transmission, pneumatic, rather than essentially duplicating details for both pneumatic and electronic systems.

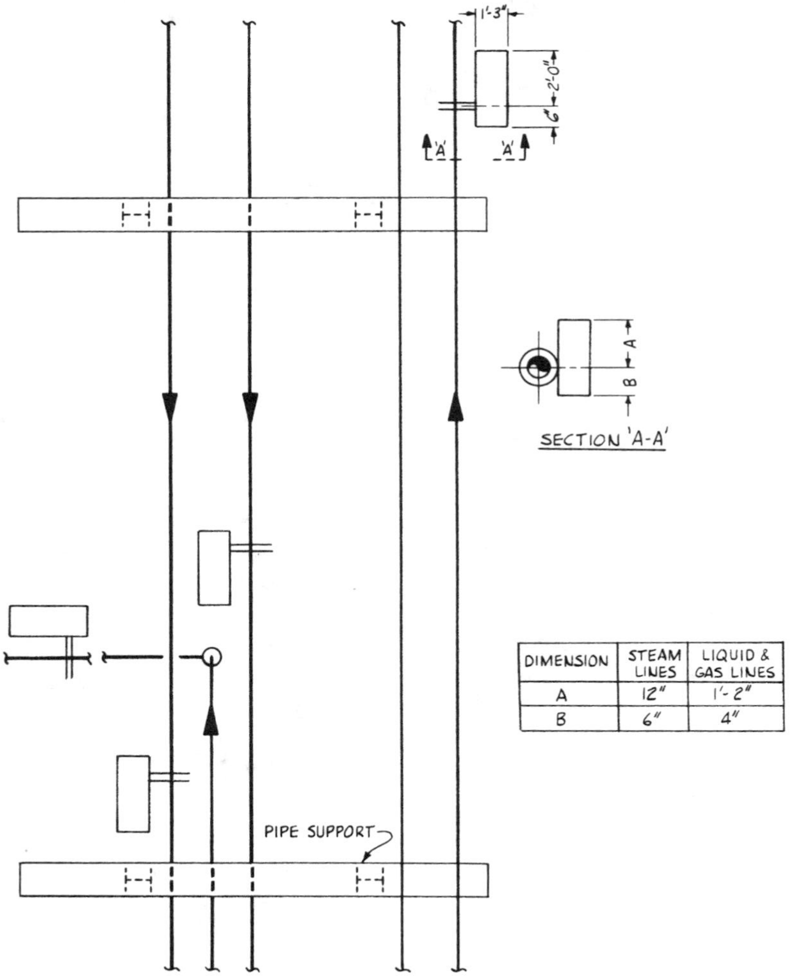

DIMENSION	STEAM LINES	LIQUID & GAS LINES
A	12"	1'-2"
B	6"	4"

Figure 7.1. Block space requirements for flow transmitters and primary elements in pipe racks.

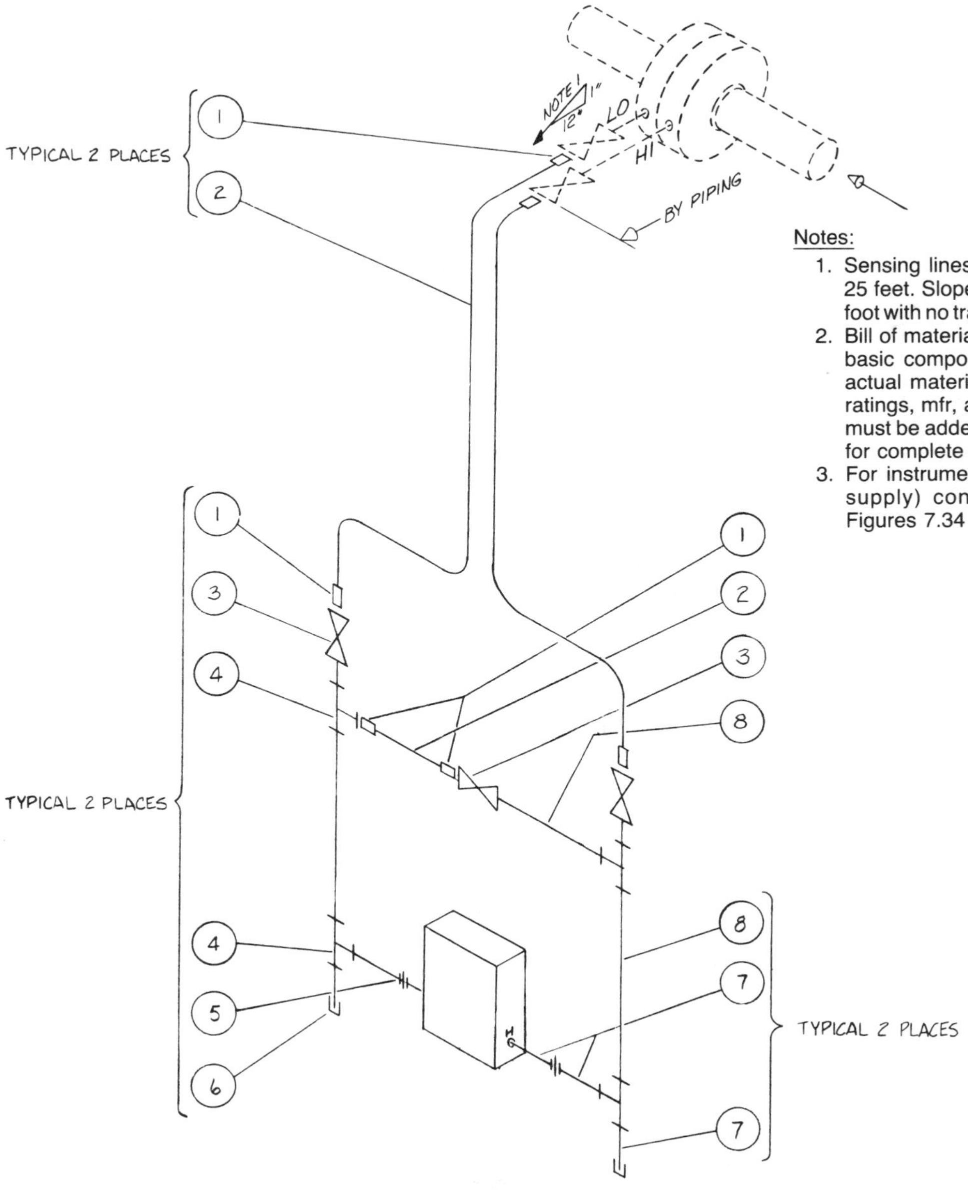

Notes:

1. Sensing lines not to exceed 25 feet. Slope at least 1" per foot with no trapped sections.
2. Bill of material describes the basic components required; actual material composition, ratings, mfr, and/or part no's must be added as necessary for complete definition.
3. For instrument air (signal & supply) connections, see Figures 7.34 & 7.35.

BILL OF MATERIAL		
ITEM	QTY	DESCRIPTION
1	6	½" T x ½" NPT MALE CONNECTOR
2	A/R	½" O.D. x .049" WALL SEAMLESS TUBING
3	3	½" NPT MxF VALVE
4	4	½" NPT TEE
5	2	½" NPT UNION
6	2	½" NPT CAP
7	6	½" NPT x 3" LG. NIPPLE TBE
8	A/R	½" NPT PIPE

Figure 7.2. Typical differential pressure-sensing instrument in liquid service—tubing connections with block and equalizing valves.

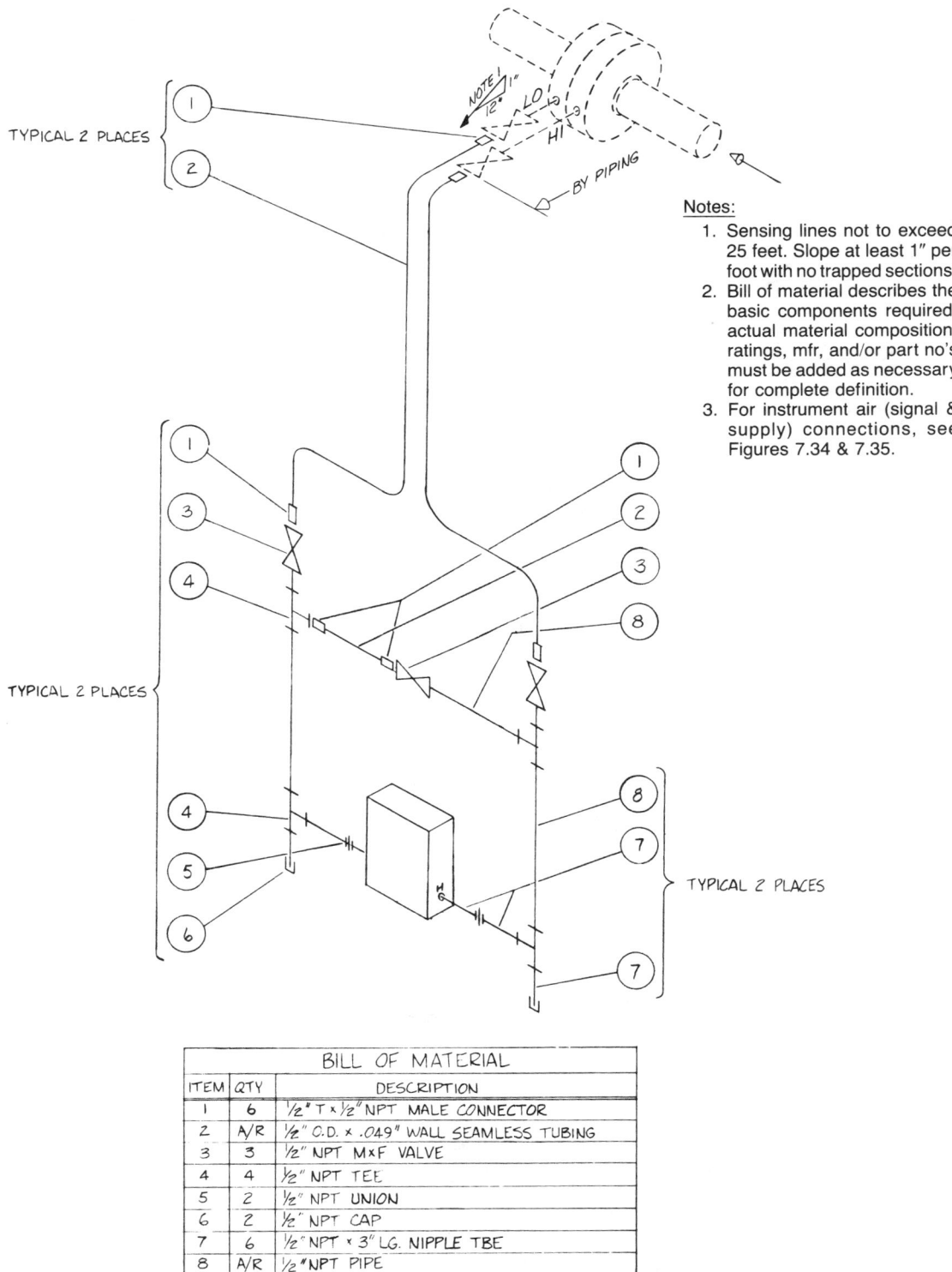

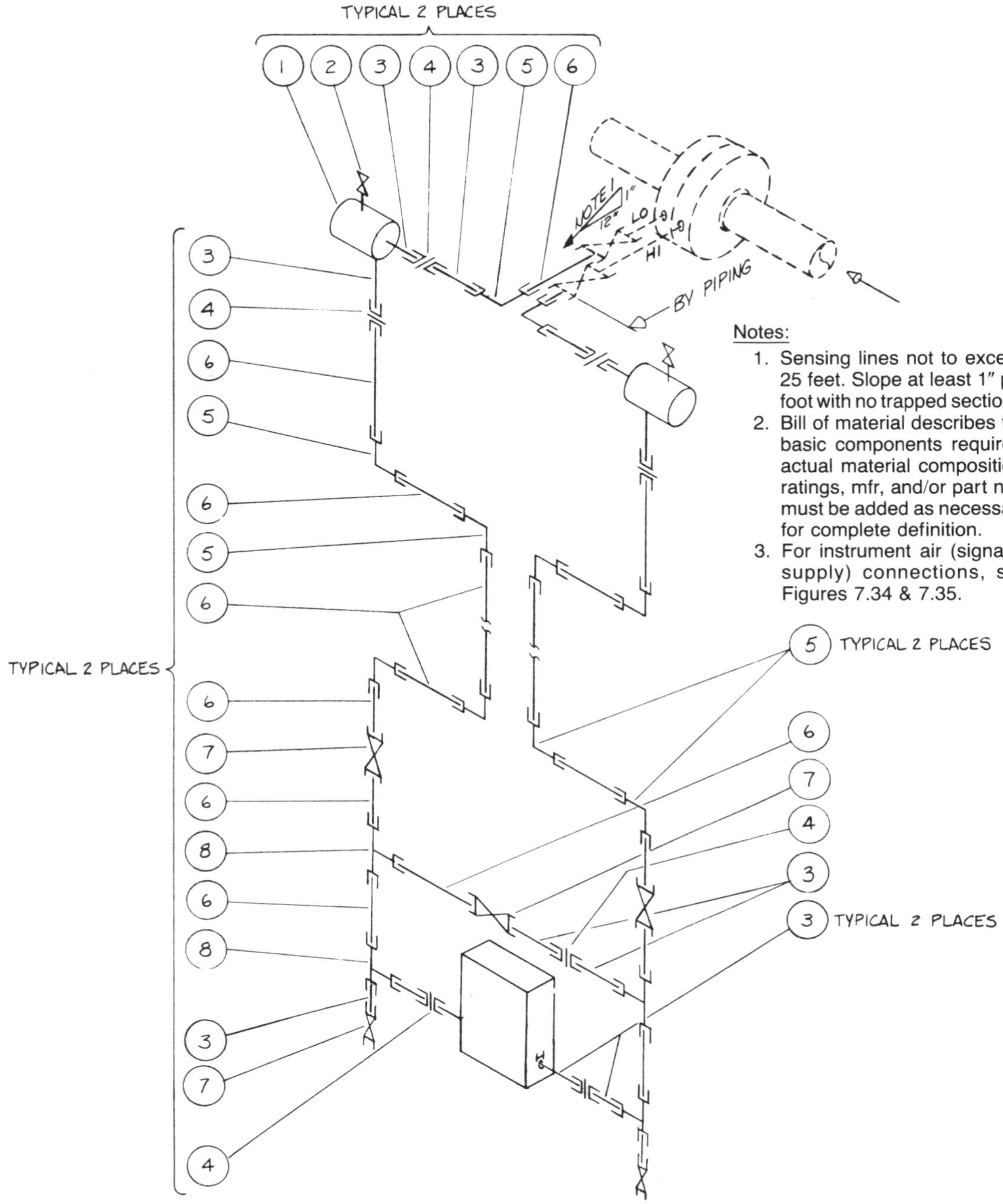

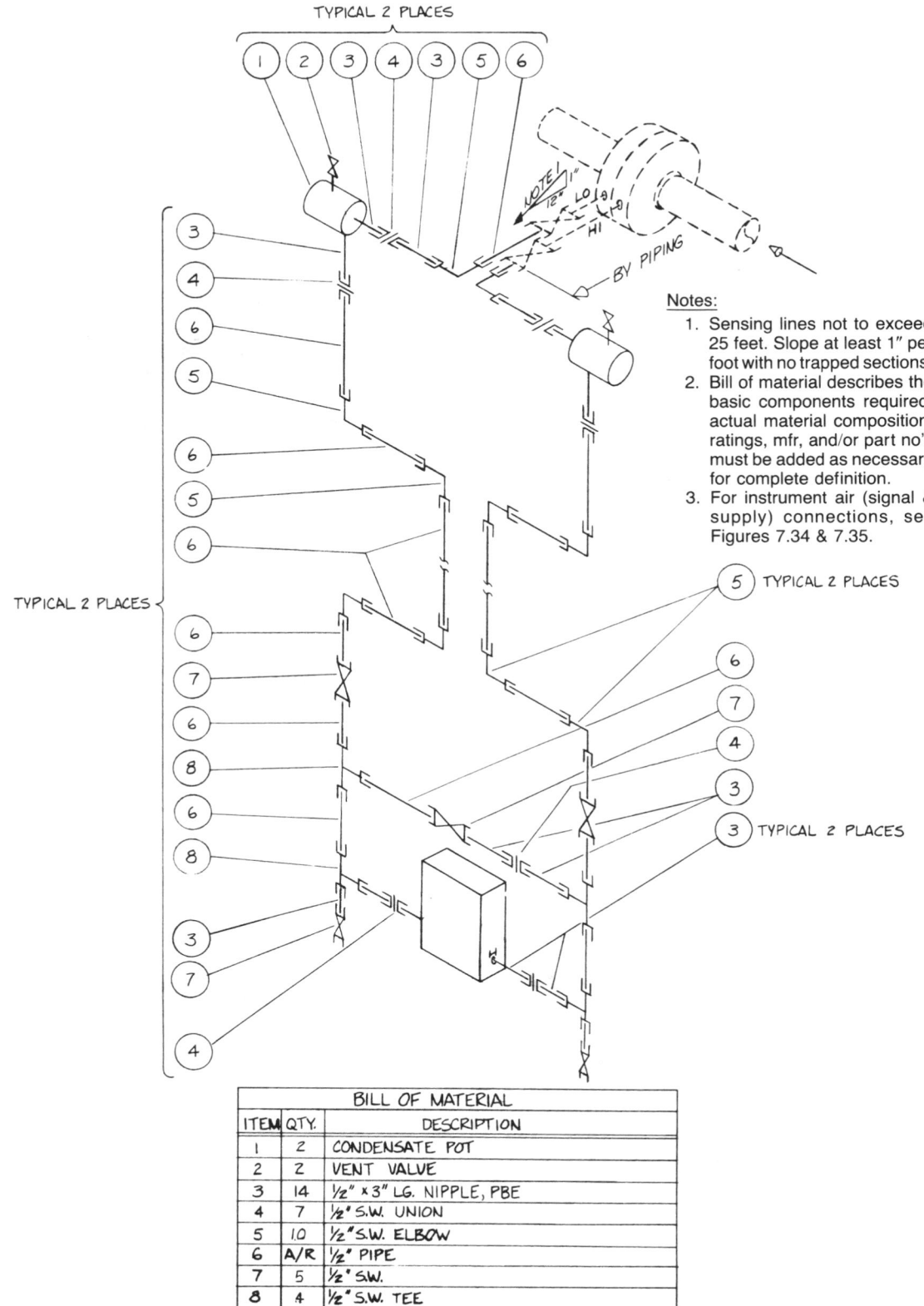

Notes:
1. Sensing lines not to exceed 25 feet. Slope at least 1" per foot with no trapped sections.
2. Bill of material describes the basic components required; actual material composition, ratings, mfr, and/or part no's must be added as necessary for complete definition.
3. For instrument air (signal & supply) connections, see Figures 7.34 & 7.35.

BILL OF MATERIAL		
ITEM	QTY.	DESCRIPTION
1	2	CONDENSATE POT
2	2	VENT VALVE
3	14	½" × 3" LG. NIPPLE, PBE
4	7	½" S.W. UNION
5	10	½" S.W. ELBOW
6	A/R	½" PIPE
7	5	½" S.W.
8	4	½" S.W. TEE

Figure 7.3. Typical differential pressure-sensing instrument in high-pressure steam or condensable vapor service—pipe (socket weld) connections with block, equalizing, and blowdown valves.

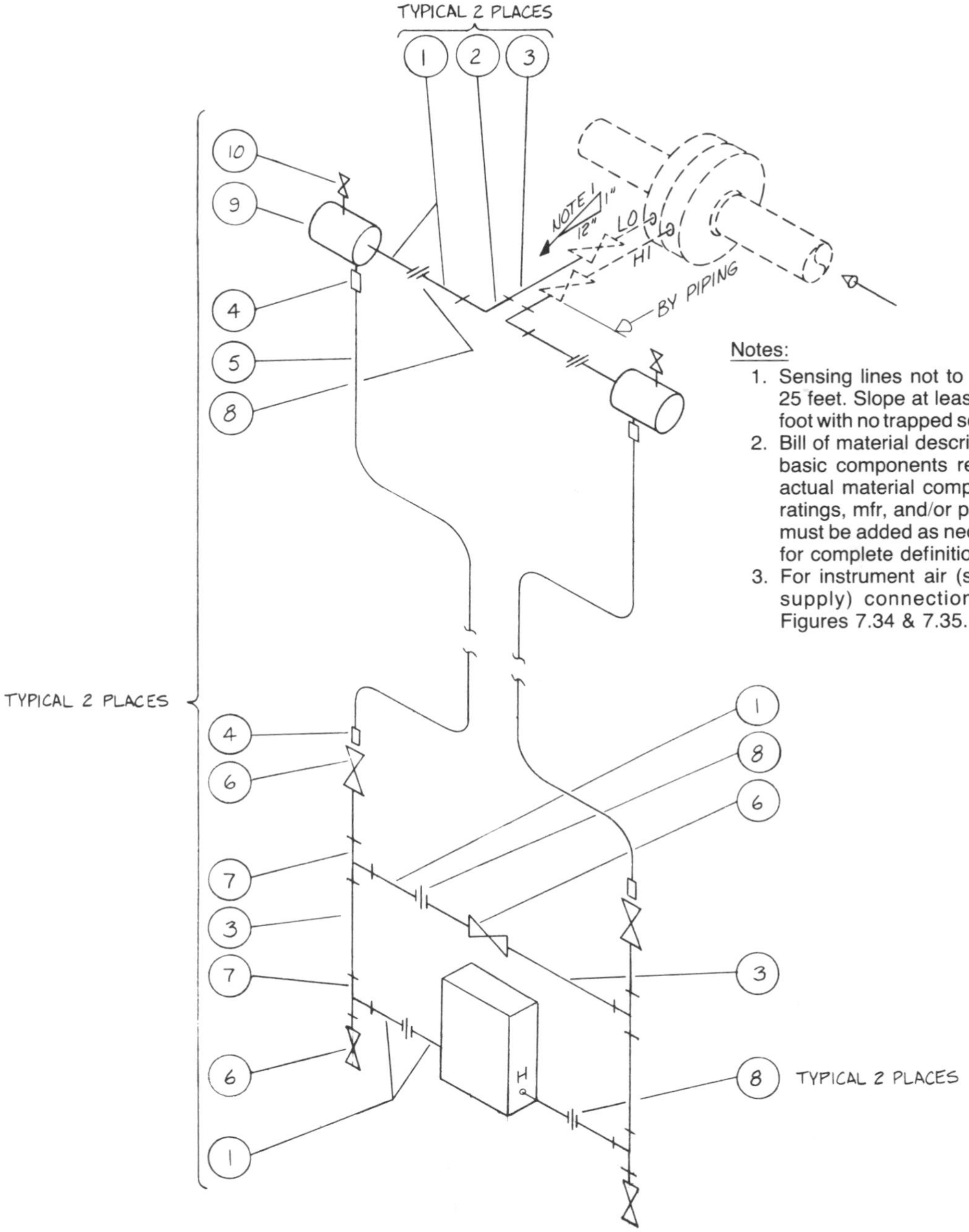

Notes:

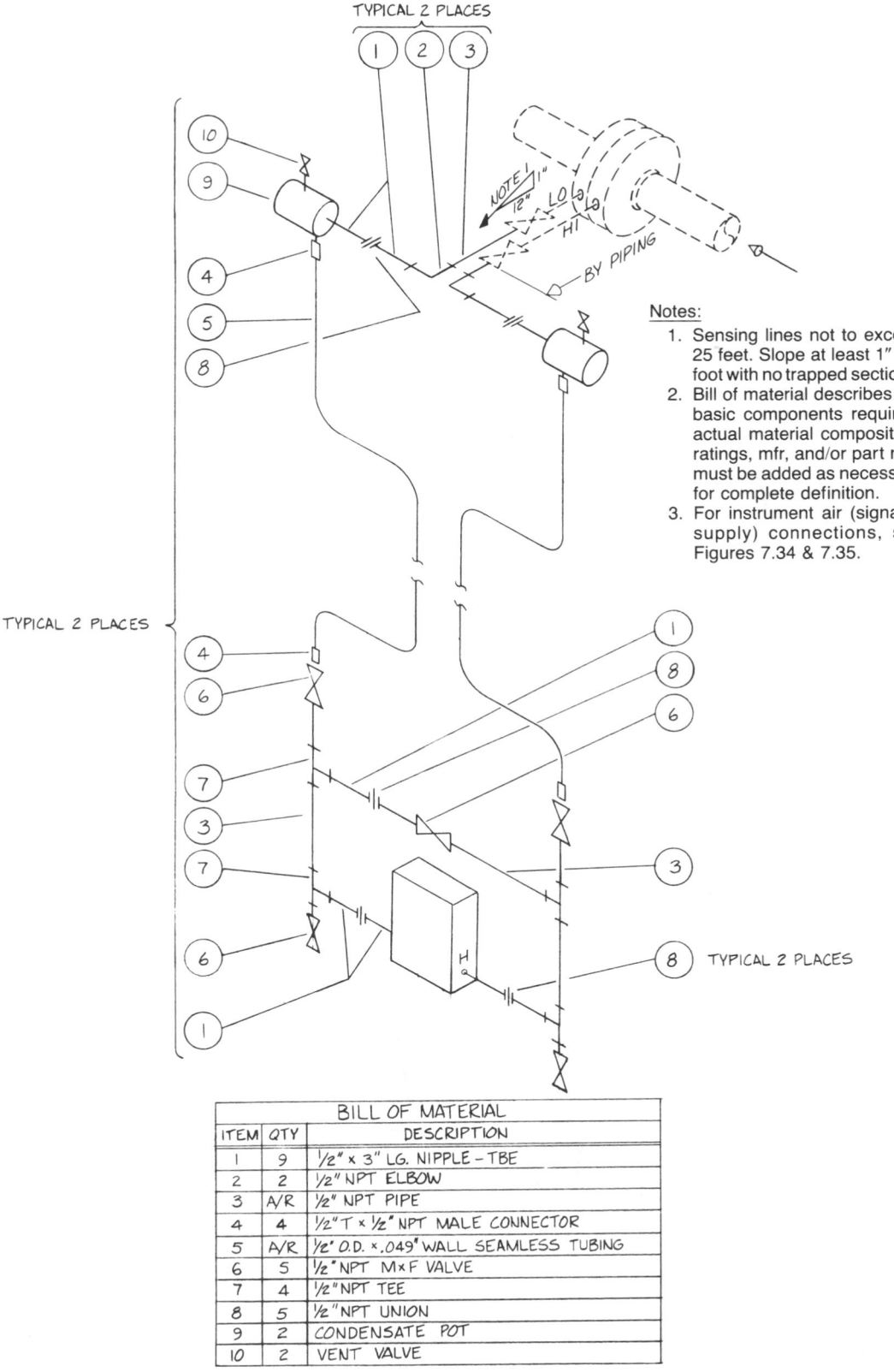

1. Sensing lines not to exceed 25 feet. Slope at least 1" per foot with no trapped sections.
2. Bill of material describes the basic components required; actual material composition, ratings, mfr, and/or part no's must be added as necessary for complete definition.
3. For instrument air (signal & supply) connections, see Figures 7.34 & 7.35.

BILL OF MATERIAL		
ITEM	QTY	DESCRIPTION
1	9	½" x 3" LG. NIPPLE - TBE
2	2	½" NPT ELBOW
3	A/R	½" NPT PIPE
4	4	½"T x ½" NPT MALE CONNECTOR
5	A/R	½" O.D. x .049" WALL SEAMLESS TUBING
6	5	½" NPT MxF VALVE
7	4	½" NPT TEE
8	5	½" NPT UNION
9	2	CONDENSATE POT
10	2	VENT VALVE

Figure 7.4. Typical differential pressure-sensing instrument in steam or condensable vapor service—tubing connections with block, equalizing, and blowdown valves.

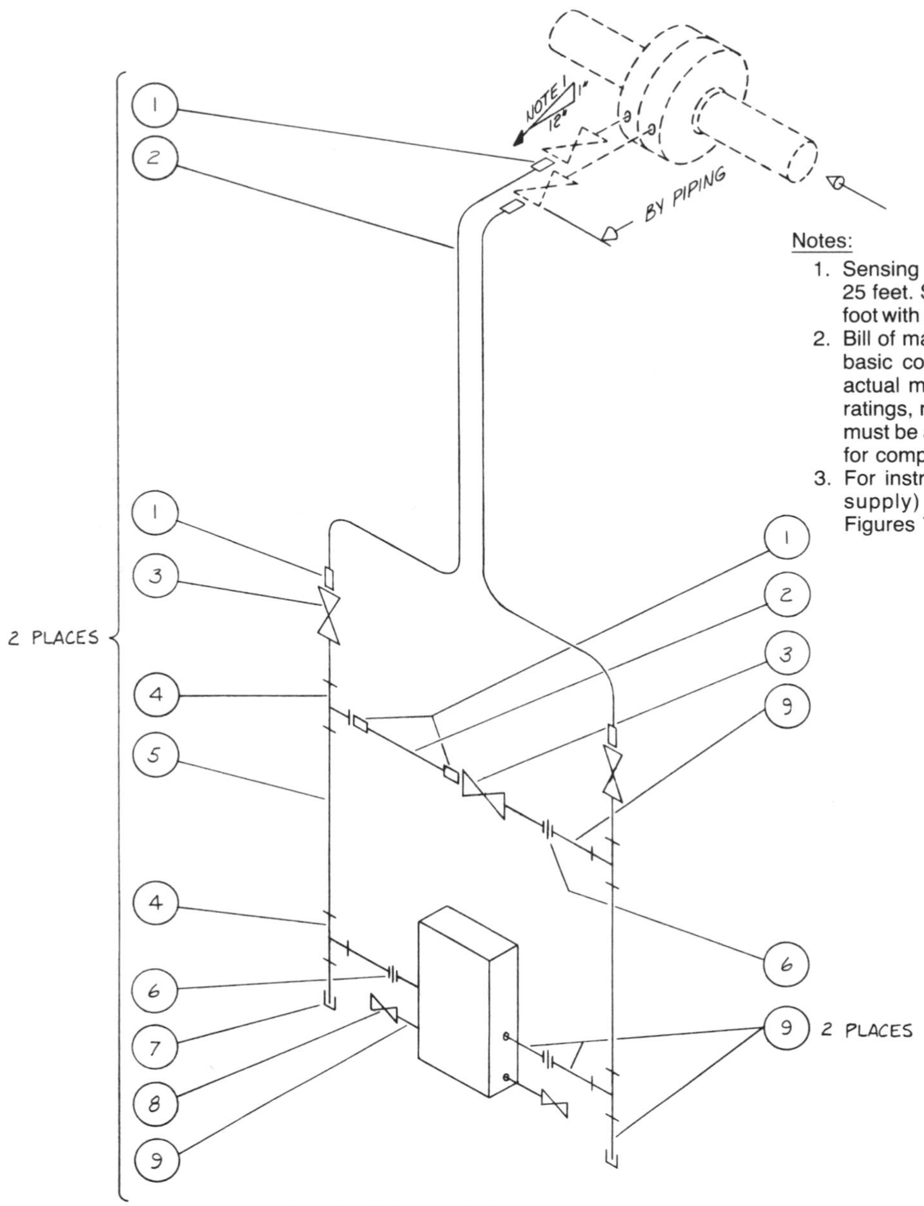

Notes:

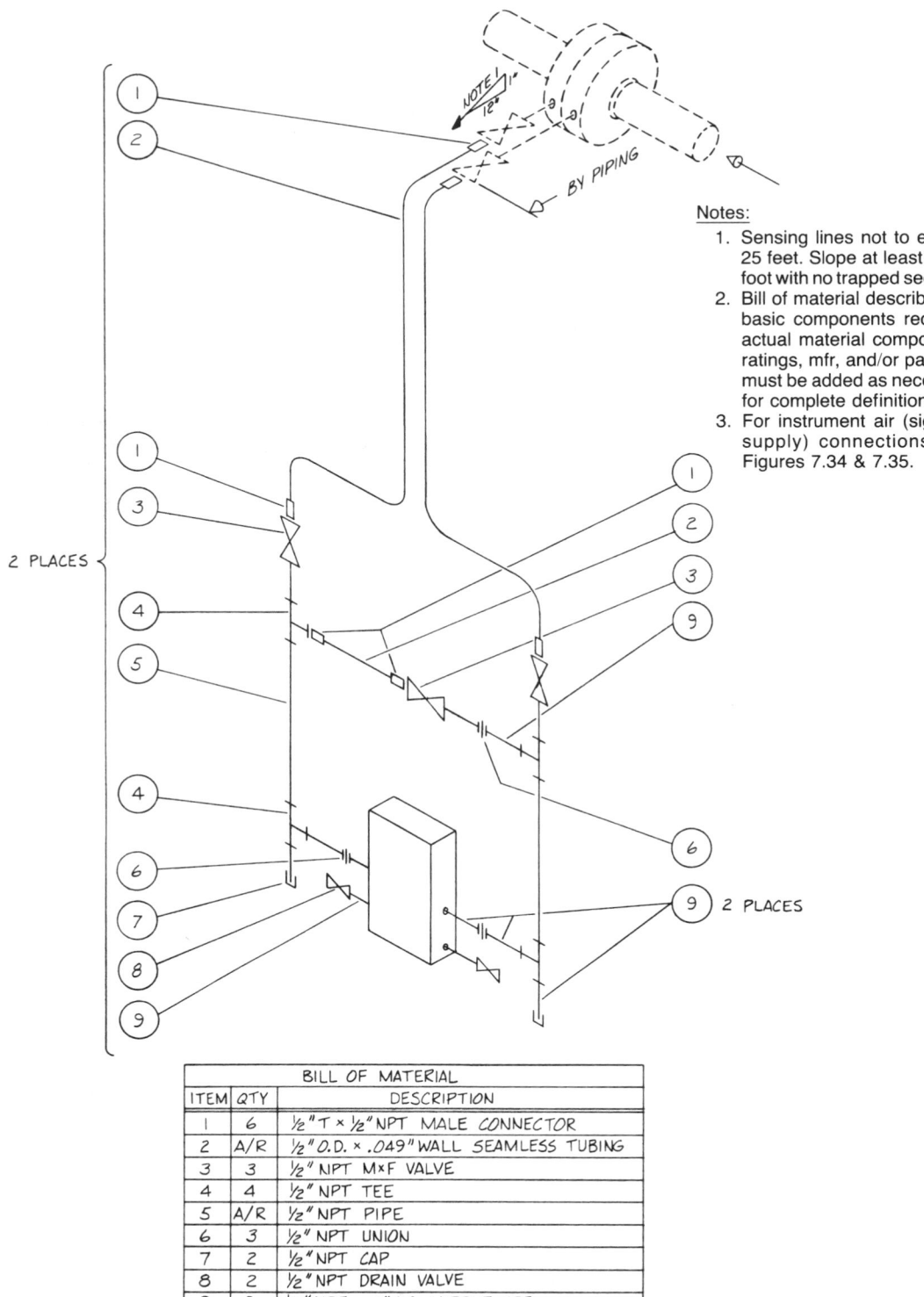

1. Sensing lines not to exceed 25 feet. Slope at least 1″ per foot with no trapped sections.
2. Bill of material describes the basic components required; actual material composition, ratings, mfr, and/or part no's must be added as necessary for complete definition.
3. For instrument air (signal & supply) connections, see Figures 7.34 & 7.35.

BILL OF MATERIAL

ITEM	QTY	DESCRIPTION
1	6	½"T × ½"NPT MALE CONNECTOR
2	A/R	½"O.D. × .049"WALL SEAMLESS TUBING
3	3	½" NPT M×F VALVE
4	4	½" NPT TEE
5	A/R	½" NPT PIPE
6	3	½" NPT UNION
7	2	½" NPT CAP
8	2	½" NPT DRAIN VALVE
9	9	½" NPT × 3" LG. NIPPLE, TBE

Figure 7.5. Typical differential pressure-sensing instrument in dry gas service—tubing connections with block, equalizing, and instrument vent valves.

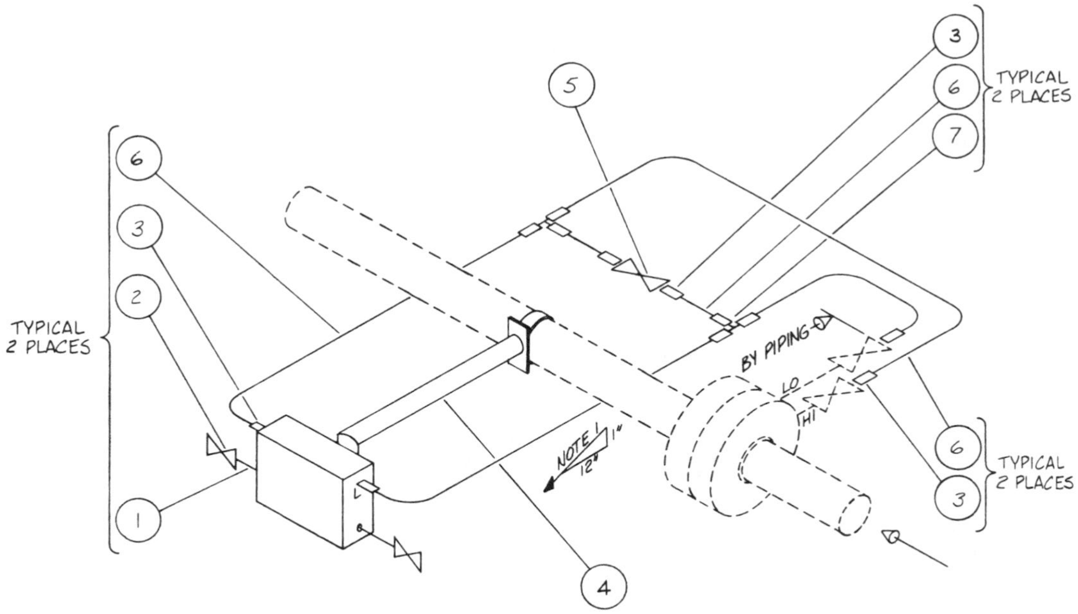

BILL OF MATERIAL		
ITEM	QTY	DESCRIPTION
1	2	½" NPT × 3" LG. NIPPLE, TBE
2	2	½" NPT DRAIN VALVE
3	6	½" T × ½" NPT MALE CONNECTOR
4	1	LEVELING SADDLE
5	1	½" NPT BLOCK VALVE
6	A/R	½" O.D. × .049" WALL SEAMLESS TUBING
7	2	½" TUBING TEE

Notes:

1. Sensing lines not to exceed 25 feet. Slope at least 1" per foot with no trapped sections.
2. Bill of material describes the basic components required; actual material composition, ratings, mfr, and/or part no's must be added as necessary for complete definition.
3. For instrument air (signal & supply) connections, see Figures 7.34 & 7.35.

Figure 7.6. Typical differential pressure-sensing flow indicator in liquid service—line mounted with tubing connections.

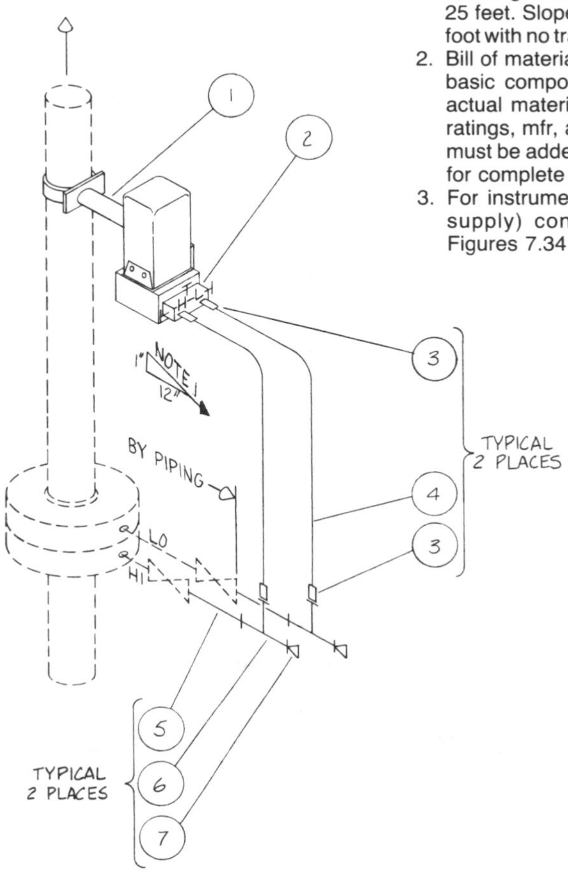

Notes:

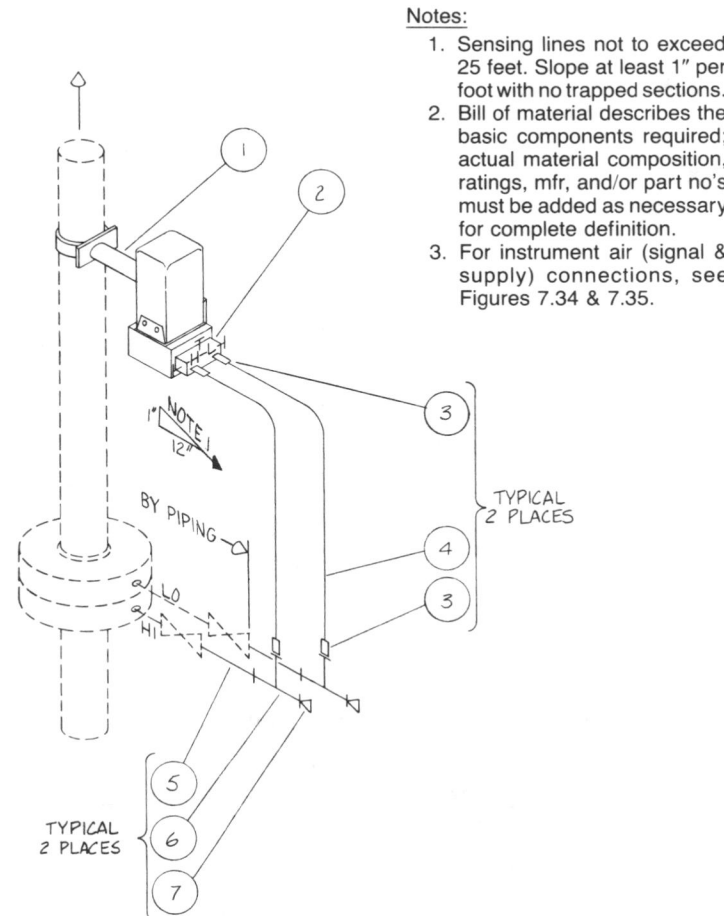

1. Sensing lines not to exceed 25 feet. Slope at least 1" per foot with no trapped sections.
2. Bill of material describes the basic components required; actual material composition, ratings, mfr, and/or part no's must be added as necessary for complete definition.
3. For instrument air (signal & supply) connections, see Figures 7.34 & 7.35.

BILL OF MATERIAL		
ITEM	QTY	DESCRIPTION
1	1	LEVELING SADDLE
2	1	3-VALVE MANIFOLD
3	4	½"T × ½" NPT MALE CONNECTOR
4	A/R	½" O.D. × .049" WALL SEAMLESS TUBING
5	2	½" NPT × 3" LG NIPPLE, TBE
6	2	½" NPT TEE
7	2	½" NPT PLUG

Figure 7.7. Typical differential pressure-sensing flow transmitter in dry gas service—vertical run, line mounted, tubing connections with rod-out provision.

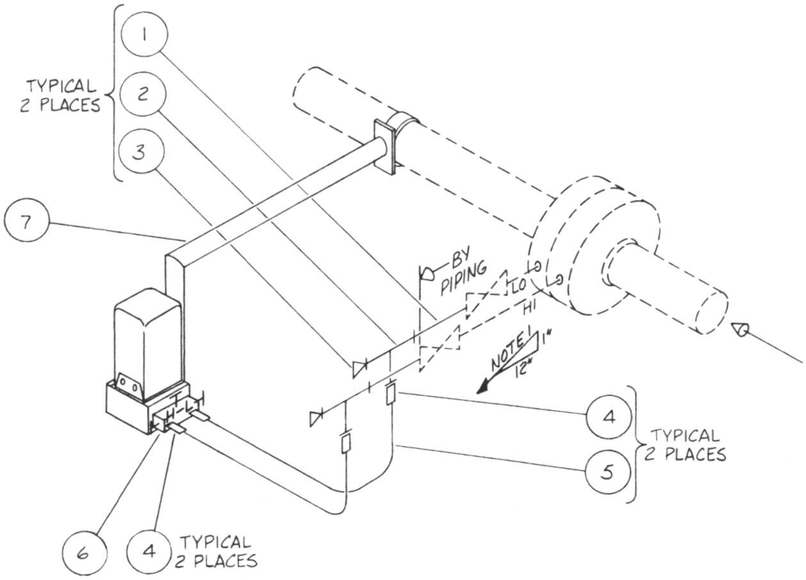

BILL OF MATERIAL		
ITEM	QTY	DESCRIPTION
1	2	½" NPT × 3" LG. NIPPLE, TBE
2	2	½" NPT TEE
3	2	½" NPT PLUG
4	4	½" T × ½" NPT MALE CONNECTOR
5	A/R	½" O.D. × .049" WALL SEAMLESS TUBING
6	1	3-VALVE MANIFOLD
7	1	LEVELING SADDLE

Notes:

1. Sensing lines not to exceed 25 feet. Slope at least 1" per foot with no trapped sections.
2. Bill of material describes the basic components required; actual material composition, ratings, mfr, and/or part no's must be added as necessary for complete definition.
3. For instrument air (signal & supply) connections, see Figures 7.34 & 7.35.

Figure 7.8. Typical differential pressure-sensing flow transmitter in liquid service—horizontal run, line mounted, tubing connections with rod-out provision.

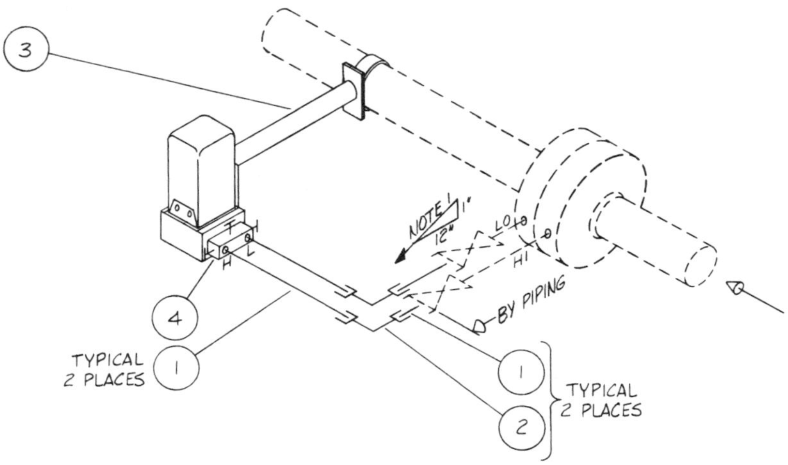

HORIZONTAL LINE

BILL OF MATERIAL		
ITEM	QTY	DESCRIPTION
1	A/R	½" PIPE
2	2	½" SW ELBOW
3	1	LEVELING SADDLE
4	1	3-VALVE MANIFOLD

Notes:

1. Sensing lines not to exceed 25 feet. Slope at least 1" per foot with no trapped sections.
2. Bill of material describes the basic components required; actual material composition, ratings, mfr, and/or part no's must be added as necessary for complete definition.
3. For instrument air (signal & supply) connections, see Figures 7.34 & 7.35.

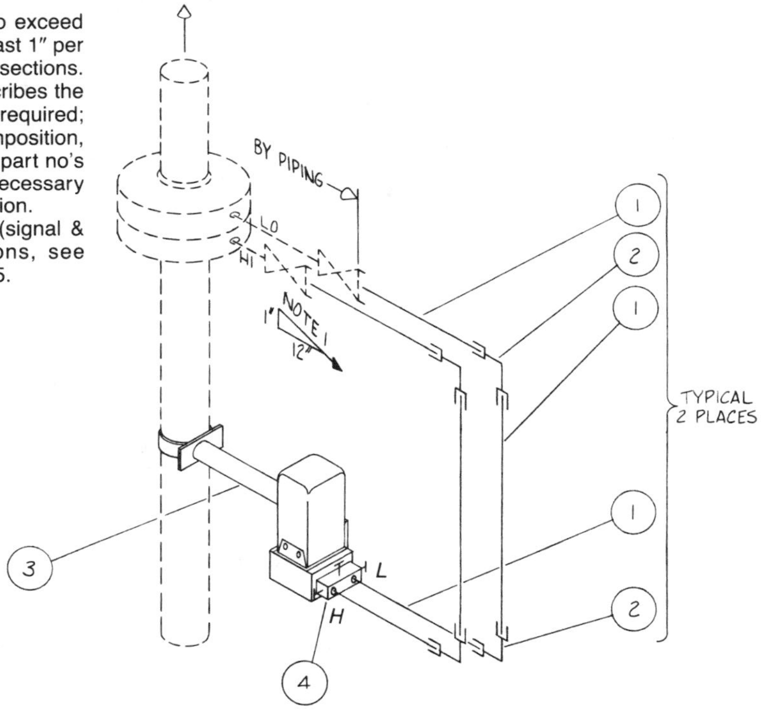

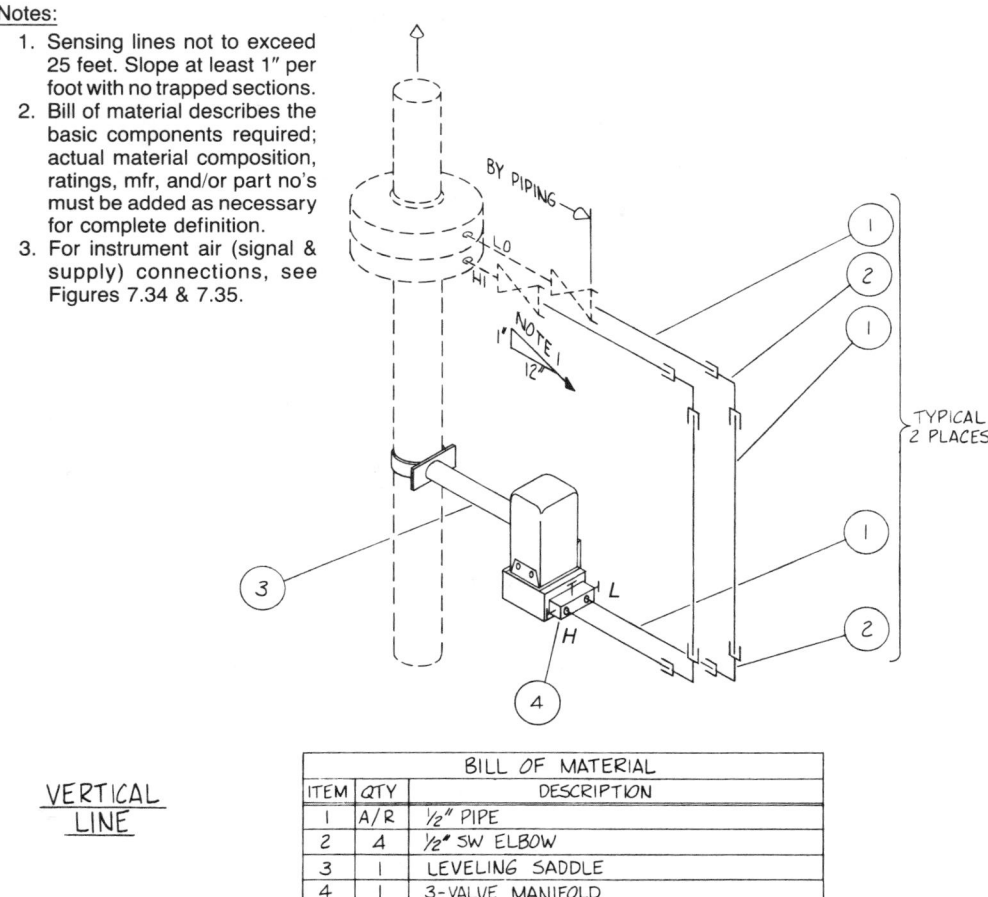

VERTICAL LINE

BILL OF MATERIAL		
ITEM	QTY	DESCRIPTION
1	A/R	½" PIPE
2	4	½" SW ELBOW
3	1	LEVELING SADDLE
4	1	3-VALVE MANIFOLD

Figure 7.9 Typical differential pressure-sensing flow transmitter in high-pressure steam service— line mounted with pipe (socket weld) connections.

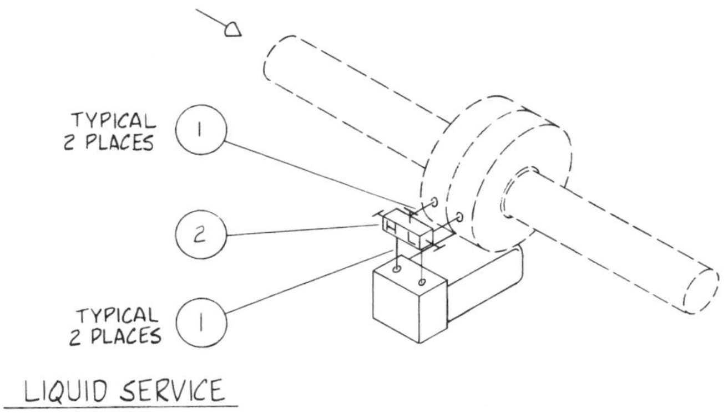

BILL OF MATERIAL		
ITEM	QTY	DESCRIPTION
1	4	½" NPT × 3" LG NIPPLE, TBE
2	1	3-VALVE MANIFOLD

Figure 7.10. Typical differential pressure-sensing flow transmitter—close-coupled in liquid service.

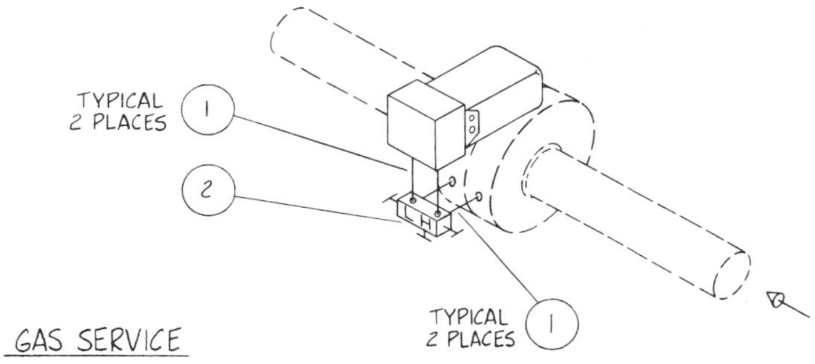

BILL OF MATERIAL		
ITEM	QTY	DESCRIPTION
1	4	½" NPT × 3" LG NIPPLE, TBE
2	1	3-VALVE MANIFOLD

Notes:

1. Bill of material describes the basic components required; actual material composition, ratings, mfr, and/or part no's must be added as necessary for complete definition.
2. For instrument air (signal & supply) connections, see Figures 7.34 & 7.35.

Figure 7.11. Typical differential pressure-sensing flow transmitter—close-coupled in gas service.

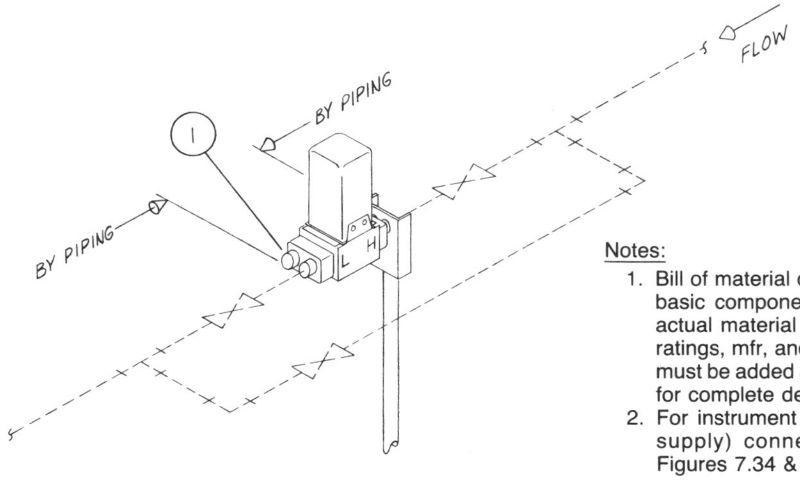

Notes:

1. Bill of material describes the basic components required; actual material composition, ratings, mfr, and/or part no's must be added as necessary for complete definition.
2. For instrument air (signal & supply) connections, see Figures 7.34 & 7.35.

BILL OF MATERIAL		
ITEM	QTY	DESCRIPTION
1	1	INTEGRAL ORIFICE MANIFOLD

Figure 7.12. Typical differential pressure-sensing transmitter with in-line integral orifice manifold assembly and by-pass.

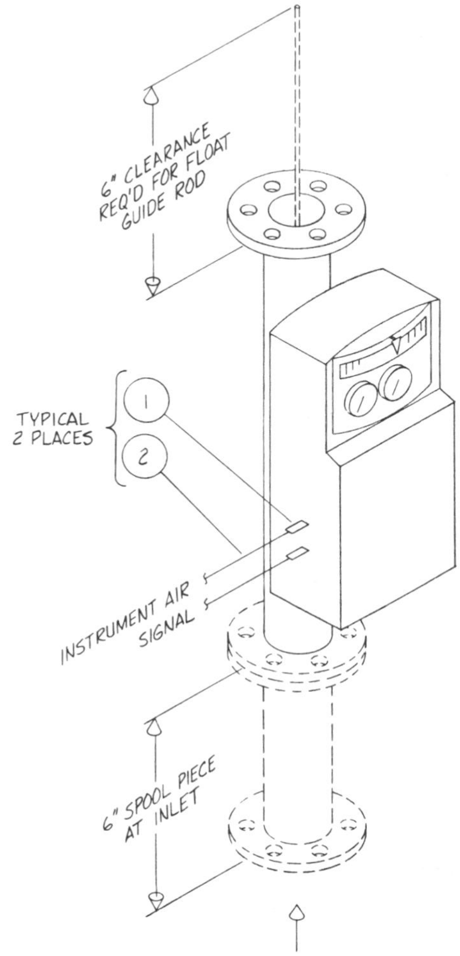

Notes:

1. Bill of material describes the basic components required; actual material composition, ratings, mfr, and/or part no's must be added as necessary for complete definition.
2. For instrument air (signal & supply) connections, see Figures 7.34 & 7.35.

BILL OF MATERIAL		
ITEM	QTY	DESCRIPTION
1	2	1/4"T × 1/4"NPT MALE CONNECTOR
2	A/R	1/4" O.D. × .030" WALL SEAMLESS COPPER TUBING, PVC COATED

Figure 7.13. Typical in-line transmitting rotameter.

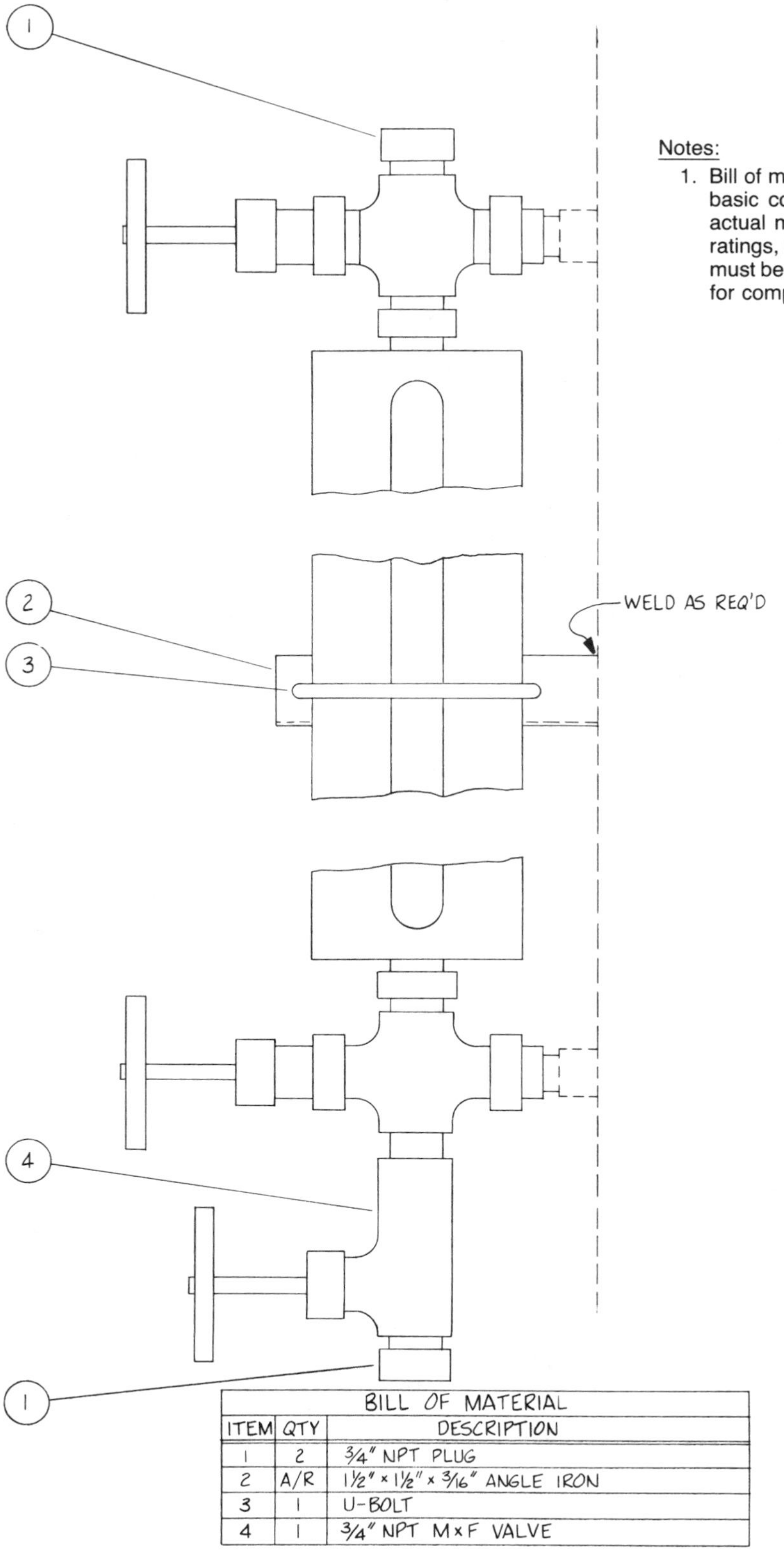

Notes:

1. Bill of material describes the basic components required; actual material composition, ratings, mfr, and/or part no's must be added as necessary for complete definition.

BILL OF MATERIAL		
ITEM	QTY	DESCRIPTION
1	2	3/4″ NPT PLUG
2	A/R	1½″ × 1½″ × 3/16″ ANGLE IRON
3	1	U-BOLT
4	1	3/4″ NPT M×F VALVE

Figure 7.14. Typical multi-section level gauge.

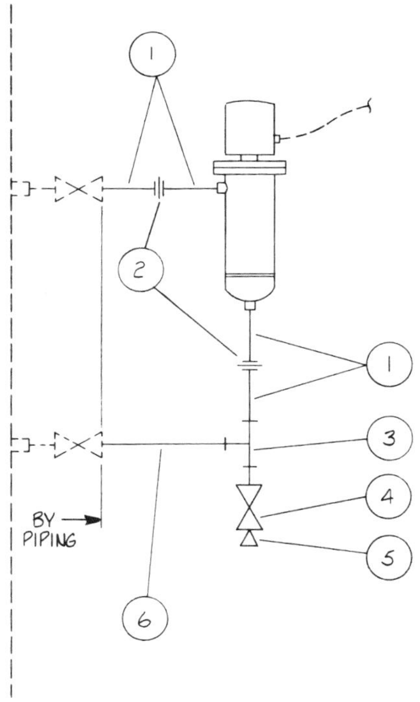

Notes:

1. Bill of material describes the basic components required; actual material composition, ratings, mfr, and/or part no's must be added as necessary for complete definition.

BILL OF MATERIAL		
ITEM	QTY	DESCRIPTION
1	4	1" NPT × 3" LG NIPPLE, TBE
2	2	1" NPT UNION
3	1	1" NPT TEE
4	1	1" NPT M×F VALVE
5	1	1" NPT PLUG
6	A/R	1" NPT PIPE

Figure 7.15. Typical level switch, screwed-pipe connections.

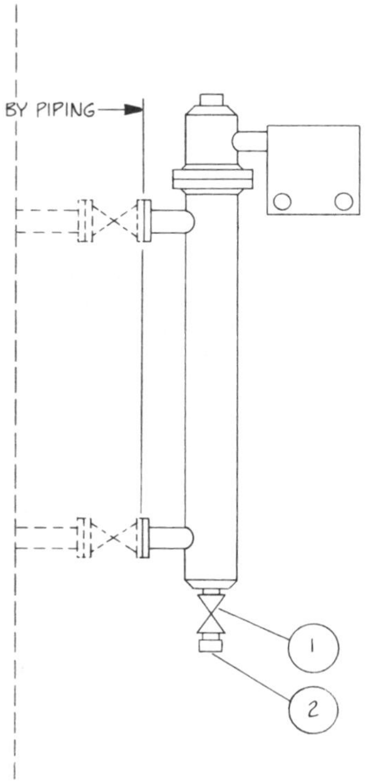

Notes:

1. Bill of material describes the basic components required; actual material composition, ratings, mfr, and/or part no's must be added as necessary for complete definition.
2. For instrument air (signal & supply) connections, see Figures 7.34 & 7.35.

BILL OF MATERIAL		
ITEM	QTY	DESCRIPTION
1	1	3/4" NPT M×F VALVE
2	1	3/4" NPT PLUG

Figure 7.16. Typical level controller—flanged connections.

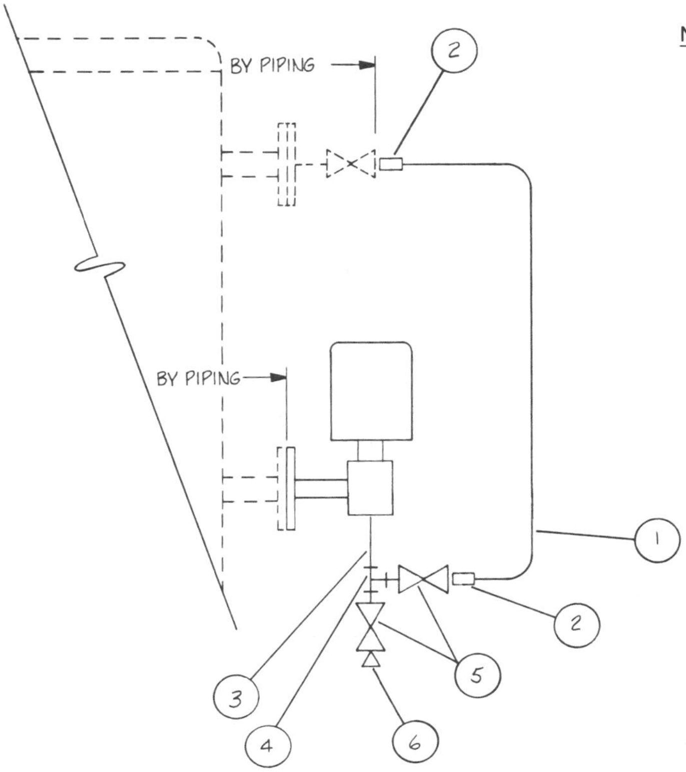

Notes:

1. Bill of material describes the basic components required; actual material composition, ratings, mfr, and/or part no's must be added as necessary for complete definition.
2. For instrument air (signal & supply) connections, see Figures 7.34 & 7.35.

BILL OF MATERIAL		
ITEM	QTY	DESCRIPTION
1	A/R	1/2" O.D. x .049" WALL SEAMLESS TUBING
2	2	1/2" T x 1/2" NPT MALE CONNECTOR
3	1	1/2" NPT x 3" LG NIPPLE, TBE
4	1	1/2" NPT TEE
5	2	1/2" NPT MxF VALVE
6	1	1/2" NPT PLUG

Figure 7.17. Typical flanged differential pressure-sensing transmitter—level service, pressure vessel with wet leg.

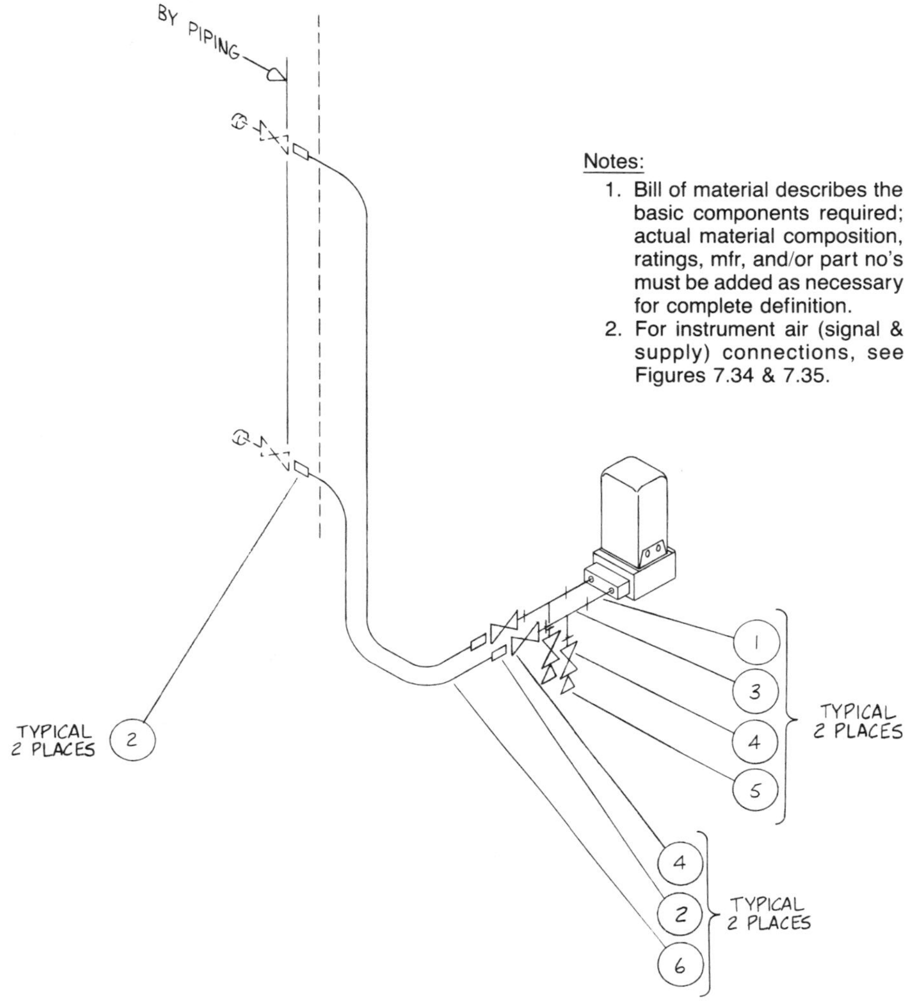

Notes:

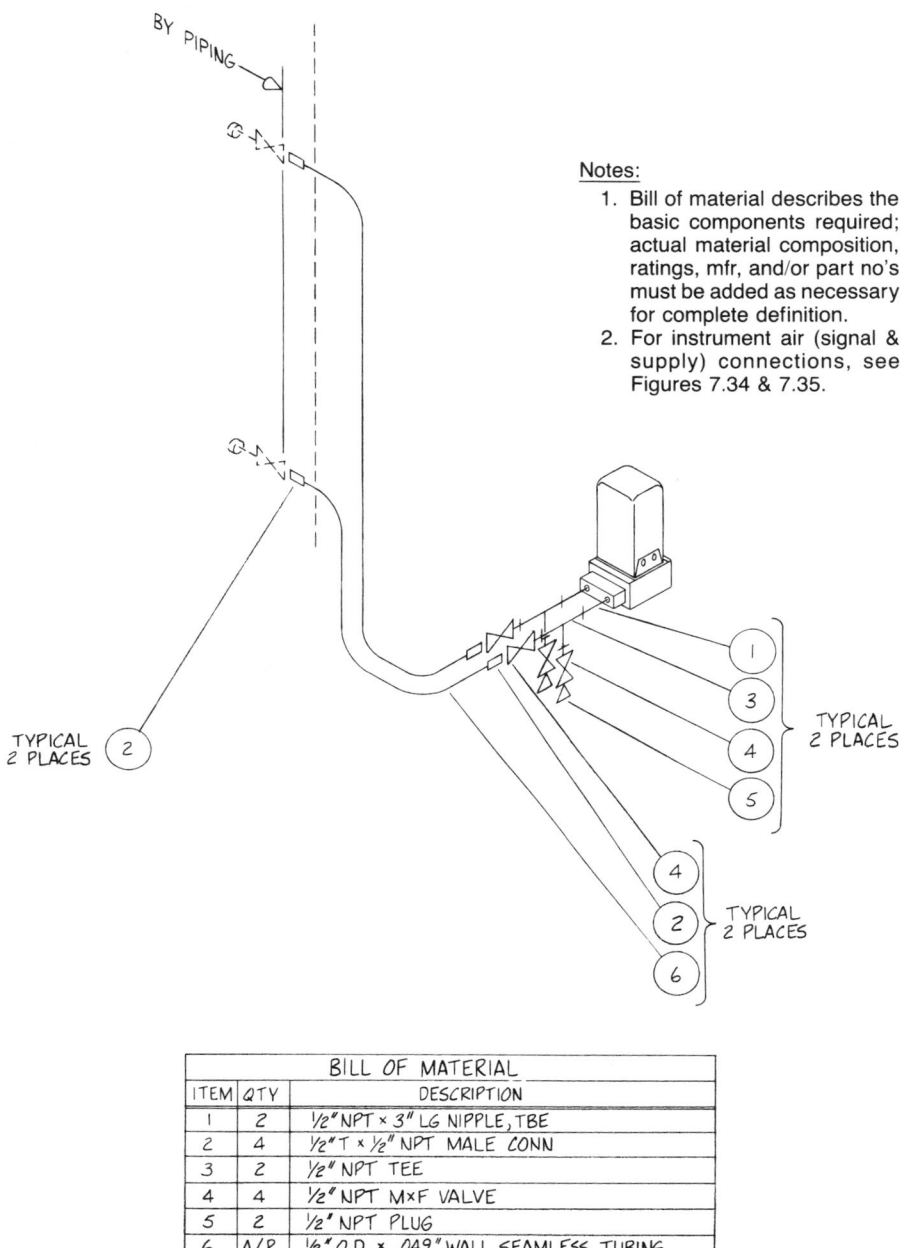

1. Bill of material describes the basic components required; actual material composition, ratings, mfr, and/or part no's must be added as necessary for complete definition.
2. For instrument air (signal & supply) connections, see Figures 7.34 & 7.35.

BILL OF MATERIAL		
ITEM	QTY	DESCRIPTION
1	2	½" NPT × 3" LG NIPPLE, TBE
2	4	½" T × ½" NPT MALE CONN
3	2	½" NPT TEE
4	4	½" NPT M×F VALVE
5	2	½" NPT PLUG
6	A/R	½" O.D. × .049" WALL SEAMLESS TUBING

Figure 7.18. Typical differential pressure-sensing transmitter—level service, pressure vessel with wet leg.

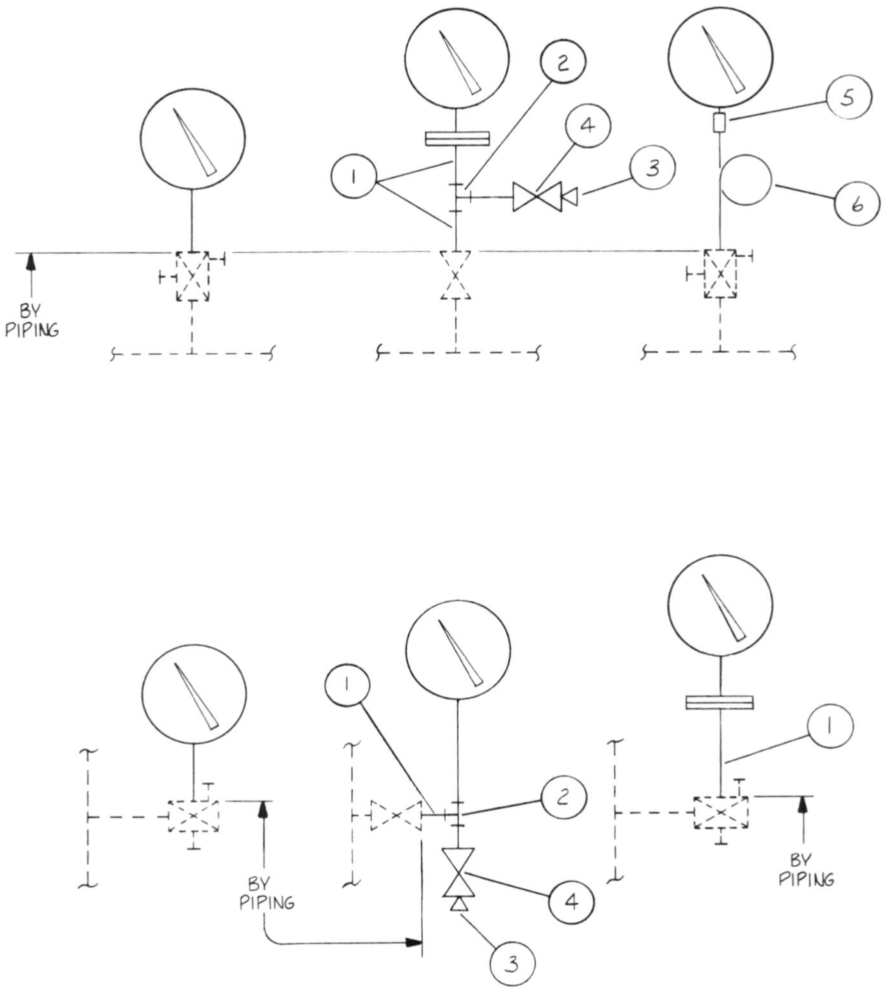

BILL OF MATERIAL		
ITEM	QTY	DESCRIPTION
1	A/R	½″ NPT × 3″ LG NIPPLE, TBE
2	A/R	½″ NPT TEE
3	A/R	½″ NPT PLUG
4	A/R	½″ NPT M×F VALVE
5	A/R	½″ NPT COUPLING
6	A/R	½″ NPT SIPHON

Notes:

1. Bill of material describes the basic components required; actual material composition, ratings, mfr, and/or part no's must be added as necessary for complete definition.

Figure 7.19. Typical pressure gauge installations.

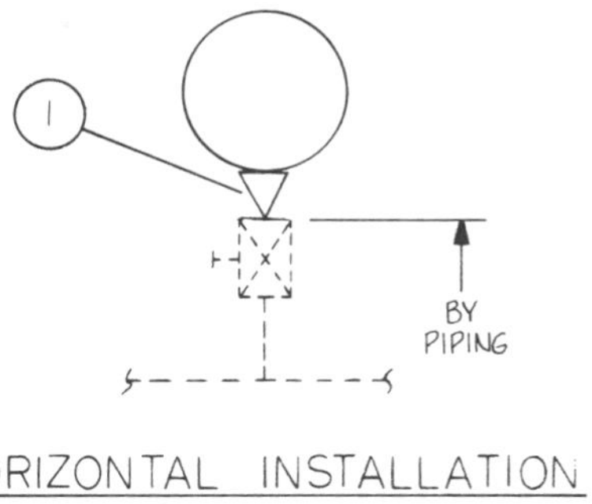

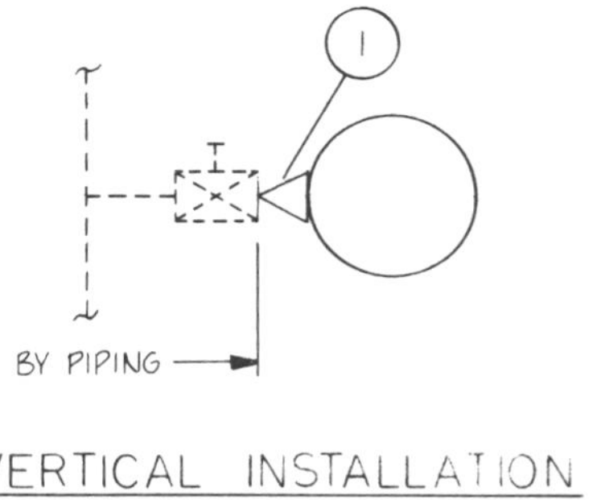

BILL OF MATERIAL		
ITEM	QTY	DESCRIPTION
1	1	½" MNPT × ¼" FNPT REDUCING BUSHING

Notes:

1. Bill of material describes the basic components required; actual material composition, ratings, mfr, and/or part no's must be added as necessary for complete definition.

Figure 7.20. Typical pressure switch installations.

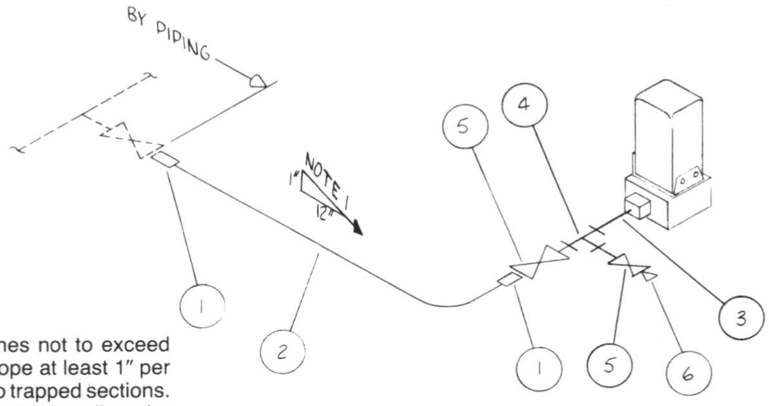

Notes:

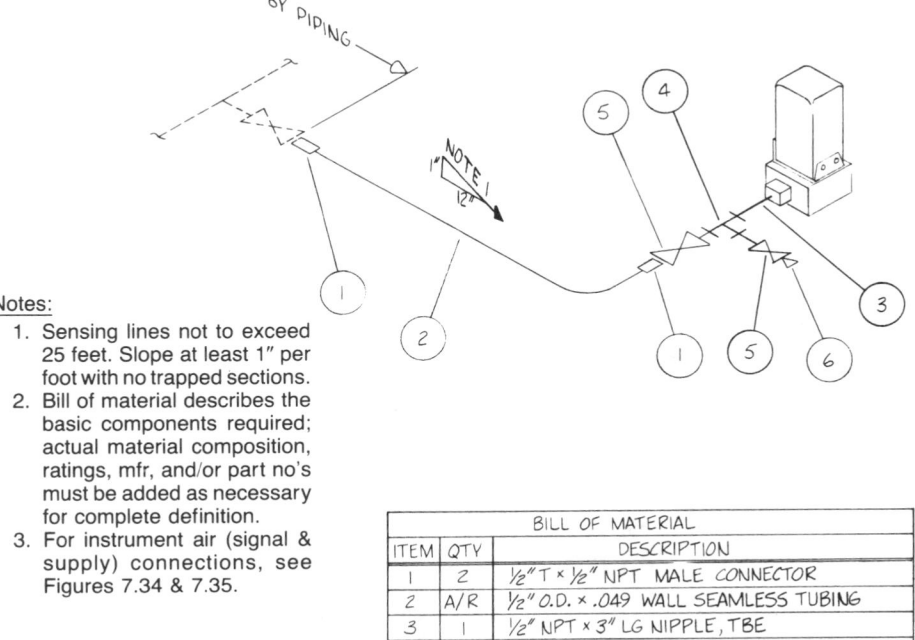

1. Sensing lines not to exceed 25 feet. Slope at least 1″ per foot with no trapped sections.
2. Bill of material describes the basic components required; actual material composition, ratings, mfr, and/or part no's must be added as necessary for complete definition.
3. For instrument air (signal & supply) connections, see Figures 7.34 & 7.35.

BILL OF MATERIAL		
ITEM	QTY	DESCRIPTION
1	2	½″T × ½″ NPT MALE CONNECTOR
2	A/R	½″ O.D. × .049 WALL SEAMLESS TUBING
3	1	½″ NPT × 3″ LG NIPPLE, TBE
4	1	½″ NPT TEE
5	2	½″ NPT M×F VALVE
6	1	½″ NPT PLUG

Figure 7.21. Typical pressure transmitter in liquid service—tubing connections.

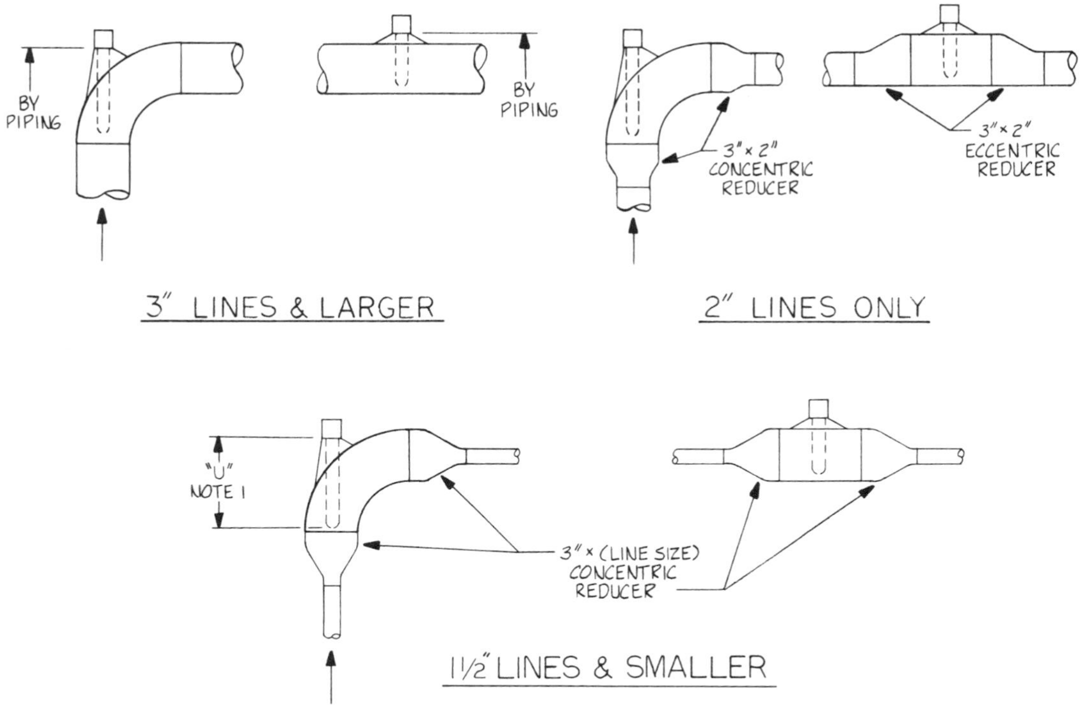

Notes:

1. "U" dimension denotes length (insertion) of thermowell from tip to, but not including, the flange or other means of attachment.

Figure 7.22. Typical thermowell installations.

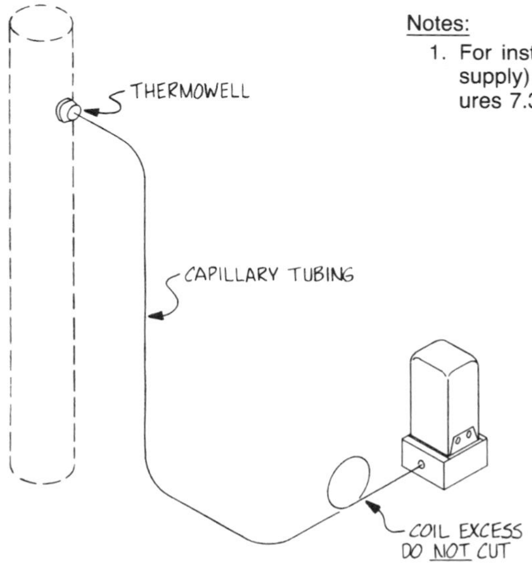

Notes:

1. For instrument air (signal & supply) connections, see figures 7.34 & 7.35.

Figure 7.23. Typical blind temperature transmitter.

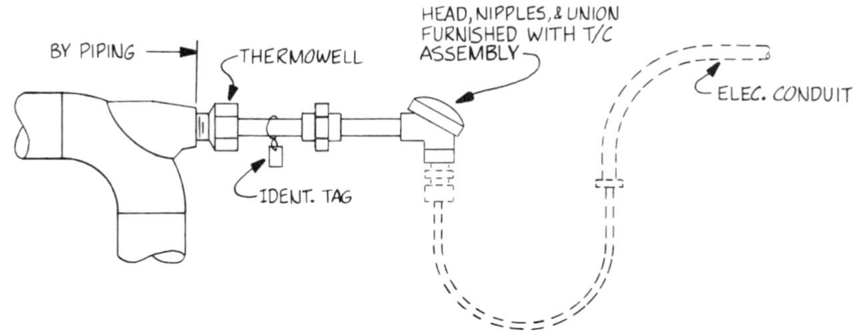

Figure 7.24. Typical thermocouple assembly.

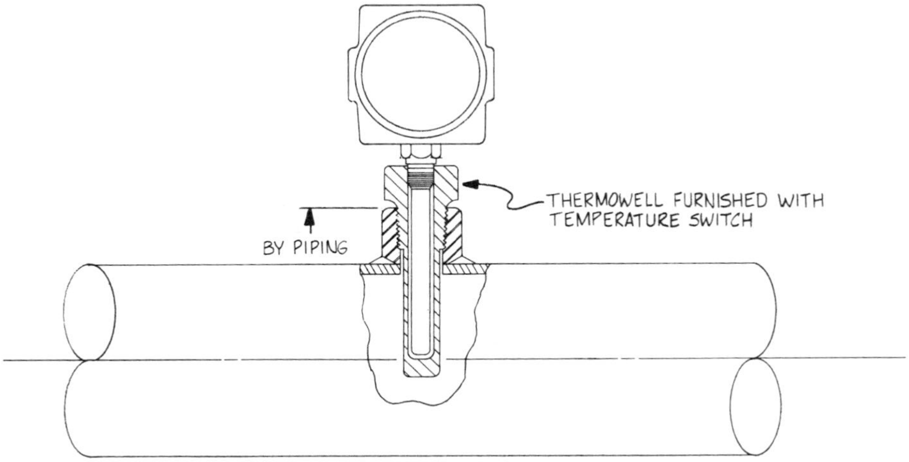

Figure 7.25. Typical temperature switch—line mounted.

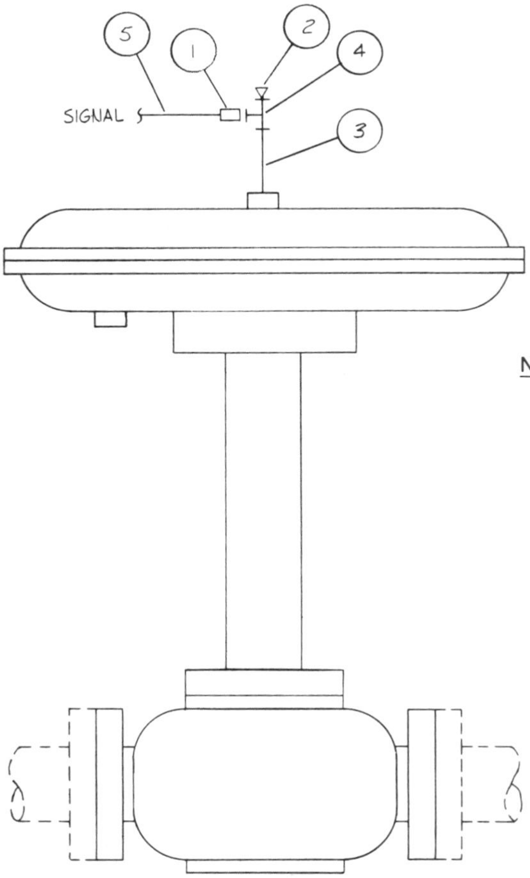

Notes:

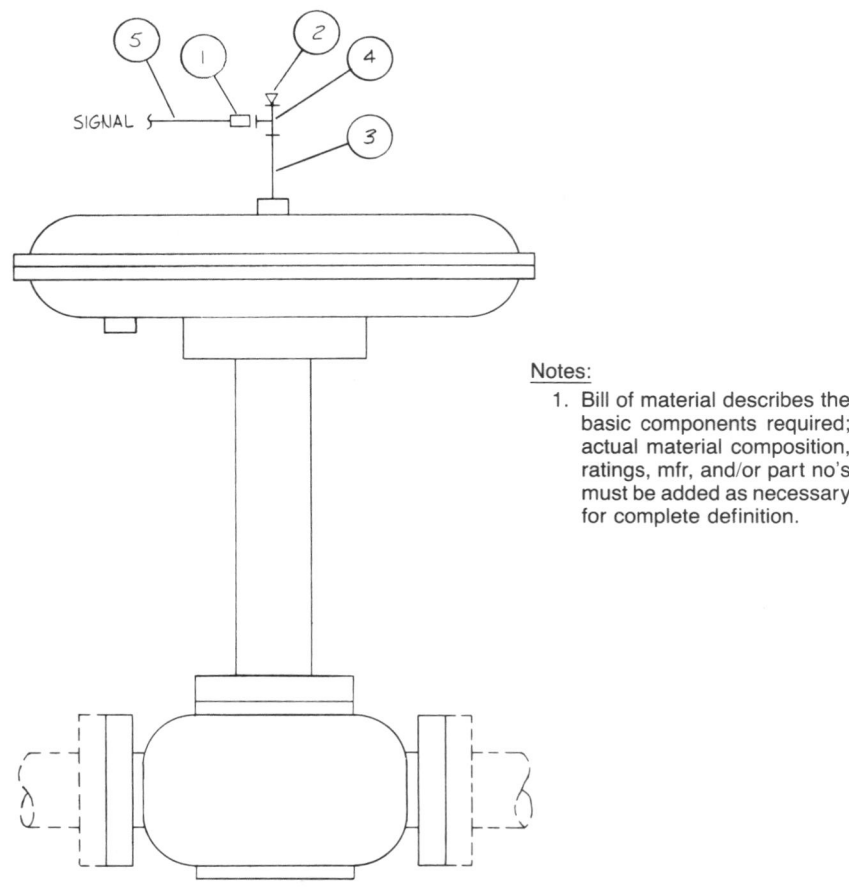

1. Bill of material describes the basic components required; actual material composition, ratings, mfr, and/or part no's must be added as necessary for complete definition.

BILL OF MATERIAL		
ITEM	QTY	DESCRIPTION
1	1	1/4" T × 1/4" NPT MALE CONNECTOR
2	1	1/4" NPT PLUG
3	1	1/4" NPT × 3" LG NIPPLE, TBE
4	1	1/4" NPT TEE
5	A/R	1/4" O.D. × .030" WALL SEAMLESS COPPER TUBING, PVC COATED

Figure 7.26. Typical diaphragm-operated control valve.

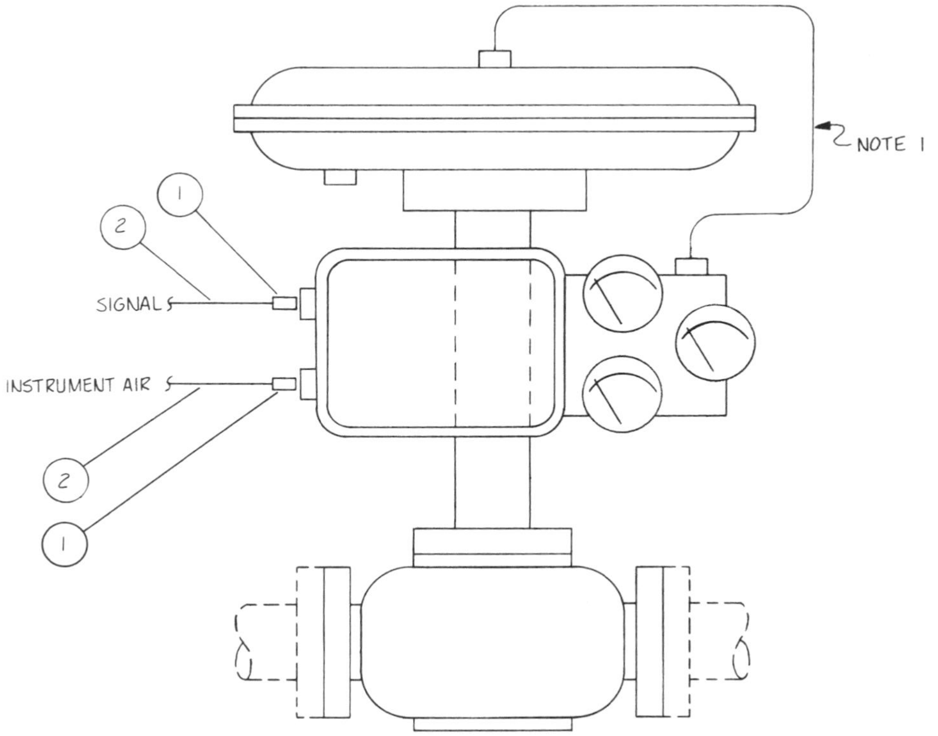

BILL OF MATERIAL		
ITEM	QTY	DESCRIPTION
1	2	¼"T × ¼" NPT MALE CONNECTOR
2	A/R	¼" O.D. × .030" WALL SEAMLESS COPPER TUBING, PVC COATED

Notes:

1. Pre-piped by valve vendor.
2. Bill of material describes the basic components required; actual material composition, ratings, mfr, and/or part no's must be added as necessary for complete definition.
3. For instrument air supply, see Figure 7.34.

Figure 7.27. Typical diaphragm-operated control valve with positioner.

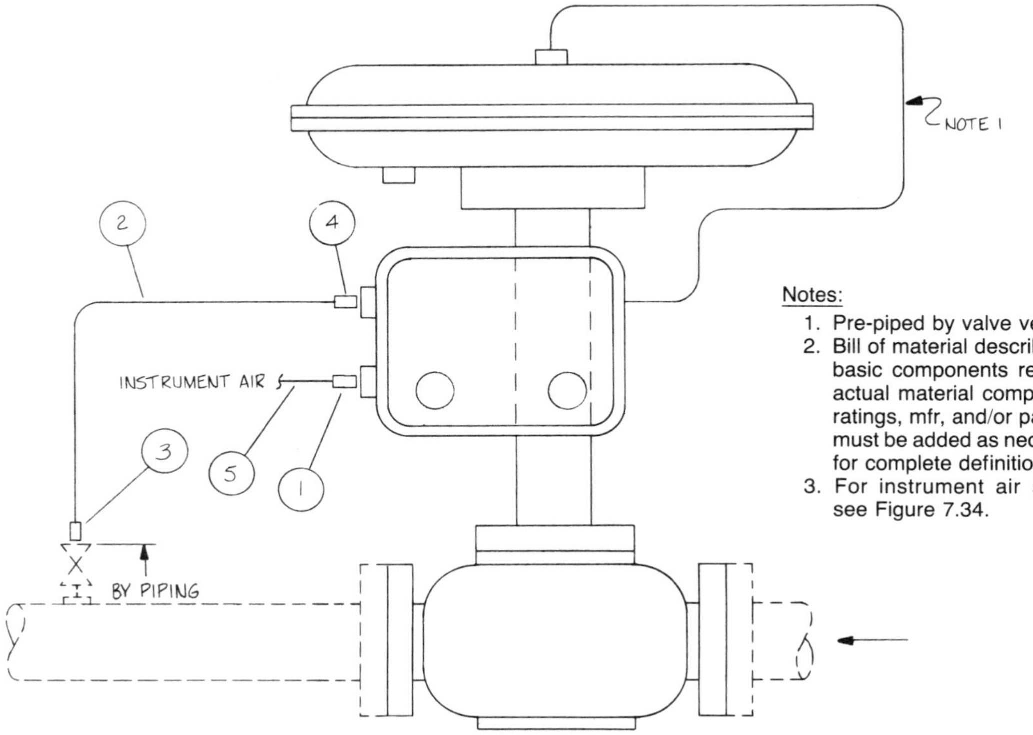

Notes:

1. Pre-piped by valve vendor.
2. Bill of material describes the basic components required; actual material composition, ratings, mfr, and/or part no's must be added as necessary for complete definition.
3. For instrument air supply, see Figure 7.34.

BILL OF MATERIAL		
ITEM	QTY	DESCRIPTION
1	1	¼″T × ¼″ NPT MALE CONNECTOR
2	A/R	½″ O.D. × .049″ WALL SEAMLESS TUBING
3	1	½″T × ½″ NPT MALE CONNECTOR
4	1	½″T × ¼″ NPT MALE CONNECTOR
5	A/R	¼″ O.D. × .030″ WALL SEAMLESS COPPER TUBING, PVC COATED

Figure 7.28. Typical diaphragm-operated control valve with pressure controller.

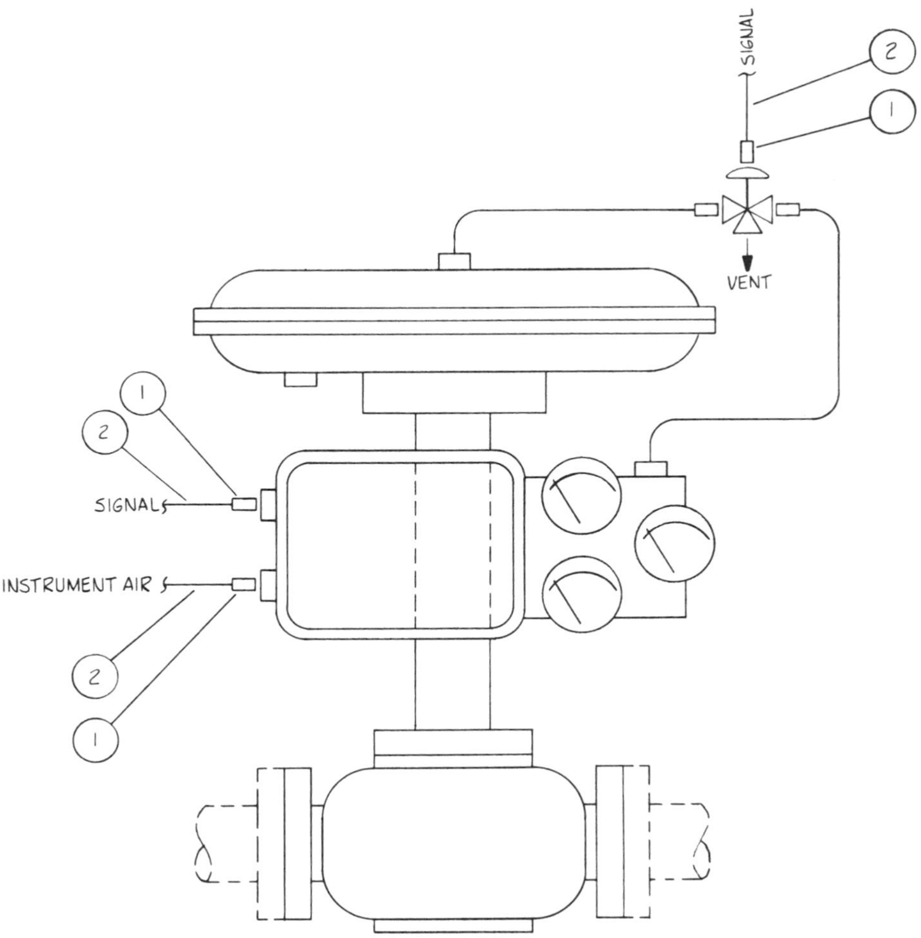

BILL OF MATERIAL		
ITEM	QTY	DESCRIPTION
1	3	1/4"T x 1/4" NPT MALE CONNECTOR
2	A/R	1/4" O.D. x .030" WALL SEAMLESS COPPER TUBING, PVC COATED

Notes:

1. Positioner and 3-way valve mounted and pre-piped by valve vendor.
2. Bill of material describes the basic components required; actual material composition, ratings, mfr, and/or part no's must be added as necessary for complete definition.
3. For instrument air supply, see Figure 7.34.

Figure 7.29. Typical diaphragm-operated control valve with positioner and three-way pneumatic switching valve for rapid diaphragm venting.

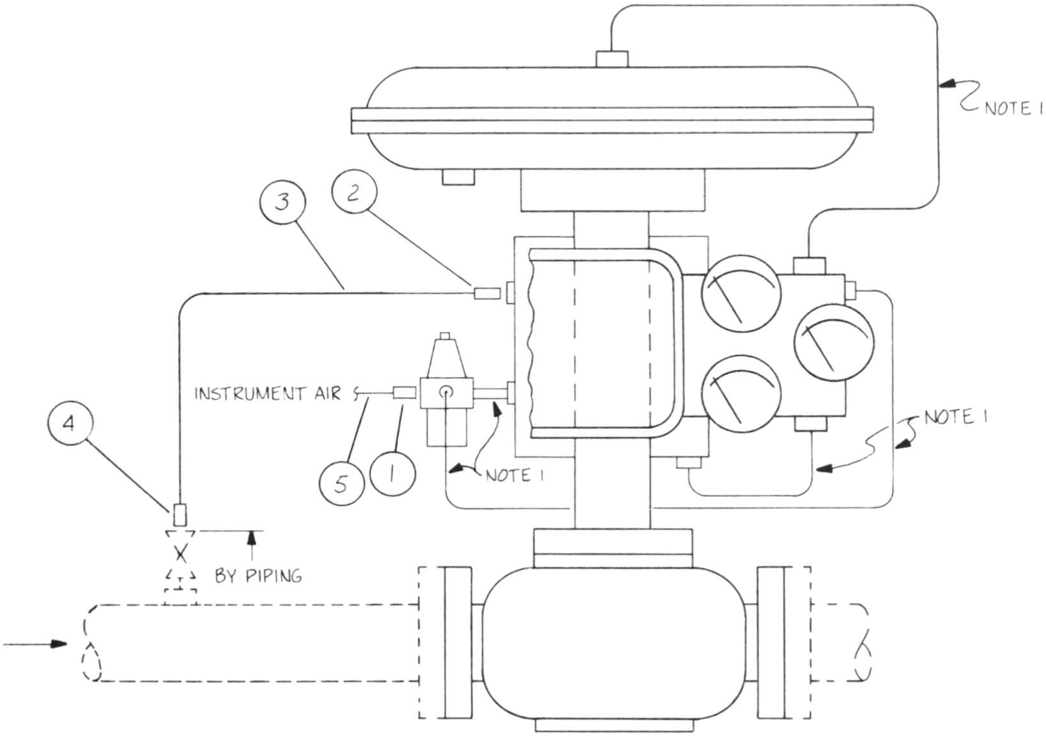

BILL OF MATERIAL		
ITEM	QTY	DESCRIPTION
1	1	¼"T × ¼"NPT MALE CONNECTOR
2	1	½"T × ¼"NPT MALE CONNECTOR
3	A/R	½"O.D. × .049" WALL SEAMLESS TUBING
4	1	½"T × ½"NPT MALE CONNECTOR
5	A/R	¼"O.D. × .030" WALL SEAMLESS COPPER TUBING, PVC COATED

Notes:

1. Supplied & pre-piped by valve vendor.
2. Bill of material describes the basic components required; actual material composition, ratings, mfr, and/or part no's must be added as necessary for complete definition.
3. For instrument air supply, see Figure 7.34. (Note that regulator shown in Figure 7.34 is not req'd, since it is supplied with valve.)

Figure 7.30. Typical diaphragm-operated control valve in back-pressure service with pressure controller and positioner.

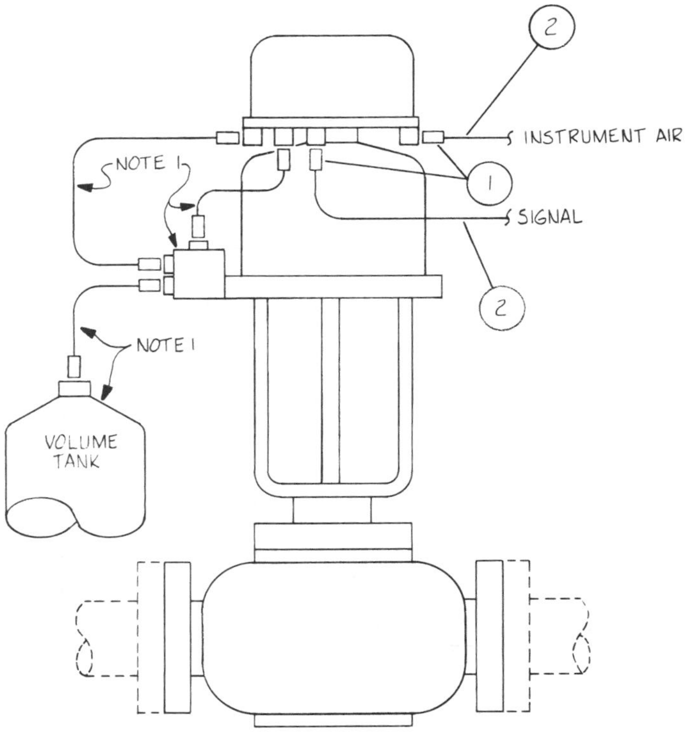

BILL OF MATERIAL		
ITEM	QTY	DESCRIPTION
1	2	1/4" T × 1/4" NPT MALE CONNECTOR
2	A/R	1/4" O.D. × .030" WALL SEAMLESS COPPER TUBING, PVC COATED

Notes:

1. Supplied & pre-piped by valve vendor.
2. Bill of material describes the basic components required; actual material composition, ratings, mfr, and/or part no's must be added as necessary for complete definition.
3. For instrument air supply, see Figure 7.34.

Figure 7.31. Typical piston-operated control valve with volume tank for automatic piston stroking upon supply pressure loss.

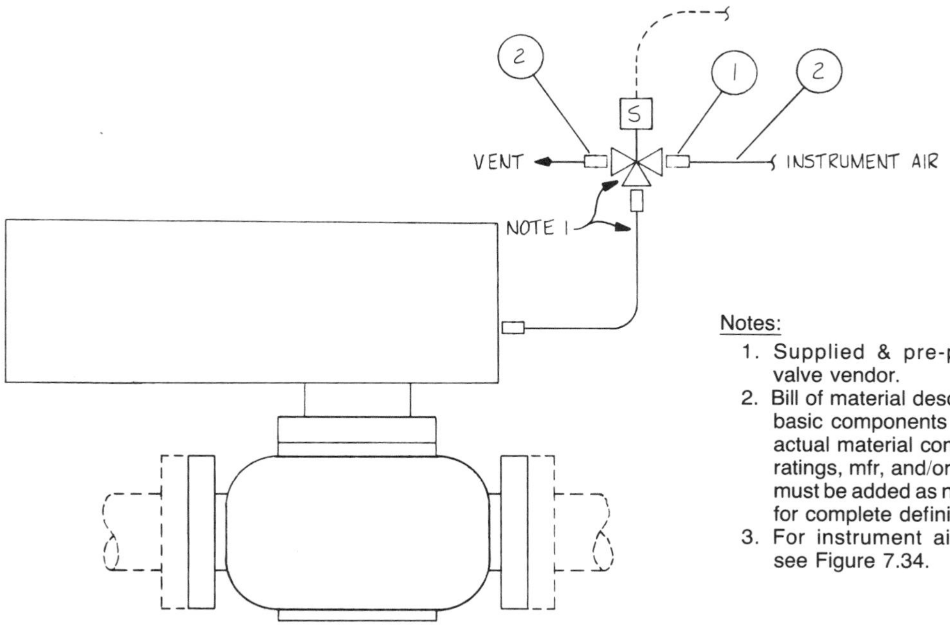

Notes:

1. Supplied & pre-piped by valve vendor.
2. Bill of material describes the basic components required; actual material composition, ratings, mfr, and/or part no's must be added as necessary for complete definition.
3. For instrument air supply, see Figure 7.34.

BILL OF MATERIAL		
ITEM	QTY	DESCRIPTION
1	1	1/4"T × 1/4" NPT MALE CONNECTOR
2	1	1/4" NPT VENT SCREEN
3	A/R	1/4" O.D. × .030" WALL SEAMLESS COPPER TUBING, PVC COATED

Figure 7.32. Typical piston-operated valve, spring return, solenoid actuated.

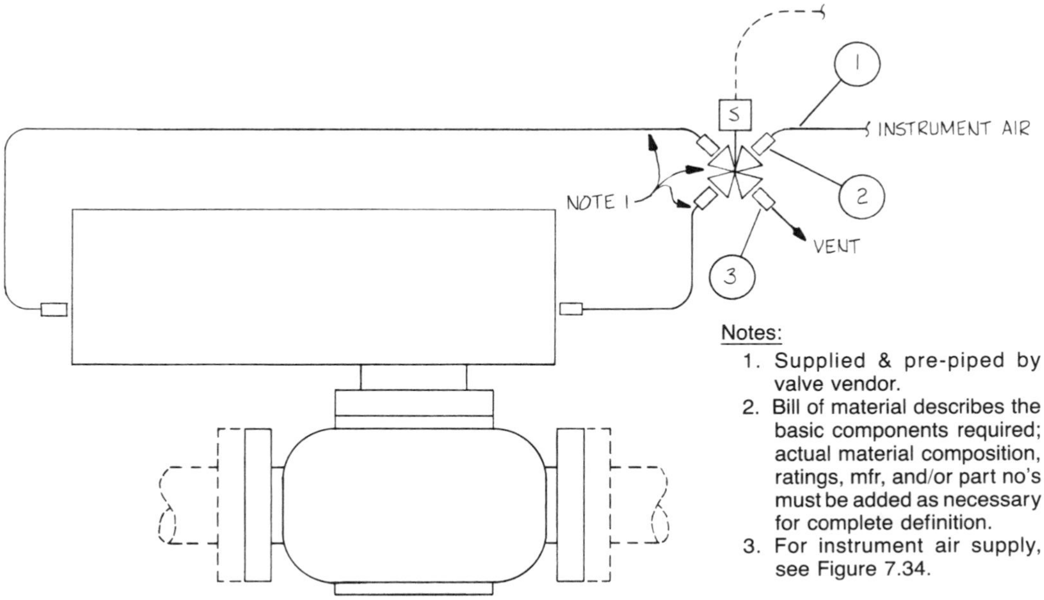

Notes:

1. Supplied & pre-piped by valve vendor.
2. Bill of material describes the basic components required; actual material composition, ratings, mfr, and/or part no's must be added as necessary for complete definition.
3. For instrument air supply, see Figure 7.34.

BILL OF MATERIAL		
ITEM	QTY	DESCRIPTION
1	A/R	1/4" O.D. × .030" WALL SEAMLESS COPPER TUBING, PVC COATED
2	1	1/4" T × 1/4" NPT MALE CONNECTOR
3	1	1/4" NPT VENT SCREEN

Figure 7.33. Typical piston-operated valve—double-acting, solenoid actuated.

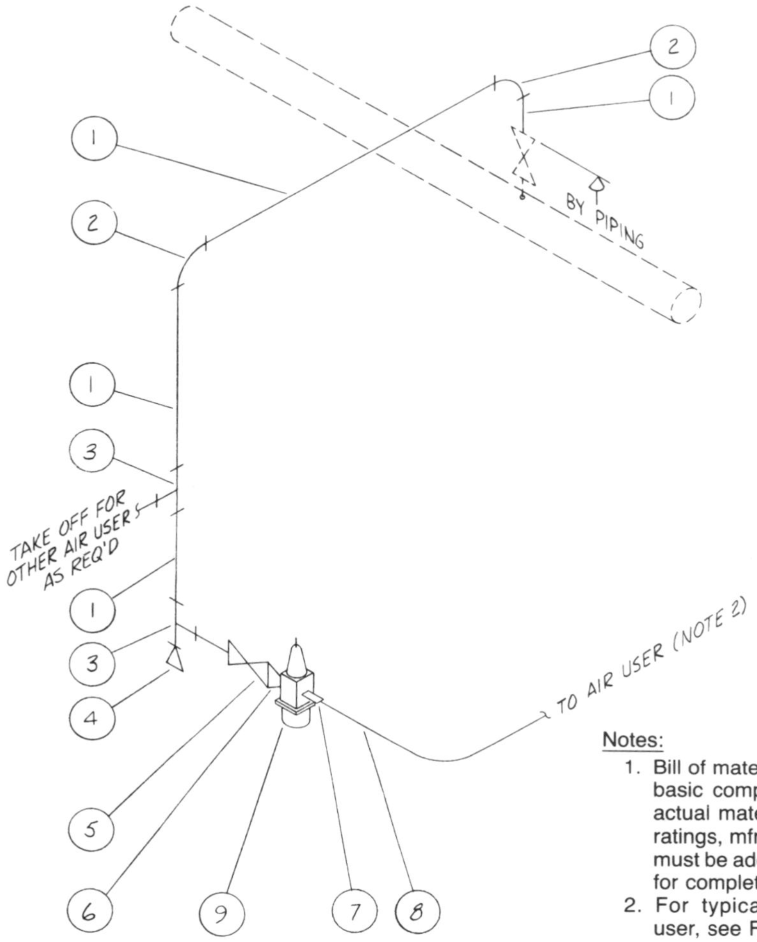

Notes:

1. Bill of material describes the basic components required; actual material composition, ratings, mfr, and/or part no's must be added as necessary for complete definition.
2. For typical instrument air user, see Figure 7.35.

BILL OF MATERIAL		
ITEM	QTY	DESCRIPTION
1	A/R	1/2" NPT PIPE
2	A/R	1/2" NPT ELBOW
3	A/R	1/2" NPT TEE
4	A/R	1/2" NPT PLUG
5	A/R	1/2" NPT M×F VALVE
6	A/R	1/2" NPT × 1/4" NPT SWAGE TBE
7	A/R	1/4" T × 1/4" NPT MALE CONNECTOR
8	A/R	1/4" O.D. × .030" WALL SEAMLESS COPPER TUBING, PVC COATED
9	A/R	FILTER REGULATOR

Figure 7.34. Typical instrument air supply.

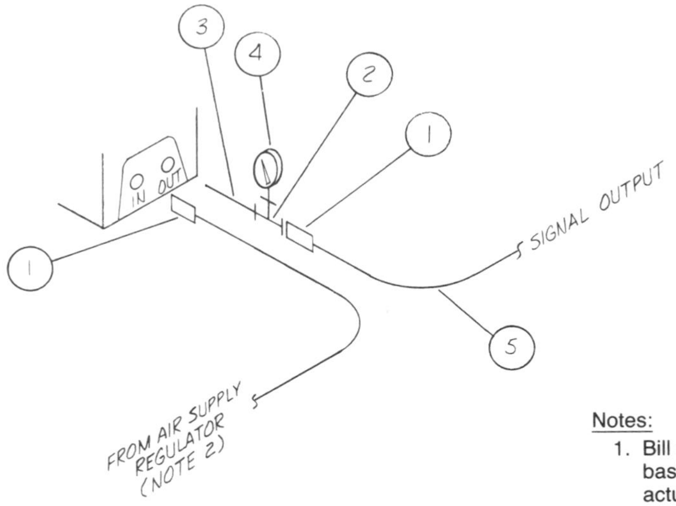

Notes:
1. Bill of material describes the basic components required; actual material composition, ratings, mfr, and/or part no's must be added as necessary for complete definition.
2. For typical air supply, see Figure 7.34.

BILL OF MATERIAL		
ITEM	QTY	DESCRIPTION
1	2	1/4" T × 1/4" NPT MALE CONNECTOR
2	1	1/4" NPT TEE
3	1	1/4" NPT × 3" LG NIPPLE, TBE
4	1	3½" RECEIVER GAUGE
5	A/R	1/4" O.D. × .030" WALL SEAMLESS COPPER TUBING, PVC COATED

Figure 7.35. Typical pneumatic instrument connections for air supply and output with receiver gauge.

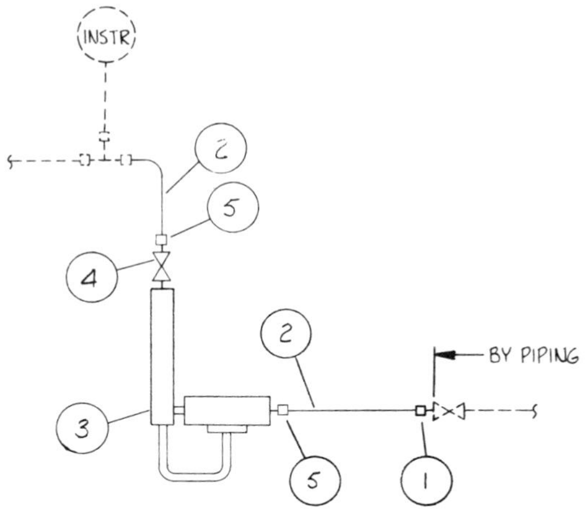

Notes:

1. Bill of material describes the basic components required; actual material composition, ratings, mfr, and/or part no's must be added as necessary for complete definition.

BILL OF MATERIAL		
ITEM	QTY	DESCRIPTION
1	1	¼" T × ½" NPT MALE CONNECTOR
2	A/R	¼" O.D. × .030" WALL SEAMLESS COPPER TUBING, PVC COATED
3	1	PURGE METER W/DIFFERENTIAL REGULATOR
4	2	¼" NPT M×F VALVE
5	3	¼" T × ¼" NPT MALE CONNECTOR

Figure 7.36. Typical instrument purging—single-line, inlet-type purge meter.

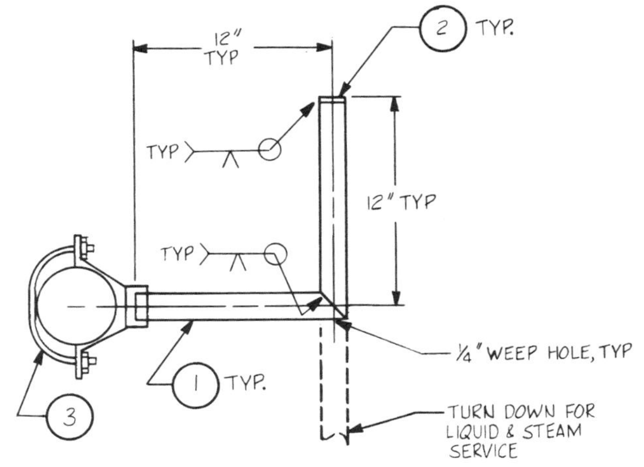

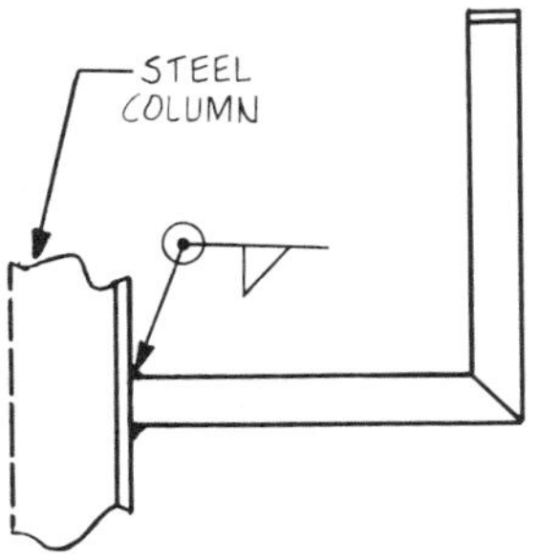

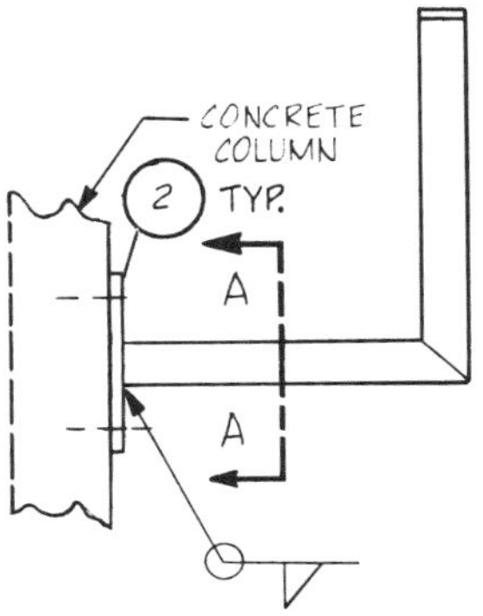

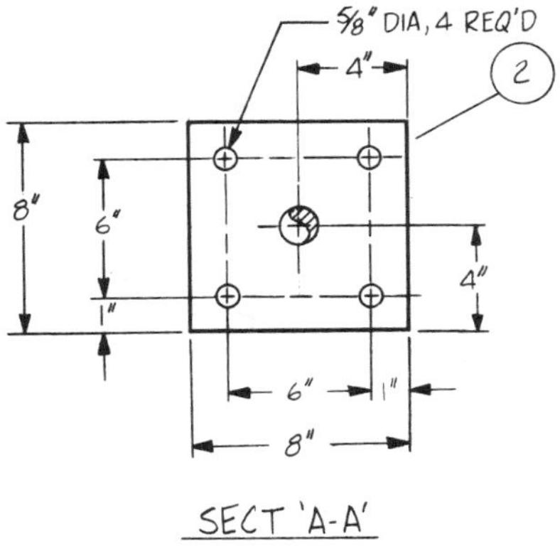

BILL OF MATERIAL		
ITEM	QTY	DESCRIPTION
1	A/R	2" NPT PIPE
2	A/R	1/4" THK CARBON STEEL PLATE
3	1	LEVELING SADDLE

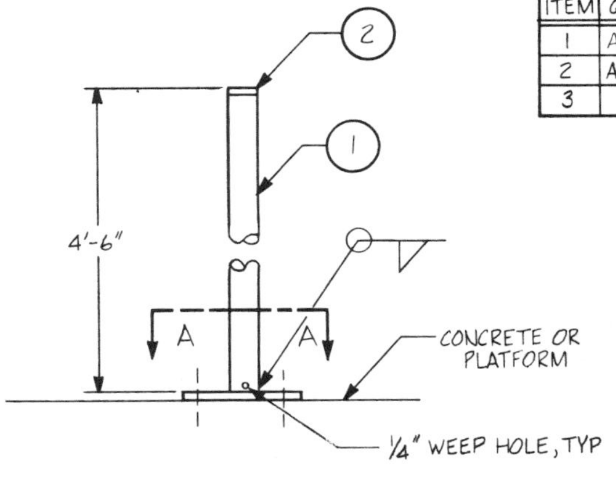

Notes:

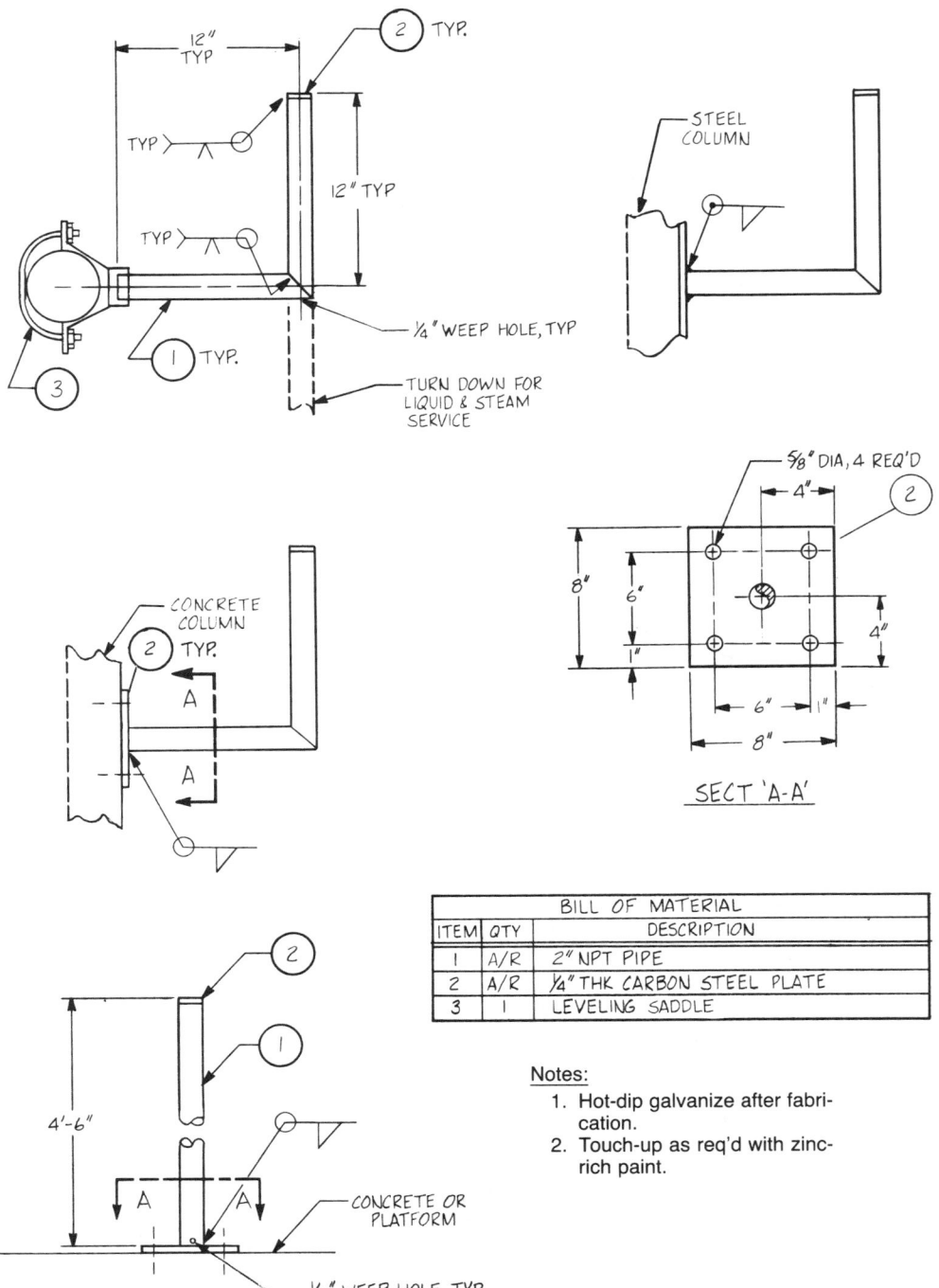

1. Hot-dip galvanize after fabrication.
2. Touch-up as req'd with zinc-rich paint.

Figure 7.37. Typical single-mount instrument supports.

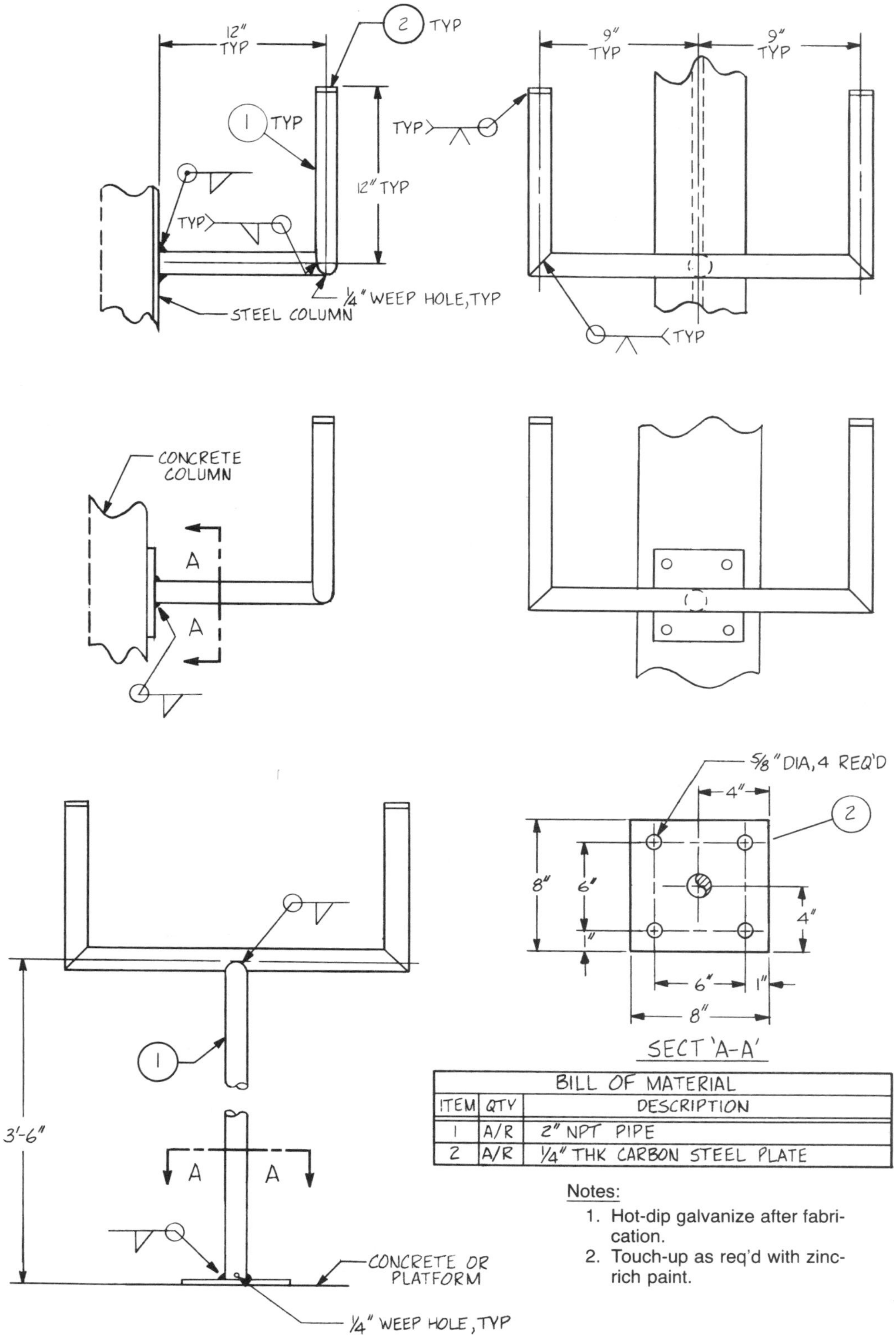

BILL OF MATERIAL		
ITEM	QTY	DESCRIPTION
1	A/R	2" NPT PIPE
2	A/R	¼" THK CARBON STEEL PLATE

Notes:

1. Hot-dip galvanize after fabrication.
2. Touch-up as req'd with zinc-rich paint.

Figure 7.38. Typical multimount instrument supports.

8 Typical Calculations

Baxter Williams
Andrew Jackson Stockton
Tom Lisch

The solution of sample problems provides valuable help for designers and engineers in industry, particularly in areas such as instrumentation where formal curriculium is lacking.

The examples which follow represent many of the typical calculations which must be performed in designing control systems in the processing industries. The step-by-step approach to solving these problems encourages a thorough analysis of the problems and provides guidelines for solving most routine operations. See Chapter 9 for electrical calculations.

Orifices

PROBLEM: Orifice Calculation for Measurement of Liquid Flow Rate

Given

Fluid:	Water
Pressure:	100 psig
Temperature:	160°F
Flow Rate:	600 gpm
Pipe Size:	6.065 inches i.d.
Meter Type:	Dry (mercury-less)
Maximum Differential:	100 inches of water
Connections:	Orifice flanges
Orifice Material:	304 S.S.

The formula[1] for calculating orifice bores for liquids is:

$$S = \frac{Q_m \; G_l}{(ND^2 \; F_a F_m \; \sqrt{G_f} \; \sqrt{h_m}) \; (F_c F_p)}, \quad \text{p. 216}$$

where

- β = ratio of orifice bore to pipe ID
- D = pipe ID, inches
- F_a = ratio of orifice bore area at flowing temperature to that at 68°F, p. 156 Spink 9th Ed.
- F_p = correction for compressibility of the liquid, p. 275.
- F_c = viscosity or Reynolds number correction, p. 225.
- F_m = manometer correction factor, p. 157
- G_f = specific gravity at flowing temperature, p. 165
- G_l = specific gravity at 60°F
- h_m = maximum differential, inches W.C. (span of instrument)
- N = a constant depending on units of measurements, p. 154
- Q_m = maximum flow rate in units corresponding with N
- S = an operating figure from which β can be obtained, p. 167

[1] Formulas and Symbols, followed by page numbers, are from Spink, L. K. *Principles and Practice of Flowmeter Engineering, 9th Edition.* Foxboro Co. Foxboro, Mass.: 1967.

The following equation term values are derived from charts on the previously referenced pages.

Establishing Values:

D = 6.065 inches
F_a = 1.002
F_m = 1.000
$F_p\sqrt{G_f}$ = 0.996
G_l = 1.000
h_m = 100 inches W.C.
N = 5.667
Q_m = 600 gpm

Calculation:

Ignoring the effects of viscosity,

$$S = \frac{600(1)}{5.667\,(6.065)^2\,(1.002)\,(1)\,(0.996)\,\sqrt{100}} = 0.28841$$

Obtain R_D = 700,000 from Figure B-2278 (p.218).

Then find
F_c = 0.980 from Figure B-2291 (p. 225)

Recalculating S,

$$S = \frac{0.28841}{0.980} = 0.29430$$

From Spink, Table 12, p. 172,

β = 0.65564
Bore = βD
= (0.65564)(6.065)
= 3.9765

PROBLEM: Orifice Calculation for Minimum Flow By-pass of Pump.

Given

Fluid:	Water
Pump Differential:	200′ water TDH (Total Dynamic Head)
Temperature:	60°F
Flow Rate:	200 gpm
Pipe Size:	1-½″ Sch. 40, C.S.
Orifice Material:	304 S.S.

The formula[2] for calculating orifice bores for liquids is

$$S = \frac{Q_m\ G_l}{ND^2F_aF_m\ \sqrt{G_f}\ \sqrt{h_m}}\quad \text{, p. 148}$$

where

β = ratio of orifice bore to pipe ID
D = pipe ID, inches
F_a = ratio of orifice bore area at flowing temperature to that at 68°F, p. 156
F_m = manometer correction factor, p. 157
G_f = specific gravity at flowing temperature, p. 158
G_l = specific gravity at 60°F
h_m = maximum differential, inches W.C.
N = a constant depending on units of measurement, p. 154
Q_m = maximum flow rate in units corresponding with N
S = an operating figure from which β can be obtained, p. 167

Establishing Values:

D = 1.610 inches
F_a = 1.000
F_m = 1.000
G_f = 1.000; $\sqrt{G_f}$ = 1.000
G_l = 1.000
h_m = 2,400 inches W.C. (200′ W.C. x 12″/ft.)
N = 5.667
Q_m = 200 gpm

Calculation:

$$S = \frac{200\,(1)}{(5.667)\,(1.61)^2\,(1)\,(1)\,(1)\,\sqrt{2{,}400}}$$

$$= 0.2779$$

From Spink, Table 13, p. 178,

β = 0.5838
Bore = βD
= 0.5838(1.61)
= 0.9399 inch

Use a plate with a 1.00 inch bore.

[2] op. cit., Spink

PROBLEM: Differential Pressure Calculation for Measurement of Liquid Flow Rate with an Integral Orifice Transmitter

Given

Flow Rate:	0.5 gpm maximum
Temperature:	100°F
Specific Gravity:	0.950 at 60°F
	0.935 at 100°F
Absolute Viscosity:	0.06 poise
Orifice Material:	304 S.S.
Orifice Bore:	0.0995 inch*

*Calculations using more than one standard bore size are often necessary to obtain the desired differential pressure range.

The formulas[3] to be used for calculating integral orifice heads are

$$C_n = k_s C_{wi} F_a F_m \sqrt{\frac{G_f}{G_l}} F_c F_p, \text{ p. 216}$$

$$R_D = \frac{0.5271\,(Q_{gph})\,(G_l)}{d\,v}, \text{ p. 191}$$

$$Q_n = C_n \sqrt{h_w}, \text{ p. 216}$$

where

C_n = flow constant in units the same as N, p. 214
C_{wi} = integral orifice flow constant for water at 60°F in gph, p. 187
d = orifice bore, in inches
F_a = ratio of orifice bore area at flowing temperature to that at 68°F, p. 156
F_c = viscosity correction, p. 256
F_m = manometer correction factor, p. 157
F_p = compressibility correction factor, p. 274
G_f = specific gravity at flowing temperature, p. 158
G_l = specific gravity at 60°F
h_w = operating differential, inches W.C.
k_s = units conversion factor, p. 154
Q_n = flow rate at h_w, units same as N
Q_{gph} = flow rate at 60°F and h_w, gph
R_D = pipe Reynolds number
v = absolute viscosity, poises

Establishing values:

C_{wi} = 2.77
d = 0.0995 inch
F_a = 1.000
F_m = 1.000
F_p = 1.000 (See Spink, p. 271; for high operating temperatures, see Figure B-2477, p. 274)
G_f = 0.935
G_l = 0.950
k_s = 0.01667 (for gpm)
Q_n = 0.5 gpm
Q_{gph} = 30 gph
v = 0.06 poise

[3]op. cit., Spink

Calculation:

$$R_D = \frac{0.5271\,(30)\,(0.950)}{0.0995\,(0.06)} = 2{,}516$$

Using an assumed value of $F_c = 1$, calculate $R_D/F_c = 2{,}516$. From Figure B-8652, Spink, p. 257, obtain a corrected value of F_c.

F_c = 1.000

Note: For Reynolds Numbers exceeding 1,500, $F_e = 1$ for integral orifices with diameters of 0.0995 inch and above. For Reynolds numbers exceeding 10,000, $F_e = 1$ for integral orifices with diameters of 0.0595 inch and below.

From Figure B-8652 –

F_c = 1.00

Substituting:

$$C_n = 0.01667\,(2.770)\,(1)\,(1)\sqrt{\frac{0.935}{0.950}}\,(1)\,(1)$$

$$= 0.047$$

$$\sqrt{h_w} = \frac{Q_n}{C_n} = \frac{0.5}{0.047} = 10.638$$

h_w = <u>113.17 inches W.C.</u>

PROBLEM: Orifice Calculation for Measurement of Steam Flow Rate (for Flange, Vena Contracta, and Pipe Taps for Orifices, and for Flow Nozzles and Venturi Tubes)

Given

Fluid:	Saturated Steam
Flow Rate:	50,000 pph
Pressure:	650 psig

Line Size:	8″ Sch. 80, C.S.
Absolute Viscosity:	0.025 CP
Differential:	100 inches W.C.
Orifice Material:	304 S.S.
Taps:	Orifice Flanges

The formulas[4] for calculating orifice bores for steam are

$$S = \frac{W_m}{359D^2 F_a F_m F_c Y \sqrt{\gamma_f} \sqrt{h_m}} \text{, p. 332}$$

$$R_D = \frac{6.32\ W_m}{Dv} \text{, p. 191}$$

where

γ_f = specific weight of steam at flowing temperature and pressure, lbm/cu. ft.
D = pipe ID, inches
F_a = ratio of orifice bore area at flowing temperature to that at 68°F, p. 156
F_c = viscosity correction, p. 256
F_m = manometer correction factor, p. 157
h_m = maximum differential, inches W.C.
R_D = pipe Reynolds Number
S = an operating figure from which β can be obtained, p. 167
v = absolute viscosity, centipoises
W_m = maximum flow rate, pph
Y = expansion factor

Establishing Values:

γ_f = 1.2020, Table 26, p. 341
D = 7.625 inches
F_a = 1.008 at 498°F, p. 156
F_m = 1.000, p. 157
h_m = 100 inches W.C.
v = 0.025 centipoises

Calculations:
Assume F_c = 1.0 and Y = 1.0, and calculate a trial value of S

$$S = \frac{50{,}000}{359\ (7.625)^2\ (1)\ (1)\ (1)\ (1)\ (1.2020)\ \sqrt{100}}$$

Obtain a value of β from Spink, Table 12, p. 171.

$$\beta = 0.55699$$

Calculate R_D at *normal* flow, or 0.7 W_m.

$$R_D = \frac{6.32\ (50{,}000)\ (0.7)}{7.625\ (0.025)}$$

$$= 1{,}160{,}393$$

From Spink, Figure B-2288, p. 223.

$$F_c \cong 0.995$$

To obtain Y, calculate a value for $\frac{(F_m)^2 h_w}{\text{psia}}$ to use in Fig. B-2503, p. 357

$$\frac{(F_m)^2 h_w}{\text{psia}} = \frac{(1)^2\ (49)}{664.7} = 0.07372$$

where h_w corresponds to 0.7 W_m, and plot a value in Figure B-2503, Spink, p. 357

$$Y = 0.999$$

Divide S by F_c times Y, and obtain a new value S'.

$$S' = \frac{S}{F_c Y} = \frac{0.19929}{(0.9925)\ (0.999)} = 0.20099$$

Obtain a new value for β from Table 12.

$$\beta' = 0.559$$

If $\frac{\beta - \beta'}{\beta} \geqslant 1\%$, use β' to obtain new values for F_c and Y, and refigure.

$$\frac{0.55699 - 0.559}{0.55699} = 0.0036 = 0.36\%$$

Therefore, $\beta = 0.55699$

$$\text{Bore} = \beta D = 0.55699\ (7.625) = 4.247 \text{ inches}$$

PROBLEM: Orifice Calculation for Measurement of Gas Flow Rate

Given

Fluid:	Nitrogen
Pressure:	200 psig
Temperature:	70°F
Specific Gravity:	0.967

[4] op. cit., Spink

Flow Rate:	2,000 scfm, max.
Differential:	100 inches W.C.
Line Size:	3" Sch. 40, C.S.
Orifice Material:	304 S.S.
Taps:	Orifice Flanges

The formula[5] to be used for calculating orifice bores for gas measurement is

$$K_o\ \beta^2 = \frac{Q_h}{338.17 D^2 F_r Y F_{pb} F_{tb} F_{tf} F_g F_{pv} F_{wv} F_a F_m \sqrt{h_w p_f}},$$

p. 428

where

β = ratio of orifice bore to pipe ID
D = pipe ID, inches
F_a = ratio of orifice bore area at flowing temperature to that at 68°F, p. 156
F_g = specific gravity correction, p. 456
F_m = manometer correction factor, p. 157
F_{pb} = pressure base correction, p. 453
F_{pv} = supercompressibility factor, p. 458
F_r = Reynolds number correction
F_{tb} = temp. base correction, p. 453
F_{tf} = flowing temp. correction, p. 454
F_{wv} = volume correction for water vapor, p. 425
h_w = operating differential pressure, inches W.C.
K_o = coefficient of discharge including velocity of approach at a hypothetical Reynolds Number of infinity, p. 391.
p_f = flowing pressure, psia
Q_h = operating flow rate, cfh, at base conditions
Y = expansion factor

Establishing Values:

D = 3.068 inches
F_a = 1.0000
F_g = 1.0168
F_m = 1.0000
F_{pb} = 1.0020 (using 14.7 psia as base pressure)
F_{pv} = 1.0520
F_{tb} = 1.0000 (using 60°F as base temperature)
F_{tf} = 0.9905
F_{wv} = 1.0, dry gas
h_w = 100 inches W.C.
p_f = 214.7 psia
Q_h = 120,000 scfh

[5] op. cit., Spink

Assume F_r = 1.0 because of low viscosity. Nitrogen has a k-value of 1.4, therefore use Fig. B-2502, p. 356, to find Y, after calculating a value for

$$\frac{(F_m)^2\ h_w}{p_f} = \frac{(1)\ 100}{214.7} = 0.466; \text{ therefore}$$

$Y \approx 0.994$

Substituting: (See box at bottom of page)

From Table 29, p. 391 of Spink, for a 3.000 inch run,

β	$K_o\beta^2$ flange taps
0.61	0.243130
	0.243815
0.62	0.252560

interpolating, β = 0.61073

Bore = βD
= 0.61073 (3.068)
= 1.8737 inches
= 1.874 inches

Level

PROBLEM: Level Transmitter Range Calculation (for Open Tank or With Equalizer Leg)

Given

Vessel:	Vertical cylindrical tank
Dimensions:	40 ft. floor-to-seam, lower level conn. 1'-6" above bottom – upper level conn. in roof
Fluid Specific Gravity:	0.87
Nitrogen Blanket:	1 inch W.C. pressure
Transmitter:	Diaphragm-type with dry equalizer leg

The formula for calculating the range, P_m, in inches W.C. is

$$P_m = L_m \times G$$

where

P_m = fluid pressure at maximum level, (L_m) in inches W.C.

$$K_o\beta^2 = \frac{120{,}000}{338.17\ (3.068)^2\ (1)\ (0.994)\ (1.0020)\ (1)\ (0.9905)\ (1.0168)\ (1.052)\ (1)\ (1)\ (1)} \times \frac{}{\sqrt{100\ (214.7)}} = 0.243815$$

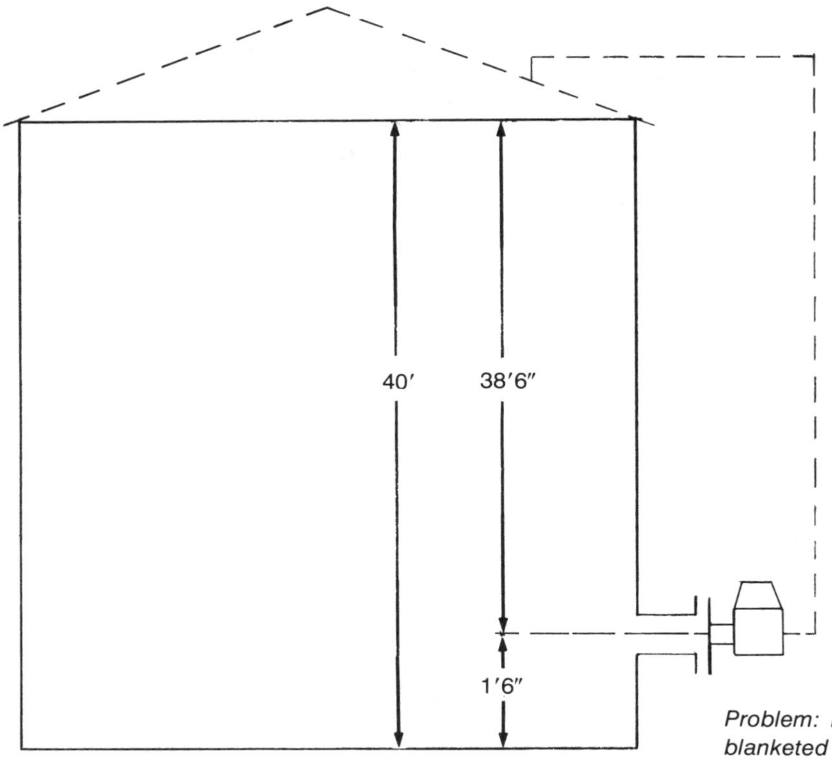

Problem: Level transmitter range calculation (pressurized or blanketed vessel) with equalizer leg.

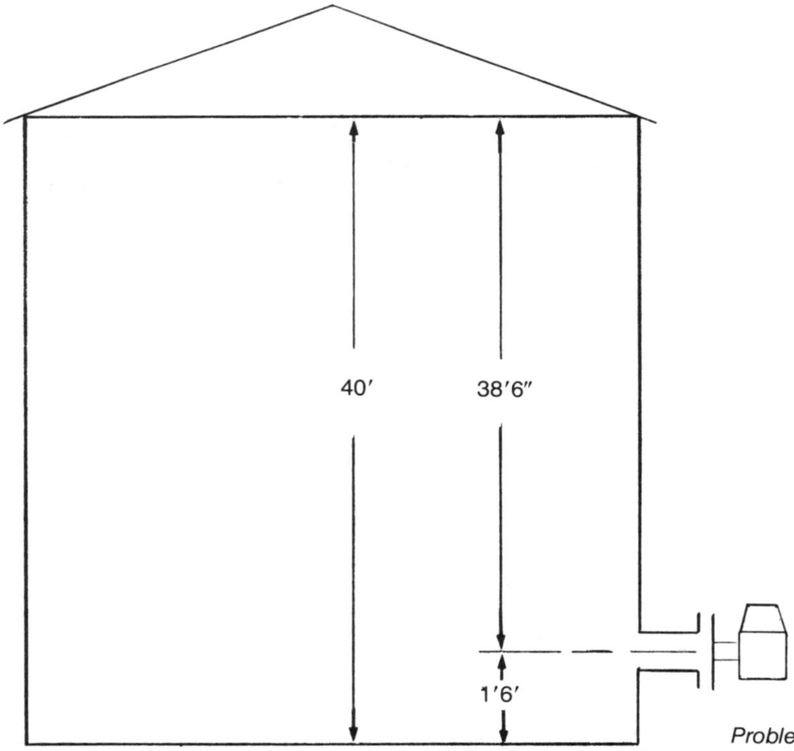

Problem: Same as above, except no equalizer leg. (for open tank)

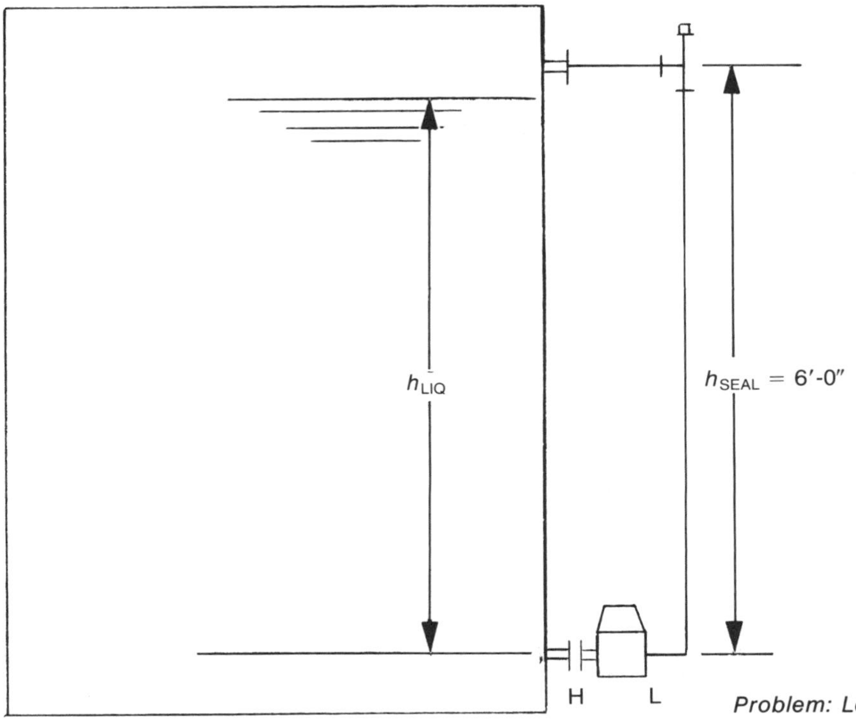

Problem: Level transmitter range calculation with seal fluid.

L_m = maximum liquid level, in inches
G = fluid specific gravity

Establishing Values:

L_m = (40′ − 1.5′) (12″/ft) = 462 inches
G = 0.87

Substituting:

P_m = 462 (0.87)
= 402 inches W.C.

PROBLEM: Same as Problem above, except no equalizer leg. (Pressurized or Blanketed Vessel)

The formula for calculating the range is

$$P_m = L_m \times G + P_{be}$$

where P_{be} = blanket pressure, inches W.C.

Substituting

P_m = 462 (0.87) + 1
= 403 inches W.C.

NOTE: The transmitter will sense 1″ W.C. when no measurable level exists. The range will be 1″ W.C. to 403″ W.C., a span of 402″.

PROBLEM: Level Transmitter Range Calculation with Seal Fluid

Given

Fluid:	A corrosive liquid
Specific gravity:	0.80
Connections:	6′-0″ apart, lowest transmitter connection at the same elevation as the lower vessel connection.
Seal fluid specific gravity:	0.975

The formulas for calculating the span are

$$\text{Span} = (P_{seal} - P_{liq.\ min.}) - (P_{seal} - P_{liq.\ max.})$$
$$P_{liq.\ max.} = h_{max} \times G_{liq.}$$
$$P_{seal} = h_{seal} \times G_{seal}$$

where

$G_{liq.}$ = specific gravity of measured liquid
G_{seal} = specific gravity of seal fluid

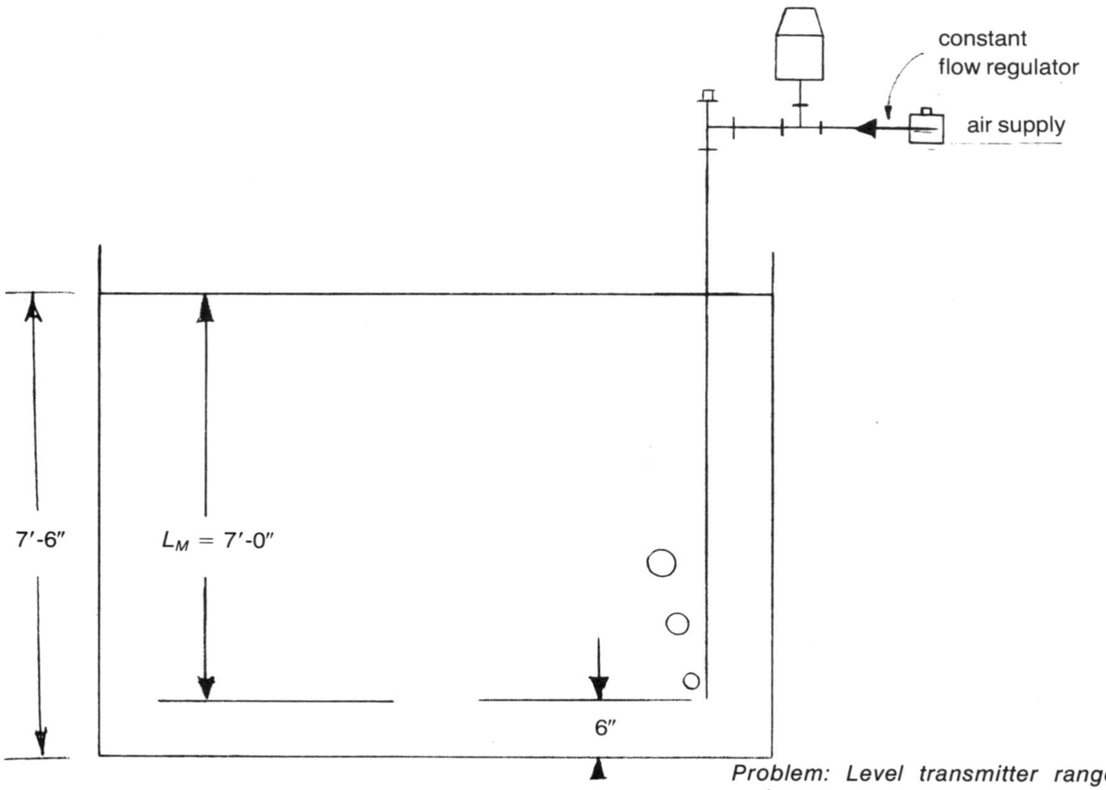

Problem: Level transmitter range calculation for bubbler action.

$h_{max.}$	=	maximum liquid level above measurement conn., inches
h_{seal}	=	seal leg length, inches
$P_{liq.\,max.}$	=	maximum pressure at measurement conn., inches W.C.
P_{seal}	=	pressure at the transmitter seal conn., inches W.C.

Establishing Values:

$G_{liq.}$ = 0.80
G_{seal} = 0.975
$h_{max.}$ = 6'-0"
h_{seal} = 6'-0"

Substituting:

$$P_{seal} = 6\,(12)\,(0.975) = 70.2 \text{ inches W.C.}$$

$$P_{liq.max.} = 6\,(12)\,(0.80) = 57.6 \text{ inches W.C.}$$

$$P_{liq.min.} = 0$$

$$\text{Span} = (70.2-0) - (70.2-57.6) = 70.2 - 12.6 = \underline{\underline{57.6 \text{ inches W.C.}}}$$

NOTE: Range is from 70.2" W.C. (level min.) to 12.6" W.C. (level max.). A transmitter with zero elevation to be specified.

PROBLEM: Level Transmitter Range Calculation for Bubbler Application

Given

Vessel:	Sump
Dimensions:	8'-0" deep
Measuring device:	Bubble pipe with end 6" above the bottom
Maximum level:	7'-6"
Liquid:	Water
Temperature:	98°C

The formula for calculating the range is

$$P_m = L_m \times G$$

where

P_m	=	fluid pressure at maximum level, (L_m) in inches W.C.
L_m	=	maximum liquid level, in inches
G	=	fluid specific gravity

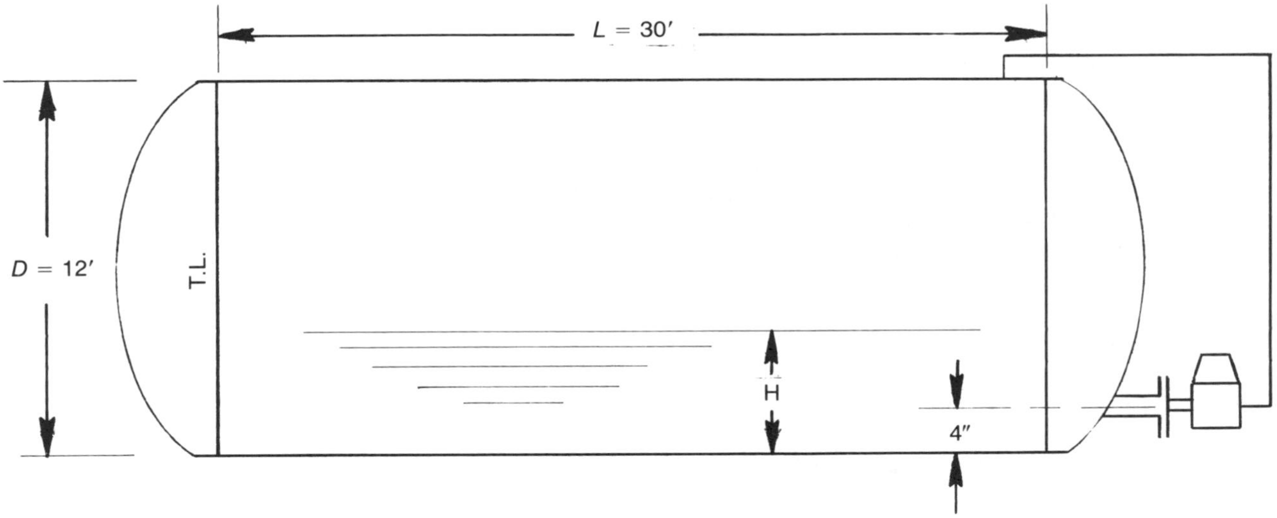

Problem: Level alarm point calculation for horizontal tank.

Establishing Values:

L_m = 7′-0″
G = 0.964 (water at 98°C)

Substituting:

$$P_m = 7\,(12)\,(0.964) = 80.976$$

Range is from 0″ W.C. to 81″ W.C.

PROBLEM: Level Alarm Point Calculation for Horizontal Tank

GIVEN

Vessel:	Horizontal cylinder with elliptical heads
Dimensions:	12′-0″ ID x 30′-0″ T-T
Connections:	4″ above bottom, and top
Alarm points:	20% and 80% capacity

Formulas used for calculations are

$$V_c = \frac{\pi L D^2}{4}$$

$$V_h = \frac{\pi D^3}{12}$$

$$V_t = V_c + V_h$$
$$V_L = K_c V_c + K_h V_h = 0.2\ V_t$$
$$V_H = K_{c2} V_c + K_{h2} V_h = 0.8\ V_t$$

where

D = ID of vessel, feet
K_c = coefficient for partial volume of horizontal cylinders (from Table 4.14, p. 261)
K_h = coefficient for partial volume of ellipsoids (from Table 4.15, p. 262)
L = T-T length of vessel, feet
V_c = total volume of the cylinder, cu. ft.
V_h = total volume of two heads, cu. ft.
V_t = total volume of the vessel, cu. ft.
V_L = partial volume at low alarm point, cu. ft.
V_H = partial volume at high alarm point, cu. ft.

Establishing Values:

D = 12 feet
L = 30 feet

Substituting

$$V_c = \frac{\pi\,(30)\,(12)^2}{4} = 3{,}392.92 \text{ cu. ft.}$$

$$V_h = \frac{\pi\,(12)^3}{12} = 452.39 \text{ cu. ft.}$$

$$V_t = 3{,}392.92 + 452.39 = 3{,}845.31 \text{ cu. ft.}$$

$$V_L = 0.2\,(3{,}845.31) = 769.06 \text{ cu. ft.}$$

$$V_H = 0.8\,(3{,}845.31) = \underline{\underline{3{,}076.25 \text{ cu. ft.}}}$$

Tables 4.14 and 4.15 make possible the calculation of capacity when level and diameter are known. Since, for this problem, capacity and diameter are known and level is to be calculated, a trial-and-error method is employed.

Assume a value of the ratio of the liquid depth with respect to the diameter *(H/D)*, and use the respective coefficients for partial volumes of both horizontal cylinders and ellipsoids (Tables 4.14 and 4.15) to check for equality with V_L in the equation.

$$V_L = K_c V_c + K_h V_h$$

$$769.06 = K_{c1}\,(3{,}392.92) + K_{h1}\,(452.39)$$

Assume $H/D = 0.260; K_{c1} = 0.206600, K_{h1} = 0.167684$

$$769.06 \neq 0.206600\,(3{,}392.92) + 0.167684\,(452.39)$$

$$\neq 776.84$$

Try $H/D = 0.255; K_{c1} = 0.201031, K_{h1} = 0.161912$

$$769.06 \neq 0.201031\,(3{,}392.92) + 0.161912\,(452.39)$$

$$\neq 755.33$$

Try $H/D = 0.258; K_{c1} = 0.204368, K_{h1} = 0.165345$

$$769.06 \cong 0.204368\,(3{,}392.92) + 0.165345\,(452.39)$$

$$\cong 768.20$$

Therefore

$$H_L = 0.258D, \text{ where } H_L = \text{low alarm level}$$
$$= 0.258\,(12)\,(12)$$
$$= 37.152$$
$$\cong 37 \text{ inches}$$

In this case,

$$H_H = D - H_L, \text{ where } H_H = \text{high alarm level}$$
$$= 12\,(12) - 37$$
$$= 107 \text{ inches}$$

The transmitter connection is 4″ above the bottom, therefore the range is 0″ to 140″ liquid. Low alarm is at 33″ liquid, or 23.571% of span. High alarm is at 103″ liquid, or 73.571% of span. Transmitter output at 20% volume is

$$P_L = 0.23571\,(15 - 3) + 3$$
$$= 5.829 \text{ psig, or } 161.9 \text{ inches W.C.}$$

Transmitter output at 80% volume is

$$P_H = 0.73571\,(15 - 3) + 3$$
$$= 11.829 \text{ psig, or } 328.6 \text{ inches W.C.}$$

Interface

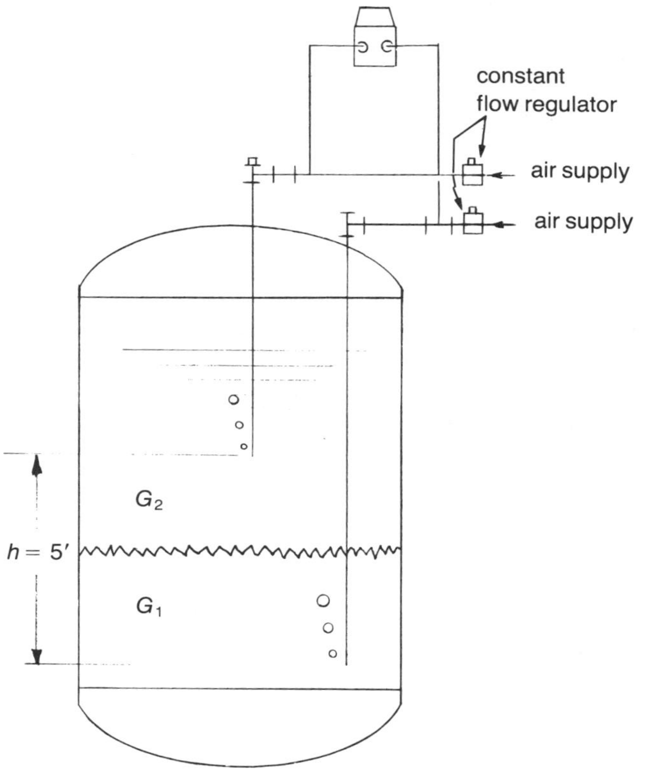

Problem: Interface measurement calculation using a d/p transmitter and two bubbler pipes.

PROBLEM: Interface Measurement Calculation Using a D/P Transmitter and Two Bubbler Pipes

Given

Measuring devices:	D/P Transmitter, two bubbler pipes
Level difference:	5′-0″
Specific gravities:	0.82 and 0.70

Formulas used for calculations are

$$P_{max.} = h \times G_1$$
$$P_{min.} = h \times G_2$$
$$S = P_{max.} - P_{min.}$$

where

G_1 = specific gravity of heavier liquid

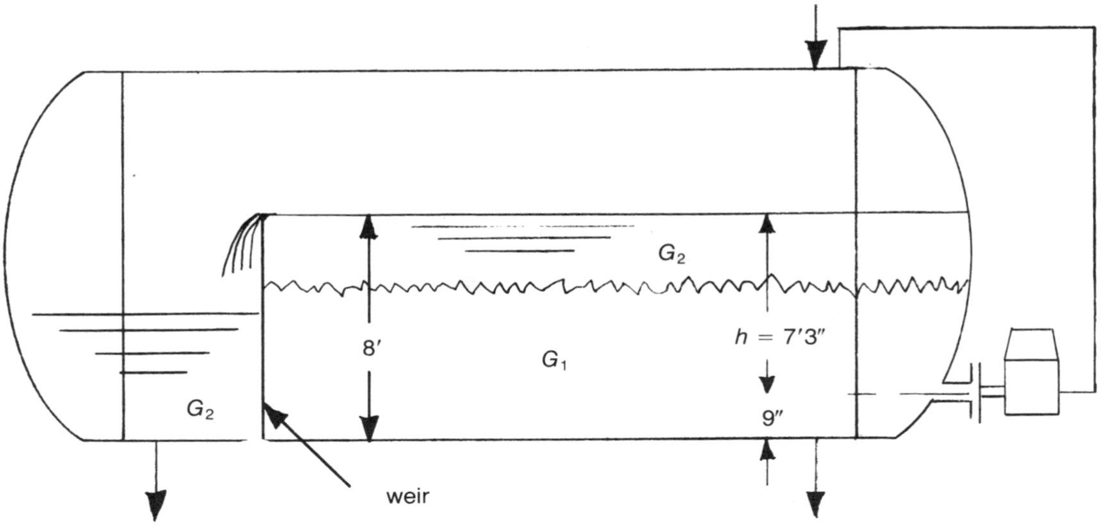

Problem: Interface measurement calculation using a d/p transmitter and a fixed weir.

G_2 = specific gravity of lighter liquid
h = level difference, inches
$P_{max.}$ = transmitter differential at maximum interface level, inches W.C.
$P_{min.}$ = transmitter differential at minimum interface level, inches W.C.
S = transmitter span, inches W.C.

Establishing Values

G_1 = 0.82
G_2 = 0.70
h = 60 inches

Substituting:

$P_{max.}$ = 60 (0.82)
= 49.2 inches W.C.
$P_{min.}$ = 60 (0.70)
= 42.0 inches W.C.
S = 49.2 - 42.0
= 7.2 inches W.C.

Transmitter range is 42.0″ W.C. to 49.2″ W.C. The zero must be suppressed to give the proper operating span.

PROBLEM: Interface Measurement Calculation Using a D/P Transmitter and a Fixed Weir

Given

Vessel: Horizontal drum with an 8′-0″ weir, constant flow over the weir
Measuring device: Flange mounted D/P transmitter
Connections: 9″ above the drum bottom, and top
Specific gravities: 0.85 and 0.60.

Formulas used for calculations are

$P_{max.} = h \times G_1$
$P_{min.} = h \times G_2$
$S = P_{max.} - P_{min.}$

where

G_1 = specific gravity of heavier liquid
G_2 = specific gravity of lighter liquid
h = level difference, inches
$P_{max.}$ = transmitter differential at maximum interface level, inches W.C.
$P_{min.}$ = transmitter differential at minimum interface level, inches W.C.
S = transmitter span, inches W.C.

Establishing Values:

G_1 = 0.85
G_2 = 0.60
h = 87 inches

Substituting:

$P_{max.}$ = 87 (0.85)
= 73.95

$$\approx 74 \text{ inches W.C.}$$
$$P_{min.} = 87\ (0.60)$$
$$= 52.2 \text{ inches W.C.}$$
$$S = 74.0 - 52.2$$
$$= \underline{\underline{21.8 \text{ inches W.C.}}}$$

Transmitter range is 52.2″ W.C. to 74″ W.C. The zero must be suppressed to give the proper operating span. Use caution to ensure that total level exceeds weir height to prevent erroneous readings.

Control Valves

PROBLEM: Control Valve Sizing Calculation for Liquid Service (Volume Basis)

Fluid:	Ethyl Alchol
Specific Gravity:	0.789
Flow Rate:	200 gpm
Sizing Differential:	15 psi

The formula used for sizing is

$$C_v = Q\sqrt{\frac{G_l}{\Delta P}} \quad \text{(Eq. 5.2, Vol. 2 2nd Ed., p. 112)}$$

where

C_v = valve flow coefficient
G_l = specific gravity at flowing conditions
ΔP = pressure drop across the valve at Q, psi
Q = maximum flow rate, gpm

Established Values:

G_l = 0.789
ΔP = 15 psi
Q = 200 gpm

Substituting:

$$C_v = 200\sqrt{\frac{0.789}{15}}$$
$$= \underline{\underline{45.9}}$$

PROBLEM: Control Valve Sizing Calculation for Steam Service

Given

Fluid:	Steam
Pressure:	450 psig
Temperature:	489°F
Flow Rate:	10,000 pph
Sizing Differential:	50 psi

The formula used for sizing is

$$C_v = \frac{W(1 + 0.0007\,T_{sh})}{2.1\sqrt{\Delta P(P_1 + P_2)}} \quad \text{(Eq. 5.5, Vol. 2, 2nd Ed., p. 112)}$$

where

C_v = valve flow coefficient
P_1 = inlet pressure, psia
ΔP = pressure drop across the valve at W, psi
T_{sh} = superheat degrees above saturation temp., °F
W = maximum flow rate, pph

Establishing Values:

P_1 = 464.7 psia
ΔP = 50 psi
T_{sh} = 29.5°F (interpolating from Table 4.43)
W = 10,000 pph

Substituting:

$$C_v = \frac{10{,}000\,[1 + 0.0007\,(29.5)]}{2.1\sqrt{50\,[464.7 + 414.7]}} = \underline{\underline{23.18}}$$

PROBLEM: Maximum Flow Rate Calculation for a Valve of Given Size at Given Conditions

Given

Fluid:	Steam
Pressure:	150 psig
Temperature:	Saturation
Valve C_v:	400, Max.
Downstream Pressure:	14 psig
Downstream Pipe Loss:	Negligible

The formula used for calculation is

$$C_v = \frac{W(1 + 0.0007\,T_{sh})}{2.1\sqrt{\Delta P(P_1 + P_2)}} \quad \text{(Eq. 5.5, Vol. 2, 2nd Ed., p. 112)}$$

Rewriting:

$$W = \frac{2.1\,C_v\sqrt{\Delta P\,(P_1 + P_2)}}{(1 + 0.007\,T_{sh})}$$

where

C_v = valve flow coefficient
P_1 = inlet pressure, psia
P_2 = discharge pressure, psia
ΔP = pressure drop across the valve at W, psi
T_{sh} = superheat degrees above saturation temp., °F
W = maximum flow rate, pph

Establishing Values:

C_v = 400
P_1 = 164.7 psia
P_2 = 28.7 psia
T_{sh} = 0°F

Since $\Delta P_{actual} > \frac{P_1}{2}$, set $\Delta P = \frac{P_1}{2}$ (P_2 in the equation also becomes $\frac{P_1}{2}$)

$$\frac{P_1}{2} = \frac{164.7}{2} = 82.35 \text{ psi}$$

Substituting:

$$W = \frac{2.1\,(400)\sqrt{82.35\,(164.7 + 82.35)}}{1 + (0)}$$
$$= \underline{\underline{119{,}813 \text{ pph}}}$$

PROBLEM: Control Valve Sizing Calculation for Gas Service (Volume Basis)

Given

Fluid:	Nitrogen
Pressure:	190 psig
Temperature:	70°F
Flow Rate:	2,000 scfm
Sizing Differential:	75 psi
Compressibility Factor:	0.994

The formula used for sizing is

$$C_v = \frac{Q}{1{,}360}\sqrt{\frac{T \times G_b \times Z}{\Delta P \times P_2}}$$ (Eq. 5.4, Vol. 2 2nd Ed., p. 112)

where

C_v = valve flow coefficient
G_b = specific gravity at base conditions
P_2 = discharge pressure, psia
ΔP = pressure drop across the valve, psi
Q = = flow rate, scfh
T = flowing temperature, °R
Z = compressibility factor

Establishing Values:

G_b = 0.967 (Table 4.26, p. 327)
P_2 = 129.7 psia
ΔP = 75 psi
Q = 120,000 scfh
T = 530°R (°R = °F + 460)
Z = 0.994

Substituting:

$$C_v = \frac{120{,}000}{1{,}360}\sqrt{\frac{530\,(0.967)\,(0.994)}{75\,(129.7)}}$$
$$= \underline{\underline{20.19}}$$

PROBLEM: Control Valve Capacity Calculation for Gas Service (Volume Basis)

Given

Fluid:	Air
Pressure:	100 psig
Temperature:	60°F
Compressibility Factor:	0.998
Discharge Pressure:	15 psig
Sizing Coefficient:	C_v = 63

The formula used for sizing is

$$C_v = \frac{Q}{1{,}360}\sqrt{\frac{T \times G_b \times Z}{\Delta P \times P_2}}$$ (Eq. 5.4, Vol. 2 2nd Ed., p. 112)

Rewriting:

$$Q = \frac{1{,}360\,C_v}{\sqrt{\dfrac{T \times G_b \times Z}{\Delta P \times P_2}}}$$

where

C_v = valve flow coefficient
G_b = specific gravity at base conditions
P_2 = discharge pressure, psia
ΔP = valve pressure drop at Q, psi
Q = flow rate, scfh
T = operating temperature, °R
Z = compressibility factor

Establishing Values:

C_v = 63
G_b = 1.0
P_1 = 114.7 psia
P_2 = 29.7 psia
T = 520°R (°R = °F + 460)
Z = 0.998

When $\Delta P > \frac{P_1}{2}$, set $\Delta P = \frac{P_1}{2} = P_2 = 57.35$ psi

Substituting:

$$Q = \frac{1{,}360(63)}{\sqrt{\frac{520(1)\,(0.963)}{57.35\ (57.35)}}}$$
$$= 219{,}583 \text{ scfh}$$
$$= \underline{\underline{3{,}660 \text{ scfm}}}$$

Ethane has a molecular weight of 30.068 (Table 4.21, p. 284)

Therefore:

$$G_b = \frac{MW_{\text{gas}}}{MW_{\text{air}}} = \frac{30.068}{28.966} = 1.038$$

To obtain Z, use Spink[6], Figure B-2518, p. 374.

T_r = 1.43, and p_r = 0.42. From Figure B-2520, p. 377, Z = 0.955.

Substituting:

$$C_v = \frac{1{,}500}{104}\sqrt{\frac{760\ (0.955)}{60\ (1.038)\ (254.7)}}$$
$$= \underline{\underline{3.085}}$$

PROBLEM: Control Valve Sizing Calculation for Gas Service (Weight Basis)

Given

Fluid:	Ethane
Pressure:	300 psig
Temperature:	300°F
Sizing Differential:	60 psi
Flow Rate:	1,500 pph

The formula used for sizing is:

$$C_v = \frac{W}{104}\sqrt{\frac{T \times Z}{\Delta P \times P_2 \times G_b}}$$ (Eq. 5.14, Vol. 2 2nd Ed., p. 115)

where

C_v = valve flow coefficient
G_b = specific gravity at base conditions
P_2 = discharge pressure, psia
ΔP = pressure drop across valve at W, psi
T = operating temperature, °R
W = flow rate, pph
Z = compressibility factor

Establishing Values:

P_2 = 254.7 psia
ΔP = 60 psi
T = 760°R (°R = °F + 460)
W = 1,500 pph

Relief Valves

PROBLEM Thermal Relief of an Exchanger That has Been Blocked in:

Given

Fluid:	60 Deg API Hydrocarbon Liq @ 300°F
Exchanger heat transfer rate:	5,000,000 Btu/Hr
Specific heat of trapped fluid:	0.7 Btu/lbs/°F
Cubical expansion factor, B:	0.0006 cuft/°F
Specific gravity, referred to H_20 = 1.0 @ 60°F:	0.74
Set pressure:	150 psig

The formula for sizing is

$$\text{gpm} = \frac{BH}{500GC}$$ (See Vol. II, 2nd Ed., Eq. 6.6, p. 141)

where:

gpm = flow rate at the flowing temperature, in U.S. gallons per minute.

B = cubical expansion coefficient per degree Fahrenheit for the liquid at the expected temperature differential. It is best to obtain this information from the process design data; however, the following are typical values for hydrocarbon liquids and for water:

3-35 deg API	0.0004
35-51 deg API	0.0005
51-64 deg API	0.0006
64-79 deg API	0.0007
79-89 deg API	0.0008
89-94 deg API	0.00085
94-100 deg API and lighter	0.0009
Water	0.0001

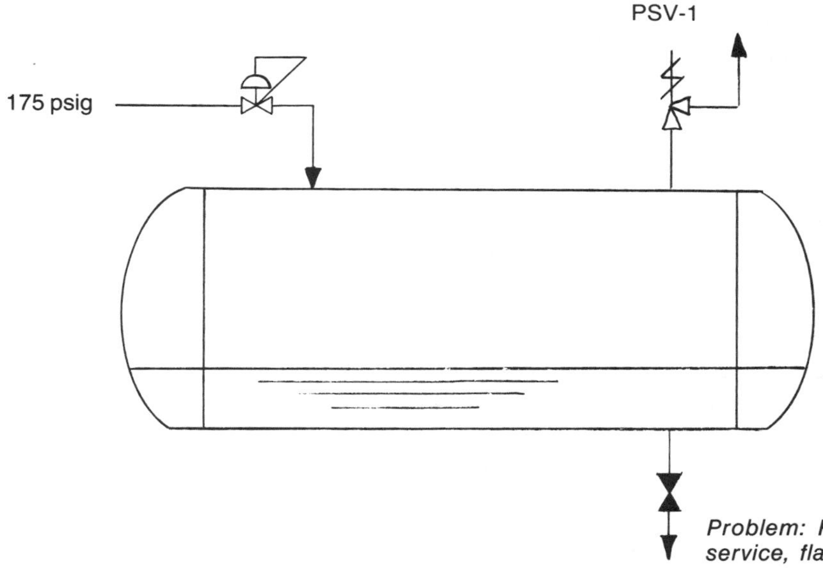

Problem: Relief valve calculation for blocked outlet, liquid service, flashing in valve.

H = total heat transfer rate, in British thermal units per hour. This should be taken as the maximum exchanger duty during operation.
G = specific gravity referred to water = 1.00 at 60°F. Compressibility of liquid is usually ignored.
C = specific heat of the trapped fluid, in British thermal units per pound per degree Fahrenheit.

Substituting:

$$\text{gpm} = \frac{0.0006\,(5{,}000{,}000)}{500\,(0.74)\,(0.7)}$$

$$= \underline{11.5}$$

$$A = \frac{\text{gpm}}{27.2\,K_p\,K_w\,K_v}\sqrt{\frac{G}{P - P_b}}$$

$$= \frac{11.5}{27.2\,(1.0)\,(1.0)\,(1.0)}\sqrt{\frac{.74}{150 - 0}}$$

$A = \underline{\underline{0.0297 \text{ sq. in.}}}$ Orifice Area Required

PROBLEM: Relief Valve Calculation for Blocked Outlet, Liquid Service, Flashing in Valve.

Given

Fluid:	Steam condensate
Normal fluid flow rate:	10,000 pph
Temperature:	350°F
Pressure:	175 psig
Vessel:	Make-up tank, operated at 120 psig. Relief valve on this tank.
Back pressure:	0 psig

Condition for sizing: Blocked outlet on make-up tank. The formulas used for sizing are:

$$A_l = \frac{gpm\ \sqrt{G}}{27.2\ K_p K_w K_v\ \sqrt{P_1 - P_b}}$$

$$A_s = \frac{W_s}{51.5\,K_1 P_1 K_b K_{sh}}$$

$$A_t = A_l + A_s$$

where

A_l = required orifice area for liquid, sq. in.
A_s = required orifice area for steam, sq. in.

A_t = total required area, sq. in.
G = liquid specific gravity at operating temperature
K = coefficient of discharge, obtainable from valve manufacturer. For many nozzle-type valves, K = 0.975.
K_p = liquid capacity correction factor, Chart 3.73, p. 230
K_{sh} = superheat correction factor, Table 4.43, p. 310

[6]op. cit., Spink

K_v = viscosity correction factor, Chart 3.75, p. 232
K_b = vapor or gas flow factor for variable back pressure
K_w = liquid flow factor for variable back pressure
P_1 = relieving pressure, psia, which is set pressure + over pressure + atmospheric pressure
p = set pressure at which relief valve is to begin opening, psig
p_b = back pressure, psig
gpm = required liquid capacity, gpm
W_s = required steam capacity, pph

Since the valve discharges into a 0.00 psig back pressure header, part of the condensate will flash into steam. This quantity is determined by the relationship

$$\text{Evap. \%} = \frac{h_{f1} - h_{f2}}{h_{fg1}} \times 100$$

From steam tables:

h_{f1} = 321.64 Btu/lbm, the heat content of water at 350°F
h_{f2} = 180.07 Btu/lbm, the heat content of water at 14.696 psia (212°F)
h_{fg1} = 970.3 Btu/lbm, the latent heat of vaporization, or heat required to vaporize a pound of water at 14.696 psia

Substituting:

$$\text{Evap. \%} = \frac{321.64 - 180.07}{970.3} \times 100$$
$$= 14.6$$

Relieving load, therefore, is 14.6% steam (1,460 pph) and 85.4% water (8,540 pph).

To determine condensate specific gravity, divide condensate density at operating conditions by the density of water at standard conditions (ρ_s = 62.356 lbm/cu. ft. at 60°F). From steam tables: v_f = 0.01799

$$\rho_l = \frac{1}{v_f} = \frac{1}{0.01799} = 55.586 \text{ lbm/cu. ft.}$$

Therefore:

$$G = \frac{\rho_l}{\rho_s} = \frac{55.586}{62.356}$$
$$= 0.891$$

Since set pressure was not given, 110% of the operating pressure will be used, provided the working pressure rating of the vessel is not exceeded. Vapor over pressure is 10% above set pressure, per ASME Code, Section VIII.

Converting water flow rate,

$$\text{gpm} = \frac{\text{pph}}{500G}$$
$$= \frac{8540}{500\,(0.891)}$$
$$= 19.16$$

Establishing Values:

G = 0.891
K = 0.975
K_p = 0.6
K_{sh} = 1.0 at saturation
K_v = 1.0 (non-viscous liquid)
K_b = 1.0 (constant back pressure)
K_w = 1.0 (constant back pressure)
P_1 = (120 + 12) (1.10) + 14.7 = 159.9 psia
p = 132 psig
p_b = 0 psig
gpm = 19.16 gpm
W_s = 1,460 pph

Substituting into the sizing formulas

$$A_l = \frac{19.16 \sqrt{0.891}}{27.2\,(0.6)\,(1)\,(1)\,\sqrt{132 - 0}}$$
$$= 0.0965 \text{ sq. in.}$$

$$A_s = \frac{1{,}460}{51.5\,(0.975)\,(159.9)\,(1)\,(1)}$$
$$= 0.1818 \text{ sq. in.}$$

But the National Board of Boiler and Pressure Vessel Inspectors has de-rated many valves for use in steam service by a factor of 10%. Therefore,

$$A_s' = \frac{A_s}{0.9}$$
$$= \frac{0.1818}{0.9}$$
$$= 0.2020 \text{ sq. in.}$$

and

$$A_t = 0.2020 + 0.0965$$
$$= \underline{\underline{0.2985 \text{ sq. in.}}}$$

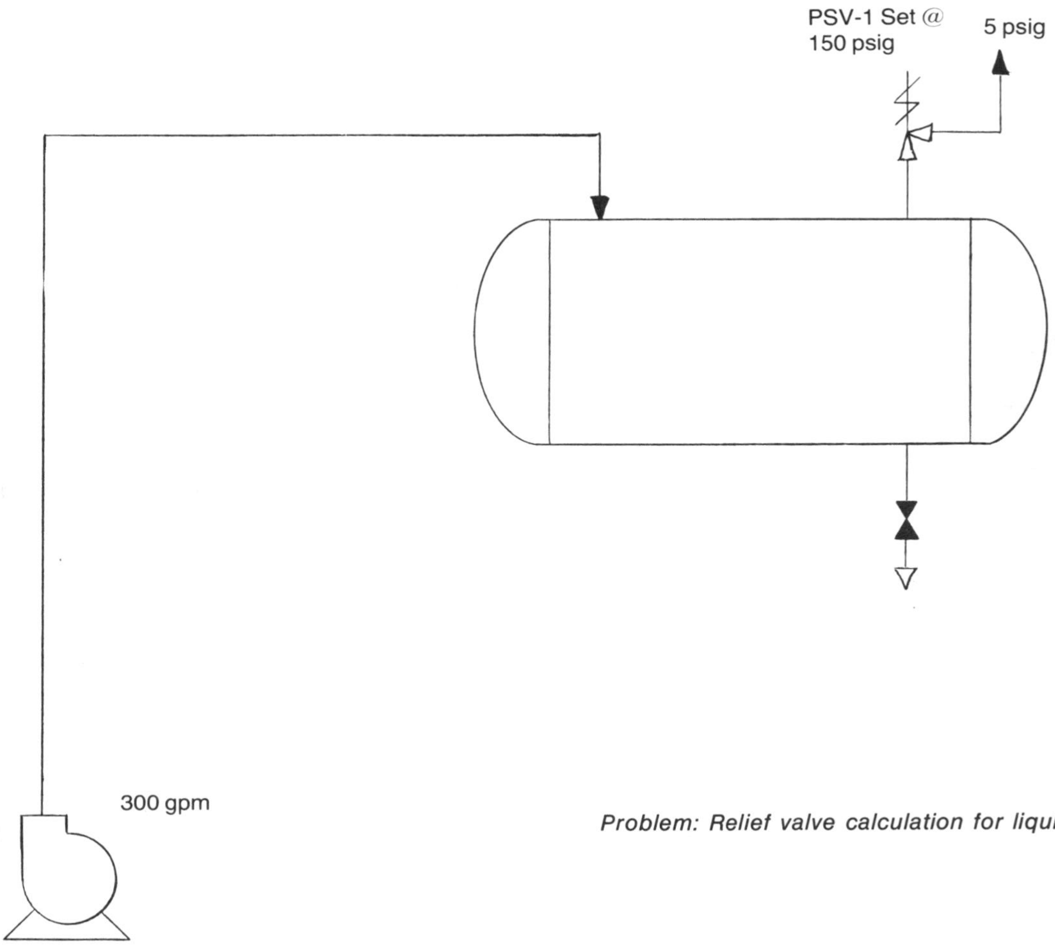

Problem: Relief valve calculation for liquid relief, non-flashing service.

PROBLEM: Relief Valve Calculation for Liquid Relief, Non-Flashing Service

Given

Fluid:	23.1°API Heavy Motor Oil
Relieving Pressure:	150 psig
Back Pressure:	5 psig
Operating Temperature:	200°F
Quantity to be Relieved:	300 gpm
Condition for Sizing:	Pumping into a Vessel with Closed Outlets

The formula for sizing non-viscous Liquid Relief is

$$A = \frac{\text{gpm}}{27.2\, K_p\, K_w\, K_v} \sqrt{\frac{G}{p - p_b}} \quad \text{(page 442)}$$

where

A = discharge area, square inches
gpm = flow rate, gallons per minute
K_p = capacity correction factor due to overpressure
K_w = capacity correction factor due to backpressure
K_v = capacity correction factor due to viscosity
G = Specific gravity of the Liquid at flowing temperature
p_1 = set pressure, psig
p_b = backpressure, psig

Checking for the Unknowns:

K_p = 0.6 at 10% overpressure (accumulation-obtained from Chart 3.73, p. 230)

K_w = 1.0 for conventional valves
K_v must be determined from a trial calculation
SG @ 60°F is determined by the relationship –

$$SG = \frac{141.5}{131.5 + \text{Deg. API}}$$
$$= \frac{141.5}{131.5 + 23.1}$$
$$= 0.915$$

G (at operating temperature – 200°F) is obtained from Chart 3.57, p. 210.

Where the specific gravity curve (0.915, interpolating between 0.90 and 0.92) intersects the operating temperature abscissa (200°F), follow the ordinate to obtain 0.865 as the Operating Specific Gravity, G.

$$G = 0.865$$

Trial Orifice Size:

$$A = \frac{\text{gpm}}{27.2\,K_p}\sqrt{\frac{G}{p_1 - p_b}}$$
$$= \frac{300}{27.2\,(0.6)}\sqrt{\frac{0.865}{150 - 5}}$$
$$= 1.42 \text{ sq. in.}$$

From any table of standard orifice sizes (see Volume 2, 2nd Ed. Table 6.1, page 132), the orifice selected would be a K orifice = 1.838 sq. in.

To determine K_v, the correction factor for viscosity, it is necessary to calculate the Reynolds Number R. Using the relationship

$$R = \frac{\text{gpm}\,(2{,}800G)}{\mu\ \sqrt{A}}$$

Where gpm, G, and A are as noted above, and

μ = is viscosity in centipoises
= 24.5 (obtained from Chart 3.59, p. 212.).

$$R = \frac{300 \times 2800 \times 0.865}{24.5\ \sqrt{1.838}}$$
$$= 21{,}876$$

From Chart 3.75, p. 232, $R = 21{,}876$ intersects the correction factor curve at 0.98.

$$K_v = 0.98$$

Using the given and derived values in the bore formula

$$A = \frac{\text{gpm}}{27.2\,K_p\,K_w\,K_v}\sqrt{\frac{G}{p - p_b}}$$

$$A = \frac{300}{27.2\,(0.6)\,(1.0)\,(0.98)}\sqrt{\frac{0.865}{150 - 5}}$$

$$= \underline{\underline{1.449 \text{ square inches}}}$$

Therefore the K orifice, at 1.838 sq. in., is adequate; and a 3″ x 4″ valve is selected.

PROBLEM: Relief Valve Sizing Calculation for Gas Expansion Due to External Fire

Given

Fluid:	Methane Gas
Operating Pressure:	711 psig
Operating Temperature:	81°F
Vessel:	Dry gas scrubber
Dimensions:	95¾″ OD by 96″ T-T. vertical
Heads	2:1 elliptical
Elevation:	3′-0″
Spring set pressure;	1,050 psig
Allowable Accumulation:	10%

Condition for sizing: Scrubber blocked-in during a fire.

The discharge areas for safety and safety relief valves on gas-containing vessels exposed to open fires can be determined by use of the following formula:

$$A = \frac{F'A_s}{\sqrt{P_1}}$$

where

A = effective discharge area of the valve, in square inches.

F' = an operating factor determined from Chart 3.72, p. 229)

A_s = exposed surface area of the vessel, in square feet.

P_1 = upstream pressure, in pounds per square inch absolute. This is the set pressure multiplied by 1.10 or 1.20 (depending on the overpressure permissible) plus the atmospheric pressure, in pounds per square inch absolute.

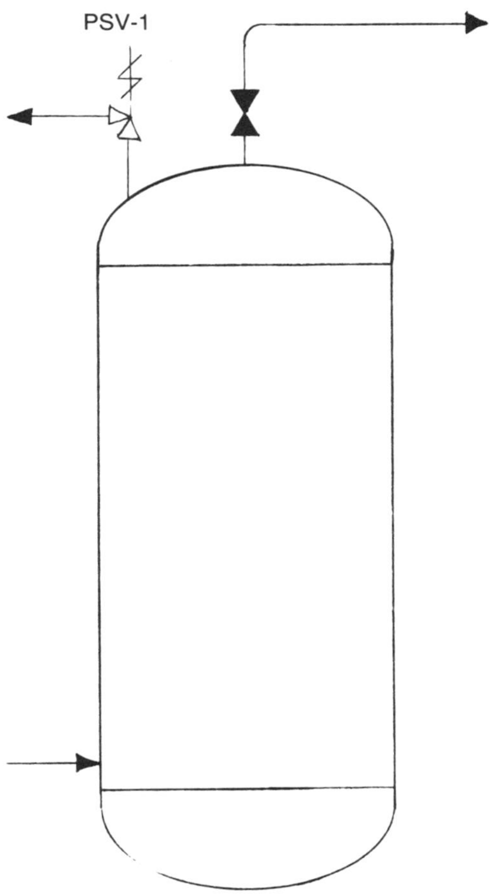

Problem: Relief valve sizing calculation for gas expansion due to external fire.

Assume the shell and both heads are engulfed in flames. The total vessel surface exposed to flame is

$$A_s = \pi DL + 2\,(1.09\ OD^2)$$
$$= 3.1416\left(\frac{95.75}{12}\right)\left(\frac{96}{12}\right) + 2(1.09)\left(\frac{95.75}{12}\right)^2$$
$$= 340 \text{ sq. ft.}$$

The relieving temperature $T_1 = \left(\frac{P_1}{P_n}\right) T_n$, where

P_1 = relieving pressure, psia
P_n = normal pressure, psia
T_n = normal temperature, °R

Substituting

$$T_1 = \frac{1{,}050\,(1.10) + 14.7}{711 + 14.7}\,(81 + 460)$$
$$= 872°\text{R } (= 412°\text{F})$$

From Table 4.32, p. 296, $k = 1.31$. And, therefore, from Chart 3.72, p. 229, $F' = 0.018$.

Substituting in the equation for orifice area,

$$A = \frac{0.018\,(340)}{\sqrt{1169.7}}$$
$$= \underline{\underline{0.1789 \text{ sq. in.}}}$$

Select a size "E" orifice having an area of 0.196 sq. in.

For purposes of sizing downstream piping, calculate the required flow rate through the valve, using the equation

$$W = \frac{CKAP_1K_b\ \sqrt{M}}{\sqrt{T_1 Z}}$$

where

A = effective discharge area, sq. in.
C = coefficient determined by the ratio of specific heats, from Chart 3.76, p. 233.
K = coefficient of discharge, obtainable from valve manufacturer. For many nozzle-type valves, K = 0.975.
K_b = back pressure correction factor at P_1 from Chart 3.78, p. 235.
M = molecular weight of the gas or vapor.
P_1 = upstream pressure, psia, which is set pressure + overpressure + atmospheric pressure.
T_1 = relieving temperature, °R
W = valve capacity, pph
Z = compressibility factor

Establishing values:

A = 0.1789 sq. in.
C = 347.91
K = 0.975 (assumed)
K_b = 1.0
M = 16, from Table 4.31, p. 295.
P_1 = 1169.7 psia
T_1 = 872°R

Critical pressure and critical temperature for methane gas are 673 psia and 344°R, respectively (from Table 4.22, p. 287). Therefore, reduced pressure is

$$P_r = \frac{P_1}{P_c} = \frac{1{,}169.7}{673} = 1.738$$

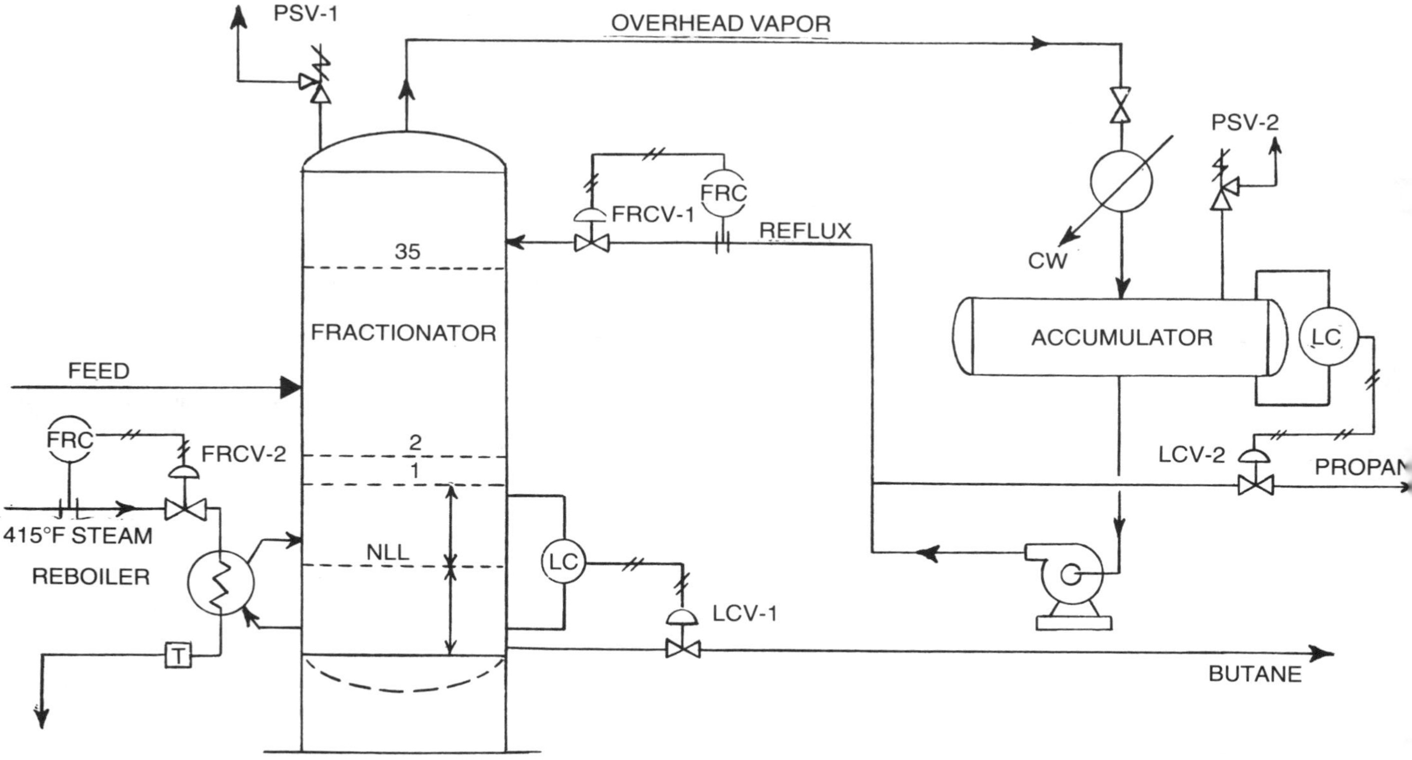

Problem: Safety relief valve sizing calculations for fractionating column.

and reduced temperature is

$$T_r = \frac{T_1}{T_{ca}} = \frac{872}{344} = 2.53$$

The compressibility factor, Z, is now obtained from Chart 3.19, p. 173.

$$Z = 1.005$$

Substituting:

$$W = \frac{347.91\,(0.975)\,(0.1789)\,(1{,}169.7)\,\sqrt{16}}{\sqrt{872}\,(1.005)}$$

$$= \underline{\underline{9{,}591 \text{ pph}}}$$

NOTE: Initial factory adjustment specifications should call for the longest blowdown possible without chattering.

PROBLEM: Safety Relief Valve Sizing Calculations for Fractionating Column.

Given

Normal Operating Pressure:	217 Psig
Normal Operating Temperatures:	
(1) Overhead	130°F
(2) Bottoms	220°F
Normal Overhead Vapor Rate:	18,000 pph
Molecular Wt. of Overhead Vapors:	46.9
PSV-1 and PSV-2 Set Pressure:	250 Psig
PSV back pressure:	0 Psig

Factionator Dimensions:

Diameter:	4′-8″ O D
Height:	55′-0″ T-T
Elevation:	6′-0″ Above Grade
Heads:	2:1 Elliptical ASME Code

Insulation:	1½″ Thick Calcium Silicate, with aluminum sheet metal cover.
Liquid Level:	4′-0″ (bot. of tower)
Tray Spacing:	2′-0″ (2½″ liquid on ea. tray.)

Reboiler:

Tube I D :	0.584 inch
Steam:	415°F saturated
Heat Duty:	2,160,000 Btu/Hr
Dimensions:	16″ OD x 16′-0″ T-T
Configuration:	See Sketch above.

Accumulator Dimensions:

Diameter:	4′-0″ O.D.
Length:	10′-0″
Heads:	2:1 elliptical ASME Code
Insulation:	None
Elevation:	10′-0″ above grade

Calculation of Contingency Relief Rates:

1. Blocked Outlet
If the block valve on the inlet to the condenser is inadvertently closed during normal operation, the total overhead vapor rate must be relieved, i.e., 18,000 pph of 46.9 *MW* vapors. The relieving pressure would be 110% of set pressure (250 x 1.1) + 14.7 = 289.7 psia. If process data is not available to provide the corresponding relief temperature, estimate it from Chart 3.56, page 209 to be 150°F. The temperature may also be estimated from Chart 3.44, page 198 by drawing a parallel vapor pressure line to propane (*MW* = 44.094).

2. Cooling Water Failure
The total overhead vapor rate must be relieved. Relieving pressures and temperatures may be calculated as in Case "A".

3. Loss of Reflux
Normally, loss of reflux because of instrument failure (FRC) will not result in an overpressure of the fractionator. However, if it is possible for loss of reflux to cause flooding of the condenser, the total overhead vapor rate must be relieved. Such a condition could be caused by: (a) LCV valve too small to remove condensed liquid within reasonable operator response time, or (b) loss of reflux pump (with no spare). Each situation must be individually evaluated.

4. Instrument Failure
Failure of FRCV-1 and LCV-2 could result in overpressure due to flooding of the condenser as discussed in Case "C". Normally, failure of LCV-1 would not cause overpressure.
If FRCV-2 were to fail in the full open position, an overpressure could be caused. Additional vapors generated would not normally be as great as the overhead rate, since a large excess of heat transfer area is not normally provided in a reboiler. If inspection of the heat exchanger and control valve specification sheets fail to confirm this, a detailed analysis of the heat transfer in the reboiler loop must be performed.

5. Split Reboiler Tube
In the example given, steam to the reboiler is on flow control. Although the tower would be upset, it is doubtful that an overpressure would occur as a result of a split reboiler tube.

6. External Fire
Calculations must now be made to find the orifice area of a valve which will relieve the vapors generated by a fire occurring around and beneath the fractionator and the overhead accumulator.
The surface area of a vessel exposed to fire which is effective in generating vapor is that area wetted by its internal liquid contents. A valid assumption in this case is that the entire fractionator below the 25′-0″ height (API recommendation) should be considered wet. Therefore, only 19′-0″ of the fractionator need be considered. However, not all of the fractionator surface is wetted, i.e. only 4′ in the bottom and 2½″ in each tray. Only the wetted portion must be included. The approximate surface area of the 2:1 elliptical head is calculated by the following formula:

$$A_h = 1.09\,(\text{O D})^2$$

The wetted area of the fractionator may be calculated as follows:

No. of trays within 25′ of grade = 6

$$A_{wet}\ (3.14 \times 4.667)\left[4.0 + (6)\left(\frac{2.5}{12}\right)\right] + 1.09 \times (4.667)^2 = 100.7 \text{ sq. ft.}$$

The surface of the reboiler may also produce vapors when fractionator bottoms liquid is in the shell rather than in the tubes. Its wetted area is (3.1416 x 1.333 x 16) = 67 sq. ft.

Use the equation $Q = 21{,}000\ FA^{0.82}$ to solve for the total amount of heat absorbed from the fire.

Where

Q = total heat absorption (input) to the wetted surface, in British thermal units per hour.
A = total wetted surface, in square feet.
F = environment factor, values of which are shown in Chart 3.65, page 222, for various types of installations.

Interpolating on Chart 3.65, page 222, the value of F is found to be 0.225.

Caution: The total heat absorption, Q, for the fractionator and from the heat absorbed by the reboiler must be computed separately. Calculations must be of the form $A_1{}^{0.82} + A_2{}^{0.82}$, *NOT* $(A_1 + A_2)^{0.82}$!

Solving for heat absorption in the fractionator from the fire

$$Q_f = 21{,}000\ (0.225)(100.7)^{0.82}$$
$$= 207{,}437 \text{ Btu/hr.}$$

Heat absorption in the reboiler from the fire

$$Q_{rf} = 21{,}000\ (0.225)\ (67)^{0.82}$$
$$= 148{,}519 \text{ Btu/hr.}$$

The total additional heat, resulting from a fire engulfing the fractionator and reboiler only, is

$$Q_t = 207{,}437 + 148{,}519 = 355{,}956 \text{ Btu/hr.}$$

Since this additional heat input results in vaporization of overhead material, calculate the additional vapors resulting from the fire (allowing 20% overpressure for fire):

P (relieving) = (250 x 1.2) + 14.7 = 314.7 psia
Estimated T (relieving) is 156°F as discussed under "A."
h_{fg} = 108 Btu/lb. (Chart 3.56, p. 209)
W = 355,956/108 = 3,300 lb./hr.

This is the relief requirement of PSV-1 for the contingency of a fire engulfing the fractionator and reboiler only.

If a fire engulfs the overhead accumulator as well as the fractionator and reboiler, additional relief should be considered for the accumulator. In this example, a block valve is located upstream of the condenser, necessitating a relief valve, PSV-2, to protect the accumulator in case of a fire while blocked in or shut down. If the block valve were omitted, one PSV could be considered for the entire system, provided the pressure drop between the source and PSV did not exceed 3% of set pressure.

For a fire engulfing the overhead accumulator, calculate the relief rate as follows:

A_{vessel} = $(3.14)(10) + (1.09)(4.0)^2 = 143$ sq. ft.
A_{wet} = (0.5) (143) = 71.5 sq. ft.
F_{bare} = 1.0
Q = $(21{,}000)(1.0)(71.5)^{0.82} = 696{,}226$ Btu/hr.
W = 696,226/108 = 6,446 lb/hr. (relief rate for PSV-2)

Summary of Contingency Relief Rates for PSV-1:

Contingency	M.W.	lb/hr	Temp., °F	Press., PSIA
A. Blocked Outlet	46.9	18,000	150	289.7
B. Cooling Water Failure	46.9	18,000	150	289.7
C. Loss of Reflux	46.9	18,000	150	289.7
D. Instrument Failure	46.9	18,000	150	289.7
E. Split Reboiler Tube	—	0	—	—
F. External Fire	46.9	3,300	156	314.7

It should be recognized that the approach described above represents a conservative, but practical solution to determining relief rates for a distillation system. Since a distillation column is a dynamic system sensitive to concentration (which varies throughout the column), temperature, pressure and many other variables, the exact relief rate for each contingency would require a very complex study of the system, which is usually not warranted.

Generally, when a distillation column is overpressured, the process temperature in the reboiler rises and the ΔT for heat transfer decreases, resulting in poorer heat transfer, so that the boil-up or overhead vapor rate would not reach the normal rate (18,000 #/hr. in the example). However, each system must be evaluated or judged on the basis of its own control scheme, process variables, etc. Usually the relief rate will not exceed the normal overhead vapor rate.

Calculation of Relief Valve Sizes:

By inspection, contingencies A, B, C and D will determine the valve size.

Safety and safety relief valves in gas or vapor service may be sized by use of one of the following formulas that are modifications of the basic ASME and API formula. Most manufacturers' sizing methods also agree with these forms.

Basic ASME and API Formula

$$W = CKAP_1 \sqrt{\frac{M}{ZT}} \qquad (1)$$

Modifications of Formula (1)

$$W = \frac{CKAP_1 K_b \sqrt{M}}{\sqrt{TZ}}$$

or (M-1)

$$A = \frac{W \sqrt{TZ}}{CKP_1 \ K_b \ \sqrt{M}}$$

$$V = \frac{6.32\, CKAP_1 K_b}{\sqrt{TZM}}$$

or (M-2)

$$A = \frac{V \sqrt{TZM}}{6.32\, CKP_1 K_b}$$

$$V = \frac{1.175\, CKAP_1 K_b}{\sqrt{TZG}}$$

or (M-3)

$$A = \frac{V \sqrt{TZG}}{1.175\, CKP_1 K_b}$$

where

W = flow through valve, in pounds per hour.

V = flow through valve, in standard cubic feet per minute at 14.7 psia and 60°F.

C = coefficient determined by the ratio of the specific heats of the gas or vapor at standard conditions. This can be obtained from Chart 3.76, p. 233.

K = coefficient of discharge, which value is obtainable from the valve manufacturer. The K for a number of nozzle-type valves is 0.975.

A = effective discharge area of the valve, in square inches.

P_1 = upstream pressure, in pounds per square inch absolute. This is the set pressure multiplied by 1.10 or 1.20 (depending on the overpressure permissible) plus the atmospheric pressure, in pounds per square inch absolute.

K_b = This can be obtained from Chart 3.78, p. 235, which applies to conventional safety relief valves, or from Chart 3.79, p. 236, which applies to balanced bellows valves. The correction factor value should be read from the curve specifically applying to the type of valve under consideration. In connection with Chart 3.79, p. 236, for set pressures lower than 50 psig, the valve manufacturer should be consulted for the proper value of correction factor K_b.

M = molecular weight of the gas or vapor
Various handbooks carry tables of molecular weights of materials, but the composition of the flowing gas or vapor is seldom the same as that listed in the tables. This value should be obtained from the process data.

T = absolute temperature of the inlet vapor, in degrees Rankine (Fahrenheit + 460).

Z = compressibility factor for the deviation of the actual gas from a perfect gas, a ratio evaluated at inlet conditions. Values of Z for various paraffin hydrocarbons can be obtained from Chart 3.12, p. 166.

G = specific gravity of gas referred to air = 1.00 at 60°F and 14.7 psia.

In Chart 3.76, p. 233, k is tabulated, where

$$k = \frac{C_p}{C_v}, \text{ i.e.,}$$

the ratio of specific heats of any ideal gas, or the ratio of specific heats of a diatomic actual gas that expands in accordance with the Perfect Gas Laws. Values of k can be determined from the properties of gases as presented in any acceptable reference work.

The tabulation in Table 4.43, p. 310, also lists n, where n = isentropic expansion coefficient of an actual gas, such as a paraffin hydrocarbon, expanding through an orifice or nozzle of a safety relief valve in which the upstream pressure is at saturation and the pressure in the throat of the orifice or nozzle is the critical flow pressure.

When k or n cannot be determined, it is suggested to let C = 315.

Establishing Values:

W = 18,000 pph (contingencies a-d), or 3,300 pph (contingency f)

C = 306.86 (from Chart 3.76, p. 233, since n = 0.93 for 46.9 MW hydrocarbon, Chart 3.80, page 230.

K = 0.975

P_1 = (250 + 10%) + 14.7 = 289.7 psia

K_b = 1.0 (from Chart 3.78, p. 235.

M = 46.9
T = 610°R
Z = 0.69 (from chart 3.12, p. 159).

To calculate orifice area,

$$A = \frac{(18{,}000)\ \sqrt{(610)\,(0.69)}}{(306.86)\,(0.975)\,(289.7)\,(1.0)\ \sqrt{46.9}}$$

$$A = \underline{\underline{0.622 \text{ sq. in.}}}$$

An "H" orifice is selected.

Rupture Disc

PROBLEM: Rupture Disc Calculation for Liquid Relief

Given

Relief rate:	6,500 gpm
Relieving specific gravity:	1.5
Liquid viscosity:	30,000 cp
Disc burst pressure:	110 psig
Backpressure developed:	0 psig

The formula used for sizing a rupture disc for liquid relief is

$$a = \frac{0.0438\ Q}{K_v} \sqrt{\frac{S}{\Delta P}}$$

where:

a = required relieving area, sq. in.
K_v = capacity correction factor due to viscosity
ΔP = pressure drop across disc at Q, psid, equal to burst pressure + accumulation (10%) – backpressure
Q = required relieving capacity, gpm
S = specific gravity at relieving temperature

Establishing values:

ΔP = 110(1.1) - 0 = 121 psid
Q = 6,500 gpm
S = 1.5

A value for K_v must be determined from a trial calculation. Using an assumed value of K_v = 1, solve for a trial required area.

$$a' = 0.0438\ (6{,}500) \sqrt{\frac{1.5}{121}}$$

$$= 31.699 \text{ sq. in.}$$

Using area a' as a minimum size, choose a pipe size whose transverse internal area is equal to or greater than this value. *Note:* If disc assembly inlet and outlet size are different, use the smaller of the two. For most low pressure services, schedule 40 pipe is used for vessel nozzles above 2″. From Table 4.34, p. 298, choose 8″ sch. 40 wrought steel pipe, whose transverse internal area, a'' = 50.027 sq. in.

Calculate a value for the Reynolds number, R, using the new area and the following formula:

$$R = \frac{\text{gpm}\ (2{,}800\ G)}{\mu\ \sqrt{a''}}$$

where

a'' = selected area, sq. in.
G = specific gravity at relieving temperature (same as S above)
gpm = required relieving capacity, gpm
R = Reynolds number
μ = absolute viscosity at relieving temperature, centipoises

Substituting:

$$R = \frac{6{,}500\ (2{,}800)\ (1.5)}{30{,}000\ \sqrt{50.027}}$$

Select a value of K_v from Chart 3.75, p. 232, of K_v = 0.65

Substituting in the original equation,

$$a = \frac{0.0438\ (6{,}500)}{0.65} \sqrt{\frac{1.5}{121}}$$

$$= \underline{\underline{48.767 \text{ sq. in.}}}$$

Since this value of a is still less than the value of a″, then a safety head with a sch. 40 bore (7.981″ ID, 50.027 sq. in area) is selected.

PROBLEM: Rupture Disc Calculation for Vapors or Gases

Given

Relief rate:	15,300 pph
Fluid molecular weight:	58
Isentropic coefficient, n:	0.94
Disc burst pressure:	150 psig
Back pressure developed:	0 psig

Relieving temperature:	177°F
Compressibility factor, Z:	0.79
Permissible overpressure:	10% of burst pressure

The formula used for sizing a rupture disc for vapors or gases is the basic ASME and API formula:

$$W = CKAP_1 \sqrt{\frac{M}{ZT}}$$

where

W = flow through valve, in pounds per hour.
C = coefficient which is determined by the ratio of the specific heats of the gas or vapor at standard conditions (see Chart 3.76, p. 233, for values).
K = coefficient of discharge.
A = effective discharge area of the valve, in square inches.
P_1 = upstream pressure, in pounds per square inch absolute. This is the set pressure multiplied by 1.10 or 1.20 (depending on the permissible overpressure) plus the atmospheric pressure, in pounds per square inch absolute.
M = molecular weight of gas or vapor.
Z = compressibility factor for the deviation of the actual gas from a perfect gas, a ratio evaluated at inlet conditions. A chart for evaluating Z is given in Chart 3.12, p. 166.
T = absolute temperature for the inlet vapor, in degrees Fahrenheit + 460.

Establishing values:

W = 15,300 pph
C = 308.11 (corresponding to $n = 0.94$ in Chart 3.76, p. 233).
K = 0.62 (taken from API Report #43-1963 by R. Solter, L. Fike, and F. Hausen)
P_1 = 150 (1.10) + 14.7 = 179.7 psia
M = 58
Z = 0.79
T = 177 + 460 = 637°R

Rewriting the basic equation to solve for area, and substituting:

$$A = \frac{W}{CKP}\sqrt{\frac{TZ}{M}} = \frac{15{,}300}{308.11\ (0.62)\ (179.7)}\sqrt{\frac{637\ (0.79)}{58}}$$

$$= \underline{\underline{1.313 \text{ sq. in.}}}$$

From Table 4.34, p. 298, the smallest commonly available pipe size would be 1½″ sch. 80, which has a transverse internal area of 1.767 sq. in.

Note: If extended lengths of upstream or downstream piping are connected to the rupture disc assembly, or if flow pressures are below that required for critical flow, the pressure drop for the complete discharge piping must be calculated. Assume rupture disc assembly as a pressure drop equivalent to 75 pipe diameters (i.e., L/D =75) and use Crane Technical Paper (Technical Paper No. 410, "Flow of Fluids through Valve, Fittings and Pipes," published by Crane Company, Chicago, Illinois.) as the basis for calculating the flow rate or size of this type system.

Miscellaneous

PROBLEM: Selection of Pneumatic Integrator Change Gear

Given

Integrator:	Foxboro Model 14A
Flow Rate:	10,000 pph, maximum
Reading Period:	Weekly
Multiplier Limitations:	Must be a multiple of 10

Solution:

Integrator Counts
(From Foxboro Catalog Sheet PS2B-5A1A, Table 1.)

Counts Per Hour	Counts Per Day[1]	Teeth in Change Gear
300	7,500	6
350	8,750	7
400	10,000	8
450	11,250	9
500	12,500	10
600	15,000	12
700	17,500	14
750	18,750	15
800	20,000	16
900	22,500	18
1000	25,000	20
1100	27,500	22
1200	30,000	24
1250	31,250	25
1300	32,500	26
1400	35,000	28
1500	37,500	30

(table continued)

Counts Per Hour	Counts Per Day[1]	Teeth in Change Gear
1600	40,000	32
1750	43,750	35
1800	45,000	36
2000	50,000	40
2100	52,500	42
2400	60,000	48
2500	62,500	50
2700	67,500	54

NOTES: 1. 25-hour day

Express the flow rate in units per hour, units per 25-hour day, and units per week.

$$F_{max} = 10{,}000 \text{ pph} = 250{,}000 \text{ ppd} = 1{,}750{,}000 \text{ lbs./wk.}$$

An integrator having the maximum number of teeth in the change gear (54) will give only 67,500 x 7 = 472,500 counts per week + 4% (calibration accuracy) = 491,400 counts per week. A complete meter turnover for a 6-digit integrator would take 14.25 days at that rate.

The proper decade meter factor multiplied by 491,400 will be equal to or greater than the value of units per week, e.g.,

$$f \times 491{,}400 \geqslant F \text{ units per week, or}$$

$$\frac{F \text{ units per week}}{f\text{, integrator factor}} \geqslant 491{,}400$$

Substituting:

$$\frac{1{,}750{,}000}{f} \geqslant 491{,}400$$

Therefore: $f = 10$

This value of f must be checked to make sure that actual values fall within the ± 4% calibration of the integrator. Approximate counts per day = F units per day ÷ f.

$$\text{Counts per day} = \frac{250{,}000}{10} = 25{,}000.$$

This value for counts per day is compared to counts per day values in the table above. If it exceeds the ± 4% limit of the nearest value, an uneven factor rather than a decade value for f must be used. In this particular case, 25,000 counts per day exactly fits an available gear.

Therefore, specify a 20-tooth change gear to give 25,000 counts per day and an integrator factor of "times 10."

PROBLEM: Meter Factor Calculation for Pneumatic Integrator

GIVEN

Integrator:	Foxboro Model 14A
Flow Rate:	600 gpm, maximum
Change Gear:	20-tooth

Solution:

Express the maximum flow rate as an hourly rate:

$$F_{max} = 600 \text{ gpm} = 36{,}000 \text{ gph}$$

The 20-tooth change gear gives 1,000 counts per hour at 15 psig input. The following equation relates the flow rate and counts per day on an hourly basis:

$$f = \frac{F_H}{C_H}$$

where

C_H = counts per hour for the change gear
f = meter factor (scale multiplier)
F_H = maximum flow rate, units per hour

Establishing Values:

C_H = 1,000 (corresponds with 20-tooth from table in Problem above.)
F_H = 36,000 gph

Substituting:

$$f = \frac{36{,}000}{1{,}000} = \underline{\underline{36}}$$

PROBLEM: Fill or Retention Time Calculation

Given

Vessel:	Rectangular sump
Dimensions:	8′ W x 10′ L x 6′ D; normal level 1′; fills to within 6″ of top.
Fill Rate:	200 gpm, maximum
Pumps:	Two, 150 gpm each.

Conditions for sizing:

a. Neither pump starts at level = 1′, and increasing.
b. One pump starts at level = 1′, and increasing.
c. Second pump starts at level = 5′-6″, and both stay on until level = 1′.

Formulas used for calculation are:

$$V_u = L \times W \times D$$
$$V_u = F \times t$$

where

D = usable depth of the sump, ft.
F = flow rate, gpm
L = sump length, ft.
t = elapsed time, min.
V_u = usable volume, cu. ft.
W = sump width, ft.

Establishing Values:

D = 4′-6″
F_a = 200 gpm (condition "a")
F_b = 50 gpm (condition "b")
F_c = 100 gpm (condition "c")
L = 10′
W = 8′

Solve for the usable capacity of the sump:

$$V_u = 10\,(8)\,(4.5) = 360 \text{ cu. ft.} = 2{,}693 \text{ gal.}$$

Substituting to find the fill time for condition "a":

$$t_a = \frac{2{,}693}{200} = \underline{\underline{13\tfrac{1}{2} \text{ min.}}}$$

Substituting to find the fill time for condition "b":

$$t_b = \frac{2{,}693}{50} = \underline{\underline{54 \text{ min.}}}$$

Substituting to find the retention time for condition "c":

$$t_c = \frac{2{,}693}{100} = \underline{\underline{27 \text{ min.}}}$$

PROBLEM: Calculate the Specific Gravity of a Mixture of Gases

Given A mixture of gases, composed of 92.2% methane, 4.9% ethane, 2.4% propane, and 0.5% isobutane, by volume.

Formulas used for calculations are:

$$m = y \times M$$
$$M_m = \Sigma m$$
$$G_m = \frac{M_m}{M_{air}}$$

where

G_m = specific gravity of the mixture
m = mass per mole of mixture, lbm
M = molecular weight of air
M_{air} = molecular weight of air
M_m = molecular weight of the mixture
y = mol. fraction, same as volume fraction

Establishing Values (Refer to Table 4.21, p. 284):

$M_{methane}$ = 16.04	$y_{methane}$ = 0.922
M_{ethane} = 30.07	y_{ethane} = 0.049
$M_{propane}$ = 44.09	$y_{propane}$ = 0.024
$M_{iso\text{-}butane}$ = 58.12	$y_{iso\text{-}butane}$ = 0.005
M_{air} = 28.966	

Substituting:

Component	y	X	M	=	m
Methane	0.922		16.04		14.789
Ethane	0.049		30.07		1.473
Propane	0.024		44.09		1.058
Iso-butane	0.005		58.12		0.291
				M_m =	17.611

$$G_m = \frac{17.611}{28.966}$$
$$= \underline{\underline{0.608}}$$

PROBLEM: Conversion of Flow Rate at Actual Conditions to Standard Conditions

GIVEN

Unit:	Plant Air compressor
Rated Capacity:	650 acfm at standard conditions
Ambient Conditions:	
Barometric Pressure:	30.07 inches of mercury
Temperature:	45°F

The formula used to relate pressure, volume, and temperature is:

$$\frac{P_a V_a}{T_a} = \frac{P_s V_s}{T_s}$$

Where

P_a = inlet pressure, psia
P_s = standard pressure, psia
T_a = inlet temperature, °R
T_s = standard temperature, °R
V_a = actual volume, cu. ft.
V_s = standard volume, cu. ft.

Since flow rate is volume per unit time, flow rate can be substituted for volume in the equation. Rewriting:

$$F_s = F_a \frac{P_a}{P_s}\left(\frac{T_s}{T_a}\right)$$

where

F_a = flow rate at inlet conditions, acfm
F_s = standard flow rate, scfm

Establishing Values:

F_a = 650 acfm
P_a = 14.769 psia (30.07″ Hg x 0.49″ psi/in. Hg.)
T_a = 505°R (460 + °F)
P_s = 14.696 psia
T_s = 520°R

Substituting:

$$F_s = 650 \frac{14.767}{14.696}\left(\frac{520}{505}\right)$$
$$= \underline{\underline{672.5\ scfm}}$$

PROBLEM: For a Gas Mixture, Determine the Molecular Weight, the Molal Heat Capacity, the Pseudo-Critical Pressure, the Pseudo-Critical Temperature, and the Compressibility Factor.

GIVEN

Fluid:	A propane and butane gas mixture
Component weight:	Propane – 18,000 lbm
	Butane – 13,488 lbm
Condition A:	289.7 psia, and 700°R
Condition B:	314.7 psia, and 705°R (assumed)

To calculate the molecular weight of the mixture, first obtain individual component molecular weights from Table 4.33, p. 297.

$M_{propane}$ = 44.094
M_{butane} = 58.120

Tabulating information and relationships,

Gas Mixture	Determination of Weight Fraction			Gas Mixture No. of Moles	Determination of Equivalent Mol. Weight	
Component Name	Individual Component W	Weight Fraction $w = \frac{W}{W_T}$	Individual Component Mol. Weight M	$n = W/M$	Mol. Fract. $y = \frac{n}{n_T}$	$y \times M$
Propane	18,000	0.5716	44.094	408.2188	0.6376	28.114
Butane	13,488	0.4284	58.120	232.0716	0.3624	21.063
TOTALS	31,488	1.000		640.2904		
				MW mix = 49.177		

To calculate molal heat capacity, pseudo-critical pressure and pseudo-critical temperature of the mixture, first obtain individual component values. From Table 4.32, p. 296, obtain critical pressures and temperatures; and from Table 4.33, p. 297, obtain values of molal heat capacity at 240°F by interpolation.

Tabulating information and relationships,

Gas Mixture		Determination of $MC_p°$ Molal Heat Capacity		Determination of Pseudo-Critical Pressure PP_c, and Temperature PT_c			
Component Name	Mol. Fract. y	Individual Component @ 240°F	$y \times MC_p°$ @ 240°F	Component Critical Pressure P_c, psia	$y \times P_c$	Component Critical Temp. T_c °R	$y \times T_c$
Propane	0.6376	21.95	13.995	617	393.4	666	424.6
Butane	0.3624	28.89	10.470	551	199.7	766	277.6
	1.0000	MC_p mix =	24.465	PP_c mix =	493.1	PT_c mix =	702.2

Using pseudo-critical values, calculate reduced values of pressure and temperature:

$$P_r = \frac{P_1}{PP_c} = \frac{289.7}{593.1} = 0.488$$

$$T_R = \frac{T_1}{PT_c} = \frac{700}{702.2} = 0.997$$

Find a value for the compressibility factor, Z, on chart 3.13, p. 167.

$Z = 0.80$

For Condition B, new values of MC_p mix are calculated. Individual component values of molal heat capacity are interpolated from Table 4.33, p. 297.

Tabulating information and relationships,

Gas Mixture		Determination of MC_p° Molal Heat Capacity	
Component Name	Mol. Fraction y	Individual Component MC_p° @ 245°F	$y \times MC_p^\circ$ @ 245°F
Propane	0.6376	22.11	14.09
Butane	0.3624	29.06	10.53
	1.0000	MC_p mix =	24.62

$$k = \frac{MC_p^\circ}{MC_p^\circ - 1.986} = \frac{24.62}{24.62 - 1.986} = 1.087$$

where

k = specific heat ratio
MC_p° = constant pressure specific heat at zero pressure

Checking the value of relieving temperature

$$T_2 = T_1 \left(\frac{P_2}{P_1}\right)^{\frac{k-1}{k}}$$

$$= 700 \left(\frac{314.7}{289.7}\right)^{\frac{1.087-1.0}{1.087}}$$

$$= 705^\circ R \ (=245^\circ F)$$

An assumed value of $T_2 = 245^\circ F$ had been used. Using MC_p values at this value of T_2, a value of MC_p mix was calculated; using MC_p mix, a value of k was calculated; and using this value of k, the relieving temperature was calculated. If the assumed value of temperature varied from the calculated value by more than 1%, a new assumed value should be used and the values recalculated.

The molecular weight, pseudo critical pressure and pseudo critical temperature of the mixture do not change.

Calculating new values of reduced pressure and temperature:

$$P_r = \frac{314.7}{593.1} = 0.5306$$

$$T_R = \frac{705}{702.2} = 1.004$$

Obtaining a value of the compressibility factor from Chart 3.19, p. 173.

$Z = 0.79$

PROBLEM: Electronic Loop Resistance Calculation

Given

An electronic loop measuring flowrate, consisting of a d/p transmitter (rated to work into a 1650 ohm external load with a 45 VDC power supply), a local indicator (10 ohms), a current-actuated trip switch (250 ohms), an integrator (250 ohms), a strip chart recorder (250 ohms), an indicating controller (250 ohms), and 1250 feet of No. 22 AWG 2-conductor copper cable.

The formula used for calculation is:

$$Rt = R_1 + R_2 + R_3 \ldots \text{etc.}$$

where:
R_t = total loop resistance, in ohms
R_1, etc. = resistance of individual instruments, in ohms

To determine cable resistance, the resistance per 1000 feet of conductor is looked up in Table 4.30, and then doubled to obtain the resistance of the pair. The value is:

$$R_{cable} = \frac{16.8 \text{ ohms}}{1000 \text{ ft}} \times 1.25 \text{ thousand ft} \times 2 = 42 \text{ ohms}$$

Substituting:

$$R_t = 10 \times 250 + 250 + 250 + 250 + 42$$
$$= 1{,}052 \text{ ohms}$$

Note that this value does not include the resistance of the transmitter, since its value varies to change the signal over a 4-20 ma span. Note also that if the value had slightly exceeded the recommended 1,650 ohms, a larger wire size might be employed to reduce the total.

9 Electrical Data

Definitions and Symbols*

NEMA Standard 250-1985
Enclosures for Electrical Equipment (1,000 Volts Maximum)

DEFINITIONS

2.1 INTRODUCTION

An enclosure is a surrounding case constructed to provide a degree of protection to personnel against incidental contact with the enclosed equipment and to provide a degree of protection to the enclosed equipment against specified environmental conditions.

A brief description of the more common types of enclosures used by the electrical industry relating to their environmental capabilities follows. Refer to the appropriate sections of this Standards Publication for more information regarding applications, features, and design tests.

NEMA Standard 3-8-1985.

Individual NEMA product standards publications or third party certification standards may contain additional requirements for product testing and performance.

Authorized Engineering Information 3-8-1985.

2.2 DEFINITIONS PERTAINING TO NONHAZARDOUS LOCATIONS

Type 1 Enclosures are intended for indoor use primarily to provide a degree of protection against contact with the enclosed equipment.

NEMA Standard 1-10-1979.

Type 2 Enclosures are intended for indoor use primarily to provide a degree of protection against limited amounts of falling water and dirt.

NEMA Standard 1-10-1979.

Type 3 Enclosures are intended for outdoor use primarily to provide a degree of protection against windblown dust, rain, sleet, and external ice formation.

NEMA Standard 1-10-1979.

Type 3R Enclosures are intended for outdoor use primarily to provide a degree of protection against falling rain, sleet, and external ice formation.

NEMA Standard 1-10-1979.

Type 3S Enclosures are intended for outdoor use primarily to provide a degree of protection against windblown dust, rain, sleet, and to provide for operation of external mechanisms when ice laden.

NEMA Standard 1-10-1979.

Type 4 Enclosures are intended for indoor or outdoor use primarily to provide a degree of protection against windblown dust and rain, splashing water, and hose-directed water.

NEMA Standard 1-10-1979.

Type 4X Enclosures are intended for indoor or outdoor use primarily to provide a degree of protection against corrosion, windblown dust and rain, splashing water, and hose-directed water.

NEMA Standard 1-10-1979.

Type 5 Enclosures are intended for indoor use primarily to provide a degree of protection against dust and falling dirt.

NEMA Standard 1-10-1979.

Type 6 Enclosures are intended for indoor or outdoor use primarily to provide a degree of protection against the entry of water during occasional temporary submersion at a limited depth.

NEMA Standard 1-10-1979.

Type 6P Enclosures are intended for indoor or outdoor use primarily to provide a degree of protection against the entry of water during prolonged submersion at a limited depth.

NEMA Standard 1-10-1979.

Type 11 Enclosures are intended for indoor use primarily to provide, by oil immersion, a degree of protection to enclosed equipment against the corrosive effects of liquids and gases.

NEMA Standard 1-10-1979.

Type 12 Enclosures are intended for indoor use primarily to provide a degree of protection against dust, falling dirt, and dripping noncorrosive liquids.

NEMA Standard 1-10-1979.

Type 12K Enclosures with knockouts are intended for indoor use primarily to provide a degree of protection against dust, falling dirt, and dripping non-corrosive liquids other than at knockouts.

NEMA Standard 1-10-1979.

Type 13 Enclosures are intended for indoor use primarily to provide a degree of protection against dust, spraying of water, oil, and noncorrosive coolant.

NEMA Standard 1-10-1979.

2.3 DEFINITIONS PERTAINING TO HAZARDOUS (CLASSIFIED) LOCATIONS

Type 7 Enclosures are for use indoors in locations classified as Class I, Groups A, B, C, or D, as defined in the *National Electrical Code*.

NEMA Standard 1-10-1979.

Type 8 Enclosures are for indoor or outdoor use in locations classified as Class I, Groups A, B, C, or D, as defined in the *National Electrical Code*.

NEMA Standard 1-10-1979.

Type 9 Enclosures are for use in indoor locations classified as Class II, Groups E or G, as defined in the *National Electrical Code*.

NEMA Standard 3-8-85.

Type 10 Enclosures are constructed to meet the applicable requirements of the Mine Safety and Health Administration.

NEMA Standard 1-10-1979.

2.4 GENERAL DEFINITIONS PERTAINING TO ENCLOSURES

Apparatus is the enclosure, the enclosed equipment, and the attached protruding accessories.

NEMA Standard 1-10-1979.

Design Tests demonstrate performance of a product designed to applicable standards; they are not intended to be production tests.

NEMA Standard 1-10-1979.

Flush Mounting means so constructed as to have a minimal front projection when set into a recessed opening and secured to a flat surface.

NEMA Standard 1-10-1979.

Hazardous (Classified) Locations are those areas which may contain hazardous (classified) materials in sufficient quantity to create an explosion. See Article 500 of the *National Electrical Code*.

NEMA Standard 3-8-1985.

Hazardous Materials are those gases, vapors, combustible dusts, fibers, or flyings which are explosive under certain conditions.

NEMA Standard 1-10-1979.

Indoor Locations are those areas which are protected from exposure to the weather.

NEMA Standard 1-10-1979.

Knockout is a portion of the wall of an enclosure so fashioned that it may be removed readily by a hammer, screwdriver, and pliers at the time of installation in order to provide a hole for the attachment of an auxiliary device or raceway, cable, or fitting.

NEMA Standard 1-10-1979.

Nonventilated means so constructed as to provide no intentional circulation of external air through the enclosure.

NEMA Standard 1-10-1979.

Oil-Resistant Gaskets are those made of material which is resistant to oil or oil fumes.

NEMA Standard 1-10-1979.

Outdoor Locations are those areas which are exposed to the weather.

NEMA Standard 1-10-1979.

Surface Mounting means so constructed as to be secured to, and projected from a flat surface.

NEMA Standard 1-10-1979.

Ventilated means so constructed as to provide for the circulation of external air through the enclosure to remove excess heat, fumes, or vapors.

NEMA Standard 1-10-1979.

NEMA ICS-1-1988

GENERAL STANDARDS FOR INDUSTRIAL CONTROL AND SYSTEMS

Part 1-100

REFERENCED STANDARDS AND DEFINITIONS

1-100.1 REFERENCED STANDARDS

In this publication, reference is made to the standards listed below:

American National Standards Institute
1430 Broadway
New York, NY 10018

C84.1-1982	*Voltage Ratings for Electric Power Systems and Equipment (60 Hz)*

American Society for Testing and Materials
1916 Race Street
Philadelphia, PA 19103

D-3638-1985	*Test Method for Comparative Tracking Index of Electrical Insulating Materials*

Institute of Electrical and Electronics Engineers
345 East 47th Street
New York, NY 10017

91-1984 (ANSI Y32.14)	*Standard Graphic Symbols for Logic Functions**
100-1984 (ANSI C42.100)	*IEEE Standard Dictionary of Electrical and Electronics Terms**
315-1975, 315A-1986 (ANSI Y32.2)	*Graphic Symbols for Electrical and Electronics Diagrams**
C37.2-1987	*Standard Electrical Power System Device Function Numbers*

International Electrotechnical Commission
1, rue de Varembe
Geneva, Switzerland

112-1979	*Method for Determining the Comparative and the Tracking Indices of Solid Insulating Materials Under Moist Conditions*
664-1980, 664A-1981	*Insulation Co-ordination Within Low-Voltage Systems, Including Clearances and Creepage Distances for Equipment*

National Electrical Manufacturers Association
2101 L Street, N.W.
Washington, DC 20037

ICS 1.1-1984	*Safety Guidelines for the Application, Installation and Maintenance of Solid State Control*

* Also available from the American National Standards Institute.

ICS 2-1983	*Industrial Control Devices, Controllers and Assemblies*
ICS 3-1983	*Industrial Systems**
ICS 4-1988	*Terminal Blocks for Industrial Use**
ICS 6-1988	*Enclosures for Industrial Controls and Systems*
LI 1-1983	*Industrial Laminated Thermosetting Products*
LI 6-1983	*Relative Temperature Indices of Industrial Thermosetting Products*
ST 1-1988	*Specialty Transformers (Except General-purpose Type)*

National Fire Protection Association
Publication Sales Department
Batterymarch Park
Quincy, MA 02269

70-87	*National Electrical Code*
70B-1987	*Electrical Equipment Maintenance*
70E-1988	*Electrical Safety Requirements for Employee Workplaces*

Underwriters Laboratories Inc.
333 Pfingsten Rd.
Northbrook, IL 60062

347-1985	*High Voltage Industrial Control Equipment**
508-1984	*Industrial Control Equipment**

* Also available from the American National Standards Institute.

The standards in this part apply to all NEMA Standards Publications for Industrial Controls and Systems unless otherwise specified. Definitions that are marked with an asterisk (*) are identical to those appearing in ANSI/IEEE Std 100, IEEE Standard Dictionary of Electrical and Electronic Terms.

The terms have been arranged in an alphabetical order that reflects common usage and are selectively cross-referenced through the grouping of commonly used nouns followed by their modifier.

Specialized definitions applying to the contents of their respective parts appear in the other Standards Publications for Industrial Controls and Systems.

1-100.2 DEFINITIONS

accelerating contactor. A contactor, other than the line or directional contactor, used primarily for the purpose of obtaining a change of accelerating torque.
NEMA Standard 9-20-1973.

accuracy, repeat. See **repeat accuracy.**

adaptive control systems. See **control system, adaptive.***

alternating current contactor. A contactor for the specific purpose of establishing and interrupting an alternating current power circuit.
NEMA Standard 3-17-1971.

ambient temperature.* The temperature of the medium such as air, water, or earth into which the heat of the equipment is dissipated.
NEMA Standard 9-20-1973.

NOTES: (A) For self-ventilated equipment, the ambient temperature is the average† temperature of the air in the immediate neighborhood of the equipment. (B) For air- or gas-cooled equipment with forced ventilation or secondary water cooling, the ambient temperature is taken as that of the ingoing air or cooling gas. (C) For self-ventilated enclosed (including oil-immersed) equipment considered as a complete unit, the ambient temperature is the average† temperature of the air outside of the enclosure in the immediate neighborhood of the equipment. (These notes are approved as Authorized Engineering Information.)

†The average of temperature readings at several locations. (This note is not a part of ANSI/IEEE Std 100 definition.)

antihunt. See **damping.**

apparatus. Includes the enclosure, the enclosed equipment, and attached appurtenances.
NEMA Standard 7-16-1969.

application tests. Those tests performed by a manufacturer to determine those operating characteristics that are not necessarily established by standards but that are of interest in the application of devices.
NEMA Standard 9-15-1971.

automatic reset. A function that operates to automatically reestablish specific conditions.
NEMA Standard 1-26-1977.

auxiliary contacts. See **contacts (auxiliary) (switching device).***

barrier.* A partition for the insulation or isolation of electric circuits or electric arcs.
NEMA Standard 5-24-1960.

block diagram. See Part 1-101.

brake, magnet. See **magnet brake.**

breaking, dynamic. See **dynamic braking.**

break rating. The value of current for which a contact assembly is rated for opening a circuit repeatedly at a specified voltage and under specified operating conditions.
NEMA Standard 5-17-1971.

bus. (1) A set of power supply leads or (2) a conductor which provides multiple connections.
NEMA Standard 5-21-1962.

capability, interrupting. See **interrupting capability.**

capability, short-time. See **short-time capability.**

carryover period. The maximum time of power loss during which a system will remain within specified limits of performance.
NEMA Standard 8-20-1966.

clearance. The shortest distance through air between conducting parts, or between a conducting part and the outer surface of the insulating enclosure considered as though metal foil were in contact with the accessible surfaces of the enclosure.
NEMA Standard 1-5-1977.

NOTE: A joint between two parts of an insulating barrier is considered to be part of the surface except where the interstices are completely filled with an insulating material (cement).
Authorized Engineering Information 1-5-1977.

combination controller—600 volts or less.† A full magnetic or semimagnetic controller with additional externally operable disconnecting means contained in a common enclosure. The disconnecting means may be a circuit breaker or a disconnect switch.
NEMA Standard 9-29-1954.

†Combination starters are specific forms of combination controllers.

NOTE: NEMA terms are shown indexed with associated definitions, identical to ANSI, IEEE, and IEC standards, while retaining the advantages of finding definitions through the groupings of commonly used nouns.

*Identical to ANSI/IEEE Std 100, *IEEE Standard Dictionary of Electrical and Electronics Terms.*

compensation, IR-drop. See **IR-drop compensation.**

conducting parts (live parts). Those parts that are designed to carry current or that are conductively connected therewith.

NEMA Standard 4-25-1928.

connection diagram. See Part 1-101.

connection diagram, wireless. See Part 1-101.

construction diagram.* See Part 1-101.

contact, electric. See **electric contact.***

contactor. A two-state (on-off) device for repeatedly establishing and interrupting an electric power circuit. Interruption is obtained by introducing a gap or a very large impedance.

NEMA Standard 9-20-1973.

contactor, accelerating. See **accelerating contactor.**

contactor, alternating current. See **alternating current contactor.**

contactor, direct current. See **direct current contactor.**

contactor, electronic. See **electronic contactor.**

contactor, low-torque. See **low-torque contactor.**

contactor, magnetic. See **magnetic contactor.***

contactor, solid-state. See **solid-state contactor.**

contacts.* Conducting parts that co-act to complete or to interrupt a circuit.

NEMA Standard 5-24-1960.

contacts (auxiliary)(switching device).* Contacts in addition to the main-circuit contacts that function with the movement of the latter.

NEMA Standard 5-24-1960.

contacts (nonoverlapping).* Combinations of two sets of contacts, actuated by a common means, each set closing in one of two positions, and so arranged that the contacts of one set open before the contacts of the other set close.

NEMA Standard 5-24-1960.

contacts (overlapping).* Combinations of two sets of contacts, actuated by a common means, each set closing in one of two positions, and so arranged that the contacts of one set open after the contacts of the other set have been closed.

NEMA Standard 5-24-1960.

continuous periodic rating.* The load that can be carried for the alternate periods of load and rest specified in the rating and repeated continuously without exceeding the specified limitation.

NEMA Standard 12-17-1947.

continuous rating.* The maximum constant load that can be carried continuously without exceeding established temperature-rise limitations under prescribed conditions of test and within the limitations of established standards.

NEMA Standard 1-5-1977.

control.* Broadly the methods and means of governing the performance of any electric apparatus, machine or system.

NEMA Standard 5-21-1962.

control apparatus.* A set of control devices used to accomplish the intended control functions.

NEMA Standard 5-21-1962.

control circuit (industrial control).* The circuit that carries the electric signals directing the performance of the controller but does not carry the main power circuit.

NEMA Standard 5-21-1962.

control-circuit devices, oiltight. See **oiltight control-circuit devices.**

control-circuit transformer.* A voltage transformer utilized to supply a voltage suitable for the operation of control devices.

NEMA Standard 5-24-1960.

control device.* An individual device used to execute a control function.

NEMA Standard 5-21-1962.

control, industrial. See **industrial control.**

control relay. See **relay.***

control relay, magnetic. See **magnetic control relay.**

control, remote. See **remote control.**

control sequence diagram. See Part 1-101.

control sequence table.* A tabulation of the connections that are made for each successive position of the controller.

NEMA Standard 12-18-1932.

NOTE: NEMA terms are shown indexed with associated definitions, identical to ANSI, IEEE, and IEC standards, while retaining the advantages of finding definitions through the groupings of commonly used nouns.

*Identical to ANSI/IEEE Std 100, *IEEE Standard Dictionary of Electrical and Electronics Terms.*

control system. A system in which deliberate guidance or manipulation is used to achieve a prescribed value of a variable.

NEMA Standard 7-7-1965.

control system, adaptive.* A control system within which automatic means are used to change the system parameters in a way intended to improve the performance of the control system.

NEMA Standard 7-7-1965.

control system diagram. See Part 1-101.

control, three-wire. See **three-wire control.***

control, two-wire. See **two-wire control.***

controller, combination—600 volts or less. See **combination controller—600 volts or less.**

controller, definite purpose. See **definite purpose controller.**

controller diagram. See Part 1-101.

controller, drum. See **drum controller.***

controller, electric. See **electric controller.***

controller, electric motor. See **electric motor controller.***

controller, electropneumatic. See **electropneumatic controller.***

controller, full-magnetic. See **full-magnetic controller.***

controller, general-purpose. See **general-purpose controller.***

controller, manual. See **manual controller.***

controller, semimagnetic. See **semimagnetic controller.***

controllers for steel-mill accessory machines. See 3-441.05 of NEMA Standards Publication No. ICS 3.

controllers for steel-mill auxiliaries. See 3-441.05 of NEMA Standards Publication No. ICS 3.

converter. A network or device for changing the form of information or energy.

NEMA Standard 5-21-1962.

corrosion-resistant. So constructed, protected or treated that corrosion will not exceed specified limits under specified test conditions.

NEMA Standard 10-21-1987.

counter. A network or device for storing integers and permitting these integers to be changed by unity or by an arbitrary integer as successive input signals are received.

NEMA Standard 5-21-1962.

counter, reversible. See **reversible counter.**

creepage distance. The shortest distance along the surface of an insulating material between two conducting parts, or between a conducting part and the outer surface of the insulating enclosure considered as though metal foil were in contact with the accessible surfaces of the enclosure.

A joint between two parts of an insulating barrier is considered to be part of the surface except where the interstices are completely filled with an insulating material (cement).

NEMA Standard 1-5-1977.

critically damped. Damping that is sufficient to prevent any overshoot of the output following an abrupt stimulus.

NEMA Standard 9-29-1960.

current relay.* A relay that functions at a predetermined value of current.

NEMA Standard 5-24-1960.

NOTE: It may be an overcurrent, undercurrent, or reverse-current relay. (This note is approved as Authorized Engineering Information.)

damped, critically. See **critically damped.**

damping. The reduction or suppression of the oscillation of a system. See critically damped.

NEMA Standard 1-25-1961.

deactivating means, interlocking (defeater). See **interlocking deactivating means (defeater).**

dead time. The time interval between the initiation of an input and the start of the resulting response. Dead time shall be expressed in seconds.

NEMA Standard 8-19-1964.

defeater. See **interlocking deactivating means (defeater).**

definite-purpose controller.* Any controller having ratings, operating characteristics, or mechanical construction for use under service conditions other than usual or for use on a definite type of application.

NEMA Standard 9-29-1960.

NOTE: NEMA terms are shown indexed with associated definitions, identical to ANSI, IEEE, and IEC standards, while retaining the advantages of finding definitions through the groupings of commonly used nouns.

*Identical to ANSI/IEEE Std 100, *IEEE Standard Dictionary of Electrical and Electronics Terms.*

delay, off. See **off delay.**

delay, on. See **on delay.**

delay, time. See **time delay.***

design test. A test that demonstrates compliance of a product design with applicable standards; it is not intended to be a production test.
NEMA Standard 7-16-1969.

diagram, block. See Part 1-101.

diagram, connection or wiring. See Part 1-101.

diagram, construction. See Part 1-101.

diagram, control sequence, or table. See Part 1-101.

diagram, control system. See Part 1-101.

diagram, illustrative. See Part 1-101.

diagram, interconnection. See Part 1-101.

diagram, logic. See Part 1-101.

diagram, one-line (single-line). See Part 1-101.

diagram, process or flow. See Part 1-101.

diagram, schematic or elementary. See Part 1-101.

diagram, wireless connection. See Part 1-101.

dielectric withstand-voltage tests.* Tests made to determine the ability of insulating materials and spacings to withstand specified overvoltages for a specified time without flashover or puncture.
NEMA Standard 1-5-1977.

NOTE: The purpose of the tests is to determine the adequacy against breakdown of insulating materials and spacings under normal or transient conditions.(This note is approved as Authorized Engineering Information.)

direct current contactor. A contactor for the specific purpose of establishing and interrupting a direct current power circuit.
NEMA Standard 3-17-1971.

disruptive discharge. The phenomena associated with the failure of insulation under electric stress; these include a collapse of voltage and the passage of current; the term applies to electrical breakdown in solid, liquid, and gaseous dielectrics, and combinations of these dielectrics.
NEMA Standard 1-5-1977.

disturbance.* An undesired input variable that may occur at any point within a feedback control system.
NEMA Standard 9-29-1960.

drawing, dimension or outline. See Part 1-101.

drift (as applied to devices). As applied to devices such as pressure switches, temperature switches, and so forth is a change in operating characteristics over a specified number of operations or time and specified environmental conditions.
NEMA Standard 1-5-1977.

drift (as applied to systems). An undesired but relatively slow change in output over a specified time with a fixed reference input and fixed load, with specified environmental conditions. The specified time is normally after the warm-up period.

Drift shall be expressed in percent of the maximum rated value of the variable being measured.
NEMA Standard 9-8-1976.

drum controller.* An electric controller that utilizes a drum switch as the main switching element.
NEMA Standard 5-24-1960.

NOTE: A drum controller usually consists of a drum switch and a resistor. (This note is approved as Authorized Engineering Information.)

drum switch.* A switch in which the electric contacts are made on segments or surfaces on the periphery of a rotating cylinder or sector, or by the operation of a rotating cam.
NEMA Standard 8-20-1966.

dynamic braking. A control function that brakes the drive by dissipating its stored energy in a resistor.
NEMA Standard 8-19-1964.

effective actuation time. The time that elapses between initial energization of the control circuit and the time the contacts of a normally open device close and remain closed or the contacts of a normally closed device open and remain open. Effective actuation time includes any contact bounce time or chatter time which occurs due to the operation of the device being tested.
NEMA Standard 1-5-1977.

electric contact.* The junction of conducting parts permitting current to flow.
NEMA Standard 5-24-1960.

NOTE: NEMA terms are shown indexed with associated definitions, identical to ANSI, IEEE, and IEC standards, while retaining the advantages of finding definitions through the groupings of commonly used nouns.
*Identical to ANSI/IEEE Std 100, *IEEE Standard Dictionary of Electrical and Electronics Terms.*

electric controller.* A device or group of devices that serves to govern in some predetermined manner the electric power delivered to the apparatus to which it is connected.

NEMA Standard 5-2-1916.

electric motor controller.* A device or group of devices that serves to govern in some predetermined manner the electric power delivered to the motor.

NEMA Standard 6-3-1956.

NOTE: An electric motor controller is distinct functionally from a simple disconnecting means whose principal purpose in a motor circuit is to disconnect the circuit, together with the motor and its controller from the source of power. See **combination controller** for an example where the two devices are combined in one piece of apparatus. (The preceding sentence is not a part of the ANSI/IEEE Std 100 definition and is approved as Authorized Engineering Information.)

electronic contactor. A contactor whose function is performed by electron tube(s) or semiconductor device(s).

NEMA Standard 9-20-1973.

electronic direct current motor controller. For the purposes of this standard, a phase-controlled rectifying system using semiconductors or tubes of the vapor- or gas-filled variety for power conversion to supply the armature circuit or the armature and shunt-field circuits of a direct current motor, to provide adjustable-speed, adjustable- and compensated-speed, or adjustable- and regulated-speed characteristics.

NEMA Standard 5-17-1971.

electropneumatic controller.* An electrically supervised controller having some or all of its basic functions performed by air pressure.

NEMA Standard 5-24-1960.

enclosure, ventilated. See **ventilated enclosure.***

failure of a control component or system. A state or condition in which a control component or system does not perform its essential function(s) when its ratings are not exceeded.

NEMA Standard 8-20-1966.

fault current. A current that results from the loss of insulation between conductors or between a conductor and ground.

NEMA Standard 3-22-1972.

fault current, low-level (as applied to a motor branch circuit). A fault current that is equal to or less than the maximum operating overload.

NEMA Standard 3-22-1972.

fault withstandability. The ability of electrical apparatus to withstand the effects of specified electrical fault current conditions without exceeding specified damage criteria.

NEMA Standard 3-17-1971.

feedback, negative. See **negative feedback.***

feedback, positive. See **positive feedback.**

full-magnetic controller.* An electric controller having all of its basic functions‡ performed by devices that are operated by electromagnets.

NEMA Standard 5-24-1960.

‡"Basic functions" usually refers to acceleration, retardation, line closing, reversing, and so forth. (This note is not a part of the ANSI/IEEE Std 100 definition and is approved as Authorized Engineering Information.)

general-purpose controller. Any controller having ratings, characteristics, and mechanical construction for use under usual service conditions in accordance with the NEMA Standards Publications for Industrial Controls and Systems.

NEMA Standard 9-29-1960.

general-use switch. A switch that is intended for use in general distribution and branch circuits. It is rated in amperes and is capable of interrupting the rated current at the rated voltage.

NEMA Standard 8-20-1966.

graphic symbol. See Part 1-101.

ground (earth).* A conducting connection, whether intentional or accidental, by which an electric circuit or equipment is connected to the earth or to some conducting body of relatively large extent that serves in place of the earth.

NEMA Standard 10-29-1979.

inch (jog). See **jog (inch).***

industrial control. Broadly the methods and means of governing the performance of an electric device, apparatus, equipment, or system used in industry.

NEMA Standard 5-21-1962.

interconnection diagram. See Part 1-101.

interlock. A device actuated by the operation of some other device with which it is directly associated, to govern succeeding operations of the same or allied devices.

Interlocks shall be permitted to be either electrical or mechanical.

NEMA Standard 5-24-1960.

NOTE: NEMA terms are shown indexed with associated definitions, identical to ANSI, IEEE, and IEC standards, while retaining the advantages of finding definitions through the groupings of commonly used nouns.

*Identical to ANSI/IEEE Std 100, *IEEE Standard Dictionary of Electrical and Electronics Terms.*

interlocking deactivating means (defeater). A manually actuated provision for temporarily rendering an interlocking device ineffective, thus permitting an operation that would otherwise be prevented.

NEMA Standard 8-20-1962.

For example, when applied to apparatus such as combination controllers (Part 2-321 of NEMA Standards Publication No. ICS 2) or control centers (Part 2-322 of NEMA Standards Publication No. ICS 2, it refers to voiding of the mechanical interlocking mechanism between the externally operable disconnect device and the enclosure doors to permit entry into the enclosure while the disconnect device is closed.

Authorized Engineering Information 8-20-1962.

intermittent periodic duty or intermittent duty. A duty in which the main contacts of an equipment remain closed for periods bearing a definite relation to the no-load periods, both periods being too short to allow the equipment to reach thermal equilibrium. See also **periodic rating.**

NEMA Standard 5-25-1988.

interrupting capability. The maximum value of current that a contact assembly is required to successfully interrupt at a specified voltage for a limited number of operations under specified conditions. The term is usually applied to abnormal or emergency conditions.

NEMA Standard 5-17-1971.

inverse time. A qualifying term applied to a relay indicating that its time of operation decreases as the magnitude of the operating quantity increases.

NEMA Standard 9-29-1960.

IR-drop compensation. A control function that compensates for voltage drop due to current.

The amount of IR-drop compensation shall be expressed in percent of rated voltage, i.e.,

$$\text{IR-drop compensation percent} = \frac{\text{Volts compensation at full-load current}}{\text{Rated full-load voltage}} \times 100$$

NEMA Standard 8-19-1964.

isolating switch.* A switch intended for isolating an electric circuit from the source of power.

NEMA Standard 8-20-1966.

NOTE: It has no interrupting rating, and it is intended to be operated only after the circuit has been opened by some other means. (This note is approved as Authorized Engineering Information.)

jog (inch).* A control function that provides for the momentary operation of a drive for the purpose of accomplishing a small movement of the driven machine.

NEMA Standard 7-12-1961.

jogging speed.* The steady-state speed that would be attained if the jogging pilot device contacts were maintained closed.

NEMA Standard 8-19-1964.

NOTE: It may be expressed either as an absolute magnitude of speed or a percentage of maximum rated speed. (This note is approved as Authorized Engineering Information.)

limit switch. See Part 2-225 of NEMA Standards Publication No. ICS 2 for definition of limit switch and associated terms.

live parts. See **conducting parts.**

loop-position control systems. See Part 3-106C of NEMA Standards Publication No. ICS 3.

low-torque contactor. A contactor used in a motor circuit in addition to a line, directional, plugging, or accelerating contactor, for the purpose of limiting stalled torque of the motor to less than 100 percent of the full load torque by inserting resistance in the circuit.

NEMA Standard 7-14-1982.

magnet brake. A friction brake that is controlled by electromagnetic means.

NEMA Standard 5-23-1919.

magnetic contactor.* A contactor actuated by electromagnetic means.

NEMA Standard 5-24-1960.

magnetic control relay. A relay that is actuated by electromagnetic means. When not otherwise qualified, the term refers to a relay intended to be operated by the opening and closing of its coil circuit, and having contacts designated for energizing or deenergizing, or both, the coils of magnetic contactors or other magnetically operated device.

NEMA Standard 7-27-1955.

make rating. The value of current for which a contact assembly is rated for closing a circuit repeatedly under specified operating conditions.

NEMA Standard 5-17-1971.

manual controller.* An electric controller having all of its basic functions performed by devices that are operated by hand.

NEMA Standard 4-25-1960.

NOTE: NEMA terms are shown indexed with associated definitions, identical to ANSI, IEEE, and IEC standards, while retaining the advantages of finding definitions through the groupings of commonly used nouns.

*Identical to ANSI/IEEE Std 100, *IEEE Standard Dictionary of Electrical and Electronics Terms.*

manual reset. A function that requires a manual operation to reestablish specific conditions.
NEMA Standard 1-26-1977.

motor-circuit switch.* A switch intended for use in a motor branch circuit.
NEMA Standard 8-20-1966.

NOTE: It is rated in horsepower, and it is capable of interrupting the maximum operating overload current of a motor of the same rating at the rated voltage. See **operating overload.** (This note is approved as Authorized Engineering Information.)

negative feedback.* A feedback signal in a direction to reduce the variable that the feedback represents.
NEMA Standard 5-16-1963.

nonoverlapping contacts. See **contacts (nonoverlapping).***

off delay. The timing period of a pneumatic time delay relay is initiated upon deenergization of its coil.
NEMA Standard 6-1-1959.

oiltight control-circuit devices. Devices such as push-button switches, pilot lights, and selector switches that are so designed that, when properly installed, they will prevent oil and coolant from entering around the operating or mounting means.
NEMA Standard 9-26-1952.

on delay. The timing period of a pneumatic time delay relay is initiated upon energization of its coil.
NEMA Standard 6-1-1959.

operate time. The elapsed time between the initial energization of the control circuit and the time the contacts of normally open device first touch or the time the contacts of a normally closed device first open.
NEMA Standard 1-5-1977.

operating overload. The overcurrent to which electric apparatus is subjected in the course of the normal operating conditions that it may encounter.
NEMA Standard 5-24-1960.

For example, those currents in excess of running current that occur for a short time as a motor is started or jogged are considered normal operating overloads for control apparatus. See 1-112.40.
Authorized Engineering Information 5-24-1960.

overcurrent relay. A relay that operates when the current through the relay during its operating period is equal to or greater than its setting.

overlapping contacts. See **contacts (overlapping).***

overload protection.* The effect of a device operative on excessive current, but not necessarily on short circuit, to cause and maintain the interruption of current flow to the device governed.
NEMA Standard 5-24-1960.

part-winding starter. A starter that applies to voltage successively to the partial sections of the primary winding of an ac motor.
NEMA Standard 8-7-1952.

period, carryover. See **carryover period.**

periodic rating.* The load that can be carried for the alternate periods of load and rest specified in the rating, the apparatus starting at approximately room temperature, and for the total time specified in the rating, without causing any of the specified limitations to be exceeded. See also **intermittent periodic duty** or **intermittent duty**.
NEMA Standard 5-25-1988.

periodic rating, continuous. See **continuous periodic rating.***

pick-up and seal voltage (magnetically operated device). The minimum voltage, suddenly applied, at which the device moves from its deenergized into its fully energized position.
NEMA Standard 9-15-1971.

pick-up voltage or current (magnetically operated device). The voltage or current, suddenly applied, at which the device starts to operate.
NEMA Standard 9-15-1971.

pilot duty rating. A generic term formerly used to indicate the ability of a control device to control other devices (See Part 2-125 of NEMA Standards Publication No. ICS 2.)
NEMA Standard 9-15-1971.

plugging.* A control function that provides braking by reversing the motor line voltage polarity or phase sequence so that the motor develops a counter-torque that exerts a retarding force.
NEMA Standard 7-12-1961.

pole (electrical switching device). A combination of mating contacts that are normally open, or normally closed, or both.
NEMA Standard 3-12-1975.

NOTE: NEMA terms are shown indexed with associated definitions, identical to ANSI, IEEE, and IEC standards, while retaining the advantages of finding definitions through the groupings of commonly used nouns.
*Identical to ANSI/IEEE Std 100, *IEEE Standard Dictionary of Electrical and Electronics Terms.*

pole, multi (electrical switching device). The term "multipole" is applied to a contact arrangement that includes two or more separate mating contact combinations; that is, two or more single-pole contact assemblies, having a common operating means.

NEMA Standard 3-12-1975.

pole, single (electrical switching device). The term "single pole" is applied to a contact arrangement in which all contacts connect in one position or another to a common contact.

NEMA Standard 3-12-1975.

positive feedback. A feedback signal in a direction to increase the variable that the feedback represents.

NEMA Standard 5-16-1963.

pressure operated switches. See Part 2-226 of NEMA Standards Publication No. ICS 2 for definitions of pressure operated switches and associated terms.

production tests. Those tests that are made at the discretion of the manufacturer on some or all production units for the purpose of maintaining quality and performance.

NEMA Standard 9-15-1971.

protection, overload. See **overload protection.***

protection, undervoltage. See **undervoltage protection.**

pulse.* A signal of relatively short duration.

NEMA Standard 1-25-1961.

pushbutton station.* A unit assembly of one or more externally operable pushbutton switches, sometimes including other pilot devices such as indicating lights or selector switches, in a suitable enclosure.

NEMA Standard 8-20-1966.

rating, break. See **break rating.**

rating, continuous. See **continuous rating.***

rating, continuous periodic. See **continuous periodic rating.***

rating (controller).* An arbitrary designation of an operating limit. It is based on power governed, the duty and service required.

NEMA Standard 1-5-1977.

NOTE: A rating is arbitrary in the sense that it must necessarily be established by definite field standards and cannot, therefore, indicate the safe operating limit under all conditions that may occur. (This note is approved as Authorized Engineering Information.)

rating (device).* The designated limit(s) of the rated operating characteristic(s).

NEMA Standard 1-5-1977.

NOTE: Such operating characteristics as current, voltage, frequency, and so forth, may be given in the rating. (This note is approved as Authorized Engineering Information.)

rating, make. See **make rating.**

rating, periodic. See **periodic rating.***

rating, pilot duty. See **pilot duty rating.**

rating, short-time. See **short-time rating.***

regeneration. The transfer of rotational energy through a motor and its control equipment back to its electrical source.

NEMA Standard 5-17-1971.

relay.* An electric device that is designed to interpret input conditions in a prescribed manner and after specified conditions are met to respond to cause contact operation or similar abrupt change in associated electric control circuits.

NEMA Standard 1-5-1977.

NOTES: (A) Inputs are usually electric, but may be mechanical, thermal, or other quantities. Limit switches and similar simple devices are not relays. (B) A relay may consist of several units, each responsive to specified inputs, the combination providing the desired performance characteristics. (These notes are approved as Authorized Engineering Information.)

relay, current. See **current relay.***

relay, magnetic control. See **magnetic control relay.**

relay, overcurrent. See **overcurrent relay.***

relay, voltage. See **voltage relay.***

release time. The time between initial deenergization of the control circuit and the initial opening of a normally open contact or the initial closing of a normally closed contact.

NEMA Standard 1-5-1977.

release, undervoltage. See **undervoltage release.**

remote control. A control function that provides for initiation or change of a control function from a remote point.

NEMA Standard 9-29-1960.

NOTE: NEMA terms are shown indexed with associated definitions, identical to ANSI, IEEE, and IEC standards, while retaining the advantages of finding definitions through the groupings of commonly used nouns.

*Identical to ANSI/IEEE Std 100, *IEEE Standard Dictionary of Electrical and Electronics Terms.*

repeat accuracy. A term used to express the degree of consistency of repeat operations under specified conditions.

NEMA Standard 3-17-1971.

reset. To restore a mechanism, storage, or device to a prescribed state.

NEMA Standard 5-21-1962.

reset, automatic. See **automatic reset.**

reset, manual. See **manual reset.**

reversible counter. A counter that will count either up or down.

NEMA Standard 5-21-1962.

schematic or elementary diagram. See Part 1-101.

sealing voltage or current. The voltage or current that is necessary to seat the armature of a magnetic circuit closing device from the position at which the contacts first touch each other.

NEMA Standard 10-27-1964.

semimagnetic controller.* An electric controller having only part of its basic functions performed by devices that are operated by electromagnets.

NEMA Standard 5-24-1960.

sequence diagram, control. See Part 1-101.

service of a controller. The specific application in which the controller is to be used, for example:

(1) General purpose.
(2) Definite purpose.
  (a) Crane and hoist.
  (b) Elevator.
  (c) Machine tool, and so forth.

NEMA Standard 9-26-1970.

short-time capability. The ability of electrical apparatus to operate within specified performance criteria when carrying electrical overloads of a specified current and time duration under specified conditions.

NEMA Standard 3-17-1971.

short-time rating.* Defines the load that can be carried for a short and definitely specified time, the machine, apparatus, or device being at approximately room temperature at the time the load is applied.

NEMA Standard 10-27-1944.

single-line diagram. See Part 1-101.

snap action. A rapid motion of the contacts from one position to another position, or their return. This action is relatively independent of the rate of travel of the actuator.

NEMA Standard 6-24-1957.

solid-state contactor. A contactor whose function is performed by semiconductor device(s).

NEMA Standard 7-16-1969.

starter, part-winding. See **part-winding starter.**

surface-mounted (type). Designed to be secured to and to project from a flat surface.

NEMA Standard 7-16-1969.

switch.* A device for opening and closing or for changing the connection of a circuit.

NEMA Standard 1-5-1977.

NOTE: A switch is understood to be manually operated, unless otherwise stated. (This note is approved as Authorized Engineering Information.)

switch, drum. See **drum switch.***

switch, general-use. See **general-use switch.**

switch, isolating. See **isolating switch.***

switch, motor-circuit. See motor-circuit switch.*

symbol, graphic. See Part 1-101.

system diagram, control. See Part 1-101.

temperature, ambient. See **ambient temperature.***

temperature operated controllers. See Part 2-227 of NEMA Standards Publication No. ICS 2 for definitions of temperature operated controllers and associated terms.

tests, design. See **design tests.**

tests, application. See **application tests.**

tests, production. See **production tests.**

three-wire control.* A control function that utilizes a momentary-contact pilot device and a holding-circuit contact to provide undervoltage protection.

NEMA Standard 1-15-1963.

time, dead. See **dead time.**

time, effective actuation. See **effective actuation time.**

NOTE: NEMA terms are shown indexed with associated definitions, identical to ANSI, IEEE, and IEC standards, while retaining the advantages of finding definitions through the groupings of commonly used nouns.

*Identical to ANSI/IEEE Std 100, *IEEE Standard Dictionary of Electrical and Electronics Terms.*

time, inverse. See **inverse time.**

time, operate. See **operate time.**

time, release. See **release time.**

time delay.* A time interval is purposely introduced in the performance of a function.

NEMA Standard 8-13-1959.

transformer, control-circuit. See **control-circuit transformer.**

transient. That part of the variation in a variable that ultimately disappears during transition from one steady-state operating condition to another.

NEMA Standard 5-16-1963.

two-wire control.* A control function that utilizes a maintained-contact type of pilot device to provide undervoltage release.

NEMA Standard 1-15-1963.

undervoltage protection (low-voltage protection). The effect of a device, operative on the reduction or failure of voltage, to cause and maintain the interruption of power to the main circuit.

NEMA Standard 1-5-1977.

NOTE: The principal objective of this device is to prevent automatic restarting of the equipment. Undervoltage or low-voltage protection devices are usually not designed to become effective at any specific degree of voltage reduction.

Authorized Engineering Information 1-5-1977.

undervoltage release (low-voltage release). The effect of a device, operative on the reduction or failure of voltage, to cause the interruption of power to the main circuit but not to prevent the re-establishment of the main circuit on return of voltage.

NEMA Standard 1-11-1984.

NOTE: Undervoltage or low-voltage releases are generally not designed to become effective at any specific degree of voltage reduction.

Authorized Engineering Information 1-11-1984.

variable operating. See Part 3-106A of NEMA Standards Publication No. ICS 3.

ventilated enclosure.* An enclosure provided with means to permit circulation of sufficient air to remove an excess of heat, fumes, or vapors.

NEMA Standard 1-5-1977.

NOTE: For outdoor applications, ventilating openings or louvers are usually filtered, screened, or restricted to limit the entrance of dust, dirt, or other foreign objects. (This note is approved as Authorized Engineering Information.)

voltage, pick-up. See **pick-up voltage or current.**

voltage, pick-up and seal. See **pick-up and seal voltage.**

voltage, sealing. See **sealing voltage or current.**

voltage relay.* A relay that functions at a predetermined value of voltage.

NEMA Standard 5-29-1960.

NOTE: It may be an overvoltage relay, an undervoltage relay, or a combination of both. (This note is approved as Authorized Engineering Information.)

wireless connection diagram. See Part 1-101.

withstandability, fault. See **fault withstandability.**

NOTE: NEMA terms are shown indexed with associated definitions, identical to ANSI, IEEE, and IEC standards, while retaining the advantages of finding definitions through the groupings of commonly used nouns.

*Identical to ANSI/IEEE Std 100, *IEEE Standard Dictionary of Electrical and Electronics Terms.*

Part 1-101

DIAGRAMS, DEVICE DESIGNATIONS, AND SYMBOLS

The standards in this part apply to all other parts in the NEMA Standards Publications for Industrial Controls and Systems unless otherwise specified. The definitions contained in Part 1-100 of this Standards Publication also apply to this part. Parts 1-101 and 1-101A are alternative standards for specifying diagrams, device designations, and symbols. Part 1-101A is intended to encourage harmonization with other national and international standards. ▲

1-101.1 IDENTIFICATION

Diagrams and drawings should be identified by one of the titles shown in 1-101.2 or by a combination of titles from 1-101.2 and 1-101.3, where applicable, but not solely by the titles given in 1-101.3.

NEMA Standard 5-17-1965.

1-101.2 TYPES OF DIAGRAMS AND DRAWINGS

1-101.2.1 Process Diagram or Flow Diagram

A process or flow diagram is a conceptual diagram of the functional interrelationship of subsystems in block or pictorial form. Process equipment such as machinery is shown for proper understanding.

1-101.2.2 Control System Diagram

A control system diagram is a conceptual diagram of the functional interrelationship of subsystems, usually in block form. It does not include the process equipment or details of circuits and device elements.

1-101.2.3 Schematic Diagram or Elementary Diagram

A schematic or elementary diagram is one which shows all circuits and device elements of an equipment and its associated apparatus or any clearly defined functional portion thereof. Such a diagram emphasizes the device elements of a circuit and their functions as distinguished from the physical arrangement of the conductors, devices or elements of a circuit system.

Circuits which function in a definite sequence should be arranged to indicate that sequence.

1-101.2.4 Control Sequence Diagram and Table

A control sequence diagram or table is a portrayal of the contact positions or connections which are made for each successive step of the control action.

1-101.2.5 Wiring Diagram or Connection Diagram

A wiring or connection diagram is one which locates and identifies electrical devices, terminals, and interconnecting wiring in an assembly. This diagram shall be (1) in a form showing interconnecting wiring by lines or indicating interconnecting wiring only by terminal designations (wireless diagram), or (2) a panel layout diagram showing the physical location of devices plus:

(a) The elementary diagram, or
(b) A wiring table, or
(c) A computer wiring chart, or
(d) A machine command tape or cards.

The term does not include mechanical drawings, commonly referred to as wiring templates, wiring assemblies, cable assemblies, etc.

1-101.2.6 Interconnecting Diagram

An interconnection diagram is one which shows only the external connections between controllers and associated machinery and equipment.

1-101.2.7 Dimension or Outline Drawing

A dimension or outline drawing (base plan, floor plan, and so forth) is one which shows the physical space and mounting requirements of a piece of equipment. It shall be permitted to also indicate ventilation requirements and space provided for connections or the location to which connections are to be made.

1-101.2.8 Construction Diagram*

A diagram that shows the physical arrangement of parts, such as wiring, buses, resistor units, and so forth. Example: A diagram showing the arrangement of grids and terminals in a grid-type resistor.

1-101.2.9 Controller Diagram*

A diagram that shows the electrical connections between the parts comprising the controller and that shows the external connections.

1-101.2.10 Illustrative Diagram*

A diagram whose principal purpose is to show the operating principle of a device or group of devices without necessarily showing actual connections or circuits. Illustrative diagrams may use pictures or symbols to illustrate or represent devices or their elements. Illustrative diagrams may be made of electric, hydraulic, pneumatic, and combination systems. They are applicable chiefly to instruction books, descriptive folders, or other media whose purpose is to explain or instruct.

* Definition from ANSI/IEEE Std 100.
▲Revised 7-11-1984.

1-101.2.11 Wireless Connection Diagram*

The general physical arrangement of devices in a control equipment and connections between these devices, terminals, and terminal boards for outgoing connections to external apparatus. Connections are shown in tabular form and not by lines. An elementary (or schematic) diagram may be included in the connection diagram.

NEMA Standard 9-15-1971.

1-101.2.12 Other Drawings

There may be additional drawings, such as conduit layout drawings, foundation drawings, and so forth.

Authorized Engineering Information 9-15-1971.

1-101.3 FORMS OF DIAGRAMS

Diagrams shall be permitted to take one or a combination of the following forms.

1-101.3.1 Block Diagram

A block diagram is made up of a group of interconnected blocks, each of which represents a device or subsystem.

1-101.3.2 One-Line Diagram (Single-Line)

A one-line or single-line diagram is one which shows, by means of single lines and graphic symbols, the course of an electrical circuit or circuits and the component devices or parts used therein. Physical relationships are usually disregarded.

1-101.3.3 Logic Diagram

A logic diagram is a particular form of one-line or single-line diagram of a logic circuit using logic symbols.

NEMA Standard 9-21-1966.

1-101.4 WIRELESS CONNECTION DIAGRAMS

1-101.4.1 Symbols

Symbols for the devices shown in connection diagrams shall be in accordance with 1-101.8. Detailed device symbols shall be made up to represent the physical arrangement of the main component parts and of the terminals to which connections are made.

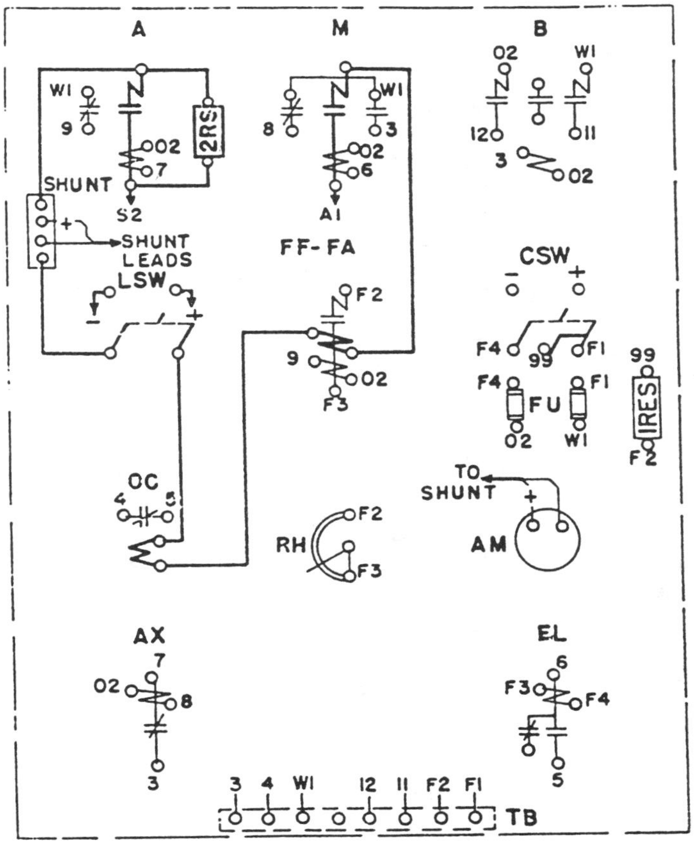

WIRE TABLE		
WIRE	CONNECT TO	
+		LSW, CSW
−		LSW, CSW
F1	TB	CSW, FU
F2	TB	IRES, RH, FF-FA
F3		FF-FA, RH, FL
F4		FL, CSW, FU
02		FU, M, A, AX, FF-FA, B
W1	TB	FU, M, A, B
3	TB	M, AX, B
4	TB	OC
5		OC, FL
6		FL, M
7		A, AX
8		M, AX
9		A, FF-FA
11	TB	B
12	TB	B
99		IRES, CSW

Figure 1-101-1
WIRELESS CONNECTION DIAGRAMS

*Definition from ANSI/IEEE Std 100.

1-101.4.2 Physical Arrangement

The physical arrangement of devices in the connection diagram shall correspond to the physical arrangement of the equipment. Physical groupings comprising control panel sections, auxiliary panels, subpanels, overhead racks, resistor compartments and the like shall be so indicated and marked. See Figure 1-101-1 for wireless connection diagram.

1-101.4.3 Device Designations

Each device in an equipment to which a connection is made shall be assigned a device designation which shall be in accordance with 1-101.5. These designations shall be distinct and there shall be no duplication of designations for the devices in any connection diagram. All designations shall correspond to those used in the elementary diagram.

1-101.4.4 Terminal Markings

Each terminal of a device to which a connection is to be made shall be assigned a distinct terminal marking. This marking shall correspond to the one used in the elementary diagram for designating the same circuit.

1-101.4.5 Control Circuit Connections

Control circuit connections shall be listed in the form of a panel wire table which shall consist of a single continuous column. This table shall list the circuit (terminal) numbers in numerical or alphabetical order, or both. Opposite each circuit (terminal) number shall be listed the designations of the devices to which the circuit will be connected. Short connections between terminals of the same device or between points on a terminal board are sometimes drawn as lines.

1-101.4.6 Power Circuit Connections

Power circuit connections shall be drawn completely by lines or shall be included in the panel wire table.

NEMA Standard 11-12-1953.

1-101.5 DEVICE DESIGNATIONS

Device designations are intended for use on diagrams in connection with graphic symbols to indicate the function of the particular device. Device designations are based on the assignment of a letter or letters to each of the fundamental functions performed by the component devices of a complete control equipment. Suitable prefix numbers or letters, or both, and suffix letters shall be permitted to be added to the basic device designations to discriminate between devices performing similar functions. When two or more basic device designations are combined, the function designation is normally given first. For example, the first control relay initiating a jog function is designated "1JCR."

Table 1-101-1
DEVICE DESIGNATIONS

Device or Function	Designation
Accelerating	A
Ammeter	AM
Braking	B
Capacitor, capacitance	C or CAP
Circuit breaker	CB
Control relay	CR
Current transformer	CT
Demand meter	DM
Diode	D
Disconnect switch	DS or DISC
Dynamic braking	DB
Field accelerating	FA
Field contactor	FC
Field decelerating	FD
Field-loss	FL
Forward	F or FWD
Frequency meter	FM
Fuse	FU
Ground protective	GP
Hoist	H
Jog	J
Limit switch	LS
Lower	L
Main contactor	M
Master control relay	MCR
Master switch	MS
Overcurrent	OC
Overload	OL
Overvoltage	OV
Plugging or potentiometer	P
Power factor meter	PFM
Pressure switch	PS
Pushbutton	PB
Reactor, reactance	X
Rectifier	REC
Resistor, resistance	R or RES
Reverse	REV
Rheostat	RH
Selector switch	SS
Silicon controlled rectifier	SCR
Solenoid valve	SV
Squirrel cage	SC
Starting contactor	S
Suppressor	SU
Tachometer generator	TACH
Terminal block or board	TB
Time-delay relay	TR
Transformer	T
Transistor	Q
Undervoltage	UV
Voltmeter	VM
Watthour meter	WHM
Wattmeter	WM

NEMA Standard 9-4-1969.

Device designations are given in alphabetical order in Table 1-101-1.

Where alternate designations are shown, care shall be taken not to use the same designation for different kinds of devices on the same drawing.

NEMA Standard 9-21-1966.

1-101.6 COIL AND CONTACT DESIGNATIONS

Table 1-101-2 shows the designations used to identify the functions of coils and contacts on complex devices. They are not a part of the device designation. If used in connection with a device designation, the two designations shall be separated by a hyphen, parenthesis, or other suitable means.

NEMA Standard 9-21-1966.

Table 1-101-2
COIL AND CONTACT DESIGNATIONS

Function	Designation
Closing coil	CC
Holding coil	HC
Latch coil	LC
Time-delay closing contacts	TC or TDC
Time-delay opening contacts	TO or TDO
Trip coil	TC
Unlatch coil	ULC

1-101.7 PREFIX NUMBERS

Prefix numbers are used with device designations to distinguish two or more devices having the same function. These numbers are assigned in an orderly fashion in agreement with the order of the relaying or switching or functional sequence, if possible.

NEMA Standard 9-21-1966.

1-101.8 GRAPHIC SYMBOLS

1-101.8.1 Definition

Graphic symbols are a shorthand means of showing graphically the function or interconnections of a circuit. They are used on schematic, one-line, or wiring diagrams. Graphic symbols are correlated with parts lists, descriptions, or instructions by means of device designations. See Table 1-101-3.

1-101.8.2 Symbol Principles

(a) The graphic symbols given in Table 1-101-3 are in general use on industrial control diagrams. Symbols not readily recognizable have been excluded. Other symbols shall be permitted to be used on industrial control diagrams provided a suitable explanation is given as to their meaning.

(b) The orientation of a symbol on a drawing, including a mirror image presentation, does not alter the meaning of the symbol.

(c) The width of a line does not affect the meaning of the symbol. A wider line shall be permitted to be used for emphasis or for power wiring in contrast to control wiring.

(d) The symbols shown in Table 1-101-3 are in their correct relative size; they were prepared on a 1/10-inch grid basis. A symbol shall be permitted to be drawn to any proportional size that suits a particular drawing, depending on the reduction or enlargement anticipated.

(e) When polarity marks are used, the sign + is positive and the − sign is negative.

(f) The arrowhead or triangle of a symbol shall be permitted to be closed, →, or open, →, unless noted.

(g) The symbol for a terminal (o) shall be permitted to be added to each point of attachment of conductors to any one of the symbols, but such added terminal symbols shall not be considered to be a part of the individual graphic symbol itself. The use of terminal symbols on all diagrams is optional. DO NOT SHOW TERMINALS ON ELEMENTARY DIAGRAMS UNLESS THOSE TERMINALS ARE ACCESSIBLE TO THE CUSTOMER!

(h) On a schematic diagram, parts of a symbol for a device, such as a relay or contactor, electron tube or transformer, shall be permitted to be separated. Each of the parts of the device then must carry the same designation.

(i) In general, the angle at which a connecting line is brought to a symbol has no particular significance unless otherwise noted.

(j) Associated or future paths and equipment shall be shown by lines composed of short dashes, - - - - - - - -

(k) An enclosure of a device or panel outline shall be permitted to be shown on a wiring or interconnection diagram as a solid line or a series of long dashes.

(m) A pictorial representation shall be permitted to be used as an alternate for any of the wiring and interconnection diagram symbols shown in this standard.

NEMA Standard 9-20-1972.

1-101.8.3 Uniformly Shaped Symbols

Rather than create new distinctively shaped symbols for new devices, a uniformly shaped symbol shall be used. The uniformly shaped symbol shall also be permitted to be used as an alternate for any distinctively shaped symbol shown in this standard. The uniformly shalped symbol is a rectangle, properly labled at the top to designate the type of device it represents and at the bottom to identify the device in a particular circuit. Terminal identification shall be immediately outside the rectangle.

NEMA Standard 5-25-1988.

Examples to illustrate the use of uniformly shaped symbols are shown in Figure 1-101-2.

Authorized Engineering Information 9-20-1972.

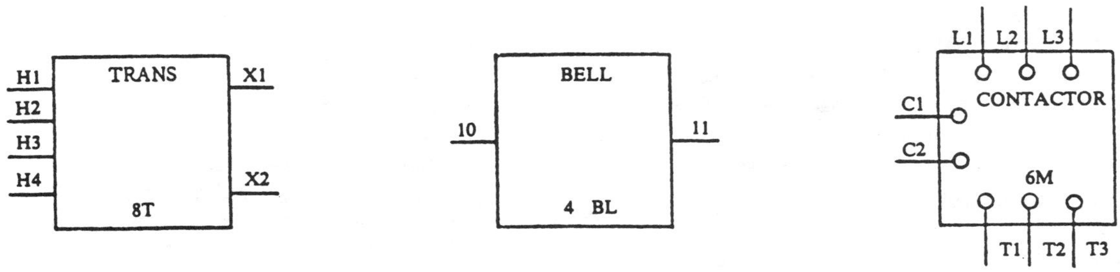

Figure 1-101-2
UNIFORMLY SHAPED SYMBOLS

1-101.8.4 Organization of Symbols

The symbols shown in Table 1-101-3 are given in alphabetical order and are in general agreement with ANSI Y32.2.

Examples identified with an asterisk have been approved as Authorized Engineering Information and not as NEMA Standards. The wiring diagram symbol for each such device will consist of the basic NEMA Standard symbols arranged to represent the particular device and will vary with the manufacturer.

NEMA Standard 9-20-1972.

When more than one symbol is shown, the first one shown is preferred. The symbols, arranged in two columns from left to right, are:

(a) For use on schematic or elementary diagrams.
(b) For use on wiring or interconnection diagrams.

Authorized Engineering Information 9-20-1972.

1-101.9 TERMINAL MARKINGS

Terminal markings used on connection diagrams for designating connections shall conform to those shown in the applicable standards of NEMA Standards Publication No. ICS 2.

NEMA Standard 11-15-1979.

1-101.10 STATIC ELEMENTS OR DEVICES

A diamond surrounding a symbol indicates a semiconductor device which has the same function as the electromechanical device represented by the symbol without the diamond.

NEMA Standard 9-20-1973.

These symbols are intended primarily for use on control circuit diagrams along with electromechanical devices.

It is recommended that, when the diamond is used for this purpose, an explanation of its meaning be stated on the diagram.

A uniformly shaped symbol with a description or a complete circuit of the device may be used if preferred.

Authorized Engineering Information 9-20-1973.

Table 1-101-3
GRAPHIC SYMBOLS

Elementary or Schematic Diagram	Connection or Interconnection Diagram
AC MACHINE See "Induction Machine" and "Wound Rotor Induction Motor."	
ADJUSTABILITY General 	none
AIR CIRCUIT BREAKER See "Circuit Breaker."	
AIR CORE If it is necessary to identify an air core, add a note adjacent to the symbol of the inductor or transformer.	
AMPLIFIER †	†
AMPLIFIER, ROTATING See "Armature with Commutator and Brushes," "Armature, Rotary Amplifier with Shorted Brushes," and "Machine, Rotating."	
ANODE OR PLATE 	none

† Use uniformly shaped symbol. See 1-101.8.

(*Continued*)

Table 1-101-3 *(Cont.)*
GRAPHIC SYMBOLS

Elementary or Schematic Diagram	Connection or Interconnection Diagram
ANNUNCIATOR	
	†
ARMATURE WITH COMMUTATOR AND BRUSHES	
	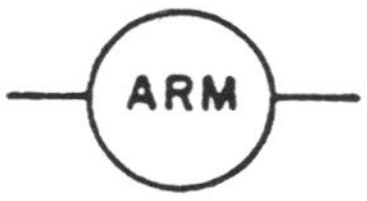
ARMATURE, ROTARY AMPLIFIER WITH SHORTED BRUSHES	
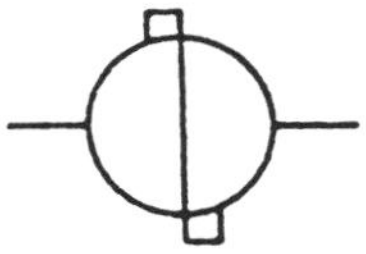	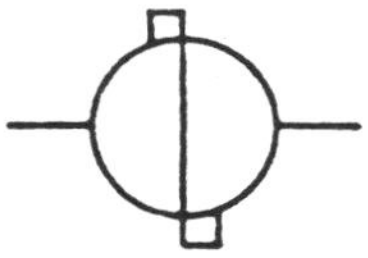
ARRESTER, LIGHTNING See also "Gap, Protective."	
	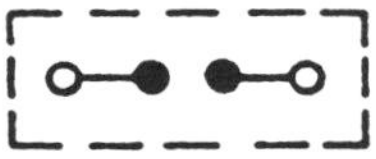 *
AUTOTRANSFORMER	
 or 	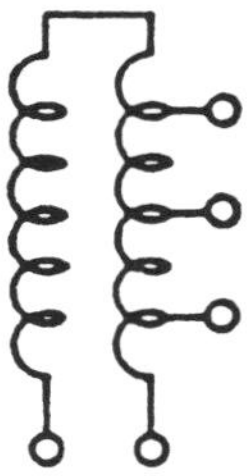 Example *

* Example only; approved as Authorized Engineering Information. See 1-101.8.
† Use uniformly shaped symbol. See 1-101.8.

(Continued)

Table 1-101-3 *(Cont.)*
GRAPHIC SYMBOLS

Elementary or Schematic Diagram	Connection or Interconnection Diagram

BATTERY
The long line is always positive, but polarity may be indicated in addition.

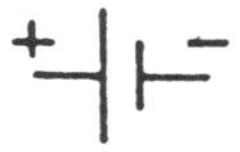

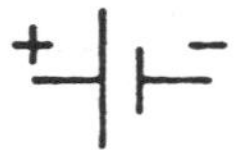

BELL

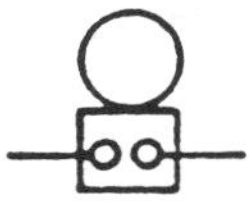

BLOWOUT COIL

BRAKE COIL

BREAKER
See "Circuit Breaker."

BRIDGE RECTIFIER
See "Rectifier."

BUZZER

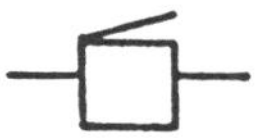

(Continued)

Table 1-101-3 ***(Cont.)***
GRAPHIC SYMBOLS

Elementary or Schematic Diagram	Connection or Interconnection Diagram
CABLE See "Conductor."	
CAPACITOR If it is necessary to identify capacitor electrodes, have the curved element represent the outside electrode in fixed paper and ceramic capacitors, the moving element in adjustable and variable capacitors, and the low-potential element in feed-through capacitors.	
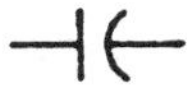	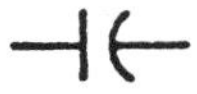 or 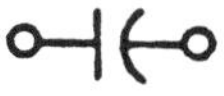
CAPACITOR, POLARIZED Show polarity of leads, + and/or −. + on straight side, − on curved side.	
CATHODE CATHODE—COLD, INCLUDING IONICALLY HEATED CATHODE	
	none
CATHODE, DIRECTLY HEATED (FILAMENTARY)	
	none
CATHODE, INDIRECTLY HEATED	
	none
CATHODE, PHOTOCATHODE	
	none
CATHODE, POOL	
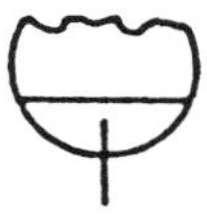	none

(Continued)

Table 1-101-3 *(Cont.)*
GRAPHIC SYMBOLS

Elementary or Schematic Diagram	Connection or Interconnection Diagram

CELL
See "Battery."

CIRCUIT BREAKER, 3 POLE NONAUTOMATIC
(Without trip units.)

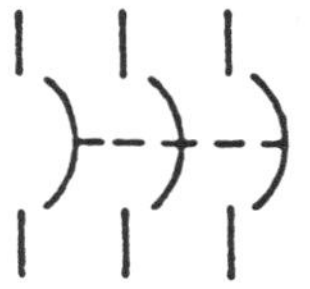

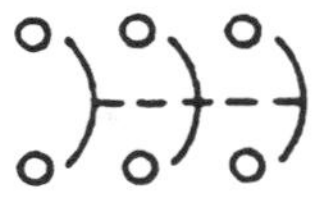

CIRCUIT BREAKER, 3 POLE WITH THERMAL OR THERMOMAGNETIC TRIP UNITS

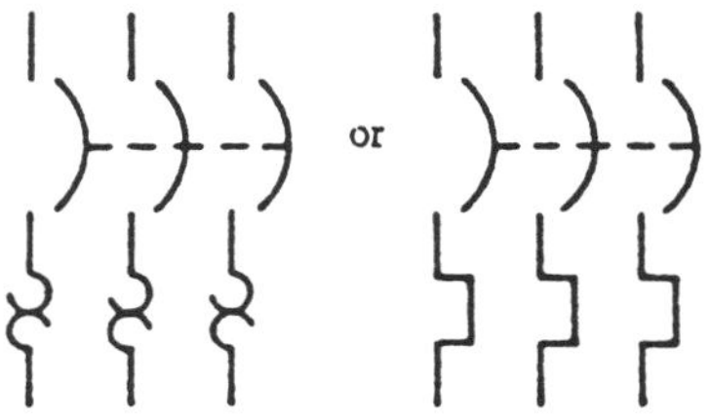

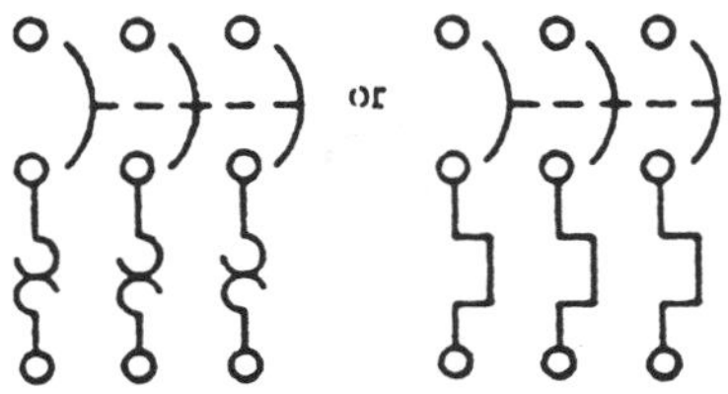

CIRCUIT BREAKER, 3 POLE WITH ONLY MAGNETIC TRIP UNITS

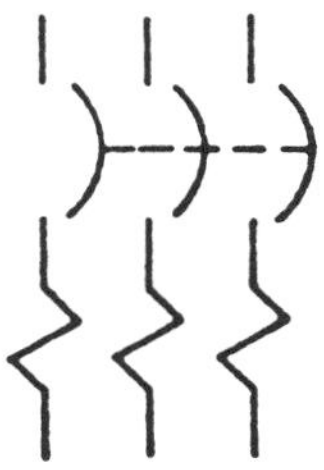

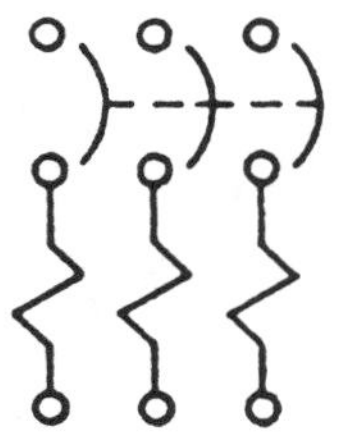

CLOCKWISE ROTATION
See "Motion, Mechanical—Rotation One Direction."

CLUTCH
See "Coil, Operating."

COIL, BLOWOUT
See "Blowout Coil."

(Continued)

Table 1-101-3 ***(Cont.)***
GRAPHIC SYMBOLS

Elementary or Schematic Diagram	Connection or Interconnection Diagram
COIL, OPERATING	
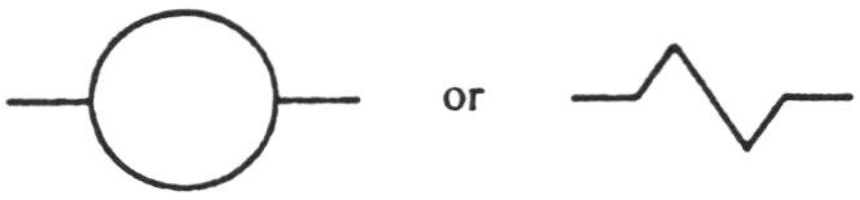	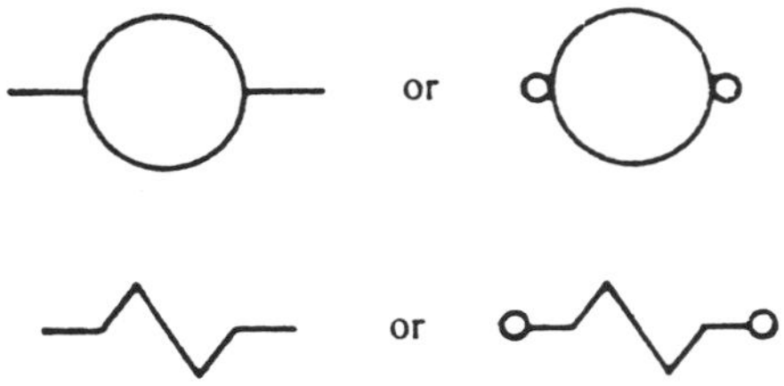
CONDUCTOR, OR CONDUCTIVE PATH	
———	———
CONDUCTOR, CROSSING OF PATHS OR CONDUCTORS NOT CONNECTED	

CONDUCTOR, JUNCTION OF CONNECTED PATHS, CONDUCTORS OR WIRES	
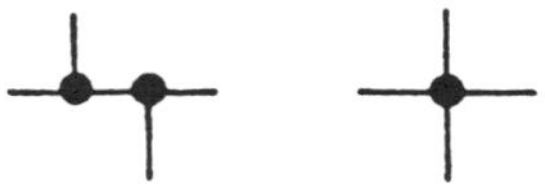	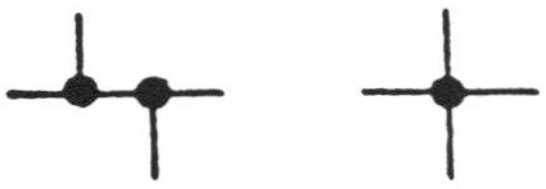
CONDUCTOR, MULTICONDUCTOR CABLE The bend of the line indicates the direction the conductor continues within the cable.	
none	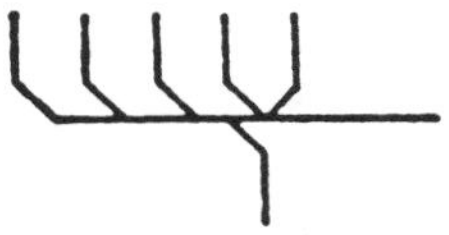
CONDUCTOR, SHIELDED SINGLE OR MULTICONDUCTOR Shield may be grounded.	
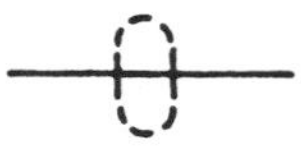	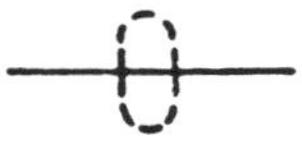
CONDUCTOR, ASSOCIATED OR FUTURE	
– – – – –	– – – – –

(Continued)

Table 1-101-3 *(Cont.)*
GRAPHIC SYMBOLS

Elementary or Schematic Diagram	Connection or Interconnection Diagram
CONNECTION, MECHANICAL	

CONNECTION, MECHANICAL WITH INTERLOCK	

CONNECTOR, DISCONNECTING DEVICE The connector symbol is not an arrowhead. It is larger and the lines are drawn at a 90° angle.	
CONNECTOR, FEMALE	

CONNECTOR, MALE	

CONNECTOR, SEPARABLE OR JACKS ENGAGED	

CONTACT, BASIC ASSEMBLIES Show a contact by a symbol indicating the circuit condition it produces when the actuating device is in the deenergized or nonoperated position. The actuating device may be of any nature, and a note may be necessary with the symbol to explain the proper point at which the contact functions, for example, the point where a contact closes or opens as a function of changing pressure, level, flow, voltage, current, etc. Where it is desirable to show contacts in the energized condition or where confusion may result, add a note.	
CONTACT, NORMALLY CLOSED—BREAK	
 or 	 or 

(*Continued*)

Table 1-101-3 *(Cont.)*
GRAPHIC SYMBOLS

Elementary or Schematic Diagram	Connection or Interconnection Diagram
CONTACT, BASIC ASSEMBLIES (Continued)	
CONTACT, NORMALLY OPEN—MAKE	
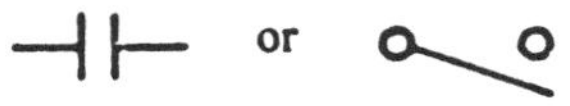	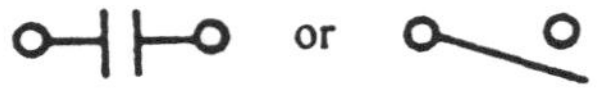
CONTACT, TRANSFER	
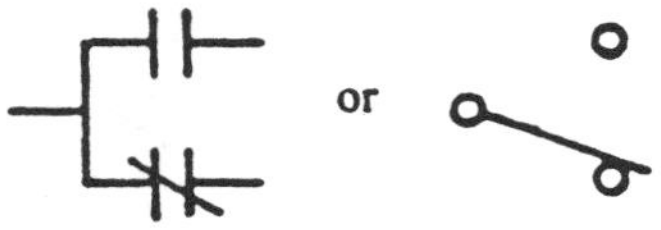	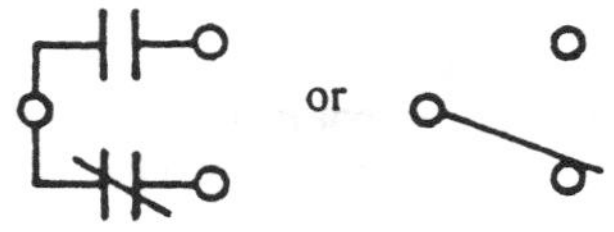
CONTACT, TIME DELAY The point of the arrow indicates the direction of switch operation in which contact action is delayed. See also 1-101.6 Time-delay closing contacts and Time-delay opening contacts.	
NORMALLY OPEN WITH TIME DELAY CLOSING (TC)	
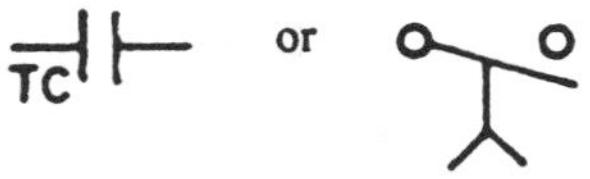	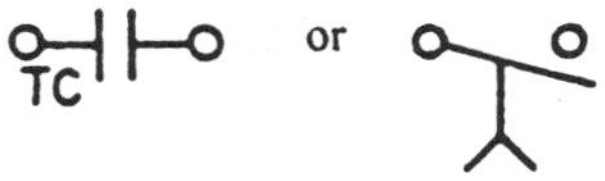
NORMALLY OPEN WITH TIME DELAY OPENING (TO)	
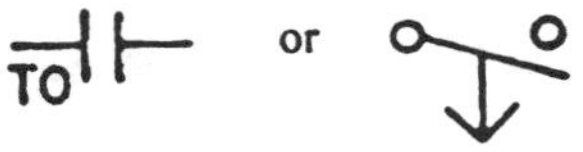	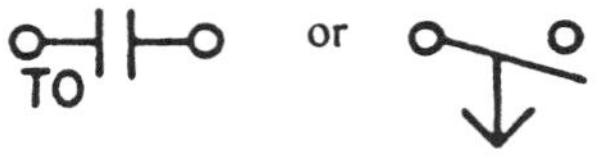
NORMALLY CLOSED WITH TIME DELAY OPENING (TO)	
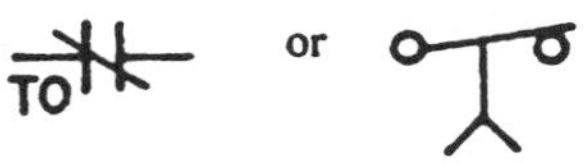	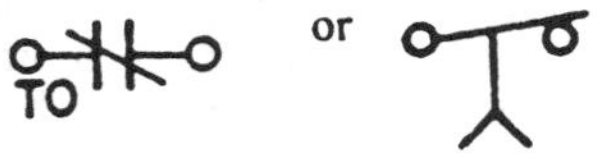
NORMALLY CLOSED WITH TIME DELAY CLOSING (TDC) OR (TC)	
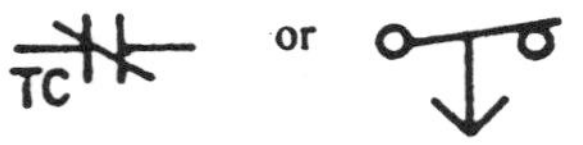	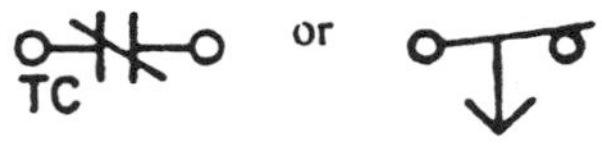
CONTACT, TIME SEQUENTIAL CLOSING	
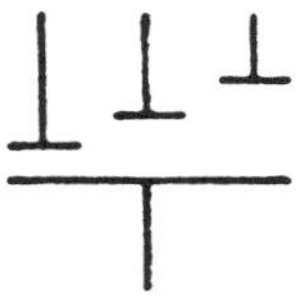	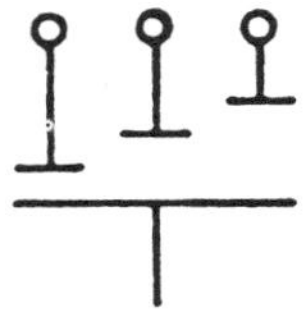

(*Continued*)

Table 1-101-3 *(Cont.)*
GRAPHIC SYMBOLS

Elementary or Schematic Diagram	Connection or Interconnection Diagram

CONTACTOR, WITH INTERLOCKS
See 1-101.8.2(h).

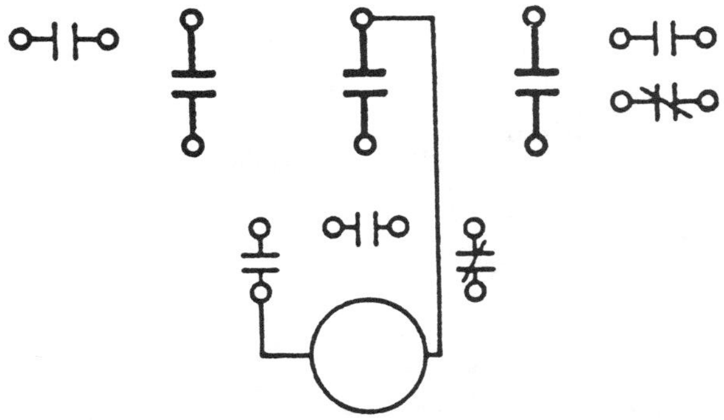

Example*

CONTROL RELAY
See 1-101.8.2(h).

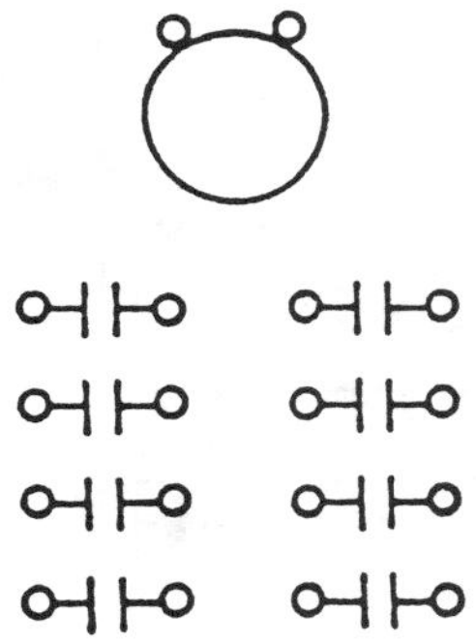

Example*

CONTROL TRANSFORMER
See "Transformer, General."

CONTROL TRANSMITTER
See "Synchro."

CORE, MAGNETIC
See "Transformer, With Magnetic Core."

*Example only; approved as Authorized Engineering Information. See 1-101.8.

(Continued)

Table 1-101-3 *(Cont.)*
GRAPHIC SYMBOLS

Elementary or Schematic Diagram	Connection or Interconnection Diagram
CURRENT TRANSFORMER See "Transformer, Current."	
DC MACHINES See "Armature," "Field," and "Machine, Rotating."	
DC SOURCE See "Battery."	
DIFFERENTIAL RECEIVER TRANSMITTER See "Synchro."	
DIODE, SEMICONDUCTOR 	‡
DIODE, TUNNEL 	‡
DIODE, UNIDIRECTIONAL BREAKDOWN (ZENER) 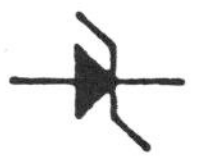	‡
DIODE, BIDIRECTIONAL BREAKDOWN 	‡
DISCONNECT SWITCH See "Switch, Single Throw."	

‡ Use pictorial representation. See 1-101.8.

(Continued)

Table 1-101-3 ***(Cont.)***
GRAPHIC SYMBOLS

Elementary or Schematic Diagram	Connection or Interconnection Diagram

DRUM SWITCH
See "Switch, Drum Switch, Sliding Contact Type."

ELECTRON TUBE COMPONENTS
See "Anode or Plate," "Cathode," "Envelope," "Grid, Including Beam-confining or Beam-forming Electrodes," "Semiconductor Device," "Triode", "Pentrode, Equipotential Cathode," and "Rectifier, General."

ENVELOPE
The envelope symbol indicates a vacuum enclosure, unless otherwise specified. A gas filled device is indicated by a dot within the envelope.

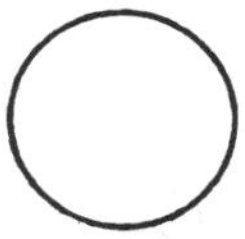 or 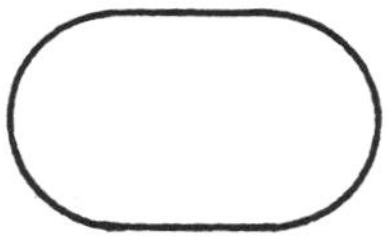

none

ENVELOPE, SPLIT

none

FIELD
See also "Winding."

FIELD, COMPENSATING OR COMMUTATING

FIELD, SERIES

FIELD, SHUNT OR SEPARATELY EXCITED

(Continued)

Table 1-101-3 *(Cont.)*
GRAPHIC SYMBOLS

Elementary or Schematic Diagram	Connection or Interconnection Diagram
FLOW-ACTUATED SWITCH See "Switch, Flow Actuated."	
FLUORESCENT LAMP See "Light, Illuminating, Fluorescent."	
FOOT-OPERATED SWITCH See "Switch, Foot Operated."	
FUSE 	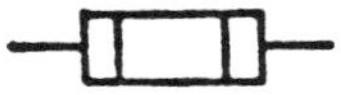 
GAP, PROTECTIVE See also "Arrester, Lightning." These triangles shall not be filled. 	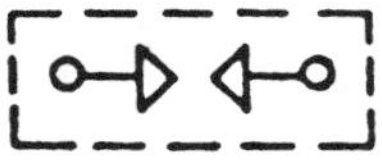
GENERATOR See "Armature with Commutator and Brushes," and "Field."	
GENERATOR FIELD See "Field."	
GRID, INCLUDING BEAM-CONFINING OR BEAM-FORMING ELECTRODES 	

(Continued)

Table 1-101-3 ***(Cont.)***
GRAPHIC SYMBOLS

Elementary or Schematic Diagram	Connection or Interconnection Diagram

GROUND, CHASSIS OR FRAME

A CHASSIS, BUS, OR FRAME CONNECTION WHICH IS INTENDED TO BE AT EARTH POTENTIAL

A CHASSIS, BUS, OR FRAME CONNECTION WHICH MAY BE AT A SUBSTANTIAL POTENTIAL WITH RESPECT TO THE EARTH OR STRUCTURE IN WHICH IT IS MOUNTED.

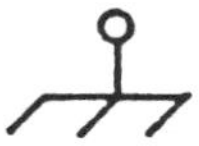

HEATER, ELECTRON TUBE

none

HORN

‡

IGNITOR, POOL TUBE
Should extend into pool.

none

INDICATING LAMP
See "Light, Indicating" and "Light, Indicating, Push-to-test."

‡ Use pictorial representation. See 1-101.8.

(*Continued*)

Table 1-101-3 *(Cont.)*
GRAPHIC SYMBOLS

Elementary or Schematic Diagram	Connection or Interconnection Diagram
INDUCTION MACHINE, SQUIRREL-CAGE INDUCTION MOTOR OR GENERATOR	
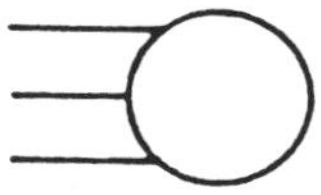	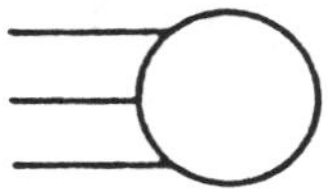
INDUCTOR See "Winding."	
INSTRUMENT (METER) See "Meter (Instrument)."	
INSTRUMENT SHUNT See "Shunt, Instrument or Relay."	
INTERLOCK, MECHANICAL See "Connection, Mechanical With Interlock."	
KNIFE, SWITCH See "Switch, Knife."	
LIGHT OR LAMP	
LIGHT, ILLUMINATING, FLUORESCENT	
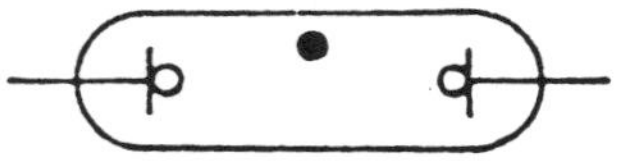	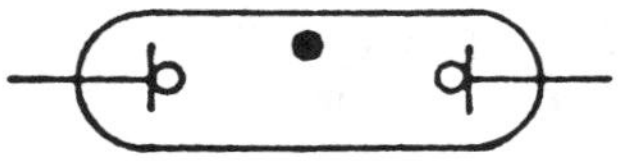
LIGHT, ILLUMINATING, INCANDESCENT	
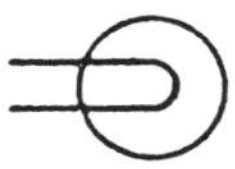	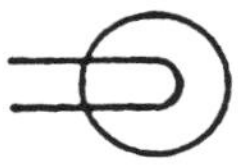

(Continued)

Table 1-101-3 *(Cont.)*
GRAPHIC SYMBOLS

Elementary or Schematic Diagram	Connection or Interconnection Diagram

LIGHT OR LAMP (Continued)

LIGHT, INDICATING

(Pilot, Signalling, or Switchboard.) To indicate characteristic, insert the specified letter or letters inside the symbol.

A —Amber	O —Orange
B —Blue	OP—Opalescent
C —Clear	P —Purple
G —Green	R —Red
NE—Neon	W —White
	Y —Yellow

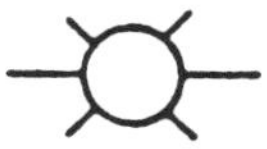

LIGHT, INDICATING, PUSH-TO-TEST

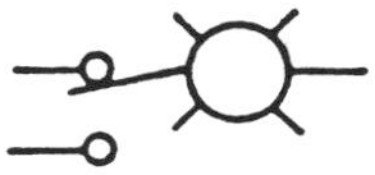

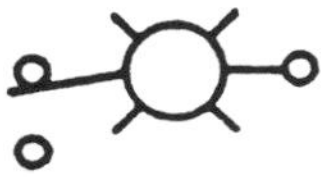

LIGHTNING ARRESTER

See "Arrester, Lightning."

LIMIT SWITCH

See "Switch, Limit."

LIQUID LEVEL SWITCH

See "Switch, Liquid Level Actuated (Float)."

MACHINE, ROTATING

See "Induction Machine, Squirrel-cage Induction Motor or Generator," "Wound-rotor Induction Motor or Induction Frequency Converter," "Armature with Commutator and Brushes" and "Field."

(Continued)

Table 1-101-3 *(Cont.)*
GRAPHIC SYMBOLS

Elementary or Schematic Diagram	Connection or Interconnection Diagram
MAGNETIC CORE OF INDUCTOR OR TRANSFORMER See "Transformer, With Magnetic Core."	
MAGNET, PERMANENT See "Permanent Magnet."	
MASTER, SWITCH See "Switch, Master or Control."	
MECHANICAL INTERLOCK See "Connection, Mechanical With Interlock."	
MERCURY-POOL TUBE Single Anode Vapor Rectifier with Ignitor. See also "Rectifier, Pool-type Cathode." 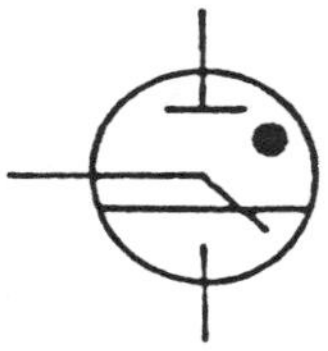	‡
METER (INSTRUMENT) To indicate the function of the meter or instrument, place the specified letter or letters within the symbol. AM—Ammeter AH—Ampere-Hour μA —Microammeter mA—Millammeter PF —Power Factor V —Voltmeter VA —Volt-Ammeter VAR —Varmeter VARH—Varhour Meter W —Wattmeter WH —Watthour Meter For other functions, use adequate description. 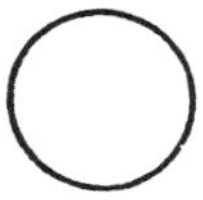	 or 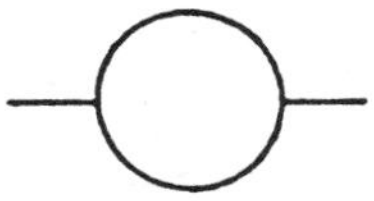

‡ Use pictorial representation. See 1-101.8.

(*Continued*)

Table 1-101-3 *(Cont.)*
GRAPHIC SYMBOLS

Elementary or Schematic Diagram	Connection or Interconnection Diagram

MOTION, MECHANICAL

TRANSLATION, ONE DIRECTION

TRANSLATION, BOTH DIRECTIONS

ROTATION, ONE DIRECTION

ROTATION, BOTH DIRECTIONS

MOTOR
See "Induction Machine, Squirrel-cage Induction Motor or Generator," "Wound-rotor Induction Motor or Induction Frequency Converter," "Armature with Commutator and Brushes," and "Field."

MOTOR FIELD
See "Field."

OVERLOAD RELAY

OVERLOAD RELAY, MAGNETIC
See "Coil, Operating" and "Contact, Normally Closed—Break."

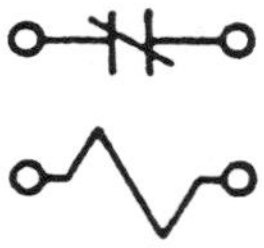

Example*

* Example only; approved as Authorized Engineering Information. See 1-101.8.

(Continued)

Table 1-101-3 *(Cont.)*
GRAPHIC SYMBOLS

Elementary or Schematic Diagram	Connection or Interconnection Diagram

OVERLOAD RELAY (Continued)

OVERLOAD RELAY, THERMAL
See "Thermal Element Actuating Device" and "Contact, Normally Closed—Break."

Example*

PENTODE, EQUIPOTENTIAL CATHODE

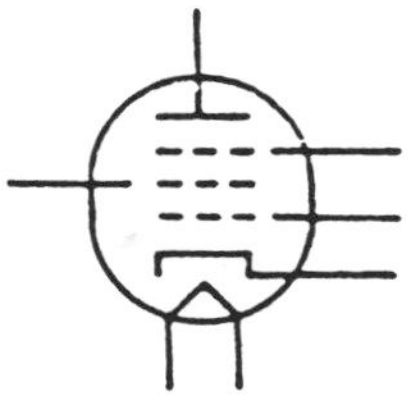

Use appropriate tube socket symbol and tube type number; see "Tube Socket."

PERMANENT MAGNET

PM or **PM —**

PHOTOSENSITIVE CELL, SEMICONDUCTOR

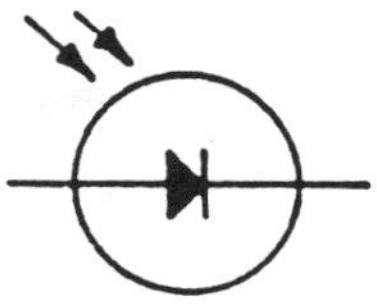

PHOTOTUBE, SINGLE UNIT, VACUUM TYPE

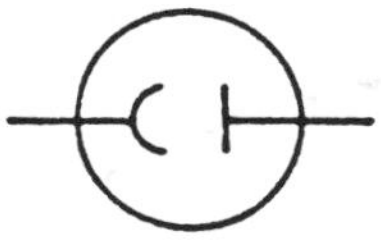

Use appropriate tube socket symbol and tube type number; see "Tube Socket."

* Example only; approved as Authorized Engineering Information. See 1-101.8.

(Continued)

Table 1-101-3 *(Cont.)*
GRAPHIC SYMBOLS

Elementary or Schematic Diagram	Connection or Interconnection Diagram
PILOT LIGHT See "Light, Indicating" and "Light, Indicating, Push-to-test."	
PLATE See "Anode or Plate."	
PLUG See "Connector, Disconnecting Device."	
POLARIZED CAPACITOR See "Capacitor, Polarized."	
POLARIZING MARKS See "Transformer, Current," "Transformer, Potential" and item 2(e) of 1-101.8.	
POTENTIOMETER See "Resistor, Adjustable" and "Resistor, Adjustable With a Section Having Negligible Resistance."	
PRESSURE OPERATED SWITCH See "Switch, Pressure or Vacuum Operated."	
PUSHBUTTON See "Switch, Pushbutton."	

(Continued)

Table 1-101-3 *(Cont.)*
GRAPHIC SYMBOLS

Elementary or Schematic Diagram	Connection or Interconnection Diagram

REACTOR
See "Winding."

REACTOR, SATURABLE CORE
Polarity may be added to direct current winding.

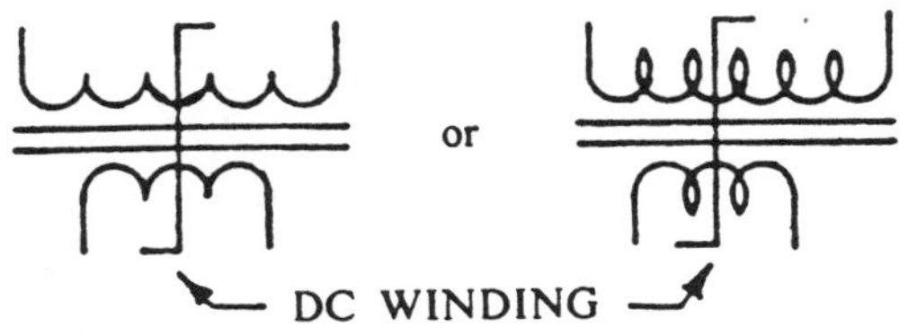

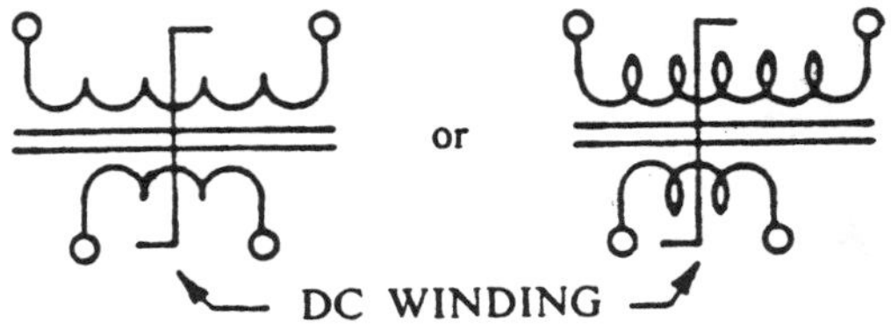

RECEIVER, SYNCHRO
See "Synchro."

RECEPTACLE
See "Connector, Disconnecting Device."

RECTIFIER, GENERAL
Triangle points in direction of forward current as indicated by a dc ammeter. Electron flow is in the opposite direction.

RECTIFIER, SEMICONDUCTOR DIODE, METALLIC OR ELECTROLYTIC
See "Diode, Semiconductor."

‡

RECTIFIER
Application: Full Wave Bridge-type.

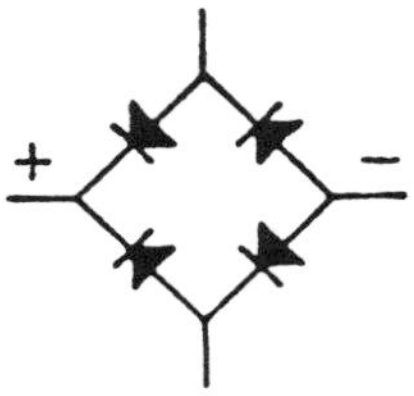

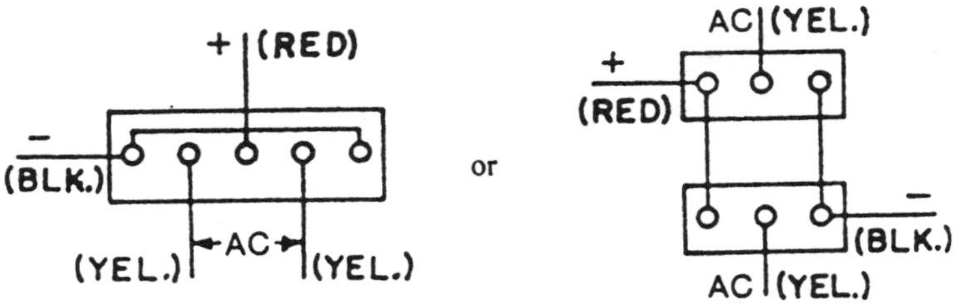

Example*

* Example only; approved as Authorized Engineering Information. See 1-101.8.
‡ Use pictorial representation. See 1-101.8.

(*Continued*)

Table 1-101-3 ***(Cont.)***
GRAPHIC SYMBOLS

Elementary or Schematic Diagram	Connection or Interconnection Diagram

RECTIFIER, GENERAL (Continued)

RECTIFIER, POOL-TYPE CATHODE
See "Mercury-pool Tube."

RECTIFIER, SILICON CONTROLLED
(Thyristor)

P-type Gate 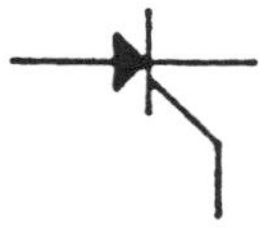 ‡

N-type Gate  ‡

REGULATING GENERATOR
See "Armature, Rotary Amplifier With Shorted Brushes."

RELAY
See "Control Relay" and "Static Relay." For stepping relay use "Switch, Stepping."

RELAY COIL
See "Coil, Operating."

RESISTOR GENERAL

 or 

RESISTOR, ADJUSTABLE
On elementary diagrams, use arrows or descriptive terms to show change in function.

 or 

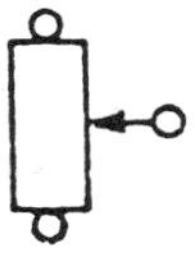 or 

‡ Use pictorial representation. See 1-101.8.

(Continued)

Table 1-101-3 *(Cont.)*
GRAPHIC SYMBOLS

Elementary or Schematic Diagram	Connection or Interconnection Diagram
RESISTOR, GENERAL (Continued)	
RESISTOR, ADJUSTABLE WITH A SECTION HAVING NEGLIGIBLE RESISTANCE Open area is resistance. The remaining portion has negligible resistance.	
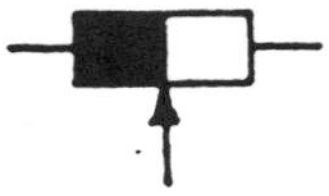 or 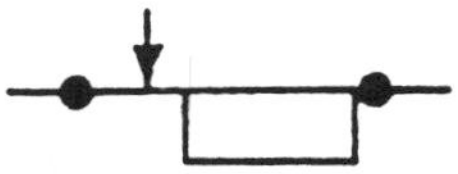	‡
RESISTOR, TAPPED	
	 or 
RHEOSTAT See "Resistor, Adjustable."	
SELECTOR SWITCH See "Switch, Selector."	
SELECTOR PUSHBUTTON See "Switch—With Pushbutton Type Contact Mechanism" and "Switch—Selector Pushbutton."	
SEMICONDUCTOR DEVICE See "Diode, Semiconductor," "Rectifier, General" and "Transistor."	

‡ Use pictorial representation. See 1-101.8.

(Continued)

Table 1-101-3 *(Cont.)*
GRAPHIC SYMBOLS

Elementary or Schematic Diagram	Connection or Interconnection Diagram

SERIES FIELD
See "Field, Series."

SHIELD, SHIELDING
Used for electrical or magnetic shielding. See "Conductor, Shielded Single or Multiconductor."

SHUNT, INSTRUMENT OR RELAY
Terminals in the rectangle are for the connection of instrument or relay.

or

SHUNT FIELD
See "Field, Shunt or Separately Excited."

SILICON CONTROLLED RECTIFIER
See "Rectifier, Silicon Controlled."

SPEED OPERATED SWITCH
See "Switch, Speed Responsive."

(Continued)

Table 1-101-3 ***(Cont.)***
GRAPHIC SYMBOLS

Elementary or Schematic Diagram	Connection or Interconnection Diagram

STARTER, WITH INTERLOCKS (EXAMPLE)

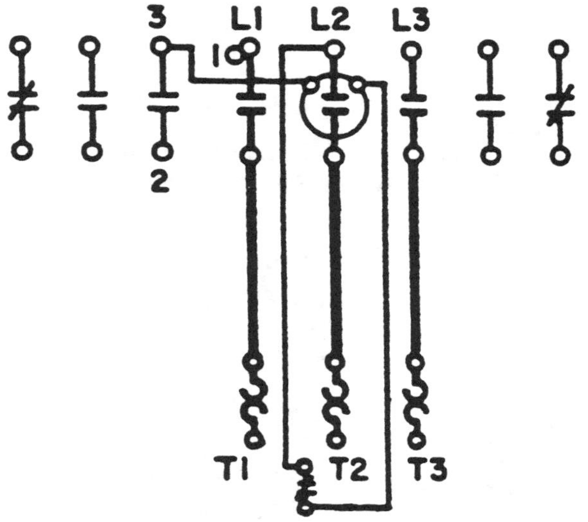

Example* ▲

STATIC ELEMENTS OR DEVICES (See 1-101.10)
Any circuit symbol enclosed in a diamond denotes a static element or device.

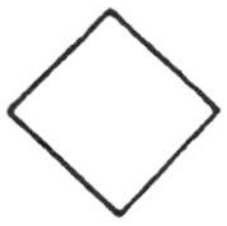

Examples

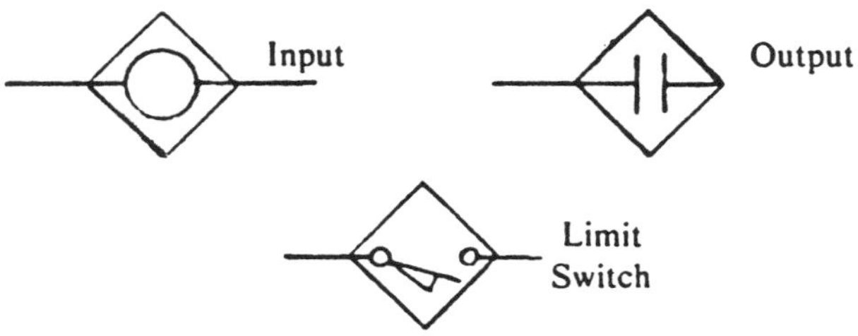

* Example only: approved as Authorized Engineering Information. See 1-101.8.
▲Revised 7-11-1984.

(*Continued*)

Table 1-101-3 *(Cont.)*
GRAPHIC SYMBOLS

Elementary or Schematic Diagram	**Connection or Interconnection Diagram**

STATIC ELEMENTS OR DEVICES (Continued)

When the input(s) and output(s) of static devices are not isolated, enclose complete device in a box. Show connection between input and output.

Examples

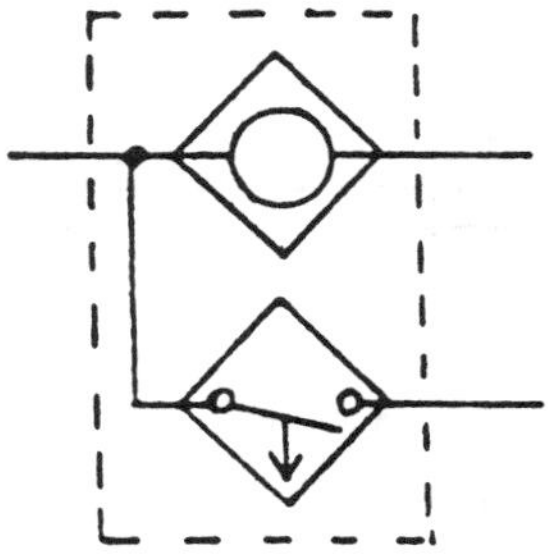

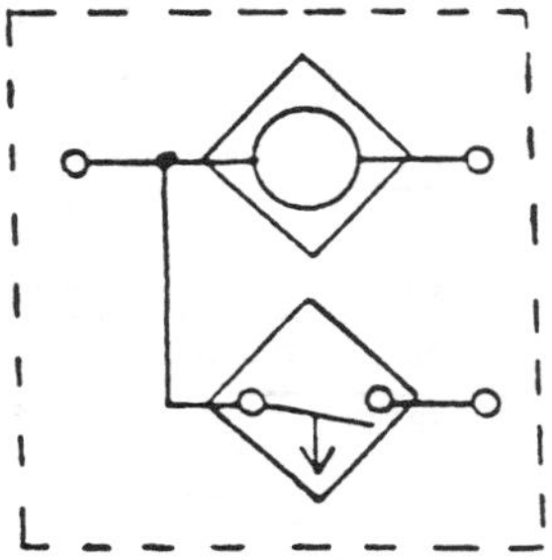

STATIC RELAY

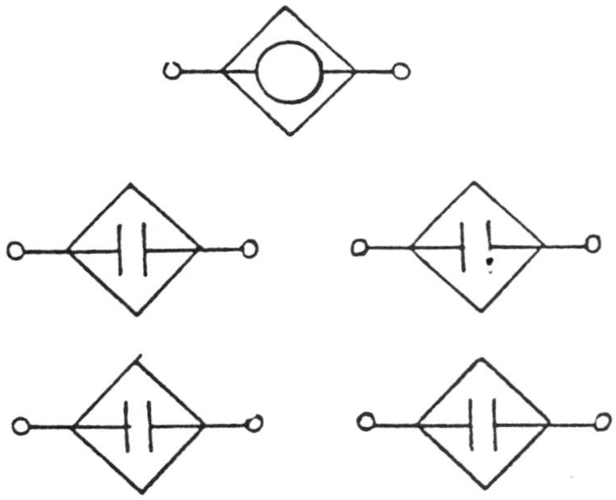

SWITCH
See also "Contact—Basic Assemblies."

SWITCH, DOUBLE THROW

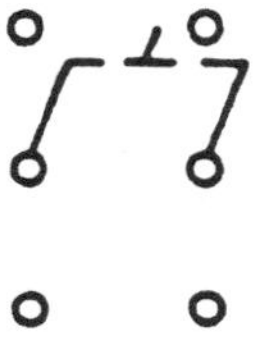

(Continued)

Table 1-101-3 *(Cont.)*
GRAPHIC SYMBOLS

Elementary or Schematic Diagram	Connection or Interconnection Diagram

SWITCH (Continued)

SWITCH, DRUM SWITCH, SLIDING CONTACT TYPE
Open rectangles are sliding contact segments.

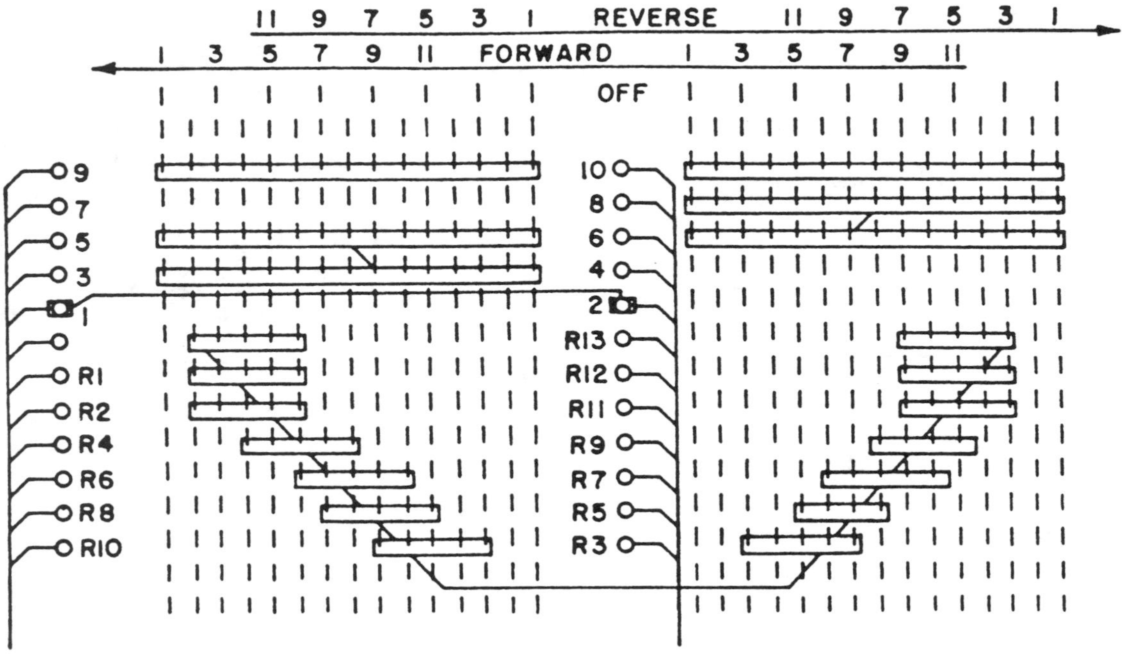

Example*

On elementary diagram, sections of the switch may be separated. Show position indications for each section.

On interconnection diagram, a uniformly shaped symbol may be used omitting segment details.

SWITCH, FLOW ACTUATED

CLOSING ON INCREASE IN FLOW

OPENING ON INCREASE IN FLOW

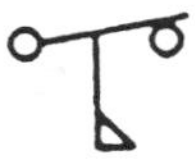

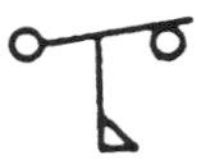

* Example only; approved as Authorized Engineering Information. See 1-101.8.

(*Continued*)

Table 1-101-3 *(Cont.)*
GRAPHIC SYMBOLS

Elementary or Schematic Diagram	Connection or Interconnection Diagram
SWITCH (Continued)	
SWITCH, FOOT OPERATED	
OPENED BY FOOT PRESSURE	

CLOSE BY FOOT PRESSURE	

SWITCH, KNIFE	
	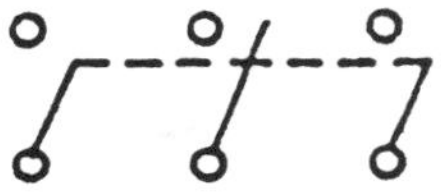
SWITCH, LIMIT As option to symbols shown below use 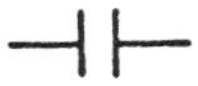 or  labeled "____LS" and describe function.	
NORMALLY OPEN	

NORMALLY OPEN, HELD CLOSED	

NORMALLY CLOSED	

NORMALLY CLOSED, HELD OPEN	

(*Continued*)

Table 1-101-3 ***(Cont.)***
GRAPHIC SYMBOLS

Elementary or Schematic Diagram	Connection or Interconnection Diagram

SWITCH (Continued)

SWITCH, LIQUID LEVEL ACTUATED (FLOAT)

CLOSING ON RISING LEVEL

OPENING ON RISING LEVEL

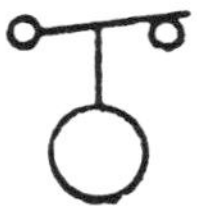

SWITCH, MASTER OR CONTROL

Show a table of contact operation on the elementary diagram. A typical table is shown below.

CONTACT	INDICATOR POSITION		
	A	B	C
1 – 2			X
3 – 4	X		
5 – 6			X
7 – 8	X		
X- INDICATES CONTACTS CLOSED			

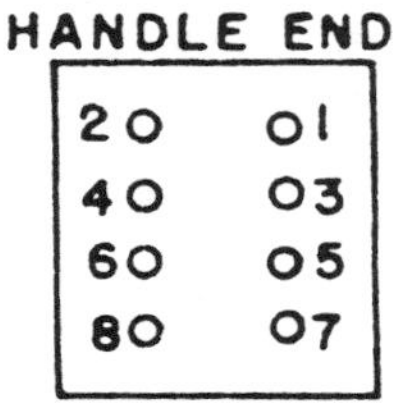

Example*

Show detached contacts (see "Contact, Basic Assemblies") elsewhere on diagram.

* Example only; approved as Authorized Engineering Information. See 1-101.8.

(Continued)

Table 1-101-3 *(Cont.)*
GRAPHIC SYMBOLS

Elementary or Schematic Diagram	Connection or Interconnection Diagram

SWITCH (Continued)

SWITCH, MASTER OR CONTROL, CAM TYPE

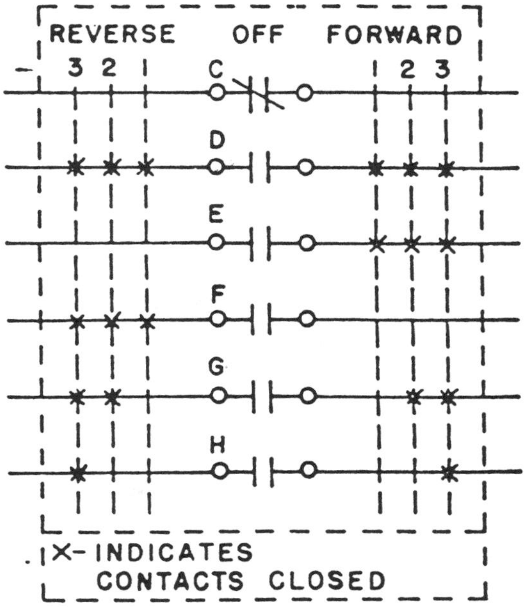

On elementary diagram, sections on the switch may be separated. Show position indications for each section

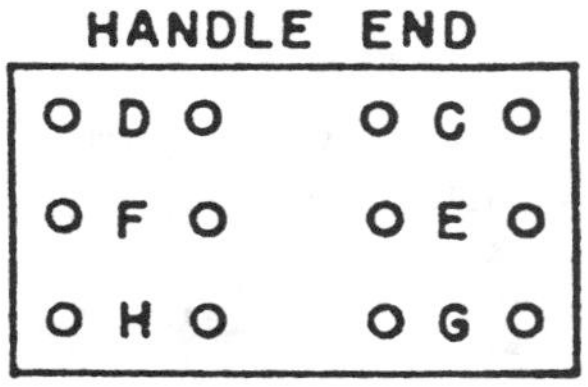

Example*

SWITCH, PRESSURE OR VACUUM OPERATED

CLOSING ON RISING PRESSURE

OPENING ON RISING PRESSURE

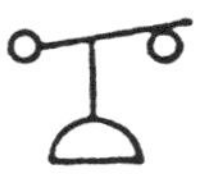

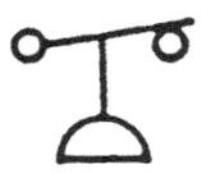

* Example only; approved as Authorized Engineering Information. See 1-101.8.

(*Continued*)

Table 1-101-3 *(Cont.)*
GRAPHIC SYMBOLS

Elementary or Schematic Diagram	Connection or Interconnection Diagram
SWITCH (Continued)	
SWITCH, PUSHBUTTON	
ILLUMINATED	
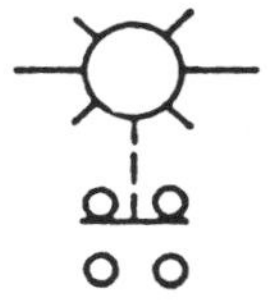	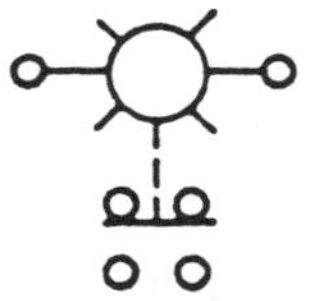
MAINTAINED, TWO CIRCUIT (Latched or not spring return)	
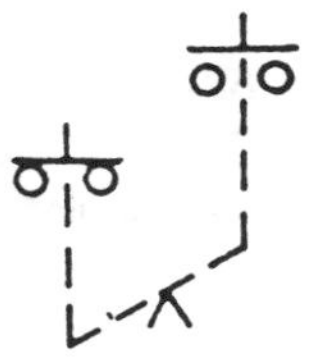	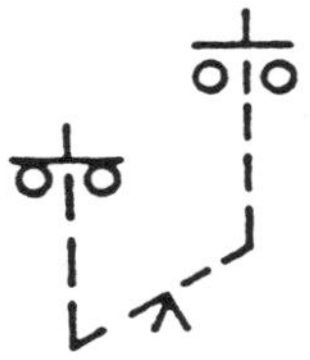
MOMENTARY OR SPRING RETURN, NORMALLY OPEN CIRCUIT CLOSING—MAKE	

MOMENTARY OR SPRING RETURN, NORMALLY CLOSED CIRCUIT OPENING—BREAK	

MOMENTARY OR SPRING RETURN, TWO CIRCUIT, NORMALLY OPEN AND NORMALLY CLOSED	

MUSHROOM HEAD Applied to two circuit pushbutton.	
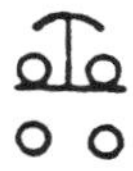	
WOBBLE STICK	

(Continued)

Table 1-101-3 *(Cont.)*
GRAPHIC SYMBOLS

Elementary or Schematic Diagram	Connection or Interconnection Diagram

SWITCH (Continued)

SWITCH, SELECTOR (Multiposition)

BREAK-BEFORE-MAKE

Nonshorting (nonbridging) during contact transfer.

or

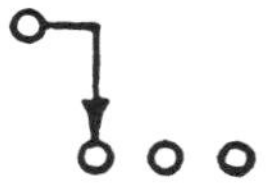

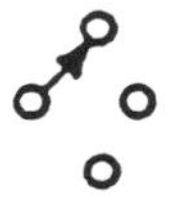

MAKE-BEFORE-BREAK

Shorting (bridging) during contact transfer.

or

SEGMENTAL CONTACT

MULTI-POINT WITH FIXED SEGMENT

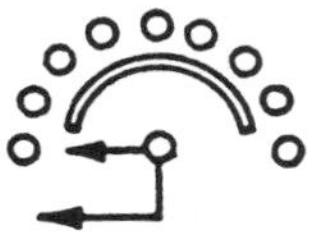

WITH PUSHBUTTON TYPE CONTACT MECHANISM

Show a table of contact operation if necessary. Point the arrow to the position label which identifies the position in which the contacts are shown.

	J	K	L
A1	X		
A2			X
B1	X		
B2			X

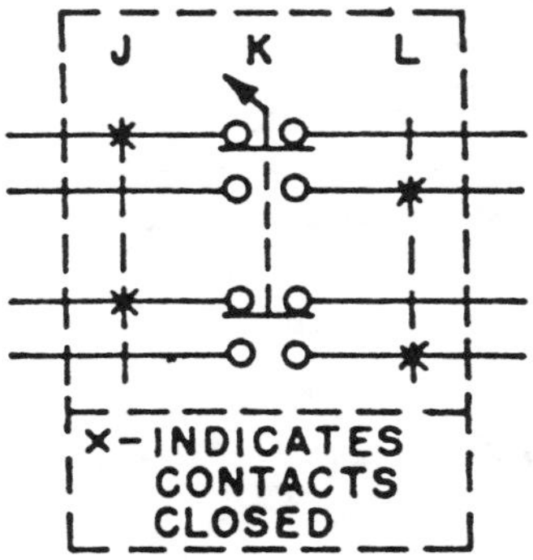

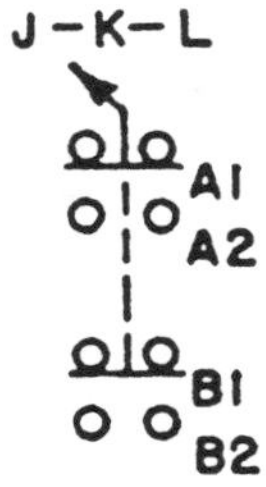

(*Continued*)

Table 1-101-3 *(Cont.)*
GRAPHIC SYMBOLS

Elementary or Schematic Diagram	Connection or Interconnection Diagram

SWITCH (Continued)

SWITCH, SELECTOR (Continued)

SELECTOR PUSHBUTTON
(Can be pushed into two or more positions and rotated into two or more positions.)

Show a table of contact operation on the diagram. Point the arrow to the position label which identifies the position in which the contacts are shown when the button is in the normal mode.

CONTACTS	SELECTOR POSITION								
	A			B			C		
	BUTTON			BUTTON			BUTTON		
	IN	NORMAL	OUT	IN	NORMAL	OUT	IN	NORMAL	OUT
1 – 2		X		X	X		X		
3 – 4	X					X		X	X
5 – 6	X	X	X		X				
7 – 8							X	X	X
X– INDICATES CONTACT CLOSED									

or

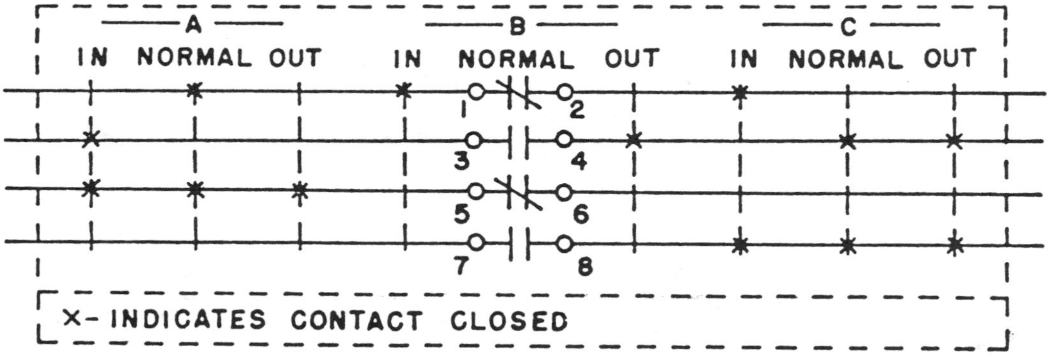

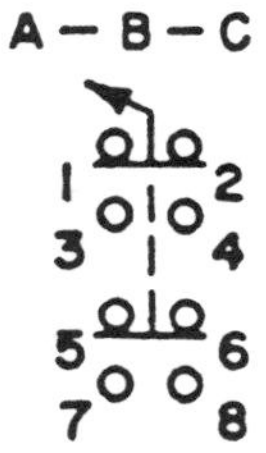

SWITCH, SINGLE THROW

(Continued)

Table 1-101-3 *(Cont.)*
GRAPHIC SYMBOLS

Elementary or Schematic Diagram	Connection or Interconnection Diagram

SWITCH (Continued)

SWITCH, SPEED RESPONSIVE
(Operated by shaft rotation) (F = Forward, R = Reverse)

ANTI-PLUGGING SWITCH
(Prevents plugging of drive)

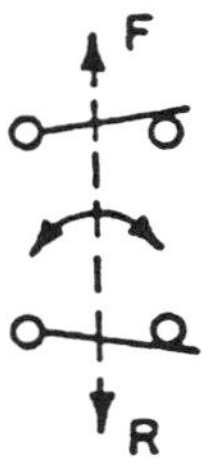

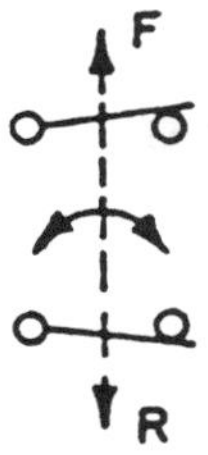

PLUGGING SWITCH
(Stops plugging action after drive has practically come to rest.)

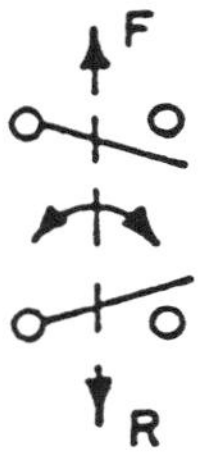

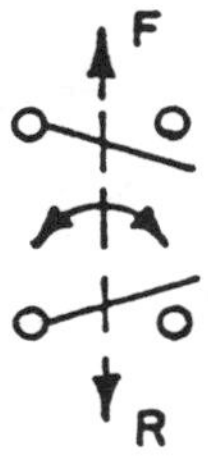

SPEED SWITCH
(Operates at a preset speed)

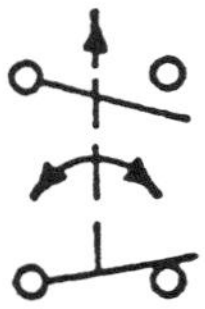

SWITCH, STEPPING

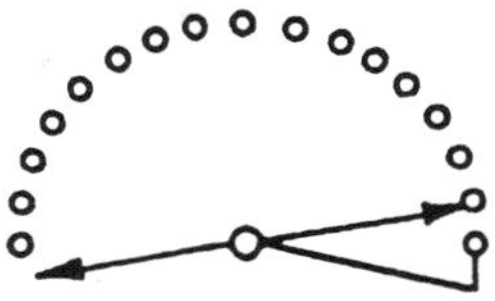

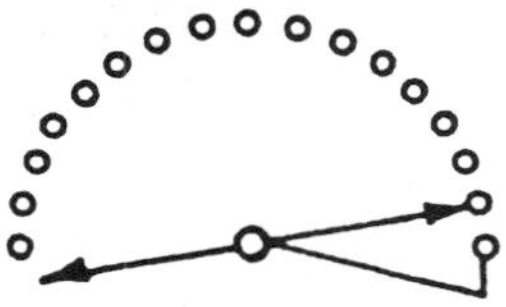

(*Continued*)

Table 1-101-3 *(Cont.)*
GRAPHIC SYMBOLS

Elementary or Schematic Diagram	Connection or Interconnection Diagram
SWITCH (Continued)	
SWITCH, TEMPERATURE ACTUATED	
CLOSING ON RISING TEMPERATURE	

OPENING ON RISING TEMPERATURE	

SWITCH, TOGGLE (Maintained position)	
SINGLE POLE	

TRANSFER, SINGLE POLE 2 POSITION	
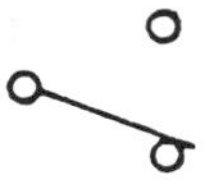	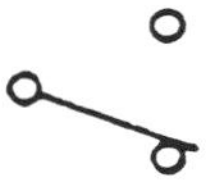
TRANSFER, SINGLE POLE 3 POSITION	
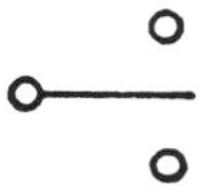	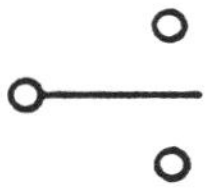
SYNCHRO, (SELSYN) Identify by note.	
CONTROL TRANSFORMER, RECEIVER, OR TRANSMITTER	
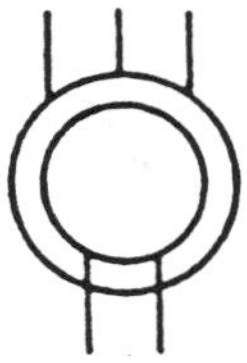	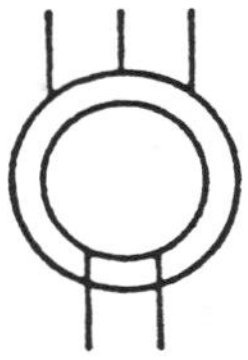

(Continued)

Table 1-101-3 *(Cont.)*
GRAPHIC SYMBOLS

Elementary or Schematic Diagram	Connection or Interconnection Diagram

SYNCHRO, (SELSYN) (Continued)

DIFFERENTIAL, RECEIVER, OR TRANSMITTER

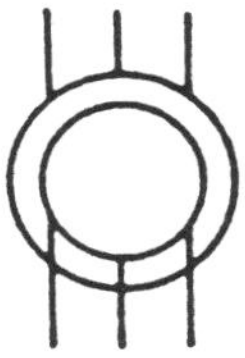

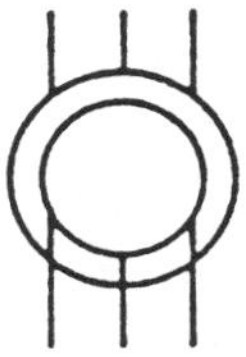

SYNCHRO, RESOLVER

*Example, 1 Phase Rotor and 2 Phase Stator.

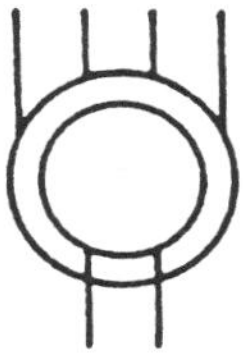

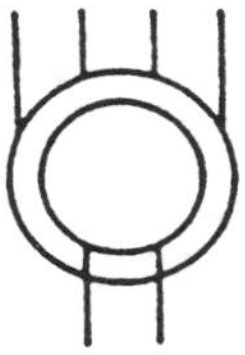

SYNCHRONOUS MOTOR, OR AC GENERATOR
Omit the field for synchronous induction, reluctance, or hysteresis motor.

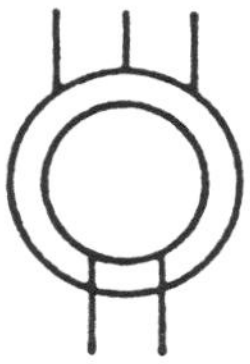 or 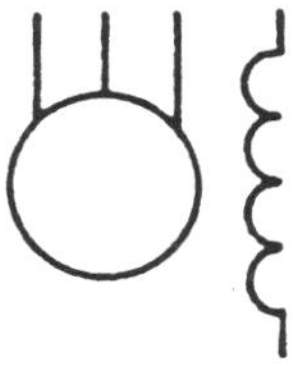

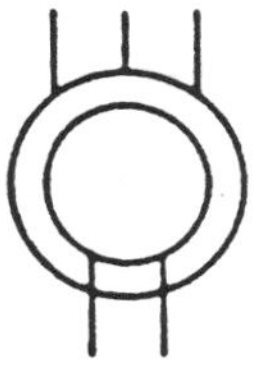 or 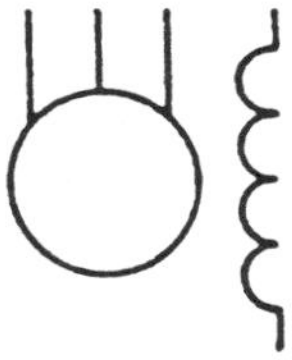

TEMPERATURE OPERATED SWITCH
See "Switch, Temperature Actuated."

TERMINAL

* Example only; approved as Authorized Engineering Information. See 1-101.8.

(*Continued*)

Table 1-101-3 ***(Cont.)***
GRAPHIC SYMBOLS

Elementary or Schematic Diagram	Connection or Interconnection Diagram
TERMINAL BOARD	
none	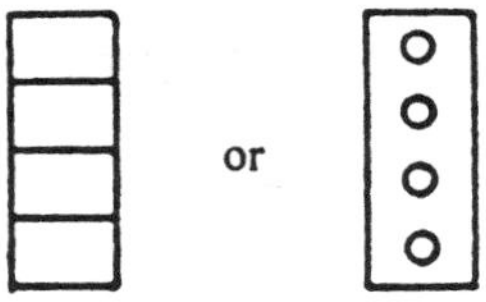
THERMAL ELEMENT ACTUATING DEVICE	
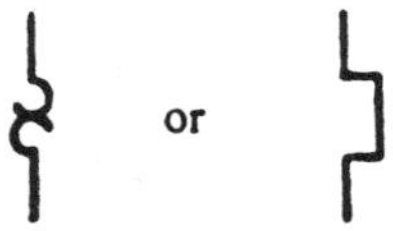	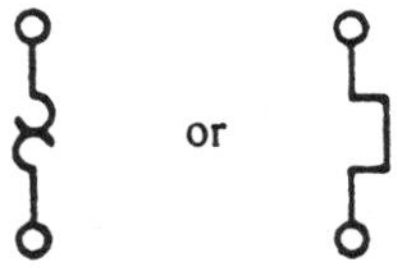
THERMOCOUPLE (Temperature measuring device)	

THERMOSTAT See "Switch, Temperature Actuated."	
THYRISTOR See "Rectifier, Silicon Controlled."	
TIME-DELAY SWITCHES See "Contact, Time Delay."	
TOGGLE SWITCH See "Switch, Toggle."	
TORQUE RECEIVER OR TRANSMITTER See "Synchro."	

(Continued)

Table 1-101-3 *(Cont.)*
GRAPHIC SYMBOLS

Elementary or Schematic Diagram	Connection or Interconnection Diagram

TRANSFORMER, GENERAL (CORE NOT SPECIFIED)

See also "Autotransformer." On elementary diagram, windings of the same transformer may be shown at different locations.

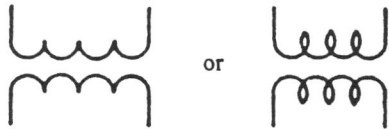

TRANSFORMER, WITH MAGNETIC CORE

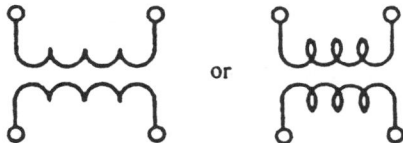

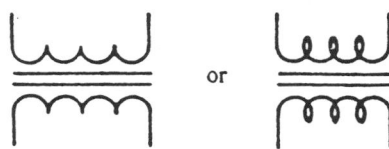

TRANSFORMER, CURRENT

With polarity mark (x) when required. Instantaneous direction of current into one polarity mark corresponds to current out of the other polarity mark.

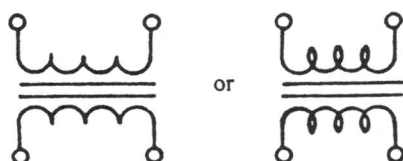

TRANSFORMER, POTENTIAL

With polarity mark (x) when required. Instantaneous direction of current into one polarity mark corresponds to current out of the other polarity mark. In lieu of polarity marking shown, use terminal identification, in accordance with 2.10 of the NEMA Standards Publication for *Specialty Transformers*, Standards Publication No. ST 1.

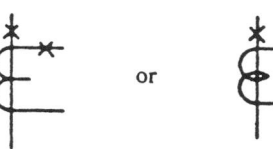

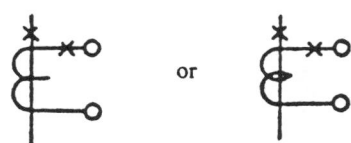

TRANSMITTER, SYNCHRO

See "Synchro."

(Continued)

Table 1-101-3 *(Cont.)*
GRAPHIC SYMBOLS

Elementary or Schematic Diagram	Connection or Interconnection Diagram
TRANSISTOR	
UNIJUNCTION TRANSISTOR P BASE 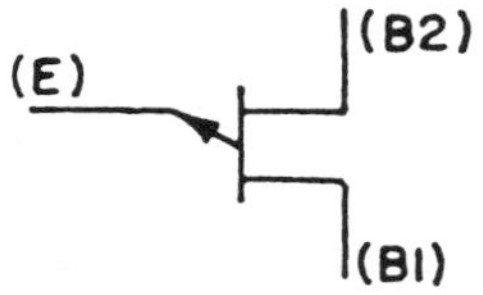	‡
UNIJUNCTION TRANSISTOR N BASE 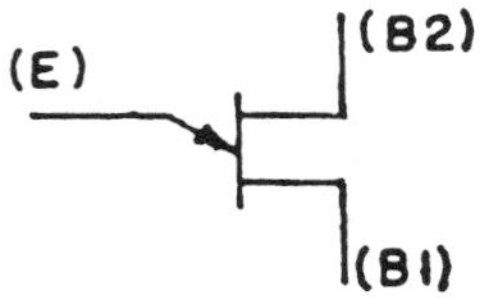	‡
PNP TRANSISTOR 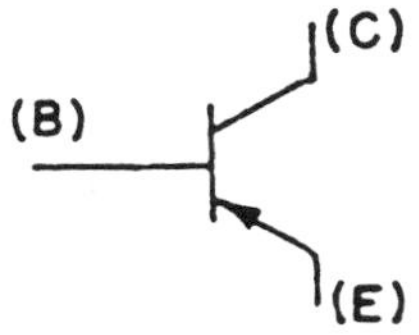	‡
NPN TRANSISTOR 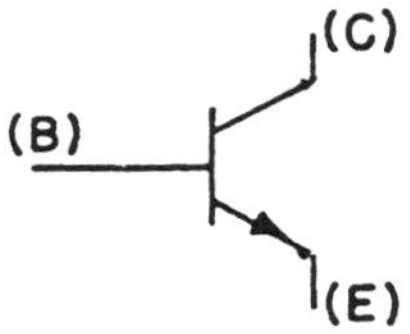	‡
TRIAC (Bidirectional Triode Thyristor) 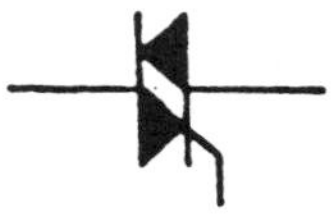	

‡ Use pictorial representation. See 1-101.8.

(Continued)

Table 1-101-3 *(Cont.)*
GRAPHIC SYMBOLS

Elementary or Schematic Diagram	Connection or Interconnection Diagram
TRIODE With directly heated filament cathode. 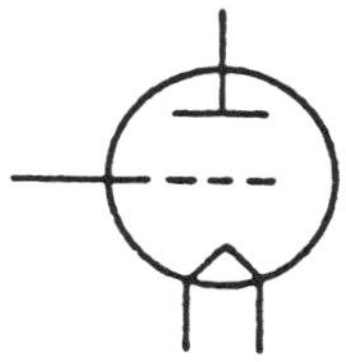	Use appropriate tube socket symbol and tube type number; see "Tube Socket."
TRIODE, TWIN EQUIPOTENTIAL CATHODE Showing use of elongated envelope. (See also "Envelope, Split.") 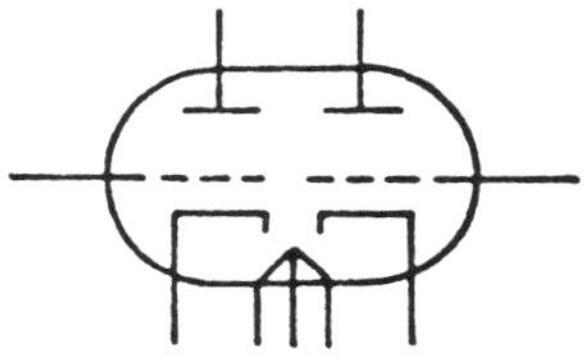	Use appropriate tube socket symbol and tube type number; see "Tube Socket."
TUBE SOCKET none	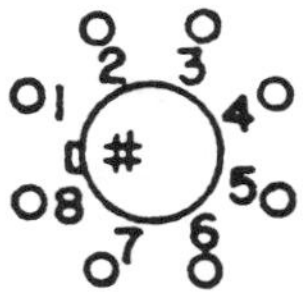 or 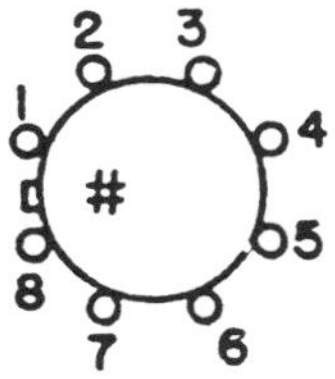 (Bottom View) Example* # Show tube type number here.

* Example only; approved as Authorized Engineering Information. See 1-101.8.

(*Continued*)

Table 1-101-3 *(Cont.)*
GRAPHIC SYMBOLS

Elementary or Schematic Diagram	Connection or Interconnection Diagram

VACUUM-OPERATED SWITCH
See "Switch, Pressure or Vacuum Operated."

VACUUM TUBE
See "Electron Tube Components."

VARIABLE
See "Adjustability" and "Resistor, Adjustable."

WINDING
See also "Field" and "Transformer, General."

 or 

 or 

WIRING
See "Conductor, or Conductive Path."

WOUND-ROTOR INDUCTION MOTOR OR
INDUCTION FREQUENCY CONVERTER

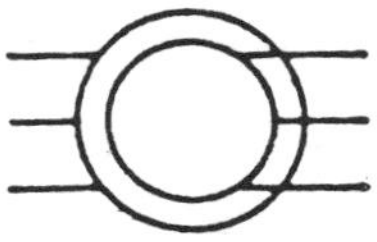

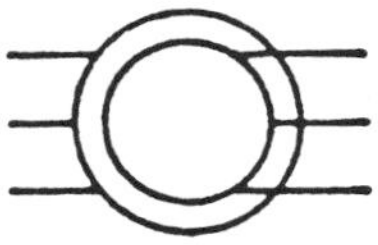

ZENER DIODE
See "Diode, Unidirectional Breakdown (Zener)" and "Diode, Bidirectional Breakdown."

Electrical Relations and Equations

Table 9.1. Electrical Conversion Table

To Convert	into	Multiply by
amperes/sq cm	amps/sq in.	6.452
amperes/sq cm	amps/sq meter	10^4
amperes/sq in.	amps/sq cm	0.1550
amperes/sq in.	amps/sq meter	1,550.0
amperes/sq meter	amps/sq cm	10^{-4}
amperes/sq meter	amps/sq in.	6.452×10^{-4}
ampere-hours	coulombs	3,600.0
ampere-hours	faradays	0.03731
ampere-turns	gilberts	1.257
ampere-turns/cm	amp-turns/in.	2.540
ampere-turns/cm	amp-turns/meter	100.0
ampere-turns/cm	gilberts/cm	1.257
ampere-turns/in.	amp-turns/cm	0.3937
ampere-turns/in.	amp-turns/meter	39.37
ampere-turns/in.	gilberts/cm	0.4950
ampere-turns/meter	amp/turns/cm	0.01
ampere-turns/meter	amp-turns/in.	0.0254
ampere-turns/meter	gilberts/cm	0.01257
Coulomb	Statcoulombs	2.998×10^4
coulombs	faradays	1.036×10^{-5}
coulombs/sq cm	coulombs/sq in.	64.52
coulombs/sq cm	coulombs/sq meter	10^4
coulombs/sq in.	coulombs/sq cm	0.1550
coulombs/sq in.	coulombs/sq meter	1,550.
coulombs/sq meter	coulombs/sq cm	10^{-4}
coulombs/sq meter	coulombs/sq in.	6.452×10^{-4}
ergs	joules	10^{-7}
ergs	kg-calories	2.389×10^{-11}
ergs	kg-meters	1.020×10^{-8}
ergs	kilowatt-hrs	0.2778×10^{-13}
ergs	watt-hours	0.2778×10^{-10}
farads	microfarads	10^4
Faraday/sec	Ampere (absolute)	9.6500×10^4
faradays	ampere-hours	26.80
faradays	coulombs	9.649×10^4
ergs/sec	kilowatts	10^{-10}
foot-pounds	kilowatt-hrs	3.766×10^{-7}
foot-pounds/min	foot-pounds/sec	0.01667
foot-pounds/min	horsepower	3.030×10^{-5}
foot-pounds/min	kilowatts	2.260×10^{-5}
gausses	lines/sq in.	6.452
gausses	webers/sq cm	10^{-8}
gausses	webers/sq in.	6.452×10^{-8}
gausses	webers/sq meter	10^{-4}
gilberts	ampere-turns	0.7958
gilberts/cm	amp-turns/cm	0.7958
gilberts/cm	amp-turns/in	2.021
gilberts/cm	amp-turns/meter	79.58
hectowatts	watts	100.0
henries	millihenries	1,000.0
horsepower	kilowatts	0.7457
horsepower	watts	745.7
horsepower (boiler)	kilowatts	9.803
horsepower-hrs	kilowatt-hrs	0.7457
International Ampere	Ampere (absolute)	.9998
International Volt	Volts (absolute)	1.0003
International volt	Joules (absolute)	$1\text{-}593 \times 10^{-19}$
International volt	Joules	9.654×10^4
joules	watt-hrs	2.778×10^{-4}
kilowatts	Btu/min	56.92
kilowatts	foot-lbs/min	4.426×10^4
kilowatts	foot-lbs/sec	737.6
kilowatts	horsepower	1.341
kilowatts	kg-calories/min	14.34
kilowatts	watts	1,000.0
kilowatt-hrs	Btu	3,413.
kilowatt-hrs	ergs	3.600×10^{13}
kilowatt-hrs	foot-lbs	2.655×10^4
kilowatt-hrs	gram-calories	859,850.
kilowatt-hrs	horsepower-hrs	1.341
kilowatt-hrs	joules	3.6×10^6
kilowatt-hrs	kg-calories	860.5
kilowatt-hrs	kg-meters	3.671×10^5
kilowatt-hrs	pounds of water evaporated from and at 212°F.	3.53
kilowatt-hrs	pounds of water raised from 62° to 212°F.	22.75
Lumen	Watt	.001496
maxwells	kilolines	0.001
maxwells	webers	10^{-8}
megalines	maxwells	10^6
megohms	microhms	10^{12}
megohms	ohms	10^6
microfarad	farads	10^{-6}
myriawatts	kilowatts	10.0
OHM (International)	OHM (absolute)	1.0005
ohms	megohms	10^{-6}
ohms	microhms	10^6
Volt/inch	Volt/cm.	.39370
Volt (absolute)	Statvolts	.003336
watts	Btu/hr	3.4129
watts	Btu/min	0.05688
watts	ergs/sec	107.
watts	foot-lbs/min	44.27
watts	foot-lbs/sec	0.7378
watts	horsepower	1.341×10^{-3}
watts	horsepower (metric)	1.360×10^{-3}
watts	kg-calories/min	0.01433
watts	kilowatts	0.001
Watts (Abs.)	B.T.U. (mean)/min.	0.056884
Watts (Abs.)	joules/sec.	1
watt-hours	Btu	3.413
watt-hours	ergs	3.60×10^{10}
watt-hours	foot-pounds	2,656.
watt-hours	gram-calories	859.85
watt-hours	horsepower-hrs	1.341×10^{-3}
watt-hours	kilogram-calories	0.8605
watt-hours	kilogram-meters	367.2
watt-hours	kilowatt-hrs	0.001
Watt (International)	Watt (absolute)	1.0002
webers	maxwells	10^8
webers	kilolines	10^5
webers/sq in.	gausses	1.550×10^7
webers/sq in.	lines/sq in.	10^8
webers/sq in.	webers/sq cm	0.1550
webers/sq in.	webers/sq meter	1,550.
webers/sq meter	gausses	10^4
webers/sq meter	lines/sq in.	6.452×10^4
webers/sq meter	webers/sq cm	10^{-4}
webers/sq meter	webers/sq in.	6.452×10^{-4}

Table 9.2.
Ohm's Law for Single-Phase and Three-Phase Power

Desired Data	Single-Phase Alternating Current	Three-Phase Alternating Current
Amperes, when kVA is known	$\frac{(kVA)(1000)}{volts}$	$\frac{(kVA)(1000)}{(1.732)(volts)}$
Amperes, when watts are known	$\frac{(watts)}{(volts)(PF)}$	$\frac{(watts)}{(volts)(PF)(1.732)}$
kVA	$\frac{(volts)(amps)}{1000}$	$\frac{(volts)(amps)(1.732)}{1000}$
watts	(volts)(amps)(PF)	(volts)(amps)(PF)(1.732)
vars	(volts)(amps)(1 – PF)	(1.732)(volts)(amps)(1 – PF)
PF (Power Factor)	$\frac{watts}{(volts)(amps)}$	$\frac{watts}{(volts)(amps)(1.732)}$
Z	$\frac{(volts)^2 (PF)}{(kVA)(1000)}$	$\frac{(volts)^2 (PF)(1.732)}{(kVA)(1000)}$
Z	$\frac{(kVA)(1000)}{(amps)^2 (PF)}$	$\frac{(kVA)(1000)}{(amps)^2 (PF)(1.732)}$

Table 9.3
Electrical Formula for Determining Amperes, Horsepower, Kilowatts, and Kilovolt-Amperes

Desired Data	Alternating Current Single Phase	*Two-Phase Four-Wire
Kilowatts	$\frac{I \times E \times PF}{1000}$	$\frac{I \times E \times 2 \times PF}{1000}$
kVA	$\frac{I \times E}{1000}$	$\frac{I \times E \times 2}{1000}$
Horsepower Output	$\frac{I \times E \times \%Eff. \times PF}{746}$	$\frac{I \times E \times 2 \times \%Eff. \times PF}{746}$
Amperes when Hp is known	$\frac{Hp \times 746}{E \times \%Eff. \times PF}$	$\frac{Hp \times 746}{2 \times E \times \%Eff. PF}$
Amperes when Kilowatts is known	$\frac{kW \times 1000}{E \times PF}$	$\frac{kW \times 1000}{2 \times E \times PF}$
Amperes when kVA is known	$\frac{kVA \times 1000}{E}$	$\frac{kVA \times 1000}{2 \times E}$
Desired Data	AC Three – Phase	DC
Kilowatts	$\frac{I \times E \times 1.73 \times PF}{1000}$	$\frac{I \times E}{1000}$
kVA	$\frac{I \times E \times 1.73}{1000}$	—
Horsepower Output	$\frac{I \times E \times 1.73 \times \%Eff. \times PF}{746}$	$\frac{I \times E \times \%Eff.}{746}$
Amperes when Hp is known	$\frac{Hp \times 746}{1.73 \times E \times \%Eff. \times PF}$	$\frac{Hp \times 746}{E \times \%Eff.}$
Amperes when Kilowatts is known	$\frac{kW \times 1000}{1.73 \times E \times PF}$	$\frac{kW \times 1000}{E}$
Amperes when kVA is known	$\frac{kVA \times 1000}{1.73 \times E}$	—

*In three-wire, two-phase circuits the current in the common conductor is 1.41 times that in either other conductor.

Note: E = Volts; I = Amperes; %Eff. = Percent Efficiency; PF = Power Factor.

Electrical Formulas

Table 9.4

$$kW = \frac{1.73 \times E \times I \times PF}{1000} \text{ (3-phase alternating current)}$$

$$kW = \frac{2 \times E \times I \times PF}{1000} \text{ (2-phase alternating current)}$$

$$kW = \frac{E \times I \times PF}{1000} \text{ (1-phase alternating current)}$$

Ohms Law (direct or non-inductive alternating current circuits):

$$I = \frac{E}{R} \quad E = IR \quad R = \frac{E}{I} \quad W = IE = \frac{E^2}{R} = I^2R$$

(Alternating inductive circuits):

$$I = \frac{E}{Z} \quad E = IZ \quad Z = \frac{E}{I}$$

Amperes per motor (current input per line):

Direct current I $= \dfrac{hp \times 746}{E \times Eff.}$

3-Phase ACI $= \dfrac{hp \times 746}{1.73 \times E \times Eff. \times PF}$

2-Phase ACI $= \dfrac{hp \times 746}{2 \times E \times Eff. \times PF}$

1-Phase ACI $= \dfrac{hp \times 746}{E \times Eff. \times PF}$

kW = 1000 watts = 1.34 hp
hp = 746 watts = 0.746 kW

$$kW = \frac{E \times I}{1000} \text{ (Direct current).}$$

kW = kVA × PF × (Alternating current)

Power Factor (PF) is the ratio of wattmeter reading (true power) to volt and ammeter readings (apparent power), and is never greater than 1.00. If true power is expressed in kilowatts (kW) and apparent power as the product of kilovolts (1000 volts) and amperes, or kVA, then

$$PF = \frac{\text{true power}}{\text{apparent power}} = \frac{\text{watts}}{\text{volts} \times \text{amperes}} = \frac{kW}{kVA}$$

For estimating purposes, power factors in any circuit can be assumed as follows: incandescent lighting circuit, no motors, 0.95 to 1.00; lighting and motors, 0.85; motors only, 0.80.

Table 9.5
Ohm's Law

Ohm's Law for Direct or Single Phase Non-inductive Alternating Current

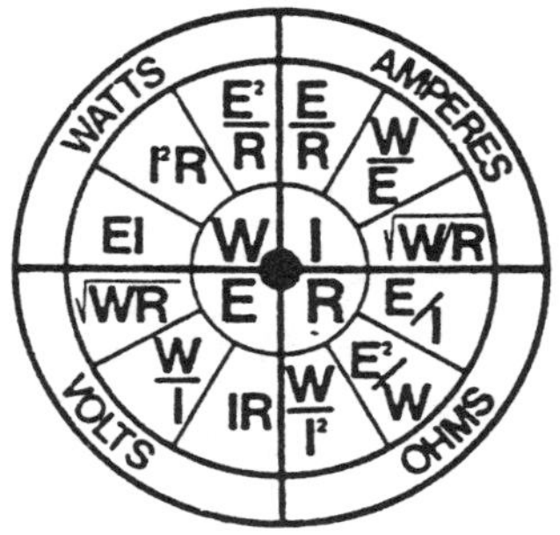

Let
I = Ampere = unit of current strength or rate of flow
E = Volt = unit of electro-motive force or electric pressure
R = Ohm = unit of resistance to flow of current
W = Watt = unit of power

Then $I = \frac{E}{R} \quad E = IR \quad R = \frac{E}{I}$

$$W = IE = \frac{E^2}{R} = I^2R$$

1,000W = 1 kW or kilowatt; this is the usual unit of measure of electric power. One kilowatt hour or kWh is the work done by one kilowatt in one hour.

For alternating current circuits, the following rules are useful for finding the power of a polyphase circuit.
Let PF = Power Factor.

3-Phase Alternating Current

$$kW = \frac{1.73 \times E \times I \times PF}{1,000}$$

2-Phase Alternating Current

$$kW = \frac{2 \times E \times I \times PF}{1,000}$$

1-Phase Alternating Current

$$kW = \frac{E \times I \times PF}{1,000}$$

1 Horsepower, hp = 746 watts
33,000 foot-pounds per minute*
2,545 heat units per hour, Btu**
2.64 lbs. water evaporated at 212°F

*1 foot-pound = raising one pound one foot
**1 Btu (British Thermal Unit) = heat required to raise the temperature of one pound of pure water one degree Fahrenheit.

Ohm's Law for D-C Circuits

The fundamental Ohm's law formulas for d-c circuits are given by,

$$I = \frac{E}{R}, \quad R = \frac{E}{I},$$

$$E = IR, \quad P = EI.$$

where

I = current in amperes,
R = resistance in ohms,
E = potential across R in volts,
P = power, in watts.

Ohms Law Formulas for D-C Circuits

Known Values	Formulas for Determining Unknown Values of . . .			
	I	**R**	**E**	**P**
I & R			IR	I^2R
I & E		$\frac{E}{I}$		EI
I & P		$\frac{P}{I^2}$	$\frac{P}{I}$	
R & E	$\frac{E}{R}$			$\frac{E^2}{R}$
R & P	$\sqrt{\frac{P}{R}}$		$\sqrt{PR}$	
E & P	$\frac{P}{E}$	$\frac{E^2}{P}$		

Ohm's Law for A-C Circuits

The fundamental Ohm's law formulas for a-c circuits are given by

$$I = \frac{E}{Z}, \quad Z = \frac{E}{I},$$

$$E = IZ, \quad P = EI \cos\theta$$

where

I = current in amperes,
Z = impedance in ohms,
E = volts across Z,
P = power in watts,
θ = phase angle in degrees

Power factor

The power-factor of any a-c circuit is equal to the true power in watts divided by the apparent power in volt-amperes which is equal to the cosine of the phase angle, and is expressed by

$$p.f. = \frac{EI\cos\theta}{EI} = \cos\theta$$

where

$p.f.$ = the circuit load power factor,
$EI\cos\theta$ = the true power in watts,
EI = the apparent power in volt-amperes,
E = the applied potential in volts,
I = load current in amperes.

Ohm's Law Formulas for A-C Circuits

Known Values	Formulas for Determining Unknown Values of . . .			
	I	**Z**	**E**	**P**
I & Z			IZ	$I^2Z\cos\theta$
I & E		$\frac{E}{I}$		$IE\cos\theta$
I & P		$\frac{P}{I^2\cos\theta}$	$\frac{P}{I\cos\theta}$	
Z & E	$\frac{E}{Z}$			$\frac{E^2\cos\theta}{Z}$
Z & P	$\sqrt{\frac{P}{Z\cos\theta}}$		$\sqrt{\frac{PZ}{\cos\theta}}$	
E & P	$\frac{P}{E\cos\theta}$	$\frac{E^2\cos\theta}{P}$		

Electrical

Resistance

In series

$$R_t = R_1 + R_2 + R_3 \ldots \text{etc.}$$

In parallel

$$R_t = \frac{1}{\frac{1}{R_1} + \frac{1}{R_2} + \frac{1}{R_3} \ldots \text{etc.}}$$

Two resistors in parallel

$$R_t = \frac{R_1 R_2}{R_1 + R_2}$$

Capacitance

In parallel

$$C_t = C_1 + C_2 + C_3 \ldots \text{etc.}$$

In series

$$C_t = \frac{1}{\frac{1}{C_1} + \frac{1}{C_2} + \frac{1}{C_3} \ldots \text{etc.}}$$

Two capacitors in series

$$C_t = \frac{C_1 C_2}{C_1 + C_2}$$

The quantity of electricity stored within a capacitor is given by

$$Q = CE$$

where

Q = the quantity stored, in coulombs,
E = the potential impressed across the condenser, in volts,
C = capacitance in farads.

The capacitance of a parallel plate capacitor is given by

$$C = 0.0885 \frac{KS(N-1)}{d}$$

where

C = capacitance in mmfd.,
K = dielectric constant,
*S = area of one plate in square centimeters,
N = number of plates,
*d = thickness of the dielectric in centimeters (same as the distance between plates).

Inductance

Self Inductance

In series

$$L_t = L_1 + L_2 + L_3 \ldots \text{etc.}$$

In parallel

$$L_t = \frac{1}{\frac{1}{L_1} + \frac{1}{L_2} + \frac{1}{L_3} \ldots \text{etc.}}$$

Two inductors in parallel

$$L_t = \frac{L_1 L_2}{L_1 + L_2}$$

Coupled Inductance

In series with fields aiding

$$L_t = L_1 + L_2 + 2M$$

In series with fields opposing

$$L_t = L_1 + L_2 - 2M$$

In parallel with fields aiding

$$L_t = \frac{1}{\frac{1}{L_1 + M} + \frac{1}{L_2 + M}}$$

*When S and d are given in inches, change constant 0.0885 to 0.224. Answer will still be in micromicrofarads.

In parallel with fields opposing.

$$L_t = \frac{1}{\frac{1}{L_1 - M} + \frac{1}{L_2 - M}}$$

where

L_t = the total inductance,
M = the mutual inductance,
L_1 and L_2 = the self inductance of the individual coils.

Mutual Inductance

The mutual inductance of two r-f coils with fields interacting, is given by

$$M = \frac{L_A - L_O}{4}$$

where

M = mutual inductance, expressed in same units as L_A and L_O,
L_A = Total inductance of coils L_1 and L_2 with fields *aiding,*
L_O = Total inductance of coils L_1 and L_2 with fields *opposing.*

Coupling coefficient

When two r-f coils are inductively coupled so as to give transformer action the coupling coefficient is expressed by

$$K = \frac{M}{\sqrt{L_1 L_2}}$$

where

K = the coupling coefficient; - (K x 10^2 = coupling coefficient in %),
M = the mutual inductance value,
L_1 and L_2 = the self-inductance of the two coils respectively, both being expressed in the same units.

Resonance

The resonant frequency, or frequency at which inductive reactance X_L equals capacitive reactance X_C, is expressed by

$$f_r = \frac{1}{2\pi \sqrt{LC}}$$

also

$$L = \frac{1}{4\pi^2 f_r^2 C}$$

and

$$C = \frac{1}{4\pi^2 f_r^2 L}$$

where

f_r = resonant frequency in Hertz,
L = inductance in henrys,
C = capacitance in farads,
2π = 6.28
$4\pi^2$ = 39.5

Reactance

of an inductance is expressed by

$$X_L = 2\pi fL$$

of a capacitance is expressed by

$$X_C = \frac{1}{2\pi fC}$$

where

X_L = inductive reactance in ohms, (known as positive reactance),
X_C = capacitive reactance in ohms, (known as negative reactance),
f = frequency in hertz,
L = inductance in henrys,
C = capacitance in farads,
2π = 6.28

Frequency

Frequency from wavelength

$$f = \frac{3 \text{ x } 10^5}{\lambda} \text{ (kilohertz)}$$

where λ = wavelength in *meters.*

$$f = \frac{3 \text{ x } 10^4}{\lambda} \text{ (megahertz)}$$

where λ = wavelength in *centimeters.*

Wavelength from frequency

$$\lambda = \frac{3 \text{ x } 10^5}{f} \text{ (meters)}$$

where f = frequency in *kilohertz.*

$$\lambda = \frac{3 \text{ x } 10^4}{f} \text{ (centimeters)}$$

where f = frequency in *megahertz.*

Impedance

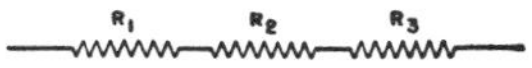

of resistance in series

$Z = R_1 + R_2 + R_3$ . . . etc.
$\theta = 0°$

of inductance alone

$Z = X_L$
$\theta = +90°$

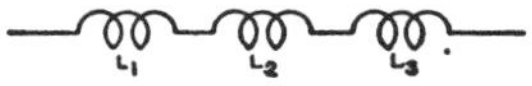

of induction in series

$Z = X_{L1} + X_{L2} + X_{L3}$ . . . etc.
$\theta = +90°$

of capacitance alone

$Z = X_C$
$\theta = -90°$

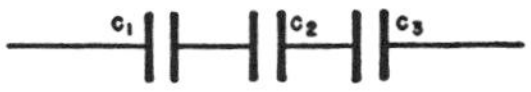

of capacitance in series

$Z = X_{C1} + X_{C2} + X_{C3}$ . . . etc.
$\theta = -90°$

or where only 2 capacitances C_1 and C_2 are involved,

$$Z = \frac{1}{2\pi f}\left(\frac{C_1 + C_2}{C_1 C_2}\right)$$
$$\theta = -90^\circ$$

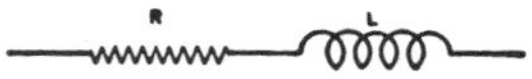

of resistance and inductance in series

$$Z = \sqrt{R^2 + X_L{}^2}$$
$$\theta = \text{arc tan}\,\frac{X_L}{R}$$

of resistance and capacitance in series

$$Z = \sqrt{R^2 + X_C{}^2}$$
$$\theta = \text{arc tan}\,\frac{X_C}{R}$$

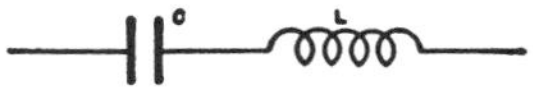

of inductance and capacitance in series

$$Z = X_L - X_C$$
$$\theta = -90^\circ \text{ when } X_L < X_C$$
$$= 0^\circ \text{ when } X_L = X_C$$
$$= +90^\circ \text{ when } X_L > X_C$$

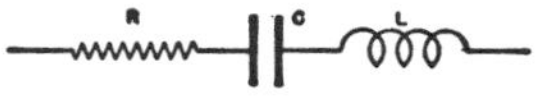

of resistance, inductance and capacitance in series

$$Z = \sqrt{R^2 + (X_L - X_C)^2}$$
$$\theta = \text{arc tan}\,\frac{X_L - X_C}{R}$$

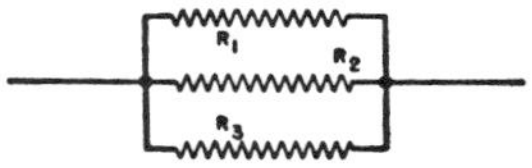

of resistance in parallel

$$Z = \cfrac{1}{\cfrac{1}{R_1} + \cfrac{1}{R_2} + \cfrac{1}{R_3} \ldots \text{etc.}}$$
$$\theta = 0^\circ$$

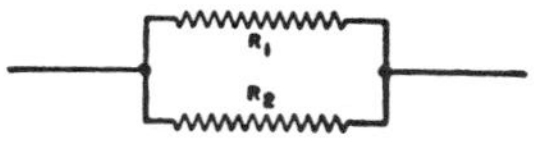

or where only 2 resistances R_1 and R_2 are involved,

$$Z = \frac{R_1 R_2}{R_1 + R_2}$$
$$\theta = 0^\circ$$

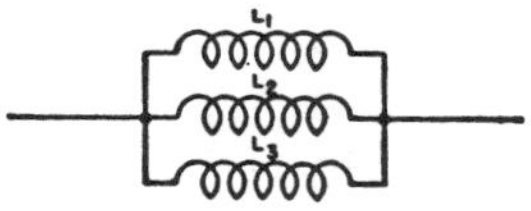

of inductance in parallel

$$Z = \cfrac{1}{\cfrac{1}{X_{L1}} + \cfrac{1}{X_{L2}} + \cfrac{1}{X_{L3}} \ldots \text{etc.}}$$
$$\theta = +90^\circ$$

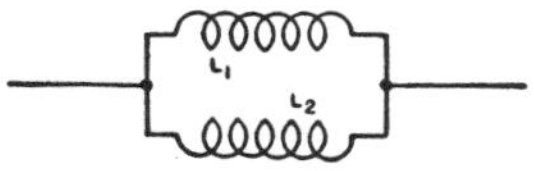

or where only 2 inductances L_1 and L_2 are involved,

$$Z = 2\pi f\left(\frac{L_1 L_2}{L_1 + L_2}\right)$$
$$\theta = +90^\circ$$

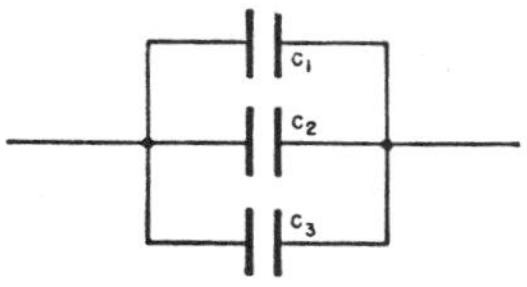

of capacitance in parallel

$$Z = \cfrac{1}{\cfrac{1}{X_{C1}} + \cfrac{1}{X_{C2}} + \cfrac{1}{X_{C3}} \ldots \text{etc.}}$$
$$\theta = -90^\circ$$

or where only 2 capacitances C_1 and C_2 are involved,

$$Z = \frac{1}{2\pi f\,(C_1 + C_2)}$$
$$\theta = -90^\circ$$

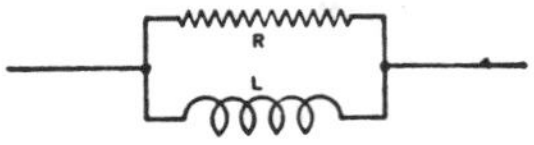

of inductance and resistance in parallel,

$$Z = \frac{RX_L}{\sqrt{R^2 + X_L{}^2}}$$

$$\theta = \text{arc tan} \frac{R}{X_L}$$

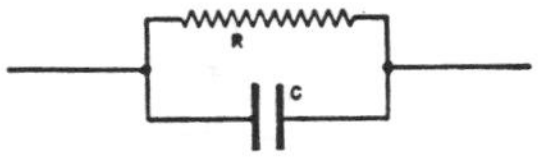

of capacitance and resistance in parallel,

$$Z = \frac{RX_C}{\sqrt{R^2 + X_C{}^2}}$$

$$\theta = - \text{ arc tan} \frac{R}{X_C}$$

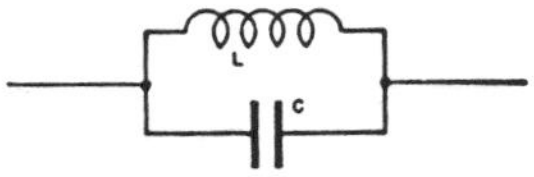

of inductance and capacitance in parallel,

$$Z = \frac{X_L X_C}{X_L - X_C}$$

$$\theta = 0^\circ \text{ when } X_L = X_C$$

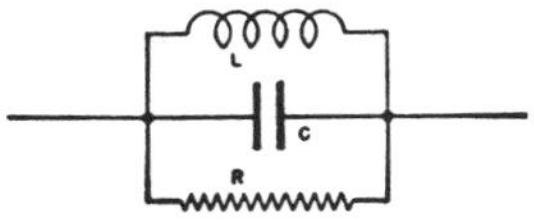

of inductance, resistance and capacitance in parallel

$$Z = \frac{RX_L X_C}{\sqrt{X_L{}^2 X_C{}^2 + (RX_L - RX_C)^2}}$$

$$\theta = \text{arc tan} \frac{RX_C - RX_L}{X_L X_C}$$

Conductance

In direct current circuits, conductance is expressed by

$$G = \frac{1}{R}$$

where

G = conductance in mhos,
R = resistance in ohms.

In d-c circuits involving resistances R_1, R_2, R_3, etc., in parallel,

the total conductance is expressed by

$$G_{\text{total}} = G_1 + G_2 + G_3 \ldots \text{etc.}$$

and the total current by

$$I_{\text{total}} = EG_{\text{total}}$$

and the amount of current in any single resistor, R_2 for example, in a parallel group, by

$$I_2 = \frac{I_{\text{total}}\, G_2}{G_1 + G_2 + G_3 \ldots \text{etc.}}$$

R, E and I in Ohm's law·formulas for d-c circuits may be expressed in terms of conductance as follows:

$$R = \frac{1}{G}, \quad E = \frac{I}{G}, \quad I = EG,$$

where

G = conductance in mhos,
R = resistance in ohms,
E = potential in volts,
I = current in amperes.

Susceptance

In an alternating current circuit, the susceptance of a series circuit is expressed by

$$B = \frac{X}{R^2 + X^2}$$

or, when the resistance is 0, susceptance becomes the reciprocal of reactance, or

$$B = \frac{1}{X}$$

where

B = susceptance in mhos

R = resistance in ohms,
X = reactance in ohms

Admittance

In an alternating current circuit, the admittance of a series circuit is expressed by

$$Y = \frac{1}{\sqrt{R^2 + X^2}}$$

Admittance is also expressed as the reciprocal of impedance, or

$$Y = \frac{1}{Z}$$

where

Y = addmittance in mhos,
R = resistance in ohms,
X = reactance in ohms,
Z = impedance in ohms.

R and X in terms of G and B

Resistance and reactance may be expressed in terms of conductance and susceptance as follows:

$$R = \frac{G}{G^2 + B^2}, \quad X = \frac{B}{G^2 + B^2}$$

G, B, Y and Z in parallel circuits

In any given a-c circuit containing a number of smaller parallel circuits only,

the effective conductance G_t is expressed by

$$G_t = G_1 + G_2 + G_3 \ldots \text{etc.},$$

and the effective susceptance B_t by

$$B_t = B_1 + B_2 + B_3 \ldots \text{etc.}$$

and the effective admittance Y_t by

$$Y_t = \sqrt{G_t^2 + B_t^2}$$

and the effective impedance Z_t by

$$Z_t = \frac{1}{\sqrt{G_t^2 + B_t^2}} \text{ or } \frac{1}{Y_t}$$

where

R = resistance in ohms,
X = reactance (capacitive or inductive) in ohms,
G = conductance in mhos,
B = susceptance in mhos,
Y = admittance in mhos,
Z = impedance in ohms.

Steady State Current Flow

In a capacitive circuit

In a capacitive circuit, where resistance loss components may be considered as negligible, the flow of current at a given alternating potential of constant frequency, is expressed by

$$I = \frac{E}{X_C} = \frac{E}{\left(\frac{1}{2\pi fC}\right)} = E(2\pi fC)$$

where

I = current in amperes,
X_C = capacitive reactance of the circuit in ohms,
E = applied potential in volts.

In an inductive circuit

In an inductive circuit, where inherent resistance and capacitance components may be so low as to be negligible, the flow of current at a given alternating potential of a constant frequency, is expressed by

$$I = \frac{E}{X_L} = \frac{E}{2\pi fL}$$

where

I = current in amperes,
X_L = inductive reactance of the circuit in ohms,
E = applied potential in volts.

Decibels

The number of db by which two power outputs P_1 and P_2 (in watts) may differ, is expressed by

$$10 \log \frac{P_1}{P_2};$$

or in terms of volts,

$$20 \log \frac{E_1}{E_2};$$

or in current,

$$20 \log \frac{I_1}{I_2}.$$

While power ratios are independent of source and load impedance values, voltage and current ratios in these formulas hold true only when the source and load impedances R_1 and R_2 are equal. In circuits where these impedances differ, voltage and current ratios are expressed by,

$$db = 20 \log \frac{E_1 \sqrt{R_2}}{E_2 \sqrt{R_1}} \text{ or, } 20 \log \frac{I_1 \sqrt{R_1}}{I_2 \sqrt{R_2}}$$

Table 9.6. DB Expressed in Watts and Volts

	Above Zero Level		Below Zero Level	
DB*	Watts	Volts	Watts	Volts
0	0.00600	1.73	6.00×10^{-3}	1.73
1	0.00755	1.94	4.77×10^{-3}	1.54
2	0.00951	2.18	3.78×10^{-3}	1.38
3	0.0120	2.45	3.01×10^{-3}	1.23
4	0.0151	2.74	2.39×10^{-3}	1.09
5	0.0190	3.08	1.90×10^{-3}	0.974
6	0.0239	3.46	1.51×10^{-3}	0.868
7	0.0301	3.88	1.20×10^{-3}	0.774
8	0.0378	4.35	9.51×10^{-4}	0.690
9	0.0477	4.88	7.55×10^{-4}	0.614
10	0.0600	5.49	6.00×10^{-4}	0.548
11	0.0755	6.14	4.77×10^{-4}	0.488
12	0.0951	6.90	3.78×10^{-4}	0.435
13	0.120	7.74	3.01×10^{-4}	0.388
14	0.151	8.68	2.39×10^{-4}	0.346
15	0.190	9.74	1.90×10^{-4}	0.308
16	0.239	10.93	1.51×10^{-4}	0.275
17	0.301	12.26	1.20×10^{-4}	0.245
18	0.378	13.76	9.51×10^{-5}	0.218
19	0.477	15.44	7.55×10^{-5}	0.194
20	0.600	17.32	6.00×10^{-5}	0.173
25	1.90	30.8	1.90×10^{-5}	0.0974
30	6.00	54.8	6.00×10^{-6}	0.0548
35	19.0	97.4	1.90×10^{-6}	0.0308
40	60.0	173.	6.00×10^{-7}	0.0173
45	190.	308.	1.90×10^{-7}	0.00974
50	600.	548.	6.00×10^{-8}	0.00548
60	6,000.	1,730.	6.00×10^{-9}	0.00173
70	60,000.	5,480.	6.00×10^{-10}	0.000548
80	600,000.	17,300.	6.00×10^{-11}	0.000173

*Zero db = 6 milliwatts into a 500 ohm load. Power ratios hold for any impedance, but voltages must be referred to an impedance load of 500 ohms.

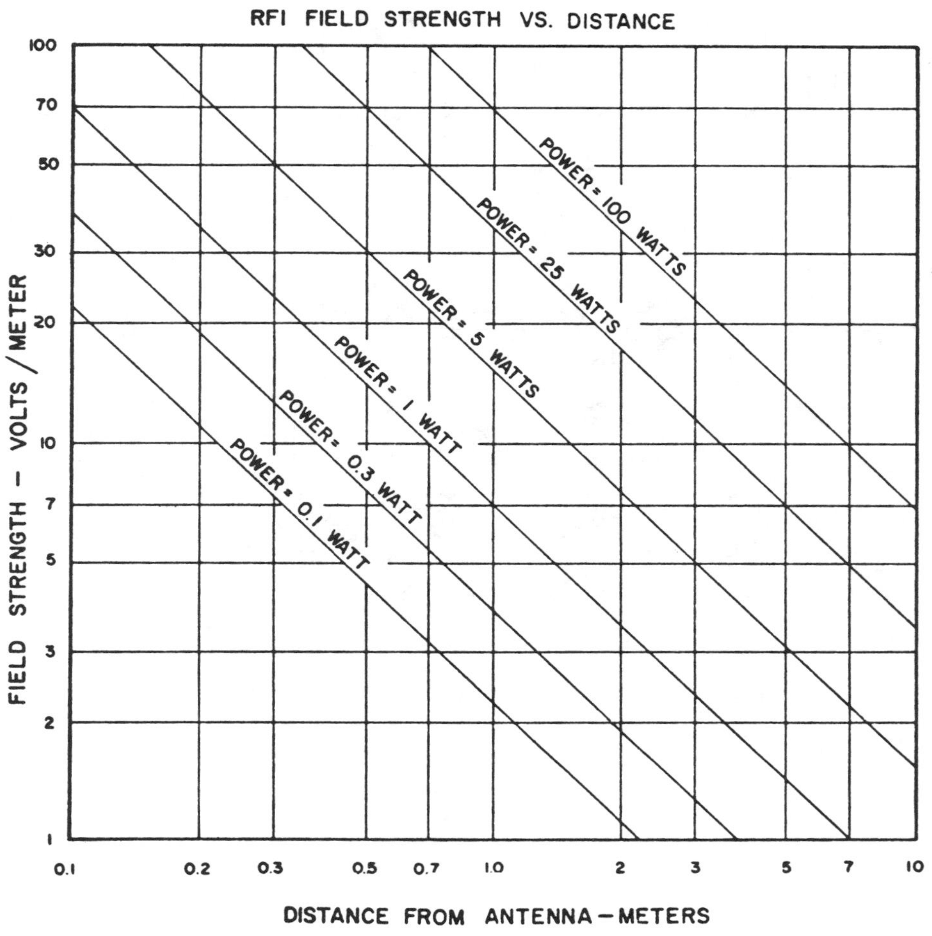

Chart 9.1. Radio frequency interference (RFI) field strength versus distance. Courtesy Transmation, Inc.

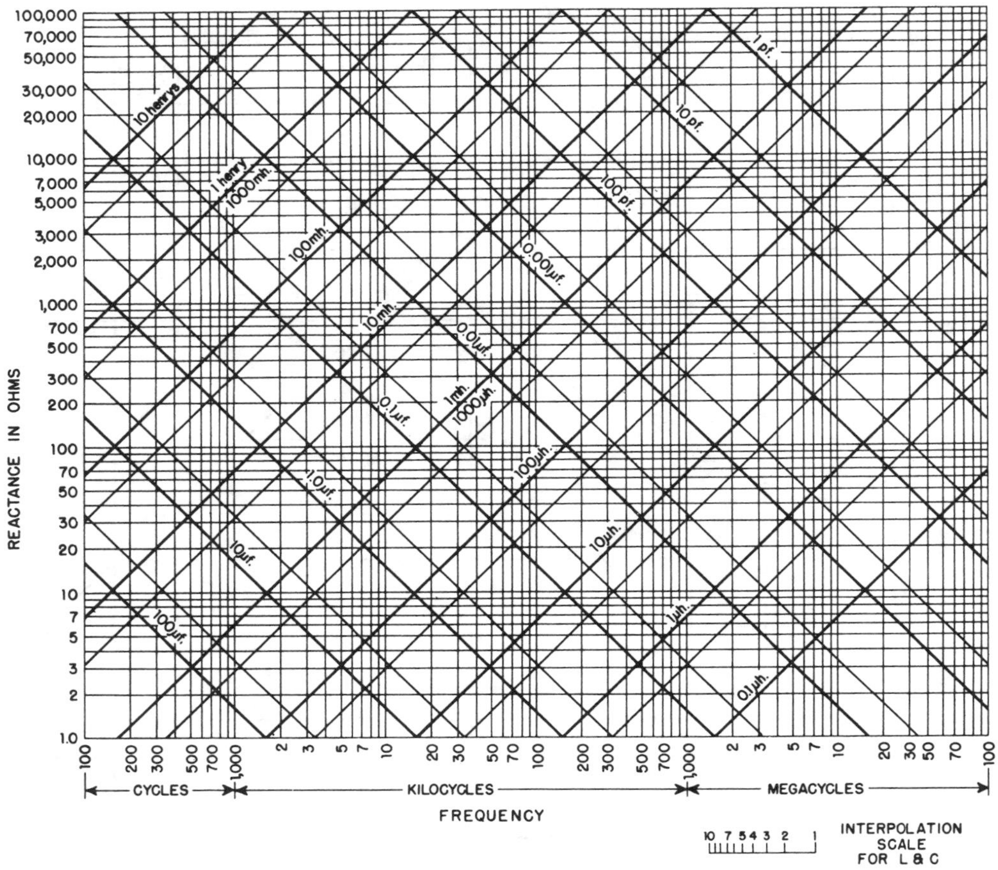

Chart 9.2. Inductive and capacitative reactance vs. frequency.

Electrical Area Classifications*

Area Classification Drawings

General

Plants that manufacture or package flammable products (vapors, gases, liquids, combustible dust, ignitible fibers, etc.) can be susceptible to explosion under abnormal conditions. Therefore special precautions must be taken to decrease the risk of explosions in these facilities. A number of organizations have contributed to the development of these precautions. For example, the American Petroleum Institute (API) has taken on the task of establishing the boundaries of various Hazardous Areas, and the NFPA and the ISA have taken on the task of establishing installation requirements, recommended practices, and standards for equipment within these boundaries.

The API. The American Petroleum Institute (API) has published recommended practices that present a basis for determining the classification of areas in Petroleum Refineries (RP 500A), Drilling Rigs and Production Facilities (RP 500B), and Pipeline Transportation Facilities (RP500C). The bases include the dimensional boundaries of hazardous areas for the many common installation configurations.

API 500A concludes that Class I, Division 1, Areas are likely to have flammable materials present under "normal" conditions and Class II, Division 2, Areas are likely to have flammable materials present only under abnormal conditions. The NFPA has developed installation requirements in accordance with this probability of ignition. The National Electrical Code (NEC) is published by the NFPA.

The NFPA. The National Fire Protection Association (NFPA) has published the following documents that define installation requirements for hazardous areas:

NFPA 30	The "Flammable and Combustible Liquid Code"
NFPA 70	The "National Electrical Code"
NFPA 325M	"Fire Hazard Properties of Flammable Liquids, Gases and Volatile Solids"
NFPA 493	The "Standards for Intrinsically Safe Apparatus and Associated Apparatus in Class I, Class II, and Class III, Division 1 Hazardous Locations." NFPA 493 provides requirements for the construction and testing of electrical apparatus, or parts of such apparatus, that are certified as intrinsically safe. NFPA 493 applies not only to apparatus or parts of apparatus in the hazardous location, but also to any parts located outside of the hazardous location (such as power supplies and recorders), where the intrinsic safety of the circuits inside the hazardous location may be influenced by the design and construction of those parts outside the hazardous location.
NFPA 496	The "Standard for Purged and Pressurized Enclosures for Electrical Equipment"
NFPA 497	The "Recommended Practice for Classification of Class I Hazardous Locations for Electrical Installations in Chemical Plants"

The NEC. Articles 500 through 517 of the National Electrical Code (NEC) establish requirements for electrical equipment and wiring in Classified Areas and provide important cross references governing electrical installations in hazardous areas.

The ISA. The Instrument Society of America (ISA) has published the following documents that define installation and equipment standards for hazardous areas:

RP12.1 "Electrical Instruments in Hazardous Atmospheres"

- Recommends max fuse size to prevent ignition;
- Discusses the classification of areas within suitable enclosures as being different from space external to the enclosure;
- Discusses the classification of instrument housings with connections to hazardous process materials; and,
- Provides various conduit seal criteria

RP12.2 "Intrinsic Safety"

RP12.3 "Explosion Proof"

RP12.4 "Purging"

RP12.5 "Sealing and Immersion"

RP12.6 "Installation of Intrinsically Safe System in Class I Hazardous Locations"

- Paragraph 4.2. Intrinsically safe and non-intrinsically safe wiring may occupy the same enclosure if they are spaced at least 2 inches apart and tied down separately.
- Paragraph 4.2. Identifying intrinsically safe wiring.
- Paragraph 4.3.1. Positive means shall be used to separate intrinsically safe wiring from non-intrinsically safe wiring inside panels.
- Paragraph 4.3.2.1. Identifying intrinsically safe wiring.
- Paragraph 5. Grounding and bonding intrinsically safe circuits.
- Paragraph 6. Intrinsically safe wiring in hazardous locations.
- Paragraph 6.7. Different intrinsically safe systems not allowed in the same multiconductor cable.

S12.4 "Instrument Purging for Reduction of Hazardous Area Classification"

- Defines the requirements for X, Y, and Z purging (X = div1 to nonhaz.; Y = div1 to div2; Z = div2 to nonhaz.).
- Ties an enclosures external case temp to fuse size and enclosure wall thickness.

S12.10 "Area Classification in Hazardous Dust Locations"

- Standard for classifying Class II locations.

*Please see the "Electrical Equipment" and "Wiring" sections in this chapter for other NEC requirements.

S12.11 "Electrical Instruments in Hazardous Dust Locations"
- Standard for electrical instrument installations in Class II locations.

S12.12 "Electrical Equipment for Use in Class I, Division 2, Hazardous Classified Locations"
- Standards for nonincendive circuits in Class I, Division 2, areas

S12.13 Part I "Performance Requirements, Combustible Gas Detectors"

Establishing Boundaries

Areas are classified according to the media that could be ignited (Class and Group designations) and according to the probability of ignition (Division designation). The boundaries of a plant's classified area can be determined by identifying all spaces in or around a process that could have ignitable products present under normal and abnormal conditions. This task has been done for various installation configurations in the API recommended practices described.

Applying these API recommended practices is a matter of knowing enough about the process to identify all places where ignitable products could exist and superimposing the area classification boundaries to these spaces. Often it is necessary to also know the process well enough to be able to identify an opening or potential opening from which ignitable products could leak into the surrounding atmosphere. This type of opening is called a "source" by API. The API documents referenced give boundary dimensions from sources.

NFPA 497 has recognized that some locations in chemical plants may appropriately be established with shorter distances and smaller areas than in refineries. NFPA 497 also distinguishes three pressure ranges and three process equipment sizes that can be applied when determining area classifications.

Introduction to API RP 500A

The following figures (9.1 through 9.9) are reproduced by permission of the American Petroleum Institute from API RP 500A. These figures are considered conservative by API in that they are based on heavier-than-air flammable gases in a hydrocarbon refining plant. Lighter-than-air gases are those gases having less than 75 percent of the density of air at standard conditions.

Refer to API RP 500A when applying these figures. References to "The Text" refers to the text in API RP 500A.

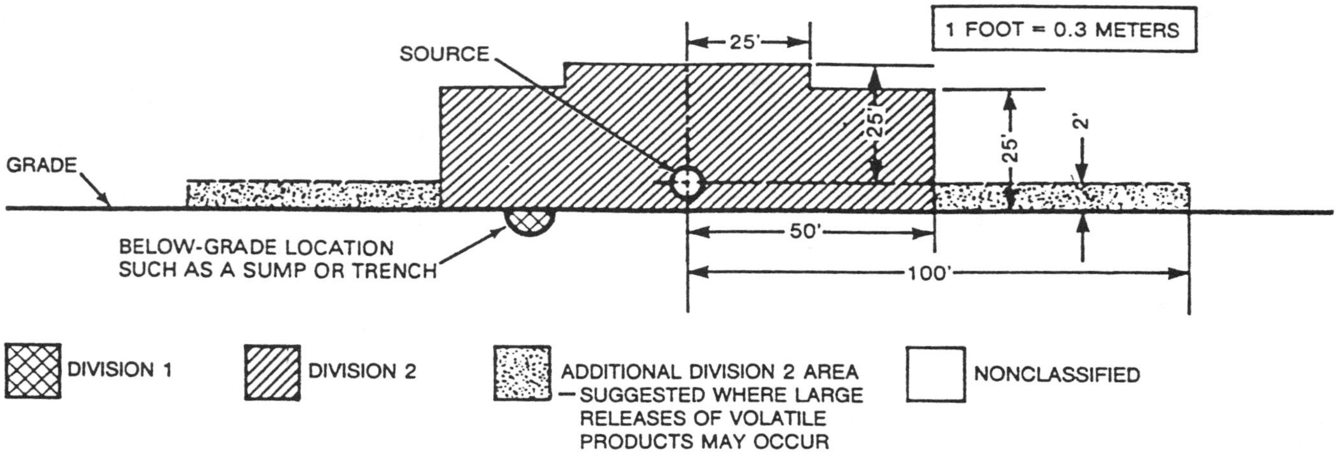

NOTE: Distances given are for typical refinery installations; they must be used with judgment, with consideration given to all factors discussed in the text. In some instances, greater or lesser distances may be justified.

Figure 9.1. Adequately ventilated process area with heavier-than-gas source (see introduction to API RP 500A) located near grade. Source: Figure 1 of API R 500A.

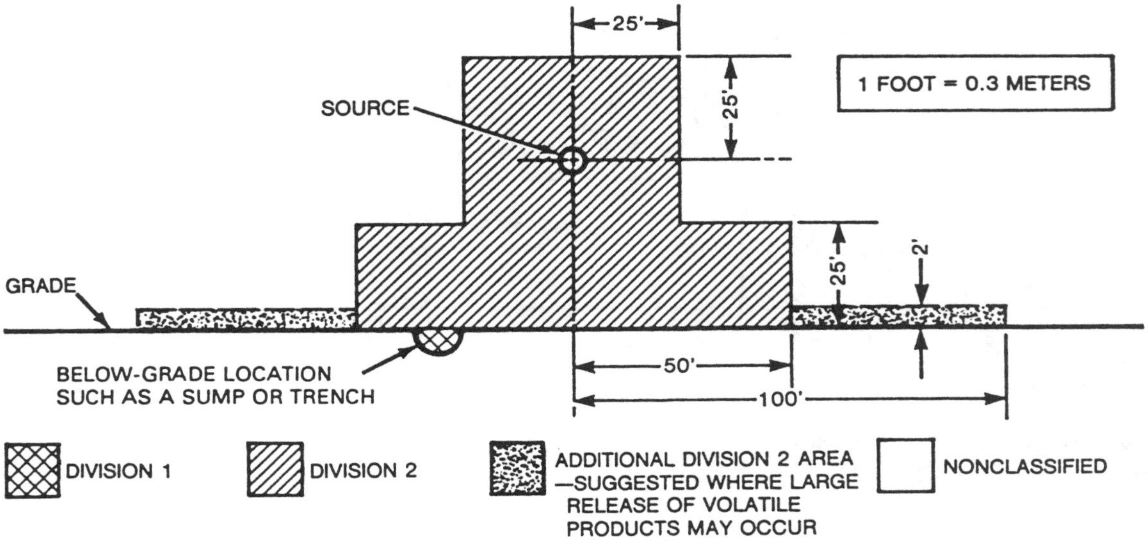

NOTE: Distances given are for typical refinery installations; they must be used with judgment, with consideration given to all factors discussed in the text. In some instances, greater or lesser distances may be justified.

Figure 9.2. Adequately ventilated process area with heavier-than-air gas source (see Introduction to API RP 500A) located above grade. Source: Figure 2 of API RP 500A.

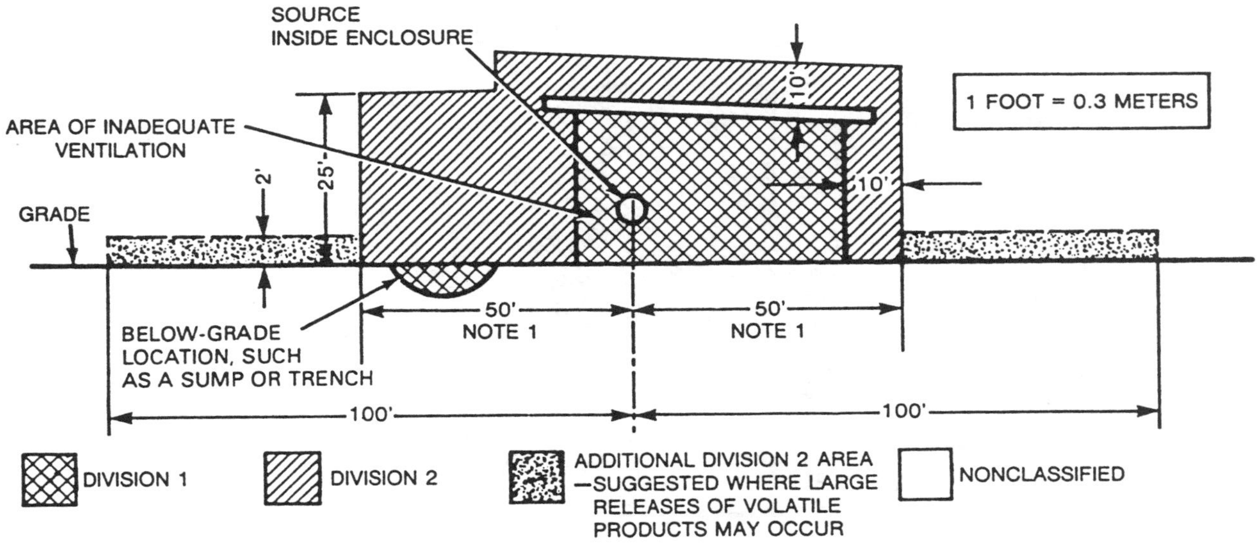

NOTES:

1. Apply horizontal distances of 50 feet from the source of vapor or 10 feet beyond the perimeter of the building, whichever is greater, except that beyond unpierced vaportight walls the area is nonclassified.
2. Distances given are for typical refinery installations; they must be used with judgment, with consideration given to all factors discussed in the text.

Figure 9.3. Inadequately ventilated process area with heavier-than-air source (see Introduction to API RP 500A). Source: Figure 3 of API RP 500A.

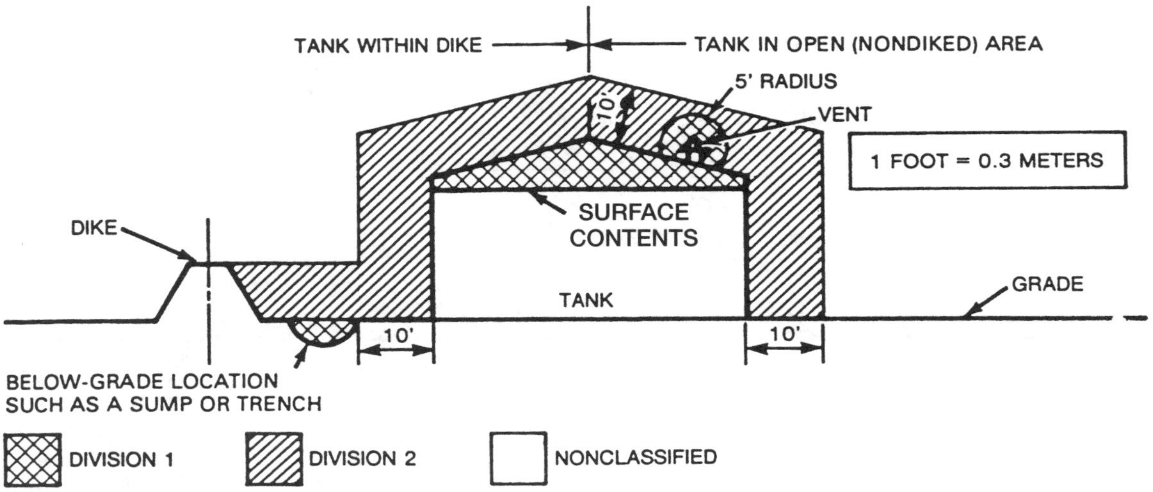

NOTES:
1. For floating-roof tanks, the area above the tank roof and within the shell is classified Division 1.
2. High filling rates or blending operations involving Class I liquids may require extending the boundaries of classified areas.
3. Distances given are for typical refinery installations; they must be used with judgment, with consideration given to all factors discussed in the text.

Figure 9.4. Refinery tank with heavier-than-air source (see Introduction to API RP 500A). Source: Figure 4 of API RP 500A.

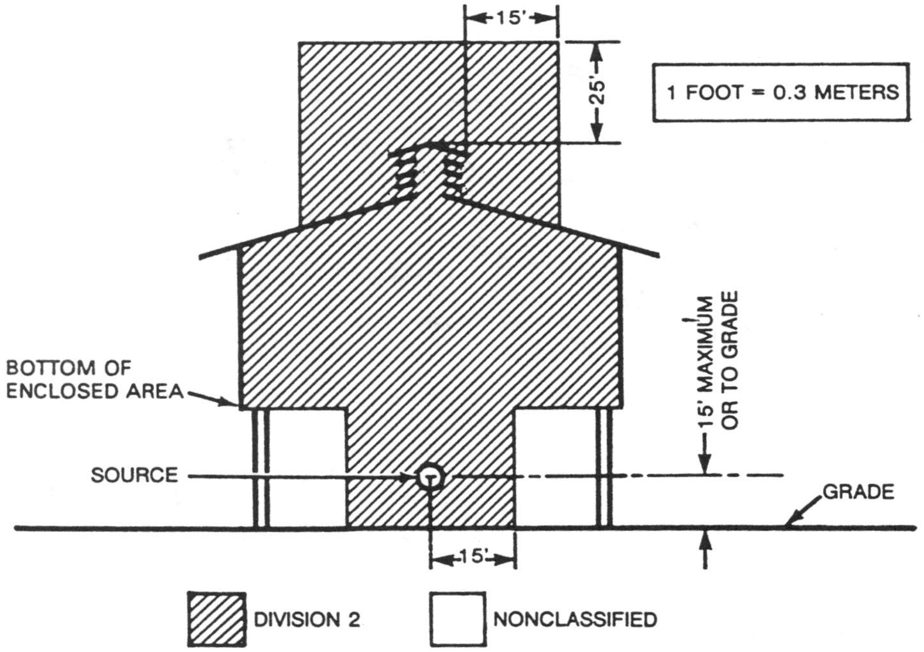

NOTE: Distances given are for typical refinery installations; they must be used with judgment, with consideration given to all factors discussed in the text.

Figure 9.5. Adequately ventilated compressor shelter with lighter-than-air gas source (see Introduction to API RP 500A). Source: Figure 5 of API RP 500A.

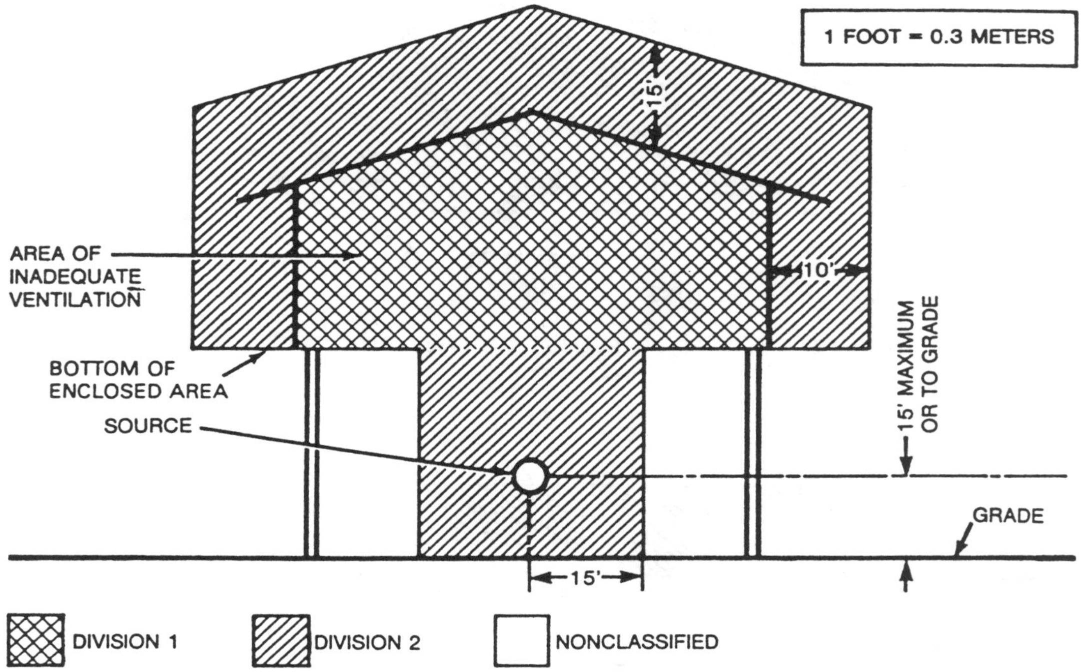

NOTE: Distances given are for typical refinery installations; they must be used with judgment, with consideration given to all factors discussed in the text.

Figure 9.6. Inadequately ventilated compressor shelter with lighter-than-air gas source (see Introduction to API RP 500A). Source: Figure 5 of API RP 500A.

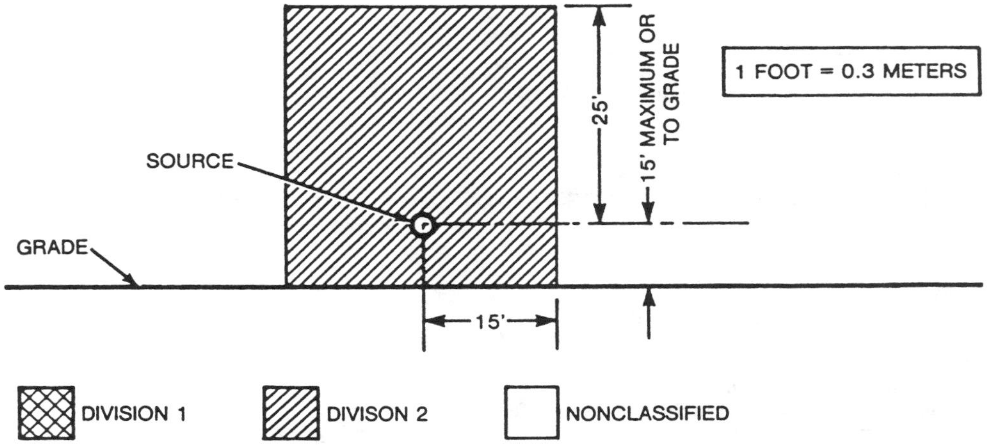

NOTE: Distances given are for typical refinery installations; they must be used with judgment, with consideration given to all factors discussed in the text.

Figure 9.7. Adequately ventilated process area with lighter-than-air gas source (see Introduction to API RP 500A). Source: Figure 5 of API RP 500A.

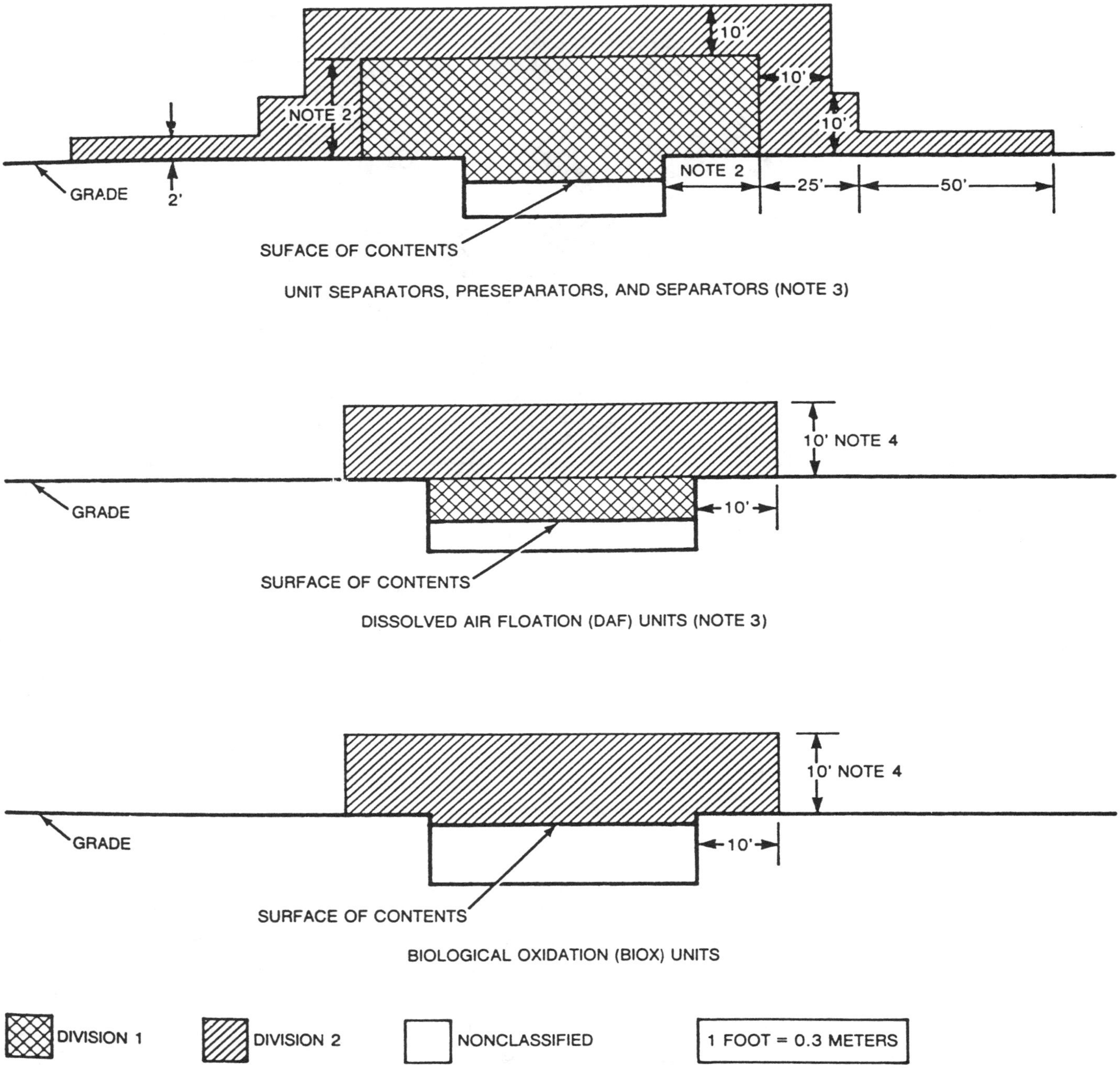

Notes:
1. The extent of the classified areas shown shall be modified as required by the proximity of other potential sources of release or of nearby obstructions, such as dikes or hills, that would impede dispersal of vapors. Distances given are for typical refinery installations; they must be used with judgment, with consideration given to all factors discussed in the text.
2. For unit separators and preseparators, 25 feet; for separators, 10 feet.
3. Applies to open top tanks or basins. Classify closed tank units per Figure 4.
4. Distance above top of basin or tank. Extend to grade for basins or tanks located above ground.

Figure 9.8. Separators, dissolved air flotation (DAF) units, and biological oxidation (BIOX) units. Source: Figure 8 of API RP 500A.

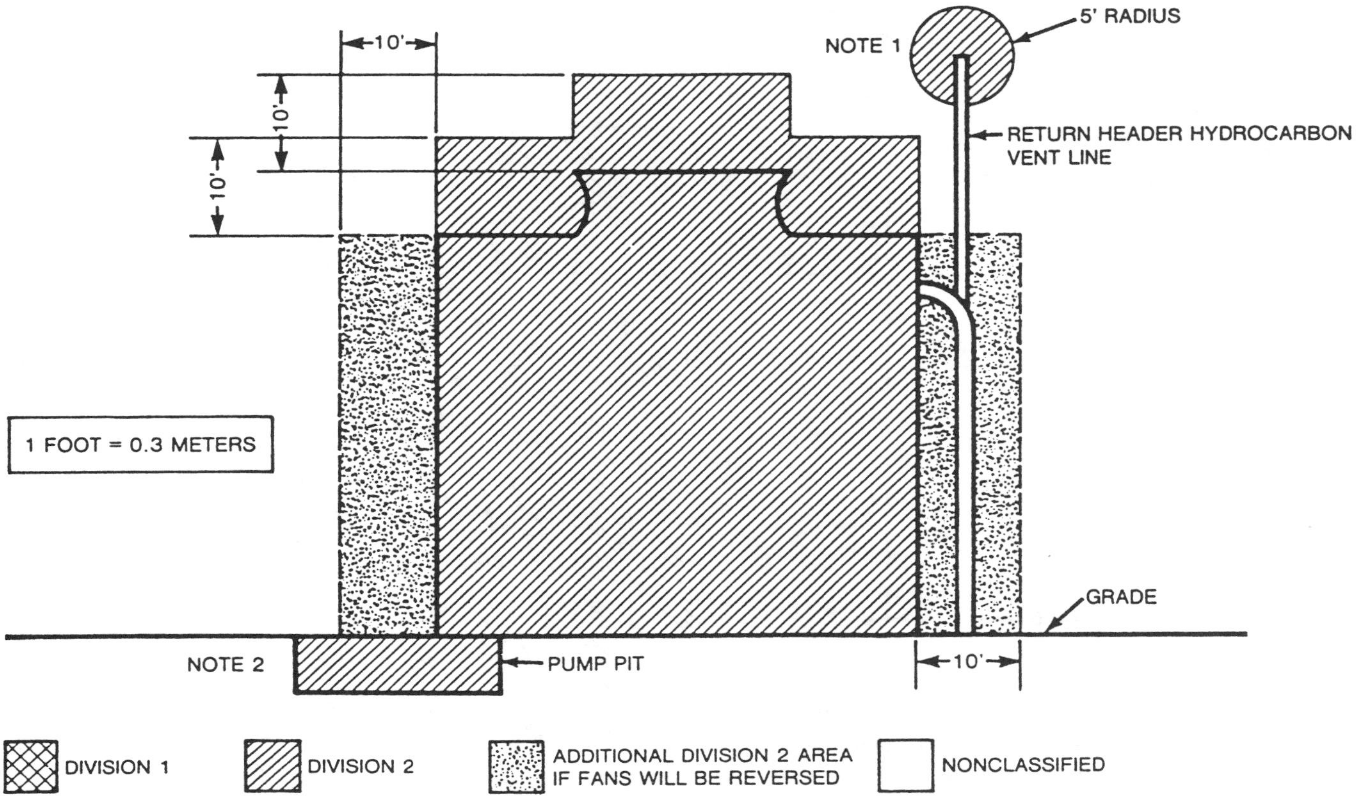

NOTES:

1. It is recommended that electrical equipment be located away from the vent area.
2. Reversal of the fans and excessive carryover from exchanger leaks are the reason for classifying the pump pit Division 2.

Figure 9.9. Mechanical draft cooling tower handling process cooling water. Source: Figure 9 of API RP 500A.

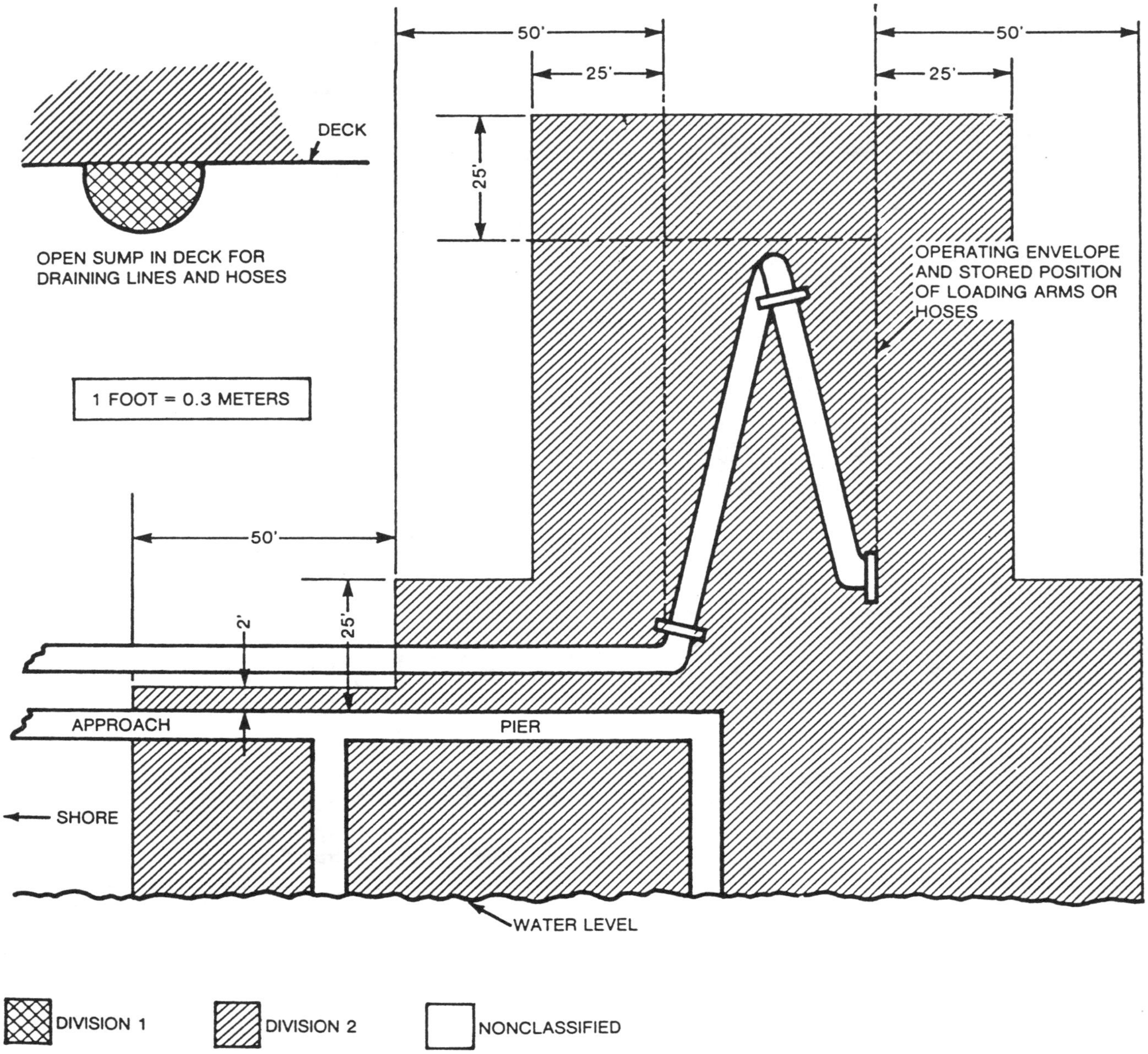

NOTES:
1. The "source of vapor" shall be the operating envelope and stored position of the outboard flange connection of the loading arm (or hose).
2. The berth area adjacent to tanker and barge cargo tanks is to be Division 2 to the following extent:
   a. 25 feet horizontally in all directions on the pier side from that portion of the hull containing cargo tanks.
   b. From the water level to 25 feet above the cargo tanks at their highest position.
3. Additional locations may have to be classified as required by the presence of other sources of flammable liquids on the berth, or by Coast Guard or other regulations.

Figure 9.10. Refinery marine terminal handling flammable liquids. Source: Figure 10 of API RP 500A.

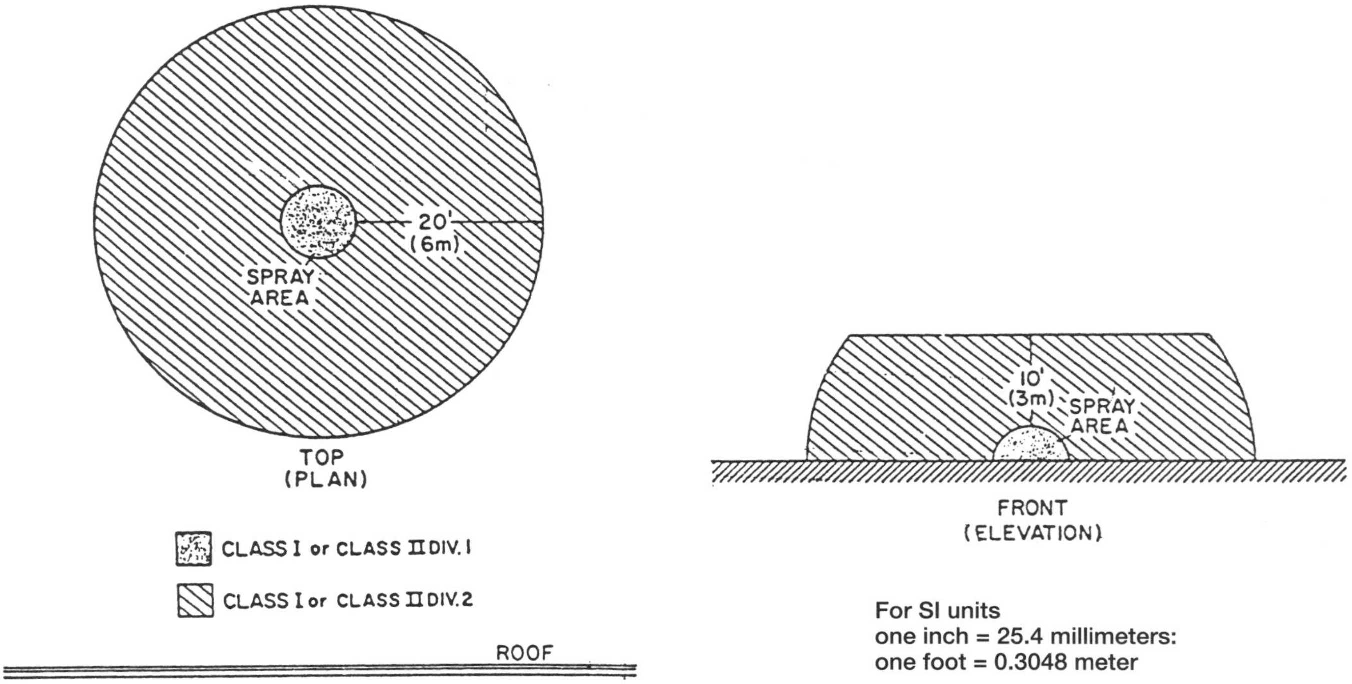

Figure 9.11. Class I or Class II, Division 2, locations adjacent to an unenclosed spray operation. Source: Figure 1, Article 516, of the 1990 NEC. Reprinted with permission from NFPA 70-1990, the National Electrical Code®, *Copyright© 1989, National Fire Protection Association, Quincy, MA 02269. This reprinted material is not the complete and official position of the National Fire Protection Association, on the referenced subject which is represented only by the standard in its entirety.*

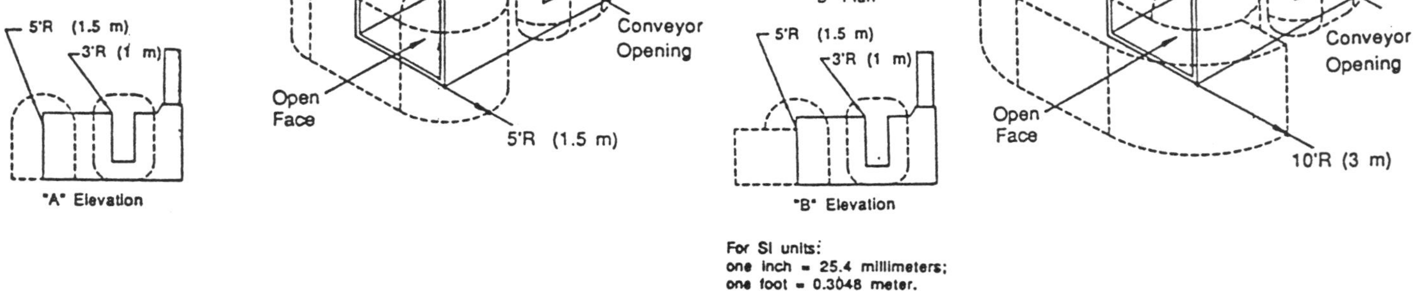

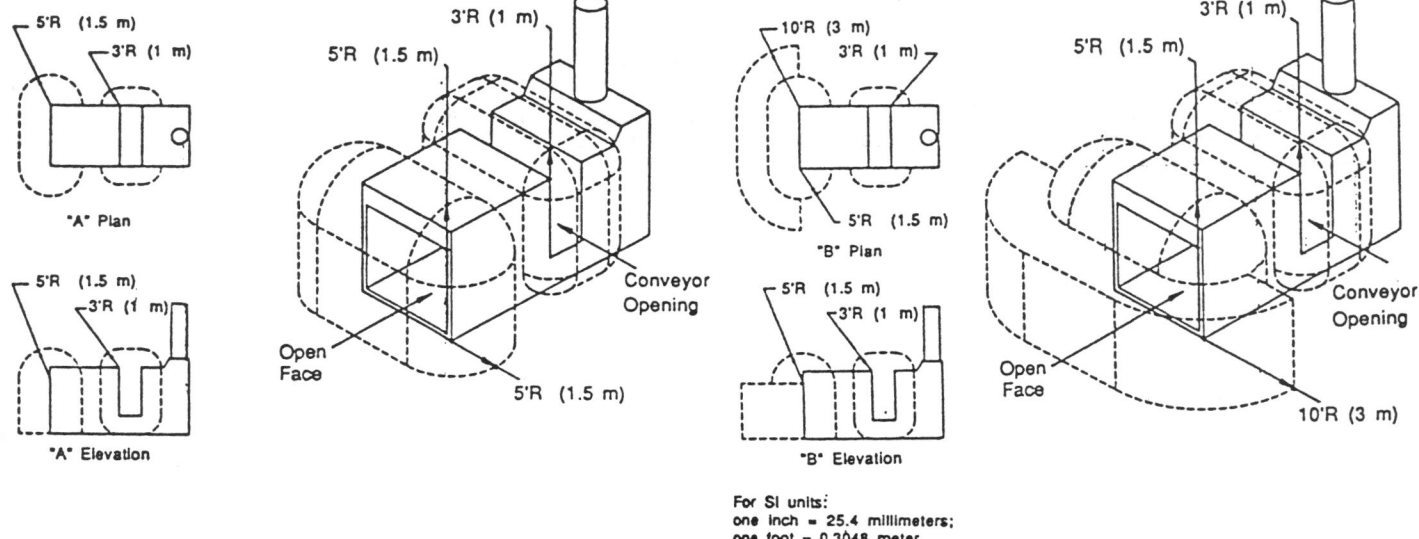

Figure 9.12. Class I or Class II, Division 2, locations adjacent to a closed top, open-face, or open-front spray booth or room. Source: Figure 2, Article 516, of the 1990 NEC. Reprinted with permission from NFPA 70-1990, the National Electrical Code®, *Copyright© 1989, National Fire Protection Association, Quincy, MA 02269. This reprinted material is not the complete and official position of the National Fire Protection Association, on the referenced subject which is represented only by the standard in its entirety.*

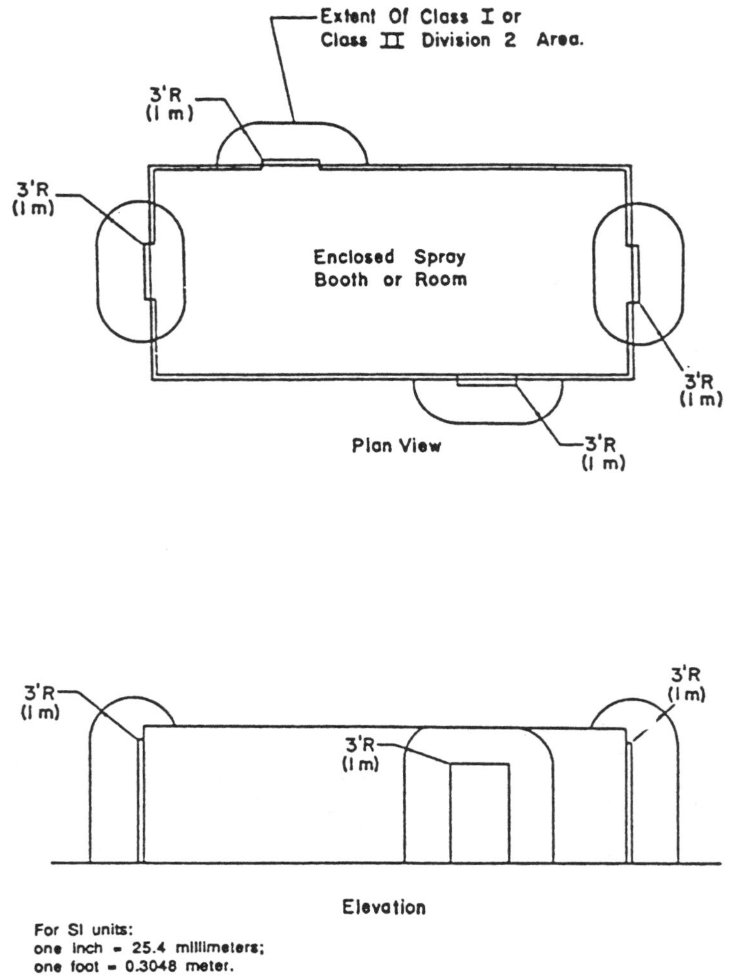

Figure 9.13. Class I or Class II, Division 2, locations adjacent to openings in an enclosed spray booth or room. Source: Figure 3, Article 516, of the 1990 NEC. Reprinted with permission from NFPA 70-1990, the National Electrical Code®, *Copyright© 1989, National Fire Protection Association, Quincy, MA 02269. This reprinted material is not the complete and official position of the National Fire Protection Association, on the referenced subject which is represented only by the standard in its entirety. Refer to the 1991 NEC when applying this information.*

Table 9.7

Class I Locations — Bulk Plants

Location	Class I Division	Extent of Classified Location
Indoor equipment installed in accordance with NFPA 30-1987, *Flammable and Combustible Liquids Code,* Section 5-3.3.2 where flammable vapor-air mixtures may exist under normal operation	1	Space within 5 feet of any edge of such equipment, extending in all directions.
	2	Space between 5 feet and 8 feet of any edge of such equipment, extending in all directions. Also, space up to 3 feet above floor or grade level within 5 feet to 25 feet horizontally from any edge of such equipment.*
Outdoor equipment ~~of~~ the type covered in NFPA 30-1987, *Flammable and Combustible Liquids Code,* Section 5-3.3.2 where flammable vapor-air mixtures may exist under normal operation	1	Space within 3 feet of any edge of such equipment, extending in all directions.
	2	Space between 3 feet and 8 feet of any edge of such equipment, extending in all directions. Also space up to 3 feet above floor or grade level within 3 feet to 10 feet horizontally from any edge of such equipment.
Tank — Aboveground**	1	Space inside dike where dike height is greater than the distance from the tank to the dike for more than 50 percent of the tank circumference.
Shell, Ends, or Roof and Dike Space	2	Within 10 feet from shell, ends, or roof of tank. Space inside dikes to level of top of dike.
Vent	1	Within 5 feet of open end of vent, extending in all directions.
	2	Space between 5 feet and 10 feet from open end of vent, extending in all directions.
Floating Roof	1	Space above the roof and within the shell.
Underground Tank Fill Opening	1	Any pit, box, or space below grade level, if any part is within a Division 1 or 2 classified location.
	2	Up to 18 inches above grade level within a horizontal radius of 10 feet from a loose fill connection, and within a horizontal radius of 5 feet from a tight fill connection.
Vent — Discharging Upward	1	Within 3 feet of open end of vent, extending in all directions.
	2	Space between 3 feet and 5 feet of open end of vent, extending in all directions.

*The release of Class I liquids may generate vapors to the extent that the entire building, and possibly a zone surrounding it, should be considered a Class I, Division 2 location.

**For Tanks—Underground, see Section 514-2 of the 1990 NEC.

Source: Table 515-2 of the 1990 NEC. Reprinted with permission from NFPA 70-1990, the National Electrical Code®*, Copyright© 1989, National Fire Protection Association, Quincy, MA 02269. This reprinted material is not the complete and official position of the National Fire Protection Association, on the referenced subject which is represented only by the standard in its entirety. Refer to the 1991 NEC when applying this information.*

Table 9.8. Class I Locations—Bulk Plants

Location	Class I Division	Extent of Classified Location
Drum and Container Filling		
Outdoors, or Indoors with Adequate Ventilation	1	Within 3 feet of vent and fill openings, extending in all directions.
	2	Space between 3 feet and 5 feet from vent or fill opening, extending in all directions. Also, up to 18 inches above floor or grade level within a horizontal radius of 10 feet from vent or fill openings.
Pumps, Bleeders, Withdrawal Fittings, Meters and Similar Devices		
Indoors	2	Within 5 feet of any edge of such devices, extending in all directions. Also up to 3 feet above floor or grade level within 25 feet horizontally from any edge of such devices.
Outdoors	2	Within 3 feet, of any edge of such devices, extending in all directions. Also up to 18 inches above grade level within 10 feet horizontally from any edge of such devices.
Pits		
Without Mechanical Ventilation	1	Entire space within pit if any part is within a Division 1 or 2 classified location.
With Adequate Mechanical Ventilation	2	Entire space within pit if any part is within a Division 1 or 2 classified location.
Containing Valves, Fittings or Piping, and not within a Division 1 or 2 Classified Location	2	Entire pit.
Drainage Ditches, Separators, Impounding Basins		
Outdoor	2	Space up to 18 inches above ditch, separator, or basin. Also up to 18 inches above grade within 15 feet horizontally from any edge.
Indoor		Same as pits.
Tank Vehicle and Tank Car*		
Loading Through Open Dome	1	Within 3 feet of edge of dome, extending in all directions.
	2	Space between 3 feet and 15 feet from edge of dome, extending in all directions.
Loading Through Bottom Connections With Atmospheric Venting	1	Within 3 feet of point of venting to atmosphere, extending in all directions.
	2	Space between 3 feet and 15 feet from point of venting to atmosphere, extending in all directions. Also up to 18 inches above grade within a horizontal radius of 10 feet from point of loading connection.
Office and Rest Rooms	Ordinary	If there is any opening to these rooms within the extent of an indoor classified location, the room shall be classified the same as if the wall, curb, or partition did not exist.

*When classifying extent of space, consideration shall be given to fact that tank cars or tank vehicles may be spotted at varying points. Therefore, the extremities of the loading or unloading positions shall be used.

(continued)

Table 9.8 continued

Class I Locations—Bulk Plants (cont.)

Location	Class I Division	Extent of Classified Location
Loading Through Closed Dome With Atmospheric Venting	1	Within 3 feet of open end of vent, extending in all directions.
	2	Space between 3 feet and 15 feet from open end of vent, extending in all directions. Also within 3 feet of edge of dome, extending in all directions.
Loading Through Closed Dome with Vapor Control	2	Within 3 feet of point of connection of both fill and vapor lines, extending in all directions.
Bottom Loading With Vapor Control Any Bottom Unloading	2	Within 3 feet of point of connections, extending in all directions. Also up to 18 inches above grade within a horizontal radius of 10 feet from point of connections.
Storage and Repair Garage for Tank Vehicles	1	All pits or spaces below floor level.
	2	Space up to 18 inches above floor or grade level for entire storage or repair garage.
Garages for Other Than Tank Vehicles	Ordinary	If there is any opening to these rooms within the extent of an outdoor classified location, the entire room shall be classified the same as the space classification at the point of the opening.
Outdoor Drum Storage	Ordinary	
Indoor Warehousing Where There Is No Flammable Liquid Transfer	Ordinary	If there is any opening to these rooms within the extent of an indoor classified location, the room shall be classified the same as if the wall, curb, or partition did not exist.
Piers and Wharves		See Figures 5-3.5.6.

For SI units: one inch = 25.4 millimeters; one foot = 0.3048 meter.

Source: Table 515-2 of the 1990 NEC. Reprinted with permission from NFPA 70-1990, the National Electrical Code®*, Copyright© 1989, National Fire Protection Association, Quincy, MA 02269. This reprinted material is not the complete and official position of the National Fire Protection Association, on the referenced subject which is represented only by the standard in its entirety. Refer to the 1991 NEC when applying this information.*

Establishing the Classification of a Hazardous Atmosphere (Based on the National Electrical Code.)

The NEC provides requirements for electrical equipment and wiring for locations where fire or explosion hazards may exist. Hazardous locations are put into three classes, as follows

Class I

Class I locations have atmospheres that contain explosive gases/vapors. These gases/vapors are divided into four groups according to flammable properties. Table 9.9 lists some Class I gases/vapors according to these groups, and shows the associated ignition temperature.

Class II

Class II locations have atmospheres that contain combustible dust that may or may not be suspended in air. Class II atmospheres are divided into three groups according to the electrical resistance of the dust. Table 9.9 describes the dust of a Class II atmosphere for each of these groups.

Class III

Class III locations have atmospheres that contain easily ignitable fibers and flyings that are not likely suspended in air in quantities sufficient to produce ignitable mixtures. Table 9.9 describes the flyings/fibers of a Class III atmosphere.

Table 9.9. Classification of Hazardous Locations (Based on the NEC®)

Class	Group	Typical Atmosphere (Ignition Temperature)
I GASES VAPORS	A	acetylene (305C, 581F)
	B	butadiene (420C, 788F)
		ethylene oxide (429C, 804F)
		hydrogen (400C, 752F)
		manufactured gases containing more than 30% hydrogen (by volume)
		propylene oxide (449C, 840F)
	C	acetaldehyde (175C, 347F)
		cyclopropane (503C, 938F)
		diethyl ether (160C, 320F)
		ethylene (450C, 842F)
		unsymmetrical dimethyl hydrazine (UDMH 1, 1-dimethyl hydrazine) (249C, 480F)
	D	acetone (465C, 869F)
		acrylonitrile (481C, 898F)
		ammonia (498C, 928F)
		benzene (560C, 1040F)
		butane (288C, 550F)
		1-butanol (butyl alcohol) (343C, 650F)
		2-butanol (secondary butyl alcohol) (450C, 761F)
		n-butyl acetate (421C, 790F)
		ethane (472C, 882F)
		ethanol (ether alcohol) (363C, 685F)
		ethyl acetate (427C, 800F)
		ethylene dichloride (413C, 775F)
		gasoline (56-60 octane: 280C, 536F) (100 octane: 471C, 880F)
		heptanes (204C, 399F)
		hexanes (225C, 437F)
		isoprene (220C, 428F)
		methane (natural gas) (537C, 999F)
		methanol (methyl alcohol) (385C, 725F)
		3-methyl-1-butanol (isoamyl alcohol) (350C, 662F)
		methyl ethyl ketone (404C, 759F)
		methyl isobutyl ketone (440C, 840F)
		2-methyl-1 propanol (isobutyl alcohol) (416C, 780F)
		2-methyl-2-propanol (tertiary butyl alcohol) (478C, 892F)
		petroleum naphtha (288C, 550F)
		octanes (206C, 403F)
		pentanes (243C, 470F)
		1-pentanol (amyl alcohol) (300C, 572F)

		propane (450C, 842F) 1-propanol (propyl alcohol) (413C, 775F) 2-propanol (isoprophyl alcohol) (399C, 750F) propylene (455C, 851F) styrene (490C, 914F) toluene (480C, 896F) vinyl acetate (402C, 756F) vinyl chloride (472C, 882F) xylenes (464 to 529C, 867 to 984F)
II DUST	E	Atmospheres containing combustible metal dusts regardless of resistivity, or other combustible dusts of similarly hazardous characteristics having resistivity of less than 10^5 ohm-centimeter.
	F	Atmospheres containing carbon black, charcoal, coal, or coke dusts that have more than 8 percent total volatile material (carbon black per ASTM D 1620; charcoal, coal, and coke dusts per ASTM D 271) or atmospheres containing these dusts sensitized by other materials so that they present an explosion hazard, and having resistivity greater than 10^2 ohm-centimeter but equal to or less than 10^8 ohm-centimeter.
	G	Atmospheres containing combustible dusts having resistivity of 10^5 ohm-centimeter or greater.
III FLYING FIBERS		Such locations usually include some parts of rayon, cotton, and other textile mills; combustible fiber manufacturing and processing plants; cotton gins and cotton-seed mills; flax-processing plants; clothing manufacturing plants; woodworking plants; and establishments and industries involving similar hazardous processes or conditions.
	NOT APPLICABLE	Easily ignitible fibers and flyings include rayon, cotton (including cotton linters and cotton waste), sisal or henequen, istle, jute, hemp, tow, cocoa fiber, oakum, baled waste kapok, Spanish moss, excelsior, and other materials of similar nature.

CAUTION: Table 9.9 is not a complete listing of requirements. Consult NFPA-497M for a more complete listing of requirements.

Electrical Properties of Materials

SOLIDS*

MATERIAL	DIELECTRIC CONSTANT	MATERIAL	DIELECTRIC CONSTANT
Acetamide	41	Paraffin	2.2
Acetanilide	2.8	Phenanthrene	2.8
Acetic Acid	4.1	Phenol	4.3
Aluminum Phosphate	6.0	Phosphorus, Red	4.1
Ammonium Bromide	7.2	Phosphorus, Yellow	3.6
Ammonium Chloride	7.0	Plastic Pellets	1.1-3.2
Antimony Trichloride	5.3	Polyethylene	4-5.0
Asbestos	4.8	Polypropylene	1.5
Asphalt	2.7	Porcelain	5-7.0
Bakelite	5.0	Potassium Aluminum Sulphate	3.8
Barium Chloride	11.0	Potassium Carbonate	5.6
Barium Chloride	9.4	Potassium Chlorate	5.1
Barium Nitrate	5.8	Potassium Chloride	5.0
Barium Sulfate	11.4	Potassium Chloronate	7.3
Calcium Carbonate	9.1	Potassium Iodide	5.6
Calcium Fluoride	7.4	Potassium Nitrate	5.0
Calcium Sulfate	5.6	Potassium Sulfate	5.9
Cellulose	4.0	Quartz	4.4
Cellulose Acetate	3.6-7.5	Resorcinol	3.2
Cement	1.5-2.1	Rice	3.5
Cereals	3-5.0	Rubber	3.0
Charcoal	1.2-1.81	Sand	3-5.0
Coke	1.1-2.2	Selinium	11.0
Cupric Oleate	2.8	Shellac	3.5
Cupric Oxide	18.1	Silver Bromide	12.2
Cupric Sulfate	10.3	Silver Chloride	11.2
Diamond	10.0	Silver Cyanide	5.6
Diphenylethane	2.7	Slate	7.0
Dolomite	8.0	Soap Powder	1.25-1.5
Ferrous Oxide	14.2	Sodium Carbonate	8.4
Fly Ash	1.9-2.6	Sodium Carbonate	5.3
Iodine	4.0	Sodium Chloride	6.1
Glass	3.7-4.2	Sodium Nitrate	5.2
Lead Acetate	2.5	Sodium Oleate	2.8
Lead Carbonate	18.1	Sodium Perchlorate	5.4
Lead Chloride	4.2	Sulphur	3.4
Lead Nomoxide	25.9	Sugar	3.0
Lead Nitrate	37.7	Sucrose	3.3
Lead Oleate	3.3	Tantalum Oxide	11.6
Lead Oxide	25.9	Thallium Chloride	46.9
Lead Sulfate	14.3	Thorium Oxide	10.6
Magnesium Oxide	9.7	P-Toluidine	3.0
Malachite	7.2	Urea	3.5
Mercuric Chloride	3.2	Zinc Sulfide	8.2
Mica	7.0	Zirconium Oxide	12.5
Napthalene	2.5	Teflon	2.0
Nylon Pellets	1.1-3.0		
Paper	2.0		

LIQUIDS*

MATERIAL	DIELECTRIC CONSTANT	MATERIAL	DIELECTRIC CONSTANT
Acetal	3.6	Acetophenone	17.3
Acetaldehyde	22.2	Acetoxime	3
Acetal Doxime	3.4	Acetyl Acetone	23.1
Acetomide	4	Acetyl Bromide	16.5
Acetanilide	2.9	Acetyl Chloride	15.8
Acetic Acid	6.1	Acetylmethyl Hexyl Ketone	27.9
Acetic Anhydride	22		
Acetone	17.7-20.7	Allyl Alcohol	21
Acetonitrile	37.5	Allyl Bromide	7

Liquids Cont.

MATERIAL	DIELECTRIC CONSTANT	MATERIAL	DIELECTRIC CONSTANT
Allyl Chloride	8.2	Bromoheptane	5.3
Allyl Iodide	6.1	Bromohexane	5.8
Allyl Isothiocyanate	17.5	Bromoform	4.4
Aluminum Bromide	3.4	Bromoisovaleric Acid	6.5
Aluminum Oleate	2.4		
Ammonia	15.5-25.0	Bromonaphthalene	5.1
Amyl Acetate	5	Bromooctadecane	3.5
Amyl Alcohol	11.2-35.5	Bromopropionic Acid	11.0
Amylamine	4.6	Bromotoluene	5.1
Amyl Benzoate	5.1	Bromotridecane	4.2
Amyl Bromide	6.3	Bromopentadecone	3.8
Amyl Chloride	6.6	Bromoundecane	4.7
Amylene	2.0	Butane	1.4
Amylene Bromide	5.6	N Butylacetate	5.1
Amyl Ether	3.1	Iso Butylacetate	5.6
Amyl Formate	5.7	Iso-Butylamine	4.5
Amyl Iodide	6.9	N-Butyl Alcohol	7.8
Amylmercaptan	4.7	Iso Butyl Alcohol	18.7-31.7
Amyl Nitrate	9.1	Butylamine	5.4
Amyl Thiocyanate	17.4	N Butyl Bromide	6.6
Aniline	5.5- 7.8	Butyl Chloral	10.0
Anisaldehyde	15.8	Butyl Chloride	9.6
Anisaldoxine	9.2	N-Butyl Formate	2.4
Anisole	4.3	N-Butyl Iodide	6.1
Antimony Pentachloride	3.2	Iso-Butyl Iodide	5.8
		Iso-Butyl Nitrate	11.9
Antimony Tribromide	20.9	Butyric Anhydride	12.0
Antimony Trichloride	33	Butyraldehyde	13.4
Antimony Triiodide	13.9	Butyric Acid	2.8
Arsenic Tribromide	9	N-Butyricacid	2.9
Arsenic Trichloride	12.4	Iso Butyric Acid	2.7
Arsenic Triiodide	7.0	Butyric Anhydride	12.9
Arsine	2.7	Butyronitrile	20.7
Asphalt	2.65	Iso Butyronitrile	20.8
Azoxyanisole	2.3	Cable Oil	2.2
Azoxybenzene	5.1	Camphanedione	16.0
Benzal Chloride	6.9	Camphene	2.7
Benzaldehyde	17.8	Camphorpinacone	3.6
Benzaldoxime	3.8	Caproic Acid	2.6
Benzene	2.3	Caprylic Acid	3.2
Benzil	13.0	Carbon Dioxide	1.6
Benzonitrile	26.0	Carbon Disulphide	2.6
Benzophenone	11.4	Carbon Tetrachloride	2.2
Benzotrichloride	7.4	Carvenone	18.4
Benzophenone	11.4-13	Carvol	11.2
Benzoylacetone	3.8	Carvone	11.0
Benzoyl Chloride	22.1	Caster Oil	2.6 4.8
Benzyl Acetate	5.0	Camphene	2.3
Benzyl Alcohol	13.0	Camphoric imide	5.5
Benzylamine	4.6	Cetyl Iodide	3.3
Benzyl Benzoate	4.8	Chloral	5.5
Benzyl Chloride	6.4	Chloracetic Acid	12.3
Benzyl Cyanide	18.3	Chlorine	2.0
Benzylethylamine	4.3	Chloroacetic Acid	21.0
Benzylmethylamine	4.4	Chloroacetone	29.8
Benzyl Salicylate	4.1	Chlorobenzene	4.7-5.9
Bornyl Acetate	4.6	Chlorocyclohexane	7.6
Boron Bromide	2.6	Chlorophetane	5.4
Boronyl Chloride	5.21	Chloroform	5.5
Bromal	7.6	Chlorohexanone oxime	3.0
Bromoacetyl Bromide	12.6		
Bromohexadeoane	3.71	Chlorohydrate	3.3
Bromine	3.1	Chloronaphthalene	5.0
Bromoaniline	13.0	0-Chlorophenol	8.2
Bromoanisole	7.0	3-Chloro-1, Dihydroxypropone	31.0
Bromobenzene	5.4		
Bromobutylene	5.8	Chlorooctane	5.0
Bromobutyric Acid	7.2	Chlorotoluene	4.7
Bromodecane	4.4	Cholestral	2.9
Bromodocosane	3.1	Chorine	1.7
Bromododecane	4.0	Chromyl Choride	2.6
Bromo 2 ethoxypentane	6.4	Cis-3-Hexene	2.0
		Cinnamaldehyde	16.9

Table 9.10 continued

Table 9.10. Dielectric constants—solids, liquids, and powders. Courtesy Endress & Hauser, Inc.

Table 9.10 continued

Liquids Cont.

MATERIAL	DIELECTRIC CONSTANT	MATERIAL	DIELECTRIC CONSTANT
Citraconic Anhydride	40.3	Diimylamine	2.5
Cocaine	3.1	Diisoamylene	2.4
Copper Oleate	2.8	Diiodoethylene	4.0
Creosol	10.6	Diiodomethane	5.3
0-Cresol	5.8	Diisoamyl	2.0
M-Cresol	5.0	Diisobutylamine	2.7
P-Cresol	5.6	Dimethoxybenzene	4.5
Cresol	5.0	Dimethylaniline	4.4
Crotonic Nitrice	28.0	Dimethyl Ethyl	11.7
Cumaldehyde	11.0	Dimethylheptane	1.9
Cumene	2.4	Dimethyl-2-hexane	2.4
Cumicaldehyde	10.7	Dimethyl Malonate	10.4
Cyanoacetic Acid	33	Dimethyl Oxalate	3.0
Cyanoethyl Acetate	19.3	Dimethylpentane	1.9
Cyanogen	2.6	Dimethylquinoxaline	2.28
Cyclohexane	2.0	Dimethyl Sulfide	6.3
Cyclohexanone oxime	3.0	Dimethyl Sulfate	55.0
Cyclohexanemethanol	9.7	Dimethyltoluidine	3.3
Cyclohexanone	18.2	M-Dinitro Benzene	2.8
Cyclohexylamine	5.3	Dinitrogen Oxide	1.6
Cyclohexylphenol	3.9	Dinitrogen Tetroxide	2.5
Cyclohexanol	15.0	Dioctyl phthalate	5.1
P-Cymene	2.3	Dipalmitin	3.5
Cymene	2.3	Dipentene	2.3
Deuterium	1.3	Diphenyl	2.5
Deuterium Oxide	78.3	Diphenylamine	3.3
Diacetoxybutone	6.64	Diphenylethane	2.3
Diallyl Sulfide	4.9	Diphenyl Ether	3.9
Dibromobenzene	8.8	Diphenytmethane	2.6
P-Dibromobenzene	4.5	Dipropylamine	2.9
Dibromobutane	5.7	Dipropyl Ketane	12.6
Dibromoheptane	5.0	Distearin	3.3
Dibromohexane	5.0	Docosane	2.0
Dibromopropane	4.3	Dodecanol	6.5
Dibromopropyl Alcohol	9.1	Dodecyne	2.1
Dibenzylamine	3.6	Dowtherm	3.3
Dibutyl phthalate	6.4	Epichlorohydrin	22.9
Dibutyl tartrate	9.4	Ethanediamine	14.2
Dichloracetic Acid	10.7	Ethanethiol	6.9
Dibutyl sebacate	4.5	Ethanethiolic Acid	13.0
Dichloracetone	14.0	Ethanol	24.3
0-Dichlorobenzene	7.5	Ethoxybenzene	4.2
Dichlorobenzene	2.8	Ethoxyethyl Acetate	7.5
1, 2-Dichloroethane	10.7	Ethoxypentane	3.6
Dichloroethane	16.7	Ethoxy-3-methylbutane	3.9
Dichlorostyrene	2.58	Ethoxytoluene	3.9
Dichlorotoluene	6.9	Ethoxynaphthalone	3.3
Dicyclohexyl adipate	4.8	Ethyl Acetate	6.4
1-Diethoxyethane	3.8	Ethyl Acetoacetate	15.9
Diethylaniline	5.5	Ethyl Acetoneoxalate	16.1
Diethyl benzalmalonate	8.0	Ethyl Acetophenoneoxalate	3.3
Diethyl di-malate	10.2	Ethyl Alcohol (See Ethanol)	
Diethyl glutarate	6.6	Ethylamine	6.3
Diethyl Ketone	17.3	Ethyl Amyl Ether	4.0
Diethyl l-malate	9.5	Ethylaniline	5.9
Diethyl Malonate	7.9	Ethyl Benzene	2.5
Diethylamine	3.7	Ethyl Benzoate	6.0
Diethyl Oxalate	8.2	Ethyl Benzoylacetate	12.8
Diethyl Oxaloacetate	6.1	Ethyl Benzoylacetoacetate	8.6
Diethyl Racemote	4.5	Ethyl Benzyl Ether	3.8
Diethyl Sebacate	5.0	Ethyl 1-Brombutyrate	8.0
Diethyl Succinate	6.6	Ethyl Bromide	4.9
Diethyl Succinosuccinate	2.5	Ethyl Bromoisobutyrate	7.9
Diethyl Sulfide	7.2	Ethyl Bromopropionate	9.4
Diethyl Sulfite	15.9	Ethyl Butyrate	5.1
Diethyl Tartrate	4.5		
Diethyl Disulfide	15.9		
Dihydrocaroone	8.7		
Dihydrocarvone	8.5		

Liquids Cont.

MATERIAL	DIELECTRIC CONSTANT	MATERIAL	DIELECTRIC CONSTANT
Ethyl Carbonate	3.1	Hydrogen Fluoride	84.0
Ethyl Chloracetate	11.6	Hydrogen Iodide	2.9
Ethyl Chloroformate	11.3	Hydrogen Peroxide	84.2
Ethyl Chloropropionate	10.1	Hydrogen Sulfide	5.8-9.3
Ethyl Cinnamate	5.3	Hydrazine	52.9
Ethyl Cyanoacetate	27.0	Indanol	7.8
Ethyl Cyclobutane	1.9	Iso-Iodohexadecane	3.5
Ethyl Dodecanoate	3.4	Iodine	118.0
Ethylene Chloride	10.5	Iodohexane	5.3
Ethylenechlorohydrin	25.0	Iodoheptane	4.9
Ethylene Cyanide	58.3	Iodomethane	7.0
Ethylenediamine	16.0	Iodotoluene	6.1
Ethylene Diamine	16.0	Isoamyl Alcohol	15.3
Ethylene Glycol	37.0	Iodloctane	4.6
Ethylene Oxide	13.9	Isoamyl Bromide	6.1
Ethyl Ether	4.7-8.1	Isoamyl Butyrate	3.9
Ethyl Ethoxybenzoate	7.1	Isoamyl Chloracetate	7.8
Ethyl Formate	8.4	Isoamyl Chloride	6.4
Ethyl Fumarate	6.5	Isoamyl Chloroformate	7.8
Ethyl Iodide	7.4	Isoamyl Iodide	5.6
Ethyl 2-Iodopropionate	8.8	Isoamyl Propionate	4.2
Ethyl Isothiocyanate	19.7	Isoamyl Salicylate	5.4
Ethyl Levulinete	12.1	Isoamyl Valerate	3.6
Ethyl Maleate	8.5	Isobutyl Acetate	5.6
Ethyl Mercaptan	8.0	Isobutyl Alcohol	18.7
Ethyl Nitrate	19.7	Isobutylamine	4.5
Ethyl Palmitate	3.2	Isobutylbenzene	2.3
Ethylpentane	1.9	Isobutyl Benzoate	5.9
Ethyl Phenylacetate	5.4	Isobutyl Bromide	6.6
Ethyl Propionate	5.7	Isobutyl Butyrate	4.0
Ethyl Salicylate	8.6	Isobutyl Chloride	7.1
Ethyl Silicate	4.1	Isobutyl Cyanide	13.3
Ethyl Stearate	2.9	Isobutyl Chloroformate	9.2
Ethyltoluene	2.2	Isobutyl Formate	6.5
Ethyl Trichloroacetate	7.8	Isobutylene Bromide	4.0
Ethyl Thiocyanate	29.6	Isobutyl Iodide	5.8
Ethyl Undecanoate	3.5	Isobutyl Nitrate	11.9
Ethyl Valerate	4.7	Isobutyl Rininoleate	4.7
Ethylbenzene	2.9	Isobutyl Valerate	3.8
Etibine	2.5	Isobutyric Acid	2.6
Eugenol	6.1	Isobutyric Anhydride	13.9
Fenchone	12.0	Isobutyronitrile	20.8
Ferric Oleate	2.6	Isocapronitrile	15.7
Ferrous Sulfate	14.2	Iso-propyl Alcohol	18.3
Fluorotoluene	4.2	Isopropylamine	5.5
Formamide	84.0	Iso-propyl Nitrate	11.5
Formic Acid	58.5	Isoquinoline	10.7
Freon 12	2.4	Isosafrol	3.4
Freon 11	3.1	Iso Valoric Acid	2.7
Freon 113	2.6	Jet Fuel (Military-JP4)	1.7
Glycerine	47.0	Kerosene	1.8
Glycerol	43.0-47.2	Lactic Acid	19.4
Glyceryl Triocetate	6.0	Lactonitrile	38.4
Glycol	35.6-41.2	Lead Oleate	3.2
Glycolic Nitrile	27.0	Lead Tetrachloride	2.8
Guaiacol	11.0	Limonene	2.3
Glucoheptitol	27.0	Linolaic Acid	2.9
Hagemannie Ester	10.6	Lonone	10.0
Heptadecanone	5.3	Malonic Nitrile	47
Heptane	1.9	Maloic Anhydride	51
Heptanone	11.9	Mannitol	3.0
Heptanoic Acid	2.5	Mandelic Nitrile	18.1
Heptyl Alcohol	6.7	Mandenitrile	17
Hexamethyldisiloxane	2.1	Menthol	3.95
Hexane	1.9	Menthonol	2.1
Hexanol	13.3	Methoxyethyl stearate	3.39
Hexanone	14.6	Mercury Diethyl	2.3
Hexyl Iodide	6.6	Mesitylene	3.4
Hexylene	2.0	Mesityl Oxide	15.4
Hydrocyanic Acid	2.3	Methal Cyanoacetate	29.4
Hydrogen Bromide	3.8	Methanol	33.6
Hydrogen Chloride	4.6	Methoxybenzene	4.33
Hydrogen Cyanide	95.4		

Table 9.10 continued

Table 9.10 continued

Liquids Cont.

MATERIAL	DIELECTRIC CONSTANT
Methoxytoluene	3.5
Methoxy-4-Methylphenol	11.0
Methyl Acetate	7.3
Methyl Acetophenoneoxalate	2.8
Methylal	2.7
Methyl Alcohol	33-56.6
Methylomine	10.5
Methylaniline	6.0
Methyl Benzoate	6.6
Methylbenzylamine	4.4
Methyl Butyl Ketone	12.4
Methyl Butyrate	5.6
Methyl Chloroacetate	12.9
Methylcyclohexanol	13.0
Methylcyclohexanone	18.0
Methylcylopentane	1.9
Methyl-1-Cyclopentanol	6.9
Methylene Iodide	5.1
Methyl Ether	5.0
Methyl Ethyl Ketone	18.4
Methyl Ethyl Ketoxime	3.4
Methyl Heptanol	5.25
Methylhexane	1.9
Methyl Kexyl Ketone	10.7
Methyl Iodide	7.1
Methyl-5 Ketocyclohexylene	24.0
Methyl Nitrobenzoate	27.0
Methyloctane	30.0
Methoxyphenol	11.0
Methyl O-Methoxybenzoate	7.8
Methyl-2, 4-pentanediol	24.4
Methyl-2-pentanone	13.1
Methylpheny Hydrazin	7.3
Methyl Propianate	5.4
Methyl Propyl Ketone	16.8
Methyl p-toluate	4.3
Methyl Salicylate	9.0
Methyl Thiocyanate	35.9
Methyl Valerate	4.3
Minerai Oil	2.1
Monomyristin	6.1
Monopalmitin	5.34
Monostearin	4.87
Napthalene	2.3
Naptholene	2.5
Napthonitrile	6.4
Naphthyl Ethyl Ether	3.2
Nitroanisole	24.0
Nitrobenzal Doxime	48.1
Nibrobenzene	26-36.1
Nitrobenzyl Alcohol	22.0
Nitroethane	19.7
Nitromethane	39.4
Nitroglycerin	19
Nitrosodimethylamine	54
Nitrosyl Bromide	13.4
Nitrosvl Chloride	18.2
Nitrotoluene	25.0
Nitrous Exide	1.6
Nonane	1.96
O-Chlorophenol	8.2
Octadecanol	3.42
Octanone	10.3
Octane	1.96
Octene	2.0
Octyl Alcohol	3.4
Octylene	4.1
Octyl Iodide	4.9
Oleic Acid	2.46
Oil, Almond	2.8
Oil, Cottonseed	3.1
Oil, Grapeseed	2.9
Oil, Lemon	2.3
Oil, Linseed	3.4
Oil, Olive	3.1
Oil, Paraffin	2.2 - 4.7
Oil, Peanut	3.0
Oil, Petroleum	2.1
Oil, Pyranol	5.3
Oil, Sesame	3.0
Oil, Sperm	3.2
Oil, Terpentine	2.2
Oil, Transformer	2.2
Oleic Acid	2.5
Palmitic Acid	2.30
Paraldehyde	14.5
Pentochlorethane	3.7
Pentane	1.8
Phenanthrene	2.8
Phentidine	7.3
Phenetole	4.5
Phenol	9.9-15.0
Phenol Ether	9.8
Phenoxyacetylene	4.76
Phenylacetaldehyde	4.8
Phenyiacetonitrile	18.0
Phenyl Acetate	6.9
Phenylacetic	3.0
Phenylethanol	13.0
Phenylethyl Acetate	4.5
Phenyl Isocyanate	8.9
Phenyl Iso-Thiocyanate	10.7
Phosgene	4.7
Phenyl-1-Iropane	2.7
Phosphine	2.5
Phosphorus	4.1
Phenylsalicylate	6.3
Phthalide	36.0
Pinacolin	12.8
Pinacone	7.4
Pinene	2.7
Piperidine	5.9
Propane	1.6
Propionaldehyde	18.9
Propionic Acid	3.1
Propionic Anhydride	18.0
Propionitrile	27.7
Propyl Acetate	6.3
Propyl Alcohol	21.8
Propyl Benzene	2.4
Propyl Bromide	7.2
Propyl Butyrate	4.3
Propyl Chloroformate	11.2
Propyl Formate	7.9
Propyl Nitrate	14.2
Propyl Propionate	4.7
Propyl Valerate	4.0
Psuedocumene	2.4
Pulegone	9.5
Pulezone	9.7
Pyridine	12.5
Quinoline	2.6-9.0
Safrol	3.1
Salicyladehyde	13.9
Santawax	2.29
Selevium	5.4
Silicon Tetrachloride	2.4
Sodium Oleate	2.7

Liquids Cont.

MATERIAL	DIELECTRIC CONSTANT
Sorbitol	33.5
Stannec Chloride	3.2
Stearic Acid	2.3
Styrene (Phenylethene)	2.4
Succinamide	2.9
Succinic Acid	2.4
Sulfur Dioxide	17.6
Sulfurous Oxychloride	9.1
Sulfur monochloride	4.8
Sulfur Trioxide	3.1
Sulfuryl Chloride	10.0
Sulphur	3.4
Sulphur Dioxide	15.6
Sulphuric Acid	84.0
Sulphuric Oxychloride	9.2
Sulphur Trioxide	3.6
Tartaric Acid	6.0-36
Terpinene	2.7
Terpineol	2.8
Tetrabromiethane	7.1
Tetrachloroethylene	2.5
Tetraethyl silicate	4.1
Tetradecanol	4.7
Tetrahydro-B-Naphthol	11.0
Tetratriacontodiene	2.82
Tetranitromethone	2.2
Thioacetic Acid	13.0
Thionyl Bromide	9.1
Thionyl Chloride	9.3
Thiophene	2.8
Thiophosphoryl Chloride	5.8
Tin Tetrachloride	2.9
Titanium Tetrachloride	2.8
Thujona	10.0
Toluene	2.4
Toluidine	6.0
Tolunitrile	18.8
Totana	5.5
Tolyl Methyl Ether	3.5
Trans-3-Hexene	2.0
Transmission Oil	2.2
Trimethyl-3-heptene	2.2
Tribromopropane	6.4
Tributylphosphate	7.9
Trichloroacetic Acid	4.5
Trichloroxoluene	6.9
Trichloroethylene	3.4
Trichloropropane	2.4
Tricresyl phosphate	6.9
Triethyl Aconitate	6.4
Triethylamine	3.2
Triethyl Aluminum	2.9
Triethyl Isoaconitate	7.2
Trifluoroactic Acid	39.0
Trifluorotoluene	9.1
Trimethylamine	2.9
Trimethylbenzene	2.2
Trimethyl Borate	8.2
Trimethylbutane	1.9
Trimethylpentane	1.9
Trimethylsulfanilic Acid	89.0
Trinitrobenzene	2.2
Tripolmitin	2.9
Triphenylmethane	2.4
Tristearin	2.7
Triolein	3.2
Undecane	2.0
Undecanone	8.4
Urea	3.5
Urethane	3.2
Valeraldehyde	11.8
Valeric Acid	2.6
Valeronitrile	17.7
Vanadium Oxybromide	3.6
Vanadium Oxychloride	3.4
Vanadium Tetrachloride	3.0
Iso-Valeric Acid	2.7
Veratrol	4.5
Vinyl Ether	3.9
Water	48-88.0
M-Xylene	2.4
Xylene	2.4
Xylenol	
Xylidine	5.0

POWDERS

MATERIAL	DIELECTRIC CONSTANT
Cement	1.5 - 2.1
Coke	1.1 - 2.2
Charcoal	1.2 - 1.81
Fly Ash	1.9 - 2.6
Plastic Grain	65 - 75
Soap Powder	1.25 - 1.5

* Values may vary considerably depending on physical size, moisture, content, density and other material characteristics.

Electrical Equipment*

Table 9.11
Comparison of Specific Applications of Enclosures for Indoor Nonhazardous Locations

Provides a Degree of Protection Against the Following Environmental Conditions	Type of Enclosure										
	1*	2*	4	4X	5	6	6P	11	12	12K	13
Incidental contact with the enclosed equipment	X	X	X	X	X	X	X	X	X	X	X
Falling dirt	X	X	X	X	X	X	X	X	X	X	X
Falling liquids and light splashing	...	X	X	X	...	X	X	X	X	X	X
Dust, lint, fibers, and flyings†	...	...	X	X	X	X	X	...	X	X	X
Hosedown and splashing water	...	...	X	X	...	X	X	...	...	...	...
Oil and coolant seepage	...	...	...	...	...	...	...	...	X	X	X
Oil or coolant spraying and splashing	...	...	...	...	...	...	...	...	...	...	X
Corrosive agents	...	...	...	X	...	...	X	X	...	...	...
Occasional temporary submersion	...	...	...	...	...	X	X	...	...	...	...
Occasional prolonged submersion	...	...	...	...	...	...	X	...	...	...	...

*These enclosures may be ventilated. However, Type 1 may not provide protection against small particles of falling dirt when ventilation is provided in the enclosure top. Consult the manufacturer. See 3.7 of NEMA Standard 250-1985.
†These fibers and flyings are nonhazardous materials and are not considered the Class III type ignitable fibers or combustible flyings. For Class III type ignitable fibers or combustible flyings see the *National Electrical Code,* Section 500-6(a).

This table is reproduced by permission of the National Electrical Manufacturers Association from NEMA Standards Publication 250-1985, "Enclosures for Electrical Equipment (1000 Volts Maximum)," copyright 1985 by NEMA.

Table 9.12
Comparison of Specific Applications of Enclosures for Outdoor Nonhazardous Locations

Provides a Degree of Protection Against the Following Environmental Conditions	Type of Enclosure						
	3	3R†	3S	4	4X	6	6P
Incidental contact with the enclosed equipment	X	X	X	X	X	X	X
Rain, snow, and sleet*	X	X	X	X	X	X	X
Sleet**	...	...	X	...	...	...	...
Windblown dust	X	...	X	X	X	X	X
Hosedown	...	...	...	X	X	X	X
Corrosive agents	...	...	...	...	X	...	X
Occasional temporary submersion	...	...	...	...	...	X	X
Occasional prolonged submersion	...	...	...	...	...	...	X

*External operating mechanisms are not required to be operable when the enclosure is ice covered.
**External operating mechanisms are operable when the enclosure is ice covered. See 6.6.2.1 of NEMA Standard 250-1985.
†These enclosures may be ventilated.
See "Definitions and Symbols" on pages 468–526 for definitions of NEMA-type numbers.

This table is reproduced by permission of the National Electrical Manufacturers Association from NEMA Standards Publication 250-1985, "Enclosures for Electrical Equipment (1000 Volts Maximum)," copyright 1985 by NEMA.

* Please see the "Area Classification" and "Wiring" sections of this chapter for other NEC requirements. Some modifications of headings and footnotes have been made by the authors to aid the reader; tabular material has not been altered.

Table 9.13

Full Load Current for Sizing 3-Phase Transformers

KVA Rating	208V	240V	480V	600V	2400V	4160V	7200V	12,470V	13,200V	13,800V
	Amperes									
3	8.34	7.23	3.61	2.89	.722	.416	.241	.139	.131	.126
6	16.6	14.4	7.2	5.8	1.44	.833	.481	.278	.262	.251
9	25.0	21.7	10.8	8.67	2.17	1.25	.723	.417	.393	.377
15	41.7	36.1	18.1	14.5	3.61	2.08	1.20	.694	.656	.628
30	83.4	72.3	36.1	28.9	7.22	41.6	2.41	1.39	1.31	1.26
45	125	108	54.2	43.4	10.8	6.25	3.61	2.08	1.96	1.88
75	208	181	90.3	72.3	18.0	10.4	6.01	3.47	3.28	3.14
112.5	313	271	135	108	27.1	15.6	9.02	5.21	4.92	4.71
150	417	361	181	145	36.1	20.8	12.0	6.94	6.56	6.28
225	625	542	271	217	54.1	31.2	18.0	10.4	9.84	9.41
300	834	723	361	289	72.2	41.6	24.0	13.9	13.1	12.6
500	1390	1204	602	482	120.2	69.3	40.0	23.1	21.8	20.8

Courtesy Warren Electric Company.

Table 9.14

Full Load Current for Sizing Single Phase Transformers

KVA Rating	120V	208V	240V	277V	480V	600V	2400V	4160V	4800V
	Amperes								
.050	.416	.240	.208	.108	.104	.083			
.075	.625	.360	.312	.270	.156	.125			
.100	.833	.480	.417	.361	.208	.167			
.150	1.25	.721	.625	.541	.313	.250			
.250	2.08	1.20	1.04	.902	.521	.417			
.500	4.16	2.40	2.08	1.80	1.04	.833			
.750	6.25	3.60	3.13	2.70	1.56	1.25			
1	8.33	4.81	4.17	3.61	2.08	1.67			
1.5	12.5	7.21	6.25	5.42	3.13	2.50			
2	16.7	9.62	8.33	7.22	4.17	3.33			
3	25.0	14.4	12.5	10.8	6.25	5.0	1.25	.721	
5	41.6	24.0	20.8	18.0	10.4	8.33	2.08	1.20	
7.5	62.5	36.1	31.3	27.1	15.6	12.5	3.13	1.80	
10	83.3	48.1	41.7	36.1	20.8	16.7	4.17	2.40	2.08
15	125	72.1	62.5	54.2	31.3	25.0	6.25	3.61	3.13
25	208	120	104	90.3	52.1	41.7	10.4	6.01	5.21
37.5	313	180	156	135	78	62.5	15.6	9.01	7.80
50	416	240	208	181	104	83.3	20.8	12.0	10.4
75	625	361	313	271	156	125	31.3	18.0	15.6
100	833	481	417	361	208	167	41.7	24.0	20.8
167	1392	803	695	603	348	278	69.6	40.1	34.8
200	1667	962	833	722	417	333	83.3	48.1	41.7
250	2083	1202	1042	903	521	417	104	60.1	52.1

Courtesy Warren Electric Company.

Table 9.15

Insulation System Rating

During recent years, the terminology used by electrical equipment manufacturers regarding insulation systems has undergone a major change. Letter designations, such as Class A, B, F and H, are obsolete. Insulation systems are now classified numerically, by temperature rating. What used to be Class A, B, F and H are now Class 105, 150, 185 and 220 respectively. The preceding designations pertain only to the rating of the insulation system. The transformer's rating has also been changed—from Class A, B, F and H, to 55°C rise, 80°C rise, 115°C rise and 150°C rise. What previously was a Class H transformer is now a 150°C rise transformer utilizing a Class 220 insulation system. The following table shows the old and new designations.

TRANSFORMER AND INSULATION SYSTEM RATINGS

Old Terminology		New Terminology			
Insulation Rating	Transformer Rating	Insulation Rating	Transformer Rating	Ambient Temperature	Hot Spot Allowance
A	A	Class 105	55°C Rise	40°C	10°C
B	B	Class 150	80°C Rise	40°C	30°C
F	F	Class 185	115°C Rise	40°C	30°C
H	H	Class 220	150°C Rise	40°C	30°C

Courtesy Warren Electric Company.

Below are connections showing three single-phase transformers banked for three-phase operation. The total KVA output is the sum of the three transformers.

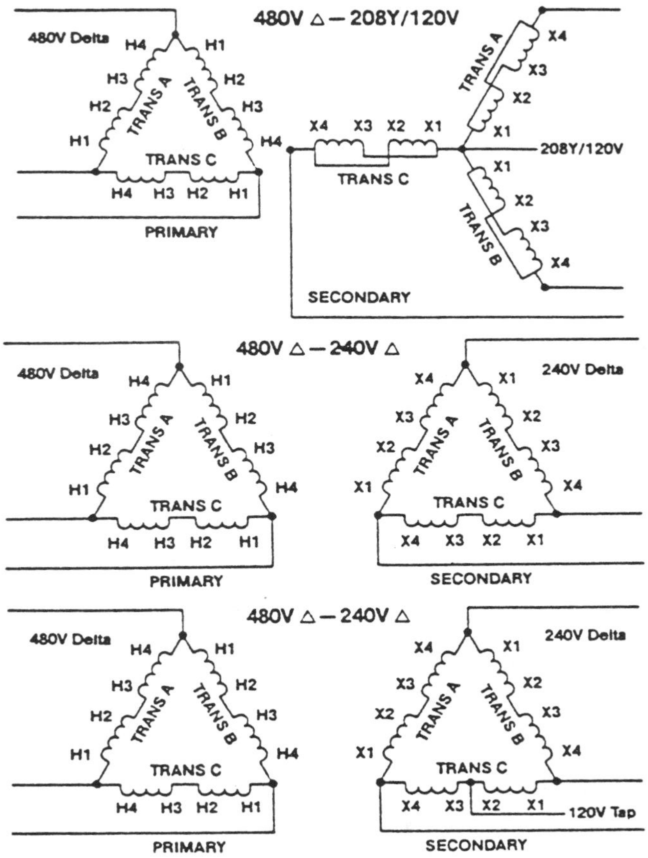

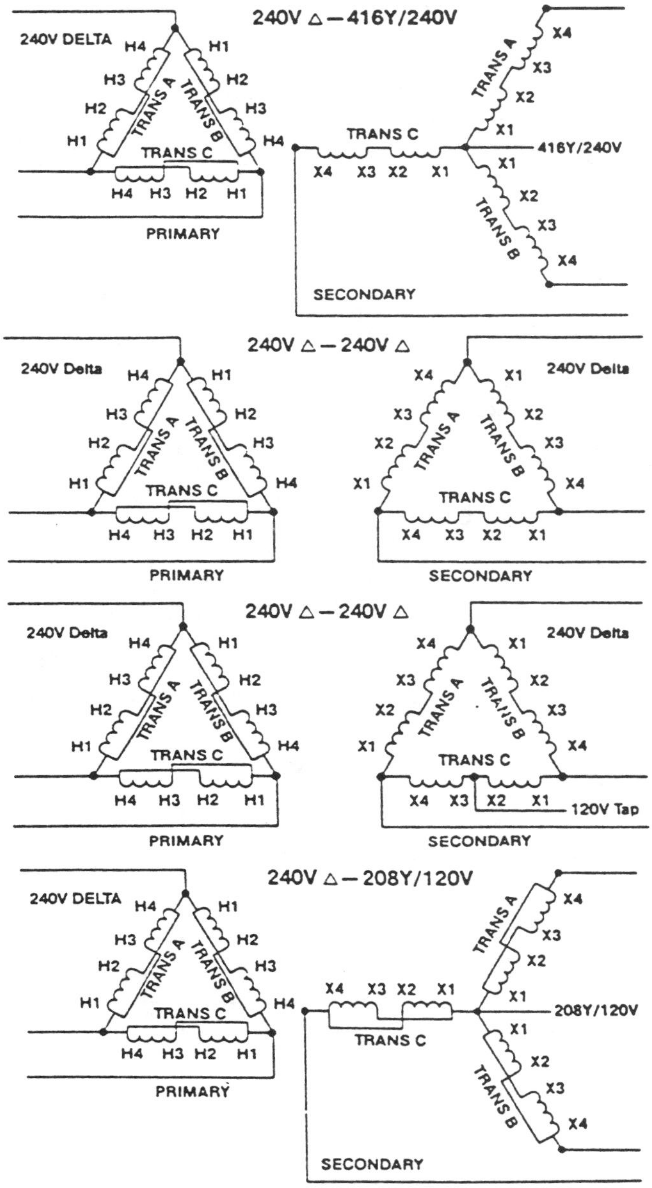

Figure 9.14. Wiring diagrams for banking single-phase transformers. Courtesy Warren Electric Company.

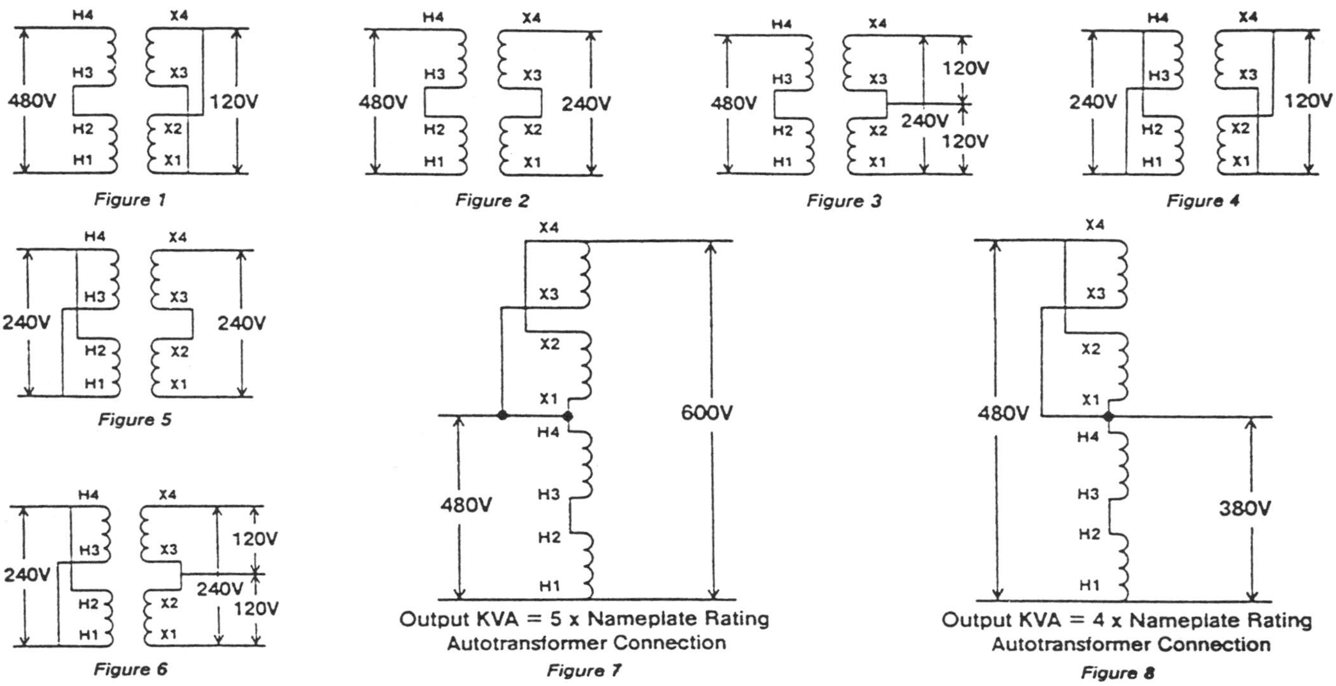

Figure 9.15. Wiring diagrams for single-phase transformers. Courtesy Warren Electric Company.

Resistor-Capacitor Color Code

Color	Significant Figure	Decimal Multiplier	Tolerance (%)	Voltage Rating*
Black	0	1	—	—
Brown	1	10	1*	100
Red	2	100	2*	200
Orange	3	1,000	3*	300
Yellow	4	10,000	4*	400
Green	5	100,000	5*	500
Blue	6	1,000,000	6*	600
Violet	7	10,000,000	7*	700
Gray	8	100,000,000	8*	800
White	9	1,000,000,000	9*	900
Gold	—	0.1	5	1000
Silver	—	0.01	10	2000
No color	—		20	500

* Applies to capacitors only.

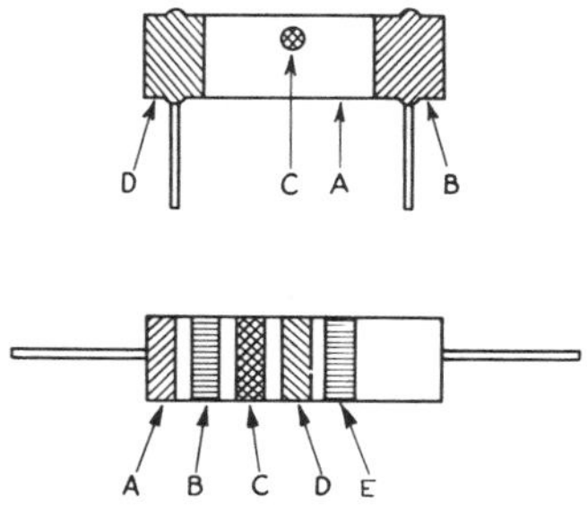

Fixed composition resistors

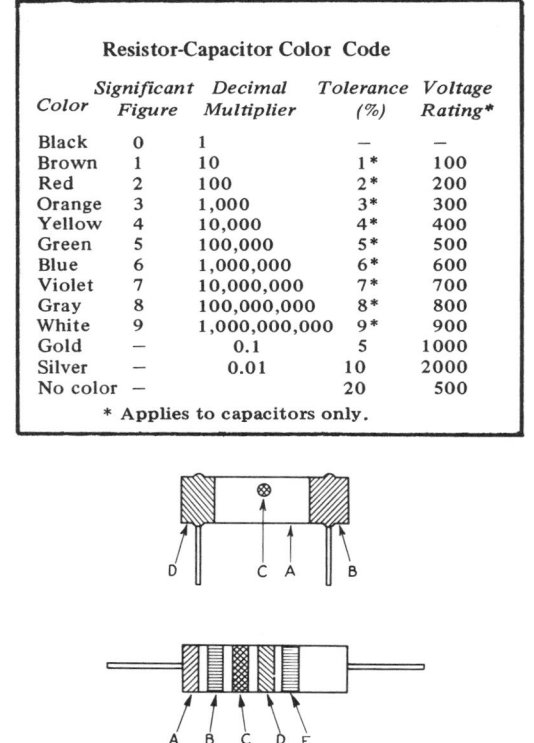

Table 9.16. Color coding of fixed composition resistors. The color code is given in the table above. The colored areas have the significance: A—first significant figure of resistance in ohms; B—second significant figure; C—decimal multiplier; D—resistance tolerance in percent. If no color is shown, the tolerance is ±20%. E—relative percent change in value per 1000 hours of operation; brown, 1%; red, 0.1%; orange, .01%; yellow, .001%.

Table 9.17. Dimensions and Percent Area of Conduit and of Tubing

		Area—Square Inches								
			Not Lead Covered			Lead Covered				
Trade Size	Internal Diameter Inches	Total 100%	2 Cond. 31%	Over 2 Cond. 40%	1 Cond. 53%	1 Cond. 55%	2 Cond. 30%	3 Cond. 40%	4 Cond. 38%	Over 4 Cond. 35%
½	.622	.30	.09	.12	.16	.17	.09	.12	.11	.11
¾	.824	.53	.16	.21	.28	.29	.16	.21	.20	.19
1	1.049	.86	.27	.34	.46	.47	.26	.34	.33	.30
1¼	1.380	1.50	.47	.60	.80	.83	.45	.60	.57	.53
1½	1.610	2.04	.63	.82	1.08	1.12	.61	.82	.78	.71
2	2.067	3.36	1.04	1.34	1.78	1.85	1.01	1.34	1.28	1.18
2½	2.469	4.79	1.48	1.92	2.54	2.63	1.44	1.92	1.82	1.68
3	3.068	7.38	2.29	2.95	3.91	4.06	2.21	2.95	2.80	2.58
3½	3.548	9.90	3.07	3.96	5.25	5.44	2.97	3.96	3.76	3.47
4	4.026	12.72	3.94	5.09	6.74	7.00	3.82	5.09	4.83	4.45
5	5.047	20.00	6.20	8.00	10.60	11.00	6.00	8.00	7.60	7.00
6	6.065	28.89	8.96	11.56	15.31	15.89	8.67	11.56	10.98	10.11

Table 9.18. Dimensions of Rubber-Covered and Thermoplastic-Covered Conductors

Size AWG kcmil	Types RFH-2, RH, RHH, *** RHW, *** SF-2 Approx. Diam. Inches	Approx. Area Sq. In.	Types TF, THW,† TW Approx. Diam. Inches	Approx. Area Sq. In.	Types TFN, THHN, THWN Approx. Diam. Inches	Approx. Area Sq. In.	Types**** FEP, FEPB, FEPW TFE, PF, PFA, PFAH, PGF, PTF, Z, ZF, ZFF Approx. Diam. Inches	Approx. Area Sq. In.	Type XHHW, ZW†† Approx. Diam. Inches	Approx. Area Sq. In.	Types KF-1, KF-2, KFF-1, KFF-2 Approx. Diam. Inches	Approx. Area Sq. In.
Col. 1	Col. 2	Col. 3	Col. 4	Col. 5	Col. 6	Col. 7	Col. 8	Col. 9	Col. 10	Col. 11	Col. 12	Col. 13
18	.146	.0167	.106	.0088	.089	.0062	.081	.0052	...	....	.065	.0033
16	.158	.0196	.118	.0109	.100	.0079	.092	.0066	...	....	.070	.0038
14	30 mils .171	.0230	.131	.0135	.105	.0087	.105 .105	.0087 .0087	...	....	.083	.0054
14	45 mils .204*	.0327*	...	....	...	....	....	....	...	....	...	....
14	....	....	.162†	.0206†	...	....	....	....	.129	.0131	...	....
12	30 mils .188	.0278	.148	.0172	.122	.0117	.121 .121	.0115 .0115	...	....	.102	.0082
12	45 mils .221*	.0384*	...	....	...	....	....	....	...	....	...	....
12	....	....	.179†	.0252†	...	....	....	....	.146	.0167	...	....
10	...... .242	.0460	.168	.0222	.153	.0184	.142 .142	.0158 .0158	...	....	.124	.0121
10	....	....	.199†	.0311†	...	....	....	....	.166	.0216	...	....
8	...... .328	.0845	.245	.0471	.218	.0373	.206 .186	.0333 .0272	...	....	...	....
8	....	....	.276†	.0598†	...	....	....	....	.241	.0456	...	....
6	.397	.1238	.323	.0819	.257	.0519	.244 .302	.0468 .0716	.282	.0625	...	....
4	.452	.1605	.372	.1087	.328	.0845	.292 .350	.0670 .0962	.328	.0845	...	....
3	.481	.1817	.401	.1263	.356	.0995	.320 .378	.0804 .1122	.356	.0995	...	....
2	.513	.2067	.433	.1473	.388	.1182	.352 .410	.0973 .1320	.388	.1182	...	....
1	.588	.2715	.508	.2027	.450	.1590	.420 ...	.1385	.450	.1590	...	....
1/0	.629	.3107	.549	.2367	.491	.1893	.462 ...	.1676	.491	.1893	...	....
2/0	.675	.3578	.595	.2781	.537	.2265	.498 ...	.1948	.537	.2265	...	....
3/0	.727	.4151	.647	.3288	.588	.2715	.560 ...	.2463	.588	.2715	...	....
4/0	.785	.4840	.705	.3904	.646	.3278	.618 ...	.3000	.646	.3278	...	....
250	.868	.5917	.788	.4877	.716	.4026	...	....	.716	.4026		
300	.933	.6837	.843	.5581	.771	.4669	...	....	.771	.4669		
350	.985	.7620	.895	.6291	.822	.5307	...	....	.822	.5307		
400	1.032	.8365	.942	.6969	.869	.5931	...	....	.869	.5931		
500	1.119	.9834	1.029	.8316	.955	.7163	...	....	.955	.7163		

Table 9.18 continued

Table 9.18. Dimensions of Rubber-Covered and Thermoplastic-Covered Conductors (continued)

Size AWG kcmil	Types RFH-2, RH, RHH, *** RHW, *** SF-2 Approx. Diam. Inches	Types RFH-2, RH, RHH, *** RHW, *** SF-2 Approx. Area Sq. In.	Types TF, THW,† TW Approx. Diam. Inches	Types TF, THW,† TW Approx. Area Sq. In.	Types TFN, THHN, THWN Approx. Diam. Inches	Types TFN, THHN, THWN Approx. Area Sq. In.	Types**** FEP, FEPB, FEPW TFE, PF, PFA, PFAH, PGF, PTF, Z, ZF, ZFF Approx. Diam. Inches	Types**** FEP, FEPB, FEPW TFE, PF, PFA, PFAH, PGF, PTF, Z, ZF, ZFF Approx. Area Sq. In.	Type XHHW, ZW†† Approx. Diam. Inches	Type XHHW, ZW†† Approx. Area Sq. In.	Types KF-1, KF-2, KFF-1, KFF-2 Approx. Diam. Inches	Types KF-1, KF-2, KFF-1, KFF-2 Approx. Area Sq. In.
Col. 1	Col. 2	Col. 3	Col. 4	Col. 5	Col. 6	Col. 7	Col. 8	Col. 9	Col. 10	Col. 11	Col. 12	Col. 13
600	1.233	1.1940	1.143	1.0261	1.058	.8791	...	....	1.073	.9043		
700	1.304	1.3355	1.214	1.1575	1.129	1.0011	...	....	1.145	1.0297		
750	1.339	1.4082	1.249	1.2252	1.163	1.0623	...	....	1.180	1.0936		
800	1.372	1.4784	1.282	1.2908	1.196	1.1234	...	....	1.210	1.1499		
900	1.435	1.6173	1.345	1.4208	1.259	1.2449	...	....	1.270	1.2668		
1000	1.494	1.7530	1.404	1.5482	1.317	1.3623	...	....	1.330	1.3893		
1250	1.676	2.2062	1.577	1.9532	...	....	...	....	1.500	1.7671		
1500	1.801	2.5475	1.702	2.2751	...	....	...	....	1.620	2.0612		
1750	1.916	2.8832	1.817	2.5930	...	....	...	....	1.740	2.3779		
2000	2.021	3.2079	1.922	2.9013	...	....	...	....	1.840	2.6590		

*The dimensions of Types RHH and RHW.

†Dimensions of THW in sizes No. 14 through No. 8 No. 6 THW and larger is the same dimension as TW.

***Dimensions of RHH and RHW without outer covering are the same as THW No. 18 through No. 10, solid: No. 8 and larger, stranded.

****In Columns 8 and 9 the values shown for sizes No. 1 through 4/0 are for TFE and Z only. The right-hand values in Columns 8 and 9 are for FEPB, Z, and ZFF only.

††No. 14 through No. 2.

Table 9.19. Aluminum Building Wire Nominal Dimensions* and Areas

Size AWG or kcmil	Bare Conductor** Number of Strands	Bare Conductor** Diam. Inches	Type THW Approx. Diam. Inches	Type THW Approx. Area Sq. In.	Type THHN Approx. Diam. Inches	Type THHN Approx. Area Sq. In.	Type XHHW Approx. Diam. Inches	Type XHHW Approx. Area Sq. In.	Size AWG or kcmil
8	7	.134	.255	.0510	—	—	.224	.0394	8
6	7	.169	.290	.0660	.240	.0452	.260	.0530	6
4	7	.213	.335	.0881	.305	.0730	.305	.0730	4
2	7	.268	.390	.1194	.360	.1017	.360	.1017	2
1	19	.299	.465	.1698	.415	.1352	.415	.1352	1
1/0	19	.336	.500	.1963	.450	.1590	.450	.1590	1/0
2/0	19	.376	.545	.2332	.495	.1924	.490	.1885	2/0
3/0	19	.423	.590	.2733	.540	.2290	.540	.2290	3/0
4/0	19	.475	.645	.3267	.595	.2780	.590	.2733	4/0
250	37	.520	.725	.4128	.670	.3525	.660	.3421	250
300	37	.570	.775	.4717	.720	.4071	.715	.4015	300
350	37	.616	.820	.5281	.770	.4656	.760	.4536	350
400	37	.659	.865	.5876	.815	.5216	.800	.5026	400
500	37	.736	.940	.6939	.885	.6151	.880	.6082	500
600	61	.813	1.050	.8659	.985	.7620	.980	.7542	600
700	61	.877	1.110	.9676	1.050	.8659	1.050	.8659	700
750	61	.908	1.150	1.0386	1.075	.9076	1.090	.9331	750
1000	61	1.060	1.285	1.2968	1.255	1.2370	1.230	1.1882	1000

* Dimensions are from industry sources

** Compact conductor per ASTM B400

Table 9.20. Nominal Diameters for Solid Copper Conductors

Conductor Size	Solid Diameters
AWG	inch
22	0.0253
20	0.0320
19	0.0359
18	0.0403
17	0.0453
16	0.0508

This material is reproduced by permission of the National Electrical Manufacturers Association from NEMA Standards Publication WC 55-1986, "Instrumentation Cables and Thermocouple Wire," copyright 1986 by NEMA.

Table 9.21. Extension Wire Elements

Type	Extension Wire	Conductor Material**
T	TX	TPX Copper TNX Constantan
J	JX	JPX Iron JNX Constantan
E	EX	EPX Chromel*, Tophel*, T-1* ENX Constantan
K	KX	KPX Chromel*, Tophel*, T-1* KNX Alumel*, Nial*, T-2*
R or S	SX	SPX Copper SNX Copper Nickel Alloy
B	BX	BPX Copper BNX Copper

*Trade Name

**Materials are listed as typical of those commercially available at present, and their listing implies no endorsement by this standard.

This material is reproduced by permission of the National Electrical Manufacturers Association from NEMA Standards Publication WC 55-1986, "Instrumentation Cables and Thermocouple Wire," copyright 1986 by NEMA.

Table 9.22. Nominal Direct Current Resistance in Ohms per 1,000 Feet at 20°C of Solid or Stranded Thermocouple Conductors

Size	Conductor Material					
(AWG)	KNX	EPX KPX	ENX JNX TNX	TPX SPX BPX BNX	SN SNX	JPX
20	173	415	287	10.1	27.4	69.9
18	111	266	184	6.39	17.5	44.6
16	68.3	164	113	4.02	10.8	27.6

This material is reproduced by permission of the National Electrical Manufacturers Assocation from NEMA Standards Publication WC 55-1986, "Instrumentation Cables and Thermocouple Wire," copyright 1986 by NEMA.

Table 9.23. Nominal Diameters of Thermocouple Wires

Conductor Size	Solid Conductor	Stranded Individual Wire	Stranded Conductor
AWG	inch	inch	inch
20	0.0320	0.0126	0.038
18	0.0403	0.0159	0.048
16	0.0508	0.0201	0.060

This material is reproduced by permission of the National Electrical Manufacturers Association form NEMA Standards Publication WC 55-1986, "Instrumentation Cables and Thermocouple Wire," copyright 1986 by NEMA.

The temperature-electromotive-force relationship for a thermocouple in general cannot be expressed by a simple equation. It is convenient, therefore, to have empirical tables giving the temperature-electromotive-force relationship for the various types of commercially available thermocouples. For any thermocouple type, a table is based on calibrations of representative thermocouples at sufficient points to yield a temperature-electromotive-force plot characteristic of the material.

These tables are taken from National Bureau of Standards Circular 561.

Electromotive force is expressed in absolute units. Temperatures are expressed on the International Temperature Scale of 1948.

IRON vs. CONSTANTAN THERMOCOUPLE

Degrees Centigrade — Reference Junction 0° C.

°C	0	1	2	3	4	5	6	7	8	9
					Millivolts					
-190	-7.66	-7.69	-7.71	-7.73	-7.76	-7.78				
-180	-7.40	-7.43	-7.46	-7.49	-7.51	-7.54	-7.56	-7.59	-7.61	-7.64
-170	-7.12	-7.15	-7.18	-7.21	-7.24	-7.27	-7.30	-7.32	-7.35	-7.38
-160	-6.82	-6.85	-6.88	-6.91	-6.94	-6.97	-7.00	-7.03	-7.06	-7.09
-150	-6.50	-6.53	-6.56	-6.60	-6.63	-6.66	-6.69	-6.72	-6.76	-6.79
-140	-6.16	-6.19	-6.22	-6.26	-6.29	-6.33	-6.36	-6.40	-6.43	-6.46
-130	-5.80	-5.84	-5.87	-5.91	-5.94	-5.98	-6.01	-6.05	-6.08	-6.12
-120	-5.42	-5.46	-5.50	-5.54	-5.58	-5.61	-5.65	-5.69	-5.72	-5.76
-110	-5.03	-5.07	-5.11	-5.15	-5.19	-5.23	-5.27	-5.31	-5.35	-5.38
-100	-4.63	-4.67	-4.71	-4.75	-4.79	-4.83	-4.87	-4.91	-4.95	-4.99
-90	-4.21	-4.25	-4.30	-4.34	-4.38	-4.42	-4.46	-4.50	-4.55	-4.59
-80	-3.78	-3.82	-3.87	-3.91	-3.96	-4.00	-4.04	-4.08	-4.13	-4.17
-70	-3.34	-3.38	-3.43	-3.47	-3.52	-3.56	-3.60	-3.65	-3.69	-3.74
-60	-2.89	-2.94	-2.98	-3.03	-3.07	-3.12	-3.16	-3.21	-3.25	-3.30
-50	-2.43	-2.48	-2.52	-2.57	-2.62	-2.66	-2.71	-2.75	-2.80	-2.84
-40	-1.96	-2.01	-2.06	-2.10	-2.15	-2.20	-2.24	-2.29	-2.34	-2.38
-30	-1.48	-1.53	-1.58	-1.63	-1.67	-1.72	-1.77	-1.82	-1.87	-1.91
-20	-1.00	-1.04	-1.09	-1.14	-1.19	-1.24	-1.29	-1.34	-1.39	-1.43
-10	-0.50	-0.55	-0.60	-0.65	-0.70	-0.75	-0.80	-0.85	-0.90	-0.95
(-)0	0.00	-0.05	-0.10	-0.15	-0.20	-0.25	-0.30	-0.35	-0.40	-0.45
(+)0	0.00	0.05	0.10	0.15	0.20	0.25	0.30	0.35	0.40	0.45
10	0.50	0.56	0.61	0.66	0.71	0.76	0.81	0.86	0.91	0.97
20	1.02	1.07	1.12	1.17	1.22	1.28	1.33	1.38	1.43	1.48
30	1.54	1.59	1.64	1.69	1.74	1.80	1.85	1.90	1.95	2.00
40	2.06	2.11	2.16	2.22	2.27	2.32	2.37	2.42	2.48	2.53
50	2.58	2.64	2.69	2.74	2.80	2.85	2.90	2.96	3.01	3.06
60	3.11	3.17	3.22	3.27	3.33	3.38	3.43	3.49	3.54	3.60
70	3.65	3.70	3.76	3.81	3.86	3.92	3.97	4.02	4.08	4.13
80	4.19	4.24	4.29	4.35	4.40	4.46	4.51	4.56	4.62	4.67
90	4.73	4.78	4.83	4.89	4.94	5.00	5.05	5.10	5.16	5.21
100	5.27	5.32	5.38	5.43	5.48	5.54	5.59	5.65	5.70	5.76
110	5.81	5.86	5.92	5.97	6.03	6.08	6.14	6.19	6.25	6.30
120	6.36	6.41	6.47	6.52	6.58	6.63	6.68	6.74	6.79	6.85
130	6.90	6.96	7.01	7.07	7.12	7.18	7.23	7.29	7.34	7.40
140	7.45	7.51	7.56	7.62	7.67	7.73	7.78	7.84	7.89	7.95
150	8.00	8.06	8.12	8.17	8.23	8.28	8.34	8.39	8.45	8.50
160	8.56	8.61	8.67	8.72	8.78	8.84	8.89	8.95	9.00	9.06
170	9.11	9.17	9.22	9.28	9.33	9.39	9.44	9.50	9.56	9.61
180	9.67	9.72	9.78	9.83	9.89	9.95	10.00	10.06	10.11	10.17
190	10.22	10.28	10.34	10.39	10.45	10.50	10.56	10.61	10.67	10.72
200	10.78	10.84	10.89	10.95	11.00	11.06	11.12	11.17	11.23	11.28
210	11.34	11.39	11.45	11.50	11.56	11.62	11.67	11.73	11.78	11.84
220	11.89	11.95	12.00	12.06	12.12	12.17	12.23	12.28	12.34	12.39
230	12.45	12.50	12.56	12.62	12.67	12.73	12.78	12.84	12.89	12.95
240	13.01	13.06	13.12	13.17	13.23	13.28	13.34	13.40	13.45	13.51
250	13.56	13.62	13.67	13.73	13.78	13.84	13.89	13.95	14.00	14.06
260	14.12	14.17	14.23	14.28	14.34	14.39	14.45	14.50	14.56	14.61
270	14.67	14.72	14.78	14.83	14.89	14.94	15.00	15.06	15.11	15.17
280	15.22	15.28	15.33	15.39	15.44	15.50	15.55	15.61	15.66	15.72
290	15.77	15.83	15.88	15.94	16.00	16.05	16.11	16.16	16.22	16.27
300	16.33	16.38	16.44	16.49	16.55	16.60	16.66	16.71	16.77	16.82
310	16.88	16.93	16.99	17.04	17.10	17.15	17.21	17.26	17.32	17.37
320	17.43	17.48	17.54	17.60	17.65	17.71	17.76	17.82	17.87	17.93
330	17.98	18.04	18.09	18.15	18.20	18.26	18.32	18.37	18.43	18.48
340	18.54	18.59	18.65	18.70	18.76	18.81	18.87	18.92	18.98	19.03

IRON vs. CONSTANTAN THERMOCOUPLE

Degrees Centigrade — Reference Junction 0° C.

°C	0	1	2	3	4	5	6	7	8	9
					Millivolts					
350	19.09	19.14	19.20	19.26	19.31	19.37	19.42	19.48	19.53	19.59
360	19.64	19.70	19.75	19.81	19.86	19.92	19.97	20.03	20.08	20.14
370	20.20	20.25	20.31	20.36	20.42	20.47	20.53	20.58	20.64	20.69
380	20.75	20.80	20.86	20.91	20.97	21.02	21.08	21.13	21.19	21.24
390	21.30	21.35	21.41	21.46	21.52	21.57	21.63	21.68	21.74	21.79
400	21.85	21.90	21.96	22.02	22.07	22.13	22.18	22.24	22.29	22.35
410	22.40	22.46	22.51	22.57	22.62	22.68	22.73	22.79	22.84	22.90
420	22.95	23.01	23.06	23.12	23.17	23.23	23.28	23.34	23.39	23.45
430	23.50	23.56	23.61	23.67	23.72	23.78	23.83	23.89	23.94	24.00
440	24.06	24.11	24.17	24.22	24.28	24.33	24.39	24.44	24.50	24.55
450	24.61	24.66	24.72	24.77	24.83	24.88	24.94	25.00	25.05	25.11
460	25.16	25.22	25.27	25.33	25.38	25.44	25.49	25.55	25.60	25.66
470	25.72	25.77	25.83	25.88	25.94	25.99	26.05	26.10	26.16	26.22
480	26.27	26.33	26.38	26.44	26.49	26.55	26.61	26.66	26.72	26.77
490	26.83	26.89	26.94	27.00	27.05	27.11	27.17	27.22	27.28	27.33
500	27.39	27.45	27.50	27.56	27.61	27.67	27.73	27.78	27.84	27.90
510	27.95	28.01	28.07	28.12	28.18	28.23	28.29	28.35	28.40	28.46
520	28.52	28.57	28.63	28.69	28.74	28.80	28.86	28.91	28.97	29.02
530	29.08	29.14	29.20	29.25	29.31	29.37	29.42	29.48	29.54	29.59
540	29.65	29.71	29.76	29.82	29.88	29.94	29.99	30.05	30.11	30.16
550	30.22	30.28	30.34	30.39	30.45	30.51	30.57	30.62	30.68	30.74
560	30.80	30.85	30.91	30.97	31.02	31.08	31.14	31.20	31.26	31.31
570	31.37	31.43	31.49	31.54	31.60	31.66	31.72	31.78	31.83	31.89
580	31.95	32.01	32.06	32.12	32.18	32.24	32.30	32.36	32.41	32.47
590	32.53	32.59	32.65	32.71	32.76	32.82	32.88	32.94	33.00	33.06
600	33.11	33.17	33.23	33.29	33.35	33.41	33.46	33.52	33.58	33.64
610	33.70	33.76	33.82	33.88	33.94	33.99	34.05	34.11	34.17	34.23
620	34.29	34.35	34.41	34.47	34.53	34.58	34.64	34.70	34.76	34.82
630	34.88	34.94	35.00	35.06	35.12	35.18	35.24	35.30	35.36	35.42
640	35.48	35.54	35.60	35.66	35.72	35.78	35.84	35.90	35.96	36.02
650	36.08	36.14	36.20	36.26	36.32	36.38	36.44	36.50	36.56	36.62
660	36.69	36.75	36.81	36.87	36.93	36.99	37.05	37.11	37.18	37.24
670	37.30	37.36	37.42	37.48	37.54	37.60	37.66	37.73	37.79	37.85
680	37.91	37.97	38.04	38.10	38.16	38.22	38.28	38.34	38.41	38.47
690	38.53	38.59	38.66	38.72	38.78	38.84	38.90	38.97	39.03	39.09
700	39.15	39.22	39.28	39.34	39.40	39.47	39.53	39.59	39.65	39.72
710	39.78	39.84	39.91	39.97	40.03	40.10	40.16	40.22	40.28	40.35
720	40.41	40.48	40.54	40.60	40.66	40.73	40.79	40.86	40.92	40.98
730	41.05	41.11	41.17	41.24	41.30	41.36	41.43	41.49	41.56	41.62
740	41.68	41.75	41.81	41.87	41.94	42.00	42.07	42.13	42.19	42.26
750	42.32	42.38	42.45	42.51	42.58	42.64	42.70	42.77	42.83	42.90
760	42.96	43.02	43.09	43.15	43.22	43.28	43.35	43.41	43.48	43.54
770	43.60	43.67	43.73	43.80	43.86	43.92	43.99	44.05	44.12	44.18
780	44.25	44.31	44.38	44.44	44.50	44.57	44.63	44.70	44.76	44.82
790	44.89	44.95	45.02	45.08	45.15	45.21	45.28	45.34	45.40	45.47
800	45.53	45.60	45.66	45.72	45.79	45.85	45.92	45.98	46.05	46.11
810	46.18	46.24	46.30	46.37	46.43	46.50	46.56	46.62	46.69	46.75
820	46.82	46.88	46.94	47.01	47.07	47.14	47.20	47.27	47.33	47.39
830	47.46	47.52	47.58	47.65	47.71	47.78	47.84	47.90	47.97	48.03
840	48.09	48.16	48.22	48.28	48.35	48.41	48.48	48.54	48.60	48.66
850	48.73	48.79	48.85	48.92	48.98	49.04	49.10	49.17	49.23	49.29
860	49.36	49.42	49.48	49.54	49.61	49.67	49.73	49.79	49.86	49.92
870	49.98	50.04								

Table 9.24 continued

Table 9.24. Thermocouple conversion tables: iron constantan (0°C), iron constantan (32°F), chromel alumel (0°C), chromel alumel (32°F), copper constantan (0°C), copper constantan (32°F), platinum—platinum + 10% rhodium (0°C), platinum—platinum + 10% rhodium (32°F), platinum—platinum + 13% rhodium (0°C), and platinum—platinum + 13% rhodium (32°F). Courtesy of Leeds and Northrup Company.

IRON vs. CONSTANTAN THERMOCOUPLE

Degrees Fahrenheit — Reference Junction 32° F.

Millivolts

°F	0	1	2	3	4	5	6	7	8	9
-310	-7.66	-7.68	-7.69	-7.70	-7.71	-7.73	-7.74	-7.75	-7.76	-7.78
-300	-7.52	-7.54	-7.55	-7.57	-7.58	-7.59	-7.61	-7.62	-7.64	-7.65
-290	-7.38	-7.39	-7.40	-7.42	-7.44	-7.45	-7.46	-7.48	-7.49	-7.51
-280	-7.22	-7.24	-7.25	-7.27	-7.28	-7.30	-7.31	-7.33	-7.34	-7.36
-270	-7.06	-7.07	-7.09	-7.11	-7.12	-7.14	-7.15	-7.17	-7.19	-7.20
-260	-6.89	-6.90	-6.92	-6.94	-6.96	-6.97	-6.99	-7.01	-7.02	-7.04
-250	-6.71	-6.73	-6.75	-6.77	-6.78	-6.80	-6.82	-6.84	-6.85	-6.87
-240	-6.53	-6.55	-6.57	-6.59	-6.61	-6.62	-6.64	-6.66	-6.68	-6.70
-230	-6.35	-6.37	-6.38	-6.40	-6.42	-6.44	-6.46	-6.48	-6.50	-6.52
-220	-6.16	-6.18	-6.19	-6.21	-6.23	-6.25	-6.27	-6.29	-6.31	-6.33
-210	-5.96	-5.98	-6.00	-6.02	-6.04	-6.06	-6.08	-6.10	-6.12	-6.14
-200	-5.76	-5.78	-5.80	-5.82	-5.84	-5.86	-5.88	-5.90	-5.92	-5.94
-190	-5.55	-5.57	-5.59	-5.61	-5.63	-5.65	-5.67	-5.70	-5.72	-5.74
-180	-5.34	-5.36	-5.38	-5.40	-5.42	-5.44	-5.46	-5.49	-5.51	-5.53
-170	-5.12	-5.14	-5.16	-5.19	-5.21	-5.23	-5.25	-5.27	-5.30	-5.32
-160	-4.90	-4.92	-4.94	-4.97	-4.99	-5.01	-5.03	-5.06	-5.08	-5.10
-150	-4.68	-4.70	-4.72	-4.74	-4.76	-4.79	-4.81	-4.83	-4.86	-4.88
-140	-4.44	-4.47	-4.49	-4.51	-4.54	-4.56	-4.58	-4.61	-4.63	-4.65
-130	-4.21	-4.23	-4.26	-4.28	-4.30	-4.33	-4.35	-4.38	-4.40	-4.42
-120	-3.97	-4.00	-4.02	-4.04	-4.07	-4.09	-4.12	-4.14	-4.16	-4.19
-110	-3.73	-3.76	-3.78	-3.81	-3.83	-3.85	-3.88	-3.90	-3.93	-3.95
-100	-3.49	-3.51	-3.54	-3.56	-3.59	-3.61	-3.64	-3.66	-3.68	-3.71
-90	-3.24	-3.27	-3.29	-3.32	-3.34	-3.36	-3.39	-3.41	-3.44	-3.46
-80	-2.99	-3.02	-3.04	-3.07	-3.09	-3.12	-3.14	-3.17	-3.19	-3.22
-70	-2.74	-2.76	-2.79	-2.81	-2.84	-2.86	-2.89	-2.92	-2.94	-2.97
-60	-2.48	-2.51	-2.53	-2.56	-2.58	-2.61	-2.64	-2.66	-2.69	-2.71
-50	-2.22	-2.25	-2.27	-2.30	-2.33	-2.35	-2.38	-2.40	-2.43	-2.46
-40	-1.96	-1.99	-2.01	-2.04	-2.06	-2.09	-2.12	-2.14	-2.17	-2.20
-30	-1.70	-1.72	-1.75	-1.78	-1.80	-1.83	-1.86	-1.88	-1.91	-1.94
-20	-1.43	-1.46	-1.48	-1.51	-1.54	-1.56	-1.59	-1.62	-1.64	-1.67
-10	-1.16	-1.19	-1.21	-1.24	-1.27	-1.29	-1.32	-1.35	-1.38	-1.40
(-)0	-.89	-.91	-.94	-.97	-1.00	-1.02	-1.05	-1.08	-1.10	-1.13
(+)0	-.89	-.86	-.83	-.80	-.78	-.75	-.72	-.70	-.67	-.64
10	-.61	-.58	-.56	-.53	-.50	-.48	-.45	-.42	-.39	-.36
20	-.34	-.31	-.28	-.25	-.22	-.20	-.17	-.14	-.11	-.09
30	-.06	-.03	.00	.03	.05	.08	.11	.14	.17	.19
40	.22	.25	.28	.31	.34	.36	.39	.42	.45	.48
50	.50	.53	.56	.59	.62	.65	.67	.70	.73	.76
60	.79	.82	.84	.87	.90	.93	.96	.99	1.02	1.04
70	1.07	1.10	1.13	1.16	1.19	1.22	1.25	1.28	1.30	1.33
80	1.36	1.39	1.42	1.45	1.48	1.51	1.54	1.56	1.59	1.62
90	1.65	1.68	1.71	1.74	1.77	1.80	1.83	1.85	1.88	1.91
100	1.94	1.97	2.00	2.03	2.06	2.09	2.12	2.14	2.17	2.20
110	2.23	2.26	2.29	2.32	2.35	2.38	2.41	2.44	2.47	2.50
120	2.52	2.55	2.58	2.61	2.64	2.67	2.70	2.73	2.76	2.79
130	2.82	2.85	2.88	2.91	2.94	2.97	3.00	3.03	3.06	3.08
140	3.11	3.14	3.17	3.20	3.23	3.26	3.29	3.32	3.35	3.38
150	3.41	3.44	3.47	3.50	3.53	3.56	3.59	3.62	3.65	3.68
160	3.71	3.74	3.77	3.80	3.83	3.86	3.89	3.92	3.95	3.98
170	4.01	4.04	4.07	4.10	4.13	4.16	4.19	4.22	4.25	4.28
180	4.31	4.34	4.37	4.40	4.43	4.46	4.49	4.52	4.55	4.58
190	4.61	4.64	4.67	4.70	4.73	4.76	4.79	4.82	4.85	4.88
200	4.91	4.94	4.97	5.00	5.03	5.06	5.09	5.12	5.15	5.18
210	5.21	5.24	5.27	5.30	5.33	5.36	5.39	5.42	5.45	5.48
220	5.51	5.54	5.57	5.60	5.63	5.66	5.69	5.72	5.75	5.78
230	5.81	5.84	5.87	5.90	5.93	5.96	5.99	6.02	6.05	6.08
240	6.11	6.14	6.17	6.20	6.24	6.27	6.30	6.33	6.36	6.39
250	6.42	6.45	6.48	6.51	6.54	6.57	6.60	6.63	6.66	6.69
260	6.72	6.75	6.78	6.81	6.84	6.87	6.90	6.93	6.96	7.00
270	7.03	7.06	7.09	7.12	7.15	7.18	7.21	7.24	7.27	7.30
280	7.33	7.36	7.39	7.42	7.45	7.48	7.51	7.54	7.58	7.61
290	7.64	7.67	7.70	7.73	7.76	7.79	7.82	7.85	7.88	7.91

IRON vs. CONSTANTAN THERMOCOUPLE

Degrees Fahrenheit — Reference Junction 32° F.

Millivolts

°F	0	1	2	3	4	5	6	7	8	9
300	7.94	7.97	8.00	8.04	8.07	8.10	8.13	8.16	8.19	8.22
310	8.25	8.28	8.31	8.34	8.37	8.40	8.44	8.47	8.50	8.53
320	8.56	8.59	8.62	8.65	8.68	8.71	8.74	8.77	8.80	8.84
330	8.87	8.90	8.93	8.96	8.99	9.02	9.05	9.08	9.11	9.14
340	9.17	9.20	9.24	9.27	9.30	9.33	9.36	9.39	9.42	9.45
350	9.48	9.51	9.54	9.58	9.61	9.64	9.67	9.70	9.73	9.76
360	9.79	9.82	9.85	9.88	9.92	9.95	9.98	10.01	10.04	10.07
370	10.10	10.13	10.16	10.19	10.22	10.25	10.28	10.32	10.35	10.38
380	10.41	10.44	10.47	10.50	10.53	10.56	10.60	10.63	10.66	10.69
390	10.72	10.75	10.78	10.81	10.84	10.87	10.90	10.94	10.97	11.00
400	11.03	11.06	11.09	11.12	11.15	11.18	11.21	11.24	11.28	11.31
410	11.34	11.37	11.40	11.43	11.46	11.49	11.52	11.55	11.58	11.62
420	11.65	11.68	11.71	11.74	11.77	11.80	11.83	11.86	11.89	11.92
430	11.96	11.99	12.02	12.05	12.08	12.11	12.14	12.17	12.20	12.23
440	12.26	12.30	12.33	12.36	12.39	12.42	12.45	12.48	12.51	12.54
450	12.57	12.60	12.64	12.67	12.70	12.73	12.76	12.79	12.82	12.85
460	12.88	12.91	12.94	12.98	13.01	13.04	13.07	13.10	13.13	13.16
470	13.19	13.22	13.25	13.28	13.31	13.34	13.38	13.41	13.44	13.47
480	13.50	13.53	13.56	13.59	13.62	13.65	13.68	13.72	13.75	13.78
490	13.81	13.84	13.87	13.90	13.93	13.96	13.99	14.02	14.05	14.08
500	14.12	14.15	14.18	14.21	14.24	14.27	14.30	14.33	14.36	14.39
510	14.42	14.45	14.48	14.52	14.55	14.58	14.61	14.64	14.67	14.70
520	14.73	14.76	14.79	14.82	14.85	14.88	14.91	14.94	14.98	15.01
530	15.04	15.07	15.10	15.13	15.16	15.19	15.22	15.25	15.28	15.31
540	15.34	15.37	15.40	15.44	15.47	15.50	15.53	15.56	15.59	15.62
550	15.65	15.68	15.71	15.74	15.77	15.80	15.84	15.87	15.90	15.93
560	15.96	15.99	16.02	16.05	16.08	16.11	16.14	16.17	16.20	16.23
570	16.26	16.30	16.33	16.36	16.39	16.42	16.45	16.48	16.51	16.54
580	16.57	16.60	16.63	16.66	16.69	16.72	16.75	16.78	16.82	16.85
590	16.88	16.91	16.94	16.97	17.00	17.03	17.06	17.09	17.12	17.15
600	17.18	17.21	17.24	17.28	17.31	17.34	17.37	17.40	17.43	17.46
610	17.49	17.52	17.55	17.58	17.61	17.64	17.68	17.71	17.74	17.77
620	17.80	17.83	17.86	17.89	17.92	17.95	17.98	18.01	18.04	18.08
630	18.11	18.14	18.17	18.20	18.23	18.26	18.29	18.32	18.35	18.38
640	18.41	18.44	18.47	18.50	18.54	18.57	18.60	18.63	18.66	18.69
650	18.72	18.75	18.78	18.81	18.84	18.87	18.90	18.94	18.97	19.00
660	19.03	19.06	19.09	19.12	19.15	19.18	19.21	19.24	19.27	19.30
670	19.34	19.37	19.40	19.43	19.46	19.49	19.52	19.55	19.58	19.61
680	19.64	19.67	19.70	19.74	19.77	19.80	19.83	19.86	19.89	19.92
690	19.95	19.98	20.01	20.04	20.07	20.10	20.13	20.16	20.20	20.23
700	20.26	20.29	20.32	20.35	20.38	20.41	20.44	20.47	20.50	20.53
710	20.56	20.59	20.62	20.66	20.69	20.72	20.75	20.78	20.81	20.84
720	20.87	20.90	20.93	20.96	20.99	21.02	21.05	21.08	21.11	21.14
730	21.18	21.21	21.24	21.27	21.30	21.33	21.36	21.39	21.42	21.45
740	21.48	21.51	21.54	21.57	21.60	21.64	21.67	21.70	21.73	21.76
750	21.79	21.82	21.85	21.88	21.91	21.94	21.97	22.00	22.03	22.06
760	22.10	22.13	22.16	22.19	22.22	22.25	22.28	22.31	22.34	22.37
770	22.40	22.43	22.46	22.49	22.52	22.55	22.58	22.62	22.65	22.68
780	22.71	22.74	22.77	22.80	22.83	22.86	22.89	22.92	22.95	22.98
790	23.01	23.04	23.08	23.11	23.14	23.17	23.20	23.23	23.26	23.29
800	23.32	23.35	23.38	23.41	23.44	23.47	23.50	23.53	23.56	23.60
810	23.63	23.66	23.69	23.72	23.75	23.78	23.81	23.84	23.87	23.90
820	23.93	23.96	23.99	24.02	24.06	24.09	24.12	24.15	24.18	24.21
830	24.24	24.27	24.30	24.33	24.36	24.39	24.42	24.45	24.48	24.52
840	24.55	24.58	24.61	24.64	24.67	24.70	24.73	24.76	24.79	24.82
850	24.85	24.88	24.91	24.94	24.98	25.01	25.04	25.07	25.10	25.13
860	25.16	25.19	25.22	25.25	25.28	25.32	25.35	25.38	25.41	25.44
870	25.47	25.50	25.53	25.56	25.59	25.62	25.65	25.68	25.72	25.75
880	25.78	25.81	25.84	25.87	25.90	25.93	25.96	25.99	26.02	26.06
890	26.09	26.12	26.15	26.18	26.21	26.24	26.27	26.30	26.33	26.36

Table 9.24 continued

IRON vs. CONSTANTAN THERMOCOUPLE
Degrees Fahrenheit — Reference Junction 32° F.

°F	0	1	2	3	4	5	6	7	8	9
					Millivolts					
900	26.40	26.43	26.46	26.49	26.52	26.55	26.58	26.61	26.64	26.67
910	26.70	26.74	26.77	26.80	26.83	26.86	26.89	26.92	26.95	26.98
920	27.02	27.05	27.08	27.11	27.14	27.17	27.20	27.23	27.26	27.30
930	27.33	27.36	27.39	27.42	27.45	27.48	27.51	27.54	27.58	27.61
940	27.64	27.67	27.70	27.73	27.76	27.80	27.83	27.86	27.89	27.92
950	27.95	27.98	28.02	28.05	28.08	28.11	28.14	28.17	28.20	28.23
960	28.26	28.30	28.33	28.36	28.39	28.42	28.45	28.48	28.52	28.55
970	28.58	28.61	28.64	28.67	28.70	28.74	28.77	28.80	28.83	28.86
980	28.89	28.92	28.96	28.99	29.02	29.05	29.08	29.11	29.14	29.18
990	29.21	29.24	29.27	29.30	29.33	29.37	29.40	29.43	29.46	29.49
1000	29.52	29.56	29.59	29.62	29.65	29.68	29.71	29.75	29.78	29.81
1010	29.84	29.87	29.90	29.94	29.97	30.00	30.03	30.06	30.10	30.13
1020	30.16	30.19	30.22	30.25	30.28	30.32	30.35	30.38	30.41	30.44
1030	30.48	30.51	30.54	30.57	30.60	30.64	30.67	30.70	30.73	30.76
1040	30.80	30.83	30.86	30.89	30.92	30.96	30.99	31.02	31.05	31.08
1050	31.12	31.15	31.18	31.21	31.24	31.28	31.31	31.34	31.37	31.40
1060	31.44	31.47	31.50	31.53	31.56	31.60	31.63	31.66	31.69	31.72
1070	31.76	31.79	31.82	31.85	31.88	31.92	31.95	31.98	32.01	32.05
1080	32.08	32.11	32.14	32.18	32.21	32.24	32.27	32.30	32.34	32.37
1090	32.40	32.43	32.47	32.50	32.53	32.56	32.60	32.63	32.66	32.69
1100	32.72	32.76	32.79	32.82	32.86	32.89	32.92	32.95	32.98	33.02
1110	33.05	33.08	33.11	33.15	33.18	33.21	33.24	33.28	33.31	33.34
1120	33.37	33.41	33.44	33.47	33.50	33.54	33.57	33.60	33.64	33.67
1130	33.70	33.73	33.76	33.80	33.83	33.86	33.90	33.93	33.96	33.99
1140	34.03	34.06	34.09	34.12	34.16	34.19	34.22	34.26	34.29	34.32
1150	34.36	34.39	34.42	34.45	34.49	34.52	34.55	34.58	34.62	34.65
1160	34.68	34.72	34.75	34.78	34.82	34.85	34.88	34.92	34.95	34.98
1170	35.01	35.05	35.08	35.11	35.15	35.18	35.21	35.25	35.28	35.31
1180	35.35	35.38	35.41	35.45	35.48	35.51	35.54	35.58	35.61	35.64
1190	35.68	35.71	35.74	35.78	35.81	35.84	35.88	35.91	35.94	35.98
1200	36.01	36.05	36.08	36.11	36.15	36.18	36.21	36.25	36.28	36.31
1210	36.35	36.38	36.42	36.45	36.48	36.52	36.55	36.58	36.62	36.65
1220	36.69	36.72	36.75	36.79	36.82	36.86	36.89	36.92	36.96	36.99
1230	37.02	37.06	37.09	37.13	37.16	37.20	37.23	37.26	37.30	37.33
1240	37.36	37.40	37.43	37.47	37.50	37.54	37.57	37.60	37.64	37.67
1250	37.71	37.74	37.78	37.81	37.84	37.88	37.91	37.95	37.98	38.02
1260	38.05	38.08	38.12	38.15	38.19	38.22	38.26	38.29	38.32	38.36
1270	38.39	38.43	38.46	38.50	38.53	38.57	38.60	38.64	38.67	38.70
1280	38.74	38.77	38.81	38.84	38.88	38.91	38.95	38.98	39.02	39.05
1290	39.08	39.12	39.15	39.19	39.22	39.26	39.29	39.33	39.36	39.40
1300	39.43	39.47	39.50	39.54	39.57	39.61	39.64	39.68	39.71	39.75
1310	39.78	39.82	39.85	39.89	39.92	39.96	39.99	40.03	40.06	40.10
1320	40.13	40.17	40.20	40.24	40.27	40.31	40.34	40.38	40.41	40.45
1330	40.48	40.52	40.55	40.59	40.62	40.66	40.69	40.73	40.76	40.80
1340	40.83	40.87	40.90	40.94	40.98	41.01	41.05	41.08	41.12	41.15
1350	41.19	41.22	41.26	41.29	41.33	41.36	41.40	41.43	41.47	41.50
1360	41.54	41.58	41.61	41.65	41.68	41.72	41.75	41.79	41.82	41.86
1370	41.90	41.93	41.97	42.00	42.04	42.07	42.11	42.14	42.18	42.22
1380	42.25	42.29	42.32	42.36	42.39	42.43	42.46	42.50	42.53	42.57
1390	42.61	42.64	42.68	42.71	42.75	42.78	42.82	42.85	42.89	42.92
1400	42.96	43.00	43.03	43.07	43.10	43.14	43.18	43.21	43.25	43.28
1410	43.32	43.35	43.39	43.43	43.46	43.50	43.53	43.57	43.60	43.64
1420	43.68	43.71	43.75	43.78	43.82	43.85	43.89	43.92	43.96	44.00
1430	44.03	44.07	44.10	44.14	44.18	44.21	44.25	44.28	44.32	44.35
1440	44.39	44.42	44.46	44.50	44.53	44.57	44.60	44.64	44.68	44.71
1450	44.75	44.78	44.82	44.85	44.89	44.93	44.96	45.00	45.03	45.07
1460	45.10	45.14	45.18	45.21	45.25	45.28	45.32	45.35	45.39	45.42
1470	45.46	45.50	45.53	45.57	45.60	45.64	45.68	45.71	45.75	45.78
1480	45.82	45.85	45.89	45.92	45.96	46.00	46.03	46.07	46.10	46.14
1490	46.18	46.21	46.25	46.28	46.32	46.35	46.39	46.42	46.46	46.50

IRON vs. CONSTANTAN THERMOCOUPLE
Degrees Fahrenheit — Reference Junction 32° F.

°F	0	1	2	3	4	5	6	7	8	9
					Millivolts					
1500	46.53	46.57	46.60	46.64	46.67	46.71	46.74	46.78	46.82	46.85
1510	46.89	46.92	46.96	47.00	47.03	47.07	47.10	47.14	47.17	47.21
1520	47.24	47.28	47.32	47.35	47.39	47.42	47.46	47.49	47.53	47.56
1530	47.60	47.63	47.67	47.70	47.74	47.78	47.81	47.85	47.88	47.92
1540	47.95	47.99	48.02	48.06	48.09	48.13	48.16	48.20	48.24	48.27
1550	48.31	48.34	48.38	48.41	48.45	48.48	48.52	48.55	48.59	48.62
1560	48.66	48.69	48.73	48.76	48.80	48.83	48.87	48.90	48.94	48.97
1570	49.01	49.04	49.08	49.11	49.15	49.18	49.22	49.25	49.29	49.32
1580	49.36	49.39	49.43	49.46	49.50	49.53	49.56	49.60	49.63	49.67
1590	49.70	49.74	49.77	49.81	49.84	49.88	49.91	49.94	49.98	50.01
1600	50.05									

Because of the known instability of iron-constantan thermocouples above 1600° F., this temperature has been used as the upper limit for the tables. We do not normally recommend the use of iron-constantan above 1500° F. except in special circumstances.

For those occasional applications requiring higher ranges the values listed below are included for reference purposes. These values are derived from the formerly used 1913 tables.

Reference Junction 32° F.

° F.	MV
1700	53.55
1800	57.07
1900	60.59
2000	64.11
2100	67.63
2200	71.15

Reference Junction 0° C.

° C.	MV
900	51.87
1000	58.21
1100	64.54
1200	70.87

Table 9.24 continued

Table 9.24 continued

CHROMEL vs. ALUMEL THERMOCOUPLE

Degrees Centigrade — Reference Junction 0° C.

Millivolts

°C	0	1	2	3	4	5	6	7	8	9
0	0.00	0.04	0.08	0.12	0.16	0.20	0.24	0.28	0.32	0.36
10	0.40	0.44	0.48	0.52	0.56	0.60	0.64	0.68	0.72	0.76
20	0.80	0.84	0.88	0.92	0.96	1.00	1.04	1.08	1.12	1.16
30	1.20	1.24	1.28	1.32	1.36	1.40	1.44	1.49	1.53	1.57
40	1.61	1.65	1.69	1.73	1.77	1.81	1.85	1.90	1.94	1.98
50	2.02	2.06	2.10	2.14	2.18	2.23	2.27	2.31	2.35	2.39
60	2.43	2.47	2.51	2.56	2.60	2.64	2.68	2.72	2.76	2.80
70	2.85	2.89	2.93	2.97	3.01	3.05	3.10	3.14	3.18	3.22
80	3.26	3.30	3.35	3.39	3.43	3.47	3.51	3.56	3.60	3.64
90	3.68	3.72	3.76	3.81	3.85	3.89	3.93	3.97	4.01	4.06
100	4.10	4.14	4.18	4.22	4.26	4.31	4.35	4.39	4.43	4.47
110	4.51	4.55	4.60	4.64	4.68	4.72	4.76	4.80	4.84	4.88
120	4.92	4.96	5.01	5.05	5.09	5.13	5.17	5.21	5.25	5.29
130	5.33	5.37	5.41	5.45	5.49	5.53	5.57	5.61	5.65	5.69
140	5.73	5.77	5.81	5.85	5.89	5.93	5.97	6.01	6.05	6.09
150	6.13	6.17	6.21	6.25	6.29	6.33	6.37	6.41	6.45	6.49
160	6.53	6.57	6.61	6.65	6.69	6.73	6.77	6.81	6.85	6.89
170	6.93	6.97	7.01	7.05	7.09	7.13	7.17	7.21	7.25	7.29
180	7.33	7.37	7.41	7.45	7.49	7.53	7.57	7.61	7.65	7.69
190	7.73	7.77	7.81	7.85	7.89	7.93	7.97	8.01	8.05	8.09
200	8.13	8.17	8.21	8.25	8.29	8.33	8.37	8.41	8.46	8.50
210	8.54	8.58	8.62	8.66	8.70	8.74	8.78	8.82	8.86	8.90
220	8.94	8.98	9.02	9.06	9.10	9.14	9.18	9.22	9.26	9.30
230	9.34	9.38	9.42	9.46	9.50	9.54	9.59	9.63	9.67	9.71
240	9.75	9.79	9.83	9.87	9.91	9.95	9.99	10.03	10.07	10.11
250	10.16	10.20	10.24	10.28	10.32	10.36	10.40	10.44	10.48	10.52
260	10.57	10.61	10.65	10.69	10.73	10.77	10.81	10.85	10.89	10.93
270	10.98	11.02	11.06	11.10	11.14	11.18	11.22	11.26	11.30	11.34
280	11.39	11.43	11.47	11.51	11.55	11.59	11.63	11.67	11.72	11.76
290	11.80	11.84	11.88	11.92	11.96	12.01	12.05	12.09	12.13	12.17
300	12.21	12.25	12.29	12.34	12.38	12.42	12.46	12.50	12.54	12.58
310	12.63	12.67	12.71	12.75	12.79	12.83	12.88	12.92	12.96	13.00
320	13.04	13.08	13.12	13.17	13.21	13.25	13.29	13.33	13.37	13.42
330	13.46	13.50	13.54	13.58	13.62	13.67	13.71	13.75	13.79	13.83
340	13.88	13.92	13.96	14.00	14.04	14.09	14.13	14.17	14.21	14.25
350	14.29	14.34	14.38	14.42	14.46	14.50	14.55	14.59	14.63	14.67
360	14.71	14.76	14.80	14.84	14.88	14.92	14.97	15.01	15.05	15.09
370	15.13	15.18	15.22	15.26	15.30	15.34	15.39	15.43	15.47	15.51
380	15.55	15.60	15.64	15.68	15.72	15.76	15.81	15.85	15.89	15.93
390	15.98	16.02	16.06	16.10	16.14	16.19	16.23	16.27	16.31	16.36
400	16.40	16.44	16.48	16.52	16.57	16.61	16.65	16.69	16.74	16.78
410	16.82	16.86	16.91	16.95	16.99	17.03	17.07	17.12	17.16	17.20
420	17.24	17.29	17.33	17.37	17.41	17.46	17.50	17.54	17.58	17.62
430	17.67	17.71	17.75	17.79	17.84	17.88	17.92	17.96	18.01	18.05
440	18.09	18.13	18.17	18.22	18.26	18.30	18.34	18.39	18.43	18.47
450	18.51	18.56	18.60	18.64	18.68	18.73	18.77	18.81	18.85	18.90
460	18.94	18.98	19.02	19.07	19.11	19.15	19.19	19.24	19.28	19.32
470	19.36	19.41	19.45	19.49	19.54	19.58	19.62	19.66	19.71	19.75
480	19.79	19.84	19.88	19.92	19.96	20.01	20.05	20.09	20.13	20.18
490	20.22	20.26	20.31	20.35	20.39	20.43	20.48	20.52	20.56	20.60
500	20.65	20.69	20.73	20.77	20.82	20.86	20.90	20.94	20.99	21.03
510	21.07	21.11	21.16	21.20	21.24	21.28	21.32	21.37	21.41	21.45
520	21.50	21.54	21.58	21.63	21.67	21.71	21.75	21.80	21.84	21.88
530	21.92	21.97	22.01	22.05	22.09	22.14	22.18	22.22	22.26	22.31
540	22.35	22.39	22.43	22.48	22.52	22.56	22.61	22.65	22.69	22.73
550	22.78	22.82	22.86	22.90	22.95	22.99	23.03	23.07	23.12	23.16
560	23.20	23.25	23.29	23.33	23.38	23.42	23.46	23.50	23.54	23.59
570	23.63	23.67	23.72	23.76	23.80	23.84	23.89	23.93	23.97	24.01
580	24.06	24.10	24.14	24.18	24.23	24.27	24.31	24.36	24.40	24.44
590	24.49	24.53	24.57	24.61	24.65	24.70	24.74	24.78	24.83	24.87

CHROMEL vs. ALUMEL THERMOCOUPLE

Degrees Centigrade — Reference Junction 0° C.

Millivolts

°C	0	1	2	3	4	5	6	7	8	9
600	24.91	24.95	25.00	25.04	25.08	25.12	25.17	25.21	25.25	25.29
610	25.34	25.38	25.42	25.47	25.51	25.55	25.59	25.64	25.68	25.72
620	25.76	25.81	25.85	25.89	25.93	25.98	26.02	26.06	26.10	26.15
630	26.19	26.23	26.27	26.32	26.36	26.40	26.44	26.48	26.53	26.57
640	26.61	26.65	26.70	26.74	26.78	26.82	26.86	26.91	26.95	26.99
650	27.03	27.07	27.12	27.16	27.20	27.24	27.28	27.33	27.37	27.41
660	27.45	27.49	27.54	27.58	27.62	27.66	27.71	27.75	27.79	27.83
670	27.87	27.92	27.96	28.00	28.04	28.08	28.13	28.17	28.21	28.25
680	28.29	28.34	28.38	28.42	28.46	28.50	28.55	28.59	28.63	28.67
690	28.72	28.76	28.80	28.84	28.88	28.93	28.97	29.01	29.05	29.10
700	29.14	29.18	29.22	29.26	29.30	29.35	29.39	29.43	29.47	29.52
710	29.56	29.60	29.64	29.68	29.72	29.77	29.81	29.85	29.89	29.93
720	29.97	30.02	30.06	30.10	30.14	30.18	30.23	30.27	30.31	30.35
730	30.39	30.44	30.48	30.52	30.56	30.60	30.65	30.69	30.73	30.77
740	30.81	30.85	30.90	30.94	30.98	31.02	31.06	31.10	31.15	31.19
750	31.23	31.27	31.31	31.35	31.40	31.44	31.48	31.52	31.56	31.60
760	31.65	31.69	31.73	31.77	31.81	31.85	31.90	31.94	31.98	32.02
770	32.06	32.10	32.15	32.19	32.23	32.27	32.31	32.35	32.39	32.43
780	32.48	32.52	32.56	32.60	32.64	32.68	32.72	32.76	32.81	32.85
790	32.89	32.93	32.97	33.01	33.05	33.09	33.13	33.18	33.22	33.26
800	33.30	33.34	33.38	33.42	33.46	33.50	33.54	33.59	33.63	33.67
810	33.71	33.75	33.79	33.83	33.87	33.91	33.95	33.99	34.04	34.08
820	34.12	34.16	34.20	34.24	34.28	34.32	34.36	34.40	34.44	34.48
830	34.53	34.57	34.61	34.65	34.69	34.73	34.77	34.81	34.85	34.89
840	34.93	34.97	35.02	35.06	35.10	35.14	35.18	35.22	35.26	35.30
850	35.34	35.38	35.42	35.46	35.50	35.54	35.58	35.63	35.67	35.71
860	35.75	35.79	35.83	35.87	35.91	35.95	35.99	36.03	36.07	36.11
870	36.15	36.19	36.23	36.27	36.31	36.35	36.39	36.43	36.47	36.51
880	36.55	36.59	36.63	36.67	36.72	36.76	36.80	36.84	36.88	36.92
890	36.96	37.00	37.04	37.08	37.12	37.16	37.20	37.24	37.28	37.32
900	37.36	37.40	37.44	37.48	37.52	37.56	37.60	37.64	37.68	37.72
910	37.76	37.80	37.84	37.88	37.92	37.96	38.00	38.04	38.08	38.12
920	38.16	38.20	38.24	38.28	38.32	38.36	38.40	38.44	38.48	38.52
930	38.56	38.60	38.64	38.68	38.72	38.76	38.80	38.84	38.88	38.92
940	38.95	38.99	39.03	39.07	39.11	39.15	39.19	39.23	39.27	39.31
950	39.35	39.39	39.43	39.47	39.51	39.55	39.59	39.63	39.67	39.71
960	39.75	39.79	39.83	39.86	39.90	39.94	39.98	40.02	40.06	40.10
970	40.14	40.18	40.22	40.26	40.30	40.34	40.38	40.41	40.45	40.49
980	40.53	40.57	40.61	40.65	40.69	40.73	40.77	40.81	40.85	40.89
990	40.92	40.96	41.00	41.04	41.08	41.12	41.16	41.20	41.24	41.28
1000	41.31	41.35	41.39	41.43	41.47	41.51	41.55	41.59	41.63	41.67
1010	41.70	41.74	41.78	41.82	41.86	41.90	41.94	41.98	42.02	42.05
1020	42.09	42.13	42.17	42.21	42.25	42.29	42.33	42.36	42.40	42.44
1030	42.48	42.52	42.56	42.60	42.63	42.67	42.71	42.75	42.79	42.83
1040	42.87	42.90	42.94	42.98	43.02	43.06	43.10	43.14	43.17	43.21
1050	43.25	43.29	43.33	43.37	43.41	43.44	43.48	43.52	43.56	43.60
1060	43.63	43.67	43.71	43.75	43.79	43.83	43.87	43.90	43.94	43.98
1070	44.02	44.06	44.10	44.13	44.17	44.21	44.25	44.29	44.33	44.36
1080	44.40	44.44	44.48	44.52	44.55	44.59	44.63	44.67	44.71	44.74
1090	44.78	44.82	44.86	44.90	44.93	44.97	45.01	45.05	45.09	45.12
1100	45.16	45.20	45.24	45.27	45.31	45.35	45.39	45.43	45.46	45.50
1110	45.54	45.58	45.62	45.65	45.69	45.73	45.77	45.80	45.84	45.88
1120	45.92	45.96	45.99	46.03	46.07	46.11	46.14	46.18	46.22	46.26
1130	46.29	46.33	46.37	46.41	46.44	46.48	46.52	46.56	46.59	46.63
1140	46.67	46.70	46.74	46.78	46.82	46.85	46.89	46.93	46.97	47.00
1150	47.04	47.08	47.12	47.15	47.19	47.23	47.26	47.30	47.34	47.38
1160	47.41	47.45	47.49	47.52	47.56	47.60	47.63	47.67	47.71	47.75
1170	47.78	47.82	47.86	47.89	47.93	47.97	48.00	48.04	48.08	48.12
1180	48.15	48.19	48.23	48.26	48.30	48.34	48.37	48.41	48.45	48.48
1190	48.52	48.56	48.59	48.63	48.67	48.70	48.74	48.78	48.81	48.85

Table 9.24 continued

Table 9.24 continued

CHROMEL vs. ALUMEL THERMOCOUPLE

Degrees Centigrade — Reference Junction 0° C.

°C	0	1	2	3	4	5	6	7	8	9
					Millivolts					
1200	48.89	48.92	48.96	49.00	49.03	49.07	49.11	49.14	49.18	49.22
1210	49.25	49.29	49.32	49.36	49.40	49.43	49.47	49.51	49.54	49.58
1220	49.62	49.65	49.69	49.72	49.76	49.80	49.83	49.87	49.90	49.94
1230	49.98	50.01	50.05	50.08	50.12	50.16	50.19	50.23	50.26	50.30
1240	50.34	50.37	50.41	50.44	50.48	50.52	50.55	50.59	50.62	50.66
1250	50.69	50.73	50.77	50.80	50.84	50.87	50.91	50.94	50.98	51.02
1260	51.05	51.09	51.12	51:16	51.19	51.23	51.27	51.30	51.34	51.37
1270	51.41	51.44	51.48	51.51	51.55	51.58	51.62	51.66	51.69	51.73
1280	51.76	51.80	51.83	51.87	51.90	51.94	51.97	52.01	52.04	52.08
1290	52.11	52.15	52.18	52.22	52.25	52.29	52.32	52.36	52.39	52.43
1300	52.46	52.50	52.53	52.57	52.60	52.64	52.67	52.71	52.74	52.78
1310	52.81	52.85	52.88	52.92	52.95	52.99	53.02	53.06	53.09	53.13
1320	53.16	53.20	53.23	53.27	53.30	53.34	53.37	53.41	53.44	53.47
1330	53.51	53.54	53.58	53.61	53.65	53.68	53.72	53.75	53.79	53.82
1340	53.85	53.89	53.92	53.96	53.99	54.03	54.06	54.10	54.13	54.16
1350	54.20	54.23	54.27	54.30	54.34	54.37	54.40	54.44	54.47	54.51
1360	54.54	54.57	54.61	54.64	54.68	54.71	54.74	54.78	54.81	54.85
1370	54.88	54.91								

CHROMEL vs. ALUMEL THERMOCOUPLE

Degrees Fahrenheit — Reference Junction 32° F.

°F	0	1	2	3	4	5	6	7	8	9
					Millivolts					
0	-0.68	-0.66	-0.64	-0.62	-0.60	-0.58	-0.56	-0.54	-0.52	-0.49
10	-0.47	-0.45	-0.43	-0.41	-0.39	-0.37	-0.34	-0.32	-0.30	-0.28
20	-0.26	-0.24	-0.22	-0.19	-0.17	-0.15	-0.13	-0.11	-0.09	-0.07
30	-0.04	-0.02	0.00	0.02	0.04	0.07	0.09	0.11	0.13	0.15
40	0.18	0.20	0.22	0.24	0.26	0.29	0.31	0.33	0.35	0.37
50	0.40	0.42	0.44	0.46	0.48	0.51	0.53	0.55	0.57	0.60
60	0.62	0.64	0.66	0.68	0.71	0.73	0.75	0.77	0.80	0.82
70	0.84	0.86	0.88	0.91	0.93	0.95	0.97	1.00	1.02	1.04
80	1.06	1.09	1.11	1.13	1.15	1.18	1.20	1.22	1.24	1.27
90	1.29	1.31	1.33	1.36	1.38	1.40	1.43	1.45	1.47	1.49
100	1.52	1.54	1.56	1.58	1.61	1.63	1.65	1.68	1.70	1.72
110	1.74	1.77	1.79	1.81	1.84	1.86	1.88	1.90	1.93	1.95
120	1.97	2.00	2.02	2.04	2.06	2.09	2.11	2.13	2.16	2.18
130	2.20	2.23	2.25	2.27	2.29	2.32	2.34	2.36	2.39	2.41
140	2.43	2.46	2.48	2.50	2.52	2.55	2.57	2.59	2.62	2.64
150	2.66	2.69	2.71	2.73	2.75	2.78	2.80	2.82	2.85	2.87
160	2.89	2.92	2.94	2.96	2.98	3.01	3.03	3.05	3.08	3.10
170	3.12	3.15	3.17	3.19	3.22	3.24	3.26	3.29	3.31	3.33
180	3.36	3.38	3.40	3.43	3.45	3.47	3.49	3.52	3.54	3.56
190	3.59	3.61	3.63	3.66	3.68	3.70	3.73	3.75	3.77	3.80
200	3.82	3.84	3.87	3.89	3.91	3.94	3.96	3.98	4.01	4.03
210	4.05	4.08	4.10	4.12	4.15	4.17	4.19	4.21	4.24	4.26
220	4.28	4.31	4.33	4.35	4.38	4.40	4.42	4.44	4.47	4.49
230	4.51	4.54	4.56	4.58	4.61	4.63	4.65	4.67	4.70	4.72
240	4.74	4.77	4.79	4.81	4.83	4.86	4.88	4.90	4.92	4.95
250	4.97	4.99	5.02	5.04	5.06	5.08	5.11	5.13	5.15	5.17
260	5.20	5.22	5.24	5.26	5.29	5.31	5.33	5.35	5.38	5.40
270	5.42	5.44	5.47	5.49	5.51	5.53	5.56	5.58	5.60	5.62
280	5.65	5.67	5.69	5.71	5.73	5.76	5.78	5.80	5.82	5.85
290	5.87	5.89	5.91	5.93	5.96	5.98	6.00	6.02	6.05	6.07
300	6.09	6.11	6.13	6.16	6.18	6.20	6.22	6.25	6.27	6.29
310	6.31	6.33	6.36	6.38	6.40	6.42	6.45	6.47	6.49	6.51
320	6.53	6.56	6.58	6.60	6.62	6.65	6.67	6.69	6.71	6.73
330	6.76	6.78	6.80	6.82	6.84	6.87	6.89	6.91	6.93	6.96
340	6.98	7.00	7.02	7.04	7.07	7.09	7.11	7.13	7.15	7.18
350	7.20	7.22	7.24	7.26	7.29	7.31	7.33	7.35	7.38	7.40
360	7.42	7.44	7.46	7.49	7.51	7.53	7.55	7.58	7.60	7.62
370	7.64	7.66	7.69	7.71	7.73	7.75	7.78	7.80	7.82	7.84
380	7.87	7.89	7.91	7.93	7.95	7.98	8.00	8.02	8.04	8.07
390	8.09	8.11	8.13	8.16	8.18	8.20	8.22	8.24	8.27	8.29
400	8.31	8.33	8.36	8.38	8.40	8.42	8.45	8.47	8.49	8.51
410	8.54	8.56	8.58	8.60	8.62	8.65	8.67	8.69	8.71	8.74
420	8.76	8.78	8.80	8.82	8.85	8.87	8.89	8.91	8.94	8.96
430	8.98	9.00	9.03	9.05	9.07	9.09	9.12	9.14	9.16	9.18
440	9.21	9.23	9.25	9.27	9.30	9.32	9.34	9.36	9.39	9.41
450	9.43	9.45	9.48	9.50	9.52	9.54	9.57	9.59	9.61	9.63
460	9.66	9.68	9.70	9.73	9.75	9.77	9.79	9.82	9.84	9.86
470	9.88	9.91	9.93	9.95	9.97	10.00	10.02	10.04	10.06	10.09
480	10.11	10.13	10.16	10.18	10.20	10.22	10.25	10.27	10.29	10.31
490	10.34	10.36	10.38	10.40	10.43	10.45	10.47	10.50	10.52	10.54
500	10.57	10.59	10.61	10.63	10.66	10.68	10.70	10.72	10.75	10.77
510	10.79	10.82	10.84	10.86	10.88	10.91	10.93	10.95	10.98	11.00
520	11.02	11.04	11.07	11.09	11.11	11.13	11.16	11.18	11.20	11.23
530	11.25	11.27	11.29	11.32	11.34	11.36	11.39	11.41	11.43	11.45
540	11.48	11.50	11.52	11.55	11.57	11.59	11.61	11.64	11.66	11.68
550	11.71	11.73	11.75	11.78	11.80	11.82	11.84	11.87	11.89	11.91
560	11.94	11.96	11.98	12.01	12.03	12.05	12.07	12.10	12.12	12.14
570	12.17	12.19	12.21	12.24	12.26	12.28	12.30	12.33	12.35	12.37
580	12.40	12.42	12.44	12.47	12.49	12.51	12.53	12.56	12.58	12.60
590	12.63	12.65	12.67	12.70	12.72	12.74	12.76	12.79	12.81	12.83

Table 9.24 continued

Table 9.24 continued

CHROMEL vs. ALUMEL THERMOCOUPLE

Degrees Fahrenheit — Reference Junction 32° F.

°F	0	1	2	3	4	5	6	7	8	9
					Millivolts					
600	12.86	12.88	12.90	12.93	12.95	12.97	13.00	13.02	13.04	13.06
610	13.09	13.11	13.13	13.16	13.18	13.20	13.23	13.25	13.27	13.30
620	13.32	13.34	13.36	13.39	13.41	13.44	13.46	13.48	13.50	13.53
630	13.55	13.57	13.60	13.62	13.64	13.67	13.69	13.71	13.74	13.76
640	13.78	13.81	13.83	13.85	13.88	13.90	13.92	13.95	13.97	13.99
650	14.02	14.04	14.06	14.09	14.11	14.13	14.15	14.18	14.20	14.22
660	14.25	14.27	14.29	14.32	14.34	14.36	14.39	14.41	14.43	14.46
670	14.48	14.50	14.53	14.55	14.57	14.60	14.62	14.64	14.67	14.69
680	14.71	14.74	14.76	14.78	14.81	14.83	14.85	14.88	14.90	14.92
690	14.95	14.97	14.99	15.02	15.04	15.06	15.09	15.11	15.13	15.16
700	15.18	15.20	15.23	15.25	15.27	15.30	15.32	15.34	15.37	15.39
710	15.41	15.44	15.46	15.48	15.51	15.53	15.55	15.58	15.60	15.62
720	15.65	15.67	15.69	15.72	15.74	15.76	15.79	15.81	15.83	15.86
730	15.88	15.90	15.93	15.95	15.98	16.00	16.02	16.05	16.07	16.09
740	16.12	16.14	16.16	16.19	16.21	16.23	16.26	16.28	16.30	16.33
750	16.35	16.37	16.40	16.42	16.45	16.47	16.49	16.52	16.54	16.56
760	16.59	16.61	16.63	16.66	16.68	16.70	16.73	16.75	16.77	16.80
770	16.82	16.84	16.87	16.89	16.92	16.94	16.96	16.99	17.01	17.03
780	17.06	17.08	17.10	17.13	17.15	17.17	17.20	17.22	17.24	17.27
790	17.29	17.31	17.34	17.36	17.39	17.41	17.43	17.46	17.48	17.50
800	17.53	17.55	17.57	17.60	17.62	17.64	17.67	17.69	17.71	17.74
810	17.76	17.78	17.81	17.83	17.86	17.88	17.90	17.93	17.95	17.97
820	18.00	18.02	18.04	18.07	18.09	18.11	18.14	18.16	18.18	18.21
830	18.23	18.25	18.28	18.30	18.33	18.35	18.37	18.40	18.42	18.44
840	18.47	18.49	18.51	18.54	18.56	18.58	18.61	18.63	18.65	18.68
850	18.70	18.73	18.75	18.77	18.80	18.82	18.84	18.87	18.89	18.91
860	18.94	18.96	18.99	19.01	19.03	19.06	19.08	19.10	19.13	19.15
870	19.18	19.20	19.22	19.25	19.27	19.29	19.32	19.34	19.36	19.39
880	19.41	19.44	19.46	19.48	19.51	19.53	19.55	19.58	19.60	19.63
890	19.65	19.67	19.70	19.72	19.75	19.77	19.79	19.82	19.84	19.86
900	19.89	19.91	19.94	19.96	19.98	20.01	20.03	20.05	20.08	20.10
910	20.13	20.15	20.17	20.20	20.22	20.24	20.27	20.29	20.32	20.34
920	20.36	20.39	20.41	20.43	20.46	20.48	20.50	20.53	20.55	20.58
930	20.60	20.62	20.65	20.67	20.69	20.72	20.74	20.76	20.79	20.81
940	20.84	20.86	20.88	20.91	20.93	20.95	20.98	21.00	21.03	21.05
950	21.07	21.10	21.12	21.14	21.17	21.19	21.21	21.24	21.26	21.28
960	21.31	21.33	21.36	21.38	21.40	21.43	21.45	21.47	21.50	21.52
970	21.54	21.57	21.59	21.62	21.64	21.66	21.69	21.71	21.73	21.76
980	21.78	21.81	21.83	21.85	21.88	21.90	21.92	21.95	21.97	21.99
990	22.02	22.04	22.07	22.09	22.11	22.14	22.16	22.18	22.21	22.23
1000	22.26	22.28	22.30	22.33	22.35	22.37	22.40	22.42	22.44	22.47
1010	22.49	22.52	22.54	22.56	22.59	22.61	22.63	22.66	22.68	22.71
1020	22.73	22.75	22.78	22.80	22.82	22.85	22.87	22.90	22.92	22.94
1030	22.97	22.99	23.01	23.04	23.06	23.08	23.11	23.13	23.16	23.18
1040	23.20	23.23	23.25	23.27	23.30	23.32	23.35	23.37	23.39	23.42
1050	23.44	23.46	23.49	23.51	23.54	23.56	23.58	23.61	23.63	23.65
1060	23.68	23.70	23.72	23.75	23.77	23.80	23.82	23.84	23.87	23.89
1070	23.91	23.94	23.96	23.99	24.01	24.03	24.06	24.08	24.10	24.13
1080	24.15	24.18	24.20	24.22	24.25	24.27	24.29	24.32	24.34	24.36
1090	24.39	24.41	24.44	24.46	24.49	24.51	24.53	24.55	24.58	24.60
1100	24.63	24.65	24.67	24.70	24.72	24.74	24.77	24.79	24.82	24.84
1110	24.86	24.89	24.91	24.93	24.96	24.98	25.01	25.03	25.05	25.08
1120	25.10	25.12	25.15	25.17	25.20	25.22	25.24	25.27	25.29	25.31
1130	25.34	25.36	25.38	25.41	25.43	25.46	25.48	25.50	25.53	25.55
1140	25.57	25.60	25.62	25.65	25.67	25.69	25.72	25.74	25.76	25.79
1150	25.81	25.83	25.86	25.88	25.91	25.93	25.95	25.98	26.00	26.02
1160	26.05	26.07	26.09	26.12	26.14	26.16	26.19	26.21	26.24	26.26
1170	26.28	26.31	26.33	26.35	26.38	26.40	26.42	26.45	26.47	26.49
1180	26.52	26.54	26.56	26.59	26.61	26.63	26.66	26.68	26.70	26.73
1190	26.75	26.77	26.80	26.82	26.85	26.87	26.89	26.91	26.94	26.96

CHROMEL vs. ALUMEL THERMOCOUPLE

Degrees Fahrenheit — Reference Junction 32° F.

°F	0	1	2	3	4	5	6	7	8	9
					Millivolts					
1200	26.98	27.01	27.03	27.06	27.08	27.10	27.12	27.15	27.17	27.20
1210	27.22	27.24	27.27	27.29	27.31	27.34	27.36	27.38	27.40	27.43
1220	27.45	27.48	27.50	27.52	27.55	27.57	27.59	27.62	27.64	27.66
1230	27.69	27.71	27.73	27.76	27.78	27.80	27.83	27.85	27.87	27.90
1240	27.92	27.94	27.97	27.99	28.01	28.04	28.06	28.08	28.11	28.13
1250	28.15	28.18	28.20	28.22	28.25	28.27	28.29	28.32	28.34	28.37
1260	28.39	28.41	28.44	28.46	28.48	28.50	28.53	28.55	28.58	28.60
1270	28.62	28.65	28.67	28.69	28.72	28.74	28.76	28.79	28.81	28.83
1280	28.86	28.88	28.90	28.93	28.95	28.97	29.00	29.02	29.04	29.07
1290	29.09	29.11	29.14	29.16	29.18	29.21	29.23	29.25	29.28	29.30
1300	29.32	29.35	29.37	29.39	29.42	29.44	29.46	29.49	29.51	29.53
1310	29.56	29.58	29.60	29.63	29.65	29.67	29.70	29.72	29.74	29.77
1320	29.79	29.81	29.84	29.86	29.88	29.91	29.93	29.95	29.97	30.00
1330	30.02	30.05	30.07	30.09	30.11	30.14	30.16	30.18	30.21	30.23
1340	30.25	30.28	30.30	30.32	30.35	30.37	30.39	30.42	30.44	30.46
1350	30.49	30.51	30.53	30.56	30.58	30.60	30.63	30.65	30.67	30.70
1360	30.72	30.74	30.77	30.79	30.81	30.83	30.86	30.88	30.90	80.93
1370	30.95	30.97	31.00	31.02	31.04	31.07	31.09	31.11	31.14	31.16
1380	31.18	31.21	31.23	31.25	31.28	31.30	31.32	31.34	31.37	31.39
1390	31.42	31.44	31.46	31.48	31.51	31.53	31.55	31.58	31.60	31.62
1400	31.65	31.67	31.69	31.72	31.74	31.76	31.78	31.81	31.83	31.85
1410	31.88	31.90	31.92	31.95	31.97	31.99	32.02	32.04	32.06	32.08
1420	32.11	32.13	32.15	32.18	32.20	32.22	32.25	32.27	32.29	32.31
1430	32.34	32.36	32.38	32.41	32.43	32.45	32.48	32.50	32.52	32.54
1440	32.57	32.59	32.61	32.64	32.66	32.68	32.70	32.73	32.75	32.77
1450	32.80	32.82	32.84	32.86	32.89	32.91	32.93	32.96	32.98	33.00
1460	33.02	33.05	33.07	33.09	33.12	33.14	33.16	33.18	33.21	33.23
1470	33.25	33.28	33.30	33.32	33.34	33.37	33.39	33.41	33.43	33.46
1480	33.48	33.50	33.53	33.55	33.57	33.59	33.62	33.64	33.66	33.69
1490	33.71	33.73	33.75	33.78	33.80	33.82	33.84	33.87	33.89	33.91
1500	33.93	33.96	33.98	34.00	34.03	34.05	34.07	34.09	34.12	34.14
1510	34.16	34.18	34.21	34.23	34.25	34.28	34.30	34.32	34.34	34.37
1520	34.39	34.41	34.43	34.46	34.48	34.50	34.53	34.55	34.57	34.59
1530	34.62	34.64	34.66	34.68	34.71	34.73	34.75	34.77	34.80	34.82
1540	34.84	34.87	34.89	34.91	34.93	34.96	34.98	35.00	35.02	35.05
1550	35.07	35.09	35.11	35.14	35.16	35.18	35.21	35.23	35.25	35.27
1560	35.29	35.32	35.34	35.36	35.39	35.41	35.43	35.45	35.48	35.50
1570	35.52	35.54	35.57	35.59	35.61	35.63	35.66	35.68	35.70	35.72
1580	35.75	35.77	35.79	35.81	35.84	35.86	35.88	35.90	35.93	35.95
1590	35.97	35.99	36.02	36.04	36.06	36.08	36.11	36.13	36.15	36.17
1600	36.19	36.22	36.24	36.26	36.29	36.31	36.33	36.35	36.37	36.40
1610	36.42	36.44	36.46	36.49	36.51	36.53	36.55	36.58	36.60	36.62
1620	36.64	36.67	36.69	36.71	36.73	36.76	36.78	36.80	36.82	36.84
1630	36.87	36.89	36.91	36.93	36.96	36.98	37.00	37.02	37.05	37.07
1640	37.09	37.11	37.14	37.16	37.18	37.20	37.23	37.25	37.27	37.29
1650	37.31	37.34	37.36	37.38	37.40	37.43	37.45	37.47	37.49	37.52
1660	37.54	37.56	37.58	37.60	37.63	37.65	37.67	37.69	37.72	37.74
1670	37.76	37.78	37.81	37.83	37.85	37.87	37.89	37.92	37.94	37.96
1680	37.98	38.01	38.03	38.05	38.07	38.09	38.12	38.14	38.16	38.18
1690	38.20	38.23	38.25	38.27	38.29	38.32	38.34	38.36	38.38	38.40
1700	38.43	38.45	38.47	38.49	38.51	38.54	38.56	38.58	38.60	38.62
1710	38.65	38.67	38.69	38.71	38.73	38.76	38.78	38.80	38.82	38.84
1720	38.87	38.89	38.91	38.93	38.95	38.98	39.00	39.02	39.04	39.06
1730	39.09	39.11	39.13	39.15	39.17	39.20	39.22	39.24	39.26	39.28
1740	39.31	39.33	39.35	39.37	39.39	39.42	39.44	39.46	39.48	39.50
1750	39.53	39.55	39.57	39.59	39.61	39.64	39.66	39.68	39.70	39.72
1760	39.75	39.77	39.79	39.81	39.83	39.86	39.88	39.90	39.92	39.94
1770	39.96	39.99	40.01	40.03	40.05	40.07	40.10	40.12	40.14	40.16
1780	40.18	40.20	40.23	40.25	40.27	40.29	40.31	40.34	40.36	40.38
1790	40.40	40.42	40.44	40.47	40.49	40.51	40.53	40.55	40.58	40.60

Table 9.24 continued

Table 9.24 continued

CHROMEL vs. ALUMEL THERMOCOUPLE

Degrees Fahrenheit — Reference Junction 32° F.

°F	0	1	2	3	4	5	6	7	8	9
					Millivolts					
1800	40.62	40.64	40.66	40.68	40.71	40.73	40.75	40.77	40.79	40.82
1810	40.84	40.86	40.88	40.90	40.92	40.95	40.97	40.99	41.01	41.03
1820	41.05	41.08	41.10	41.12	41.14	41.16	41.18	41.21	41.23	41.25
1830	41.27	41.29	41.31	41.34	41.36	41.38	41.40	41.42	41.45	41.47
1840	41.49	41.51	41.53	41.55	41.57	41.60	41.62	41.64	41.66	41.68
1850	41.70	41.73	41.75	41.77	41.79	41.81	41.83	41.85	41.88	41.90
1860	41.92	41.94	41.96	41.99	42.01	42.03	42.05	42.07	42.09	42.11
1870	42.14	42.16	42.18	42.20	42.22	42.24	42.26	42.29	42.31	42.33
1880	42.35	42.37	42.39	42.42	42.44	42.46	42.48	42.50	42.52	42.55
1890	42.57	42.59	42.61	42.63	42.65	42.67	42.69	42.72	42.74	42.76
1900	42.78	42.80	42.82	42.84	42.87	42.89	42.91	42.93	42.95	42.97
1910	42.99	43.01	43.04	43.06	43.08	43.10	43.12	43.14	43.17	43.19
1920	43.21	43.23	43.25	43.27	43.29	43.31	43.34	43.36	43.38	43.40
1930	43.42	43.44	43.47	43.49	43.51	43.53	43.55	43.57	43.59	43.61
1940	43.63	43.66	43.68	43.70	43.72	43.74	43.76	43.78	43.81	43.83
1950	43.85	43.87	43.89	43.91	43.93	43.95	43.98	44.00	44.02	44.04
1960	44.06	44.08	44.10	44.13	44.15	44.17	44.19	44.21	44.23	44.25
1970	44.27	44.30	44.32	44.34	44.36	44.38	44.40	44.42	44.44	44.47
1980	44.49	44.51	44.53	44.55	44.57	44.59	44.61	44.63	44.66	44.68
1990	44.70	44.72	44.74	44.76	44.78	44.80	44.82	44.85	44.87	44.89
2000	44.91	44.93	44.95	44.97	44.99	45.01	45.03	45.06	45.08	45.10
2010	45.12	45.14	45.16	45.18	45.20	45.22	45.24	45.27	45.29	45.31
2020	45.33	45.35	45.37	45.39	45.41	45.43	45.45	45.48	45.50	45.52
2030	45.54	45.56	45.58	45.60	45.62	45.64	45.66	45.69	45.71	45.73
2040	45.75	45.77	45.79	45.81	45.83	45.85	45.87	45.90	45.92	45.94
2050	45.96	45.98	46.00	46.02	46.04	46.06	46.08	46.11	46.13	46.15
2060	46.17	46.19	46.21	46.23	46.25	46.27	46.29	46.31	46.33	46.36
2070	46.38	46.40	46.42	46.44	46.46	46.48	46.50	46.52	46.54	46.56
2080	46.58	46.60	46.63	46.65	46.67	46.69	46.71	46.73	46.75	46.77
2090	46.79	46.81	46.83	46.85	46.87	46.90	46.92	46.94	46.96	46.98
2100	47.00	47.02	47.04	47.06	47.08	47.10	47.12	47.14	47.17	47.19
2110	47.21	47.23	47.25	47.27	47.29	47.31	47.33	47.35	47.37	47.39
2120	47.41	47.43	47.45	47.47	47.49	47.52	47.54	47.56	47.58	47.60
2130	47.62	47.64	47.66	47.68	47.70	47.72	47.74	47.76	47.78	47.80
2140	47.82	47.84	47.86	47.89	47.91	47.93	47.95	47.97	47.99	48.01
2150	48.03	48.05	48.07	48.09	48.11	48.13	48.15	48.17	48.19	48.21
2160	48.23	48.25	48.27	48.29	48.32	48.34	48.36	48.38	48.40	48.42
2170	48.44	48.46	48.48	48.50	48.52	48.54	48.56	48.58	48.60	48.62
2180	48.64	48.66	48.68	48.70	48.72	48.74	48.76	48.79	48.81	48.83
2190	48.85	48.87	48.89	48.91	48.93	48.95	48.97	48.99	49.01	49.03
2200	49.05	49.07	49.09	49.11	49.13	49.15	49.17	49.19	49.21	49.23
2210	49.25	49.27	49.29	49.31	49.33	49.35	49.37	49.39	49.41	49.43
2220	49.45	49.47	49.49	49.51	49.53	49.55	49.57	49.59	49.61	49.63
2230	49.65	49.67	49.69	49.71	49.73	49.76	49.78	49.80	49.82	49.84
2240	49.86	49.88	49.90	49.92	49.94	49.96	49.98	50.00	50.02	50.04
2250	50.06	50.08	50.10	50.12	50.14	50.16	50.18	50.20	50.22	50.24
2260	50.26	50.28	50.30	50.32	50.34	50.36	50.38	50.40	50.42	50.44
2270	50.46	50.48	50.50	50.52	50.54	50.56	50.57	50.59	50.61	50.63
2280	50.65	50.67	50.69	50.71	50.73	50.75	50.77	50.79	50.81	50.83
2290	50.85	50.87	50.89	50.91	50.93	50.95	50.97	50.99	51.01	51.03
2300	51.05	51.07	51.09	51.11	51.13	51.15	51.17	51.19	51.21	51.23
2310	51.25	51.27	51.29	51.31	51.33	51.35	51.37	51.39	51.41	51.43
2320	51.45	51.47	51.48	51.50	51.52	51.54	51.56	51.58	51.60	51.62
2330	51.64	51.66	51.68	51.70	51.72	51.74	51.76	51.78	51.80	51.82
2340	51.84	51.86	51.88	51.90	51.92	51.94	51.96	51.98	52.00	51.01
2350	52.03	52.05	52.07	52.09	52.11	52.13	52.15	52.17	52.19	52.21
2360	52.23	52.25	52.27	52.29	52.31	52.33	52.35	52.37	52.39	52.41
2370	52.42	52.44	52.46	52.48	52.50	52.52	52.54	52.56	52.58	52.60
2380	52.62	52.64	52.66	52.68	52.70	52.72	52.74	52.76	52.77	52.79
2390	52.81	52.83	52.85	52.87	52.89	52.91	52.93	52.95	52.97	52.99

CHROMEL vs. ALUMEL THERMOCOUPLE

Degrees Fahrenheit — Reference Junction 32° F.

°F	0	1	2	3	4	5	6	7	8	9
					Millivolts					
2400	53.01	53.03	53.05	53.07	53.08	53.10	53.12	53.14	53.16	53.18
2410	53.20	53.22	53.24	53.26	53.28	53.30	53.32	53.34	53.35	53.37
2420	53.39	53.41	53.43	53.45	53.47	53.49	53.51	53.53	53.55	53.57
2430	53.59	53.60	53.62	53.64	53.66	53.68	53.70	53.72	53.74	53.76
2440	53.78	53.80	53.82	53.83	53.85	53.87	53.89	53.91	53.93	53.95
2450	53.97	53.99	54.01	54.03	54.04	54.06	54.08	54.10	54.12	54.14
2460	54.16	54.18	54.20	54.22	54.24	54.25	54.27	54.29	54.31	54.33
2470	54.35	54.37	54.39	54.41	54.43	54.44	54.46	54.48	54.50	54.52
2480	54.54	54.56	54.58	54.60	54.62	54.63	54.65	54.67	54.69	54.71
2490	54.73	54.75	54.77	54.79	54.81	54.82	54.84	54.86	54.88	54.90

Table 9.24 continued

COPPER vs. CONSTANTAN THERMOCOUPLE
Degrees Centigrade — Reference Junction 0° C.

°C	0	1	2	3	4	5	6	7	8	9
					Millivolts					
-190	-5.379	-5.395	-5.411	....	....	....	....	....	....	....
-180	-5.205	-5.223	-5.241	-5.258	-5.276	-5.294	-5.311	-5.328	-5.345	-5.362
-170	-5.018	-5.037	-5.056	-5.075	-5.094	-5.113	-5.132	-5.150	-5.169	-5.187
-160	-4.817	-4.838	-4.858	-4.878	-4.899	-4.919	-4.939	-4.959	-4.978	-4.998
-150	-4.603	-4.625	-4.647	-4.669	-4.690	-4.712	-4.733	-4.754	-4.775	-4.796
-140	-4.377	-4.400	-4.423	-4.446	-4.469	-4.492	-4.514	-4.537	-4.559	-4.581
-130	-4.138	-4.162	-4.187	-4.211	-4.235	-4.259	-4.283	-4.307	-4.330	-4.354
-120	-3.887	-3.912	-3.938	-3.964	-3.989	-4.014	-4.039	-4.064	-4.089	-4.114
-110	-3.624	-3.651	-3.678	-3.704	-3.730	-3.757	-3.783	-3.809	-3.835	-3.861
-100	-3.349	-3.377	-3.405	-3.432	-3.460	-3.488	-3.515	-3.542	-3.570	-3.597
-90	-3.062	-3.091	-3.120	-3.149	-3.178	-3.207	-3.235	-3.264	-3.292	-3.320
-80	-2.764	-2.794	-2.824	-2.854	-2.884	-2.914	-2.944	-2.974	-3.003	-3.033
-70	-2.455	-2.486	-2.518	-2.549	-2.580	-2.611	-2.642	-2.672	-2.703	-2.733
-60	-2.135	-2.167	-2.200	-2.232	-2.264	-2.296	-2.328	-2.360	-2.392	-2.423
-50	-1.804	-1.838	-1.871	-1.905	-1.938	-1.971	-2.004	-2.037	-2.070	-2.103
-40	-1.463	-1.498	-1.532	-1.567	-1.601	-1.635	-1.669	-1.703	-1.737	-1.771
-30	-1.112	-1.148	-1.183	-1.218	-1.254	-1.289	-1.324	-1.359	-1.394	-1.429
-20	-0.751	-0.788	-0.824	-0.860	-0.897	-0.933	-0.969	-1.005	-1.041	-1.076
-10	-0.380	-0.417	-0.455	-0.492	-0.530	-0.567	-0.604	-0.641	-0.678	-0.714
(-)0	0.000	-0.038	-0.077	-0.115	-0.153	-0.191	-0.229	-0.267	-0.305	-0.343
(+)0	0.000	0.038	0.077	0.116	0.154	0.193	0.232	0.271	0.311	0.350
10	0.389	0.429	0.468	0.508	0.547	0.587	0.627	0.667	0.707	0.747
20	0.787	0.827	0.868	0.908	0.949	0.990	1.030	1.071	1.112	1.153
30	1.194	1.235	1.277	1.318	1.360	1.401	1.443	1.485	1.526	1.568
40	1.610	1.652	1.694	1.737	1.779	1.821	1.864	1.907	1.949	1.992
50	2.035	2.078	2.121	2.164	2.207	2.250	2.293	2.336	2.380	2.423
60	2.467	2.511	2.555	2.599	2.643	2.687	2.731	2.775	2.820	2.864
70	2.908	2.953	2.997	3.042	3.087	3.132	3.177	3.222	3.267	3.312
80	3.357	3.402	3.448	3.493	3.539	3.584	3.630	3.676	3.722	3.767
90	3.813	3.859	3.906	3.952	3.998	4.044	4.091	4.138	4.184	4.230
100	4.277	4.324	4.371	4.418	4.465	4.512	4.559	4.606	4.654	4.701
110	4.749	4.796	4.843	4.891	4.939	4.987	5.035	5.083	5.131	5.179
120	5.227	5.275	5.323	5.372	5.420	5.469	5.518	5.566	5.615	5.663
130	5.712	5.761	5.810	5.859	5.908	5.957	6.007	6.056	6.105	6.155
140	6.204	6.254	6.303	6.353	6.403	6.453	6.503	6.553	6.603	6.653
150	6.703	6.753	6.803	6.853	6.904	6.954	7.004	7.055	7.106	7.157
160	7.208	7.258	7.309	7.360	7.411	7.462	7.513	7.565	7.616	7.667
170	7.719	7.770	7.822	7.874	7.926	7.978	8.029	8.080	8.132	8.184
180	8.236	8.288	8.340	8.392	8.445	8.497	8.549	8.601	8.654	8.707
190	8.759	8.812	8.864	8.917	8.970	9.023	9.076	9.129	9.182	9.235
200	9.288	9.341	9.394	9.448	9.501	9.555	9.608	9.662	9.715	9.769
210	9.823	9.877	9.931	9.985	10.039	10.093	10.147	10.201	10.255	10.309
220	10.363	10.417	10.471	10.526	10.580	10.635	10.689	10.744	10.799	10.854
230	10.909	10.963	11.018	11.073	11.128	11.183	11.238	11.293	11.348	11.403
240	11.459	11.514	11.569	11.624	11.680	11.735	11.791	11.847	11.903	11.959
250	12.015	12.071	12.126	12.182	12.238	12.294	12.350	12.406	12.462	12.518
260	12.575	12.631	12.688	12.744	12.800	12.857	12.913	12.970	13.027	13.083
270	13.140	13.197	13.254	13.311	13.368	13.425	13.482	13.539	13.596	13.653
280	13.710	13.768	13.825	13.882	13.939	13.997	14.055	14.112	14.170	14.227
290	14.285	14.343	14.400	14.458	14.515	14.573	14.631	14.689	14.747	14.805
300	14.864	14.922	14.980	15.038	15.096	15.155	15.213	15.271	15.330	15.388
310	15.447	15.506	15.564	15.623	15.681	15.740	15.799	15.858	15.917	15.976
320	16.035	16.094	16.153	16.212	16.271	16.330	16.389	16.449	16.508	16.567
330	16.626	16.685	16.745	16.804	16.864	16.924	16.983	17.043	17.102	17.162
340	17.222	17.281	17.341	17.401	17.461	17.521	17.581	17.641	17.701	17.761
350	17.821	17.881	17.941	18.002	18.062	18.123	18.183	18.243	18.304	18.364
360	18.425	18.485	18.546	18.607	18.667	18.727	18.788	18.849	18.910	18.971
370	19.032	19.093	19.154	19.215	19.276	19.337	19.398	19.459	19.520	19.581
380	19.642	19.704	19.765	19.827	19.888	19.949	20.011	20.072	20.134	20.195
390	20.257	20.318	20.380	20.442	20.504	20.565	20.627	20.688	20.750	20.812

COPPER vs. CONSTANTAN THERMOCOUPLE
Degrees Fahrenheit — Reference Junction 32° F.

°F	0	1	2	3	4	5	6	7	8	9
					Millivolts					
-310	-5.379	-5.388	-5.397	-5.406	....	....	....	....	....	....
-300	-5.284	-5.294	-5.303	-5.313	-5.322	-5.332	-5.341	-5.351	-5.360	-5.370
-290	-5.185	-5.195	-5.205	-5.215	-5.225	-5.235	-5.245	-5.254	-5.264	-5.274
-280	-5.081	-5.092	-5.102	-5.113	-5.124	-5.134	-5.144	-5.154	-5.165	-5.175
-270	-4.974	-4.985	-4.996	-5.007	-5.018	-5.029	-5.039	-5.050	-5.060	-5.071
-260	-4.863	-4.874	-4.885	-4.897	-4.908	-4.919	-4.930	-4.941	-4.952	-4.963
-250	-4.747	-4.759	-4.770	-4.782	-4.794	-4.805	-4.817	-4.829	-4.840	-4.851
-240	-4.627	-4.640	-4.652	-4.664	-4.676	-4.688	-4.700	-4.712	-4.724	-4.735
-230	-4.504	-4.517	-4.529	-4.542	-4.554	-4.566	-4.579	-4.591	-4.603	-4.615
-220	-4.377	-4.390	-4.403	-4.415	-4.428	-4.441	-4.454	-4.466	-4.479	-4.492
-210	-4.246	-4.259	-4.272	-4.286	-4.299	-4.312	-4.325	-4.338	-4.351	-4.364
-200	-4.111	-4.125	-4.138	-4.151	-4.165	-4.179	-4.192	-4.206	-4.219	-4.232
-190	-3.972	-3.986	-4.000	-4.014	-4.028	-4.042	-4.056	-4.069	-4.083	-4.097
-180	-3.829	-3.844	-3.858	-3.873	-3.887	-3.901	-3.915	-3.929	-3.944	-3.958
-170	-3.684	-3.698	-3.713	-3.727	-3.742	-3.757	-3.771	-3.786	-3.800	-3.815
-160	-3.533	-3.548	-3.564	-3.579	-3.594	-3.609	-3.624	-3.639	-3.654	-3.669
-150	-3.380	-3.396	-3.411	-3.426	-3.441	-3.457	-3.472	-3.488	-3.503	-3.518
-140	-3.223	-3.238	-3.254	-3.270	-3.286	-3.301	-3.317	-3.333	-3.349	-3.365
-130	-3.062	-3.078	-3.094	-3.110	-3.127	-3.143	-3.159	-3.175	-3.191	-3.207
-120	-2.897	-2.914	-2.931	-2.947	-2.964	-2.980	-2.997	-3.013	-3.030	-3.046
-110	-2.730	-2.747	-2.764	-2.781	-2.797	-2.814	-2.831	-2.847	-2.864	-2.881
-100	-2.559	-2.577	-2.594	-2.611	-2.628	-2.645	-2.662	-2.679	-2.696	-2.713
-90	-2.385	-2.402	-2.420	-2.437	-2.455	-2.472	-2.490	-2.507	-2.525	-2.542
-80	-2.207	-2.225	-2.243	-2.260	-2.278	-2.296	-2.314	-2.332	-2.349	-2.367
-70	-2.026	-2.044	-2.063	-2.081	-2.099	-2.117	-2.135	-2.153	-2.171	-2.189
-60	-1.842	-1.860	-1.879	-1.897	-1.916	-1.934	-1.953	-1.971	-1.989	-2.008
-50	-1.654	-1.673	-1.692	-1.711	-1.729	-1.748	-1.767	-1.786	-1.804	-1.823
-40	-1.463	-1.482	-1.502	-1.521	-1.540	-1.559	-1.578	-1.597	-1.616	-1.635
-30	-1.270	-1.289	-1.308	-1.328	-1.347	-1.367	-1.386	-1.406	-1.425	-1.444
-20	-1.072	-1.092	-1.112	-1.132	-1.152	-1.171	-1.191	-1.210	-1.230	-1.250
-10	-0.872	-0.893	-0.913	-0.933	-0.953	-0.973	-0.993	-1.013	-1.033	-1.053
(-)0	-0.670	-0.690	-0.710	-0.730	-0.751	-0.771	-0.792	-0.812	-0.832	-0.852
(+)0	-0.670	-0.649	-0.629	-0.608	-0.588	-0.567	-0.546	-0.526	-0.505	-0.484
10	-0.463	-0.442	-0.421	-0.401	-0.380	-0.359	-0.339	-0.318	-0.297	-0.275
20	-0.254	-0.233	-0.212	-0.191	-0.170	-0.149	-0.128	-0.107	-0.085	-0.064
30	-0.042	-0.021	0.000	0.021	0.042	0.064	0.086	0.107	0.129	0.150
40	0.171	0.193	0.215	0.236	0.258	0.280	0.302	0.324	0.346	0.367
50	0.389	0.411	0.433	0.455	0.477	0.499	0.521	0.543	0.565	0.587
60	0.609	0.631	0.654	0.676	0.698	0.720	0.743	0.765	0.787	0.809
70	0.832	0.854	0.877	0.899	0.922	0.944	0.967	0.990	1.012	1.035
80	1.057	1.080	1.103	1.126	1.148	1.171	1.194	1.217	1.240	1.263
90	1.286	1.309	1.332	1.355	1.378	1.401	1.424	1.448	1.471	1.494
100	1.517	1.540	1.563	1.587	1.610	1.633	1.657	1.680	1.704	1.727
110	1.751	1.774	1.798	1.821	1.845	1.869	1.893	1.916	1.940	1.963
120	1.987	2.011	2.035	2.059	2.083	2.107	2.131	2.154	2.178	2.202
130	2.226	2.250	2.274	2.298	2.322	2.346	2.370	2.394	2.418	2.443
140	2.467	2.491	2.516	2.540	2.565	2.589	2.614	2.638	2.663	2.687
150	2.711	2.736	2.760	2.785	2.810	2.835	2.859	2.884	2.908	2.933
160	2.958	2.982	3.007	3.032	3.057	3.082	3.107	3.132	3.157	3.182
170	3.207	3.232	3.257	3.282	3.307	3.332	3.357	3.382	3.407	3.433
180	3.458	3.483	3.508	3.534	3.559	3.584	3.610	3.635	3.661	3.686
190	3.712	3.737	3.762	3.787	3.813	3.839	3.864	3.890	3.915	3.941
200	3.967	3.993	4.018	4.044	4.070	4.096	4.122	4.148	4.174	4.199
210	4.225	4.251	4.277	4.303	4.329	4.355	4.381	4.408	4.434	4.460
220	4.486	4.512	4.538	4.564	4.590	4.617	4.643	4.670	4.696	4.722
230	4.749	4.775	4.801	4.827	4.854	4.880	4.907	4.934	4.960	4.987
240	5.014	5.040	5.067	5.094	5.120	5.147	5.174	5.200	5.227	5.254

Table 9.24 continued

Table 9.24 continued

COPPER vs. CONSTANTAN THERMOCOUPLE

Degrees Fahrenheit — Reference Junction 32° F.

Millivolts

°F	0	1	2	3	4	5	6	7	8	9
250	5.280	5.307	5.334	5.361	5.388	5.415	5.442	5.469	5.496	5.523
260	5.550	5.577	5.604	5.631	5.658	5.685	5.712	5.739	5.766	5.794
270	5.821	5.848	5.875	5.903	5.930	5.957	5.985	6.012	6.040	6.067
280	6.094	6.122	6.149	6.177	6.204	6.232	6.259	6.287	6.314	6.342
290	6.370	6.397	6.425	6.453	6.481	6.508	6.536	6.564	6.592	6.620
300	6.647	6.675	6.703	6.731	6.759	6.786	6.814	6.842	6.870	6.898
310	6.926	6.954	6.982	7.010	7.038	7.066	7.095	7.123	7.151	7.180
320	7.208	7.236	7.264	7.292	7.321	7.349	7.377	7.405	7.434	7.462
330	7.491	7.519	7.548	7.576	7.605	7.633	7.661	7.690	7.719	7.747
340	7.776	7.805	7.834	7.862	7.891	7.920	7.949	7.978	8.006	8.035
350	8.064	8.092	8.120	8.149	8.178	8.207	8.236	8.265	8.294	8.323
360	8.352	8.381	8.410	8.439	8.468	8.497	8.526	8.555	8.584	8.613
370	8.642	8.672	8.701	8.730	8.759	8.788	8.818	8.847	8.876	8.905
380	8.935	8.964	8.994	9.023	9.052	9.082	9.111	9.141	9.170	9.200
390	9.229	9.259	9.288	9.317	9.347	9.376	9.406	9.436	9.466	9.495
400	9.525	9.555	9.584	9.614	9.644	9.674	9.703	9.733	9.763	9.793
410	9.823	9.853	9.883	9.913	9.943	9.973	10.003	10.033	10.063	10.093
420	10.123	10.153	10.183	10.213	10.243	10.273	10.303	10.333	10.363	10.393
430	10.423	10.453	10.483	10.514	10.544	10.574	10.604	10.635	10.665	10.695
440	10.726	10.756	10.787	10.817	10.848	10.878	10.909	10.939	10.969	11.000
450	11.030	11.061	11.091	11.122	11.152	11.183	11.214	11.244	11.275	11.305
460	11.336	11.366	11.397	11.428	11.459	11.490	11.520	11.551	11.581	11.612
470	11.643	11.674	11.704	11.735	11.766	11.797	11.828	11.859	11.891	11.922
480	11.953	11.984	12.015	12.046	12.077	12.108	12.138	12.170	12.201	12.232
490	12.263	12.294	12.325	12.356	12.387	12.418	12.450	12.481	12.512	12.543
500	12.575	12.606	12.637	12.669	12.700	12.732	12.763	12.794	12.825	12.857
510	12.888	12.919	12.951	12.983	13.014	13.046	13.077	13.108	13.140	13.172
520	13.203	13.235	13.267	13.298	13.330	13.362	13.393	13.425	13.457	13.488
530	13.520	13.552	13.583	13.615	13.647	13.678	13.710	13.742	13.774	13.806
540	13.838	13.869	13.901	13.933	13.965	13.997	14.029	14.061	14.093	14.125
550	14.157	14.189	14.221	14.253	14.285	14.317	14.349	14.381	14.413	14.445
560	14.477	14.509	14.541	14.573	14.605	14.637	14.670	14.702	14.734	14.766
570	14.799	14.831	14.864	14.896	14.928	14.961	14.993	15.025	15.057	15.090
580	15.122	15.155	15.187	15.219	15.252	15.284	15.317	15.349	15.382	15.414
590	15.447	15.480	15.512	15.545	15.577	15.610	15.642	15.675	15.707	15.740
600	15.773	15.806	15.838	15.871	15.904	15.937	15.969	16.002	16.035	16.068
610	16.101	16.133	16.166	16.199	16.232	16.264	16.297	16.330	16.363	16.396
620	16.429	16.462	16.495	16.528	16.560	16.593	16.626	16.659	16.692	16.725
630	16.758	16.791	16.824	16.857	16.890	16.924	16.957	16.990	17.023	17.056
640	17.089	17.122	17.155	17.189	17.222	17.255	17.288	17.321	17.354	17.388
650	17.421	17.454	17.488	17.521	17.554	17.588	17.621	17.654	17.688	17.721
660	17.754	17.788	17.821	17.854	17.888	17.921	17.955	17.988	18.022	18.055
670	18.089	18.123	18.156	18.190	18.223	18.257	18.290	18.324	18.357	18.391
680	18.425	18.458	18.492	18.526	18.560	18.593	18.627	18.660	18.694	18.727
690	18.761	18.795	18.829	18.863	18.896	18.930	18.964	18.998	19.032	19.066
700	19.100	19.134	19.168	19.201	19.235	19.269	19.303	19.337	19.371	19.405
710	19.439	19.473	19.506	19.540	19.574	19.608	19.642	19.676	19.711	19.745
720	19.779	19.813	19.847	19.881	19.915	19.949	19.983	20.018	20.052	20.086
730	20.120	20.154	20.188	20.223	20.257	20.291	20.325	20.359	20.394	20.428
740	20.463	20.497	20.531	20.565	20.599	20.634	20.668	20.702	20.736	20.771
750	20.805	20.840	20.874	....	....	....	....	....	....	....

PLAT. vs. PLAT. +10% RHODIUM THERMOCOUPLE

Degrees Centigrade — Reference Junction 0° C.

Millivolts

°C	0	1	2	3	4	5	6	7	8	9
0	0.000	0.005	0.011	0.016	0.022	0.028	0.033	0.039	0.044	0.050
10	.056	.061	.067	.073	.078	.084	.090	.096	.102	.107
20	.113	.119	.125	.131	.137	.143	.149	.155	.161	.167
30	.173	.179	.185	.191	.198	.204	.210	.216	.222	.229
40	.235	.241	.247	.254	.260	.266	.273	.279	.286	.292
50	.299	.305	.312	.318	.325	.331	.338	.344	.351	.357
60	.364	.371	.377	.384	.391	.397	.404	.411	.418	.425
70	.431	.438	.445	.452	.459	.466	.473	.479	.486	.493
80	.500	.507	.514	.521	.528	.535	.543	.550	.557	.564
90	.571	.578	.585	.593	.600	.607	.614	.621	.629	.636
100	.643	.651	.658	.665	.673	.680	.687	.694	.702	.709
110	.717	.724	.732	.739	.747	.754	.762	.769	.777	.784
120	.792	.800	.807	.815	.823	.830	.838	.845	.853	.861
130	.869	.876	.884	.892	.900	.907	.915	.923	.931	.939
140	.946	.954	.962	.970	.978	.986	.994	1.002	1.009	1.017
150	1.025	1.033	1.041	1.049	1.057	1.065	1.073	1.081	1.089	1.097
160	1.106	1.114	1.122	1.130	1.138	1.146	1.154	1.162	1.170	1.179
170	1.187	1.195	1.203	1.211	1.220	1.228	1.236	1.244	1.253	1.261
180	1.269	1.277	1.286	1.294	1.302	1.311	1.319	1.327	1.336	1.344
190	1.352	1.361	1.369	1.377	1.386	1.394	1.403	1.411	1.419	1.428
200	1.436	1.445	1.453	1.462	1.470	1.479	1.487	1.496	1.504	1.513
210	1.521	1.530	1.538	1.547	1.555	1.564	1.573	1.581	1.590	1.598
220	1.607	1.615	1.624	1.633	1.641	1.650	1.659	1.667	1.676	1.685
230	1.693	1.702	1.710	1.719	1.728	1.736	1.745	1.754	1.763	1.771
240	1.780	1.789	1.798	1.806	1.815	1.824	1.833	1.841	1.850	1.859
250	1.868	1.877	1.885	1.894	1.903	1.912	1.921	1.930	1.938	1.947
260	1.956	1.965	1.974	1.983	1.992	2.001	2.009	2.018	2.027	2.036
270	2.045	2.054	2.063	2.072	2.081	2.090	2.099	2.108	2.117	2.126
280	2.135	2.144	2.153	2.162	2.171	2.180	2.189	2.198	2.207	2.216
290	2.225	2.234	2.243	2.252	2.261	2.271	2.280	2.289	2.298	2.307
300	2.316	2.325	2.334	2.343	2.353	2.362	2.371	2.380	2.389	2.398
310	2.408	2.417	2.426	2.435	2.444	2.453	2.463	2.472	2.481	2.490
320	2.499	2.509	2.518	2.527	2.536	2.546	2.555	2.564	2.573	2.583
330	2.592	2.601	2.610	2.620	2.629	2.638	2.648	2.657	2.666	2.676
340	2.685	2.694	2.704	2.713	2.722	2.731	2.741	2.750	2.760	2.769
350	2.778	2.788	2.797	2.806	2.816	2.825	2.834	2.844	2.853	2.863
360	2.872	2.881	2.891	2.900	2.910	2.919	2.929	2.938	2.947	2.957
370	2.966	2.976	2.985	2.995	3.004	3.014	3.023	3.032	3.042	3.051
380	3.061	3.070	3.080	3.089	3.099	3.108	3.118	3.127	3.137	3.146
390	3.156	3.165	3.175	3.184	3.194	3.203	3.213	3.222	3.232	3.241
400	3.251	3.261	3.270	3.280	3.289	3.299	3.308	3.318	3.327	3.337
410	3.347	3.356	3.366	3.375	3.385	3.394	3.404	3.414	3.423	3.433
420	3.442	3.452	3.462	3.471	3.481	3.490	3.500	3.510	3.519	3.529
430	3.539	3.548	3.558	3.567	3.577	3.587	3.596	3.606	3.616	3.625
440	3.635	3.645	3.654	3.664	3.674	3.683	3.693	3.703	3.712	3.722
450	3.732	3.741	3.751	3.761	3.771	3.780	3.790	3.800	3.809	3.819
460	3.829	3.839	3.848	3.858	3.868	3.878	3.887	3.897	3.907	3.917
470	3.926	3.936	3.946	3.956	3.965	3.975	3.985	3.995	4.004	4.014
480	4.024	4.034	4.044	4.053	4.063	4.073	4.083	4.093	4.103	4.112
490	4.122	4.132	4.142	4.152	4.162	4.171	4.181	4.191	4.201	4.211
500	4.221	4.230	4.240	4.250	4.260	4.270	4.280	4.290	4.300	4.310
510	4.319	4.329	4.339	4.349	4.359	4.369	4.379	4.389	4.399	4.409
520	4.419	4.428	4.438	4.448	4.458	4.468	4.478	4.488	4.498	4.508
530	4.518	4.528	4.538	4.548	4.558	4.568	4.578	4.588	4.598	4.608
540	4.618	4.628	4.638	4.648	4.658	4.668	4.678	4.688	4.698	4.708
550	4.718	4.728	4.738	4.748	4.758	4.768	4.778	4.788	4.798	4.808
560	4.818	4.828	4.839	4.849	4.859	4.869	4.879	4.889	4.899	4.909
570	4.919	4.929	4.939	4.950	4.960	4.970	4.980	4.990	5.000	5.010
580	5.020	5.031	5.041	5.051	5.061	5.071	5.081	5.091	5.102	5.112
590	5.122	5.132	5.142	5.152	5.163	5.173	5.183	5.193	5.203	5.214

Table 9.24 continued

Table 9.24 continued

PLAT. vs. PLAT. +10% RHODIUM THERMOCOUPLE
Degrees Fahrenheit — Reference Junction 32° F.

°F	0	1	2	3	4	5	6	7	8	9
					Millivolts					
1200	5.726	5.732	5.738	5.744	5.749	5.755	5.761	5.767	5.773	5.778
1210	5.784	5.790	5.796	5.801	5.807	5.813	5.819	5.824	5.830	5.836
1220	5.842	5.847	5.853	5.859	5.865	5.871	5.876	5.882	5.888	5.894
1230	5.899	5.905	5.911	5.917	5.923	5.928	5.934	5.940	5.946	5.951
1240	5.957	5.963	5.969	5.975	5.980	5.986	5.992	5.998	6.004	6.009
1250	6.015	6.021	6.027	6.033	6.038	6.044	6.050	6.056	6.062	6.067
1260	6.073	6.079	6.085	6.091	6.096	6.102	6.108	6.114	6.120	6.126
1270	6.131	6.137	6.143	6.149	6.155	6.161	6.166	6.172	6.178	6.184
1280	6.190	6.196	6.201	6.207	6.213	6.219	6.225	6.231	6.236	6.242
1290	6.248	6.254	6.260	6.266	6.271	6.277	6.283	6.289	6.295	6.301
1300	6.307	6.312	6.318	6.324	6.330	6.336	6.342	6.348	6.353	6.359
1310	6.365	6.371	6.377	6.383	6.389	6.394	6.400	6.406	6.412	6.418
1320	6.424	6.430	6.436	6.441	6.447	6.453	6.459	6.465	6.471	6.477
1330	6.483	6.488	6.494	6.500	6.506	6.512	6.518	6.524	6.530	6.536
1340	6.542	6.547	6.553	6.559	6.565	6.571	6.577	6.583	6.589	6.595
1350	6.601	6.606	6.612	6.618	6.624	6.630	6.636	6.642	6.648	6.654
1360	6.660	6.666	6.671	6.677	6.683	6.689	6.695	6.701	6.707	6.713
1370	6.719	6.725	6.731	6.737	6.743	6.749	6.754	6.760	6.766	6.772
1380	6.778	6.784	6.790	6.796	6.802	6.808	6.814	6.820	6.826	6.832
1390	6.838	6.844	6.850	6.855	6.861	6.867	6.873	6.879	6.885	6.891
1400	6.897	6.903	6.909	6.915	6.921	6.927	6.933	6.939	6.945	6.951
1410	6.957	6.963	6.969	6.975	6.981	6.987	6.993	6.999	7.005	7.011
1420	7.017	7.023	7.029	7.034	7.040	7.046	7.052	7.058	7.064	7.070
1430	7.076	7.082	7.088	7.094	7.100	7.106	7.112	7.118	7.124	7.130
1440	7.136	7.142	7.148	7.154	7.160	7.166	7.172	7.178	7.184	7.190
1450	7.196	7.202	7.208	7.214	7.220	7.226	7.233	7.239	7.245	7.251
1460	7.257	7.263	7.269	7.275	7.281	7.287	7.293	7.299	7.305	7.311
1470	7.317	7.323	7.329	7.335	7.341	7.347	7.353	7.359	7.365	7.371
1480	7.377	7.383	7.389	7.395	7.401	7.407	7.414	7.420	7.426	7.432
1490	7.438	7.444	7.450	7.456	7.462	7.468	7.474	7.480	7.486	7.492
1500	7.498	7.504	7.510	7.517	7.523	7.529	7.535	7.541	7.547	7.553
1510	7.559	7.565	7.571	7.577	7.583	7.589	7.596	7.602	7.608	7.614
1520	7.620	7.626	7.632	7.638	7.644	7.650	7.656	7.662	7.669	7.675
1530	7.681	7.687	7.693	7.699	7.705	7.711	7.717	7.723	7.730	7.736
1540	7.742	7.748	7.754	7.760	7.766	7.772	7.778	7.785	7.791	7.797
1550	7.803	7.809	7.815	7.821	7.827	7.834	7.840	7.846	7.852	7.858
1560	7.864	7.870	7.876	7.882	7.889	7.895	7.901	7.907	7.913	7.919
1570	7.925	7.932	7.938	7.944	7.950	7.956	7.962	7.968	7.975	7.981
1580	7.987	7.993	7.999	8.005	8.012	8.018	8.024	8.030	8.036	8.042
1590	8.048	8.055	8.061	8.067	8.073	8.079	8.085	8.092	8.098	8.104
1600	8.110	8.116	8.122	8.129	8.135	8.141	8.147	8.153	8.159	8.166
1610	8.172	8.178	8.184	8.190	8.197	8.203	8.209	8.215	8.221	8.228
1620	8.234	8.240	8.246	8.252	8.258	8.265	8.271	8.277	8.283	8.289
1630	8.296	8.302	8.308	8.314	8.320	8.327	8.333	8.339	8.345	8.352
1640	8.358	8.364	8.370	8.376	8.383	8.389	8.395	8.401	8.407	8.414
1650	8.420	8.426	8.432	8.439	8.445	8.451	8.457	8.464	8.470	8.476
1660	8.482	8.488	8.495	8.501	8.507	8.513	8.520	8.526	8.532	8.538
1670	8.545	8.551	8.557	8.563	8.570	8.576	8.582	8.588	8.595	8.601
1680	8.607	8.613	8.620	8.626	8.632	8.638	8.645	8.651	8.657	8.663
1690	8.670	8.676	8.682	8.689	8.695	8.701	8.707	8.714	8.720	8.726
1700	8.732	8.739	8.745	8.751	8.758	8.764	8.770	8.776	8.783	8.789
1710	8.795	8.802	8.808	8.814	8.820	8.827	8.833	8.839	8.846	8.852
1720	8.858	8.864	8.871	8.877	8.883	8.890	8.896	8.902	8.909	8.915
1730	8.921	8.927	8.934	8.940	8.946	8.953	8.959	8.965	8.972	8.978
1740	8.984	8.991	8.997	9.003	9.010	9.016	9.022	9.029	9.035	9.041
1750	9.048	9.054	9.060	9.067	9.073	9.079	9.086	9.092	9.098	9.105
1760	9.111	9.117	9.124	9.130	9.136	9.143	9.149	9.155	9.162	9.168
1770	9.174	9.181	9.187	9.193	9.200	9.206	9.212	9.219	9.225	9.232
1780	9.238	9.244	9.251	9.257	9.263	9.270	9.276	9.282	9.289	9.295
1790	9.302	9.308	9.314	9.321	9.327	9.333	9.340	9.346	9.353	9.359

PLAT. vs. PLAT. +10% RHODIUM THERMOCOUPLE
Degrees Fahrenheit — Reference Junction 32° F.

°F	0	1	2	3	4	5	6	7	8	9
					Millivolts					
1800	9.365	9.372	9.378	9.384	9.391	9.397	9.404	9.410	9.416	9.423
1810	9.429	9.436	9.442	9.448	9.455	9.461	9.468	9.474	9.480	9.487
1820	9.493	9.500	9.506	9.512	9.519	9.525	9.532	9.538	9.544	9.551
1830	9.557	9.564	9.570	9.576	9.583	9.589	9.596	9.602	9.609	9.615
1840	9.621	9.628	9.634	9.641	9.647	9.654	9.660	9.666	9.673	9.679
1850	9.686	9.692	9.699	9.705	9.711	9.718	9.724	9.731	9.737	9.744
1860	9.750	9.757	9.763	9.769	9.776	9.782	9.789	9.795	9.802	9.808
1870	9.815	9.821	9.828	9.834	9.840	9.847	9.853	9.860	9.866	9.873
1880	9.879	9.886	9.892	9.899	9.905	9.912	9.918	9.925	9.931	9.937
1890	9.944	9.950	9.957	9.963	9.970	9.976	9.983	9.989	9.996	10.002
1900	10.009	10.015	10.022	10.028	10.035	10.041	10.048	10.054	10.061	10.067
1910	10.074	10.080	10.087	10.093	10.100	10.106	10.113	10.119	10.126	10.132
1920	10.139	10.145	10.152	10.158	10.165	10.171	10.178	10.184	10.191	10.197
1930	10.204	10.210	10.217	10.223	10.230	10.237	10.243	10.250	10.256	10.263
1940	10.269	10.276	10.282	10.289	10.295	10.302	10.308	10.315	10.321	10.328
1950	10.334	10.341	10.348	10.354	10.361	10.367	10.374	10.380	10.387	10.393
1960	10.400	10.406	10.413	10.420	10.426	10.433	10.439	10.446	10.452	10.459
1970	10.465	10.472	10.478	10.485	10.492	10.498	10.505	10.511	10.518	10.524
1980	10.531	10.538	10.544	10.551	10.557	10.564	10.570	10.577	10.583	10.590
1990	10.597	10.603	10.610	10.616	10.623	10.629	10.636	10.643	10.649	10.656
2000	10.662	10.669	10.675	10.682	10.689	10.695	10.702	10.708	10.715	10.722
2010	10.728	10.735	10.741	10.748	10.754	10.761	10.768	10.774	10.781	10.787
2020	10.794	10.801	10.807	10.814	10.820	10.827	10.834	10.840	10.847	10.853
2030	10.860	10.866	10.873	10.880	10.886	10.893	10.899	10.906	10.913	10.919
2040	10.926	10.932	10.939	10.946	10.952	10.959	10.966	10.972	10.979	10.985
2050	10.992	10.999	11.005	11.012	11.018	11.025	11.032	11.038	11.045	11.051
2060	11.058	11.065	11.071	11.078	11.085	11.091	11.098	11.104	11.111	11.118
2070	11.124	11.131	11.137	11.144	11.151	11.157	11.164	11.171	11.177	11.184
2080	11.190	11.197	11.204	11.210	11.217	11.224	11.230	11.237	11.243	11.250
2090	11.257	11.263	11.270	11.277	11.283	11.290	11.296	11.303	11.310	11.316
2100	11.323	11.330	11.336	11.343	11.350	11.356	11.363	11.369	11.376	11.383
2110	11.389	11.396	11.403	11.409	11.416	11.423	11.429	11.436	11.443	11.449
2120	11.456	11.462	11.469	11.476	11.482	11.489	11.496	11.502	11.509	11.516
2130	11.522	11.529	11.536	11.542	11.549	11.556	11.562	11.569	11.575	11.582
2140	11.589	11.595	11.602	11.609	11.615	11.622	11.629	11.635	11.642	11.649
2150	11.655	11.662	11.669	11.675	11.682	11.689	11.695	11.702	11.709	11.715
2160	11.722	11.729	11.735	11.742	11.749	11.755	11.762	11.769	11.775	11.782
2170	11.789	11.795	11.802	11.809	11.815	11.822	11.829	11.835	11.842	11.848
2180	11.855	11.862	11.868	11.875	11.882	11.888	11.895	11.902	11.908	11.915
2190	11.922	11.928	11.935	11.942	11.949	11.955	11.962	11.969	11.975	11.982
2200	11.989	11.995	12.002	12.009	12.015	12.022	12.029	12.035	12.042	12.049
2210	12.055	12.062	12.069	12.075	12.082	12.089	12.095	12.102	12.109	12.115
2220	12.122	12.129	12.135	12.142	12.149	12.155	12.162	12.169	12.175	12.182
2230	12.189	12.196	12.202	12.209	12.216	12.222	12.229	12.236	12.242	12.249
2240	12.256	12.262	12.269	12.276	12.282	12.289	12.296	12.302	12.309	12.316
2250	12.322	12.329	12.336	12.342	12.349	12.356	12.363	12.369	12.376	12.383
2260	12.389	12.396	12.403	12.409	12.416	12.423	12.429	12.436	12.443	12.449
2270	12.456	12.463	12.470	12.476	12.483	12.490	12.496	12.503	12.510	12.516
2280	12.523	12.530	12.536	12.543	12.550	12.556	12.563	12.570	12.577	12.583
2290	12.590	12.597	12.603	12.610	12.617	12.623	12.630	12.637	12.643	12.650
2300	12.657	12.663	12.670	12.677	12.684	12.690	12.697	12.704	12.710	12.717
2310	12.724	12.730	12.737	12.744	12.750	12.757	12.764	12.770	12.777	12.784
2320	12.790	12.797	12.804	12.810	12.817	12.824	12.830	12.837	12.844	12.851
2330	12.857	12.864	12.871	12.877	12.884	12.891	12.897	12.904	12.911	12.917
2340	12.924	12.931	12.937	12.944	12.951	12.957	12.964	12.971	12.977	12.984
2350	12.991	12.997	13.004	13.011	13.018	13.024	13.031	13.038	13.044	13.051
2360	13.058	13.064	13.071	13.078	13.084	13.091	13.098	13.104	13.111	13.118
2370	13.124	13.131	13.138	13.144	13.151	13.158	13.164	13.171	13.178	13.184
2380	13.191	13.198	13.204	13.211	13.218	13.224	13.231	13.238	13.244	13.251
2390	13.258	13.265	13.271	13.278	13.285	13.291	13.298	13.305	13.311	13.318

Table 9.24 continued

Table 9.24 continued

PLAT. vs. PLAT. +10% RHODIUM THERMOCOUPLE

Degrees Centigrade — Reference Junction 0° C.

°C	0	1	2	3	4	5	6	7	8	9
					Millivolts					
600	5.224	5.234	5.244	5.254	5.265	5.275	5.285	5.295	5.306	5.316
610	5.326	5.336	5.346	5.357	5.367	5.377	5.388	5.398	5.408	5.418
620	5.429	5.439	5.449	5.459	5.470	5.480	5.490	5.501	5.511	5.521
630	5.532	5.542	5.552	5.563	5.573	5.583	5.593	5.604	5.614	5.624
640	5.635	5.645	5.655	5.666	5.676	5.686	5.697	5.707	5.717	5.728
650	5.738	5.748	5.759	5.769	5.779	5.790	5.800	5.811	5.821	5.831
660	5.842	5.852	5.862	5.873	5.883	5.894	5.904	5.914	5.925	5.935
670	5.946	5.956	5.967	5.977	5.987	5.998	6.008	6.019	6.029	6.040
680	6.050	6.060	6.071	6.081	6.092	6.102	6.113	6.123	6.134	6.144
690	6.155	6.165	6.176	6.186	6.197	6.207	6.218	6.228	6.239	6.249
700	6.260	6.270	6.281	6.291	6.302	6.312	6.323	6.333	6.344	6.355
710	6.365	6.376	6.386	6.397	6.407	6.418	6.429	6.439	6.450	6.460
720	6.471	6.481	6.492	6.503	6.513	6.524	6.534	6.545	6.556	6.566
730	6.577	6.588	6.598	6.609	6.619	6.630	6.641	6.651	6.662	6.673
740	6.683	6.694	6.705	6.715	6.726	6.737	6.747	6.758	6.769	6.779
750	6.790	6.801	6.811	6.822	6.833	6.844	6.854	6.865	6.876	6.886
760	6.897	6.908	6.919	6.929	6.940	6.951	6.962	6.972	6.983	6.994
770	7.005	7.015	7.026	7.037	7.047	7.058	7.069	7.080	7.091	7.102
780	7.112	7.123	7.134	7.145	7.156	7.166	7.177	7.188	7.199	7.210
790	7.220	7.231	7.242	7.253	7.264	7.275	7.286	7.296	7.307	7.318
800	7.329	7.340	7.351	7.362	7.372	7.383	7.394	7.405	7.416	7.427
810	7.438	7.449	7.460	7.470	7.481	7.492	7.503	7.514	7.525	7.536
820	7.547	7.558	7.569	7.580	7.591	7.602	7.613	7.623	7.634	7.645
830	7.656	7.667	7.678	7.689	7.700	7.711	7.722	7.733	7.744	7.755
840	7.766	7.777	7.788	7.799	7.810	7.821	7.832	7.843	7.854	7.865
850	7.876	7.887	7.898	7.910	7.921	7.932	7.943	7.954	7.965	7.976
860	7.987	7.998	8.009	8.020	8.031	8.042	8.053	8.064	8.076	8.087
870	8.098	8.109	8.120	8.131	8.142	8.153	8.164	8.176	8.187	8.198
880	8.209	8.220	8.231	8.242	8.254	8.265	8.276	8.287	8.298	8.309
890	8.320	8.332	8.343	8.354	8.365	8.376	8.388	8.399	8.410	8.421
900	8.432	8.444	8.455	8.466	8.477	8.488	8.500	8.511	8.522	8.533
910	8.545	8.556	8.567	8.578	8.590	8.601	8.612	8.623	8.635	8.646
920	8.657	8.668	8.680	8.691	8.702	8.714	8.725	8.736	8.747	8.759
930	8.770	8.781	8.793	8.804	8.815	8.827	8.838	8.849	8.861	8.872
940	8.883	8.895	8.906	8.917	8.929	8.940	8.951	8.963	8.974	8.986
950	8.997	9.008	9.020	9.031	9.042	9.054	9.065	9.077	9.088	9.099
960	9.111	9.122	9.134	9.145	9.157	9.168	9.179	9.191	9.202	9.214
970	9.225	9.236	9.248	9.260	9.271	9.282	9.294	9.305	9.317	9.328
980	9.340	9.351	9.363	9.374	9.386	9.397	9.409	9.420	9.432	9.443
990	9.455	9.466	9.478	9.489	9.501	9.512	9.524	9.535	9.547	9.559
1000	9.570	9.582	9.593	9.605	9.616	9.628	9.639	9.651	9.663	9.674
1010	9.686	9.697	9.709	9.720	9.732	9.744	9.755	9.767	9.779	9.790
1020	9.802	9.813	9.825	9.837	9.848	9.860	9.871	9.883	9.895	9.906
1030	9.918	9.930	9.941	9.953	9.965	9.976	9.988	10.000	10.011	10.023
1040	10.035	10.046	10.058	10.070	10.082	10.093	10.105	10.117	10.128	10.140
1050	10.152	10.163	10.175	10.187	10.199	10.210	10.222	10.234	10.246	10.257
1060	10.269	10.281	10.293	10.304	10.316	10.328	10.340	10.351	10.363	10.375
1070	10.387	10.399	10.410	10.422	10.434	10.446	10.458	10.469	10.481	10.493
1080	10.505	10.517	10.528	10.540	10.552	10.564	10.576	10.587	10.599	10.611
1090	10.623	10.635	10.647	10.658	10.670	10.682	10.694	10.706	10.718	10.729
1100	10.741	10.753	10.765	10.777	10.789	10.801	10.812	10.824	10.836	10.848
1110	10.860	10.872	10.884	10.896	10.907	10.919	10.931	10.943	10.955	10.967
1120	10.979	10.991	11.003	11.014	11.026	11.038	11.050	11.062	11.074	11.086
1130	11.098	11.110	11.122	11.133	11.145	11.157	11.169	11.181	11.193	11.205
1140	11.217	11.229	11.241	11.253	11.265	11.277	11.289	11.300	11.312	11.324
1150	11.336	11.348	11.360	11.372	11.384	11.396	11.408	11.420	11.432	11.444
1160	11.456	11.468	11.480	11.492	11.504	11.516	11.528	11.540	11.552	11.564
1170	11.575	11.587	11.599	11.611	11.623	11.635	11.647	11.659	11.671	11.683
1180	11.695	11.707	11.719	11.731	11.743	11.755	11.767	11.779	11.791	11.803
1190	11.815	11.827	11.839	11.851	11.863	11.875	11.887	11.899	11.911	11.923

PLAT. vs. PLAT. +10% RHODIUM THERMOCOUPLE

Degrees Centigrade — Reference Junction 0° C.

°C	0	1	2	3	4	5	6	7	8	9
					Millivolts					
1200	11.935	11.947	11.959	11.971	11.983	11.995	12.007	12.019	12.031	12.043
1210	12.055	12.067	12.079	12.091	12.103	12.115	12.127	12.139	12.151	12.163
1220	12.175	12.187	12.200	12.212	12.224	12.236	12.248	12.260	12.272	12.284
1230	12.296	12.308	12.320	12.332	12.344	12.356	12.368	12.380	12.392	12.404
1240	12.416	12.428	12.440	12.452	12.464	12.476	12.488	12.500	12.512	12.524
1250	12.536	12.548	12.560	12.573	12.585	12.597	12.609	12.621	12.633	12.645
1260	12.657	12.669	12.681	12.693	12.705	12.717	12.729	12.741	12.753	12.765
1270	12.777	12.789	12.801	12.813	12.825	12.837	12.849	12.861	12.873	12.885
1280	12.897	12.909	12.921	12.933	12.945	12.957	12.969	12.981	12.993	13.005
1290	13.018	13.030	13.042	13.054	13.066	13.078	13.090	13.102	13.114	13.126
1300	13.138	13.150	13.162	13.174	13.186	13.198	13.210	13.222	13.234	13.246
1310	13.258	13.270	13.282	13.294	13.306	13.318	13.330	13.342	13.354	13.366
1320	13.378	13.390	13.402	13.414	13.426	13.438	13.450	13.462	13.474	13.486
1330	13.498	13.510	13.522	13.534	13.546	13.558	13.570	13.582	13.594	13.606
1340	13.618	13.630	13.642	13.654	13.666	13.678	13.690	13.702	13.714	13.726
1350	13.738	13.750	13.762	13.774	13.786	13.798	13.810	13.822	13.834	13.846
1360	13.858	13.870	13.882	13.894	13.906	13.918	13.930	13.942	13.954	13.966
1370	13.978	13.990	14.002	14.014	14.026	14.038	14.050	14.062	14.074	14.086
1380	14.098	14.110	14.122	14.133	14.145	14.157	14.169	14.181	14.193	14.205
1390	14.217	14.229	14.241	14.253	14.265	14.277	14.289	14.301	14.313	14.325
1400	14.337	14.349	14.361	14.373	14.385	14.397	14.409	14.421	14.433	14.445
1410	14.457	14.469	14.481	14.493	14.504	14.516	14.528	14.540	14.552	14.564
1420	14.576	14.588	14.600	14.612	14.624	14.636	14.648	14.660	14.672	14.684
1430	14.696	14.708	14.720	14.732	14.744	14.755	14.767	14.779	14.791	14.803
1440	14.815	14.827	14.839	14.851	14.863	14.875	14.887	14.899	14.911	14.923
1450	14.935	14.946	14.958	14.970	14.982	14.994	15.006	15.018	15.030	15.042
1460	15.054	15.066	15.078	15.090	15.102	15.113	15.125	15.137	15.149	15.161
1470	15.173	15.185	15.197	15.209	15.221	15.233	15.245	15.256	15.268	15.280
1480	15.292	15.304	15.316	15.328	15.340	15.352	15.364	15.376	15.387	15.399
1490	15.411	15.423	15.435	15.447	15.459	15.471	15.483	15.495	15.507	15.518
1500	15.530	15.542	15.554	15.566	15.578	15.590	15.602	15.614	15.625	15.637
1510	15.649	15.661	15.673	15.685	15.697	15.709	15.721	15.732	15.744	15.756
1520	15.768	15.780	15.792	15.804	15.816	15.827	15.839	15.851	15.863	15.875
1530	15.887	15.899	15.911	15.922	15.934	15.946	15.958	15.970	15.982	15.994
1540	16.006	16.017	16.029	16.041	16.053	16.065	16.077	16.089	16.100	16.112
1550	16.124	16.136	16.148	16.160	16.171	16.183	16.195	16.207	16.219	16.231
1560	16.243	16.254	16.266	16.278	16.290	16.302	16.314	16.325	16.337	16.349
1570	16.361	16.373	16.385	16.396	16.408	16.420	16.432	16.444	16.456	16.467
1580	16.479	16.491	16.503	16.515	16.527	16.538	16.550	16.562	16.574	16.586
1590	16.597	16.609	16.621	16.633	16.645	16.657	16.668	16.680	16.692	16.704
1600	16.716	16.727	16.739	16.751	16.763	16.775	16.786	16.798	16.810	16.822
1610	16.834	16.845	16.857	16.869	16.881	16.893	16.904	16.916	16.928	16.940
1620	16.952	16.963	16.975	16.987	16.999	17.010	17.022	17.034	17.046	17.058
1630	17.069	17.081	17.093	17.105	17.116	17.128	17.140	17.152	17.163	17.175
1640	17.187	17.199	17.211	17.222	17.234	17.246	17.258	17.269	17.281	17.293
1650	17.305	17.316	17.328	17.340	17.352	17.363	17.375	17.387	17.398	17.410
1660	17.422	17.434	17.446	17.457	17.469	17.481	17.492	17.504	17.516	17.528
1670	17.539	17.551	17.563	17.575	17.586	17.598	17.610	17.621	17.633	17.645
1680	17.657	17.668	17.680	17.692	17.704	17.715	17.727	17.739	17.750	17.762
1690	17.774	17.785	17.797	17.809	17.821	17.832	17.844	17.856	17.867	17.879
1700	17.891	17.902	17.914	17.926	17.938	17.949	17.961	17.973	17.984	17.996
1710	18.008	18.019	18.031	18.043	18.054	18.066	18.078	18.089	18.101	18.113
1720	18.124	18.136	18.148	18.159	18.171	18.183	18.194	18.206	18.218	18.229
1730	18.241	18.253	18.264	18.276	18.288	18.299	18.311	18.323	18.334	18.346
1740	18.358	18.369	18.381	18.393	18.404	18.416	18.427	18.439	18.451	18.462
1750	18.474	18.486	18.497	18.509	18.520	18.532	18.544	18.555	18.567	18.579
1760	18.590	18.602	18.613	18.625	18.637	18.648	18.660	18.672	18.683	18.695

Table 9.24 continued

Table 9.24 continued

PLAT. vs. PLAT. +10% RHODIUM THERMOCOUPLE

Degrees Fahrenheit — Reference Junction 32° F.

°F	0	1	2	3	4	5	6	7	8	9
					Millivolts					
0	-0.093	-0.090	-0.087	-0.085	-0.082	-0.079	-0.076	-0.073	-0.071	-0.068
10	-.065	-.062	-.059	-.056	-.053	-.051	-.048	-.045	-.042	-.039
20	-.036	-.033	-.030	-.027	-.024	-.021	-.018	-.015	-.012	-.009
30	-.006	-.003	.000	.003	.006	.009	.012	.015	.018	.021
40	.024	.028	.031	.034	.037	.040	.043	.046	.049	.052
50	.056	.059	.062	.065	.068	.071	.075	.078	.081	.084
60	.087	.091	.094	.097	.100	.104	.107	.110	.113	.117
70	.120	.123	.126	.130	.133	.136	.140	.143	.146	.150
80	.153	.156	.160	.163	.166	.170	.173	.176	.180	.183
90	.187	.190	.193	.197	.200	.204	.207	.211	.214	.218
100	.221	.224	.228	.231	.235	.238	.242	.245	.249	.252
110	.256	.259	.263	.266	.270	.274	.277	.281	.284	.288
120	.291	.295	.299	.302	.306	.309	.313	.317	.320	.324
130	.327	.331	.335	.338	.342	.346	.349	.353	.357	.360
140	.364	.368	.371	.375	.379	.383	.386	.390	.394	.397
150	.401	.405	.409	.412	.416	.420	.424	.428	.431	.435
160	.439	.443	.447	.450	.454	.458	.462	.466	.469	.473
170	.477	.481	.485	.489	.493	.496	.500	.504	.508	.512
180	.516	.520	.524	.528	.532	.535	.539	.543	.547	.551
190	.555	.559	.563	.567	.571	.575	.579	.583	.587	.591
200	.595	.599	.603	.607	.611	.615	.619	.623	.627	.631
210	.635	.639	.643	.647	.651	.655	.659	.664	.668	.672
220	.676	.680	.684	.688	.692	.696	.700	.705	.709	.713
230	.717	.721	.725	.729	.734	.738	.742	.746	.750	.754
240	.758	.763	.767	.771	.775	.779	.784	.788	.792	.796
250	.800	.805	.809	.813	.817	.822	.826	.830	.834	.839
260	.843	.847	.851	.856	.860	.864	.869	.873	.877	.881
270	.886	.890	.894	.899	.903	.907	.912	.916	.920	.925
280	.929	.933	.938	.942	.946	.951	.955	.959	.964	.968
290	.973	.977	.981	.986	.990	.994	.999	1.003	1.008	1.012
300	1.017	1.021	1.025	1.030	1.034	1.039	1.043	1.048	1.052	1.056
310	1.061	1.065	1.070	1.074	1.079	1.083	1.088	1.092	1.097	1.101
320	1.106	1.110	1.115	1.119	1.124	1.128	1.132	1.137	1.142	1.146
330	1.151	1.155	1.160	1.164	1.169	1.173	1.178	1.182	1.187	1.191
340	1.196	1.200	1.205	1.210	1.214	1.219	1.223	1.228	1.232	1.237
350	1.242	1.246	1.251	1.255	1.260	1.264	1.269	1.274	1.278	1.283
360	1.287	1.292	1.297	1.301	1.306	1.311	1.315	1.320	1.324	1.329
370	1.334	1.338	1.343	1.348	1.352	1.357	1.362	1.366	1.371	1.376
380	1.380	1.385	1.390	1.394	1.399	1.404	1.408	1.413	1.418	1.422
390	1.427	1.432	1.436	1.441	1.446	1.450	1.455	1.460	1.465	1.469
400	1.474	1.479	1.483	1.488	1.493	1.498	1.502	1.507	1.512	1.516
410	1.521	1.526	1.531	1.535	1.540	1.545	1.550	1.554	1.559	1.564
420	1.569	1.573	1.578	1.583	1.588	1.593	1.597	1.602	1.607	1.612
430	1.616	1.621	1.626	1.631	1.636	1.640	1.645	1.650	1.655	1.660
440	1.664	1.669	1.674	1.679	1.684	1.688	1.693	1.698	1.703	1.708
450	1.712	1.717	1.722	1.727	1.732	1.736	1.741	1.746	1.751	1.756
460	1.761	1.765	1.770	1.775	1.780	1.785	1.790	1.795	1.799	1.804
470	1.809	1.814	1.819	1.824	1.829	1.833	1.838	1.843	1.848	1.853
480	1.858	1.863	1.868	1.873	1.877	1.882	1.887	1.892	1.897	1.902
490	1.907	1.912	1.917	1.922	1.927	1.931	1.936	1.941	1.946	1.951
500	1.956	1.961	1.966	1.971	1.976	1.981	1.986	1.991	1.996	2.000
510	2.005	2.010	2.015	2.020	2.025	2.030	2.035	2.040	2.045	2.050
520	2.055	2.060	2.065	2.070	2.075	2.080	2.085	2.090	2.095	2.100
530	2.105	2.110	2.115	2.120	2.125	2.130	2.135	2.140	2.145	2.150
540	2.155	2.160	2.165	2.170	2.175	2.180	2.185	2.190	2.195	2.200
550	2.205	2.210	2.215	2.220	2.225	2.230	2.235	2.240	2.245	2.250
560	2.255	2.260	2.265	2.270	2.276	2.281	2.286	2.291	2.296	2.301
570	2.306	2.311	2.316	2.321	2.326	2.331	2.336	2.341	2.346	2.351
580	2.357	2.362	2.367	2.372	2.377	2.382	2.387	2.392	2.397	2.402
590	2.407	2.413	2.418	2.423	2.428	2.433	2.438	2.443	2.448	2.453

PLAT. vs. PLAT. +10% RHODIUM THERMOCOUPLE

Degrees Fahrenheit — Reference Junction 32° F.

°F	0	1	2	3	4	5	6	7	8	9
					Millivolts					
600	2.458	2.464	2.469	2.474	2.479	2.484	2.489	2.494	2.499	2.505
610	2.510	2.515	2.520	2.525	2.530	2.535	2.540	2.546	2.551	2.556
620	2.561	2.566	2.571	2.576	2.582	2.587	2.592	2.597	2.602	2.607
630	2.613	2.618	2.623	2.628	2.633	2.638	2.644	2.649	2.654	2.659
640	2.664	2.669	2.675	2.680	2.685	2.690	2.695	2.700	2.706	2.711
650	2.716	2.721	2.726	2.731	2.737	2.742	2.747	2.752	2.757	2.763
660	2.768	2.773	2.778	2.783	2.789	2.794	2.799	2.804	2.809	2.815
670	2.820	2.825	2.830	2.836	2.841	2.846	2.851	2.856	2.862	2.867
680	2.872	2.877	2.882	2.888	2.893	2.898	2.903	2.909	2.914	2.919
690	2.924	2.930	2.935	2.940	2.945	2.951	2.956	2.961	2.966	2.972
700	2.977	2.982	2.987	2.992	2.998	3.003	3.008	3.014	3.019	3.024
710	3.029	3.035	3.040	3.045	3.050	3.056	3.061	3.066	3.071	3.077
720	3.082	3.087	3.092	3.098	3.102	3.108	3.114	3.119	3.124	3.129
730	3.135	3.140	3.145	3.150	3.156	3.161	3.166	3.172	3.177	3.182
740	3.188	3.193	3.198	3.203	3.209	3.214	3.219	3.225	3.230	3.235
750	3.240	3.246	3.251	3.256	3.262	3.267	3.272	3.278	3.283	3.288
760	3.293	3.299	3.304	3.309	3.315	3.320	3.325	3.331	3.336	3.341
770	3.347	3.352	3.357	3.363	3.368	3.373	3.378	3.384	3.389	3.394
780	3.400	3.405	3.410	3.416	3.421	3.426	3.432	3.437	3.442	3.448
790	3.453	3.458	3.464	3.469	3.474	3.480	3.485	3.490	3.496	3.501
800	3.506	3.512	3.517	3.522	3.528	3.533	3.539	3.544	3.549	3.555
810	3.560	3.565	3.571	3.576	3.581	3.587	3.592	3.597	3.603	3.608
820	3.614	3.619	3.624	3.630	3.635	3.640	3.646	3.651	3.656	3.662
830	3.667	3.673	3.678	3.683	3.689	3.694	3.699	3.705	3.710	3.716
840	3.721	3.726	3.732	3.737	3.743	3.748	3.753	3.759	3.764	3.769
850	3.775	3.780	3.786	3.791	3.796	3.802	3.807	3.813	3.818	3.823
860	3.829	3.834	3.840	3.845	3.850	3.856	3.861	3.867	3.872	3.878
870	3.883	3.888	3.894	3.899	3.905	3.910	3.915	3.921	3.926	3.932
880	3.937	3.943	3.948	3.953	3.959	3.964	3.970	3.975	3.981	3.986
890	3.991	3.997	4.002	4.008	4.013	4.019	4.024	4.030	4.035	4.040
900	4.046	4.051	4.057	4.062	4.068	4.073	4.079	4.084	4.089	4.095
910	4.100	4.106	4.111	4.117	4.122	4.128	4.133	4.139	4.144	4.149
920	4.155	4.160	4.166	4.171	4.177	4.182	4.188	4.193	4.199	4.204
930	4.210	4.215	4.221	4.226	4.232	4.237	4.243	4.248	4.254	4.259
940	4.264	4.270	4.275	4.281	4.286	4.292	4.297	4.303	4.308	4.314
950	4.319	4.325	4.330	4.336	4.341	4.347	4.352	4.358	4.363	4.369
960	4.374	4.380	4.385	4.391	4.396	4.402	4.408	4.413	4.419	4.424
970	4.430	4.435	4.441	4.446	4.452	4.457	4.463	4.468	4.474	4.479
980	4.485	4.490	4.496	4.501	4.507	4.512	4.518	4.524	4.529	4.535
990	4.540	4.546	4.551	4.557	4.562	4.568	4.573	4.579	4.584	4.590
1000	4.596	4.601	4.607	4.612	4.618	4.623	4.629	4.634	4.640	4.646
1010	4.651	4.657	4.662	4.668	4.673	4.679	4.685	4.690	4.696	4.701
1020	4.707	4.712	4.718	4.724	4.729	4.735	4.740	4.746	4.751	4.757
1030	4.763	4.768	4.774	4.779	4.785	4.790	4.796	4.802	4.807	4.813
1040	4.818	4.824	4.830	4.835	4.841	4.846	4.852	4.858	4.863	4.869
1050	4.874	4.880	4.886	4.891	4.897	4.902	4.908	4.914	4.919	4.925
1060	4.930	4.936	4.942	4.947	4.953	4.959	4.964	4.970	4.975	4.981
1070	4.987	4.992	4.998	5.004	5.009	5.015	5.020	5.026	5.032	5.037
1080	5.043	5.049	5.054	5.060	5.066	5.071	5.077	5.082	5.088	5.094
1090	5.099	5.105	5.111	5.116	5.122	5.128	5.133	5.139	5.145	5.150
1100	5.156	5.162	5.167	5.173	5.178	5.184	5.190	5.195	5.201	5.207
1110	5.212	5.218	5.224	5.229	5.235	5.241	5.246	5.252	5.258	5.264
1120	5.269	5.275	5.281	5.286	5.292	5.298	5.303	5.309	5.315	5.320
1130	5.326	5.332	5.337	5.343	5.349	5.354	5.360	5.366	5.372	5.377
1140	5.383	5.389	5.394	5.400	5.406	5.411	5.417	5.423	5.429	5.434
1150	5.440	5.446	5.451	5.457	5.463	5.469	5.474	5.480	5.486	5.491
1160	5.497	5.503	5.509	5.514	5.520	5.526	5.532	5.537	5.543	5.549
1170	5.555	5.560	5.566	5.572	5.577	5.583	5.589	5.595	5.600	5.606
1180	5.612	5.617	5.623	5.629	5.635	5.640	5.646	5.652	5.658	5.663
1190	5.669	5.675	5.681	5.686	5.692	5.698	5.704	5.709	5.715	5.721

Table 9.24 continued

Table 9.24 continued

PLAT. vs. PLAT. +10% RHODIUM THERMOCOUPLE
Degrees Fahrenheit — Reference Junction 32° F.

°F	0	1	2	3	4	5	6	7	8	9
					Millivolts					
2400	13.325	13.331	13.338	13.345	13.351	13.358	13.365	13.371	13.378	13.385
2410	13.391	13.398	13.405	13.411	13.418	13.425	13.431	13.438	13.445	13.451
2420	13.458	13.465	13.471	13.478	13.485	13.491	13.498	13.505	13.511	13.518
2430	13.525	13.531	13.538	13.545	13.551	13.558	13.565	13.571	13.578	13.585
2440	13.591	13.598	13.605	13.611	13.618	13.625	13.631	13.638	13.645	13.651
2450	13.658	13.665	13.671	13.678	13.685	13.691	13.698	13.705	13.711	13.718
2460	13.725	13.731	13.738	13.745	13.751	13.758	13.765	13.771	13.778	13.785
2470	13.791	13.798	13.805	13.811	13.818	13.825	13.831	13.838	13.845	13.851
2480	13.858	13.865	13.871	13.878	13.885	13.891	13.898	13.905	13.911	13.918
2490	13.924	13.931	13.938	13.944	13.951	13.958	13.964	13.971	13.978	13.984
2500	13.991	13.998	14.004	14.011	14.018	14.024	14.031	14.038	14.044	14.051
2510	14.058	14.064	14.071	14.078	14.084	14.091	14.098	14.104	14.111	14.118
2520	14.124	14.131	14.137	14.144	14.151	14.157	14.164	14.171	14.177	14.184
2530	14.191	14.197	14.204	14.211	14.217	14.224	14.231	14.237	14.244	14.251
2540	14.257	14.264	14.271	14.277	14.284	14.290	14.297	14.304	14.310	14.317
2550	14.324	14.330	14.337	14.344	14.350	14.357	14.364	14.370	14.377	14.384
2560	14.390	14.397	14.403	14.410	14.417	14.423	14.430	14.437	14.443	14.450
2570	14.457	14.463	14.470	14.477	14.483	14.490	14.497	14.503	14.510	14.516
2580	14.523	14.530	14.536	14.543	14.550	14.556	14.563	14.570	14.576	14.583
2590	14.589	14.596	14.603	14.609	14.616	14.623	14.629	14.636	14.643	14.649
2600	14.656	14.663	14.669	14.676	14.682	14.689	14.696	14.702	14.709	14.716
2610	14.722	14.729	14.736	14.742	14.749	14.755	14.762	14.769	14.775	14.782
2620	14.789	14.795	14.802	14.809	14.815	14.822	14.828	14.835	14.842	14.848
2630	14.855	14.862	14.868	14.875	14.881	14.888	14.895	14.901	14.908	14.915
2640	14.921	14.928	14.935	14.941	14.948	14.954	14.961	14.968	14.974	14.981
2650	14.988	14.994	15.001	15.007	15.014	15.021	15.027	15.034	15.041	15.047
2660	15.054	15.060	15.067	15.074	15.080	15.087	15.094	15.100	15.107	15.113
2670	15.120	15.127	15.133	15.140	15.147	15.153	15.160	15.166	15.173	15.180
2680	15.186	15.193	15.200	15.206	15.213	15.219	15.226	15.233	15.239	15.246
2690	15.253	15.259	15.266	15.272	15.279	15.286	15.292	15.299	15.305	15.312
2700	15.319	15.325	15.332	15.339	15.345	15.352	15.358	15.365	15.372	15.378
2710	15.385	15.391	15.398	15.405	15.411	15.418	15.425	15.431	15.438	15.444
2720	15.451	15.458	15.464	15.471	15.477	15.484	15.491	15.497	15.504	15.510
2730	15.517	15.524	15.530	15.537	15.544	15.550	15.557	15.563	15.570	15.577
2740	15.583	15.590	15.596	15.603	15.610	15.616	15.623	15.629	15.636	15.643
2750	15.649	15.656	15.662	15.669	15.676	15.682	15.689	15.695	15.702	15.709
2760	15.715	15.722	15.728	15.735	15.742	15.748	15.755	15.761	15.768	15.775
2770	15.781	15.788	15.794	15.801	15.808	15.814	15.821	15.827	15.834	15.841
2780	15.847	15.854	15.860	15.867	15.874	15.880	15.887	15.893	15.900	15.907
2790	15.913	15.920	15.926	15.933	15.940	15.946	15.953	15.959	15.966	15.973
2800	15.979	15.986	15.992	15.999	16.006	16.012	16.019	16.025	16.032	16.038
2810	16.045	16.052	16.058	16.065	16.071	16.078	16.085	16.091	16.098	16.104
2820	16.111	16.117	16.124	16.131	16.137	16.144	16.150	16.157	16.164	16.170
2830	16.177	16.183	16.190	16.196	16.203	16.210	16.216	16.223	16.229	16.236
2840	16.243	16.249	16.256	16.262	16.269	16.275	16.282	16.289	16.295	16.302
2850	16.308	16.315	16.322	16.328	16.335	16.341	16.348	16.354	16.361	16.368
2860	16.374	16.381	16.387	16.394	16.400	16.407	16.414	16.420	16.427	16.433
2870	16.440	16.446	16.453	16.460	16.466	16.473	16.479	16.486	16.492	16.499
2880	16.506	16.512	16.519	16.525	16.532	16.538	16.545	16.552	16.558	16.565
2890	16.571	16.578	16.584	16.591	16.597	16.604	16.611	16.617	16.624	16.630
2900	16.637	16.643	16.650	16.657	16.663	16.670	16.676	16.683	16.689	16.696
2910	16.702	16.709	16.716	16.722	16.729	16.735	16.742	16.748	16.755	16.761
2920	16.768	16.775	16.781	16.788	16.794	16.801	16.807	16.814	16.820	16.827
2930	16.834	16.840	16.847	16.853	16.860	16.866	16.873	16.879	16.886	16.893
2940	16.899	16.906	16.912	16.919	16.925	16.932	16.938	16.945	16.952	16.958
2950	16.965	16.971	16.978	16.984	16.991	16.997	17.004	17.010	17.017	17.023
2960	17.030	17.037	17.043	17.050	17.056	17.063	17.069	17.076	17.082	17.089
2970	17.095	17.102	17.109	17.115	17.122	17.128	17.135	17.141	17.148	17.154
2980	17.161	17.167	17.174	17.180	17.187	17.194	17.200	17.207	17.213	17.220
2990	17.226	17.233	17.239	17.246	17.252	17.259	17.265	17.272	17.278	17.285

PLAT. vs. PLAT. +10% RHODIUM THERMOCOUPLE
Degrees Fahrenheit — Reference Junction 32° F.

°F	0	1	2	3	4	5	6	7	8	9
					Millivolts					
3000	17.292	17.298	17.305	17.311	17.318	17.324	17.331	17.337	17.344	17.350
3010	17.357	17.363	17.370	17.376	17.383	17.389	17.396	17.402	17.409	17.416
3020	17.422	17.429	17.435	17.442	17.448	17.455	17.461	17.468	17.474	17.481
3030	17.487	17.494	17.500	17.507	17.513	17.520	17.526	17.533	17.539	17.546
3040	17.552	17.559	17.565	17.572	17.578	17.585	17.592	17.598	17.605	17.611
3050	17.618	17.624	17.631	17.637	17.644	17.650	17.657	17.663	17.670	17.676
3060	17.683	17.689	17.696	17.702	17.709	17.715	17.722	17.728	17.735	17.741
3070	17.748	17.754	17.761	17.767	17.774	17.780	17.787	17.793	17.800	17.806
3080	17.813	17.819	17.826	17.832	17.839	17.845	17.852	17.858	17.865	17.871
3090	17.878	17.884	17.891	17.897	17.904	17.910	17.917	17.923	17.930	17.936
3100	17.943	17.949	17.956	17.962	17.669	17.975	17.982	17.988	17.995	18.001
3110	18.008	18.014	18.021	18.027	18.034	18.040	18.047	18.053	18.060	18.066
3120	18.073	18.079	18.086	18.092	18.098	18.105	18.111	18.118	18.124	18.131
3130	18.137	18.144	18.150	18.157	18.163	18.170	18.176	18.183	18.189	18.196
3140	18.202	18.209	18.215	18.222	18.228	18.235	18.241	18.248	18.254	18.260
3150	18.267	18.273	18.280	18.286	18.293	18.299	18.306	18.312	18.319	18.325
3160	18.332	18.338	18.345	18.351	18.358	18.364	18.371	18.377	18.383	18.390
3170	18.396	18.403	18.409	18.416	18.422	18.429	18.435	18.442	18.448	18.455
3180	18.461	18.468	18.474	18.480	18.487	18.493	18.500	18.506	18.513	18.519
3190	18.526	18.532	18.539	18.545	18.551	18.558	18.564	18.571	18.577	18.584
3200	18.590	18.597	18.603	18.610	18.616	18.622	18.629	18.635	18.642	18.648
3210	18.655	18.661	18.668	18.674	18.681	18.687				

Table 9.24 continued

Table 9.24 continued

PLAT. vs. PLAT. +13% RHODIUM THERMOCOUPLE

Degrees Centigrade — Reference Junction 0° C.

°C	0	1	2	3	4	5	6	7	8	9
					Millivolts					
0	0.000	0.005	0.011	0.016	0.022	0.027	0.033	0.038	0.043	0.049
10	.055	.061	.066	.072	.078	.083	.089	.095	.101	.106
20	.112	.118	.124	.130	.136	.142	.148	.154	.160	.166
30	.172	.178	.184	.190	.196	.203	.209	.215	.221	.228
40	.234	.240	.246	.252	.259	.265	.272	.278	.285	.291
50	.298	.304	.311	.317	.324	.330	.337	.343	.350	.357
60	.363	.370	.377	.383	.390	.397	.403	.410	.417	.424
70	.431	.438	.445	.451	.458	.465	.472	.479	.486	.493
80	.500	.507	.514	.521	.528	.536	.543	.550	.557	.565
90	.572	.579	.586	.594	.601	.609	.616	.623	.631	.638
100	.645	.653	.660	.668	.675	.683	.690	.698	.705	.713
110	.721	.728	.736	.744	.752	.759	.767	.775	.782	.790
120	.798	.805	.813	.821	.829	.837	.845	.853	.861	.869
130	.877	.885	.893	.901	.909	.917	.925	.933	.941	.949
140	.957	.966	.974	.982	.990	.998	1.006	1.014	1.022	1.031
150	1.039	1.047	1.055	1.063	1.072	1.080	1.088	1.096	1.104	1.112
160	1.121	1.129	1.138	1.146	1.154	1.163	1.171	1.179	1.188	1.196
170	1.205	1.213	1.222	1.231	1.239	1.247	1.256	1.265	1.273	1.282
180	1.290	1.298	1.307	1.316	1.324	1.333	1.342	1.351	1.359	1.368
190	1.377	1.386	1.395	1.403	1.412	1.420	1.429	1.438	1.447	1.456
200	1.465	1.473	1.482	1.491	1.500	1.509	1.517	1.526	1.535	1.544
210	1.553	1.562	1.571	1.580	1.589	1.598	1.607	1.616	1.625	1.634
220	1.643	1.652	1.661	1.670	1.679	1.688	1.697	1.706	1.715	1.725
230	1.734	1.743	1.752	1.761	1.770	1.779	1.788	1.798	1.807	1.816
240	1.826	1.835	1.844	1.853	1.863	1.872	1.881	1.890	1.900	1.909
250	1.918	1.928	1.937	1.946	1.956	1.965	1.974	1.984	1.993	2.002
260	2.012	2.021	2.031	2.040	2.050	2.059	2.068	2.078	2.087	2.097
270	2.107	2.116	2.126	2.135	2.145	2.154	2.164	2.173	2.183	2.192
280	2.202	2.211	2.221	2.231	2.240	2.250	2.259	2.269	2.279	2.288
290	2.298	2.308	2.317	2.327	2.337	2.346	2.356	2.366	2.375	2.385
300	2.395	2.405	2.415	2.424	2.434	2.444	2.454	2.464	2.473	2.483
310	2.493	2.503	2.513	2.522	2.532	2.542	2.552	2.562	2.572	2.581
320	2.591	2.601	2.611	2.621	2.631	2.641	2.650	2.660	2.670	2.680
330	2.690	2.700	2.710	2.720	2.730	2.740	2.750	2.760	2.770	2.780
340	2.790	2.800	2.810	2.820	2.830	2.840	2.850	2.860	2.870	2.880
350	2.890	2.900	2.910	2.820	2.930	2.940	2.950	2.961	2.971	2.981
360	2.991	3.001	3.011	3.021	3.031	3.041	3.051	3.062	3.072	3.082
370	3.092	3.102	3.112	3.122	3.133	3.143	3.153	3.163	3.173	3.183
380	3.194	3.204	3.214	3.224	3.234	3.245	3.255	3.265	3.276	3.286
390	3.296	3.306	3.317	3.327	3.337	3.347	3.358	3.368	3.378	3.389
400	3.399	3.409	3.420	3.430	3.440	3.451	3.461	3.471	3.481	3.492
410	3.502	3.512	3.523	3.533	3.544	3.554	3.565	3.575	3.586	3.596
420	3.607	3.617	3.627	3.638	3.648	3.659	3.669	3.680	3.690	3.701
430	3.712	3.722	3.732	3.743	3.753	3.764	3 774	3.785	3.796	3.806
440	3.817	3.827	3.838	3.848	3.859	3.870	3.880	3.891	3.901	3.912
450	3.923	3.933	3.944	3.954	3.965	3.976	3.987	3.997	4.008	4.018
460	4.029	4.039	4.050	4.060	4.071	4.081	4.092	4.102	4.113	4.123
470	4.134	4.145	4.156	4.166	4.177	4.187	4.198	4.209	4.219	4.230
480	4.241	4.251	4.262	4.273	4.283	4.294	4.305	4.315	4.326	4.337
490	4.348	4.358	4.369	4.380	4.390	4.401	4.412	4.422	4.433	4.444
500	4.455	4.466	4.477	4.488	4.498	4.509	4.520	4.531	4.542	4.552
510	4.563	4.574	4.585	4.596	4.607	4.618	4.629	4.640	4.651	4.662
520	4.672	4.683	4.694	4.705	4.716	4.727	4.738	4.749	4.760	4.771
530	4.782	4.793	4.804	4.815	4.826	4.837	4.848	4.859	4.870	4.881
540	4.893	4.904	4.915	4.926	4.937	4.948	4.959	4.970	4.981	4.992
550	5.004	5.015	5.026	5.037	5.048	5.059	5.070	5.081	5.092	5.104
560	5.115	5.126	5.137	5.148	5.159	5.170	5.182	5.193	5.204	5.215
570	5.226	5.238	5.249	5.260	5.271	5.282	5.293	5.304	5.316	5.327
580	5.338	5.349	5.360	5.371	5.383	5.394	5.405	5.416	5.428	5.439
590	5.450	5.461	5.472	5.484	5.495	5.507	5.518	5.529	5.540	5.551

PLAT. vs. PLAT. +13% RHODIUM THERMOCOUPLE

Degrees Centigrade — Reference Junction 0° C.

°C	0	1	2	3	4	5	6	7	8	9
					Millivolts					
600	5.563	5.574	5.586	5.597	5.609	5.620	5.631	5.642	5.654	5.665
610	5.677	5.688	5.700	5.711	5.723	5.734	5.746	5.757	5.769	5.780
620	5.792	5.803	5.814	5.826	5.837	5.849	5.861	5.872	5.883	5.895
630	5.907	5.918	5.930	5.941	5.952	5.964	5.976	5.987	5.999	6.010
640	6.022	6.033	6.044	6.056	6.068	6.079	6.091	6.102	6.114	6.126
650	6.137	6.149	6.160	6.171	6.183	6.194	6.206	6.218	6.229	6.240
660	6.252	6.264	6.275	6.287	6.299	6.310	6.321	6.333	6.344	6.356
670	6.368	6.380	6.391	6.403	6.415	6.427	6.438	6.450	6.461	6.473
680	6.485	6.497	6.508	6.520	6.532	6.544	6.555	6.567	6.579	6.590
690	6.602	6.614	6.626	6.637	6.649	6.661	6.672	6.684	6.696	6.708
700	6.720	6.732	6.744	6.756	6.768	6.779	6.791	6.803	6.815	6.827
710	6.838	6.850	6.862	6.874	6.886	6.898	6.910	6.922	6.834	6.946
720	6.957	6.969	6.981	6.993	7.005	7.017	7.029	7.040	7.052	7.064
730	7.076	7.088	7.100	7.112	7.124	7.136	7.147	7.159	7.171	7.183
740	7.195	7.207	7.219	7.231	7.243	7.255	7.267	7.279	7.291	7.303
750	7.315	7.327	7.339	7.351	7.364	7.376	7.388	7.400	7.412	7.424
760	7.436	7.448	7.460	7.472	7.485	7.497	7.509	7.521	7.533	7.545
770	7.557	7.570	7.582	7.594	7.606	7.618	7.631	7.643	7.655	7.667
780	7.679	7.692	7.704	7.716	7.728	7.740	7.752	7.765	7.777	7.789
790	7.801	7.814	7.826	7.838	7.850	7.863	7.875	7.888	7.900	7.912
800	7.924	7.936	7.949	7.961	7.973	7.986	7.998	8.010	8.022	8.035
810	8.047	8.059	8.071	8.084	8.096	8.109	8.121	8.134	8.146	8.158
820	8.170	8.182	8.195	8.208	8.220	8.232	8.245	8.257	8.269	8.281
830	8.294	8.306	8.319	8.331	8.343	8.356	8.369	8.381	8.394	8.406
840	8.419	8.431	8.444	8.456	8.469	8.481	8.494	8.506	8.519	8.531
850	8.544	8.556	8.569	8.581	8.594	8.606	8.619	8.631	8.644	8.656
860	8.669	8.681	8.694	8.706	8.719	8.732	8.744	8.757	8.769	8.782
870	8.795	8.807	8.820	8.832	8.845	8.858	8.870	8.883	8.895	8.908
880	8.921	8.933	8.946	8.959	8.971	8.984	8.996	9.009	9.021	9.034
890	9.047	9.060	9.072	9.085	9.098	9.111	9.123	9.136	9.149	9.161
900	9.175	9.188	9.200	9.213	9.226	9.239	9.251	9.264	9.277	9.290
910	9.303	9.316	9.328	9.341	9.354	9.367	9.379	9.392	9.405	9.418
920	9.431	9.444	9.456	9.469	9.482	9.495	9.508	9.520	9.533	9.546
930	9.559	9.572	9.585	9.598	9.610	9.623	9.636	9.649	9.661	9.674
940	9.687	9.700	9.713	9.726	9.739	9.752	9.765	9.778	9.790	9.803
950	9.816	9.829	9.842	9.855	9.868	9.881	9.894	9.907	9.920	9.933
960	9.946	9.960	9.973	9.986	9.999	10.012	10.025	10.038	10.051	10.064
970	10.077	10.090	10.103	10.116	10.130	10.143	10.156	10.169	10.182	10.195
980	10.208	10.221	10.234	10.247	10.260	10.274	10.287	10.300	10.313	10.326
990	10.339	10.352	10.366	10.379	10.392	10.405	10.419	10.432	10.445	10.458
1000	10.471	10.484	10.497	10.510	10.523	10.537	10.550	10.563	10.576	10.589
1010	10.603	10.616	10.629	10.642	10.655	10.669	10.682	10.695	10.709	10.722
1020	10.735	10.748	10.761	10.775	10.788	10.801	10.815	10.828	10.841	10.855
1030	10.869	10.882	10.895	10.909	10.922	10.936	10.949	10.963	10.976	10.989
1040	11.003	11.016	11.030	11.043	11.057	11.070	11.084	11.097	11.111	11.124
1050	11.138	11.151	11.165	11.178	11.191	11.205	11.219	11.232	11.246	11.259
1060	11.273	11.286	11.300	11.313	11.327	11.340	11.354	11.367	11.381	11.394
1070	11.408	11.421	11.435	11.449	11.463	11.476	11.490	11.504	11.517	11.531
1080	11.544	11.558	11.571	11.585	11.599	11.613	11.626	11.640	11.654	11.667
1090	11.681	11.694	11.708	11.722	11.736	11.749	11.763	11.776	11.790	11.803
1100	11.817	11.830	11.844	11.858	11.871	11.885	11.899	11.913	11.926	11.940
1110	11.954	11.967	11.981	11.994	12.008	12.022	12.035	12.049	12.063	12.077
1120	12.090	12.104	12.118	12.131	12.145	12.159	12.173	12.186	12.200	12.214
1130	12.227	12.241	12.254	12.268	12.282	12.296	12.310	12.323	12.337	12.351
1140	12.365	12.378	12.392	12.406	12.420	12.434	12.447	12.461	12.475	12.489
1150	12.503	12.516	12.530	12.544	12.558	12.572	12.585	12.599	12.613	12.627
1160	12.641	12.654	12.668	12.682	12.696	12.710	12.723	12.737	12.751	12.765
1170	12.779	12.792	12.806	12.820	12.834	12.848	12.861	12.875	12.889	12.903
1180	12.917	12.931	12.944	12.958	12.972	12.986	13.000	13.014	13.028	13.042
1190	13.055	13.069	13.083	13.097	13.111	13.125	13.139	13.152	13.166	13.180

Table 9.24 continued

Table 9.24 continued

PLAT. vs. PLAT. +13% RHODIUM THERMOCOUPLE
Degrees Centigrade — Reference Junction 0° C.

°C	0	1	2	3	4	5	6	7	8	9
					Millivolts					
1200	13.193	13.207	13.221	13.235	13.249	13.263	13.277	13.291	13.305	13.319
1210	13.332	13.346	13.360	13.374	13.388	13.402	13.416	13.429	13.443	13.457
1220	13.471	13.485	13.499	13.513	13.526	13.540	13.554	13.568	13.582	13.596
1230	13.610	13.624	13.638	13.652	13.666	13.679	13.693	13.707	13.721	13.735
1240	13.749	13.763	13.777	13.791	13.805	13.818	13.832	13.846	13.860	13.874
1250	13.888	13.902	13.916	13.930	13.943	13.957	13.971	13.985	13.999	14.013
1260	14.027	14.041	14.055	14.069	14.082	14.096	14.110	14.124	14.138	14.152
1270	14.165	14.179	14.193	14.207	14.221	14.235	14.249	14.263	14.277	14.291
1280	14.304	14.318	14.332	14.346	14.360	14.374	14.388	14.402	14.416	14.430
1290	14.443	14.457	14.471	14.485	14.499	14.513	14.527	14.541	14.555	14.569
1300	14.582	14.596	14.610	14.624	14.638	14.652	14.666	14.680	14.694	14.707
1310	14.721	14.735	14.749	14.763	14.777	14.791	14.804	14.818	14.832	14.846
1320	14.860	14.874	14.888	14.901	14.915	14.929	14.943	14.957	14.971	14.985
1330	14.999	15.013	15.026	15.040	15.054	15.068	15.082	15.096	15.110	15.124
1340	15.138	15.151	15.165	15.179	15.193	15.207	15.221	15.234	15.248	15.262
1350	15.276	15.290	15.304	15.318	15.331	15.345	15.359	15.373	15.387	15.401
1360	15.415	15.429	15.443	15.456	15.470	15.484	15.498	15.512	15.526	15.540
1370	15.553	15.567	15.581	15.595	15.609	15.623	15.637	15.651	15.665	15.679
1380	15.692	15.706	15.720	15.734	15.748	15.761	15.775	15.789	15.803	15.817
1390	15.831	15.845	15.859	15.873	15.886	15.900	15.914	15.928	15.942	15.956
1400	15.969	15.983	15.997	16.011	16.025	16.039	16.053	16.067	16.081	16.095
1410	16.108	16.122	16.136	16.150	16.164	16.178	16.192	16.206	16.219	16.233
1420	16.247	16.261	16.275	16.289	16.303	16.317	16.330	16.344	16.358	16.372
1430	16.386	16.400	16.414	16.427	16.441	16.455	16.469	16.483	16.497	16.511
1440	16.524	16.538	16.552	16.566	16.580	16.594	16.608	16.621	16.635	16.649
1450	16.663	16.677	16.691	16.705	16.719	16.733	16.746	16.760	16.774	16.788
1460	16.802	16.816	16.830	16.844	16.858	16.872	16.885	16.899	16.913	16.927
1470	16.940	16.954	16.968	16.982	16.996	17.010	17.024	17.037	17.051	17.065
1480	17.079	17.092	17.106	17.120	17.134	17.148	17.161	17.175	17.189	17.203
1490	17.217	17.230	17.244	17.258	17.272	17.286	17.299	17.313	17.327	17.341
1500	17.355	17.368	17.382	17.396	17.410	17.424	17.437	17.451	17.465	17.479
1510	17.493	17.506	17.520	17.534	17.547	17.561	17.575	17.589	17.603	17.617
1520	17.631	17.644	17.658	17.672	17.686	17.699	17.713	17.726	17.740	17.754
1530	17.768	17.781	17.795	17.809	17.823	17.837	17.850	17.864	17.878	17.892
1540	17.906	17.919	17.933	17.947	17.960	17.974	17.988	18.002	18.016	18.029
1550	18.043	18.056	18.070	18.084	18.098	18.111	18.125	18.139	18.152	18.166
1560	18.179	18.193	18.207	18.220	18.234	18.248	18.261	18.275	18.289	18.303
1570	18.316	18.330	18.344	18.357	18.371	18.385	18.399	18.412	18.426	18.440
1580	18.453	18.467	18.481	18.494	18.508	18.522	18.536	18.549	18.563	18.576
1590	18.590	18.604	18.618	18.631	18.645	18.659	18.672	18.686	18.700	18.714
1600	18.727	18.741	18.754	18.768	18.782	18.796	18.810	18.823	18.836	18.850
1610	18.864	18.878	18.891	18.905	18.919	18.932	18.946	18.960	18.973	18.987
1620	19.001	19.014	19.028	19.042	19.056	19.069	19.083	19.096	19.110	19.124
1630	19.137	19.150	19.164	19.178	19.191	19.205	19.219	19.232	19.246	19.260
1640	19.273	19.287	19.300	19.314	19.328	19.341	19.355	19.369	19.382	19.396
1650	19.409	19.423	19.437	19.450	19.464	19.477	19.491	19.504	19.518	19.531
1660	19.545	19.559	19.573	19.586	19.600	19.614	19.627	19.641	19.654	19.668
1670	19.682	19.695	19.709	19.722	19.736	19.750	19.763	19.777	19.790	19.804
1680	19.818	19.831	19.845	19.859	19.873	19.886	19.900	19.913	19.927	19.940
1690	19.954	19.967	19.981	19.994	20.008	20.022	20.035	20.049	20.062	20.076

PLAT. vs. PLAT. +13% RHODIUM THERMOCOUPLE
Degrees Fahrenheit — Reference Junction 32° F.

°F	0	1	2	3	4	5	6	7	8	9
					Millivolts					
0	-0.092	-0.089	-0.086	-0.084	-0.081	-0.078	-0.075	-0.072	-0.070	-0.067
10	-.064	-.061	-.058	-.055	-.052	-.050	-.047	-.044	-.041	-.038
20	-.035	-.032	-.029	-.026	-.023	-.021	-.018	-.015	-.012	-.009
30	-.006	-.003	.000	.003	.006	.009	.012	.015	.018	.021
40	.024	.027	.030	.033	.036	.039	.042	.045	.048	.052
50	.055	.058	.061	.064	.068	.071	.074	.077	.080	.083
60	.086	.090	.093	.096	.099	.103	.106	.109	.112	.116
70	.119	.122	.126	.129	.132	.135	.139	.142	.145	.149
80	.152	.155	.159	.162	.165	.169	.172	.175	.179	.182
90	.186	.189	.192	.196	.199	.203	.206	.210	.213	.217
100	.220	.224	.227	.230	.234	.237	.241	.244	.248	.251
110	.255	.258	.262	.265	.269	.272	.276	.280	.284	.287
120	.291	.294	.298	.301	.305	.308	.312	.316	.319	.323
130	.327	.330	.334	.337	.341	.345	.349	.352	.356	.359
140	.363	.367	.370	.374	.378	.381	.385	.389	.393	.397
150	.400	.404	.408	.411	.415	.419	.423	.427	.431	.435
160	.438	.442	.446	.450	.453	.457	.461	.465	.469	.473
170	.476	.480	.484	.488	.492	.496	.500	.504	.508	.512
180	.516	.520	.524	.528	.532	.536	.540	.544	.548	.552
190	.556	.560	.564	.568	.572	.576	.580	.584	.588	.592
200	.596	.600	.604	.608	.612	.616	.620	.625	.629	.633
210	.637	.641	.645	.649	.653	.657	.662	.666	.670	.674
220	.678	.683	.687	.691	.695	.700	.704	.708	.712	.716
230	.721	.725	.729	.734	.738	.742	.746	.750	.755	.759
240	.763	.767	.772	.776	.780	.785	.789	.793	.798	.802
250	.807	.811	.815	.820	.824	.828	.833	.837	.842	.846
260	.850	.855	.859	.863	.868	.872	.877	.881	.886	.890
270	.894	.899	.904	.908	.912	.917	.921	.926	.930	.935
280	.939	.944	.948	.953	.957	.962	.966	.971	.975	.980
290	.984	.989	.993	.998	1.002	1.007	1.011	1.016	1.020	1.025
300	1.030	1.034	1.039	1.043	1.048	1.052	1.057	1.061	1.066	1.071
310	1.075	1.080	1.084	1.089	1.094	1.098	1.103	1.107	1.112	1.117
320	1.121	1.126	1.130	1.135	1.140	1.144	1.149	1.153	1.158	1.163
330	1.167	1.172	1.176	1.181	1.186	1.191	1.195	1.200	1.205	1.210
340	1.214	1.219	1.223	1.228	1.233	1.238	1.242	1.247	1.252	1.257
350	1.261	1.266	1.271	1.276	1.280	1.285	1.290	1.295	1.300	1.304
360	1.309	1.314	1.319	1.323	1.328	1.333	1.338	1.343	1.348	1.352
370	1.357	1.362	1.367	1.372	1.377	1.381	1.386	1.391	1.396	1.401
380	1.406	1.410	1.415	1.420	1.425	1.430	1.435	1.440	1.445	1.450
390	1.455	1.460	1.465	1.470	1.475	1.480	1.484	1.489	1.494	1.499
400	1.504	1.509	1.514	1.519	1.524	1.529	1.533	1.438	1.543	1.548
410	1.553	1.558	1.563	1.568	1.573	1.578	1.583	1.588	1.593	1.598
420	1.603	1.608	1.613	1.618	1.623	1.628	1.633	1.638	1.643	1.648
430	1.653	1.658	1.663	1.668	1.673	1.678	1.683	1.688	1.693	1.698
440	1.703	1.708	1.713	1.719	1.724	1.729	1.734	1.739	1.744	1.749
450	1.754	1.759	1.764	1.769	1.774	1.779	1.785	1.790	1.795	1.800
460	1.805	1.811	1.816	1.821	1.826	1.831	1.836	1.841	1.846	1.851
470	1.856	1.862	1.867	1.872	1.877	1.882	1.887	1.892	1.898	1.903
480	1.908	1.913	1.918	1.924	1.929	1.934	1.939	1.944	1.950	1.955
490	1.960	1.965	1.970	1.976	1.981	1.986	1.991	1.996	2.002	2.007
500	2.012	2.017	2.023	2.028	2.033	2.038	2.044	2.049	2.054	2.059
510	2.065	2.070	2.075	2.081	2.086	2.091	2.096	2.101	2.107	2.112
520	2.117	2.123	2.128	2.133	2.139	2.144	2.149	2.154	2.160	2.165
530	2.170	2.176	2.181	2.186	2.192	2.197	2.202	2.207	2.213	2.218
540	2.223	2.229	2.234	2.239	2.245	2.250	2.255	2.261	2.266	2.271
550	2.277	2.282	2.287	2.293	2.298	2.303	2.308	2.314	2.319	2.325
560	2.330	2.335	2.341	2.346	2.352	2.357	2.363	2.368	2.373	2.379
570	2.384	2.389	2.395	2.401	2.406	2.412	2.417	2.423	2.428	2.433
580	2.438	2.444	2.449	2.455	2.460	2.466	2.471	2.477	2.482	2.487
590	2.493	2.498	2.504	2.509	2.515	2.520	2.526	2.531	2.537	2.542

Table 9.24 continued

Table 9.24 continued

PLAT. vs. PLAT. +13% RHODIUM THERMOCOUPLE

Degrees Fahrenheit — Reference Junction 32° F.

°F	0	1	2	3	4	5	6	7	8	9
					Millivolts					
600	2.547	2.553	2.558	2.564	2.569	2.575	2.580	2.586	2.591	2.597
610	2.602	2.608	2.613	2.619	2.624	2.630	2.635	2.641	2.646	2.652
620	2.657	2.663	2.668	2.674	2.679	2.685	2.690	2.696	2.701	2.707
630	2.712	2.718	2.723	2.729	2.734	2.740	2.746	2.751	2.757	2.762
640	2.768	2.773	2.779	2.784	2.790	2.796	2.801	2.807	2.812	2.818
650	2.823	2.829	2.834	2.840	2.846	2.851	2.857	2.862	2.868	2.873
660	2.879	2.884	2.890	2.896	2.901	2.907	2.912	2.918	2.923	2.929
670	2.935	2.940	2.946	2.952	2.957	2.963	2.968	2.974	2.979	2.985
680	2.991	2.997	3.002	3.008	3.013	3.019	3.024	3.030	3.036	3.041
690	3.047	3.053	3.058	3.064	3.069	3.075	3.081	3.087	3.092	3.098
700	3.103	3.109	3.115	3.120	3.126	3.132	3.137	3.143	3.148	3.154
710	3.160	3.166	3.171	3.177	3.182	3.188	3.194	3.199	3.205	3.211
720	3.217	3.222	3.228	3.234	3.239	3.245	3.251	3.256	3.262	3.268
730	3.273	3.279	3.285	3.291	3.296	3.302	3.308	3.313	3.319	3.325
740	3.330	3.336	3.342	3.348	3.353	3.359	3.365	3.370	3.376	3.382
750	3.387	3.393	3.399	3.405	3.411	3.416	3.422	3.428	3.433	3.439
760	3.445	3.451	3.456	3.462	3.468	3.473	3.479	3.485	3.491	3.497
770	3.502	3.508	3.514	3.519	3.525	3.531	3.537	3.543	3.549	3.554
780	3.560	3.566	3.572	3.577	3.583	3.589	3.595	3.601	3.607	3.612
790	3.618	3.624	3.630	3.635	3.641	3.647	3.653	3.659	3.665	3.671
800	3.677	3.682	3.688	3.694	3.700	3.706	3.712	3.718	3.723	3.729
810	3.735	3.741	3.746	3.752	3.758	3.764	3.770	3.776	3.782	3.788
820	3.794	3.799	3.805	3.811	3.817	3.823	3.829	3.835	3.841	3.846
830	3.852	3.858	3.864	3.870	3.876	3.882	3.888	3.894	3.899	3.905
840	3.911	3.917	3.923	3.929	3.935	3.941	3.946	3.952	3.958	3.964
850	3.970	3.976	3.982	3.988	3.994	3.999	4.005	4.011	4.017	4.023
860	4.029	4.035	4.041	4.047	4.052	4.058	4.064	4.070	4.075	4.081
870	4.087	4.093	4.099	4.105	4.111	4.116	4.122	4.128	4.134	4.140
880	4.146	4.152	4.158	4.164	4.169	4.175	4.181	4.187	4.193	4.199
890	4.205	4.211	4.217	4.223	4.229	4.235	4.241	4.246	4.252	4.258
900	4.264	4.270	4.276	4.282	4.288	4.294	4.300	4.306	4.312	4.318
910	4.324	4.330	4.336	4.342	4.348	4.354	4.360	4.366	4.372	4.378
920	4.384	4.389	4.395	4.401	4.407	4.413	4.419	4.425	4.431	4.437
930	4.443	4.449	4.455	4.461	4.467	4.473	4.479	4.485	4.491	4.497
940	4.503	4.509	4.515	4.521	4.527	4.533	4.539	4.545	4.551	4.557
950	4.563	4.569	4.575	4.581	4.587	4.593	4.599	4.605	4.612	4.618
960	4.624	4.630	4.636	4.642	4.648	4.654	4.660	4.666	4.672	4.679
970	4.685	4.691	4.697	4.703	4.709	4.715	4.721	4.727	4.733	4.740
980	4.746	4.752	4.758	4.764	4.770	4.776	4.782	4.788	4.794	4.801
990	4.807	4.613	4.819	4.825	4.831	4.837	4.844	4.850	4.856	4.862
1000	4.868	4.874	4.881	4.887	4.893	4.899	4.905	4.911	4.917	4.924
1010	4.930	4.936	4.942	4.948	4.954	4.960	4.966	4.972	4.979	4.985
1020	4.991	4.998	5.004	5.010	5.016	5.022	5.028	5.034	5.041	5.047
1030	5.053	5.059	5.066	5.072	5.078	5.084	5.090	5.096	5.102	5.109
1040	5.115	5.121	5.127	5.133	5.139	5.146	5.152	5.158	5.164	5.170
1050	5.176	5.182	5.189	5.195	5.201	5.208	5.214	5.220	5.226	5.232
1060	5.238	5.244	5.251	5.257	5.263	5.270	5.276	5.282	5.288	5.294
1070	5.301	5.307	5.313	5.319	5.326	5.332	5.338	5.344	5.351	5.357
1080	5.363	5.369	5.376	5.382	5.388	5.394	5.401	5.407	5.413	5.419
1090	5.426	5.432	5.438	5.444	5.450	5.457	5.463	5.469	5.476	5.482
1100	5.488	5.494	5.501	5.507	5.513	5.519	5.526	5.532	5.538	5.544
1110	5.551	5.557	5.563	5.570	5.576	5.582	5.589	5.595	5.601	5.607
1120	5.614	5.620	5.626	5.633	5.639	5.645	5.652	5.658	5.664	5.671
1130	5.677	5.684	5.690	5.696	5.703	5.709	5.716	5.722	5.728	5.734
1140	5.741	5.747	5.753	5.760	5.766	5.773	5.779	5.786	5.792	5.798
1150	5.805	5.811	5.817	5.824	5.830	5.837	5.843	5.849	5.856	5.862
1160	5.869	5.875	5.881	5.888	5.894	5.901	5.907	5.913	5.920	5.926
1170	5.933	5.939	5.945	5.952	5.958	5.964	5.971	5.977	5.983	5.990
1180	5.996	6.003	6.009	6.015	6.022	6.028	6.035	6.041	6.047	6.054
1190	6.060	6.067	6.073	6.079	6.086	6.092	6.099	6.105	6.111	6.118

PLAT. vs. PLAT. +13% RHODIUM THERMOCOUPLE

Degrees Fahrenheit — Reference Junction 32° F.

°F	0	1	2	3	4	5	6	7	8	9
					Millivolts					
1200	6.125	6.131	6.137	6.143	6.150	6.156	6.163	6.169	6.175	6.182
1210	6.188	6.195	6.201	6.207	6.214	6.220	6.227	6.233	6.239	6.246
1220	6.252	6.259	6.265	6.272	6.278	6.285	6.291	6.298	6.304	6.310
1230	6.317	6.323	6.329	6.336	6.342	6.349	6.355	6.362	6.368	6.375
1240	6.381	6.388	6.394	6.401	6.407	6.414	6.420	6.427	6.433	6.440
1250	6.446	6.453	6.459	6.466	6.472	6.479	6.485	6.492	6.498	6.505
1260	6.511	6.518	6.524	6.531	6.537	6.544	6.550	6.557	6.563	6.570
1270	6.577	6.583	6.589	6.596	6.602	6.609	6.616	6.622	6.629	6.635
1280	6.642	6.648	6.655	6.661	6.668	6.674	6.681	6.687	6.694	6.701
1290	6.707	6.714	6.720	6.727	6.733	6.740	6.746	6.753	6.759	6.766
1300	6.773	6.779	6.786	6.792	6.799	6.805	6.812	6.818	6.825	6.832
1310	6.838	6.845	6.851	6.858	6.865	6.871	6.877	6.884	6.891	6.898
1320	6.904	6.911	6.917	6.924	6.931	6.937	6.943	6.950	6.957	6.964
1330	6.970	6.977	6.983	6.990	6.997	7.003	7.010	7.017	7.023	7.030
1340	7.037	7.043	7.049	7.056	7.063	7.069	7.076	7.083	7.089	7.096
1350	7.103	7.109	7.116	7.123	7.129	7.136	7.143	7.149	7.155	7.162
1360	7.169	7.175	7.182	7.189	7.195	7.202	7.209	7.215	7.222	7.229
1370	7.235	7.242	7.249	7.255	7.262	7.269	7.275	7.282	7.289	7.295
1380	7.302	7.309	7.315	7.322	7.329	7.336	7.342	7.349	7.356	7.362
1390	7.369	7.376	7.382	7.389	7.396	7.403	7.409	7.416	7.423	7.429
1400	7.436	7.443	7.449	7.456	7.463	7.470	7.477	7.483	7.490	7.497
1410	7.503	7.510	7.517	7.523	7.530	7.537	7.544	7.551	7.557	7.564
1420	7.571	7.578	7.585	7.591	7.598	7.605	7.611	7.618	7.625	7.632
1430	7.639	7.645	7.652	7.659	7.665	7.672	7.679	7.686	7.693	7.699
1440	7.706	7.713	7.720	7.727	7.733	7.740	7.747	7.754	7.761	7.767
1450	7.774	7.781	7.788	7.795	7.801	7.808	7.815	7.822	7.829	7.835
1460	7.842	7.849	7.856	7.863	7.870	7.877	7.884	7.891	7.897	7.904
1470	7.911	7.918	7.924	7.931	7.938	7.945	7.952	7.959	7.965	7.972
1480	7.979	7.986	7.993	7.999	8.006	8.013	8.020	8.027	8.033	8.040
1490	8.047	8.054	8.061	8.068	8.075	8.081	8.089	8.095	8.102	8.109
1500	8.116	8.123	8.129	8.136	8.143	8.150	8.157	8.163	8.170	8.177
1510	8.184	8.191	8.198	8.205	8.212	8.218	8.225	8.232	8.239	8.246
1520	8.253	8.260	8.267	8.274	8.281	8.287	8.294	8.301	8.308	8.315
1530	8.322	8.329	8.336	8.343	8.350	8.356	8.363	8.370	8.377	8.384
1540	8.391	8.398	8.405	8.412	8.419	8.426	8.433	8.439	8.446	8.453
1550	8.460	8.467	8.474	8.481	8.488	8.495	8.502	8.509	8.516	8.523
1560	8.530	8.537	8.544	8.551	8.558	8.565	8.571	8.578	8.585	8.592
1570	8.599	8.606	8.613	8.620	8.627	8.634	8.641	8.648	8.655	8.662
1580	8.669	8.676	8.683	8.690	8.697	8.704	8.711	8.718	8.725	8.732
1590	8.739	8.746	8.753	8.760	8.767	8.774	8.781	8.788	8.795	8.802
1600	8.809	8.816	8.823	8.830	8.837	8.844	8.851	8.858	8.865	8.872
1610	8.879	8.866	8.893	8.900	8.907	8.914	8.921	8.928	8.935	8.942
1620	8.949	8.956	8.963	8.970	8.977	8.984	8.991	8.998	9.005	9.012
1630	9.019	9.026	9.033	9.040	9.047	9.054	9.061	9.068	9.075	9.082
1640	9.090	9.097	9.104	9.111	9.118	9.125	9.132	9.139	9.146	9.153
1650	9.161	9.168	9.175	9.182	9.189	9.196	9.203	9.210	9.218	9.225
1660	9.232	9.239	9.246	9.253	9.260	9.267	9.274	9.281	9.289	9.296
1670	9.303	9.310	9.317	9.324	9.331	9.338	9.345	9.353	9.360	9.367
1680	9.374	9.381	9.388	9.395	9.402	9.409	9.416	9.424	9.431	9.438
1690	9.445	9.452	9.459	9.466	9.474	9.481	9.488	9.495	9.502	9.509
1700	9.516	9.523	9.531	9.538	9.545	9.552	9.559	9.566	9.573	9.580
1710	9.587	9.594	9.602	9.609	9.616	9.623	9.630	9.637	9.644	9.651
1720	9.659	9.666	9.673	9.680	9.687	9.694	9.701	9.709	9.716	9.723
1730	9.730	9.737	9.744	9.751	9.759	9.766	9.773	9.780	9.787	9.794
1740	9.802	9.809	9.816	9.823	9.830	9.838	9.845	9.852	9.859	9.866
1750	9.874	9.881	9.888	9.895	9.902	9.910	9.917	9.924	9.931	9.939
1760	9.946	9.953	9.961	9.968	9.975	9.982	9.990	9.997	10.004	10.012
1770	10.019	10.026	10.034	10.041	10.048	10.056	10.063	10.070	10.077	10.084
1780	10.092	10.099	10.106	10.114	10.121	10.129	10.136	10.143	10.150	10.157
1790	10.164	10.172	10.179	10.186	10.194	10.201	10.208	10.215	10.223	10.230

Table 9.24 continued

Table 9.24 continued

PLAT. vs. PLAT. +13% RHODIUM THERMOCOUPLE
Degrees Fahrenheit — Reference Junction 32° F.

°F	0	1	2	3	4	5	6	7	8	9
					Millivolts					
1800	10.237	10.244	10.251	10.259	10.266	10.274	10.281	10.288	10.296	10.303
1810	10.310	10.318	10.325	10.332	10.339	10.347	10.354	10.361	10.369	10.376
1820	10.383	10.391	10.398	10.405	10.412	10.420	10.427	10.434	10.441	10.449
1830	10.456	10.464	10.471	10.478	10.485	10.493	10.500	10.507	10.514	10.522
1840	10.529	10.537	10.544	10.551	10.559	10.566	10.574	10.581	10.588	10.596
1850	10.603	10.610	10.618	10.625	10.632	10.639	10.647	10.654	10.661	10.669
1860	10.676	10.683	10.691	10.698	10.705	10.712	10.720	10.727	10.735	10.742
1870	10.749	10.757	10.764	10.771	10.779	10.786	10.794	10.801	10.809	10.816
1880	10.823	10.831	10.839	10.846	10.854	10.861	10.869	10.876	10.884	10.891
1890	10.898	10.906	10.914	10.921	10.929	10.936	10.944	10.951	10.959	10.966
1900	10.973	10.981	10.988	10.996	11.003	11.011	11.018	11.026	11.033	11.040
1910	11.048	11.055	11.063	11.070	11.078	11.085	11.093	11.100	11.108	11.115
1920	11.122	11.130	11.138	11.145	11.153	11.160	11.168	11.175	11.183	11.190
1930	11.197	11.205	11.213	11.220	11.228	11.235	11.243	11.250	11.258	11.265
1940	11.273	11.280	11.288	11.295	11.303	11.310	11.318	11.325	11.333	11.340
1950	11.348	11.355	11.363	11.371	11.379	11.385	11.393	11.401	11.408	11.416
1960	11.424	11.431	11.439	11.446	11.454	11.461	11.468	11.476	11.484	11.492
1970	11.499	11.507	11.515	11.522	11.529	11.537	11.544	11.552	11.560	11.568
1980	11.575	11.582	11.590	11.598	11.605	11.613	11.620	11.628	11.636	11.643
1990	11.651	11.658	11.666	11.674	11.681	11.689	11.696	11.704	11.712	11.719
2000	11.726	11.734	11.742	11.749	11.757	11.765	11.772	11.779	11.787	11.795
2010	11.802	11.810	11.817	11.825	11.832	11.840	11.848	11.855	11.863	11.871
2020	11.878	11.885	11.893	11.901	11.908	11.916	11.924	11.931	11.938	11.946
2030	11.954	11.961	11.969	11.976	11.984	11.992	11.999	12.007	12.014	12.022
2040	12.029	12.037	12.045	12.052	12.060	12.068	12.075	12.082	12.090	12.098
2050	12.105	12.113	12.121	12.128	12.136	12.144	12.151	12.159	12.166	12.174
2060	12.182	12.189	12.197	12.205	12.212	12.220	12.227	12.235	12.243	12.250
2070	12.258	12.265	12.273	12.281	12.288	12.296	12.304	12.312	12.319	12.327
2080	12.335	12.342	12.350	12.358	12.365	12.373	12.381	12.388	12.396	12.403
2090	12.411	12.419	12.427	12.434	12.442	12.450	12.458	12.465	12.473	12.480
2100	12.488	12.495	12.503	12.511	12.518	12.526	12.534	12.541	12.549	12.557
2110	12.564	12.572	12.579	12.587	12.595	12.602	12.610	12.618	12.625	12.633
2120	12.641	12.648	12.656	12.664	12.672	12.679	12.687	12.695	12.702	12.710
2130	12.718	12.725	12.733	12.741	12.748	12.756	12.764	12.772	12.779	12.787
2140	12.795	12.802	12.810	12.818	12.825	12.833	12.841	12.848	12.856	12.864
2150	12.871	12.879	12.887	12.894	12.902	12.909	12.917	12.925	12.932	12.940
2160	12.948	12.955	12.963	12.971	12.978	12.986	12.994	13.002	13.009	13.017
2170	13.025	13.032	13.040	13.048	13.055	13.063	13.071	13.078	13.086	13.094
2180	13.102	13.109	13.117	13.125	13.132	13.140	13.148	13.155	13.163	13.170
2190	13.178	13.186	13.193	13.201	13.208	13.216	13.224	13.232	13.239	13.247
2200	13.255	13.263	13.270	13.278	13.285	13.293	13.301	13.309	13.316	13.324
2210	13.332	13.340	13.347	13.355	13.363	13.371	13.378	13.386	13.394	13.402
2220	13.409	13.417	13.425	13.432	13.440	13.448	13.455	13.463	13.471	13.479
2230	13.486	13.494	13.502	13.509	13.517	13.525	13.532	13.540	13.548	13.556
2240	13.564	13.571	13.579	13.587	13.595	13.602	13.610	13.618	13.625	13.633
2250	13.641	13.648	13.656	13.664	13.672	13.679	13.687	13.695	13.702	13.710
2260	13.718	13.725	13.733	13.741	13.749	13.756	13.764	13.772	13.779	13.787
2270	13.795	13.802	13.810	13.818	13.826	13.833	13.841	13.849	13.857	13.865
2280	13.872	13.880	13.888	13.895	13.903	13.911	13.918	13.926	13.934	13.942
2290	13.949	13.957	13.965	13.972	13.980	13.988	13.995	14.003	14.011	14.019
2300	14.027	14.034	14.042	14.050	14.058	14.065	14.073	14.081	14.088	14.096
2310	14.104	14.111	14.119	14.127	14.135	14.142	14.150	14.158	14.165	14.173
2320	14.181	14.188	14.196	14.204	14.212	14.219	14.227	14.235	14.242	14.250
2330	14.258	14.265	14.273	14.281	14.288	14.296	14.304	14.312	14.319	14.327
2340	14.335	14.342	14.350	14.358	14.366	14.374	14.382	14.389	14.397	14.405
2350	14.412	14.420	14.428	14.435	14.443	14.451	14.459	14.467	14.475	14.482
2360	14.490	14.498	14.505	14.513	14.521	14.528	14.536	14.544	14.552	14.560
2370	14.567	14.575	14.583	14.591	14.598	14.606	14.614	14.621	14.629	14.637
2380	14.644	14.652	14.660	14.668	14.675	14.683	14.691	14.698	14.706	14.714
2390	14.721	14.729	14.737	14.745	14.752	14.760	14.768	14.775	14.783	14.791

PLAT. vs. PLAT. +13% RHODIUM THERMOCOUPLE
Degrees Fahrenheit — Reference Junction 32° F.

°F	0	1	2	3	4	5	6	7	8	9
					Millivolts					
2400	14.798	14.806	14.814	14.822	14.829	14.837	14.845	14.852	14.860	14.868
2410	14.875	14.883	14.891	14.898	14.906	14.914	14.922	14.929	14.937	14.945
2420	14.952	14.960	14.968	14.975	14.983	14.991	14.999	15.006	15.014	15.022
2430	15.029	15.037	15.045	15.052	15.060	15.068	15.076	15.084	15.091	15.099
2440	15.107	15.115	15.122	15.130	15.138	15.145	15.153	15.161	15.168	15.176
2450	15.184	15.192	15.199	15.207	15.215	15.222	15.230	15.238	15.245	15.253
2460	15.261	15.268	15.276	15.284	15.292	15.299	15.307	15.315	15.322	15.330
2470	15.338	15.345	15.353	15.361	15.369	15.377	15.385	15.392	15.400	15.408
2480	15.415	15.423	15.431	15.438	15.446	15.454	15.462	15.469	15.477	15.484
2490	15.492	15.500	15.508	15.515	15.523	15.531	15.538	15.546	15.553	15.561
2500	15.568	15.576	15.584	15.592	15.599	15.607	15.615	15.623	15.630	15.638
2510	15.645	15.653	15.661	15.668	15.676	15.684	15.692	15.700	15.707	15.715
2520	15.722	15.730	15.738	15.745	15.753	15.761	15.769	15.777	15.785	15.792
2530	15.800	15.808	15.815	15.823	15.831	15.838	15.846	15.854	15.862	15.869
2540	15.877	15.885	15.892	15.900	15.908	15.915	15.923	15.931	15.938	15.946
2550	15.954	15.962	15.969	15.977	15.985	15.992	16.000	16.008	16.015	16.023
2560	16.031	16.039	16.046	16.054	16.062	16.070	16.078	16.085	16.093	16.101
2570	16.108	16.116	16.124	16.132	16.139	16.147	16.155	16.163	16.170	16.178
2580	16.185	16.193	16.201	16.208	16.216	16.224	16.232	16.240	16.247	16.255
2590	16.263	16.271	16.278	16.286	16.294	16.301	16.309	16.317	16.325	16.332
2600	16.340	16.348	16.355	16.363	16.371	16.378	16.386	16.394	16.402	16.409
2610	16.417	16.425	16.432	16.440	16.448	16.455	16.463	16.471	16.478	16.486
2620	16.494	16.502	16.509	16.517	16.524	16.532	16.540	16.548	16.556	16.564
2630	16.571	16.579	16.586	16.594	16.602	16.610	16.618	16.625	16.633	16.641
2640	16.648	16.656	16.663	16.671	16.679	16.687	16.695	16.702	16.710	16.718
2650	16.725	16.733	16.741	16.749	16.756	16.764	16.772	16.780	16.788	16.795
2660	16.802	16.810	16.818	16.826	16.834	16.842	16.849	16.857	16.865	16.872
2670	16.880	16.887	16.895	16.903	16.911	16.918	16.926	16.933	16.941	16.949
2680	16.957	16.964	16.972	16.979	16.987	16.995	17.002	17.010	17.018	17.025
2690	17.033	17.041	17.048	17.056	17.064	17.072	17.079	17.087	17.095	17.102
2700	17.110	17.118	17.125	17.133	17.141	17.148	17.756	17.163	17.171	17.179
2710	17.186	17.194	17.202	17.209	17.217	17.225	17.232	17.240	17.248	17.255
2720	17.263	17.271	17.278	17.286	17.294	17.301	17.309	17.317	17.325	17.332
2730	17.340	17.347	17.355	17.363	17.370	17.378	17.385	17.393	17.401	17.406
2740	17.416	17.424	17.432	17.439	17.447	17.455	17.462	17.470	17.478	17.485
2750	17.493	17.500	17.508	17.516	17.524	17.532	17.539	17.546	17.554	17.562
2760	17.569	17.577	17.585	17.592	17.600	17.608	17.615	17.623	17.631	17.638
2770	17.646	17.654	17.662	17.669	17.677	17.685	17.692	17.700	17.708	17.715
2780	17.723	17.731	17.738	17.746	17.753	17.761	17.768	17.776	17.784	17.792
2790	17.799	17.807	17.814	17.822	17.830	17.837	17.845	17.852	17.860	17.868
2800	17.875	17.882	17.890	17.898	17.906	17.913	17.921	17.928	17.936	17.944
2810	17.951	17.958	17.966	17.974	17.982	17.989	17.997	18.004	18.012	18.020
2820	18.027	18.035	18.043	18.050	18.058	18.065	18.073	18.080	18.088	18.096
2830	18.103	18.111	18.119	18.126	18.134	18.141	18.149	18.156	18.164	18.172
2840	18.179	18.187	18.195	18.202	18.210	18.218	18.225	18.233	18.240	18.248
2850	18.255	18.263	18.271	18.278	18.286	18.294	18.301	18.309	18.316	18.324
2860	18.332	18.339	18.347	18.355	18.362	18.370	18.377	18.385	18.392	18.400
2870	18.408	18.415	18.423	18.431	18.438	18.446	18.453	18.461	18.468	18.476
2880	18.484	18.492	18.499	18.507	18.514	18.522	18.529	18.537	18.545	18.552
2890	18.560	18.568	18.575	18.583	18.590	18.598	18.605	18.613	18.621	18.628
2900	18.636	18.644	18.651	18.659	18.666	18.674	18.681	18.689	18.697	18.705
2910	18.712	18.720	18.727	18.735	18.743	18.750	18.758	18.765	18.773	18.781
2920	18.788	18.796	18.803	18.811	18.819	18.826	18.834	18.842	18.849	18.857
2930	18.864	18.872	18.879	18.887	18.895	18.902	18.910	18.918	18.925	18.932
2940	18.940	18.948	18.955	18.963	18.971	18.978	18.986	18.993	19.001	19.008
2950	19.016	19.024	19.031	19.039	19.046	19.054	19.062	19.069	19.077	19.084
2960	19.092	19.099	19.107	19.115	19.122	19.129	19.137	19.145	19.152	19.160
2970	19.168	19.175	19.182	19.190	19.198	19.205	19.213	19.220	19.228	19.235
2980	19.243	19.250	19.258	19.265	19.273	19.281	19.288	19.295	19.303	19.311
2990	19.318	19.326	19.333	19.341	19.348	19.356	19.364	19.371	19.378	19.386

Table 9.24 continued

Table 9.24 continued

PLAT. vs. PLAT. +13% RHODIUM THERMOCOUPLE
Degrees Fahrenheit **Reference Junction 32° F.**

°F	0	1	2	3	4	5	6	7	8	9
					Millivolts					
3000	19.394	19.402	19.409	19.417	19.424	19.432	19.439	19.447	19.454	19.462
3010	19.470	19.477	19.485	19.492	19.500	19.508	19.515	19.523	19.530	19.538
3020	19.545	19.553	19.561	19.568	19.576	19.583	19.591	19.598	19.606	19.614
3030	19.621	19.628	19.636	19.644	19.651	19.659	19.667	19.674	19.682	19.689
3040	19.697	19.704	19.712	19.720	19.727	19.735	19.742	19.750	19.758	19.765
3050	19.773	19.780	19.788	19.795	19.803	19.811	19.818	19.826	19.833	19.841
3060	19.848	19.856	19.864	19.871	19.878	19.886	19.894	19.902	19.909	19.916
3070	19.924	19.932	19.939	19.947	19.954	19.962	19.969	19.977	19.984	19.992
3080	19.999	20.007	20.014	20.022	20.029	20.037	20.044	20.052	20.059	20.067
3090	20.075	20.082	20.090	20.097	20.105	20.112	20.120	20.127	20.135	20.142

Table 9.25. Schedule for Establishing Maximum Direct Current Resistance per Unit Length of Completed Cable

Cable Type	Maximum DC Resistance
Single Conductors	Table 9.25 Value Plus 2% [R max = R × 1.02]
Multiple Conductor Cables	Table 9.25 Value Plus 2% One of the following: 2% — One layer of Conductors [R max = R × 1.02 × 1.02] 3% — More than One Layer of Conductors [R max = R × 1.02 × 1.03] 4% — Pairs or Other Precabled Units [R max = R × 1.02 × 1.04] 5% — More than One Layer of Pairs or Other Precabled Units [R max = R × 1.02 × 1.05]

This material is reproduced by permission of the National Electrical Manufacturers Association from NEMA Standards Publication WC 55-1986, "Instrumentation Cables and Thermocouple Wire," copyright 1986 by NEMA.

Table 9.26. Nominal Direct Current Resistance in Ohms per 1,000 Feet at 25°C of Solid and Concentric Lay Stranded Copper Conductors

Conductor Size	Solid Copper		Stranded Copper		
AWG	uncoated	coated	uncoated	coated	
			Class B,C	Class B	Class C
22	16.5	17.2	16.7	17.9	18.1
20	10.3	10.7	10.5	11.1	11.3
19	8.20	8.52	8.33	8.83	8.96
18	6.51	6.76	6.67	7.07	7.14
17	5.15	5.35	5.21	5.52	5.64
16	4.10	4.26	4.18	4.43	4.44

This material is reproduced by permission of the National Electrical Manufacturers Association from NEMA Standards Publication WC 55-1986, "Instrumentation Cables and Thermocouple Wire," copyright 1986 by NEMA.

Table 9.27
Tinned Solid Copper Wire

Size AWG	Nominal Diameter Inches	Maximum Breaking Strength Lbs.*	Max. d-c @ 20 C Ohms/M Ft.*	Weight Lbs./M Ft.*
40	.0031		11.6×10^2	.0291
39	.0035		909	.0371
38	.0040		696	.0484
37	.0045		550	.0613
36	.0050		445	.0757
35	.0056		355	.0949
34	.0063		280	.120
33	.0071		221	.153
32	.0080		174	.194
31	.0089		140	.240
30	.0100		111	.303
29	.0113		86.3	.387
28	.0126		69.4	.481
27	.0142		54.6	.610
26	.0159		43.6	.765
25	.0179		34.4	.970
24	.0201		26.7	1.22
23	.0226	15.4	21.1	1.55
22	.0253	19.4	16.8	1.94
21	.0285	24.6	13.3	2.46
20	.0320	31.0	10.5	3.10
19	.0359	39.0	8.37	3.90
18	.0403	49.1	6.64	4.92
17	.0453	62.0	5.26	6.21
16	.0508	78.0	4.18	7.81
15	.0571	98.6	3.31	9.87
14	.0641	124.0	2.62	12.4
13	.0720	157.0	2.08	15.7
12	.0808	197.0	1.65	19.8
11	.0907	249.0	1.31	24.9
10	.1019	314.0	1.039	31.43
9	.1144	380.3	.8156	39.61
8	.1285	479.8	.6464	49.98
7	.1443	605.1	.5126	63.03
6	.1620	762.6	.4067	79.44
5	.1819	961.5	.3226	100.2
4	.2043	1213.0	.2557	126.3
3	.2294	1529.0	.2028	159.3
2	.2576	1928.0	.1609	200.9
1	.2893	2432.0	.1275	253.3

*Calculated per ASTM B258-51T.

Table 9.27. Tinned solid copper wire. Courtesy Rome Cable Corp.

Wiring*

Table 9.28
Minimum Enclosure Wall Thickness for Manufacturers' Standard Gauge

CARBON STEEL		STAINLESS STEEL	
Manufacturers' Standard Gauge	UL and NEMA Minimum Thickness	Manufacturers' Standard Gauge	UL and NEMA Thickness
10	.123	10	.123
11	.108	11	.108
12	.093	12	.093
13	.080	13	.080
14	.067	14	.067
15	.060	15	.060
16	.053	16	.053
17	.047	17	.047
18	.042	18	.042

Table 9.29

Maximum Cord- and Plug-Connected Load to Receptacle When Two or More Receptacles are Connected to a Branch Circuit

Circuit Rating Amperes	Receptacle Rating Amperes	Maximum Load Amperes
15 or 20	15	12
20	20	16
30	30	24

Source: Table 210-21 (b)(2) and 210-21 (b)(3) of the 1990 NEC.

Receptacle Ratings for Various Size Circuits When Two or More Receptacles are Connected to a Branch Circuit

Circuit Rating Amperes	Receptacle Rating Amperes
15	Not over 15
20	15 or 20
30	30
40	40 or 50
50	50

*Please see the "Electrical Area Classifications" and "Electrical Equipment" sections of this chapter for other NEC requirements. Some notes have been added to original NEC material by the authors to aid the reader. Refer to the 1991 NEC when applying this information

Table 9.30
Summary of Branch-Circuit Requirements for Circuits Having Two or More Outlets

(Type FEP, FEPB, SA, TW, RH, RHW, RHH, THHN, THW, THWN, and XHHW conductors in raceway or cable.)

CIRCUIT RATING	15 Amp	20 Amp	30 Amp	40 Amp	50 Amp
CONDUCTORS (Min. Size)					
Circuit Wires*	14	12	10	8	6
Taps	14	14	14	12	12
Fixture Wires and Cords	Refer to Section 240-4				
OVERCURRENT PROTECTION	15 Amp	20 Amp	30 Amp	40 Amp	50 Amp
OUTLET DEVICES:					
Lampholders Permitted	Any Type	Any Type	Heavy Duty	Heavy Duty	Heavy Duty
Receptacle Rating**	15 Max. Amp	15 or 20 Amp	30 Amp	40 or 50 Amp	50 Amp
MAXIMUM LOAD	15 Amp	20 Amp	30 Amp	40 Amp	50 Amp
PERMISSIBLE LOAD	Refer to Section 210-23(a)	Refer to Section 210-23(a)	Refer to Section 210-23(b)	Refer to Section 210-23(c)	Refer to Section 210-23(c)

*These gages are for copper conductors.
**For receptacle rating of cord-connected electric-discharge lighting fixtures, see Section 410-30(c) of the 1990 NEC.

Source: Table 210-24 of the 1990 NEC. Reprinted with permission. Please see full credit line on page 385.

Table 9.31
Minimum Size of Grounding Electrode Conductor for AC Systems

Size of Largest Service-Entrance Conductor or Equivalent Area for Parallel Conductors: Copper	Size of Largest Service-Entrance Conductor or Equivalent Area for Parallel Conductors: Aluminum or Copper-Clad Aluminum	Size of Grounding Electrode Conductor: Copper	Size of Grounding Electrode Conductor: *Aluminum or Copper-Clad Aluminum
2 or smaller	1/0 or smaller	8	6
1 or 1/0	2/0 or 3/0	6	4
2/0 or 3/0	4/0 or 250 kcmil	4	2
Over 3/0 thru 350 kcmil	Over 250 kcmil thru 500 kcmil	2	1/0
Over 350 kcmil thru 600 kcmil	Over 500 kcmil thru 900 kcmil	1/0	3/0
Over 600 kcmil thru 1100 kcmil	Over 900 kcmil thru 1750 kcmil	2/0	4/0
Over 1100 kcmil	Over 1750 kcmil	3/0	250 kcmil

Where multiple sets of service-entrance conductors are used as permitted in Section 230-40, Exception No. 2 of the 1990 NEC, the equivalent size of the largest service-entrance conductor shall be determined by the largest sum of the areas of the corresponding conductors of each set.

Where there are no service-entrance conductors, the grounding electrode conductor size shall be determined by the equivalent size of the largest service-entrance conductor required for the load to be served.

*See installation restrictions in Section 250-92(a) of the 1990 NEC.
(FPN): See Section 250-23 (b) of the 1990 NEC.
Note: See exception in paragraph 250-94 of the 1990 NEC

Source: Table 250-94 of the 1990 NEC. Reprinted with permission. Please see full credit line on page 585.

Table 9.32
Minimum Size Equipment Grounding Conductors for Grounding Raceway and Equipment

Rating or Setting of Automatic Overcurrent Device in Circuit Ahead of Equipment, Conduit, etc., Not Exceeding (Amperes)	Size	
	Copper Wire No.	Aluminum or Copper-Clad Aluminum Wire No.*
15	14	12
20	12	10
30	10	8
40	10	8
60	10	8
100	8	6
200	6	4
300	4	2
400	3	1
500	2	1/0
600	1	2/0
800	1/0	3/0
1000	2/0	4/0
1200	3/0	250 kcmil
1600	4/0	350 "
2000	250 kcmil	400 "
2500	350 "	600 "
3000	400 "	600 "
4000	500 "	800 "
5000	700 "	1200 "
6000	800 "	1200 "

Note: See exceptions in the 1990 NEC.

Source: Table 250-95 of the 1990 NEC. Reprinted with permission. Please see full credit line on page 585.

Table 9.33
Ampacities of Insulated Conductors Rated 0–2,000 Volts, 60° to 90°C (140° to 194°F), Not More Than Three Conductors in Raceway or Cable or Earth (Directly Buried), Based on Ambient Temperature of 30°C (86°F)

Size	Temperature Rating of Conductor. See Table 310-13.								Size
	60°C (140°F)	75°C (167°F)	85°C (185°F)	90°C (194°F)	60°C (140°F)	75°C (167°F)	85°C (185°F)	90°C (194°F)	
AWG kcmil	Types †TW, †UF	Types †FEPW, †RH, †RHW, †THHW, †THW, †THWN, †XHHW, †USE, †ZW	Type V	Types TA, TBS,SA SIS, †FEP, †FEPB, †RHH, †THHN, †THHW, †XHHW	Types †TW, †UF	Types †RH, †RHW, †THHW, †THW, †THWN, †XHHW †USE	Type V	Types TA, TBS, SA, SIS, †RHH, †THHW, †THHN, †XHHW	AWG kcmil
	COPPER				ALUMINUM OR COPPER-CLAD ALUMINUM				
18	....	....	....	14	....	....	....	....	....
16	....	....	18	18	....	....	....	....	....
14	20†	20†	25	25†	....	....	....	....	....
12	25†	25†	30	30†	20†	20†	25	25†	12
10	30	35†	40	40†	25	30†	30	35†	10
8	40	50	55	55	30	40	40	45	8
6	55	65	70	75	40	50	55	60	6
4	70	85	95	95	55	65	75	75	4
3	85	100	110	110	65	75	85	85	3
2	95	115	125	130	75	90	100	100	2
1	110	130	145	150	85	100	110	115	1
1/0	125	150	165	170	100	120	130	135	1/0
2/0	145	175	190	195	115	135	145	150	2/0
3/0	165	200	215	225	130	155	170	175	3/0
4/0	195	230	250	260	150	180	195	205	4/0
250	215	255	275	290	170	205	220	230	250
300	240	285	310	320	190	230	250	255	300
350	260	310	340	350	210	250	270	280	350
400	280	335	365	380	225	270	295	305	400
500	320	380	415	430	260	310	335	350	500
600	355	420	460	475	285	340	370	385	600
700	385	460	500	520	310	375	405	420	700
750	400	475	515	535	320	385	420	435	750
800	410	490	535	555	330	395	430	450	800
900	435	520	565	585	355	425	465	480	900
1000	455	545	590	615	375	445	485	500	1000
1250	495	590	640	665	405	485	525	545	1250
1500	520	625	680	705	435	520	565	585	1500
1750	545	650	705	735	455	545	595	615	1750
2000	560	665	725	750	470	560	610	630	2000

AMPACITY CORRECTION FACTORS

Ambient Temp. °C	For ambient temperatures other than 30°C (86°F), multiply the ampacities shown above by the appropriate factor shown below.								Ambient Temp. °F
21–25	1.08	1.05	1.04	1.04	1.08	1.05	1.04	1.04	70–77
26–30	1.00	1.00	1.00	1.00	1.00	1.00	1.00	1.00	79–86
31–35	.91	.94	.95	.96	.91	.94	.95	.96	88–95
36–40	.82	.88	.90	.91	.82	.88	.90	.91	97–104
41–45	.71	.82	.85	.87	.71	.82	.85	.87	106–113
46–50	.58	.75	.80	.82	.58	.75	.80	.82	115–122
51–55	.419	.67	.74	.76	.41	.67	.74	.76	124–131
56–60	....	.58	.67	.71	....	.58	.67	.71	133–140
61–70	....	.33	.52	.58	....	.33	.52	.58	142–158
71–80	....	....	.30	.41	....	....	.30	.41	160–176

†Unless otherwise specifically permitted elsewhere in this Code, the overcurrent protection for conductor types marked with an obelisk (†) shall not exceed 15 amperes for 14 AWG, 20 amperes for 12 AWG, and 30 amperes for 10 AWG copper; or 15 amperes for 12 AWG and 25 amperes for 10 AWG aluminum and copper-clad aluminum after any correction factors for ambient temperature and number of conductors have been applied.

Notes: The following notes apply to Table 9.33. See the 1990 NEC for exceptions and clarification.

Table 9.33 continued

1. Ampacity Adjustment Factors.

(a) More than Three Conductors in a Raceway or Cable. Where the number of conductors in a raceway or cable exceeds three, the ampacities shall be reduced as shown in the following table:

Number of Conductors	Column A Percent of Values in Tables as Adjusted for Ambient Temperature if Necessary	Number of Conductors	Column B** Percent of Values in Tables as Adjusted for Ambient Temperature if Necessary
4 through 6	80	4 through 6	80
7 through 9	70	7 through 9	70
10 through 24*	70	10 through 20	50
25 through 42*	60	21 through 30	45
43 and above*	50	31 through 40	40
		41 through 60	35

*These factors include the effects of a load diversity of 50 percent.
**No diversity.
(FPN): Column A is based on the following formula:

$$A_2 = \sqrt{0.5\frac{N}{E}} \times (A_1) \text{ where}$$

A_1 = Table ampacity multiplied by factor from Note 8(a)
N = Total number of conductors used to obtain factor from Note 8(a)
E = Desired number of energized conductors
A_2 = Ampacity limit for energized conductors

Where single conductors or multiconductor cables are stacked or bundled longer than 24 inches (610 mm) without maintaining spacing and are not installed in raceways, the ampacity of each conductor shall be reduced as shown in the above table.

Exception No. 1: When conductors of different systems, as provided in Section 300-3 of the 1990 NEC, are installed in a common raceway or cable the derating factors shown above shall apply to the number of power and lighting (Articles 210, 215, 220, and 230) conductors only.

Exception No. 2: For conductors installed in cable trays, the provisions of Section 318-11 of the 1990 NEC shall apply.

Exception No. 3: Derating factors shall not apply to conductors in nipples having a length not exceeding 24 inches (610 mm).

Exception No. 4: Derating factors shall not apply to underground conductors entering or leaving an outdoor trench if those conductors have physical protection in the form of rigid metal conduit, intermediate metal conduit or rigid nonmetallic conduit having a length not exceeding 10 feet (3.05 m) above grade and the number of conductors does not exceed 4.

(b) More than One Conduit, Tube, or Raceway. Spacing between conduits, tubing, or raceways shall be maintained.

2. Overcurrent Protection. Where the standard ratings and settings of overcurrent devices do not correspond with the ratings and settings allowed for conductors, the next higher standard rating and setting shall be permitted.

Exception: As limited in Section 240-3 of the 1990 NEC.

3. Neutral Conductor.

(a) A neutral conductor which carries only the unbalanced current from other conductors, as in the case of normally balanced circuits of three or more conductors, shall not be counted when applying the provisions of Note 1.

(b) In a 3-wire circuit consisting of 2-phase wires and the neutral of a 4-wire, 3-phase wye-connected system, a common conductor carries approximately the same current as the other conductors and shall be counted when applying the provisions of Note 1.

(c) On a 4-wire, 3-phase wye circuit where the major portion of the load consists of electric-discharge lighting, data processing, or similar equipment, there are harmonic currents present in the neutral conductor and the neutral shall be considered to be a current-carrying conductor.

4. Grounding or Bonding Conductor. A grounding or bonding conductor shall not be counted when applying the provisions of Note 1.

Source: Table 310-16 and associated notes of the 1990 NEC. Reprinted with permission. Please see full credit line on page 585.

Table 9.34
Allowable Cable Fill Area for Multiconductor Cables in Ladder, Ventilated Trough, or Solid Bottom Cable Trays for Cables Rated 2,000 Volts or Less

Inside Width of Cable Tray (Inches)	Maximum Allowable Fill Area in Square Inches for Multiconductor Cables			
	Ladder or Ventilated Trough Cable Trays, Section 318-9(a)		Solid Bottom Cable Trays, Section 318-9(c)	
	Column 1 Applicable for Section 318-9(a) (2) Only (Square inches)	Column 2* Applicable for Section 318-9(a)(3) Only (Square inches)	Column 3 Applicable for Section 318-9(c) (2) Only (Square inches)	Column 4* Applicable for Section 318-9(c) (3) Only (Square inches)
6	7	7—(1.2 Sd)**	5.5	5.5—Sd**
12	14	14—(1.2 Sd)	11.0	11.0—Sd
18	21	21—(1.2 Sd)	16.5	16.5—Sd
24	28	28—(1.2 Sd)	22.0	22.0—Sd
30	35	35—(1.2 Sd)	27.5	27.5—Sd
36	42	42—(1.2 Sd)	33.0	33.0—Sd

For SI units: one square inch = 645 square millimeters.

*The maximum allowable fill areas in Columns 2 and 4 shall be computed. For example, the maximum allowable fill, in square inches, for a 6-inch (152-mm) wide cable tray in Column 2 shall be: 7 minus (1.2 multiplied by Sd).

**The term Sd in Columns 2 and 4 is equal to the sum of the diameters, in inches, of all 4/0 AWG and larger multiconductor cables in the same cable tray with smaller cables.

Note: All section numbers refer to the 1990 NEC.

Source: Table 318-9 of the 1990 NEC. Reprinted with permission. Please see full credit line on page 585.

Table 9.35
Minimum Radius of the Curve of the Inner Edge of Any Field Conduit Bent (Inches)

Size of Conduit (In.)	Conductors Without Lead Sheath (In.)	Conductors With Lead Sheath (In.)
½	4	6
¾	5	8
1	6	11
1¼	8	14
1½	10	16
2	12	21
2½	15	25
3	18	31
3½	21	36
4	24	40
5	30	50
6	36	61

For SI units: (Radius) one inch = 25.4 millimeters.

Source: Table 346-10 of the 1990 NEC. Reprinted with permission. Please see full credit line on page 585.

Table 9.36
Exceptions for the Field Bends for Conductors Without Lead Sheath and Made With Single Operation (One Shot) Bending Machine

Size of Conduit (In.)	Radius to Center of Conduit (In.)
½	4
¾	4½
1	5¾
1¼	7¼
1½	8¼
2	9½
2½	10½
3	13
3½	15
4	16
5	24
6	30

For SI units: (Radius) one inch = 25.4 millimeters.

Source: Table 346-10 of the 1990 NEC. Reprinted with permission. Please see full credit line on page 585.

Table 9.37
Supports for Rigid Metal Conduit if Made With Threaded Couplings*

Conduit Size (Inches)	Maximum Distance Between Rigid Metal Conduit Supports (Feet)
½–¾	10
1	12
1¼–1½	14
2–2½	16
3 and larger	20

For SI units: (Supports) one foot = 0.3048 meter.

*Provided such supports prevent transmission of stresses to termination where conduit is deflected between supports.

Source: Table 346-12 of the 1990 NEC. Reprinted with permission. Please see full credit line on page 585.

Table 9.38
Support of Rigid Nonmetallic Conduit

Conduit Size (Inches)	Maximum Spacing Between Supports (Feet)
½-1	3
1¼-2	5
2½-3	6
3½-5	7
6	8

For SI units: (Supports) one foot = 0.3048 meter.

Note: See other support requirements in the 1990 NEC.

Source: Table 347-8 of the 1990 NEC. Reprinted with permission. Please see full credit line on page 585.

Table 9.39
Minimum Radii for Fixed Bends (From Inside of Bend) That Is Bent for Installation Purposes and Is Not Flexed or Bent After Installation

Trade Size	Minimum Radii
⅜ inch	3½ inches
½ inch	4 inches
¾ inch	5 inches

For SI units: (Radii) one inch = 25.4 millimeters.

Source: Table 349-20 of the 1990 NEC. Reprinted with permission. Please see full credit line on page 585.

Table 9.40
Minimum Radii for Flexing Use (From Inside of Bend) That Will Be Infrequently Flexed After Installation

Trade Size	Minimum Radii
⅜ inch	10 inches
½ inch	12½ inches
¾ inch	17½ inches

For SI units: (Radii) one inch = 25.4 millimeters.

Source: Table 349-20(a) of the 1990 NEC. Reprinted with permission. Please see full credit line on page 585.

Table 9.41
Maximum Number of Conductors Permitted in Standard Boxes

Metal Boxes

Box Dimension, Inches Trade Size or Type	Min. Cu. In. Cap.	Maximum Number of Conductors No. 18	No. 16	No. 14	No. 12	No. 10	No. 8	No. 6
4 x 1¼ Round or Octagonal	12.5	8	7	6	5	5	4	2
4 x 1½ Round or Octagonal	15.5	10	8	7	6	6	5	3
4 x 2⅛ Round or Octagonal	21.5	14	12	10	9	8	7	4
4 x 1¼ Square	18.0	12	10	9	8	7	6	3
4 x 1½ Square	21.0	14	12	10	9	8	7	4
4 x 2⅛ Square	30.3	20	17	15	13	12	10	6
4¹¹/₁₆ x 1¼ Square	25.5	17	14	12	11	10	8	5
4¹¹/₁₆ x 1½ Square	29.5	19	16	14	13	11	9	5
4¹¹/₁₆ x 2⅛ Square	42.0	28	24	21	18	16	14	8
3 x 2 x 1½ Device	7.5	5	4	3	3	3	2	1
3 x 2 x 2 Device	10.0	6	5	5	4	4	3	2
3 x 2 x 2¼ Device	10.5	7	6	5	4	4	3	2
3 x 2 x 2½ Device	12.5	8	7	6	5	5	4	2
3 x 2 x 2¾ Device	14.0	9	8	7	6	5	4	2
3 x 2 x 3½ Device	18.0	12	10	9	8	7	6	3
4 x 2⅛ x 1½ Device	10.3	6	5	5	4	4	3	2
4 x 2⅛ x 1⅞ Device	13.0	8	7	6	5	5	4	2
4 x 2⅛ x 2⅛ Device	14.5	9	8	7	6	5	4	2
3¾ x 2 x 2½ Masonry Box/Gang	14.0	9	8	7	6	5	4	2
3¾ x 2 x 3½ Masonry Box/Gang	21.0	14	12	10	9	8	7	4
FS—Minimum Internal Depth 1¾ Single Cover/Gang	13.5	9	7	6	6	5	4	2
FD—Minimum Internal Depth 2⅜ Single Cover/Gang	18.0	12	10	9	8	7	6	3
FS—Minimum Internal Depth 1¾ Multiple Cover/Gang	18.0	12	10	9	8	7	6	3
FD—Minimum Internal Depth 2⅜ Multiple Cover/Gang	24.0	16	13	12	10	9	8	4

Volume Required per Conductor

Size of Conductor	Free Space Within Box for Each Conductor
No. 18	1.5 cubic inches
No. 16	1.75 cubic inches
No. 14	2. cubic inches
No. 12	2.25 cubic inches
No. 10	2.5 cubic inches
No. 8	3. cubic inches
No. 6	5. cubic inches

Note: See Article 370 of the 1990 NEC for requirements of boxes or conduit bodies used as junction boxes.

Source: Table 370-6(a) of the 1990 NEC. Reprinted with permission. Please see full credit line on page 585.

Table 9.42
Minimum Wire Bending Space at Terminals and Minimum Width of Wiring Gutters in Inches*

AWG or Circular-Mil Size of Wire	Wires per Terminal 1	2	3	4	5
14-10	Not Specified	—	—	—	—
8-6	1½	—	—	—	—
4-3	2	—	—	—	—
2	2½	—	—	—	—
1	3	—	—	—	—
1/0-2/0	3½	5	7	—	—
3/0-4/0	4	6	8	—	—
250 kcmil	4½	6	8	10	—
300-350 kcmil	5	8	10	12	
400-500 kcmil	6	8	10	12	14
600-700 kcmil	8	10	12	14	16
750-900 kcmil	8	12	14	16	18
1000-1250 kcmil	10	—	—	—	—
1500-2000 kcmil	12	—	—	—	—

For SI units: one inch = 25.4 millimeters.

Bending space at terminals shall be measured in a straight line from the end of the lug or wire connector (in the direction that the wire leaves the terminal) to the wall, barrier, or obstruction.

*For conductors that do not enter or leave the enclosure through the wall opposite their terminals.

Note: Conductors shall not be deflected within a cabinet or cutout box unless a gutter having a width in accordance with the above table is provided.

Source: Table 373-6(a) of the 1990 NEC. Reprinted with permission. Please see full credit line on page 585.

Table 9.43
Minimum Wire Bending Space at Terminals in Inches*

Wire Size (AWG or kcmil)	Wires per Terminal 1	2	3	4 or More
14-10	Not Specified	—	—	—
8	1½	—	—	—
6	2	—	—	—
4	3	—	—	—
3	3	—	—	—
2	3½	—	—	—
1	4½	—	—	—
1/0	5½	5½	7	—
2/0	6	6	7½	—
3/0	6½ (½)	6½ (½)	8	
4/0	7 (1)	7½ (1½)	8½ (½)	—
250	8½ (2)	8½ (2)	9 (1)	10
300	10 (3)	10 (2)	11 (1)	12
350	12 (3)	12 (3)	13 (3)	14 (2)
400	13 (3)	13 (3)	14 (3)	15 (3)
500	14 (3)	14 (3)	15 (3)	16 (3)
600	15 (3)	16 (3)	18 (3)	19 (3)
700	16 (3)	18 (3)	20 (3)	22 (3)
750	17 (3)	19 (3)	22 (3)	24 (3)
800	18	20	22	24
900	19	22	24	24
1000	20	—	—	—
1250	22	—	—	—
1500	24	—	—	—
1750	24	—	—	—
2000	24	—	—	—

*For conductors that enter or leave the enclosure through the wall opposite their terminals.

For SI units: one inch = 25.4 millimeters.

Bending space at terminals shall be measured in a straight line from the end of the lug or wire connector in a direction perpendicular to the enclosure wall.

For removable and lay-in wire terminals intended for only one wire, bending space shall be permitted to be reduced by the number of inches shown in parentheses.

Source: Table 373-6(b) of the 1990 NEC. Reprinted with permission. Please see full credit line on page 585.

Table 9.44
Minimum Spacings Between Bare Metal Parts, Busbars, Etc.

	Opposite Polarity Where Mounted on the Same Surface	Opposite Polarity Where Held Free in Air	Live Parts to Ground*
Not over 125 volts, nominal	¾ inch	½ inch	½ inch
Not over 250 volts, nominal	1¼ inch	¾ inch	½ inch
Not over 600 volts, nominal	2 inches	1 inch	1 inch

For SI units: one inch = 25.4 millimeters.

*For spacing between live parts and doors of cabinets, see Section 373-11(a) (1), (2), and (3) of the 1990 NEC.

Source: Table 384-36 of the 1990 NEC. Reprinted with permission. Please see full credit line on page 585.

Table 9.45
Ampacity of Flexible Cords and Cables with not More than Three Conductors

[Based on Ambient Temperature of 30°C (86°F).

Size AWG	Thermoset Types TS / Thermoplastic Types TPT, TST	Thermoset Types C, E, EO, PD, S, SJ, SJO, SJOO, SO, SOO, SP-1, SP-2, SP-3, SRD, SV, SVO, SVOO / Thermoplastic Types ET, ETT, ETLB, SE, SEO, SJE, SJEO, SJT, SJTO, SJTOO, SPE-1, SPE-2, SPE-3, SPT-1, SPT-2, SPT-3, ST, STO, STOO, SRDE, SRDT, SVE, SVEO, SVT, SVTO, SVTOO		Types AFS, AFSJ, HPD, HPN, HS, HSJ, HSJO, HSO	Asbestos Types AFC*, AFPD*, AFPO
		A†	B†		
27**	0.5	..	..	..	..
20	..	5***	7***	..	..
18	..	7	10	10	6
17	..	..	12	..	..
16	..	10	13	15	8
15	..	..	..	17	..
14	..	15	18	20	17
12	..	20	25	30	23
10	..	25	30	35	28
8	..	35	40	..	..
6	..	45	55	..	..
4	..	60	70	..	..
2	..	80	95	..	..

* These types are used almost exclusively in fixtures where they are exposed to high temperatures and ampere ratings are assigned accordingly.
** Tinsel cord.
*** Elevator cables only.
† The ampacities under subheading A apply to 3-conductor cords and other multiconductor cords connected to utilization equipment so that only 3 conductors are current carrying. The ampacities under subheading B apply to 2-conductor cords and other multiconductor cords connected to utilization equipment so that only 2 conductors are current carrying.

NOTE. Ultimate Insulation Temperature. In no case shall conductors be associated together in such a way with respect to the kind of circuit, the wiring method used, or the number of conductors that the limiting temperature of the conductors will be exceeded.

Note: The ampacity of each conductor shall be reduced from the three conductor ratings shown in Table 9.45 as follows:

Column A		Column B**	
Number of Conductors	Percent of Values in Tables 400-5(A) and 400-5(B)	Number of Conductors	Percent of Values in Tables 400-5(A) and 400-5(B)
4 through 6	80	4 through 6	80
7 through 9	70	7 through 9	70
10 through 24*	70	10 through 20	50
25 through 42*	60	21 through 30	45
43 and above*	50	31 through 40	40
		41 through 60	35

* Column A-These factors include the effects of a load diversity of 50 percent.
**Column B-No Diversity.

(FPN): Column A is based on the following formula:

$A_2 = \sqrt{0.5\frac{N}{E}} \times (A_1)$ where

A_1 = Table ampacity multiplied by factor from table above.
N = Total number of conductors used to obtain factor from table above.
E = Desired number of energized conductors.
A_2 = Ampacity limit for energized conductors.

Source: Table 400-5 (a) of the 1990 NEC. Reprinted with permission. Please see full credit line on page 585.

Table 9.46
Minimum Size and Number of Conductors Required in Cables Used for Power-Limited Fire Protective Signaling Circuits

Size	Minimum Number of Conductors in Cable
26 AWG	10
24 AWG	6
22 AWG	4
19 AWG	2
16 AWG or larger	1

Source: Table 760-51 of the 1990 NEC. Reprinted with permission. Please see full credit line on page 585.

Table 9.47
Communication Wires and Cables that are Approved for Installation in Buildings

Cable Marking	Type	Reference
MPP	Multipurpose plenum cable	Sections 800-51(f) and 800-53(e)
CMP	Communications plenum cable	Sections 800-51(a) and 800-53(a)
MPR	Multipurpose riser cable	Sections 800-51(f) and 800-53(e)
CMR	Communications riser cable	Sections 800-51(b) and 800-53(b)
MP	Multipurpose cable	Sections 800-51(f) and 800-53(e)
CM	Communications cable	Sections 800-51(c) and 800-53(c)
CMX	Communications cable, limited use	Sections 800-51(d) and 800-53(c), Exceptions No. 1, 2, 3, and 4
CMUC	Undercarpet communications cable	800-51(e) and 800-53(c), Exception No. 5

(FPN No. 1): Cable types are listed in descending order of fire resistance rating.

(FPN No. 2): See the referenced sections for permitted uses.

Note: All section numbers refer to the 1990 NEC.

Source: Table 800-50 of the 1990 NEC. Reprinted with permission. Please see full credit line on page 585.

Table 9.48
Acceptable Substitutes for Approved Communication Wires and Cables Installed in Buildings

Cable Type	Permitted Substitutions
MPP	None
CMP	MPP
MPR	MPP
CMR	MPP, CMP, MPR
MP	MPP, MPR
CM	MPP, CMP, MPR, CMR, MP
CMX	MPP, CMP, MPR, CMR, MP, CM

(FPN): For the use of communications cable and multipurpose cable in place of Class 2 or Class 3 cables, see Section 725-53(f); and for the use of communications cable and multipurpose cable in place of power-limited fire protective signaling cables, see Section 760-53(d). The permitted cable substitutions are illustrated in Figure 800-53. (All section and figure numbers refer to the 1990 NEC.)

Source: Table 800-53 of the 1990 NEC. Reprinted with permission. Please see full credit line on page 585.

Table 9.49
Size of Receiving-Station, Wire-Strung Outdoor Antenna Conductors

Material	Minimum Size of Conductors — When Maximum Open Span Length Is Less than 35 feet	35 feet to 150 feet	Over 150 feet
Aluminum alloy, hard-drawn copper	19	14	12
Copper-clad steel, bronze, or other high-strength material	20	17	14

For SI units: one foot = 0.3048 meter.

Source: Table 810-16(a) of the 1990 NEC. Reprinted with permission. Please see full credit line on page 585.

Table 9.50
Size of Amateur Station Outdoor Antenna Conductors for Transmitting and Receiving Stations

Material	Minimum Size of Conductors — Where Maximum Open Span Length Is Less Than 150 feet	Over 150 feet
Hard-drawn copper	14	10
Copper-clad steel, bronze or other high-strength material	14	12

For SI units: one foot = 0.3048 meter.

Source: Table 810-52 of the 1990 NEC. Reprinted with permission. Please see full credit line on page 585.

Table 9.51
Acceptable Cable Markings for Coaxial Cables Installed Within Buildings Listed for that Purpose

Cable Marking	Type	Reference
CATVP	CATV plenum cable	Sections 820-51(a) and 820-53(a)
CATVR	CATV riser cable	Sections 820-51(b) and 820-53(b)
CATV	CATV cable	Sections 820-51(c) and 820-53(c)
CATVX	CATV cable, limited use	Sections 820-51(d) and 820-53(c), Exceptions No. 1, 2, and 3

(FPN No. 1): Cable types are listed in descending order of fire-resistance rating.

(FPN No. 2): See the referenced sections for listing requirements and permitted uses.

Note: All section numbers refer to the 1990 NEC.

Source: Table 820-50 of the 1990 NEC. Reprinted with permission. Please see full credit line on page 585.

Table 9.52
Substitutions for Coaxial Cables Installed in Buildings Listed for that Purpose

Cable Type	Permitted Substitutions
CATVP	None
CATVR	CATVP
CATV	CATVP, CATVR
CATVX	CATVP, CATVR, CATV

Source: Table 820-53 of the 1990 NEC. Reprinted with permission. Please see full credit line on page 585.

Notes to Tables

1. Tables 9.53–9.55 apply only to complete conduit or tubing systems and are not intended to apply to sections of conduit or tubing used to protect exposed wiring from physical damage.

2. Equipment grounding or bonding conductors, when installed, shall be included when calculating conduit or tubing fill. The actual dimensions of the equipment grounding or bonding conductor (insulated or bare) shall be used in the calculation.

3. When conduit or tubing nipples having a maximum length not to exceed 24 inches (610 mm) are installed between boxes, cabinets, and similar enclosures, the nipple shall be permitted to be filled to 60 percent of its total cross-sectional area. See Article 310 of the 1990 NEC for other requirements.

4. For conductors not included, such as compact or multiconductor cables, the actual dimensions shall be used.

5. The maximum allowable percent of cross section of conduit and tubing for conductors is as follows:

Number of Conductors	1	2	3	4	Over 4
All conductor types except lead-covered	53	31	40	40	40
Lead-covered conductors	55	30	40	38	35

6. A multiconductor cable of two or more conductors shall be treated as a single conductor cable for calculating percentage conduit fill area. For cables that have elliptical cross section, the cross-sectional area calculation shall be based on using the major diameter of the ellipse as a circle diameter.

Table 9.53
Maximum Number of Identically Sized Conductors in Trade Sizes of Conduit or Tubing (Based on Table 1, Chapter 9, of the 1990 NEC)

Conduit Trade Size (Inches)		½	¾	1	1¼	1½	2	2½	3	3½	4	5	6
Type Letters	Conductor Size AWG, kcmil												
TW, XHHW (14 through 8)	14	9	15	25	44	60	99	142					
	12	7	12	19	35	47	78	111	171				
	10	5	9	15	26	36	60	85	131	176			
	8	2	4	7	12	17	28	40	62	84	108		
RHW and RHH (without outer covering), THW	14	6	10	16	29	40	65	93	143	192			
	12	4	8	13	24	32	53	76	117	157			
	10	4	6	11	19	26	43	61	95	127	163		
	8	1	3	5	10	13	22	32	49	66	85	133	
TW,	6	1	2	4	7	10	16	23	36	48	62	97	141
	4	1	1	3	5	7	12	17	27	36	47	73	106
THW,	3	1	1	2	4	6	10	15	23	31	40	63	91
	2	1	1	2	4	5	9	13	20	27	34	54	78
	1		1	1	3	4	6	9	14	19	25	39	57
FEPB (6 through 2), RHW and RHH (without outer covering)	1/0		1	1	2	3	5	8	12	16	21	33	49
	2/0		1	1	1	3	5	7	10	14	18	29	41
	3/0		1	1	1	2	4	6	9	12	15	24	35
	4/0			1	1	1	3	5	7	10	13	20	29
	250			1	1	1	2	4	6	8	10	16	23
	300			1	1	1	2	3	5	7	9	14	20
	350				1	1	1	3	4	6	8	12	18
	400				1	1	1	2	4	5	7	11	16
	500				1	1	1	1	3	4	6	9	14
	600					1	1	1	3	4	5	7	11
	700					1	1	1	2	3	4	7	10
	750					1	1	1	2	3	4	6	9

Note: This table is for concentric stranded conductors only.

Source: Tables 3B and 3C of the 1990 NEC. Reprinted with permission. Please see full credit line on page 585.

Table 9.54
Maximum Number of Conductors in Trade Sizes of Conduit or Tubing (Based on Table 1, Chapter 9, of the 1990 NEC)

Type Letters	Conduit Trade Size (Inches) / Conductor Size AWG, kcmil	½	¾	1	1½	1½	2	2½	3	3½	4	5	6
	14	13	24	39	69	94	154						
	12	10	18	29	51	70	114	164					
THWN	10	6	11	18	32	44	73	104	160				
	8	3	5	9	16	22	36	51	79	106	136		
THHN													
FEP (14 through 2)	6	1	4	6	11	15	26	37	57	76	98	154	
FEPB (14 through 8)	4	1	2	4	7	9	16	22	35	47	60	94	137
PFA (14 through 4/0)	3	1	1	3	6	8	13	19	29	39	51	80	116
PFAH (14 through 4/0)	2	1	1	3	5	7	11	16	25	33	43	67	97
Z (14 through 4/0)	1		1	1	3	5	8	12	18	25	32	50	72
XHHW (4 through	1/0		1	1	3	4	7	10	15	21	27	42	61
500 kcmil)	2/0		1	1	2	3	6	8	13	17	22	35	51
	3/0		1	1	1	3	5	7	11	14	18	29	42
	4/0		1	1	1	2	4	6	9	12	15	24	35
	250			1	1	1	3	4	7	10	12	20	28
	300			1	1	1	3	4	6	8	11	17	24
	350			1	1	1	2	3	5	7	9	15	21
	400				1	1	1	3	5	6	8	13	19
	500				1	1	1	2	4	5	7	11	16
	600				1	1	1	1	3	4	5	9	13
	700					1	1	1	3	4	5	8	11
	750					1	1	1	2	3	4	7	11
XHHW	6	1	3	5	9	13	21	30	47	63	81	128	185
	600				1	1	1	1	3	4	5	9	13
	700					1	1	1	3	4	5	7	11
	750					1	1	1	2	3	4	7	10
	14	3	6	10	18	25	41	58	90	121	155		
	12	3	5	9	15	21	35	50	77	103	132		
RHW	10	2	4	7	13	18	29	41	64	86	110		
	8	1	2	4	7	9	16	22	35	47	60	94	137
RHH	6	1	1	2	5	6	11	15	24	32	41	64	93
	4	1	1	1	3	5	8	12	18	24	31	50	72
(with	3	1	1	1	3	4	7	10	16	22	28	44	63
outer	2		1	1	3	4	6	9	14	19	24	38	56
covering)	1		1	1	1	3	5	7	11	14	18	29	42
	1/0		1	1	1	2	4	6	9	12	16	25	37
	2/0			1	1	1	3	5	8	11	14	22	32
	3/0			1	1	1	3	4	7	9	12	19	28
	4/0			1	1	1	2	4	6	8	10	16	24
	250				1	1	1	3	5	6	8	13	19
	300				1	1	1	3	4	5	7	11	17
	350				1	1	1	2	4	5	6	10	15
	400				1	1	1	1	3	4	6	9	14
	500				1	1	1	1	3	4	5	8	11
	600					1	1	1	2	3	4	6	9
	700					1	1	1	1	3	3	6	8
	750						1	1	1	3	3	5	8

Note: This table is for concentric stranded conductors only.

Source: Table 3A of the 1990 NEC. Reprinted with permission. Please see full credit line on page 585.

Table 9.55
Maximum Number of Compact Conductors in Trade Sizes of Conduit or Tubing

Conductor Size AWG or kcmil	1 in. THW	1 in. THHN	1 in. XHHW	1¼ in. THW	1¼ in. THHN	1¼ in. XHHW	1½ in. THW	1½ in. THHN	1½ in. XHHW	2 in. THW	2 in. THHN	2 in. XHHW	2½ in. THW	2½ in. THHN	2½ in. XHHW	3 in. THW	3 in. THHN	3 in. XHHW	3½ in. THW	3½ in. THHN	3½ in. XHHW	4 in. THW	4 in. THHN	4 in. XHHW	Conductor Size AWG or kcmil
6	5	7	6	9	13	11	12	18	15																6
4	4	4	4	7	8	8	9	11	11	15	18	18													4
2	3	3	3	5	6	6	7	8	8	11	13	13													2
1				3	4	4	5	6	6	8	10	10	11	14	14										1
1/0				3	3	3	4	5	5	7	8	8	9	12	12										1/0
2/0					3	3	3	4	4	5	7	7	8	10	10	12	15	15							2/0
3/0							3	3	3	5	6	6	7	8	8	10	13	13							3/0
4/0								3	3	4	5	5	6	7	7	9	10	10	12	14	14				4/0
250										3	4	4	4	5	5	7	8	8	9	11	11				250
300										3	3	3	4	4	4	6	7	7	8	9	10	10	12	12	300
350											3	3	3	4	4	5	6	6	7	8	8	9	11	11	350
400													3	3	4	5	5	6	6	7	8	8	9	10	400
500														3	3	4	4	5	5	6	6	7	8	8	500
600																3	4	4	4	5	5	6	6	7	600
700																3	3	3	4	4	4	5	6	6	700
750																3	3	3	4	4	4	5	5	5	750
1000																			3	3	3	4	4	4	1000

Source: Table 5B of the 1990 NEC. Reprinted with permission. Please see full credit line on page 585.

Table 9.56
Voltage Drop Calculations

To find volts lost:
Multiply current (amperes) by the distance (feet in one conductor) by the figure in table for the kind of system and size of wire used.
Place a decimal point in front of the last 6 figures.
The result is the number of volts lost.
Note: For A-c 3 Phase Current, Voltage Drop obtained is phase-to-phase.

Wire Size	Power Factor %	AC Single Phase	AC Three Phase	DC
14 Awg	100	5880	5090	5880
	90	5360	4640	
	80	4790	4150	
	70	4230	3660	
	60	3650	3160	
12 Awg	100	3690	3190	3690
	90	3380	2930	
	80	3030	2620	
	70	2680	2320	
	60	2320	2010	
10 Awg	100	2320	2010	2820
	90	2150	1861	
	80	1935	1675	
	70	1718	1487	
	60	1497	1296	
8 Awg	100	1462	1265	1462
	90	1373	1189	
	80	1248	1081	
	70	1117	969	
	60	981	849	
6 Awg	100	918	795	918
	90	882	764	
	80	812	703	
	70	734	636	
	60	653	565	
4 Awg	100	578	501	578
	90	571	494	
	80	533	462	
	70	489	423	
	60	440	381	
0 Awg	100	233	202	229
	90	257	222	
	80	252	218	
	70	241	209	
	60	227	196	
00 Awg	100	187	162	181
	90	213	184	
	80	212	183	
	70	206	178	
	60	196	169	
000 Awg	100	149	129	144
	90	179	155	
	80	181	156	
	70	177	153	
	60	171	148	
0000 Awg	100	121	104	114
	90	152	131	
	80	156	135	
	70	155	134	
	60	151	131	
250 MCM	100	102	89	97
	90	136	117	
	80	143	123	
	70	143	124	
	60	141	122	

Wire Size	Power Factor %	AC Single Phase	AC Three Phase	DC
300 MCM	100	86	75	81
	90	121	104	
	80	128	111	
	70	131	113	
	60	130	113	
350 MCM	100	74	64	69
	90	109	95	
	80	118	102	
	70	122	105	
	60	122	106	
400 MCM	100	66	57	60
	90	101	88	
	80	111	96	
	70	115	99	
	60	116	101	
500 MCM	100	54	47	48
	90	89	78	
	80	99	86	
	70	105	91	
	60	108	93	
600 MCM	100	47	41	40
	90	83	72	
	80	93	81	
	70	99	86	
	60	103	89	
700 MCM	100	41	36	34
	90	77	67	
	80	88	76	
	70	94	82	
	60	98	85	
750 MCM	100	39	34	32
	90	75	65	
	80	86	75	
	70	93	81	
	60	97	84	
800 MCM	100	37	32	30
	90	73	63	
	80	84	73	
	70	91	79	
	60	95	82	
900 MCM	100	34	29	27
	90	69	61	
	80	81	71	
	70	88	77	
	60	93	81	
1000 MCM	100	31	27	24
	90	67	58	
	80	79	68	
	70	86	75	
	60	91	78	

Courtesy of Warren Electric Company.

Table 9.57
Characteristics of Commonly Used Transmission Lines

Type of Line	Z_0 *Ohms*	*Vel.* *%*	*pF/ per ft.*	*OD*	*Attenuation in dB per 100 feet* 3.5	7	14	21	28	50	144	420
RG58/A-AU	53	66	28.5	0.195	0.68	1.0	1.5	1.9	2.2	3.1	5.7	10.4
RG58 Foam Diel.	50	79	25.4	0.195	0.52	0.8	1.1	1.4	1.7	2.2	4.1	7.1
RG59/A-AU	73	66	21.0	0.242	0.64	0.90	1.3	1.6	1.8	2.4	4.2	7.2
RG59 Foam Diel.	75	79	16.9	0.242	0.48	0.70	1.0	1.2	1.4	2.0	3.4	6.1
RG8/A-AU	52	66	29.5	0.405	0.30	0.45	0.66	0.83	0.98	1.35	2.5	4.8
RG8 Foam Diel.	50	80	25.4	0.405	0.27	0.44	0.62	0.76	0.90	1.2	2.2	3.9
RG11/A-AU	75	66	20.5	0.405	0.38	0.55	0.80	0.98	1.15	1.55	2.8	4.9
Aluminum Jacket, Foam Diel.[1]												
3/8 inch	50	81	25.0	–	–	–	0.36	0.48	0.54	0.75	1.3	2.5
1/2 inch	50	81	25.0	–	–	–	0.27	0.35	0.40	0.55	1.0	1.8
3/8 inch	75	81	16.7	–	–	–	0.43	0.51	0.60	0.80	1.4	2.6
1/2 inch	75	81	16.7	–	–	–	0.34	0.40	0.48	0.60	1.2	1.9
Open-wire[2]	–	97	–		0.03	0.05	0.07	0.08	0.10	0.13	0.25	–
300-ohm Twin-lead	300	82	5.8		0.18	0.28	0.41	0.52	0.60	0.85	1.55	2.8
300-ohm tubular	300	80	4.6		0.07	0.25	0.39	0.48	0.53	0.75	1.3	1.9
Open-wire, TV type												
1/2 inch	400	95			0.028	0.05	0.09	0.13	0.17	0.30	0.75	–
1 inch	450	95			0.028	0.05	0.09	0.13	0.17	0.30	0.75	–

1 Polyfoam dielectric type line information courtesy of Times Wire and Cable Co.

2 Attenuation of open-wire line based on No. 12 conductors, neglecting radiation.

Table 9.58. Extension Cord Requirements

It is important to use extension cords of adequate current carrying capacity when utilizing a power plant to operate portable electric tools. Undersized cords result in excessive voltage drops and additional power plant loading. This also causes excessive heating of the portable tool because voltage drop reduces tool capacity.

Ampere Rating	Wire Gauge @ Cord Length 50 ft.	100 ft.	150 ft.
2	18	18	18
3	18	18	18
4	16	16	16
5	16	16	16
6	16	16	14
8	16	14	12
10	16	14	12
12	14	14	12
14	14	12	10
16	12	12	10

Courtesy of Warren Electric Company.

Table 9.59. Portable Cord Diameters

Sizes of Round Flexible Cord Types SO & SJO

AWG	Mustang 90[1]	Vutron 90[2]	Neoprene 60[3]	Royal 40[4]
2 CONDUCTORS				
18 SO	.345	.345	.385	.360
18 SJO	.295	.300	.300	.300
16 SO	.370	.400	.400	.380
16 SJO	.325	.325	.325	.320
14 SO	.505	.527	.527	.512
14 SJO	—	—	.355	.350
12 SO	.580	.597	.597	.585
12 SJO	—	—	.430	.416
10 SO	.630	.637	.637	.640
8 SO	—	—	.700	.715
6 SO	—	—	.820	.800
3 CONDUCTORS				
18 SO	.365	.400	.400	.375
18 SJO	.320	.330	.330	.310
16 SO	.390	.425	.425	.393
16 SJO	.340	.355	.355	.331
14 SO	.530	.557	.557	.538
14 SJO	—	—	.375	.370
12 SO	.605	.632	.632	.611
12 SJO	—	—	.455	.440
10 SO	.660	.687	.687	.670
8 SO	—	—	.750	.782
6 SO	—	—	.885	.890
4 CONDUCTORS				
18 SO	.395	.430	.430	.406
18 SJO	.340	.355	.355	.340
16 SO	.420	.480	.385	.427
16 SJO	.380	.385	.480	.367
14 SO	.575	.602	.602	.584
14 SJO	—	—	.415	.405
12 SO	.655	.662	.662	.662
12 SJO	—	—	.500	.485
10 SO	.715	.742	.742	.730
8 SO	—	—	.820	.840
6 SO	—	—	.975	.900

[1] *Mustang 90 American Insulated Wire*
[2] *Vutron 90° Carol Cable*
[3] *Neoprene 60° Carol Cable*
[4] *Royal 90° Royal Electric*

Courtesy of Warren Electric Company.

10 Typical Electrical Hardware and Installation Details

The pages in this chapter include reprints from the catalog of the Warren Electric Company, an electrical supplier, illustrating the variety of parts required to install instrumentation power and control. The catalog part numbers have been eliminated to avoid confusion should this the supplier's part numbers change in the future. The dimensions shown in this chapter are typical, but may vary between manufacturers. The construction materials are also typical, but may also vary between manufacturers. Installation details have been provided courtesy of Parsons S.I.P., Inc.*

*The material on pages 606–650 has been modified by the authors and is used courtesy of the Warren Electric Company, Houston, Texas. The material on pages 651–676 are used courtesy of Parsons S.I.P. Inc., Houston, Texas.

Conduit

Rigid Steel Conduit

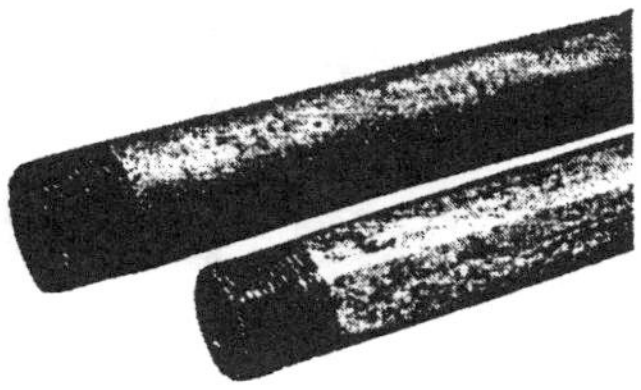

Trade Size (inches)	Min. Wgt. per 100 ft. (10 ft. lengths w/couplings)	NEMA Color Code Thread Protectors	Feet per Lift	Wt. per Crane Lift
½	80	Black	2500	2000 lbs.
¾	110	Red	2000	2200 lbs.
1	164	Blue	1250	2050 lbs.
1¼	210	Red	1020	2142 lbs.
1½	258	Black	1020	2631 lbs.
2	343	Blue	500	1715 lbs.
2½	543	Black	250	1357 lbs.
3	714	Blue	250	1785 lbs.
3½	856	Black	200	1712 lbs.
4	1000	Blue	200	2000 lbs.
5	1344	Blue	150	2016 lbs.
6	1798	Blue	100	1798 lbs.

Galvanized Steel Couplings

Pipe Size (In.)	Std. Ctn. (Qty.)	Wt. per 100	Wt. Per Ctn.
½	100	12	12
¾	50	18	9
1	30	30	9
1¼	25	37	9
1½	25	52	13
2	20	72	15
2½	24	170	41
3	16	210	34
3½	12	340	41
4	10	300	30
5	Bulk	475	Bulk
6	Bulk	765	Bulk

90°-45° Rigid Conduit Elbow

Pipe Size (In.)	Min. UL Radius (In.)	Offset	Straight	Wt. Per 100	Wt. Per Ctn.	Std. Ctn. (Qty.)
½	4	6⅜	1¾	71	35	50
¾	4½	7 7/16	2¼	110	55	50
1	5¾	9¼	2¼	198	49	25
1¼	7¼	10⅞	3	310	62	20
1½	8¼	12 13/16	3½	446	67	15
2	9½	14 9/16	4	675	67	10
2½	10½	18 5/16	6	1350	675	50
3	13	21½	6	2099	735	35
3½	15	24½	8	2908	727	25
4	16	25½	7	3519	880	25
5	24	36	10½	6720	Bulk	Bulk
6	30	46¼	11	11523	Bulk	Bulk

Plasti-Bond Rigid Steel Conduit

Special Products For Special Problems

In 1961, ROBROY introduced Plastibond, the first galvanized, rigid steel conduit to incorporate a permanently-bonded PVC coating. And, since a conduit system must protect against corrosion at every point to be effective, we also developed the first conduit coupling with PVC pressure-sealing sleeves, and the first bonded PVC coated access fittings with PVC pressure-sealing sleeves.

ROBROY further recognized that, in some severely corrosive atmospheres, contaminants can still find their way inside the electrical conduit system and corrode the components from within. In response, we developed and introduced Plasti-Bond 2 products. The same as standard Plasti-Bond on the outside, Plasti-Bond 2 has the extra protection of a hard, fusion-bonded phenolic coating on the inside to prevent internal corrosion.

Size	Feet Per Lift	Lbs. Per C Ft.
½"	2,500	85
¾"	2,000	112
1"	1,250	164
1¼"	1,000	217
1½	800	268
2"	600	358
2½"	350	546
3"	300	708
3½"	250	851
4"	200	1,009
5"	150	1,337
6"	100	1,993

PVC Coated Couplings

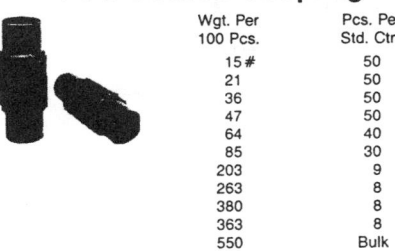

Wgt. Per 100 Pcs.	Pcs. Per Std. Ctn.
15#	50
21	50
36	50
47	50
64	40
85	30
203	9
263	8
380	8
363	8
550	Bulk
850	Bulk

Rigid Aluminum Conduit

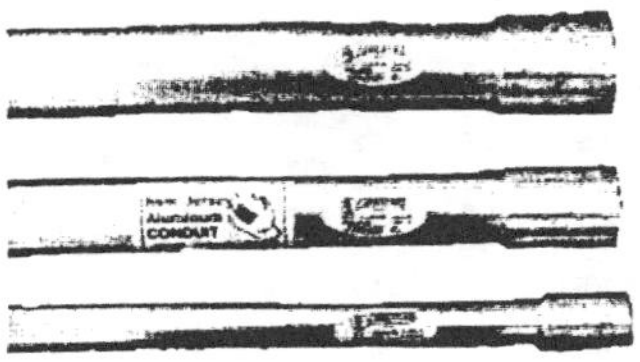

Conduit Size Inches	Weight Per 100 ft.	Pieces Per Bundle	Shipping Bundle Pieces	Shipping Bundle Feet	Shipping Bundle Pounds
½	28.1	10	440	4400	1236
¾	37.4	10	320	3200	1197
1	54.5	10	210	2100	1145
1¼	71.6	5	160	1600	1146
1½	88.7	5	140	1400	1242
2	118.5	3	81	810	960
2½	187.5	1	75	750	1406
3	246.3	1	50	500	1232
3½	295.6	1	36	360	1064
4	350.2	1	32	320	1121
5	478.9	1	34	340	1628
6	630.4	1	14	140	883

PVC Coated Elbows

STANDARD RADIUS ELBOWS

Size	Wgt. Per. 100 Pcs.	Pcs. Per Std. Ctn.
½"	83	25
¾"	120	25
1"	208	20
1¼"	330	20
1½"	453	10
2"	690	10
2½"	1,309	50
3"	1,938	50
3½"	2,875	40
4"	3,407	28
5"	6,788	Bulk
6"	11,300	Bulk

Aluminum Couplings

Pipe Size (In.)	Wt. per 100	Wt. Per Ctn.	Std. Ctn. (Qty.)
½	6	6	100
¾	9	5	50
1	13	4	30
1¼	19	5	25
1½	23	6	25
2	35	7	20
2½	68	16	24
3	91	15	16
3½	108	13	12
4	142	14	10
5	242	Bulk	Bulk
6	320	Bulk	Bulk

Galvanized Steel Conduit Nipples

Length Pipe Size	Length	Close Std. Ctn.	Close Lbs. 100	1½" Std. Ctn.	1½" Lbs. 100	2" Std. Ctn.	2" Lbs. 100	2½" Std. Ctn.	2½" Lbs. 100	3" Std. Ctn.	3" Lbs. 100	3½" Std. Ctn.	3½" Lbs. 100	4" Std. Ctn.	4" Lbs. 100	5" Std. Ctn.	5" Lbs. 100	6" Std. Ctn.	6" Lbs. 100	8" Std. Ctn.	8" Lbs. 100	10" Std. Ctn.	10" Lbs. 100	12" Std. Ctn.	12" Lbs. 100
½"	1⅛	25	6	25	8	25	12	25	15	25	19	25	22	25	26	25	33	25	40	25	54	25	68	25	82
¾"	1⅜"	25	9	25	15	25	14	25	19	25	24	25	28	25	34	25	43	25	52	25	73	25	89	25	109
1"	1½"	25	16			25	22	25	28	25	36	25	43	25	49	25	64	25	78	25	109	25	138	25	166
1¼"	1⅝"	25	22			25	28	25	37	25	47	25	55	25	66	25	84	25	100	10	136	10	176	10	216
1½"	1¾"	25	28			25	34	25	44	25	56	25	68	25	80	25	103	25	122	10	170	10	216	10	260
2"	2"	25	44					25	59	25	72	25	88	25	103	25	132	25	160	10	220	10	285	10	335
2½"	2½"	12	84							12	100	12	120	12	150	12	197	12	240	6	329	6	422	6	505
3"	2⅝"	12	118							12	100	12	157	12	200	12	260	12	300	6	411	6	528	6	630
3½"	2¾"	B	160											12	240	12	320	12	373	6	510	6	655	6	785
4"	2⅞"	B	180											12	285	12	380	12	440	6	600	6	775	6	925
5"	3"	B	240													B	480	B	600	B	825	B	1055	B	1260
6"	3⅛"	B	350													B	660	B	820	B	1125	B	1440	B	1720

Aluminum Conduit Nipples

Length Pipe Size	Length	Close Std. Ctn.	Close Lbs. 100	1½" Std. Ctn.	1½" Lbs. 100	2" Std. Ctn.	2" Lbs. 100	2½" Std. Ctn.	2½" Lbs. 100	3" Std. Ctn.	3" Lbs. 100	3½" Std. Ctn.	3½" Lbs. 100	4" Std. Ctn.	4" Lbs. 100	5" Std. Ctn.	5" Lbs. 100	6" Std. Ctn.	6" Lbs. 100	8" Std. Ctn.	8" Lbs. 100	10" Std. Ctn.	10" Lbs. 100	12" Std. Ctn.	12" Lbs. 100
½"	1⅛	25	2	25	3	25	4	25	5	25	7	25	8	25	9	25	12	25	14	25	19	25	24	25	29
¾"	1⅜"	25	3	25	3	25	5	25	7	25	8	25	10	25	12	25	15	25	18	25	25	25	31	25	38
1"	1½"	25	4			25	7	25	8	25	12	25	14	25	16	25	21	25	26	10	36	10	45	10	55
1¼"	1⅝"	25	7			25	10	25	13	25	16	25	20	25	23	25	29	25	36	10	49	10	62	10	75
1½"	1¾"	25	10			25	12	25	16	25	20	25	24	25	28	25	36	25	44	10	59	10	74	10	90
2"	2"	25	16					25	22	25	27	25	32	25	38	25	48	25	58	10	78	10	100	10	120
2½"	2½"	12	22							12	30	12	39	12	47	12	64	12	80	6	114	6	148	6	180
3"	2⅝"	12	34							12	42	12	53	12	64	12	86	12	108	6	152	6	196	6	240
3½"	2¾"	12	42											12	75	12	101	12	128	6	180	6	234	6	286
4"	2⅞"	12	51											12	87	12	118	12	149	6	212	6	274	6	336
5"	3"	B	71													B	155	B	197	B	281	B	366	B	450
6"	3⅛"	B	105													B	208	B	263	B	372	B	481	B	591

90°-45° Aluminum Elbows

Pipe Size (in.)	Min. UL Radius (in.)	Offset	Straight	Weight Per 100	Weight Per Ctn.	Std. Ctn. Qty.
½	4	6⅜	1¾	25	13	50
¾	4½	7 7/16	2¼	39	20	50
1	5¾	9¼	2¼	68	17	25
1¼	7¼	10⅞	3¼	107	21	20
1½	8¼	12 13/16	3½	152	23	15
2	9½	14 9/16	4	242	25	10
2½	10½	18 5/16	6¼	480	240	50
3	13	21½	6	749	262	35
3½	15½	24½	8	1035	259	25
4	16	25½	7	1255	314	25
5	24	36	10½	2399	Bulk	Bulk
6	30	46¼	11	4124	Bulk	Bulk

Electrical Metallic Tubing

Article 348—Electrical Metallic Tubing

Use (348-1)—The use of electrical metallic tubing shall be permitted for both exposed and concealed work. Electrical metallic tubing shall not be used: (1) where, during installation or afterward, it will be subject to severe physical damage; (2) where protected from corrosion solely by enamel; (3) in cinder concrete or cinder fill where subject to permanent moisture unless protected on all sides by a layer of noncinder concrete at least 2 inches thick or unless the tubing is at least 18 inches under the fill. Where practicable, dissimilar metals in contact anywhere in the system shall be avoided to eliminate the possibility of galvanic action

Exception: Aluminum fittings and enclosures are permitted to be used with steel electrical metallic tubing.

Ferrous or nonferrous electrical metallic tubing, elbows, couplings and fittings shall be permitted to be installed in concrete, in direct contact with the earth, or in areas subject to severe corrosive influences when protected by corrosion protection and judged suitable for the condition.

Trade Size (In.)	Wt. Per 100 Ft.	Feet Per Lift	Wt. Per Crane Lift (Lbs.)
½	29	7,000	2030
¾	45	5,000	2250
1	68	3,000	2040
1¼	95	2,000	1900
1½	119	1,750	2082
2	144	1,200	1728
2½	215	510	1096
3	260	440	1144
3½	325	370	1202
4	400	240	960

90°-45° EMT Elbows

Pipe Size (In.)	Min. UL Radius (In.)	Offset	Wt. per 100	Wt. per Ctn.	Std. Ctn. (Qty.)
½	4	6 5/16	27	14	50
¾	4½	7½	47	24	50
1	5¾	8⅞	84	21	25
1¼	7¼	11	157	31	20
1½	8¼	12 13/16	215	32	15
2	9½	14¼	299	30	10
2½	10½	18 5/16	589	294	50
3	13	21½	836	418	50
3½	15	24½	1181	413	35
4	16	25½	1433	501	35

PVC Schedule 80 Conduit

Carlon PLUS 80 is designed for applications where an extra heavy wall conduit is needed such as outside installations subject to severe physical abuse, pole risers, bridge crossings, and other similar conditions. Carlon C-2000 compound combines high physical properties with high chemical resistance and impact damage resistance. Complies with UL651 and NEMA TC-2.

Nominal Size	Weight Per 100'	Feet Per Bundle 10'	Feet Per Bundle 20'
½"	21	100	200
¾"	29	100	200
1"	42	100	200
1¼"	59	50	100
1½	71	50	100
2"	99	10	20
2½	150	10	20
3"	209	10	20
4"	306	10	20
5"	424	10	20

PVC Schedule 40 Conduit

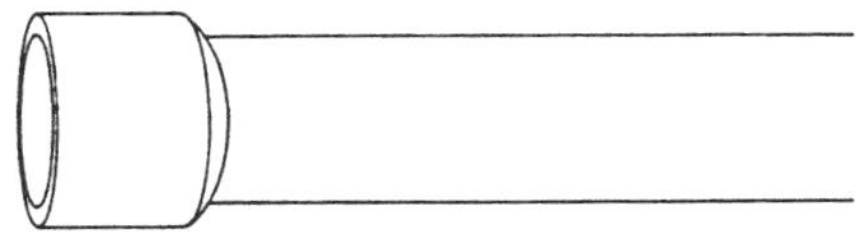

Carlon Plus 40, a UL-listed non-metallic Schedule 40 conduit, has superior weathering characteristics and lower smoke and HCL emission in fire conditions.

Manufactured from Carlon's new compound formulation, Carflex 300, Carlon Plus 40 meets and exceeds all industry and UL requirements (UL Standards 651, NEMA TC2-1978 and Federal Specification WC-1094A). It provides lower expansion and contraction, higher modulus strength and is more rigid than other non-metallic conduits.

Nominal Size	Weight Per 100′	Feet Per Bundle 10′	Feet Per Bundle 20′
½″	17	100	200
¾″	23	100	200
1″	34	100	200
1¼″	46	50	100
1½	55	50	100
2″	73	50	100
2½	124	10	20
3″	162	10	20
3½″	195	10	20
4″	231	10	20
5″	312	10	20
6″	406	10	20

PVC Couplings

STANDARD COUPLINGS

Size	Std. Ctn. Qty.	Std. Ctn. Wgt. (Lbs.)	Size	Std. Ctn. Qty.	Std. Ctn. Wgt. (Lbs.)
½	150	4	2½	20	7
¾	100	4	3	25	14
1	50	4	3½	20	13
1¼	30	3	4	15	13
1½	25	4	5	8	10
2	30	5	6	5	9

PVC 90° and 45° Elbows

Elbows—Non-Metallic

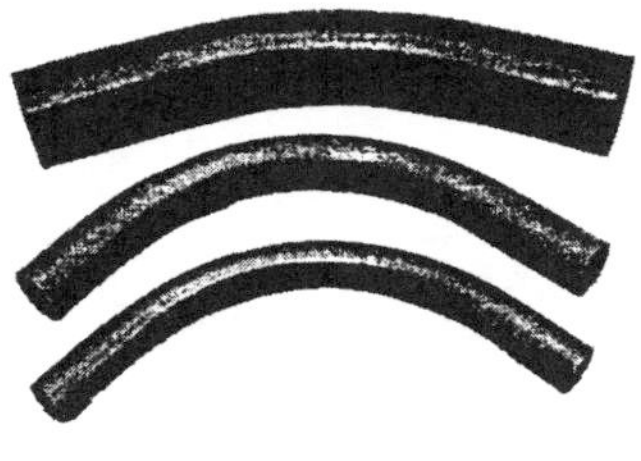

90° ELBOWS

Size	Std. Ctn. Qty.	Std. Ctn. Wt. (Lbs.)
½	50	7
¾	25	5
1	25	9
1¼	20	12
1½	25	20
2	20	24
2½	10	23
3	30	108
3½	25	125
4	25	150
5	8	96
6	8	144

45° ELBOWS

Size	Std. Ctn. Qty.	Std. Ctn. Wt. (Lbs.)
½	50	5
¾	25	3
1	20	5
1¼	20	8
1½	20	10
2	20	14
2½	20	30
3	30	66
3½	25	75
4	25	95
5	8	56
6	8	88

Steel Greenfield

Trade Size	Standard Coil Length		Approx. Wt. 100′ of Aluminum Alflex®	Comparable Wt. for Steel Flex
5/16″	100′-250′		13	1½″
⅝″	100′-250′	1000′	18	1⅜″
½″	100′	1000′	28	2⅛″
¾	100′	500′-1000′	33	3″
1″	50′		52	3¾″
1¼″	50′		65	4¼″
1½″″	25′		80	3¾″
2″	25′		100	5⅝″
2½″″	25′		150	7¾″
3″	25′		190	11½″
3½″″	25′		220	13″
4″	25′		270	18″

Aluminum Greenfield

Trade Size	Standard Coil Length	Approx. Wt. 100′ of Aluminum Alflex®	Comparable Wt. For Steel Flex
⅜″	100′	8.5	25.5
½″	100′	16.0	48.0
¾″	100′	20.5	59.5
1″	50′	35.0	102.0
1¼″	50′	43.0	125.0
1½″	25′	55.5	162.5
2″	25′	73.0	212.5
2½″	25′	93.0	263.0
3″	25′	107.5	313.0
3½″	25′	124.5	363.0
4″	25′	142.0	413.5

Type E.F.

Galvanized steel core, thermo-plastic cover. Extra flexible typical applications: vibration, movement crossover connections tight bends.

Electrical Trade Size	Approx. Inside Bend Dia. (inches)	Standard Carton Lgth. (ft)	Standard Carton Wt. (lbs)
¼	3½	500	73
5/16	4	500	78
⅜	4	250	53
½	5	200	57
¾	6	175	73
1	8	100	57
1¼	9	100	78
1½	11	50	60
2	14	50	77
2½	19	50	103
3	30	25	72
3½	32	25	91
4	34	25	101
5	44	25	169
6	60	25	231

Type U.A.

Galvanized steel core, thermo-plastic cover. UL listed, NEC article 351A integral copper ground wire. Typical applications where U.L. listing is required, where grounding is necessary.

Electrical Trade Size	Approx. Inside Bend Dia. (inches)	Standard Carton Lgth. (ft)	Standard Carton Wt. (lbs)
⅜	6	200	59
½	7	200	71
¾	10	150	76
1	12	100	87
1¼	14	50	59
1½	11	50	64
2	14	50	83
2½	19	50	104
3	30	25	73
4	34	25	100

Type O.R.

Galvanized steel core, cord packing, special oil resistant thermoplastic cover, resists oil degradation. Typical applications, exposure to cutting oils.

Electrical Trade Size	Approx. Inside Bend Dia. (inches)	Standard Carton Lgth. (ft)	Standard Carton Wt. (lbs)
⅜	4	250	53
½	5	200	57
¾	6	175	73
1	8	100	57
1¼	9	100	78
1½	11	50	60
2	14	50	77
2½	19	50	103
3	30	25	72
4	34	25	101
5	44	25	169
6	60	25	231

Type C.N.-P

Smooth inner core, nylon reinforcing, rugged outer jacket. U.L. listed NEC. article 351 B non-metallic, extra rugged abrasion resistant meets JIC requirements. Typical applications where abrasion or physical abuse are factors, constant flexing.

Electrical Trade Size	Standard Carton Lgth. (ft)	Standard Carton Wt. (lbs)
⅜	250	39
½	200	45
¾	175	60
1	100	49
1¼	100	70
1½	50	48
2	50	75

Conduit/Cable Fittings

Chase Nipples

Chase® liquidtight cord connectors feature an insulated throat and beveled neoprene bushing to provide insulation protection and strain relief for cord applications. They are UL listed, CSA certified. Series 2631.

Size	Description	Unit Qty.	Std. Pkg.	Wgt. Per 100
CHASE® NIPPLES NYLON INSULATED				
½"	Steel or malleable	25	100	4
¾"	iron (steel thru ½").	25	100	7
1"	Temperature rating	25	100	13
1¼"	105°C. Use Chase	10	50	21
1½"	nipples to connect	10	50	29
2"	conduit coupling to	5	50	52
2½"	box knockout, with a	5	20	83
3"	locknut to hold two	2	10	125
3½"	boxes back to back or	1	5	140
4"	side by side, with a	1	5	180
5"	locknut to connect fix-		1	300
6"	ture housings in continuous runs.		1	340
CHASE® NIPPLES				
⅜"		100	1000	3
½"		100	500	4
¾"		100	200	7
1"		25	100	12
1¼"		25	100	20
1½"		10	50	28
2"		5	50	50
2½"		5	20	80
3"		2	10	120
3½"		1	5	140
4"		1	5	180
5"		1	5	300
6"			1	340
CHASE® NIPPLES Copper-free aluminum alloy. aluminum: less than .5% copper				
½"		100	500	2
¾"		100	200	3
1"		25	100	4
1¼"		25	100	7
1½"		10	50	10
2"		5	50	18
2½"		5	20	28
3"		2	10	42
3½"		1	5	50
4"		1	5	62
5"		1	5	100
6"			1	130

Buna N Sealing Ring

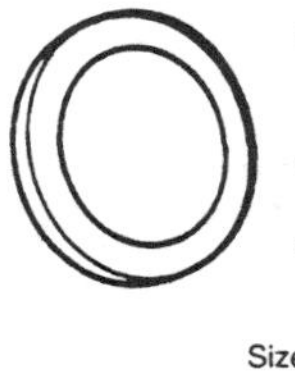

- Eliminates need for fitting on rigid conduit when used with double locknuts
- Provides watertight and dusttight seal
- Stainless steel retainer

Size	Unit Qty.	Std. Pkg.	Wgt. Per 100
½"	50	100	1
¾"	25	50	1
1"	25	50	2½
1¼"	5	25	3
1½"	5	25	3½
2"	5	25	5
2½"	1	5	8
3"	1	5	12½
4"	1	5	16

Bullet Hubs

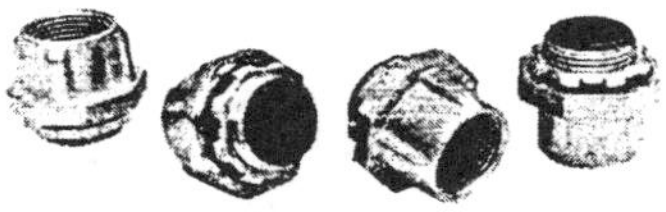

Bullet® Hub connectors for rigid metal conduit eliminate welding. They are simply installed with a wrench, and will lower your installed cost. Series 370. They are also available with a PVC coating for added protection against corrosive environments. Series 485.

Size	Description	Unit Qty.	Std. Pkg.	Wgt. Per 100
BULLET® HUB CONNECTORS NYLON INSULATED				
½"	Steel or malleable iron	25	100	20
¾"	(steel through 1¼").	25	50	26
1"	With neoprene "O" ring,	5	25	44
1¼"	provides a watertight	5	25	75
1½"	threaded hub on enclo-	2	10	100
2"	sures. U.L. listed 105°C.	1	5	138
2½"	Dura-Plate™ finish.	1	5	200
3"		1	5	375
3½"			1	425
4"			1	500
5"			1	
6"			1	
BULLET® HUB CONNECTORS—ALUMINUM				
½"		25	100	15
¾"		25	50	20
1"		5	25	36
1¼"		5	25	60
1½"		2	10	80
2"		1	5	110
2½"		1	5	160
3"		1	5	300
3½"			1	340
4"			1	400

Note: suitable or hazardous locations use per Class 1, Div. 2; Class II, Div. 1 & 2; Class III, Div. 1 & 2 NEC 501-4(b); 502-4(a) (2); 503-3(a).

Rigid Metal Threadless Couplings

Size	Unit Qty.	Std. Pkg.	Wgt. Per 100
½"	25	100	22
¾"	25	50	30
1"	5	25	52
1¼"	5	25	112
1½"	2	10	140
2"	1	5	220
2½"	1	5	480
3"		1	700
3½"		1	800
4"		1	900

Rigid Metal Offset Nipples and Connectors

Size	Unit Qty.	Std. Pkg.	Wgt. Per 100
OFFSET NIPPLES			
½"	25	100	12
¾"	10	100	16
1"	5	50	28
1¼"	5	25	36
1½"	5	10	44
2"	1	5	70
OFFSET CONNECTORS			
½"	25	100	12
¾"	10	100	20
1"	5	50	34

EMT Connectors

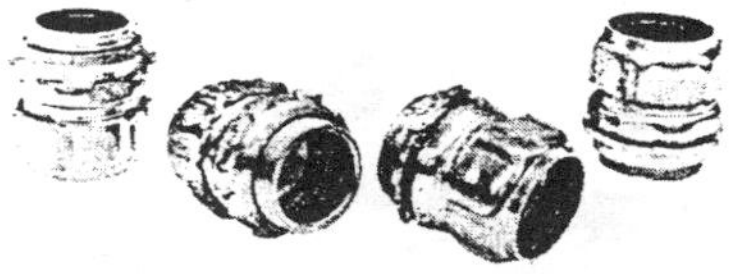

Size	Unit Qty.	Std. Pkg.	Wgt. Per 100
E.M.T. CONNECTORS NYLON INSULATED			
½"	50	500	10
¾"	25	250	14
1"	10	100	23
1¼"	5	25	46
1½"	2	10	59
2"	2	10	80
E.M.T. CONNECTORS			
½"	50	500	10
¾"	25	250	17
1"	10	100	28
1¼"	5	25	56
1½"	2	10	70
2"	2	10	120

E.M.T. Couplings

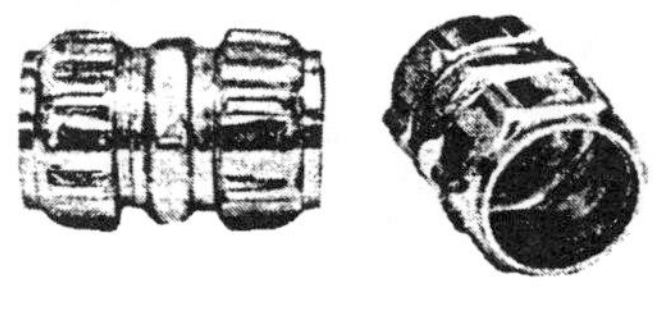

Size	Unit Qty.	Std. Pkg.	Wgt. Per 100
½"	50	500	11
¾"	25	250	19
1"	10	100	28
1¼"	5	25	88
1½"	2	10	90
2"	2	10	120

E.M.T. Offset Connectors

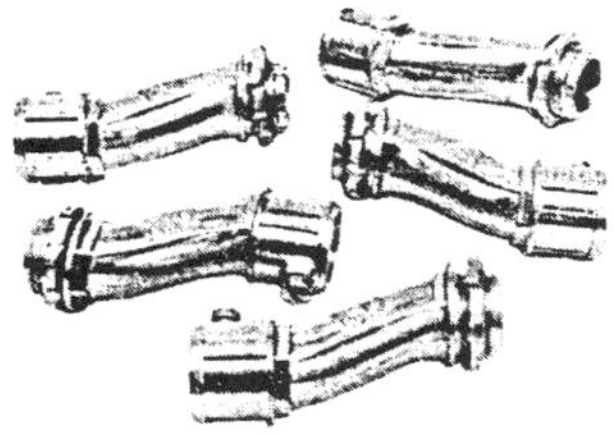

Size	Unit Qty.	Std. Pkg.	Wgt. Per 100
½"	25	100	14
¾"	10	100	22
1"	5	50	28

EMT Short Elbows

Size	Unit Qty.	Std. Pkg.	Wgt. Per 100
INSULATED SHORT ELBOWS			
½"	10	20	20
¾"	10	20	32
1"	5	15	56
1¼"	2	4	80
1½"	2	4	160
2"	2	4	230
SHORT ELBOWS			
½"	25	50	20
¾"	25	50	32
1"	5	25	56
1¼"	2	10	80
1½"	2	10	160
2"	2	4	230

Flexible Metal Conduit Fittings

Cable Opening Max.	Cable Opening Min.	Trade Size	K.O. Size	Unit Qty.	Std. Pkg.	Wgt. Per 100
FLEXIBLE METAL CONDUIT CONNECTORS TITE-BITE STYLE—NYLON INSULATED						
.656	.437	¾"	½"	50	500	7
.937	.750	½"	½"	25	100	15
1.125	.906	¾"	¾"	10	50	23
1.468	1.250	1"	1"	5	25	45
1.750	1.562	1¼"	1¼"	5	10	62
2.031	1.812	1½"	1½"	5	10	102
2.500	2.312	2"	2"	1	5	130
3.062	2.812	2½"	2½"	1	5	230
3.562	3.312	3"	3"	1	5	270
4.060	3.620	3½"	3½"		1	
4.560	4.120	4	4		1	
5.500	4.600	5	5		1	
NON INSULATED						
.656	.437	⅜"	½"	100	1000	7
.937	.750	½"	½"	25	100	21
1.093	.906	¾"	¾"	25	100	22
1.468	1.250	1"	1"	10	25	40
1.750	1.562	1¼"	1¼"	5	10	60
2.031	1.812	1½"	1½"	5	10	100
2.500	2.312	2"	2"	1	5	120
3.000	2.812	2½"	2½"	1	5	200
3.562	3.312	3"	3"	1	5	260

Flexible Metal Conduit Connectors

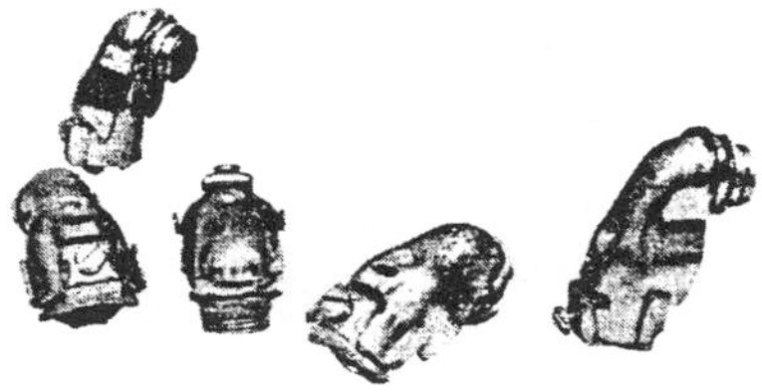

Cable Opening Max.	Cable Opening Min.	Trade Size	K.O. Size	Unit Qty.	Std. Pkg.	Wgt. Per 100
FLEXIBLE METAL CONDUIT CONNECTORS 90° ANGLE-NYLON NON INSULATED						
.656	.437	⅜"	½"	25	100	15
.937	.750	½"	½"	23	100	25
1.093	.906	¾"	¾"	10	50	41
1.720	1.250	1"	1"	5	25	62
1.750	1.562	1¼"	1¼"	5	10	135
2.031	1.812	1½"	1½"	5	10	205
2.500	2.312	2"	2"	1	5	280
3.062	2.812	2½"	2½"	1	5	430
3.562	3.312	3"	3"	1	5	580
4.060	3.620	3½"	3½"		1	
4.560	4.120	4"	4"		1	
TITE-BITE® CONNECTORS 90° ANGLE						
.656	.437	⅜"	½"	50	100	13
.937	.750	½"	½"	25	100	27
1.093	.906	¾"	¾"	10	50	34
1.720	1.250	1"	1"	5	25	60
1.750	1.562	1¼"	1¼"	5	10	110
2.031	1.812	1½"	1½"	5	10	170
2.500	2.312	2"	2"	1	5	280
3.000	2.812	2½"	2½"	1	5	500
3.562	3.312	3"	3"	1	5	800

Flexible Metal Conduit Connectors

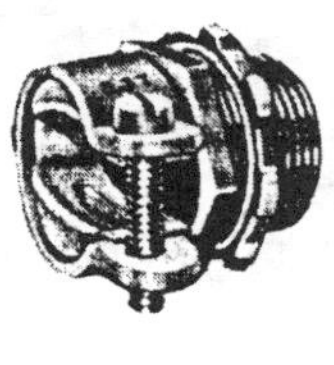

Cable Opening Max.	Cable Opening Min.	Trade Size	K.O. Size	Unit Qty.	Std. Pkg.	Wgt. Per 100
FLEXIBLE METAL CONDUIT CONNECTORS SQUEEZE STYLE						
.531	.437	5/16"	⅜"	50	100	5
.585	.455	⅜"	½"	100	1000	6
.938	.812	½"	½"	50	100	13
1.094	.938	¾"	¾"	25	100	19
1.375	1.250	1"	1"	10	25	32
1.656	1.500	1¼"	1¼"	5	10	40
1.906	1.688	1½"	1½"	5	10	70
2.500	2.313	2"	2"	5	10	100
3.000	2.812	2½"	2½"	1	5	180
3.563	3.312	3"	3"	1	5	200

Flexible Metal Conduit Connectors

Cable Opening Max.	Cable Opening Min.	Trade Size	K.O. Size	Unit Qty.	Std. Pkg.	Wgt. Per 100
90° ANGLE FLEXIBLE METAL CONDUIT CONNECTORS SQUEEZE TYPE						
.656	.406	⅜"	½"	50	100	15
.813	.688	⅜"	½"	25	100	23
.937	.813	½"	½"	25	100	21
1.000	.875	¾"	¾"	25	50	32
1.125	1.000	¾"	¾"	10	50	28
1.406	1.187	1"	1"	5	25	52
1.656	1.375	1¼"	1¼"	5	10	100
1.875	1.625	1½"	1½"	5	10	180
2.500	2.125	2"	2"	1	5	300
45° ANGLE						
.656	.406	⅜"	½"	25	100	16
.937	.813	½"	½"	25	100	21
1.125	1.000	¾"	¾"	25	50	26

Flexible Metal Conduit Connectors—2-Screw Type

Cable Opening Max.	Cable Opening Min.	Trade Size	K.O. Size	Unit Quan.	Std. Pkg.	Wgt. Per 100
.656	.390	⅜"	½"	100	500	6
.937	.500	½"	½"	50	100	13

Liquidtight Flexible Metal Conduit Fittings Straight

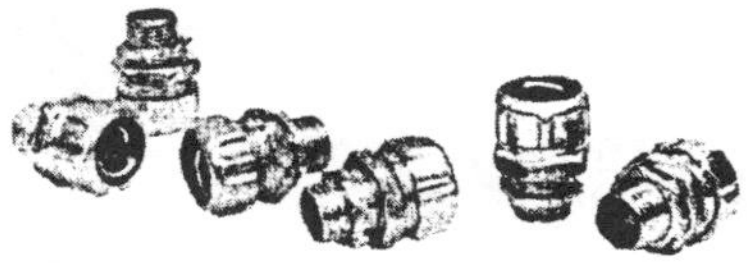

Standard liquidtight fittings provide a dependable connection for liquidtight flexible metal conduit. They feature a double beveled sealing ring which gives you a plastic-to-plastic seal to prevent conduit pullout and maintain the liquidtight integrity of the connection. They are UL listed and CSA certified. Series 5331. Also available with PVC coating for added protection against corrosive environments. Series 3321.

Hub Size	Conduit Size	Unit Qty.	Std. Pkg.	Wgt. Per 100
NYLON INSULATED STRAIGHT CONNECTORS				
¼"	¼"	25	100	7
⅜"	5/16""	25	100	10
½"	⅜"	25	100	14
½"	½"	25	100	18
¾"	¾"	25	50	28
1"	1"	10	50	51
1¼"	1¼"	5	25	92
1½"	1½"	2	10	105
2"	2"	1	5	160
2½"	2½"	1	5	312
3"	3"	1	5	375
4"	4"		1	612
5"	5"		1	2800
6"	6"		1	4000

STRAIGHT CONNECTORS

½"	⅜"	25	100	13
½"	½"	25	100	17
¾"	¾"	25	50	24
1"	1"	10	50	41
1¼"	1¼"	5	25	84
1½"	1½"	2	10	105
2"	2"	1	5	160
2½"	2½"	1	5	312
3"	3"	1	5	375
4"	4"		1	612
5"	5"		1	2800

Liquidtight Flexible Metal Conduit Fittings 45° and 90° Elbow

Hub Size	Conduit Size	Unit Qty.	Std. Pkg.	Wgt. Per 100
NYLON INSULATED 45° ANGLE CONNECTORS				
½"	⅜"	25	50	20
½"	½"	25	50	28
¾"	¾"	10	50	40
1"	1"	5	25	60
1¼"	1¼"	5	25	112
1½"	1½"	2	10	160
2"	2"	1	5	230
2½"	2½"		1	600
3"	3"		1	875
4"	4"		1	1225
45° ANGLE CONNECTORS				
½"	⅜"	25	50	18
½"	½"	25	50	28
¾"	¾"	10	50	40
1"	1"	5	25	68
1¼"	1¼"	5	25	100
1½"	1½"	2	10	160
2"	2"	1	5	230
2½"	2½"		1	600
3"	3"		1	875
4"	4"		1	1225
NYLON INSULATED 90° ANGLE CONNECTORS				
½"	⅜"	25	50	28
½"	½"	25	50	32
¾"	¾"	10	50	44
1"	1"	5	25	68
1¼"	1¼"	5	25	120
1½"	1½"	2	10	160
2"	2"	1	5	230
2½"	2½"		1	862
3"	3"		1	1287
4"	4"		1	2181
90° ANGLE CONNECTORS				
½"	⅜"	25	50	18
½"	½"	25	50	28
¾"	¾"	10	50	40
1"	1"	5	25	68
1¼"	1¼"	5	25	100
1½"	1½"	2	10	160
2"	2"	1	5	230
2½"	2½"		1	862
3"	3"		1	1287
4"	4"		1	2181

Nonmetallic Flexible Liquidtight

Conduit Size	Unit Qty.	Std. Pkg.	Wgt. Per 100
STRAIGHT CONNECTOR			
½"	20	100	8
¾"	10	50	13
1"	5	25	20
1¼"	5	25	38
90° ANGLE CONNECTOR			
½"	10	50	11
¾"	10	50	16
1"	5	25	26
1¼"	5	10	48

Nonmetallic Sheath Cable and Flexible Cord Connectors

Cable Opening Max.	Min.	K.O. Size	Unit Qty.	Std. Pkg.	Wgt. Per 100
TITE-BITE® CONNECTORS 90° ANGLE-NYLON INSULATED					
.590	.250	½"	100	500	6
.750	.530	¾"	50	250	11
.990	.690	1"	25	100	28
1.320	.850	1¼"	10	20	50
1.515	.930	1½"	5	20	60
1.980	1.150	2"	5	10	90
TWO-SCREW CONNECTORS					
.656	.250	½"	100	1000	6
.820	.530	¾"	50	250	11
.070	.690	1"	25	100	28
1.385	.850	1¼"	10	20	50
1.530	.930	1½"	5	20	60
1.980	1.150	2"	5	10	90
2.375	1.500	2½"	1	5	120
2.875	1.750	3"		1	200
3.375	2.250	3½"		1	320
3.875	2.500	4"		1	440

Xtraflex Conduit System

The Xtraflex™ System conduit and tubing gives you the flexibility, toughness, environmental protection and cost effectiveness of non metallic—plus the unique ability to utilize not just one, but three families of Thomas & Betts fittings. This means you can meet many machine or industrial requirements with wiring protection that's easy to install, liquidtight and so flexible it will fit into cramped spaces.

Both the Xtraflex™ LTC Conduit and the Xtraflex™ EFC Tubing are liquidtight and oiltight when used in conjunction with T&B's APC series all plastic liquidtight connecters.

The Xtraflex™ System's APC fittings, tubing and conduit are manufactured from weather resistant and corrosion resistant thermoplastic materials making them suitable for applications in outdoor or corrosive environments.

Conduit Size	Unit Qty.	Std. Pkg.
STRAIGHT ALL PLASTIC LIQUIDTIGHT CONNECTOR		
⅜"	25	100
½"	25	100
¾"	25	50
1"	10	50
1¼"	5	25
1½"	2	10
2"	1	5
90° ELBOW ALL PLASTIC LIQUIDTIGHT CONNECTOR		
⅜"	25	100
½"	25	100
¾"	25	50
1"	10	50
SMOOTH LIQUIDTIGHT NON-METALLIC CONDUIT TYPE 2		
⅜"		100'
½"		100'
¾"		100'
1"		100'
1¼"		100'
1½"		50'
2"		50'
CORRUGATED FLEXIBLE NON-METALLIC TUBING		
⅜"		100'
½"		100'
¾"		100'
1"		100'
1¼"		100'
1½"		50'
2"		50'

Metallic Bushings

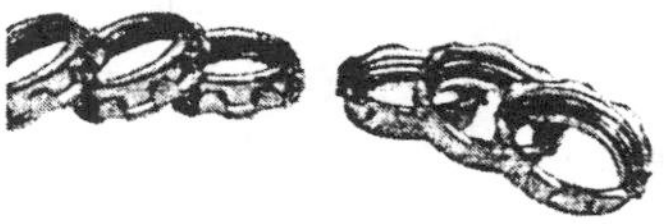

Size	Unit Qty.	Std. Pkg.	Wgt. Per 100
LIC BUSHINGS			
⅜"	100	500	1
½"	100	1000	2
¾"	100	1000	3
1"	50	500	8
1¼"	50	200	11
1½"	25	100	15
2"	25	50	18
2½"	10	30	36
3"	5	25	48
3½"	5	10	76
4"	5	10	100
4½"	2	4	140
5"	2	4	170
6"	1	2	250
INSULATED METALLIC BUSHING—150°C RATED			
½"	50	100	6
¾"	50	100	8
1"	10	40	12
1¼"	5	20	20
1½"	5	20	26
2"	5	10	35
2½"	5	10	64
3"	1	5	73
3½"	1	5	98
4"	1	5	110
NYLON INSULATED METALLIC BUSHINGS COPPER-FREE ALUMINUM			
½"	50	100	1
¾"	25	50	2
1"	10	40	3
1¼"	5	20	4
1½"	5	20	6
2"	5	10	7
2½"	5	10	13
3"	1	5	18
3½"	1	5	28
4"	1	5	38
4½"		1	52
5"		1	60
6"		1	90

Size	Unit Qty.	Std. Pkg.	Wgt. Per 100
METALLIC BUSHINGS—COPPER-FREE ALUMINUM			
½"	100	1000	1
¾"	100	1000	2
1"	50	500	3
1¼"	50	200	4
1½"	25	100	6
2"	25	50	7
2½"	10	30	13
3"	5	25	18
3½"	5	10	28
4"	5	10	38
5"	2	4	60
6"	1	2	90

Plastic Bushings

Size	Unit Qty.	Std. Pkg.	Wgt. Per 100
INSULATED ALL PLASTIC BUSHINGS—105°C RATED			
½"	100	400	1
¾"	100	400	1
1"	50	200	2
1¼"	25	100	3
1½"	25	50	4
2"	20	40	5
2½"	10	20	8
3"	10	20	10
3½"	5	10	13
4"	5	10	15
4½"	1	5	18
5"	1	2	21
6"	1	2	25
INSULATED ALL PLASTIC BUSHING—150°C RATED			
½"	100	400	1.5
¾"	100	400	2
1"	50	200	3
1¼"	25	100	4
1½"	25	100	5
2"	20	40	6
2½"	10	20	9
3"	10	20	11
3½"	5	10	15
4"	5	10	17
4½"	4	8	20
5"	1	2	23
6"	1	2	28

Locknuts

Size	Unit Qty.	Std. Pkg.	Wgt. Per 100
STEEL* (through 2")			
¼"	100	500	1
⅜"	100	500	1
½"	100	1000	1
¾"	100	1000	2
1"	50	500	4
1¼"	50	200	7
1½"	50	100	9
2"	25	50	14
MALLEABLE IRON* (2½" and above)			
2½"	15	30	26
3"	5	25	40
3½"	5	10	44
4"	5	10	52
4½"	2	4	73
5"	2	4	90
6"	2	4	100

COPPER-FREE ALUMINUM ALLOY			
½"	100	1000	½
¾"	100	1000	1
1"	50	500	2
1¼"	50	200	3
1½"	50	100	4
2"	25	50	5
2½"	15	30	9
3"	5	25	14
3½"	5	10	15
4"	5	10	18
4½"	2	4	25
5"	2	4	30
6"	2	4	35

X Series Outlet Bodies—Explosion Proof

Hub Size
½
¾
1
½
¾
1
½
¾
1

Y Type Capped Elbows

90° Capped Elbow Hub Size
½
¾
1
1¼

Reducers

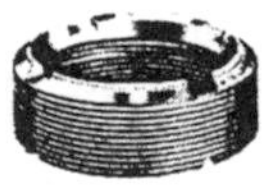

R-21 thru R-65 Cond. Size		R-71 thru R-109 Cond. Size	
½-⅜	2 -¾	3 - ¾	3½-2½
¾-½	2-1	3 -1	3½-3
1 -½	2-1¼	3 -1¼	4 - ½
1 -¾	2-1½	3 -1½	4 - ¾
1¼-½	2½-½	3 -2	4 -1
1¼-¾	2½-¾	3 -2½	4 -1¼
1¼-1	2½-1	3½- ½	4 -1½
1½-½	2½-1¼	3½- ¾	4 -2
1½-¾	2½-1½	3 -1	4 -2½
1½-1	2½-2	3½-1¼	4 -3
1½-1¼	3-½	3½-1½	4 -3½
2 -½		3½-2	

Cord Grips—CGB

CGB, cord and cable fittings with neoprene bushing are for use with portable cords and Types MV (unarmored), PLTC, SE (round), TC and UF cables. CGB cord and cable fittings are installed to:

- Provide means for passing a cord, cable (armored or unarmored) or flexible conduit into an enclosure, through a bulkhead or into a rigid conduit
- Form a seal for cord or unarmored round cables
- Form a secure connection or termination for flexible cord, cable (armored or unarmored), or flexible conduit

Male Thread NTP Trade Size	Cord Range Diameter		
⅜	.125 to .250	⅜	.250 to .375
⅜	.250 to .375	⅜	.375 to .500
⅜	.375 to .437	½	.125 to .250
⅜	.125 to .250	½	.250 to .375

Male Thread NTP Trade Size	Cord Range Diameter		
½	.375 to .437	1	.875 to 1.000
½	.125 to .250	1	.875 to 1.000
½	.250 to .375	1	1.000 to 1.188
½	.375 to .500	1	1.188 to 1.375
½	.500 to .625	1¼	1.000 to 1.188
½	.625 to .750	1¼	1.188 to 1.375
½	.750 to .875	1¼	1.375 to 1.625
¾	.125 to .250	1¼	1.625 to 1.875
¾	.250 to .375	1½	.875 to 1.000
¾	.375 to .500	1½	1.000 to 1.188
¾	.500 to .625	1½	1.188 to 1.375
¾	.625 to .750	1½	1.375 to 1.625
¾	.750 to .875	1½	1.625 to 1.875
¾	.875 to 1.000	2	1.375 to 1.625
1	.250 to .375	2	1.625 to 1.875
1	.375 to .500	2	1.875 to 2.188
1	.500 to .625	2	2.188 to 2.500
1	.625 to .750	2½	1.375 to 1.625
1	.750 to .875	2½	1.625 to 1.875
1	.750 to .875		

Cord Grips—CGF Series

- CGF series for use with Types AC, MC and MV unjacketed armored cables
- CGFP series for use with portable cords and Types MV, PLTC, SE (round), TC and UF jacketed unarmored cables
- CFFJ series for use with Types MC and MV jacketed armored cables
- CGFR series for use with Types MC and MV unjacketed armored solid core cables

CG cord and cable fittings are installed to:

- Provide means for passing armored, metal clad, jacketed or unjacketed and unarmored cable through a bulkhead or into an enclosure
- Form a mechanically gripping water and/or oiltight termination
- Provide grounding continuity of cable armor and flexible conduit

SG series sealing gaskets can be used with locknuts to provide a watertight seal in slip holes of sheet metal structures and boxes.

RSMP series terminating hub plates can be used on panels, junction boxes and bulkheads.

With Neoprene Bushing for Wet Locations Seal and Lead Bushing with Armor Stop for Armor Grip

Male Thread Size	Cable O.D. Range	Armor O.D. Range	Form
½	.625– .875	.500– .625	C-B
¾	.625– .875	.625– .813	C
¾	.750–1.000	.625– .813	C
¾	.750–1.000	.813–1.000	C
1	.625– .875	.625– .813	C
1	.750–1.000	.625– .813	C
1	.750–1.000	.813–1.000	C
1	.875–1.188	.625– .813	D-C
1	1.000–1.375	.813–1.000	D-C
1¼	1.000–1.375	1.000–1.375	D
1½	1.375–1.625	1.000–1.375	E-D
2	1.375–1.625	1.375–1.875	E
2	1.625–1.875	1.375–1.875	E
2	1.875–2.188	1.375–1.875	F-E
2½	1.875–2.188	1.875–2.188	F
2½	2.188–2.500	1.875–2.188	F
2½	2.188–2.500	2.188–2.500	F
3	1.875–2.188	1.875–2.188	F
3	2.188–2.500	1.875–2.188	F
3	2.188–2.500	2.188–2.500	F
3	2.500–3.000	2.188–2.500	G-F
3½	2.500–3.000	2.500–3.000	G
3½	3.000–3.500	2.500–3.000	G
3½	3.000–3.500	3.000–3.500	G
4	2.500–3.000	2.500–3.000	G
4	3.000–3.500	2.500–3.000	G
4	3.000–3.500	3.000–3.500	G
4	3.500–4.250	3.000–3.500	H-G
5	3.500–4.250	3.500–4.250	H

Cable Fittings—CGSJ Series

CGS series cable fittings are installed to:

- Provide means for passing armored, metal clad, jacketed or unjacketed and unarmored cables through a bulkhead or enclosure in hazardous areas. These terminators are suitable for use in Class I, Groups C, D, Div. 2 locations only when Chico A sealing compound is used to make the seal in the sealing fitting
- Sealing fitting of CGSC and CGSP terminator may be filled with hot potting compound for use in wet, non-hazardous locations
- Form a mechanical grip and wator and/or oil-resistant termination
- Provide ground continuity of cable armor and flexible conduit RSMP series terminating hub plates can be used on panels, junction boxes and bulkheads

CGSJ

Male Thread Size	Cable O.D. Range	Armor O.D. Range	Form	Potting Head/Sealing Fitting Internal Volume (Cubic Inches) Horiz.	Ver.
½	.625– .875	.500– .625	C-B	¾	1¼
¾	.625– .875	.625– .813	C	1	1½
¾	.750–1.000	.625– .813	C	1	1½
¾	.750–1.000	.813–1.000	C	1	1½
1	.625– .875	.625– .813	C	1¾	2¾
1	.750–1.000	.625– .813	C	1¾	2¾
1	.750–1.000	.813–1.000	C	1¾	2¾
1	.875–1.188	.625– .813	D-C	1¾	2¾
1	1.000–1.375	.813–1.000	D-C	1¾	2¾
1¼	1.000–1.375	1.000–1.375	D	6	8
1½	1.375–1.625	1.000–1.375	E-D	8½	10¾
2	1.375–1.625	1.375–1.875	E	16½	20
2	1.625–1.875	1.375–1.875	E	16½	20
2	1.875–2.188	1.375–1.875	F-E	16½	20
2½	1.875–2.188	1.875–2.188	F	24	35
2½	2.188–2.500	1.875–2.188	F	24	35
2½	2.188–2.500	2.188–2.500	F	24	35
3	1.875–2.188	1.875–2.188	F	47½	57
3	2.188–2.500	1.875–2.188	F	47½	57
3	2.188–2.500	2.188–2.500	F	47½	57
3	2.500–3.000	2.188–2.500	G-F	47½	57
3½	2.500–3.000	2.500–3.000	G	82	75
3½	3.000–3.500	2.500–3.000	G	82	75
3½	3.000–3.500	3.000–3.500	G	82	75
4	2.500–3.000	2.500–3.000	G	119	105
4	3.000–3.500	2.500–3.000	G	119	105
4	3.000–3.500	3.000–3.500	G	119	105
4	3.500–4.250	3.000–3.500	H-G	119	105
5	3.500–4.250	3.500–4.250	H	208	186

Liquidtight Conduit and Fittings

Easy to install, easy to assemble. New Carflex/Compuflex liquidtight conduit and fittings are totally non-metallic, non-conductive, corrosion resistant, faster and easier to install. When used with Carflex/Compuflex non-metallic liquidtight flexible c onduits, the fittings provide completely non-metallic systems which are functionally superior, yet competitive with metallic systems.

Nom. Size	Ctn. Length	Reel** Length	Wgt. Per 100′
LIQUIDTIGHT CONDUIT*			
⅜"	100′	1000	12.8
½"	100′	1000	13.3
¾"	100′	1000	18.8
1"	100′	500	25.9
1¼"	100′	200	33.9
1½"	100′	150	44.6
2"	100′	100	61.7
LIQUIDTIGHT FOR COMPUTER APPLICATIONS			
⅜"	100′	1000	12.8
½"	100′	1000	13.3
¾"	100′	1000	18.8
1"	100′	500	25.9
1¼"	100′	200	33.9
1½"	100′	150	44.6
2"	100′	100	61.7

Item	Pkg. Qty.
STRAIGHT FITTINGS U.L. LISTED	
½" straight	50
¾" straight	50
1" straight	25
1¼" straight	5
1½" straight	5
90° FITTINGS U.L. LISTED	
½" 90°	50
¾" 90°	50
1" 90°	25
1¼" 90°	5
1½" 90°	5
45° FITTINGS U.L. LISTED	
½" 45°	25
¾" 45°	25
1" 45°	25
1¼" 45°	5
1½" 45°	5

Flexible Couplings

Application

EC couplings are used:

- In hazardous areas where a flexible member is required in a conduit system to accomplish difficult bends, or to allow for movement or vibration of connected equipment or units

Features

- Rugged design to withstand explosive pressure (Class I)
- Mechanical abuse
- Liquid-tight for wet locations
- For use where lack of space makes use of rigid conduit difficult
- Wire duct liner in sizes ½" to 1½" insulates against grounds and burn-through from short circuit
- No bonding jumpers required, metallic braid provides continuous electrical path
- ECGJH combination has two threaded male end fittings
- ECLK combination has one female union and one male threaded end fitting

ECGJH

Flexible Length	Size	Flexible Length	Size
4"	½	21"	4
4"	¾	24"	½
6"	½	24"	¾
6"	¾	24"	1
6"	1	24"	1¼
8"	½	24"	1½
8"	¾	24"	2
8"	1	24"	2½
10"	½	24"	3
10"	¾	24"	4
10"	1	27"	½
12"	½	27"	¾
12"	¾	27"	1
12"	1	27"	1¼
12"	1	27"	1¼
12"	1¼	27"	1½
12"	1½	27"	2
12"	2	27"	2½
12"	2½	27"	3
12"	3	27"	4
12"	4	30"	½
15"	½	30"	¾
15"	¾	30"	1
15"	1	30"	1¼
15"	1¼	30"	1½
15"	1½	30"	2
15"	2	30"	2½
15"	2½	30"	3
15"	3	30"	4
15"	4	33"	½
18"	½	33"	¾
18"	¾	33"	1
18"	1	33"	1¼
18"	1¼	33"	1½
18"	1½	33"	2
18"	2	33"	2½
18"	2½	33"	3
18"	3	33"	4
18"	4	36"	½
21"	½	36"	¾
21"	¾	36"	1
21"	1	36"	1¼
21"	1¼	36"	1½
21"	1½	36"	2
21"	2	36"	2½
21"	2½	36"	3
21"	3	36"	4

Flexible Couplings—ECLK

ECLK

Flexible Length	Size	Flexible Length	Size
4"	½	21"	4
4"	¾	24"	½
6"	½	24"	¾
6"	¾	24"	1
6"	1	24"	1¼
8"	½	24"	1½
8"	¾	24"	2
8"	1	24"	2½
10"	½	24"	3
10"	¾	24"	4
10"	1	27"	½
12"	½	27"	¾
12"	¾	27"	1
12"	1	27"	1¼
12"	1¼	27"	1½
12"	1½	27"	2
12"	2	27"	2½
12"	2½	27"	3
12"	3	27"	4
12"	4	30"	½
15"	½	30"	¾
15"	¾	30"	1
15"	1	30"	1¼
15"	1¼	30"	1½
15"	1½	30"	2
15"	2	30"	2½
15"	2½	30"	3
15"	3	30"	4
15"	4	33"	½
18"	½	33"	¾
18"	¾	33"	1
18"	1	33"	1¼
18"	1¼	33"	1½
18"	1½	33"	2
18"	2	33"	2½
18"	2½	33"	3
18"	3	33"	4
18"	4	36"	½
21"	½	36"	¾
21"	¾	36"	1
21"	1	36"	1¼
21"	1¼	36"	1½
21"	1½	36"	2
21"	2	36"	2½
21"	2½	36"	3
21"	3	36"	4

GUA

GUA series conduit outlet bodies are installed within hazardous area conduit systems to:

- Protect conductors in threaded rigid conduit
- Act as pull and splice boxes
- Interconnect lengths of conduit
- Change conduit direction
- Provide access to conductors for maintenance and future system changes
- Act as mounting outlets for fixtures (with appropriate covers)
- Act as sealing fittings (with appropriate covers)

Compliances

- NEC:
  Class I, Groups C, D
  Class II, Groups E, F, G
  Class III
- UL Standard: 886
- ANSI Standard: C33.27

Options

(Suffix to be added to Cat. No.)

- SA: Bodies–copper-free aluminum
- WOD: Covers–cast Feraloy-zinc electroplate and aluminum cellulose lacquer
- Corro-free™ epoxy powder coat—information available on request
- Hot-dip galvanize—Information on request
- S302: "O" ring gaskets for blank covers watertight

Hub Size	Cover Opening Cover
½	2
¾	2
½	3
¾	3
1	3
1¼	3⅝
1½	5

GUAB

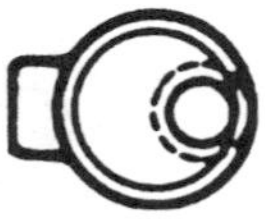

Hub Size	Cover Opening Dia.
½	2
¾	2
½	3
¾	3
1	3
1¼	3⅝
1½	5
2	5

GUAC

Hub Size	Cover Opening Dia.
½	2
¾	2
½	3
¾	3
1	3
1¼	3⅝
1¼	5
1½	5
2	5

GUAD

Hub Size	Cover Opening Dia.
½	2
¾	2
½	3
¾	3
1	3
1¼	5

GUAL

Hub Size	Cover Opening Dia.
½	2
¾	2
½	3
¾	3
1	3
1¼	3⅝
1¼	5
1½	5
2	5

GUAM

Hub Size	Cover Opening Dia.
½	2
¾	2
½	3
¾	3
1¼	3⅝
2	5

GUAN

Hub Size	Cover Opening Dia.
½	2
¾	2
½	3
¾	3
1	3
1¼	3⅝
1½	5
2	5

GUAT

Hub Size	Cover Opening Dia.
½	2
¾	2
½	3
¾	3
1	3
1	3⅝
1¼	3⅝
1¼	5
1½	5
2	5

GUAW

Hub Size	Cover Opening Dia.
½	2
¾	2
½	3
¾	3

GUAX

Hub Size	Cover Opening Dia.
½	2
¾	2
½	3
¾	3
1	3
1	3⅝
1¼	3⅝
1¼	5
1½	5
2	5

GUA Series Covers

Threaded covers, canopies and extensions are used:

- Interchangeably on GUA and GUF series bodies
- To conceal opening in plaster when enclosure body is mounted flush (flush cover)
- To provide a seal in hazardous areas (sealing cover).
- To mount pendant lighting fixtures such as EVA listed in lighting section (fixture canopy)
- To mount EVA pendant lighting fixtures on cover which is then screwed into outlet body without twisting conductors (union hub cover)
- To mount pendant lighting fixtures on cover which is then screwed into outlet body as above, for wiring after fixture stem is installed (nipple cover)
- To provide means of increasing outlet body depth (threaded extension)

GUAX EXTENSION*

Size ¾

Cover Opening Dia.	Depth
DOME COVER	
2	2
3	2
3 ⅝	2
3 ⅝	4
5	4
5	10
SURFACE COVER	
2	
3	
3⅝	
5	
FLUSH COVER	
2	
3	
SEALING COVER	
2	
3	
3⅝	
5	

Cover Opening Dia.	Fixt. Stem Size
NIPPLE COVER	
2	½
3	½
3	¾
FIXTURE COVER—UNION HUB TYPE	
3	½
3	¾
FIXTURE CANOPY	
3	½
3	¾
3	1
3	1¼

GUA THREADED EXTENSION

Cover Opening Dia.	Ext. Depth
3	1¼

* *For flush mounted GUA and GUF series with 3" cover opening to make one or more exposed extensions. Furnished with 3 pipe plugs.*

Conduit Seals—Vertical Only

EYS sealing fittings:

- Restrict the passage of gases, vapors or flames from one portion of the electrical installation to another at atmospheric pressure and normal ambient temperatures
- Limit explosions to the sealed-off enclosure
- Limit precompression or "pressure piling" in conduit systems

Sealing fittings are required:

- At each entrance to an enclosure housing an arcing or sparking device when used in Class I, Division 1 and 2 hazardous locations. To be located as close as practicable and, in no case, more than 18" from such enclosures
- At each entrance of 2" size or larger to an enclosure or fitting housing terminals, splices or taps when used in Class I, Division 1 hazardous locations. To be located as close as practicable and, in no case, more than 18" from such enclosures
- In conduit systems when leaving the Class I, Division 1 or Division 2 hazardous locations
- In cable systems when the cables either do not have a gas/vaportight continous sheath or are capable of transmitting gases or vapors through the cable core when these cables leave the Class I, Division 1 or Division 2 hazardous locations

Compliances

NEC:

- EYS1-3, 16-36, 11-31, 116-316:
  Class I, Groups A, B, C, D
  Class II, Groups E, F, G
- EYS41-61:
  Class I, Groups B, C, D
  Class II, Groups E, F, G
- EYS29, 4-014, 46-0146; EZS1-8, 16-86:
  Class I, Groups C, D
  Class II, Groups E, F, G
- UL: Standard: 886

FOR SEALING IN VERTICAL POSITIONS ONLY

Hub Size	Female Hub Cat. No.	Male & Female Hub Cat. No.	Approx. Internal Volume in Cubic Inches Vertical	Horizontal
½	EYS1	EYS16	1	1
¾	EYS2	EYS26	2	2
1	EYS3	EYS36	3¾	3¾

Conduit Seals—Vertical or Horizontal

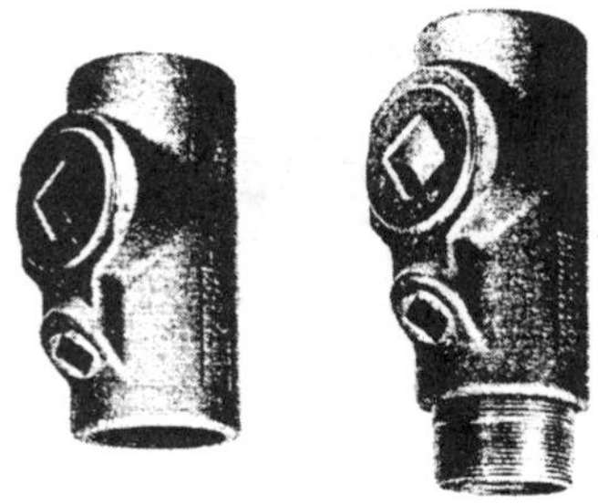

FOR SEALING IN VERTICAL OR HORIZONTAL POSITIONS

Hub Size	Approx. Internal Volume in Cubic Inches Vertical	Horizontal
½	1	1
¾	2	2
1	3¾	3¾
1¼	8	6
1½	10¾	8½
2	20	16½

Conduit Seals—Vertical or Horizontal Class I, Groups C, D Class II, Groups E, F, G

Hub Size	Approximate Internal Volume in Cubic Inches Vertical	Horizontal
1¼	8	6
1½	10¾	8½
2	20	16½
2½	35	24
3	57	47½
3½	75	82
4	105	119
5	186	208
6	320	339

Chico Sealing Compound

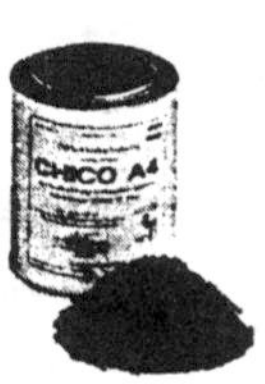

Chico X Fiber: forms a dam between the integral bushing of the sealing fitting and the end of the conduit and around the electrical conductors entering the hub.

Chico A Sealing Compound: forms a seal around each electrical conductor and between them and inside of the sealing fitting to restrict the passage of gases, vapors or flames through the sealing fitting at atmospheric pressure and at normal ambient temperatures.

CHICO A—SEALING COMPOUND

Net Weight	Vol. Cu. In.
9 oz.	13
1 lb.	23
1 lb.	23
2 lb.	46
2 lb.	46
5 lb.	115
10 lb.	230

CHICO X—FIBER

Net Weight
1 oz.
2 oz.
4 oz.
8 oz.
1 lb.

Chart for Approximate Amount of Fiber Per Hub

Hub Size	Ozs. Required	Hub Size	Ozs. Required
½	1/32	2½	1½
¾	1/16	3	2
1	⅛	3½	3
1¼	¼	4	4½
1½	½	5	7
2	1	6	10

Unions

UNF—FEMALE

Size
½
¾ to ½
¾
1
1¼
1¼
1½
1½
2
2
2½
3
3½
4
5
6

Unions

UNL—90° ANGLE

Size
½ to ½
¾ female to ½ male
½ female to ¾ male
¾ to ¾

Unions

UNY and UNF unions are installed in threaded thickwall conduit systems:

- UNY—to connect conduit to a conduit fitting, junction box or device enclosure
- UNF—to connect conduit to conduit, or to provide a means for future modification of the conduit system

UNL unions are used in conduit and fitting installations when entrance angle is between 90° and 180°.

UNY—MALE

Size
½
½ female to ¾ male
¾
1
1¼
1¼
1½
1½
2
2
2½
3
3½
4
5
6

OE Series Condulets

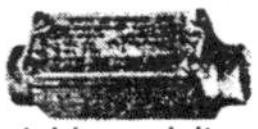

OE series are installed in conduit systems within hazardous areas to:

- Protect conductors in threaded rigid conduit
- Act as pulling and splice fittings
- Interconnect lengths of conduit
- Change direction of conduit
- Provide access for maintenance and future system changes
- NEC:
  Class I, Groups C, D
  Class II, Groups E, F, G
  Class III
- UL Standard: 886
- ANSI Standard: C33.27

Hub Size

OEC
½
¾

OELL
½
¾

OET
½
¾

OELB
½
¾

OELR
½
¾

Type C

Conduit outlet bodies are installed in conduit systems to:

- Act as pull outlets for conductors being installed
- Provide openings for making splices and taps in conductors
- Act as mounting outlets for lighting fixtures and wiring devices
- Connect conduit sections
- Provide taps for branch conduit runs
- Make 90° bends in conduit runs
- Provide for access to conductors for maintenance and future system changes
- Form 7 Condulet outlet bodies approach conduit in size for neat, compact installations. Form 8 and Mark 9 bodies provide more room for heavier conductors
- Many shapes and sizes are available for rigid and thin wall (EMT) conduit
- Conduit hubs have tapered threads and feature integral bushings for protection of wire insulation. Form 7 has exclusive wedge-nut cover attachment to provide clear, unobstructed cover opening

Standard Materials

- Form 7, Form 8 outlet bodies—Feraloy
- Mark 9 outlet bodies—copper-free aluminum, ½" to 2" die-cast, 2½" to 4" sand cast

Standard Finishes

- Form 7, Form 8 outlet bodies—zinc electroplate with aluminum lacquer
- Mark 9 outlet bodies—aluminum lacquer

Hub Size
½
¾
1
1¼
1½
2
2½
3
3½
4

Type E

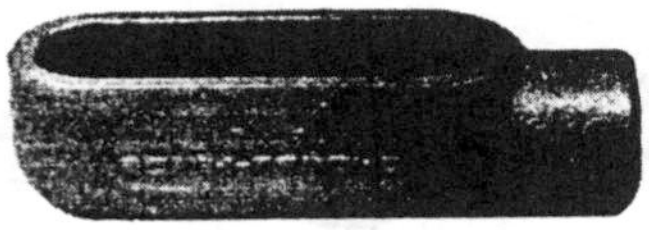

Hub Size
½
¾
1

Type L

Hub Size
½
¾
1
1¼
1½
2

Double faced—may be used as LL or LR—has 2 openings, one of which is furnished with a blank sheet steel cover

Type LB

Hub Size
½
¾
1
1¼
1½
2
2½
3
3½
4

Type LL

Hub Size
½
¾
1
1¼
1½
2
2½
3
3½
4

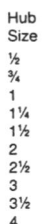

Type BC

Mogul bodies are installed in conduit systems to:

- Act as pull outlets for conductors that are stiff, due to large size or type of insulation
- Provide the longer openings needed when pulling large conductors
- Prevent sharp bends and kinks in large conductors (protects insulation during installation)
- Provide ample openings for splices and taps
- Provide access to wiring for maintenance and future system changes

Size	Size
1	2½
1¼	3
1½	3½
2	4

Type BLB

Size
1
1¼
1½
2
2½
3
3½
4

Type BT

Size
1
1¼
1½
2
2½
3
3½
4

Type BUB

Size
1
1¼
1½
2
2½
3
3½
4

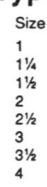

Type B Series Covers and Gaskets

BLANK COVERS—FERALOY®

Size	Without Gasket Cat. No.	With Round Neoprene Gasket Cat. No.
1 or 1¼	BG47	BG48
1½ or 2	BG67	BG68
2½ or 3	BG87	BG88
3½ or 4	BG97	BG98

CORK GASKETS—FOR USE W/BG47, BG67, BG87 & BG97

Size
1 or 1¼
1½ or 2
2½ or 3
3½ or 4

Corrosion-Resistant Cord Grips

NCGB corrosion-resistant cord and cable fittings:

- Are used with Type SO portable cords and Types MV, PLTC, SE (round), TC and UF (round) jacketed unarmored cable
- Are used indoors and outdoors for ordinary locations where wet, salty or harsh atmospheres rust or corrode the electrical system components
- Provide a means for passing a cord or cable into an enclosure, through bulkheads or into rigid conduit
- Forms a seal for a non-slip connection or termination
- Provide a raintight, dust-tight and oil-resistant seal

N.P.T. Trade Size	Cable O.D. Range
½	.25 to .42
½	.40 to .57
¾	.64 to .78
1	.76 to .91
1	.89 to 1.03

Type LR

Hub Size	Style Form 7	Form 8	Mark 9
½			
¾			
1			
1¼			
1½			
2			
2½			
3			
3½			
4			

Type T

Hub Size	Style Form 7*	Form 8	Mark 9
½			
¾			
1			
1¼			
1½			
2			
2½			
3			
3½			
4			

Type TB

Hub Size	Style Form 7	Form 8	Mark 9
½			
¾			
1			
1¼			
1½			
2			

Type X

Hub Size	Style Form 7	Form 8	Mark 9
½			
¾			
1			
1¼			
1½			
2			

Blank Covers

- Form 7: Wedge nut design facilitates installation and removal. Nuts are held captive in cover. Covers can be used with or without gaskets
- Form 8: Four cover screws provided in 1½" and larger sizes to provide tight cover and gasket assembly. Feraloy covers have dome shapes for added strength and extra wiring room
- Mark 9: Self-retaining cover screws

Size	Sheet Steel Form 7	Sheet Steel Form 8	Sheet Alum. Mark 9	Feraloy Form 7	Feraloy Form 8	Composition Form 7
½						
¾						
1						
1¼						
1½						
2						
2½						
3						
3½						
4						

Gaskets

Solid gaskets:

- Are used with blank covers
- For Mark 9, can be converted to open type gaskets by tearing out center section along scored lines—½" to 2" sizes
- For Form 7 are used with blank sheet steel and Feraloy® covers

Open gaskets:

- Are used with wire hole and wiring device covers
- For Form 7 are used with blank composition, wire hole, and wiring device covers
- For Mark 9—2½" to 4" sizes

Size	Form 7	Form 8	Mark 9
SOLID GASKETS—NEOPRENE			
½			
¾			
1			
1¼			
1½			
2			
2½			
3			
3½			
4			
OPEN GASKETS—NEOPRENE			
½			
¾			
1			
1¼			
1½			
2			
2½			
3			
3½			
4			

Service Entrance Elbow NEMA 7

LBH conduit outlet bodies are installed inhazardous areas to:

- Act as pull outlets especially for conductors that are stiff due to large size or type of insulation
- Act as pull outlets especially for conductors that are stiff due to large size or type of insulation
- Make 90° bends in conduit system, allowing straight pull in either direction
- Provide for conduit service entrance to buildings

- Provide for conductor entrance to motors
- Provide access to wiring for maintenance and future system changes

Size	Size
½	2
¾	2½
1	3
1¼	3½
1½	4

SLB—Service Entrance Elbows

SLB elbows are installed in conduit systems to:

- Act as service entrance elbows between service entrance and vertical weatherhead conduit runs
- Make 90° bends in conduit systems where space is limited
- Act as pull outlets
- Provide access to conductors for maintenance and future system changes

Size	Size
½	1¼
¾	1½
1	2

Pulling Elbow NEMA 7-9

LBY elbows are installed in conduit systems within hazardous areas to:

- Make 90° bends in conduit systems where space is limited
- Act as pull outlets
- Provide access to conductors for maintenance and future system changes

Size	Size
½	1
¾	1¼

LBY—Pulling Elbows

LBY elbows have:

- Maximum volume for bends within a compact overall size
- Screw-on cover for ease of installation and removal
- Cover openings on an angle, permitting conductors to be pulled straight through either hub
- Taper tapped hubs and integral bushing for standard threaded conduit

Size	Size
½	1¼
¾	1½
1	

Type LBD

LBD bodies are installed at 90° bends in rigid conduit to:

- Act as pull outlets for conductors that are stiff due to large size or type of insulation
- Make 90° bends in conduit system, allowing straight pull in either direction
- Provide for conduit service entrance to buildings
- Provide for conductor entrance to motors

- Provide access to wiring for maintenance and future system changes

Size
½
¾
1
1¼
1½
2
2½
3
3½
4
5
6

Elbows—45° Female

Size
½
¾
1
1¼
1½
2
2½
3
3½
4

Elbows—90° Female

Size
½
¾
1
1¼
1½
2
2½

Elbows—90° Male

EL elbows are installed in conduit run or in box or fitting hub to change direction in threaded rigid conduit run by 90°, or when terminating at a box or fitting.

Size
½
¾
1
1¼

Elbows—90° Male and Female

Size
½
¾
1
1¼

Reducer Couplings

REC reducers connect two different sizes of conduit together or are used to replace a coupling and reducer in an installation.

Large Hub Size	Small Hub Size
¾	½
1	½
1	¾
1¼	¾
1¼	1
1½	¾
1½	1
1½	1¼
2	¾
2	1
2	1¼
2	1½
2½	1½
3	2
3½	2½
4	3
5	4

Reducers—Plugs

- RE reducers are used in threaded heavy wall conduit systems
- RE reduces conduit hubs to a smaller size
- PLG plugs are used for closing threaded conduit hubs

RE

Size	
½–⅛	2½–1
½–¾	2½–1¼
½–⅜	2½–1½
¾–½	2½–2
1–½	3–1
1–¾	3–1¼
1¼–½	3–1½
1¼–¾	3–2
1¼–1	3–2½
1½–½	3½–2
1½–¾	3½–2½
1½–1	3½–3
1½–1¼	4–2
2–½	4–2½
2–¾	4–3
2–1	4–3½
2–1¼	5–4
2–1½	6–5

PLG

Size	
Recessed	
½	2
¾	2½
1	3
1¼	3½
1½	4
Square Head	
½	2
¾	2½
1	3
1¼	3½
1½	4

Breathers and Drains

Applications

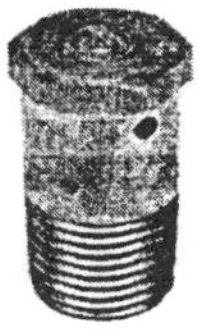

- ECD drains and breathers are installed in enclosures or conduit systems to:
  Provide ventilation to minimize condensation;
  Drain accumulated condensate
- At least one breather should be used with each drain
- Breather is installed in top of enclosure or upper section of conduit system
- A "standard" drain is installed in bottom of enclosure or in lower section of conduit system
- "Universal" breather or drain functions as a breather when mounted at the top of an enclosure, or as a drain when mounted in the bottom of an enclosure
- "Combination" breather and drain is used in those applications hwere the use of a top mounted breather is not practical due to limited space; or in offshore and marine installations where moisture may enter the enclosure through the breather located on top of enclosure
- Drains and breathers are installed in hubs or drilled and tapped openings

ECD "STANDARD"—DRAIN AND BREATHER

Male Thread Size	
¼	½

ECD "UNIVERSAL"—DRAIN OR BREATHER

¼	⅜
⅜	½

ECD "COMBINATION"—DRAIN AND BREATHER

½	

Cable Tray Conduit Clamps—LCC

LCC cable tray conduit clamps are used for installation on cable tray side rails with inside flanges (requiring inside tray mounting) and outside flanges; LCCF clamps are for use exclusively on inside flanges.

LCC/LCCF cable tray conduit clamps:

- Provide a means of clamping metal conduit (rigid steel or aluminum, IMC and EMT) to cable tray to provide for the exit of power and/or control cables from tray
- Provide a means to firmly bond exit conduit to cable tray for best grounding continuity
- Provide strong mechanical support for exit conduits and cables
- Can be used indoors or outdoors, wherever cable tray systems are installed
- Facilitate the safe exit of cables from tray–insure protection of cables from damage

Conduit Size	
½	2
¾	2½
1	3
1¼	3½
1½	4

XD Expansion Coupling

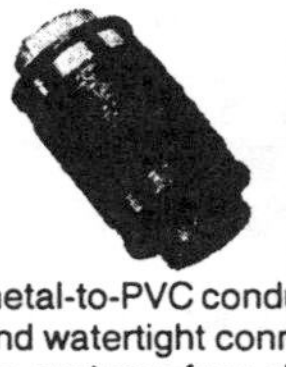

XD couplings can be installed indoors, outdoors, buried underground, or embedded in concrete in non-hazardous areas. XD's are used with standard rigid conduit or PVC rigid conduit. (PVC requires rigid metal conduit nipples and rigid metal-to-PVC conduit adapters.) XD's provide a flexible and watertight connection for protection of conduit wiring systems from damage due to movement.

Typical applications include:

- Underground conduit feeder runs
- Runs between sections of concrete subject to relative movement
- Runs between fixed structures
- Conduit entrances in high-rise buildings
- Bridges
- Marinas, docks, piers

Hub Size	Hub Size
1	3
1¼	3½
1½	4
2	5
2½	6

HEAVY DUTY CONTACT BLOCK RATINGS

Voltage	Inrush	Breaking	Carrying
120 VAC	60	6	10
240 VAC	30	3	10
480 VAC	15	1.5	10
600 VAC	12	1.2	10

Contact Arrangement

Hub Size	Normal Position	Alarm Activated Position
¾" / 1"		
¾" / 1"		
¾" / 1"		

Adapters

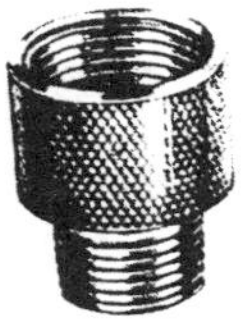

Conduit Size

Male ⅜
Female ½
Male ½
Female ¾

Bell Reducers

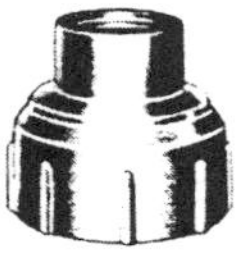

Conduit Size

Female to Female
1¼-½
1¼-¾
1½-½
1½-¾

Close Up Plugs

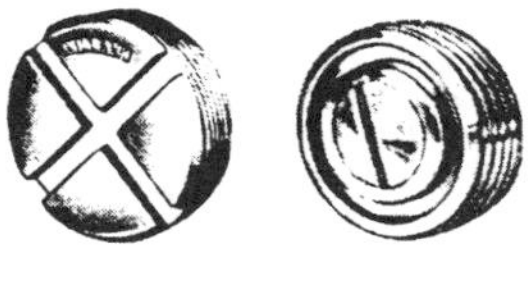

Conduit Size	
⅜	2
½	2½
¾	3
1	3½
1¼	4
1½	

Sealing Compound

SC Series Sealing Compound is a magnesite base cement used extensively for sealing conduit to prevent the spread of explosive gases. It is non-shrinking and a secure seal is formed. SC Series resists acids, water, oil, etc. Listed by Underwriters' Laboratories, Inc., for use with Killark EY Series sealing fittings.

Approximately one ounce of Killark sealing compound is needed per cubic inch of space to be filled.

Size Package	
	4 oz.
	8 oz.
	1 lb.

Conduit Seals

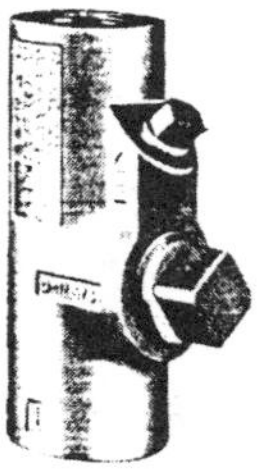

ENY Series (For vertical or horizontal conduit

EYS Series (For vertical or horizontal conduit)

EY Series (For vertical conduit)

Hub Size	Female Top and Bottom ENY	EYS	EY	Male Top Female Bottom ENY	EYS	EY	Female Top and Bottom with Drain EY
½	ENY-1	EYS-1	EY-1	ENY-1-T	EYS-1-T	EY-1-T	
¾	ENY-2	EYS-2	EY-2	ENY-2-T	EYS-2-T	EY-2-T	
1	ENY-3	EYS-3	EY-3	ENY-3-T	EYS-3-T	EY-3-T	EYD-3
1¼	ENY-4	EYS-4	EY-4	ENY-4-T	EYS-4-T	EY-4-T	EYD-4
1½	ENY-5	EYS-5	EY-5	ENY-5-T	EYS-5-T	EY-5-T	EYD-5
2	ENY-6	EYS-6	EY-6	ENY-6-T	EYS-6-T	EY-6-T	EYD-6
2½		EYS-7	EY-7		EYS-7-T	EY-7-T	EYD-7
3		EYS-8	EY-8		EYS-8-T	EY-8-T	EYD-8
3½		EYS-9	EY-9		EYS-9-T	EY-9-T	EYD-9
4		EYS-0	EY-0		EYS-0-T	EY-0-T	EYD-0

Sealing Fiber

Fiber is made from an environmentally safe, non-asbestos material.

Size Package

2 oz.
4 oz.
8 oz.
1 lb.

EYMF Series Capped Elbow

Hub Size

½
¾

GE Series Accessories

	Cover Type
—	Blank
2"	Dome
3"	Dome
4"	Dome
5"	Dome
—	Gasket
½	Hub
¾	Hub
—	Lens
—	Sealing

Sealing Cover

GE Series Outlet Bodies

(Threaded Hubs—covers included.) Gaskets available—sizes ½" thru 2"

GEM Series (2¹³⁄₁₆ Dia.) GEC Series (3¹¹⁄₁₆ Dia.)

GES Series (5³⁄₃₂ Dia.) GEJ Series (6¼ Dia.)

Hub Type	Hub Size
N	½, ¾, 1
E	½, ¾, 1, 1½, 2
C	½, ¾, 1, 1¼, 1½, 2
L	½, ¾, 1, 1¼, 1½, 2
LA	½, ¾
T	½, ¾, 1, 1¼, 1½, 2
TA	½, ¾
X	½, ¾, 1, 1¼, 1½, 2
XA	½, ¾, 1, 1¼
LB	½, ¾, 1, 1¼, 1½, 2
CA	½, ¾, —, 1¼

"GRR" Series Explosion-Proof Junction Box

3½" diameter, 2¼" deep, with blank screw cover, 5 threaded hubs and 3 close-up plugs.

Hub
½"
¾
1

Swivel Elbows

Swivel elbows have a range from 90° to 180°.

	Hub Size
Both Hubs	½
Female	¾
One Hub Female	½
One Hub Male	¾

GU Series Unions

Hazardous Locations
Class I, Groups C, D
Class II, Groups E, F, G
Class III

	Dimensions	
Hub Size	Diameter	Length
STYLE-FEMALE AND FEMALE		
½	1½	1 23/32
¾	1 11/16	1 23/32
1	2	2 23/64
1¼	2⅜	2 3/32
1½	2⅝	2 3/32
2	3¼	2 5/16
2½	4½	3 3/32
3	5⅜	3 1/16
3½	6¼	3 9/16
4	6¼	3 9/16
STYLE-MALE AND FEMALE		
½	1½	2 3/16
¾	1 11/16	2 13/64
1	2	2 41/64
1¼	2⅜	2 45/64
1½	2⅝	2 23/32
2	3¼	2 31/32
2½	4½	4 9/32
3	5⅗	4¼
3½	6¼	4½
4	6¼	4½

90° Angle Union

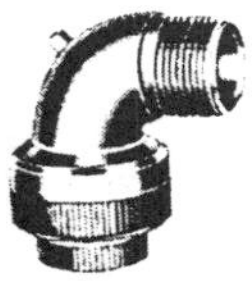

Hub Size
½
¾

Elbows

Type	Hub Size
MALE/FEMALE	
MF	½
90°	¾
FEMALE/FEMALE	
	½
	½
FF	¾
90°	1
	1¼
FF	½
45°	¾

MALE/MALE	
MF	½
45°	¾
MM	½
90°	¾
MALE/FEMALE B-L TYPE	
	½
BL	¾
90°	1
	1¼

Drain and Breather

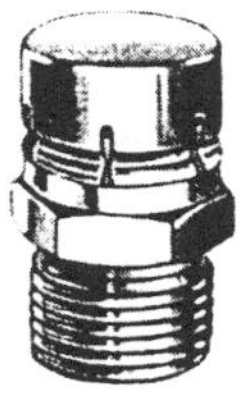

KDB-1
Killark KDB-1 may be used as a combination drain or breather.

Thread Lubricant

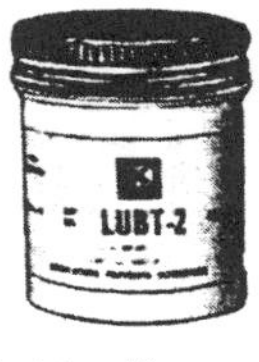

LUBG is a general purpose lubricant to be used in temperatures ranging from 0° to 125°F.

LUBT is a high quality lubricant to be used in temperatures ranging from −40° to +500° F. It is recommended to be used on hazardous location lighting fixtures.

Container Size
6 oz.
12 oz.
2 oz.

"OC" Electrolets

Size
½
¾
1
1¼
1½

"OLB" Electrolets

Size
½
¾
1
1¼
1½
2
2½
3

"OLL" and "OLR" Electrolets

Size	Size
½	½
¾	¾
1	1
1¼	1¼
1½	1½
2	2
2½	2½
3	3

"OLRL" Electrolets

Size
½
¾
1
1¼
1½
2
2½
3

"OT" Electrolets

Size
½
¾
1
1¼
1½
2

"OTB" Electrolets

Size
½
¾
1
1¼
1½
2

"OX" Electrolets

Size
½
¾
1
1¾
1½
2

"OL" Covers

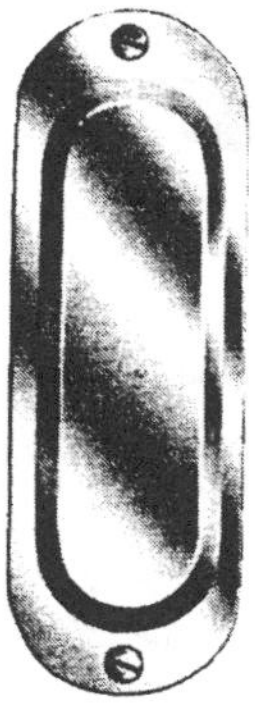

Size
½
¾
1
1¼-1½
2
2½-3

"OLK" Gaskets

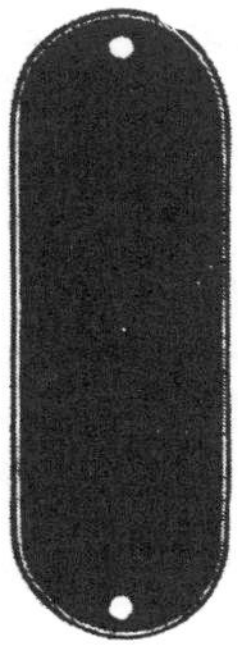

Size
½
¾
1
1¼-1½
2
2½-3

Cable Terminators

Cable Terminators and Sealing Fittings are used on conduit ends and cable ends to effectively seal the cable and conduit. They also provide a positive method for grounding lead sheath.

Where Installed:

- Wherever the jacket or sheath is removed from multi-conductor cable
- Wherever it is desirable that the ends of conduit and cables be sealed against the entrance of water, damp or corrosive atmospheres, hot or cold air or dust
- Wherever it is desirable to seal the entrance of exposed cable into cabinets, switchboards or terminal boxes

Specify:

- Size of conduit
- Number of conductors
- Diameter over insulation of individual conductors

For rubber or plastic insulated cables in conduit. Use with one or more single or multiple conductor cables

	Max. Diameter of Wire Permitted—Inches				Approx.
Conduit Size	1 Wire	1 Wires	2 Wires	3 Wires	Comp'd Req'd. Pints
1¼	1.02	.55	.50	44	⅛
1½	1.20	.63	.59	.51	⅛
2″	1.53	.81	.73	.65	¼
2½	1.83	.97	.93	.78	½
3″	2.28	1.21	1.16	.97	¾
3½	2.65	1.40	1.34	1.12	1
4″	3.00	1.58	1.52	1.27	1½
4½″	3.35	1.77	1.71	1.43	2
5″	3.75	1.99	1.91	1.60	2¾
6″	4.50	2.39	2.30	1.92	4

Insulating Bushings Type A

- Features: High impact thermosetting phenolic bushings, with a 150°C U.L. Temperature Rating, that will not soften melt or burn
- Standard Sizes: ½″ thru 6″
- Type A: Specification Bushing for threaded Rigid and IMC.
- Use: To insulate end of threaded rigid and IMC

TYPE A

Trade Size Inches	Wt. Per 100	Ctn.	Std. Pkg.
½	1.3	100	500
¾	1.8	100	500
1	3.5	50	200
1¼	6.0	25	125
1½	7.0	25	100
2	10.5	20	100
2½	14.4	10	40
3	19.0	10	40
3½	25.0	5	25
4	28.0	5	20
5	48.0	—	10
6	63.0	—	10

Insulating Bushing Type BB

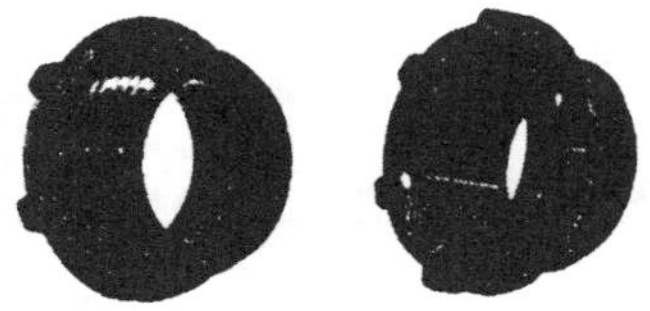

Use: To insulate cables passing through metal boxes or troughs.

TYPE BB

Trade Size Inches	Wt. Per 1100	Ctn.	Std. Pkg.
½	1.2	100	500
¾	2.2	50	500
1	3.1	50	200
1¼	5.3	25	125
1½	7.0	25	100
2	11.0	20	100
2½	19.0	10	50
3	23.0	10	50
3½	27.0	5	25
4	30.0	5	25
5	35.0	—	15
6	45.0	—	10

Beam Clamps

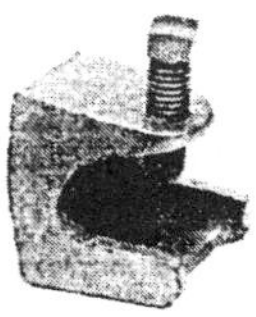

- Material/Finish: malleable iron, zinc plated. Steel Bolt—case hardened
- Optional finish: hot dip galvanized and/or mechanically galvanized
- Standard Sizes: ¾″ thru 2½″

Trade Size Inches	Wt. Per 100	Ctn.	Std. Pkg.
¾	12	50	250
1	17	25	100
1	22	25	100
1	23	25	100
1½	50	10	50
2	96	5	25
2½	168	5	25

Edge Type Hot Dip Galvanized

Designed to secure conduit or EMT perpendicular to a beam channel or angle support.

Trade Size Inches	Weight Per 100	Pkg.
½	65	50
¾	74	50
1	79	50
1¼	93	50
1½	100	50
2	132	50
2½	177	25
3	229	25

Parallel Type Hot Dip Galvanized

Designed to secure conduit or EMT parallel to a beam channel or angle support.

Trade Size Inches	Weight Per 100	Std. Pkg.
½	55	50
¾	59	50
1	65	50
1¼	72	50
1½	77	50
2	108	50
2½	155	25
3	184	25
3½	213	25
4	220	25

Right Angle Type Zinc Plated or Hot Dip Galvanized

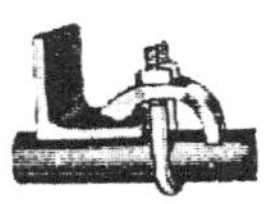

Designated to secure conduit or EMT across a beam, channel or angle support.

Trade Size Inches	Weight Per 100	Std. Pkg.
RIGHT ANGLE TYPE ZINC PLATED		
½	37	50
¾	41	50
1	45	50
1¼	52	50
1½	57	50
2	78	50
2½	90	25
3	101	25
3½	110	25
4	128	25
5	—	5
6	—	5

RIGHT ANGLE TYPE HOT DIP GALVANIZED

½	37	50
¾	41	50
1	45	50
1¼	52	50
1½	57	50
2	78	50
2½	90	25
3	101	25
3½	110	25
4	128	25
5	—	5
6	—	5

Cable and Accessories

Ty-Rap® Cable Ties

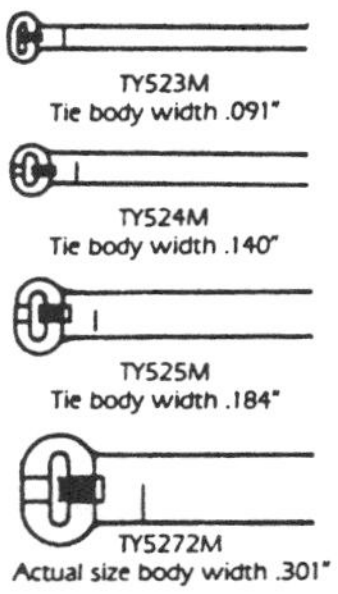

In the years since T&B introduced the first nylon cable ties for electrical wiring, hundreds of new products have been added making it the broadest line for electrical construction and industrial applications.

The TY-RAP® tying system incorporates a complete line of innovative products designed to increase productivity and lower the installed cost of wire bundling. These products include a full range of cable ties, identification ties, clamps, bases, harnessing aids and installing tools with both manual and automatic features. Utilization of the TY-RAP® tying system for all wiring jobs, large and small, will speed and improve harnessing and wire bundling while saving you money.

SELF-LOCKING TY-RAP® CABLE TIES

Reference No.	Dia. Range	Tie Body Width (in.)	Length (in.)	Min. Tensile Holding Strength (lbs.)	Unit Qty.	Std. Pkg.
TY523M	1/16"–5/8"	.091	3.62	18	100	1000
TY52315M	1/16"–1½"	.091	7.00	18	100	500
TY5232M	1/16"–2"	.091	8.00	18	100	500
TY5234M	1/16"–4"	.091	14.00	18	100	500
TY524M	1/16"–1⅛"	.140	5.50	40	100	1000
TY5242M	1/16"–2"	.140	8.19	40	100	500
TY526M	1/16"–3"	.140	11.08	40	100	500
TY5244M	1/16"–4"	.140	14.50	40	100	500
TY525M	1/16"–1¾"	.184	7.31	50	100	1000
TY5253M	1/16"–3"	.184	11.41	50	100	500
TY528M	1/16"–4"	.184	14.19	50	100	500
TY5272M	1/16"–2"	.301	8.00	120	50	250
TY527M	3/16"–3½"	.301	13.38	120	50	500
TY529M	3/16"–9"	.301	30.00	120	50	500

Ty-Rap Cable Straps and Clamps

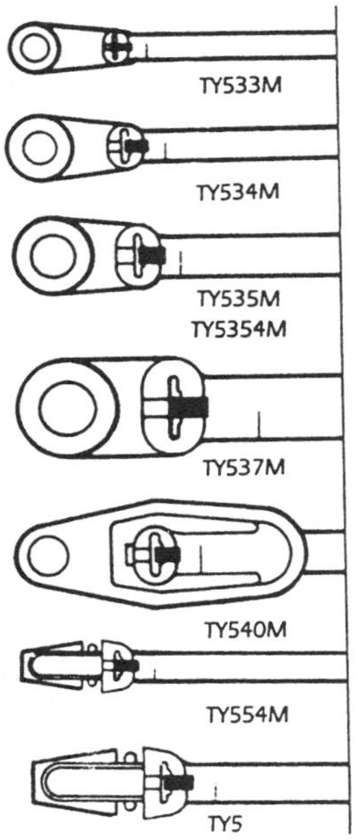

Because they're adjustable, TY-RAP clamps and straps eliminate a large inventory of plastic and metal fasteners. Various style mounting heads let you choose the method of fastening you desire: bolts, push-in, nails.

Reference No.	Wire Bundle Range	Holding Stacks	Mount Hole	Width	Length	Unit Qty.	Std. Pkg.
TY-533M	1/16"–5/8"	18	#4	.091"	4"	100	500
TY-534M	1/16"–1⅛"	40	#8	.140"	6"	100	500
TY-535M	13/16"–1¾"	50	#10	.184"	7.8"	100	500
TY-5354M	1 1/16"–4"	50	#10	.184"	14.6"	100	500
TY-537M	3/16"–3½"	120	¼"	.301"	13.9"	50	500
TY540M	3/16"–1½""	50	#10	.184"	8.5"	100	500
TY-554M	1/16"–1¼"	30	3/16"	.140"	5.96"	100	500
TY-538M	⅛"–1¾"	50	¼"	.184"	7.8"	100	500

Self-locking Ty-Rap® Identification Ties

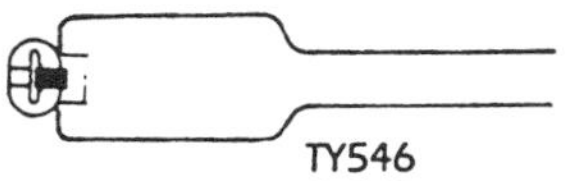

TY546
TY548

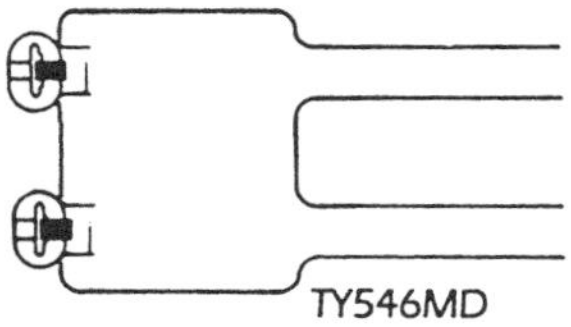

TY546MD

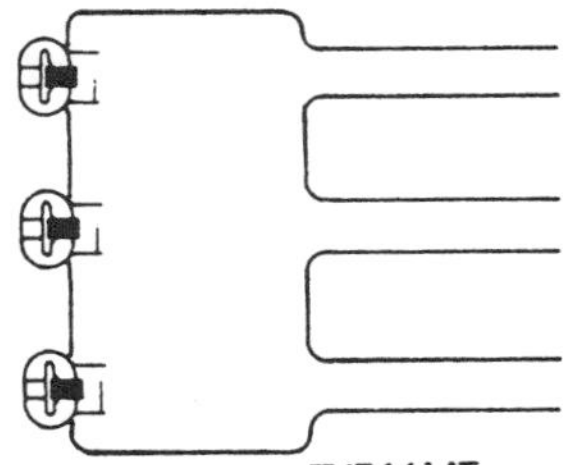

TY546MT

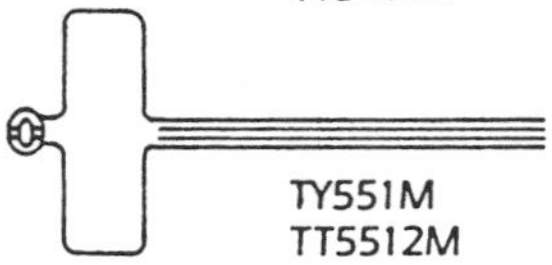

TY551M
TT5512M

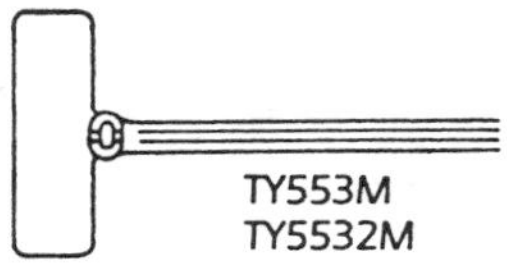

TY553M
TY5532M

Self-locking TY-RAP identification ties are available in single, double, and triple configurations. They may be marked with the special T & B black or red marking pen which produces a permanent legend, or they can be heat stamped. Because the ties serve the dual purpose of tying and identifying, you will achieve extra economies. They can be applied with the standard TY-RAP tying tools used for regular ties.

TY-RAP® IDENTIFICATION CABLE TIES

Reference No.	Wire Bundle	Marking Pad Size	Tensile Str.	Width	Length	Unit Qty.	Std. Pkg.	Wt. Per 100
TY546M	3/8"-1 3/4"	1/2" × 7/8"	50	.184"	7.25"	100	500	4
TY548M	3/4"-4"	1/2" × 2 1/8"	50	.184"	14.2"	100	500	5
T546MD	3/8"-1 3/4"	1 11/64" × 15/16"	50	.184"	7.25"	50	250	8
TY546MT	3/8"-1 3/4"	1 13/16" × 15/16"	50	.184"	7.25"	100	200	12
TY551M	3/16"-5/8"	1" × 5/16"	18	.091"	3.6"	50	500	3
TY5512M	3/16"-2"	1" × 5/16"	18	.091"	8.3"	100	500	4
TY553M	1/16"-5/8"	13/16" × 23/64"	18	.091"	4"	50	500	3
TY5532M*	1/16"-2"	13/16" × 23/64"	18	.091"	8.4"	100	500	4

RHH-RHW-USE

Okonite-Okolon is Okonite's trade name for its ethylene propylene-propylene, chlorosulfonated polyethylene composite insulation. The combination of the two materials provides a dielectric which requires no outer covering and has excellent resistance to heat, mechanical abuse, flame, weathering, most oils, acids and alkalies.

Okonite-Okolon 600 Volt Power and Control Cables are recommended for use in all low voltage circuits where continuity of service is the prime consideration. These cables may be installed in wet or dry locations, indoors or outdoors, in raceways, underground ducts, cable tray (size 250 kcmil and larger), directly buried in the earth, or lashed to a messenger for aerial installation.

Conductor Size - AWG or kcmil	No. of Strands	Composite Insulation Thickness - mils	Approx. O.D. (inches)	Approx. Ship Weight - lbs./M'	ICEA Ampacity	NEC Ampacity 75C Wet	NEC Ampacity 90C Dry
14	1	45	.16	27	24	20	25
14	7	45	.17	28	24	20	25
12	1	45	.18	36	30	25	30
12	7	45	.19	37	30	25	30
10	1	45	.20	50	42	35	40
10	7	45	.21	55	42	35	40
9	19	45	.23	65	48	35	40
8	7	60	.27	85	55	50	55
6	7	75	.34	135	75	65	75
4	7	75	.39	195	97	85	95
2	7	75	.45	285	130	115	130
1	19	100	.54	387	156	130	150
1/0	19	100	.58	460	179	150	170
2/0	19	100	.62	560	204	175	195
3/0	19	100	.68	696	242	200	225
4/0	19	100	.73	840	278	230	260
250	37	130	.84	1040	317	255	299
350	37	130	.94	1405	384	310	350
500	37	130	1.07	1950	477	380	430
750	61	145	1.28	2870	498	475	535
1000	61	145	1.45	3810	689	545	615

600 Volt Power and Control Tray Cable—Type FMR

Okonite-FMR® is Okonite's trade name for its heat, moisture, flame and chemical resistant, mechanically rugged ethylene-propylene insulating compound.

Okonite-FMR insulation is an evolutionary development of our ethylene-propylene rubber technology.

The new properties of Okonite-FMR insulation substantially enhance the well known features of ethylene-propylene rubber insulations.

The Okoseal® (PVC) jacket is mechanically rugged and has excellent resistance to most chemicals.

Okonite-FMR Okoseal Type TC tray cable is permitted for use on power, lighting, control, and signal circuits; indoors or outdoors; in cable trays, raceways, direct burial in the ground, or where supported in outdoor locations by a messenger wire; for Class I circuits as permitted in Article 725 of the NEC; and in cable trays in Class I, Division 2 hazardous locations in industrial establishments where the conditions of maintenance and supervision assure that only qualified persons will service the installation.

Conductor Size (AWG/ kcmil)	Number of Conductors	Approx. O.D. (in)	Approx. Ship Weight (lbs/ 1000′)	Tray Ampacity 90C Dry	Tray Ampacity 75C Wet	Ampacity 90C Dry	Ampacity 75C Wet
	2	0.39	100			25	20
	3	0.41	120			25	20
	4	0.45	145			20	16
	5	0.48	170			20	16
14(7x)	7	0.52	210	25	20		
	9	0.64	285			17.5	14
	12	0.72	355				
	19	0.83	505				
	37	1.15	950			15	12
	2	0.42	120			30	25
	3	0.45	150			30	25
	4	0.49	185			24	20
	5	0.53	220			24	20
12(7x)	7	0.61	295	30	25		
	9	0.71	375			21	17.5
	12	0.79	475				
	19	0.96	725				
	37	1.27	1305			18	15
	2	0.47	155			40	35
	3	0.50	200			40	35
	4	0.58	265			32	28
10(7x)	5	0.63	315	40	35	32	28
	7	0.68	405				
	9	0.79	515			28	24.5
	12	0.93	700				
	3	1.52	335			55	50
8(7x)	3	1.52	365	55	50		
	4	1.52	415			44	40
	4	1.52	450				

Type PLTC Instrumentation Cable

Okonite Type P-OS (Pair/triads—Overall Shield) instrumentation cables are designed for use as instrumentation, process control and computer cables in Class II or III Power-Limited Circuits as defined in NEC Article 725. They are suitable for installation in wet or dry locations with conductor operating temperature up to 105C. They may be installed in cable tray, in any raceway, as open runs of cable. UL Listed Type PLTC is authorized for use above grade in Class I, Division 2 hazardous locations.

The overall shield eliminates most of the static interference from the electric field radiated by power cables and other electrical equipment.

For dc service in wet locations X-Olene insulation having an overall aluminum C-L-X armor construction is recommended.

Conductors: Bare soft annealed copper, Class B, 7-strand concentric per ASTM B-8.
Insulation: Flame-retardant Okoseal (PVC) per UL Subject 13, 15 mils nominal thickness, 105C temperature rating.
Cable Shield: 1.35 mil blue double faced aluminum/synthetic polymer backed tape overlapped to provide 100% coverage, and a 7-strand tinned copper drain wire, two sizes smaller than the conductor.
Jacket: Black, flame-retardant Okoseal per UL Subject 13. A rip cord is laid longitudinally under the jacket to facilitate removal.
UL Listed Type PLTC (Power-Limited Tray Cable) and Power-Limited Circuit Cable for use in Class II or III circuits in accordance with Article 725 of the National Electrical Code.

Size (Awg)	Number of Pairs	Number of Triads	Cable O.D. (Inches)	Ship Weight - (Lbs./1000′)
22	1		.20	23
22		1	.21	28
20	1		.22	28
20		1	.23	34
18	1		.24	34
18		1	.25	42
16	1		.26	43
16		1	.28	55

PLTC Armored Instrumentation Cable Multiple Pairs or Triads

MULTIPLE PAIRS OR TRIADS—OVERALL SHIELD; 300 VOLTS, 105C RATING (FOR CABLE TRAY USE)

Strand Size (AWG)	No. of Pairs	No. of Triads	Cable O.D. (in.)	Approx. Ship Weight (lbs/1000′)
	2		0.68	220
	4		0.76	295
	6		0.85	345
	8		0.90	380
	10		0.95	420
	12		0.99	495
		4	0.76	320
		8	0.95	435
20(7x)		12	1.13	595
		16	1.21	700
		24	1.48	1035
	2		0.76	290
	4		0.76	325
	6		0.90	390
	8		0.99	490
	10		1.08	550
	12		1.08	580
	16		1.17	665
	20		1.21	750
18(7x)	24		1.40	935
		4	0.85	375
		8	1.08	575
		12	1.21	725
		16	1.35	920
		24	1.65	1285
	2		0.76	310
	4		0.85	380
	6		1.03	520
	8		1.08	580
	10		1.17	650
	12		1.21	720
	16		1.26	820
	20		1.40	1020
16(7x)	24		1.53	1210
		4	0.95	450
		8	1.17	695
		12	1.40	980
		16	1.53	1205
		24	1.82	1705
		36	2.05	2335

	2		0.49	135
	4		0.54	205
	6		0.67	300
	8		0.72	360
	10		0.84	455
	12		0.87	510
	16		0.98	685
	20		1.09	815
16(7x)	24		1.21	955
		4	0.60	260
		8	0.80	480
		12	0.99	705
		16	1.09	875
		24	1.37	1285

PLTC Armored Instrumentation Cable Single Pair or Triad

C-L-X Type P-OS (Pair/triad—Overall Shield) instrumentation cables are designed for use as instrumentation, process control and computer cables in Class II or III Power-Limited Circuits as defined in NEC Article 725. They are suitable for installation in wet or dry locations with conductor operating temperature up to 105C. They may be installed in cable tray, in any raceway, as open runs of cable. UL Type PLTC is authorized for use above grade in Class I, Division 2 hazardous locations.

The overall shield eliminates most of the static interference from the electric field radiated by power cables and other electrical equipment. The C-L-X sheath provides the physical protection against mechanical damage as required in NEC Section 725-18 as well as complete protection against moisture or gases entering the cable. For dc service in wet locations X-Olene insulation having an overall aluminum C-L-X armor construction is recommended.

Number of Pairs	Number of Triads	Conductors: 16 AWG Inner Jacket O.D. (Inches)	C-L-X O.D. (Inches)	Cable O.D. (Inches)	Ship Weight - Lbs./1000′
1		.26	.49	.60	210
	1	.28	.49	.60	217

Shielded Pairs—Overall Shield Type PLTC Instrumentation Cable

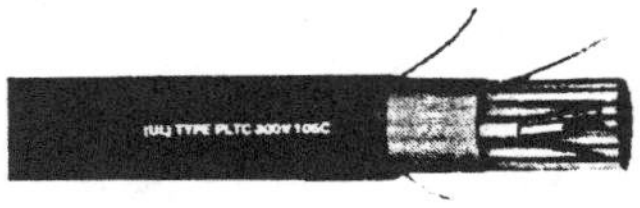

Okonite Type SP-OS (Shielded Pairs/triads - Overall Shield) instrumentation cables are designed for use as instrumentation, process control and computer cables. They are suitable for installation in wet or dry locations with conductor operating temperature up to 105C. They may be installed in cable tray, in any raceway, as open runs of cable. UL Type PLTC is authorized for use above grade in Class I, Division 2 hazardous locations.

The isolated individual shields over each pair, when properly grounded, prevent crosstalk or capacitive coupling between adjacent pairs which occurs with ac signals, particularly the pulse type. The overall shield eliminates most of the static interference from the electrical field radiated by power cables and other electrical equipment. For dc service in wet locations X-Olene insulation having an aluminum C-L-X armor construction is recommended.

Strand Size (AWG)	No. in Pairs	No. of Triads	Cable O.D. (in.)	Approx. Ship Weight (lbs/1000′)
	2		0.38	80
	4		0.44	115
	6		0.52	165
	8		0.56	195
	10		0.67	255
	12		0.70	285
	16		0.77	365
	20		0.85	430
20(7x)	24		0.96	545
		4	0.49	145
		8	0.62	250
		12	0.77	375
		16	0.85	460
		24	1.07	680
	2		0.44	105
	4		0.49	150
	6		0.58	205
	8		0.65	275
	10		0.75	350
	12		0.78	390
	16		0.86	475
	20		0.97	615
18(7x)	24		1.08	705
		4	0.54	195
		8	0.71	345
		12	0.86	490
		16	0.97	655
		24	1.20	910

Shielded Pairs—Overall Shield CLX Type PLTC Instrumentation Cable

C-L-X Type SP-OS (Shielded Pairs/triads—Overall Shield) instrumentation cables are designed for use as instrumentation, process control and computer cables. They are suitable for installation in wet or dry locations with conductor operating temperature up to 105C. They may be installed in cable tray, in any raceway, as open runs of cable. UL Type PLTC is authorized for use above grade in Class I, Division 2 hazardous location.

The isolated individual shields over each pair, when properly grounded, prevent crosstalk or capacitive coupling between adjacent pairs which occurs with ac signals, particularly the pulse type.

The overall shield eliminates most of the static interference from the electric field radiated by power cables and other electrical equipment.

The C-L-X sheath provides the physical protection against mechanical damage as required in NEC Section 725-18 as well as complete protection against moisture or gases entering the cable.

For dc service in wet locations X-Olene insulation having an overall aluminum C-L-X armor construction is recommended.

Strand Size (AWG)	No. of Pairs	No. of Triads	Cable O.D. (in.)	Approx. Ship Weight (lbs/1000′)
	2		0.72	240
	4		0.76	310
	6		0.85	365
	8		0.90	405
	10		1.03	515
	12		1.08	555
	16		1.17	635
	20		1.21	715
20(7x)	24		1.40	890
		4	0.76	335
		8	0.99	500
		12	1.17	645
		16	1.21	745
		24	1.48	1105
	2		0.78	300
	4		0.82	340
	6		0.95	425
	8		0.99	525
	10		1.13	605
	12		1.17	655
	16		1.26	770
	20		1.40	960
18(7x)	24		1.48	1135
		4	0.90	405
		8	1.08	610
		12	1.26	785
		16	1.40	1095
		24	1.65	1385
	2		0.76	325
	4		0.90	415
	6		1.03	555
	8		1.08	630
	10		1.21	740
	12		1.26	810
	16		1.40	1035
	20		1.53	1245
16(7x)	24		1.69	1515

Bare Copper Conductor

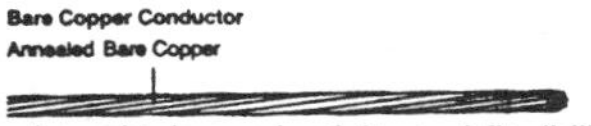

Copper conductors are used in overhead electrical transmission and distribution systems as well as for grounding electrical systems, where high conductivity and flexibility is required.

Solid bare copper wires for use as feed stock for manufactured products are available in various grades, tempers and sizes.

Conductors

Bare copper conductors, solid or stranded, are available in various tempers and sizes.

Soft Temper

Solid ASTM B2: 26 AWG—4/0 AWG; Stranded ASTM B8: 8AWG—2000 MCM

Medium Hard Temper

Solid ASTM B2: 18 AWG—4/0 AWG; Stranded ASTM B8: 8 AWG—2000 MCM

Hard Temper

Solid ASTM B1: 26 AWG—4/0 AWG; Stranded ASTM B8: 8 AWG—2000 MCM.

Coated copper conductors, solid or stranded, are available in various sizes but in soft temper only.

Tinned ASTM B33

Alloy Coated ASTM B189

Size AWG or MCM	Number of Strands	Net Weight Lbs./MFT
26	Solid	.765
25	Solid	.970
24	Solid	1.22
23	Solid	1.55
22	Solid	1.94
21	Solid	2.46
20	Solid	3.10
19	Solid	3.90
18	Solid	4.92
17	Solid	6.21
16	Solid	7.81
15	Solid	9.87
14	Solid	12.40
13	Solid	15.69
12	Solid	19.80
11	Solid	24.90
10	Solid	31.43
9	Solid	39.61
8	Solid	49.98
8	7	50.97
6	Solid	79.44
6	7	81.05
5	Solid	100.2
5	7	102.2
4	Solid	126.3
4	3	127.6
4	7	128.9
3	Solid	159.3
3	3	160.9
3	7	162.5
2	Solid	200.9
2	3	202.9
2	7	204.9
1	Solid	253.3
1	3	255.9
1	7	258.4
1/0	Solid	319.5
1/0	7	325.8
1/0	12	325.8
1/0	19	325.8
2/0	Solid	402.8
2/0	7	410.9
2/0	12	410.9
2/0	19	410.9
3/0	Solid	507.8
3/0	7	518.1
3/0	12	518.1
3/0	19	518.1
4/0	Solid	640.5
4/0	7	653.3
4/0	12	653.3
4/0	19	653.3
250	12	771.9
250	19	771.9
250	37	771.9
300	12	926.3
300	19	926.3
300	37	926.3
350	12	1081
350	19	1081
350	37	1081
400	19	1235
400	37	1235
450	19	1389
450	37	1389
500	19	1544
500	37	1544
600	37	1853
600	61	1853
700	37	2161
700	61	2161
750	37	2316
750	61	2316
800	37	2470
800	61	2470
900	37	2779
900	61	2779
1000	37	3088
1000	61	3088
1250	61	3859
1250	91	3859
1500	61	4631
1500	91	4631
1750	91	5403
2000	91	6175

XHHW—600V—90°C

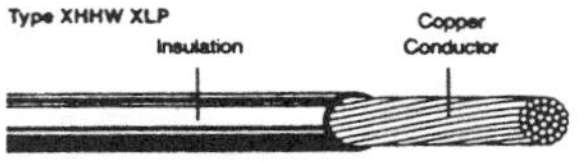

Anaconda Type XHHW wire may be used for power or control circuits, in recognized raceways, in wet or dry locations and in the presence of oil, in accordance with the National Electrical Code (NEC).

Bare copper per ASTM B8: sizes 14 AWG—1000 MCM

Insulation—Cross-linked polyethylene, thermosetting XOP.

Temperature ratings: Wet, 75C; Dry, 90C

Colors—1. Black; 2. White; 3. Red; 4. Blue; 5. Yellow; 6. Green; 7. Orange; 8. Brown

Size AWG or MCM	Number of Strands	Overall Diameter (in.)	Net Weight Lbs./MFT	Stock Colors
14	7	0.14	19	8
12	7	0.16	29	8
10	7	0.18	45	7
8	7	0.24	73	3
6	7	0.28	99	1
4	7	0.33	152	1
2	7	0.39	234	1
1	19	0.45	296	1
1/0	19	0.49	368	1
2/0	19	0.54	459	1
3/0	19	0.59	573	1
4/0	19	0.65	716	1
250	37	0.71	847	1
300	37	0.77	1009	1
350	37	0.82	1171	1
400	37	0.85	1322	1
500	37	0.95	1653	1
600	61	1.04	1974	1
750	61	1.16	2465	1
1000	61	1.31	3260	1

Type THHN/THWN

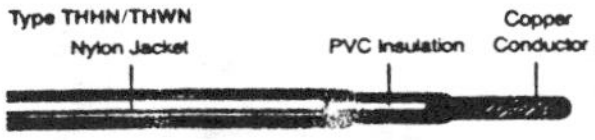

PVC Nylon; 600 Volts

14-10 AWG solid, 14 AWG-1000 MCM stranded bare copper conductor, PVC insulation, nylon jacket.

Maximum Conductor Temperature: Continuous Operation

THHN 90°C dry THWN 75°C wet or dry

MTW 90°C dry 60°C in oil

AWM 105°C dry 80°C in oil

Features—UL listed as THWN or THHN for general purpose wiring. Small diameter. Pulls easily into conduit. Excellent oil, gasoline, chemical resistance. 250 MCM and larger pass ribbon burner flame test—listed For CT Use. Stranded items UL listed as MTW.

Applications—General purpose wiring in conduit in accordance with the NEC. Stranded cable suitable for use as Machine Tool Wire. Sizes 250 MCM and larger suitable for use in cable trays.

Size AWG	No. of Strands	Ampacity 75°C	Ampacity 90°C	Overall Dia. Inches	Net Weight Lbs/Mft
14	Solid	15	15	.107	16
14	19	15	15	.116	17
12	Solid	20	20	.124	24
12	19	20	20	.135	25
10	Solid	30	30	.155	38
10	19	30	30	.169	40
8	Solid	45	50	.203	61
8	19	45	50	.222	65
6	19	65	70	.27	98
4	19	85	90	.34	157
2	19	115	120	.40	240
1	19	130	140	.46	305
1/0	19	150	155	.50	379
2/0	19	175	185	.55	470
3/0	19	200	210	.60	585
4/0	19	230	235	.66	726
250	37	255	270	.73	865
300	37	285	300	.78	1030
350	37	310	325	.83	1190
400	37	335	360	.88	1355
500	37	380	405	.97	1680
750	61	475	500	1.17	2500

Thermocouple Extension Cable

Single and Multiple Pairs—Overall Shield
300 Volts 105°C Rating Type PLTC

Multiple Shielded Pairs

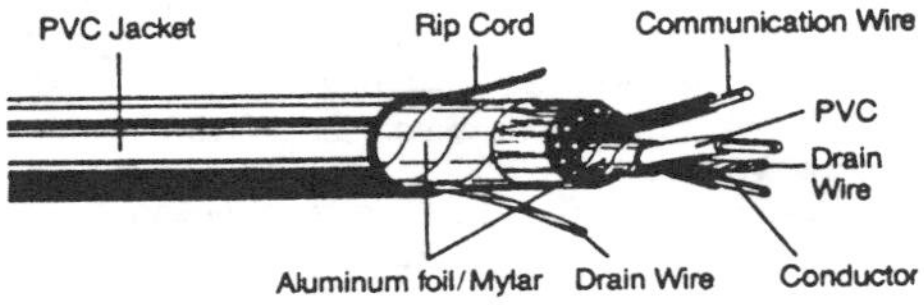

Overall shielded thermocouple extension cables are for use on circuits where shielding from ambient interference is required but shielding between pairs is not essential. They are suitable for installation in wet or dry locations and conductor temperatures up to 105° centigrade. They may be installed in cable tray, rigid metal conduit, intermediate metallic conduit, electrical metallic conduit (EMT), or other raceways as approved by the appropriate authority.

ANSI Type	Alloys Positive	Alloys Negative	Insulation Colors Positive	Insulation Colors Negative	Jacket Color	Temperature Range C	Limits of Error
TYPE DESIGNATIONS							
EX	Chromel	Constantan	Purple	Red	Purple	0 to +200	±1.7°C
JX	Iron	Constantan	White	Red	Black	0 to +200	±2.2°C
KX	Chromel	Alumel	Yellow	Red	Yellow	0 to +200	±2.2°C
TX	Copper	Constantan	Blue	Red	Blue	60 to +100	±1.0°C

Note: The red (negative) in each pair is sequentially numbered for identification.

ANSI Type	Size AWG	No. Pairs	Jacket Mils	Overall Diameter
EX	20	4	40	0.360
EX	20	8	50	0.460
EX	20	12	50	0.540
EX	20	24	60	.780
EX	16	1	35	0.250
JX	20	4	40	0.360
JX	20	8	50	0.470
JX	20	12	50	0.540
JX	20	16	60	0.620
JX	20	24	60	0.730
JX	20	36	70	0.840
JX	16	1	35	0.250
KX	20	8	50	0.470
KX	20	12	50	0.540
KX	20	16	60	0.620
KX	20	24	60	0.730
KX	20	36	70	0.840
KX	16	1	35	0.250

Instrumentation Cable

Single Pair or Triad Shielded
300 Volt 105°C Power Limited Tray Cable

Single Pair or Triad—Shielded

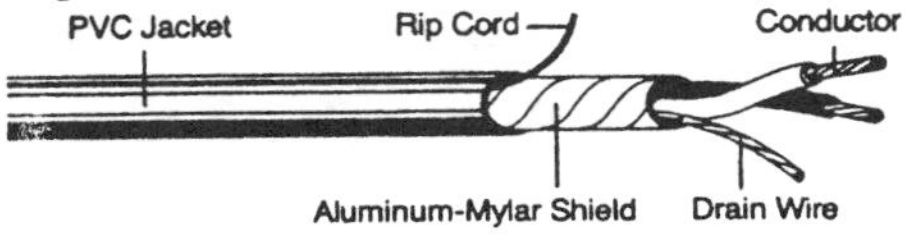

Applications

Overall shielded instrumentation cables are for use in instrumentation, computer, and control applications where signals are transmitted in excess of 100 millivolts except in areas where exceptional high voltage or current interferes. They are suitable for installation in wet or dry locations and conductor temperatures up to 105° centigrade. They may be installed in cable tray, rigid metal conduit, intermediate metallic conduit, electrical metallic tubing (EMT), or other raceways as approved by the appropriate authority.

Size AWG	No. of Conductors	Overall Diameter Inches
20	2	.230
18	2	.250
18	3	.260
16	2	.265
16	3	.280

Coaxial Cables

Description	Trade & UL Type Number	AWG (Stranding) [Dia. in mm] Nom. D.C.R.	Nominal O.D. Inch	No. of Shields and Material Nom. D.C.R.	Nom. Imp. (ohms)	Nominal Capacitance pF/ft.	Nominal Attenuation MHz	db/100 ft.
58/U JAN-C-17A	8240 60C	20 (Solid) [.81] bare 10.1Ω/M' 33.1Ω/km	195	1 tinned copper 4.1Ω/M 95% shield coverage	53.5	28.5 Black PVC jacket	200	6.8
59/U Type 100% Sweep Tested 5-300 MHz	9259 1354 60C	22 (7×30) [.76] Bare copper 15.0Ω/M' 49.2Ω/km	.242	1 bare copper 2.6Ω/M' 95% shield coverage	75	17.3 Black PVC jacket. For CCTV applications.	200	4.5
62A/U Type	9269 1478 60C	22 (Solid) [.64] bare copper covered steel 41.2Ω/M' 135.2Ω/km	.242	1 bare copper 2.6Ω/M' 95% shield coverage	93	13.5 Black PVC jacket.	200	3.8

Fiber Optic Cables

In all types of cable structures, the individual optical fibers are the signal transmission media acting much the same as individual optical wave guides. The fibers have an all dielectric structure consisting of a central circular transparent core region which propagates the optical radiation and an outer cladding layer that completes the guiding structure. For low loss transmission the core is typically silica glass while the cladding may be glass or polymer material (PCS, polymer clad silica). To achieve high signal bandwidth capabilities, the core region has a varying or graded refractive index.

No. of Fibers	Bandwidth MHz-km	Outer Diameter Inch	Wt. Lbs./1000'	Minimum Bend Radius Inches Installation	Minimum Bend Radius Inches Long-Term Application	Strength Member	Outer Jacket
300 MICRON/440 MICRON (CORE/CLAD) PVC JACKET							
1	20	.150	8	4	2	Kevlar	PVC
2	20	.150 × .311	16	4	2	Kevlar	PVC
4	20	.315	30	6	4	Kevlar	PVC
6	20	.315	30	6	4	Kevlar	PVC
8	20	.394	52	7	5	Kevlar	PVC
10	20	.551	107	8	6	Kevlar	PVC
12	20	.551	107	8	6	Kevlar	PVC
18	20	.551	107	8	6	Kevlar	PVC
200 MICRON/380 MICRON (CORE/CLAD) PVC JACKET							
1	25	.150	8	4	2	Kevlar	PVC
2	25	.150 × .311	16	4	2	Kevlar	PVC
4	25	.315	30	6	4	Kevlar	PVC
6	25	.315	30	6	4	Kevlar	PVC
8	25	.394	52	7	5	Kevlar	PVC
10	25	.551	107	8	6	Kevlar	PVC
12	25	.551	107	8	6	Kevlar	PVC
18	25	.551	107	8	6	Kevlar	PVC

Electrical Supports

Aluminum Ladder—4″ Deep

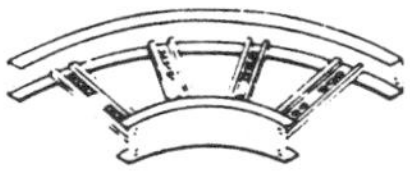
Horizontal 90° Elbow

Vertical Inside 90° Elbow

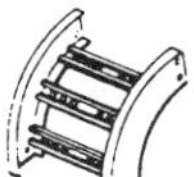
Vertical Outside 90° Elbow

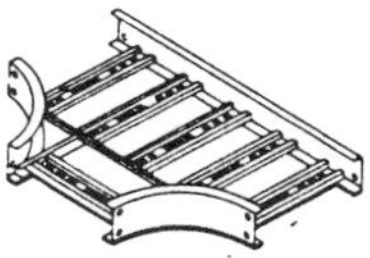
Horizontal Tee

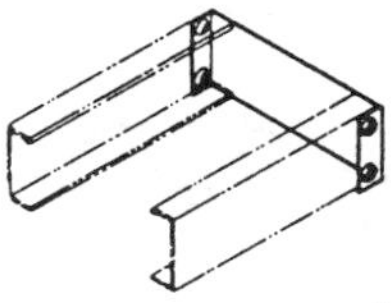
End Plate

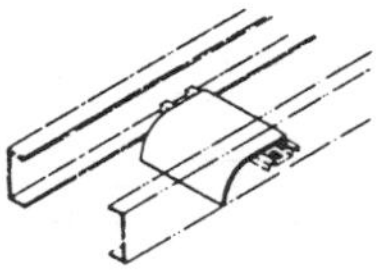
Dropout

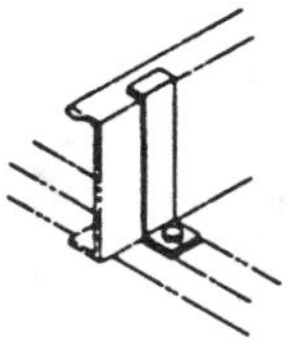
Z-Type Hold Down Clip

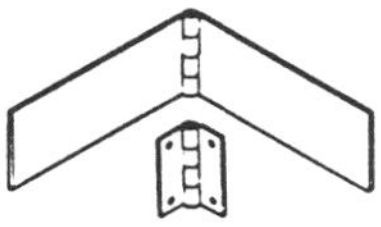
Horizontal Adj. Splice

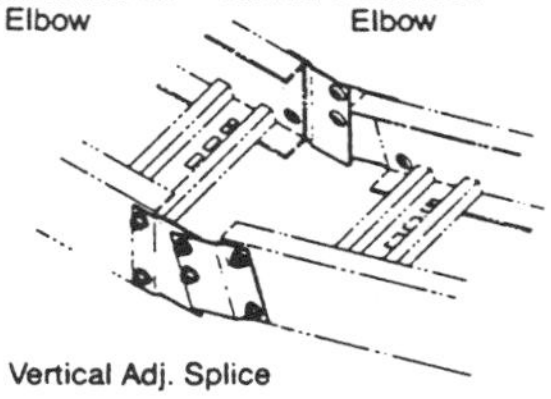
Vertical Adj. Splice

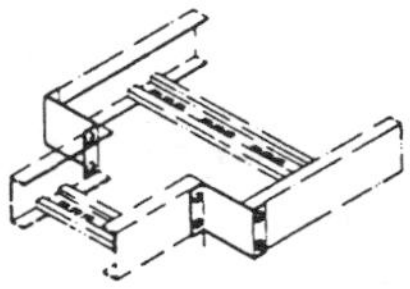
Reducing Splice

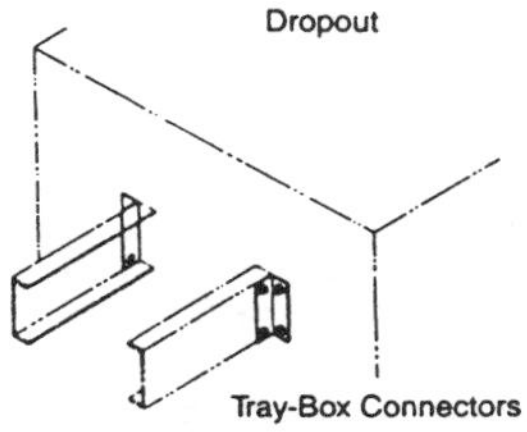
Tray-Box Connectors

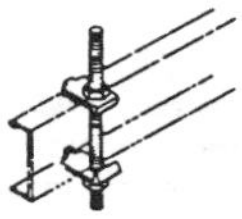
Tray Hanger Clamp

Aluminum ladder cable tray with nine inch "dougle rung" spacing provides support for many types of cables. These rugged, lightweight trays offer four inch cable fill depth and excellent load carrying capabilities. (NEMA Class 12B for cable loads up to 75 lbs.; Class 12C for loads up to 100 lbs.) Fittings are available in 12 & 24 in. radius. Splices furnished with all tray sections.

SPLICES & ACCESSORIES

Description
Standard Splice-Extra Pair
Horizontal Adjustable Splice—Set
Vertical Adjustable Splice—Set
6" Reducing Splice
12" Reducing Splice
Tray-Box Connector—Pair
Z-Type Hold Down Clip—Pair
Square Hold Down Clip—Pair
Hanger Clamp—Pair

TRAY AND FITTINGS

Description	12" Wide Tray	24" Wide Tray
Straight Section		
Class 12C Ladder		
Class 12B Ladder		
Horizontal 90° Elbow		
12" Radius		
24" Radius		
Vertical Outside 90°		
12" Radius		
24" Radius		
Vertical Inside 90°		
12" Radius		
24" Radius		
Tee Fitting		
12" Radius		
24" Radius		
Accessories		
End Plate		
Dropout		

Cable Tray Selector Chart

This selector chart shows the full line of Square D cable tray products. The maximum cable load (in lbs/linear ft.) is given for each tray when installed on various support spans.

Overall Height	Load Depth	Support Span (Ft.) 6	8	10	12	14	16	18	20	Max. NEMA Class	
ALUMINUM CABLE TRAYS											
3⅝"	3"	200#	113.	72.	50#					12A	Ladder Type Trough Type Solid Bottom
5⅝"†	4"†	340#	191.	122.	85#					12B†	Ladder Type† Trough Type Solid Bottom
4⅝"†	4"†	400#	225.	144.	100#					12C†	Ladder Type† Trough Type Solid Bottom
4⅝"	4"				139#	102.	78.	62	50#	20A	Ladder Type Trough Type Solid Bottom
6"	5⅜"	300#	170.	108#	75#					12B	Ladder Type Trough Type Solid Bottom
6"	5⅜"				147#	108.	83.	65.	53#	20A	Ladder Type Trough Type Solid Bottom
6"	5⅜"				214#	157.	120.	95.	77#	20B	Ladder Type Trough Type Solid Bottom

Overall Height	Load Depth	Support Span (Ft.) 6	8	10	12	14	16	18	20	Max. NEMA Class	
STEEL CABLE TRAYS											
3⅝"	3"	204#	115.	73.	51#					12A	
4⅝"	4"	300#	170.	108.	75#					12B	Ladder Type Trough Type Solid Bottom
4⅝"	4"				139#	102.	78.	62.	50#	20A	Ladder Type Trough Type Solid Bottom
6"	5⅜"	332#	187.	120.	83#					12B	Ladder Type Trough Type Solid Bottom
6"	5⅜"				231#	169.	130.	102.	83#	20B	Ladder Type Trough Type Solid Bottom
CHANNEL TRAYS											
−4⅝"	+ 6" W.—										Aluminum Type Galvanized Steel

Trapeze Hangers

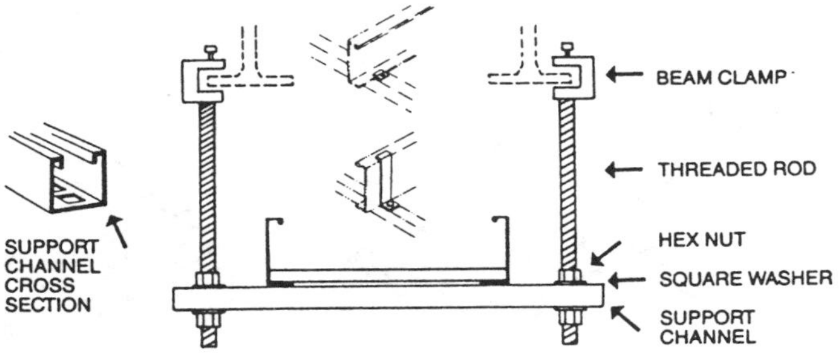

The most versatile supports for cable trays. The main element is the mill-galvanized steel slotted Support Channel. It is supplied as pre-cut sections or in 10 ft. lengths for fieldcutting. All elements sized for ½ in. hardware. All components ordered separately.

COMPONENTS

Description
Beam Clamp-(Rated 600#)
Threaded Rod-6 ft. lg.
Threaded Rod-10 ft. lg.
Hex Nut-100 pcs.
Sq. Washer-100 pcs.
Bolt 1½" lg. 100 pcs.
Strut Nut-100 pcs.
Pre-Cut Channels:
For 12" tray-20" long
For 24" tray-32" long
Full Length-10 ft. long

continued

Units can easily be assembled by the user; factory assembly, custom colors, words and private labeling are accepted on orders of sufficient quantity.

Applications

Visalert units are recommended for use in industrial plants, hospitals, schools, apartment or office buildings, and any other situation where effective visual warning is needed. All models are UL-listed for indoor applications; and model VALS is listed for outdoor use when used with the gasket supplied. Units can be flush- or surface-mounted on 350 and 450 horns, or 50GC speakers. Models VAL and VALF can also be mounted on 4" bells with appropriate mounting accessories.

SPECIFICATIONS

Voltage		Current	Flash Energy
*VAL-VALF	6 VDC high power	500mA	—
	12 VDC high power	700mA	—
	24 VDC high power	340mA	—
*VALS	24 VDC high power	100mA	1.60 candela-sec.
VALS	12 VDC low power	75mA	.25 candela-sec.
*VALS	120 VAC high power	30mA	1.60 candela-sec.

Voltage		Current (Amps)	dB level
18-31 VDC	High-power strobe With 0.8W horn	.135	90 dB min.
	With 1.5W horn	.163	93 dB min.
	With 6W horn	.350	99 dB min.
120 VAC	High-power horn/ strobe	.210	100 dB min.

Square Channel Washer

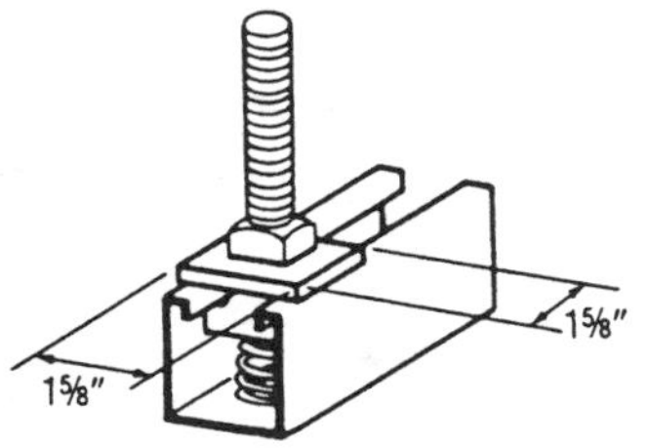

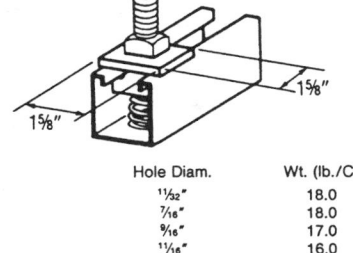

Hole Diam.	Wt. (lb./C)
11/32"	18.0
7/16"	18.0
9/16"	17.0
11/16"	16.0
13/16"	16.0

Globe Strut

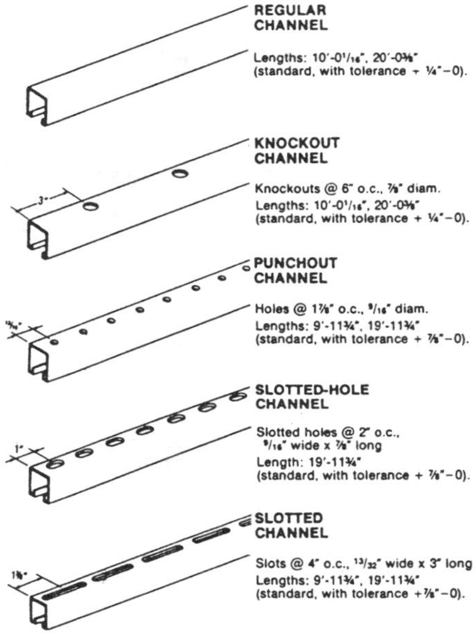

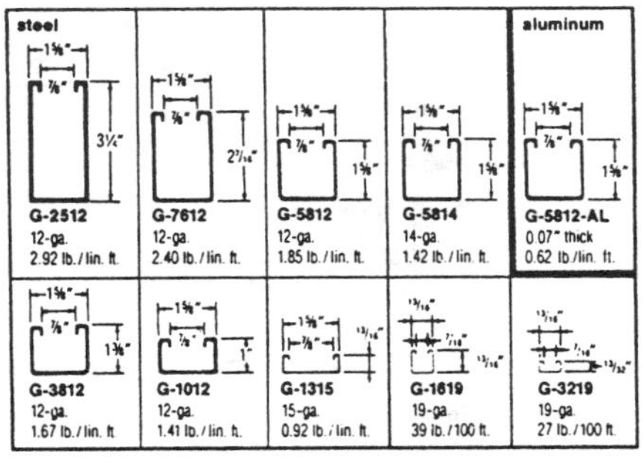

Globe Strut Channels are available in either a basic solid continuous type or six variations—combination, slotted, slotted-hole, double-slotted, knockout, punchout and concrete inserts—to meet specific uses. A specially designed lock nut and cap screw is used with appropriate fittings to assemble channels in any position or in any combination with other channel types.

Conduit Clamps

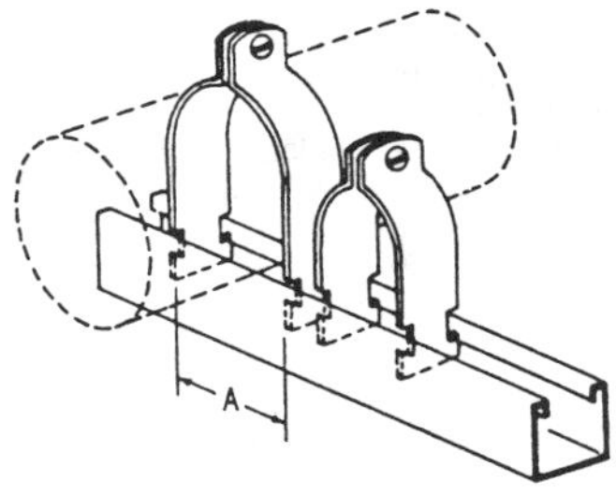

Clamps for Standard Pipe and Rigid Conduit

STEEL AND ALUMINUM CLAMPS

Nom. Pipe Size (I.D.)	Dim. A	Wt. (lb./C) Steel	Alum.
⅜"	675"	10.0	—
½"	840"	11.0	5.0
¾"	1.050"	15.0	6.0
1"	1.315"	16.0	7.0
1¼"	1.660"	21.0	8.0
1½"	1.900"	32.0	12.0
2"	2.375"	34.0	14.0
2½"	2.875"	36.0	16.0
3"	3.500"	51.0	17.0
3½"	4.000"	57.0	19.0
4"	4.500"	61.0	21.0
5"	5.563"	75.0	23.0
6"	6.625"	95.0	26.0
7"	7.625"	110.0	—
8"	8.625"	117.0	—

Clamps for Thin Wall Conduit

STEEL CLAMPS

Nom. Pipe Size (I.D.)	Dimension A (in.)	Wt. (lb./C)
⅜"	.577	11.0
½"	.706	12.0
¾"	.922	12.0
1"	1.163	17.0
1¼"	1.510	18.0
1½"	1.740	32.0
2"	2.197	33.0

Standard/Universal Channel Lock Nuts

Thread Size (diam.–no./in.)	Wt. (lb./C) Without Spring G-7612, G-5814 G-5812, G-3814	G-1012 G-1315	Universal (of 1½", 1⅝" widths)		With Spring G-5812/14 G-3812	G-1012 G-1315	Universal (of 1½", 1⅝" widths)
#10-24	6.5	—	—	—	7.3	6.3	—
¼"-20	6.5	—	6.0	6.0	7.3	7.3	6.5
5/16"-18	6.2	—	—	—	10.3	10.3	—
⅜"-16	9.3	—	7.5	7.6	9.6	9.2	8.0
½"-13	11.3	8.0	8.0	12.3	11.3	8.3	8.5
⅝"-11	8.7	10.3	—	14.8	13.5	10.3	—
¾"-10	11.3	9.0	—	13.3	12.6	9.3	—

Channel Splice

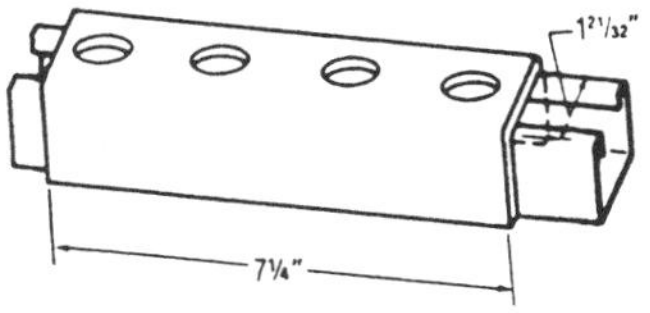

	Wt. (lb./C)
2-HOLE SPLICE CHANNELS	
	121.0
	76.0
3-HOLE SPLICE CHANNELS	
	116.0
	174.0
4-HOLE SPLICE CHANNELS	
	344.0
	233.0
	157.0

90° Angle Fittings

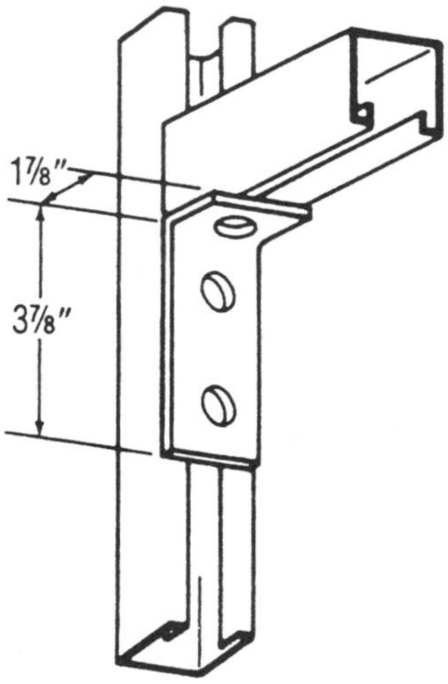

3 Hole Corner Angle
4 Hole Corner Angle
3 Hole Shelf Gusset Angle
4 Hole Webbed Corner Angle

Pierced-Angle Beam Clamps

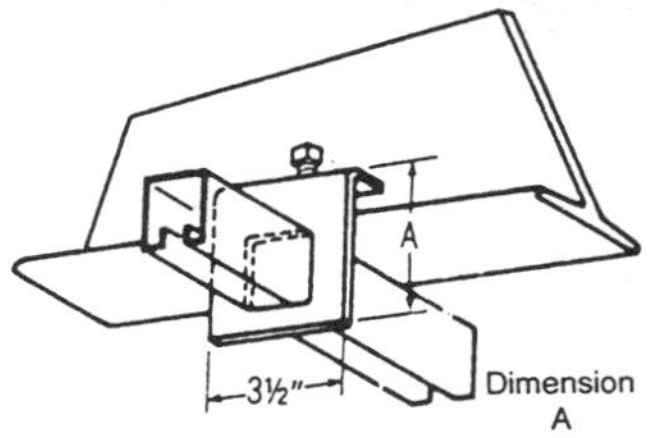

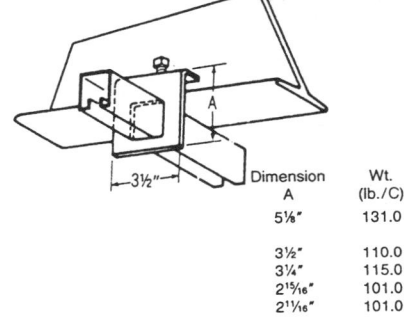

Dimension A	Wt. (lb./C)
5⅛"	131.0
3½"	110.0
3¼"	115.0
2¹⁵⁄₁₆"	101.0
2¹¹⁄₁₆"	101.0

U-Bolt Clamps

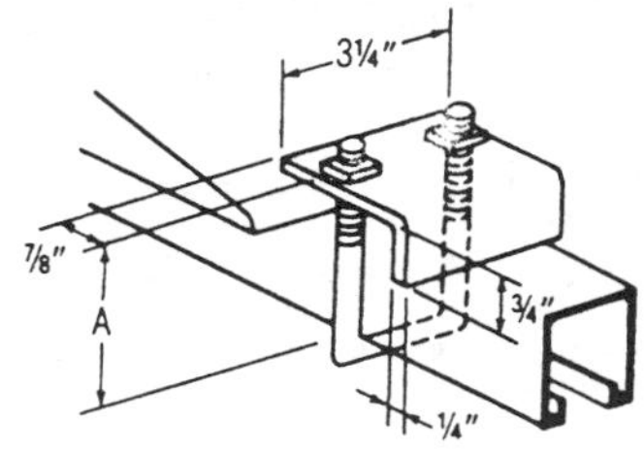

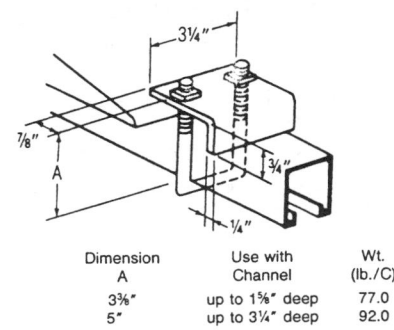

Dimension A	Use with Channel	Wt. (lb./C)
3⅜"	up to 1⅝" deep	77.0
5"	up to 3¼" deep	92.0

Hardware

Size (diam. × lgth.)	Wt. (lb./C)
HEX HEAD CAP SCREWS	
¼ × ⅝"	1.4
¼ × ¾"	1.4
¼ × 1"	2.0
⅜ × ¾"	3.4
⅜ × 1"	4.1
⅜ × 1¼"	5.3
⅜ × 1½"	6.0
⅜ × 2"	7.3
⅜ × 2¼"	8.0
½ × 1"	8.3
½ × 1¼"	9.3
½ × 1½"	11.0
½ × 1¾"	12.4
½ × 2"	13.3
½ × 2¼"	14.9
½ × 2½"	16.3
5/16 × 1¾"	3.6
SLOTTED HEX HEAD CAP SCREWS	
¼ × ¾"	1.7
5/16 × 1"	2.6
5/16 × 1¼"	3.0
⅜ × 1¼"	5.3
FLAT HEAD MACHINE SCREWS	
¼ × ⅝"	1.0
¼ × ¾"	1.5
⅜ × 2"	6.0
SET SCREWS	
⅜ × 1½"	4.3
⅜ × 2"	6.1
½ × 1½"	8.5
½ × 2"	10.3
ROUND HEAD MACHINE SCREWS	
¼ × ½"	1.0
¼ × ¾"	1.4
¼ × 1"	1.7
⅜ × 1"	4.1
⅜ × 1¼"	4.7
⅜ × 1½"	5.3
HEX NUTS	
¼"-20	0.6
5/16"-18	1.3
⅜"-16	1.6
½"-13	4.3
⅝"-11	7.3
FLAT WASHERS	
¼"	0.6
5/16"	1.2
⅜"	1.5
½"	3.3
⅝"	3.9
LOCK WASHERS	
¼"	0.3
5/16"	0.4
⅜"	0.7
½"	1.3

STEEL HEX COUPLING NUTS

Thread Size	Dimension B	Wt. (lb./C)
¼"-20	⅞"	4.0
⅜"-16	1¾"	11.3
½"-13	1¾"	11.6
⅝"-11	2⅛"	17.3
¾"-10	2¼"	29.3

HANGER ROD WITH CONTINUOUS THREAD

Size & Lgth.	Allowable Load* (lb.)	Wt. (lb./C)
¼"-20 × 72"	670	12.0
¼"-20 × 120"	670	12.0
¼"-20 × 144"	670	12.0
⅜"-16 × 72"	1580	29.0
⅜"-16 × 120"	1580	29.0
⅜"-16 × 144"	1580	29.0
½"-13 × 72"	2770	54.0
½"-13 × 120"	2770	54.0
½"-13 × 144"	2770	54.0
⅝"-11 × 72"	4500	83.0
¾"-10 × 72"	6500	125.0

The Kindorf Channel Erector System

Kindorf is your practical method of support for electrical and mechanical systems. Roll-formed from 12 or 14 gauge steel. Finish is Galv-Krōm—a superior galvanized finish plus zinc chromate—protects the steel against rust and corrosion. Complies with Nema Standard ML1-1975. Available in 10 ft. and 20 ft. lengths.

All channels have ⅞" continuous slot and are produced with solid bases, bolt holes, slots or KO's.

Wt. Lbs. 100 Ft.	Description	Gauge
76	Solid Base	14
71	Holes	14
80	Slots*	14
168	Solid Base	12
107	Solid Base	14
160	Holes	12
100	Holes	14
154	Slots	12
168	KO's	12
107	KO's	14
196	Solid Base	12
188	Holes	12
191	KO's	12
194	Slots**	12
285	Solid Base	12
277	Holes	12
270	KO's	12
58	Solid Base	12
56	Holes	12

Holes—17/32" holes on 1½" centers
*Slots—*3" slots on 4" centers **one 2¼" slots on 4' centers*
KO's—½" KO's on 6" centers

Single Bolt Pipe Straps C105 & C106 Series

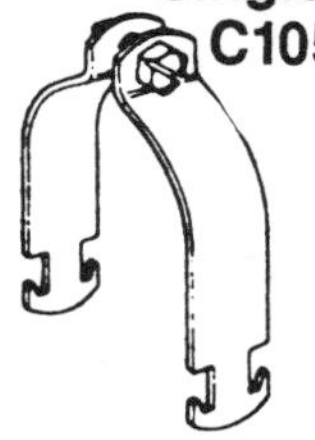

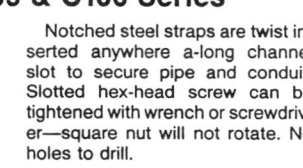

Notched steel straps are twist inserted anywhere a-long channel slot to secure pipe and conduit. Slotted hex-head screw can be tightened with wrench or screwdriver—square nut will not rotate. No holes to drill.

C-105 PIPE AND RIGID CONDUIT STRAPS—STEEL

Conduit or Pipe Size In.	Conduit or Pipe O.D. In.	Gauge or In. Thick	Wt. Per 100 Lbs.
¼	0.540	14 ga.	13
⅜	0.675	14 ga.	12
½	0.840	14 ga.	13
¾	1.050	14 ga.	15
1	1.315	14 ga.	17
1¼	1.660	14 ga.	19
1½	1.900	12 ga.	28
2	2.375	12 ga.	31
2½	2.875	12 ga.	36
3	3.500	12 ga.	42
3½	4.000	⅛"	56
4	4.500	⅛"	64
5	5.563	⅛"	76
6	6.625	⅛"	89
8	8.625	⅛"	114

C-106 E.M.T. STRAPS—STEEL

EMT Size In.	EMT O.D.	Gauge or In. Thick.	Wt. Lbs. Per 100
⅜	0.577	14 ga.	13
½	0.706	14 ga.	14
¾	0.922	14 ga.	15
1	1.163	14 ga.	16
1¼	1.510	14 ga.	19
1½	1.740	14 ga.	20
2	2.197	14 ga.	22
2½	2.875	12 ga.	36
3	3.500	12 ga.	42
3½	4.000	⅛"	56
4	4.500	⅛"	64

Conduit & Pipe Hangers

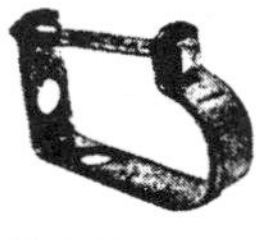

C-149 LAY-IN PIPE HANGERS

Pipe Size In.	Max. Recom. Load in Lbs.	Wt. Lbs. Per 100
½	400	17
¾	400	19
1	400	22
1¼	400	26
1½	400	26
2	400	31
2½	500	66
3	500	75
3½	500	84
4	800	178
5	800	199
6	1000	225
8	1100	250

6-H SERIES CONDUIT AND PIPE HANGERS

Conduit Size, In. Rigid	EMT
⅜—½	½
¾	¾
1	1
—	1¼
1¼	1½
1½	—
2	2
2½	—
3	—
3½	—

E-231 and E-232 Series Rod Hangers

The E-231 clamp may be suspended from a recessed nut or in a "straight-through" manner which permits vertical position of rod.

The E-232 clamp is the E-231 with a swing connector. For ⅜" & ½" rod.

For Rod Inches	Wt. Lbs. Per 100
⅜	32
½	56
⅜	48
½	80

E-760 Channel to Beam Clamp

Secures channel to beam. Fits all I-beams where flange edge does not exceed .8" thickness. Load rating of each clamp 1,275 lbs. with a safety factor of 5⅛" steel, ⅜" U-bolt.

Standard finish: Galv-Krōm. Two required.

Dimension 'A'	Wt. in. lbs./C
3¼"	77
4¾"	88

500 Series Beam Clamps

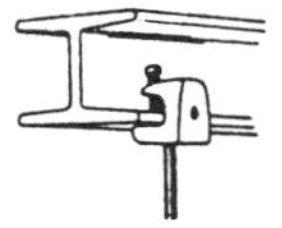

For ¼" through ½" Rod. Furnished with tapped holes in base and back for bolt or hanger rod.

Jaw Opening Inches	Hole Tapped	Weight Lbs. Per 100
⅝	¼-20	15
⅝	10-24	15
15/16	¼-20	23
15/16	10-24	23
⅞	5/16-18	47
1	⅜-16	95
1	½-13	164
⅞	Two 10-24 & two ⅜" untapped	78
1⅛	Two 10-24 & two ½-13	108
1⅜	½-13	165
2⅛	½-13	184

Angle Clamps

Malleable iron. All parts are electrogalvanized including the threads.

Cat. No.	Size In.	Wt. Lbs. Per 100
RC CLAMP (For mounting pipe or conduit at right angle to beam)		
	⅜	25
	½	41
	¾	46
	1	51
	1¼	54
	1½	58
	2	86
	2½	109
	3	118
	3½	134
	4	144
EC CLAMP (For mounting pipe or conduit vertically across beam edge)		
	½	68
	¾	76
	1	86
	1¼	98
	1½	105
	2	141
	2½	188
	3	232
PC CLAMPS (For mounting pipe or conduit parallel to beam)		
	⅜	31
	½	54
	¾	57
	1	64
	1¼	81
	1½	92
	2	116
	2½	148
	3	178
	3½	209
	4	224

Steel Nuts

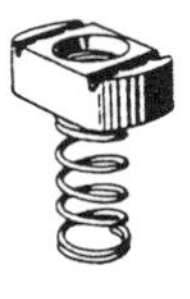

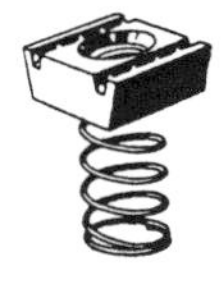

- B-910 Series for use with all channels—no spring.
- B-911 Series for use with 1½" deep channels—with spring.
- B-912 Series for use with ¾" deep channels—with spring.

Size In.	Size In.
¼	⅜
5/16	½
⅜	¼
½	5/16
¼	⅜
5/16	½

Hex Head Cap Screw

For connections of fittings to channel. Less nuts.

HEX NUT

Size In.
¼
5/16
⅜
½

CAP SCREW

Size In.	Size In.
½ × ¾	⅜ × ¾
½ × 1	⅜ × 1
½ × 1¼	⅜ × 1¼
½ × 1½	⅜ × 1½
½ × 1¾	⅜ × 1¾
½ × 2	⅜ × 2
½ × 2¼	⅜ × 2¼
½ × 2½	⅜ × 2½
½ × 3	⅜ × 2¾
½ × 3½	⅜ × 3
½ × 4	⅜ × 3½
—	⅜ × 4

H-193 Hanger Rod

Continuous thread. Precision made with true, free running threads—Suffix indicates rod size and length.

Dia. In.	Wt. Lbs. Per 100 Ft.
¼	36
¼	73
¼	124
¼	148
⅜	88
⅜	172
⅜	293
⅜	348
½	162
½	313
½	530
½	648

H-286 U Bolts

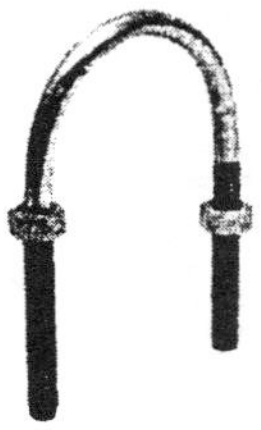

'U' Bolts to support, anchor or guide pipe lines. Furnished with one hex nut per leg. Galv-krom finish.

Size In.	Wt. Lbs. Per 100
⅜	7
½	13
¾	15
1	16
1¼	17
1½	18
2	32
2½	34
3	38
3½	40
4	46

Electrical Termination

Ka-Lug—Type KA

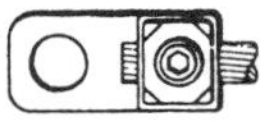

One-Hole tongue. For copper conductors, cable to flat. Low cost lug of high conductivity copper alloy, compactly designed, with Allen Head screw, for easy terminating at service switch, terminal block, etc. where space and cost are the important factors. Each lug takes a convenient range of conductors.

Cat. No.	Copper Conductor Range Min.	Max.
KA8C	14 Sol.	8 Str.
KA4C	14 Sol.	4 Str.
KA25	4 Str.	1/0 Str.
KA28	1 Str.	4/0 Str.
KA34	4/0 Str.	500 MCM

Ground Connector - Type Gar

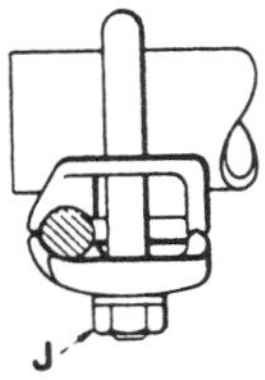

Double-purpose ground connector–joins ground bus to rod or pipe either parallel or at right angles. Easily installed by tightening two nuts on U-bolt.

Accomodates wide range of cable.

Conductor Tube I.P.S.	Rod	Cable	J
¼	½	8 Sol.-4 Str. 4 Sol.-2/0 Str. 2/0 Sol.-250	⅜
⅜	⅝-¾	8 Sol.-4 Str. 4 Sol. 2/0 Str. 2/0 Sol.-250	
		300-500	½
½-¾	⅞-1	8 Sol.-4Str. 4 Sol. 2/0 Str. 2/0 Sol.-250	⅜
		300-500	½
1	1⅛-1¼	8 Sol.-4 Str. 4 Sol. 2/0 Str. 2/0 Sol.-250	⅜
		300-500	½
1¼	1⅜-1½	8 Sol.-4 Str. 4 Sol. 2/0 Str. 2/0 Sol.-250	⅜
		300-500	½
1½	1⅝-1⅞	8 Sol.-4 Str. 4 Sol. 2/0 Str. 2/0 Sol.-250	⅜
		300-500	½
2	2-2⅜	8 Sol.-4 Str. 4 Sol. 2/0 Str. 2/0 Sol.-250	⅜
		300-500	½
2½	2½-2⅞	8 Sol.-4 Str. 4 Sol. 2/0 Str. 2/0 Sol.-250	⅜
		300-500	½
3	3-3½	8 Sol.-4 Str. 4 Sol. 2/0 Str. 2/0 Sol.-250	⅜
		300-500	½
3½	3½-4	8 Sol.-4 Str. 4 Sol. 2/0 Str. 2/0 Sol.-250	⅜
		300-500	½
4	4-4½	8 Sol.-4 Str. 4 Sol. 2/0 Str. 2/0 Sol.-250	⅜
		300-500	½
5	—	8 Sol.-4 Str. 4 Sol.-2/0 Str.	⅜
		2/0 Sol.-250 300-500	½

Scrulug—Type KPA

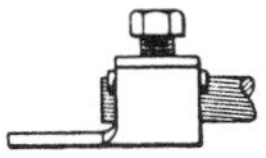

For low cost indoor terminal connections. Installed with screwdriver or pliers. Compact construction, especially useful on panelboards,in fuse boxes, safety switches, etc.

Conductor Range Minimum	Maximum
14 Sol.	8 Str.
14 Sol.	4 Str.
4 Str.	1/0 Str.
1/0 Str.	4/0 Str.
4/0 Str.	500 MCM

Type KC, K2C Servit Post For Copper Cable to Flat

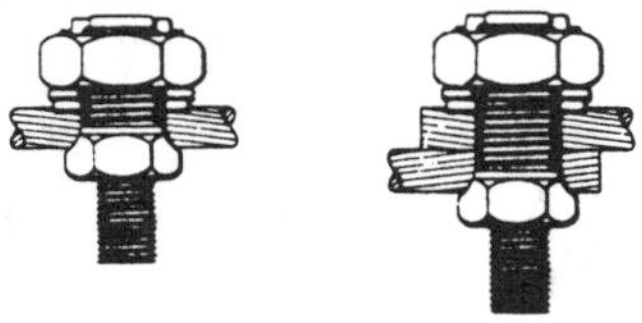

Servit Post–Type KC

Servit Post used to ground one or two cables to steel structures, fence posts, transformers. Also used to tap one or two cables from bus bar. One wrench installation.

Conductor Stranded	Conductor Solid	Stud Diameter
12-9	12-8	¼-20
10-7	10-6	¼-20
10-5	10-4	5/16-18
10-3	10-2	⅜-16
8-2	10-1	⅜-16
2-1/0	2-2/0	½-13
2-2/0	2-3/0	½-13
1-4/0	1-4/0	⅝-11
1-350	—	⅝-11
3/0-500	—	¾-10

Servit—Type KS

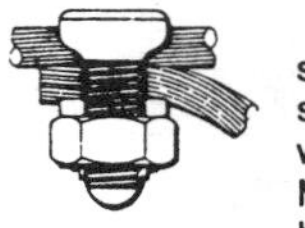

Type KS designed for making service connections quickly and surely. Units for accommodating wire sizes up to and including 500 MCM are made of forged Durium, a high-copper alloy, which eliminates danger of corrosion and insures good contact between conductors. Each size of Servit is designed to accommodate at least two conductors of maximum diameter, but it will close down on a single maximum conductor or its equivalent.

For Copper Cable or Wire

Maximum Conductor Size	
10 Sol.	12 Str.
8 Sol.	8 Str.
6 Sol.	7 Str.
6 Sol.	7 Str.
4 Sol.	5 Str.
4 Sol.	5 Str.
2 Sol.	3 Str.
2 Sol.	3 Str.
1 Sol.	1 Str.
2/0 Sol.	1/0 Str.
3/0 Sol.	2/0 Str.
4/0 Sol.	3/0 Str.
4/0 Sol.	250 Mem.
—	350 Mem.
—	500 Mem.
—	750 Mem.
—	1000 Mem.

Servit—Type KSU

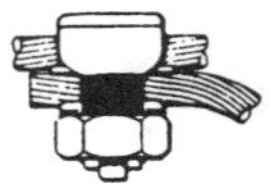

Universal copper alloy Servit, tin-plated for all cimbinations of aluminum and copper conductors. Single tin-plated, copper spacer separates dissimilar metals.

For Aluminum and Copper Conductors

	Max. Conductor Steel		
ACSR	Sol.	Str.	Copper or Alum.
—	8	—	7 Str.
6	6	8	5 Str.
4	4	6	3 Str.
2	—	4	2 Str.
1	—	—	1/0 Str.
1/0	—	—	2/0 Str.
4/0	—	—	250 Mem.

OKLIP—Type KVS

Compact, two piece, high strength, high copper alloy Burndy OKLIP recommended for heavy duty connections. Neoprene rings hold bolts in place during installation. Longer peened bolt, permits swivel action for easier installation. Installed with ordinary wrench.

Conductor Range
Two 2/0 Str. cables max.
Two 4/0 Str. cables max.
Two 350 Mem cables max.
Two 500 Mem cables max.
Two 800 Mem cables max.
Two 1000 Mem cables max.

Universal Oklip—Type KVSU

Compact, tin-plated, high strength, high copper alloy two-piece connector with spacer and tin-plated Durium hardware. Recommended for heavy duty connections. Spacer separates dissimilar conductors and provides long contact length.

Conductor Run	Conductor Tap
2 Str.-2/0 Str.	6 Str.-2/0 Str.
1/0 Str.-4/0 Str.	6 Str.-4/0 Str.
250-350	6 Str.-350
400-500	4 Str.-500
400/800	4/0-800
500/1000	4/0-1000

Qiklug—Type QA For Copper Cable

	Copper Cond. Range	Stud Size In.
ONE HOLE		
	14 Sol. to 8 Str.	10
	8 Str. to 4 Str.	¼
	4 Str. to 1 Str.	¼
	1/0 to 2/0	⅜
	3/0 to 4/0	⅜
	250 to 350	½
	400 to 500	½
	600 to 800	⅝
	850 to 1000	⅝
TWO HOLE		
	14 Sol. to 8 Str.	10
	8 Str. to 4 Str.	¼
	4 Str. to 1 Str.	5/16
	1/0 to 2/0	⅜
	1/0 to 2/0	½
	3/0 to 4/0	½
	250 to 350	½
	400 to 500	½
	600 to 800	½
	850 to 1000	½

Qiklug—Types QA-B and QA-2B

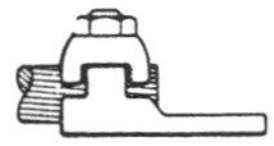

A heavy duty terminal for rapid termination of wires and cables. The Qiklug accommodates a range of cable sizes, and is supplied with one, two or more holes in the terminal tongue. Also furnished with tongues and drilling to order.

	Copper Conductor Range Min.	Max.
TYPE QA-B		
	14 Sol.	8 Str.
	8 Str.	4 Str.
	4 Str.	1 Str.
	1/0 Str.	2/0 Str.
	3/0 Str.	4/0 Str.
	250	350
	400	500
	600	800
	850	1000
TYPE QA-2		
	14 Sol.	8 Str.
	8 Str.	4 Str.
	4 Str.	1 Str.
	1/0 Str.	2/0 Str.
	1/0 Str.	2/0 Str.
	3/0 Str.	4/0 Str.
	250	350
	400	500
	600	800
	850	1000

Hylug Type YA

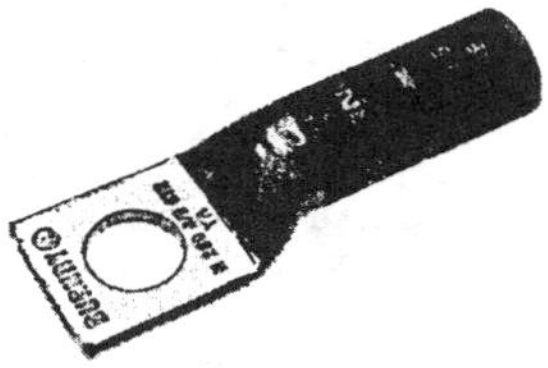

Long-barreled connectors for terminating cable sizes from No. 6 Str. to 2000 MCM.

The long barrels permit double indentations on sizes above 250 MCM for added mechanical strength in heavy duty applications.

	Copper Cable Size	Stud Size In.
ONE HOLE		
	6 Str.	¼
	4 Str.	¼
	3 Str.	5/16
	2 Str.	5/16
	1 Str.	5/16
	1/0 Str.	5/16
	2/0 Str.	⅜
	3/0 Str.	½
	4/0 Str.	½
	250 Mem.	½
	300 Mem.	½
	350 Mem.	½
	400 Mem.	⅝
	500 Mem.	⅝
	600 Mem.	⅝
	750 Mem.	⅝
	1000 Mem.	⅝

continued

continued from previous page

	Copper Cable Size	Stud Size In.
TWO HOLE		
	6 Str.	¼
	6 Str.	½
	4 Str.	¼
	4 Str.	½
	2 Str.	5/16
	2 Str.	½
	1 Str.	5/16
	1 Str.	½
	1/0 Str.	5/16
	1/0 Str.	½
	2/0 Str.	½
	3/0 Str.	½
	4/0 Str.	½
	250 Mem.	½
	300 Mem.	½
	350 Mem.	½
	400 Mem.	½
	500 Mem.	½
	600 Mem.	½
	750 Mem.	½
	800 Mem.	½
	1000 Mem.	½
	1500 Mem.	½
	2000 Mem.	½
FOUR HOLE		
	1000 Mem.	½
	1500 Mem.	½
	2000 Mem.	½

Hylug Type YA-L

Copper compression terminals for regular-duty applications. Inspection hole ensures proper insertion of cable. Tin-plated. Marked with die index and color code.

	Conductor Size Commercial AWG	Screw Sizes
INSTALLED W/HYTOOL NO. MR4C, MY29-3, Y35 or Y39		
	#6 Str.	¼
		5/16
	#4 Str.	¼
		⅜
INSTALLED W/HYTOOL NO. MY29-3, Y35 OR Y39		
		¼
	#2 Str.	5/16
		⅜
	#1 Str.	5/16
		5/16
	1/0 Str.	⅜
	2/0 Str.	⅜
	3/0 Str.	⅜
	4/0 Str.	⅜
	250 Mem.	½
INSTALLED W/HYTOOL NO. Y35 OR Y39		
	300 Mem.	½
	350 Mem.	½
	400 Mem.	⅝
	500 Mem.	⅝

Hylug® Type YAV

	Conductor Sizes Commercial AWG	Screw Sizes
INSTALLED w/HYTOOL NO. Y10M, Y10D OR Y14MV		
	#22–#18 Sol.	4–6
	#22–#18 Str.	8–10
		4–6
	#20–#12 Sol.	8–10
	#20–#14 Str.	¼
		4–6
	#11–#9 Sol.	8–10
	#12–#10 Str.	¼
		5/16
		⅜
INSTALLED w/HYTOOL NO. MR4C, MY29-3 OR Y34A		
		8–10
	6 Sol.	¼
	8 Str.	5/16
		⅜

Copper CRIMPIT Type YC-C

Range-taking copper compression tap connector. C-shaped.

Cable Range Run	Tap
12 Sol.-10 Str.	12 Sol.-10 Str.
8 Sol-8 Str.	10 Sol.-8 Str.
6 sol.-4 Str.	8 Sol.-8 Str.
6 Sol-4 Str.	6 Sol.-6 Str.
6 sol.-4 Str.	4 Sol.-4 Str.
2 Sol.-2 Str.	8 Sol.-4 Str.
2 Sol.-2 Str.	2 Sol.-2 Str.
1/0 Str.-2/0 Str.	8 Sol.-2 Str.
1/0 Str.-2/0 Str.	1/0 Str.-2/0 Str.
3/0 Str.-4/0 Str.	6 Sol.-2 Str.
3/0 Str.-4/0 Str.	1/0 Str. 2/0 Str.
3/0 Str.-4/0 Str.	3/0 Str.4/0 Str.

Hylink Type YS

Long barreled Hylinks for cables from No. 6 Str. to 2000 MCM. Cable stops insure proper insertion of cable. Double indentation in each half of the barrel on sizes above 250 MCM provides mechanical strength for heavy duty applications.

Cable	Color Code
6 Str.	Blue
5 Str.	—
4 Str.	Gray
3 Str.	White
2 Str.	Brown
1 Str.	Green
1/0 Str.	Pink
2/0 Str.	Black
3/0 Str.	Orange
4/0 Str.	Purple
250	Yellow
300	White
350	Red
400	Blue
500	Brown
600	Green
750	Black
1000	White
1500	Green
2000	Brown

Aluminum Hylink—Type YS-A

Recommended for use on aluminum wire. These splices are of seamless tubular construction, tin-plated to resist corrosion and factory filled with Penetrox compound. They are clearly marked with the catalog number, wire size and type, stud size, Die Index and Color Code for easy identification; they are also knurled for the proper number and location of crimps.

Cable
12 Str.
10 Sol.
8 Str.
6 Str.
4 Str.
2 Str.
1 Str.
1/0 Str.
2/0 Str.
3/0 Str.
4/0 Str.
250
300
350

Hylink® Type YS-L

Short-barrel copper compression splices for regular-duty applications. Center stop ensures proper insertion of cable. Tin-plated. Marked with die index and color code.

	Conductor Sizes Commercial AWG
INSTALLED W/HYTOOL NO. MR4C, MY29-3, Y35 OR Y39	
	#6 Str.
	#4 Str.
INSTALLED W/HYTOOL NO. MY29-3, Y35 OR Y39	
	#2 Str.
	#1 Str.
	1/0 Str.
	2/0 Str.
	3/0 Str.
	4/0 Str.
	250 Mem.
INSTALLED W/HYTOOL NO. Y35 OR Y39	
	300 Mem.
	350 Mem.
	400 Mem.
	500 Mem.

Penetrox Joint Compound

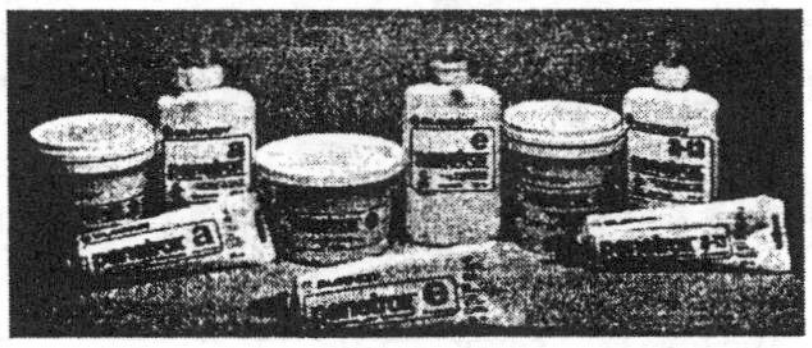

PENETROX compounds help in breaking through oxide films to establish intimate contact between conductor strands and connector. PENETROX A is recommended for all bare aluminum outdoor applications. PENETROX A13 is recommended for all insulated aluminum conductors. PENETROX E is recommended for use on copper-to-copper applications.

	Container Description
PENETROX A	
	3 fl. oz. tube
	4 fl. oz. squeeze bottle
	8 fl. oz. squeeze bottle
	1 pt. tub
	1 qt. tub
	1 gal. tub
	5 gal. tub
	1 lb. cartridge
PENETROX A 13	
	3 fl. oz. tube
	4 fl. oz. squeeze bottle
	8 fl. oz. squeeze bottle
	1 pt. tub
	1 qt. tub
	1 gal. tub
PENETROX E	
	3 fl. oz. tube
	4 fl. oz. squeeze bottle
	8 fl. oz. squeeze bottle
	1 pt. tub
	1 qt. tub

Edge Type Conduit Support (ET)

Size	Standard Package	Wt. Per 100 Pcs.
½"	50	67
¾"	50	71
1	25	85
1¼"	25	91
1½"	25	115
2"	25	135
2½"	10	183
3"	10	230

Parallel Conduit Supports (PAR)

Size	Standard Package	Wt. Per 100 Pcs.
⅜"	50	32
½"	50	58
¾"	50	64
1"	50	70
1¼"	25	72
1½"	25	93
2"	25	100
2½"	25	135
3"	10	155
3½"	10	190
4"	10	205

Right Angle Conduit Supports (RA)

As is well known to the electrical trade "Korns" Conduit Supports are specially designed malleable iron clamps with a sharp bite contact point which upon tightening grips a structural steel member firmly to insure adequate grip against slippage under vibration, temperature changes and other adverse installation conditions. These hangers likewise, as built by "Korns", provide an almost equally solid grip of the conduit.

As the type names Right Angle, Parallel, and Edge Type suggest they carry conduit respectively at right angle to, parallel to, and vertical to the edge of a structural member as illustrated.

The Right Angle and Parallel in sizes ⅜" through 4" are usable on rigid conduit, standard pipe or thin wall conduit. Edge Type in sizes ½" through 3" is usable on rigid conduit or standard pipe.

Size	Std. Pkg.	Wt. Per 100 Pcs.
⅜"	50	25
½"	50	40
¾"	50	43
1"	50	48
1¼"	50	53
1½"	50	58
2"	50	85
2½"	25	106
3"	25	110
3½"	25	128
4"	25	140

Electric Fuses

Small Dimension Fuses

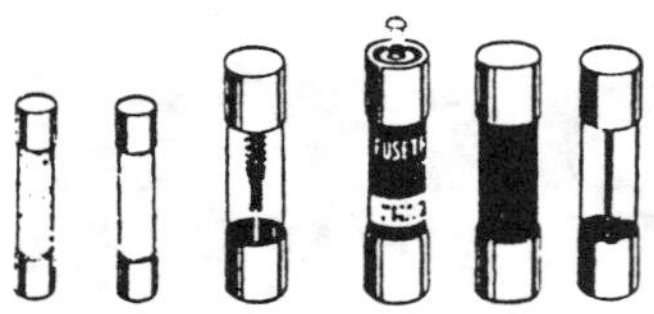

A select group of widely used small-dimension fuses are included in this catalog.

The ampere rating of fuse selected is dependent upon:

a. 1. Overload and short-circuit protection. Generally, select fuse ampere rating at 125% of the full load amperes.
   2. Short-circuit protection only. Select fuse ampere rating at 150% to 300% of equipment or circuit rating.

b. Ambient temperature affects the current carrying capacity of fuses.
   For circuit protection, the voltage rating of the fuses must be equal to, or greater than the voltage of the circuit in which the fuse is applied.

Time Current Characteristics

The fuse time current characteristic should be compatible with the time-current characteristic of the load and the time current characteristic of the circuit components to be protected.

a. Select a dual-element, time-delay or time-delay fuse where high inrush or starting loads are present as with motors, solenoids, or control transformers. (Usually sized at 125% of full load amperes.)
b. Select non-time-delay fuses for resistive currents or other currents where no transients or surges are encountered. (Usually sized at 125% of full load amperes.)
c. Select a limiter or non-time-delay fuse where short-circuit protection only is required. (Usually sized at 150% to 300% of circuit ampere rating.)
d. Select very fast-acting fuses to protect very low energy withstand components, such as semi-conductors.
e. Test the selected fuse in the intended circuit under all normal circuit conditions that may include transient, inrush, or any other non-steady-state currents.

Ferrule Fuses

5 mm × 20 mm (FOR FOREIGN EQUIPMENT)

Fuse Size (Amps)	Quick-Acting GDA Ceramic. Amperes interrupting rating (AIC) 1500A. IEC Sheet 1.	GDB Glass. 35 AIC or 10 times Amp rating whichever is larger. IEC Sheet 2.	Time-Lag GDC Glass. For surge currents. 35 AIC or 10 times Amp rating whichever is larger. IEC Sheet 3.	Fuse Size (Amps)	Non-Time Delay GMA Glass. Formerly GJU.
	Meet IEC Specification 127 (except GDA 315 ma and 400 ma); SEMKO 104 approval; CEE4 Certification.				
32ma	—	250V	250V	1/32	250V
40ma	—	250V	250V	1/20	250V
50ma	250V	250V	250V	1/16	250V
63ma	250V	250V	250V	1/10	250V
80ma	250V	250V	250V	1/8	250V
100ma	250V	250V	250V	2/10	250V
125ma	250V	250V	250V	1/4	250V
160ma	250V	250V	250V	3/10	250V
200ma	250V	250V	250V	4/10	250V
250ma	250V	250V	250V	1/2	250V
315ma	250V	250V	250V	6/10	250V
400ma	250V	250V	250V	7/10	250V
500ma	250V	250V	250V	3/4	250V
630ma	250V	250V	250V	8/10	250V
800ma	250V	250V	250V	1	250V
1a	250V	250V	250V	1 2/10	250V
1.25a	250V	250V	250V	1 1/2	250V
1.6a	250V	250V	250V	1 6/10	250V
2a	250V	250V	250V	2	250V
2.5a	250V	250V	250V	2 1/2	250V
3.15a	250V	250V	250V	3	250V
4a	250V	250V	250V	3 1/2	250V
5a	250V	250V	250V	4	250V
6.3a	250V	250V	250V	5	250V
8a	—	250V	—	6	250V
10a	—	250V	—	7	125V
—	—	—	—	8	125V
—	—	—	—	10	125V
—	—	—	—	15	125V

Small Dimension Fuses

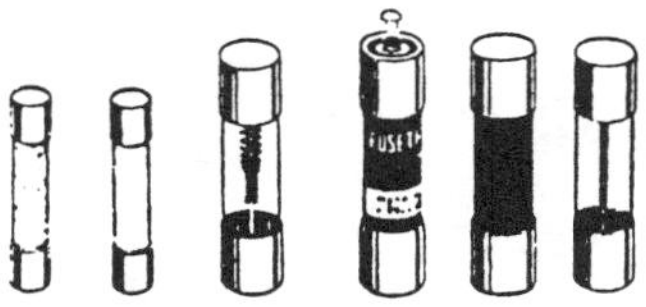

Ferrule Fuses

Fuse Size (Amps)	1/4" × 1" Non-Time-Delay AGX Glass. For Instrument, electronic and small appliance circuits; MKG—low resistance; 1/16 & 1/8 A, 250V.	1/4" × 1 1/4" Non-Time-Delay AGC Glass. For instrument, electronic and small appliance circuits.	GLH Glass. For instrument, electronic and small appliance circuits.	MGB Glass. For instrument, electronic and small appliance circuits.	MTH Glass. For instrument, electronic and small appliance circuits.	ABC Ceramic. For instrument, electronic and small appliance circuits.
1/500	250V[U]	250V[UC]	—	—	—	—
1/200	250V[U]	250V[UC]	—	—	—	—
1/100	250V[U]	250V[UC]	—	—	—	—
1/32	250V[U]	250V[UC] (also 1/20)	—	—	—	—
1/16	250V	250V	—	250V	—	—
1/10	250V[U]	250V	—	—	—	—
1/8	250V[U]	250V[UC]	—	250V	—	—
15/100	—	250V[UC]	—	—	—	—
175/1000	—	250V[UC]	—	—	—	—
2/10	250V[U]	250V[UC]	—	—	—	—
3/16	250V[U]	250V[UC]	—	—	—	—
1/4	250V[U]	250V[UC]	—	—	—	250V[U]
3/10	250V[U]	250V[UC]	—	—	—	—
3/8	250V[U]	250V[UC]	—	—	—	—
4/10	250V[U]	250V[UC] (also 45/100)	—	—	—	—
1/2	250V[U]	250V[UC]	—	—	—	250V[U]
6/10	—	250V[UC]	—	—	—	—
3/4	250V[U]	250V[UC]	—	—	—	250V[U]
8/10	—	250V[UC]	—	—	—	—
1	250V[U]	250V[UC]	—	—	—	250V[U]
1 2/10	—	—	—	—	—	—
1 1/8	—	—	—	—	—	—
1 1/4	250V[U]	250V[UC]	—	—	—	—
1 4/10	—	—	—	—	—	—
1 1/2	250V[U]	250V[UC]	—	—	—	250V[U]
1 6/10	—	250V[UC] (also 1 3/4)	—	—	—	—
1 8/10	—	250V[UC]	—	—	—	—
2	250V[U]	250V[UC]	—	—	—	250V[U]
2 1/4	—	250V[UC]	—	—	—	—
2 1/2	125V[U]	250V[UC]	—	—	—	250V[U]
2 8/10	—	—	—	—	—	—
3	125V[U]	250V[UC]	—	—	—	250V[U]
3 2/10	—	32V[R]	—	—	—	—
3 1/2	—	—	—	—	—	—
4	125V	32V[R]	—	—	250V[UC]	250V[U]
4 1/2	—	—	—	—	—	—
5	125V[U]	32V[R]	—	—	250V[UC]	250V[U]
6 4/10	—	—	—	—	—	—
6	32V[U]	32V[R]	—	—	250V[UC]	250V[U]
6 1/4	—	32V[R]	—	—	—	—
7	32V[U]	32V[R]	125V[U]	—	250V[UC]	250V[U]
7 1/2	—	32V[R]	—	—	—	—
8	32V[U]	32V[R]	125V[U]	—	250V[UC]	250V[U]
9	—	32V[R]	—	—	—	—
10	32V[U]	32V[R]	125V[U]	—	—	250V[U]
12	—	32V[R]	—	—	—	250V[U]
15	32V[U]	32V[R]	—	—	—	250V[U]
20	32V[U]	32V[R]	—	—	—	250V[R]
25	32V	32V[R]	—	—	—	125V[R]
30	32V	32V[R]	—	—	—	125V

Small Dimension Fuses

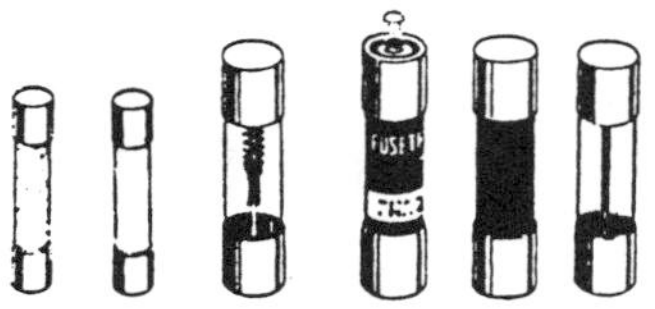

Ferrule Fuses

Fuse Size (Amps)	¼″ × 1¼″ (Continued) Very Fast Acting GBB Ceramic. For protection of semiconductors. (See Section 7.)	Dual-Element Time-Delav MDL Glass. For circuits with high inrush currents.	MDQ Glass. For circuits with high inrush currents.	MDX Glass. For circuits with high inrush currents.	MDA Ceramic. For circuits with high inrush currents.	Time-Delay MSL Glass. Spiral wound, single element for moderate inrush I. Superior to "slow-blow" type. 16-18 sec. delay at 200% load. (MSV has radial leads.)
1/500	—	—	—	—	—	
1/200	—	—	—	—	—	—
1/100	—	250V^{UC}	—	—	250V	—
1/32	—	250V^{UC}	—	—	250V	—
1/16	—	250V^{UC}	—	—	250V^{R}	—
1/10	—	250V^{UC}	—	—	250V^{R}	—
1/8	60V^{R}	250V^{UC}	—	—	250V^{R}	—
15/100	—	250V^{UC}	—	—	250V^{R}	—
175/1000	—	250V^{UC}	—	—	250V^{R}	—
2/10	—	250V^{UC}	—	—	250V^{R}	—
3/16	—	250V^{UC}	—	—	250V^{R}	—
1/4	60V^{R}	250V^{UC}	—	—	250V^{R}	250V (MSL/MSV)^{UC}
3/10	—	250V^{UC}	—	—	250V^{R}	250V (MSL)^{UC}
3/8	—	250V^{UC}	—	—	250V^{R}	250V (MSL/MSV)^{UC}
4/10	—	250V^{UC}	—	—	250V^{R}	250V (MSL)
1/2	60V^{R}	250V^{UC}	—	—	250V^{R}	250V (MSL/MSV)^{UC}
6/10	—	250V^{UC} (Also 7/10)	—	—	250V^{R}	250V (MSL)^{UC}
3/4	60V^{R}	250V^{UC}	—	—	250V^{R}	250V (MSL)^{UC}
8/10	—	250V^{UC}	—	—	250V^{R}	250V (MSL)^{uc}
1	60V^{R}	250V^{UC}	—	—	250V^{R}	250V (MSL/MSV)^{UC}
1 2/10	—	125V^{uc}	—	—	—	—
1 1/8	—	—	—	—	—	—
1 1/4	60V^{R}	125V^{uc}	—	250V^{U}	250V^{R}	250V (MSL)^{UC}
1 4/10	—	—	—	—	—	—
1 1/2	60V^{R}	125V^{UC}	—	250V^{U}	250V^{R}	250V (MSL/MSV)^{UC}
1 6/10	—	125V^{UC}	—	250V^{U}	250V^{R}	250V (MSL)^{UC}
1 8/10	—	125V^{UC}	—	250V^{U}	—	—
2	60V^{R}	60V^{R}	—	250V^{U}	250V^{R}	250V (MSL/MSV)^{UC}
2 1/4	—	125V^{UC}	—	—	—	—
2 1/2	—	125V^{UC}	250V^{U}	125V^{UC}	250V^{R}	250V (MSL/MSV)^{UC}
2 8/10	—	125V^{UC}	250V^{U}	—	250V^{R}	—
3	60V^{R}	32V	250V^{U}	125V^{UC}	250V^{R}	250V (MSL/MSV)^{UC}
3 2/10	—	32V	250V^{U}	125V^{UC}	250V^{R}	250V (MSL/MSV)^{UC}
3 1/2	—	—	—	—	—	—
4	60V^{R}	32V	250V^{U}	125V^{UC}	250V^{R}	250V (MSL/MSV)^{UC}
4 1/2	—	—	—	—	—	—
5	60V^{R}	32V	250V^{U}	125V^{UC}	250V^{R}	250V (MSL/MSV)^{UC}
5 6/10	—	—	—	—	—	—
6	60V^{R}	32V	250V^{U}	—	250V^{R}	250V (MSL)^{UC}
6 1/4	—	32V	250V^{U}	125V^{UC}	250V^{R}	250V (MSL/MSV)^{UC}
7	60V^{R}	32V	250V^{U}	125V^{UC}	250V^{R}	250V (MSL/MSV)^{UC}
7 1/2	—	32V^{R}	—	—	—	—
8	60V^{R}	32V^{R}	—	—	250V^{R}	250V (MSL/MSV)^{UC}
9	60V^{R}	32V^{R}	—	—	—	—
10	60V^{R}	32V^{R}	—	—	250V^{R}	—
12	60V^{R}	32V^{R}	—	—	250V^{R}	—
15	60V^{R}	32V^{R}	—	—	250V^{R}	—
20	60V^{R}	32V^{R}	—	—	250V^{R}	—
25	60V^{R}	32V^{R}	—	—	125V	—
30	60V^{R}	32V^{R}	—	—	125V	—

Small Dimension Fuses

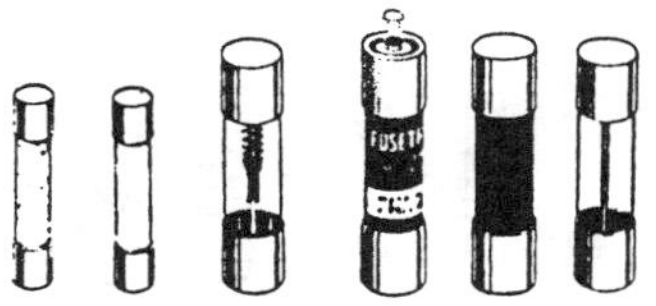

Ferrule Fuses

Fuse Size (Amps)	Non-Time-Delay BBS Fibre. For control, gaseous vapor, and electronic circuits.	13/32" × 1 3/8" KTQ Fibre. For control, gaseous vapor, and electronic circuits. Slightly more delay than BBS for transient I.	13/32" × 1 1/2" Non-Time-Delay AGU Glass. Formerly 5AG.	BAF Laminated.	BAN Fibre. Formerly 5AB.	KTK And KLM Melamine. Limitron fast-acting. For control, gaseous vapor, or circuits having high fault I. 20900,000 AIC; current limiting. (KLM's have d-c rating of 500V.)
1/500	—	—	—	—	—	—
1/200	—	—	—	—	—	—
1/100	—	—	—	—	—	—
1/32	—	—	—	—	—	—
1/16	—	—	—	—	—	—
1/10	—	—	—	—	—	600V[UC]
1/8	—	—	—	—	—	600V[UC]
15/100	—	—	—	—	—	—
175/1000	—	—	—	—	—	—
2/10	600V[UC]	—	—	—	—	600V[UC]
3/16	—	—	—	—	—	—
1/4	—	—	—	—	—	600V[UC]
3/10	—	—	—	—	—	600V[UC]
3/8	—	—	—	—	—	—
4/10	600V[UC]	—	—	—	—	—
1/2	600V[UC]	—	—	250V[UC]	—	600V[UC]
6/10	600V[UC]	—	—	—	—	—
3/4	600V[UC]	—	—	—	—	600V[UC]
8/10	600V[UC]	—	—	—	—	—
1	600V[UC]	600V	250V[U]	250V[UC]	250V	600V[UC]
1 2/10	—	—	—	—	—	—
1 1/8	—	—	—	—	—	—
1 1/4	—	—	—	—	—	—
1 4/10	—	—	—	—	—	—
1 1/2	600V[UC]	—	—	250V[UC]	—	600V[UC]
1 6/10	600V[UC]	—	—	—	—	—
1 8/10	600V[UC]	—	—	—	—	—
2	600V[UC]	600V	250V[U]	250V[UC]	250V	600V[UC]
2 1/4	—	—	—	—	—	—
2 1/2	—	—	—	250V[UC]	—	600V[UC]
2 8/10	—	—	—	—	—	—
3	600V[UC]	600V	250V[U]	250V[UC]	250V	600V[UC]
3 2/10	—	—	—	—	—	—
3 1/2	—	—	—	—	—	600V[UC]
4	600V[UC]	600V[R]	32V	250V[UC]	250V	600V[UC]
4 1/2	—	—	—	—	—	—
5	600V[UC]	250V[UC]	32V	250V[UC]	250V	600V[UC]
5 6/10	—	—	—	—	—	—
6	600V	600V[R]	—	250V[UC]	250V	600V[UC]
6 1/4	—	—	—	250V[UC]	—	—
7	600V	—	—	250V[UC]	—	600V[UC]
7 1/2	—	—	—	—	—	—
8	600V	—	32V	250V[UC]	250V	600C[UC]
9	—	—	—	250V[UC]	—	600V[UC]
10	600V	—	32V	250V[UC]	250V	600V[UC]
12	—	—	—	250V[UC]	250V	600V[UC]
15	600V	—	32V	250V[UC]	250V	600V[UC]
20	600V	—	32V	125V	250V	600V[U]
25	600V	—	32V	125V	250V	600V[U]
30	600V	—	32V	125V	250V	600V[U]

Small Dimension Fuses

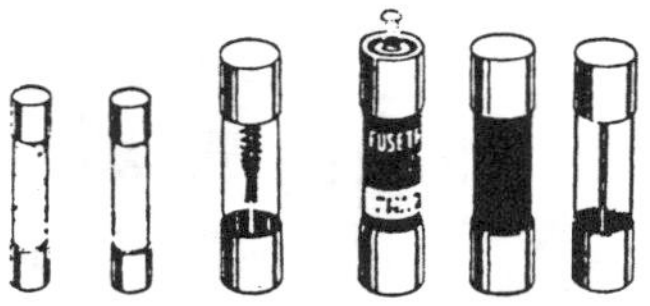

Ferrule Fuses

Fuse Size (Amps)	13/32" × 1 1/2" Non-Time-Delay KTK-R KTK with rejection feature. U.L. Class CC. Branch circuit fuse. (See Section 1-15.	Dual-Element Time-Delay FNM Fusetron. Fibre. For circuits with high inrush currents.	FNQ Fibre. For motor control transformers & other circuits with inrush I's. 10,000 AIC. Duel element to 3 2/10A; single element above 3 2/10 A.	Time-Delay FNW Fibre. For motor control transformers & other circuits with inrush I's. 10,000 AIC.	Pin Indicating Fuses 1/4" × 1 1/4" Non-Time-Delay GBA Fibre. Highly visible red pin ejects when fuse opens.	GLD Fibre. Pin silver-plated for positive signal activation when fuse opens.
1/500	—	—	—	—	—	—
1/200	—	—	—	—	—	—
1/100	—	—	—	—	—	—
1/32	—	—	—	—	—	—
1/16	—	—	—	—	—	—
1/10	600V^{UC}	250V^{UC}	500V^{UC}	—	—	—
1/8	600V^{UC}	—	500V^{UC}	—	—	—
15/100	—	250V^{UC}	500V^{UC}	—	—	—
175/1000	—	—	—	—	—	—
2/10	600V^{UC}	250V^{UC}	500V^{UC}	—	—	—
3/16	—	—	500V^{UC}	—	—	—
1/4	600V^{UC}	250V^{UC}	500V^{UC}	—	—	—
3/10	600V^{UC}	250V^{UC}	500V^{UC}	—	—	—
3/8	—	—	—	—	—	—
4/10	600V^{UC}	250V^{UC}	500V^{UC}	—	—	—
1/2	600V^{U}	250V^{UC}	500V^{UC}	—	125V^{U}	125V^{UC}
6/10	600V^{UC}	250V^{UC}	500V^{UC}	—	—	—
3/4	600V^{UC}	—	—	—	125V^{U}	125V^{UC}
8/10	—	250V^{UC}	500V^{UC}	—	—	—
1	600V^{UC}	250V^{UC}	500V^{UC}	—	125V^{U}	125V^{UC}
1 2/10	—	—	—	—	—	—
1 1/8	—	250V^{UC}	500V^{UC}	—	—	—
1 1/4	—	250V^{UC}	500V^{UC}	—	—	—
1 4/10	—	250V^{UC}	—	—	—	—
1 1/2	600V^{UC}	250V^{UC}	500V^{UC}	—	125V^{U}	125V^{UC}
1 6/10	—	250V^{UC}	500V^{UC}	—	—	—
1 8/10	—	250V^{UC}	—	—	—	—
2	600V^{UC}	250V^{UC}	500V^{UC}	—	125V^{U}	125V^{UC}
2 1/4	—	250V^{UC}	500V^{UC}	—	—	—
2 1/2	—	250V^{UC}	500V^{UC}	—	—	—
2 8/10	—	250V^{UC}	—	—	—	—
3	600V^{UC}	250V^{UC}	500V^{UC}	—	125V^{U}	125V^{UC}
3 2/10	—	250V^{UC}	500V^{UC}	—	—	—
3 1/2	—	250V^{UC}	500V^{UC}	—	—	—
4	600V^{UC}	250V^{UC}	500V^{UC}	—	125V^{U}	125V^{UC}
4 1/2	—	250V^{UC}	500V^{UC}	—	—	—
5	600V^{UC}	250V^{UC}	500V^{UC}	—	125V^{U}	125V^{UC}
5 6/10	—	250V^{UC}	500V^{UC}	—	—	—
6	600V^{UC}	250V^{UC}	500V^{UC}	—	125V	125V
6 1/4	—	250V^{UC}	500V^{UC}	—	—	—
7	600V^{UC}	250V^{UC}	500V^{UC}	—	—	—
7 1/2	—	—	—	—	—	—
8	600V^{UC}	250V^{UC}	500V^{UC}	—	125V	125V
9	600V^{UC}	250V^{UC}	500V^{UC}	—	—	—
10	600V^{UC}	250V^{UC}	500V^{UC}	—	32V	32V
12	600V^{UC}	125V^{UC}	500V^{UC} (Also 14)	250V^{U}	32V	32V
15	600V^{UC}	125V^{UC}	500V^{UC}	250V^{U}	32V	32V
20	600V^{UC}	32V	500V^{U}	250V^{U}	—	—
25	600V^{UC}	32V	500V^{U}	250V^{U}	—	—
30	600V^{UC}	32V	500V^{U}	250V^{U}	—	—

"O" Ring Gaskets for GUB Series

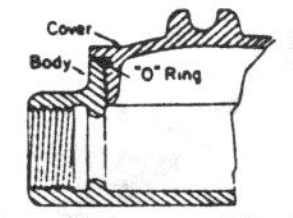

Standard explosion-proof and dust-ignition-proof GUB series are recommended for locations requiring NEMA 3 enclosures and, when furnished with "O" ring gaskets, for locations requiring NEMA 4 (watertight) enclosures. As illustrated in the drawing, the external "O" ring gasket in no way impairs the effectiveness ofthe explosion-proof and dust-ignition-proof threaded joint

When specifying this Neoprene gasket, add suffix S302 to the Cat. No.

Nominal Diameter of Cover Opening
3⅝
3⅝
5½
7
9⅝
9⅝
7
7

Drilling and Tapping GUB Boxes

MAXIMUM SIZE AND NO. OF DRILLED AND TAPPED HOLES

No. of Holes Top & Bottom				No. of Holes Each Side				No. of Holes Back		
1	2	3	4	1	2	3	4	1	2	4
2	2	1	¾	2	2	1	¾	¾	¾	¾
2	2	1	¾	2	2	1	¾	2	2	2
¾	¾	¾		¾	¾	¾	¾	2	2	2
2	2	1½	1	2	2	2	1	4	3	3
2	2	1½	1	2	2	2	1	4	2	2
2	2	2	1½	2	2	2	2	4	4	4
4	4	3	2	4	4	3½	2	4	4	4

Division 2—1 Pole Circuit Breakers

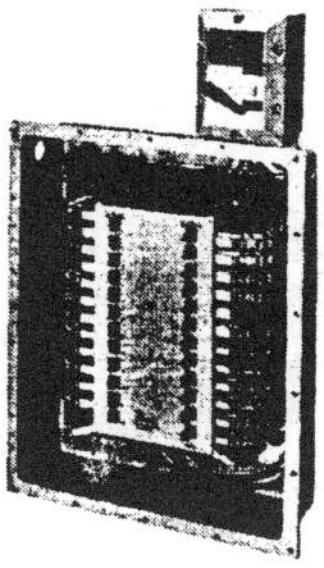

Panelboards are available with 15, 20 or 30 ampere circuit breakers. To order a panelboard with all breakers of the same rating, add the desired rating as a suffix to the Cat. No. For example, the 12 circuit D2PB1512 panelboard with all the circuit breakers rated 20 amperes would be ordered as D2PB1512-20.

Panelboards can be furnished with an assortment of breaker ratings. To order, the quantities and ampere ratings are added as suffixes to the Cat. No. For example, the 12 circuit D2PB1512 with six 15 ampere, four 20 ampere and two 30 ampere circuit breakers would be ordered as D2PB1512-615-420-230.

The D2PB with a main breaker is available up to 100 amps. To order D2PB with main breaker, add the appropriate suffix. Example: D2PB1512-15 with 3 pole, 100 amp main circuit breaker would be ordered as D2PB1512-15-3M100. If 2 pole main is required, change the number 3 to 2. If a lower trip rating than 100 is required, the suffix will change accordingly.

1-POLE CIRCUIT BREAKERS

No. of Single-Pole Breakers	Panel Size	Wiring System 4 Mains: 3-Wire Branches: 2-Wire Solid Neutral Main Lug Size	Wiring System 5 Mains: 4-Wire, 3-Phase Branches: 2-Wire, Solid Neutral Main Lug Size
6	1	1/0	1/0
8	1	1/0	1/0
10	1	1/0	1/0
12	1	1/0	1/0
12	2	4/0	4/0
14	2	4/0	4/0
16	2	4/0	4/0
18	2	4/0	4/0
20	2	4/0	4/0
22	2	4/0	4/0
24	2	4/0	4/0

Division 2 2 Pole Circuit Breakers

TWO-POLE CIRCUIT BREAKERS

No. of 2-Pole Breaking	Panel Size	Wiring System 3 Mains: 3-wire Branches: 3-wire Solid Neutral Main Lug Size	Wiring System 8 Mains: 4-wire, 3-Phase Branches: 3-wire, 1 Pha-se Solid Neutral Main Lug Size
4	1	1/0	1/0
5	1	1/0	1/0
6	1	1/0	1/0
6	2	4/0	4/0
7	2	4/0	4/0
8	2	4/0	4/0
9	2	4/0	4/0
10	2	4/0	4/0
11	2	4/0	4/0
12	2	4/0	4/0

Circuit Breakers

Where D2PB panelboards have been ordered with less than the maximum number of circuit breakers, breakers can easily be added in the field. Circuit breaker assemblies for field addition or replacement are listed below; they consist of the breaker itself in its factory sealed Class I, Division 2, Groups C, D enclosure and necessary hardware.

CIRCUIT BREAKER ASSEMBLIES

Ampere Rating
15
20
30

QP Plug-In BQ Bolt On

QP Plug-In

BQ Bolt-on

QP, BQ Breakers—10,000 Amperes I.R.

Ampere Rating	
1-POLE	120/240 VOLT AC
15	
20	
30	
40	
50	
2-POLE[1]	120/240 VOLT AC
15	
20	
30	
40	
50	
60	
70	
90	
100	
125	
2-POLE[1,2]	240 VOLT AC
15	
20	
30	
40	
50	
60	
70	
90	
100	
3-POLE[1,2]	240 VOLT AC
15	
20	
30	
40	
50	
60	
70	
90	
100	

ACCESSORIES
Description
2-Pole Tie Handle
Handle Block Lock-Off Device
Padlocking Device 1, 2, or 3 Pole
Filler Plate

QPH, BQH Breakers—22,000 Amperes, I.R. (Main Lug Panels and 400 Ampere Main Breaker Panel)

Ampere Rating	Max. I.R. Symmetrical Amperes	
1-POLE		120/240 VOLT AC
15		
20		
25		
30		
40	22,000	
50		
55		
60		
70		
2-POLE COMMON TRIP		120/240 VOLT AC
15		
20		
25		
30		
40		
50	22,000	
60		
70		
90		
100		
3-POLE COMMON TRIP		240 VOLT AC
15		
20		
30		
40		
50	22,000	
60		
70		
90		
100		

Ground Fault Interrupters

Ampere Rating	Max. I.R. Symmetrical Amperes	Plug-In	Bolt-On
GFI 1-POLE			
	QPF		
15			
20	10,000		
30			
GFI 1-POLE			
	QPHF		
15			
20	22,000		
30			

HQP, HBQ Breakers—65,000 Amperes I.R. (Main Lug Panels Only)

Ampere Rating	HQP Plug-In	HBQ Bolt-On
1-POLE		120/240 VOLT AC
15		
20		
30		
40		
50		
2-POLE[1]		120/240 VOLT AC
15		
20		
30		
40		
50		
60		
70		
90		
100		
3-POLE[1,2]		240 VOLT AC
15		
20		
30		
40		
50		
60		
70		
90		
100		
SPECIAL GAS STATION, 2 WIRE[1]		120/240 VOLT AC
15		
20		
30		
GAS STATION, 3 WIRE[1]		120/240 VOLT AC
15		
20		
30		

[1] *Common trip.*
[2] *Suitable for 3 phase, 3 wire systems.*

Enclosures

FLB Circuit Breaker Enclosures

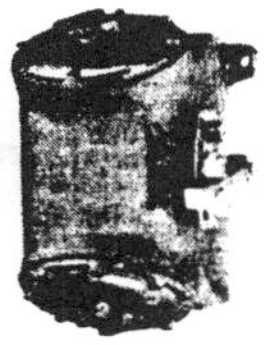

FLB circuit breakers and enclosures are used:

- For service entrance, feeder or branch circuit protection for lighting, heating, appliance and motor circuits
- In areas made hazardous due to the presence of flammable vapors, gases or combustible dusts
- In damp, wet or corrosive locations
- Indoors or outdoors at petroleum refineries, chemical and petrochemical plants and other process industry facilities where similar hazards exist
- To provide disconnect means, short circuit protection and thermal time delay overload protection

Electrical Rating Ranges

- 50, 100 and 225 ampere frame sizes

Compliances

- NEC: Class I, Groups C, D
  Class II, Groups E, F, G
  Class III
- NEMA: 3, 7CD, 9EFG, 12
- UL Standard: 698

120 VAC/125 VDC, 240 VAC/250 VDC—NON-INTERCHANGEABLE TRIP

Circuit Breaker Poles	Hub Size	Circuit Breaker Amps
1	¾	15 20 30 40 50
2	1	15 20 30 40 50
	1½	70 90 100
3	1¼	15 20 30 40 50
	1½	70 90 100

480 VAC/250 VDC—100 AMPERE FRAME SIZE WITH NON-INTERCHANGEABLE TRIP, 480 VAC MAX.

Circuit Breaker Poles	Hub Size	Circuit Breaker Amps
2	1	15 20 30 40 50
	1½	70 90 100
3	1¼	15 20 30 40 50
	1½	70 90 100

Krydon Hubs

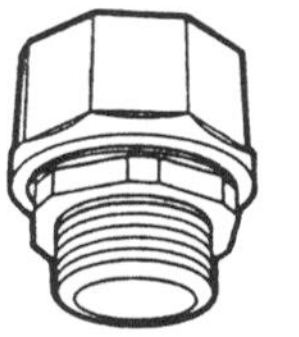

Krydon® material hubs for conduit entrances, in sizes ½″ through 3″ are available for factory or field installation in all enclosures made of Krydon material. For factory installation, send drawing showing sizes and locations of hubs. Furnished with locknuts and gaskets to assure weathertightness.

Conduit Size	Conduit Size
½	1½
¾	2
1	2½
1¼	3

NJB-NCE-NCS-Krydon Accessories

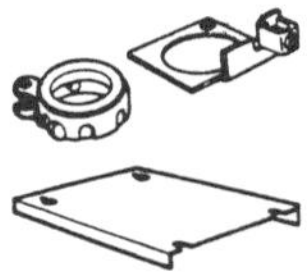

Grounding Plates (½″ through 1″) and insulated bushings (½″ through 3″) permit use of the conduit as the grounding circuit. Both types have set screws and ground-wire terminals

- Mounting plates, made of steel with zinc chromate finish. Can be custom drilled to your specifications or supplied blank.

GROUNDING PLATES AND GROUNDING BUSHINGS

Conduit Size
½
¾
1
1¼
1½
2
2½
3

MOUNTING PLATES

Fits Enclosure
3⅛ × 6¾
3¼ × 9
3¼ × 11
5 × 9
7 × 10
7 × 14
10 × 18
10 × 24
14 × 26
24 × 26

EJB Junction Boxes

EJB junction boxes are used in threaded rigid conduit systems in hazardous areas:

- As a junction or pull box
- To provide enclosures for splices and branch circuit taps
- For housing terminal blocks, relays andother electrical devices
- As mounting boxes for multi-device control panels with EMP barrel assemblies
- Indoors or outdoors in damp, wet, dusty, corrosive, hazardous locations
- Where exposure to frequent or heavy rain, water, spray, moisture, and humidity is common; such as: offshore drilling facilities, cooling towers, coal preparation and handling facilities and sewage and waste water treatment plant (use EJB-N4)
- In areas which are hazardous due to the presence of hydrogen or gases and vapors of equivalent hazard such as found in process industries, missile bases and gas manufacturing plants

Compliances

- NEC: Class I, Group D (Styles A, B1, B2)
  Class 1, Groups B, C, D (Style C only)
  Class 11, Groups E, F, G
  Class 111
- UL Standard: 886
- UL Raintight
- NEMA 3, 4 (optional for Style C only), 7, 9

STYLES A, B1, B2

	Nominal Inside Dimensions			
	Sides (aa)	Top & Bottom (bb)	Depth	Style
	4	6	4	A
	6	8	4	A
	6	8	6	A
	8	8	6	B1
	8	12	4	B2
	4	16	4	B2
	8	10	6	BR
	8	13	8	B1
	9	16	6	B1
	12	16	8	B1
	12	23	6	B1
	8	15	10	B1
ALUMINUM				
	4	6	4	A
	6	8	4	A
	6	8	6	A
	8	8	6	B1
	8	12	4	B2
	8	10	6	B2
	12	16	8	B1

EJB Style C Junction Boxes

- External flange design—wide unobstructed cover opening provides a completely accessible interior for wiring and electrical equipment
- Square corners of enclosure body provide maximum interior space and area for conduit openings
- Flat cover provides additional space for mounting a greater number of control devices
- Special neoprene cover gasket provides a water seal to meet NEMA 4 requirements and provides superior protection for enclosure equipment water/corrosion (use EJB-N4)
- Detachable mounting feet provide mounting flexibility. No need to replace enclosure if mounting feet are broken
- Optional stainless steel hinges provide convenient and easy access for inspection, maintenance and system changes

Standard Materials

- Body and cover—copper-free aluminum
  (suffix-SA items and Style C)
  Felaroy® (Styles A, B1, B2)
- Gasket—neoprene
- Cover screws—steel
- Hinges—stainless steel

STYLE C

Nominal Inside Dimensions			
Sides (aa)	Top & bottom (bb)	Depth	Style
12	12	5	C
12	12	6	C
12	12	8	C
16	16	6	C
16	16	8	C
12	18	6	C
12	18	8	C
12	24	8	C
12	24	10	C
18	24	8	C
18	24	10	C
24	24	8	C
24	24	10	C
12	36	8	C
18	36	8	C
18	36	10	C

GU, GUE, GUB Junction Boxes

Application

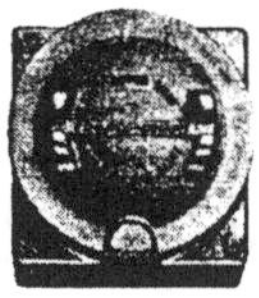

GU, GUE, GUB series junction boxes are used in threaded rigid conduit systems in hazardous areas:

- To function as a splice box, pull box or equipment and device enclosure
- Indoors and outdoors

Features

- Threaded construction throughout permits use in hazardous areas
- Wide variety of conduit hub arrangements
- Hubs provide integral bushings
- Covers are sealed with optional "O" ring gasket when watertight enclosures are required

Compliances

- NEC:
  GU—Class I, Groups C, D
  Class II, Groups E, F, G
  Class III
  GUE and GUB—Class I, Group D
  Class II, Groups E, F, G
  Class III
  GUE, GUB01, GUB02, GUB03, GUB06—also available for Class I, Group B hazardous areas (see listings)
- UL Standard: 886
- Fed. Spec.: W-C-583b

JUNCTION BOXES WITHOUT HUBS

GU: 4⅛″ × 4⅛″ × 3⁷⁄₁₆″ 3⅝″ cover opening	GUE: 4⅝″ × 4⅝″ × 4⅝″ 3⅝″ cover opening
GUB01: 6½″ × 7″ × 5¾″ 5½″ cover opening	GUB02: 8″ × 10″ × 5⅞″ 7″ cover opening
GUB03: 11″ × 12″ × 8¹³⁄₁₆″ 9⅝″ cover opening	GUB06: 8½″ × 10″ × 6⅞″ 7″ cover opening
GUB04: 11″ × 12″ × 8¹¹⁄₁₆″ 9⅝″ cover opening	GUB08: 8½″ × 10″ × 6¹³⁄₁₆″ 7″ cover opening
GUB01110: 14″ × 18″ × 13½″ 12¼″ cover opening	GUB15151: 19″ × 21″ × 16⅝″ 16¾″ cover opening

GUB Junction Boxes

Application

- GUB series junction boxes for drilling and tapping conduit holes are used in threaded rigid conduit systems in hazardous areas:
- To function as a splice box, pull box or equipment and device enclosures
- Indoors and outdoors

Features

- Threaded construction throughout permits use in hazardous areas
- Bodies have thick walls so they can be drilled and tapped for 5 threads to meet NEC requirements for Class I hazardous areas
- Covers are sealed with optional "O" ring gasket when watertight enclosures are required

Compliances

- NEC: Class I, Group D
  Class II, Groups E,,F,G
  Class III
- UL Standard: 886
- Fed. Spec.: W-C-583b

JUNCTION BOXES WITHOUT DRILLED & TAPPED HOLES

GUB220:	8″ × 10″ × 5⅞″	7″ cover opening
GUB260:	8½″ × 10″ × 6⅞″	7″ cover opening
GUB03:	11″ × 12″ × 8¹³⁄₁₆″	9⅝″ cover opening
GUB01110:	14″ × 18″ × 13½″	12¼″ cover opening
GUB15151:	19″ × 21″ × 16⅝″	16¾″ cover opening
GUB04:	11″ × 12″ × 8¹¹⁄₁₆″	9⅝″ cover opening
GUB08:	8½″ × 10″ × 6¹³⁄₁₆″	7″ cover opening

Brazed Hubs For GU, GUE, GUB Boxes

MAXIMUM SIZE HUBS THAT CAN BE USED

		No. of Hubs on Top & Bottom				No. of Hubs on Each Side				No. of Hubs on Back		
	Type of Hub	1	2	3	4	1	2	3	4	1	2	4
GU	Threaded	1¼	1¼	½		1¼	1¼	½		1¼	1	1
	Union	1	1			1	1			1½	1	1
GUE	Threaded	2 (2)	1½ (¾)	¾		2 (2)	1½	¾		1¼ (1¼)	1	1
	Union	2	1	½		2	1	½		1¼	1	1
GUB01	Threaded	2½ (2)	2 (1¼)	1 (¾)	½	2½ (2)	2	1	¾	4 (2)	2½	2
	Union	2½	1½	1	½	2½	1½	1	½	4	1½	1½
GUB02	Threaded	2½ (2)	2½ (2)	1¼(1)	1	2½ (2)	2½	2	1	4 (2)	2½	2½
	Union	2½	2	1	¾	2½	2	1½	1	4	2	2
GUB06	Threaded	3	2½	1½	1	3	2½	2	1	4	2½	2½
	Union	3	2	1	¾	3	2	1½	1	4	2	2
GUB08	Threaded	3	2	1	½	3	2	1½	1	4	2	2
	Union	3	1½	1	½	3	1½	1	1	4	1½	1½
GUB03	Threaded	4 (2)	3½ (2)	2 (1¼)	1¼ (1)	4 (2)	3½	2	1¼	4 (2)	3½	3½
	Union	4	3	1½	1	4	3	2	¼	4	3	3
GUB04	Threaded	4	3	1½	1	4	3	2	1¼	4	3	3
	Union	4	2½	1½	1	4	2½	1½	1	4	2½	2½
GUB01110	Threaded	4	4	3	2	4	4	3½	2½	4	4	4
	Union	4	4	2½	1½	4	4	3	2	4	4	4
GUB15151	Threaded	4	4	3½	2½	4	4	4	3	4	4	4
	Union	4	4	3	2	4	4	3½	2½	4	4	4

Covers for GUB Boxes

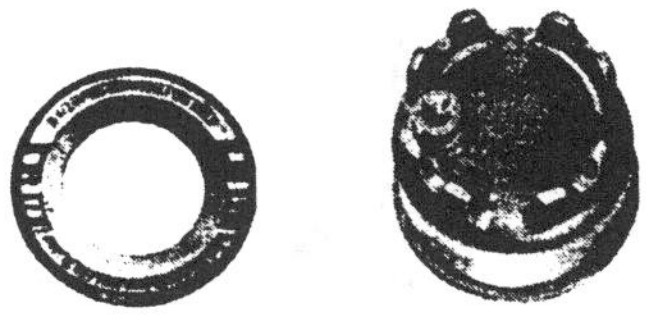

Application

GUB threaded covers are used with GUB boxes in control systems in hazardous areas:

- Indoors and outdoors
- In three categories:

  Flat—for normal use; furnished with standard GUB boxes

  Glass window—to provide visibility of meter indications when used to enclose meters

  Domed—for increasing volume of GUB to make it easier to splice and pull large conductors

Features

- Domed—more suitable for use when splices of heavy conductors are made and enclosed, since the conductors may be pulled in with the ends outside the box. After the splices are made, they do not have to be crowded back into the box
- Glass window—has maximum diameter glassto give best visibility. In selecting, the diameter of the meter face should match or be slightly smaller than window diameter

Compliances

- NEC: GUB714, GUB7110:
  Class I, Groups B,C,D
  Class II, Groups E,F,G
  Class III

  All other covers:
  Class I, Group D
  Class II, Groups E,F,G
  Class III

Body Size	Flat Cover	Glass Window Cover	Dome Cover Nominal Depth
GUB01			4
			10
GUBO2			6
GUB06			17
GUB08			24
GUB03			10
GUB04			17
GUB01110			5
			16
			22
GUB15151			17

Note: GUB covers are suitable for use in hazardous areas only when used with appropriate GUB series enclosures.

GLX Lens Cover Assemblies

For installation in XB, and XJB.

Cat. No.	Lens Diameter
GLX-275	2⁵⁄₁₆
GLX-300	2⁹⁄₁₆
GLX-375	3⁹⁄₁₆
GLX-537	4⅞
GLFX-537	4⅞
GLX-600	5¹³⁄₁₆
GLX-775	7⁷⁄₁₆

Enclosure Size Inside Dimensions (Inches) Width × Length × Depth	Enclosure Size Inside Dimensions (Inches) Width × Length × Depth
EXB- 6 × 6 × 4	EXB-16 × 16 × 6
EXB- 8 × 8 × 6	EXB-16 × 16 × 8
EXB- 8 × 10 × 6	EXB-18 × 24 × 8
EXB-12 × 12 × 6	EXB-18 × 24 × 10
EXB-12 × 12 × 8	EXB-18 × 36 × 8
EXB-12 × 18 × 6	EXB-18 × 36 × 10
EXB-12 × 18 × 8	EXB-24 × 24 × 8
EXB-12 × 24 × 6	EXB-24 × 24 × 10
EXB-12 × 24 × 8	

EXB Series Explosion Proof Junction Box "Quantum"

Innovative Waterproofing
Waterproof gasket is located to the inside of the cover bolts. Prevents water from seeping into enclosure through bolt holes.

Copper-free Aluminum Alloy
Killark K-27® (less than 0.4% copper). High strength. Lightweight. Corrosion resistant. Attractive finish.

Bi-directional Mounting Lugs
Lugs can be positioned in two directions to aid installation in restricted areas.

Hinges Standard
Provides maximum surface protection to flame path surface during installation and servicing.

Fewer Bolts
Revolutionary design eliminates corner bolts. Some models have as many as eight fewer bolts than competitive enclosures. Saves maintenance and installation time.

Recessed Cover Flange
New cover design accepts prying instrument while protecting ground flame path. Cover opens easier.

Flat Covers
Provides maximum mounting area.

Class I, Groups B, C, D
Class II, Groups E, F, G
Class III
NEMA 3, 4, 7, 9

XB Series Explosion-Proof Junction Box

Cast Aluminum
NEMA 7D, 9EFG, Class I, Group D, Class II, Groups E, F and G, Class III, Division 1 and 2 Areas

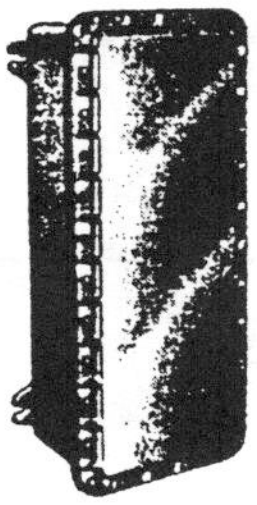

XB·Series

Cat. No.	Inside Box Size W × L × D	Maximum Conduit Size	Mounting Dimensions A	Mounting Dimensions B
XB-444	4 × 4 × 4	2	2	5⅞
XB-464	4 × 6 × 4	2	4	5⅞
XB-664	6 × 6 × 4	2	4	7⅝
XB-684	6 × 8 × 4	2	6	7⅞
XB-6186	6 × 18 × 6	3	12	8⅜
XB-6366	6 × 36 × 6	4	30	8⅜
XB-884	8 × 8 × 4	2	6	10⁷⁄₁₆
XB-886	8 × 8 × 6	2	6	10⁷⁄₁₆
XB-8106	8 × 10 × 6	2	6	10½
XB-10126	10 × 12 × 6	2	6	12½
XB-12126	12 × 12 × 6	2	6	14½
XB-12128	12 × 12 × 8	4	6	14½
XB-12168	12 × 16 × 8	4	12	14½
XB-12245	12 × 24 × 5	3	19½	14¾
XB-12248	12 × 24 × 8	4	19½	14⅜
XB-12328	12 × 32 × 8	4	27½	14⅜
XB-123610	12 × 36 × 10	4	31½	14⅜
XB-14146	14 × 14 × 6	3	9¾	17¹¹⁄₁₆
XB-14286	14 × 28 × 6	3½	22½	17¾
XB-14246	18 × 24 × 6	3	18⅜	21½
XB-20248	20 × 24 × 8	4	18⅜	23½

XJB Series Explosion-Proof Junction Box

Cast Aluminum
NEMA 7D, 9EFG, Class I, Group D,
Class II, Groups E, F and G
Class III, Division I and 2 Areas

XJB Series Inside Box Size W × L × D	Maximum Conduit Size	Mounting Dimensions A	B
5 × 10 × 4	1	5 3/4	7 7/32
5 × 10 × 5	1 1/2	5 3/4	7 3/16
5 × 10 × 6	1 1/2	5 3/4	7 7/32
5 × 13 × 3	1 1/2	8 1/4	7 7/32
5 × 13 × 5	1 1/2	8 1/4	7 1/4
5 × 13 × 6	2	8 1/4	7 1/4
5 × 18 × 3	1 1/2	13 1/4	7 1/4
5 × 18 × 6	2	13 1/4	7 1/4
6 × 10 × 4	1	5 1/2	8 9/32
6 × 10 × 5	1 1/2	5 1/2	8 5/32
6 × 10 × 6	1 1/2	5 1/2	8 5/32
8 × 10 × 4	1	6	10 9/16
8 × 10 × 5	1 1/2	6	10 9/16
8 × 10 × 6	1 1/2	6	10 9/16
8 × 13 × 6	2	8 1/2	10 9/16
8 × 15 × 6	2	11 1/2	10 9/16
8 × 15 × 8	4	11 1/2	10 9/16
8 × 15 × 10	4	11 1/2	10 9/16
8 × 18 × 6	2	13 1/2	10 9/16
8 × 18 × 8	4	13 1/2	10 9/16
8 × 18 × 10	4	13 1/2	10 9/16
8 × 21 × 6	2	16 1/2	10 5/8
10 × 16 × 6	1 1/2	11 1/2	12 5/8
10 × 24 × 6	1 1/2	19 1/2	12 5/8

XJB Series Inside Box Size W × L × D	Maximum Conduit Size	Mounting Dimensions A	B
12 × 19 × 6	1 1/2	14 1/2	14 1/8
12 × 19 × 8	3 1/2	14 1/2	14 1/8
12 × 19 × 10	3 1/2	14 1/2	14 1/8
12 × 24 × 3	1 1/2	19 1/2	14 7/16
12 × 24 × 5	3	19 1/2	14 5/8
12 × 24 × 6	1 1/2	19 1/2	14 7/16
12 × 24 × 8	3	19 1/2	14 5/8
12 × 24 × 10	3	19 1/2	14 5/8
12 × 32 × 3	2	27 1/2	14 1/2
12 × 32 × 5	3 1/2	27 1/2	14 9/16
12 × 32 × 6	2	27 1/2	14 1/2
12 × 32 × 8	3 1/2	27 1/2	14 9/16
12 × 32 × 10	3 1/2	27 1/2	14 9/16
12 × 32 × 11	4	27 1/2	14 3/8
12 × 32 × 13	4	27 1/2	14 3/8
12 × 40 × 6	2	35 1/2	14 9/16
12 × 40 × 8	4	35 1/2	14 9/16
12 × 40 × 10	4	35 1/2	14 9/16
16 × 16 × 6	1 1/2	11 1/2	18 7/8
16 × 16 × 8	4	11 1/2	18 7/8
16 × 16 × 10	4	11 1/2	18 7/8
16 × 24 × 6	2	19 1/2	18 9/16
16 × 24 × 8	4	19 1/2	18 9/16
16 × 24 × 10	4	19 1/2	18 9/16

XJB Series Inside Box Size W × L × D	Maximum Conduit Size	Mounting Dimensions A	B
16 × 32 × 6	2	27 1/2	18 5/8
16 × 32 × 8	4	27 1/2	18 5/8
16 × 32 × 10	4	27 1/2	18 5/8
16 × 40 × 6	2	35 1/2	18 3/4
16 × 40 × 8	3 1/2	35 1/2	18 3/4
16 × 40 × 10	3 1/2	35 1/2	18 3/4
16 × 48 × 4	3	43 9/16	18 1/8
16 × 48 × 6	4	43 9/16	18 1/8
16 × 48 × 8	2 1/2	43 9/16	18 1/8
16 × 48 × 10	4	43 9/16	18 1/8
16 × 48 × 12	4	43 9/16	18 1/8
20 × 24 × 6	2	19 1/2	22 13/32
20 × 24 × 8	4	19 1/2	22 17/32
20 × 24 × 10	4	19 1/2	22 17/32
20 × 32 × 6	2	28 3/8	23 3/8
20 × 32 × 8	4	27 1/2	21 7/8
20 × 32 × 10	4	27 1/2	21 7/8
20 × 40 × 10	4	36	22 3/4
24 × 24 × 6	2	19 1/2	26 7/16
24 × 24 × 8	4	19 1/2	26 7/16
24 × 24 × 10	4	19 1/2	26 7/16
24 × 24 × 11	4	19 1/2	26 7/16
24 × 24 × 13	4	19 1/2	26 7/16
24 × 24 × 16	4	19 1/2	26 7/16

"GR" Series Explosion-Proof Junction Boxes

Compact rectangular enclosures with threaded screw cover.

W/Blank Cover
W/Lens Cover
Lens Diameter
W/2" Dome Cover
W/3" Dome Cover
W/4" Dome cover

W/5" Dome Cover
W/6" Dome Cover
W/8" Dome Cover
W/12" Dome Cover
W/16" Dome Cover

DIMENSIONS								
Length	4 5/8	5 7/16	6 5/8	6 7/8	9 7/8	10 1/2	12	12
Width	4 5/8	5 7/16	5 1/2	6 7/8	7 7/8	8 7/8	11	11
Depth	4	3 1/4	4 7/8	5 3/4	6	6 11/16	7 1/4	15 3/4
Opening Dia.	4	3 3/16	4	5 3/8	7 3/8	8	9 3/4	9 3/4
Max. Cond. Size	1 1/2	2	2	2	2	2	4	4
Mtg. Length	4	—	5 3/4	5 7/8	9 1/8	9 1/2	10 3/4	8 1/4
Mtg. Width	6	—	6 1/2	8 1/8	9 1/8	10 1/2	12 1/8	12 1/8

"GR" Series Junction Box Accessories

Item Description
Blank Cover
Lens Cover
Flanged Lens Cover
Gasket
Dome Cover 2"
3"
4"
5"
6"
8"
12"
16"
Flat Pan
L Type Pan (Blank)
(Listed with (2" dome)
Reference to (3" dome)
Cover Type) (4" dome)
(5" dome)
(6" dome)
(8" dome)
(12" dome)
(16" dome)

"GRD" Series Explosion-Proof Junction Box

Non-Rusting Aluminum
Threaded Hubs—Won't Spark
5 7/16" × 5 7/16" × 4 13/16"
O.A. Dimens. Inc. Cover

"GRR" Series Explosion-Proof Junction Box

3 1/2" diameter, 2 1/4" deep, with blank screw cover, 5 threaded hubs and 3 close-up plugs.

Hub
1/2"
3/4
1

"GRS" Series Explosion-Proof Junction Boxes

Cast aluminum with cover

Furnished	Size
w/7 3/4" hubs	4 1/8 × 2 5/16"
Furnished w/8 3/4" hubs, inc. one 3/4" flush hub in back center	4 5/8 × 3 5/32"

"GRSS" Series Explosion-Proof Junction Boxes

4 5/8" square, 3 9/32" deep including cover. 7 threaded hubs and 4 close-up plugs. One hub is in back.

Hub
1/2"
3/4
1

NEMA 3R Screw Cover

Designed for use as wiring boxes and junction boxes. Provide protection in outdoor installations against rain, sleet and snow, or indoors against dripping water.

Construction

- 16 gauge, 14 gauge, or 12 gauge galvanized steel.
- Drip shield top and seamfree sides, front, and back protection from rain, snow or sleet.
- Slip-on removable cover fastened with screws along bottom edge.
- Door handles included on all 12 gauge enclosures
- Embossed mounting holes on back of enclosure.
- Knockouts in bottom if "C" dimension is less than 8 inches.
- Provisions for padlocking.
- No gasketing.

Enclosure Size A × B × C	Enclosure Size A × B × C
6 × 4 × 4	24 × 12 × 6
6 × 6 × 4	24 × 15 × 6
8 × 6 × 4	24 × 18 × 6
8 × 8 × 4	12 × 12 × 8
8 × 12 × 4	18 × 12 × 8
10 × 8 × 4	18 × 15 × 8
10 × 10 × 4	18 × 18 × 8
12 × 6 × 4	24 × 18 × 8
12 × 8 × 4	24 × 24 × 8
12 × 12 × 4	30 × 24 × 8
15 × 12 × 4	12 × 12 × 10
8 × 6 × 6	16 × 16 × 10
8 × 8 × 6	18 × 18 × 12
10 × 8 × 6	24 × 18 × 12
10 × 10 × 6	24 × 24 × 12
12 × 8 × 6	36 × 24 × 12
12 × 10 × 6	36 × 30 × 12
12 × 12 × 6	42 × 36 × 12
12 × 18 × 6	48 × 36 × 12
15 × 12 × 6	24 × 24 × 16
18 × 12 × 6	30 × 30 × 16
18 × 15 × 6	36 × 36 × 16
18 × 18 × 6	

NEMA 4X Fiberglass Enclosures

Designed to house electrical and electronic controls, instruments, and components in highly corrosive environments indoors or outdoors. Ideal for use in petro-chemical plants, sewage plants, food processing areas, packing plants, electro-plating plants, etc.

Enclosure Size A × B × C
20.19 × 16.25 × 6.00
20.19 × 20.25 × 6.00
24.25 × 20.25 × 6.00
30.25 × 24.25 × 6.00
20.19 × 16.25 × 8.00
20.19 × 20.25 × 8.00
24.25 × 20.25 × 8.00
24.25 × 24.25 × 8.00
30.25 × 24.25 × 8.00
36.25 × 30.25 × 8.00
20.19 × 16.25 × 10.00
20.19 × 20.25 × 10.00
24.25 × 20.25 × 10.00
30.25 × 24.25 × 10.00
20.19 × 16.25 × 12.00
20.19 × 20.25 × 12.00
24.25 × 24.25 × 12.00
30.25 × 24.25 × 12.00
36.25 × 30.25 × 12.00
36.25 × 36.25 × 12.00
48.25 × 36.25 × 12.00
60.25 × 36.25 × 12.00
30.25 × 24.25 × 16.00
48.25 × 36.25 × 16.00
60.25 × 36.25 × 16.00

NEMA 4X Stainless Steel

Designed to house electrical and electronic controls, instruments, and components in areas which may be regularly hosed down or are otherwise very wet. Designed for use in areas where serious corrosion problems exist. Suitable for use outdoors, or in dairies, breweries, petro-chemical plants, sewage plants, food processing areas, packing plants, and similar installations.

STANDARD SIZES

Enclosure Size A × B × C
16 × 12 × 6
16 × 16 × 6
20 × 16 × 6
20 × 20 × 6
24 × 20 × 6
16 × 12 × 8
20 × 16 × 8
24 × 16 × 8
24 × 20 × 8
24 × 24 × 6
24 × 30 × 8
30 × 20 × 8
30 × 24 × 8
30 × 30 × 8
36 × 24 × 8
36 × 30 × 8
42 × 36 × 8
48 × 36 × 8
20 × 16 × 10
24 × 20 × 10
36 × 24 × 10
36 × 30 × 10
42 × 30 × 10
24 × 24 × 12
30 × 24 × 12
36 × 30 × 12
48 × 36 × 12
60 × 36 × 12
30 × 24 × 16
36 × 30 × 16
48 × 36 × 16

Small NEMA 4X Fiberglass Enclosures

Designed for use as electrical junction boxes, terminal wiring boxes, instrument housings, and electrical control enclosures in highly corrosive environments indoors or outdoors. Ideal for use in most petro-chemical plants, sewage plants, food processing areas, packing plants, electro-plating plants, etc.

SCREW COVER ENCLOSURES STANDARD SIZES

ENCLOSURES W/QUICK-RELEASE LATCHES, STD. SIZES

"CHNF" Continuous Hinge Clamp Cover

Designed for use in areas which may be regularly hosed down or are otherwise very wet. Suitable for use in dairies, breweries, and similar installations.

Construction

- 16 gauge or 14 gauge steel.
- Seams continuously welded and ground smooth, no holes or knockouts.
- Stainless steel clamps on three sides of cover assure watertight seal.
- Heavy gauge continuous hinge.
- Cover removed by pulling stainless steel hinge pin.
- Oil-resistant gasket and adhesive.
- Weldnuts provided for mounting OPTIONAL panels and terminal kits in size 6″ × 4″ and larger.

- Specify side to be hinged when ordering special boxes.

Gauge	Box Size A × B × C
16	4 × 4 × 3
16	6 × 4 × 3
14	8 × 6 × 3.50
16	6 × 6 × 4
14	10 × 8 × 4
14	12 × 10 × 5
14	10 × 8 × 6
14	12 × 12 × 6
14	14 × 12 × 6
14	16 × 14 × 6

"CH" Continuous Hinge Clamp Cover

Designed for use as electrical junction boxes, terminal wiring boxes, instrument housings, and electrical control enclosures. Provide protection from dust, dirt, oil, and water.

Gauge	Box Size A × B × C
16	4 × 4 × 3
16	6 × 4 × 3
14	8 × 6 × 3.50
14	8 × 6 × 3.50
16	4 × 4 × 4
16	6 × 4 × 4
16	6 × 6 × 4
14	8 × 8 × 4
14	10 × 8 × 4
14	10 × 8 × 4
14	12 × 6 × 4
14	12 × 10 × 5
14	12 × 10 × 5
14	8 × 6 × 6
14	10 × 8 × 6
14	10 × 10 × 6
14	12 × 12 × 6
14	14 × 8 × 6
14	14 × 12 × 6
14	14 × 12 × 6
14	16 × 10 × 6
14	16 × 14 × 6
14	16 × 14 × 6
14	12 × 10 × 8
14	14 × 12 × 8
14	16 × 14 × 8
14	16 × 14 × 10

JIC/NEMA 4-4X Stainless Steel

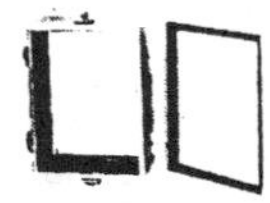

Designed for use in areas which may be regularly hosed down or are otherwise very wet. Suitable for outdoors, or in dairies, breweries, and similar installations. Also designed for use in areas where serious corrosion problems exist.

Gau.	Box. Size A × B × C	Stainless Steel Panel	Steel Panel
14	4 × 4 × 3		
14	6 × 4 × 4		
14	8 × 6 × 4		
14	6 × 6 × 4		
14	10 × 8 × 4		
14	12 × 10 × 6		
14	14 × 12 × 6		
14	16 × 14 × 6		

NEMA 12 Single Door

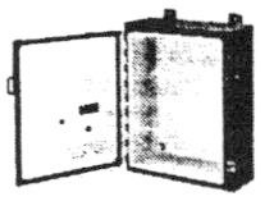

Designed to house electrical and electronic controls, instruments, and components. Provide protection from dust, dirt, oil, and water.

Construction

- 14 gauge steel
- Seams continuously welded and ground smooth, no holes or knockouts
- Door and body stiffeners in larger enclosures
- Rolled lip around three sides of door and all sides of enclosure opening excludes liquids and contaminants
- Door clamps are quick and easy to operate
- Door removed by pulling heavy gauge continuous hinge pin
- Hasp and staple for padlocking
- Removable and reversible print pocket
- Oil-resistant gasket attached with oil-resistant adhesive and held in place with steel retaining strips
- Collar studs provided for mounting optional panels

Enclosure Size A × B × C	Enclosure Size A × B × C
12 × 24 × 6	16 × 12 × 10
16 × 12 × 6	20 × 16 × 10
16 × 16 × 6	20 × 20 × 10
16 × 20 × 6	24 × 12 × 10
20 × 12 × 6	24 × 20 × 10
20 × 16 × 6	24 × 24 × 10
20 × 20 × 6	30 × 20 × 10
20 × 24 × 6	30 × 24 × 10
24 × 12 × 6	36 × 24 × 10
24 × 16 × 6	36 × 30 × 10
24 × 20 × 6	42 × 30 × 10
24 × 24 × 6	42 × 36 × 10
30 × 16 × 6	48 × 30 × 10
30 × 20 × 6	48 × 36 × 10
30 × 24 × 6	60 × 36 × 10
36 × 24 × 6	20 × 16 × 12
36 × 30 × 6	24 × 20 × 12
12 × 24 × 8	24 × 24 × 12
16 × 12 × 8	30 × 24 × 12
16 × 16 × 8	30 × 30 × 12
16 × 20 × 8	36 × 24 × 12
20 × 12 × 8	36 × 30 × 12
20 × 16 × 8	36 × 36 × 12
20 × 20 × 8	42 × 30 × 12
20 × 24 × 8	42 × 36 × 12
24 × 12 × 8	48 × 36 × 12
24 × 16 × 8	60 × 36 × 12
24 × 20 × 8	72 × 30 × 12
24 × 24 × 8	72 × 36 × 12
24 × 30 × 8	24 × 20 × 16
30 × 20 × 8	24 × 24 × 16
30 × 24 × 8	30 × 24 × 16
30 × 30 × 8	36 × 30 × 16
30 × 36 × 8	42 × 36 × 16
36 × 24 × 6	48 × 36 × 16
36 × 30 × 8	60 × 36 × 16
36 × 36 × 8	72 × 30 × 16
42 × 24 × 8	30 × 24 × 20
42 × 30 × 8	36 × 30 × 20
42 × 36 × 8	48 × 36 × 20
48 × 24 × 8	60 × 36 × 20
48 × 30 × 8	72 × 30 × 20
48 × 36 × 8	30 × 24 × 24
60 × 36 × 8	72 × 30 × 24

Drawn Screw Cover Box

Designed for use as electrical junction boxes, terminal wiring boxes, and instrument housings. Suitable for use with liquid-tight flexible conduit. Provides protection from dust, dirt, oil, and water.

Box Size A × B × C
3.50 × 4.00 × 2.75
5.75 × 4.00 × 2.75
8.00 × 4.00 × 2.75
10.25 × 4.00 × 2.75
3.50 × 4.00 × 3.75
5.75 × 4.00 × 3.75
8.00 × 4.00 × 3.75
10.25 × 4.00 × 3.75

NEMA 12 Two Door Floor Mounted

Designed to house electrical and electronic controls, instruments, and components. Provide protection from dust, dirt, oil, and water.

FINISH—White enamel inside and gray prime outside over phosphatized surfaces. OPTIONAL panels are white enamel.

Construction

- Catalog number A-544208LP is fabricated from 14 gauge steel. All other enclosures are 12 gauge steel (10 gauge back when A=72.06).
- Seams continuously welded and ground smooth, no holes or knockouts.
- Strong, rigid construction with body stiffeners.
- Gasketed overlapping doors eliminate need for center post.
- 3-point latching mechanism operated by oil-tight key-locking handle.
- Latch rods have rollers for easier door closing.
- Heavy gauge continuous hinges support each door.
- Removable print pocket.
- 12-inch floor stands.
- Heavy duty lifting eyes.
- Panel supports.
- Oil-resistant door gasket attached with oil-resistant adhesive and held in place with steel retaining strips.
- Collar studs provided for mounting optional panels.

STANDARD SIZES

Enclosure Size A × B × C	Panel Size	Enclosure Size A × B × C	Panel Size
54.00 × 42.00 × 8.00	50 × 38	60.06 × 60.06 × 16.06	56 × 56
60.06 × 48.06 × 8.06	56 × 44	72.06 × 60.06 × 16.06	68 × 56
60.06 × 48.06 × 10.06	56 × 44	72.06 × 72.06 × 16.06	68 × 68
60.06 × 60.06 × 10.06	56 × 56	60.06 × 48.06 × 20.06	56 × 44
72.06 × 60.06 × 10.06	68 × 56	72.06 × 60.06 × 20.06	68 × 56
72.06 × 72.06 × 10.06	68 × 68	72.06 × 72.06 × 20.06	68 × 68
60.06 × 48.06 × 12.06	56 × 44	60.06 × 48.06 × 24.06	56 × 44
60.06 × 60.06 × 12.06	56 × 56	60.06 × 60.06 × 24.06	56 × 56
72.06 × 60.06 × 12.06	68 × 56	72.06 × 60.06 × 24.06	68 × 56
72.06 × 72.06 × 12.06	68 × 68	72.06 × 72.06 × 24.06	68 × 68
60.06 × 48.06 × 16.06	56 × 44		

NEMA 12 Enclosures for Allen-Bradley Disconnects

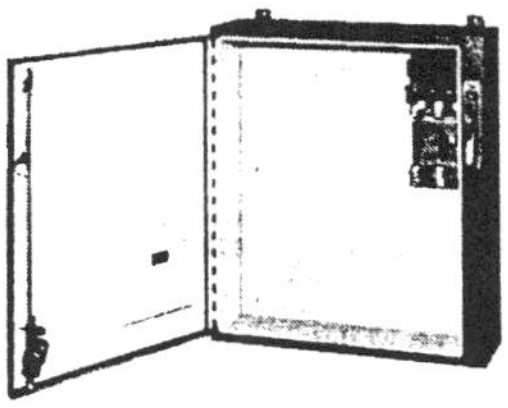

Designed to house Allen-Bradley Bulletin 1494F flange-mounted disconnect switches and Bulletin 1494D flange-mounted operators for circuit breakers. (See page 85 for single door enclosures which receive Allen-Bradley Bulletin 1494V disconnect devices.) Provide protection from dust, dirt, oil, and water.

Construction

- 14 gauge steel
- Seams continuously welded and ground smooth
- Door and body stiffeners in larger enclosures for extra rigidity
- Rolled lip around all sides of enclosure opening excludes liquids and contaminants
- Door removed by pulling continuous hinge pin
- Removable print pocket
- Oil-resistant door gasket attached with oil-resistant adhesive and held in place with steel retaining strips
- Collar studs provided for mounting optional panels
- Holes provided in body and door for mounting disconnect switch, operating mechanism, and door for mounting disconnect switch, operating mechanism, and door hardware
- Furnished with enclosure:
  - Guide brackets and door catch welded to door
  - Locking bar(s) for door hardware
  - Spacers and screws to support disconnect to panel
  - Instructions to locate and install disconnects

NOTE: Disconnect switch and operating mechanism are not furnished with the enclosure

Door hardware is not furnished with the enclosure. Enclosures with an "A" dimension under 40 inches require 2-point door hardware (one Allen-Bradley catalog number 1494F-L1). Enclosures with an "A" dimension of 40 inches or more require 3-point door hardware (one Allen-Bradley catalog number 1494F-L2 and one catalog number 1494F-L3). Exception: Enclosure catalog numbers A-48AB3812A2LP and A-60AB3812A2LP require one Allen-Bradley catalog number 1494-L4 3-point door hardware.

Enclosure Size A × B × C	H
20 × 21.38 × 8	—
24 × 21.38 × 8	—
24 × 25.38 × 8	—
30 × 21.38 × 8	—
30 × 25.38 × 8	—
36 × 25.38 × 8	—
36 × 31.38 × 8	9.88
42 × 31.38 × 8	9.88
42 × 37.38 × 8	9.88
48 × 37.38 × 8	11.38
60 × 37.38 × 8	14.38
24 × 21.38 × 10	—
24 × 25.38 × 10	—
30 × 21.38 × 10	—
30 × 25.38 × 10	—
36 × 25.38 × 10	—
36 × 31.38 × 10	9.00
42 × 31.38 × 10	9.88
42 × 37.38 × 10	9.88
48 × 37.38 × 10	11.38
60 × 37.38 × 10	14.38
36 × 31.38 × 12	9.00
42 × 31.38 × 12	9.88
42 × 37.38 × 12	9.88
48 × 37.38 × 12	11.38
60 × 37.38 × 12	14.38
48 × 37.38 × 16	11.38
60 × 37.38 × 16	14.38
48 × 37.50 × 12	11.38
60 × 37.50 × 12	14.38

Screw Cover Pull Boxes

Designed for use as junction boxes and pull boxes. For flush installations, flush covers and screw cover pull boxes must be ordered separately.

WITH KNOCKOUTS

Enamel	Galvanized
4 × 4 × 4	4 × 4 × 4
6 × 4 × 4	6 × 4 × 4
6 × 4 × 4	6 × 6 × 4
8 × 6 × 4	8 × 6 × 4
8 × 8 × 4	8 × 8 × 4
10 × 8 × 4	10 × 8 × 4
10 × 10 × 4	10 × 10 × 4
12 × 6 × 4	—
12 × 8 × 4	12 × 8 × 4
12 × 10 × 4	12 × 10 × 4
12 × 12 × 4	12 × 12 × 4
15 × 12 × 4	15 × 12 × 4
15 × 15 × 4	—
18 × 12 × 4	18 × 12 × 4
18 × 15 × 4	—
6 × 6 × 6	6 × 6 × 6
8 × 6 × 6	8 × 6 × 6
8 × 8 × 6	8 × 8 × 6
10 × 8 × 6	10 × 8 × 6
10 × 10 × 6	10 × 10 × 6
12 × 8 × 6	12 × 8 × 6
12 × 10 × 6	12 × 10 × 6
12 × 12 × 6	12 × 12 × 6
15 × 12 × 6	15 × 12 × 6
15 × 15 × 6	15 × 15 × 6
18 × 12 × 6	18 × 12 × 6
18 × 15 × 6	18 × 15 × 6
18 × 18 × 6	18 × 18 × 6
24 × 18 × 6	24 × 18 × 6
24 × 24 × 6	24 × 24 × 6
30 × 24 × 6	30 × 24 × 6
6 × 6 × 4	6 × 6 × 4
8 × 6 × 4	8 × 6 × 4
8 × 8 × 4	8 × 8 × 4
10 × 8 × 4	10 × 8 × 4
10 × 10 × 4	10 × 10 × 4
12 × 8 × 4	12 × 8 × 4
12 × 10 × 4	12 × 10 × 4
12 × 12 × 4	12 × 12 × 4
	15 × 12 × 4
6 × 6 × 6	6 × 6 × 6
8 × 8 × 6	8 × 8 × 6
10 × 8 × 6	—
10 × 10 × 6	10 × 10 × 6
12 × 12 × 6	12 × 12 × 6
15 × 12 × 6	15 × 12 × 6
18 × 12 × 6	18 × 12 × 6
18 × 18 × 6	18 × 18 × 6
24 × 18 × 6	24 × 18 × 6
24 × 24 × 6	24 × 24 × 6
24 × 18 × 8	—
24 × 24 × 8	24 × 24 × 8
30 × 24 × 8	—
30 × 30 × 8	30 × 30 × 8
36 × 24 × 8	—
36 × 30 × 8	—
36 × 36 × 12	36 × 36 × 12

Designline Electronic Enclosures

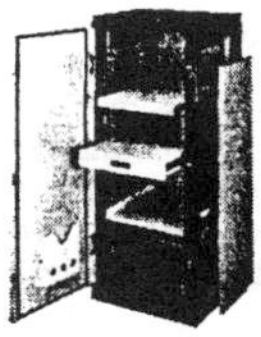

Designed to house electrical and electronic controls, instruments, and components. Provide protection from dust, dirt, oil, and water. Standard enclosure assemblies or standard frame assemblies and optional accessories are available to meet the requirements of machine tool centers, security systems, electronic systems, automated mass production centers, telecommunications, environmental control systems, and industrial process control centers. Smooth, blended surfaces and attractive finish complement other electronic equipment.

Enclosure Size A × B × C
73.91 × 24.12 × 24.12
73.91 × 24.12 × 29.12
73.91 × 29.12 × 24.12
73.91 × 29.12 × 29.12
56.41 × 24.12 × 24.12
56.41 × 24.12 × 29.12

Push Button Enclosures

Designed to hold all standard brands of oil-tight pushbuttons, switches, and pilot lights. Provide protection against dirt, dust, oil, and water.

Holes	Enclosure Size A × B × C
1	3.50 × 3.25 × 2.75
2	5.75 × 3.25 × 2.75
0	5.75 × 3.25 × 2.75
3	8.00 × 3.25 × 2.75
0	8.00 × 3.25 × 2.75
4	10.25 × 3.25 × 2.75
4	7.25 × 6.25 × 3.00
5	12.50 × 3.25 × 2.75
6	14.75 × 3.25 × 2.75
6	9.50 × 6.25 × 3.00
7	18.00 × 3.25 × 2.75
8	20.25 × 3.25 × 2.75
9	22.50 × 3.25 × 2.75
9	9.50 × 8.50 × 3.00
10	24.75 × 3.25 × 2.75
12	11.75 × 8.50 × 3.00
16	11.75 × 10.75 × 3.00
20	14.00 × 10.75 × 3.00
25	14.00 × 13.00 × 3.00

Designline Instrument Enclosures

Designed to house a variety of electrical and electronic controls, instruments, and components in indoor environments where dust, dirt, oil, water, or other contaminants may be encountered. Extra-deep cover is ideal for mounting meters or gauges. Smooth, blended surfaces and attractive finish complement hi-tech electronic equipment.

Enclosure Size A × B × C
6 × 6 × 4
8 × 6 × 4
10 × 8 × 4
12 × 10 × 4
6 × 6 × 6
8 × 6 × 6
10 × 8 × 6
12 × 10 × 6
12 × 10 × 8

Designline NEMA 13

Designed to house electrical and electronic controls, instruments, and components in indoor areas where dirt, dust, oil, water, or other contaminants are present. Smooth, blended surfaces and attractive finish complement hi-tech electronic equipment.

Enclosure Size A × B × C
12 × 12 × 6
16 × 12 × 6
16 × 16 × 6
20 × 16 × 6
20 × 20 × 6
20 × 16 × 6
20 × 20 × 6
24 × 20 × 8
24 × 24 × 8
30 × 24 × 8
30 × 30 × 8
36 × 24 × 8
36 × 30 × 8
30 × 24 × 12
36 × 36 × 12
42 × 36 × 12

JIC Wiring Troughs

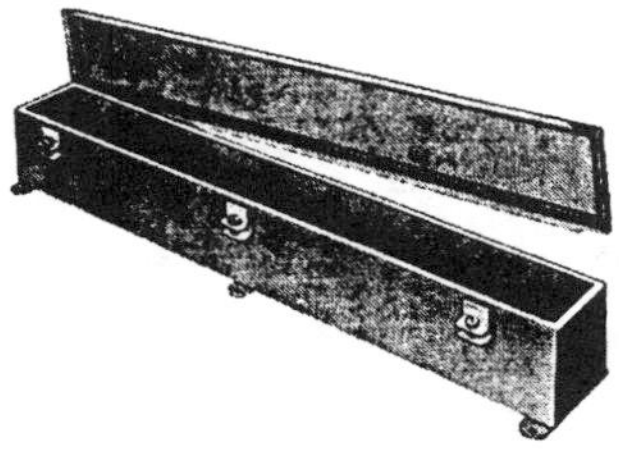

Designed to protect wiring from dust, dirt, oil, and water. Available in sizes: 2.50" × 2.50", 4" × 4", and 6" × 6".

Construction

- 14 gauge steel covers and bodies (catalog numbers F-2212 and F-2218 are 16 gauge steel).
- Seams continuously welded and ground smooth, no holes or knockouts.
- Lift-off hinges and external screw clamps assure complete seal between covers and bodies.
- Screw clamps are quick and easy to operate.
- Chain holds cover to body when cover is removed.
- Oil-resistant gasket and adhesive.

Trough A × B × C
2.50 × 2.50 × 12
2.50 × 2.50 × 18
2.50 × 2.50 × 24
4.00 × 4.00 × 12
4.00 × 4.00 × 18
4.00 × 4.00 × 24
4.00 × 4.00 × 36
4.00 × 4.00 × 48
6.00 × 6.00 × 12
6.00 × 6.00 × 18
6.00 × 6.00 × 24
6.00 × 6.00 × 36
6.00 × 6.00 × 48

Transformers

Hevi-Duty Low Voltage General Purpose Transformers

Since no vaults are required for installation, they can be located near the load, eliminating the need for long expensive low voltage feeder lines.

The enclosures may be supplied either ventilated or totally enclosed for indoor or outdoor use.

Common applications are resistive loads such as lighting and heating. Industrial uses include providing power for manufacturing machinery such as conveyors, shears, ovens, presses and welding equipment.

Units are available for both indoor and outdoor application.

- Low voltage general purpose transformers are used primarily for step down applications from 600 to 480 volts primary to 120, 208, 240 or 277 volts secondary.

Single Phase

KVA	Design Style
240 × 480V Primary 120/240V Secondary	
NO TAPS	
.050	A
.075	A
.100	A
.150	A
.250	A
.500	B
.750	B
1.0	B
1.5	B
2.0	B
3.0	B
5.0	B
7.5	B
10	B
WITH 2-2½% FCAN & 4-2½% FCBN TAPS	
15	C
25	C
37.5	C
50	C
75	C
100	C
167	C
480V Primary 120/240V Secondary	
WITH 2-5% FCBN TAPS	
3	B
5	B
7.5	B
10	B

Three Phase

KVA	Design Style
480V Delta Primary 208Y/120V Secondary	
WITH 2-5% FCBN TAPS	
3	B
6	B
9	B
15	B
WITH 2-2½% FCAN & 4-2½% FCBN TAPS	
15	C
25	C
30	C
37.5	C
45	C
50	C
75	C
112.5	C
150	C
225	C
300	C
500	C
480V Delta Primary 240V Delta Secondary/120 CT	
WITH 2-2½% FCAN & 4-2½% FCBN TAPS	
15	C
30	C
45	C
75	C
112.5	C

Buck-Boost Transformers

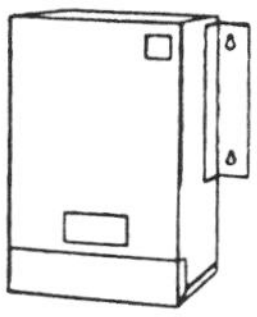

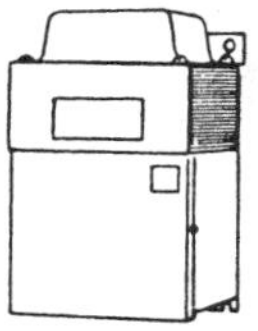

Hevi-Duty Buck-Boost Transformers are insulating transformers with low voltage secondary windings. By field connecting the primary and secondary windings in accordance with recommended instructions, the transformer is "field converted" to an autotransformer which will buck or boost the supply voltage a small amount, usually no more than ±20%. Autotransformers are inherently much more economical to purchase and operate compared to isolation transformers of equivalent rating. This is why buck-boost transformers are the most practical and economical solution for changing line voltages small amounts. Most common applications are boosting 208V. to 230V., 230V. to 277V. and vice-versa. When a buck-boost transformer is wired as an autotransformer, only a fraction of the load KVA is actually transformed. In an autotransformer circuit, the majority of the load KVA passes directly from the supply to the load. This is why a buck-boost transformer can supply a load which has much larger KVA rating than the nameplate indicates.

KVA/ Voltage	120 × 240 12/24	120 × 240 16/32	240 × 480 24/48	Approx. Weight (Lbs.)
.050				3
.100				3⅞
.150				5
.250				7
.500				22
.750				27
1.0				28
1.5				40
2.0				46
3.0				62
5.0				100

Encapsulated Transformers for Hazardous Locations

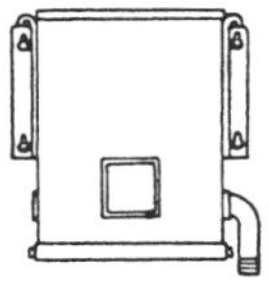

There is a growing need today for local power transformation in hazardous location areas where liquid-filled and standard dry type transformers are not suitable. Such applications are where volatile gases, dust, or liquids are present in dangerous concentrations. In these instances, a spark or excess heat could cause a fire or explosion. These transformers are designed to preclude the need for large, expensive explosion-proof boxes to cover the transformer. The basic unit is a standard Hevi-Duty type HS or HT low voltage general purpose unit. However, in these designs the normal wiring compartment is not air filled. Rather, all wiring leads for normal connections and taps are carried out of the unit through rigid conduit. Then the connection chamber is completely filled with encapsulant, excluding all air. The leads are of sufficient length to allow connection by the installer in an adjacent explosion-proof conduit box.

KVA	Volt. Code	Primary Volts	Secondary Volts
	Voltage Codes		
SINGLE PHASE			
.500			
.750	1	240 × 480	120/240
1	6	120 × 240	120/240
1.5	7	208	120/240
2	9	277	120/240
3	10	600	120/240
5			
7.5			
10			
15			
25			

KVA	Volt. Code	Primary Volts	Secondary Volts
	Voltage Codes		
THREE PHASE			
3			
6	1	480	208Y/120
9	4	480	240
15	6	240	208Y/120
30	7	600	208Y/120
45	8	600	240

Installation Details

S.I.P. Engineering, Inc. Houston, Texas	ENGINEERING STANDARDS ELECTRICAL INSTR. INTRINSICALLY SAFE AND THERMOCOUPLE (GENERAL PURPOSE)	ENGR. STD. D-I01 PAGE: 1 OF 2

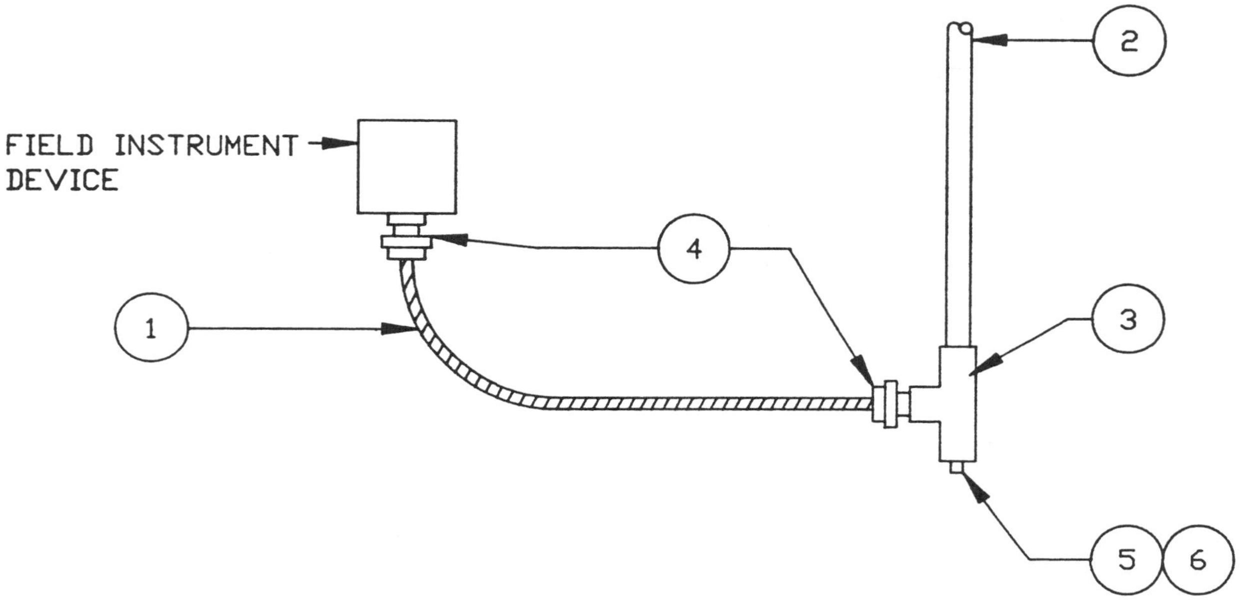

DETAIL #*

INSTR. CONNECTION

(PLAN DWG. #E-XXXXX)

NOTES:

1. CATALOG NUMBERS AND VENDOR DESIGNATIONS ARE FOR REFERENCE ONLY. SUBSTITUTION OF ANY APPROVED EQUAL MAY BE MADE.
2. GENERAL PURPOSE & CLASS I, DIV. 2.
3. FOR LIST OF MATERIALS, SEE SH. 2 OF 2.

APPROVED ____ DATE 1-04-91 ____ REVISION 00 ____ DATE 1-04-91

PC ACAD: DI01

S.I.P. Engineering, Inc.
Houston, Texas

ENGINEERING STANDARDS
ELECTRICAL
INSTR. INTRINSICALLY SAFE
AND THERMOCOUPLE (GENERAL PURPOSE)

ENGR. STD. D-I01

PAGE 2 OF 2

LIST OF MATERIAL

ITEM NO.	CODE		QTY.	DESCRIPTION
1	FLX	UA	AR	FLEXIBLE CONDUIT, TYPE UA
2	CND	**	AR	CONDUIT
3	T	**	1	CONDUIT FITTING, TYPE T
4	CONN ST	UA	2	CONNECTOR, SEALTITE, STRAIGHT
5	RCC		1	REDUCER
6	DRAIN		1	DRAIN, C-H ECD

** MATERIAL COMPATIBLE WITH CONDUIT REQUIREMENTS

APPROVED DATE 1-04-91 REVISION 00 DATE 1-04-91

PC ACAD: DI01

S.I.P. Engineering, Inc.
Houston, Texas

ENGINEERING STANDARDS
ELECTRICAL
INSTR. INTRINSICALLY SAFE
AND THERMOCOUPLE CONNECTION

ENGR. STD.	D-I02
PAGE:	1 OF 2

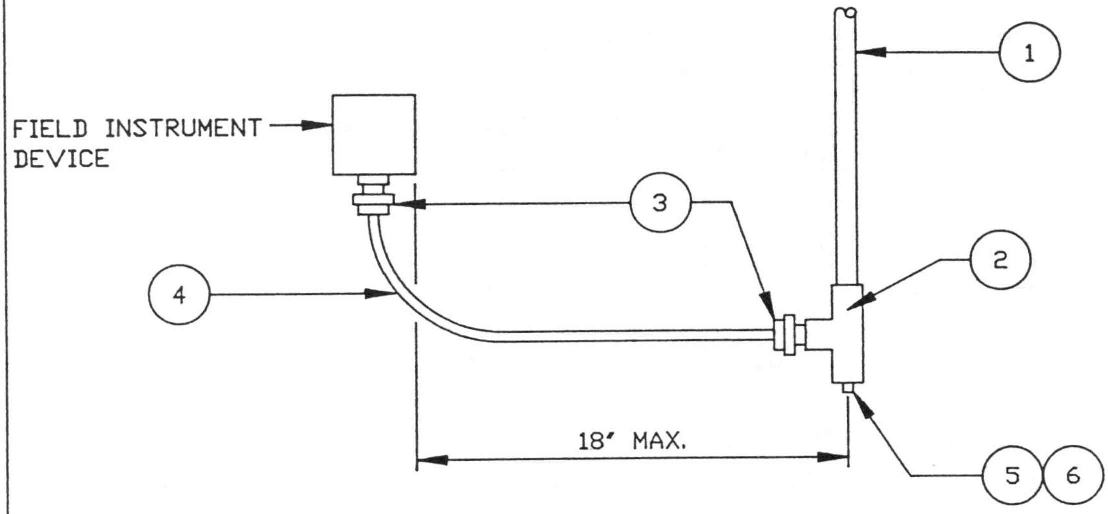

DETAIL #*

INSTR. CONNECTION
(PLAN DWG. #E-XXXXX)

NOTES:

1. CATALOG NUMBERS AND VENDOR DESIGNATIONS ARE FOR REFERENCE ONLY. SUBSTITUTION OF ANY APPROVED EQUAL MAY BE MADE.
2. FOR LIST OF MATERIALS, SEE SH. 2 OF 2.

APPROVED DATE 1-04-91 REVISION 00 DATE 1-04-91

PC ACAD: DI02

S.I.P. Engineering, Inc.
Houston, Texas

ENGINEERING STANDARDS
ELECTRICAL
INSTR. INTRINSICALLY SAFE
AND THERMOCOUPLE CONNECTION

ENGR. STD.
D-I02
PAGE: 2 OF 2

LIST OF MATERIAL

ITEM NO.	CODE	QTY.	DESCRIPTION
1	CND **	AR	CONDUIT
2	T **	1	CONDUIT FITTING, TYPE T
3	CONNCGB **	2	CABLE TERMINATOR, TYPE CGB
4	CABLE T/C **	AR	CABLE, PVC INSULATED
5	RCC	1	REDUCER
6	DRAIN	1	DRAIN, C-H ECD

** MATERIAL COMPATIBLE WITH CONDUIT REQUIREMENTS

APPROVED DATE 1-04-91 REVISION 00 DATE 1-04-91

PC ACAD: D102

S.I.P. Engineering, Inc.
Houston, Texas

ENGINEERING STANDARDS
ELECTRICAL
INSTR. NOT INTRINSICALLY SAFE
CONNECTION (CLASS 1, DIV. 2)

ENGR. STD.
D-I03
PAGE: 1 OF 2

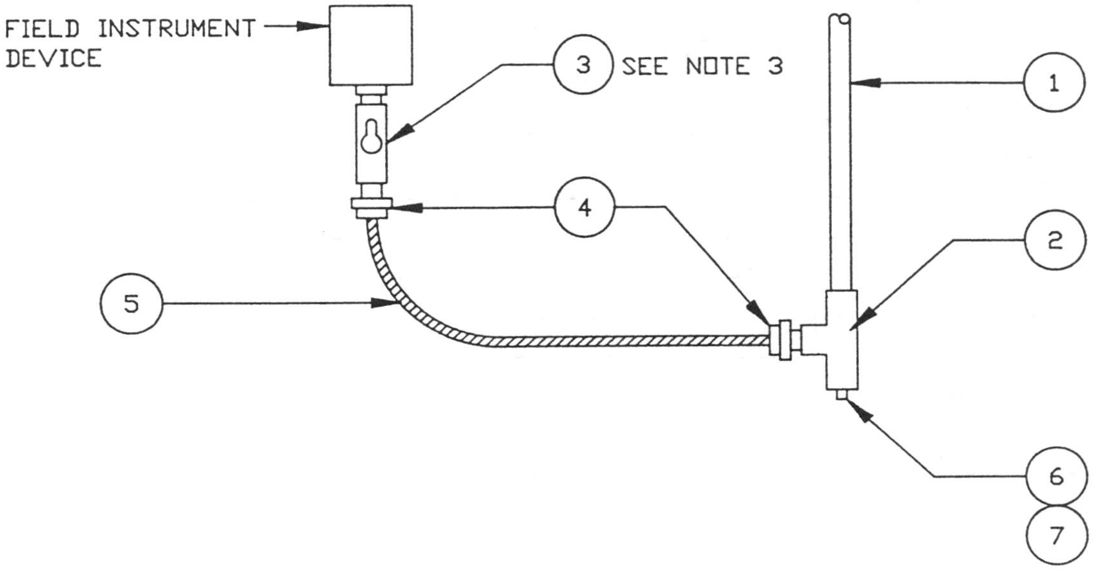

DETAIL #*

INSTR. CONNECTION
(PLAN DWG. #E-XXXXX)

NOTES:

1. CATALOG NUMBERS AND VENDOR DESIGNATIONS ARE FOR REFERENCE ONLY. SUBSTITUTION OF ANY APPROVED EQUAL MAY BE MADE.
2. HAZARDOUS, CLASS 1, DIV. 2.
3. SEAL REQUIRED ONLY IF ARCHING CONTACTS ARE SPECIFIED.
4. FOR LIST OF MATERIALS, SEE SH. 2 OF 2.

APPROVED DATE 1-04-91 REVISION 00 DATE 1-04-91

PC ACAD: D103

S.I.P. Engineering, Inc.
Houston, Texas

ENGINEERING STANDARDS
ELECTRICAL
INSTR. NOT INTRINSICALLY SAFE
CONNECTION (CLASS 1, DIV. 2)

ENGR. STD. D-I03

PAGE 2 OF 2

LIST OF MATERIAL

ITEM NO.	CODE		QTY.	DESCRIPTION
1	CND	**	AR	CONDUIT
2	T	**	1	CONDUIT FITTING, TYPE T
3	EYD	**	1	SEAL, DRAIN
4	CONN ST	UA	2	CONNECTOR, SEALTITE, STRAIGHT
5	FLX	UA	AR	FLEXIBLE CONDUIT, TYPE UA
6	RE		1	REDUCER
7	DRAIN		1	DRAIN, C-H ECD

** MATERIAL COMPATIBLE WITH CONDUIT REQUIREMENTS

APPROVED DATE 1-04-91 REVISION 00 DATE 1-04-91

PC ACAD: DI03

S.I.P. Engineering, Inc.
Houston, Texas

ENGINEERING STANDARDS
ELECTRICAL
INSTR. CONNECTION LOCATED IN
CLASS 1, DIV. 1

ENGR. STD.
D-I04
PAGE 1 OF 2

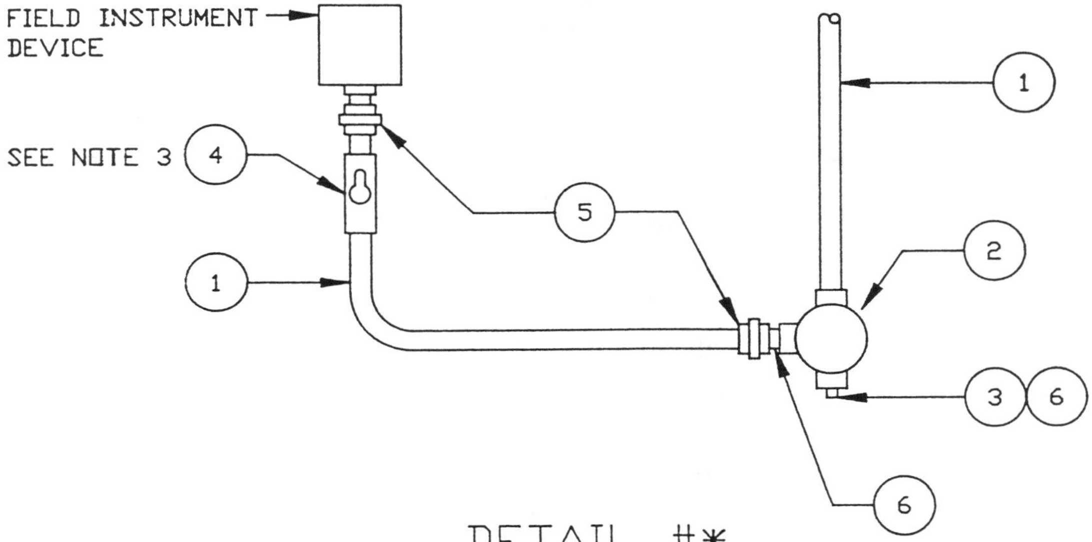

DETAIL #*

INSTR. CONNECTION
(PLAN DWG. #E-XXXXX)

NOTES:

1. CATALOG NUMBERS AND VENDOR DESIGNATIONS ARE FOR REFERENCE ONLY. SUBSTITUTION OF ANY APPROVED EQUAL MAY BE MADE.
2. HAZARDOUS, CLASS I, DIV. 1.
3. SEAL REQUIRED ON ALL DEVICES LOCATED IN HAZARDOUS AREA.
4. FOR LIST OF MATERIALS, SEE SH. 2 OF 2.

APPROVED ____ DATE 1-04-91 REVISION 00 DATE 1-04-91

PC ACAD\ I104

S.I.P. Engineering, Inc. Houston, Texas	ENGINEERING STANDARDS ELECTRICAL INSTR. CONNECTION LOCATED IN CLASS 1, DIV. 1	ENGR. STD. D-I04 PAGE: 2 OF 2

LIST OF MATERIAL

ITEM NO.	CODE		QTY.	DESCRIPTION
1	CND	**	AR	CONDUIT
2	GUAT XP	**	1	CONDUIT FITTING, EXPLOSION-PROOF, TYPE GUA
3	ECD	SST	1	EXPLOSION-PROOF BREATHER/DRAIN
4	EYD	**	1	SEAL, DRAIN
5	UNY	**	2	UNION
6	RE	**	2	REDUCER

** MATERIAL COMPATIBLE WITH CONDUIT REQUIREMENTS

APPROVED DATE 1-04-91 REVISION 00 DATE 1-04-91

PC ACAD: D104

S.I.P. Engineering, Inc.
Houston, Texas

ENGINEERING STANDARDS
ELECTRICAL
INSTR. JUNCTION BOX ASSEMBLY
WITH BACK PANEL (16 PR.)

ENGR. STD.
D-JB01A
PAGE:

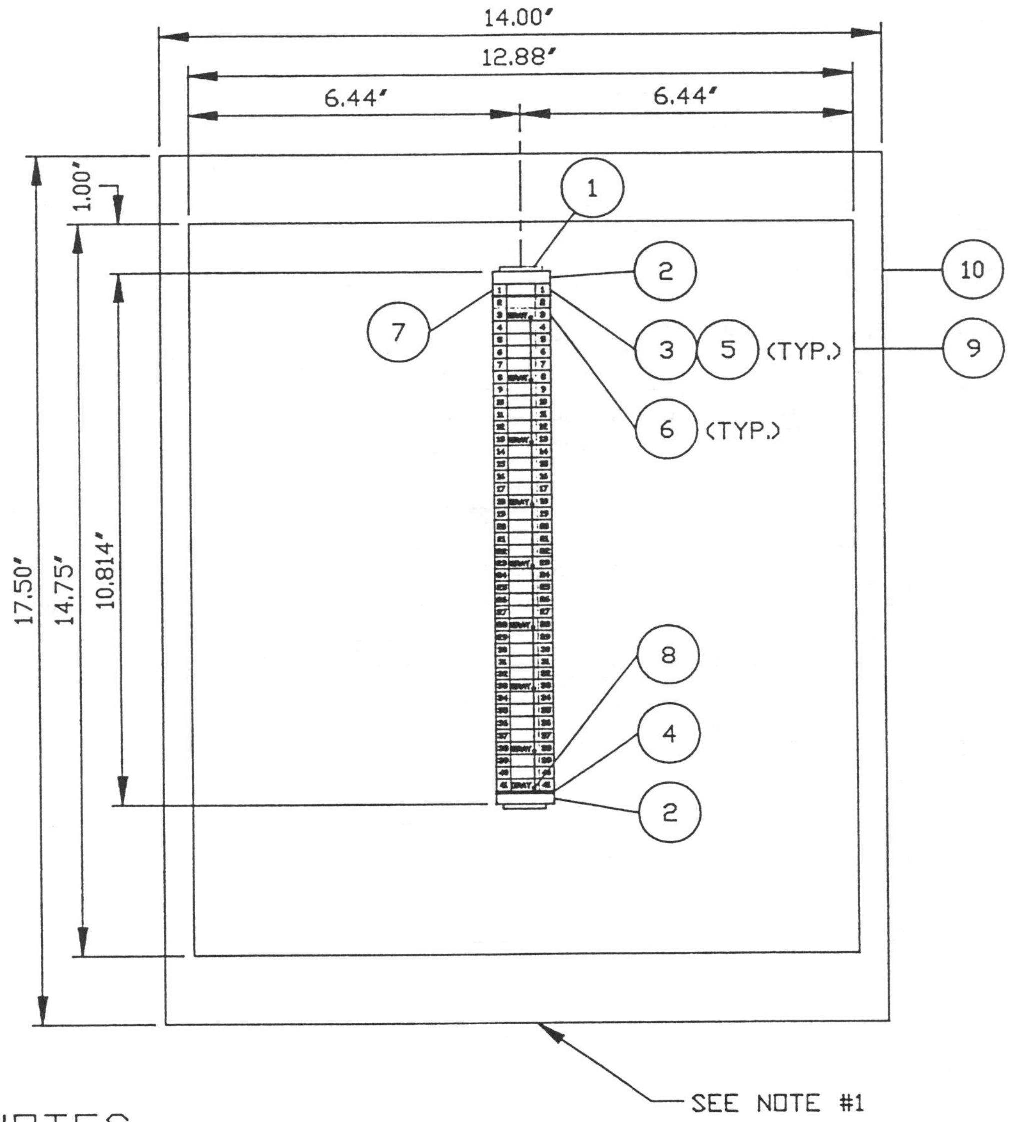

NOTES:

1. DRILL 1/8" DIAMETER WEEP HOLE IN THE BOTTOM REAR CORNER OF JUNCTION BOX FOR DRAINAGE.

*NOTE TO CAD OPERATORS:

1. SCALE THIS STANDARD UP FROM A REFERENCE DIMENSION OF 5.500" TO 17.500" TO GET FULL SCALE DRAWING.
2. DIMENSIONS AND BALLOON NUMBER CALL OUTS ARE DRAWN LARGER THAN ACTUAL DRAWING FOR CLARITY.

APPROVED DATE 1-04-91 REVISION 00 DATE 1-04-91

PC ACAD: D-JB01A

S.I.P. Engineering, Inc.
Houston, Texas

ENGINEERING STANDARDS
ELECTRICAL
INSTR. JUNCTION BOX ASSEMBLY
WITH BACK PANEL (16 PR.)

ENGR. STD.
D-JB01A
PAGE:

BILL OF MATERIAL			
ITEM	CATALOG	QTY.	DESCRIPTION
1	0801733	1	PHOENIX NS35/7,5 MOUNTING RAIL
			(CUT LENGTH AS REQ'D.)
2	1201442	2	PHOENIX E/UK END CLAMP
3	3003017	41	PHOENIX UK-4 TERMINAL BLOCK
4	3003020	1	PHOENIX D-UK4/10 END COVER
5	1001037	64	PHOENIX DST-6 CLEAR MARKER CARRIERS
6	1001079	18	PHOENIX DST-6 GRAY MARKER CARRIERS
7	1001121	2	PHOENIX ES6:1-50 INSERT STRIP
8	0202303	1	PHOENIX LB BRIDGE W/ SCREWS, TYPE LB100-6
			(CUT LENGTH AS REQ'D.)
9	A-16P14	1	HOFFMAN BACK PANEL
10	A-18149-JFGQR	1	HOFFMAN NEMA TYPE 4X FIBERGLASS ENCLOSURE
			W/ QUICK RELEASE LATCHES

APPROVED DATE 1-04-91 REVISION 00 DATE 1-04-91

PC ACAD: D.JB01A

S.I.P. Engineering, Inc.
Houston, Texas

ENGINEERING STANDARDS
ELECTRICAL
INSTR. JUNCTION BOX ASSEMBLY
WITHOUT BACK PANEL (12 PR.)

ENGR. STD.
D-JB01B
PAGE:

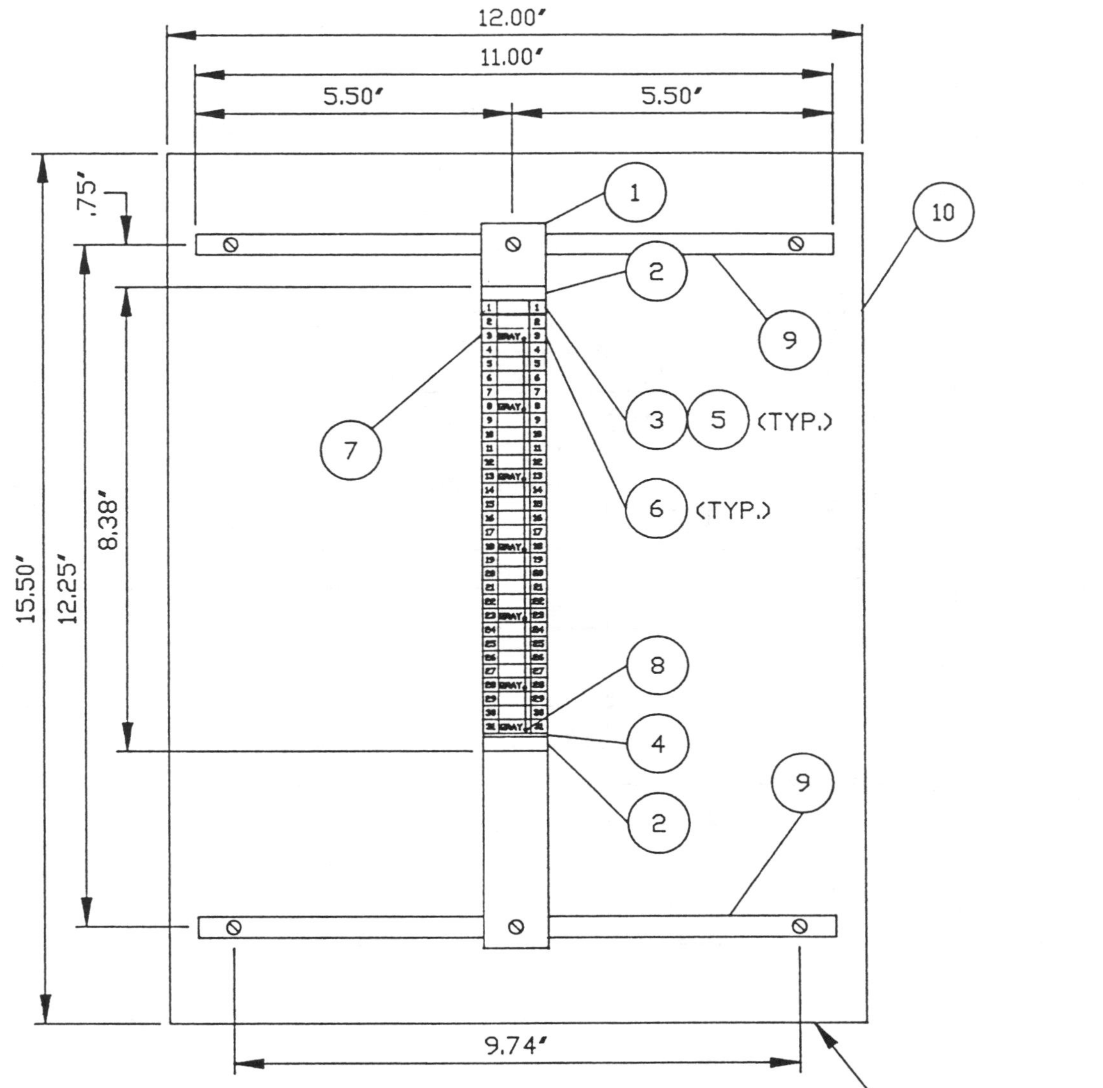

NOTES:

1. DRILL 1/8" DIAMETER WEEP HOLE IN THE BOTTOM REAR CORNER OF JUNCTION BOX FOR DRAINAGE.

*NOTE TO CAD OPERATORS:

1. SCALE THIS STANDARD UP FROM A REFERENCE DIMENSION OF 5.500" TO 15.500" TO GET FULL SCALE DRAWING.
2. DIMENSIONS AND BALLOON NUMBER CALL OUTS ARE DRAWN LARGER THAN ACTUAL DRAWING FOR CLARITY.

APPROVED ___ DATE 1-04-91 REVISION 00 DATE 1-04-91

S.I.P. Engineering, Inc.
Houston, Texas

ENGINEERING STANDARDS
ELECTRICAL
INSTR. JUNCTION BOX ASSEMBLY
WITHOUT BACK PANEL (12 PR.)

ENGR. STD. D-JB01B
PAGE: 3 OF 3

BILL OF MATERIAL

ITEM	CATALOG	QTY.	DESCRIPTION
1	0801733	1	PHOENIX NS35/7,5 MOUNTING RAIL
			(CUT LENGTH AS REQ'D.)
2	1201442	2	PHOENIX E/UK END CLAMP
3	3003017	31	PHOENIX UK-4 TERMINAL BLOCK
4	3003020	1	PHOENIX D-UK4/10 END COVER
5	1001037	48	PHOENIX DST-6 CLEAR MARKER CARRIERS
6	1001079	14	PHOENIX DST-6 GRAY MARKER CARRIERS
7	1001121	2	PHOENIX ES6:1-50 INSERT STRIP
8	0202303	1	PHOENIX LB BRIDGE W/ SCREWS, TYPE LB100-6
			(CUT LENGTH AS REQ'D.)
9	A-12JTMA	1	HOFFMAN BRACKET ASSEMBLY
10	A-16128-JFGQR	1	HOFFMAN NEMA TYPE 4X FIBERGLASS ENCLOSURE
			W/ QUICK RELEASE LATCHES

APPROVED DATE 1-04-91 REVISION 00 DATE 1-04-91

PC ACAD: DJB01B

S.I.P. Engineering, Inc.
Houston, Texas

ENGINEERING STANDARDS
ELECTRICAL
INSTR. JUNCTION BOX ASSEMBLY
WITH BACK PANEL (36 PR.)

ENGR. STD.
D-JB02A
PAGE: 1 OF 3

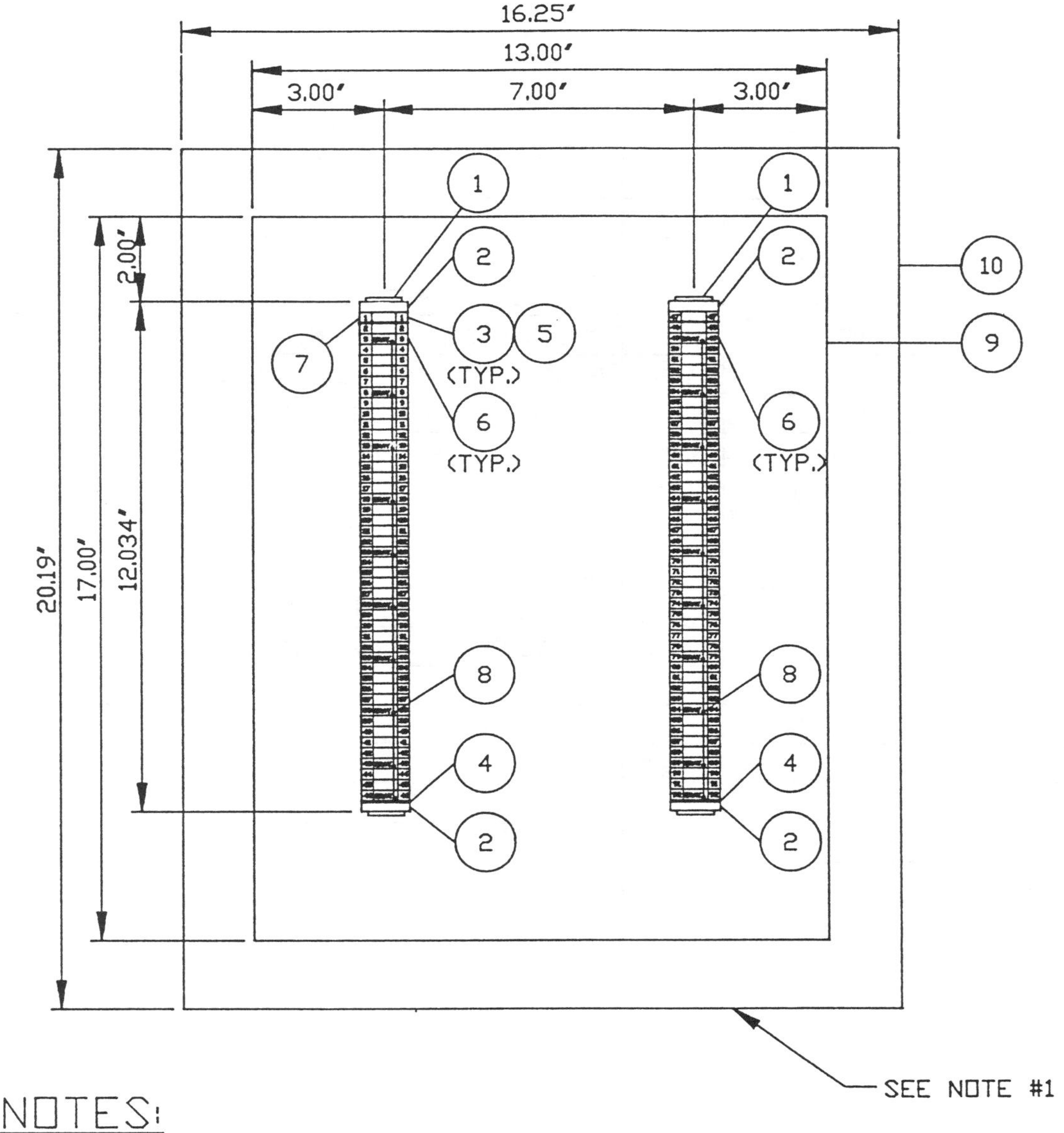

NOTES:

1. DRILL 1/8″ DIAMETER WEEP HOLE IN THE BOTTOM REAR CORNER OF JUNCTION BOX FOR DRAINAGE.

*NOTE TO CAD OPERATORS:

1. SCALE THIS STANDARD UP FROM A REFERENCE DIMENSION OF 5.500″ TO 20.190″ TO GET FULL SCALE DRAWING.
2. DIMENSIONS AND BALLOON NUMBER CALL OUTS ARE DRAWN LARGER THAN ACTUAL DRAWING FOR CLARITY.

APPROVED DATE 1-04-91 REVISION 00 DATE 1-04-91

PC ACAD: D.JB02A

S.I.P. Engineering, Inc.
Houston, Texas

ENGINEERING STANDARDS
ELECTRICAL
INSTR. JUNCTION BOX ASSEMBLY
WITH BACK PANEL (36 PR.)

ENGR. STD.
D-JB02A
PAGE: 3 OF 3

BILL OF MATERIAL

ITEM	CATALOG	QTY.	DESCRIPTION
1	0801733	1	PHOENIX NS35/7,5 MOUNTING RAIL
			(CUT LENGTH AS REQ'D.)
2	1201442	4	PHOENIX E/UK END CLAMP
3	3003017	92	PHOENIX UK-4 TERMINAL BLOCK
4	3003020	2	PHOENIX D-UK4/10 END COVER
5	1001037	144	PHOENIX DST-6 CLEAR MARKER CARRIERS
6	1001079	40	PHOENIX DST-6 GRAY MARKER CARRIERS
7	1001121	2	PHOENIX ES6:1-100 INSERT STRIP
8	0202303	2	PHOENIX LB BRIDGE W/ SCREWS, TYPE LB100-6
			(CUT LENGTH AS REQ'D.)
9	A-20P16	1	HOFFMAN BACK PANEL
10	A-20H1608-GQRLP	1	HOFFMAN NEMA TYPE 4X FIBERGLASS ENCLOSURE
			W/ QUICK RELEASE LATCHES

APPROVED ____ DATE 1-04-91 REVISION 00 DATE 1-04-91

P.C. ACAD: DJB02A

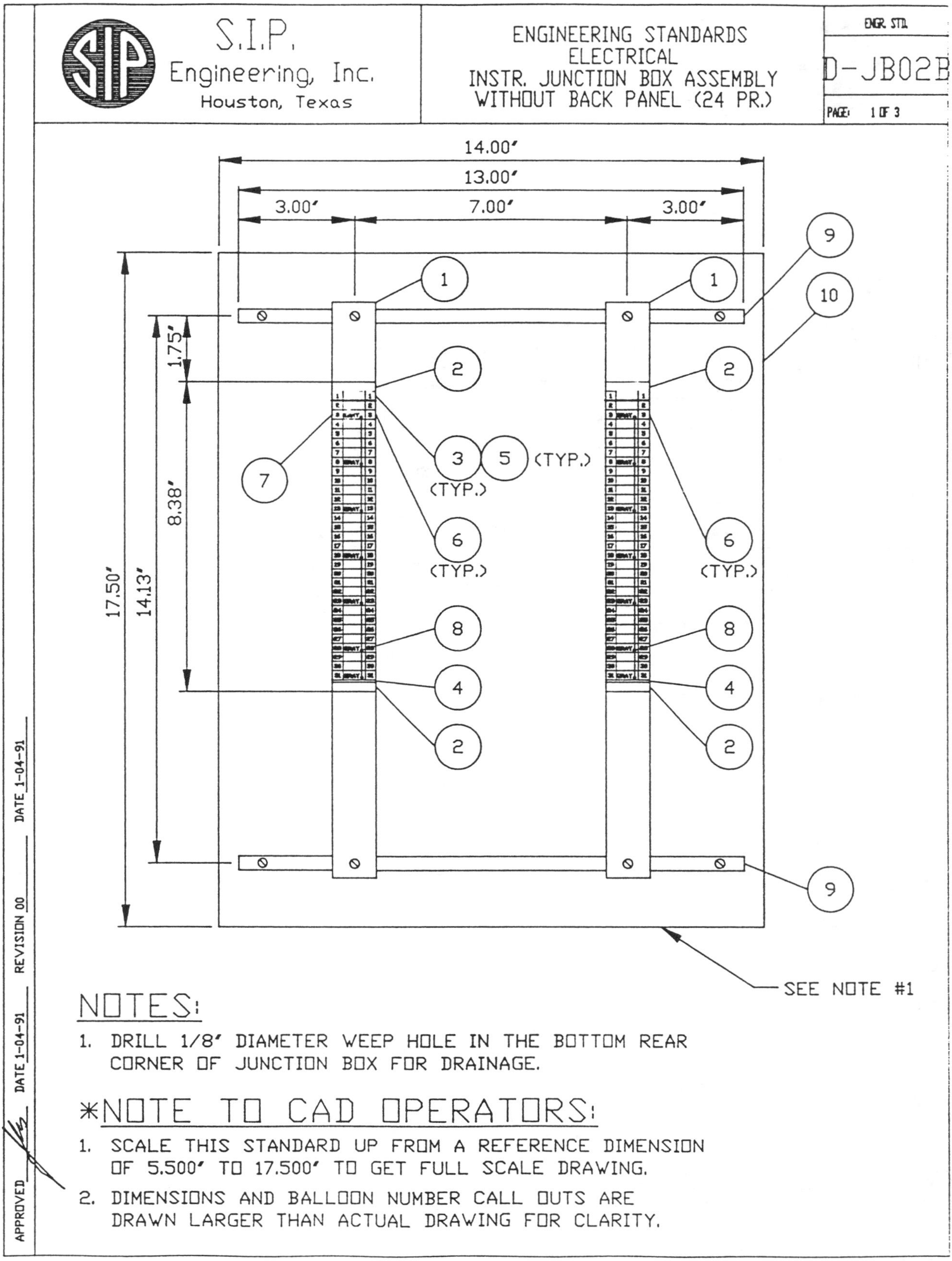

NOTES:

1. DRILL 1/8' DIAMETER WEEP HOLE IN THE BOTTOM REAR CORNER OF JUNCTION BOX FOR DRAINAGE.

*NOTE TO CAD OPERATORS:

1. SCALE THIS STANDARD UP FROM A REFERENCE DIMENSION OF 5.500' TO 17.500' TO GET FULL SCALE DRAWING.
2. DIMENSIONS AND BALLOON NUMBER CALL OUTS ARE DRAWN LARGER THAN ACTUAL DRAWING FOR CLARITY.

S.I.P. Engineering, Inc.
Houston, Texas

ENGINEERING STANDARDS
ELECTRICAL
INSTR. JUNCTION BOX ASSEMBLY
WITHOUT BACK PANEL (24 PR.)

ENGR. STD. D-JB02B

PAGE: 3 OF 3

BILL OF MATERIAL

ITEM	CATALOG	QTY.	DESCRIPTION
1	0801733	1	PHOENIX NS35/7,5 MOUNTING RAIL
			(CUT LENGTH AS REQ'D.)
2	1201442	4	PHOENIX E/UK END CLAMP
3	3003017	62	PHOENIX UK-4 TERMINAL BLOCK
4	3003020	2	PHOENIX D-UK4/10 END COVER
5	1001037	96	PHOENIX DST-6 CLEAR MARKER CARRIERS
6	1001079	28	PHOENIX DST-6 GRAY MARKER CARRIERS
7	1001121	2	PHOENIX ES6:1-175 INSERT STRIP
8	0202303	2	PHOENIX LB BRIDGE W/ SCREWS, TYPE LB100-6
			(CUT LENGTH AS REQ'D.)
9	A-14JTMA	1	HOFFMAN BRACKET ASSEMBLY
10	A-18149-JFGQR	1	HOFFMAN NEMA TYPE 4X FIBERGLASS ENCLOSURE
			W/ QUICK RELEASE LATCHES

APPROVED ____ DATE 1-04-91 REVISION 00 DATE 1-04-91

PC ACAD: D.JBOPR

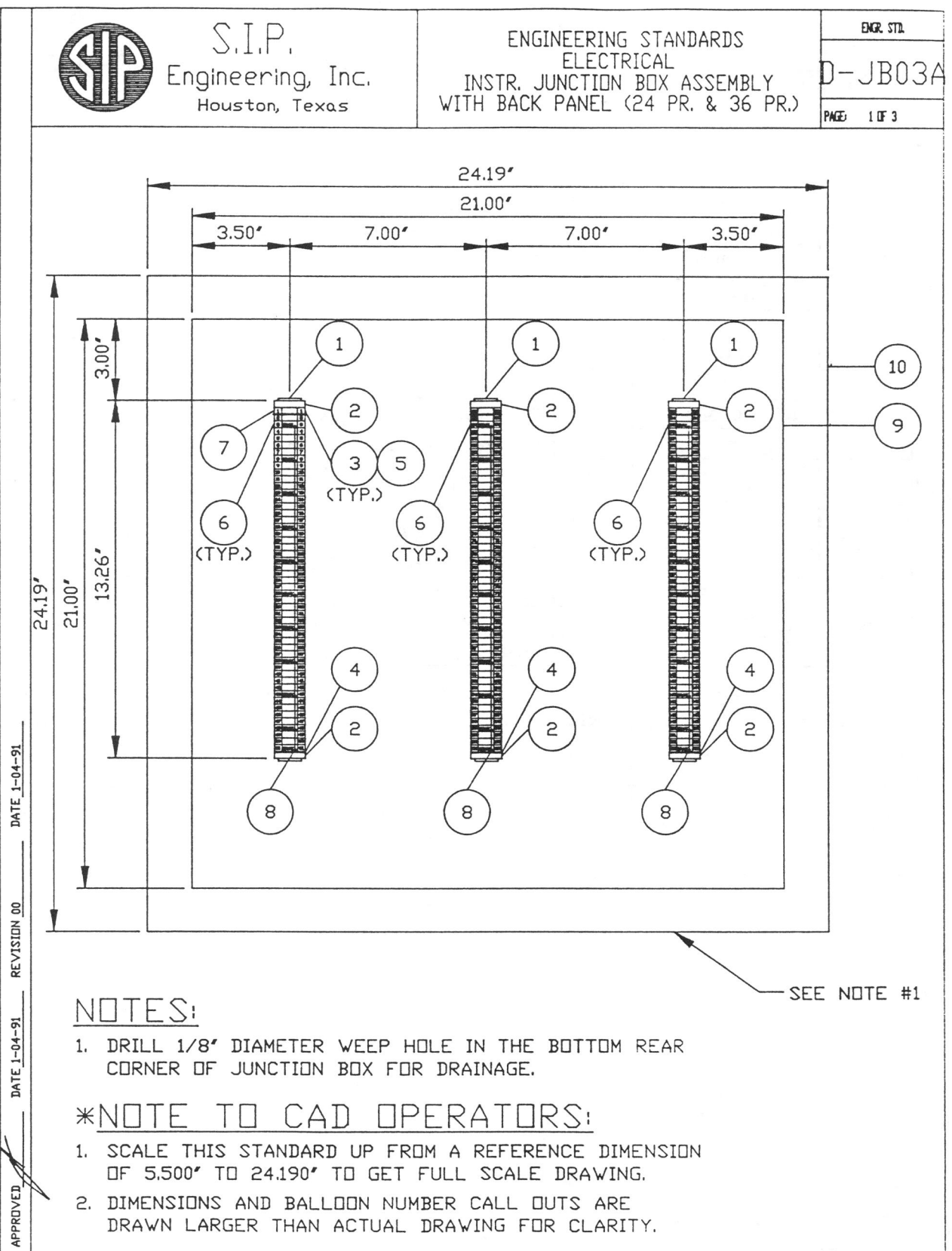

S.I.P. Engineering, Inc.
Houston, Texas

ENGINEERING STANDARDS
ELECTRICAL
INSTR. JUNCTION BOX ASSEMBLY
WITH BACK PANEL (24 PR. & 36 PR.)

ENGR. STD. D-JB03A
PAGE: 1 OF 3

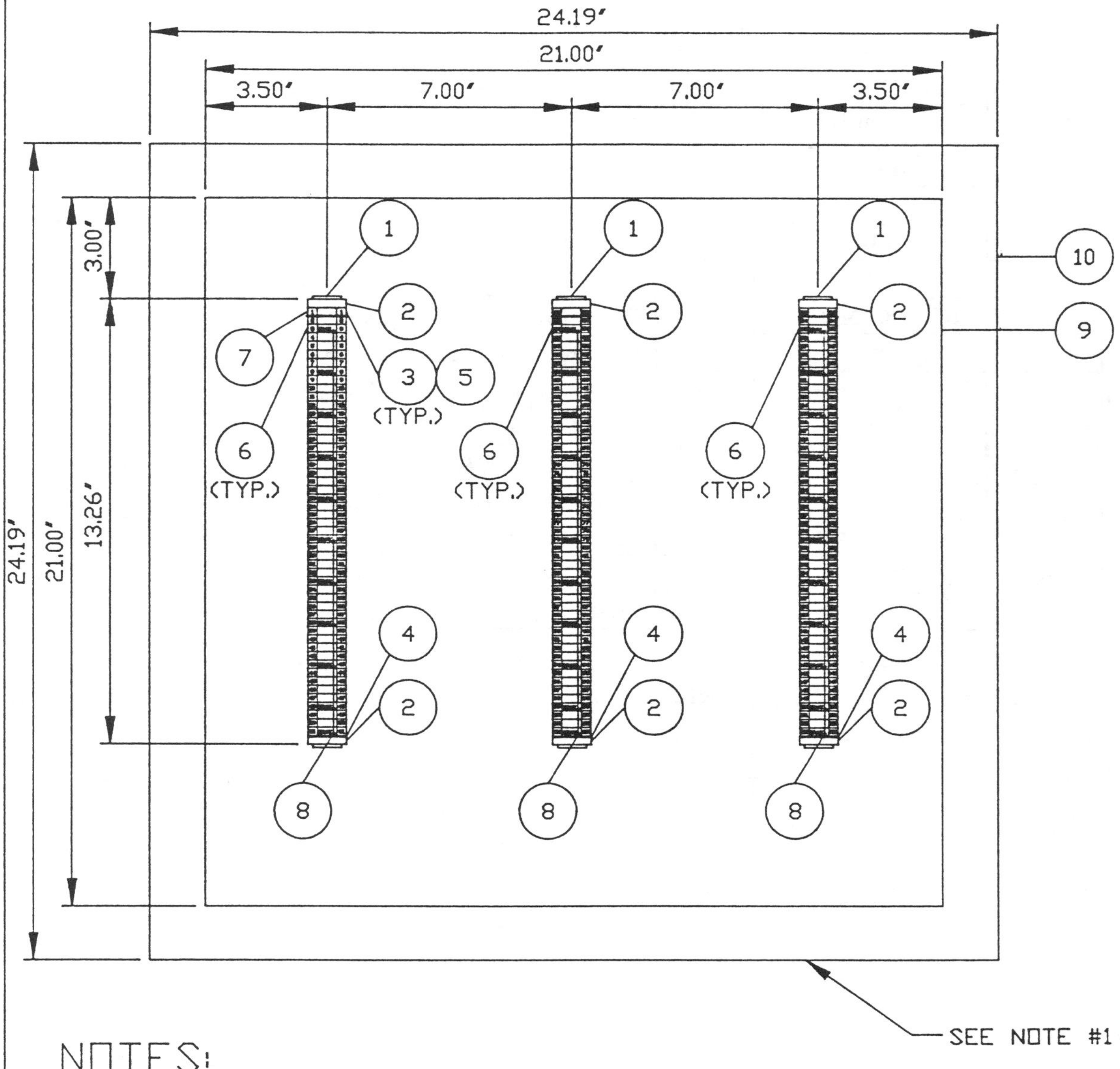

NOTES:

1. DRILL 1/8″ DIAMETER WEEP HOLE IN THE BOTTOM REAR CORNER OF JUNCTION BOX FOR DRAINAGE.

*NOTE TO CAD OPERATORS:

1. SCALE THIS STANDARD UP FROM A REFERENCE DIMENSION OF 5.500″ TO 24.190″ TO GET FULL SCALE DRAWING.
2. DIMENSIONS AND BALLOON NUMBER CALL OUTS ARE DRAWN LARGER THAN ACTUAL DRAWING FOR CLARITY.

APPROVED ____ DATE 1-04-91 ____ REVISION 00 ____ DATE 1-04-91

S.I.P. Engineering, Inc.
Houston, Texas

ENGINEERING STANDARDS
ELECTRICAL
INSTR. JUNCTION BOX ASSEMBLY
WITH BACK PANEL (24 PR. & 36 PR.)

ENGR. STD.
D-JB03A
PAGE: 3 OF 3

BILL OF MATERIAL

ITEM	CATALOG	QTY.	DESCRIPTION
1	0801733	2	PHOENIX NS35/7,5 MOUNTING RAIL
			(CUT LENGTH AS REQ'D.)
2	1201442	6	PHOENIX E/UK END CLAMP
3	3003017	153	PHOENIX UK-4 TERMINAL BLOCK
4	3003020	3	PHOENIX D-UK4/10 END COVER
5	1001037	240	PHOENIX DST-6 CLEAR MARKER CARRIERS
6	1001079	66	PHOENIX DST-6 GRAY MARKER CARRIERS
7	1001121	2	PHOENIX ES6:1-175 INSERT STRIP
8	0202303	3	PHOENIX LB BRIDGE W/ SCREWS, TYPE LB100-6
			(CUT LENGTH AS REQ'D.)
9	A-24P24	1	HOFFMAN BACK PANEL
10	A-24H2408-GQRLP	1	HOFFMAN NEMA TYPE 4X FIBERGLASS ENCLOSURE
			W/ QUICK RELEASE LATCHES

APPROVED DATE 1-04-91 REVISION 00 DATE 1-04-91

PC ACAD

S.I.P. Engineering, Inc.
Houston, Texas

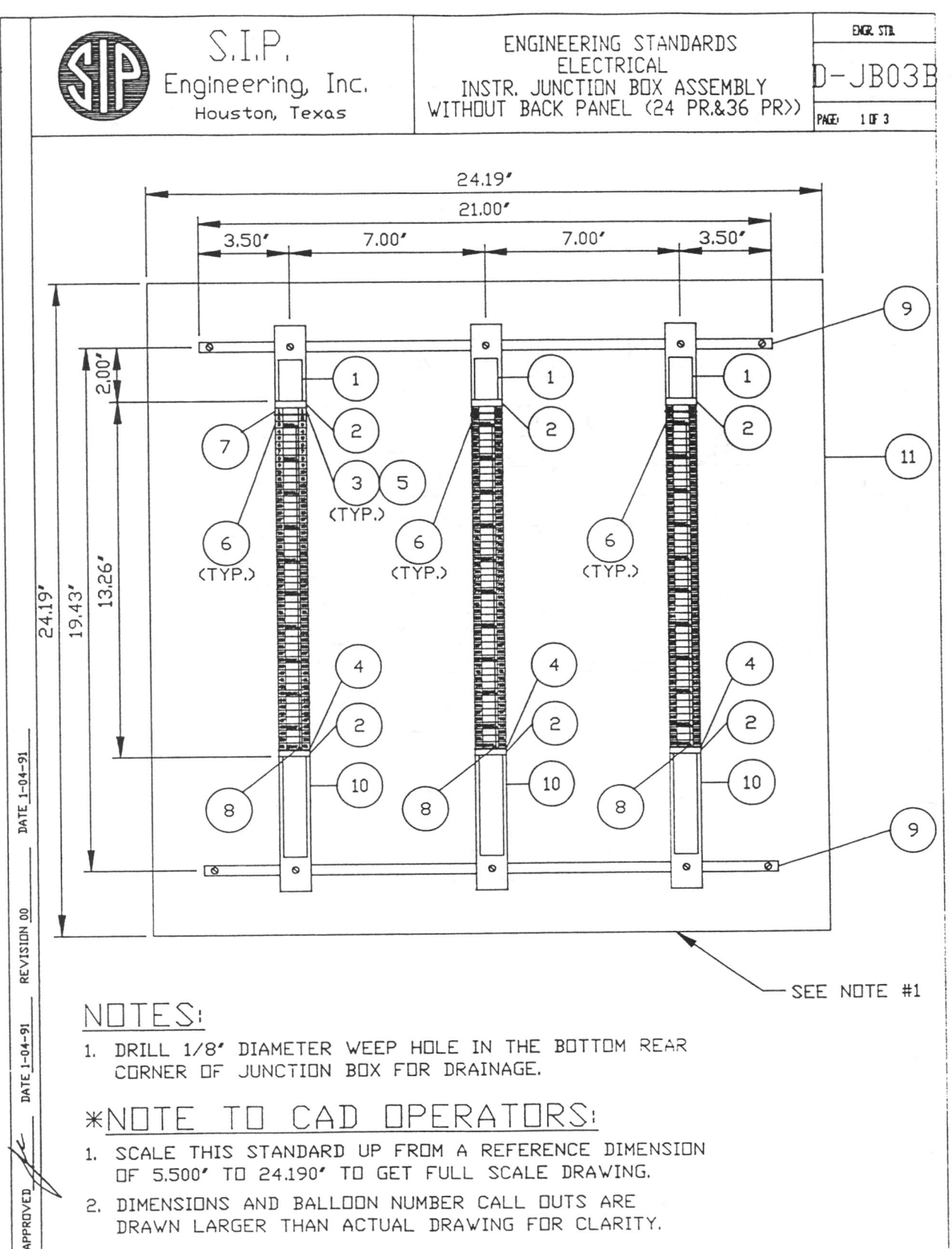

ENGINEERING STANDARDS
ELECTRICAL
INSTR. JUNCTION BOX ASSEMBLY
WITHOUT BACK PANEL (24 PR.&36 PR.)

ENGR. STD.
D-JB03B
PAGE: 1 OF 3

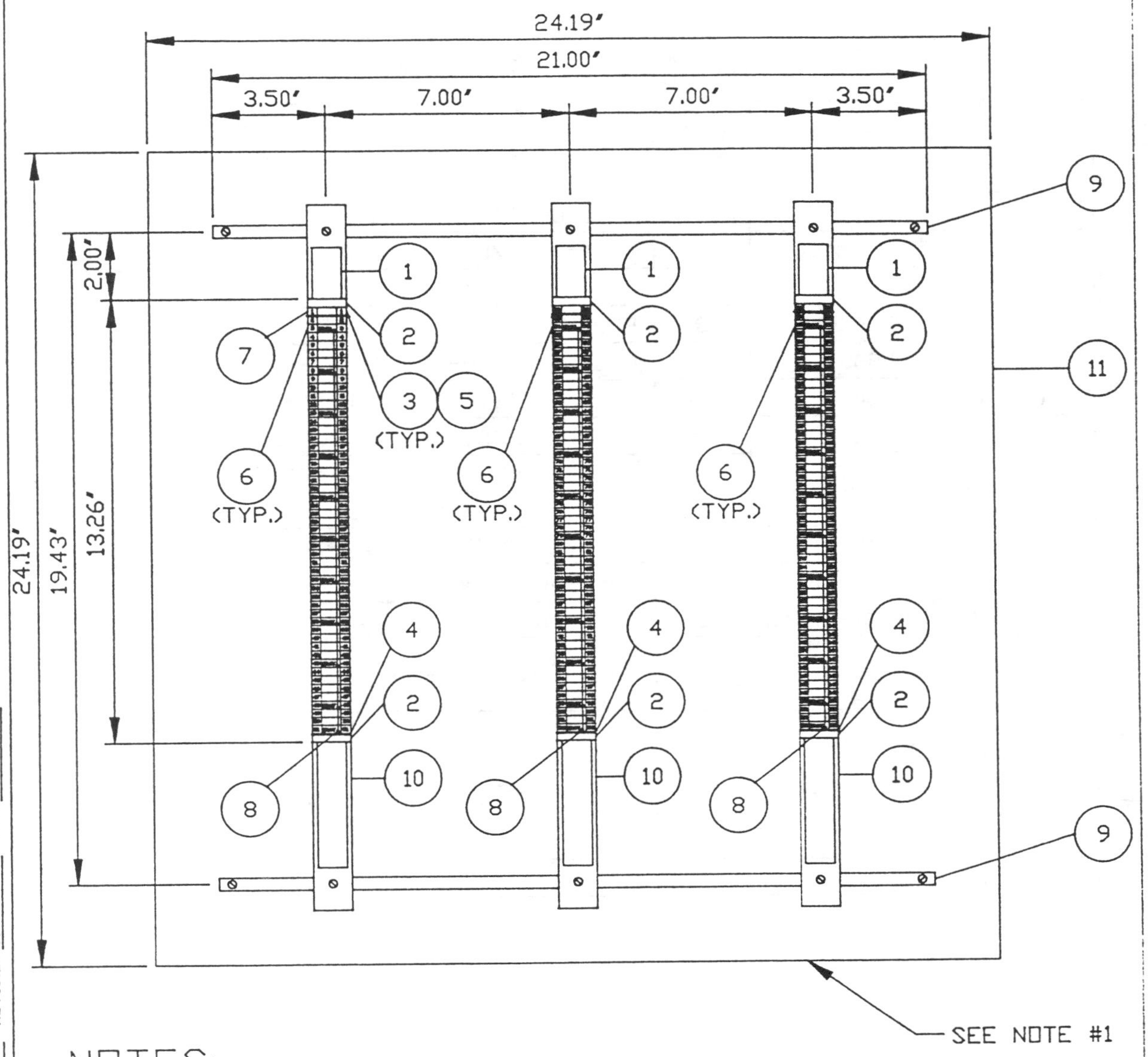

NOTES:

1. DRILL 1/8″ DIAMETER WEEP HOLE IN THE BOTTOM REAR CORNER OF JUNCTION BOX FOR DRAINAGE.

*NOTE TO CAD OPERATORS:

1. SCALE THIS STANDARD UP FROM A REFERENCE DIMENSION OF 5.500″ TO 24.190″ TO GET FULL SCALE DRAWING.
2. DIMENSIONS AND BALLOON NUMBER CALL OUTS ARE DRAWN LARGER THAN ACTUAL DRAWING FOR CLARITY.

APPROVED DATE 1-04-91 REVISION 00 DATE 1-04-91

PC ACAD: DJB03B

S.I.P. Engineering, Inc.
Houston, Texas

ENGINEERING STANDARDS
ELECTRICAL
INSTR. JUNCTION BOX ASSEMBLY
WITHOUT BACK PANEL (24 PR.&36 PR.)

ENGR. STD. D-JB03B
PAGE 3 OF 3

BILL OF MATERIAL

ITEM	CATALOG	QTY.	DESCRIPTION
1	0801733	2	PHOENIX NS35/7,5 MOUNTING RAIL
			(CUT LENGTH AS REQ'D.)
2	1201442	6	PHOENIX E/UK END CLAMP
3	3003017	153	PHOENIX UK-4 TERMINAL BLOCK
4	3003020	3	PHOENIX D-UK4/10 END COVER
5	1001037	240	PHOENIX DST-6 CLEAR MARKER CARRIERS
6	1001079	66	PHOENIX DST-6 GRAY MARKER CARRIERS
7	1001121	2	PHOENIX ES6:1-175 INSERT STRIP
8	0202303	2	PHOENIX LB BRIDGE W/ SCREWS, TYPE LB100-6
			(CUT LENGTH AS REQ'D.)
9	A-24NTMA2	1	HOFFMAN BRACKET ASSEMBLY
10	A-24T	3	HOFFMAN TERMINAL STRAP
11	A-24H2408-GQRLP	1	HOFFMAN NEMA TYPE 4X FIBERGLASS ENCLOSURE
			W/ QUICK RELEASE LATCHES

APPROVED – DATE 1-04-91 – REVISION 00 – DATE 1-04-91

PC ACAD: D-JB03B

S.I.P. Engineering, Inc.
Houston, Texas

ENGINEERING STANDARDS
ELECTRICAL
ELECT. 600V. POWER JUNCTION BOX
19 CONDUCTOR

ENGR. STD. D-JB04

PAGE: 1 OF 3

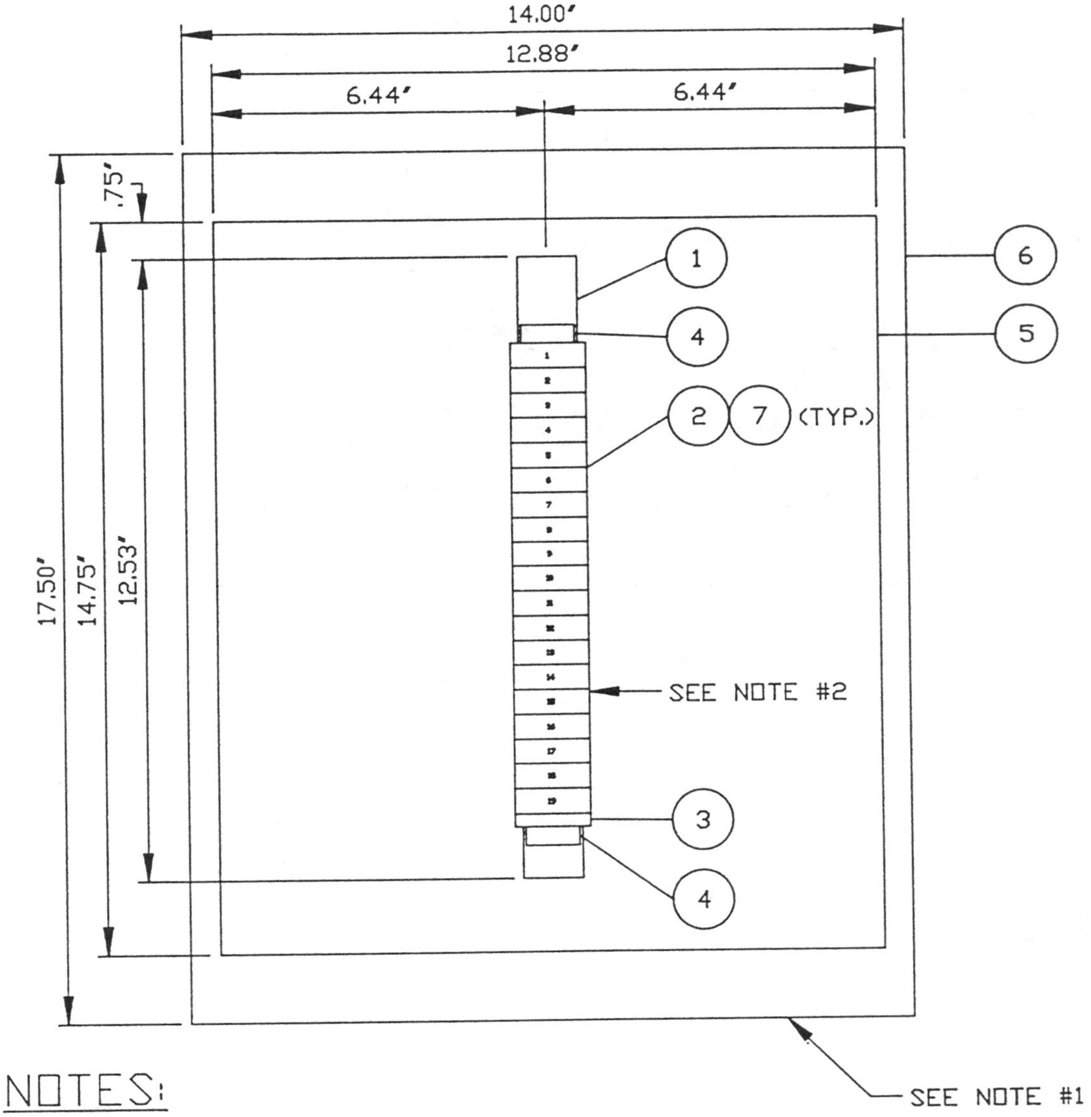

NOTES:

1. DRILL 1/8" DIAMETER WEEP HOLE IN THE BOTTOM REAR CORNER OF JUNCTION BOX FOR DRAINAGE.
2. BUCHANAN TERMINALS RATED FOR #18 THRU #4 WIRE SIZES.

*NOTE TO CAD OPERATORS:

1. SCALE THIS STANDARD UP FROM A REFERENCE DIMENSION OF 5.500" TO 17.500" TO GET FULL SCALE DRAWING.
2. DIMENSIONS AND BALLOON NUMBER CALL OUTS ARE DRAWN LARGER THAN ACTUAL DRAWING FOR CLARITY.

APPROVED ____ DATE 1-04-91 REVISION 00 DATE 1-04-91

PC ACAD: DJB04

S.I.P. Engineering, Inc.
Houston, Texas

ENGINEERING STANDARDS
ELECTRICAL
ELECT. 600V. POWER JUNCTION BOX
19 CONDUCTOR

ENGR. STD.
D-JB04
PAGE: 3 OF 3

BILL OF MATERIAL

ITEM	CATALOG	QTY.	DESCRIPTION
1	67	1	BUCHANAN ALUM. MOUNTING CHANNEL
			(CUT LENGTH AS REQ'D.)
2	P0242	19	BUCHANAN HEAVY DUTY TERMINAL BLOCK
3	P0250	1	BUCHANAN END SECTION
4	68	2	BUCHANAN UNIVERSAL CLAMP
5	A-16P14	1	HOFFMAN BACK PANEL
6	A-18149-JFGQR	1	HOFFMAN NEMA TYPE 4X FIBERGLASS ENCLOSURE
			W/ QUICK RELEASE LATCHES
7	50	1	BUCHANAN VINYL MARKING STRIP

APPROVED ____ DATE 1-04-91 REVISION 00 DATE 1-04-91

PC ACAD: D.JB04

S.I.P. Engineering, Inc.
Houston, Texas

ENGINEERING STANDARDS
ELECTRICAL
ELECT. 600V. POWER JUNCTION BOX
37 CONDUCTOR

ENGR. STD.
D-JB05
PAGE: 1 OF 3

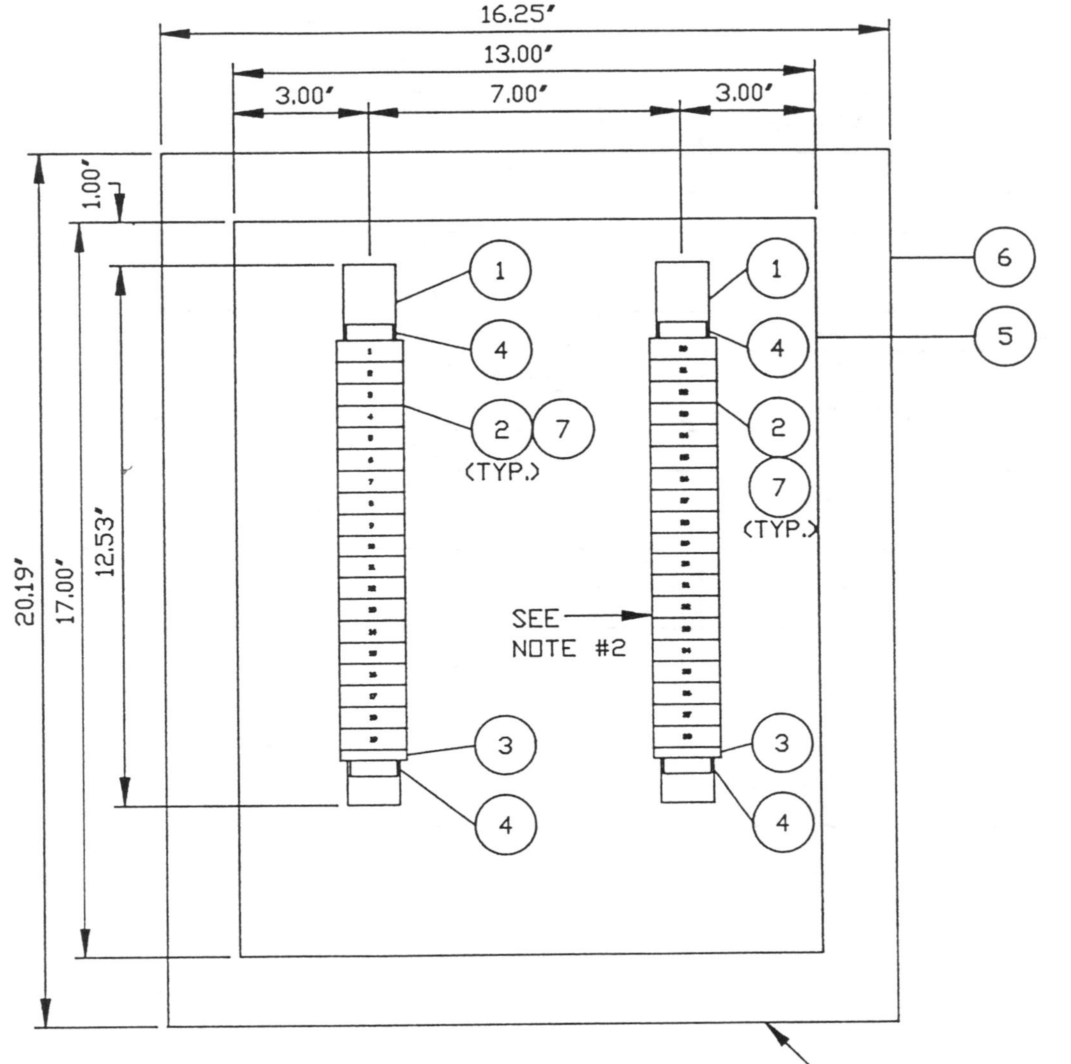

NOTES:

1. DRILL 1/8" DIAMETER WEEP HOLE IN THE BOTTOM REAR CORNER OF JUNCTION BOX FOR DRAINAGE.
2. BUCHANAN TERMINALS RATED FOR #18 THRU #4 WIRE SIZES.

*NOTE TO CAD OPERATORS:

1. SCALE THIS STANDARD UP FROM A REFERENCE DIMENSION OF 5.500" TO 20.190" TO GET FULL SCALE DRAWING.
2. DIMENSIONS AND BALLOON NUMBER CALL OUTS ARE DRAWN LARGER THAN ACTUAL DRAWING FOR CLARITY.

APPROVED DATE 1-04-91 REVISION 00 DATE 1-04-91

PC ACAD: DJB05

S.I.P. Engineering, Inc.
Houston, Texas

ENGINEERING STANDARDS
ELECTRICAL
ELECT. 600V. POWER JUNCTION BOX
37 CONDUCTOR

ENGR. STD.
D-JB05
PAGE: 3 OF 3

BILL OF MATERIAL

ITEM	CATALOG	QTY.	DESCRIPTION
1	67	1	BUCHANAN ALUM. MOUNTING CHANNEL
			(CUT LENGTH AS REQ'D.)
2	P0242	38	BUCHANAN HEAVY DUTY TERMINAL BLOCK
3	P0250	2	BUCHANAN END SECTION
4	68	4	BUCHANAN UNIVERSAL CLAMP
5	A-20P16	1	HOFFMAN BACK PANEL
6	A-20H1608-GQRLP	1	HOFFMAN NEMA TYPE 4X FIBERGLASS ENCLOSURE
			W/ QUICK RELEASE LATCHES
7	50	1	BUCHANAN VINYL MARKING STRIP

APPROVED ____ DATE 1-04-91 ____ REVISION 00 ____ DATE 1-04-91

PC ACAD: D.JB05

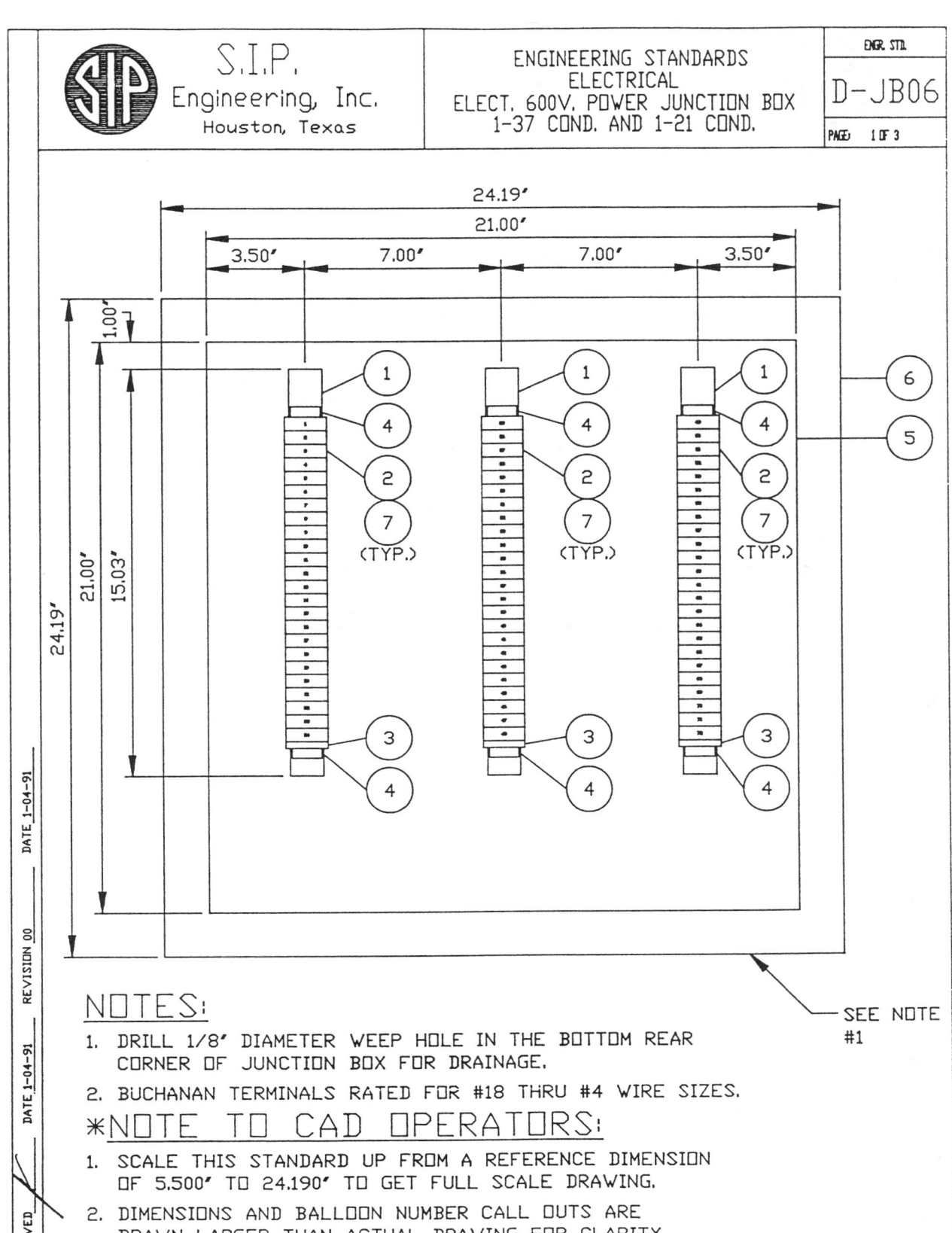

S.I.P.
Engineering, Inc.
Houston, Texas

ENGINEERING STANDARDS
ELECTRICAL
ELECT. 600V. POWER JUNCTION BOX
1-37 COND. AND 1-21 COND.

ENGR. STD.
D-JB06
PAGE: 1 OF 3

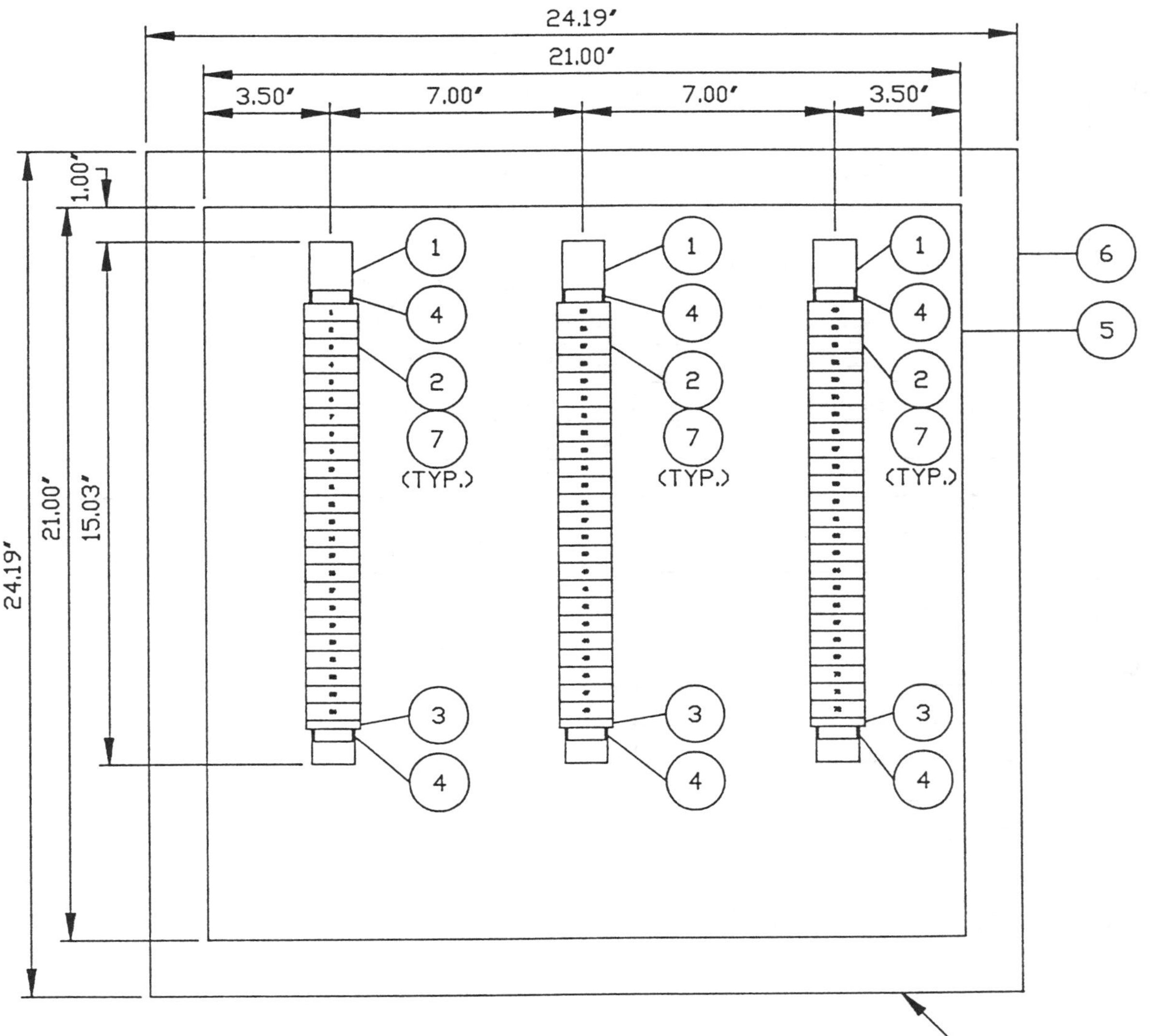

NOTES:

1. DRILL 1/8" DIAMETER WEEP HOLE IN THE BOTTOM REAR CORNER OF JUNCTION BOX FOR DRAINAGE.
2. BUCHANAN TERMINALS RATED FOR #18 THRU #4 WIRE SIZES.

*NOTE TO CAD OPERATORS:

1. SCALE THIS STANDARD UP FROM A REFERENCE DIMENSION OF 5.500" TO 24.190" TO GET FULL SCALE DRAWING.
2. DIMENSIONS AND BALLOON NUMBER CALL OUTS ARE DRAWN LARGER THAN ACTUAL DRAWING FOR CLARITY.

APPROVED ____ DATE 1-04-91 ____ REVISION 00 ____ DATE 1-04-91

PC ACAD: DJB06

S.I.P. Engineering, Inc.
Houston, Texas

ENGINEERING STANDARDS
ELECTRICAL
ELECT. 600V. POWER JUNCTION BOX
1-37 COND. AND 1-21 COND.

ENGR. STD.
D-JB06
PAGE: 3 OF 3

BILL OF MATERIAL

ITEM	CATALOG	QTY.	DESCRIPTION
1	67	1	BUCHANAN ALUM. MOUNTING CHANNEL
			(CUT LENGTH AS REQ'D.)
2	P0242	72	BUCHANAN HEAVY DUTY TERMINAL BLOCK
3	P0250	3	BUCHANAN END SECTION
4	68	6	BUCHANAN UNIVERSAL CLAMP
5	A-24P24	1	HOFFMAN BACK PANEL
6	A-24H2408-GQRLP	1	HOFFMAN NEMA TYPE 4X FIBERGLASS ENCLOSURE
			W/ QUICK RELEASE LATCHES
7	50	1	BUCHANAN VINYL MARKING STRIP

APPROVED ____ DATE 1-04-91 REVISION 00 DATE 1-04-91

PC ACAD: D.JB06

Index